Annals of Mathematics Studies

Number 170

Higher Topos Theory

Jacob Lurie

PRINCETON UNIVERSITY PRESS

PRINCETON AND OXFORD

2009

Copyright © 2009 by Princeton University Press

Published by Princeton University Press
41 William Street, Princeton, New Jersey 08540

In the United Kingdom: Princeton University Press
6 Oxford Street, Woodstock, Oxfordshire OX20 1TW

All Rights Reserved

Library of Congress Cataloging-in-Publication Data

Lurie, Jacob, 1977-
    Higher topos theory / Jacob Lurie.
        p. cm. (Annals of mathematics studies ; no. 170)
    Includes bibliographical references and index.
    ISBN 978-0-691-14048-3 (hardcover : alk. paper) – ISBN 978-0-691-14049-0
(pbk. : alk. paper) 1. Toposes. 2. Categories (Mathematics) I. Title.
    QA169.L87 2009
    512'.62–dc22   2008038170

British Library Cataloging-in-Publication Data is available

This book has been composed in Times in LATEX

The publisher would like to acknowledge the authors of this volume for
providing the camera-ready copy from which this book was printed.

press.princeton.edu

Printed and bound by CPI Group (UK) Ltd, Croydon, CR0 4YY

Contents

Preface

Let X be a nice topological space (for example, a CW complex). One goal of algebraic topology is to study the topology of X by means of algebraic invariants, such as the singular cohomology groups $H^n(X; G)$ of X with coefficients in an abelian group G. These cohomology groups have proven to be an extremely useful tool largely because they enjoy excellent formal properties (which have been axiomatized by Eilenberg and Steenrod, see [26]) and because they tend to be very computable. However, the usual definition of $H^n(X; G)$ in terms of singular G-valued cochains on X is perhaps somewhat unenlightening. This raises the following question: can we understand the cohomology group $H^n(X; G)$ in more conceptual terms?

As a first step toward answering this question, we observe that $H^n(X; G)$ is a *representable* functor of X. That is, there exists an *Eilenberg-MacLane space* $K(G, n)$ and a universal cohomology class $\eta \in H^n(K(G, n); G)$ such that, for any nice topological space X, pullback of η determines a bijection

$$[X, K(G, n)] \to H^n(X; G).$$

Here $[X, K(G, n)]$ denotes the set of homotopy classes of maps from X to $K(G, n)$. The space $K(G, n)$ can be characterized up to homotopy equivalence by the above property or by the formula

$$\pi_k K(G, n) \simeq \begin{cases} * & \text{if } k \neq n \\ G & \text{if } k = n. \end{cases}$$

In the case $n = 1$, we can be more concrete. An Eilenberg-MacLane space $K(G, 1)$ is called a *classifying space* for G and is typically denoted by BG. The universal cover of BG is a contractible space EG, which carries a free action of the group G by covering transformations. We have a quotient map $\pi : EG \to BG$. Each fiber of π is a discrete topological space on which the group G acts simply transitively. We can summarize the situation by saying that EG is a G-*torsor* over the classifying space BG. For every continuous map $X \to BG$, the fiber product $\widetilde{X} : EG \times_{BG} X$ has the structure of a G-torsor on X: that is, it is a space endowed with a free action of G and a homeomorphism $\widetilde{X}/G \simeq X$. This construction determines a map from $[X, BG]$ to the set of isomorphism classes of G-torsors on X. If X is a sufficiently well-behaved space (such as a CW complex), then this map is a bijection. We therefore have (at least) three different ways of thinking about a cohomology class $\eta \in H^1(X; G)$:

(1) As a G-valued singular cocycle on X, which is well-defined up to coboundaries.

(2) As a continuous map $X \to BG$, which is well-defined up to homotopy.

(3) As a G-torsor on X, which is well-defined up to isomorphism.

These three points of view are equivalent if the space X is sufficiently nice. However, they are generally quite different from one another. The singular cohomology of a space X is constructed using continuous maps from simplices Δ^k into X. If there are not many maps *into* X (for example, if every path in X is constant), then we cannot expect singular cohomology to tell us very much about X. The second definition uses maps from X into the classifying space BG, which (ultimately) relies on the existence of continuous real-valued functions on X. If X does not admit many real-valued functions, then the set of homotopy classes $[X, BG]$ is also not a very useful invariant. For such spaces, the third approach is the most powerful: there is a good theory of G-torsors on an arbitrary topological space X.

There is another reason for thinking about $\mathrm{H}^1(X; G)$ in the language of G-torsors: it continues to make sense in situations where the traditional ideas of topology break down. If $\widetilde{X}$ is a G-torsor on a topological space X, then the projection map $\widetilde{X} \to X$ is a local homeomorphism; we may therefore identify $\widetilde{X}$ with a sheaf of sets $\mathcal{F}$ on X. The action of G on $\widetilde{X}$ determines an action of G on $\mathcal{F}$. The sheaf $\mathcal{F}$ (with its G-action) and the space $\widetilde{X}$ (with its G-action) determine each other up to canonical isomorphism. Consequently, we can formulate the definition of a G-torsor in terms of the category $\mathrm{Shv}_{\mathrm{Set}}(X)$ of sheaves of sets on X without ever mentioning the topological space X itself. The same definition makes sense in any category which bears a sufficiently strong resemblance to the category of sheaves on a topological space: for example, in any *Grothendieck topos*. This observation allows us to construct a theory of torsors in a variety of nonstandard contexts, such as the étale topology of algebraic varieties (see [2]).

Describing the cohomology of X in terms of the sheaf theory of X has still another advantage, which comes into play even when the space X is assumed to be a CW complex. For a general space X, isomorphism classes of G-torsors on X are classified not by the singular cohomology $\mathrm{H}^1_{\mathrm{sing}}(X; G)$ but by the sheaf cohomology $\mathrm{H}^1_{\mathrm{sheaf}}(X; \mathcal{G})$ of X with coefficients in the constant sheaf $\mathcal{G}$ associated to G. This sheaf cohomology is defined more generally for *any* sheaf of groups $\mathcal{G}$ on X. Moreover, we have a conceptual interpretation of $\mathrm{H}^1_{\mathrm{sheaf}}(X; \mathcal{G})$ in general: it classifies $\mathcal{G}$-torsors on X (that is, sheaves $\mathcal{F}$ on X which carry an action of $\mathcal{G}$ and locally admit a $\mathcal{G}$-equivariant isomorphism $\mathcal{F} \simeq \mathcal{G}$) up to isomorphism. The general formalism of sheaf cohomology is extremely useful, even if we are interested only in the case where X is a nice topological space: it includes, for example, the theory of cohomology with coefficients in a local system on X.

Let us now attempt to obtain a similar interpretation for cohomology classes $\eta \in \mathrm{H}^2(X; G)$. What should play the role of a G-torsor in this case?

To answer this question, we return to the situation where X is a CW complex, so that η can be identified with a continuous map $X \to K(G, 2)$. We can think of $K(G, 2)$ as the classifying space of a group: not the discrete group G but instead the classifying space BG (which, if built in a sufficiently careful way, comes equipped with the structure of a topological abelian group). Namely, we can identify $K(G, 2)$ with the quotient E/BG, where E is a contractible space with a free action of BG. Any cohomology class $\eta \in \mathrm{H}^2(X; G)$ determines a map $X \to K(G, 2)$ (which is well-defined up to homotopy), and we can form the pullback $\widetilde{X} = E \times_{BG} X$. We now think of $\widetilde{X}$ as a torsor over X: not for the discrete group G but instead for its classifying space BG.

To complete the analogy with our analysis in the case $n = 1$, we would like to interpret the fibration $\widetilde{X} \to X$ as defining some kind of sheaf $\mathcal{F}$ on the space X. This sheaf $\mathcal{F}$ should have the property that for each $x \in X$, the stalk $\mathcal{F}_x$ can be identified with the fiber $\widetilde{X}_x \simeq BG$. Since the space BG is not discrete (or homotopy equivalent to a discrete space), the situation cannot be adequately described in the usual language of set-valued sheaves. However, the classifying space BG is *almost* discrete: since the homotopy groups $\pi_i BG$ vanish for $i > 1$, we can recover BG (up to homotopy equivalence) from its fundamental groupoid. This suggests that we might try to think about $\mathcal{F}$ as a "groupoid-valued sheaf" on X, or a *stack* (in groupoids) on X.

Remark. The condition that each stalk $\mathcal{F}_x$ be equivalent to a classifying space BG can be summarized by saying that $\mathcal{F}$ is a *gerbe* on X: more precisely, it is a gerbe banded by the constant sheaf $\mathcal{G}$ associated to G. We refer the reader to [31] for an explanation of this terminology and a proof that such gerbes are indeed classified by the sheaf cohomology group $\mathrm{H}^2_{\mathrm{sheaf}}(X; \mathcal{G})$.

For larger values of n, even the language of stacks is not sufficient to describe the nature of the sheaf $\mathcal{F}$ associated to the fibration $\widetilde{X} \to X$. To address the situation, Grothendieck proposed (in his infamous letter to Quillen; see [35]) that there should be a theory of n-*stacks* on X for every integer $n \geq 0$. Moreover, for every sheaf of abelian groups $\mathcal{G}$ on X, the cohomology group $\mathrm{H}^{n+1}_{\mathrm{sheaf}}(X; \mathcal{G})$ should have an interpretation as classifying a special type of n-stack: namely, the class of n-gerbes banded by $\mathcal{G}$ (for a discussion in the case $n = 2$, we refer the reader to [13]; we will treat the general case in §7.2.2). In the special case where the space X is a point (and where we restrict our attention to n-stacks in groupoids), the theory of n-stacks on X should recover the classical homotopy theory of n-*types*: that is, CW complexes Z such that the homotopy groups $\pi_i(Z, z)$ vanish for $i > n$ (and every base point $z \in Z$). More generally, we should think of an n-stack (in groupoids) on a general space X as a "sheaf of n-types" on X.

When $n = 0$, an n-stack on a topological space X is simply a sheaf of sets on X. The collection of all such sheaves can be organized into a category $\mathrm{Shv}_{\mathrm{Set}}(X)$, and this category is a prototypical example of a *Grothendieck topos*. The main goal of this book is to obtain an analogous understanding of the situation for $n > 0$. More precisely, we would like answers to the following questions:

($Q1$) Given a topological space X, what should we mean by a "sheaf of n-types" on X?

($Q2$) Let $\mathrm{Shv}_{\leq n}(X)$ denote the collection of all sheaves of n-types on X. What sort of a mathematical object is $\mathrm{Shv}_{\leq n}(X)$?

($Q3$) What special features (if any) does $\mathrm{Shv}_{\leq n}(X)$ possess?

Our answers to questions ($Q2$) and ($Q3$) may be summarized as follows (our answer to ($Q1$) is more elaborate, and we will avoid discussing it for the moment):

($A2$) The collection $\mathrm{Shv}_{\leq n}(X)$ has the structure of an ∞-*category*.

($A3$) The ∞-category $\mathrm{Shv}_{\leq n}(X)$ is an example of an $(n+1)$-*topos*: that is, an ∞-category which satisfies higher-categorical analogues of Giraud's axioms for Grothendieck topoi (see Theorem 6.4.1.5).

Remark. Grothendieck's vision has been realized in various ways thanks to the work of a number of mathematicians (most notably Brown, Joyal, and Jardine; see for example [41]), and their work can also be used to provide answers to questions ($Q1$) and ($Q2$) (for more details, we refer the reader to §6.5.2). Question ($Q3$) has also been addressed (at least in the limiting case $n = \infty$) by Toën and Vezzosi (see [78]) and in unpublished work of Rezk.

To provide more complete versions of the answers ($A2$) and ($A3$), we will need to develop the language of *higher category theory*. This is generally regarded as a technical and forbidding subject, but fortunately we will only need a small fragment of it. More precisely, we will need a theory of $(\infty, 1)$-*categories*: higher categories $\mathcal{C}$ for which the k-morphisms of $\mathcal{C}$ are required to be invertible for $k > 1$. In Chapter 1, we will present such a theory: namely, one can define an ∞-*category* to be a simplicial set satisfying a weakened version of the Kan extension condition (see Definition 1.2.4; simplicial sets satisfying this condition are also called *weak Kan complexes* or *quasi-categories* in the literature). Our intention is that Chapter 1 can be used as a short "user's guide" to ∞-categories: it contains many of the basic definitions and explains how many ideas from classical category theory can be extended to the ∞-categorical context. To simplify the exposition, we have deferred many proofs until later chapters, which contain a more thorough account of the theory. The hope is that Chapter 1 will be useful to readers who want to get the flavor of the subject without becoming overwhelmed by technical details.

In Chapter 2 we will shift our focus slightly: rather than study individual examples of ∞-categories, we consider *families* of ∞-categories $\{\mathcal{C}_D\}_{D \in \mathcal{D}}$ parametrized by the objects of another ∞-category $\mathcal{D}$. We might expect such a family to be given by a map of ∞-categories $p : \mathcal{C} \to \mathcal{D}$: given such a map, we can then define each $\mathcal{C}_D$ to be the fiber product $\mathcal{C} \times_\mathcal{D} \{D\}$. This definition behaves poorly in general (for example, the fibers $\mathcal{C}_D$ need not be ∞-categories), but it behaves well if we make suitable assumptions on the map p. Our goal in Chapter 2 is to study some of these assumptions in detail and to show that they lead to a good *relative* version of higher category theory.

One motivation for the theory of ∞-categories is that it arises naturally in addressing questions like $(Q2)$ above. More precisely, given a collection of mathematical objects $\{\mathcal{F}_\alpha\}$ whose definition has a homotopy-theoretic flavor (like n-stacks on a topological space X), one can often organize the collection $\{\mathcal{F}_\alpha\}$ into an ∞-category (in other words, there exists an ∞-category $\mathcal{C}$ whose vertices correspond to the objects $\mathcal{F}_\alpha$). Another important example is provided by higher category theory itself: the collection of all ∞-categories can itself be organized into a (very large) ∞-category, which we will denote by $\mathcal{C}\mathrm{at}_\infty$. Our goal in Chapter 3 is to study $\mathcal{C}\mathrm{at}_\infty$ and to show that it can be characterized by a universal property: namely, functors $\chi : \mathcal{D} \to \mathcal{C}\mathrm{at}_\infty$ are classified (up to equivalence) by certain kinds of fibrations $\mathcal{C} \to \mathcal{D}$ (see Theorem 3.2.0.1 for a more precise statement). Roughly speaking, this correspondence assigns to a fibration $\mathcal{C} \to \mathcal{D}$ the functor χ given by the formula $\chi(D) = \mathcal{C} \times_\mathcal{D} \{D\}$.

Classically, category theory is a useful tool not so much because of the light it sheds on any particular mathematical discipline but instead because categories are so ubiquitous: mathematical objects in many different settings (sets, groups, smooth manifolds, and so on) can be organized into categories. Moreover, many elementary mathematical concepts can be described in purely categorical terms and therefore make sense in each of these settings. For example, we can form products of sets, groups, and smooth manifolds: each of these notions can simply be described as a Cartesian product in the relevant category. Cartesian products are a special case of the more general notion of *limit*, which plays a central role in classical category theory. In Chapter 4, we will make a systematic study of limits (and the dual theory of colimits) in the ∞-categorical setting. We will also introduce the more general theory of *Kan extensions*, in both absolute and relative versions; this theory plays a key technical role throughout the later parts of the book.

In some sense, the material of Chapters 1 through 4 of this book should be regarded as purely formal. Our main results can be summarized as follows: there exists a reasonable theory of ∞-categories, and it behaves in more or less the same way as the theory of ordinary categories. Many of the ideas that we introduce are straightforward generalizations of their ordinary counterparts (though proofs in the ∞-categorical setting often require a bit of dexterity in manipulating simplicial sets), which will be familiar to mathematicians who are acquainted with ordinary category theory (as presented, for example, in [52]). In Chapter 5, we introduce ∞-categorical analogues of more sophisticated concepts from classical category theory: presheaves, Pro-categories and Ind-categories, accessible and presentable categories, and localizations. The main theme is that most of the ∞-categories which appear "in nature" are large but are nevertheless determined by small subcategories. Taking careful advantage of this fact will allow us to deduce a number of pleasant results, such as an ∞-categorical version of the adjoint functor theorem (Corollary 5.5.2.9).

In Chapter 6 we come to the heart of the book: the study of *∞-topoi*, the ∞-categorical analogues of Grothendieck topoi. The theory of ∞-topoi is our answer to the question (Q3) in the limiting case $n = \infty$ (we will also study the analogous notion for finite values of n). Our main result is an analogue of Giraud's theorem, which asserts the equivalence of "extrinsic" and "intrinsic" approaches to the subject (Theorem 6.1.0.6). Roughly speaking, an ∞-topos is an ∞-category which "looks like" the ∞-category of all homotopy types. We will show that this intuition is justified in the sense that it is possible to reconstruct a large portion of classical homotopy theory inside an arbitrary ∞-topos. In other words, an ∞-topos is a world in which one can "do" homotopy theory (much as an ordinary topos can be regarded as a world in which one can "do" other types of mathematics).

In Chapter 7 we will discuss some relationships between our theory of ∞-topoi and ideas from classical topology. We will show that, if X is a paracompact space, then the ∞-topos of "sheaves of homotopy types" on X can be interpreted in terms of the classical homotopy theory of spaces *over X*. Another main theme is that various ideas from geometric topology (such as dimension theory and shape theory) can be described naturally using the language of ∞-topoi. We will also formulate and prove "nonabelian" generalizations of classical cohomological results, such as Grothendieck's vanishing theorem for the cohomology of Noetherian topological spaces and the proper base change theorem.

Prerequisites and Suggested Reading

We have made an effort to keep this book as self-contained as possible. The main prerequisite is familiarity with the classical homotopy theory of simplicial sets (good references include [56] and [32]; we have also provided a very brief review in §A.2.7). The remaining material that we need is either described in the appendix or developed in the body of the text. However, our exposition of this background material is often somewhat terse, and the reader might benefit from consulting other treatments of the same ideas. Some suggestions for further reading are listed below.

Warning. The list of references below is woefully incomplete. We have not attempted, either here or in the body of the text, to give a comprehensive survey of the literature on higher category theory. We have also not attempted to trace all of the ideas presented to their origins or to present a detailed history of the subject. Many of the topics presented in this book have appeared elsewhere or belong to the mathematical folklore; it should not be assumed that uncredited results are due to the author.

- **Classical Category Theory:** Large portions of this book are devoted to providing ∞-categorical generalizations of the basic notions of category theory. A good reference for many of the concepts we use is MacLane's book [52] (see also [1] and [54] for some of the more advanced material of Chapter 5).

- **Classical Topos Theory:** Our main goal in this book is to describe an ∞-categorical version of topos theory. Familiarity with classical topos theory is not strictly necessary (we will define all of the relevant concepts as we need them) but will certainly be helpful. Good references include [2] and [53].

- **Model Categories:** Quillen's theory of model categories provides a useful tool for studying specific examples of ∞-categories, including the theory of ∞-categories itself. We will summarize the theory of model categories in §A.2; more complete references include [40], [38], and [32].

- **Higher Category Theory:** There are many approaches to the theory of higher categories, some of which look quite different from the approach presented in this book. Several other possibilities are presented in the survey article [48]. More detailed accounts can be found in [49], [71], and [75].

  In this book, we consider only $(\infty, 1)$-*categories*: that is, higher categories in which all k-morphisms are assumed to be invertible for $k > 1$. There are a number of convenient ways to formalize this idea: via simplicial categories (see, for example, [21] and [7]), via Segal categories ([71]), via complete Segal spaces ([64]), or via the theory of ∞-categories presented in this book (other references include [43], [44], [60], and [10]). The relationship between these various approaches is described in [8], and an axiomatic framework which encompasses all of them is described in [76].

- **Higher Topos Theory:** The idea of studying a topological space X via the theory of sheaves of n-types (or n-*stacks*) on X goes back at least to Grothendieck ([35]) and has been taken up a number of times in recent years. For small values of n, we refer the reader to [31], [74], [13], [45], and [61]. For the case $n = \infty$, we refer the reader to [14], [41], [39], and [77]. A very readable introduction to some of these ideas can be found in [4].

  Higher topos theory itself can be considered an abstraction of this idea: rather than studying sheaves of n-types on a particular topological space X, we instead study general n-categories with the same formal properties. This idea has been implemented in the work of Toën and Vezzosi (see [78] and [79]), resulting in a theory which is essentially equivalent to the one presented in this book. (A rather different variation on this idea in the case $n = 2$ can be also be found in [11].)

The subject has also been greatly influenced by the unpublished ideas of Charles Rezk.

TERMINOLOGY

Here are a few comments on some of the terminology which appears in this book:

- The word *topos* will always mean *Grothendieck* topos.

- We let Set_Δ denote the category of simplicial sets. If J is a linearly ordered set, we let Δ^J denote the simplicial set given by the nerve of J, so that the collection of n-simplices of Δ^J can be identified with the collection of all nondecreasing maps $\{0, \ldots, n\} \to J$. We will frequently apply this notation when J is a subset of $\{0, \ldots, n\}$; in this case, we can identify Δ^J with a subsimplex of the standard n-simplex Δ^n (at least if $J \neq \emptyset$; if $J = \emptyset$, then Δ^J is empty).

- We will refer to a category $\mathcal{C}$ as *accessible* or *presentable* if it is *locally accessible* or *locally presentable* in the terminology of [54].

- Unless otherwise specified, the term ∞-*category* will be used to indicate a higher category in which all n-morphisms are invertible for $n > 1$.

- We will study higher categories using Joyal's theory of *quasi-categories*. However, we do not always follow Joyal's terminology. In particular, we will use the term ∞-*category* to refer to what Joyal calls a *quasi-category* (which are, in turn, the same as the *weak Kan complex* of Boardman and Vogt); we will use the terms *inner fibration* and *inner anodyne map* where Joyal uses *mid-fibration* and *mid-anodyne map*.

- Let $n \geq 0$. We will say that a homotopy type X (described by either a topological space or a Kan complex) is *n-truncated* if the homotopy groups $\pi_i(X, x)$ vanish for every point $x \in X$ and every $i > n$. By convention, we say that X is (-1)-truncated if it is either empty or (weakly) contractible, and (-2)-truncated if X is (weakly) contractible.

- Let $n \geq 0$. We will say that a homotopy type X (described either by a topological space or a Kan complex) is *n-connective* if X is nonempty and the homotopy groups $\pi_i(X, x)$ vanish for every point $x \in X$ and every integer $i < n$. By convention, we will agree that every homotopy type X is (-1)-connective.

- More generally, we will say that a map of homotopy types $f : X \to Y$ is n-truncated (n-connective) if the homotopy fibers of f are n-truncated (n-connective).

Remark. For $n \geq 1$, a homotopy type X is n-connective if and only if it is $(n-1)$-connected (in the usual terminology). In particular, X is 1-connective if and only if it is path-connected.

Warning. In this book, we will often be concerned with sheaves on a topological space X (or some Grothendieck site) taking values in an ∞-category $\mathcal{C}$. The most "universal" case is that in which $\mathcal{C}$ is the ∞-category of $\mathcal{S}$ of spaces. Consequently, the term "sheaf on X" without any other qualifiers will typically refer to a sheaf of spaces on X rather than a sheaf of sets on X. We will see that the collection of all $\mathcal{S}$-valued sheaves on X can be organized into an ∞-category, which we denote by $\mathrm{Shv}(X)$. In particular, $\mathrm{Shv}(X)$ will not denote the ordinary category of set-valued sheaves on X; if we need to consider this latter object, we will denote it by $\mathrm{Shv}_{\mathrm{Set}}(X)$.

ACKNOWLEDGEMENTS

This book would never have come into existence without the advice and encouragement of many people. In particular, I would like to thank Vigleik Angeltveit, Rex Cheung, Vladimir Drinfeld, Matt Emerton, John Francis, Andre Henriques, Nori Minami, James Parson, Steven Sam, David Spivak, and James Wallbridge for many suggestions and corrections which have improved the readability of this book; Andre Joyal, who was kind enough to share with me a preliminary version of his work on the theory of quasi-categories; Charles Rezk, for explaining to me a very conceptual reformulation of the axioms for ∞-topoi (which we will describe in §6.1.3); Bertrand Toën and Gabriele Vezzosi, for many stimulating conversations about their work (which has considerable overlap with the material treated here); Mike Hopkins, for his advice and support throughout my time as a graduate student; Max Lieblich, for offering encouragement during early stages of this project; Josh Nichols-Barrer, for sharing with me some of his ideas about the foundations of higher category theory. I would also like to thank my copyeditor Carol Dean and my editors Kathleen Cioffi, Vickie Kearn, and Anna Pierrehumbert at Princeton Univesity Press for helping to make this the best book that it can be. Finally, I would like to thank the American Institute of Mathematics for supporting me throughout the (seemingly endless) process of writing and revising this work.

Higher Topos Theory

Chapter One

An Overview of Higher Category Theory

This chapter is intended as a general introduction to higher category theory. We begin with what we feel is the most intuitive approach to the subject using *topological categories*. This approach is easy to understand but difficult to work with when one wishes to perform even simple categorical constructions. As a remedy, we will introduce the more suitable formalism of ∞-*categories* (called *weak Kan complexes* in [10] and *quasi-categories* in [43]), which provides a more convenient setting for adaptations of sophisticated category-theoretic ideas. Our goal in §1.1.1 is to introduce both approaches and to explain why they are equivalent to one another. The proof of this equivalence will rely on a crucial result (Theorem 1.1.5.13) which we will prove in §2.2.

Our second objective in this chapter is to give the reader an idea of how to work with the formalism of ∞-categories. In §1.2, we will establish a vocabulary which includes ∞-categorical analogues (often direct generalizations) of most of the important concepts from ordinary category theory. To keep the exposition brisk, we will postpone the more difficult proofs until later chapters of this book. Our hope is that, after reading this chapter, a reader who does not wish to be burdened with the details will be able to understand (at least in outline) some of the more conceptual ideas described in Chapter 5 and beyond.

1.1 FOUNDATIONS FOR HIGHER CATEGORY THEORY

1.1.1 Goals and Obstacles

Recall that a *category* $\mathcal{C}$ consists of the following data:

(1) A collection $\{X, Y, Z, \ldots\}$ whose members are the *objects* of $\mathcal{C}$. We typically write $X \in \mathcal{C}$ to indicate that X is an object of $\mathcal{C}$.

(2) For every pair of objects $X, Y \in \mathcal{C}$, a set $\mathrm{Hom}_{\mathcal{C}}(X, Y)$ of *morphisms* from X to Y. We will typically write $f : X \to Y$ to indicate that $f \in \mathrm{Hom}_{\mathcal{C}}(X, Y)$ and say that *f is a morphism from X to Y*.

(3) For every object $X \in \mathcal{C}$, an *identity morphism* $\mathrm{id}_X \in \mathrm{Hom}_{\mathcal{C}}(X, X)$.

(4) For every triple of objects $X, Y, Z \in \mathcal{C}$, a composition map

$$\mathrm{Hom}_{\mathcal{C}}(X, Y) \times \mathrm{Hom}_{\mathcal{C}}(Y, Z) \to \mathrm{Hom}_{\mathcal{C}}(X, Z).$$

Given morphisms $f : X \to Y$ and $g : Y \to Z$, we will usually denote the image of the pair (f, g) under the composition map by gf or $g \circ f$.

These data are furthermore required to satisfy the following conditions, which guarantee that composition is unital and associative:

(5) For every morphism $f : X \to Y$, we have $\mathrm{id}_Y \circ f = f = f \circ \mathrm{id}_X$ in $\mathrm{Hom}_{\mathcal{C}}(X, Y)$.

(6) For every triple of composable morphisms

$$W \xrightarrow{f} X \xrightarrow{g} Y \xrightarrow{h} Z,$$

we have an equality $h \circ (g \circ f) = (h \circ g) \circ f$ in $\mathrm{Hom}_{\mathcal{C}}(W, Z)$.

The theory of categories has proven to be a valuable organization tool in many areas of mathematics. Mathematical structures of virtually any type can be viewed as the objects of a suitable category $\mathcal{C}$, where the morphisms in $\mathcal{C}$ are given by structure-preserving maps. There is a veritable legion of examples of categories which fit this paradigm:

- The category $\mathcal{S}$et whose objects are sets and whose morphisms are maps of sets.

- The category $\mathcal{G}$rp whose objects are groups and whose morphisms are group homomorphisms.

- The category $\mathcal{T}$op whose objects are topological spaces and whose morphisms are continuous maps.

- The category $\mathcal{C}$at whose objects are (small) categories and whose morphisms are functors. (Recall that a functor F from $\mathcal{C}$ to $\mathcal{D}$ is a map which assigns to each object $C \in \mathcal{C}$ another object $FC \in \mathcal{D}$, and to each morphism $f : C \to C'$ in $\mathcal{C}$ a morphism $F(f) : FC \to FC'$ in $\mathcal{D}$, so that $F(\mathrm{id}_C) = \mathrm{id}_{FC}$ and $F(g \circ f) = F(g) \circ F(f)$.)

- $\cdots$

In general, the existence of a morphism $f : X \to Y$ in a category $\mathcal{C}$ reflects some relationship that exists between the objects $X, Y \in \mathcal{C}$. In some contexts, these relationships themselves become basic objects of study and can be fruitfully organized into categories:

Example 1.1.1.1. Let $\mathcal{G}$rp be the category whose objects are groups and whose morphisms are group homomorphisms. In the theory of groups, one is often concerned only with group homomorphisms *up to conjugacy*. The relation of conjugacy can be encoded as follows: for every pair of groups $G, H \in \mathcal{G}$rp, there is a category $\mathrm{Map}(G, H)$ whose objects are group homomorphisms from G to H (that is, elements of $\mathrm{Hom}_{\mathcal{G}\mathrm{rp}}(G, H)$), where a morphism from $f : G \to H$ to $f' : G \to H$ is an element $h \in H$ such that $hf(g)h^{-1} = f'(g)$ for all $g \in G$. Note that two group homomorphisms $f, f' : G \to H$ are conjugate if and only if they are isomorphic when viewed as objects of $\mathrm{Map}(G, H)$.

Example 1.1.1.2. Let X and Y be topological spaces and let $f_0, f_1 : X \to Y$ be continuous maps. Recall that a *homotopy* from f_0 to f_1 is a continuous map $f : X \times [0, 1] \to Y$ such that $f|X \times \{0\}$ coincides with f_0 and $f|X \times \{1\}$ coincides with f_1. In algebraic topology, one is often concerned not with the category Top of topological spaces but with its *homotopy category*: that is, the category obtained by identifying those pairs of morphisms $f_0, f_1 : X \to Y$ which are homotopic to one another. For many purposes, it is better to do something a little bit more sophisticated: namely, one can form a category $\mathrm{Map}(X, Y)$ whose objects are continuous maps $f : X \to Y$ and whose morphisms are given by (homotopy classes of) homotopies.

Example 1.1.1.3. Given a pair of categories $\mathcal{C}$ and $\mathcal{D}$, the collection of all functors from $\mathcal{C}$ to $\mathcal{D}$ is itself naturally organized into a category $\mathrm{Fun}(\mathcal{C}, \mathcal{D})$, where the morphisms are given by *natural transformations*. (Recall that, given a pair of functors $F, G : \mathcal{C} \to \mathcal{D}$, a natural transformation $\alpha : F \to G$ is a collection of morphisms $\{\alpha_C : F(C) \to G(C)\}_{C \in \mathcal{C}}$ which satisfy the following condition: for every morphism $f : C \to C'$ in $\mathcal{C}$, the diagram

$$
\begin{array}{ccc}
F(C) & \xrightarrow{F(f)} & F(C') \\
\downarrow{\alpha_C} & & \downarrow{\alpha_{C'}} \\
G(C) & \xrightarrow{G(f)} & G(C')
\end{array}
$$

commutes in $\mathcal{D}$.)

In each of these examples, the objects of interest can naturally be organized into what is called a 2-*category* (or *bicategory*): we have not only a collection of objects and a notion of morphisms between objects but also a notion of morphisms between morphisms, which are called 2-*morphisms*. The vision of higher category theory is that there should exist a good notion of n-category for all $n \geq 0$ in which we have not only objects, morphisms, and 2-morphisms but also k-morphisms for all $k \leq n$. Finally, in some sort of limit we might hope to obtain a theory of ∞-categories, where there are morphisms of all orders.

Example 1.1.1.4. Let X be a topological space and $0 \leq n \leq \infty$. We can extract an n-category $\pi_{\leq n} X$ (roughly) as follows. The objects of $\pi_{\leq n} X$ are the points of X. If $x, y \in X$, then the morphisms from x to y in $\pi_{\leq n} X$ are given by continuous paths $[0, 1] \to X$ starting at x and ending at y. The 2-morphisms are given by homotopies of paths, the 3-morphisms by homotopies between homotopies, and so forth. Finally, if $n < \infty$, then two n-morphisms of $\pi_{\leq n} X$ are considered to be the same if and only if they are homotopic to one another.

If $n = 0$, then $\pi_{\leq n} X$ can be identified with the set $\pi_0 X$ of path components of X. If $n = 1$, then our definition of $\pi_{\leq n} X$ agrees with the usual definition for the fundamental groupoid of X. For this reason, $\pi_{\leq n} X$ is often called the *fundamental n-groupoid of X*. It is called an n-*groupoid* (rather than a mere

n-category) because every k-morphism of $\pi_{\leq k}X$ has an inverse (at least up to homotopy).

There are many approaches to realizing the theory of higher categories. We might begin by defining a 2-category to be a "category enriched over Cat." In other words, we consider a collection of objects together with a *category* of morphisms $\mathrm{Hom}(A, B)$ for any two objects A and B and composition *functors* $c_{ABC} : \mathrm{Hom}(A, B) \times \mathrm{Hom}(B, C) \to \mathrm{Hom}(A, C)$ (to simplify the discussion, we will ignore identity morphisms for a moment). These functors are required to satisfy an associative law, which asserts that for any quadruple (A, B, C, D) of objects, the diagram

$$
\begin{array}{ccc}
\mathrm{Hom}(A, B) \times \mathrm{Hom}(B, C) \times \mathrm{Hom}(C, D) & \longrightarrow & \mathrm{Hom}(A, C) \times \mathrm{Hom}(C, D) \\
\downarrow & & \downarrow \\
\mathrm{Hom}(A, B) \times \mathrm{Hom}(B, D) & \longrightarrow & \mathrm{Hom}(A, D)
\end{array}
$$

commutes; in other words, one has an *equality* of functors

$$
c_{ACD} \circ (c_{ABC} \times 1) = c_{ABD} \circ (1 \times c_{BCD})
$$

from $\mathrm{Hom}(A, B) \times \mathrm{Hom}(B, C) \times \mathrm{Hom}(C, D)$ to $\mathrm{Hom}(A, D)$. This leads to the definition of a *strict 2-category*.

At this point, we should object that the definition of a strict 2-category violates one of the basic philosophical principles of category theory: one should never demand that two functors F and F' be equal to one another. Instead one should postulate the existence of a natural isomorphism between F and F'. This means that the associative law should not take the form of an equation but of additional structure: a collection of isomorphisms $\gamma_{ABCD} : c_{ACD} \circ (c_{ABC} \times 1) \simeq c_{ABD} \circ (1 \times c_{BCD})$. We should further demand that the isomorphisms γ_{ABCD} be functorial in the quadruple (A, B, C, D) and satisfy certain higher associativity conditions, which generalize the "Pentagon axiom" described in §A.1.3. After formulating the appropriate conditions, we arrive at the definition of a *weak 2-category*.

Let us contrast the notions of strict 2-category and weak 2-category. The former is easier to define because we do not have to worry about the higher associativity conditions satisfied by the transformations γ_{ABCD}. On the other hand, the latter notion seems more natural if we take the philosophy of category theory seriously. In this case, we happen to be lucky: the notions of strict 2-category and weak 2-category turn out to be equivalent. More precisely, any weak 2-category is equivalent (in the relevant sense) to a strict 2-category. The choice of definition can therefore be regarded as a question of aesthetics.

We now plunge onward to 3-categories. Following the above program, we might define a *strict 3-category* to consist of a collection of objects together with strict 2-categories $\mathrm{Hom}(A, B)$ for any pair of objects A and B, together with a strictly associative composition law. Alternatively, we could seek a definition of *weak 3-category* by allowing $\mathrm{Hom}(A, B)$ to be a weak

2-category, requiring associativity only up to natural 2-isomorphisms, which satisfy higher associativity laws up to natural 3-isomorphisms, which in turn satisfy still higher associativity laws of their own. Unfortunately, it turns out that these notions are *not* equivalent.

Both of these approaches have serious drawbacks. The obvious problem with weak 3-categories is that an explicit definition is extremely complicated (see [33], where a definition is given along these lines), to the point where it is essentially unusable. On the other hand, strict 3-categories have the problem of not being the correct notion: most of the weak 3-categories which occur in nature are not equivalent to strict 3-categories. For example, the fundamental 3-groupoid of the 2-sphere S^2 cannot be described using the language of strict 3-categories. The situation only gets worse (from either point of view) as we pass to 4-categories and beyond.

Fortunately, it turns out that major simplifications can be introduced if we are willing to restrict our attention to ∞-categories in which most of the higher morphisms are invertible. From this point forward, we will use the term (∞, n)-*category* to refer to ∞-categories in which all k-morphisms are invertible for $k > n$. The ∞-categories described in Example 1.1.1.4 (when $n = \infty$) are all $(\infty, 0)$-categories. The converse, which asserts that every $(\infty, 0)$-category has the form $\pi_{\leq \infty} X$ for some topological space X, is a generally accepted principle of higher category theory. Moreover, the ∞-groupoid $\pi_{\leq \infty} X$ encodes the entire homotopy type of X. In other words, $(\infty, 0)$-categories (that is, ∞-categories in which *all* morphisms are invertible) have been extensively studied from another point of view: they are essentially the same thing as "spaces" in the sense of homotopy theory, and there are many equivalent ways to describe them (for example, we can use CW complexes or simplicial sets).

Convention 1.1.1.5. We will sometimes refer to $(\infty, 0)$-categories as ∞-*groupoids* and $(\infty, 2)$-categories as ∞-*bicategories*. Unless we specify otherwise, the generic term "∞-category" will refer to an $(\infty, 1)$-category.

In this book, we will restrict our attention almost entirely to the theory of ∞-categories (in which we have only invertible n-morphisms for $n \geq 2$). Our reasons are threefold:

(1) Allowing noninvertible n-morphisms for $n > 1$ introduces a number of additional complications to the theory at both technical and conceptual levels. As we will see throughout this book, many ideas from category theory generalize to the ∞-categorical setting in a natural way. However, these generalizations are not so straightforward if we allow noninvertible 2-morphisms. For example, one must distinguish between strict and lax fiber products, even in the setting of "classical" 2-categories.

(2) For the applications studied in this book, we will not need to consider (∞, n)-categories for $n > 2$. The case $n = 2$ is of some relevance

because the collection of (small) ∞-categories can naturally be viewed as a (large) ∞-bicategory. However, we will generally be able to exploit this structure in an ad hoc manner without developing any general theory of ∞-bicategories.

(3) For $n > 1$, the theory of (∞, n)-categories is most naturally viewed as a special case of *enriched* (higher) category theory. Roughly speaking, an n-category can be viewed as a category enriched over $(n-1)$-categories. As we explained above, this point of view is inadequate because it requires that composition satisfies an associative law up to equality, while in practice the associativity holds only up to isomorphism or some weaker notion of equivalence. In other words, to obtain the correct definition we need to view the collection of $(n-1)$-categories as an n-category, not as an ordinary category. Consequently, the naive approach is circular: though it does lead to a good theory of n-categories, we can make sense of it only if the theory of n-categories is already in place.

Thinking along similar lines, we can view an (∞, n)-category as an ∞-category which is *enriched over* $(\infty, n-1)$-*categories*. The collection of $(\infty, n-1)$-categories is itself organized into an (∞, n)-category $\mathrm{Cat}_{(\infty, n-1)}$, so at a first glance this definition suffers from the same problem of circularity. However, because the associativity properties of composition are required to hold up to *equivalence*, rather than up to arbitrary natural transformation, the noninvertible k-morphisms in $\mathrm{Cat}_{(\infty, n-1)}$ are irrelevant for $k > 1$. One can define an (∞, n)-category to be a category enriched over $\mathrm{Cat}_{(\infty, n-1)}$, where the latter is regarded as an ∞-category by discarding noninvertible k-morphisms for $2 \leq k \leq n$. In other words, the naive inductive definition of higher category theory is reasonable *provided that we work in the ∞-categorical setting from the outset.* We refer the reader to [75] for a definition of n-categories which follows this line of thought.

The theory of *enriched* ∞-categories is a useful and important one but will not be treated in this book. Instead we refer the reader to [50] for an introduction using the same language and formalism we employ here.

Though we will not need a theory of (∞, n)-categories for $n > 1$, the case $n = 1$ is the main subject matter of this book. Fortunately, the above discussion suggests a definition. Namely, an ∞-category $\mathcal{C}$ should consist of a collection of objects and an ∞-groupoid $\mathrm{Map}_{\mathcal{C}}(X, Y)$ for every pair of objects $X, Y \in \mathcal{C}$. These ∞-groupoids can be identified with topological spaces, and should be equipped with an associative composition law. As before, we are faced with two choices as to how to make this precise: do we require associativity on the nose or only up to (coherent) homotopy? Fortunately, the answer turns out to be irrelevant: as in the theory of 2-categories, any ∞-category with a coherently associative multiplication can be replaced by an equivalent ∞-category with a strictly associative multiplication. We are led to the following:

Definition 1.1.1.6. A *topological category* is a category which is enriched over $\mathcal{CG}$, the category of compactly generated (and weakly Hausdorff) topological spaces. The category of topological categories will be denoted by $\mathcal{Cat}_{\text{top}}$.

More explicitly, a topological category $\mathcal{C}$ consists of a collection of objects together with a (compactly generated) topological space $\text{Map}_{\mathcal{C}}(X, Y)$ for any pair of objects $X, Y \in \mathcal{C}$. These mapping spaces must be equipped with an associative composition law given by continuous maps

$$\text{Map}_{\mathcal{C}}(X_0, X_1) \times \text{Map}_{\mathcal{C}}(X_1, X_2) \times \cdots \times \text{Map}_{\mathcal{C}}(X_{n-1}, X_n) \to \text{Map}_{\mathcal{C}}(X_0, X_n)$$

(defined for all $n \geq 0$). Here the product is taken in the category of compactly generated topological spaces.

Remark 1.1.1.7. The decision to work with compactly generated topological spaces, rather than arbitrary spaces, is made in order to facilitate the comparison with more combinatorial approaches to homotopy theory. This is a purely technical point which the reader may safely ignore.

It is possible to use Definition 1.1.1.6 as a foundation for higher category theory: that is, to *define* an ∞-category to be a topological category. However, this approach has a number of technical disadvantages. We will describe an alternative (though equivalent) formalism in the next section.

1.1.2 ∞-Categories

Of the numerous formalizations of higher category theory, Definition 1.1.1.6 is the quickest and most transparent. However, it is one of the most difficult to actually work with: many of the basic constructions of higher category theory give rise most naturally to $(\infty, 1)$-categories for which the composition of morphisms is associative only up to (coherent) homotopy (for several examples of this phenomenon, we refer the reader to §1.2). In order to remain in the world of topological categories, it is necessary to combine these constructions with a "straightening" procedure which produces a strictly associative composition law. Although it is always possible to do this (see Theorem 2.2.5.1), it is much more technically convenient to work from the outset within a more flexible theory of $(\infty, 1)$-categories. Fortunately, there are many candidates for such a theory, including the theory of Segal categories ([71]), the theory of complete Segal spaces ([64]), and the theory of model categories ([40], [38]). To review all of these notions and their interrelationships would involve too great a digression from the main purpose of this book. However, the frequency with which we will encounter sophisticated categorical constructions necessitates the use of *one* of these more efficient approaches. We will employ the theory of *weak Kan complexes*, which goes back to Boardman-Vogt ([10]). These objects have subsequently been studied more extensively by Joyal ([43], [44]), who calls them *quasi-categories*. We will simply call them ∞-*categories*.

To get a feeling for what an ∞-category $\mathcal{C}$ should be, it is useful to consider two extreme cases. If *every* morphism in $\mathcal{C}$ is invertible, then $\mathcal{C}$ is equivalent to the fundamental ∞-groupoid of a topological space X. In this case, higher category theory reduces to classical homotopy theory. On the other hand, if $\mathcal{C}$ has no nontrivial n-morphisms for $n > 1$, then $\mathcal{C}$ is equivalent to an ordinary category. A general formalism must capture the features of both of these examples. In other words, we need a class of mathematical objects which can behave both like categories and like topological spaces. In §1.1.1, we achieved this by "brute force": namely, we directly amalgamated the theory of topological spaces and the theory of categories by considering topological categories. However, it is possible to approach the problem more directly using the theory of *simplicial sets*. We will assume that the reader has some familiarity with the theory of simplicial sets; a brief review of this theory is included in §A.2.7, and a more extensive introduction can be found in [32].

The theory of simplicial sets originated as a combinatorial approach to homotopy theory. Given any topological space X, one can associate a simplicial set $\mathrm{Sing}\,X$, whose n-simplices are precisely the continuous maps $|\Delta^n| \to X$, where $|\Delta^n| = \{(x_0, \ldots, x_n) \in [0,1]^{n+1} | x_0 + \ldots + x_n = 1\}$ is the standard n-simplex. Moreover, the topological space X is *determined*, up to weak homotopy equivalence, by $\mathrm{Sing}\,X$. More precisely, the singular complex functor $X \mapsto \mathrm{Sing}\,X$ admits a left adjoint, which carries every simplicial set K to its *geometric realization* $|K|$. For every topological space X, the counit map $|\mathrm{Sing}\,X| \to X$ is a weak homotopy equivalence. Consequently, if one is only interested in studying topological spaces up to weak homotopy equivalence, one might as well work with simplicial sets instead.

If X is a topological space, then the simplicial set $\mathrm{Sing}\,X$ has an important property, which is captured by the following definition:

Definition 1.1.2.1. Let K be a simplicial set. We say that K is a *Kan complex* if, for any $0 \le i \le n$ and any diagram of solid arrows

there exists a dotted arrow as indicated rendering the diagram commutative. Here $\Lambda_i^n \subseteq \Delta^n$ denotes the ith horn, obtained from the simplex Δ^n by deleting the interior and the face opposite the ith vertex.

The singular complex of any topological space X is a Kan complex: this follows from the fact that the horn $|\Lambda_i^n|$ is a retract of the simplex $|\Delta^n|$ in the category of topological spaces. Conversely, any Kan complex K "behaves like" a space: for example, there are simple combinatorial recipes for extracting homotopy groups from K (which turn out be isomorphic to the homotopy groups of the topological space $|K|$). According to a theorem of Quillen (see [32] for a proof), the singular complex and geometric realization provide mutually inverse equivalences between the homotopy category of CW complexes and the homotopy category of Kan complexes.

The formalism of simplicial sets is also closely related to category theory. To any category $\mathcal{C}$, we can associate a simplicial set $N(\mathcal{C})$ called the *nerve* of $\mathcal{C}$. For each $n \geq 0$, we let $N(\mathcal{C})_n = \mathrm{Map}_{\mathrm{Set}_\Delta}(\Delta^n, N(\mathcal{C}))$ denote the set of all functors $[n] \to \mathcal{C}$. Here $[n]$ denotes the linearly ordered set $\{0, \ldots, n\}$, regarded as a category in the obvious way. More concretely, $N(\mathcal{C})_n$ is the set of all composable sequences of morphisms

$$C_0 \xrightarrow{f_1} C_1 \xrightarrow{f_2} \cdots \xrightarrow{f_n} C_n$$

having length n. In this description, the face map d_i carries the above sequence to

$$C_0 \xrightarrow{f_1} \cdots \xrightarrow{f_{i-1}} C_{i-1} \xrightarrow{f_{i+1} \circ f_i} C_{i+1} \xrightarrow{f_{i+2}} \cdots \xrightarrow{f_n} C_n$$

while the degeneracy s_i carries it to

$$C_0 \xrightarrow{f_1} \cdots \xrightarrow{f_i} C_i \xrightarrow{\mathrm{id}_{C_i}} C_i \xrightarrow{f_{i+1}} C_{i+1} \xrightarrow{f_{i+2}} \cdots \xrightarrow{f_n} C_n.$$

It is more or less clear from this description that the simplicial set $N(\mathcal{C})$ is just a fancy way of encoding the structure of $\mathcal{C}$ as a category. More precisely, we note that the category $\mathcal{C}$ can be recovered (up to isomorphism) from its nerve $N(\mathcal{C})$. The objects of $\mathcal{C}$ are simply the *vertices* of $N(\mathcal{C})$: that is, the elements of $N(\mathcal{C})_0$. A morphism from C_0 to C_1 is given by an edge $\phi \in N(\mathcal{C})_1$ with $d_1(\phi) = C_0$ and $d_0(\phi) = C_1$. The identity morphism from an object C to itself is given by the degenerate simplex $s_0(C)$. Finally, given a diagram $C_0 \xrightarrow{\phi} C_1 \xrightarrow{\psi} C_2$, the edge of $N(\mathcal{C})$ corresponding to $\psi \circ \phi$ may be uniquely characterized by the fact that there exists a 2-simplex $\sigma \in N(\mathcal{C})_2$ with $d_2(\sigma) = \phi$, $d_0(\sigma) = \psi$, and $d_1(\sigma) = \psi \circ \phi$.

It is not difficult to characterize those simplicial sets which arise as the nerve of a category:

Proposition 1.1.2.2. *Let K be a simplicial set. Then the following conditions are equivalent:*

(1) *There exists a small category $\mathcal{C}$ and an isomorphism $K \simeq N(\mathcal{C})$.*

(2) *For each $0 < i < n$ and each diagram*

there exists a unique *dotted arrow rendering the diagram commutative.*

Proof. We first show that $(1) \Rightarrow (2)$. Let K be the nerve of a small category $\mathcal{C}$ and let $f_0 : \Lambda_i^n \to K$ be a map of simplicial sets, where $0 < i < n$. We wish to show that f_0 can be extended uniquely to a map $f : \Delta^n \to K$. For $0 \leq k \leq n$, let $X_k \in \mathcal{C}$ be the image of the vertex $\{k\} \subseteq \Lambda_i^n$. For $0 < k \leq n$, let $g_k : X_{k-1} \to X_k$ be the morphism in $\mathcal{C}$ determined by the restriction $f_0 | \Delta^{\{k-1,k\}}$. The composable chain of morphisms

$$X_0 \xrightarrow{g_1} X_1 \xrightarrow{g_2} \cdots \xrightarrow{g_n} X_n$$

determines an n-simplex $f : \Delta^n \to K$. We will show that f is the desired solution to our extension problem (the uniqueness of this solution is evident: if $f' : \Delta^n \to K$ is any other map with $f' | \Lambda_i^n = f_0$, then f' must correspond to the same chain of morphisms in $\mathcal{C}$, so that $f' = f$). It will suffice to prove the following for every $0 \leq j \leq n$:

$(*_j)$ If $j \neq i$, then

$$f | \Delta^{\{0,\ldots,j-1,j+1,\ldots,n\}} = f_0 | \Delta^{\{0,\ldots,j-1,j+1,\ldots,n\}}.$$

To prove $(*_j)$, it will suffice to show that f and f_0 have the same restriction to $\Delta^{\{k,k'\}}$, where k and k' are adjacent elements of the linearly ordered set $\{0,\ldots,j-1,j+1,\ldots,n\} \subseteq [n]$. If k and k' are adjacent in $[n]$, then this follows by construction. In particular, $(*)$ is automatically satisfied if $j = 0$ or $j = n$. Suppose instead that $k = j - 1$ and $k' = j + 1$, where $0 < j < n$. If $n = 2$, then $j = 1 = i$ and we obtain a contradiction. We may therefore assume that $n > 2$, so that either $j - 1 > 0$ or $j + 1 < n$. Without loss of generality, $j - 1 > 0$, so that $\Delta^{\{j-1,j+1\}} \subseteq \Delta^{\{1,\ldots,n\}}$. The desired conclusion now follows from $(*_0)$.

We now prove the converse. Suppose that the simplicial set K satisfies (2); we claim that K is isomorphic to the nerve of a small category $\mathcal{C}$. We construct the category $\mathcal{C}$ as follows:

(i) The objects of $\mathcal{C}$ are the vertices of K.

(ii) Given a pair of objects $x, y \in \mathcal{C}$, we let $\mathrm{Hom}_{\mathcal{C}}(x, y)$ denote the collection of all edges $e : \Delta^1 \to K$ such that $e | \{0\} = x$ and $e | \{1\} = y$.

(iii) Let x be an object of $\mathcal{C}$. Then the identity morphism id_x is the edge of K defined by the composition

$$\Delta^1 \to \Delta^0 \xrightarrow{e} K.$$

(iv) Let $f : x \to y$ and $g : y \to z$ be morphisms in $\mathcal{C}$. Then f and g together determine a map $\sigma_0 : \Lambda_1^2 \to K$. In view of condition (2), the map σ_0 can be extended uniquely to a 2-simplex $\sigma : \Delta^2 \to K$. We define the composition $g \circ f$ to be the morphism from x to z in $\mathcal{C}$ corresponding to the edge given by the composition

$$\Delta^1 \simeq \Delta^{\{0,2\}} \subseteq \Delta^2 \xrightarrow{\sigma} K.$$

We first claim that $\mathcal{C}$ is a category. To prove this, we must verify the following axioms:

(a) For every object $y \in \mathcal{C}$, the identity id_y is a unit with respect to composition. In other words, for every morphism $f : x \to y$ in $\mathcal{C}$ and every morphism $g : y \to z$ in $\mathcal{C}$, we have $\mathrm{id}_y \circ f = f$ and $g \circ \mathrm{id}_y = g$. These equations are "witnessed" by the 2-simplices $s_1(f), s_0(g) \in \mathrm{Hom}_{\mathrm{Set}_\Delta}(\Delta^2, K)$.

(b) Composition is associative. That is, for every sequence of composable morphisms
$$w \xrightarrow{f} x \xrightarrow{g} y \xrightarrow{h} z,$$
we have $h \circ (g \circ f) = (h \circ g) \circ f$. To prove this, let us first choose 2-simplices σ_{012} and σ_{123} as indicated below:

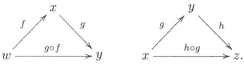

Now choose a 2-simplex σ_{023} corresponding to a diagram

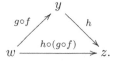

These three 2-simplices together define a map $\tau_0 : \Lambda_2^3 \to K$. Since K satisfies condition (2), we can extend τ_0 to a 3-simplex $\tau : \Delta^3 \to K$. The composition
$$\Delta^2 \simeq \Delta^{\{0,1,3\}} \subseteq \Delta^3 \xrightarrow{\tau} K$$
corresponds to the diagram

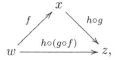

which witnesses the associativity axiom $h \circ (g \circ f) = (h \circ g) \circ f$.

It follows that $\mathcal{C}$ is a well-defined category. By construction, we have a canonical map of simplicial sets $\phi : K \to \mathrm{N}\,\mathcal{C}$. To complete the proof, it will suffice to show that ϕ is an isomorphism. We will prove, by induction on $n \geq 0$, that ϕ induces a bijection $\mathrm{Hom}_{\mathrm{Set}_\Delta}(\Delta^n, K) \to \mathrm{Hom}_{\mathrm{Set}_\Delta}(\Delta^n, \mathrm{N}\,\mathcal{C})$. For $n = 0$ and $n = 1$, this is obvious from the construction. Assume therefore that $n \geq 2$ and choose an integer i such that $0 < i < n$. We have a commutative diagram

$$\begin{array}{ccc}
\mathrm{Hom}_{\mathrm{Set}_\Delta}(\Delta^n, K) & \longrightarrow & \mathrm{Hom}_{\mathrm{Set}_\Delta}(\Delta^n, \mathrm{N}\,\mathcal{C}) \\
\downarrow & & \downarrow \\
\mathrm{Hom}_{\mathrm{Set}_\Delta}(\Lambda_i^n, K) & \longrightarrow & \mathrm{Hom}_{\mathrm{Set}_\Delta}(\Lambda_i^n, \mathrm{N}\,\mathcal{C}).
\end{array}$$

Since K and $N\,\mathcal{C}$ both satisfy (2) (for $N\,\mathcal{C}$, this follows from the first part of the proof), the vertical maps are bijective. It will therefore suffice to show that the lower horizontal map is bijective, which follows from the inductive hypothesis.   □

We note that condition (2) of Proposition 1.1.2.2 is very similar to Definition 1.1.2.1. However, it is different in two important respects. First, it requires the extension condition only for *inner* horns Λ_i^n with $0 < i < n$. Second, the asserted condition is stronger in this case: not only does any map $\Lambda_i^n \to K$ extend to the simplex Δ^n, but the extension is unique.

Remark 1.1.2.3. It is easy to see that it is not reasonable to expect condition (2) of Proposition 1.1.2.2 to hold for *outer* horns Λ_i^n where $i \in \{0, n\}$. Consider, for example, the case where $i = n = 2$ and where K is the nerve of a category $\mathcal{C}$. Giving a map $\Lambda_2^2 \to K$ corresponds to supplying the solid arrows in the diagram

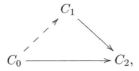

and the extension condition would amount to the assertion that one could always find a dotted arrow rendering the diagram commutative. This is true in general only when the category $\mathcal{C}$ is a *groupoid*.

We now see that the notion of a simplicial set is a flexible one: a simplicial set K can be a good model for an ∞-groupoid (if K is a Kan complex) or for an ordinary category (if it satisfies the hypotheses of Proposition 1.1.2.2). Based on these observations, we might expect that some more general class of simplicial sets could serve as models for ∞-categories in general.

Consider first an arbitrary simplicial set K. We can try to envision K as a generalized category whose objects are the vertices of K (that is, the elements of K_0) and whose morphisms are the edges of K (that is, the elements of K_1). A 2-simplex $\sigma : \Delta^2 \to K$ should be thought of as a diagram

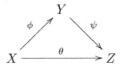

together with an identification (or homotopy) between θ and $\psi \circ \phi$ which witnesses the "commutativity" of the diagram. (In higher category theory, commutativity is not merely a condition: the homotopy $\theta \simeq \psi \circ \phi$ is an additional datum.) Simplices of larger dimension may be thought of as verifying the commutativity of certain higher-dimensional diagrams.

Unfortunately, for a general simplicial set K, the analogy outlined above is not very strong. The essence of the problem is that, though we may refer to the 1-simplices of K as morphisms, there is in general no way to compose them. Taking our cue from the example of $N(\mathcal{C})$, we might say that a morphism $\theta : X \to Z$ is a composition of morphisms $\phi : X \to Y$ and $\psi : Y \to Z$ if there exists a 2-simplex $\sigma : \Delta^2 \to K$ as in the diagram indicated above. We must now consider two potential difficulties: the desired 2-simplex σ may not exist, and if it does it exist it may not be unique, so that we have more than one choice for the composition θ.

The existence requirement for σ can be formulated as an extension condition on the simplicial set K. We note that a composable pair of morphisms (ψ, ϕ) determines a map of simplicial sets $\Lambda^2_1 \to K$. Thus, the assertion that σ can always be found may be formulated as a extension property: any map of simplicial sets $\Lambda^2_1 \to K$ can be extended to Δ^2, as indicated in the following diagram:

The uniqueness of θ is another matter. It turns out to be unnecessary (and unnatural) to require that θ be uniquely determined. To understand this point, let us return to the example of the fundamental groupoid of a topological space X. This is a category whose objects are the points $x \in X$. The morphisms between a point $x \in X$ and a point $y \in X$ are given by continuous paths $p : [0,1] \to X$ such that $p(0) = x$ and $p(1) = y$. Two such paths are considered to be equivalent if there is a homotopy between them. Composition in the fundamental groupoid is given by concatenation of paths. Given paths $p, q : [0,1] \to X$ with $p(0) = x$, $p(1) = q(0) = y$, and $q(1) = z$, the composite of p and q should be a path joining x to z. There are many ways of obtaining such a path from p and q. One of the simplest is to define

$$r(t) = \begin{cases} p(2t) & \text{if } 0 \leq t \leq \frac{1}{2} \\ q(2t - 1) & \text{if } \frac{1}{2} \leq t \leq 1. \end{cases}$$

However, we could just as well use the formula

$$r'(t) = \begin{cases} p(3t) & \text{if } 0 \leq t \leq \frac{1}{3} \\ q(\frac{3t-1}{2}) & \text{if } \frac{1}{3} \leq t \leq 1 \end{cases}$$

to define the composite path. Because the paths r and r' are homotopic to one another, it does not matter which one we choose.

The situation becomes more complicated if we try to think 2-categorically. We can capture more information about the space X by considering its *fundamental 2-groupoid*. This is a 2-category whose objects are the points of X, whose morphisms are paths between points, and whose 2-morphisms are given by homotopies between paths (which are themselves considered modulo homotopy). In order to have composition of morphisms unambiguously defined, we would have to choose some formula once and for all. Moreover, there is no particularly compelling choice; for example, neither of the formulas written above leads to a strictly associative composition law.

The lesson to learn from this is that in higher-categorical situations, we should not necessarily ask for a uniquely determined composition of two morphisms. In the fundamental groupoid example, there are many choices for a composite path, but all of them are homotopic to one another. Moreover, in keeping with the philosophy of higher category theory, *any* path which is homotopic to the composite should be just as good as the composite itself. From this point of view, it is perhaps more natural to view composition as a relation than as a function, and this is very efficiently encoded in the formalism of simplicial sets: a 2-simplex $\sigma : \Delta^2 \to K$ should be viewed as "evidence" that $d_0(\sigma) \circ d_2(\sigma)$ is homotopic to $d_1(\sigma)$.

Exactly what conditions on a simplicial set K will guarantee that it behaves like a higher category? Based on the above argument, it seems reasonable to require that K satisfy an extension condition with respect to certain horn inclusions Λ^n_i, as in Definition 1.1.2.1. However, as we observed in Remark 1.1.2.3, this is reasonable only for the inner horns where $0 < i < n$, which appear in the statement of Proposition 1.1.2.2.

Definition 1.1.2.4. An ∞-*category* is a simplicial set K which has the following property: for any $0 < i < n$, any map $f_0 : \Lambda^n_i \to K$ admits an extension $f : \Delta^n \to K$.

Definition 1.1.2.4 was first formulated by Boardman and Vogt ([10]). They referred to ∞-catgories as *weak Kan complexes*, motivated by the obvious analogy with Definition 1.1.2.1. Our terminology places more emphasis on the analogy with the characterization of ordinary categories given in Proposition 1.1.2.2: we require the same extension conditions but drop the uniqueness assumption.

Example 1.1.2.5. Any Kan complex is an ∞-category. In particular, if X is a topological space, then we may view its singular complex $\mathrm{Sing}\, X$ as an ∞-category: this is one way of defining the fundamental ∞-groupoid $\pi_{\leq \infty} X$ of X introduced informally in Example 1.1.1.4.

Example 1.1.2.6. The nerve of any category is an ∞-category. We will occasionally abuse terminology by identifying a category $\mathcal{C}$ with its nerve $\mathrm{N}(\mathcal{C})$; by means of this identification, we may view ordinary category theory as a special case of the study of ∞-categories.

The weak Kan condition of Definition 1.1.2.4 leads to a very elegant and powerful version of higher category theory. This theory has been developed by Joyal in [43] and [44] (where simplicial sets satisfying the condition of Definition 1.1.2.4 are called *quasi-categories*) and will be used throughout this book.

Notation 1.1.2.7. Depending on the context, we will use two different notations in connection with simplicial sets. When emphasizing their role as

∞-categories, we will often denote them by calligraphic letters such as $\mathcal{C}$, $\mathcal{D}$, and so forth. When casting simplicial sets in their different (though related) role as representatives of homotopy types, we will employ capital Roman letters. To avoid confusion, we will also employ the latter notation when we wish to contrast the theory of ∞-categories with some other other approach to higher category theory, such as the theory of topological categories.

1.1.3 Equivalences of Topological Categories

We have now introduced two approaches to higher category theory: one based on topological categories and one based on simplicial sets. These two approaches turn out to be equivalent to one another. However, the equivalence itself needs to be understood in a higher-categorical sense. We take our cue from classical homotopy theory, in which we can take the basic objects to be either topological spaces or simplicial sets. It is not true that every Kan complex is isomorphic to the singular complex of a topological space or that every CW complex is homeomorphic to the geometric realization of a simplicial set. However, both of these statements become true if we replace the words "isomorphic to" by "homotopy equivalent to." We would like to formulate a similar statement regarding our approaches to higher category theory. The first step is to find a concept which replaces homotopy equivalence. If $F : \mathcal{C} \to \mathcal{D}$ is a functor between topological categories, under what circumstances should we regard F as an equivalence (so that $\mathcal{C}$ and $\mathcal{D}$ really represent the same higher category)?

The most naive answer is that F should be regarded as an equivalence if it is an isomorphism of topological categories. This means that F induces a bijection between the objects of $\mathcal{C}$ and the objects of $\mathcal{D}$, and a homeomorphism $\mathrm{Map}_{\mathcal{C}}(X, Y) \to \mathrm{Map}_{\mathcal{D}}(F(X), F(Y))$ for every pair of objects $X, Y \in \mathcal{C}$. However, it is immediately obvious that this condition is far too strong; for example, in the case where $\mathcal{C}$ and $\mathcal{D}$ are ordinary categories (which we may view also as topological categories where all morphism sets are endowed with the discrete topology), we recover the notion of an isomorphism between categories. This notion does not play an important role in category theory. One rarely asks whether or not two categories are isomorphic; instead, one asks whether or not they are equivalent. This suggests the following definition:

Definition 1.1.3.1. A functor $F : \mathcal{C} \to \mathcal{D}$ between topological categories is a *strong equivalence* if it is an equivalence in the sense of enriched category theory. In other words, F is a strong equivalence if it induces homeomorphisms $\mathrm{Map}_{\mathcal{C}}(X, Y) \to \mathrm{Map}_{\mathcal{D}}(F(X), F(Y))$ for every pair of objects $X, Y \in \mathcal{C}$, and every object of $\mathcal{D}$ is isomorphic (in $\mathcal{D}$) to $F(X)$ for some $X \in \mathcal{C}$.

The notion of strong equivalence between topological categories has the virtue that, when restricted to ordinary categories, it reduces to the usual notion of equivalence. However, it is still not the right definition: for a pair of objects X and Y of a higher category $\mathcal{C}$, the morphism space $\mathrm{Map}_{\mathcal{C}}(X, Y)$ should itself be well-defined only up to homotopy equivalence.

Definition 1.1.3.2. Let $\mathcal{C}$ be a topological category. The *homotopy category* h$\mathcal{C}$ is defined as follows:

- The objects of h$\mathcal{C}$ are the objects of $\mathcal{C}$.

- If $X, Y \in \mathcal{C}$, then we define $\mathrm{Hom}_{\mathrm{h}\mathcal{C}}(X, Y) = \pi_0 \mathrm{Map}_{\mathcal{C}}(X, Y)$.

- Composition of morphisms in h$\mathcal{C}$ is induced from the composition of morphisms in $\mathcal{C}$ by applying the functor π_0.

Example 1.1.3.3. Let $\mathcal{C}$ be the topological category whose objects are CW complexes, where $\mathrm{Map}_{\mathcal{C}}(X, Y)$ is the set of continuous maps from X to Y, equipped with the (compactly generated version of the) compact-open topology. We will denote the homotopy category of $\mathcal{C}$ by $\mathcal{H}$ and refer to $\mathcal{H}$ as the *homotopy category of spaces*.

There is a second construction of the homotopy category $\mathcal{H}$ which will play an important role in what follows. First, we must recall a bit of terminology from classical homotopy theory.

Definition 1.1.3.4. A map $f : X \to Y$ between topological spaces is said to be a *weak homotopy equivalence* if it induces a bijection $\pi_0 X \to \pi_0 Y$, and if for every point $x \in X$ and every $i \geq 1$, the induced map of homotopy groups

$$\pi_i(X, x) \to \pi_i(Y, f(x))$$

is an isomorphism.

Given a space $X \in \mathcal{CG}$, classical homotopy theory ensures the existence of a CW complex X' equipped with a weak homotopy equivalence $\phi : X' \to X$. Of course, X' is not uniquely determined; however, it is unique up to canonical homotopy equivalence, so that the assignment

$$X \mapsto [X] = X'$$

determines a functor $\theta : \mathcal{CG} \to \mathcal{H}$. By construction, θ carries weak homotopy equivalences in $\mathcal{CG}$ to isomorphisms in $\mathcal{H}$. In fact, θ is universal with respect to this property. In other words, we may describe $\mathcal{H}$ as the category obtained from $\mathcal{CG}$ by formally inverting all weak homotopy equivalences. This is one version of Whitehead's theorem, which is usually stated as follows: every weak homotopy equivalence between CW complexes admits a homotopy inverse.

We can now improve upon Definition 1.1.3.2 slightly. We first observe that the functor $\theta : \mathcal{CG} \to \mathcal{H}$ preserves products. Consequently, we can apply the construction of Remark A.1.4.3 to convert any topological category $\mathcal{C}$ into a category enriched over $\mathcal{H}$. We will denote this $\mathcal{H}$-enriched category by h$\mathcal{C}$ and refer to it as the *homotopy category* of $\mathcal{C}$. More concretely, the homotopy category h$\mathcal{C}$ may be described as follows:

(1) The objects of $h\mathcal{C}$ are the objects of $\mathcal{C}$.

(2) For $X, Y \in \mathcal{C}$, we have

$$\mathrm{Map}_{h\mathcal{C}}(X, Y) = [\mathrm{Map}_{\mathcal{C}}(X, Y)].$$

(3) The composition law on $h\mathcal{C}$ is obtained from the composition law on $\mathcal{C}$ by applying the functor $\theta : \mathcal{CG} \to \mathcal{H}$.

Remark 1.1.3.5. If $\mathcal{C}$ is a topological category, we have now defined $h\mathcal{C}$ in two different ways: first as an ordinary category and later as a category enriched over $\mathcal{H}$. These two definitions are compatible with one another in the sense that $h\mathcal{C}$ (regarded as an ordinary category) is the underlying category of $h\mathcal{C}$ (regarded as an $\mathcal{H}$-enriched category). This follows immediately from the observation that for every topological space X, there is a canonical bijection $\pi_0 X \simeq \mathrm{Map}_{\mathcal{H}}(*, [X])$.

If $\mathcal{C}$ is a topological category, we may imagine that $h\mathcal{C}$ is the object which is obtained by forgetting the topological morphism spaces of $\mathcal{C}$ and remembering only their (weak) homotopy types. The following definition codifies the idea that these homotopy types should be "all that really matter."

Definition 1.1.3.6. Let $F : \mathcal{C} \to \mathcal{D}$ be a functor between topological categories. We will say that F is a *weak equivalence*, or simply an *equivalence*, if the induced functor $h\mathcal{C} \to h\mathcal{D}$ is an equivalence of $\mathcal{H}$-enriched categories.

More concretely, a functor F is an equivalence if and only if the following conditions are satisfied:

- For every pair of objects $X, Y \in \mathcal{C}$, the induced map

$$\mathrm{Map}_{\mathcal{C}}(X, Y) \to \mathrm{Map}_{\mathcal{D}}(F(X), F(Y))$$

  is a weak homotopy equivalence of topological spaces.

- Every object of $\mathcal{D}$ is isomorphic in $h\mathcal{D}$ to $F(X)$ for some $X \in \mathcal{C}$.

Remark 1.1.3.7. A morphism $f : X \to Y$ in $\mathcal{D}$ is said to be an *equivalence* if the induced morphism in $h\mathcal{D}$ is an isomorphism. In general, this is much weaker than the condition that f be an isomorphism in $\mathcal{D}$; see Proposition 1.2.4.1.

It is Definition 1.1.3.6 which gives the correct notion of equivalence between topological categories (at least, when one is using them to describe higher category theory). We will agree that all relevant properties of topological categories are invariant under this notion of equivalence. We say that two topological categories are *equivalent* if there is an equivalence between them, or more generally if there is a chain of equivalences joining them. Equivalent topological categories should be regarded as interchangeable for all relevant purposes.

Remark 1.1.3.8. According to Definition 1.1.3.6, a functor $F : \mathcal{C} \to \mathcal{D}$ is an equivalence if and only if the induced functor $h\mathcal{C} \to h\mathcal{D}$ is an equivalence. In other words, the homotopy category $h\mathcal{C}$ (regarded as a category which is enriched over $\mathcal{H}$) is an invariant of $\mathcal{C}$ which is sufficiently powerful to detect equivalences between ∞-categories. This should be regarded as analogous to the more classical fact that the homotopy groups $\pi_i(X, x)$ of a CW complex X are homotopy invariants which detect homotopy equivalences between CW complexes (by Whitehead's theorem). However, it is important to remember that $h\mathcal{C}$ does not determine $\mathcal{C}$ up to equivalence, just as the homotopy type of a CW complex is not determined by its homotopy groups.

1.1.4 Simplicial Categories

In the previous sections we introduced two very different approaches to the foundations of higher category theory: one based on topological categories, the other on simplicial sets. In order to prove that they are equivalent to one another, we will introduce a third approach which is closely related to the first but shares the combinatorial flavor of the second.

Definition 1.1.4.1. A *simplicial category* is a category which is enriched over the category Set_Δ of simplicial sets. The category of simplicial categories (where morphisms are given by simplicially enriched functors) will be denoted by Cat_Δ.

Remark 1.1.4.2. Every simplicial category can be regarded as a simplicial object in the category Cat. Conversely, a simplicial object of Cat arises from a simplicial category if and only if the underlying simplicial set of objects is constant.

Like topological categories, simplicial categories can be used as models of higher category theory. If $\mathcal{C}$ is a simplicial category, then we will generally think of the simplicial sets $\mathrm{Map}_\mathcal{C}(X, Y)$ as encoding homotopy types or ∞-groupoids.

Remark 1.1.4.3. If $\mathcal{C}$ is a simplicial category with the property that each of the simplicial sets $\mathrm{Map}_\mathcal{C}(X, Y)$ is an ∞-category, then we may view $\mathcal{C}$ itself as a kind of ∞-bicategory. We will not use this interpretation of simplicial categories in this book. Usually we will consider only *fibrant* simplicial categories; that is, simplicial categories for which the mapping objects $\mathrm{Map}_\mathcal{C}(X, Y)$ are Kan complexes.

The relationship between simplicial categories and topological categories is easy to describe. Let Set_Δ denote the category of simplicial sets and $\mathcal{CG}$ the category of compactly generated Hausdorff spaces. We recall that there exists a pair of adjoint functors

$$\mathrm{Set}_\Delta \overset{|\ |}{\underset{\mathrm{Sing}}{\rightleftarrows}} \mathcal{CG}$$

which are called the *geometric realization* and *singular complex* functors, respectively. Both of these functors commute with finite products. Consequently, if $\mathcal{C}$ is a simplicial category, we may define a topological category $|\mathcal{C}|$ in the following way:

- The objects of $|\mathcal{C}|$ are the objects of $\mathcal{C}$.

- If $X, Y \in \mathcal{C}$, then $\mathrm{Map}_{|\mathcal{C}|}(X, Y) = |\mathrm{Map}_{\mathcal{C}}(X, Y)|$.

- The composition law for morphisms in $|\mathcal{C}|$ is obtained from the composition law on $\mathcal{C}$ by applying the geometric realization functor.

Similarly, if $\mathcal{C}$ is a topological category, we may obtain a simplicial category $\mathrm{Sing}\,\mathcal{C}$ by applying the singular complex functor to each of the morphism spaces individually. The singular complex and geometric realization functors determine an adjunction between $\mathcal{C}\mathrm{at}_\Delta$ and $\mathcal{C}\mathrm{at}_{\mathrm{top}}$. This adjunction should be understood as determining an equivalence between the theory of simplicial categories and the theory of topological categories. This is essentially a formal consequence of the fact that the geometric realization and singular complex functors determine an equivalence between the homotopy theory of topological spaces and the homotopy theory of simplicial sets. More precisely, we recall that a map $f : S \to T$ of simplicial sets is said to be a *weak homotopy equivalence* if the induced map $|S| \to |T|$ of topological spaces is a weak homotopy equivalence. A theorem of Quillen (see [32] for a proof) asserts that the unit and counit morphisms

$$S \to \mathrm{Sing}\,|S|$$

$$|\mathrm{Sing}\,X| \to X$$

are weak homotopy equivalences for every (compactly generated) topological space X and every simplicial set S. It follows that the category obtained from $\mathcal{CG}$ by inverting weak homotopy equivalences (of spaces) is equivalent to the category obtained from Set_Δ by inverting weak homotopy equivalences. We use the symbol $\mathcal{H}$ to denote either of these (equivalent) categories.

If $\mathcal{C}$ is a simplicial category, we let $\mathrm{h}\mathcal{C}$ denote the $\mathcal{H}$-enriched category obtained by applying the functor $\mathrm{Set}_\Delta \to \mathcal{H}$ to each of the morphism spaces of $\mathcal{C}$. We will refer to $\mathrm{h}\mathcal{C}$ as the *homotopy category of* $\mathcal{C}$. We note that this is the same notation that was introduced in §1.1.3 for the homotopy category of a topological category. However, there is little risk of confusion: the above remarks imply the existence of canonical isomorphisms

$$\mathrm{h}\mathcal{C} \simeq \mathrm{h}|\mathcal{C}|$$

$$\mathrm{h}\mathcal{D} \simeq \mathrm{h}\mathrm{Sing}\,\mathcal{D}$$

for every simplicial category $\mathcal{C}$ and every topological category $\mathcal{D}$.

Definition 1.1.4.4. A functor $\mathcal{C} \to \mathcal{C}'$ between simplicial categories is an *equivalence* if the induced functor $\mathrm{h}\mathcal{C} \to \mathrm{h}\mathcal{C}'$ is an equivalence of $\mathcal{H}$-enriched categories.

In other words, a functor $\mathcal{C} \to \mathcal{C}'$ between simplicial categories is an equivalence if and only if the geometric realization $|\mathcal{C}| \to |\mathcal{C}'|$ is an equivalence of topological categories. In fact, one can say more. It follows easily from the preceding remarks that the unit and counit maps

$$\mathcal{C} \to \mathrm{Sing}\,|\mathcal{C}|$$

$$|\mathrm{Sing}\,\mathcal{D}| \to \mathcal{D}$$

induce *isomorphisms* between homotopy categories. Consequently, if we are working with topological or simplicial categories *up to equivalence*, we are always free to replace a simplicial category $\mathcal{C}$ by $|\mathcal{C}|$ or a topological category $\mathcal{D}$ by $\mathrm{Sing}\,\mathcal{D}$. In this sense, the notions of topological category and simplicial category are equivalent, and either can be used as a foundation for higher category theory.

1.1.5 Comparing ∞-Categories with Simplicial Categories

In §1.1.4, we introduced the theory of simplicial categories and explained why (for our purposes) it is equivalent to the theory of topological categories. In this section, we will show that the theory of simplicial categories is also closely related to the theory of ∞-categories. Our discussion requires somewhat more elaborate constructions than were needed in the previous sections; a reader who does not wish to become bogged down in details is urged to skip ahead to §1.2.1.

We will relate simplicial categories with simplicial sets by means of the *simplicial nerve functor*

$$\mathrm{N} : \mathcal{C}\mathrm{at}_\Delta \to \mathcal{S}\mathrm{et}_\Delta,$$

originally introduced by Cordier (see [16]). The nerve of an ordinary category $\mathcal{C}$ is characterized by the formula

$$\mathrm{Hom}_{\mathcal{S}\mathrm{et}_\Delta}(\Delta^n, \mathrm{N}(\mathcal{C})) = \mathrm{Hom}_{\mathcal{C}\mathrm{at}}([n], \mathcal{C});$$

here $[n]$ denotes the linearly ordered set $\{0, \ldots, n\}$ regarded as a category. This definition makes sense also when $\mathcal{C}$ is a simplicial category but is clearly not very interesting: it makes no use of the simplicial structure on $\mathcal{C}$. In order to obtain a more interesting construction, we need to replace the ordinary category $[n]$ by a suitable "thickening," a simplicial category which we will denote by $\mathfrak{C}[\Delta^n]$.

Definition 1.1.5.1. Let J be a finite nonempty linearly ordered set. The simplicial category $\mathfrak{C}[\Delta^J]$ is defined as follows:

- The objects of $\mathfrak{C}[\Delta^J]$ are the elements of J.

- If $i, j \in J$, then

$$\mathrm{Map}_{\mathfrak{C}[\Delta^J]}(i, j) = \begin{cases} \emptyset & \text{if } j < i \\ \mathrm{N}(P_{i,j}) & \text{if } i \leq j. \end{cases}$$

Here $P_{i,j}$ denotes the partially ordered set $\{I \subseteq J : (i, j \in I) \wedge (\forall k \in I)[i \leq k \leq j]\}$.

- If $i_0 \leq i_1 \leq \cdots \leq i_n$, then the composition

$$\mathrm{Map}_{\mathfrak{C}[\Delta^J]}(i_0, i_1) \times \cdots \times \mathrm{Map}_{\mathfrak{C}[\Delta^J]}(i_{n-1}, i_n) \to \mathrm{Map}_{\mathfrak{C}[\Delta^J]}(i_0, i_n)$$

is induced by the map of partially ordered sets

$$P_{i_0, i_1} \times \cdots \times P_{i_{n-1}, i_n} \to P_{i_0, i_n}$$

$$(I_1, \dots, I_n) \mapsto I_1 \cup \cdots \cup I_n.$$

In order to help digest Definition 1.1.5.1, let us analyze the structure of the topological category $|\mathfrak{C}[\Delta^n]|$. The objects of this category are elements of the set $[n] = \{0, \dots, n\}$. For each $0 \leq i \leq j \leq n$, the topological space $\mathrm{Map}_{|\mathfrak{C}[\Delta^n]|}(i, j)$ is homeomorphic to a cube; it may be identified with the set of all functions $p : \{k \in [n] : i \leq k \leq j\} \to [0, 1]$ which satisfy $p(i) = p(j) = 1$. The morphism space $\mathrm{Map}_{|\mathfrak{C}[\Delta^n]|}(i, j)$ is empty when $j < i$, and composition of morphisms is given by concatenation of functions.

Remark 1.1.5.2. Let us try to understand better the simplicial category $\mathfrak{C}[\Delta^n]$ and its relationship to the ordinary category $[n]$. These categories have the same objects: the elements of $\{0, \dots, n\}$. In the category $[n]$, there is a unique morphism $q_{ij} : i \to j$ whenever $i \leq j$. By virtue of the uniqueness, these elements satisfy $q_{jk} \circ q_{ij} = q_{ik}$ for $i \leq j \leq k$.

In the simplicial category $\mathfrak{C}[\Delta^n]$, there is a vertex $p_{ij} \in \mathrm{Map}_{\mathfrak{C}[\Delta^n]}(i, j)$ for each $i \leq j$, given by the element $\{i, j\} \in P_{ij}$. We note that $p_{jk} \circ p_{ij} \neq p_{ik}$ (except in degenerate cases where $i = j$ or $j = k$). Instead, the collection of all compositions

$$p_{i_n i_{n-1}} \circ p_{i_{n-1} i_{n-2}} \circ \cdots \circ p_{i_1 i_0},$$

where $i = i_0 < i_1 < \cdots < i_{n-1} < i_n = j$ constitute all of the different vertices of the cube $\mathrm{Map}_{\mathfrak{C}[\Delta^n]}(i, j)$. The weak contractibility of $\mathrm{Map}_{\mathfrak{C}[\Delta^n]}(i, j)$ expresses the idea that although these compositions do not coincide, they are all canonically homotopic to one another. We observe that there is a (unique) functor $\mathfrak{C}[\Delta^n] \to [n]$ which is the identity on objects. This functor is an equivalence of simplicial categories. We can summarize the situation informally as follows: the simplicial category $\mathfrak{C}[\Delta^n]$ is a thickened version of $[n]$ where we have dropped the strict associativity condition

$$q_{jk} \circ q_{ij} = q_{ik}$$

and instead have imposed associativity only up to (coherent) homotopy. (We can formulate this idea more precisely by saying that $\mathfrak{C}[\Delta^\bullet]$ is a cofibrant replacement for $[\bullet]$ with respect to a suitable model structure on the category of cosimplicial objects of Cat_Δ.)

The construction $J \mapsto \mathfrak{C}[\Delta^J]$ is functorial in J, as we now explain.

Definition 1.1.5.3. Let $f : J \to J'$ be a monotone map between linearly ordered sets. The simplicial functor $\mathfrak{C}[f] : \mathfrak{C}[\Delta^J] \to \mathfrak{C}[\Delta^{J'}]$ is defined as follows:

- For each object $i \in \mathfrak{C}[\Delta^J]$, $\mathfrak{C}[f](i) = f(i) \in \mathfrak{C}[\Delta^{J'}]$.

- If $i \leq j$ in J, then the map $\mathrm{Map}_{\mathfrak{C}[\Delta^J]}(i, j) \to \mathrm{Map}_{\mathfrak{C}[\Delta^{J'}]}(f(i), f(j))$ induced by f is the nerve of the map

$$P_{i,j} \to P_{f(i),f(j)}$$

$$I \mapsto f(I).$$

Remark 1.1.5.4. Using the notation of Remark 1.1.5.2, we note that Definition 1.1.5.3 has been rigged so that the functor $\mathfrak{C}[f]$ carries the vertex $p_{ij} \in \mathrm{Map}_{\mathfrak{C}[\Delta^J]}(i, j)$ to the vertex $p_{f(i)f(j)} \in \mathrm{Map}_{\mathfrak{C}[\Delta^{J'}]}(f(i), f(j))$.

It is not difficult to check that the construction described in Definition 1.1.5.3 is well-defined, and compatible with composition in f. Consequently, we deduce that $\mathfrak{C}$ determines a functor

$$\Delta \to \mathcal{C}at_\Delta$$

$$\Delta^n \mapsto \mathfrak{C}[\Delta^n],$$

which we may view as a cosimplicial object of $\mathcal{C}at_\Delta$.

Definition 1.1.5.5. Let $\mathcal{C}$ be a simplicial category. The *simplicial nerve* $\mathrm{N}(\mathcal{C})$ is the simplicial set described by the formula

$$\mathrm{Hom}_{\mathcal{S}et_\Delta}(\Delta^n, \mathrm{N}(\mathcal{C})) = \mathrm{Hom}_{\mathcal{C}at_\Delta}(\mathfrak{C}[\Delta^n], \mathcal{C}).$$

If $\mathcal{C}$ is a topological category, we define the *topological nerve* $\mathrm{N}(\mathcal{C})$ of $\mathcal{C}$ to be the simplicial nerve of $\mathrm{Sing}\, \mathcal{C}$.

Remark 1.1.5.6. If $\mathcal{C}$ is a simplicial (topological) category, we will often abuse terminology by referring to the simplicial (topological) nerve of $\mathcal{C}$ simply as the *nerve* of $\mathcal{C}$.

Warning 1.1.5.7. Let $\mathcal{C}$ be a simplicial category. Then $\mathcal{C}$ can be regarded as an ordinary category by ignoring all simplices of positive dimension in the mapping spaces of $\mathcal{C}$. The simplicial nerve of $\mathcal{C}$ does *not* coincide with the nerve of this underlying ordinary category. Our notation is therefore potentially ambiguous. We will adopt the following convention: whenever $\mathcal{C}$ is a simplicial category, $\mathrm{N}(\mathcal{C})$ will denote the *simplicial* nerve of $\mathcal{C}$ unless we specify otherwise. Similarly, if $\mathcal{C}$ is a topological category, then the topological nerve of $\mathcal{C}$ does not generally coincide with the nerve of the underlying category; the notation $\mathrm{N}(\mathcal{C})$ will be used to indicate the topological nerve unless otherwise specified.

Example 1.1.5.8. Any ordinary category $\mathcal{C}$ may be considered as a simplicial category by taking each of the simplicial sets $\mathrm{Hom}_{\mathcal{C}}(X, Y)$ to be *constant*. In this case, the set of simplicial functors $\mathfrak{C}[\Delta^n] \to \mathcal{C}$ may be identified with the set of functors from $[n]$ into $\mathcal{C}$. Consequently, the simplicial nerve of $\mathcal{C}$ agrees with the ordinary nerve of $\mathcal{C}$ as defined in §1.1.2. Similarly, the ordinary nerve of $\mathcal{C}$ can be identified with the topological nerve of $\mathcal{C}$, where $\mathcal{C}$ is regarded as a topological category with discrete morphism spaces.

In order to get a feel for what the nerve of a topological category $\mathcal{C}$ looks like, let us explicitly describe its low-dimensional simplices:

- The 0-simplices of $N(\mathcal{C})$ may be identified with the objects of $\mathcal{C}$.

- The 1-simplices of $N(\mathcal{C})$ may be identified with the morphisms of $\mathcal{C}$.

- To give a map from the boundary of a 2-simplex into $N(\mathcal{C})$ is to give a diagram (not necessarily commutative)

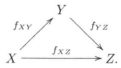

To give a 2-simplex of $N(\mathcal{C})$ having this specified boundary is equivalent to giving a path from f_{XZ} to $f_{YZ} \circ f_{XY}$ in $\mathrm{Map}_{\mathcal{C}}(X, Z)$.

The category $\mathcal{C}\mathrm{at}_{\Delta}$ of simplicial categories admits (small) colimits. Consequently, by formal nonsense, the functor $\mathfrak{C} : \Delta \to \mathcal{C}\mathrm{at}_{\Delta}$ extends uniquely (up to unique isomorphism) to a colimit-preserving functor $\mathrm{Set}_{\Delta} \to \mathcal{C}\mathrm{at}_{\Delta}$, which we will denote also by $\mathfrak{C}$. By construction, the functor $\mathfrak{C}$ is left adjoint to the simplicial nerve functor N. For each simplicial set S, we can view $\mathfrak{C}[S]$ as the simplicial category "freely generated" by S: every n-simplex $\sigma : \Delta^n \to S$ determines a functor $\mathfrak{C}[\Delta^n] \to \mathfrak{C}[S]$, which we can think of as a homotopy coherent diagram $[n] \to \mathfrak{C}[S]$.

Example 1.1.5.9. Let A be a partially ordered set. The simplicial category $\mathfrak{C}[N\,A]$ can be constructed using the following generalization of Definition 1.1.5.1:

- The objects of $\mathfrak{C}[N\,A]$ are the elements of A.

- Given a pair of elements $a, b \in A$, the simplicial set $\mathrm{Map}_{\mathfrak{C}[N\,A]}(a, b)$ can be identified with $N\,P_{a,b}$, where $P_{a,b}$ denotes the collection of linearly ordered subsets $S \subseteq A$ with least element a and largest element b, partially ordered by inclusion.

- Given a sequence of elements $a_0, \ldots, a_n \in A$, the composition map

$$\mathrm{Map}_{\mathfrak{C}[N\,A]}(a_0, a_1) \times \cdots \times \mathrm{Map}_{\mathfrak{C}[N\,A]}(a_{n-1}, a_n) \to \mathrm{Map}_{\mathfrak{C}[N\,A]}(a_0, a_n)$$

  is induced by the map of partially ordered sets

$$P_{a_0,a_1} \times \cdots \times P_{a_{n-1},a_n} \to P_{a_0,a_n}$$

$$(S_1, \ldots, S_n) \mapsto S_1 \cup \cdots \cup S_n.$$

Proposition 1.1.5.10. *Let $\mathcal{C}$ be a simplicial category having the property that, for every pair of objects $X, Y \in \mathcal{C}$, the simplicial set $\mathrm{Map}_{\mathcal{C}}(X, Y)$ is a Kan complex. Then the simplicial nerve $N(\mathcal{C})$ is an ∞-category.*

Proof. We must show that if $0 < i < n$, then $N(\mathcal{C})$ has the right extension property with respect to the inclusion $\Lambda_i^n \subseteq \Delta^n$. Rephrasing this in the language of simplicial categories, we must show that $\mathcal{C}$ has the right extension property with respect to the simplicial functor $\mathfrak{C}[\Lambda_i^n] \to \mathfrak{C}[\Delta^n]$. To prove this, we make use of the following observations concerning $\mathfrak{C}[\Lambda_i^n]$, which we view as a simplicial subcategory of $\mathfrak{C}[\Delta^n]$:

- The objects of $\mathfrak{C}[\Lambda_i^n]$ are the objects of $\mathfrak{C}[\Delta^n]$: that is, elements of the set $[n]$.

- For $0 \leq j \leq k \leq n$, the simplicial set $\mathrm{Map}_{\mathfrak{C}[\Lambda_i^n]}(j,k)$ coincides with $\mathrm{Map}_{\mathfrak{C}[\Delta^n]}(j,k)$ unless $j = 0$ and $k = n$ (note that this condition fails if $i = 0$ or $i = n$).

Consequently, every extension problem

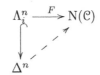

is equivalent to

$$\mathrm{Map}_{\mathfrak{C}[\Lambda_i^n]}(0,n) \longrightarrow \mathrm{Map}_{\mathcal{C}}(F(0), F(n))$$
$$\downarrow \qquad \qquad \qquad \qquad \nearrow$$
$$\mathrm{Map}_{\mathfrak{C}[\Delta^n]}(0,n).$$

Since the simplicial set on the right is a Kan complex by assumption, it suffices to verify that the left vertical map is anodyne. This follows by inspection: the simplicial set $\mathrm{Map}_{\mathfrak{C}[\Delta^n]}(0,n)$ can be identified with the cube $(\Delta^1)^{\{1,\dots,n-1\}}$. Under this identification, $\mathrm{Map}_{\mathfrak{C}[\Lambda_i^n]}(0,n)$ corresponds to the simplicial subset of $(\Delta^1)^{\{1,\dots,n-1\}}$ obtained by removing the interior of the cube together with one of its faces. $\square$

Remark 1.1.5.11. The proof of Proposition 1.1.5.10 actually provides a slightly stronger result: if $F : \mathcal{C} \to \mathcal{D}$ is a functor between simplicial categories which induces Kan fibrations $\mathrm{Map}_{\mathcal{C}}(C, C') \to \mathrm{Map}_{\mathcal{D}}(F(C), F(C'))$ for every pair of objects $C, C' \in \mathcal{C}$, then the associated map $N(\mathcal{C}) \to N(\mathcal{D})$ is an inner fibration of simplicial sets (see Definition 2.0.0.3).

Corollary 1.1.5.12. *Let $\mathcal{C}$ be a topological category. Then the topological nerve $N(\mathcal{C})$ is an ∞-category.*

Proof. This follows immediately from Proposition 1.1.5.10 (note that the singular complex of any topological space is a Kan complex). $\square$

We now cite the following theorem, which will be proven in §2.2.4 and refined in §2.2.5:

Theorem 1.1.5.13. *Let $\mathcal{C}$ be a topological category and let $X, Y \in \mathcal{C}$ be objects. Then the counit map*

$$|\operatorname{Map}_{\mathfrak{C}[N(\mathcal{C})]}(X, Y)| \to \operatorname{Map}_{\mathcal{C}}(X, Y)$$

is a weak homotopy equivalence of topological spaces.

Using Theorem 1.1.5.13, we can explain why the theory of ∞-categories is equivalent to the theory of topological categories (or equivalently, simplicial categories). The adjoint functors N and $|\mathfrak{C}[\bullet]|$ are not mutually inverse equivalences of categories. However, they *are* homotopy inverse to one another. To make this precise, we need to introduce a definition.

Definition 1.1.5.14. Let S be a simplicial set. The *homotopy category* hS is defined to be the homotopy category $h\mathfrak{C}[S]$ of the simplicial category $\mathfrak{C}[S]$. We will often view hS as a category enriched over the homotopy category $\mathcal{H}$ of spaces via the construction of §1.1.4: that is, for every pair of vertices $x, y \in S$, we have $\operatorname{Map}_{hS}(x, y) = [\operatorname{Map}_{\mathfrak{C}[S]}(x, y)]$. A map $f : S \to T$ of simplicial sets is a *categorical equivalence* if the induced map $hS \to hT$ is an equivalence of $\mathcal{H}$-enriched categories.

Remark 1.1.5.15. In [44], Joyal uses the term "weak categorical equivalence" for what we have called a categorical equivalence, and reserves the term "categorical equivalence" for a stronger notion of equivalence.

Remark 1.1.5.16. We have introduced the term "categorical equivalence," rather than simply "equivalence" or "weak equivalence," in order to avoid confusing the notion of categorical equivalence of simplicial sets with the (more classical) notion of weak homotopy equivalence of simplicial sets.

Remark 1.1.5.17. It is immediate from the definition that $f : S \to T$ is a categorical equivalence if and only if $\mathfrak{C}[S] \to \mathfrak{C}[T]$ is an equivalence (of simplicial categories) if and only if $|\mathfrak{C}[S]| \to |\mathfrak{C}[T]|$ is an equivalence (of topological categories).

We now observe that the adjoint functors $(|\mathfrak{C}[\bullet]|, N)$ determine an equivalence between the theory of simplicial sets (up to categorical equivalence) and that of topological categories (up to equivalence). In other words, for any topological category $\mathcal{C}$ the counit map $|\mathfrak{C}[N(\mathcal{C})]| \to \mathcal{C}$ is an equivalence of topological categories, and for any simplicial set S the unit map $S \to N|\mathfrak{C}[S]|$ is a categorical equivalence of simplicial sets. In view of Remark 1.1.5.17, the second assertion is a formal consequence of the first. Moreover, the first assertion is merely a reformulation of Theorem 1.1.5.13.

Remark 1.1.5.18. The reader may at this point object that we have obtained a comparison between the theory of topological categories and the theory of simplicial sets but that not every simplicial set is an ∞-category. However, every simplicial set is categorically equivalent to an ∞-category. In fact, Theorem 1.1.5.13 implies that every simplicial set S is categorically equivalent to the nerve of the topological category $|\mathfrak{C}[S]|$, which is an ∞-category (Corollary 1.1.5.12).

1.2 THE LANGUAGE OF HIGHER CATEGORY THEORY

One of the main goals of this book is to demonstrate that many ideas from classical category theory can be adapted to the setting of higher categories. In this section, we will survey some of the simplest examples.

1.2.1 The Opposite of an ∞-Category

If $\mathcal{C}$ is an ordinary category, then the opposite category $\mathcal{C}^{op}$ is defined in the following way:

- The objects of $\mathcal{C}^{op}$ are the objects of $\mathcal{C}$.

- For $X, Y \in \mathcal{C}$, we have $\mathrm{Hom}_{\mathcal{C}^{op}}(X, Y) = \mathrm{Hom}_{\mathcal{C}}(Y, X)$. Identity morphisms and composition are defined in the obvious way.

This definition generalizes without change to the setting of topological or simplicial categories. Adapting this definition to the setting of ∞-categories requires a few additional words. We may define more generally the *opposite* of a simplicial set S as follows: for any finite nonempty linearly ordered set J, we set $S^{op}(J) = S(J^{op})$, where J^{op} denotes the same set J endowed with the opposite ordering. More concretely, we have $S_n^{op} = S_n$, but the face and degeneracy maps on S^{op} are given by the formulas

$$(d_i : S_n^{op} \to S_{n-1}^{op}) = (d_{n-i} : S_n \to S_{n-1})$$

$$(s_i : S_n^{op} \to S_{n+1}^{op}) = (s_{n-i} : S_n \to S_{n+1}).$$

The formation of opposite categories is fully compatible with all of the constructions we have introduced for passing back and forth between different models of higher category theory.

It is clear from the definition that a simplicial set S is an ∞-category if and only if its opposite S^{op} is an ∞-category: for $0 < i < n$, S has the extension property with respect to the horn inclusion $\Lambda_i^n \subseteq \Delta^n$ if and only if S^{op} has the extension property with respect to the horn inclusion $\Lambda_{n-i}^n \subseteq \Delta^n$.

1.2.2 Mapping Spaces in Higher Category Theory

If X and Y are objects of an ordinary category $\mathcal{C}$, then one has a well-defined set $\mathrm{Hom}_{\mathcal{C}}(X, Y)$ of morphisms from X to Y. In higher category theory, one has instead a morphism *space* $\mathrm{Map}_{\mathcal{C}}(X, Y)$. In the setting of topological or simplicial categories, this morphism space (either a topological space or a simplicial set) is an inherent feature of the formalism. It is less obvious how to define $\mathrm{Map}_{\mathcal{C}}(X, Y)$ in the setting of ∞-categories. However, it is at least clear what to do on the level of the homotopy category.

Definition 1.2.2.1. Let S be a simplicial set containing vertices x and y and let $\mathcal{H}$ denote the homotopy category of spaces. We define $\mathrm{Map}_S(x, y) =$

$\text{Map}_{\text{h}S}(x, y) \in \mathcal{H}$ to be the object of $\mathcal{H}$ representing the space of maps from x to y in S. Here $\text{h}S$ denotes the homotopy category of S regarded as a $\mathcal{H}$-enriched category (Definition 1.1.5.14).

Warning 1.2.2.2. Let S be a simplicial set. The notation $\text{Map}_S(X, Y)$ has two *very* different meanings. When X and Y are vertices of S, then our notation should be interpreted in the sense of Definition 1.2.2.1, so that $\text{Map}_S(X, Y)$ is an object of $\mathcal{H}$. If X and Y are objects of $(\text{Set}_\Delta)_{/S}$, then we instead let $\text{Map}_S(X, Y)$ denote the simplicial mapping object

$$Y^X \times_{S^X} \{\phi\} \in \text{Set}_\Delta,$$

where ϕ denotes the structural morphism $X \to S$. We trust that it will be clear from the context which of these two definitions applies in a given situation.

We now consider the following question: given a simplicial set S containing a pair of vertices x and y, how can we compute $\text{Map}_S(x, y)$? We have defined $\text{Map}_S(x, y)$ as an object of the homotopy category $\mathcal{H}$, but for many purposes it is important to choose a simplicial set M which represents $\text{Map}_S(x, y)$. The most obvious candidate for M is the simplicial set $\text{Map}_{\mathfrak{C}[S]}(x, y)$. The advantages of this definition are that it works in all cases (that is, S does not need to be an ∞-category) and comes equipped with an associative composition law. However, the construction of the simplicial set $\text{Map}_{\mathfrak{C}[S]}(x, y)$ is quite complicated. Furthermore, $\text{Map}_{\mathfrak{C}[S]}(x, y)$ is usually not a Kan complex, so it can be difficult to extract algebraic invariants like homotopy groups even when a concrete description of its simplices is known.

In order to address these shortcomings, we will introduce another simplicial set which represents the homotopy type $\text{Map}_S(x, y) \in \mathcal{H}$, at least when S is an ∞-category. We define a new simplicial set $\text{Hom}_S^{\text{R}}(x, y)$, the space of *right morphisms* from x to y, by letting $\text{Hom}_{\text{Set}_\Delta}(\Delta^n, \text{Hom}_S^{\text{R}}(x, y))$ denote the set of all $z : \Delta^{n+1} \to S$ such that $z|\Delta^{\{n+1\}} = y$ and $z|\Delta^{\{0, \dots, n\}}$ is a constant simplex at the vertex x. The face and degeneracy operations on $\text{Hom}_S^{\text{R}}(x, y)_n$ are defined to coincide with corresponding operations on S_{n+1}.

We first observe that when S is an ∞-category, $\text{Hom}_S^{\text{R}}(x, y)$ really is a "space":

Proposition 1.2.2.3. *Let $\mathcal{C}$ be an ∞-category containing a pair of objects x and y. The simplicial set $\text{Hom}_{\mathcal{C}}^{\text{R}}(x, y)$ is a Kan complex.*

Proof. It is immediate from the definition that if $\mathcal{C}$ is a ∞-category, then $M = \text{Hom}_{\mathcal{C}}^{\text{R}}(x, y)$ satisfies the Kan extension condition for every horn inclusion $\Lambda_i^n \subseteq \Delta^n$, where $0 < i \leq n$. This implies that M is a Kan complex (Proposition 1.2.5.1). $\square$

Remark 1.2.2.4. If S is a simplicial set and $x, y, z \in S_0$, then there is no obvious composition law

$$\text{Hom}_S^{\text{R}}(x, y) \times \text{Hom}_S^{\text{R}}(y, z) \to \text{Hom}_S^{\text{R}}(x, z).$$

We will later see that if S is an ∞-category, then there is a composition law which is well-defined up to a contractible space of choices. The absence of a canonical choice for a composition law is the main drawback of $\operatorname{Hom}_S^R(x, y)$ in comparison with $\operatorname{Map}_{\mathfrak{C}[S]}(x, y)$. The main goal of §2.2 is to show that if S is an ∞-category, then there is a (canonical) isomorphism between $\operatorname{Hom}_S^R(x, y)$ and $\operatorname{Map}_{\mathfrak{C}[S]}(x, y)$ in the homotopy category $\mathcal{H}$. In particular, we will conclude that $\operatorname{Hom}_S^R(x, y)$ represents $\operatorname{Map}_S(x, y)$ whenever S is an ∞-category.

Remark 1.2.2.5. The definition of $\operatorname{Hom}_S^R(x, y)$ is not self-dual: that is, $\operatorname{Hom}_{S^{op}}^R(x, y) \neq \operatorname{Hom}_S^R(y, x)$ in general. Instead, we define $\operatorname{Hom}_S^L(x, y) = \operatorname{Hom}_{S^{op}}^R(y, x)^{op}$, so that $\operatorname{Hom}_S^L(x, y)_n$ is the set of all $z \in S_{n+1}$ such that $z|\Delta^{\{0\}} = x$ and $z|\Delta^{\{1, \dots, n+1\}}$ is the constant simplex at the vertex y.

Although the simplicial sets $\operatorname{Hom}_S^L(x, y)$ and $\operatorname{Hom}_S^R(x, y)$ are generally not isomorphic to one another, they are homotopy equivalent whenever S is an ∞-category. To prove this, it is convenient to define a third, self-dual, space of morphisms: let $\operatorname{Hom}_S(x, y) = \{x\} \times_S S^{\Delta^1} \times_S \{y\}$. In other words, to give an n-simplex of $\operatorname{Hom}_S(x, y)$, one must give a map $f : \Delta^n \times \Delta^1 \to S$ such that $f|\Delta^n \times \{0\}$ is constant at x and $f|\Delta^n \times \{1\}$ is constant at y. We observe that there exist natural inclusions

$$\operatorname{Hom}_S^R(x, y) \hookrightarrow \operatorname{Hom}_S(x, y) \hookleftarrow \operatorname{Hom}_S^L(x, y),$$

which are induced by retracting the cylinder $\Delta^n \times \Delta^1$ onto certain maximal-dimensional simplices. We will later show (Corollary 4.2.1.8) that these inclusions are homotopy equivalences provided that S is an ∞-category.

1.2.3 The Homotopy Category

For every ordinary category $\mathcal{C}$, the nerve $\operatorname{N}(\mathcal{C})$ is an ∞-category. Informally, we can describe the situation as follows: the nerve functor is a fully faithful inclusion from the bicategory of categories to the ∞-bicategory of ∞-categories. Moreover, this inclusion has a left adjoint:

Proposition 1.2.3.1. *The nerve functor* $\mathcal{C}\mathrm{at} \to \operatorname{Set}_\Delta$ *is right adjoint to the functor* $\mathrm{h} \colon \operatorname{Set}_\Delta \to \mathcal{C}\mathrm{at}$, *which associates to every simplicial set S its homotopy category* $\mathrm{h}S$ *(here we ignore the $\mathcal{H}$-enrichment of $\mathrm{h}S$).*

Proof. Let us temporarily distinguish between the nerve functor $\operatorname{N} : \mathcal{C}\mathrm{at} \to \operatorname{Set}_\Delta$ and the simplicial nerve functor $\operatorname{N'} : \mathcal{C}\mathrm{at}_\Delta \to \operatorname{Set}_\Delta$. These two functors are related by the fact that N can be written as a composition

$$\mathcal{C}\mathrm{at} \overset{i}{\subseteq} \mathcal{C}\mathrm{at}_\Delta \overset{\operatorname{N'}}{\to} \operatorname{Set}_\Delta .$$

The functor $\pi_0 : \operatorname{Set}_\Delta \to \operatorname{Set}$ is a left adjoint to the inclusion functor $\operatorname{Set} \to \operatorname{Set}_\Delta$, so the functor

$$\mathcal{C}\mathrm{at}_\Delta \to \mathcal{C}\mathrm{at}$$

$$\mathcal{C} \mapsto \mathrm{h}\mathcal{C}$$

is left adjoint to i. It follows that $N = N' \circ i$ has a left adjoint, given by the composition

$$\mathrm{Set}_\Delta \xrightarrow{\mathfrak{C}[\bullet]} \mathrm{Cat}_\Delta \xrightarrow{\mathrm{h}} \mathrm{Cat},$$

which coincides with the homotopy category functor $\mathrm{h} : \mathrm{Set}_\Delta \to \mathrm{Cat}$ by definition. □

Remark 1.2.3.2. The formation of the homotopy category is literally left adjoint to the inclusion $\mathrm{Cat} \subseteq \mathrm{Cat}_\Delta$. The analogous assertion is not quite true in the setting of topological categories because the functor $\pi_0 : \mathcal{CG} \to \mathrm{Set}$ is a left adjoint only when restricted to locally path-connected spaces.

Warning 1.2.3.3. If $\mathcal{C}$ is a simplicial category, then we do not necessarily expect that $\mathrm{h}\mathcal{C} \simeq \mathrm{hN}(\mathcal{C})$. However, this is always the case when $\mathcal{C}$ is *fibrant* in the sense that every simplicial set $\mathrm{Map}_\mathcal{C}(X, Y)$ is a Kan complex.

Remark 1.2.3.4. If S is a simplicial set, Joyal ([44]) refers to the category $\mathrm{h}S$ as the *fundamental category* of S. This is motivated by the observation that if S is a Kan complex, then $\mathrm{h}S$ is the fundamental groupoid of S in the usual sense.

Our objective for the remainder of this section is to obtain a more explicit understanding of the homotopy category $\mathrm{h}S$ of a simplicial set S. Proposition 1.2.3.1 implies that $\mathrm{h}S$ admits the following presentation by generators and relations:

- The objects of $\mathrm{h}S$ are the vertices of S.

- For every edge $\phi : \Delta^1 \to S$, there is a morphism $\overline{\phi}$ from $\phi(0)$ to $\phi(1)$.

- For each $\sigma : \Delta^2 \to S$, we have $\overline{d_0(\sigma)} \circ \overline{d_2(\sigma)} = \overline{d_1(\sigma)}$.

- For each vertex x of S, the morphism $\overline{s_0 x}$ is the identity id_x.

If S is an ∞-category, there is a much more satisfying construction of the category $\mathrm{h}S$. We will describe this construction in detail since it nicely illustrates the utility of the weak Kan condition of Definition 1.1.2.4.

Let $\mathcal{C}$ be an ∞-category. We will construct a category $\pi(\mathcal{C})$ (which we will eventually show to be equivalent to the homotopy category $\mathrm{h}\mathcal{C}$). The objects of $\pi(\mathcal{C})$ are the vertices of $\mathcal{C}$. Given an edge $\phi : \Delta^1 \to \mathcal{C}$, we shall say that ϕ has *source* $C = \phi(0)$ and *target* $C' = \phi(1)$ and write $\phi : C \to C'$. For each object C of $\mathcal{C}$, we let id_C denote the degenerate edge $s_0(C) : C \to C$.

Let $\phi : C \to C'$ and $\phi' : C \to C'$ be a pair of edges of $\mathcal{C}$ having the same source and target. We will say that ϕ and ϕ' are *homotopic* if there is a 2-simplex $\sigma : \Delta^2 \to \mathcal{C}$, which we depict as follows:

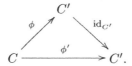

In this case, we say that σ is a *homotopy* between ϕ and ϕ'.

Proposition 1.2.3.5. *Let $\mathcal{C}$ be an ∞-category and let C and C' be objects of $\pi(\mathcal{C})$. Then the relation of homotopy is an equivalence relation on the edges joining C to C'.*

Proof. Let $\phi : \Delta^1 \to \mathcal{C}$ be an edge. Then $s_1(\phi)$ is a homotopy from ϕ to itself. Thus homotopy is a reflexive relation.

Suppose next that $\phi, \phi', \phi'' : C \to C'$ are edges with the same source and target. Let σ be a homotopy from ϕ to ϕ', and σ' a homotopy from ϕ to ϕ''. Let $\sigma'' : \Delta^2 \to \mathcal{C}$ denote the constant map at the vertex C'. We have a commutative diagram

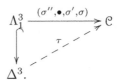

Since $\mathcal{C}$ is an ∞-category, there exists a 3-simplex $\tau : \Delta^3 \to \mathcal{C}$ as indicated by the dotted arrow in the diagram. It is easy to see that $d_1(\tau)$ is a homotopy from ϕ' to ϕ''.

As a special case, we can take $\phi = \phi''$; we then deduce that the relation of homotopy is symmetric. It then follows immediately from the above that the relation of homotopy is also transitive. $\square$

Remark 1.2.3.6. The definition of homotopy that we have given is not evidently self-dual; in other words, it is not immediately obvious that a homotopic pair of edges $\phi, \phi' : C \to C'$ of an ∞-category $\mathcal{C}$ remain homotopic when regarded as edges in the opposite ∞-category $\mathcal{C}^{op}$. To prove this, let σ be a homotopy from ϕ to ϕ' and consider the commutative diagram

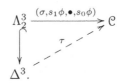

The assumption that $\mathcal{C}$ is an ∞-category guarantees a 3-simplex τ rendering the diagram commutative. The face $d_2\tau$ may be regarded as a homotopy from ϕ' to ϕ in $\mathcal{C}^{op}$.

We can now define the morphism sets of the category $\pi(\mathcal{C})$: given vertices X and Y of $\mathcal{C}$, we let $\mathrm{Hom}_{\pi(\mathcal{C})}(X, Y)$ denote the set of homotopy classes of edges $\phi : X \to Y$ in $\mathcal{C}$. For each edge $\phi : \Delta^1 \to \mathcal{C}$, we let $[\phi]$ denote the corresponding morphism in $\pi(\mathcal{C})$.

We define a composition law on $\pi(\mathcal{C})$ as follows. Suppose that X, Y, and Z are vertices of $\mathcal{C}$ and that we are given edges $\phi : X \to Y$, $\psi : Y \to Z$. The pair (ϕ, ψ) determines a map $\Lambda_1^2 \to \mathcal{C}$. Since $\mathcal{C}$ is an ∞-category, this map extends to a 2-simplex $\sigma : \Delta^2 \to \mathcal{C}$. We now define $[\psi] \circ [\phi] = [d_1\sigma]$.

Proposition 1.2.3.7. *Let $\mathcal{C}$ be an ∞-category. The composition law on $\pi(\mathcal{C})$ is well-defined. In other words, the homotopy class $[\psi] \circ [\phi]$ does not depend on the choice of ψ representing $[\psi]$, the choice of ϕ representing $[\phi]$, or the choice of the 2-simplex σ.*

Proof. We begin by verifying the independence of the choice of σ. Suppose that we are given two 2-simplices $\sigma, \sigma' : \Delta^2 \to \mathcal{C}$, satisfying

$$d_0 \sigma = d_0 \sigma' = \psi$$

$$d_2 \sigma = d_2 \sigma' = \phi.$$

Consider the diagram

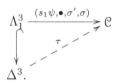

Since $\mathcal{C}$ is an ∞-category, there exists a 3-simplex τ as indicated by the dotted arrow. It follows that $d_1 \tau$ is a homotopy from $d_1 \sigma$ to $d_1 \sigma'$.

We now show that $[\psi] \circ [\phi]$ depends only on ψ and ϕ only up to homotopy. In view of Remark 1.2.3.6, the assertion is symmetric with respect to ψ and ϕ; it will therefore suffice to show that $[\psi] \circ [\phi]$ does not change if we replace ϕ by a morphism ϕ' which is homotopic to ϕ. Let σ be a 2-simplex with $d_0 \sigma = \psi$ and $d_2 \sigma = \phi$, and let σ' be a homotopy from ϕ to ϕ'. Consider the diagram

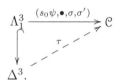

Again, the hypothesis that $\mathcal{C}$ is an ∞-category guarantees the existence of a 3-simplex τ as indicated in the diagram. Let $\sigma'' = d_1 \tau$. Then $[\psi] \circ [\phi'] = [d_1 \sigma']$. But $d_1 \sigma = d_1 \sigma'$ by construction, so that $[\psi] \circ [\phi] = [\psi] \circ [\phi']$, as desired. $\square$

Proposition 1.2.3.8. *If $\mathcal{C}$ is an ∞-category, then $\pi(\mathcal{C})$ is a category.*

Proof. Let C be a vertex of $\mathcal{C}$. We first verify that $[\mathrm{id}_C]$ is an identity with respect to the composition law on $\pi(\mathcal{C})$. For every edge $\phi : C' \to C$ in $\mathcal{C}$, the 2-simplex $s_1(\phi)$ verifies the equation

$$[\mathrm{id}_C] \circ [\phi] = [\phi].$$

This proves that id_C is a left identity; the dual argument (Remark 1.2.3.6) shows that $[\mathrm{id}_C]$ is a right identity.

The only other thing we need to check is the associative law for composition in $\pi(\mathcal{C})$. Suppose we are given a composable sequence of edges

$$C \xrightarrow{\phi} C' \xrightarrow{\phi'} C'' \xrightarrow{\phi''} C'''.$$

Choose 2-simplices $\sigma, \sigma', \sigma'' : \Delta^2 \to \mathcal{C}$ corresponding to diagrams

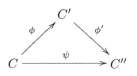

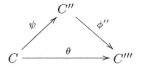

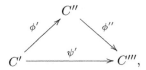

respectively. Then $[\phi'] \circ [\phi] = [\psi]$, $[\phi''] \circ [\psi] = [\theta]$, and $[\phi''] \circ [\phi'] = [\psi']$. Consider the diagram

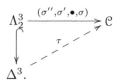

Since $\mathcal{C}$ is an ∞-category, there exists a 3-simplex τ rendering the diagram commutative. Then $d_2(\tau)$ verifies the equation $[\psi'] \circ [\phi] = [\theta]$, so that

$$([\phi''] \circ [\phi']) \circ [\phi] = [\theta] = [\phi''] \circ [\psi] = [\phi''] \circ ([\phi'] \circ [\phi]),$$

as desired. $\square$

We now show that if $\mathcal{C}$ is an ∞-category, then $\pi(\mathcal{C})$ is naturally equivalent (in fact, isomorphic) to $h\mathcal{C}$.

Proposition 1.2.3.9. *Let $\mathcal{C}$ be an ∞-category. There exists a unique functor $F : h\mathcal{C} \to \pi(\mathcal{C})$ with the following properties:*

(1) *On objects, F is the identity map.*

(2) *For every edge ϕ of $\mathcal{C}$, $F(\overline{\phi}) = [\phi]$.*

Moreover, F is an isomorphism of categories.

Proof. The existence and uniqueness of F follows immediately from our presentation of h$\mathcal{C}$ by generators and relations. It is obvious that F is bijective on objects and surjective on morphisms. To complete the proof, it will suffice to show that F is faithful.

We first show that every morphism $f : x \to y$ in h$\mathcal{C}$ may be written as $\overline{\phi}$ for some $\phi \in \mathcal{C}$. Since the morphisms in h$\mathcal{C}$ are generated by morphisms having the form $\overline{\phi}$ under composition, it suffices to show that the set of such morphisms contains all identity morphisms and is stable under composition. The first assertion is clear since $\overline{s_0 x} = \mathrm{id}_x$. For the second, we note that if $\phi : x \to y$ and $\phi' : y \to z$ are composable edges, then there exists a 2-simplex $\sigma : \Delta^2 \to \mathcal{C}$ which we may depict as follows:

Thus $\overline{\phi'} \circ \overline{\phi} = \overline{\psi}$.

Now suppose that $\phi, \phi' : x \to y$ are such that $[\phi] = [\phi']$; we wish to show that $\overline{\phi} = \overline{\phi'}$. By definition, there exists a homotopy $\sigma : \Delta^2 \to \mathcal{C}$ joining ϕ and ϕ'. The existence of σ entails the relation

$$\mathrm{id}_y \circ \overline{\phi} = \overline{\phi'}$$

in the homotopy category hS, so that $\overline{\phi} = \overline{\phi'}$, as desired. $\square$

1.2.4 Objects, Morphisms, and Equivalences

As in ordinary category theory, we may speak of *objects* and *morphisms* in a higher category $\mathcal{C}$. If $\mathcal{C}$ is a topological (or simplicial) category, these should be understood literally as the objects and morphisms in the underlying category of $\mathcal{C}$. We may also apply this terminology to ∞-categories (or even more general simplicial sets): if S is a simplicial set, then the *objects* of S are the vertices $\Delta^0 \to S$, and the *morphisms* of S are the edges $\Delta^1 \to S$. A morphism $\phi : \Delta^1 \to S$ is said to have *source* $X = \phi(0)$ and *target* $Y = \phi(1)$; we will often denote this by writing $\phi : X \to Y$. If $X : \Delta^0 \to S$ is an object of S, we will write $\mathrm{id}_X = s_0(X) : X \to X$ and refer to this as the *identity morphism* of X.

If $f, g : X \to Y$ are two morphisms in a higher category $\mathcal{C}$, then f and g are *homotopic* if they determine the same morphism in the homotopy category h$\mathcal{C}$. In the setting of ∞-categories, this coincides with the notion of homotopy introduced in the previous section. In the setting of topological categories, this simply means that f and g lie in the same path component of $\mathrm{Map}_{\mathcal{C}}(X, Y)$. In either case, we will sometimes indicate this relationship between f and g by writing $f \simeq g$.

A morphism $f : X \to Y$ in an ∞-category $\mathcal{C}$ is said to be an *equivalence* if it determines an isomorphism in the homotopy category h$\mathcal{C}$. We say that X and Y are *equivalent* if there is an equivalence between them (in other words, if they are isomorphic as objects of h$\mathcal{C}$).

If $\mathcal{C}$ is a topological category, then the requirement that a morphism $f :$ $X \to Y$ be an equivalence is quite a bit weaker than the requirement that f be an isomorphism. In fact, we have the following:

Proposition 1.2.4.1. *Let $f : X \to Y$ be a morphism in a topological category. The following conditions are equivalent:*

(1) *The morphism f is an equivalence.*

(2) *The morphism f has a homotopy inverse $g : Y \to X$: that is, a morphism g such that $f \circ g \simeq \mathrm{id}_Y$ and $g \circ f \simeq \mathrm{id}_X$.*

(3) *For every object $Z \in \mathcal{C}$, the induced map $\mathrm{Map}_{\mathcal{C}}(Z, X) \to \mathrm{Map}_{\mathcal{C}}(Z, Y)$ is a homotopy equivalence.*

(4) *For every object $Z \in \mathcal{C}$, the induced map $\mathrm{Map}_{\mathcal{C}}(Z, X) \to \mathrm{Map}_{\mathcal{C}}(Z, Y)$ is a weak homotopy equivalence.*

(5) *For every object $Z \in \mathcal{C}$, the induced map $\mathrm{Map}_{\mathcal{C}}(Y, Z) \to \mathrm{Map}_{\mathcal{C}}(X, Z)$ is a homotopy equivalence.*

(6) *For every object $Z \in \mathcal{C}$, the induced map $\mathrm{Map}_{\mathcal{C}}(Y, Z) \to \mathrm{Map}_{\mathcal{C}}(X, Z)$ is a weak homotopy equivalence.*

Proof. It is clear that (2) is merely a reformulation of (1). We will show that $(2) \Rightarrow (3) \Rightarrow (4) \Rightarrow (1)$; the implications $(2) \Rightarrow (5) \Rightarrow (6) \Rightarrow (1)$ follow using the same argument.

To see that (2) implies (3), we note that if g is a homotopy inverse to f, then composition with g gives a map $\mathrm{Map}_{\mathcal{C}}(Z, Y) \to \mathrm{Map}_{\mathcal{C}}(Z, X)$ which is homotopy inverse to composition with f. It is clear that (3) implies (4). Finally, if (4) holds, then we note that X and Y represent the same functor on $h\mathcal{C}$ so that f induces an isomorphism between X and Y in $h\mathcal{C}$. $\qquad\square$

Example 1.2.4.2. Let $\mathcal{C}$ be the category of CW complexes which we regard as a topological category by endowing each of the sets $\mathrm{Hom}_{\mathcal{C}}(X, Y)$ with the (compactly generated) compact open topology. A pair of objects $X, Y \in \mathcal{C}$ are equivalent (in the sense defined above) if and only if they are homotopy equivalent (in the sense of classical topology).

If $\mathcal{C}$ is an ∞-category (topological category, simplicial category), then we shall write $X \in \mathcal{C}$ to mean that X is an object of $\mathcal{C}$. We will generally understand that all meaningful properties of objects are invariant under equivalence. Similarly, all meaningful properties of morphisms are invariant under homotopy and under composition with equivalences.

In the setting of ∞-categories, there is a very useful characterization of equivalences which is due to Joyal.

Proposition 1.2.4.3 (Joyal [44]). *Let $\mathcal{C}$ be an ∞-category and $\phi : \Delta^1 \to \mathcal{C}$ a morphism of $\mathcal{C}$. Then ϕ is an equivalence if and only if, for every $n \geq 2$ and every map $f_0 : \Lambda^n_0 \to \mathcal{C}$ such that $f_0 | \Delta^{\{0,1\}} = \phi$, there exists an extension of f_0 to Δ^n.*

The proof requires some ideas which we have not yet introduced and will be given in §2.1.2.

1.2.5 ∞-Groupoids and Classical Homotopy Theory

Let $\mathcal{C}$ be an ∞-category. We will say that $\mathcal{C}$ is an ∞-*groupoid* if the homotopy category $h\mathcal{C}$ is a groupoid: in other words, if every morphism in $\mathcal{C}$ is an equivalence. In §1.1.1, we asserted that the theory of ∞-groupoids is equivalent to classical homotopy theory. We can now formulate this idea in a very precise way:

Proposition 1.2.5.1 (Joyal [43]). *Let $\mathcal{C}$ be a simplicial set. The following conditions are equivalent:*

(1) *The simplicial set $\mathcal{C}$ is an ∞-category, and its homotopy category $h\mathcal{C}$ is a groupoid.*

(2) *The simplicial set $\mathcal{C}$ satisfies the extension condition for all horn inclusions $\Lambda_i^n \subseteq \Delta^n$ for $0 \leq i < n$.*

(3) *The simplicial set $\mathcal{C}$ satisfies the extension condition for all horn inclusions $\Lambda_i^n \subseteq \Delta^n$ for $0 < i \leq n$.*

(4) *The simplicial set $\mathcal{C}$ is a Kan complex; in other words, it satisfies the extension condition for all horn inclusions $\Lambda_i^n \subseteq \Delta^n$ for $0 \leq i \leq n$.*

Proof. The equivalence (1) $\Leftrightarrow$ (2) follows immediately from Proposition 1.2.4.3. Similarly, the equivalence (1) $\Leftrightarrow$ (3) follows by applying Proposition 1.2.4.3 to $\mathcal{C}^{op}$. We conclude by observing that (4) $\Leftrightarrow$ (2) $\wedge$ (3). □

Remark 1.2.5.2. The assertion that we can identify ∞-groupoids with spaces is less obvious in other formulations of higher category theory. For example, suppose that $\mathcal{C}$ is a topological category whose homotopy category $h\mathcal{C}$ is a groupoid. For simplicity, we will assume furthermore that $\mathcal{C}$ has a single object X. We may then identify $\mathcal{C}$ with the topological monoid $M = \operatorname{Hom}_{\mathcal{C}}(X, X)$. The assumption that $h\mathcal{C}$ is a groupoid is equivalent to the assumption that the discrete monoid $\pi_0 M$ is a group. In this case, one can show that the unit map $M \to \Omega B M$ is a weak homotopy equivalence, where BM denotes the classifying space of the topological monoid M. In other words, up to equivalence, specifying $\mathcal{C}$ (together with the object X) is equivalent to specifying the space BM (together with its base point).

Informally, we might say that the inclusion functor i from Kan complexes to ∞-categories exhibits the ∞-category of (small) ∞-groupoids as a full subcategory of the ∞-bicategory of (small) ∞-categories. Conversely, every ∞-category $\mathcal{C}$ has an "underlying" ∞-groupoid, which is obtained by discarding the noninvertible morphisms of $\mathcal{C}$:

Proposition 1.2.5.3 ([44]). *Let* $\mathcal{C}$ *be an* ∞-*category. Let* $\mathcal{C}' \subseteq \mathcal{C}$ *be the largest simplicial subset of* $\mathcal{C}$ *having the property that every edge of* $\mathcal{C}'$ *is an equivalence in* $\mathcal{C}$. *Then* $\mathcal{C}'$ *is a Kan complex. It may be characterized by the following universal property: for any Kan complex* K, *the induced map* $\mathrm{Hom}_{\mathrm{Set}_\Delta}(K, \mathcal{C}') \to \mathrm{Hom}_{\mathrm{Set}_\Delta}(K, \mathcal{C})$ *is a bijection.*

Proof. It is straightforward to check that $\mathcal{C}'$ is an ∞-category. Moreover, if f is a morphism in $\mathcal{C}'$, then f has a homotopy inverse $g \in \mathcal{C}$. Since g is itself an equivalence in $\mathcal{C}$, we conclude that g belongs to $\mathcal{C}'$ and is therefore a homotopy inverse to f in $\mathcal{C}'$. In other words, every morphism in $\mathcal{C}'$ is an equivalence, so that $\mathcal{C}'$ is a Kan complex by Proposition 1.2.5.1. To prove the last assertion, we observe that if K is an ∞-category, then any map of simplicial sets $\phi : K \to \mathcal{C}$ carries equivalences in K to equivalences in $\mathcal{C}$. In particular, if K is a Kan complex, then ϕ factors (uniquely) through $\mathcal{C}'$. $\square$

We can describe the situation of Proposition 1.2.5.3 by saying that $\mathcal{C}'$ *is the largest Kan complex contained in* $\mathcal{C}$. The functor $\mathcal{C} \mapsto \mathcal{C}'$ is right adjoint to the inclusion functor from Kan complexes to ∞-categories. It is easy to see that this right adjoint is an invariant notion: that is, a categorical equivalence of ∞-categories $\mathcal{C} \to \mathcal{D}$ induces a homotopy equivalence $\mathcal{C}' \to \mathcal{D}'$ of Kan complexes.

Remark 1.2.5.4. It is easy to give analogous constructions in the case of topological or simplicial categories. For example, if $\mathcal{C}$ is a topological category, then we can define $\mathcal{C}'$ to be another topological category with the same objects as $\mathcal{C}$, where $\mathrm{Map}_{\mathcal{C}'}(X, Y) \subseteq \mathrm{Map}_{\mathcal{C}}(X, Y)$ is the subspace consisting of equivalences in $\mathrm{Map}_{\mathcal{C}}(X, Y)$, equipped with the subspace topology.

Remark 1.2.5.5. We will later introduce a relative version of the construction described in Proposition 1.2.5.3, which applies to certain families of ∞-categories (Corollary 2.4.2.5).

Although the inclusion functor from Kan complexes to ∞-categories does not literally have a left adjoint, it does have such an in a higher-categorical sense. This left adjoint is computed by any "fibrant replacement" functor (for the usual model structure) from Set_Δ to itself, for example, the functor $S \mapsto \mathrm{Sing}\,|S|$. The unit map $u : S \to \mathrm{Sing}\,|S|$ is always a weak homotopy equivalence but generally not a categorical equivalence. For example, if S is an ∞-category, then u is a categorical equivalence if and only if S is a Kan complex. In general, $\mathrm{Sing}\,|S|$ may be regarded as the ∞-groupoid obtained from S by freely adjoining inverses to all the morphisms in S.

Remark 1.2.5.6. The inclusion functor i and its homotopy-theoretic left adjoint may also be understood using the formalism of *localizations of model categories*. In addition to its usual model category structure, the category Set_Δ of simplicial sets may be endowed with the *Joyal model structure*, which we will define in §2.2.5. These model structures have the same cofibrations (in both cases, the cofibrations are simply the monomorphisms of simplicial sets). However, the Joyal model structure has fewer weak equivalences (categorical equivalences rather than weak homotopy equivalences) and consequently more fibrant objects (all ∞-categories rather than only Kan complexes). It follows that the usual homotopy theory of simplicial sets is a localization of the homotopy theory of ∞-categories. The identity functor from $\mathcal{S}et_\Delta$ to itself determines a Quillen adjunction between these two homotopy theories, which plays the role of i and its left adjoint.

1.2.6 Homotopy Commutativity versus Homotopy Coherence

Let $\mathcal{C}$ be an ∞-category (topological category, simplicial category). To a first approximation, working in $\mathcal{C}$ is like working in its homotopy category $h\mathcal{C}$: up to equivalence, $\mathcal{C}$ and $h\mathcal{C}$ have the same objects and morphisms. The main difference between $h\mathcal{C}$ and $\mathcal{C}$ is that in $\mathcal{C}$ one must not ask whether or not morphisms are *equal*; instead one should ask whether or not they are *homotopic*. If so, the homotopy itself is an additional datum which we will need to consider. Consequently, the notion of a commutative diagram in $h\mathcal{C}$, which corresponds to a *homotopy commutative* diagram in $\mathcal{C}$, is quite unnatural and usually needs to be replaced by the more refined notion of a *homotopy coherent* diagram in $\mathcal{C}$.

To understand the problem, let us suppose that $F : \mathcal{J} \to \mathcal{H}$ is a functor from an ordinary category $\mathcal{J}$ into the homotopy category of spaces $\mathcal{H}$. In other words, F assigns to each object $X \in \mathcal{J}$ a space (say, a CW complex) $F(X)$, and to each morphism $\phi : X \to Y$ in $\mathcal{J}$ a continuous map of spaces $F(\phi) : F(X) \to F(Y)$ (well-defined up to homotopy), such that $F(\phi \circ \psi)$ is homotopic to $F(\phi) \circ F(\psi)$ for any pair of composable morphisms ϕ, ψ in $\mathcal{J}$. In this situation, it may or may not be possible to *lift* F to an actual functor $\widetilde{F}$ from $\mathcal{J}$ to the ordinary category of topological spaces such that $\widetilde{F}$ induces a functor $\mathcal{J} \to \mathcal{H}$ which is naturally isomorphic to F. In general, there are obstructions to both the existence and the uniqueness of the lifting $\widetilde{F}$, even up to homotopy. To see this, let us suppose for a moment that $\widetilde{F}$ exists, so that there exist homotopies $k_\phi : \widetilde{F}(\phi) \simeq F(\phi)$. These homotopies determine *additional* data on F: namely, one obtains a canonical homotopy $h_{\phi,\psi}$ from $F(\phi \circ \psi)$ to $F(\phi) \circ F(\psi)$ by composing

$$F(\phi \circ \psi) \simeq \widetilde{F}(\phi \circ \psi) = \widetilde{F}(\phi) \circ \widetilde{F}(\psi) \simeq F(\phi) \circ F(\psi).$$

The functor F to the homotopy category $\mathcal{H}$ should be viewed as a first approximation to $\widetilde{F}$; we obtain a second approximation when we take into account the homotopies $h_{\phi,\psi}$. These homotopies are not arbitrary: the associativity of composition gives a relationship between $h_{\phi,\psi}, h_{\psi,\theta}, h_{\phi,\psi\circ\theta}$, and $h_{\phi\circ\psi,\theta}$, for a composable triple of morphisms (ϕ, ψ, θ) in $\mathcal{J}$. This relationship may be formulated in terms of the existence of a certain higher homotopy, which is once again canonically determined by $\widetilde{F}$ (and the homotopies k_ϕ). To obtain the next approximation to $\widetilde{F}$, we should take these higher homotopies into account and formulate the associativity properties that *they* enjoy, and so on. Roughly speaking, a *homotopy coherent* diagram in $\mathcal{C}$ is a functor $F : \mathcal{J} \to h\mathcal{C}$ together with all of the extra data that would be available if we were able to lift F to a functor $\widetilde{F} : \mathfrak{I} \to \mathcal{C}$.

The distinction between homotopy commutativity and homotopy coherence is arguably the *main* difficulty in working with higher categories. The idea of homotopy coherence is simple enough and can be made precise in the setting of a general topological category. However, the amount of data required to specify a homotopy coherent diagram is considerable, so the concept is quite difficult to employ in practical situations.

Remark 1.2.6.1. Let $\mathfrak{I}$ be an ordinary category and let $\mathcal{C}$ be a topological category. Any functor $F : \mathfrak{I} \to \mathcal{C}$ determines a homotopy coherent diagram in $\mathcal{C}$ (with all of the homotopies involved being constant). For many topological categories $\mathcal{C}$, the converse fails: not every homotopy-coherent diagram in $\mathcal{C}$ can be obtained in this way, even up to equivalence. In these cases, it is the notion of *homotopy coherent* diagram which is fundamental; a homotopy coherent diagram should be regarded as "just as good" as a strictly commutative diagram for ∞-categorical purposes. As evidence for this, we remark that given an equivalence $\mathcal{C}' \to \mathcal{C}$, a strictly commutative diagram $F : \mathfrak{I} \to \mathcal{C}$ cannot always be lifted to a strictly commutative diagram in $\mathcal{C}'$; however, it can always be lifted (up to equivalence) to a homotopy coherent diagram in $\mathcal{C}'$.

One of the advantages of working with ∞-categories is that the definition of a homotopy coherent diagram is easy to formulate. We can simply define a homotopy coherent diagram in an ∞-category $\mathcal{C}$ to be a map of simplicial sets $f : N(\mathfrak{I}) \to \mathcal{C}$. The restriction of f to simplices of low dimension encodes the induced map on homotopy categories. Specifying f on higher-dimensional simplices gives precisely the "coherence data" that the above discussion calls for.

Remark 1.2.6.2. Another possible approach to the problem of homotopy coherence is to restrict our attention to simplicial (or topological) categories $\mathcal{C}$ in which every homotopy coherent diagram is equivalent to a strictly commutative diagram. For example, this is always true when $\mathcal{C}$ arises from a simplicial model category (Proposition 4.2.4.4). Consequently, in the framework of model categories, it is possible to ignore the theory of homotopy coherent diagrams and work with strictly commutative diagrams instead. This approach is quite powerful, particularly when combined with the observation that every simplicial category $\mathcal{C}$ admits a fully faithful embedding into a simplicial model category (for example, one can use a simplicially enriched version of the Yoneda embedding). This idea can be used to show that every homotopy coherent diagram in $\mathcal{C}$ can be "straightened" to a commutative diagram, possibly after replacing $\mathcal{C}$ by an equivalent simplicial category (for a more precise version of this statement, we refer the reader to Corollary 4.2.4.7).

1.2.7 Functors Between Higher Categories

The notion of a homotopy coherent diagram in an higher category $\mathcal{C}$ is a special case of the more general notion of a functor $F : \mathcal{J} \to \mathcal{C}$ between higher categories (specifically, it is the special case in which $\mathcal{J}$ is assumed to be an ordinary category). Just as the collection of all ordinary categories forms a bicategory (with functors as morphisms and natural transformations as 2-morphisms), the collection of all ∞-categories can be organized into an ∞-bicategory. In particular, for any ∞-categories $\mathcal{C}$ and $\mathcal{C}'$, we expect to be able to construct an ∞-category $\mathrm{Fun}(\mathcal{C}, \mathcal{C}')$ of functors from $\mathcal{C}$ to $\mathcal{C}'$.

In the setting of topological categories, the construction of an appropriate mapping object $\mathrm{Fun}(\mathcal{C}, \mathcal{C}')$ is quite difficult. The naive guess is that $\mathrm{Fun}(\mathcal{C}, \mathcal{C}')$ should be a category of topological functors from $\mathcal{C}$ to $\mathcal{C}'$: that is, functors which induce continuous maps between morphism spaces. However, we saw in §1.2.6 that this notion is generally too rigid, even in the special case where $\mathcal{C}$ is an ordinary category.

Remark 1.2.7.1. Using the language of model categories, one might say that the problem is that not every topological category is *cofibrant*. If $\mathcal{C}$ is a *cofibrant* topological category (for example, if $\mathcal{C} = |\mathfrak{C}[S]|$, where S is a simplicial set), then the collection of topological functors from $\mathcal{C}$ to $\mathcal{C}'$ is large enough to contain representatives for every ∞-categorical functor from $\mathcal{C}$ to $\mathcal{C}'$. Most ordinary categories are not cofibrant when viewed as topological categories. More importantly, the property of being cofibrant is not stable under products, so that naive attempts to construct a mapping object $\mathrm{Fun}(\mathcal{C}, \mathcal{C}')$ need not give the correct answer even when $\mathcal{C}$ itself is assumed cofibrant (if $\mathcal{C}$ is cofibrant, then we are guaranteed to have "enough" topological functors $\mathcal{C} \to \mathcal{C}'$ to represent all functors between the underlying ∞-categories but not necessarily enough natural transformations between them; note that the product $\mathcal{C} \times [1]$ is usually not cofibrant, even in the simplest nontrivial case where $\mathcal{C} = [1]$.) This is arguably the most important technical disadvantage of the theory of topological (or simplicial) categories as an approach to higher category theory.

The construction of functor categories is much easier to describe in the framework of ∞-categories. If $\mathcal{C}$ and $\mathcal{D}$ are ∞-categories, then we can simply define a *functor* from $\mathcal{C}$ to $\mathcal{D}$ to be a map $p : \mathcal{C} \to \mathcal{D}$ of simplicial sets.

Notation 1.2.7.2. Let $\mathcal{C}$ and $\mathcal{D}$ be simplicial sets. We let $\mathrm{Fun}(\mathcal{C}, \mathcal{D})$ denote the simplicial set $\mathrm{Map}_{\mathrm{Set}_\Delta}(\mathcal{C}, \mathcal{D})$ parametrizing maps from $\mathcal{C}$ to $\mathcal{D}$. We will use this notation only when $\mathcal{D}$ is an ∞-category (the simplicial set $\mathcal{C}$ will often, but not always, be an ∞-category as well). We will refer to $\mathrm{Fun}(\mathcal{C}, \mathcal{D})$ as the ∞-*category of functors from $\mathcal{C}$ to $\mathcal{D}$* (see Proposition 1.2.7.3 below). We will refer to morphisms in $\mathrm{Fun}(\mathcal{C}, \mathcal{D})$ as *natural transformations* of functors, and equivalences in $\mathrm{Fun}(\mathcal{C}, \mathcal{D})$ as *natural equivalences*.

Proposition 1.2.7.3. *Let K be an arbitrary simplicial set.*

(1) *For every ∞-category $\mathcal{C}$, the simplicial set $\mathrm{Fun}(K, \mathcal{C})$ is an ∞-category.*

(2) *Let* $\mathcal{C} \to \mathcal{D}$ *be a categorical equivalence of ∞-categories. Then the induced map* $\mathrm{Fun}(K, \mathcal{C}) \to \mathrm{Fun}(K, \mathcal{D})$ *is a categorical equivalence.*

(3) *Let* $\mathcal{C}$ *be an ∞-category and* $K \to K'$ *a categorical equivalence of simplicial sets. Then the induced map* $\mathrm{Fun}(K', \mathcal{C}) \to \mathrm{Fun}(K, \mathcal{C})$ *is a categorical equivalence.*

The proof makes use of the Joyal model structure on Set_Δ and will be given in §2.2.5.

1.2.8 Joins of ∞-Categories

Let $\mathcal{C}$ and $\mathcal{C}'$ be ordinary categories. We will define a new category $\mathcal{C} \star \mathcal{C}'$, called the *join* of $\mathcal{C}$ and $\mathcal{C}'$. An object of $\mathcal{C} \star \mathcal{C}'$ is either an object of $\mathcal{C}$ or an object of $\mathcal{C}'$. The morphism sets are given as follows:

$$\mathrm{Hom}_{\mathcal{C} \star \mathcal{C}'}(X, Y) = \begin{cases} \mathrm{Hom}_{\mathcal{C}}(X, Y) & \text{if } X, Y \in \mathcal{C} \\ \mathrm{Hom}_{\mathcal{C}'}(X, Y) & \text{if } X, Y \in \mathcal{C}' \\ \emptyset & \text{if } X \in \mathcal{C}', Y \in \mathcal{C} \\ * & \text{if } X \in \mathcal{C}, Y \in \mathcal{C}'. \end{cases}$$

Composition of morphisms in $\mathcal{C} \star \mathcal{C}'$ is defined in the obvious way.

The join construction described above is often useful when discussing diagram categories, limits, and colimits. In this section, we will introduce a generalization of this construction to the ∞-categorical setting.

Definition 1.2.8.1. If S and S' are simplicial sets, then the simplicial set $S \star S'$ is defined as follows: for each nonempty finite linearly ordered set J, we set

$$(S \star S')(J) = \coprod_{J = I \cup I'} S(I) \times S'(I'),$$

where the union is taken over all decompositions of J into disjoint subsets I and I', satisfying $i < i'$ for all $i \in I$, $i' \in I'$. Here we allow the possibility that either I or I' is empty, in which case we agree to the convention that $S(\emptyset) = S'(\emptyset) = *$.

More concretely, we have

$$(S \star S')_n = S_n \cup S'_n \cup \bigcup_{i+j=n-1} S_i \times S'_j.$$

The join operation endows Set_Δ with the structure of a monoidal category (see §A.1.3). The identity for the join operation is the empty simplicial set $\emptyset = \Delta^{-1}$. More generally, we have natural isomorphisms $\phi_{ij} : \Delta^{i-1} \star \Delta^{j-1} \simeq \Delta^{(i+j)-1}$ for all $i, j \geq 0$.

Remark 1.2.8.2. The operation $\star$ is essentially determined by the isomorphisms ϕ_{ij}, together with its behavior under the formation of colimits: for any fixed simplicial set S, the functors

$$T \mapsto T \star S$$

$$T \mapsto S \star T$$

commute with colimits when regarded as functors from Set_Δ to the under-category $(\text{Set}_\Delta)_{S/}$ of simplicial sets *under* S.

Passage to the nerve carries joins of categories into joins of simplicial sets. More precisely, for every pair of categories $\mathcal{C}$ and $\mathcal{C}'$, there is a canonical isomorphism

$$N(\mathcal{C} \star \mathcal{C}') \simeq N(\mathcal{C}) \star N(\mathcal{C}').$$

(The existence of this isomorphism persists when we allow $\mathcal{C}$ and $\mathcal{C}'$ to be simplicial or topological categories and apply the appropriate generalization of the nerve functor.) This suggests that the join operation on simplicial sets is the appropriate ∞-categorical analogue of the join operation on categories.

We remark that the formation of joins does not commute with the functor $\mathfrak{C}[\bullet]$. However, the simplicial category $\mathfrak{C}[S \star S']$ contains $\mathfrak{C}[S]$ and $\mathfrak{C}[S']$ as full (topological) subcategories and contains no morphisms from objects of $\mathfrak{C}[S']$ to objects of $\mathfrak{C}[S]$. Consequently, there is unique map $\phi : \mathfrak{C}[S \star S'] \to \mathfrak{C}[S] \star \mathfrak{C}[S']$ which reduces to the identity on $\mathfrak{C}[S]$ and $\mathfrak{C}[S']$. We will later show that ϕ is an equivalence of simplicial categories (Corollary 4.2.1.4).

We conclude by recording a pleasant property of the join operation:

Proposition 1.2.8.3 (Joyal [44]). *If S and S' are ∞-categories, then $S \star S'$ is an ∞-category.*

Proof. Let $p : \Lambda_i^n \to S \star S'$ be a map, with $0 < i < n$. If p carries Λ_i^n entirely into $S \subseteq S \star S'$ or into $S' \subseteq S \star S'$, then we deduce the existence of an extension of p to Δ^n using the assumption that S and S' are ∞-categories. Otherwise, we may suppose that p carries the vertices $\{0, \ldots, j\}$ into S, and the vertices $\{j+1, \ldots, n\}$ into S'. We may now restrict p to obtain maps

$$\Delta^{\{0,\ldots,j\}} \to S$$

$$\Delta^{\{j+1,\ldots,n\}} \to S',$$

which together determine a map $\Delta^n \to S \star S'$ extending p. $\square$

Notation 1.2.8.4. Let K be a simplicial set. The *left cone* $K^\triangleleft$ is defined to be the join $\Delta^0 \star K$. Dually, the *right cone* $K^\triangleright$ is defined to be the join $K \star \Delta^0$. Either cone contains a distinguished vertex (belonging to Δ^0), which we will refer to as the *cone point*.

1.2.9 Overcategories and Undercategories

Let $\mathcal{C}$ be an ordinary category and $X \in \mathcal{C}$ an object. The *overcategory* $\mathcal{C}_{/X}$ is defined as follows: the objects of $\mathcal{C}_{/X}$ are morphisms $Y \to X$ in $\mathcal{C}$ having target X. Morphisms are given by commutative triangles

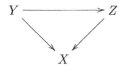

and composition is defined in the obvious way.

One can rephrase the definition of the overcategory as follows. Let $[0]$ denote the category with a single object possessing only an identity morphism. Then specifying an object $X \in \mathcal{C}$ is equivalent to specifying a functor $x : [0] \to \mathcal{C}$. The overcategory $\mathcal{C}_{/X}$ may then be described by the following universal property: for any category $\mathcal{C}'$, we have a bijection

$$\mathrm{Hom}(\mathcal{C}', \mathcal{C}_{/X}) \simeq \mathrm{Hom}_x(\mathcal{C}' \star [0], \mathcal{C}),$$

where the subscript on the right hand side indicates that we consider only those functors $\mathcal{C}' \star [0] \to \mathcal{C}$ whose restriction to $[0]$ coincides with x.

Our goal in this section is to generalize the construction of overcategories to the ∞-categorical setting. Let us begin by working in the framework of topological categories. In this case, there is a natural candidate for the relevant overcategory. Namely, if $\mathcal{C}$ is a topological category containing an object X, then the overcategory $\mathcal{C}_{/X}$ (defined as above) has the structure of a topological category where each morphism space $\mathrm{Map}_{\mathcal{C}_{/X}}(Y, Z)$ is topologized as a subspace of $\mathrm{Map}_{\mathcal{C}}(Y, Z)$ (here we are identifying an object of $\mathcal{C}_{/X}$ with its image in $\mathcal{C}$). This topological category is usually *not* a model for the correct ∞-categorical slice construction. The problem is that a morphism in $\mathcal{C}_{/X}$ consists of a commutative triangle

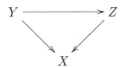

of objects over X. To obtain the correct notion, we should also allow triangles which commute only up to homotopy.

Remark 1.2.9.1. In some cases, the naive overcategory $\mathcal{C}_{/X}$ is a good approximation to the correct construction: see Lemma 6.1.3.13.

In the setting of ∞-categories, Joyal has given a much simpler description of the desired construction (see [43]). This description will play a vitally important role throughout this book. We begin by noting the following:

Proposition 1.2.9.2 ([43]). *Let S and K be simplicial sets, and $p : K \to S$ an arbitrary map. There exists a simplicial set $S_{/p}$ with the following universal property:*

$$\mathrm{Hom}_{\mathrm{Set}_\Delta}(Y, S_{/p}) = \mathrm{Hom}_p(Y \star K, S),$$

where the subscript on the right hand side indicates that we consider only those morphisms $f : Y \star K \to S$ such that $f|K = p$.

Proof. One defines $(S_{/p})_n$ to be $\mathrm{Hom}_p(\Delta^n \star K, S)$. The universal property holds by definition when Y is a simplex. It holds in general because both sides are compatible with the formation of colimits in Y. $\square$

Let $p : K \to S$ be as in Proposition 1.2.9.2. If S is an ∞-category, we will refer to $S_{/p}$ as an *overcategory* of S or as the ∞-*category of objects of S over p*. The following result guarantees that the operation of passing to overcategories is well-behaved:

Proposition 1.2.9.3. *Let $p : K \to \mathcal{C}$ be a map of simplicial sets and suppose that $\mathcal{C}$ is an ∞-category. Then $\mathcal{C}_{/p}$ is an ∞-category. Moreover, if $q : \mathcal{C} \to \mathcal{C}'$ is a categorical equivalence of ∞-categories, then the induced map $\mathcal{C}_{/p} \to \mathcal{C}'_{/qp}$ is a categorical equivalence as well.*

The proof requires a number of ideas which we have not yet introduced and will be postponed (see Proposition 2.1.2.2 for the first assertion, and §2.4.5 for the second).

Remark 1.2.9.4. Let $\mathcal{C}$ be an ∞-category. In the particular case where $p : \Delta^n \to \mathcal{C}$ classifies an n-simplex $\sigma \in \mathcal{C}_n$, we will often write $\mathcal{C}_{/\sigma}$ in place of $\mathcal{C}_{/p}$. In particular, if X is an object of $\mathcal{C}$, we let $\mathcal{C}_{/X}$ denote the overcategory $\mathcal{C}_{/p}$, where $p : \Delta^0 \to \mathcal{C}$ has image X.

Remark 1.2.9.5. Let $p : K \to \mathcal{C}$ be a map of simplicial sets. The preceding discussion can be dualized, replacing $Y \star K$ by $K \star Y$; in this case we denote the corresponding simplicial set by $\mathcal{C}_{p/}$, which (if $\mathcal{C}$ is an ∞-category) we will refer to as an *undercategory* of $\mathcal{C}$. In the special case where $K = \Delta^n$ and p classifies a simplex $\sigma \in \mathcal{C}_n$, we will also write $\mathcal{C}_{\sigma/}$ for $\mathcal{C}_{p/}$; in particular, we will write $\mathcal{C}_{X/}$ when X is an object of $\mathcal{C}$.

Remark 1.2.9.6. If $\mathcal{C}$ is an ordinary category and $X \in \mathcal{C}$, then there is a canonical isomorphism $\mathrm{N}(\mathcal{C})_{/X} \simeq \mathrm{N}(\mathcal{C}_{/X})$. In other words, the overcategory construction for ∞-categories can be regarded as a *generalization* of the relevant construction from classical category theory.

1.2.10 Fully Faithful and Essentially Surjective Functors

Definition 1.2.10.1. Let $F : \mathcal{C} \to \mathcal{D}$ be a functor between topological categories (simplicial categories, simplicial sets). We will say that F is *essentially surjective* if the induced functor $hF : h\mathcal{C} \to h\mathcal{D}$ is essentially surjective (that is, if every object of $\mathcal{D}$ is equivalent to $F(X)$ for some $X \in \mathcal{C}$).

We will say that F is *fully faithful* if hF is a fully faithful functor of $\mathcal{H}$-enriched categories. In other words, F is fully faithful if and only if, for every pair of objects $X, Y \in \mathcal{C}$, the induced map $\mathrm{Map}_{h\mathcal{C}}(X, Y) \to \mathrm{Map}_{h\mathcal{D}}(F(X), F(Y))$ is an isomorphism in the homotopy category $\mathcal{H}$.

Remark 1.2.10.2. Because Definition 1.2.10.1 makes reference only to the homotopy categories of $\mathcal{C}$ and $\mathcal{D}$, it is invariant under equivalence and under operations which pass between the various models for higher category theory that we have introduced.

Just as in ordinary category theory, a functor F is an equivalence if and only if it is fully faithful and essentially surjective.

1.2.11 Subcategories of ∞-Categories

Let $\mathcal{C}$ be an ∞-category and let $(h\mathcal{C})' \subseteq h\mathcal{C}$ be a subcategory of its homotopy category. We can then form a pullback diagram of simplicial sets:

$$
\begin{array}{ccc}
\mathcal{C}' & \longrightarrow & \mathcal{C} \\
\downarrow & & \downarrow \\
N(h\mathcal{C})' & \longrightarrow & N(h\mathcal{C}).
\end{array}
$$

We will refer to $\mathcal{C}'$ as the *subcategory of $\mathcal{C}$ spanned by* $(h\mathcal{C})'$. In general, we will say that a simplicial subset $\mathcal{C}' \subseteq \mathcal{C}$ is a *subcategory* of $\mathcal{C}$ if it arises via this construction.

Remark 1.2.11.1. We use the term "subcategory," rather than "sub-∞-category," in order to avoid awkward language. The terminology is not meant to suggest that $\mathcal{C}'$ is itself a category or is isomorphic to the nerve of a category.

In the case where $(h\mathcal{C})'$ is a full subcategory of $h\mathcal{C}$, we will say that $\mathcal{C}'$ is a *full subcategory* of $\mathcal{C}$. In this case, $\mathcal{C}'$ is determined by the set $\mathcal{C}'_0$ of those objects $X \in \mathcal{C}$ which belong to $\mathcal{C}'$. We will then say that $\mathcal{C}'$ is the *full subcategory of $\mathcal{C}$ spanned by* $\mathcal{C}'_0$.

It follows from Remark 1.2.2.4 that the inclusion $\mathcal{C}' \subseteq \mathcal{C}$ is fully faithful. In general, any fully faithful functor $f : \mathcal{C}'' \to \mathcal{C}$ factors as a composition

$$
\mathcal{C}'' \xrightarrow{f'} \mathcal{C}' \xrightarrow{f''} \mathcal{C},
$$

where f' is an equivalence of ∞-categories and f'' is the inclusion of the full subcategory $\mathcal{C}' \subseteq \mathcal{C}$ spanned by the set of objects $f(\mathcal{C}''_0) \subseteq \mathcal{C}_0$.

1.2.12 Initial and Final Objects

If $\mathcal{C}$ is an ordinary category, then an object $X \in \mathcal{C}$ is said to be *final* if $\operatorname{Hom}_\mathcal{C}(Y, X)$ consists of a single element for every $Y \in \mathcal{C}$. Dually, an object $X \in \mathcal{C}$ is *initial* if it is final when viewed as an object of $\mathcal{C}^{op}$. The goal of this section is to generalize these definitions to the ∞-categorical setting.

If $\mathcal{C}$ is a topological category, then a candidate definition immediately presents itself: we could ignore the topology on the morphism spaces and consider those objects of $\mathcal{C}$ which are final when $\mathcal{C}$ is regarded as an ordinary category. This requirement is unnaturally strong. For example, the category $\mathcal{CG}$ of compactly generated Hausdorff spaces has a final object: the topological space $*$, consisting of a single point. However, there are objects of $\mathcal{CG}$ which are equivalent to $*$ (any contractible space) but not isomorphic to $*$ (and therefore not final objects of $\mathcal{CG}$, at least in the classical sense). Since any reasonable ∞-categorical notion is stable under equivalence, we need to find a weaker condition.

Definition 1.2.12.1. Let $\mathcal{C}$ be a topological category (simplicial category, simplicial set). An object $X \in \mathcal{C}$ is *final* if it is final in the homotopy category h$\mathcal{C}$, regarded as a category enriched over $\mathcal{H}$. In other words, X is final if and only if for each $Y \in \mathcal{C}$, the mapping space $\mathrm{Map}_{h\mathcal{C}}(Y, X)$ is weakly contractible (that is, a final object of $\mathcal{H}$).

Remark 1.2.12.2. Since Definition 1.2.12.1 makes reference only to the homotopy category h$\mathcal{C}$, it is invariant under equivalence and under passing between the various models for higher category theory.

In the setting of ∞-categories, it is convenient to employ a slightly more sophisticated definition, which we borrow from [43].

Definition 1.2.12.3. Let $\mathcal{C}$ be a simplicial set. A vertex X of $\mathcal{C}$ is *strongly final* if the projection $\mathcal{C}_{/X} \to \mathcal{C}$ is a trivial fibration of simplicial sets.

In other words, a vertex X of $\mathcal{C}$ is strongly final if and only if any map $f_0 : \partial \Delta^n \to \mathcal{C}$ such that $f_0(n) = X$ can be extended to a map $f : \Delta^n \to S$.

Proposition 1.2.12.4. *Let $\mathcal{C}$ be an ∞-category containing an object Y. The object Y is strongly final if and only if, for every object $X \in \mathcal{C}$, the Kan complex $\mathrm{Hom}_{\mathcal{C}}^{\mathrm{R}}(X, Y)$ is contractible.*

Proof. The "only if" direction is clear: the space $\mathrm{Hom}_{\mathcal{C}}^{\mathrm{R}}(X, Y)$ is the fiber of the projection $p : \mathcal{C}_{/Y} \to \mathcal{C}$ over the vertex X. If p is a trivial fibration, then the fiber is a contractible Kan complex. Since p is a right fibration (Proposition 2.1.2.1), the converse holds as well (Lemma 2.1.3.4). $\square$

Corollary 1.2.12.5. *Let $\mathcal{C}$ be a simplicial set. Every strongly final object of $\mathcal{C}$ is a final object of $\mathcal{C}$. The converse holds if $\mathcal{C}$ is an ∞-category.*

Proof. Let $[0]$ denote the category with a single object and a single morphism. Suppose that Y is a strongly final vertex of $\mathcal{C}$. Then there exists a retraction of $\mathcal{C}^{\triangleright}$ onto $\mathcal{C}$ carrying the cone point to Y. Consequently, we obtain a retraction of ($\mathcal{H}$-enriched) homotopy categories from h$\mathcal{C} \star [0]$ to h$\mathcal{C}$ carrying the unique object of $[0]$ to Y. This implies that Y is final in h$\mathcal{C}$, so that Y is a final object of $\mathcal{C}$.

To prove the converse, we note that if $\mathcal{C}$ is an ∞-category, then the Kan complex $\mathrm{Hom}_{\mathcal{C}}^{\mathrm{R}}(X, Y)$ represents the homotopy type $\mathrm{Map}_{\mathcal{C}}(X, Y) \in \mathcal{H}$; by Proposition 1.2.12.4 this space is contractible for all X if and only if Y is strongly final. $\square$

Remark 1.2.12.6. The above discussion dualizes in an evident way, so that we have a notion of *initial* objects of an ∞-category $\mathcal{C}$.

Example 1.2.12.7. Let $\mathcal{C}$ be an ordinary category containing an object X. Then X is a final (initial) object of the ∞-category N($\mathcal{C}$) if and only if it is a final (initial) object of $\mathcal{C}$ in the usual sense.

Remark 1.2.12.8. Definition 1.2.12.3 is only natural in the case where $\mathcal{C}$ is an ∞-category. For example, if $\mathcal{C}$ is not an ∞-category, then the collection of strongly final vertices of $\mathcal{C}$ need not be stable under equivalence.

An ordinary category $\mathcal{C}$ may have more than one final object, but any two final objects are uniquely isomorphic to one another. In the setting of ∞-categories, an analogous statement holds but is slightly more complicated because the word "unique" needs to be interpreted in a homotopy-theoretic sense:

Proposition 1.2.12.9 (Joyal). *Let $\mathcal{C}$ be a ∞-category and let $\mathcal{C}'$ be the full subcategory of $\mathcal{C}$ spanned by the final vertices of $\mathcal{C}$. Then $\mathcal{C}'$ either is empty or is a contractible Kan complex.*

Proof. We wish to prove that every map $p : \partial \Delta^n \to \mathcal{C}'$ can be extended to an n-simplex of $\mathcal{C}'$. If $n = 0$, this is possible unless $\mathcal{C}'$ is empty. For $n > 0$, the desired extension exists because p carries the nth vertex of $\partial \Delta^n$ to a final object of $\mathcal{C}$. $\qquad\square$

1.2.13 Limits and Colimits

An important consequence of the distinction between homotopy commutativity and homotopy coherence is that the appropriate notions of limit and colimit in a higher category $\mathcal{C}$ do not coincide with the notions of limit and colimit in the homotopy category $h\mathcal{C}$ (where limits and colimits often do not exist). Limits and colimits in $\mathcal{C}$ are often referred to as *homotopy limits* and *homotopy colimits* to avoid confusing them with ordinary limits and colimits.

Homotopy limits and colimits can be defined in a topological category, but the definition is rather complicated. We will review a few special cases here and discuss the general definition in the Appendix (§A.2.8).

Example 1.2.13.1. Let $\{X_\alpha\}$ be a family of objects in a topological category $\mathcal{C}$. A *homotopy product* $X = \prod_\alpha X_\alpha$ is an object of $\mathcal{C}$ equipped with morphisms $f_\alpha : X \to X_\alpha$ which induce a weak homotopy equivalence

$$\mathrm{Map}_{\mathcal{C}}(Y, X) \to \prod_\alpha \mathrm{Map}_{\mathcal{C}}(Y, X_\alpha)$$

for every object $Y \in \mathcal{C}$.

Passing to path components and using the fact that π_0 commutes with products, we deduce that

$$\mathrm{Hom}_{h\mathcal{C}}(Y, X) \simeq \prod_\alpha \mathrm{Hom}_{h\mathcal{C}}(Y, X_\alpha),$$

so that any product in $\mathcal{C}$ is also a product in $h\mathcal{C}$. In particular, the object X is determined up to canonical isomorphism in $h\mathcal{C}$.

In the special case where the index set is empty, we recover the notion of a final object of $\mathcal{C}$: an object X for which each of the mapping spaces $\mathrm{Map}_{\mathcal{C}}(Y, X)$ is weakly contractible.

Example 1.2.13.2. Given two morphisms $\pi : X \to Z$ and $\psi : Y \to Z$ in a topological category $\mathcal{C}$, let us define $\mathrm{Map}_{\mathcal{C}}(W, X \times_Z^h Y)$ to be the space consisting of points $p \in \mathrm{Map}_{\mathcal{C}}(W, X)$ and $q \in \mathrm{Map}_{\mathcal{C}}(W, Y)$ together with a path $r : [0,1] \to \mathrm{Map}_{\mathcal{C}}(W, Z)$ joining $\pi \circ p$ to $\psi \circ q$. We endow $\mathrm{Map}_{\mathcal{C}}(W, X \times_Z^h Y)$ with the obvious topology, so that $X \times_Z^h Y$ can be viewed as a presheaf of topological spaces on $\mathcal{C}$. A *homotopy fiber product for X and Y over Z* is an object of $\mathcal{C}$ which represents this presheaf up to weak homotopy equivalence. In other words, it is an object $P \in \mathcal{C}$ equipped with a point $p \in \mathrm{Map}_{\mathcal{C}}(P, X \times_Z^h Y)$ which induces weak homotopy equivalences $\mathrm{Map}_{\mathcal{C}}(W, P) \to \mathrm{Map}_{\mathcal{C}}(W, X \times_Z^h Y)$ for every $W \in \mathcal{C}$.

We note that if there exists a fiber product (in the ordinary sense) $X \times_Z Y$ in the category $\mathcal{C}$, then this ordinary fiber product admits a (canonically determined) map to the homotopy fiber product (if the homotopy fiber product exists). This map need not be an equivalence, but it is an equivalence in many good cases. We also note that a homotopy fiber product P comes equipped with a map to the fiber product $X \times_Z Y$ taken in the category $h\mathcal{C}$ (if this fiber product exists); this map is usually not an isomorphism.

Remark 1.2.13.3. Homotopy limits and colimits in general may be described in relation to homotopy limits of topological spaces. The homotopy limit X of a diagram of objects $\{X_\alpha\}$ in an arbitrary topological category $\mathcal{C}$ is determined, up to equivalence, by the requirement that there exists a natural weak homotopy equivalence

$$\mathrm{Map}_{\mathcal{C}}(Y, X) \simeq \mathrm{holim}\{\mathrm{Map}_{\mathcal{C}}(Y, X_\alpha)\}.$$

Similarly, the homotopy colimit of the diagram is characterized by the existence of a natural weak homotopy equivalence

$$\mathrm{Map}_{\mathcal{C}}(X, Y) \simeq \mathrm{holim}\{\mathrm{Map}_{\mathcal{C}}(X_\alpha, Y)\}.$$

For a more precise discussion, we refer the reader to Remark A.3.3.13.

In the setting of ∞-categories, limits and colimits are quite easy to define:

Definition 1.2.13.4 (Joyal [43]). Let $\mathcal{C}$ be an ∞-category and let $p : K \to \mathcal{C}$ be an arbitrary map of simplicial sets. A *colimit* for p is an initial object of $\mathcal{C}_{p/}$, and a *limit* for p is a final object of $\mathcal{C}_{/p}$.

Remark 1.2.13.5. According to Definition 1.2.13.4, a colimit of a diagram $p : K \to \mathcal{C}$ is an object of $\mathcal{C}_{p/}$. We may identify this object with a map $\overline{p} : K^{\triangleright} \to \mathcal{C}$ extending p. In general, we will say that a map $\overline{p} : K^{\triangleright} \to \mathcal{C}$ is a *colimit diagram* if it is a colimit of $p = \overline{p}|K$. In this case, we will also abuse terminology by referring to $\overline{p}(\infty) \in \mathcal{C}$ as a *colimit of p*, where ∞ denotes the cone point of $K^{\triangleright}$.

If $p : K \to \mathcal{C}$ is a diagram, we will sometimes write $\varinjlim(p)$ to denote a colimit of p (considered either as an object of $\mathcal{C}_{p/}$ or of $\mathcal{C}$), and $\varprojlim(p)$ to denote a limit of p (as either an object of $\mathcal{C}_{/p}$ or an object of $\mathcal{C}$). This notation is slightly abusive since $\varinjlim(p)$ is not uniquely determined by p. This phenomenon is familiar in classical category theory: the colimit of a diagram is not unique but is determined up to canonical isomorphism. In the ∞-categorical setting, we have a similar uniqueness result: Proposition 1.2.12.9 implies that the collection of candidates for $\varinjlim(p)$, if nonempty, is parametrized by a contractible Kan complex.

Remark 1.2.13.6. In §4.2.4, we will show that Definition 1.2.13.4 agrees with the classical theory of homotopy (co)limits when we specialize to the case where $\mathcal{C}$ is the nerve of a topological category.

Remark 1.2.13.7. Let $\mathcal{C}$ be an ∞-category, $\mathcal{C}' \subseteq \mathcal{C}$ a full subcategory, and $p : K \to \mathcal{C}'$ a diagram. Then $\mathcal{C}'_{p/} = \mathcal{C}' \times_{\mathcal{C}} \mathcal{C}_{p/}$. In particular, if p has a colimit in $\mathcal{C}$ and that colimit belongs to $\mathcal{C}'$, then the same object may be regarded as a colimit for p in $\mathcal{C}'$.

Let $f : \mathcal{C} \to \mathcal{C}'$ be a map between ∞-categories. Let $p : K \to \mathcal{C}$ be a diagram in $\mathcal{C}$ having a colimit $x \in \mathcal{C}_{p/}$. The image $f(x) \in \mathcal{C}'_{fp/}$ may or may not be a colimit for the composite map $f \circ p$. If it is, we will say that f *preserves* the colimit of the diagram p. Often we will apply this terminology not to a particular diagram p but to some class of diagrams: for example, we may speak of maps f which preserve coproducts, pushouts, or filtered colimits (see §4.4 for a discussion of special classes of colimits). Similarly, we may ask whether or not a map f preserves the limit of a particular diagram or various families of diagrams.

We conclude this section by giving a simple example of a colimit-preserving functor.

Proposition 1.2.13.8. *Let $\mathcal{C}$ be an ∞-category and let $q : T \to \mathcal{C}$ and $p : K \to \mathcal{C}_{/q}$ be two diagrams. Let p_0 denote the composition of p with the projection $\mathcal{C}_{/q} \to \mathcal{C}$. Suppose that p_0 has a colimit in $\mathcal{C}$. Then*

(1) *The diagram p has a colimit in $\mathcal{C}_{/q}$, and that colimit is preserved by the projection $\mathcal{C}_{/q} \to \mathcal{C}$.*

(2) *An extension $\widetilde{p} : K^{\triangleright} \to \mathcal{C}_{/q}$ is a colimit of p if and only if the composition*

$$K^{\triangleright} \to \mathcal{C}_{/q} \to \mathcal{C}$$

is a colimit of p_0.

Proof. We first prove the "if" direction of (2). Let $\widetilde{p} : K^{\triangleright} \to \mathcal{C}_{/q}$ be such that the composite map $\widetilde{p}_0 : K^{\triangleright} \to \mathcal{C}$ is a colimit of p_0. We wish to show that $\widetilde{p}$ is a colimit of p. We may identify $\widetilde{p}$ with a map $K \star \Delta^0 \star T \to \mathcal{C}$. For this, it suffices to show that for any inclusion $A \subseteq B$ of simplicial sets, it is possible to solve the lifting problem depicted in the following diagram:

$$(K \star B \star T) \coprod_{K \star A \star T} (K \star \Delta^0 \star A \star T) \longrightarrow \mathcal{C}$$

$$\downarrow$$

$$K \star \Delta^0 \star B \star T.$$

Because $\widetilde{p}_0$ is a colimit of p_0, the projection

$$\mathcal{C}_{\widetilde{p}_0/} \to \mathcal{C}_{p_0/}$$

is a trivial fibration of simplicial sets and therefore has the right lifting property with respect to the inclusion $A \star T \subseteq B \star T$.

We now prove (1). Let $\widetilde{p}_0 : K^{\triangleright} \to \mathcal{C}$ be a colimit of p_0. Since the projection $\mathcal{C}_{\widetilde{p}_0/} \to \mathcal{C}_{p_0/}$ is a trivial fibration, it has the right lifting property with respect to T: this guarantees the existence of an extension $\widetilde{p} : K^{\triangleright} \to \mathcal{C}$ lifting $\widetilde{p}_0$. The preceding analysis proves that $\widetilde{p}$ is a colimit of p.

Finally, the "only if" direction of (2) follows from (1) since any two colimits of p are equivalent. □

1.2.14 Presentations of ∞-Categories

Like many other types of mathematical structures, ∞-categories can be described by generators and relations. In particular, it makes sense to speak of a *finitely presented* ∞-category $\mathcal{C}$. Roughly speaking, $\mathcal{C}$ is finitely presented if it has finitely many objects and its morphism spaces are determined by specifying a finite number of generating morphisms, a finite number of relations among these generating morphisms, a finite number of relations among the relations, and so forth (a finite number of relations in all).

Example 1.2.14.1. Let $\mathcal{C}$ be the free higher category generated by a single object X and a single morphism $f : X \to X$. Then $\mathcal{C}$ is a finitely presented ∞-category with a single object and $\mathrm{Hom}_{\mathcal{C}}(X, X) = \{1, f, f^2, \ldots\}$ is infinite and discrete. In particular, we note that the finite presentation of $\mathcal{C}$ does not guarantee finiteness properties of the morphism spaces.

Example 1.2.14.2. If we identify ∞-groupoids with spaces, then giving a presentation for an ∞-groupoid corresponds to giving a cell decomposition of the associated space. Consequently, the finitely presented ∞-groupoids correspond precisely to the finite cell complexes.

Example 1.2.14.3. Suppose that $\mathcal{C}$ is a higher category with only two objects X and Y, that X and Y have contractible endomorphism spaces, and that $\mathrm{Hom}_{\mathcal{C}}(X, Y)$ is empty. Then $\mathcal{C}$ is completely determined by the morphism space $\mathrm{Hom}_{\mathcal{C}}(Y, X)$, which may be arbitrary. In this case, $\mathcal{C}$ is finitely presented if and only if $\mathrm{Hom}_{\mathcal{C}}(Y, X)$ is a finite cell complex (up to homotopy equivalence).

The idea of giving a presentation for an ∞-category is very naturally encoded in Joyal's model structure on the category of simplicial sets, which we will discuss in §2.2.4. This model structure can be described as follows:

- The fibrant objects of Set_{Δ} are precisely the ∞-categories.

- The weak equivalences in Set_{Δ} are precisely those maps $p : S \to S'$ which induce equivalences $\mathfrak{C}[S] \to \mathfrak{C}[S']$ of simplicial categories.

If S is an arbitrary simplicial set, we can choose a "fibrant replacement" for S: that is, a categorical equivalence $S \to \mathcal{C}$, where $\mathcal{C}$ is an ∞-category. For example, we can take $\mathcal{C}$ to be the nerve of the topological category $|\mathfrak{C}[S]|$. The ∞-category $\mathcal{C}$ is well-defined up to equivalence, and we may regard it as an ∞-category "generated by" S. The simplicial set S itself can be thought of as a "blueprint" for building $\mathcal{C}$. We may view S as generated from the empty (simplicial) set by adjoining nondegenerate simplices. Adjoining a 0-simplex to S has the effect of adding an object to the ∞-category $\mathcal{C}$, and adjoining a 1-simplex to S has the effect of adjoining a morphism to $\mathcal{C}$. Higher-dimensional simplices can be thought of as encoding relations among the morphisms.

1.2.15 Set-Theoretic Technicalities

In ordinary category theory, one frequently encounters categories in which the collection of objects is too large to form a set. Generally speaking, this does not create any difficulties so long as we avoid doing anything which is obviously illegal (such as considering the "category of all categories" as an object of itself).

The same issues arise in the setting of higher category theory and are in some sense even more of a nuisance. In ordinary category theory, one generally allows a category $\mathcal{C}$ to have a proper class of objects but still requires $\operatorname{Hom}_{\mathcal{C}}(X, Y)$ to be a *set* for fixed objects $X, Y \in \mathcal{C}$. The formalism of ∞-categories treats objects and morphisms on the same footing (they are both simplices of a simplicial set), and it is somewhat unnatural (though certainly possible) to directly impose the analogous condition; see §5.4.1 for a discussion.

There are several means of handling the technical difficulties inherent in working with large objects (in either classical or higher category theory):

(1) One can employ some set-theoretic device that enables one to distinguish between "large" and "small". Examples include:

   – Assuming the existence of a sufficient supply of (Grothendieck) universes.

   – Working in an axiomatic framework which allows both sets and *classes* (collections of sets which are possibly too large for themselves to be considered sets).

   – Working in a standard set-theoretic framework (such as Zermelo-Frankel) but incorporating a theory of classes through some ad hoc device. For example, one can define a class to be a collection of sets which is defined by some formula in the language of set theory.

(2) One can work exclusively with small categories, and mirror the distinction between large and small by keeping careful track of relative sizes.

(3) One can simply ignore the set-theoretic difficulties inherent in discussing large categories.

Needless to say, approach (2) yields the most refined information. However, it has the disadvantage of burdening our exposition with an additional layer of technicalities. On the other hand, approach (3) will sometimes be inadequate because we will need to make arguments which play off the distinction between a large category and a small subcategory which determines it. Consequently, we shall officially adopt approach (1) for the remainder of this book. More specifically, we assume that for every cardinal κ_0, there exists a strongly inaccessible cardinal $\kappa \geq \kappa_0$. We then let $\mathcal{U}(\kappa)$ denote the collection of all sets having rank $< \kappa$, so that $\mathcal{U}(\kappa)$ is a *Grothendieck universe*: in other words, $\mathcal{U}(\kappa)$ satisfies all of the usual axioms of set theory. We will refer to a mathematical object as *small* if it belongs to $\mathcal{U}(\kappa)$ (or is isomorphic to such an object), and *essentially small* if it is equivalent (in whatever relevant sense) to a small object. Whenever it is convenient to do so, we will choose another strongly inaccessible cardinal $\kappa' > \kappa$ to obtain a larger Grothendieck universe $\mathcal{U}(\kappa')$ in which $\mathcal{U}(\kappa)$ becomes small.

For example, an ∞-category $\mathcal{C}$ is essentially small if and only if it satisfies the following conditions:

- The set of isomorphism classes of objects in the homotopy category $h\mathcal{C}$ has cardinality $< \kappa$.

- For every morphism $f : X \to Y$ in $\mathcal{C}$ and every $i \geq 0$, the homotopy set $\pi_i(\mathrm{Hom}_{\mathcal{C}}^{\mathrm{R}}(X, Y), f)$ has cardinality $< \kappa$.

For a proof and further discussion, we refer the reader to §5.4.1.

Remark 1.2.15.1. The existence of the strongly inaccessible cardinal κ cannot be proven from the standard axioms of set theory, and the assumption that κ exists cannot be proven consistent with the standard axioms for set theory. However, it should be clear that assuming the existence of κ is merely the most convenient of the devices mentioned above; none of the results proven in this book will depend on this assumption in an essential way.

1.2.16 The ∞-Category of Spaces

The category of sets plays a central role in classical category theory. The main reason is that *every* category $\mathcal{C}$ is enriched over sets: given a pair of objects $X, Y \in \mathcal{C}$, we may regard $\mathrm{Hom}_{\mathcal{C}}(X, Y)$ as an object of Set. In the higher-categorical setting, the proper analogue of Set is the ∞-category $\mathcal{S}$ of *spaces*, which we will now introduce.

Definition 1.2.16.1. Let $\mathcal{K}an$ denote the full subcategory of Set_Δ spanned by the collection of Kan complexes. We will regard $\mathcal{K}an$ as a simplicial category. Let $\mathcal{S} = \mathrm{N}(\mathcal{K}an)$ denote the (simplicial) nerve of $\mathcal{K}an$. We will refer to $\mathcal{S}$ as the ∞-*category of spaces*.

Remark 1.2.16.2. For every pair of objects $X, Y \in \mathcal{K}an$, the simplicial set $\mathrm{Map}_{\mathcal{K}an}(X, Y) = Y^X$ is a Kan complex. It follows that $\mathcal{S}$ is an ∞-category (Proposition 1.1.5.10).

Remark 1.2.16.3. There are many other ways to construction a suitable "∞-category of spaces." For example, we could instead define S to be the (topological) nerve of the category of CW complexes and continuous maps. All that really matters is that we have a ∞-category which is equivalent to $S = N(\mathcal{K}an)$. We have selected Definition 1.2.16.1 for definiteness and to simplify our discussion of the Yoneda embedding in §5.1.3.

Remark 1.2.16.4. We will occasionally need to distinguish between large and small spaces. In these contexts, we will let S denote the ∞-category of small spaces (defined by taking the simplicial nerve of the category of small Kan complexes), and $\widehat{S}$ the ∞-category of large spaces (defined by taking the simplicial nerve of the category of *all* Kan complexes). We observe that S is a large ∞-category and that $\widehat{S}$ is even bigger.

Chapter Two

Fibrations of Simplicial Sets

Many classes of morphisms which play an important role in the homotopy theory of simplicial sets can be defined by their lifting properties (we refer the reader to §A.1.2 for a brief discussion and a summary of the terminology employed below).

Example 2.0.0.1. A morphism $p : X \to S$ of simplicial sets which has the right lifting property with respect to every horn inclusion $\Lambda^n_i \subseteq \Delta^n$ is called a *Kan fibration*. A morphism $i : A \to B$ which has the left lifting property with respect to every Kan fibration is said to be *anodyne*.

Example 2.0.0.2. A morphism $p : X \to S$ of simplicial sets which has the right lifting property with respect to every inclusion $\partial\Delta^n \subseteq \Delta^n$ is called a *trivial fibration*. A morphism $i : A \to B$ has the *left* lifting property with respect to every trivial Kan fibration if and only if it is a *cofibration*: that is, if and only if i is a monomorphism of simplicial sets.

By definition, a simplicial set S is a ∞-category if it has the extension property with respect to all horn inclusions $\Lambda^n_i \subseteq \Delta^n$ with $0 < i < n$. As in classical homotopy theory, it is convenient to introduce a *relative* version of this condition.

Definition 2.0.0.3 (Joyal). A morphism $f : X \to S$ of simplicial sets is

- a *left fibration* if f has the right lifting property with respect to all horn inclusions $\Lambda^n_i \subseteq \Delta^n$, $0 \leq i < n$.

- a *right fibration* if f has the right lifting property with respect to all horn inclusions $\Lambda^n_i \subseteq \Delta^n$, $0 < i \leq n$.

- an *inner fibration* if f has the right lifting property with respect to all horn inclusions $\Lambda^n_i \subseteq \Delta^n$, $0 < i < n$.

A morphism of simplicial sets $i : A \to B$ is

- *left anodyne* if i has the left lifting property with respect to all left fibrations.

- *right anodyne* if i has the left lifting property with respect to all right fibrations.

- *inner anodyne* if i has the left lifting property with respect to all inner fibrations.

Remark 2.0.0.4. Joyal uses the terms "mid-fibration" and "mid-anodyne morphism" for what we have chosen to call *inner fibrations* and *inner anodyne morphisms*.

The purpose of this chapter is to study the notions of fibration defined above, which are basic tools in the theory of ∞-categories. In §2.1, we study the theory of right (left) fibrations $p : X \to S$, which can be viewed as the ∞-categorical analogue of *categories (co)fibered in groupoids* over S. We will apply these ideas in §2.2 to show that the theory of ∞-categories is equivalent to the theory of simplicial categories.

There is also an analogue of the more general theory of (co)fibered categories whose fibers are not necessarily groupoids: this is the theory of *(co)Cartesian fibrations*, which we will introduce in §2.4. Cartesian and co-Cartesian fibrations are both examples of inner fibrations, which we will study in §2.3.

Remark 2.0.0.5. To help orient the reader, we summarize the relationship between many of the classes of fibrations which we will study in this book. If $f : X \to S$ is a map of simplicial sets, then we have the following implications:

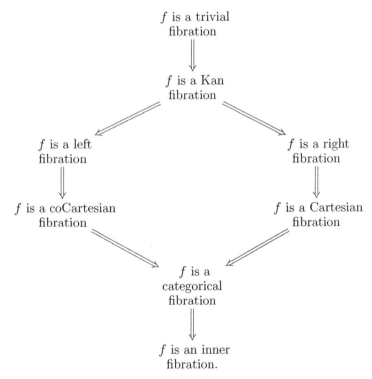

In general, none of these implications is reversible.

Remark 2.0.0.6. The small object argument (Proposition A.1.2.5) shows that every map $X \to Z$ of simplicial sets admits a factorization

$$X \xrightarrow{p} Y \xrightarrow{q} Z,$$

where p is anodyne (left anodyne, right anodyne, inner anodyne, a cofibration) and q is a Kan fibration (left fibration, right fibration, inner fibration, trivial fibration).

Remark 2.0.0.7. The theory of left fibrations (left anodyne maps) is *dual* to the theory of right fibrations (right anodyne maps): a map $S \to T$ is a left fibration (left anodyne map) if and only if the induced map $S^{op} \to T^{op}$ is a right fibration (right anodyne map). Consequently, we will generally confine our remarks in §2.1 to the case of left fibrations; the analogous statements for right fibrations will follow by duality.

2.1 LEFT FIBRATIONS

In this section, we will study the class of *left fibrations* between simplicial sets. We begin in §2.1.1 with a review of some classical category theory: namely, the theory of categories cofibered in groupoids (over another category). We will see that the theory of left fibrations is a natural ∞-categorical generalization of this idea. In §2.1.2, we will show that the class of left fibrations is stable under various important constructions, such as the formation of slice ∞-categories.

It follows immediately from the definition that every Kan fibration of simplicial sets is a left fibration. The converse is false in general. However, it is possible to give a relatively simple criterion for testing whether or not a left fibration $f : X \to S$ is a Kan fibration. We will establish this criterion in §2.1.3 and deduce some of its consequences.

The classical theory of Kan fibrations has a natural interpretation in the language of model categories: a map $p : X \to S$ is a Kan fibration if and only if X is a fibrant object of $(\mathrm{Set}_\Delta)_{/S}$, where the category $(\mathrm{Set}_\Delta)_{/S}$ is equipped with its usual model structure. There is a similar characterization of left fibrations: a map $p : X \to S$ is a left fibration if and only if X is a fibrant object of $(\mathrm{Set}_\Delta)_{/S}$ with respect to a certain model structure which we will refer to as the *covariant model structure*. We will define the covariant model structure in §2.1.4 and give an overview of its basic properties.

2.1.1 Left Fibrations in Classical Category Theory

Before beginning our study of left fibrations, let us recall a bit of classical category theory. Let $\mathcal{D}$ be a small category and suppose we are given a functor

$$\chi : \mathcal{D} \to \mathcal{G}\mathrm{pd},$$

where $\mathcal{G}\mathrm{pd}$ denotes the category of groupoids (where the morphisms are given by functors). Using the functor χ, we can extract a new category $\mathcal{C}_\chi$ via the classical *Grothendieck construction*:

- The objects of $\mathcal{C}_\chi$ are pairs (D, η), where $D \in \mathcal{D}$ and η is an object of the groupoid $\chi(D)$.

- Given a pair of objects $(D, \eta), (D', \eta') \in \mathcal{C}_\chi$, a morphism from (D, η) to (D', η') in $\mathcal{C}_\chi$ is given by a pair (f, α), where $f : D \to D'$ is a morphism in $\mathcal{D}$ and $\alpha : \chi(f)(\eta) \simeq \eta'$ is an isomorphism in the groupoid $\chi(D')$.

- Composition of morphisms is defined in the obvious way.

There is an evident forgetful functor $F : \mathcal{C}_\chi \to \mathcal{D}$, which carries an object $(D, \eta) \in \mathcal{C}_\chi$ to the underlying object $D \in \mathcal{D}$. Moreover, it is possible to reconstruct χ from the category $\mathcal{C}_\chi$ (together with the forgetful functor F) at least up to equivalence; for example, if D is an object of $\mathcal{D}$, then the groupoid $\chi(D)$ is canonically equivalent to the fiber product $\mathcal{C}_\chi \times_\mathcal{D} \{D\}$. Consequently, the Grothendieck construction sets up a dictionary which relates functors $\chi : \mathcal{D} \to \mathcal{G}\mathrm{pd}$ with categories $\mathcal{C}_\chi$ admitting a functor $F : \mathcal{C}_\chi \to \mathcal{D}$. However, this dictionary is not perfect; not every functor $F : \mathcal{C} \to \mathcal{D}$ arises via the Grothendieck construction described above. To clarify the situation, we recall the following definition:

Definition 2.1.1.1. Let $F : \mathcal{C} \to \mathcal{D}$ be a functor between categories. We say that $\mathcal{C}$ *is cofibered in groupoids over* $\mathcal{D}$ if the following conditions are satisfied:

(1) For every object $C \in \mathcal{C}$ and every morphism $\eta : F(C) \to D$ in $\mathcal{D}$, there exists a morphism $\widetilde{\eta} : C \to \widetilde{D}$ such that $F(\widetilde{\eta}) = \eta$.

(2) For every morphism $\eta : C \to C'$ in $\mathcal{C}$ and every object $C'' \in \mathcal{C}$, the map

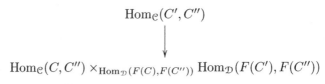

$$\mathrm{Hom}_\mathcal{C}(C', C'')$$
$$\downarrow$$
$$\mathrm{Hom}_\mathcal{C}(C, C'') \times_{\mathrm{Hom}_\mathcal{D}(F(C), F(C''))} \mathrm{Hom}_\mathcal{D}(F(C'), F(C''))$$

is bijective.

Example 2.1.1.2. Let $\chi : \mathcal{D} \to \mathcal{G}\mathrm{pd}$ be a functor from a category $\mathcal{D}$ to the category of groupoids. Then the forgetful functor $\mathcal{C}_\chi \to \mathcal{D}$ exhibits $\mathcal{C}_\chi$ as fibered in groupoids over $\mathcal{D}$.

Example 2.1.1.2 admits a converse: suppose we begin with a category $\mathcal{C}$ fibered in groupoids over $\mathcal{D}$. Then, for every every object $D \in \mathcal{D}$, the fiber $\mathcal{C}_D = \mathcal{C} \times_\mathcal{D} \{D\}$ is a groupoid. Moreover, for every morphism $f : D \to D'$ in $\mathcal{D}$, it is possible to construct a functor $f_! : \mathcal{C}_D \to \mathcal{C}_{D'}$ as follows: for each $C \in \mathcal{C}_D$, choose a morphism $\overline{f} : C \to C'$ covering the map $D \to D'$ and set $f_!(C) = C'$. The map $\overline{f}$ may not be uniquely determined, but it is unique up to isomorphism and depends functorially on C. Consequently, we obtain a functor $f_!$, which is well-defined up to isomorphism. We can then try to define a functor $\chi : \mathcal{D} \to \mathcal{G}\mathrm{pd}$ by the formulas

$$D \mapsto \mathcal{C}_D$$

$$f \mapsto f_!.$$

Unfortunately, this does not quite work: since the functor $f_!$ is determined only up to canonical isomorphism by f, the identity $(f \circ g)_! = f_! \circ g_!$ holds only up to canonical isomorphism rather than up to equality. This is merely a technical inconvenience; it can be addressed in (at least) two ways:

- The groupoid $\chi(D) = \mathcal{C} \times_{\mathcal{D}} \{D\}$ can be described as the category of functors G fitting into a commutative diagram

$$
\begin{array}{ccc}
 & & \mathcal{C} \\
 & \overset{G}{\nearrow} & \Big\downarrow F \\
\{D\} & \longrightarrow & \mathcal{D} .
\end{array}
$$

  If we replace the one-point category $\{D\}$ with the overcategory $\mathcal{D}_{D/}$ in this definition, then we obtain a groupoid equivalent to $\chi(D)$ which depends on D in a strictly functorial fashion.

- Without modifying the definition of $\chi(D)$, we can realize χ as a functor from $\mathcal{D}$ to an appropriate *bicategory* of groupoids.

We may summarize the above discussion informally by saying that the Grothendieck construction establishes an equivalence between functors $\chi : \mathcal{D} \to \mathcal{G}\mathrm{pd}$ and categories fibered in groupoids over $\mathcal{D}$.

The theory of left fibrations should be regarded as an ∞-categorical generalization of Definition 2.1.1.1. As a preliminary piece of evidence for this assertion, we offer the following:

Proposition 2.1.1.3. *Let $F : \mathcal{C} \to \mathcal{D}$ be a functor between categories. Then $\mathcal{C}$ is cofibered in groupoids over $\mathcal{D}$ if and only if the induced map $\mathrm{N}(F) : \mathrm{N}(\mathcal{C}) \to \mathrm{N}(\mathcal{D})$ is a left fibration of simplicial sets.*

Proof. Proposition 1.1.2.2 implies that $\mathrm{N}(F)$ is an inner fibration. It follows that $\mathrm{N}(F)$ is a left fibration if and only if it has the right lifting property with respect to $\Lambda_0^n \subseteq \Delta^n$ for all $n > 0$. When $n = 1$, the relevant lifting property is equivalent to (1) of Definition 2.1.1.1. When $n = 2$ ($n = 3$), the relevant lifting property is equivalent to the surjectivity (injectivity) of the map described in (2). For $n > 3$, the relevant lifting property is automatic (since a map $\Lambda_0^n \to S$ extends *uniquely* to Δ^n when S is isomorphic to the nerve of a category). $\qquad\square$

Let us now consider the structure of a general left fibration $p : X \to S$. In the case where S consists of a single vertex, Proposition 1.2.5.1 asserts that p is a left fibration if and only if X is a Kan complex. Since the class of left fibrations is stable under pullback, we deduce that for *any* left fibration $p : X \to S$ and any vertex s of S, the fiber $X_s = X \times_S \{s\}$ is a Kan complex (which we can think of as the ∞-categorical analogue of a groupoid). Moreover, these Kan complexes are related to one another. More precisely, suppose that $f : s \to s'$ is an edge of the simplicial set S and consider the inclusion $i : X_s \simeq X_s \times \{0\} \subseteq X_s \times \Delta^1$. In §2.1.2, we will prove that i is left anodyne (Corollary 2.1.2.7). It follows that we can solve the lifting problem

$$
\begin{array}{ccc}
\{0\} \times X_s & \hookrightarrow & X \\
\downarrow & \nearrow & \downarrow p \\
\Delta^1 \times X_s & \xrightarrow{\quad} \Delta^1 \xrightarrow{f} & S.
\end{array}
$$

Restricting the dotted arrow to $\{1\} \times X_s$, we obtain a map $f_! : X_s \to X_{s'}$. Of course, $f_!$ is not unique, but it is uniquely determined up to homotopy.

Lemma 2.1.1.4. *Let $q : X \to S$ be a left fibration of simplicial sets. The assignment*

$$s \in S_0 \mapsto X_s$$

$$f \in S_1 \mapsto f_!$$

determines a (covariant) functor from the homotopy category hS *into the homotopy category* $\mathcal{H}$ *of spaces.*

Proof. Let $f : s \to s'$ be an edge of S. We note the following characterization of the morphism $f_!$ in $\mathcal{H}$. Let K be any simplicial set and suppose we are given homotopy classes of maps $\eta \in \operatorname{Hom}_{\mathcal{H}}(K, X_s)$, $\eta' \in \operatorname{Hom}_{\mathcal{H}}(K, X_{s'})$. Then $\eta' = f_! \circ \eta$ if and only if there exists a map $p : K \times \Delta^1 \to X$ such that $q \circ p$ is given by the composition

$$K \times \Delta^1 \to \Delta^1 \xrightarrow{f} S,$$

η is the homotopy class of $p|K \times \{0\}$, and η' is the homotopy class of $p|K \times \{1\}$.

Now consider any 2-simplex $\sigma : \Delta^2 \to S$, which we will depict as

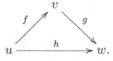

We note that the inclusion $X_u \times \{0\} \subseteq X_u \times \Delta^2$ is left anodyne (Corollary 2.1.2.7). Consequently there exists a map $p : X_u \times \Delta^2 \to X$ such that $p|X_u \times \{0\}$ is the inclusion $X_u \subseteq X$ and $q \circ p$ is the composition $X_u \times \Delta^2 \to \Delta^2 \xrightarrow{\sigma} S$. Then $f_! \simeq p|X_u \times \{1\}$, $h_! = p|X_u \times \{2\}$, and the map $p|X_u \times \Delta^{\{1,2\}}$ verifies the equation

$$h_! = g_! \circ f_!$$

in $\operatorname{Hom}_{\mathcal{H}}(X_u, X_w)$. $\qquad\square$

We can summarize the situation informally as follows. Fix a simplicial set S. To give a left fibration $q : X \to S$, one must specify a Kan complex X_s for each "object" of S, a map $f_! : X_s \to X_{s'}$ for each "morphism" $f : s \to s'$ of S, and "coherence data" for these morphisms for each higher-dimensional simplex of S. In other words, giving a left fibration ought to be more or less equivalent to giving a functor from S to the ∞-category $\mathcal{S}$ of spaces. Lemma 2.1.1.4 can be regarded as a weak version of this assertion; we will prove something considerably more precise in §2.1.4 (see Theorem 2.2.1.2).

We close this section by establishing two simple properties of left fibrations, which will be needed in the proof of Proposition 1.2.4.3:

Proposition 2.1.1.5. *Let $p : \mathcal{C} \to \mathcal{D}$ be a left fibration of ∞-categories and let $f : X \to Y$ be a morphism in $\mathcal{C}$ such that $p(f)$ is an equivalence in $\mathcal{D}$. Then f is an equivalence in $\mathcal{C}$.*

Proof. Let $\bar{g}$ be a homotopy inverse to $p(f)$ in $\mathcal{D}$ so that there exists a 2-simplex of $\mathcal{D}$ depicted as follows:

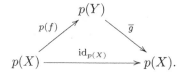

Since p is a left fibration, we can lift this to a diagram

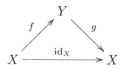

in $\mathcal{C}$. It follows that $g \circ f \simeq \mathrm{id}_X$, so that f admits a left homotopy inverse. Since $p(g) = \bar{g}$ is an equivalence in $\mathcal{D}$, the same argument proves that g has a left homotopy inverse. This left homotopy inverse must coincide with f since f is a right homotopy inverse to g. Thus f and g are homotopy inverse in the ∞-category $\mathcal{C}$, so that f is an equivalence, as desired. □

Proposition 2.1.1.6. *Let $p : \mathcal{C} \to \mathcal{D}$ be a left fibration of ∞-categories, let Y be an object of $\mathcal{C}$, and let $\bar{f} : \overline{X} \to p(Y)$ be an equivalence in $\mathcal{D}$. Then there exists a morphism $f : X \to Y$ in $\mathcal{C}$ such that $p(f) = \bar{f}$ (automatically an equivalence in view of Proposition 2.1.1.5).*

Proof. Let $\bar{g} : p(Y) \to \overline{X}$ be a homotopy inverse to $\bar{f}$ in $\mathcal{C}$. Since p is a left fibration, there exists a morphism $g : Y \to X$ such that $\bar{g} = p(g)$. Since $\bar{f}$ and $\bar{g}$ are homotopy inverse to one another, there exists a 2-simplex of $\mathcal{D}$ which we can depict as follows:

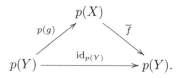

Applying the assumption that p is a left fibration once more, we can lift this to a diagram

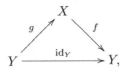

which proves the existence of f.   □

2.1.2 Stability Properties of Left Fibrations

The purpose of this section is to show that left fibrations of simplicial sets exist in abundance. Our main results are Proposition 2.1.2.1 (which is our basic source of examples for left fibrations) and Corollary 2.1.2.9 (which asserts that left fibrations are stable under the formation of functor categories).

Let $\mathcal{C}$ be an ∞-category and let $\mathcal{S}$ denote the ∞-category of spaces. One can think of a functor from $\mathcal{C}$ to $\mathcal{S}$ as a "cosheaf of spaces" on $\mathcal{C}$. By analogy with ordinary category theory, one might expect that the basic example of such a cosheaf would be the cosheaf corepresented by an object C of $\mathcal{C}$; roughly speaking this should be given by the functor

$$D \mapsto \operatorname{Map}_{\mathcal{C}}(C, D).$$

As we saw in §2.1.1, it is natural to guess that such a functor can be encoded by a left fibration $\widetilde{\mathcal{C}} \to \mathcal{C}$. There is a natural candidate for $\widetilde{\mathcal{C}}$: the undercategory $\mathcal{C}_{C/}$. Note that the fiber of the map

$$f : \mathcal{C}_{C/} \to \mathcal{C}$$

over the object $D \in \mathcal{C}$ is the Kan complex $\operatorname{Hom}^{\mathrm{L}}_{\mathcal{C}}(C, D)$. The assertion that f is a left fibration is a consequence of the following more general result:

Proposition 2.1.2.1 (Joyal). *Suppose we are given a diagram of simplicial sets*

$$K_0 \subseteq K \xrightarrow{p} X \xrightarrow{q} S,$$

where q is an inner fibration. Let $r = q \circ p : K \to S$, $p_0 = p|K_0$, and $r_0 = r|K_0$. Then the induced map

$$X_{p/} \to X_{p_0/} \times_{S_{r_0/}} S_{r/}$$

is a left fibration. If the map q is already a left fibration, then the induced map

$$X_{/p} \to X_{/p_0} \times_{S_{/r_0}} S_{/r}$$

is a left fibration as well.

Proposition 2.1.2.1 immediately implies the following half of Proposition 1.2.9.3, which we asserted earlier without proof:

Corollary 2.1.2.2 (Joyal). *Let $\mathcal{C}$ be an ∞-category and $p : K \to \mathcal{C}$ an arbitrary diagram. Then the projection $\mathcal{C}_{p/} \to \mathcal{C}$ is a left fibration. In particular, $\mathcal{C}_{p/}$ is itself an ∞-category.*

Proof. Apply Proposition 2.1.2.1 in the case where $X = \mathcal{C}$, $S = *$, $A = \emptyset$, and $B = K$. □

We can also use Proposition 2.1.2.1 to prove Proposition 1.2.4.3, which was stated without proof in §1.2.4.

Proposition. *Let $\mathcal{C}$ be an ∞-category and $\phi : \Delta^1 \to \mathcal{C}$ a morphism of $\mathcal{C}$. Then ϕ is an equivalence if and only if, for every $n \geq 2$ and every map $f_0 : \Lambda_0^n \to \mathcal{C}$ such that $f_0|\Delta^{\{0,1\}} = \phi$, there exists an extension of f_0 to Δ^n.*

Proof. Suppose first that ϕ is an equivalence and let f_0 be as above. To find the desired extension of f_0, we must produce the dotted arrow in the associated diagram

$$
\begin{array}{ccc}
\{0\} & \longrightarrow & \mathcal{C}_{/\Delta^{n-2}} \\
\downarrow & \nearrow & \downarrow q \\
\Delta^1 & \xrightarrow{\phi'} & \mathcal{C}_{/\partial\Delta^{n-2}}.
\end{array}
$$

The projection map $p : \mathcal{C}_{/\partial\Delta^{n-2}} \to \mathcal{C}$ is a right fibration (Proposition 2.1.2.1). Since ϕ' is a preimage of ϕ under p, Proposition 2.1.1.5 implies that ϕ' is an equivalence. Because q is a right fibration (Proposition 2.1.2.1 again), the existence of the dotted arrow follows from Proposition 2.1.1.6.

We now prove the converse. Let $\phi : X \to Y$ be a morphism in $\mathcal{C}$ and consider the map $\Lambda_0^2 \to \mathcal{C}$ indicated in the following diagram:

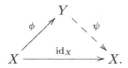

The assumed extension property ensures the existence of the dotted morphism $\psi : Y \to X$ and a 2-simplex σ which verifies the identity $\psi \circ \phi \simeq \mathrm{id}_X$. We now consider the map

$$
\tau_0 : \Lambda_0^3 \xrightarrow{(\bullet, s_0\phi, s_1\psi, \sigma)} \mathcal{C}.
$$

Once again, our assumption allows us to extend τ_0 to a 3-simplex $\tau : \Delta^3 \to \mathcal{C}$, and the face $d_0\tau$ verifies the identity $\phi \circ \psi = \mathrm{id}_Y$. It follows that ψ is a homotopy inverse to ϕ, so that ϕ is an equivalence in $\mathcal{C}$. □

We now turn to the proof of Proposition 2.1.2.1. It is an easy consequence of the following more basic observation:

Lemma 2.1.2.3 (Joyal [44]). *Let $f : A_0 \subseteq A$ and $g : B_0 \subseteq B$ be inclusions of simplicial sets. Suppose either that f is right anodyne or that g is left anodyne. Then the induced inclusion*

$$h : (A_0 \star B) \coprod_{A_0 \star B_0} (A \star B_0) \subseteq A \star B$$

is inner anodyne.

Proof. We will prove that h is inner anodyne whenever f is right anodyne; the other assertion follows by a dual argument.

Consider the class of *all* morphisms f for which the conclusion of the lemma holds (for any inclusion g). This class of morphisms is weakly saturated; to prove that it contains all right anodyne morphisms, it suffices to show that it contains each of the inclusions $f : \Lambda^n_j \subseteq \Delta^n$ for $0 < j \leq n$. We may therefore assume that f is of this form.

Now consider the collection of all inclusions g for which h is inner anodyne (where f is now fixed). This class of morphisms is also weakly saturated; to prove that it contains all inclusions, it suffices to show that the lemma holds when g is of the form $\partial \Delta^m \subseteq \Delta^m$. In this case, h can be identified with the inclusion $\Lambda^{n+m+1}_j \subseteq \Delta^{n+m+1}$, which is inner anodyne because $0 < j \leq n < n + m + 1$. $\square$

The following result can be proven by exactly the same argument:

Lemma 2.1.2.4 (Joyal). *Let $f : A_0 \to A$ and $g : B_0 \to B$ be inclusions of simplicial sets. Suppose that f is left anodyne. Then the induced inclusion*

$$(A_0 \star B) \coprod_{A_0 \star B_0} (A \star B_0) \subseteq A \star B$$

is left anodyne.

Proof of Proposition 2.1.2.1. After unwinding the definitions, the first assertion follows from Lemma 2.1.2.3 and the second from Lemma 2.1.2.4. $\square$

For future reference, we record the following counterpart to Proposition 2.1.2.1:

Proposition 2.1.2.5 (Joyal). *Let $\pi : S \to T$ be an inner fibration, $p : B \to S$ a map of simplicial sets, $i : A \subseteq B$ an inclusion of simplicial sets, $p_0 = p|A$, $p' = \pi \circ p$, and $p'_0 = \pi \circ p_0 = p'|A$. Suppose either that i is right anodyne or that π is a left fibration. Then the induced map*

$$\phi : S_{p/} \to S_{p_0/} \times_{T_{p'_0/}} T_{p'/}$$

is a trivial Kan fibration.

Proof. Consider the class of all cofibrations $i : A \to B$ for which ϕ is a trivial fibration for *every* inner fibration (right fibration) $p : S \to T$. It is not difficult to see that this is a weakly saturated class of morphisms; thus,

it suffices to consider the case where $A = \Lambda_i^m$ and $B = \Delta^m$ for $0 < i \leq m$
$(0 \leq i \leq m)$.

Let $q : \partial \Delta^n \to S_{p/}$ be a map and suppose we are given an extension of
$\phi \circ q$ to Δ^n. We wish to find a compatible extension of q. Unwinding the
definitions, we are given a map

$$r : (\Delta^m \star \partial \Delta^n) \coprod_{\Lambda_i^m \star \partial \Delta^n} (\Lambda_i^m \star \Delta^n) \to S,$$

which we wish to extend to $\Delta^m \star \Delta^n$ in a manner that is compatible with a
given extension $\Delta^m \star \Delta^n \to T$ of the composite map $\pi \circ r$. The existence of
such an extension follows immediately from the assumption that p has the
right lifting property with respect to the horn inclusion $\Lambda_i^{n+m+1} \subseteq \Delta^{n+m+1}$.

$\square$

The remainder of this section is devoted to the study of the behavior of
left fibrations under exponentiation. Our goal is to prove an assertion of the
following form: if $p : X \to S$ is a left fibration of simplicial sets, then so
is the induced map $X^K \to S^K$, for every simplicial set K (this is a special
case of Corollary 2.1.2.9 below). This is an easy consequence of the following
characterization of left anodyne maps, which is due to Joyal:

Proposition 2.1.2.6 (Joyal [44]). *The following collections of morphisms
generate the same weakly saturated class of morphisms of* Set_Δ:

(1) *The collection* A_1 *of all horn inclusions* $\Lambda_i^n \subseteq \Delta^n$, $0 \leq i < n$.

(2) *The collection* A_2 *of all inclusions*

$$(\Delta^m \times \{0\}) \coprod_{\partial \Delta^m \times \{0\}} (\partial \Delta^m \times \Delta^1) \subseteq \Delta^m \times \Delta^1.$$

(3) *The collection* A_3 *of all inclusions*

$$(S' \times \{0\}) \coprod_{S \times \{0\}} (S \times \Delta^1) \subseteq S' \times \Delta^1,$$

where $S \subseteq S'$.

Proof. Let $S \subseteq S'$ be as in (3). Working cell by cell on S', we deduce that
every morphism in A_3 can be obtained as an iterated pushout of morphisms
belonging to A_2. Conversely, A_2 is contained in A_3, which proves that they
generate the same weakly saturated collection of morphisms.

To proceed with the proof, we must first introduce a bit of notation. The
$(n+1)$-simplices of $\Delta^n \times \Delta^1$ are indexed by order-preserving maps

$$[n+1] \to [0, \ldots, n] \times [0, 1].$$

We let σ_k denote the map

$$\sigma_k(m) = \begin{cases} (m, 0) & \text{if } m \leq k \\ (m-1, 1) & \text{if } m > k. \end{cases}$$

We will also denote by σ_k the corresponding $(n+1)$-simplex of $\Delta^n \times \Delta^1$. We note that $\{\sigma_k\}_{0 \leq k \leq n}$ are precisely the nondegenerate $(n+1)$-simplices of $\Delta^n \times \Delta^1$.

We define a collection $\{X(k)\}_{0 \leq k \leq n+1}$ of simplicial subsets of $\Delta^n \times \Delta^1$ by descending induction on k. We begin by setting

$$X(n+1) = (\Delta^n \times \{0\}) \coprod_{\partial \Delta^n \times \{0\}} (\partial \Delta^n \times \Delta^1).$$

Assuming that $X(k+1)$ has been defined, we let $X(k) \subseteq \Delta^n \times \Delta^1$ be the union of $X(k+1)$ and the simplex σ_k (together with all the faces of σ_k). We note that this description exhibits $X(k)$ as a pushout

$$X(k+1) \coprod_{\Lambda_k^{n+1}} \Delta^{n+1}$$

and also that $X(0) = \Delta^n \times \Delta^1$. It follows that each step in the chain of inclusions

$$X(n+1) \subseteq X(n) \subseteq \cdots \subseteq X(1) \subseteq X(0)$$

is contained in the class of morphisms generated by A_1, so that the inclusion $X(n+1) \subseteq X(0)$ is generated by A_1.

To complete the proof, we show that each inclusion in A_1 is a retract of an inclusion in A_3. More specifically, the inclusion $\Lambda_i^n \subseteq \Delta^n$ is a retract of

$$(\Delta^n \times \{0\}) \coprod_{\Lambda_i^n \times \{0\}} (\Lambda_i^n \times \Delta^1) \subseteq \Delta^n \times \Delta^1$$

so long as $0 \leq i < n$. We will define the relevant maps

$$\Delta^n \xrightarrow{j} \Delta^n \times \Delta^1 \xrightarrow{r} \Delta^n$$

and leave it to the reader to verify that they are compatible with the relevant subobjects. The map j is simply the inclusion $\Delta^n \simeq \Delta^n \times \{1\} \subseteq \Delta^n \times \Delta^1$. The map r is induced by a map of partially ordered sets, which we will also denote by r. It may be described by the formulas

$$r(m,0) = \begin{cases} m & \text{if } m \neq i+1 \\ i & \text{if } m = i+1 \end{cases}$$

$$r(m,1) = m.$$

$\square$

Corollary 2.1.2.7. *Let* $i : A \to A'$ *be left anodyne and let* $j : B \to B'$ *be a cofibration. Then the induced map*

$$(A \times B') \coprod_{A \times B} (A' \times B) \to A' \times B'$$

is left anodyne.

Proof. This follows immediately from Proposition 2.3.2.1, which characterizes the class of left anodyne maps as the class generated by A_3 (which is stable under smash products with any cofibration). $\square$

Remark 2.1.2.8. A basic fact in the homotopy theory of simplicial sets is that the analogue of Corollary 2.1.2.7 also holds for the class of *anodyne* maps of simplicial sets. Since the class of anodyne maps is generated (as a weakly saturated class of morphisms) by the class of left anodyne maps and the class of right anodyne maps, this classical fact follows from Corollary 2.1.2.7 (together with the dual assertion concerning right anodyne maps).

Corollary 2.1.2.9. *Let $p : X \to S$ be a left fibration and let $i : A \to B$ be any cofibration of simplicial sets. Then the induced map $q : X^B \to X^A \times_{S^A} S^B$ is a left fibration. If i is left anodyne, then q is a trivial Kan fibration.*

Corollary 2.1.2.10 (Homotopy Extension Lifting Property)**.** *Let $p : X \to S$ be a map of simplicial sets. Then p is a left fibration if and only if the induced map*

$$X^{\Delta^1} \to X^{\{0\}} \times_{S^{\{0\}}} S^{\Delta^1}$$

is a trivial Kan fibration of simplicial sets.

For future use, we record the following criterion for establishing that a morphism is left anodyne:

Proposition 2.1.2.11. *Let $p : X \to S$ be a map of simplicial sets, let $s : S \to X$ be a section of p, and let $h \in \mathrm{Hom}_S(X \times \Delta^1, X)$ be a (fiberwise) simplicial homotopy from $s \circ p = h|X \times \{0\}$ to $\mathrm{id}_X = h|X \times \{1\}$. Then s is left anodyne.*

Proof. Consider a diagram

$$
\begin{array}{ccc}
S & \xrightarrow{\ g\ } & Y \\
{\scriptstyle s}\downarrow & {\overset{f}{\nearrow}} & \downarrow{\scriptstyle q} \\
X & \xrightarrow[\ g'\ ]{} & Z
\end{array}
$$

where q is a left fibration. We must show that it is possible to find a map f rendering the diagram commutative. Define $F_0 : (S \times \Delta^1) \coprod_{S \times \{0\}} (X \times \{0\})$ to be the composition of g with the projection onto S. Now consider the diagram

$$
\begin{array}{ccc}
(S \times \Delta^1) \coprod_{S \times \{0\}} (X \times \{0\}) & \xrightarrow{\ F_0\ } & Y \\
\downarrow & {\overset{F}{\nearrow}} & \downarrow{\scriptstyle q} \\
X \times \Delta^1 & \xrightarrow[\ g' \circ h\ ]{} & Z.
\end{array}
$$

Since q is a left fibration and the left vertical map is left anodyne, it is possible to supply the dotted arrow F as indicated. Now we observe that $f = F|X \times \{1\}$ has the desired properties. $\square$

2.1.3 A Characterization of Kan Fibrations

Let $p : X \to S$ be a left fibration of simplicial sets. As we saw in §2.1.1, p determines for each vertex s of S a Kan complex X_s, and for each edge $f : s \to s'$ a map of Kan complexes $f_! : X_s \to X_{s'}$ (which is well-defined up to homotopy). If p is a Kan fibration, then the same argument allows us to construct a map $X_{s'} \to X_s$, which is a homotopy inverse to $f_!$. Our goal in this section is to prove the following converse:

Proposition 2.1.3.1. *Let $p : S \to T$ be a left fibration of simplicial sets. The following conditions are equivalent:*

(1) *The map p is a Kan fibration.*

(2) *For every edge $f : t \to t'$ in T, the map $f_! : S_t \to S_{t'}$ is an isomorphism in the homotopy category $\mathcal{H}$ of spaces.*

Lemma 2.1.3.2. *Let $p : S \to T$ be a left fibration of simplicial sets. Suppose that S and T are Kan complexes and that p is a homotopy equivalence. Then p induces a surjection from S_0 to T_0.*

Proof. Fix a vertex $t \in T_0$. Since p is a homotopy equivalence, there exists a vertex $s \in S_0$ and an edge e joining $p(s)$ to t. Since p is a left fibration, this edge lifts to an edge $e' : s \to s'$ in S. Then $p(s') = t$. $\square$

Lemma 2.1.3.3. *Let $p : S \to T$ be a left fibration of simplicial sets. Suppose that T is a Kan complex. Then p is a Kan fibration.*

Proof. We note that the projection $S \to *$, being a composition of left fibrations $S \to T$ and $T \to *$, is a left fibration, so that S is also a Kan complex. Let $A \subseteq B$ be an anodyne inclusion of simplicial sets. We must show that the map $p : S^B \to S^A \times_{T^A} T^B$ is surjective on vertices. Since S and T are Kan complexes, the maps $T^B \to T^A$ and $S^B \to S^A$ are trivial fibrations. It follows that p is a homotopy equivalence and a left fibration. Now we simply apply Lemma 2.1.3.2. $\square$

Lemma 2.1.3.4. *Let $p : S \to T$ be a left fibration of simplicial sets. Suppose that for every vertex $t \in T$, the fiber S_t is contractible. Then p is a trivial Kan fibration.*

Proof. It will suffice to prove the analogous result for *right* fibrations (we do this in order to keep the notation we use below consistent with that employed in the proof of Proposition 2.1.2.6).

Since p has nonempty fibers, it has the right lifting property with respect to the inclusion $\emptyset = \partial \Delta^0 \subseteq \Delta^0$. Let $n > 0$, let $f : \partial \Delta^n \to S$ be any map, and let $g : \Delta^n \to T$ be an extension of $p \circ f$. We must show that there exists an extension $\widetilde{f} : \Delta^n \to S$ with $g = p \circ \widetilde{f}$.

Pulling back via the map G, we may suppose that $T = \Delta^n$ and g is the identity map, so that S is an ∞-category. Let t denote the initial vertex of

T. There is a unique map $g' : \Delta^n \times \Delta^1 \to T$ such that $g'|\Delta^n \times \{1\} = g$ and $g'|\Delta^n \times \{0\}$ is constant at the vertex t.

Since the inclusion $\partial \Delta^n \times \{1\} \subseteq \partial \Delta^n \times \Delta^1$ is right anodyne, there exists an extension f' of f to $\partial \Delta^n \times \Delta^1$ which covers $g'|\partial \Delta^n \times \Delta^1$. To complete the proof, it suffices to show that we can extend f' to a map $\widetilde{f'} : \Delta^n \times \Delta^1 \to S$ (such an extension is automatically compatible with g' in view of our assumptions that $T = \Delta^n$ and $n > 0$). Assuming this has been done, we simply define $\widetilde{f} = \widetilde{f'}|\Delta^n \times \{1\}$.

Recall the notation of the proof of Proposition 2.1.2.6 and filter the simplicial set $\Delta^n \times \Delta^1$ by the simplicial subsets

$$X(n+1) \subseteq \cdots \subseteq X(0) = \Delta^n \times \Delta^1.$$

We extend the definition of f' to $X(m)$ by a descending induction on m. When $m = n + 1$, we note that $X(n + 1)$ is obtained from $\partial \Delta^n \times \Delta^1$ by adjoining the interior of the simplex $\partial \Delta^n \times \{0\}$. Since the boundary of this simplex maps entirely into the contractible Kan complex S_t, it is possible to extend f' to $X(n+1)$.

Now suppose the definition of f' has been extended to $X(i+1)$. We note that $X(i)$ is obtained from $X(i + 1)$ by pushout along a horn inclusion $\Lambda_i^{n+1} \subseteq \Delta^{n+1}$. If $i > 0$, then the assumption that S is an ∞-category guarantees the existence of an extension of f' to $X(i)$. When $i = 0$, we note that f' carries the initial edge of σ_0 into the fiber S_t. Since S_t is a Kan complex, f' carries the initial edge of σ_0 to an equivalence in S, and the desired extension of f' exists by Proposition 1.2.4.3. $\qquad\square$

Proof of Proposition 2.1.3.1. Suppose first that (1) is satisfied and let $f : t \to t'$ be an edge in T. Since p is a right fibration, the edge f induces a map $f^* : S_{t'} \to S_t$ which is well-defined up to homotopy. It is not difficult to check that the maps f^* and $f_!$ are homotopy inverse to one another; in particular, $f_!$ is a homotopy equivalence. This proves that (1) $\Rightarrow$ (2).

Assume now that (2) is satisfied. A map of simplicial sets is a Kan fibration if and only if it is both a right fibration and a left fibration; consequently, it will suffice to prove that p is a right fibration. According to Corollary 2.1.2.10, it will suffice to show that

$$q : S^{\Delta^1} \to S^{\{1\}} \times_{T^{\{1\}}} T^{\Delta^1}$$

is a trivial Kan fibration. Corollary 2.1.2.9 implies that q is a left fibration. By Lemma 2.1.3.4, it suffices to show that the fibers of q are contractible.

Fix an edge $f : t \to t'$ in T. Let X denote the simplicial set of sections of the projection $S \times_T \Delta^1 \to \Delta^1$, where Δ^1 maps into T via the edge f. Consider the fiber $q' : X \to S_{t'}$ of q over the edge f. Since q and q' have the same fibers (over points of $S^{\{1\}} \times_{T^{\{1\}}} T^{\Delta^1}$ whose second projection is the edge f), it will suffice to show that q' is a trivial fibration for every choice of f.

Consider the projection $r : X \to S_t$. Since p is a left fibration, r is a trivial fibration. Because S_t is a Kan complex, so is X. Lemma 2.1.3.3 implies that

q' is a Kan fibration. We note that $f_!$ is obtained by choosing a section of r and then composing with q'. Consequently, assumption (2) implies that q' is a homotopy equivalence and thus a trivial fibration, which completes the proof. $\square$

Remark 2.1.3.5. Lemma 2.1.3.4 is an immediate consequence of Proposition 2.1.3.1 since any map between contractible Kan complexes is a homotopy equivalence. Lemma 2.1.3.3 also follows immediately (if T is a Kan complex, then its homotopy category is a groupoid, so that *any* functor $hT \to \mathcal{H}$ carries edges of T to invertible morphisms in $\mathcal{H}$).

2.1.4 The Covariant Model Structure

In §2.1.2, we saw that a left fibration $p : X \to S$ determines a functor χ from hS to the homotopy category $\mathcal{H}$, carrying each vertex s to the fiber $X_s = X \times_S \{s\}$. We would like to formulate a more precise relationship between left fibrations over S and functors from S into spaces. For this, it is convenient to employ Quillen's language of model categories. In this section, we will show that the category $(\text{Set}_\Delta)_{/S}$ can be endowed with the structure of a simplicial model category whose fibrant objects are precisely the left fibrations $X \to S$. In §2.2, we will describe an ∞-categorical version of the Grothendieck construction which is implemented by a right Quillen functor

$$(\text{Set}_\Delta)^{\mathfrak{C}[S]} \to (\text{Set}_\Delta)_{/S},$$

which we will eventually prove to be a Quillen equivalence (Theorem 2.2.1.2).

Warning 2.1.4.1. We will assume throughout this section that the reader is familiar with the theory of model categories as presented in §A.2. We will also assume familiarity with the model structure on the category Cat_Δ of simplicial categories (see §A.3.2).

Definition 2.1.4.2. Let $f : X \to S$ be a map of simplicial sets. The *left cone* of f is the simplicial set $S \coprod_X X^\triangleleft$. We will denote the left cone of f by $C^\triangleleft(f)$. Dually, we define the *right cone* of f to be the simplicial set $C^\triangleright(f) = S \coprod_X X^\triangleright$.

Remark 2.1.4.3. Let $f : X \to S$ be a map of simplicial sets. There is a canonical monomorphism of simplicial sets $S \to C^\triangleleft(f)$. We will generally identify S with its image under this monomorphism and thereby regard S as a simplicial subset of $C^\triangleleft(f)$. We note that there is a unique vertex of $C^\triangleleft(f)$ which does not belong to S. We will refer to this vertex as the *cone point* of $C^\triangleleft(f)$.

Example 2.1.4.4. Let S be a simplicial set and let id_S denote the identity map from S to itself. Then $C^\triangleleft(\text{id}_S)$ and $C^\triangleright(\text{id}_S)$ can be identified with $S^\triangleleft$ and $S^\triangleright$, respectively.

Definition 2.1.4.5. Let S be a simplicial set. We will say that a map $f : X \to Y$ in $(\text{Set}_\Delta)_{/S}$ is a

(C) *covariant cofibration* if it is a monomorphism of simplicial sets.

(W) *covariant equivalence* if the induced map

$$X^{\triangleleft} \coprod_X S \to Y^{\triangleleft} \coprod_Y S$$

is a categorical equivalence.

(F) *covariant fibration* if it has the right lifting property with respect to every map which is both a covariant cofibration and a covariant equivalence.

Lemma 2.1.4.6. *Let S be a simplicial set. Then every left anodyne map in $(\mathrm{Set}_{\Delta})_{/S}$ is a covariant equivalence.*

Proof. By general nonsense, it suffices to prove the result for a generating left anodyne inclusion of the form $\Lambda_i^n \subseteq \Delta^n$, where $0 \leq i < n$. In other words, we must show any map

$$i : (\Lambda_i^n)^{\triangleleft} \coprod_{\Lambda_i^n} S \to (\Delta^n)^{\triangleleft} \coprod_{\Delta^n} S$$

is a categorical equivalence. We now observe that i is a pushout of the inner anodyne inclusion $\Lambda_{i+1}^{n+1} \subseteq \Delta^{n+1}$. □

Proposition 2.1.4.7. *Let S be a simplicial set. The covariant cofibrations, covariant equivalences, and covariant fibrations determine a left proper combinatorial model structure on $(\mathrm{Set}_{\Delta})_{/S}$.*

Proof. It suffices to show that conditions (1), (2), and (3) of Proposition A.2.6.13 are met. We consider each in turn:

(1) The class (W) of weak equivalences is perfect. This follows from Corollary A.2.6.12 since the functor $X \mapsto X^{\triangleleft} \coprod_X S$ commutes with filtered colimits.

(2) It is clear that the class (C) of cofibrations is generated by a set. We must show that weak equivalences are stable under pushouts by cofibrations. In other words, suppose we are given a pushout diagram

$$\begin{array}{ccc} X & \xrightarrow{j} & Y \\ \downarrow{\scriptstyle i} & & \downarrow \\ X' & \xrightarrow{j'} & Y' \end{array}$$

in $(\mathrm{Set}_{\Delta})_{/S}$, where i is a covariant cofibration and j is a covariant equivalence. We must show that j' is a covariant equivalence. We obtain a pushout diagram in $\mathcal{C}\mathrm{at}_{\Delta}$:

$$\begin{array}{ccc} \mathfrak{C}[X^{\triangleleft} \coprod_X S] & \longrightarrow & \mathfrak{C}[Y^{\triangleleft} \coprod_Y S] \\ \downarrow & & \downarrow \\ \mathfrak{C}[(X')^{\triangleleft} \coprod_{X'} S] & \longrightarrow & \mathfrak{C}[(Y')^{\triangleleft} \coprod_{Y'} S]. \end{array}$$

This diagram is homotopy coCartesian because $\mathcal{C}\mathrm{at}_\Delta$ is a left proper model category. Since the upper horizontal map is an equivalence, so is the bottom horizontal map; thus j' is a covariant equivalence.

(3) We must show that a map $p : X \to Y$ in Set_Δ, which has the right lifting property with respect to every map in (C), belongs to (W). We note in that case that p is a trivial Kan fibration and therefore admits a section $s : Y \to X$. We will show that p and s induce mutually inverse isomorphisms between $\mathcal{C}[X^\triangleleft \coprod_X S]$ and $\mathcal{C}[Y^\triangleleft \coprod_Y S]$ in the homotopy category $h\mathcal{C}\mathrm{at}_\Delta$; it will then follow that p is a covariant equivalence.

Let $f : X \to X$ denote the composition $s \circ p$; we wish to show that the map $\mathcal{C}[X^\triangleleft \coprod_X S]$ induced by f is equivalent to the identity in $h\mathcal{C}\mathrm{at}_\Delta$. We observe that f is homotopic to the identity id_X via a homotopy $h : \Delta^1 \times X \to X$. It will therefore suffice to show that h is a covariant equivalence. But h admits a left inverse

$$X \simeq \{0\} \times X \subseteq \Delta^1 \times X,$$

which is left anodyne (Corollary 2.1.2.7) and therefore a covariant equivalence by Lemma 2.1.4.6.

$\square$

Proposition 2.1.4.8. *The category* $(\mathrm{Set}_\Delta)_{/S}$ *is a simplicial model category (with respect to the covariant model structure and the natural simplicial structure).*

Proof. We will deduce this from Proposition A.3.1.7. The only nontrivial point is to verify that for any $X \in (\mathrm{Set}_\Delta)_{/S}$, the projection $X \times \Delta^n \to X$ is a covariant equivalence. But this map has a section $X \times \{0\} \to X \times \Delta^n$, which is left anodyne and therefore a covariant equivalence (Proposition 2.1.4.9). $\square$

We will refer to the model structure of Proposition 2.1.4.7 as the *covariant model structure* on $(\mathrm{Set}_\Delta)_{/S}$. We will prove later that the covariantly fibrant objects of $(\mathrm{Set}_\Delta)_{/S}$ are precisely the left fibrations $X \to S$ (Corollary 2.2.3.12). For the time being, we will be content to make a much weaker observation:

Proposition 2.1.4.9. *Let S be a simplicial set.*

(1) *Every left anodyne map in $(\mathrm{Set}_\Delta)_{/S}$ is a trivial cofibration with respect to the covariant model structure.*

(2) *Every covariant fibration in $(\mathrm{Set}_\Delta)_{/S}$ is a left fibration of simplicial sets.*

(3) *Every fibrant object of $(\mathrm{Set}_\Delta)_{/S}$ determines a left fibration $X \to S$.*

Proof. Assertion (1) follows from Lemma 2.1.4.6, and the implications $(1) \Rightarrow (2) \Rightarrow (3)$ are obvious. $\square$

Our next result expresses the idea that the covariant model structure on $(\text{Set}_\Delta)_{/S}$ depends functorially on S:

Proposition 2.1.4.10. *Let $j : S \to S'$ be a map of simplicial sets. Let $j_! : (\text{Set}_\Delta)_{/S} \to (\text{Set}_\Delta)_{/S'}$ be the forgetful functor (given by composition with j) and let $j^* : (\text{Set}_\Delta)_{/S'} \to (\text{Set}_\Delta)_{/S}$ be its right adjoint, which is given by the formula*

$$j^* X' = X' \times_{S'} S.$$

Then we have a Quillen adjunction

$$(\text{Set}_\Delta)_{/S} \underset{j^*}{\overset{j_!}{\rightleftarrows}} (\text{Set}_\Delta)_{/S'}$$

(with the covariant model structures).

Proof. It is clear that $j_!$ preserves cofibrations. For $X \in (\text{Set}_\Delta)_S$, the pushout diagram

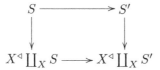

is a homotopy pushout (with respect to the Joyal model structure). Thus $j_!$ preserves covariant equivalences. It follows that $(j_!, j^*)$ is a Quillen adjunction. $\square$

Remark 2.1.4.11. Let $j : S \to S'$ be as in Proposition 2.1.4.10. If j is a categorical equivalence, then the Quillen adjunction $(j_!, j^*)$ is a Quillen equivalence. This follows from Theorem 2.2.1.2 and Proposition A.3.3.8.

Remark 2.1.4.12. Let S be a simplicial set. The covariant model structure on $(\text{Set}_\Delta)_{/S}$ is usually not self-dual. Consequently, we may define a new model structure on $(\text{Set}_\Delta)_{/S}$ as follows:

(C) A map f in $(\text{Set}_\Delta)_{/S}$ is a *contravariant cofibration* if it is a monomorphism of simplicial sets.

(W) A map f in $(\text{Set}_\Delta)_{/S}$ is a *contravariant equivalence* if f^{op} is a covariant equivalence in $(\text{Set}_\Delta)_{/S^{op}}$.

(F) A map f in $(\text{Set}_\Delta)_{/S}$ is a *contravariant fibration* if f^{op} is a covariant fibration in $(\text{Set}_\Delta)_{/S^{op}}$.

We will refer to this model structure on $(\text{Set}_\Delta)_{/S}$ as the *contravariant* model structure. Propositions 2.1.4.8, 2.1.4.9, and 2.1.4.10 have evident analogues in the contravariant setting.

2.2 SIMPLICIAL CATEGORIES AND ∞-CATEGORIES

For every topological category $\mathcal{C}$ and every pair of objects $X, Y \in \mathcal{C}$, Theorem 1.1.5.13 asserts that the counit map

$$u : |\operatorname{Map}_{\mathfrak{C}[\mathrm{N}(\mathcal{C})]}(X, Y)| \to \operatorname{Map}_{\mathcal{C}}(X, Y)$$

is a weak homotopy equivalence of topological spaces. This result is the main ingredient needed to establish the equivalence between the theory of topological categories and the theory of ∞-categories. The goal of this section is to give a proof of Theorem 1.1.5.13 and to develop some of its consequences.

We first replace Theorem 1.1.5.13 by a statement about *simplicial* categories. Consider the composition

$$\operatorname{Map}_{\mathfrak{C}[\mathrm{N}(\mathcal{C})]}(X, Y) \xrightarrow{v} \operatorname{Sing} \operatorname{Map}_{|\mathfrak{C}[\mathrm{N}(\mathcal{C})]|}(X, Y) \xrightarrow{\operatorname{Sing}(u)} \operatorname{Sing} \operatorname{Map}_{\mathcal{C}}(X, Y).$$

Classical homotopy theory ensures that v is a weak homotopy equivalence. Moreover, u is a weak homotopy equivalence of topological spaces if and only if $\operatorname{Sing}(u)$ is a weak homotopy equivalence of simplicial sets. Consequently, u is a weak homotopy equivalence of topological spaces if and only if $\operatorname{Sing}(u) \circ v$ is a weak homotopy equivalence of simplicial sets. It will therefore suffice to prove the following *simplicial* analogue of Theorem 1.1.5.13:

Theorem 2.2.0.1. *Let $\mathcal{C}$ be a fibrant simplicial category (that is, a simplicial category in which each mapping space $\operatorname{Map}_{\mathcal{C}}(x, y)$ is a Kan complex) and let $x, y \in \mathcal{C}$ be a pair of objects. The counit map*

$$u : \operatorname{Map}_{\mathfrak{C}[\mathrm{N}(\mathcal{C})]}(x, y) \to \operatorname{Map}_{\mathcal{C}}(x, y)$$

is a weak homotopy equivalence of simplicial sets.

The proof will be given in §2.2.4 (see Proposition 2.2.4.1). Our strategy is as follows:

(1) We will show that, for every simplicial set S, there is a close relationship between *right fibrations* $S' \to S$ and *simplicial presheaves* $\mathcal{F} : \mathfrak{C}[S]^{op} \to \operatorname{Set}_{\Delta}$. This relationship is controlled by the straightening and unstraightening functors which we introduce in §2.2.1.

(2) Suppose that S is an ∞-category. Then, for each object $y \in S$, the projection $S_{/y} \to S$ is a right fibration, which corresponds to a simplicial presheaf $\mathcal{F} : \mathfrak{C}[S]^{op} \to \operatorname{Set}_{\Delta}$. This simplicial presheaf $\mathcal{F}$ is related to $S_{/y}$ in two different ways:

  (*i*) As a simplicial presheaf, $\mathcal{F}$ is weakly equivalent to the functor $x \mapsto \operatorname{Map}_{\mathfrak{C}[S]}(x, y)$.

  (*ii*) For each object x of S, there is a canonical homotopy equivalence $\mathcal{F}(x) \to S_{/y} \times_S \{x\} \simeq \operatorname{Hom}^{\mathrm{R}}_S(x, y)$. Here the Kan complex $\operatorname{Hom}^{\mathrm{R}}_S(x, y)$ is defined as in §1.2.2.

(3) Combining observations (i) and (ii), we will conclude that the mapping spaces $\mathrm{Hom}^{\mathrm{R}}_S(x,y)$ are homotopy equivalent to the corresponding mapping spaces $\mathrm{Hom}_{\mathfrak{C}[S]}(x,y)$.

(4) In the special case where S is the nerve of a fibrant simplicial category $\mathfrak{C}$, there is a canonical map $\mathrm{Hom}_{\mathfrak{C}}(x,y) \to \mathrm{Hom}^{\mathrm{R}}_S(x,y)$, which we will show to be a homotopy equivalence in §2.2.2.

(5) Combining (3) and (4), we will obtain a canonical isomorphism

$$\mathrm{Map}_{\mathfrak{C}}(x,y) \simeq \mathrm{Map}_{\mathfrak{C}[\mathrm{N}(\mathfrak{C})]}(x,y)$$

in the homotopy category of spaces. We will then show that this isomorphism is induced by the unit map appearing in the statement of Theorem 2.2.0.1.

We will conclude this section with §2.2.5, where we apply Theorem 2.2.0.1 to construct the *Joyal model structure* on Set_Δ and to establish a more refined version of the equivalence between ∞-categories and simplicial categories.

2.2.1 The Straightening and Unstraightening Constructions (Unmarked Case)

In §2.1.1, we asserted that a left fibration $X \to S$ can be viewed as a functor from S into a suitable ∞-category of Kan complexes. Our goal in this section is to make this idea precise. For technical reasons, it will be somewhat more convenient to phrase our results in terms of the dual theory of *right* fibrations $X \to S$. Given any functor $\phi : \mathfrak{C}[S]^{op} \to \mathfrak{C}$ between simplicial categories, we will define an *unstraightening functor* $\mathrm{Un}_\phi : \mathrm{Set}^{\mathfrak{C}}_\Delta \to (\mathrm{Set}_\Delta)_{/S}$. If $\mathcal{F} : \mathfrak{C} \to \mathrm{Set}_\Delta$ is a diagram taking values in Kan complexes, then the associated map $\mathrm{Un}_\phi\, \mathcal{F} \to S$ is a right fibration whose fiber at a point $s \in S$ is homotopy equivalent to the Kan complex $\mathcal{F}(\phi(s))$.

Fix a simplicial set S, a simplicial category $\mathfrak{C}$, and a functor $\phi : \mathfrak{C}[S] \to \mathfrak{C}^{op}$. Given an object $X \in (\mathrm{Set}_\Delta)_{/S}$, let v denote the cone point of $X^{\triangleright}$. We can view the simplicial category

$$\mathcal{M} = \mathfrak{C}[X^{\triangleright}] \coprod_{\mathfrak{C}[X]} \mathfrak{C}^{op}$$

as a correspondence from $\mathfrak{C}^{op}$ to $\{v\}$, which we can identify with a simplicial functor

$$St_\phi X : \mathfrak{C} \to \mathrm{Set}_\Delta .$$

This functor is described by the formula

$$(St_\phi X)(C) = \mathrm{Map}_{\mathcal{M}}(C,v).$$

We may regard St_ϕ as a functor from $(\mathrm{Set}_\Delta)_{/S}$ to $(\mathrm{Set}_\Delta)^{\mathfrak{C}}$. We refer to St_ϕ as the *straightening functor* associated to ϕ. In the special case where $\mathfrak{C} = \mathfrak{C}[S]^{op}$ and ϕ is the identity map, we will write St_S instead of St_ϕ.

By the adjoint functor theorem (or by direct construction), the straightening functor St_ϕ associated to $\phi : \mathfrak{C}[S] \to \mathcal{C}^{op}$ has a right adjoint, which we will denote by Un_ϕ and refer to as the *unstraightening functor*. We now record the obvious functoriality properties of this construction.

Proposition 2.2.1.1. (1) *Let $p : S' \to S$ be a map of simplicial sets, $\mathcal{C}$ a simplicial category, and $\phi : \mathfrak{C}[S] \to \mathcal{C}^{op}$ a simplicial functor, and let $\phi' : \mathfrak{C}[S'] \to \mathcal{C}^{op}$ denote the composition $\phi \circ \mathfrak{C}[p]$. Let $p_! : (\mathrm{Set}_\Delta)_{/S'} \to (\mathrm{Set}_\Delta)_{/S}$ denote the forgetful functor given by composition with p. There is a natural isomorphism of functors*

$$St_\phi \circ p_! \simeq St_{\phi'}$$

from $(\mathrm{Set}_\Delta)_{/S'}$ to $\mathrm{Set}_\Delta^{\mathcal{C}}$.

(2) *Let S be a simplicial set, $\pi : \mathcal{C} \to \mathcal{C}'$ a simplicial functor between simplicial categories, and $\phi : \mathfrak{C}[S] \to \mathcal{C}^{op}$ a simplicial functor. Then there is a natural isomorphism of functors*

$$St_{\pi^{op} \circ \phi} \simeq \pi_! \circ St_\phi$$

from $(\mathrm{Set}_\Delta)_{/S}$ to $\mathrm{Set}_\Delta^{\mathcal{C}'}$. Here $\pi_! : \mathrm{Set}_\Delta^{\mathcal{C}} \to \mathrm{Set}_\Delta^{\mathcal{C}'}$ is the left adjoint to the functor $\pi^* : \mathrm{Set}_\Delta^{\mathcal{C}'} \to \mathrm{Set}_\Delta^{\mathcal{C}}$ given by composition with π.

Our main result is the following:

Theorem 2.2.1.2. *Let S be a simplicial set, $\mathcal{C}$ a simplicial category, and $\phi : \mathfrak{C}[S] \to \mathcal{C}^{op}$ a simplicial functor. The straightening and unstraightening functors determine a Quillen adjunction*

$$(\mathrm{Set}_\Delta)_{/S} \underset{\mathrm{Un}_\phi}{\overset{St_\phi}{\rightleftarrows}} \mathrm{Set}_\Delta^{\mathcal{C}},$$

where $(\mathrm{Set}_\Delta)_{/S}$ is endowed with the contravariant model structure and $\mathrm{Set}_\Delta^{\mathcal{C}}$ with the projective model structure. If ϕ is an equivalence of simplicial categories, then $(St_\phi, \mathrm{Un}_\phi)$ is a Quillen equivalence.

Proof. It is easy to see that St_ϕ preserves cofibrations and weak equivalences, so that the pair $(St_\phi, \mathrm{Un}_\phi)$ is a Quillen adjunction. The real content of Theorem 2.2.1.2 is the final assertion. Suppose that ϕ is an equivalence of simplicial categories; then we wish to show that $(St_\phi, \mathrm{Un}_\phi)$ is a Quillen equivalence. We will prove this result in §2.2.3 as a consequence of Proposition 2.2.3.11. $\square$

2.2.2 Straightening Over a Point

In this section, we will study the behavior of the straightening functor St_X in the case where the simplicial set $X = \{x\}$ consists of a single vertex. In this case, we can view St_X as a colimit-preserving functor from the category of simplicial sets to itself. We begin with a few general remarks about such functors.

Let Δ denote the category of combinatorial simplices and Set_Δ the category of simplicial sets, so that Set_Δ may be identified with the category of presheaves of sets on Δ. If $\mathcal{C}$ is *any* category which admits small colimits, then any functor $f : \Delta \to \mathcal{C}$ extends to a colimit-preserving functor $F : \mathrm{Set}_\Delta \to \mathcal{C}$ (which is unique up to unique isomorphism). We may regard f as a cosimplicial object $C^\bullet$ of $\mathcal{C}$. In this case, we shall denote the functor F by

$$S \mapsto |S|_{C^\bullet}.$$

Remark 2.2.2.1. Concretely, one constructs $|S|_{C^\bullet}$ by taking the disjoint union of $S_n \times C^n$ and making the appropriate identifications along the "boundaries." In the language of category theory, the geometric realization is given by the *coend*

$$\int_{[n] \in \Delta} S_n \times C^n.$$

The functor $S \mapsto |S|_{C^\bullet}$ has a right adjoint which we shall denote by $\mathrm{Sing}_{C^\bullet}$. It may be described by the formula

$$\mathrm{Sing}_{C^\bullet}(X)_n = \mathrm{Hom}_{\mathcal{C}}(C^n, X).$$

Example 2.2.2.2. Let $\mathcal{C}$ be the category $\mathcal{CG}$ of compactly generated Hausdorff spaces and let $C^\bullet$ be the cosimplicial space defined by

$$C^n = \{(x_0, \ldots, x_n) \in [0,1]^{n+1} : x_0 + \cdots + x_n = 1\}.$$

Then $|S|_{C^\bullet}$ is the usual *geometric realization* $|S|$ of the simplicial set S, and $\mathrm{Sing}_{C^\bullet} = \mathrm{Sing}$ is the functor which assigns to each topological space X its singular complex.

Example 2.2.2.3. Let $\mathcal{C}$ be the category Set_Δ and let $C^\bullet$ be the *standard simplex* (the cosimplicial object of Set_Δ given by the Yoneda embedding):

$$C^n = \Delta^n.$$

Then $||_{C^\bullet}$ and $\mathrm{Sing}_{C^\bullet}$ are both (isomorphic to) the identity functor on Set_Δ.

Example 2.2.2.4. Let $\mathcal{C} = \mathcal{C}at$ and let $f : \Delta \to \mathcal{C}at$ be the functor which associates to each finite nonempty linearly ordered set J the corresponding category. Then $\mathrm{Sing}_{C^\bullet} = \mathrm{N}$ is the functor which associates to each category its nerve, and $||_{C^\bullet}$ associates to each simplicial set S the homotopy category $\mathrm{h}S$ as defined in §1.2.3.

Example 2.2.2.5. Let $\mathcal{C} = \mathcal{C}at_\Delta$ and let $C^\bullet$ be the cosimplicial object of $\mathcal{C}$ given in Definitions 1.1.5.1 and 1.1.5.3. Then $\mathrm{Sing}_{C^\bullet}$ is the simplicial nerve functor, and $||_{C^\bullet}$ is its left adjoint

$$S \mapsto \mathfrak{C}[S].$$

Let us now return to the case of the straightening functor St_X, where $X = \{x\}$ consists of a single vertex. The above remarks show that we can identify St_X with the geometric realization functor $||_{Q^\bullet} : \mathrm{Set}_\Delta \to \mathrm{Set}_\Delta$ for some cosimplicial object $Q^\bullet$ in Set_Δ. To describe $Q^\bullet$ more explicitly, let us first define a cosimplicial simplicial set $J^\bullet$ by the formula

$$J^n = (\Delta^n \star \{y\}) \coprod_{\Delta^n} \{x\}.$$

The cosimplical simplicial set $Q^\bullet$ can then be described by the formula $Q^n = \mathrm{Map}_{\mathfrak{C}[J^n]}(x, y)$.

In order to proceed with our analysis, we need to understand better the cosimplicial object $Q^\bullet$ of Set_Δ. It admits the following description:

- For each $n \geq 0$, let $P_{[n]}$ denote the partially ordered set of *nonempty* subsets of $[n]$, and $K_{[n]}$ the simplicial set $\mathrm{N}(P)$ (which may be identified with a simplicial subset of the $(n+1)$-cube $(\Delta^1)^{n+1}$). The simplicial set Q^n is obtained by collapsing, for each $0 \leq i \leq n$, the subset

$$(\Delta^1)^{\{j:0\leq j<i\}} \times \{1\} \times (\Delta^1)^{\{j:i<j\leq n\}} \subseteq K_{[n]}$$

  to its quotient $(\Delta^1)^{\{j:i<j\leq n\}}$.

- A map $f : [n] \to [m]$ determines a map $P_f : P_{[n]} \to P_{[m]}$ by setting $P_f(I) = f(I)$. The map P_f in turn induces a map of simplicial sets $K_{[n]} \to K_{[m]}$, which determines a map of quotients $Q^n \to Q^m$ when f is order-preserving.

Remark 2.2.2.6. Let $\mathfrak{Q}^\bullet = |Q^\bullet|$ denote the cosimplicial space obtained by applying the (usual) geometric realization functor to $Q^\bullet$. The space $\mathfrak{Q}^n$ may be described as a quotient of the cube of all functions $p : [n] \to [0,1]$ satisfying $p(0) = 1$. This cube is to be divided by the following equivalence relation: $p \simeq p'$ if there exists a nonnegative integer $i \leq n$ such that $p|\{i, \ldots n\} = p'|\{i, \ldots, n\}$ and $p(i) = p'(i) = 1$.

Each $\mathfrak{Q}^n$ is homeomorphic to an n-simplex, and these homeomorphisms may be chosen to be compatible with the face maps of the cosimplicial space $\mathfrak{Q}^\bullet$. However, $\mathfrak{Q}^\bullet$ is not isomorphic to the standard simplex because it has very different degeneracies. For example, the product of the degeneracy mappings $\mathfrak{Q}^n \to (\mathfrak{Q}^1)^n$ is not injective for $n \geq 2$.

Our goal for the remainder of this section is to study the functors $\mathrm{Sing}_{Q^\bullet}$ and $||_{Q^\bullet}$ and to prove that they are "close" to the identity functor. More precisely, there is a map $\pi : Q^\bullet \to \Delta^\bullet$ of cosimplicial objects of Set_Δ. It is induced by a map $K_{[n]} \to \Delta^n$, which is the nerve of the map of partially ordered sets $P_{[n]} \to [n]$ that carries each nonempty subset of $[n]$ to its largest element.

Proposition 2.2.2.7. *Let S be a simplicial set. Then the map $p_S : |S|_{Q^\bullet} \to S$ induced by π is a weak homotopy equivalence.*

Proof. Consider the collection A of simplicial sets S for which the assertion of Proposition 2.2.2.7 holds. Since A is stable under filtered colimits, it will suffice to prove that every simplicial set S having only finitely many nondegenerate simplices belongs to A. We prove this by induction on the dimension n of S and the number of nondegenerate simplices of S of dimension n. If $S = \emptyset$, there is nothing to prove; otherwise we may write

$$S \simeq S' \coprod_{\partial \Delta^n} \Delta^n$$

$$|S|_{Q\bullet} \simeq |S'|_{Q\bullet} \coprod_{|\partial \Delta^n|_{Q\bullet}} |\Delta^n|_{Q\bullet}.$$

Since both of these pushouts are homotopy pushouts, it suffices to show that $p_{S'}$, $p_{\partial \Delta^n}$, and p_{Δ^n} are weak homotopy equivalences. For $p_{S'}$ and $p_{\partial \Delta^n}$, this follows from the inductive hypothesis; for p_{Δ^n}, we need only observe that both Δ^n and $|\Delta^n|_{Q\bullet} = Q^n$ are weakly contractible. $\square$

Remark 2.2.2.8. The strategy used to prove Proposition 2.2.2.7 will reappear frequently throughout this book: it allows us to prove theorems about arbitrary simplicial sets by reducing to the case of simplices.

Proposition 2.2.2.9. *The adjoint functors* $\mathrm{Set}_\Delta \overset{\|_{Q\bullet}}{\underset{\mathrm{Sing}_{Q\bullet}}{\rightleftarrows}} \mathrm{Set}_\Delta$ *determine a Quillen equivalence from the category* Set_Δ *(endowed with the Kan model structure) to itself.*

Proof. We first show that the functors $(\|_{Q\bullet}, \mathrm{Sing}_{Q\bullet})$ determine a Quillen adjunction from Set_Δ to itself. For this, it suffices to prove that the functor $S \mapsto |S|_{Q\bullet}$ preserves cofibrations and weak equivalences. The case of cofibrations is easy, and the second case follows from Proposition 2.2.2.7. To complete the proof, it will suffice to show that the left derived functor $L\|_{Q\bullet}$ determines an equivalence from the homotopy category $\mathcal{H}$ to itself. This follows immediately from Proposition 2.2.2.7, which implies that $L\|_{Q\bullet}$ is equivalent to the identity functor. $\square$

Corollary 2.2.2.10. *Let X be a Kan complex. Then the counit map*

$$v : |\mathrm{Sing}_{Q\bullet} X|_{Q\bullet} \to X$$

is a weak homotopy equivalence.

Remark 2.2.2.11. Let S be a simplicial set containing a vertex s. Let $\mathcal{C}$ be a simplicial category, $\phi : \mathfrak{C}[S]^{op} \to \mathcal{C}$ a simplicial functor, and $C = \phi(s) \in \mathcal{C}$. For every simplicial functor $\mathcal{F} : \mathcal{C} \to \mathrm{Set}_\Delta$, there is a canonical isomorphism

$$(\mathrm{Un}_\phi \mathcal{F}) \times_S \{s\} \simeq \mathrm{Sing}_{Q\bullet} \mathcal{F}(C).$$

In particular, we have a canonical map from $\mathcal{F}(C)$ to the fiber $(\mathrm{Un}_\phi \mathcal{F})_s$, which is a homotopy equivalence if $\mathcal{F}(C)$ is fibrant.

Remark 2.2.2.12. Let $\mathcal{C}$ and $\mathcal{C}'$ be simplicial categories. Given a pair of simplicial functors $\mathcal{F} : \mathcal{C} \to \mathrm{Set}_\Delta$, $\mathcal{F}' : \mathcal{C}' \to \mathrm{Set}_\Delta$, we let $\mathcal{F} \boxtimes \mathcal{F}' : \mathcal{C} \times \mathcal{C}' \to \mathrm{Set}_\Delta$ denote the functor described by the formula

$$(\mathcal{F} \boxtimes \mathcal{F}')(C, C') = \mathcal{F}(C) \times \mathcal{F}'(C').$$

Given a pair of simplicial functors $\phi : \mathfrak{C}[S]^{op} \to \mathcal{C}$, $\phi' : \mathfrak{C}[S']^{op} \to \mathcal{C}'$, we let $\phi \boxtimes \phi'$ denote the induced map $\mathfrak{C}[S \times S'] \to \mathcal{C} \times \mathcal{C}'$. We observe that there is a canonical isomorphism of functors

$$\mathrm{Un}_{\phi \boxtimes \phi'}(\mathcal{F} \boxtimes \mathcal{F}') \simeq \mathrm{Un}_\phi(\mathcal{F}) \times \mathrm{Un}_{\phi'}(\mathcal{F}').$$

Restricting our attention to the case where $S' = \Delta^0$ and ϕ' is an isomorphism, we obtain an isomorphism

$$\mathrm{Un}_\phi(\mathcal{F} \boxtimes K) \simeq \mathrm{Un}_\phi(\mathcal{F}) \times \mathrm{Sing}_{Q^\bullet} K$$

for every simplicial set K. In particular, for every pair of functors $\mathcal{F}, \mathcal{G} \in \mathrm{Set}_\Delta^{\mathcal{C}}$, we have a chain of maps

$$
\begin{aligned}
\mathrm{Hom}_{\mathrm{Set}_\Delta}(K, \mathrm{Map}_{\mathrm{Set}_\Delta^{\mathcal{C}}}(\mathcal{F}, \mathcal{G})) &\simeq \mathrm{Hom}_{\mathrm{Set}_\Delta^{\mathcal{C}}}(\mathcal{F} \boxtimes K, \mathcal{G}) \\
&\to \mathrm{Hom}_{(\mathrm{Set}_\Delta)_{/S}}(\mathrm{Un}_\phi(\mathcal{F} \boxtimes K), \mathrm{Un}_\phi \mathcal{G}) \\
&\simeq \mathrm{Hom}_{(\mathrm{Set}_\Delta)_{/S}}(\mathrm{Un}_\phi(\mathcal{F}) \times \mathrm{Sing}_{Q^\bullet} K, \mathrm{Un}_\phi \mathcal{G}) \\
&\to \mathrm{Hom}_{(\mathrm{Set}_\Delta)_{/S}}(\mathrm{Un}_\phi(\mathcal{F}) \times K, \mathrm{Un}_\phi \mathcal{G}) \\
&\simeq \mathrm{Hom}_{\mathrm{Set}_\Delta}(K, \mathrm{Map}_{(\mathrm{Set}_\Delta)_{/S}}(\mathrm{Un}_\phi(\mathcal{F}), \mathrm{Un}_\phi(\mathcal{G}))).
\end{aligned}
$$

This construction is natural in K and therefore determines a map of simplicial sets

$$\mathrm{Map}_{\mathrm{Set}_\Delta^{\mathcal{C}}}(\mathcal{F}, \mathcal{G}) \to \mathrm{Map}_{(\mathrm{Set}_\Delta)_{/S}}(\mathrm{Un}_\phi \mathcal{F}, \mathrm{Un}_\phi \mathcal{G}).$$

Together, these maps endow the unstraightening functor Un_ϕ with the structure of a *simplicial* functor from $\mathrm{Set}_\Delta^{\mathcal{C}}$ to $(\mathrm{Set}_\Delta)_{/S}$.

The cosimplicial object $Q^\bullet$ of Set_Δ will play an important role in our proof of Theorem 1.1.5.13. To explain this, let us suppose that $\mathcal{C}$ is a simplicial category and $S = \mathrm{N}(\mathcal{C})$ is its simplicial nerve. For every pair of vertices $\overline{x}, \overline{y} \in S$, we can consider the right mapping space $\mathrm{Hom}_S^{\mathrm{R}}(\overline{x}, \overline{y})$. By definition, giving an n-simplex of $\mathrm{Hom}_S^{\mathrm{R}}(\overline{x}, \overline{y})$ is equivalent to giving a map of simplicial sets $J^n \to S$, which carries x to $\overline{x}$ and y to $\overline{y}$. Using the identification $S \simeq \mathrm{N}(\mathcal{C})$, we see that this is equivalent to giving a map $\mathfrak{C}[J^n]$ into $\mathcal{C}$, which again carries x to $\overline{x}$ and y to $\overline{y}$. This is simply the data of a map of simplicial sets $Q^n \to \mathrm{Map}_{\mathcal{C}}(\overline{x}, \overline{y})$. Moreover, this identification is natural with respect to $[n]$; we therefore have the following result:

Proposition 2.2.2.13. *Let $\mathcal{C}$ be a simplicial category and let $X, Y \in \mathcal{C}$ be two objects. There is a natural isomorphism of simplicial sets*

$$\mathrm{Hom}_{\mathrm{N}(\mathcal{C})}^{\mathrm{R}}(X, Y) \simeq \mathrm{Sing}_{Q^\bullet} \mathrm{Map}_{\mathcal{C}}(X, Y).$$

2.2.3 Straightening of Right Fibrations

Our goal in this section is to prove Theorem 2.2.1.2, which asserts that the Quillen adjunction

$$(\text{Set}_\Delta)_{/S} \underset{\text{Un}_\phi}{\overset{St_\phi}{\rightleftarrows}} \text{Set}_\Delta^{\mathcal{C}}$$

is a Quillen equivalence when $\phi : \mathcal{C}[S] \to \mathcal{C}^{op}$ is an equivalence of simplicial categories. We first treat the case where S is a simplex.

Lemma 2.2.3.1. *Let n be a nonnegative integer, let $[n]$ denote the linearly ordered set $\{0, \ldots, n\}$, regarded as a (discrete) simplicial category, and let $\phi : \mathcal{C}[\Delta^n] \to [n]$ be the canonical functor. Then the Quillen adjunction*

$$(\text{Set}_\Delta)_{/\Delta^n} \underset{\text{Un}_\phi}{\overset{St_\phi}{\rightleftarrows}} \text{Set}_\Delta^{[n]}$$

is a Quillen equivalence.

Proof. It follows from the definition of the contravariant model structure that the left derived functor $LSt_\phi : \mathrm{h}(\text{Set}_\Delta)_{/\Delta^n} \to \mathrm{hSet}_\Delta^{[n]}$ is conservative. It will therefore suffice to show that the counit map $LSt_\phi \circ R\,\text{Un}_\phi \to \text{id}$ is an isomorphism of functors from $\mathrm{hSet}_\Delta^{[n]}$ to itself. For this, we must show that if $\mathcal{F} : [n] \to \text{Set}_\Delta$ is projectively fibrant, then the counit map

$$St_\phi \,\text{Un}_\phi \,\mathcal{F} \to \mathcal{F}$$

is an equivalence in $\text{Set}_\Delta^{[n]}$. In other words, we may assume that $\mathcal{F}(i)$ is a Kan complex for $i \in [n]$, and we wish to prove that each of the induced maps

$$v_i : (St_\phi \,\text{Un}_\phi \,\mathcal{F})(i) \to \mathcal{F}(i)$$

is a weak homotopy equivalence of simplicial sets.

Let $\psi : [n] \to [1]$ be defined by the formula

$$\psi'(j) = \begin{cases} 0 & \text{if } 0 \le j \le i \\ 1 & \text{otherwise.} \end{cases}$$

Then, for every object $X \in (\text{Set}_\Delta)_{/\Delta^n}$, we have isomorphisms

$$(St_\phi X)(i) \simeq (St_{\psi \circ \phi} X)(0) \simeq |X \times_{\Delta^n} \Delta^{\{n-i,\ldots,n\}}|_{Q^\bullet},$$

where the twisted geometric realization functor $||_{Q^\bullet}$ is as defined in §2.2.2. Taking $X = \text{Un}_\phi\,\mathcal{F}$, we see that v_i fits into a commutative diagram

$$\begin{array}{ccc}
|X \times_{\Delta^n} \{n-i\}|_{Q^\bullet} & \xrightarrow{\sim} & |\,\text{Sing}_{Q^\bullet}\,\mathcal{F}(i)|_{Q^\bullet} \\
\Big\downarrow & & \Big\downarrow \\
|X \times_{\Delta^n} \Delta^{\{n-i,\ldots,n\}}|_{Q^\bullet} & \xrightarrow{v_i} & \mathcal{F}(i).
\end{array}$$

Here the upper horizontal map is an isomorphism supplied by Corollary 2.2.2.11, and the right vertical map is a weak homotopy equivalence by

Proposition 2.2.2.10. Consequently, to prove that the map v_i is a weak homotopy equivalence, it will suffice to show that the left vertical map is a weak homotopy equivalence. In view of Proposition 2.2.2.7, it will suffice to show that the inclusion

$$X \times_{\Delta^n} \{n - i\} \subseteq X \times_{\Delta^n} \Delta^{\{n-i,\dots,n\}}$$

is a weak homotopy equivalence. In fact, $X \times_{\Delta^n} \{n - i\}$ is a deformation retract of $X \times_{\Delta^n} \Delta^{\{n-i,\dots,n\}}$: this follows from the observation that the projection $X \to \Delta^n$ is a right fibration (Proposition 2.1.4.9).   □

It will be convenient to restate Lemma 2.2.3.1 in a slightly modified form. First, we need to introduce a bit of terminology.

Definition 2.2.3.2. Suppose we are given a commutative diagram of simplicial sets

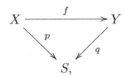

where p and q are right fibrations. We will say that f is a *pointwise equivalence* if, for each vertex $s \in S$, the induced map $X_s \to Y_s$ is a homotopy equivalence of Kan complexes.

Remark 2.2.3.3. In the situation of Definition 2.2.3.2, the following conditions are equivalent:

(a) The map f is a pointwise equivalence of right fibrations over S.

(b) The map f is a contravariant equivalence in $(\mathrm{Set}_\Delta)_{/S}$.

(c) The map f is a categorical equivalence of simplicial sets.

The equivalence $(a) \Leftrightarrow (b)$ follows from Corollary 2.2.3.13 (see below), and the equivalence $(a) \Leftrightarrow (c)$ from Proposition 3.3.1.5.

Lemma 2.2.3.4. *Let $S' \subseteq S$ be simplicial sets. Let $p : X \to S$ be any map and let $q : Y \to S$ be a right fibration. Let $X' = X \times_S S'$ and let $Y' = Y \times_S S'$. The restriction map*

$$\phi : \mathrm{Map}_{(\mathrm{Set}_\Delta)_{/S}}(X, Y) \to \mathrm{Map}_{(\mathrm{Set}_\Delta)_{/S'}}(X', Y')$$

is a Kan fibration.

Proof. We first show that ϕ is a right fibration. It will suffice to show that ϕ has the right lifting property with respect to every right anodyne inclusion $A \subseteq B$. This follows from the fact that q has the right lifting property with respect to the induced inclusion

$$i : (B \times S') \coprod_{A \times S'} (A \times S) \subseteq B \times S$$

since i is again right anodyne (Corollary 2.1.2.7).

Applying the preceding argument to the inclusion $\emptyset \subseteq S'$, we deduce that the projection map

$$\mathrm{Map}_{(\mathrm{Set}_\Delta)_{/S'}}(X', Y') \to \Delta^0$$

is a right fibration. Proposition 1.2.5.1 implies that $\mathrm{Map}_{(\mathrm{Set}_\Delta)_{/S'}}(X', Y')$ is a Kan complex. Lemma 2.1.3.3 now implies that ϕ is a Kan fibration, as desired.   □

Lemma 2.2.3.5. *Let $\mathcal{U}$ be a collection of simplicial sets. Suppose that the following conditions are satisfied:*

(i) *The collection $\mathcal{U}$ is stable under isomorphism. That is, if $S \in \mathcal{U}$ and $S' \simeq S$, then $S' \in \mathcal{U}$.*

(ii) *The collection $\mathcal{U}$ is stable under the formation of disjoint unions.*

(iii) *Every simplex Δ^n belongs to $\mathcal{U}$.*

(iv) *Suppose we are given a pushout diagram*

$$
\begin{array}{ccc}
X & \longrightarrow & X' \\
\downarrow{\scriptstyle f} & & \downarrow \\
Y & \longrightarrow & Y'
\end{array}
$$

in which X, X', and Y belong to $\mathcal{U}$. If the map f is a monomorphism, then Y' belongs to $\mathcal{U}$.

(v) *Suppose we are given a sequence of monomorphisms of simplicial sets*

$$X(0) \to X(1) \to \cdots$$

If each $X(i)$ belongs to $\mathcal{U}$, then the colimit $\varinjlim X(i)$ belongs to $\mathcal{U}$.

Then every simplicial set belongs to $\mathcal{U}$.

Proof. Let S be a simplicial set; we wish to show that $S \in \mathcal{U}$. In view of (v), it will suffice to show that each skeleton $\mathrm{sk}^n S$ belongs to $\mathcal{U}$. We may therefore assume that S is finite dimensional. We now proceed by induction on the dimension n of S. Let A denote the set of nondegenerate n-simplexes of S, so that we have a pushout diagram

$$
\begin{array}{ccc}
\coprod_{\alpha \in A} \partial \Delta^n & \longrightarrow & \mathrm{sk}^{n-1} S \\
\downarrow & & \downarrow \\
\coprod_{\alpha \in A} \Delta^n & \longrightarrow & S.
\end{array}
$$

Invoking assumption (iv), we are reduced to proving that the simplicial sets $\mathrm{sk}^{n-1} S$, $\coprod_{\alpha \in A} \partial \Delta^n$, and $\coprod_{\alpha \in A} \Delta^n$ belong to $\mathcal{U}$. For the first two this follows from the inductive hypothesis, and for the last it follows from assumptions (ii) and (iii).   □

Lemma 2.2.3.6. *Suppose we are given a commutative diagram of simplicial sets*

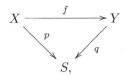

where p and q are right fibrations. The following conditions are equivalent:

(a) *The map f is a pointwise equivalence.*

(b) *The map f is an equivalence in the simplicial category $(\mathrm{Set}_\Delta)_{/S}$ (that is, f admits a homotopy inverse).*

(c) *For every object $A \in (\mathrm{Set}_\Delta)_{/S}$, composition with f induces a homotopy equivalence of Kan complexes*

$$\mathrm{Map}_{(\mathrm{Set}_\Delta)_{/S}}(A, X) \to \mathrm{Map}_{(\mathrm{Set}_\Delta)_{/S}}(A, Y).$$

Proof. The implication $(b) \Rightarrow (a)$ is clear (any homotopy inverse to f determines homotopy inverses for the maps $f_s : X_s \to Y_s$ for each vertex $s \in S$), and the implication $(c) \Rightarrow (b)$ follows from Proposition 1.2.4.1. We will prove that $(a) \Rightarrow (c)$. Let $\mathcal{U}$ denote the collection of all simplicial sets A such that, for *every* map $A \to S$, composition with f induces a homotopy equivalence of Kan complexes

$$\mathrm{Map}_{(\mathrm{Set}_\Delta)_{/S}}(A, X) \to \mathrm{Map}_{(\mathrm{Set}_\Delta)_{/S}}(A, Y).$$

We will show that $\mathcal{U}$ satisfies the hypotheses of Lemma 2.2.3.5 and therefore contains *all* simplicial sets. Conditions (i) and (ii) are obvious, and conditions (iv) and (v) follow from Lemma 2.2.3.4. It will therefore suffice to show that every simplex Δ^n belongs to $\mathcal{U}$. For every map $\Delta^n \to S$, we have a commutative diagram

$$
\begin{array}{ccc}
\mathrm{Map}_{(\mathrm{Set}_\Delta)_{/S}}(\Delta^n, X) & \longrightarrow & \mathrm{Map}_{(\mathrm{Set}_\Delta)_{/S}}(\Delta^n, Y) \\
\downarrow & & \downarrow \\
\mathrm{Map}_{(\mathrm{Set}_\Delta)_{/S}}(\{n\}, X) & \longrightarrow & \mathrm{Map}_{(\mathrm{Set}_\Delta)_{/S}}(\{n\}, Y).
\end{array}
$$

Since the inclusion $\{n\} \subseteq \Delta^n$ is right anodyne, the vertical maps are trivial Kan fibrations. It will therefore suffice to show that the bottom horizontal map is a homotopy equivalence, which follows immediately from (a). $\square$

Lemma 2.2.3.7. *Let $\phi : \mathfrak{C}[\Delta^n] \to [n]$ be as in Lemma 2.2.3.1. Suppose we are given a right fibration $X \to \Delta^n$, a projectively fibrant diagram $\mathcal{F} \in \mathrm{Set}_\Delta^{[n]}$, and a weak equivalence of diagrams $\alpha : \mathrm{St}_\phi X \to \mathcal{F}$. Then the adjoint map $X \to \mathrm{Un}_\phi \mathcal{F}$ is a pointwise equivalence of left fibrations over Δ^n.*

Proof. For $0 \leq i \leq n$, let $X(i) = X \times_{\Delta^n} \Delta^{\{n-i,\dots,n\}} \subseteq X$. We observe that $(St_\phi X)(i)$ is canonically isomorphic to the twisted geometric realization $|X(i)|_{Q^\bullet}$, where $Q^\bullet$ is defined as in §2.2.2.2. Since $X \to \Delta^n$ is a right fibration, the fiber $X \times_{\Delta^n} \{i\}$ is a deformation retract of $X(i)$. Using Proposition 2.2.2.7, we conclude that the induced inclusion $|X \times_{\Delta^n} \{n-i\}|_{Q^\bullet} \to |X(i)|_{Q^\bullet}$ is a weak homotopy equivalence. Since α is a weak equivalence, we get weak equivalences $|X \times_{\Delta^n} \{n-i\}|_{Q^\bullet} \to \mathcal{F}(i)$ for each $0 \leq i \leq n$. Using Proposition 2.2.2.9, we deduce that the adjoint maps $X \times_{\Delta^n} \{n - i\} \to \mathrm{Sing}_{Q^\bullet} \mathcal{F}(i)$ are again weak homotopy equivalences. The desired result now follows from the observation that $\mathrm{Sing}_{Q^\bullet} \mathcal{F}(i) \simeq (\mathrm{Un}_\phi \mathcal{F}) \times_{\Delta^n} \{n - i\}$ (Remark 2.2.2.11). $\quad\square$

Notation 2.2.3.8. For every simplicial set S, we let $\mathrm{RFib}(S)$ denote the full subcategory of $(\mathrm{Set}_\Delta)_{/S}$ spanned by those maps $X \to S$ which are right fibrations.

Proposition 2.1.4.9 implies that if $p : X \to S$ exhibits X as a fibrant object of the contravariant model category $(\mathrm{Set}_\Delta)_{/S}$, then p is a right fibration. We will prove the converse below (Corollary 2.2.3.12). For the moment, we will be content with the following weaker result:

Lemma 2.2.3.9. *For every integer $n \geq 0$, the inclusion $i : (\mathrm{Set}_\Delta)^\circ_{/\Delta^n} \subseteq \mathrm{RFib}(\Delta^n)$ is an equivalence of simplicial categories.*

Proof. It is clear that i is fully faithful. To prove that i is essentially surjective, consider any left fibration $X \to \Delta^n$. Let $\phi : \mathfrak{C}[\Delta^n] \to [n]$ be defined as in Lemma 2.2.3.1 and choose a weak equivalence $St_\phi X \to \mathcal{F}$, where $\mathcal{F} \in \mathrm{Set}_\Delta^{[n]}$ is a projectively fibrant diagram. Lemma 2.2.3.7 implies that the adjoint map $X \to \mathrm{Un}_\phi \mathcal{F}$ is a pointwise equivalence of right fibrations in Δ^n and therefore a homotopy equivalence in $\mathrm{RFib}(\Delta^n)$ (Lemma 2.2.3.6). It now suffices to observe that $\mathrm{Un}_\phi \mathcal{F} \in (\mathrm{Set}_\Delta)^\circ_{/\Delta^n}$. $\quad\square$

Lemma 2.2.3.10. *For each integer $n \geq 0$, the unstraightening functor $\mathrm{Un}_{\Delta^n} : (\mathrm{Set}_\Delta^{\mathfrak{C}[\Delta^n]})^\circ \to \mathrm{RFib}(\Delta^n)$ is an equivalence of simplicial categories.*

Proof. In view of Lemma 2.2.3.9 and Proposition A.3.1.10, it will suffice to show that the Quillen adjunction $(St_{\Delta^n}, \mathrm{Un}_{\Delta^n})$ is a Quillen equivalence. This follows immediately from Lemma 2.2.3.1 and Proposition A.3.3.8. $\quad\square$

Proposition 2.2.3.11. *For every simplicial set S, the unstraightening functor Un_S induces an equivalence of simplicial categories*

$$(\mathrm{Set}_\Delta^{\mathfrak{C}[S]^{op}})^\circ \to \mathrm{RFib}(S).$$

Proof. For each simplicial set S, let $(\mathrm{Set}_\Delta^{\mathfrak{C}[S]^{op}})_f$ denote the category of projectively fibrant objects of $\mathrm{Set}_\Delta^{\mathfrak{C}[S]^{op}}$ and let W_S be the class of weak equivalences in $(\mathrm{Set}_\Delta^{\mathfrak{C}[S]^{op}})_f$. Let W'_S be the collection of pointwise equivalences in

RFib(S). We have a commutative diagram of simplicial categories:

$$
\begin{array}{ccc}
(\mathrm{Set}_\Delta^{\mathcal{C}[S]^{op}})^\circ & \xrightarrow{\ \mathrm{Un}_S\ } & \mathrm{RFib}(S) \\
\downarrow & & \downarrow{\psi_S} \\
(\mathrm{Set}_\Delta^{\mathcal{C}[S]^{op}})_f[W_S^{-1}] & \xrightarrow{\ \phi_S\ } & \mathrm{RFib}[W'^{-1}_S]
\end{array}
$$

(see Notation A.3.5.1). Lemma A.3.6.17 implies that the left vertical map is an equivalence. Using Lemma 2.2.3.6 and Remark A.3.2.14, we deduce that the right vertical map is also an equivalence. It will therefore suffice to show that ϕ_S is an equivalence.

Let $\mathcal{U}$ denote the collection of simplicial sets S for which ϕ_S is an equivalence. We will show that $\mathcal{U}$ satisfies the hypotheses of Lemma 2.2.3.5 and therefore contains every simplicial set S. Conditions (i) and (ii) are obviously satisfied, and condition (iii) follows from Lemma 2.2.3.10 and Proposition A.3.1.10. We will verify condition (iv); the proof of (v) is similar.

Applying Corollary A.3.6.18, we deduce:

$(*)$ The functor $S \mapsto (\mathrm{Set}_\Delta^{\mathcal{C}[S]^{op}})_f[W_S^{-1}]$ carries homotopy colimit diagrams indexed by a partially ordered set to homotopy limit diagrams in Cat_Δ.

Suppose we are given a pushout diagram

$$
\begin{array}{ccc}
X & \longrightarrow & X' \\
\downarrow{f} & & \downarrow \\
Y & \longrightarrow & Y'
\end{array}
$$

in which $X, X', Y \in \mathcal{U}$, where f is a cofibration. We wish to prove that $Y' \in \mathcal{U}$. We have a commutative diagram

$$
\begin{array}{ccccc}
(\mathrm{Set}_\Delta^{\mathcal{C}[Y']^{op}})_f[W_{Y'}^{-1}] & \xrightarrow{\phi_{Y'}} & \mathrm{RFib}(Y')[W'^{-1}_{Y'}] & \xrightarrow{\ u\ } & \mathrm{RFib}(Y)[W'^{-1}_Y] \\
& & \downarrow{v} \quad \searrow^{w} & & \downarrow \\
& & \mathrm{RFib}(X')[W'^{-1}_{X'}] & \longrightarrow & \mathrm{RFib}(X)[W'^{-1}_X].
\end{array}
$$

Using $(*)$ and Corollary A.3.2.28, we deduce that $\phi_{Y'}$ is an equivalence if and only if, for every pair of objects $x, y \in \mathrm{RFib}(Y')[W'^{-1}_{Y'}]$, the diagram of simplicial sets

$$
\begin{array}{ccc}
\mathrm{Map}_{\mathrm{RFib}(Y')[W'^{-1}_{Y'}]}(x,y) & \longrightarrow & \mathrm{Map}_{\mathrm{RFib}(Y)[W'^{-1}_Y]}(u(x), u(y)) \\
\downarrow & & \downarrow \\
\mathrm{Map}_{\mathrm{RFib}(X')[W'^{-1}_{X'}]}(v(x), v(y)) & \longrightarrow & \mathrm{Map}_{\mathrm{RFib}(X)[W'^{-1}_X]}(w(x), w(y))
\end{array}
$$

is homotopy Cartesian. Since $\psi_{Y'}$ is a weak equivalence of simplicial categories, we may assume without loss of generality that $x = \psi_{Y'}(\overline{x})$ and that

$y = \psi_{Y'}(\bar{y})$ for some $\bar{x}, \bar{y} \in (\mathrm{Set}_\Delta)^\circ_{/Y'}$. It will therefore suffice to prove that the equivalent diagram

$$
\begin{array}{ccc}
\mathrm{Map}_{\mathrm{RFib}(Y')}(\bar{x},\bar{y}) & \longrightarrow & \mathrm{Map}_{\mathrm{RFib}(Y)}(\bar{u}(\bar{x}),\bar{u}(\bar{y})) \\
\downarrow & & \downarrow \\
\mathrm{Map}_{\mathrm{RFib}(X')}(\bar{v}(\bar{x}),\bar{v}(\bar{y})) & \xrightarrow{\;g\;} & \mathrm{Map}_{\mathrm{RFib}(X)}(\bar{w}(\bar{x}),\bar{w}(\bar{y}))
\end{array}
$$

is homotopy Cartesian. But this diagram is a pullback square, and the map g is a Kan fibration by Lemma 2.2.3.4. $\square$

We can now complete the proof of Theorem 2.2.1.2. Suppose that $\phi :$ $\mathfrak{C}[S] \to \mathfrak{C}^{op}$ is an equivalence of simplicial categories; we wish to show that the adjoint functors $(St_\phi, \mathrm{Un}_\phi)$ determine a Quillen equivalence between $(\mathrm{Set}_\Delta)_{/S}$ and $\mathrm{Set}_\Delta^{\mathfrak{C}}$. Using Proposition A.3.3.8, we can reduce to the case where ϕ is an isomorphism. In view of Proposition A.3.1.10, it will suffice to show that Un_ϕ induces an equivalence of simplicial categories $(\mathrm{Set}_\Delta^{\mathfrak{C}[S]^{op}})^\circ \to$ $(\mathrm{Set}_\Delta)^\circ_{/S}$, which follows immediately from Proposition 2.2.3.11.

Corollary 2.2.3.12. *Let $p : X \to S$ be a map of simplicial sets. The following conditions are equivalent:*

(1) *The map p is a right fibration.*

(2) *The map p exhibits X as a fibrant object of $(\mathrm{Set}_\Delta)_{/S}$ (with respect to the contravariant model structure).*

Proof. The implication (2) $\Rightarrow$ (1) follows from Proposition 2.1.4.9. For the converse, let us suppose that p is a right fibration. Proposition 2.2.3.11 implies that the unstraightening functor $\mathrm{Un}_S : (\mathrm{Set}_\Delta^{\mathfrak{C}[S]^{op}})^\circ \to \mathrm{Fun}^{\mathrm{R}}(S)$ is essentially surjective. Since Un_S factors through the inclusion $i : (\mathrm{Set}_\Delta)^\circ_{/S} \subseteq$ $\mathrm{Fun}^{\mathrm{R}}(S)$, we deduce that i is essentially surjective. Consequently, we can choose a simplicial homotopy equivalence $f : X \to Y$ in $(\mathrm{Set}_\Delta)_{/S}$, where Y is fibrant. Let g be a homotopy inverse to X so that there exists a homotopy $h : X \times \Delta^1 \to X$ from id_X to $g \circ f$.

To prove that X is fibrant, we must show that every lifting problem

$$
\begin{array}{ccc}
A & \xrightarrow{\;e_0\;} & X \\
{\scriptstyle j}\downarrow & \nearrow^{e} & \downarrow{\scriptstyle p} \\
B & \longrightarrow & S
\end{array}
$$

has a solution, provided that j is a trivial cofibration in the contravariant model category $(\mathrm{Set}_\Delta)_{/S}$. Since Y is fibrant, the map $f \circ e_0$ can be extended to a map $\bar{e} : B \to Y$ in $(\mathrm{Set}_\Delta)_{/S}$. Let $e' = g \circ \bar{e}$. The maps $\bar{e}$ and $h \circ (e_0 \times \mathrm{id}_{\Delta^1})$ determine another lifting problem

$$
\begin{array}{ccc}
(A \times \Delta^1) \coprod_{A \times \{1\}} (B \times \{1\}) & \longrightarrow & X \\
{\scriptstyle j'}\downarrow & \nearrow^{E} & \downarrow{\scriptstyle p} \\
B \times \Delta^1 & \longrightarrow & S.
\end{array}
$$

Proposition 2.1.2.6 implies that j' is right anodyne. Since p is a right fibration, there exists an extension E as indicated in the diagram. The restriction $e = E|B \times \{0\}$ is then a solution the original problem.   □

Corollary 2.2.3.13. *Suppose we are given a diagram of simplicial sets*

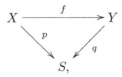

where p and q are right fibrations. Then f is a contravariant equivalence in $(\mathcal{S}et_\Delta)_{/S}$ *if and only if f is a pointwise equivalence.*

Proof. Since $(\mathcal{S}et_\Delta)_{/S}$ is a simplicial model category, this follows immediately from Corollary 2.2.3.12 and Lemma 2.2.3.6.   □

Corollary 2.2.3.12 admits the following generalization:

Corollary 2.2.3.14. *Suppose we are given a diagram of simplicial sets*

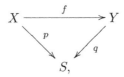

where p and q are right fibrations. Then f is a contravariant fibration in $(\mathcal{S}et_\Delta)_{/S}$ *if and only if f is a right fibration.*

Proof. The map f admits a factorization

$$X \xrightarrow{f'} X' \xrightarrow{f''} Y,$$

where f' is a contravariant equivalence and f'' is a contravariant fibration (in $(\mathcal{S}et_\Delta)_{/S}$). Proposition 2.1.4.9 implies that f'' is a right fibration, so the composition $q \circ f''$ is a right fibration. Invoking Corollary 2.2.3.13, we conclude that for every vertex $s \in S$, the map f' induces a homotopy equivalence of fibers $X_s \to X'_s$. Consider the diagram

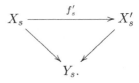

The vertical maps in this diagram are right fibrations between Kan complexes and therefore Kan fibrations (Lemma 2.1.3.3). Since f_s is a homotopy equivalence, we conclude that the induced map of fibers $f'_y : X_y \to X'_y$ is a homotopy equivalence for each vertex $y \in Y$. Invoking Lemma 2.2.3.6, we deduce that f' is an equivalence in the simplicial category $(\mathcal{S}et_\Delta)_{/Y}$.

We can now repeat the proof of Corollary 2.2.3.12. Let g be a homotopy inverse to f' in the simplicial category $(\mathrm{Set}_\Delta)_{/Y}$ and let $h : X \times \Delta^1 \to X$ be a homotopy from id_X to $g \circ f'$ (which projects to the identity on Y). To prove that f is a covariant fibration, we must show that every lifting problem

$$
\begin{array}{ccc}
A & \xrightarrow{\ e_0\ } & X \\
{\scriptstyle j}\downarrow & \nearrow^{e} & \downarrow{\scriptstyle f} \\
B & \longrightarrow & Y
\end{array}
$$

has a solution provided that j is a trivial cofibration in the contravariant model category $(\mathrm{Set}_\Delta)_{/S}$. Since f'' is a contravariant fibration, the map $f' \circ e_0$ can be extended to a map $\overline{e} : B \to X'$ in $(\mathrm{Set}_\Delta)_{/Y}$. Let $e' = g \circ \overline{e}$. The maps $\overline{e}$ and $h \circ (e_0 \times \mathrm{id}_{\Delta^1})$ determine another lifting problem

$$
\begin{array}{ccc}
(A \times \Delta^1) \coprod_{A \times \{1\}} (B \times \{1\}) & \longrightarrow & X \\
{\scriptstyle j'}\downarrow & \nearrow^{E} & \downarrow{\scriptstyle f} \\
B \times \Delta^1 & \longrightarrow & Y.
\end{array}
$$

Proposition 2.1.2.6 implies that j' is right anodyne. Since f is a right fibration, there exists an extension E as indicated in the diagram. The restriction $e = E|B \times \{0\}$ is then a solution to the original problem. $\square$

We conclude this section with one more result which will be useful in studying the Joyal model structure on Set_Δ. Suppose that $f : X \to S$ is any map of simplicial sets and that $\{s\}$ is a vertex of S. Let $Q^\bullet$ denote the cosimplicial object of Set_Δ defined in §2.2.2. Then we have a canonical map

$$
|X_s|_{Q^\bullet} \simeq (St_{\{s\}} X_s)(s) \to (St_S X)(s).
$$

Proposition 2.2.3.15. *Suppose that $f : X \to S$ is a right fibration of simplicial sets. Then for each vertex s of S, the canonical map $\phi : |X_s|_{Q^\bullet} \to (St_S X)(s)$ is a weak homotopy equivalence of simplicial sets.*

Proof. Choose a weak equivalence $St_S X \to \mathfrak{F}$, where $\mathfrak{F} : \mathfrak{C}[S]^{op} \to \mathrm{Set}_\Delta$ is a projectively fibrant diagram. Theorem 2.2.1.2 implies that the adjoint map $X \to \mathrm{Un}_S(\mathfrak{F})$ is a contravariant equivalence in $(\mathrm{Set}_\Delta)_{/S}$. Applying the "only if" direction of Corollary 2.2.3.12, we conclude that each of the induced maps

$$
X_s \to (\mathrm{Un}_S \mathfrak{F})_s \simeq \mathrm{Sing}_{Q^\bullet} \mathfrak{F}(s)
$$

is a homotopy equivalence of Kan complexes. Using Proposition 2.2.2.9, we deduce that the adjoint map $|X_s|_{Q^\bullet} \to \mathfrak{F}(s)$ is a weak homotopy equivalence. It follows from the two-out-of-three property that ϕ is also a weak homotopy equivalence. $\square$

2.2.4 The Comparison Theorem

Let S be an ∞-category containing a pair of objects x and y and let $Q^\bullet$ denote the cosimplicial object of Set_Δ described in §2.2.2. We have a canonical map of simplicial sets

$$f : |\operatorname{Hom}_S^{\mathrm{R}}(x,y)|_{Q^\bullet} \to \operatorname{Map}_{\mathcal{C}[S]}(x,y).$$

Moreover, in the special case where S is the nerve of a fibrant simplicial category $\mathcal{C}$, the composition

$$|\operatorname{Hom}_S^{\mathrm{R}}(x,y)|_{Q^\bullet} \xrightarrow{f} \operatorname{Map}_{\mathcal{C}[S]}(x,y) \to \operatorname{Map}_{\mathcal{C}}(x,y)$$

can be identified with the counit map

$$|\operatorname{Sing}_{Q^\bullet} \operatorname{Map}_{\mathcal{C}}(x,y)|_{Q^\bullet} \to \operatorname{Map}_{\mathcal{C}}(X,Y)$$

and is therefore a weak equivalence (Proposition 2.2.2.10). Consequently, we may reformulate Theorem 2.2.0.1 in the following way:

Proposition 2.2.4.1. *Let S be an ∞-category containing a pair of objects x and y. Then the natural map*

$$f : |\operatorname{Hom}_S^{\mathrm{R}}(x,y)|_{Q^\bullet} \to \operatorname{Map}_{\mathcal{C}[S]}(x,y)$$

is a weak homotopy equivalence of simplicial sets.

Proof. Let $C = S_{/y}^{\triangleright} \coprod_{S_{/y}} S$ and let v denote the image in C of the cone point of $S_{/y}^{\triangleright}$. There is a canonical projection $\pi : C \to S$, which induces a map of simplicial sets

$$f'' : (St_S S_{/y})(x) \to \operatorname{Map}_{\mathcal{C}[S]}(x,y).$$

The map f can be identified with the composition $f'' \circ f'$, where f' is the map

$$|\operatorname{Hom}_S^{\mathrm{R}}(x,y)|_{Q^\bullet} \simeq (St_{\{x\}} S_{/y} \times_S \{x\})(x) \to (St_S S_{/y})(x).$$

Since the projection $S_{/y} \to S$ is a right fibration, the map f' is a weak homotopy equivalence (Proposition 2.2.3.15). It will therefore suffice to show that f'' is a weak homotopy equivalence. To see this, we consider the commutative diagram

$$
\begin{array}{ccc}
 & (St_S S_{/y})(x) & \\
 \overset{g}{\nearrow} & & \overset{f''}{\searrow} \\
(St_S\{y\})(x) & \xrightarrow{\quad h \quad} & \operatorname{Map}_{\mathcal{C}[S]}(x,y).
\end{array}
$$

The inclusion $i : \{y\} \subseteq S_{/y}$ is a retract of the inclusion

$$(S_{/y} \times \{1\}) \coprod_{\{y\} \times \{1\}} (\{\mathrm{id}_y\} \times \Delta^1) \subseteq S_{/y} \times \Delta^1,$$

which is right anodyne by Corollary 2.1.2.7. It follows that i is a contravariant equivalence in $(\mathrm{Set}_\Delta)_{/S}$ (Proposition 2.1.4.9), so the map g is a weak homotopy equivalence of simplicial sets. Since the map h is an isomorphism, the map f'' is also a weak homotopy equivalence by virtue of the two-out-of-three property. $\qquad\square$

2.2.5 The Joyal Model Structure

The category of simplicial sets can be endowed with a model structure for which the fibrant objects are precisely the ∞-categories. The original construction of this model structure is due to Joyal, who uses purely combinatorial arguments ([44]). In this section, we will exploit the relationship between simplicial categories and ∞-categories to give an alternative description of this model structure. Our discussion will make use of a model structure on the category Cat_Δ of simplicial categories, which we review in §A.3.2.

Theorem 2.2.5.1. *There exists a left proper combinatorial model structure on the category of simplicial sets with the following properties:*

(C) *A map $p : S \to S'$ of simplicial sets is a cofibration if and only if it is a monomorphism.*

(W) *A map $p : S \to S'$ is a categorical equivalence if and only if the induced simplicial functor $\mathfrak{C}[S] \to \mathfrak{C}[S']$ is an equivalence of simplicial categories.*

Moreover, the adjoint functors $(\mathfrak{C}, \mathrm{N})$ determine a Quillen equivalence between Set_Δ (with the model structure defined above) and Cat_Δ.

Our proof will make use of the theory of *inner anodyne* maps of simplicial sets, which we will study in detail in §2.3. We first establish a simple lemma.

Lemma 2.2.5.2. *Every inner anodyne map $f : A \to B$ of simplicial sets is a categorical equivalence.*

Proof. It will suffice to prove that if f is inner anodyne, then the associated map $\mathfrak{C}[f]$ is a trivial cofibration of simplicial categories. The collection of all morphisms f for which this statement holds is weakly saturated (Definition A.1.2.2). Consequently, we may assume that f is an inner horn inclusion $\Lambda_i^n \subseteq \Delta^n$, $0 < i < n$. We now explicitly describe the map $\mathfrak{C}[f]$:

- The objects of $\mathfrak{C}[\partial \Lambda_i^n]$ are the objects of $\mathfrak{C}[\Delta^n]$: namely, elements of the linearly ordered set $[n] = \{0, \ldots, n\}$.

- For $0 \leq j \leq k \leq n$, the simplicial set $\mathrm{Map}_{\mathfrak{C}[\Lambda_i^n]}(j, k)$ is equal to $\mathrm{Map}_{\mathfrak{C}[\Delta^n]}(j, k)$ unless $j = 0$ and $k = n$. In the latter case,

$$\mathrm{Map}_{\mathfrak{C}[\Lambda_i^n]}(j, k) = K \subseteq (\Delta^1)^{n-1} \simeq \mathrm{Map}_{\mathfrak{C}[\Delta^n]}(j, k),$$

  where K is the simplicial subset of the cube $(\Delta^1)^{n-1}$ obtained by removing the interior and a single face.

We observe that $\mathfrak{C}[f]$ is a pushout of the inclusion $\mathcal{E}_K \subseteq \mathcal{E}_{(\Delta^1)^{n-1}}$ (see §A.3.2 for an explanation of this notation). It now suffices to observe that the inclusion $K \subseteq (\Delta^1)^{n-1}$ is a trivial fibration of simplicial sets (with respect to the usual model structure on Set_Δ). $\qquad\square$

Proof of Theorem 2.2.5.1. We first show that $\mathfrak{C}$ carries cofibrations of simplicial sets to cofibrations of simplicial categories. Since the class of all cofibrations of simplicial sets is generated by the inclusions $\partial \Delta^n \subseteq \Delta^n$, it suffices to show that each map $\mathfrak{C}[\partial \Delta^n] \to \mathfrak{C}[\Delta^n]$ is a cofibration of simplicial categories. If $n = 0$, then the inclusion $\mathfrak{C}[\partial \Delta^n] \subseteq \mathfrak{C}[\Delta^n]$ is isomorphic to the inclusion $\emptyset \subseteq *$ of simplicial categories, which is a cofibration. In the case where $n > 0$, we make use of the following explicit description of $\mathfrak{C}[\partial \Delta^n]$ as a subcategory of $\mathfrak{C}[\Delta^n]$:

- The objects of $\mathfrak{C}[\partial \Delta^n]$ are the objects of $\mathfrak{C}[\Delta^n]$: namely, elements of the linearly ordered set $[n] = \{0, \dots, n\}$.

- For $0 \leq j \leq k \leq n$, the simplicial set $\operatorname{Hom}_{\mathfrak{C}[\partial \Delta^n]}(j, k)$ is equal to $\operatorname{Hom}_{\mathfrak{C}[\Delta^n]}(j, k)$ unless $j = 0$ and $k = n$. In the latter case, the simplicial set $\operatorname{Hom}_{\mathfrak{C}[\partial \Delta^n]}(j, k)$ consists of the boundary of the cube

$$(\Delta^1)^{n-1} \simeq \operatorname{Hom}_{\mathfrak{C}[\Delta^n]}(j, k).$$

In particular, the inclusion $\mathfrak{C}[\partial \Delta^n] \subseteq \mathfrak{C}[\Delta^n]$ is a pushout of the inclusion $\mathcal{E}_{\partial(\Delta^1)^{n-1}} \subseteq \mathcal{E}_{(\Delta^1)^{n-1}}$, which is a cofibration of simplicial categories (see §A.3.2 for an explanation of our notation).

We now declare that a map $p : S \to S'$ of simplicial sets is a *categorical fibration* if it has the right lifting property with respect to all maps which are cofibrations and categorical equivalences. We now claim that the cofibrations, categorical equivalences, and categorical fibrations determine a left proper combinatorial model structure on $\operatorname{Set}_\Delta$. To prove this, it will suffice to show that the hypotheses of Proposition A.2.6.13 are satisfied:

(1) The class of categorical equivalences in $\operatorname{Set}_\Delta$ is perfect. This follows from Corollary A.2.6.12 since the functor $\mathfrak{C}$ preserves filtered colimits and the class of equivalences between simplicial categories is perfect.

(2) The class of categorical equivalences is stable under pushouts by cofibrations. Since $\mathfrak{C}$ preserves cofibrations, this follows immediately from the left properness of $\operatorname{Cat}_\Delta$.

(3) A map of simplicial sets which has the right lifting property with respect to *all* cofibrations is a categorical equivalence. In other words, we must show that if $p : S \to S'$ is a trivial fibration of simplicial sets, then the induced functor $\mathfrak{C}[p] : \mathfrak{C}[S] \to \mathfrak{C}[S']$ is an equivalence of simplicial categories.

Since p is a trivial fibration, it admits a section $s : S' \to S$. It is clear that $\mathfrak{C}[p] \circ \mathfrak{C}[s]$ is the identity; it therefore suffices to show that

$$\mathfrak{C}[s] \circ \mathfrak{C}[p] : \mathfrak{C}[S] \to \mathfrak{C}[S]$$

is homotopic to the identity.

Let K denote the simplicial set $\operatorname{Map}_{S'}(S, S)$. Then K is a contractible Kan complex containing vertices x and y which classify $s \circ p$ and id_S.

We note the existence of a natural "evaluation map" $e : K \times S \to S$ such that $s \circ p = e \circ (\{x\} \times \mathrm{id}_S)$, $\mathrm{id}_S = e \circ (\{y\} \times \mathrm{id}_S)$. It therefore suffices to show that the functor $\mathfrak{C}$ carries $\{x\} \times \mathrm{id}_S$ and $\{y\} \times \mathrm{id}_S$ into homotopic morphisms. Since both of these maps section the projection $K \times S \to S$, it suffices to show that the projection $\mathfrak{C}[K \times S] \to \mathfrak{C}[S]$ is an equivalence of simplicial categories. Replacing S by $S \times K$ and S' by S, we are reduced to the special case where $S = S' \times K$ and K is a contractible Kan complex.

By the small object argument, we can find an inner anodyne map $S' \to V$, where V is an ∞-category. The corresponding map $S' \times K \to V \times K$ is also inner anodyne (Proposition 2.3.2.1), so the maps $\mathfrak{C}[S'] \to \mathfrak{C}[V]$ and $\mathfrak{C}[S' \times K] \to \mathfrak{C}[V \times K]$ are both trivial cofibrations (Lemma 2.2.5.2). It follows that we are free to replace S' by V and S by $V \times K$. In other words, we may suppose that S' is an ∞-category (and now we will have no further need of the assumption that S is isomorphic to the product $S' \times K$).

Since p is surjective on vertices, it is clear that $\mathfrak{C}[p]$ is essentially surjective. It therefore suffices to show that for every pair of vertices $x, y \in S_0$, the induced map of simplicial sets $\mathrm{Map}_{\mathfrak{C}[S]}(x, y) \to \mathrm{Map}_{\mathfrak{C}[S']}(p(x), p(y))$ is a weak homotopy equivalence. Using Propositions 2.2.4.1 and 2.2.2.7, it suffices to show that the map $\mathrm{Hom}_S^R(x, y) \to \mathrm{Hom}_{S'}^R(p(x), p(y))$ is a weak homotopy equivalence. This map is obviously a trivial fibration if p is a trivial fibration.

By construction, the functor $\mathfrak{C}$ preserves weak equivalences. We verified above that $\mathfrak{C}$ preserves cofibrations as well. It follows that the adjoint functors $(\mathfrak{C}, \mathrm{N})$ determine a Quillen adjunction

$$\mathrm{Set}_\Delta \underset{\mathrm{N}}{\overset{\mathfrak{C}}{\rightleftarrows}} \mathcal{C}\mathrm{at}_\Delta .$$

To complete the proof, we wish to show that this Quillen adjunction is a Quillen equivalence. According to Proposition A.2.5.1, we must show that for every simplicial set S and every *fibrant* simplicial category $\mathcal{C}$, a map

$$u : S \to \mathrm{N}(\mathcal{C})$$

is a categorical equivalence if and only if the adjoint map

$$v : \mathfrak{C}[S] \to \mathcal{C}$$

is an equivalence of simplicial categories. We observe that v factors as a composition

$$\mathfrak{C}[S] \xrightarrow{\mathfrak{C}[u]} \mathfrak{C}[\mathrm{N}(\mathcal{C})] \xrightarrow{w} \mathcal{C} .$$

By definition, u is a categorical equivalence if and only if $\mathfrak{C}[u]$ is an equivalence of simplicial categories. We now conclude by observing that the counit map w is an equivalence of simplicial categories (Theorem 2.2.0.1). $\square$

We now establish a few pleasant properties enjoyed by the Joyal model structure on Set_Δ. We first note that every object of Set_Δ is cofibrant; in particular, the Joyal model structure is *left proper* (Proposition A.2.4.2).

Remark 2.2.5.3. The Joyal model structure is *not* right proper. To see this, we note that the inclusion $\Lambda_1^2 \subseteq \Delta^2$ is a categorical equivalence, but it does not remain so after pulling back via the fibration $\Delta^{\{0,2\}} \subseteq \Delta^2$.

Corollary 2.2.5.4. *Let $f : A \to B$ be a categorical equivalence of simplicial sets and K an arbitrary simplicial set. Then the induced map $A \times K \to B \times K$ is a categorical equivalence.*

Proof. Choose an inner anodyne map $B \to Q$, where Q is an ∞-category. Then $B \times K \to Q \times K$ is also inner anodyne, hence a categorical equivalence (Lemma 2.2.5.2). It therefore suffices to prove that $A \times K \to Q \times K$ is a categorical equivalence. In other words, we may suppose to begin with that B is an ∞-category.

Now choose a factorization $A \xrightarrow{f'} R \xrightarrow{f''} B$, where f' is an inner anodyne map and f'' is an inner fibration. Since B is an ∞-category, R is an ∞-category. The map $A \times K \to R \times K$ is inner anodyne (since f' is) and therefore a categorical equivalence; consequently, it suffices to show that $R \times K \to B \times K$ is a categorical equivalence. In other words, we may reduce to the case where A is also an ∞-category.

Choose an inner anodyne map $K \to S$, where S is an ∞-category. Then $A \times K \to A \times S$ and $B \times K \to B \times S$ are both inner anodyne and therefore categorical equivalences. Thus, to prove that $A \times K \to B \times K$ is a categorical equivalence, it suffices to show that $A \times S \to B \times S$ is a categorical equivalence. In other words, we may suppose that K is an ∞-category.

Since A, B, and K are ∞-categories, we have canonical isomorphisms

$$\mathrm{h}(A \times K) \simeq \mathrm{h}A \times \mathrm{h}K \qquad \mathrm{h}(B \times K) \simeq \mathrm{h}B \times \mathrm{h}K.$$

It follows that $A \times K \to B \times K$ is essentially surjective provided that f is essentially surjective. Furthermore, for any pair of vertices $(a, k), (a', k') \in (A \times K)_0$, we have

$$\mathrm{Hom}_{A \times K}^{\mathrm{R}}((a, k), (a', k')) \simeq \mathrm{Hom}_A^{\mathrm{R}}(a, a') \times \mathrm{Hom}_K^{\mathrm{R}}(k, k')$$

$$\mathrm{Hom}_{B \times K}^{\mathrm{R}}((f(a), k), (f(a'), k')) \simeq \mathrm{Hom}_B^{\mathrm{R}}(f(a), f(a')) \times \mathrm{Hom}_K^{\mathrm{R}}(k, k').$$

This shows that $A \times K \to B \times K$ is fully faithful provided that f is fully faithful, which completes the proof. $\square$

Remark 2.2.5.5. Since every inner anodyne map is a categorical equivalence, it follows that every categorical fibration $p : X \to S$ is a inner fibration (see Definition 2.0.0.3). The converse is false in general; however, it is true when S is a point. In other words, the fibrant objects for the Joyal model structure on Set_Δ are precisely the ∞-categories. The proof will be given in §2.4.6 as Theorem 2.4.6.1. We will assume this result for the remainder of the section. No circularity will result from this because the proof of Theorem 2.4.6.1 will not use any of the results proven below.

The functor $\mathfrak{C}[\bullet]$ does not generally commute with products. However, Corollary 2.2.5.4 implies that $\mathfrak{C}$ commutes with products in the following weak sense:

Corollary 2.2.5.6. *Let S and S' be simplicial sets. The natural map*
$$\mathfrak{C}[S \times S'] \to \mathfrak{C}[S] \times \mathfrak{C}[S']$$
is an equivalence of simplicial categories.

Proof. Suppose first that there are fibrant simplicial categories $\mathcal{C}$, $\mathcal{C}'$ with $S = \mathrm{N}(\mathcal{C})$, $S' = \mathrm{N}(\mathcal{C}')$. In this case, we have a diagram
$$\mathfrak{C}[S \times S'] \xrightarrow{f} \mathfrak{C}[S] \times \mathfrak{C}[S'] \xrightarrow{g} \mathcal{C} \times \mathcal{C}'.$$
By the two-out-of-three property, it suffices to show that g and $g \circ f$ are equivalences. Both of these assertions follow immediately from the fact that the counit map $\mathfrak{C}[\mathrm{N}(\mathcal{D})] \to \mathcal{D}$ is an equivalence for *any* fibrant simplicial category $\mathcal{D}$ (Theorem 2.2.5.1).

In the general case, we may choose categorical equivalences $S \to T$, $S' \to T'$, where T and T' are nerves of fibrant simplicial categories. Since $S \times S' \to T \times T'$ is a categorical equivalence, we reduce to the case treated above. $\square$

Let K be a fixed simplicial set and let $\mathcal{C}$ be a simplicial set which is fibrant with respect to the Joyal model structure. Then $\mathcal{C}$ has the extension property with respect to all inner anodyne maps and is therefore an ∞-category. It follows that $\mathrm{Fun}(K, \mathcal{C})$ is also an ∞-category. We might call two morphisms $f, g : K \to \mathcal{C}$ *homotopic* if they are equivalent when viewed as objects of $\mathrm{Fun}(K, \mathcal{C})$. On the other hand, the general theory of model categories furnishes another notion of homotopy: f and g are *left homotopic* if the map
$$f \coprod g : K \coprod K \to \mathcal{C}$$
can be extended over a mapping cylinder I for K.

Proposition 2.2.5.7. *Let $\mathcal{C}$ be a ∞-category and K an arbitrary simplicial set. A pair of morphisms $f, g : K \to \mathcal{C}$ are homotopic if and only if they are left-homotopic.*

Proof. Choose a contractible Kan complex S containing a pair of distinct vertices, x and y. We note that the inclusion
$$K \coprod K \simeq K \times \{x, y\} \subseteq K \times S$$
exhibits $K \times S$ as a mapping cylinder for K. It follows that f and g are left homotopic if and only if the map $f \coprod g : K \coprod K \to \mathcal{C}$ admits an extension to $K \times S$. In other words, f and g are left homotopic if and only if there exists a map $h : S \to \mathcal{C}^K$ such that $h(x) = f$ and $h(y) = g$. We note that any such map factors through Z, where $Z \subseteq \mathrm{Fun}(K, \mathcal{C})$ is the largest Kan complex contained in $\mathcal{C}^K$. Now, by classical homotopy theory, the map h exists if and only if f and g belong to the same path component of Z. It is clear that this holds if and only if f and g are equivalent when viewed as objects of the ∞-category $\mathrm{Fun}(K, \mathcal{C})$. $\square$

We are now in a position to prove Proposition 1.2.7.3, which was asserted without proof in §1.2.7. We first recall the statement.

Proposition. *Let K be an arbitrary simplicial set.*

(1) *For every ∞-category $\mathcal{C}$, the simplicial set $\mathrm{Fun}(K, \mathcal{C})$ is an ∞-category.*

(2) *Let $\mathcal{C} \to \mathcal{D}$ be a categorical equivalence of ∞-categories. Then the induced map $\mathrm{Fun}(K, \mathcal{C}) \to \mathrm{Fun}(K, \mathcal{D})$ is a categorical equivalence.*

(3) *Let $\mathcal{C}$ be an ∞-category and $K \to K'$ a categorical equivalence of simplicial sets. Then the induced map $\mathrm{Fun}(K', \mathcal{C}) \to \mathrm{Fun}(K, \mathcal{C})$ is a categorical equivalence.*

Proof. We first prove (1). To show that $\mathrm{Fun}(K, \mathcal{C})$ is an ∞-category, it suffices to show that it has the extension property with respect to every inner anodyne inclusion $A \subseteq B$. This is equivalent to the assertion that $\mathcal{C}$ has the right lifting property with respect to the inclusion $A \times K \subseteq B \times K$. But $\mathcal{C}$ is an ∞-category and $A \times K \subseteq B \times K$ is inner anodyne (Corollary 2.3.2.4).

Let hSet_{Δ} denote the homotopy category of Set_{Δ} taken with respect to the Joyal model structure. For each simplicial set X, we let $[X]$ denote the same simplicial set considered as an object of hSet_{Δ}. For every pair of objects $X, Y \in \mathrm{Set}_{\Delta}$, $[X \times Y]$ is a product of $[X]$ and $[Y]$ in hSet_{Δ}. This is a general fact when X and Y are fibrant; in the general case, we choose fibrant replacements $X \to X'$, $Y \to Y'$ and apply the fact that the canonical map $X \times Y \to X' \times Y'$ is a categorical equivalence (Corollary 2.2.5.4).

If $\mathcal{C}$ is an ∞-category, then $\mathcal{C}$ is a fibrant object of Set_{Δ} (Theorem 2.4.6.1). Proposition 2.2.5.7 allows us to identify $\mathrm{Hom}_{\mathrm{hSet}_{\Delta}}([X], [\mathcal{C}])$ with the set of equivalence classes of objects in the ∞-category $\mathrm{Fun}(X, \mathcal{C})$. In particular, we have canonical bijections

$$\mathrm{Hom}_{\mathrm{hSet}_{\Delta}}([X] \times [K], [\mathcal{C}]) \simeq \mathrm{Hom}_{\mathrm{hSet}_{\Delta}}([X \times K], [\mathcal{C}])$$
$$\simeq \mathrm{Hom}_{\mathrm{hSet}_{\Delta}}([X], [\mathrm{Fun}(K, \mathcal{C})]).$$

It follows that $[\mathrm{Fun}(K, \mathcal{C})]$ is determined up to canonical isomorphism by $[K]$ and $[\mathcal{C}]$ (more precisely, it is an *exponential* $[\mathcal{C}]^{[K]}$ in the homotopy category hSet_{Δ}), which proves (2) and (3). $\qquad\square$

Our description of the Joyal model structure on Set_{Δ} is different from the definition given in [44]. Namely, Joyal defines a map $f : A \to B$ to be a *weak categorical equivalence* if, for every ∞-category $\mathcal{C}$, the induced map

$$\mathrm{hFun}(B, \mathcal{C}) \to \mathrm{hFun}(A, \mathcal{C})$$

is an equivalence (of ordinary categories). To prove that our definition agrees with his, it will suffice to prove the following.

Proposition 2.2.5.8. *Let $f : A \to B$ be a map of simplicial sets. Then f is a categorical equivalence if and only if it is a weak categorical equivalence.*

Proof. Suppose first that f is a categorical equivalence. If $\mathcal{C}$ is an arbitrary ∞-category, Proposition 1.2.7.3 implies that the induced map $\mathrm{Fun}(B, \mathcal{C}) \to \mathrm{Fun}(A, \mathcal{C})$ is a categorical equivalence, so that $\mathrm{hFun}(B, \mathcal{C}) \to \mathrm{hFun}(A, \mathcal{C})$ is an equivalence of categories. This proves that f is a weak categorical equivalence.

Conversely, suppose that f is a weak categorical equivalence. We wish to show that f induces an isomorphism in the homotopy category of Set_Δ with respect to the Joyal model structure. It will suffice to show that for any fibrant object $\mathcal{C}$, f induces a bijection $[B, \mathcal{C}] \to [A, \mathcal{C}]$, where $[X, \mathcal{C}]$ denotes the set of homotopy classes of maps from X to $\mathcal{C}$. By Proposition 2.2.5.7, $[X, \mathcal{C}]$ may be identified with the set of isomorphism classes of objects in the category $\mathrm{hFun}(X, \mathcal{C})$. By assumption, f induces an equivalence of categories $\mathrm{hFun}(B, \mathcal{C}) \to \mathrm{hFun}(A, \mathcal{C})$ and therefore a bijection on isomorphism classes of objects. $\qquad\square$

Remark 2.2.5.9. The proof of Proposition 1.2.7.3 makes use of Theorem 2.4.6.1, which asserts that the (categorically) fibrant objects of Set_Δ are precisely the ∞-categories. Joyal proves the analogous assertion for his model structure in [44]. We remark that one cannot formally deduce Theorem 2.4.6.1 from Joyal's result since we *need* Theorem 2.4.6.1 to prove that Joyal's model structure coincides with the one we have defined above. On the other hand, our approach *does* give a new proof of Joyal's theorem.

Remark 2.2.5.10. Proposition 2.2.5.8 permits us to define the Joyal model structure without reference to the theory of simplicial categories (this is Joyal's original point of view [44]). Our approach is less elegant but allows us to easily compare the theory of ∞-categories with other models of higher category theory, such as simplicial categories. There is another approach to obtaining comparison results, due to Toën. In [76], he shows that if $\mathcal{C}$ is a model category equipped with a cosimplicial object $C^\bullet$ satisfying certain conditions, then $\mathcal{C}$ is (canonically) Quillen equivalent to Rezk's category of complete Segal spaces. Toën's theorem applies in particular when $\mathcal{C}$ is the category of simplicial sets and $C^\bullet$ is the "standard simplex" $C^n = \Delta^n$. In fact, Set_Δ is in some sense universal with respect to this property because it is generated by $C^\bullet$ under colimits and the class of categorical equivalences is dictated by Toën's axioms. We refer the reader to [76] for details.

2.3 INNER FIBRATIONS

In this section, we will study the theory of *inner fibrations* between simplicial sets. The meaning of this notion is somewhat difficult to explain because it has no counterpart in classical category theory: Proposition 1.1.2.2 implies that *every* functor between ordinary categories $\mathcal{C} \to \mathcal{D}$ induces an inner fibration of nerves $\mathrm{N}(\mathcal{C}) \to \mathrm{N}(\mathcal{D})$.

In the case where S is a point, a map $p : X \to S$ is an inner fibration if and only if X is an ∞-category. Moreover, the class of inner fibrations is stable under base change: if

$$\begin{array}{ccc} X' & \longrightarrow & X \\ \downarrow{\scriptstyle p'} & & \downarrow{\scriptstyle p} \\ S' & \longrightarrow & S \end{array}$$

is a pullback diagram of simplicial sets and p is an inner fibration, then so is p'. It follows that if $p : X \to S$ is an arbitrary inner fibration, then each fiber $X_s = X \times_S \{s\}$ is an ∞-category. We may therefore think of p as encoding a family of ∞-categories parametrized by S. However, the fibers X_s depend functorially on s only in a very weak sense.

Example 2.3.0.1. Let $F : \mathcal{C} \to \mathcal{C}'$ be a functor between ordinary categories. Then the map $N(\mathcal{C}) \to N(\mathcal{C}')$ is an inner fibration. Yet the fibers $N(\mathcal{C})_C = N(\mathcal{C} \times_{\mathcal{C}'} \{C\})$ and $N(\mathcal{C})_D = N(\mathcal{C} \times_{\mathcal{C}'} \{D\})$ over objects $C, D \in \mathcal{C}'$ can have wildly different properties even if C and D are isomorphic objects of $\mathcal{C}'$.

In order to describe how the different fibers of an inner fibration are related to one another, we will introduce the notion of a *correspondence* between ∞-categories. We review the classical theory of correspondences in §2.3.1 and explain how to generalize this theory to the ∞-categorical setting.

In §2.3.2, we will prove that the class of inner anodyne maps is stable under smash products with arbitrary cofibrations between simplicial sets. As a consequence, we will deduce that the class of inner fibrations (and hence the class of ∞-categories) is stable under the formation of mapping spaces.

In §2.3.3, we will study the theory of *minimal* inner fibrations, a generalization of Quillen's theory of minimal Kan fibrations. In particular, we will define a class of minimal ∞-categories and show that every ∞-category $\mathcal{C}$ is (categorically) equivalent to a minimal ∞-category $\mathcal{C}'$, where $\mathcal{C}'$ is well-defined up to (noncanonical) isomorphism. We will apply this theory in §2.3.4 to develop a theory of n-categories for $n < \infty$.

2.3.1 Correspondences

Let $\mathcal{C}$ and $\mathcal{C}'$ be categories. A *correspondence* from $\mathcal{C}$ to $\mathcal{C}'$ is a functor

$$M : \mathcal{C}^{op} \times \mathcal{C}' \to \mathrm{Set}.$$

If M is a correspondence from $\mathcal{C}$ to $\mathcal{C}'$, we can define a new category $\mathcal{C} \star^M \mathcal{C}'$ as follows. An object of $\mathcal{C} \star^M \mathcal{C}'$ is either an object of $\mathcal{C}$ or an object of $\mathcal{C}'$. For morphisms, we take

$$\mathrm{Hom}_{\mathcal{C} \star^M \mathcal{C}'}(X, Y) = \begin{cases} \mathrm{Hom}_{\mathcal{C}}(X, Y) & \text{if } X, Y \in \mathcal{C} \\ \mathrm{Hom}_{\mathcal{C}'}(X, Y) & \text{if } X, Y \in \mathcal{C}' \\ M(X, Y) & \text{if } X \in \mathcal{C}, Y \in \mathcal{C}' \\ \emptyset & \text{if } X \in \mathcal{C}', Y \in \mathcal{C}. \end{cases}$$

Composition of morphisms is defined in the obvious way, using the composition laws in $\mathcal{C}$ and $\mathcal{C}'$ and the functoriality of $M(X, Y)$ in X and Y.

Remark 2.3.1.1. In the special case where $F : \mathcal{C}^{op} \times \mathcal{C}' \to$ Set is the constant functor taking the value $*$, the category $\mathcal{C} \star^F \mathcal{C}'$ coincides with the ordinary join $\mathcal{C} \star \mathcal{C}'$.

For any correspondence $M : \mathcal{C} \to \mathcal{C}'$, there is an obvious functor $F : \mathcal{C} \star^M \mathcal{C}' \to [1]$ (here $[1]$ denotes the linearly ordered set $\{0, 1\}$ regarded as a category in the obvious way) which is uniquely determined by the condition that $F^{-1}\{0\} = \mathcal{C}$ and $F^{-1}\{1\} = \mathcal{C}'$. Conversely, given any category $\mathcal{M}$ equipped with a functor $F : \mathcal{M} \to [1]$, we can *define* $\mathcal{C} = F^{-1}\{0\}$, $\mathcal{C}' = F^{-1}\{1\}$, and a correspondence $M : \mathcal{C} \to \mathcal{C}'$ by the formula $M(X, Y) = \text{Hom}_{\mathcal{M}}(X, Y)$. We may summarize the situation as follows:

Fact 2.3.1.2. *Giving a pair of categories $\mathcal{C}$, $\mathcal{C}'$ and a correspondence between them is equivalent to giving a category $\mathcal{M}$ equipped with a functor $\mathcal{M} \to [1]$.*

Given this reformulation, it is clear how to generalize the notion of a correspondence to the ∞-categorical setting.

Definition 2.3.1.3. Let $\mathcal{C}$ and $\mathcal{C}'$ be ∞-categories. A *correspondence* from $\mathcal{C}$ to $\mathcal{C}'$ is a ∞-category $\mathcal{M}$ equipped with a map $F : \mathcal{M} \to \Delta^1$ and identifications $\mathcal{C} \simeq F^{-1}\{0\}$, $\mathcal{C}' \simeq F^{-1}\{1\}$.

Remark 2.3.1.4. Let $\mathcal{C}$ and $\mathcal{C}'$ be ∞-categories. Fact 2.3.1.2 generalizes to the ∞-categorical setting in the following way: there is a canonical bijection between equivalence classes of correspondences from $\mathcal{C}$ to $\mathcal{C}'$ and equivalence classes of functors $\mathcal{C}^{op} \times \mathcal{C}' \to \mathcal{S}$, where $\mathcal{S}$ denotes the ∞-category of spaces. In fact, it is possible to prove a more precise result (a Quillen equivalence between certain model categories), but we will not need this.

To understand the relevance of Definition 2.3.1.3, we note the following:

Proposition 2.3.1.5. *Let $\mathcal{C}$ be an ordinary category and let $p : X \to N(\mathcal{C})$ be a map of simplicial sets. Then p is an inner fibration if and only if X is an ∞-category.*

Proof. This follows from the fact that any map $\Lambda^n_i \to N(\mathcal{C})$, $0 < i < n$, admits a *unique* extension to Δ^n. $\qquad\square$

It follows readily from the definition that an arbitrary map of simplicial sets $p : X \to S$ is an inner fibration if and only if the fiber of p over any simplex of S is an ∞-category. In particular, an inner fibration p associates to each vertex s of S an ∞-category X_s and to each edge $f : s \to s'$ in S a correspondence between the ∞-categories X_s and $X_{s'}$. Higher-dimensional simplices give rise to what may be thought of as compatible "chains" of correspondences.

Roughly speaking, we might think of an inner fibration $p : X \to S$ as a functor from S into some kind of ∞-category of ∞-categories where the morphisms are given by correspondences. However, this description is not quite accurate because the correspondences are required to "compose" only in a weak sense. To understand the issue, let us return to the setting of *ordinary* categories. If $\mathcal{C}$ and $\mathcal{C}'$ are two categories, then the correspondences from $\mathcal{C}$ to $\mathcal{C}'$ themselves constitute a category, which we may denote by $M(\mathcal{C}, \mathcal{C}')$. There is a natural "composition" defined on correspondences. If we view an object $F \in M(\mathcal{C}, \mathcal{C}')$ as a functor $\mathcal{C}^{op} \times \mathcal{C}' \to \mathcal{S}$et and $G \in M(\mathcal{C}', \mathcal{C}'')$, then we can define $(G \circ F)(C, C'')$ to be the coend

$$\int_{C' \in \mathcal{C}'} F(C, C') \times G(C', C'').$$

If we view F as determining a category $\mathcal{C} \star^F \mathcal{C}'$ and G as determining a category $\mathcal{C}' \star^G \mathcal{C}''$, then $\mathcal{C} \star^{G \circ F} \mathcal{C}''$ is obtained by forming the pushout

$$(\mathcal{C} \star^F \mathcal{C}') \coprod_{\mathcal{C}'} (\mathcal{C}' \star^G \mathcal{C}'')$$

and then discarding the objects of $\mathcal{C}'$.

Now, giving a category equipped with a functor to [2] is equivalent to giving a triple of categories $\mathcal{C}$, $\mathcal{C}'$, $\mathcal{C}''$ together with correspondences $F \in M(\mathcal{C}, \mathcal{C}')$, $G \in M(\mathcal{C}', \mathcal{C}'')$, $H \in M(\mathcal{C}, \mathcal{C}'')$, and a map $\alpha : G \circ F \to H$. But the map α need not be an isomorphism. Consequently, the above data cannot literally be interpreted as a functor from [2] into a category (or even a higher category) in which the morphisms are given by correspondences.

If $\mathcal{C}$ and $\mathcal{C}'$ are categories, then a correspondence from $\mathcal{C}$ to $\mathcal{C}'$ can be regarded as a kind of generalized functor from $\mathcal{C}$ to $\mathcal{C}'$. More specifically, for any functor $f : \mathcal{C} \to \mathcal{C}'$, we can define a correspondence M_f by the formula $M_f(X, Y) = \operatorname{Hom}_{\mathcal{C}'}(f(X), Y)$. This construction gives a fully faithful embedding $\operatorname{Map}_{\mathcal{C}at}(\mathcal{C}, \mathcal{C}') \to M(\mathcal{C}, \mathcal{C}')$. Similarly, any functor $g : \mathcal{C}' \to \mathcal{C}$ determines a correspondence M_g given by the formula

$$M_g(X, Y) = \operatorname{Hom}_{\mathcal{C}}(X, g(Y));$$

we observe that $M_f \simeq M_g$ if and only if the functors f and g are adjoint to one another.

If an inner fibration $p : X \to S$ corresponds to a "functor" from S to a higher category of ∞-categories with morphisms given by correspondences, then some special class of inner fibrations should correspond to functors from S into an ∞-category of ∞-categories with morphisms given by actual functors. This is indeed the case, and the appropriate notion is that of a (*co*)*Cartesian fibration* which we will study in §2.4.

2.3.2 Stability Properties of Inner Fibrations

Let $\mathcal{C}$ be an ∞-category and K an arbitrary simplicial set. In §1.2.7, we asserted that $\operatorname{Fun}(K, \mathcal{C})$ is an ∞-category (Proposition 1.2.7.3). In the course of the proof, we invoked certain stability properties of the class of inner anodyne maps. The goal of this section is to establish the required properties and deduce some of their consequences. Our main result is the following analogue of Proposition 2.1.2.6:

Proposition 2.3.2.1 (Joyal [44]). *The following collections all generate the same class of morphisms of* Set_Δ:

(1) *The collection* A_1 *of all horn inclusions* $\Lambda_i^n \subseteq \Delta^n$, $0 < i < n$.

(2) *The collection* A_2 *of all inclusions*

$$(\Delta^m \times \Lambda_1^2) \coprod_{\partial \Delta^m \times \Lambda_1^2} (\partial \Delta^m \times \Delta^2) \subseteq \Delta^m \times \Delta^2.$$

(3) *The collection* A_3 *of all inclusions*

$$(S' \times \Lambda_1^2) \coprod_{S \times \Lambda_1^2} (S \times \Delta^2) \subseteq S' \times \Delta^2,$$

where $S \subseteq S'$.

Proof. We will employ the strategy that we used to prove Proposition 2.1.2.6, though the details are slightly more complicated. Working cell by cell, we conclude that every morphism in A_3 belongs to the weakly saturated class of morphisms generated by A_2. We next show that every morphism in A_1 is a retract of a morphism belonging to A_3. More precisely, we will show that for $0 < i < n$, the inclusion $\Lambda_i^n \subseteq \Delta^n$ is a retract of the inclusion

$$(\Delta^n \times \Lambda_1^2) \coprod_{\Lambda_i^n \times \Lambda_1^2} (\Lambda_i^n \times \Delta^2) \subseteq \Delta^n \times \Delta^2.$$

To prove this, we embed Δ^n into $\Delta^n \times \Delta^2$ via the map of partially ordered sets $s : [n] \to [n] \times [2]$ given by

$$s(j) = \begin{cases} (j,0) & \text{if } j < i \\ (j,1) & \text{if } j = i \\ (j,2) & \text{if } j > i \end{cases}$$

and consider the retraction $\Delta^n \times \Delta^2 \to \Delta^n$ given by the map

$$r : [n] \times [2] \to [n]$$

$$r(j,k) = \begin{cases} j & \text{if } j < i, k = 0 \\ j & \text{if } j > i, k = 2 \\ i & \text{otherwise.} \end{cases}$$

We now show that every morphism in A_2 is inner anodyne (that is, it lies in the weakly saturated class of morphisms generated by A_1). Choose $m \geq 0$. For each $0 \leq i \leq j < m$, we let σ_{ij} denote the $(m+1)$-simplex of $\Delta^m \times \Delta^2$ corresponding to the map

$$f_{ij} : [m+1] \to [m] \times [2]$$

$$f_{ij}(k) = \begin{cases} (k,0) & \text{if } 0 \leq k \leq i \\ (k-1,1) & \text{if } i+1 \leq k \leq j+1 \\ (k-1,2) & \text{if } j+2 \leq k \leq m+1. \end{cases}$$

For each $0 \leq i \leq j \leq m$, we let τ_{ij} denote the $(m+2)$-simplex of $\Delta^m \times \Delta^2$ corresponding to the map

$$g_{ij} : [m+2] \to [m] \times [2]$$

$$g_{ij}(k) = \begin{cases} (k, 0) & \text{if } 0 \leq k \leq i \\ (k-1, 1) & \text{if } i+1 \leq k \leq j+1 \\ (k-2, 2) & \text{if } j+2 \leq k \leq m+2. \end{cases}$$

Let $X(0) = (\Delta^m \times \Lambda_1^2) \coprod_{\partial \Delta^m \times \Lambda_1^2} (\partial \Delta^m \times \Delta^2)$. For $0 \leq j < m$, we let

$$X(j+1) = X(j) \cup \sigma_{0j} \cup \cdots \cup \sigma_{jj}.$$

We have a chain of inclusions

$$X(j) \subseteq X(j) \cup \sigma_{0j} \subseteq \cdots \subset X(j) \cup \sigma_{0j} \cup \cdots \cup \sigma_{jj} = X(j+1),$$

each of which is a pushout of a morphism in A_1 and therefore inner anodyne. It follows that each inclusion $X(j) \subseteq X(j+1)$ is inner anodyne. Set $Y(0) = X(m)$ so that the inclusion $X(0) \subseteq Y(0)$ is inner anodyne. We now set $Y(j+1) = Y(j) \cup \tau_{0j} \cup \cdots \cup \tau_{jj}$ for $0 \leq j \leq m$. As before, we have a chain of inclusions

$$Y(j) \subseteq Y(j) \cup \tau_{0j} \subseteq \cdots \subseteq Y_j \cup \tau_{0j} \cup \cdots \cup \tau_{jj} = Y(j+1),$$

each of which is a pushout of a morphism belonging to A_1. It follows that each inclusion $Y(j) \subseteq Y(j+1)$ is inner anodyne. By transitivity, we conclude that the inclusion $X(0) \subseteq Y(m+2)$ is inner anodyne. We conclude the proof by observing that $Y(m+2) = \Delta^m \times \Delta^2$. $\qquad\square$

Corollary 2.3.2.2 (Joyal [44]). *A simplicial set $\mathcal{C}$ is an ∞-category if and only if the restriction map*

$$\text{Fun}(\Delta^2, \mathcal{C}) \to \text{Fun}(\Lambda_1^2, \mathcal{C})$$

is a trivial fibration.

Proof. By Proposition 2.3.2.1, $\mathcal{C} \to *$ is an inner fibration if and only if S has the extension property with respect to each of the inclusions in the class A_2. $\qquad\square$

Remark 2.3.2.3. In §1.1.2, we asserted that the main function of the weak Kan condition on a simplicial set $\mathcal{C}$ is that it allows us to compose the edges of $\mathcal{C}$. We can regard Corollary 2.3.2.2 as an affirmation of this philosophy: the class of ∞-categories $\mathcal{C}$ is characterized by the requirement that one can compose morphisms in $\mathcal{C}$, and the composition is well-defined up to a contractible space of choices.

Corollary 2.3.2.4 (Joyal [44]). *Let $i : A \to A'$ be an inner anodyne map of simplicial sets and let $j : B \to B'$ be a cofibration. Then the induced map*

$$(A \times B') \coprod_{A \times B} (A' \times B) \to A' \times B'$$

is inner anodyne.

Proof. This follows immediately from Proposition 2.3.2.1, which characterizes the class of inner anodyne maps as the class generated by A_3 (which is
stable under smash products with any cofibration).   □

Corollary 2.3.2.5 (Joyal [44]). *Let $p : X \to S$ be an inner fibration and
let $i : A \to B$ be any cofibration of simplicial sets. Then the induced map
$q : X^B \to X^A \times_{S^A} S^B$ is an inner fibration. If i is inner anodyne, then q is
a trivial fibration. In particular, if X is an ∞-category, then so is X^B for
any simplicial set B.*

2.3.3 Minimal Fibrations

One of the aims of homotopy theory is to understand the classification of
spaces up to homotopy equivalence. In the setting of simplicial sets, this
problem admits an attractive formulation in terms of Quillen's theory of
minimal Kan complexes. Every Kan complex X is homotopy equivalent to
a minimal Kan complex, and a map $X \to Y$ of minimal Kan complexes is a
homotopy equivalence if and only if it is an isomorphism. Consequently, the
classification of Kan complexes up to homotopy equivalence is equivalent to
the classification of *minimal* Kan complexes up to isomorphism. Of course, in
practical terms, this is not of much use for solving the classification problem.
Nevertheless, the theory of minimal Kan complexes (and, more generally,
minimal Kan fibrations) is a useful tool in the homotopy theory of simplicial
sets. The purpose of this section is to describe a generalization of the theory
of minimal models in which Kan fibrations are replaced by inner fibrations.
An exposition of this theory can also be found in [44].

We begin by introducing a bit of terminology. Suppose we are given a
commutative diagram

$$
\begin{array}{ccc}
A & \xrightarrow{\ u\ } & X \\
{\scriptstyle i}\downarrow & \nearrow & \downarrow{\scriptstyle p} \\
B & \xrightarrow[\ v\ ]{} & S
\end{array}
$$

of simplicial sets, where p is an inner fibration, and suppose also that we
have a pair $f, f' : B \to X$ of candidates for the dotted arrow which render
the diagram commutative. We will say that f and f' are *homotopic relative
to A over S* if they are equivalent when viewed as objects in the ∞-category
given by the fiber of the map

$$X^B \to X^A \times_{S^A} S^B.$$

Equivalently, f and f' are homotopic relative to A over S if there exists a
map $F : B \times \Delta^1 \to X$ such that $F|B \times \{0\} = f, F|B \times \{1\} = f', p \circ F = v \circ \pi_B$,
$F \circ (i \times \mathrm{id}_{\Delta^1}) = u \circ \pi_A$, and $F|\{b\} \times \Delta^1$ is an equivalence in the ∞-category
$X_{v(b)}$ for every vertex b of B.

Definition 2.3.3.1. Let $p : X \to S$ be an inner fibration of simplicial sets.
We will say that p is *minimal* if $f = f'$ for every pair of maps $f, f' : \Delta^n \to X$
which are homotopic relative to $\partial \Delta^n$ over S.

We will say that an ∞-category $\mathcal{C}$ is *minimal* if the associated inner fibration $\mathcal{C} \to *$ is minimal.

Remark 2.3.3.2. In the case where p is a Kan fibration, Definition 2.3.3.1 recovers the usual notion of a minimal Kan fibration. We refer the reader to [32] for a discussion of minimal fibrations in this more classical setting.

Remark 2.3.3.3. Let $p : X \to \Delta^n$ be an inner fibration. Then X is an ∞-category. Moreover, p is a minimal inner fibration if and only if X is a minimal ∞-category. This follows from the observation that for any pair of maps $f, f' : \Delta^m \to X$, a homotopy between f and f' is automatically compatible with the projection to Δ^n.

Remark 2.3.3.4. If $p : X \to S$ is a minimal inner fibration and $T \to S$ is an arbitrary map of simplicial sets, then the induced map $X_T = X \times_S T \to T$ is a minimal inner fibration. Conversely, if $p : X \to S$ is an inner fibration and if $X \times_S \Delta^n \to \Delta^n$ is minimal for *every* map $\sigma : \Delta^n \to S$, then p is minimal. Consequently, for many purposes the study of minimal inner fibrations reduces to the study of minimal ∞-categories.

Lemma 2.3.3.5. *Let $\mathcal{C}$ be a minimal ∞-category and let $f : \mathcal{C} \to \mathcal{C}$ be a functor which is homotopic to the identity. Then f is a monomorphism of simplicial sets.*

Proof. Choose a homotopy $h : \Delta^1 \times \mathcal{C} \to \mathcal{C}$ from $\mathrm{id}_{\mathcal{C}}$ to f. We prove by induction on n that the map f induces an injection from the set of n-simplices of $\mathcal{C}$ to itself. Let $\sigma, \sigma' : \Delta^n \to \mathcal{C}$ be such that $f \circ \sigma = f \circ \sigma'$. By the inductive hypothesis, we deduce that $\sigma | \partial \Delta^n = \sigma' | \partial \Delta^n = \sigma_0$. Consider the diagram

$$(\Delta^2 \times \partial \Delta^n) \coprod_{\Lambda_2^2 \times \partial \Delta^n} (\Lambda_2^2 \times \Delta^n) \xrightarrow{\quad G_0 \quad} \mathcal{C}$$

$$\downarrow \qquad \qquad \nearrow_{G}$$

$$\Delta^2 \times \Delta^n,$$

where $G_0 | \Lambda_2^2 \times \Delta^n$ is given by amalgamating $h \circ (\mathrm{id}_{\Delta^1} \times \sigma)$ with $h \circ (\mathrm{id}_{\Delta^1} \times \sigma')$ and $G_0 | \Delta^2 \times \partial \Delta^n$ is given by the composition

$$\Delta^2 \times \partial \Delta^n \to \Delta^1 \times \partial \Delta^n \xrightarrow{\sigma_0} \Delta^1 \times \mathcal{C} \xrightarrow{h} \mathcal{C}.$$

Since $h | \Delta^1 \times \{X\}$ is an equivalence for every object $X \in \mathcal{C}$, Proposition 2.4.1.8 implies the existence of the map G indicated in the diagram. The restriction $G | \Delta^1 \times \Delta^n$ is a homotopy between σ and σ' relative to $\partial \Delta^n$. Since $\mathcal{C}$ is minimal, we deduce that $\sigma = \sigma'$. $\qquad\square$

Lemma 2.3.3.6. *Let $\mathcal{C}$ be a minimal ∞-category and let $f : \mathcal{C} \to \mathcal{C}$ be a functor which is homotopic to the identity. Then f is an isomorphism of simplicial sets.*

Proof. Choose a homotopy $h : \Delta^1 \times \mathcal{C} \to \mathcal{C}$ from $\mathrm{id}_\mathcal{C}$ to f. We prove by induction on n that the map f induces a *bijection* from the set of n-simplices of $\mathcal{C}$ to itself. The injectivity follows from Lemma 2.3.3.5, so it will suffice to prove the surjectivity. Choose an n-simplex $\sigma : \Delta^n \to \mathcal{C}$. By the inductive hypothesis, we may suppose that $\sigma | \partial \Delta^n = f \circ \sigma_0'$ for some map $\sigma_0' : \partial \Delta^n \to \mathcal{C}$. Consider the diagram

$$(\Delta^1 \times \partial \Delta^n) \coprod_{\{1\} \times \partial \Delta^n} (\{1\} \times \Delta^n) \xrightarrow{\quad G_0 \quad} \mathcal{C}$$

with G (dashed) down to $\Delta^1 \times \Delta^n$,

where $G_0 | \Delta^1 \times \partial \Delta^n = h \circ (\mathrm{id}_{\Delta^1} \times \sigma_0')$ and $G_0 | \{1\} \times \Delta^n = \sigma$. If $n > 0$, then the existence of the map G as indicated in the diagram follows from Proposition 2.4.1.8; if $n = 0$, it is obvious. Now let $\sigma' = G | \{0\} \times \Delta^n$. To complete the proof, it will suffice to show that $f \circ \sigma' = \sigma$.

Consider now the diagram

$$(\Lambda_0^2 \times \Delta^n) \coprod_{\Lambda_0^2 \times \partial \Delta^n} (\Delta^2 \times \partial \Delta^n) \xrightarrow{\quad H_0 \quad} \mathcal{C}$$

with H (dashed) down to Δ^2,

where $H_0 | \Delta^{\{0,1\}} \times \Delta^n = h \circ (\mathrm{id}_{\Delta^1} \times \sigma')$, $H_0 | \Delta^{\{1,2\}} \times \Delta^n = G$, and $H_0 | (\Delta^2 \times \partial \Delta^n)$ is given by the composition

$$\Delta^2 \times \partial \Delta^n \to \Delta^1 \times \partial \Delta^n \xrightarrow{\sigma_0'} \Delta^1 \times \mathcal{C} \xrightarrow{h} \mathcal{C}.$$

The existence of the dotted arrow H follows once again from Proposition 2.4.1.8. The restriction $H | \Delta^{\{1,2\}} \times \Delta^n$ is a homotopy from $f \circ \sigma'$ to σ relative to $\partial \Delta^n$. Since $\mathcal{C}$ is minimal, we conclude that $f \circ \sigma' = \sigma$, as desired. $\quad \square$

Proposition 2.3.3.7. *Let $f : \mathcal{C} \to \mathcal{D}$ be an equivalence of minimal ∞-categories. Then f is an isomorphism.*

Proof. Since f is a categorical equivalence, it admits a homotopy inverse $g : \mathcal{D} \to \mathcal{C}$. Now apply Lemma 2.3.3.6 to the compositions $f \circ g$ and $g \circ f$. $\quad \square$

The following result guarantees a good supply of minimal ∞-categories:

Proposition 2.3.3.8. *Let $p : X \to S$ be an inner fibration of simplicial sets. Then there exists a retraction $r : X \to X$ onto a simplicial subset $X' \subseteq X$ with the following properties:*

(1) *The restriction $p | X' : X' \to S$ is a minimal inner fibration.*

(2) *The retraction r is compatible with the projection p in the sense that $p \circ r = p$.*

(3) *The map r is homotopic over S to id_X relative to X'.*

(4) *For every map of simplicial sets $T \to S$, the induced inclusion*

$$X' \times_S T \subseteq X \times_S T$$

is a categorical equivalence.

Proof. For every $n \geq 0$, we define a relation on the set of n-simplices of X: given two simplices $\sigma, \sigma' : \Delta^n \to X$, we will write $\sigma \sim \sigma'$ if σ is homotopic to σ' relative to $\partial \Delta^n$. We note that $\sigma \sim \sigma'$ if and only if $\sigma | \partial \Delta^n = \sigma' | \partial \Delta^n$ and σ is equivalent to σ', where both are viewed as objects in the ∞-category given by a fiber of the map

$$X^{\Delta^n} \to X^{\partial \Delta^n} \times_{S^{\partial \Delta^n}} S^{\Delta^n}.$$

Consequently, $\sim$ is an equivalence relation.

Suppose that σ and σ' are both degenerate and $\sigma \sim \sigma'$. From the equality $\sigma | \partial \Delta^n = \sigma' | \partial \Delta^n$, we deduce that $\sigma = \sigma'$. Consequently, there is at most one degenerate n-simplex of X in each $\sim$-class. Let $Y(n) \subseteq X_n$ denote a set of representatives for the $\sim$-classes of n-simplices in X, which contains all degenerate simplices. We now define the simplicial subset $X' \subseteq X$ recursively as follows: an n-simplex $\sigma : \Delta^n \to X$ belongs to X' if $\sigma \in Y(n)$ and $\sigma | \partial \Delta^n$ factors through X'.

Let us now prove (1). To show that $p | X'$ is an inner fibration, it suffices to prove that every lifting problem of the form

$$
\begin{array}{ccc}
\Lambda^n_i & \xrightarrow{\;s\;} & X' \\
\big\downarrow{\scriptstyle\cap} & \nearrow^{\sigma} & \big\downarrow \\
\Delta^n & \longrightarrow & S,
\end{array}
$$

with $0 < i < n$ has a solution f in X'. Since p is an inner fibration, this lifting problem has a solution $\sigma' : \Delta^n \to X$ in the original simplicial set X. Let $\sigma'_0 = d_i \sigma : \Delta^{n-1} \to X$ be the induced map. Then $\sigma'_0 | \partial \Delta^{n-1}$ factors through X'. Consequently, σ'_0 is homotopic (over S and relative to $\partial \Delta^{n-1}$) to some map $\sigma_0 : \Delta^{n-1} \to X'$. Let $g_0 : \Delta^1 \times \Delta^{n-1} \to X$ be a homotopy from σ'_0 to σ_0 and let $g_1 : \Delta^1 \times \partial \Delta^n \to X$ be the result of amalgamating g_0 with the identity homotopy from s to itself. Let $\sigma_1 = g_1 | \{1\} \times \partial \Delta^n$. Using Proposition 2.4.1.8, we deduce that g_1 extends to a homotopy from σ' to some other map $\sigma'' : \Delta^n \to X$ with $\sigma'' | \partial \Delta^n = \sigma_1$. It follows that σ'' is homotopic over S relative to $\partial \Delta^n$ to a map $\sigma : \Delta^n \to X$ with the desired properties. This proves that $p | X'$ is an inner fibration. It is immediate from the construction that $p | X'$ is minimal.

We now verify (2) and (3) by constructing a map $h : X \times \Delta^1 \to X$ such that $h | X \times \{0\}$ is the identity, $h | X \times \{1\}$ is a retraction $r : X \to X$ with image X', and h is a homotopy over S and relative to X'. Choose an exhaustion of X by a transfinite sequence of simplicial subsets

$$X' = X^0 \subseteq X^1 \subseteq \cdots,$$

where each X^α is obtained from

$$X^{<\alpha} = \bigcup_{\beta < \alpha} X^\beta$$

by adjoining a single nondegenerate simplex, provided that such a simplex exists. We construct $h_\alpha = h|X^\alpha \times \Delta^1$ by induction on α. By the inductive hypothesis, we may suppose that we have already defined $h_{<\alpha} = h|X^{<\alpha} \times \Delta^1$. If $X = X^{<\alpha}$, then we are done. Otherwise, we can write $X^\alpha = X^{<\alpha} \coprod_{\partial \Delta^n} \Delta^n$ corresponding to some nondegenerate simplex $\tau : \Delta^n \to X$, and it suffices to define $h_\alpha|\Delta^n \times \Delta^1$. If τ factors through X', we define $h_\alpha|\Delta^n \times \Delta^1$ to be the composition

$$\Delta^n \times \Delta^1 \to \Delta^n \overset{\sigma}{\to} X.$$

Otherwise, we use Proposition 2.4.1.8 to deduce the existence of the dotted arrow h' in the diagram

$$(\Delta^n \times \{0\}) \coprod_{\partial \Delta^n \times \{0\}} (\partial \Delta^n \times \Delta^1) \xrightarrow{(\tau, h_{<\alpha})} X$$

with h_0 and $p \circ \sigma$, p,

$$\Delta^n \times \Delta^1 \xrightarrow{p \circ \sigma} S.$$

Let $\tau' = h'|\Delta^n \times \{1\}$. Then $\tau'|\partial \Delta^n$ factors through X'. It follows that there is a homotopy $h'' : \Delta^n \times \Delta^{\{1,2\}} \to X$ from τ' to τ'', which is over S and relative to $\partial \Delta^n$ and such that τ'' factors through X'. Now consider the diagram

$$(\Delta^n \times \Lambda_1^2) \coprod_{\partial \Delta^n \times \Lambda_1^2} (\partial \Delta^n \times \Delta^2) \xrightarrow{H_0} X$$

with H and p,

$$\Delta^n \times \Delta^2 \longrightarrow S,$$

where $H_0|\Delta^n \times \Delta^{\{0,1\}} = h'$, $H_0|\Delta^n \times \Delta^{\{1,2\}} = h''$, and $H_0|\partial \Delta^n \times \Delta^2$ is given by the composition

$$\partial \Delta^n \times \Delta^2 \to \partial \Delta^n \times \Delta^1 \overset{h_{<\alpha}}{\to} X.$$

Using the fact that p is an inner fibration, we deduce that there exists a dotted arrow H rendering the diagram commutative. We may now define $h_\alpha|\Delta^n \times \Delta^1 = H|\Delta^n \times \Delta^{\{0,2\}}$; it is easy to see that this extension has all the desired properties.

We now prove (4). Using Proposition 3.2.2.8, we can reduce to the case where $T = \Delta^n$. Without loss of generality, we can replace S by $T = \Delta^n$, so that X and X' are ∞-categories. The above constructions show that $r : X \to X'$ is a homotopy inverse of the inclusion $i : X' \to X$, so that i is an equivalence, as desired. $\square$

We conclude by recording a property of minimal ∞-categories which makes them very useful for certain applications.

Proposition 2.3.3.9. *Let* $\mathcal{C}$ *be a minimal* ∞*-category and let* $\sigma : \Delta^n \to \mathcal{C}$ *be an n-simplex of* $\mathcal{C}$ *such that* $\sigma|\Delta^{\{i,i+1\}} = \mathrm{id}_C : C \to C$ *is a degenerate edge. Then* $\sigma = s_i\sigma_0$ *for some* $\sigma_0 : \Delta^{n-1} \to \mathcal{C}$.

Proof. We work by induction on n. Let $\sigma_0 = d_{i+1}\sigma$ and let $\sigma' = s_i\sigma_0$. We will prove that $\sigma = \sigma'$. Our first goal is to prove that $\sigma|\partial\Delta^n = \sigma'|\partial\Delta^n$; in other words, that $d_j\sigma = d_j\sigma'$ for $0 \leq j \leq n$. If $j = i+1$, this is obvious; if $j \notin \{i, i+1\}$, then it follows from the inductive hypothesis. Let us consider the case $i = j$, and set $\sigma_1 = d^i\sigma$. We need to prove that $\sigma_0 = \sigma_1$. The argument above establishes that $\sigma_0|\partial\Delta^{n-1} = \sigma_1|\partial\Delta^{n-1}$. Since $\mathcal{C}$ is minimal, it will suffice to show that σ_0 and σ_1 are homotopic relative to $\partial\Delta^{n-1}$. We now observe that

$$\left(s_{n-1}\sigma_0, s_{n-2}\sigma_0, \ldots, s_{i+1}\sigma_0, \sigma, s_{i-1}\sigma_1, \ldots, s_0\sigma_1\right)$$

provides the desired homotopy $\Delta^{n-1} \times \Delta^1 \to \mathcal{C}$.

Since σ and σ' coincide on $\partial\Delta^n$, to prove that $\sigma = \sigma'$ it will suffice to prove that σ and σ' are homotopic relative to $\partial\Delta^n$. We now observe that

$$\left(s_n\sigma', \ldots, s_{i+2}\sigma', s_i\sigma', s_i\sigma, s_{i-1}\sigma, \ldots, s_0\sigma\right)$$

is a homotopy $\Delta^n \times \Delta^1 \to \mathcal{C}$ with the desired properties. $\square$

We can interpret Proposition 2.3.3.9 as asserting that in a minimal ∞-category $\mathcal{C}$, composition is "strictly unital." For example, in the special case where $n = 2$ and $i = 1$, Proposition 2.3.3.9 asserts that if $f : X \to Y$ is a morphism in an ∞-category $\mathcal{C}$, then f is the *unique* composition $\mathrm{id}_Y \circ f$.

2.3.4 *n*-Categories

The theory of ∞-categories can be regarded as a generalization of classical category theory: if $\mathcal{C}$ is an ordinary category, then its nerve $\mathrm{N}(\mathcal{C})$ is an ∞-category which determines $\mathcal{C}$ up to canonical isomorphism. Moreover, Proposition 1.1.2.2 provides a precise characterization of those ∞-categories which can be obtained from ordinary categories. In this section, we will explain how to specialize the theory of ∞-categories to obtain a theory of n-categories for every nonnegative integer n. (However, the ideas described here are appropriate for describing only those n-categories which have only invertible k-morphisms for $k \geq 2$.)

Before we can give the appropriate definition, we need to introduce a bit of terminology. Let $f, f' : K \to \mathcal{C}$ be two diagrams in an ∞-category $\mathcal{C}$ and suppose that $K' \subseteq K$ is a simplicial subset such that $f|K' = f'|K' = f_0$. We will say that f and f' are *homotopic relative to* K' if they are equivalent when viewed as objects of the ∞-category $\mathrm{Fun}(K, \mathcal{C}) \times_{\mathrm{Fun}(K',\mathcal{C})} \{f_0\}$. Equivalently, f and f' are homotopic relative to K' if there exists a homotopy

$$h : K \times \Delta^1 \to \mathcal{C}$$

with the following properties:

(*i*) The restriction $h|K' \times \Delta^1$ coincides with the composition

$$K' \times \Delta^1 \to K' \xrightarrow{f_0} \mathcal{C}.$$

(*ii*) The restriction $h|K \times \{0\}$ coincides with f.

(*iii*) The restriction $h|K \times \{1\}$ coincides with f'.

(*iv*) For every vertex x of K, the restriction $h|\{x\} \times \Delta^1$ is an equivalence in $\mathcal{C}$.

We observe that if K' contains every vertex of K, then condition (*iv*) follows from condition (*i*).

Definition 2.3.4.1. Let $\mathcal{C}$ be a simplicial set and let $n \geq -1$ be an integer. We will say that $\mathcal{C}$ is an *n-category* if it is an ∞-category and the following additional conditions are satisfied:

(1) Given a pair of maps $f, f' : \Delta^n \to \mathcal{C}$, if f and f' are homotopic relative to $\partial \Delta^n$, then $f = f'$.

(2) Given $m > n$ and a pair of maps $f, f' : \Delta^m \to \mathcal{C}$, if $f|\partial \Delta^m = f'|\partial \Delta^m$, then $f = f'$.

It is sometimes convenient to extend Definition 2.3.4.1 to the case where $n = -2$: we will say that a simplicial set $\mathcal{C}$ is a (-2)-*category* if it is a final object of Set_Δ: in other words, if it is isomorphic to Δ^0.

Example 2.3.4.2. Let $\mathcal{C}$ be a (-1)-category. Using condition (2) of Definition 2.3.4.1, one shows by induction on m that $\mathcal{C}$ has at most one m-simplex. Consequently, we see that up to isomorphism there are precisely two (-1)-categories: $\Delta^{-1} \simeq \emptyset$ and Δ^0.

Example 2.3.4.3. Let $\mathcal{C}$ be a 0-category and let $X = \mathcal{C}_0$ denote the set of objects of $\mathcal{C}$. Let us write $x \leq y$ if there is a morphism ϕ from x to y in $\mathcal{C}$. Since $\mathcal{C}$ is an ∞-category, this relation is reflexive and transitive. Moreover, condition (2) of Definition 2.3.4.1 guarantees that the morphism ϕ is unique if it exists. If $x \leq y$ and $y \leq x$, it follows that the morphisms relating x and y are mutually inverse equivalences. Condition (1) then implies that $x = y$. We deduce that $(X, \leq)$ is a partially ordered set. It follows from Proposition 2.3.4.5 below that the map $\mathcal{C} \to \mathrm{N}(X)$ is an isomorphism.

Conversely, it is easy to see that the nerve of any partially ordered set $(X, \leq)$ is a 0-category in the sense of Definition 2.3.4.1. Consequently, the full subcategory of Set_Δ spanned by the 0-categories is equivalent to the category of partially ordered sets.

Remark 2.3.4.4. Let $\mathcal{C}$ be an n-category and let $m > n + 1$. Then the restriction map

$$\theta : \mathrm{Hom}_{\mathrm{Set}_\Delta}(\Delta^m, \mathcal{C}) \to \mathrm{Hom}_{\mathrm{Set}_\Delta}(\partial \Delta^m, \mathcal{C})$$

is bijective. If $n = -1$, this is clear from Example 2.3.4.2; let us therefore suppose that $n \geq 0$, so that $m \geq 2$. The injectivity of θ follows immediately from part (2) of Definition 2.3.4.1. To prove the surjectivity, we consider an arbitrary map $f_0 : \partial \Delta^m \to \mathcal{C}$. Let $f : \Delta^m \to \mathcal{C}$ be an extension of $f_0 | \Lambda_1^m$ (which exists since $\mathcal{C}$ is an ∞-category and $0 < 1 < m$). Using condition (2) again, we deduce that $\theta(f) = f_0$.

The following result shows that, in the case where $n = 1$, Definition 2.3.4.1 recovers the usual definition of a category:

Proposition 2.3.4.5. *Let S be a simplicial set. The following conditions are equivalent:*

(1) *The unit map $u : S \to \mathrm{N}(\mathrm{h}S)$ is an isomorphism of simplicial sets.*

(2) *There exists a small category $\mathcal{C}$ and an isomorphism $S \simeq \mathrm{N}(\mathcal{C})$ of simplicial sets.*

(3) *The simplicial set S is a 1-category.*

Proof. The implications (1) $\Rightarrow$ (2) $\Rightarrow$ (3) are clear. Let us therefore assume that (3) holds and show that $f : S \to \mathrm{N}(\mathrm{h}S)$ is an isomorphism. We will prove, by induction on n, that the map u is bijective on n-simplices.

For $n = 0$, this is clear. If $n = 1$, the surjectivity of u obvious. To prove the injectivity, we note that if $f(\phi) = f(\psi)$, then the edges ϕ and ψ are homotopic in S. A simple application of condition (2) of Definition 2.3.4.1 then shows that $\phi = \psi$.

Now suppose $n > 1$. The injectivity of u on n-simplices follows from condition (3) of Definition 2.3.4.1 and the injectivity of u on $(n-1)$-simplices. To prove the surjectivity, let us suppose we are given a map $s : \Delta^n \to \mathrm{N}(\mathrm{h}S)$. Choose $0 < i < n$. Since u is bijective on lower-dimensional simplices, the map $s | \Lambda_i^n$ factors uniquely through S. Since S is an ∞-category, this factorization extends to a map $\widetilde{s} : \Delta^n \to S$. Since $\mathrm{N}(\mathrm{h}S)$ is the nerve of a category, a pair of maps from Δ^n into $\mathrm{N}(\mathrm{h}S)$ which agree on Λ_i^n must be the same. We deduce that $u \circ \widetilde{s} = s$, and the proof is complete. $\square$

Remark 2.3.4.6. The condition that an ∞-category $\mathcal{C}$ be an n-category is not invariant under categorical equivalence. For example, if $\mathcal{D}$ is a category with several objects, all of which are uniquely isomorphic to one another, then $\mathrm{N}(\mathcal{D})$ is categorically equivalent to Δ^0 but is not a (-1)-category. Consequently, there can be no intrinsic characterization of the class of n-categories itself. Nevertheless, there does exist a convenient description for the class of ∞-categories which are *equivalent* to n-categories; see Proposition 2.3.4.18.

Our next goal is to establish that the class of n-categories is stable under the formation of functor categories. In order to do so, we need to reformulate Definition 2.3.4.1 in a more invariant manner. Recall that for any simplicial set X, the *n-skeleton* $\mathrm{sk}^n X$ is defined to be the simplicial subset of X generated by all the simplices of X having dimension $\leq n$.

Proposition 2.3.4.7. *Let $\mathcal{C}$ be an ∞-category and let $n \geq -1$. The following are equivalent:*

(1) *The ∞-category $\mathcal{C}$ is an n-category.*

(2) *For every simplicial set K and every pair of maps $f, f' : K \to \mathcal{C}$ such that $f | \operatorname{sk}^n K$ and $f' | \operatorname{sk}^n K$ are homotopic relative to $\operatorname{sk}^{n-1} K$, we have $f = f'$.*

Proof. The implication (2) $\Rightarrow$ (1) is obvious. Suppose that (1) is satisfied and let $f, f' : K \to \mathcal{C}$ be as in the statement of (2). To prove that $f = f'$, it suffices to show that f and f' agree on every nondegenerate simplex of K. We may therefore reduce to the case where $K = \Delta^m$. We now work by induction on m. If $m < n$, there is nothing to prove. In the case where $m = n$, the assumption that $\mathcal{C}$ is an n-category immediately implies that $f = f'$. If $m > n$, the inductive hypothesis implies that $f | \partial \Delta^m = f' | \partial \Delta^m$, so that (1) implies that $f = f'$. $\qquad\square$

Corollary 2.3.4.8. *Let $\mathcal{C}$ be an n-category and X a simplicial set. Then $\operatorname{Fun}(X, \mathcal{C})$ is an n-category.*

Proof. Proposition 1.2.7.3 asserts that $\operatorname{Fun}(X, \mathcal{C})$ is an ∞-category. We will show that $\operatorname{Fun}(X, \mathcal{C})$ satisfies condition (2) of Proposition 2.3.4.7. Suppose we are given a pair of maps $f, f' : K \to \operatorname{Fun}(X, \mathcal{C})$ such that $f | \operatorname{sk}^n K$ and $f' | \operatorname{sk}^n K$ are homotopic relative to $f | \operatorname{sk}^{n-1} K$. We wish to show that $f = f'$. We may identify f and f' with maps $F, F' : K \times X \to \mathcal{C}$. Since $\mathcal{C}$ is an n-category, to prove that $F = F'$ it suffices to show that $F | \operatorname{sk}^n(K \times X)$ and $F' | \operatorname{sk}^n(K \times X)$ are homotopic relative to $\operatorname{sk}^{n-1}(K \times X)$. This follows at once because $\operatorname{sk}^p(K \times X) \subseteq (\operatorname{sk}^p K) \times X$ for every integer p. $\qquad\square$

When $n = 1$, Proposition 1.1.2.2 asserts that the class of n-categories can be characterized by the uniqueness of certain horn fillers. We now prove a generalization of this result.

Proposition 2.3.4.9. *Let $n \geq 1$ and let $\mathcal{C}$ be an ∞-category. Then $\mathcal{C}$ is an n-category if and only if it satisfies the following condition:*

* *For every $m > n$ and every diagram*

  where $0 < i < m$, there exists a unique *dotted arrow f as indicated which renders the diagram commutative.*

Proof. Suppose first that $\mathcal{C}$ is an n-category. Let $f, f' : \Delta^m \to \mathcal{C}$ be two maps with $f | \Lambda_i^m = f' | \Lambda_i^m$, where $0 < i < m$ and $m > n$. We wish to prove that $f = f'$. Since Λ_i^m contains the $(n-1)$-skeleton of Δ^m, it will suffice (by Proposition 2.3.4.7) to show that f and f' are homotopic relative to Λ_i^m. This follows immediately from the fact that the inclusion $\Lambda_i^m \subseteq \Delta^m$ is a categorical equivalence.

Now suppose that every map $f_0 : \Lambda_i^m \to \mathcal{C}$, where $0 < i < m$ and $n < m$, extends uniquely to an m-simplex of $\mathcal{C}$. We will show that $\mathcal{C}$ satisfies conditions (1) and (2) of Definition 2.3.4.1. Condition (2) is obvious: if $f, f' : \Delta^m \to \mathcal{C}$ are two maps which coincide on $\partial \Delta^m$, then they coincide on Λ_1^m and are therefore equal to one another (here we use the fact that $m > 1$ because of our assumption that $n \geq 1$). Condition (1) is a bit more subtle. Suppose that $f, f' : \Delta^n \to \mathcal{C}$ are homotopic via a homotopy $h : \Delta^n \times \Delta^1 \to \mathcal{C}$ which is constant on $\partial \Delta^n \times \Delta^1$. For $0 \leq i \leq n$, let σ_i denote the $(n+1)$-simplex of $\mathcal{C}$ obtained by composing h with the map

$$[n+1] \to [n] \times [1]$$

$$j \mapsto \begin{cases} (j, 0) & \text{if } j \leq i \\ (j-1, 1) & \text{if } j > i. \end{cases}$$

If $i < n$, then we observe that $\sigma_i | \Lambda_{i+1}^{n+1}$ is equivalent to the restriction $(s_i d_i \sigma_i) | \Lambda_{i+1}^{n+1}$. Applying our hypothesis, we conclude that $\sigma_i = s_i d_i \sigma_i$, so that $d_i \sigma_i = d_{i+1} \sigma_i$. A dual argument establishes the same equality for $0 < i$. Since $n > 0$, we conclude that $d_i \sigma_i = d_{i+1} \sigma_i$ for all i. Consequently, we have a chain of equalities

$$f' = d_0 \sigma_0 = d_1 \sigma_0 = d_1 \sigma_1 = d_2 \sigma_1 = \cdots = d_n \sigma_n = d_{n+1} \sigma_n = f,$$

so that $f' = f$, as desired. $\square$

Corollary 2.3.4.10. *Let $\mathcal{C}$ be an n-category and let $p : K \to \mathcal{C}$ be a diagram. Then $\mathcal{C}_{/p}$ is an n-category.*

Proof. If $n \leq 0$, this follows easily from Examples 2.3.4.2 and 2.3.4.3. We may therefore suppose that $n \geq 1$. Proposition 1.2.9.3 implies that $\mathcal{C}_{/p}$ is an ∞-category. According to Proposition 2.3.4.9, it suffices to show that for every $m > n$, $0 < i < m$, and every map $f_0 : \Lambda_i^m \to \mathcal{C}_{/p}$, there exists a *unique* map $f : \Delta^m \to \mathcal{C}_{/p}$ extending f. Equivalently, we must show that there is a unique map g rendering the diagram

$$\begin{array}{ccc} \Lambda_i^m \star K & \xrightarrow{g_0} & \mathcal{C} \\ \big\downarrow & \nearrow{\scriptstyle g} & \\ \Delta^m \star K & & \end{array}$$

commutative. The existence of g follows from the fact that $\mathcal{C}_{/p}$ is an ∞-category. Suppose that $g' : \Delta^m \star K \to \mathcal{C}$ is another map which extends g_0. Proposition 1.1.2.2 implies that $g' | \Delta^m = g | \Delta^m$. We conclude that g and g' coincide on the n-skeleton of $\Delta^m \star K$. Since $\mathcal{C}$ is an n-category, we deduce that $g = g'$, as desired. $\square$

We conclude this section by introducing a construction which allows us to pass from an arbitrary ∞-category $\mathcal{C}$ to its "underlying" n-category by discarding information about k-morphisms for $k > n$. In the case where $n = 1$, we have already introduced the relevant construction: we simply replace $\mathcal{C}$ by (the nerve of) its homotopy category.

Notation 2.3.4.11. Let $\mathcal{C}$ be an ∞-category and let $n \geq 1$. For every simplicial set K, let $[K, \mathcal{C}]_n \subseteq \mathrm{Fun}(\mathrm{sk}^n K, \mathcal{C})$ be the subset consisting of those diagrams $\mathrm{sk}^n K \to \mathcal{C}$ which extend to the $(n + 1)$-skeleton of K (in other words, the image of the restriction map $\mathrm{Fun}(\mathrm{sk}^{n+1} K, \mathcal{C}) \to \mathrm{Fun}(\mathrm{sk}^n K, \mathcal{C})$). We define an equivalence relation $\sim$ on $[K, \mathcal{C}]_n$ as follows: given two maps $f, g : \mathrm{sk}^n K \to \mathcal{C}$, we write $f \sim g$ if f and g are homotopic relative to $\mathrm{sk}^{n-1} K$.

Proposition 2.3.4.12. *Let $\mathcal{C}$ be an ∞-category and $n \geq 1$.*

(1) *There exists a simplicial set $\mathrm{h}_n\mathcal{C}$ with the following universal mapping property:* $\mathrm{Fun}(K, \mathrm{h}_n\mathcal{C}) = [\mathrm{K}, \mathcal{C}]_\mathrm{n}/ \sim.$

(2) *The simplicial set $\mathrm{h}_n\mathcal{C}$ is an n-category.*

(3) *If $\mathcal{C}$ is an n-category, then the natural map $\theta : \mathcal{C} \to \mathrm{h}_n\mathcal{C}$ is an isomorphism.*

(4) *For every n-category $\mathcal{D}$, composition with θ induces an isomorphism of simplicial sets*

$$\psi : \mathrm{Fun}(\mathrm{h}_n\mathcal{C}, \mathcal{D}) \to \mathrm{Fun}(\mathcal{C}, \mathcal{D}).$$

Proof. To prove (1), we begin by defining $\mathrm{h}_n\mathcal{C}([m]) = [\Delta^m, \mathcal{C}]_n/ \sim$, so that the desired universal property holds by definition whenever K is a simplex. Unwinding the definitions, to check the universal property for a general simplicial set K we must verify the following fact:

(∗) Given two maps $f, g : \partial \Delta^{n+1} \to \mathcal{C}$ which are homotopic relative to $\mathrm{sk}^{n-1} \Delta^{n+1}$, if f extends to an $(n + 1)$-simplex of $\mathcal{C}$, then g extends to an $(n + 1)$-simplex of $\mathcal{C}$.

This follows easily from Proposition A.2.3.1.

We next show that $\mathrm{h}_n\mathcal{C}$ is an ∞-category. Let $\eta_0 : \Lambda_i^m \to \mathrm{h}_n\mathcal{C}$ be a morphism, where $0 < i < m$. We wish to show that η_0 extends to an m-simplex $\eta : \Delta^m \to \mathcal{C}$. If $m \leq n + 2$, then $\Lambda_i^m = \mathrm{sk}^{n+1} \Lambda_i^m$, so that η_0 can be written as a composition

$$\Lambda_i^m \to \mathcal{C} \xrightarrow{\theta} \mathrm{h}_n\mathcal{C} .$$

The existence of η now follows from our assumption that $\mathcal{C}$ is an ∞-category. If $m > n+2$, then $\mathrm{Hom}_{\mathrm{Set}_\Delta}(\Lambda_i^m, \mathrm{h}_n\mathcal{C}) \simeq \mathrm{Hom}_{\mathrm{Set}_\Delta}(\Delta^m, \mathrm{h}_n\mathcal{C})$ by construction, so there is nothing to prove.

We next prove that $\mathrm{h}_n\mathcal{C}$ is an n-category. It is clear from the construction that for $m > n$, any two m-simplices of $\mathrm{h}_n\mathcal{C}$ with the same boundary must coincide. Suppose next that we are given two maps $f, f' : \Delta^n \to h_n \mathcal{C}$ which are homotopic relative to $\partial \Delta^n$. Let $F : \Delta^n \times \Delta^1 \to h_n \mathcal{C}$ be a homotopy from f to f'. Using $(*)$, we deduce that F is the image under θ of a map $\widetilde{F} : \Delta^n \times \Delta^1 \to h_n \mathcal{C}$, where $\widetilde{F}| \partial \Delta^n \times \Delta^1$ factors through the projection $\partial \Delta^n \times \Delta^1 \to \partial \Delta^n$. Since $n > 0$, we conclude that $\widetilde{F}$ is a homotopy from $\widetilde{F}|\Delta^n \times \{0\}$ to $\widetilde{F}|\Delta^n \times \{1\}$, so that $f = f'$. This completes the proof of (2).

To prove (3), let us suppose that $\mathcal{C}$ is an n-category; we prove by induction on m that the map $\mathcal{C} \to h_n \mathcal{C}$ is bijective on m-simplices. For $m < n$, this is clear. When $m = n$, it follows from part (1) of Definition 2.3.4.1. When $m = n + 1$, surjectivity follows from the construction of $h_n \mathcal{C}$ and injectivity from part (2) of Definition 2.3.4.1. For $m > n + 1$, we have a commutative diagram

$$
\begin{array}{ccc}
\operatorname{Hom}_{\mathrm{Set}_\Delta}(\Delta^m, \mathcal{C}) & \longrightarrow & \operatorname{Hom}_{\mathrm{Set}_\Delta}(\Delta^m, h_n \mathcal{C}) \\
\downarrow & & \downarrow \\
\operatorname{Hom}_{\mathrm{Set}_\Delta}(\partial \Delta^m, \mathcal{C}) & \longrightarrow & \operatorname{Hom}_{\mathrm{Set}_\Delta}(\partial \Delta^m, h_n \mathcal{C}),
\end{array}
$$

where the bottom horizontal map is an isomorphism by the inductive hypothesis, the left vertical map is an isomorphism by construction, and the right vertical map is an isomorphism by Remark 2.3.4.4; it follows that the upper horizontal map is an isomorphism as well.

To prove (4), we observe that if $\mathcal{D}$ is an n-category, then the composition

$$\operatorname{Fun}(\mathcal{C}, \mathcal{D}) \to \operatorname{Fun}(h_n \mathcal{C}, h_n \mathcal{D}) \simeq \operatorname{Fun}(h_n \mathcal{C}, \mathcal{D})$$

is an inverse to ϕ, where the second isomorphism is given by (3). $\square$

Remark 2.3.4.13. The construction of Proposition 2.3.4.12 does not quite work if $n \leq 0$ because there may exist equivalences in $h_n \mathcal{C}$ which do not arise from equivalences in $\mathcal{C}$. However, it is a simple matter to give an alternative construction in these cases which satisfies conditions (2), (3), and (4); we leave the details to the reader.

Remark 2.3.4.14. In the case $n = 1$, the ∞-category $h_1 \mathcal{C}$ constructed in Proposition 2.3.4.12 is isomorphic to the nerve of the homotopy category $h\mathcal{C}$.

We now apply the theory of minimal ∞-categories (§2.3.3) to obtain a characterization of the class of ∞-categories which are *equivalent* to n-categories. First, we need a definition from classical homotopy theory.

Definition 2.3.4.15. Let $k \geq -1$ be an integer. A Kan complex X is k-*truncated* if, for every $i > k$ and every point $x \in X$, we have

$$\pi_i(X, x) \simeq *.$$

By convention, we will also say that X is (-2)-*truncated* if X is contractible.

Remark 2.3.4.16. If X and Y are homotopy equivalent Kan complexes, then X is k-truncated if and only if Y is k-truncated. In other words, we may view k-truncatedness as a condition on objects in the homotopy category $\mathcal{H}$ of spaces.

Example 2.3.4.17. A Kan complex X is (-1)-truncated if it is either empty or contractible. It is 0-truncated if the natural map $X \to \pi_0 X$ is a homotopy equivalence (equivalently, X is 0-truncated if it is homotopy equivalent to a discrete space).

Proposition 2.3.4.18. *Let $\mathcal{C}$ be an ∞-category and let $n \geq -1$. The following conditions are equivalent:*

(1) *There exists a minimal model $\mathcal{C}' \subseteq \mathcal{C}$ such that $\mathcal{C}'$ is an n-category.*

(2) *There exists a categorical equivalence $\mathcal{D} \simeq \mathcal{C}$, where $\mathcal{D}$ is an n-category.*

(3) *For every pair of objects $X, Y \in \mathcal{C}$, the mapping space $\mathrm{Map}_{\mathcal{C}}(X, Y) \in \mathcal{H}$ is $(n-1)$-truncated.*

Proof. It is clear that (1) implies (2). Suppose next that (2) is satisfied; we will prove (3). Without loss of generality, we may replace $\mathcal{C}$ by $\mathcal{D}$ and thereby assume that $\mathcal{C}$ is an n-category. If $n = -1$, the desired result follows immediately from Example 2.3.4.2. Choose $m \geq n$ and an element $\eta \in \pi_m(\mathrm{Map}_{\mathcal{C}}(X, Y), f)$. We can represent η by a commutative diagram of simplicial sets

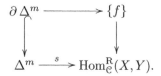

We can identify s with a map $\Delta^{m+1} \to \mathcal{C}$ whose restriction to $\partial \Delta^{m+1}$ is specified. Since $\mathcal{C}$ is an n-category, the inequality $m + 1 > n$ shows that s is uniquely determined. This proves that $\pi_m(\mathrm{Map}_{\mathcal{C}}(X, Y), f) \simeq *$, so that (3) is satisfied.

To prove that (3) implies (1), it suffices to show that if $\mathcal{C}$ is a *minimal* ∞-category which satisfies (3), then $\mathcal{C}$ is an n-category. We must show that the conditions of Definition 2.3.4.1 are satisfied. The first of these conditions follows immediately from the assumption that $\mathcal{C}$ is minimal. For the second, we must show that if $m > n$ and $f, f' : \partial \Delta^m \to \mathcal{C}$ are such that $f | \partial \Delta^m = f' | \partial \Delta^m$, then $f = f'$. Since $\mathcal{C}$ is minimal, it suffices to show that f and f' are homotopic relative to $\partial \Delta^m$. We will prove that there is a map $g : \Delta^{m+1} \to \mathcal{C}$ such that $d_{m+1} g = f$, $d_m g = f'$, and $d_i g = d_i s_m f = d_i s_m f'$ for $0 \leq i < m$. Then the sequence $(s_0 f, s_1 f, \ldots, s_{m-1} f, g)$ determines a map $\Delta^m \times \Delta^1 \to \mathcal{C}$ which gives the desired homotopy between f and f' (relative to $\partial \Delta^m$).

To produce the map g, it suffices to solve the lifting problem depicted in the diagram

Choose a fibrant simplicial category $\mathcal{D}$ and an equivalence of ∞-categories $\mathcal{C} \to N(\mathcal{D})$. According to Proposition A.2.3.1, it will suffice to prove that we can solve the associated lifting problem

$$
\begin{array}{ccc}
\mathfrak{C}[\partial \Delta^{m+1}] & \xrightarrow{G_0} & \mathcal{D} \\
\big\downarrow & \nearrow{\scriptstyle G} & \\
\mathfrak{C}[\Delta^{m+1}]. &
\end{array}
$$

Let X and Y denote the initial and final vertices of Δ^{m+1}, regarded as objects of $\mathfrak{C}[\partial \Delta^{m+1}]$. Note that G_0 determines a map

$$
e_0 : \partial(\Delta^1)^m \simeq \operatorname{Map}_{\mathfrak{C}[\partial \Delta^{m+1}]}(X, Y) \to \operatorname{Map}_{\mathcal{D}}(G_0(X), G_0(Y))
$$

and that giving the desired extension G is equivalent to extending e_0 to a map

$$
e : (\Delta^1)^m \simeq \operatorname{Map}_{\mathfrak{C}[\Delta^{m+1}]}(X, Y) \to \operatorname{Map}_{\mathcal{D}}(G_0(X), G_0(Y)).
$$

The obstruction to constructing e lies in $\pi_{m-1}(\operatorname{Map}_{\mathcal{D}}(G_0(X), G_0(Y)), p)$ for an appropriately chosen base point p. Since $(m-1) > (n-1)$, condition (3) implies that this homotopy set is trivial, so that the desired extension can be found. $\qquad\square$

Corollary 2.3.4.19. *Let X be a Kan complex. Then X is (categorically) equivalent to an n-category if and only if it is n-truncated.*

Proof. For $n = -2$ this is obvious. If $n \geq -1$, this follows from characterization (3) of Proposition 2.3.4.18 and the following observation: a Kan complex X is n-truncated if and only if, for every pair of vertices $x, y \in X_0$, the Kan complex .

$$
\{x\} \times_X X^{\Delta^1} \times_X \{y\}
$$

of paths from x to y is $(n-1)$-truncated. $\qquad\square$

Corollary 2.3.4.20. *Let $\mathcal{C}$ be an ∞-category and K a simplicial set. Suppose that, for every pair of objects $C, D \in \mathcal{C}$, the space $\operatorname{Map}_{\mathcal{C}}(C, D)$ is n-truncated. Then the ∞-category $\operatorname{Fun}(K, \mathcal{C})$ has the same property.*

Proof. This follows immediately from Proposition 2.3.4.18 and Corollary 2.3.4.8 because the functor

$$
\mathcal{C} \mapsto \operatorname{Fun}(K, \mathcal{C})
$$

preserves categorical equivalences between ∞-categories. $\qquad\square$

2.4 CARTESIAN FIBRATIONS

Let $p : X \to S$ be an inner fibration of simplicial sets. Each fiber of p is an ∞-category, and each edge $f : s \to s'$ of S determines a correspondence between the fibers X_s and $X_{s'}$. In this section, we would like to study the case in which each of these correspondences is associated to a functor $f^* : X_{s'} \to X_s$. Roughly speaking, we can attempt to construct f^* as follows: for each vertex $y \in X_{s'}$, we choose an edge $\widetilde{f} : x \to y$ lifting f, and set $f^*y = x$. However, this recipe does not uniquely determine x, even up to equivalence, since there might be many different choices for $\widetilde{f}$. To get a good theory, we need to make a good choice of $\widetilde{f}$. More precisely, we should require that $\widetilde{f}$ be a p-*Cartesian* edge of X. In §2.4.1, we will introduce the definition of p-Cartesian edges and study their basic properties. In particular, we will see that a p-Cartesian edge $\widetilde{f}$ is determined up to equivalence by its target y and its image in S. Consequently, if there is a sufficient supply of p-Cartesian edges of X, then we can use the above prescription to define the functor $f^* : X_{s'} \to X_s$. This leads us to the notion of a *Cartesian fibration*, which we will study in §2.4.2.

In §2.4.3, we will establish a few basic stability properties of the class of Cartesian fibrations (we will discuss other results of this type in Chapter 3 after we have developed the language of marked simplicial sets). In §2.4.4, we will show that if $p : \mathcal{C} \to \mathcal{D}$ is a Cartesian fibration of ∞-categories, then we can reduce many questions about $\mathcal{C}$ to similar questions about the base $\mathcal{D}$ and about the fibers of p. This technique has many applications, which we will discuss in §2.4.5 and §2.4.6. Finally, in §2.4.7, we will study the theory of *bifibrations*, which is useful for constructing examples of Cartesian fibrations.

2.4.1 Cartesian Morphisms

Let $\mathcal{C}$ and $\mathcal{C}'$ be ordinary categories and let $M : \mathcal{C}^{op} \times \mathcal{C}' \to \mathrm{Set}$ be a correspondence between them. Suppose that we wish to know whether or not M arises as the correspondence associated to some functor $g : \mathcal{C}' \to \mathcal{C}$. This is the case if and only if, for each object $C' \in \mathcal{C}'$, we can find an object $C \in \mathcal{C}$ and a point $\eta \in M(C, C')$ having the property that the "composition with η" map

$$\psi : \mathrm{Hom}_{\mathcal{C}}(D, C) \to M(D, C')$$

is bijective for all $D \in \mathcal{C}$. Note that η may be regarded as a morphism in the category $\mathcal{C} \star^M \mathcal{C}'$. We will say that η is a *Cartesian* morphism in $\mathcal{C} \star^M \mathcal{C}'$ if ψ is bijective for each $D \in \mathcal{C}$. The purpose of this section is to generalize this notion to the ∞-categorical setting and to establish its basic properties.

Definition 2.4.1.1. Let $p : X \to S$ be an inner fibration of simplicial sets. Let $f : x \to y$ be an edge in X. We shall say that f is p-*Cartesian* if the induced map

$$X_{/f} \to X_{/y} \times_{S_{/p(y)}} S_{/p(f)}$$

is a trivial Kan fibration.

Remark 2.4.1.2. Let $\mathcal{M}$ be an ordinary category, let $p : N(\mathcal{M}) \to \Delta^1$ be a map (automatically an inner fibration), and let $f : x \to y$ be a morphism in $\mathcal{M}$ which projects isomorphically onto Δ^1. Then f is p-Cartesian in the sense of Definition 2.4.1.1 if and only if it is Cartesian in the classical sense.

We now summarize a few of the formal properties of Definition 2.4.1.1:

Proposition 2.4.1.3. (1) *Let $p : X \to S$ be an isomorphism of simplicial sets. Then every edge of X is p-Cartesian.*

(2) *Suppose we are given a pullback diagram*

$$
\begin{array}{ccc}
X' & \xrightarrow{\;q\;} & X \\
\downarrow{\scriptstyle p'} & & \downarrow{\scriptstyle p} \\
S' & \longrightarrow & S
\end{array}
$$

of simplicial sets, where p (and therefore also p') is an inner fibration. Let f be an edge of X'. If $q(f)$ is p-Cartesian, then f is p'-Cartesian.

(3) *Let $p : X \to Y$ and $q : Y \to Z$ be inner fibrations and let $f : x' \to x$ be an edge of X such that $p(f)$ is q-Cartesian. Then f is p-Cartesian if and only if f is $(q \circ p)$-Cartesian.*

Proof. Assertions (1) and (2) follow immediately from the definition. To prove (3), we consider the commutative diagram

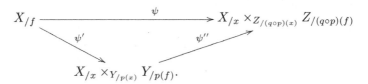

The map ψ'' is a pullback of

$$
Y_{/p(f)} \to Y_{/p(x)} \times_{Z_{/(q \circ p)(x)}} Z_{/(q \circ p)(f)}
$$

and therefore a trivial fibration in view of our assumption that $p(f)$ is q-Cartesian. If ψ' is a trivial fibration, it follows that ψ is a trivial fibration as well, which proves the "only if" direction of (3).

For the converse, suppose that ψ is a trivial fibration. Proposition 2.1.2.1 implies that ψ' is a right fibration. According to Lemma 2.1.3.4, it will suffice to prove that the fibers of ψ' are contractible. Let t be a vertex of $X_{/x} \times_{Y_{/p(x)}} Y_{/p(f)}$ and let $K = (\psi'')^{-1}\{\psi''(t)\}$. Since ψ'' is a trivial fibration, K is a contractible Kan complex. Since ψ is a trivial fibration, the simplicial set $(\psi')^{-1}K = \psi^{-1}\{\psi''(t)\}$ is also a contractible Kan complex. It follows that the fiber of ψ' over the point t is weakly contractible, as desired. $\square$

Remark 2.4.1.4. Let $p : X \to S$ be an inner fibration of simplicial sets. Unwinding the definition, we see that an edge $f : \Delta^1 \to X$ is p-Cartesian if and only if for every $n \geq 2$ and every commutative diagram

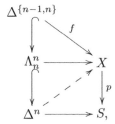

there exists a dotted arrow as indicated, rendering the diagram commutative.

In particular, we note that Proposition 1.2.4.3 may be restated as follows:

($*$) Let $\mathcal{C}$ be a ∞-category and let $p : \mathcal{C} \to \Delta^0$ denote the projection from $\mathcal{C}$ to a point. A morphism ϕ of $\mathcal{C}$ is p-Cartesian if and only if ϕ is an equivalence.

In fact, it is possible to strengthen assertion ($*$) as follows:

Proposition 2.4.1.5. *Let $p : \mathcal{C} \to \mathcal{D}$ be an inner fibration between ∞-categories and let $f : C \to C'$ be a morphism in $\mathcal{C}$. The following conditions are equivalent:*

(1) *The morphism f is an equivalence in $\mathcal{C}$.*

(2) *The morphism f is p-Cartesian, and $p(f)$ is an equivalence in $\mathcal{D}$.*

Proof. Let q denote the projection from $\mathcal{D}$ to a point. We note that both (1) and (2) imply that $p(f)$ is an equivalence in $\mathcal{D}$ and therefore q-Cartesian by ($*$). The equivalence of (1) and (2) now follows from ($*$) and the third part of Proposition 2.4.1.3. □

Corollary 2.4.1.6. *Let $p : \mathcal{C} \to \mathcal{D}$ be an inner fibration between ∞-categories. Every identity morphism of $\mathcal{C}$ (in other words, every degenerate edge of $\mathcal{C}$) is p-Cartesian.*

We now study the behavior of Cartesian edges under composition.

Proposition 2.4.1.7. *Let $p : \mathcal{C} \to \mathcal{D}$ be an inner fibration between simplicial sets and let $\sigma : \Delta^2 \to \mathcal{C}$ be a 2-simplex of $\mathcal{C}$, which we will depict as a diagram*

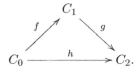

Suppose that g is p-Cartesian. Then f is p-Cartesian if and only if h is p-Cartesian.

Proof. We wish to show that the map

$$i_0 : \mathcal{C}_{/h} \to \mathcal{C}_{/C_2} \times_{\mathcal{D}_{/p(C_2)}} \mathcal{D}_{/p(h)}$$

is a trivial fibration if and only if

$$i_1 : \mathcal{C}_{/f} \to \mathcal{C}_{/C_1} \times_{\mathcal{D}_{/p(C_1)}} \mathcal{D}_{/p(f)}$$

is a trivial fibration. The dual of Proposition 2.1.2.1 implies that both maps are right fibrations. Consequently, by (the dual of) Lemma 2.1.3.4, it suffices to show that the fibers of i_0 are contractible if and only if the fibers of i_1 are contractible.

For any simplicial subset $B \subseteq \Delta^2$, let $X_B = \mathcal{C}_{/\sigma|B} \times_{\mathcal{D}_{\sigma|B}} \mathcal{D}_{/\sigma}$. We note that X_B is functorial in B in the sense that an inclusion $A \subseteq B$ induces a map $j_{A,B} : X_B \to X_A$ (which is a right fibration, again by Proposition 2.1.2.1). Observe that $j_{\Delta^{\{2\}}, \Delta^{\{0,2\}}}$ is the base change of i_0 by the map $\mathcal{D}_{/p(\sigma)} \to \mathcal{D}_{/p(h)}$ and that $j_{\Delta^{\{1\}}, \Delta^{\{0,1\}}}$ is the base change of i_1 by the map $\mathcal{D}_{/\sigma} \to \mathcal{D}_{/p(f)}$. The maps

$$\mathcal{D}_{/p(f)} \leftarrow \mathcal{D}_{/p(\sigma)} \to \mathcal{D}_{/p(h)}$$

are both surjective on objects (in fact, both maps have sections). Consequently, it suffices to prove that $j_{\Delta^{\{1\}}, \Delta^{\{0,1\}}}$ has contractible fibers if and only if $j_{\Delta^{\{2\}}, \Delta^{\{0,2\}}}$ has contractible fibers. Now we observe that the compositions

$$X_{\Delta^2} \to X_{\Delta^{\{0,2\}}} \to X_{\Delta^{\{2\}}}$$

$$X_{\Delta^2} \to X_{\Lambda_1^2} \to X_{\Delta^{\{1,2\}}} \to X_{\Delta^{\{2\}}}$$

coincide. By Proposition 2.1.2.5, $j_{A,B}$ is a trivial fibration whenever the inclusion $A \subseteq B$ is left anodyne. We deduce that $j_{\Delta^{\{2\}}, \Delta^{\{0,2\}}}$ is a trivial fibration if and only if $j_{\Delta^{\{1,2\}}, \Lambda_1^2}$ is a trivial fibration. Consequently, it suffices to show that $j_{\Delta^{\{1,2\}}, \Lambda_1^2}$ is a trivial fibration if and only if $j_{\Delta^{\{1\}}, \Delta^{\{0,1\}}}$ is a trivial fibration.

Since $j_{\Delta^{\{1,2\}}, \Lambda_1^2}$ is a pullback of $j_{\Delta^{\{1\}}, \Delta^{\{0,1\}}}$, the "if" direction is obvious. For the converse, it suffices to show that the natural map

$$\mathcal{C}_{/g} \times_{\mathcal{D}_{/p(g)}} \mathcal{D}_{/p(\sigma)} \to \mathcal{C}_{/C_1} \times_{\mathcal{D}_{/p(C_1)}} \mathcal{D}_{/p(\sigma)}$$

is surjective on vertices. But this map is a trivial fibration because the inclusion $\{1\} \subseteq \Delta^{\{1,2\}}$ is left anodyne.   □

Our next goal is to reformulate the notion of a Cartesian morphism in a form which will be useful later. For convenience of notation, we will prove this result in a dual form. If $p : X \to S$ is an inner fibration and f is an edge of X, we will say that f is *p-coCartesian* if it is Cartesian with respect to the morphism $p^{op} : X^{op} \to S^{op}$.

Proposition 2.4.1.8. *Let $p : Y \to S$ be an inner fibration of simplicial sets and let $e : \Delta^1 \to Y$ be an edge. Then e is p-coCartesian if and only if for each $n \geq 1$ and each diagram*

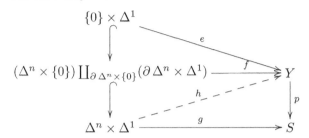

there exists a map h as indicated, rendering the diagram commutative.

Proof. Let us first prove the "only if" direction. We recall a bit of the notation used in the proof of Proposition 2.1.2.6; in particular, the filtration

$$X(n+1) \subseteq \cdots \subseteq X(0) = \Delta^n \times \Delta^1$$

of $\Delta^n \times \Delta^1$. We construct $h|X(m)$ by descending induction on m. To begin, we set $h|X(n+1) = f$. Now, for each m the space $X(m)$ is obtained from $X(m+1)$ by pushout along a horn inclusion $\Lambda_m^{n+1} \subseteq \Delta^{m+1}$. If $m > 0$, the desired extension exists because p is an inner fibration. If $m = 0$, the desired extension exists because of the hypothesis that e is a p-coCartesian edge.

We now prove the "if" direction. Suppose that e satisfies the condition in the statement of Proposition 2.4.1.8. We wish to show that e is p-coCartesian. In other words, we must show that for every $n \geq 2$ and every diagram

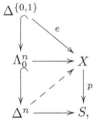

there exists a dotted arrow as indicated, rendering the diagram commutative. Replacing S by Δ^n and Y by $Y \times_S \Delta^n$, we may reduce to the case where S is an ∞-category. We again make use of the notation (and argument) employed in the proof of Proposition 2.1.2.6. Namely, the inclusion $\Lambda_0^n \subseteq \Delta^n$ is a retract of the inclusion

$$(\Lambda_0^n \times \Delta^1) \coprod_{\Lambda_0^n \times \{0\}} (\Delta^n \times \{0\}) \subseteq \Delta^n \times \Delta^1.$$

The retraction is implemented by maps

$$\Delta^n \xrightarrow{j} \Delta^n \times \Delta^1 \xrightarrow{r} \Delta^n,$$

which were defined in the proof of Proposition 2.1.2.6. We now set $F = f \circ r$, $G = g \circ r$.

Let $K = \Delta^{\{1,2,\ldots,n\}} \subseteq \Delta^n$. Then

$$F|(\partial K \times \Delta^1) \coprod_{\partial K \times \{0\}} (K \times \Delta^1)$$

carries $\{1\} \times \Delta^1$ into e. By assumption, there exists an extension of F to $K \times \Delta^1$ which is compatible with G. In other words, there exists a compatible extension F' of F to

$$\partial \Delta^n \times \Delta^1 \coprod_{\partial \Delta^n \times \{0\}} \Delta^n \times \{0\}.$$

Moreover, F' carries $\{0\} \times \Delta^1$ to a degenerate edge; such an edge is automatically coCartesian (this follows from Corollary 2.4.1.6 because S is an ∞-category), and therefore there exists an extension of F' to all of $\Delta^n \times \Delta^1$ by the first part of the proof. $\qquad\square$

Remark 2.4.1.9. Let $p : X \to S$ be an inner fibration of simplicial sets, let x be a vertex of X, and let $\overline{f} : \overline{x}' \to p(x)$ be an edge of S ending at $p(x)$. There may exist many p-Cartesian edges $f : x' \to x$ of X with $p(f) = \overline{f}$. However, there is a sense in which any two edges having the same target x are equivalent to one another. Namely, any p-Cartesian edge $f : x' \to x$ lifting $\overline{f}$ can be regarded as a final object of the ∞-category $X_{/x} \times_{S_{/p(x)}} \{\overline{f}\}$ and is therefore determined up to equivalence by $\overline{f}$ and x.

We now spell out the meaning of Definition 2.4.1.1 in the setting of simplicial categories.

Proposition 2.4.1.10. *Let* $F : \mathcal{C} \to \mathcal{D}$ *be a functor between simplicial categories. Suppose that* $\mathcal{C}$ *and* $\mathcal{D}$ *are fibrant and that for every pair of objects* $C, C' \in \mathcal{C}$, *the associated map*

$$\mathrm{Map}_{\mathcal{C}}(C, C') \to \mathrm{Map}_{\mathcal{D}}(F(C), F(C'))$$

is a Kan fibration. Then the following assertions hold:

(1) *The associated map* $q : \mathrm{N}(\mathcal{C}) \to \mathrm{N}(\mathcal{D})$ *is an inner fibration between* ∞-*categories.*

(2) *A morphism* $f : C' \to C''$ *in* $\mathcal{C}$ *is* q-*Cartesian if and only if, for every object* $C \in \mathcal{C}$, *the diagram of simplicial sets*

$$
\begin{array}{ccc}
\mathrm{Map}_{\mathcal{C}}(C, C') & \longrightarrow & \mathrm{Map}_{\mathcal{C}}(C, C'') \\
\downarrow & & \downarrow \\
\mathrm{Map}_{\mathcal{D}}(F(C), F(C')) & \longrightarrow & \mathrm{Map}_{\mathcal{D}}(F(C), F(C''))
\end{array}
$$

is homotopy Cartesian.

Proof. Assertion (1) follows from Remark 1.1.5.11. Let f be a morphism in $\mathcal{C}$. By definition, $f : C' \to C''$ is q-Cartesian if and only if

$$\theta : \mathrm{N}(\mathcal{C})_{/f} \to \mathrm{N}(\mathcal{C})_{/C''} \times_{\mathrm{N}(\mathcal{D})_{/F(C'')}} \mathrm{N}(\mathcal{D})_{/F(f)}$$

is a trivial fibration. Since θ is a right fibration between right fibrations over $\mathcal{C}$, f is q-Cartesian if and only if for every object $C \in \mathcal{C}$, the induced map

$$\theta_C : \{C\} \times_{N(\mathcal{C})} N(\mathcal{C})_{/f} \to \{C\} \times_{N(\mathcal{C})} N(\mathcal{C})_{/C''} \times_{N(\mathcal{D})_{/F(C'')}} N(\mathcal{D})_{/F(f)}$$

is a homotopy equivalence of Kan complexes. This is equivalent to the assertion that the diagram

$$
\begin{array}{ccc}
N(\mathcal{C})_{/f} \times_{\mathcal{C}} \{C\} & \longrightarrow & N(\mathcal{C})_{/C''} \times_{N(\mathcal{C})} \{C\} \\
\downarrow & & \downarrow \\
N(\mathcal{D})_{/F(f)} \times_{N(\mathcal{D})} \{F(C)\} & \longrightarrow & N(\mathcal{D})_{/F(C'')} \times_{N(\mathcal{D})} \{F(C)\}
\end{array}
$$

is homotopy Cartesian. In view of Theorem 1.1.5.13, this diagram is equivalent to the diagram of simplicial sets

$$
\begin{array}{ccc}
\mathrm{Map}_{\mathcal{C}}(C, C') & \longrightarrow & \mathrm{Map}_{\mathcal{C}}(C, C'') \\
\downarrow & & \downarrow \\
\mathrm{Map}_{\mathcal{D}}(F(C), F(C')) & \longrightarrow & \mathrm{Map}_{\mathcal{D}}(F(C), F(C'')).
\end{array}
$$

This proves (2).   □

In some contexts, it will be convenient to consider a slightly larger class of edges:

Definition 2.4.1.11. Let $p : X \to S$ be an inner fibration and let $e : \Delta^1 \to X$ be an edge. We will say that e is *locally p-Cartesian* if it is a p'-Cartesian edge of the fiber product $X \times_S \Delta^1$, where $p' : X \times_S \Delta^1 \to \Delta^1$ denotes the projection.

Remark 2.4.1.12. Suppose we are given a pullback diagram

$$
\begin{array}{ccc}
X' & \overset{f}{\longrightarrow} & X \\
{\scriptstyle p'}\downarrow & & \downarrow{\scriptstyle p} \\
S' & \longrightarrow & S
\end{array}
$$

of simplicial sets, where p (and therefore also p') is an inner fibration. An edge e of X' is locally p'-Cartesian if and only if its image $f(e)$ is locally p-Cartesian.

We conclude with a somewhat technical result which will be needed in §3.1.1:

Proposition 2.4.1.13. *Let $p : X \to S$ be an inner fibration of simplicial sets and let $f : x \to y$ be an edge of X. Suppose that there is a 3-simplex $\sigma : \Delta^3 \to X$ such that $d_1\sigma = s_0 f$ and $d_2\sigma = s_1 f$. Suppose furthermore that there exists a p-Cartesian edge $\widetilde{f} : \widetilde{x} \to y$ such that $p(\widetilde{f}) = p(f)$. Then f is p-Cartesian.*

Proof. We have a diagram of simplicial sets

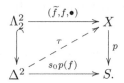

Because $\widetilde{f}$ is p-Cartesian, there exists a map τ rendering the diagram commutative. Let $g = d_2(\tau)$, which we regard as a morphism $x \to \widetilde{x}$ in the ∞-category $X_{p(x)} = X \times_S \{p(x)\}$. We will show that g is an equivalence in $X_{p(x)}$. It will follow that g is p-Cartesian and that f, being a composition of p-Cartesian edges, is p-Cartesian (Proposition 2.4.1.7).

Now consider the diagram

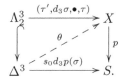

The map τ' exists since p is an inner fibration. Let $g' = d_1\tau'$. We will show that $g' : \widetilde{x} \to x$ is a homotopy inverse to g in the ∞-category $X_{p(x)}$.

Using τ and τ', we construct a new diagram

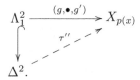

Since p is an inner fibration, we deduce the existence of $\theta : \Delta^3 \to X$, rendering the diagram commutative. The simplex $d_2(\theta)$ exhibits id_x as a composition $g' \circ g$ in the ∞-category $X_{p(s)}$. It follows that g' is a left homotopy inverse to g.

We now have a diagram

$$
\begin{array}{ccc}
\Lambda_1^2 & \xrightarrow{(g,\bullet,g')} & X_{p(x)} \\
\Big\uparrow & \nearrow^{\tau''} & \\
\Delta^2. & &
\end{array}
$$

The indicated 2-simplex τ'' exists since $X_{p(x)}$ is an ∞-category and exhibits $d_1(\tau'')$ as a composition $g \circ g'$. To complete the proof, it will suffice to show that $d_1(\tau'')$ is an equivalence in $X_{p(x)}$.

Consider the diagrams

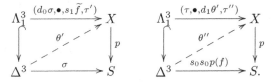

Since p is an inner fibration, there exist 3-simplices $\theta', \theta'' : \Delta^3 \to X$ with the indicated properties. The 2-simplex $d_1(\theta'')$ identifies $d_1(\tau'')$ as a map between two p-Cartesian lifts of $p(f)$; it follows that $d_1(\tau'')$ is an equivalence, which completes the proof.   □

2.4.2 Cartesian Fibrations

In this section, we will introduce the study of *Cartesian fibrations* between simplicial sets. The theory of Cartesian fibrations is a generalization of the theory of right fibrations studied in §2.1. Recall that if $f : X \to S$ is a right fibration of simplicial sets, then the fibers $\{X_s\}_{s \in S}$ are Kan complexes, which depend in a (contravariantly) functorial fashion on the choice of vertex $s \in S$. The condition that f be a Cartesian fibration has a similar flavor: we still require that X_s depend functorially on s but weaken the requirement that X_s be a Kan complex; instead, we merely require that it be an ∞-category.

Definition 2.4.2.1. We will say that a map $p : X \to S$ of simplicial sets is a *Cartesian fibration* if the following conditions are satisfied:

(1) The map p is an inner fibration.

(2) For every edge $f : x \to y$ of S and every vertex $\widetilde{y}$ of X with $p(\widetilde{y}) = y$, there exists a p-Cartesian edge $\widetilde{f} : \widetilde{x} \to \widetilde{y}$ with $p(\widetilde{f}) = f$.

We say that p is a *coCartesian fibration* if the opposite map $p^{op} : X^{op} \to S^{op}$ is a Cartesian fibration.

If a general inner fibration $p : X \to S$ associates to each vertex $s \in S$ an ∞-category X_s and to each edge $s \to s'$ a correspondence from X_s to $X_{s'}$, then p is Cartesian if each of these correspondences arises from a (canonically determined) functor $X_{s'} \to X_s$. In other words, a Cartesian fibration with base S ought to be roughly the same thing as a contravariant functor from S into an ∞-category of ∞-categories, where the morphisms are given by *functors*. One of the main goals of Chapter 3 is to give a precise formulation (and proof) of this assertion.

Remark 2.4.2.2. Let $F : \mathcal{C} \to \mathcal{C}'$ be a functor between (ordinary) categories. The induced map of simplicial sets $N(F) : N(\mathcal{C}) \to N(\mathcal{C}')$ is automatically an inner fibration; it is Cartesian if and only if F is a *fibration* of categories in the sense of Grothendieck.

The following formal properties follow immediately from the definition:

Proposition 2.4.2.3.   (1) *Any isomorphism of simplicial sets is a Cartesian fibration.*

(2) *The class of Cartesian fibrations between simplicial sets is stable under base change.*

(3) *A composition of Cartesian fibrations is a Cartesian fibration.*

Recall that an ∞-category $\mathcal{C}$ is a Kan complex if and only if every morphism in $\mathcal{C}$ is an equivalence. We now establish a relative version of this statement:

Proposition 2.4.2.4. *Let $p : X \to S$ be an inner fibration of simplicial sets. The following conditions are equivalent:*

(1) *The map p is a Cartesian fibration, and every edge in X is p-Cartesian.*

(2) *The map p is a right fibration.*

(3) *The map p is a Cartesian fibration, and every fiber of p is a Kan complex.*

Proof. In view of Remark 2.4.1.4, the assertion that every edge of X is p-Cartesian is equivalent to the assertion that p has the right lifting property with respect to $\Lambda_n^n \subseteq \Delta^n$ for all $n \geq 2$. The requirement that p be a Cartesian fibration further imposes the right lifting property with respect to $\Lambda_1^1 \subseteq \Delta^1$. This proves that $(1) \Leftrightarrow (2)$.

Suppose that (2) holds. Since we have established that (2) implies (1), we know that p is Cartesian. Furthermore, we have already seen that the fibers of a right fibration are Kan complexes. Thus (2) implies (3).

We complete the proof by showing that (3) implies that every edge $f : x \to y$ of X is p-Cartesian. Since p is a Cartesian fibration, there exists a p-Cartesian edge $f' : x' \to y$ with $p(f') = p(f)$. Since f' is p-Cartesian, there exists a 2-simplex $\sigma : \Delta^2 \to X$ which we may depict as a diagram

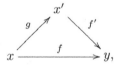

where $p(\sigma) = s_0 p(f)$. Then g lies in the fiber $X_{p(x)}$ and is therefore an equivalence (since $X_{p(x)}$ is a Kan complex). It follows that f is equivalent to f' as objects of $X_{/y} \times_{S_{/p(y)}} \{p(f)\}$, so that f is p-Cartesian, as desired. $\square$

Corollary 2.4.2.5. *Let $p : X \to S$ be a Cartesian fibration. Let $X' \subseteq X$ consist of all those simplices σ of X such that every edge of σ is p-Cartesian. Then $p|X'$ is a right fibration.*

Proof. We first show that $p|X'$ is an inner fibration. It suffices to show that $p|X'$ has the right lifting property with respect to every horn inclusion Λ_i^n, $0 < i < n$. If $n > 2$, then this follows immediately from the fact that p has the appropriate lifting property. If $n = 2$, then we must show that if $f : \Delta^2 \to X$ is such that $f|\Lambda_1^2$ factors through X', then f factors through X'. This follows immediately from Proposition 2.4.1.7.

We now wish to complete the proof by showing that p is a right fibration. According to Proposition 2.4.2.4, it suffices to prove that every edge of X' is $p|X'$-Cartesian. This follows immediately from the characterization given in Remark 2.4.1.4 because every edge of X' is p-Cartesian (when regarded as an edge of X). $\square$

In order to verify that certain maps are Cartesian fibrations, it is often convenient to work in a slightly more general setting.

Definition 2.4.2.6. A map $p : X \to S$ of simplicial sets is a *locally Cartesian fibration* if it is an inner fibration and, for every edge $\Delta^1 \to S$, the pullback $X \times_S \Delta^1 \to \Delta^1$ is a Cartesian fibration.

In other words, an inner fibration $p : X \to S$ is a locally Cartesian fibration if and only if, for every vertex $x \in X$ and every edge $e : s \to p(x)$ in S, there exists a locally p-Cartesian edge $\overline{s} \to x$ which lifts e.

Let $p : X \to S$ be an inner fibration of simplicial sets. It is clear that every p-Cartesian morphism of X is locally p-Cartesian. Moreover, Proposition 2.4.1.7 implies that the class of p-Cartesian edges of X is stable under composition. Then following result can be regarded as a sort of converse:

Lemma 2.4.2.7. *Let $p : X \to S$ be a locally Cartesian fibration of simplicial sets and let $f : x' \to x$ be an edge of X. The following conditions are equivalent:*

(1) *The edge e is p-Cartesian.*

(2) *For every 2-simplex σ*

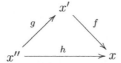

in X, the edge g is locally p-Cartesian if and only if the edge h is locally p-Cartesian.

(3) *For every 2-simplex σ*

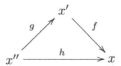

in X, if g is locally p-Cartesian, then h is locally p-Cartesian.

Proof. We first show that (1) $\Rightarrow$ (2). Pulling back via the composition $p \circ \sigma : \Delta^2 \to S$, we can reduce to the case where $S = \Delta^2$. In this case, g is locally p-Cartesian if and only if it is p-Cartesian, and likewise for h. We now conclude by applying Proposition 2.4.1.7.

The implication (2) $\Rightarrow$ (3) is obvious. We conclude by showing that (3) $\Rightarrow$ (1). We must show that $\eta : X_{/f} \to X_{/x} \times_{S_{/p(x)}} S_{/p(f)}$ is a trivial fibration. Since η is a right fibration, it will suffice to show that the fiber of η over any vertex is contractible. Any such vertex determines a map $\sigma : \Delta^2 \to S$ with $\sigma | \Delta^{\{1,2\}} = p(f)$. Pulling back via σ, we may suppose that $S = \Delta^2$.

It will be convenient to introduce a bit of notation: for every map $q : K \to X$ let $Y_{/q} \subseteq X_{/q}$ denote the full simplicial subset spanned by those vertices of $X_{/q}$ which map to the initial vertex of S. We wish to show that the natural map $Y_{/f} \to Y_{/x}$ is a trivial fibration. By assumption, there exists a locally p-Cartesian morphism $g : x'' \to x'$ in X covering the edge $\Delta^{\{0,1\}} \subseteq S$. Since X is an ∞-category, there exists a 2-simplex $\tau : \Delta^2 \to X$ with $d_2(\tau) = g$ and $d_0(\tau) = f$. Then $h = d_1(\tau)$ is a composite of f and g, and assumption (3) guarantees that h is locally p-Cartesian. We have a commutative diagram

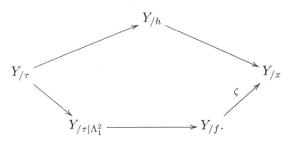

Moreover, all of the maps in this diagram are trivial fibrations except possibly ζ, which is known to be a right fibration. It follows that ζ is a trivial fibration as well, which completes the proof. $\qquad\square$

In fact, we have the following:

Proposition 2.4.2.8. *Let $p : X \to S$ be a locally Cartesian fibration. The following conditions are equivalent:*

(1) *The map p is a Cartesian fibration.*

(2) *Given a 2-simplex*

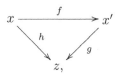

if f and g are locally p-Cartesian, then h is locally p-Cartesian.

(3) *Every locally p-Cartesian edge of X is p-Cartesian.*

Proof. The equivalence (2) $\Leftrightarrow$ (3) follows from Lemma 2.4.2.7, and the implication (3) $\Rightarrow$ (1) is obvious. To prove that (1) $\Rightarrow$ (3), let us suppose that $e : x \to y$ is a locally p-Cartesian edge of X. Choose a p-Cartesian edge $e' : x' \to y$ lifting $p(e)$. The edges e and e' are both p'-Cartesian in $X' = X \times_S \Delta^1$, where $p' : X' \to \Delta^1$ denotes the projection. It follows that e and e' are equivalent in X' and therefore also equivalent in X. Since e' is p-Cartesian, we deduce that e is p-Cartesian as well. $\qquad\square$

Remark 2.4.2.9. If $p : X \to S$ is a locally Cartesian fibration, then we can associate to every edge $s \to s'$ of S a functor $X_{s'} \to X_s$, which is well-defined

up to homotopy. A 2-simplex

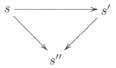

determines a triangle of ∞-categories

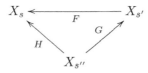

which commutes up to a (generally noninvertible) natural transformation $\alpha : F \circ G \to H$. Proposition 2.4.2.8 implies that p is a Cartesian fibration if and only if every such natural transformation is an equivalence of functors.

Corollary 2.4.2.10. *Let* $p : X \to S$ *be an inner fibration of simplicial sets. Then* p *is Cartesian if and only if every pullback* $X \times_S \Delta^n \to \Delta^n$ *is a Cartesian fibration for* $n \leq 2$.

One advantage the theory of locally Cartesian fibrations holds over the theory of Cartesian fibrations is the following "fiberwise" existence criterion:

Proposition 2.4.2.11. *Suppose we are given a commutative diagram of simplicial sets*

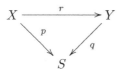

satisfying the following conditions:

(1) *The maps* p *and* q *are locally Cartesian fibrations, and* r *is an inner fibration.*

(2) *The map* r *carries locally* p-*Cartesian edges of* X *to locally* q-*Cartesian edges of* Y.

(3) *For every vertex* s *of* S, *the induced map* $r_s : X_s \to Y_s$ *is a locally Cartesian fibration.*

Then r *is a locally Cartesian fibration. Moreover, an edge* e *of* X *is locally* r-*Cartesian if and only if there exists a 2-simplex* σ

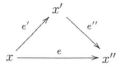

with the following properties:

(i) *In the simplicial set S, we have $p(\sigma) = s^0(p(e))$.*

(ii) *The edge e'' is locally p-Cartesian.*

(iii) *The edge e' is locally $r_{p(x)}$-Cartesian.*

Proof. Suppose we are given a vertex $x'' \in X$ and an edge $e_0 : y \to p(x'')$ in Y. It is clear that we can construct a 2-simplex σ in X satisfying (i) through (iii), with $p(e) = q(e_0)$. Moreover, σ is uniquely determined up to equivalence. We will prove that e is locally r-Cartesian. This will prove that r is a locally Cartesian fibration and the "if" direction of the final assertion. The converse will then follow from the uniqueness (up to equivalence) of locally r-Cartesian lifts of a given edge (with specified terminal vertex).

To prove that e is locally r-Cartesian, we are free to pull back by the edge $p(e) : \Delta^1 \to S$ and thereby reduce to the case $S = \Delta^1$. Then p and q are Cartesian fibrations. Since e'' is p-Cartesian and $r(e'')$ is q-Cartesian, Proposition 2.4.1.3 implies that e'' is r-Cartesian. Remark 2.4.1.12 implies that e' is locally p-Cartesian. It follows from Lemma 2.4.2.7 that e is locally p-Cartesian as well. $\qquad\square$

Remark 2.4.2.12. The analogue of Proposition 2.4.2.11 for Cartesian fibrations is false.

2.4.3 Stability Properties of Cartesian Fibrations

In this section, we will prove the class of Cartesian fibrations is stable under the formation of overcategories and undercategories. Since the definition of a Cartesian fibration is not self-dual, we must treat these results separately, using slightly different arguments (Propositions 2.4.3.2 and 2.4.3.3). We begin with the following simple lemma.

Lemma 2.4.3.1. *Let $A \subseteq B$ be an inclusion of simplicial sets. Then the inclusion*

$$(\{1\} \star B) \coprod_{\{1\}\star A} (\Delta^1 \star A) \subseteq \Delta^1 \star B$$

is inner anodyne.

Proof. Working by transfinite induction, we may reduce to the case where B is obtained from A by adjoining a single nondegenerate simplex and therefore to the universal case $B = \Delta^n$, $A = \partial \Delta^n$. Now the inclusion in question is isomorphic to $\Lambda_1^{n+2} \subseteq \Delta^{n+2}$. $\qquad\square$

Proposition 2.4.3.2. *Let $p : \mathcal{C} \to \mathcal{D}$ be a Cartesian fibration of simplicial sets and let $q : K \to \mathcal{C}$ be a diagram. Then*

(1) *The induced map $p' : \mathcal{C}_{/q} \to \mathcal{D}_{/pq}$ is a Cartesian fibration.*

(2) *An edge f of $\mathcal{C}_{/q}$ is p'-Cartesian if and only if the image of f in $\mathcal{C}$ is p-Cartesian.*

Proof. Proposition 2.1.2.5 implies that p' is an inner fibration. Let us call an edge f of $\mathcal{C}_{q/}$ *special* if its image in $\mathcal{C}$ is p-Cartesian. To complete the proof, we will verify the following assertions:

(i) Given a vertex $\overline{q} \in \mathcal{C}_{/q}$ and an edge $\widetilde{f} : \overline{r}' \to p'(\overline{q})$, there exists a special edge $f : \overline{r} \to \overline{q}$ with $p'(f) = \widetilde{f}$.

(ii) Every special edge of $\mathcal{C}_{/q}$ is p'-Cartesian.

To prove (i), let $\widetilde{f}'$ denote the image of $\widetilde{f}$ in $\mathcal{D}$ and c the image of $\overline{q}$ in $\mathcal{C}$. Using the assumption that p is a coCartesian fibration, we can choose a p-coCartesian edge $f' : c \to d$ lifting $\widetilde{f}'$. To extend this data to the desired edge f of $\mathcal{C}_{/q}$, it suffices to solve the lifting problem depicted in the diagram

$$
\begin{array}{ccc}
(\{1\} \star K) \coprod_{\{1\}} \Delta^1 & \longrightarrow & \mathcal{C} \\
\Big\downarrow{\scriptstyle i} & \nearrow & \Big\downarrow{\scriptstyle p} \\
\Delta^1 \star K & \longrightarrow & \mathcal{D} .
\end{array}
$$

This lifting problem has a solution because p is an inner fibration and i is inner anodyne (Lemma 2.4.3.1).

To prove (ii), it will suffice to show that if $n \geq 2$, then any lifting problem of the form

$$
\begin{array}{ccc}
\Lambda^n_n \star K & \overset{g}{\longrightarrow} & \mathcal{C} \\
\Big\downarrow & \overset{G}{\nearrow} & \Big\downarrow{\scriptstyle p} \\
\Delta^n \star K & \longrightarrow & \mathcal{D}
\end{array}
$$

has a solution provided that $e = g(\Delta^{\{n-1,n\}})$ is a p-Cartesian edge of $\mathcal{C}$. Consider the set P of pairs $(K', G_{K'})$, where $K' \subseteq K$ and $G_{K'}$ fits in a commutative diagram

$$
\begin{array}{ccc}
(\Lambda^n_n \star K) \coprod_{\Lambda^n_n \star K'} (\Delta^n \star K') & \overset{G_{K'}}{\longrightarrow} & \mathcal{C} \\
\Big\downarrow & & \Big\downarrow{\scriptstyle p} \\
\Delta^n \star K & \longrightarrow & \mathcal{D} .
\end{array}
$$

Because e is p-Cartesian, there exists an element $(\emptyset, G_\emptyset) \in P$. We regard P as partially ordered, where $(K', G_{K'}) \leq (K'', G_{K''})$ if $K' \subseteq K''$ and $G_{K'}$ is a restriction of $G_{K''}$. Invoking Zorn's lemma, we deduce the existence of a maximal element $(K', G_{K'})$ of P. If $K' = K$, then the proof is complete. Otherwise, it is possible to enlarge K' by adjoining a single nondegenerate m-simplex of K. Since $(K', G_{K''})$ is maximal, we conclude that the associated lifting problem

$$
\begin{array}{ccc}
(\Lambda^n_n \star \Delta^m) \coprod_{\Lambda^n_n \star \partial \Delta^m} (\Delta^n \star \partial \Delta^m) & \longrightarrow & \mathcal{C} \\
\Big\downarrow & \overset{\sigma}{\nearrow} & \Big\downarrow{\scriptstyle p} \\
\Delta^n \star \Delta^m & \longrightarrow & \mathcal{D}
\end{array}
$$

has no solution. The left vertical map is equivalent to the inclusion $\Lambda_{n+1}^{n+m+1} \subseteq \Delta^{n+m+1}$, which is inner anodyne. Since p is an inner fibration by assumption, we obtain a contradiction. $\qquad\square$

Proposition 2.4.3.3. *Let $p : \mathcal{C} \to \mathcal{D}$ be a coCartesian fibration of simplicial sets and let $q : K \to \mathcal{C}$ be a diagram. Then*

(1) *The induced map $p' : \mathcal{C}_{/q} \to \mathcal{D}_{/pq}$ is a coCartesian fibration.*

(2) *An edge f of $\mathcal{C}_{/q}$ is p'-coCartesian if and only if the image of f in $\mathcal{C}$ is p-coCartesian.*

Proof. Proposition 2.1.2.5 implies that p' is an inner fibration. Let us call an edge f of $\mathcal{C}_{/q}$ *special* if its image in $\mathcal{C}$ is p-coCartesian. To complete the proof, it will suffice to verify the following assertions:

(i) Given a vertex $\overline{q} \in \mathcal{C}_{/q}$ and an edge $\widetilde{f} : p'(\overline{q}) \to \overline{r}'$, there exists a special edge $f : \overline{q} \to \overline{r}$ with $p'(f) = \widetilde{f}$.

(ii) Every special edge of $\mathcal{C}_{/q}$ is p'-coCartesian.

To prove (i), we begin with a commutative diagram

$$
\begin{array}{ccc}
\Delta^0 \star K & \xrightarrow{\ \overline{q}\ } & \mathcal{C} \\
\uparrow & & \downarrow \\
\Delta^1 \star K & \xrightarrow{\ \widetilde{f}\ } & \mathcal{D}.
\end{array}
$$

Let $C \in \mathcal{C}$ denote the image under $\overline{q}$ of the cone point of $\Delta^0 \star K$ and choose a p-coCartesian morphism $u : C \to C'$ lifting $\widetilde{f}|\Delta^1$. We now consider the collection P of all pairs (L, f_L), where L is a simplicial subset of K and f_L is a map fitting into a commutative diagram

$$
\begin{array}{ccc}
(\Delta^0 \star K) \coprod_{\Delta^0 \star L} (\Delta^1 \star L) & \xrightarrow{\ f_L\ } & \mathcal{C} \\
\uparrow & & \downarrow \\
\Delta^1 \star K & \xrightarrow{\ \widetilde{f}\ } & \mathcal{D},
\end{array}
$$

where $f_L|\Delta^1 = u$ and $f_L|\Delta^0 \star K = \overline{q}$. We partially order the set P as follows: $(L, f_L) \le (L', f_{L'})$ if $L \subseteq L'$ and f_L is equal to the restriction of $f_{L'}$. The partially ordered set P satisfies the hypotheses of Zorn's lemma and therefore contains a maximal element (L, f_L). If $L \ne K$, then we can choose a simplex $\sigma : \Delta^n \to K$ of minimal dimension which does not belong to L. By maximality, we obtain a diagram in which the indicated dotted arrow cannot be supplied. This is a contradiction since the upper horizontal map carries the initial edge of Λ_0^{n+2} to a p-coCartesian edge of $\mathcal{C}$. It follows that $L = K$, and we may take $f = f_L$. This completes the proof of (i).

The proof of (ii) is similar. Suppose we are given $n \geq 2$ and let

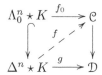

be a commutative diagram, where $f_0|K = q$ and $f_0|\Delta^{\{0,1\}}$ is a p-coCartesian edge of $\mathcal{C}$. We wish to prove the existence of the dotted arrow f indicated in the diagram. As above, we consider the collection P of all pairs (L, f_L), where L is a simplicial subset of K and f_L extends f_0 and fits into a commutative diagram

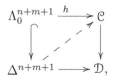

We partially order P as follows: $(L, f_L) \leq (L', f_{L'})$ if $L \subseteq L'$ and f_L is a restriction of $f_{L'}$. Using Zorn's lemma, we conclude that P contains a maximal element (L, f_L). If $L \neq K$, then we can choose a simplex $\sigma : \Delta^m \to K$ which does not belong to L, where m is as small as possible. Invoking the maximality of (L, f_L), we obtain a diagram

$$\Lambda_0^{n+m+1} \xrightarrow{h} \mathcal{C}$$
$$\downarrow \qquad \qquad \downarrow$$
$$\Delta^{n+m+1} \longrightarrow \mathcal{D},$$

where the indicated dotted arrow cannot be supplied. However, the map h carries the initial edge of Δ^{n+m+1} to a p-coCartesian edge of $\mathcal{C}$, so we obtain a contradiction. It follows that $L = K$, so that we can take $f = f_L$ to complete the proof. $\qquad \square$

2.4.4 Mapping Spaces and Cartesian Fibrations

Let $p : \mathcal{C} \to \mathcal{D}$ be a functor between ∞-categories and let X and Y be objects of $\mathcal{C}$. Then p induces a map

$$\phi : \mathrm{Map}_{\mathcal{C}}(X, Y) \to \mathrm{Map}_{\mathcal{D}}(p(X), p(Y)).$$

Our goal in this section is to understand the relationship between the fibers of p and the *homotopy* fibers of ϕ.

Lemma 2.4.4.1. *Let $p : \mathcal{C} \to \mathcal{D}$ be an inner fibration of ∞-categories and let $X, Y \in \mathcal{C}$. The induced map $\phi : \mathrm{Hom}^{\mathrm{R}}_{\mathcal{C}}(X, Y) \to \mathrm{Hom}^{\mathrm{R}}_{\mathcal{D}}(p(X), p(Y))$ is a Kan fibration.*

Proof. Since p is an inner fibration, the induced map $\widetilde{\phi} : \mathcal{C}_{/X} \to \mathcal{D}_{/p(X)} \times_{\mathcal{D}} \mathcal{C}$ is a right fibration by Proposition 2.1.2.1. We note that ϕ is obtained from $\widetilde{\phi}$ by restricting to the fiber over the vertex Y of $\mathcal{C}$. Thus ϕ is a right fibration; since the target of ϕ is a Kan complex, ϕ is a Kan fibration by Lemma 2.1.3.3. $\qquad\square$

Suppose the conditions of Lemma 2.4.4.1 are satisfied. Let us consider the problem of computing the fiber of ϕ over a vertex $\overline{e} : p(X) \to p(Y)$ of $\mathrm{Hom}^{\mathrm{R}}_{\mathcal{D}}(X, Y)$. Suppose that there is a p-Cartesian edge $e : X' \to Y$ lifting $\overline{e}$. By definition, we have a trivial fibration

$$\psi : \mathcal{C}_{/e} \to \mathcal{C}_{/Y} \times_{\mathcal{D}_{/p(Y)}} \mathcal{D}_{/\overline{e}}.$$

Consider the 2-simplex $\sigma = s_1(\overline{e})$ regarded as a vertex of $\mathcal{D}_{/\overline{e}}$. Passing to the fiber, we obtain a trivial fibration

$$F \to \phi^{-1}(e),$$

where F denotes the fiber of $\mathcal{C}_{/e} \to \mathcal{D}_{/\overline{e}} \times_{\mathcal{D}} \mathcal{C}$ over the point (σ, X). On the other hand, we have a trivial fibration $\mathcal{C}_{/e} \to \mathcal{D}_{/\overline{e}} \times_{\mathcal{D}_{/p(X)}} \mathcal{C}_{/X'}$ by Proposition 2.1.2.5. Passing to the fiber again, we obtain a trivial fibration $F \to \mathrm{Hom}^{\mathrm{R}}_{\mathcal{C}_{p(X)}}(X, X')$. We may summarize the situation as follows:

Proposition 2.4.4.2. *Let $p : \mathcal{C} \to \mathcal{D}$ be an inner fibration of ∞-categories. Let $X, Y \in \mathcal{C}$, let $\overline{e} : p(X) \to p(Y)$ be a morphism in $\mathcal{D}$, and let $e : X' \to Y$ be a locally p-Cartesian morphism of $\mathcal{C}$ lifting $\overline{e}$. Then in the homotopy category $\mathcal{H}$ of spaces, there is a fiber sequence*

$$\mathrm{Map}_{\mathcal{C}_{p(X)}}(X, X') \to \mathrm{Map}_{\mathcal{C}}(X, Y) \to \mathrm{Map}_{\mathcal{D}}(p(X), p(Y)).$$

Here the fiber is taken over the point classified by $\overline{e} : p(X) \to p(Y)$.

Proof. The edge $\overline{e}$ defines a map $\Delta^1 \to \mathcal{D}$. Note that the fiber of the Kan fibration $\mathrm{Hom}^{\mathrm{R}}_{\mathcal{C}}(X, Y) \to \mathrm{Hom}^{\mathrm{R}}_{\mathcal{D}}(pX, pY)$ does not change if we replace p by the induced projection $\mathcal{C} \times_{\mathcal{D}} \Delta^1 \to \Delta^1$. We may therefore assume without loss of generality that e is p-Cartesian, and the desired result follows from the above analysis. $\qquad\square$

A similar assertion can be taken as a characterization of Cartesian morphisms:

Proposition 2.4.4.3. *Let $p : \mathcal{C} \to \mathcal{D}$ be an inner fibration of ∞-categories and let $f : Y \to Z$ be a morphism in $\mathcal{C}$. The following are equivalent:*

(1) *The morphism f is p-Cartesian.*

(2) *For every object X of $\mathcal{C}$, composition with f gives rise to a homotopy Cartesian diagram*

$$\begin{array}{ccc} \mathrm{Map}_{\mathcal{C}}(X,Y) & \longrightarrow & \mathrm{Map}_{\mathcal{C}}(X,Z) \\ \downarrow & & \downarrow \\ \mathrm{Map}_{\mathcal{D}}(p(X),p(Y)) & \longrightarrow & \mathrm{Map}_{\mathcal{D}}(p(X),p(Z)). \end{array}$$

Proof. Let $\phi : \mathcal{C}_{/f} \to \mathcal{C}_{/Z} \times_{\mathcal{D}_{/p(Z)}} \mathcal{D}_{/p(f)}$ be the canonical map; then (1) is equivalent to the assertion that ϕ is a trivial fibration. According to Proposition 2.1.2.1, ϕ is a right fibration. Thus, ϕ is a trivial fibration if and only if the fibers of ϕ are contractible Kan complexes. For each object $X \in \mathcal{C}$, let

$$\phi_X : \mathcal{C}_{/f} \times_{\mathcal{C}} \{X\} \to \mathcal{C}_{/Z} \times_{\mathcal{D}_{/p(Z)}} \mathcal{D}_{/p(f)} \times_{\mathcal{C}} \{X\}$$

be the induced map. Then ϕ_X is a right fibration between Kan complexes and therefore a Kan fibration; it has contractible fibers if and only if it is a homotopy equivalence. Thus (1) is equivalent to the assertion that ϕ_X is a homotopy equivalence for every object X of $\mathcal{C}$.

We remark that (2) is somewhat imprecise: although all the maps in the diagram are well-defined in the homotopy category $\mathcal{H}$ of spaces, we need to represent this by a commutative diagram in the category of simplicial sets before we can ask whether or not the diagram is homotopy Cartesian. We therefore rephrase (2) more precisely: it asserts that the diagram of Kan complexes

$$\begin{array}{ccc} \mathcal{C}_{/f} \times_{\mathcal{C}} \{X\} & \longrightarrow & \mathcal{C}_{/Z} \times_{\mathcal{C}} \{X\} \\ \downarrow & & \downarrow \\ \mathcal{D}_{/p(f)} \times_{\mathcal{D}} \{p(X)\} & \longrightarrow & \mathcal{D}_{/p(Z)} \times_{\mathcal{D}} \{p(X)\} \end{array}$$

is homotopy Cartesian. Lemma 2.4.4.1 implies that the right vertical map is a Kan fibration, so the homotopy limit in question is given by the fiber product

$$\mathcal{C}_{/Z} \times_{\mathcal{D}_{/p(Z)}} \mathcal{D}_{/p(f)} \times_{\mathcal{C}} \{X\}.$$

Consequently, assertion (2) is also equivalent to the condition that ϕ_X be a homotopy equivalence for every object $X \in \mathcal{C}$. $\qquad\square$

Corollary 2.4.4.4. *Suppose we are given maps $\mathcal{C} \xrightarrow{p} \mathcal{D} \xrightarrow{q} \mathcal{E}$ of ∞-categories such that both q and $q \circ p$ are locally Cartesian fibrations. Suppose that p carries locally $(q \circ p)$-Cartesian edges of $\mathcal{C}$ to locally q-Cartesian edges of $\mathcal{D}$ and that for every object $Z \in \mathcal{E}$, the induced map $\mathcal{C}_Z \to \mathcal{D}_Z$ is a categorical equivalence. Then p is a categorical equivalence.*

Proof. Proposition 2.4.4.2 implies that p is fully faithful. If Y is any object of $\mathcal{D}$, then Y is equivalent in the fiber $\mathcal{D}_{q(Y)}$ to the image under p of some vertex of $\mathcal{C}_{q(Y)}$. Thus p is essentially surjective, and the proof is complete. $\qquad\square$

Corollary 2.4.4.5. *Let* $p : \mathcal{C} \to \mathcal{D}$ *be a Cartesian fibration of ∞-categories. Let* $q : \mathcal{D}' \to \mathcal{D}$ *be a categorical equivalence of ∞-categories. Then the induced map* $q' : \mathcal{C}' = \mathcal{D}' \times_{\mathcal{D}} \mathcal{C} \to \mathcal{C}$ *is a categorical equivalence.*

Proof. Proposition 2.4.4.2 immediately implies that q' is fully faithful. We claim that q' is essentially surjective. Let X be any object of $\mathcal{C}$. Since q is fully faithful, there exists an object y of T' and an equivalence $\bar{e} : q(Y) \to p(X)$. Since p is Cartesian, we can choose a p-Cartesian edge $e : Y' \to X$ lifting $\bar{e}$. Since e is p-Cartesian and $p(e)$ is an equivalence, e is an equivalence. By construction, the object Y' of S lies in the image of q'. $\qquad\square$

Corollary 2.4.4.6. *Let* $p : \mathcal{C} \to \mathcal{D}$ *be a Cartesian fibration of ∞-categories. Then p is a categorical equivalence if and only if p is a trivial fibration.*

Proof. The "if" direction is clear. Suppose then that p is a categorical equivalence. We first claim that p is surjective on objects. The essential surjectivity of p implies that for each $Y \in \mathcal{D}$, there is an equivalence $Y \to p(X)$ for some object X of $\mathcal{C}$. Since p is Cartesian, this equivalence lifts to a p-Cartesian edge $\widetilde{Y} \to X$ of S, so that $p(\widetilde{Y}) = Y$.

Since p is fully faithful, the map $\mathrm{Map}_{\mathcal{C}}(X, X') \to \mathrm{Map}_{\mathcal{D}}(p(X), p(X'))$ is a homotopy equivalence for any pair of objects $X, X' \in \mathcal{C}$. Suppose that $p(X) = p(X')$. Then, applying Proposition 2.4.4.2, we deduce that $\mathrm{Map}_{\mathcal{C}_{p(X)}}(X, X')$ is contractible. It follows that the ∞-category $\mathcal{C}_{p(X)}$ is nonempty with contractible morphism spaces; it is therefore a contractible Kan complex. Proposition 2.4.2.4 now implies that p is a right fibration. Since p has contractible fibers, it is a trivial fibration by Lemma 2.1.3.4. $\quad\square$

We have already seen that if an ∞-category S has an initial object, then that initial object is essentially unique. We now establish a relative version of this result.

Lemma 2.4.4.7. *Let* $p : \mathcal{C} \to \mathcal{D}$ *be a Cartesian fibration of ∞-categories and let C be an object of $\mathcal{C}$. Suppose that $D = p(C)$ is an initial object of $\mathcal{D}$ and that C is an initial object of the ∞-category $\mathcal{C}_D = \mathcal{C} \times_{\mathcal{D}} \{D\}$. Then C is an initial object of $\mathcal{C}$.*

Proof. Let C' be any object of $\mathcal{C}$ and set $D' = p(C')$. Since D is an initial object of $\mathcal{D}$, the space $\mathrm{Map}_{\mathcal{D}}(D, D')$ is contractible. In particular, there exists a morphism $f : D \to D'$ in $\mathcal{D}$. Let $\widetilde{f} : \widetilde{D} \to C'$ be a p-Cartesian lift of f. According to Proposition 2.4.4.2, there exists a fiber sequence in the homotopy category $\mathcal{H}$:

$$\mathrm{Map}_{\mathcal{C}_D}(C, \widetilde{D}) \to \mathrm{Map}_{\mathcal{C}}(C, C') \to \mathrm{Map}_{\mathcal{D}}(D, D').$$

Since the first and last spaces in this sequence are contractible, we deduce that $\mathrm{Map}_{\mathcal{C}}(C, C')$ is contractible as well, so that C is an initial object of $\mathcal{C}$. $\qquad\square$

Lemma 2.4.4.8. *Suppose we are given a diagram of simplicial sets*

$$
\begin{array}{ccc}
\partial \Delta^n & \xrightarrow{f_0} & X \\
\downarrow & \nearrow^{f} & \downarrow{p} \\
\Delta^n & \xrightarrow{g} & S,
\end{array}
$$

where p is a Cartesian fibration and $n > 0$. Suppose that $f_0(0)$ is an initial object of the ∞-category $X_{g(0)} = X \times_S \{g(0)\}$. Then there exists a map $f : \Delta^n \to S$ as indicated by the dotted arrow in the diagram, which renders the diagram commutative.

Proof. Pulling back via g, we may replace S by Δ^n and thereby reduce to the case where S is an ∞-category and $g(0)$ is an initial object of S. It follows from Lemma 2.4.4.7 that $f_0(v)$ is an initial object of S, which implies the existence of the desired extension f. □

Proposition 2.4.4.9. *Let $p : X \to S$ be a Cartesian fibration of simplicial sets. Assume that for each vertex s of S, the ∞-category $X_s = X \times_S \{s\}$ has an initial object.*

(1) *Let $X' \subseteq X$ denote the full simplicial subset of X spanned by those vertices x which are initial objects of $X_{p(x)}$. Then $p|X'$ is a trivial fibration of simplicial sets.*

(2) *Let $\mathcal{C} = \mathrm{Map}_S(S, X)$ be the ∞-category of sections of p. An arbitrary section $q : S \to X$ is an initial object of $\mathcal{C}$ if and only if q factors through X'.*

Proof. Since every fiber X_s has an initial object, the map $p|X'$ has the right lifting property with respect to the inclusion $\emptyset \subseteq \Delta^0$. If $n > 0$, then Lemma 2.4.4.8 shows that $p|X'$ has the right lifting property with respect to $\partial \Delta^n \subseteq \Delta^n$. This proves (1). In particular, we deduce that there exists a map $q : S \to X'$ which is a section of p. In view of the uniqueness of initial objects, (2) will follow if we can show that q is an initial object of $\mathcal{C}$. Unwinding the definitions, we must show that for $n > 0$, any lifting problem

$$
\begin{array}{ccc}
S \times \partial \Delta^n & \xrightarrow{f} & X \\
\downarrow & \nearrow & \downarrow{q} \\
S \times \Delta^n & \longrightarrow & S
\end{array}
$$

can be solved provided that $f|S \times \{0\} = q$. The desired extension can be constructed simplex by simplex using Lemma 2.4.4.8. □

2.4.5 Application: Invariance of Undercategories

Our goal in this section is to complete the proof of Proposition 1.2.9.3 by proving the following assertion:

(∗) Let $p : \mathcal{C} \to \mathcal{D}$ be an equivalence of ∞-categories and let $j : K \to \mathcal{C}$ be a diagram. Then the induced map

$$\mathcal{C}_{j/} \to \mathcal{D}_{pj/}$$

is a categorical equivalence.

We will need a lemma.

Lemma 2.4.5.1. *Let $p : \mathcal{C} \to \mathcal{D}$ be a fully faithful map of ∞-categories and let $j : K \to \mathcal{C}$ be any diagram in $\mathcal{C}$. Then, for any object x of $\mathcal{C}$, the map of Kan complexes*

$$\mathcal{C}_{j/} \times_{\mathcal{C}} \{x\} \to \mathcal{D}_{pj/} \times_{\mathcal{D}} \{p(x)\}$$

is a homotopy equivalence.

Proof. For any map $r : K' \to K$ of simplicial sets, let $C_r = \mathcal{C}_{jr/} \times_{\mathcal{C}} \{x\}$ and $D_r = \mathcal{D}_{pjr/} \times_{\mathcal{D}} \{p(x)\}$.

Choose a transfinite sequence of simplicial subsets K_α of K such that $K_{\alpha+1}$ is the result of adjoining a single nondegenerate simplex to K_α and $K_\lambda = \bigcup_{\alpha < \lambda} K_\alpha$ whenever λ is a limit ordinal (we include the case where $\lambda = 0$, so that $K_0 = \emptyset$). Let $i_\alpha : K_\alpha \to K$ denote the inclusion. We claim the following:

(1) For every ordinal α, the map $\phi_\alpha : C_{i_\alpha} \to D_{i_\alpha}$ is a homotopy equivalence of simplicial sets.

(2) For every pair of ordinals $\beta \leq \alpha$, the maps $C_{i_\alpha} \to C_{i_\beta}$ and $D_{i_\alpha} \to D_{i_\beta}$ are Kan fibrations of simplicial sets.

We prove both of these claims by induction on α. When $\alpha = 0$, (2) is obvious and (1) follows since both sides are isomorphic to Δ^0. If α is a limit ordinal, (2) is again obvious, while (1) follows from the fact that both C_{i_α} and D_{i_α} are obtained as the inverse limits of transfinite sequences of fibrations, and the map ϕ_α is an inverse limit of maps which are individually homotopy equivalences.

Assume that $\alpha = \beta + 1$ is a successor ordinal, so that $K_\alpha \simeq K_\beta \coprod_{\partial \Delta^n} \Delta^n$. Let $f : \Delta^n \to K_\alpha$ be the induced map, so that

$$C_{i_\alpha} = C_{i_\beta} \times_{C_{f|\partial \Delta^n}} C_f$$

$$D_{i_\alpha} = D_{i_\beta} \times_{D_{f|\partial \Delta^n}} D_f.$$

We note that the projections $C_f \to D_{f|\partial \Delta^n}$ and $C_f \to D_{f|\partial \Delta^n}$ are left fibrations by Proposition 2.1.2.1 and therefore Kan fibrations by Lemma 2.1.3.3. This proves (2) since the class of Kan fibrations is stable under pullback. We also note that the pullback diagrams defining X_{i_α} and Y_{i_α} are also homotopy pullback diagrams. Thus, to prove that ϕ_α is a homotopy equivalence, it suffices to show that ϕ_β and the maps

$$C_{f|\partial \Delta^n} \to D_{f|\partial \Delta^n}$$

$$C_f \to D_f$$

are homotopy equivalences. In other words, we may reduce to the case where K is a *finite* complex.

We now work by induction on the dimension of K. Suppose that the dimension of K is n and that the result is known for all simplicial sets having smaller dimensions. Running through the above argument again, we can reduce to the case where $K = \Delta^n$. Let v denote the final vertex of Δ^n. By Proposition 2.1.2.5, the maps

$$C_j \to C_{j|\{v\}}$$

$$D_j \to D_{j|\{v\}}$$

are trivial fibrations. Thus, it suffices to consider the case where K is a single point $\{v\}$. In this case, we have $C_j = \operatorname{Hom}_{\mathcal{C}}^{\mathrm{L}}(j(v), x)$ and $Y_j = \operatorname{Hom}_{\mathcal{D}}^{\mathrm{L}}(p(j(v)), p(x))$. It follows that the map ϕ is a homotopy equivalence since p is assumed to be fully faithful. $\qquad\square$

Proof of (∗). Let $p : \mathcal{C} \to \mathcal{D}$ be a categorical equivalence of ∞-categories and $j : K \to \mathcal{C}$ any diagram. We have a factorization

$$\mathcal{C}_{j/} \xrightarrow{f} \mathcal{D}_{pj/} \times_{\mathcal{D}} \mathcal{C} \xrightarrow{g} \mathcal{D}_{pj/}.$$

Lemma 2.4.5.1 implies that $\mathcal{C}_{j/}$ and $\mathcal{D}_{pj/} \times_{\mathcal{D}} \mathcal{C}$ are fiberwise equivalent left fibrations over $\mathcal{C}$, so that f is a categorical equivalence by Corollary 2.4.4.4 (we note that the map f automatically carries coCartesian edges to coCartesian edges because *all* edges of the target $\mathcal{D}_{pj/} \times_{\mathcal{D}} \mathcal{C}$ are coCartesian). The map g is a categorical equivalence by Corollary 2.4.4.5. It follows that $g \circ f$ is a categorical equivalence, as desired. $\qquad\square$

2.4.6 Application: Categorical Fibrations over a Point

Our main goal in this section is to prove the following result:

Theorem 2.4.6.1. *Let $\mathcal{C}$ be a simplicial set. Then $\mathcal{C}$ is fibrant for the Joyal model structure if and only if $\mathcal{C}$ is an ∞-category.*

The proof will require a few technical preliminaries.

Lemma 2.4.6.2. *Let $p : \mathcal{C} \to \mathcal{D}$ be a categorical equivalence of ∞-categories and let $m \geq 2$ be an integer. Suppose we are given maps $f_0 : \partial \Delta^{\{1,\dots,m\}} \to \mathcal{C}$ and $h_0 : \Lambda_0^m \to \mathcal{D}$ with $h_0| \partial \Delta^{\{1,\dots,m\}} = p \circ f_0$. Suppose further that the restriction of h to $\Delta^{\{0,1\}}$ is an equivalence in $\mathcal{D}$. Then there exist maps $f : \Delta^{\{1,\dots,m\}} \to \mathcal{C}$, $h : \Delta^m \to \mathcal{D}$ such that $h|\Delta^{\{1,\dots,n\}} = p \circ f$, $f_0 = f| \partial \Delta^{\{1,\dots,m\}}$, and $h_0 = h|\Lambda_0^m$.*

Proof. We may regard h_0 as a point of the simplicial set $\mathcal{D}_{/p \circ f_0}$. Since p is a categorical equivalence, Proposition 1.2.9.3 implies that $p' : \mathcal{C}_{/f_0} \to \mathcal{D}_{/p \circ f_0}$ is a categorical equivalence. It follows that h_0 lies in the essential image of p'. Consider the linearly ordered set $\{0 < 0' < 1 < \cdots < n\}$ and the corresponding simplex $\Delta^{\{0,0',\ldots,n\}}$. By hypothesis, we can extend f_0 to a map $f_0' : \Lambda_{0'}^{\{0',\ldots,m\}} \to \mathcal{C}$ and h_0 to a map $h_0' : \Delta^{\{0,0'\}} \star \partial\,\Delta^{\{1,\ldots,m\}} \to \mathcal{D}$ such that $h_0'|\Delta^{\{0,0'\}}$ is an equivalence and $h_0'|\Lambda_0^{\{0',\ldots,m\}} = p \circ f_0'$.

Since $h_0'|\Delta^{\{0,0'\}}$ and $h_0'|\Delta^{\{0,1\}}$ are both equivalences in $\mathcal{D}$, we deduce that $h_0'|\Delta^{\{0',1\}}$ is an equivalence in $\mathcal{D}$. Since p is a categorical equivalence, it follows that $f_0'|\Delta^{\{0',1\}}$ is an equivalence in $\mathcal{C}$. Proposition 1.2.4.3 implies that f_0' extends to a map $f' : \Delta^{\{0',\ldots,m\}} \to \mathcal{C}$. The union of $p \circ f'$ and h_0' determines a map $\Lambda_{0'}^{\{0,0',\ldots,m\}} \to \mathcal{D}$; since $\mathcal{D}$ is an ∞-category, this extends to a map $h' : \Delta^{\{0,0',\ldots,m\}} \to \mathcal{D}$. We may now take $f = f'|\Delta^{\{1,\ldots,m\}}$ and $h = h'|\Delta^m$. $\square$

Lemma 2.4.6.3. *Let $p : \mathcal{C} \to \mathcal{D}$ be a categorical equivalence of ∞-categories and let $A \subseteq B$ be an inclusion of simplicial sets. Let $f_0 : A \to \mathcal{C}$, $g : B \to \mathcal{D}$ be any maps and let $h_0 : A \times \Delta^1 \to \mathcal{D}$ be an equivalence from $g|A$ to $p \circ f_0$. Then there exists a map $f : B \to \mathcal{C}$ and an equivalence $h : B \times \Delta^1 \to \mathcal{D}$ from g to $p \circ f$ such that $f_0 = f|A$ and $h_0 = h|A \times \Delta^1$.*

Proof. Working simplex by simplex with the inclusion $A \subseteq B$, we may reduce to the case where $B = \Delta^n$, $A = \partial\,\Delta^n$. If $n = 0$, the existence of the desired extensions is a reformulation of the assumption that p is essentially surjective. Let us assume therefore that $n \geq 1$.

We consider the task of constructing $h : \Delta^n \times \Delta^1 \to \mathcal{D}$. Consider the filtration

$$X(n+1) \subseteq \cdots \subseteq X(0) = \Delta^n \times \Delta^1$$

described in the proof of Proposition 2.1.2.6. We note that the value of h on $X(n+1)$ is uniquely prescribed by h_0 and g. We extend the definition of h to $X(i)$ by descending induction on i. We note that $X(i) \simeq X(i+1) \coprod_{\Lambda_k^{n+1}} \Delta^{n+1}$. For $i > 0$, the existence of the required extension is guaranteed by the assumption that $\mathcal{D}$ is an ∞-category. Since $n \geq 1$, Lemma 2.4.6.2 allows us to extend h over the simplex σ_0 and to define f so that the desired conditions are satisfied. $\square$

Lemma 2.4.6.4. *Let $\mathcal{C} \subseteq \mathcal{D}$ be an inclusion of simplicial sets which is also a categorical equivalence. Suppose further that $\mathcal{C}$ is an ∞-category. Then $\mathcal{C}$ is a retract of $\mathcal{D}$.*

Proof. Enlarging $\mathcal{D}$ by an inner anodyne extension if necessary, we may suppose that $\mathcal{D}$ is an ∞-category. We now apply Lemma 2.4.6.3 in the case where $A = \mathcal{C}$, $B = \mathcal{D}$. $\square$

Proof of Theorem 2.4.6.1. The "only if" direction has already been established (Remark 2.2.5.5). For the converse, we must show that if $\mathcal{C}$ is an ∞-category, then $\mathcal{C}$ has the extension property with respect to every inclusion of simplicial sets $A \subseteq B$ which is a categorical equivalence. Fix any map $A \to \mathcal{C}$. Since the Joyal model structure is left proper, the inclusion

$\mathcal{C} \subseteq \mathcal{C} \coprod_A B$ is a categorical equivalence. We now apply Lemma 2.4.6.4 to conclude that $\mathcal{C}$ is a retract of $\mathcal{C} \coprod_A B$. □

We can state Theorem 2.4.6.1 as follows: if S is a point, then $p : X \to S$ is a categorical fibration (in other words, a fibration with respect to the Joyal model structure on $\mathcal{S}$) if and only if it is an inner fibration. However, the class of inner fibrations does *not* coincide with the class of categorical fibrations in general. The following result describes the situation when T is an ∞-category:

Corollary 2.4.6.5 (Joyal). *Let* $p : \mathcal{C} \to \mathcal{D}$ *be a map of simplicial sets, where* $\mathcal{D}$ *is an* ∞-category. *Then* p *is a categorical fibration if and only if the following conditions are satisfied:*

(1) *The map* p *is an inner fibration.*

(2) *For every equivalence* $f : D \to D'$ *in* $\mathcal{D}$ *and every object* $C \in \mathcal{C}$ *with* $p(C) = D$, *there exists an equivalence* $\overline{f} : C \to C'$ *in* $\mathcal{C}$ *with* $p(\overline{f}) = f$.

Proof. Suppose first that p is a categorical fibration. Then (1) follows immediately (since the inclusions $\Lambda^n_i \subseteq \Delta^n$ are categorical equivalences for $0 < i < n$). To prove (2), we let $\mathcal{D}^0$ denote the largest Kan complex contained in $\mathcal{D}$, so that the edge f belongs to $\mathcal{D}$. There exists a contractible Kan complex K containing an edge $\widetilde{f} : \widetilde{D} \to \widetilde{D}'$ and a map $q : K \to \mathcal{D}$ such that $q(\widetilde{f}) = f$. Since the inclusion $\{\widetilde{D}\} \subseteq K$ is a categorical equivalence, our assumption that p is a categorical fibration allows us to lift q to a map $\widetilde{q} : K \to \mathcal{C}$ such that $\widetilde{q}(\widetilde{D}) = C$. We can now take $\overline{f} = \widetilde{q}(\widetilde{f})$; since $\widetilde{f}$ is an equivalence in K, $\overline{f}$ is an equivalence in $\mathcal{C}$.

Now suppose that (1) and (2) are satisfied. We wish to show that p is a categorical fibration. Consider a lifting problem

$$
\begin{array}{ccc}
A & \xrightarrow{g_0} & \mathcal{C} \\
{\scriptstyle i}\downarrow & {\scriptstyle g}\nearrow & \downarrow{\scriptstyle p} \\
B & \xrightarrow{h} & \mathcal{D},
\end{array}
$$

where i is a cofibration and a categorical equivalence; we wish to show that there exists a morphism g as indicated which renders the diagram commutative. We first observe that condition (1), together with our assumption that $\mathcal{D}$ is an ∞-category, guarantees that $\mathcal{C}$ is an ∞-category. Applying Theorem 2.4.6.1, we can extend g_0 to a map $g' : B \to \mathcal{C}$ (not necessarily satisfying $h = p \circ g'$). The maps h and $p \circ g'$ have the same restriction to A. Let

$$
H_0 : (B \times \partial \Delta^1) \coprod_{A \times \partial \Delta^1} (A \times \Delta^1) \to \mathcal{D}
$$

be given by $(p \circ g', h)$ on $B \times \partial \Delta^1$ and by the composition

$$
A \times \Delta^1 \to A \subseteq B \xrightarrow{h} \mathcal{D}
$$

on $A \times \Delta^1$. Applying Theorem 2.4.6.1 once more, we deduce that H_0 extends to a map $H : B \times \Delta^1 \to \mathcal{D}$. The map H carries $\{a\} \times \Delta^1$ to an equivalence in $\mathcal{D}$ for every vertex a of A. Since the inclusion $A \subseteq B$ is a categorical equivalence, we deduce that H carries $\{b\} \times \Delta^1$ to an equivalence for every $b \in B$.

Let

$$G_0 : (B \times \{0\}) \coprod_{A \times \{0\}} (A \times \Delta^1) \to \mathcal{C}$$

be the composition of the projection to B with the map g'. We have a commutative diagram

$$
\begin{array}{ccc}
(B \times \{0\}) \coprod_{A \times \{0\}} (A \times \Delta^1) & \xrightarrow{\ G_0\ } & \mathcal{C} \\
\downarrow & \overset{G}{\nearrow} & \downarrow{\scriptstyle p} \\
B \times \Delta^1 & \xrightarrow{\ \ H\ \ } & \mathcal{D} \, .
\end{array}
$$

To complete the proof, it will suffice to show that we can supply a map G as indicated, rendering the diagram commutative; in this case, we can solve the original lifting problem by defining $g = G|B \times \{1\}$.

We construct the desired extension G working simplex by simplex on B. We start by applying assumption (2) to construct the map $G|\{b\} \times \Delta^1$ for every vertex b of B (that does not already belong to A); moreover, we ensure that $G|\{b\} \times \Delta^1$ is an equivalence in $\mathcal{C}$.

To extend G_0 to simplices of higher dimension, we encounter lifting problems of the type

$$
\begin{array}{ccc}
(\Delta^n \times \{0\}) \coprod_{\partial \Delta^n \times \{0\}} (\partial \Delta^n \times \Delta^1) & \xrightarrow{\ e\ } & \mathcal{C} \\
\downarrow & \nearrow & \downarrow{\scriptstyle p} \\
\Delta^n \times \Delta^1 & \xrightarrow{\quad\quad\quad} & \mathcal{D} \, .
\end{array}
$$

According to Proposition 2.4.1.8, these lifting problems can be solved provided that e carries $\{0\} \times \Delta^1$ to a p-coCartesian edge of $\mathcal{C}$. This follows immediately from Proposition 2.4.1.5. $\qquad\square$

2.4.7 Bifibrations

As we explained in §2.1.2, left fibrations $p : X \to S$ can be thought of as *covariant* functors from S into an ∞-category of spaces. Similarly, right fibrations $q : Y \to T$ can be thought of as *contravariant* functors from T into an ∞-category of spaces. The purpose of this section is to introduce a convenient formalism which encodes covariant and contravariant functoriality simultaneously.

Remark 2.4.7.1. The theory of bifibrations will not play an important role in the remainder of the book. In fact, the only result from this section that we will actually use is Corollary 2.4.7.12, whose statement makes no mention of bifibrations. A reader who is willing to take Corollary 2.4.7.12 on faith, or supply an alternative proof, may safely omit the material covered in this section.

Definition 2.4.7.2. Let S, T, and X be simplicial sets and let $p : X \to S \times T$ be a map. We shall say that p is a *bifibration* if it is an inner fibration having the following properties:

- For every $n \geq 1$ and every diagram of solid arrows

$$
\begin{array}{ccc}
\Lambda_0^n & \longrightarrow & X \\
\downarrow & \nearrow & \downarrow \\
\Delta^n & \xrightarrow{\ f\ } & S \times T
\end{array}
$$

  such that $\pi_T \circ f$ maps $\Delta^{\{0,1\}} \subseteq \Delta^n$ to a degenerate edge of T, there exists a dotted arrow as indicated, rendering the diagram commutative. Here π_T denotes the projection $S \times T \to T$.

- For every $n \geq 1$ and every diagram of solid arrows

$$
\begin{array}{ccc}
\Lambda_n^n & \longrightarrow & X \\
\downarrow & \nearrow & \downarrow \\
\Delta^n & \xrightarrow{\ f\ } & S \times T
\end{array}
$$

  such that $\pi_S \circ f$ maps $\Delta^{\{n-1,n\}} \subseteq \Delta^n$ to a degenerate edge of T, there exists a dotted arrow as indicated, rendering the diagram commutative. Here π_S denotes the projection $S \times T \to S$.

Remark 2.4.7.3. The condition that p be a bifibration is not a condition on p alone but also refers to a decomposition of the codomain of p as a product $S \times T$. We note also that the definition is not symmetric in S and T: instead, $p : X \to S \times T$ is a bifibration if and only if $p^{op} : X^{op} \to T^{op} \times S^{op}$ is a bifibration.

Remark 2.4.7.4. Let $p : X \to S \times T$ be a map of simplicial sets. If $T = *$, then p is a bifibration if and only if it is a *left* fibration. If $S = *$, then p is a bifibration if and only if it is a *right* fibration.

Roughly speaking, we can think of a bifibration $p : X \to S \times T$ as a bifunctor from $S \times T$ to an ∞-category of spaces; the functoriality is covariant in S and contravariant in T.

Lemma 2.4.7.5. *Let $p : X \to S \times T$ be a bifibration of simplicial sets. Suppose that S is an ∞-category. Then the composition $q = \pi_T \circ p$ is a Cartesian fibration of simplicial sets. Furthermore, an edge e of X is q-Cartesian if and only if $\pi_S(p(e))$ is an equivalence.*

Proof. The map q is an inner fibration because it is a composition of inner fibrations. Let us say that an edge $e : x \to y$ of X is *quasi-Cartesian* if $\pi_S(p(e))$ is degenerate in S. Let $y \in X_0$ be any vertex of X and let $\bar{e} : \bar{x} \to q(y)$ be an edge of S. The pair $(\bar{e}, s_0 q(y))$ is an edge of $S \times T$ whose projection to T is degenerate; consequently, it lifts to a (quasi-Cartesian) edge $e : x \to y$ in X. It is immediate from Definition 2.4.7.2 that any quasi-Cartesian edge of X is q-Cartesian. Thus q is a Cartesian fibration.

Now suppose that e is a q-Cartesian edge of X. Then e is equivalent to a quasi-Cartesian edge of X; it follows easily that $\pi_S(p(e))$ is an equivalence. Conversely, suppose that $e : x \to y$ is an edge of X and that $\pi_S(p(e))$ is an equivalence. We wish to show that e is q-Cartesian. Choose a quasi-Cartesian edge $e' : x' \to y$ with $q(e') = q(e)$. Since e' is q-Cartesian, there exists a simplex $\sigma \in X_2$ with $d_0\sigma = e'$, $d_1\sigma = e$, and $q(\sigma) = s_0 q(e)$. Let $f = d_2(\sigma)$, so that $\pi_S(p(e')) \circ \pi_S(p(f)) \simeq \pi_S p(e)$ in the ∞-category S. We note that f lies in the fiber $X_{q(x)}$, which is left fibered over S; since f maps to an equivalence in S, it is an equivalence in $X_{q(x)}$. Consequently, f is q-Cartesian, so that $e = e' \circ f$ is q-Cartesian as well. $\square$

Proposition 2.4.7.6. *Let $X \xrightarrow{p} Y \xrightarrow{q} S \times T$ be a diagram of simplicial sets. Suppose that q and $q \circ p$ are bifibrations and that p induces a homotopy equivalence $X_{(s,t)} \to Y_{(s,t)}$ of fibers over each vertex (s,t) of $S \times T$. Then p is a categorical equivalence.*

Proof. By means of a standard argument (see the proof of Proposition 2.2.2.7), we may reduce to the case where S and T are simplices; in particular, we may suppose that S and T are ∞-categories. Fix $t \in T_0$ and consider the map of fibers $p_t : X_t \to Y_t$. Both sides are left fibered over $S \times \{t\}$, so that p_t is a categorical equivalence by (the dual of) Corollary 2.4.4.4. We may then apply Corollary 2.4.4.4 again (along with the characterization of Cartesian edges given in Lemma 2.4.7.5) to deduce that p is a categorical equivalence. $\square$

Proposition 2.4.7.7. *Let $p : X \to S \times T$ be a bifibration, let $f : S' \to S$, $g : T' \to T$ be categorical equivalences between ∞-categories, and let $X' = X \times_{S \times T} (S' \times T')$. Then the induced map $X' \to X$ is a categorical equivalence.*

Proof. We will prove the result assuming that f is an isomorphism. A dual argument will establish the result when g is an isomorphism and applying the result twice we will deduce the desired statement for arbitrary f and g.

Given a map $i : A \to S$, let us say that i is *good* if the induced map $X \times_{S \times T} (A \times T') \to X \times_{S \times T} (A \times T')$ is a categorical equivalence. We wish to show that the identity map $S \to S$ is good; it will suffice to show that *all* maps $A \to S$ are good. Using the argument of Proposition 2.2.2.7, we can reduce to showing that every map $\Delta^n \to S$ is good. In other words, we may assume that $S = \Delta^n$, and in particular that S is an ∞-category. By Lemma 2.4.7.5, the projection $X \to T$ is a Cartesian fibration. The desired result now follows from Corollary 2.4.4.5. $\square$

We next prove an analogue of Lemma 2.4.6.3.

Lemma 2.4.7.8. *Let $X \xrightarrow{p} Y \xrightarrow{q} S \times T$ satisfy the hypotheses of Proposition 2.4.7.6. Let $A \subseteq B$ be a cofibration of simplicial sets over $S \times T$. Let $f_0 : A \to X$, $g : B \to Y$ be morphisms in $(\mathrm{Set}_\Delta)_{/S \times T}$ and let $h_0 : A \times \Delta^1 \to Y$ be a homotopy (again over $S \times T$) from $g|A$ to $p \circ f_0$.*

Then there exists a map $f : B \to X$ (of simplicial sets over $S \times T$) and a homotopy $h : B \times \Delta^1 \to T$ (over $S \times T$) from g to $p \circ f$ such that $f_0 = f|A$ and $h_0 = h|A \times \Delta^1$.

Proof. Working simplex by simplex with the inclusion $A \subseteq B$, we may reduce to the case where $B = \Delta^n$, $A = \partial \Delta^n$. If $n = 0$, we may invoke the fact that p induces a surjection $\pi_0 X_{(s,t)} \to \pi_0 Y_{(s,t)}$ on each fiber. Let us assume therefore that $n \geq 1$. Without loss of generality, we may pull back along the maps $B \to S$, $B \to T$ and reduce to the case where S and T are simplices.

We consider the task of constructing $h : \Delta^n \times \Delta^1 \to T$. We now employ the filtration

$$X(n+1) \subseteq \cdots \subseteq X(0)$$

described in the proof of Proposition 2.1.2.6. We note that the value of h on $X(n+1)$ is uniquely prescribed by h_0 and g. We extend the definition of h to $X(i)$ by descending induction on i. We note that $X(i) \simeq X(i+1) \coprod_{\Lambda_k^{n+1}} \Delta^{n+1}$. For $i > 0$, the existence of the required extension is guaranteed by the assumption that Y is inner-fibered over $S \times T$.

We note that, in view of the assumption that S and T are simplices, any extension of h over the simplex σ_0 is automatically a map *over $S \times T$*. Since S and T are ∞-categories, Proposition 2.4.7.6 implies that p is a categorical equivalence of ∞-categories; the existence of the desired extension of h (and the map f) now follows from Lemma 2.4.6.2. $\qquad\square$

Proposition 2.4.7.9. *Let $X \xrightarrow{p} Y \xrightarrow{q} S \times T$ satisfy the hypotheses of Proposition 2.4.7.6. Suppose that p is a cofibration. Then there exists a retraction $r : Y \to X$ (as a map of simplicial sets over $S \times T$) such that $r \circ p = \mathrm{id}_X$.*

Proof. Apply Lemma 2.4.7.8 in the case where $A = X$ and $B = Y$. $\qquad\square$

Let $q : \mathcal{M} \to \Delta^1$ be an inner fibration, which we view as a correspondence from $\mathcal{C} = q^{-1}\{0\}$ to $\mathcal{D} = q^{-1}\{1\}$. Evaluation at the endpoints of Δ^1 induces maps $\mathrm{Map}_{\Delta^1}(\Delta^1, \mathcal{M}) \to \mathcal{C}$, $\mathrm{Map}_{\Delta^1}(\Delta^1, \mathcal{M}) \to \mathcal{D}$.

Proposition 2.4.7.10. *For every inner fibration $q : \mathcal{M} \to \Delta^1$ as above, the map $p : \mathrm{Map}_{\Delta^1}(\Delta^1, \mathcal{M}) \to \mathcal{C} \times \mathcal{D}$ is a bifibration.*

Proof. We first show that p is an inner fibration. It suffices to prove that q has the right lifting property with respect to

$$(\Lambda_i^n \times \Delta^1) \coprod_{\Lambda_i^n \times \partial \Delta^1} (\Delta^n \times \partial \Delta^1) \subseteq \Delta^n \times \Delta^1$$

for any $0 < i < n$. But this is a smash product of $\partial \Delta^1 \subseteq \Delta^1$ with the inner anodyne inclusion $\Lambda_i^n \subseteq \Delta^n$.

To complete the proof that p is a bifibration, we verify that for every $n \geq 1$, $f_0 : \Lambda_n^0 \to X$, and $g : \Delta^n \to S \times T$ with $g|\Lambda_0^n = p \circ f_0$, if $(\pi_S \circ g)|\Delta^{\{0,1\}}$ is degenerate, then there exists $f : \Delta^n \to X$ with $g = p \circ f$ and $f_0 = f|\Lambda_0^n$. (The dual assertion, regarding extensions of maps $\Lambda_n^n \to X$, is verified in the same way.) The pair (f_0, g) may be regarded as a map

$$h_0 : (\Delta^n \times \{0,1\}) \coprod_{\Lambda_0^n \times \{0,1\}} (\Lambda_0^n \times \Delta^1) \to \mathcal{M},$$

and our goal is to prove that h_0 extends to a map $h : \Delta^n \times \Delta^1 \to \mathcal{M}$.

Let $\{\sigma_i\}_{0 \leq i \leq n}$ be the maximal-dimensional simplices of $\Delta^n \times \Delta^1$, as in the proof of Proposition 2.1.2.6. We set

$$K(0) = (\Delta^n \times \{0,1\}) \coprod_{\Lambda_0^n \times \{0,1\}} (\Lambda_0^n \times \Delta^1)$$

and, for $0 \leq i \leq n$, let $K(i+1) = K(i) \cup \sigma_i$. We construct maps $h_i : K_i \to \mathcal{M}$, with $h_i = h_{i+1}|K_i$, by induction on i. We note that for $i < n$, $K(i+1) \simeq K(i) \coprod_{\Lambda_{i+1}^{n+1}} \Delta^{n+1}$, so that the desired extension exists by virtue of the assumption that $\mathcal{M}$ is an ∞-category. If $i = n$, we have instead an isomorphism $\Delta^n \times \Delta^1 = K(n+1) \simeq K(n) \coprod_{\Lambda_0^{n+1}} \Delta^{n+1}$. The desired extension of h_n can be found using Proposition 1.2.4.3 because $h_0|\Delta^{\{0,1\}} \times \{0\}$ is an equivalence in $\mathcal{C} \subseteq \mathcal{M}$ by assumption. $\square$

Corollary 2.4.7.11. *Let $\mathcal{C}$ be an ∞-category. Evaluation at the endpoints gives a bifibration $\operatorname{Fun}(\Delta^1, \mathcal{C}) \to \mathcal{C} \times \mathcal{C}$.*

Proof. Apply Proposition 2.4.7.10 to the correspondence $\mathcal{C} \times \Delta^1$. $\square$

Corollary 2.4.7.12. *Let $f : \mathcal{C} \to \mathcal{D}$ be a functor between ∞-categories. The projection*

$$p : \operatorname{Fun}(\Delta^1, \mathcal{D}) \times_{\operatorname{Fun}(\{1\}, \mathcal{D})} \mathcal{C} \to \operatorname{Fun}(\{0\}, \mathcal{D})$$

is a Cartesian fibration. Moreover, a morphism of $\operatorname{Fun}(\Delta^1, \mathcal{D}) \times_{\operatorname{Fun}(\{1\}, \mathcal{D})} \mathcal{C}$ is p-Cartesian if and only if its image in $\mathcal{C}$ is an equivalence.

Proof. Combine Corollary 2.4.7.11 with Lemma 2.4.7.5. $\square$

Chapter Three

The ∞-Category of ∞-Categories

The power of category theory lies in its role as a unifying language for mathematics: nearly every class of mathematical structures (groups, manifolds, algebraic varieties, and so on) can be organized into a category. This language is somewhat inadequate in situations where the structures need to be classified up to some notion of equivalence less rigid than isomorphism. For example, in algebraic topology one wishes to study topological spaces up to homotopy equivalence; in homological algebra one wishes to study chain complexes up to quasi-isomorphism. Both of these examples are most naturally described in terms of higher category theory (for example, the theory of ∞-categories developed in this book).

Another source of examples arises in category theory itself. In classical category theory, it is generally regarded as unnatural to ask whether two categories are isomorphic; instead, one asks whether or not they are equivalent. The same phenomenon arises in higher category theory. Throughout this book, we generally regard two ∞-categories $\mathcal{C}$ and $\mathcal{D}$ as the same if they are categorically equivalent, even if they are not isomorphic to one another as simplicial sets. In other words, we are not interested in the *ordinary* category of ∞-categories (a full subcategory of Set_Δ) but in an underlying ∞-category which we now define.

Definition 3.0.0.1. The simplicial category $\mathcal{C}\mathrm{at}_\infty^\Delta$ is defined as follows:

(1) The objects of $\mathcal{C}\mathrm{at}_\infty^\Delta$ are (small) ∞-categories.

(2) Given ∞-categories $\mathcal{C}$ and $\mathcal{D}$, we define $\mathrm{Map}_{\mathcal{C}\mathrm{at}_\infty^\Delta}(\mathcal{C}, \mathcal{D})$ to be the largest Kan complex contained in the ∞-category $\mathrm{Fun}(\mathcal{C}, \mathcal{D})$.

We let $\mathcal{C}\mathrm{at}_\infty$ denote the simplicial nerve $\mathrm{N}(\mathcal{C}\mathrm{at}_\infty^\Delta)$. We will refer to $\mathcal{C}\mathrm{at}_\infty$ as the ∞-*category of (small)* ∞-*categories.*

Remark 3.0.0.2. By construction, $\mathcal{C}\mathrm{at}_\infty$ arises as the nerve of a simplicial category $\mathcal{C}\mathrm{at}_\infty^\Delta$, where composition is strictly associative. This is one advantage of working with ∞-categories: the correct notion of functor is encoded by simply considering maps of simplicial sets (rather than homotopy coherent diagrams, say), so there is no difficulty in composing them.

Remark 3.0.0.3. The mapping spaces in $\mathcal{C}\mathrm{at}_\infty^\Delta$ are Kan complexes, so that $\mathcal{C}\mathrm{at}_\infty$ is an ∞-category (Proposition 1.1.5.10) as suggested by the terminology.

Remark 3.0.0.4. By construction, the objects of $\mathcal{C}at_\infty$ are ∞-categories, morphisms are given by functors, and 2-morphisms are given by *homotopies* between functors. In other words, $\mathcal{C}at_\infty$ discards all information about non-invertible natural transformations between functors. If necessary, we could retain this information by forming an ∞-*bicategory* of (small) ∞-categories. We do not wish to become involved in any systematic discussion of ∞-bicategories, so we will be content to consider only $\mathcal{C}at_\infty$.

Our goal in this chapter is to study the ∞-category $\mathcal{C}at_\infty$. For example, we would like to show that $\mathcal{C}at_\infty$ admits limits and colimits. There are two approaches to proving this assertion. We can attack the problem directly by giving an explicit construction of the limits and colimits in question: see §3.3.3 and §3.3.4. Alternatively, we can try to realize $\mathcal{C}at_\infty$ as the ∞-category underlying a (simplicial) model category $\mathbf{A}$ and deduce the existence of limits and colimits in $\mathcal{C}at_\infty$ from the existence of homotopy limits and homotopy colimits in $\mathbf{A}$ (Corollary 4.2.4.8). The objects of $\mathcal{C}at_\infty$ can be identified with the fibrant-cofibrant objects of Set_Δ with respect to the Joyal model structure. However, we cannot apply Corollary 4.2.4.8 directly because the Joyal model structure on Set_Δ is not compatible with the (usual) simplicial structure. We will remedy this difficulty by introducing the category Set_Δ^+ of *marked* simplicial sets. We will explain how to endow Set_Δ^+ with the structure of a *simplicial* model category in such a way that there is an equivalence of simplicial categories $\mathcal{C}at_\infty^\Delta \simeq (\mathrm{Set}_\Delta^+)^\circ$. This will allow us to identify $\mathcal{C}at_\infty$ with the ∞-category underlying Set_Δ^+, so that Corollary 4.2.4.8 can be invoked.

We will introduce the formalism of marked simplicial sets in §3.1. In particular, we will explain the construction of a model structure not only on Set_Δ^+ itself but also for the category $(\mathrm{Set}_\Delta^+)_{/S}$ of marked simplicial sets *over* a given simplicial set S. The fibrant objects of $(\mathrm{Set}_\Delta^+)_{/S}$ can be identified with Cartesian fibrations $X \to S$, which we can think of as contravariant functors from S into $\mathcal{C}at_\infty$. In §3.2, we will justify this intuition by introducing the *straightening* and *unstraightening* functors which will allow us to pass back and forth between Cartesian fibrations over S and functors from S^{op} to $\mathcal{C}at_\infty$. This correspondence has applications both to the study of Cartesian fibrations and to the study of the ∞-category $\mathcal{C}at_\infty$; we will survey some of these applications in §3.3.

Remark 3.0.0.5. In the later chapters of this book, it will be necessary to undertake a systematic study of ∞-categories which are not small. For this purpose, we introduce the following notational conventions: $\mathcal{C}at_\infty$ will denote the simplicial nerve of the category of *small* ∞-categories, while $\widehat{\mathcal{C}at_\infty}$ denotes the simplicial nerve of the category of ∞-categories which are not necessarily small.

3.1 MARKED SIMPLICIAL SETS

The Joyal model structure on Set_Δ is a powerful tool in the study of ∞-categories. Nevertheless, in *relative* situations it is somewhat inconvenient. Roughly speaking, a categorical fibration $p : X \to S$ determines a family of ∞-categories X_s parametrized by the vertices s of S. However, we are generally more interested in those cases where X_s can be regarded as a functor of s. As we explained in §2.4.2, this naturally translates into the assumption that p is a Cartesian (or coCartesian) fibration. According to Proposition 3.3.1.7, every Cartesian fibration is a categorical fibration, but the converse is false. Consequently, it is natural to try to endow $(\mathrm{Set}_\Delta)_{/S}$ with some *other* model structure in which the fibrant objects are precisely the Cartesian fibrations over S.

Unfortunately, this turns out to be an unreasonable demand. In order to have a model category, we need to be able to form fibrant replacements: in other words, we need the ability to enlarge an arbitrary map $p : X \to S$ into a commutative diagram

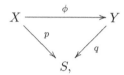

where q is a Cartesian fibration *generated by* p. A question arises: for which edges f of X should $\phi(f)$ be a q-Cartesian edge of Y? This sort of information is needed for the construction of Y; consequently, we need a formalism in which certain edges of X have been distinguished.

Definition 3.1.0.1. A *marked simplicial set* is a pair $(X, \mathcal{E})$, where X is a simplicial set and $\mathcal{E}$ is a set of edges of X which contains every degenerate edge. We will say that an edge of X will be called *marked* if it belongs to $\mathcal{E}$.

A morphism $f : (X, \mathcal{E}) \to (X', \mathcal{E}')$ of marked simplicial sets is a map $f : X \to X'$ having the property that $f(\mathcal{E}) \subseteq \mathcal{E}'$. The category of marked simplicial sets will be denoted by Set_Δ^+.

Every simplicial set S may be regarded as a marked simplicial set, usually in many different ways. The two extreme cases deserve special mention: if S is a simplicial set, we let $S^\sharp = (S, S_1)$ denote the marked simplicial set in which *every* edge of S has been marked and let $S^\flat = (S, s_0(S_0))$ denote the marked simplicial set in which only the degenerate edges of S have been marked.

Notation 3.1.0.2. Let S be a simplicial set. We let $(\mathrm{Set}_\Delta^+)_{/S}$ denote the category of marked simplicial sets equipped with a map to S (which might otherwise be denoted as $(\mathrm{Set}_\Delta^+)_{/S^\sharp}$).

Our goal in this section is to study the theory of marked simplicial sets and, in particular, to endow each $(\mathrm{Set}_\Delta^+)_{/S}$ with the structure of a model category. We will begin in §3.1.1 by introducing the notion of a *marked anodyne* morphism in Set_Δ^+. In §3.1.2, we will establish a basic stability property of the class of marked anodyne maps, which implies the stability of Cartesian fibrations under exponentiation (Proposition 3.1.2.1). In §3.1.3, we will introduce the *Cartesian model structure* on $(\mathrm{Set}_\Delta^+)_{/S}$ for every simplicial set S. In §3.1.4, we will study these model categories; in particular, we will see that each $(\mathrm{Set}_\Delta^+)_{/S}$ is a *simplicial* model category whose fibrant objects are precisely the Cartesian fibrations $X \to S$ (with Cartesian edges of X marked). Finally, we will conclude with §3.1.5, where we compare the Cartesian model structure on $(\mathrm{Set}_\Delta^+)_{/S}$ with other model structures considered in this book (such as the Joyal and contravariant model structures).

3.1.1 Marked Anodyne Morphisms

In this section, we will introduce the class of *marked anodyne* morphisms in Set_Δ^+. The definition is chosen so that the condition that a map $\overline{X} \to \overline{S}$ have the right lifting property with respect to all marked anodyne morphisms is closely related to the condition that the underlying map of simplicial sets $X \to S$ be a Cartesian fibration (we refer the reader to Proposition 3.1.1.6 for a more precise statement). The theory of marked anodyne maps is a technical device which will prove useful when we discuss the Cartesian model structure in §3.1.3: every marked anodyne morphism is a trivial cofibration with respect to the Cartesian model structure, but not conversely. In this respect, the class of marked anodyne morphisms of Set_Δ^+ is analogous to the class of inner anodyne morphisms of Set_Δ.

Definition 3.1.1.1. The class of *marked anodyne* morphisms in Set_Δ^+ is the smallest weakly saturated (see §A.1.2) class of morphisms with the following properties:

(1) For each $0 < i < n$, the inclusion $(\Lambda_i^n)^\flat \subseteq (\Delta^n)^\flat$ is marked anodyne.

(2) For every $n > 0$, the inclusion

$$(\Lambda_n^n, \mathcal{E} \cap (\Lambda_n^n)_1) \subseteq (\Delta^n, \mathcal{E})$$

is marked anodyne, where $\mathcal{E}$ denotes the set of all degenerate edges of Δ^n together with the final edge $\Delta^{\{n-1,n\}}$.

(3) The inclusion

$$(\Lambda_1^2)^\sharp \coprod_{(\Lambda_1^2)^\flat} (\Delta^2)^\flat \to (\Delta^2)^\sharp$$

is marked anodyne.

(4) For every Kan complex K, the map $K^\flat \to K^\sharp$ is marked anodyne.

Remark 3.1.1.2. The definition of a marked simplicial set is self-dual. However, Definition 3.1.1.1 is not self-dual: if $A \to B$ is marked anodyne, then the opposite morphism $A^{op} \to B^{op}$ need not be marked anodyne. This reflects the fact that the theory of Cartesian fibrations is not self-dual.

Remark 3.1.1.3. In part (4) of Definition 3.1.1.1, it suffices to allow K to range over a set of representatives for all isomorphism classes of Kan complexes with only countably many simplices. Consequently, we deduce that the class of marked anodyne morphisms in Set_Δ^+ is of small generation, so that the small object argument applies (see §A.1.2). We will refine this observation further: see Corollary 3.1.1.8 below.

Remark 3.1.1.4. In Definition 3.1.1.1, we are free to replace (1) by

(1′) For every inner anodyne map $A \to B$ of simplicial sets, the induced map $A^\flat \to B^\flat$ is marked anodyne.

Proposition 3.1.1.5. *Consider the following classes of morphisms in* Set_Δ^+:

(2) *All inclusions*

$$(\Lambda_n^n, \mathcal{E} \cap (\Lambda_n^n)_1) \subseteq (\Delta^n, \mathcal{E}),$$

where $n > 0$ and $\mathcal{E}$ denotes the set of all degenerate edges of Δ^n together with the final edge $\Delta^{\{n-1,n\}}$.

(2′) *All inclusions*

$$((\partial\,\Delta^n)^\flat \times (\Delta^1)^\sharp) \coprod_{(\partial\,\Delta^n)^\flat \times \{1\}^\sharp} ((\Delta^n)^\flat \times \{1\}^\sharp) \subseteq (\Delta^n)^\flat \times (\Delta^1)^\sharp.$$

(2″) *All inclusions*

$$(A^\flat \times (\Delta^1)^\sharp) \coprod_{A^\flat \times \{1\}^\sharp} (B^\flat \times \{1\}^\sharp) \subseteq B^\flat \times (\Delta^1)^\sharp,$$

where $A \subseteq B$ is an inclusion of simplicial sets.

The classes (2′) and (2″) generate the same weakly saturated class of morphisms of Set_Δ^+ which contains the weakly saturated class generated by (2). Conversely, the weakly saturated class of morphisms generated by (1) and (2) from Definition 3.1.1.1 contains (2′) and (2″).

Proof. To see that each of the morphisms specified in (2″) is contained in the weakly saturated class generated by (2′), it suffices to work simplex by simplex with the inclusion $A \subseteq B$. The converse is obvious since the class of morphisms of type (2′) is contained in the class of morphisms of type (2″). To see that the weakly saturated class generated by (2″) contains (2), it suffices to show that every morphism in (2) is a retract of a morphism in (2″). For this, we consider maps

$$\Delta^n \xrightarrow{j} \Delta^n \times \Delta^1 \xrightarrow{r} \Delta^n.$$

Here j is the composition of the identification $\Delta^n \simeq \Delta^n \times \{0\}$ with the inclusion $\Delta^n \times \{0\} \subseteq \Delta^n \times \Delta^1$, and r may be identified with the map of partially ordered sets

$$r(m,i) = \begin{cases} n & \text{if } m = n-1, i = 1 \\ m & \text{otherwise.} \end{cases}$$

Now we simply observe that j and r exhibit the inclusion

$$(\Lambda_n^n, \mathcal{E} \cap (\Lambda_n^n)_0) \subseteq (\Delta^n, \mathcal{E})$$

as a retract of

$$((\Lambda_n^n)^\flat \times (\Delta^1)^\sharp) \coprod_{(\Lambda_n^n)^\flat \times \{1\}^\sharp} ((\Delta^n)^\flat \times \{1\}^\sharp) \subseteq (\Delta^n)^\flat \times (\Delta^1)^\sharp.$$

To complete the proof, we must show that each of the inclusions

$$((\partial \Delta^n)^\flat \times (\Delta^1)^\sharp) \coprod_{(\partial \Delta^n)^\flat \times \{1\}^\sharp} ((\Delta^n)^\flat \times \{1\}^\sharp) \subseteq (\Delta^n)^\flat \times (\Delta^1)^\sharp$$

of type $(2')$ belongs to the weakly saturated class generated by (1) and (2). To see this, consider the filtration

$$Y_{n+1} \subseteq \cdots \subseteq Y_0 = \Delta^n \times \Delta^1$$

which is the *opposite* of the filtration defined in the proof of Proposition 2.1.2.6. We let $\mathcal{E}_i$ denote the class of all edges of Y_i which are marked in $(\Delta^n)^\flat \times (\Delta^1)^\sharp$. It will suffice to show that each inclusion $f_i : (Y_{i+1}, \mathcal{E}_{i+1}) \subseteq (Y_i, \mathcal{E}_i)$ lies in the weakly saturated class generated by (1) and (2). For $i \neq 0$, the map f_i is a pushout of $(\Lambda_{n+1-i}^{n+1})^\flat \subseteq (\Delta^{n+1})^\flat$. For $i = 0$, f_i is a pushout of

$$(\Lambda_{n+1}^{n+1}, \mathcal{E} \cap (\Lambda_{n+1}^{n+1})_1) \subseteq (\Delta^{n+1}, \mathcal{E}),$$

where $\mathcal{E}$ denotes the set of all degenerate edges of Δ^{n+1}, together with $\Delta^{\{n,n+1\}}$. $\square$

We now characterize the class of marked anodyne maps:

Proposition 3.1.1.6. *A map $p : X \to S$ in Set_Δ^+ has the right lifting property with respect to all marked anodyne maps if and only if the following conditions are satisfied:*

(A) *The map p is an inner fibration of simplicial sets.*

(B) *An edge e of X is marked if and only if $p(e)$ is marked and e is p-Cartesian.*

(C) *For every object y of X and every marked edge $\bar{e} : \bar{x} \to p(y)$ in S, there exists a marked edge $e : x \to y$ of X with $p(e) = \bar{e}$.*

Proof. We first prove the "only if" direction. Suppose that p has the right lifting property with respect to all marked anodyne maps. By considering maps of the form (1) from Definition 3.1.1.1, we deduce that (A) holds. Considering (2) in the case $n = 0$, we deduce that (C) holds. Considering (2) for $n > 0$, we deduce that every marked edge of X is p-Cartesian. For the converse, let us suppose that $e : x \to y$ is a p-Cartesian edge of X and that $p(e)$ is marked in S. Invoking (C), we deduce that there exists a marked edge $e' : x' \to y$ with $p(e) = p(e')$. Since e' is Cartesian, we can find a 2-simplex σ of X with $d_0(\sigma) = e'$, $d_1(\sigma) = e$, and $p(\sigma) = s_1 p(e)$. Then $d_2(\sigma)$ is an equivalence between x and x' in the ∞-category $X_{p(x)}$. Let K denote the largest Kan complex contained in $X_{p(x)}$. Since p has the right lifting property with respect to $K^\flat \to K^\sharp$, we deduce that every edge of K is marked; in particular, $d_2(\sigma)$ is marked. Since p has the right lifting property with respect to the morphism described in (3) of Definition 3.1.1.1, we deduce that $d_1(\sigma) = e$ is marked.

Now suppose that p satisfies the hypotheses of the proposition. We must show that p has the right lifting property with respect to the classes of morphisms (1), (2), (3), and (4) of Definition 3.1.1.1. For (1), this follows from the assumption that p is an inner fibration. For (2), this follows from (C) and from the assumption that every marked edge is p-Cartesian. For (3), we are free to replace S by $(\Delta^2)^\sharp$; then p is a Cartesian fibration over an ∞-category S, and we can apply Proposition 2.4.1.7 to deduce that the class of p-Cartesian edges is stable under composition.

Finally, for (4), we may replace S by $K^\sharp$; then S is a Kan complex and p is a Cartesian fibration, so the p-Cartesian edges of X are precisely the equivalences in X. Since K is a Kan complex, any map $K \to X$ carries the edges of K to equivalences in X. □

By Quillen's small object argument, we deduce that a map $j : A \to B$ in Set_Δ^+ is marked anodyne if and only if it has the left lifting property with respect to all morphisms $p : X \to S$ satisfying the hypotheses of Proposition 3.1.1.6. From this, we deduce:

Corollary 3.1.1.7. *The inclusion*

$$ i : (\Lambda_2^2)^\sharp \coprod_{(\Lambda_2^2)^\flat} (\Delta^2)^\flat \hookrightarrow (\Delta^2)^\sharp $$

is marked anodyne.

Proof. It will suffice to show that i has the left lifting property with respect to any of the morphisms $p : X \to S$ described in Proposition 3.1.1.6. Without loss of generality, we may replace S by $(\Delta^2)^\sharp$; we now apply Proposition 2.4.1.7. □

The following somewhat technical corollary will be needed in §3.1.3:

Corollary 3.1.1.8. *In Definition 3.1.1.1, we can replace the class of morphisms (4) by*

(4') *the map $j : A^\flat \to (A, s_0 A_0 \cup \{f\})$, where A is the quotient of Δ^3 which corepresents the functor*

$$ \mathrm{Hom}_{\mathrm{Set}_\Delta}(A, X) = \{\sigma \in X_3, e \in X_1 : d_1\sigma = s_0 e, d_2\sigma = s_1 e\} $$

and $f \in A_1$ is the image of $\Delta^{\{0,1\}} \subseteq \Delta^3$ in A.

Proof. We first show that for every Kan complex K, the map $i : K^\flat \to K^\sharp$ lies in the weakly saturated class of morphisms generated by $(4')$. We note that i can be obtained as an iterated pushout of morphisms having the form $K^\flat \to (K, s_0 K_0 \cup \{e\})$, where e is an edge of K. It therefore suffices to show that there exists a map $p : A \to K$ such that $p(f) = e$. In other words, we must prove that there exists a 3-simplex $\sigma : \Delta^3 \to K$ with $d_1 \sigma = s_0 e$ and $d_2 \sigma = s_1 e$. This follows immediately from the Kan extension condition.

To complete the proof, it will suffice to show that the map j is marked anodyne. To do so, it suffices to prove that for any diagram

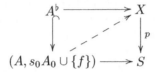

for which p satisfies the conditions of Proposition 3.1.1.6, there exists a dotted arrow as indicated, rendering the diagram commutative. This is simply a reformulation of Proposition 2.4.1.13. $\square$

Definition 3.1.1.9. Let $p : X \to S$ be a Cartesian fibration of simplicial sets. We let $X^\natural$ denote the marked simplicial set $(X, \mathcal{E})$, where $\mathcal{E}$ is the set of p-Cartesian edges of X.

Remark 3.1.1.10. Our notation is slightly abusive because $X^\natural$ depends not only on X but also on the map $X \to S$.

Remark 3.1.1.11. According to Proposition 3.1.1.6, a map $(Y, \mathcal{E}) \to S^\sharp$ has the right lifting property with respect to all marked anodyne maps if and only if the underlying map $Y \to S$ is a Cartesian fibration and $(Y, \mathcal{E}) = Y^\natural$.

We conclude this section with the following easy result, which will be needed later:

Proposition 3.1.1.12. *Let $p : X \to S$ be an inner fibration of simplicial sets, let $f : A \to B$ be a marked anodyne morphism in Set_Δ^+, let $q : B \to X$ be map of simplicial sets which carries each marked edge of B to a p-Cartesian edge of X, and set $q_0 = q \circ f$. Then the induced map*

$$X_{/q} \to X_{/q_0} \times_{S_{/pq_0}} S_{/pq}$$

is a trivial fibration of simplicial sets.

Proof. It is easy to see that the class of all morphisms f of Set_Δ^+ which satisfy the desired conclusion is weakly saturated. It therefore suffices to prove that this class contains a collection of generators for the weakly saturated class of marked anodyne morphisms. If f induces a left anodyne map on the underlying simplicial sets, then the desired result is automatic. It therefore suffices to consider the case where f is the inclusion

$$(\Lambda_n^n, \mathcal{E} \cap (\Lambda_n^n)_1) \subseteq (\Delta^n, \mathcal{E})$$

as described in (2) of Definition 3.1.1.1. In this case, a lifting problem

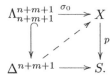

can be reformulated as an equivalent lifting problem

$$\Lambda^{n+m+1}_{n+m+1} \xrightarrow{\sigma_0} X$$
$$\downarrow \qquad \qquad \downarrow p$$
$$\Delta^{n+m+1} \longrightarrow S.$$

This lifting problem admits a solution since the hypothsis on q guarantees that σ_0 carries $\Delta^{\{n+m,n+m+1\}}$ to a p-Cartesian edge of X. □

3.1.2 Stability Properties of Marked Anodyne Morphisms

Our main goal in this section is to prove the following stability result:

Proposition 3.1.2.1. *Let $p : X \to S$ be a Cartesian fibration of simplicial sets and let K be an arbitrary simplicial set. Then*

(1) *The induced map $p^K : X^K \to S^K$ is a Cartesian fibration.*

(2) *An edge $\Delta^1 \to X^K$ is p^K-Cartesian if and only if, for every vertex k of K, the induced edge $\Delta^1 \to X$ is p-Cartesian.*

We could easily have given an ad hoc proof of this result in §2.4.3. However, we have opted instead to give a proof using the language of marked simplicial sets.

Definition 3.1.2.2. A morphism $(X, \mathcal{E}) \to (X', \mathcal{E}')$ in Set^+_Δ is a *cofibration* if the underlying map $X \to X'$ of simplicial sets is a cofibration.

The main ingredient we will need to prove Proposition 3.1.2.1 is the following:

Proposition 3.1.2.3. *The class of marked anodyne maps in Set^+_Δ is stable under smash products with arbitrary cofibrations. In other words, if $f : X \to X'$ is marked anodyne and $g : Y \to Y'$ is a cofibration, then the induced map*

$$(X \times Y') \coprod_{X \times Y} (X' \times Y) \to X' \times Y'$$

is marked anodyne.

Proof. The argument is tedious but straightforward. Without loss of generality we may suppose that f belongs either to the class $(2')$ of Proposition 3.1.1.5, or to one of the classes specified in (1), (3), or (4) of Definition 3.1.1.1. The class of cofibrations is generated by the inclusions $(\partial \Delta^n)^\flat \subseteq (\Delta^n)^\flat$ and $(\Delta^1)^\flat \subseteq (\Delta^1)^\sharp$; thus we may suppose that $g : Y \to Y'$ is one of these maps. There are eight cases to consider:

(A1) Let f be the inclusion $(\Lambda_i^n)^\flat \subseteq (\Delta^n)^\flat$ and let g be the inclusion $(\partial \Delta^n)^\flat \to (\Delta^n)^\flat$, where $0 < i < n$. Since the class of inner anodyne maps between simplicial sets is stable under smash products with inclusions, the smash product of f and g is marked anodyne (see Remark 3.1.1.4).

(A2) Let f be the inclusion $(\Lambda_i^n)^\flat \to (\Delta^n)^\flat$, and let g be the map $(\Delta^1)^\flat \to (\Delta^1)^\sharp$, where $0 < i < n$. Then the smash product of f and g is an isomorphism (since Λ_i^n contains all vertices of Δ^n).

(B1) Let f be the inclusion

$$(\{1\}^\sharp \times (\Delta^n)^\flat) \coprod_{\{1\}^\sharp \times (\partial \Delta^n)^\flat} ((\Delta^1)^\sharp \times (\partial \Delta^n)^\flat) \subseteq (\Delta^1)^\sharp \times (\Delta^n)^\flat$$

and let g be the inclusion $(\partial \Delta^n)^\flat \to (\Delta^n)^\flat$. Then the smash product of f and g belongs to the class $(2'')$ of Proposition 3.1.1.5.

(B2) Let f be the inclusion

$$(\{1\}^\sharp \times (\Delta^n)^\flat) \coprod_{\{1\}^\sharp \times (\partial \Delta^n)^\flat} ((\Delta^1)^\sharp \times (\partial \Delta^n)^\flat) \subseteq (\Delta^1)^\sharp \times (\Delta^n)^\flat$$

and let g denote the map $(\Delta^1)^\flat \to (\Delta^1)^\sharp$. If $n > 0$, then the smash product of f and g is an isomorphism. If $n = 0$, then the smash product may be identified with the map $(\Delta^1 \times \Delta^1, \mathcal{E}) \to (\Delta^1 \times \Delta^1)^\sharp$, where $\mathcal{E}$ consists of all degenerate edges together with $\{0\} \times \Delta^1$, $\{1\} \times \Delta^1$, and $\Delta^1 \times \{1\}$. This map may be obtained as a composition of two marked anodyne maps: the first is of type (3) in Definition 3.1.1.1 (adjoining the "diagonal" edge to $\mathcal{E}$), and the second is the map described in Corollary 3.1.1.7 (adjoining the edge $\Delta^1 \times \{0\}$ to $\mathcal{E}$).

(C1) Let f be the inclusion

$$(\Lambda_1^2)^\sharp \coprod_{(\Lambda_1^2)^\flat} (\Delta^2)^\flat \to (\Delta^2)^\sharp$$

and let g be the inclusion $(\partial \Delta^n)^\flat \subseteq (\Delta^n)^\flat$. Then the smash product of f and g is an isomorphism for $n > 0$ and is isomorphic to f for $n = 0$.

(C2) Let f be the inclusion

$$(\Lambda_1^2)^\sharp \coprod_{(\Lambda_1^2)^\flat} (\Delta^2)^\flat \to (\Delta^2)^\sharp$$

and let g be the canonical map $(\Delta^1)^\flat \to (\Delta^1)^\sharp$. Then the smash product of f and g is a pushout of the map f.

(D1) Let f be the map $K^\flat \to K^\sharp$, where K is a Kan complex, and let g be the inclusion $(\partial \Delta^n)^\flat \subseteq (\Delta^n)^\flat$. Then the smash product of f and g is an isomorphism for $n > 0$, and isomorphic to f for $n = 0$.

(D2) Let f be the map $K^\flat \to K^\sharp$, where K is a Kan complex, and let g be the map $(\Delta^1)^\flat \to (\Delta^1)^\sharp$. The smash product of f and g can be identified with the inclusion

$$(K \times \Delta^1, \mathcal{E}) \subseteq (K \times \Delta^1)^\sharp,$$

where $\mathcal{E}$ denotes the class of all edges $e = (e', e'')$ of $K \times \Delta^1$ for which either $e' : \Delta^1 \to K$ or $e'' : \Delta^1 \to \Delta^1$ is degenerate. This inclusion can be obtained as a transfinite composition of pushouts of the map

$$(\Lambda_1^2)^\sharp \coprod_{(\Lambda_1^2)^\flat} (\Delta^2)^\flat \to (\Delta^2)^\sharp.$$

$\square$

We now return to our main objective:

Proof of Proposition 3.1.2.1. Since p is a Cartesian fibration, it induces a map $X^\natural \to S^\sharp$ which has the right lifting property with respect to all marked anodyne maps. By Proposition 3.1.2.3, the induced map

$$(X^\natural)^{K^\flat} \to (S^\sharp)^{K^\flat} = (S^K)^\sharp$$

has the right lifting property with respect to all marked anodyne morphisms. The desired result now follows from Remark 3.1.1.11. $\square$

3.1.3 The Cartesian Model Structure

Let S be a simplicial set. Our goal in this section is to introduce the *Cartesian model structure* on the category $(\mathrm{Set}_\Delta^+)_{/S}$ of marked simplicial sets over S. We will eventually show that the fibrant objects of $(\mathrm{Set}_\Delta^+)_{/S}$ correspond precisely to Cartesian fibrations $X \to S$ and that they encode (contravariant) functors from S into the ∞-category Cat_∞.

The category Set_Δ^+ is *Cartesian-closed*; that is, for any two objects $X, Y \in \mathrm{Set}_\Delta^+$, there exists an internal mapping object Y^X equipped with an "evaluation map" $Y^X \times X \to Y$ which induces bijections

$$\mathrm{Hom}_{\mathrm{Set}_\Delta^+}(Z, Y^X) \to \mathrm{Hom}_{\mathrm{Set}_\Delta^+}(Z \times X, Y)$$

for every $Z \in \mathrm{Set}_\Delta^+$. We let $\mathrm{Map}^\flat(X, Y)$ denote the underlying simplicial set of Y^X and $\mathrm{Map}^\sharp(X, Y) \subseteq \mathrm{Map}^\flat(X, Y)$ the simplicial subset consisting of all simplices $\sigma \subseteq \mathrm{Map}^\flat(X, Y)$ such that every edge of σ is a marked edge of Y^X. Equivalently, we may describe these simplicial sets by the mapping properties

$$\mathrm{Hom}_{\mathrm{Set}_\Delta}(K, \mathrm{Map}^\flat(X, Y)) \simeq \mathrm{Hom}_{\mathrm{Set}_\Delta^+}(K^\flat \times X, Y)$$

$$\mathrm{Hom}_{\mathrm{Set}_\Delta}(K, \mathrm{Map}^\sharp(X, Y)) \simeq \mathrm{Hom}_{\mathrm{Set}_\Delta^+}(K^\sharp \times X, Y).$$

If X and Y are objects of $(\mathrm{Set}_\Delta^+)_{/S}$, then we let $\mathrm{Map}_S^\sharp(X, Y) \subseteq \mathrm{Map}^\sharp(X, Y)$ and $\mathrm{Map}_S^\flat(X, Y) \subseteq \mathrm{Map}^\flat(X, Y)$ denote the simplicial subsets classifying those maps which are compatible with the projections to S.

Remark 3.1.3.1. If $X \in (\mathrm{Set}_\Delta^+)_{/S}$ and $p : Y \to S$ is a Cartesian fibration, then $\mathrm{Map}_S^\flat(X, Y^\natural)$ is an ∞-category and $\mathrm{Map}_S^\sharp(X, Y^\natural)$ is the largest Kan complex contained in $\mathrm{Map}_S^\flat(X, Y^\natural)$.

Lemma 3.1.3.2. *Let $f : \mathcal{C} \to \mathcal{D}$ be a functor between ∞-categories. The following are equivalent:*

(1) *The functor f is a categorical equivalence.*

(2) *For every simplicial set K, the induced map $\mathrm{Fun}(K, \mathcal{C}) \to \mathrm{Fun}(K, \mathcal{D})$ is a categorical equivalence.*

(3) *For every simplicial set K, the functor $\mathrm{Fun}(K, \mathcal{C}) \to \mathrm{Fun}(K, \mathcal{D})$ induces a homotopy equivalence from the largest Kan complex contained in $\mathrm{Fun}(K, \mathcal{C})$ to the largest Kan complex contained in $\mathrm{Fun}(K, \mathcal{D})$.*

Proof. The implications $(1) \Rightarrow (2) \Rightarrow (3)$ are obvious. Suppose that (3) is satisfied. Let $K = \mathcal{D}$. According to (3), there exists an object x of $\mathrm{Fun}(K, \mathcal{C})$ whose image in $\mathrm{Fun}(K, \mathcal{D})$ is equivalent to the identity map $K \to \mathcal{D}$. We may identify x with a functor $g : \mathcal{D} \to \mathcal{C}$ having the property that $f \circ g$ is homotopic to the identity $\mathrm{id}_\mathcal{D}$. It follows that g also has the property asserted by (3), so the same argument shows that there is a functor $f' : \mathcal{C} \to \mathcal{D}$ such that $g \circ f'$ is homotopic to $\mathrm{id}_\mathcal{C}$. It follows that $f \circ g \circ f'$ is homotopic to both f and f', so that f is homotopic to f'. Thus g is a homotopy inverse to f, which proves that f is an equivalence. $\square$

Proposition 3.1.3.3. *Let S be a simplicial set and let $p : X \to Y$ be a morphism in $(\mathrm{Set}_\Delta^+)_{/S}$. The following are equivalent:*

(1) *For every Cartesian fibration $Z \to S$, the induced map*

$$\mathrm{Map}_S^\flat(Y, Z^\natural) \to \mathrm{Map}_S^\flat(X, Z^\natural)$$

is an equivalence of ∞-categories.

(2) *For every Cartesian fibration $Z \to S$, the induced map*

$$\mathrm{Map}_S^\sharp(Y, Z^\natural) \to \mathrm{Map}_S^\sharp(X, Z^\natural)$$

is a homotopy equivalence of Kan complexes.

Proof. Since $\mathrm{Map}_S^\sharp(M, Z^\natural)$ is the largest Kan complex contained in the ∞-category $\mathrm{Map}_S^\flat(M, Z^\natural)$, it is clear that (1) implies (2). Suppose that (2) is satisfied and let $Z \to S$ be a Cartesian fibration. We wish to show that

$$\mathrm{Map}_S^\flat(Y, Z^\natural) \to \mathrm{Map}_S^\flat(X, Z^\natural)$$

is an equivalence of ∞-categories. According to Lemma 3.1.3.2, it suffices to show that

$$\mathrm{Map}_S^\flat(Y, Z^\natural)^K \to \mathrm{Map}_S^\flat(X, Z^\natural)^K$$

induces a homotopy equivalence on the maximal Kan complexes contained in each side. Let $Z(K) = Z^K \times_{S^K} S$. Proposition 3.1.2.1 implies that $Z(K) \to S$ is a Cartesian fibration and that there is a natural identification

$$\mathrm{Map}_S^\flat(M, Z(K)^\natural) \simeq \mathrm{Map}_S^\flat(M, Z(K)^\natural).$$

We observe that the largest Kan complex contained in the right hand side is $\mathrm{Map}_S^\sharp(M, Z(K)^\natural)$. On the other hand, the natural map

$$\mathrm{Map}_S^\sharp(Y, Z(K)^\natural) \to \mathrm{Map}_S^\sharp(X, Z(K)^\natural)$$

is a homotopy equivalence by assumption (2).   □

We will say that a map $X \to Y$ in $(\mathcal{S}\mathrm{et}_\Delta^+)_{/S}$ is a *Cartesian equivalence* if it satisfies the equivalent conditions of Proposition 3.1.3.3.

Remark 3.1.3.4. Let $f : X \to Y$ be a morphism in $(\mathcal{S}\mathrm{et}_\Delta^+)_{/S}$ which is *marked anodyne* when regarded as a map of marked simplicial sets. Since the smash product of f with any inclusion $A^\flat \subseteq B^\flat$ is also marked anodyne, we deduce that the map

$$\phi : \mathrm{Map}_S^\flat(Y, Z^\natural) \to \mathrm{Map}_S^\flat(X, Z^\natural)$$

is a trivial fibration for *every* Cartesian fibration $Z \to S$. Consequently, f is a Cartesian equivalence.

Let S be a simplicial set and let $X, Y \in (\mathcal{S}\mathrm{et}_\Delta^+)_{/S}$. We will say a pair of morphisms $f, g : X \to Y$ are *strongly homotopic* if there exists a contractible Kan complex K and a map $K \to \mathrm{Map}_S^\flat(X, Y)$ whose image contains both of the vertices f and g. If $Y = Z^\natural$, where $Z \to S$ is a Cartesian fibration, then this simply means that f and g are equivalent when viewed as objects of the ∞-category $\mathrm{Map}_S^\flat(X, Y)$.

Proposition 3.1.3.5. *Let $X \xrightarrow{p} Y \xrightarrow{q} S$ be a diagram of simplicial sets, where both q and $q \circ p$ are Cartesian fibrations. The following assertions are equivalent:*

(1) *The map p induces a Cartesian equivalence $X^\natural \to Y^\natural$ in $(\mathcal{S}\mathrm{et}_\Delta^+)_{/S}$.*

(2) *There exists a map $r : Y \to X$ which is a strong homotopy inverse to p, in the sense that $p \circ r$ and $r \circ p$ are both strongly homotopic to the identity.*

(3) *The map p induces a categorical equivalence $X_s \to Y_s$ for each vertex s of S.*

Proof. The equivalence between (1) and (2) is easy, as is the assertion that (2) implies (3). It therefore suffices to show that (3) implies (2). We will construct r and a homotopy from $r \circ p$ to the identity. It then follows that the map r satisfies (3), so the same argument will show that r has a right homotopy inverse; by general nonsense this right homotopy inverse will automatically be homotopic to p, and the proof will be complete.

Choose a transfinite sequence of simplicial subsets $S(\alpha) \subseteq S$, where each $S(\alpha)$ is obtained from $\bigcup_{\beta < \alpha} S(\beta)$ by adjoining a single nondegenerate simplex (if such a simplex exists). We construct $r_\alpha : Y \times_S S(\alpha) \to X$ and an equivalence $h_\alpha : (X \times_S S(\alpha)) \times \Delta^1 \to X \times_S S(\alpha)$ from $r_\alpha \circ p$ to the identity by induction on α. By this device we may reduce to the case where $S = \Delta^n$, and the maps

$$r^0 : Y' \to X$$

$$h^0 : X' \times \Delta^1 \to X$$

are already specified, where $Y' = Y \times_{\Delta^n} \partial \Delta^n \subseteq Y$ and $X' = X \times_{\Delta^n} \partial \Delta^n \subseteq X$. We may regard r' and h' together as defining a map $\psi_0 : Z' \to X$, where

$$Z' = Y' \coprod_{X' \times \{0\}} (X' \times \Delta^1) \coprod_{X' \times \{1\}} X.$$

Let $Z = Y \coprod_{X \times \{0\}} X \times \Delta^1$; then our goal is to solve the lifting problem depicted in the diagram

in such a way that ψ carries $\{x\} \times \Delta^1$ to an equivalence in X for every object x of X. We note that this last condition is vacuous for $n > 0$.

If $n = 0$, the problem amounts to constructing a map $Y \to X$ which is homotopy inverse to p: this is possible in view of the assumption that p is a categorical equivalence. For $n > 0$, we note that any map $\phi : Z \to X$ extending ϕ_0 is automatically compatible with the projection to S (since S is a simplex and Z' contains all vertices of Z). Since the inclusion $Z' \subseteq Z$ is a cofibration between cofibrant objects in the model category Set_Δ (with the Joyal model structure) and X is a ∞-category (since q is an inner fibration and Δ^n is an ∞-category), Proposition A.2.3.1 asserts that it is sufficient to show that the extension ϕ exists up to homotopy. Since Corollary 2.4.4.4 implies that p is an equivalence, we are free to replace the inclusion $Z' \subseteq Z$ with the weakly equivalent inclusion

$$(X \times \{1\}) \coprod_{X \times_{\Delta^n} \partial \Delta^n \times \Delta^1} (X \times_{\Delta^n} \partial \Delta^n \times \{1\}) \subseteq X \times \Delta^1.$$

Since ϕ_0 carries $\{x\} \times \Delta^1$ to a $(q \circ p)$-Cartesian edge of X, for every vertex x of X, the existence of ϕ follows from Proposition 3.1.1.5.   □

Lemma 3.1.3.6. *Let S be a simplicial set, let $i : X \to Y$ be a cofibration in $(\mathrm{Set}_\Delta^+)_{/S}$, and let $Z \to S$ be a Cartesian fibration. Then the associated map $p : \mathrm{Map}_S^\sharp(Y, Z^\natural) \to \mathrm{Map}_S^\sharp(X, Z^\natural)$ is a Kan fibration.*

Proof. Let $A \subseteq B$ be an anodyne inclusion of simplicial sets. We must show that p has the right lifting property with respect to p. Equivalently, we must show that $Z^\flat \to S$ has the right lifting property with respect to the inclusion

$$(B^\sharp \times X) \coprod_{A^\sharp \times X} (A^\sharp \times Y) \subseteq B^\sharp \times Y.$$

This follows from Proposition 3.1.2.3 since the inclusion $A^\sharp \subseteq B^\sharp$ is marked anodyne. ☐

Proposition 3.1.3.7. *Let S be a simplicial set. There exists a left proper combinatorial model structure on $(\mathrm{Set}_\Delta^+)_{/S}$ which may be described as follows:*

(C) *The cofibrations in $(\mathrm{Set}_\Delta^+)_{/S}$ are those morphisms $p : X \to Y$ in $(\mathrm{Set}_\Delta^+)_{/S}$ which are cofibrations when regarded as morphisms of simplicial sets.*

(W) *The weak equivalences in $(\mathrm{Set}_\Delta^+)_{/S}$ are the Cartesian equivalences.*

(F) *The fibrations in $(\mathrm{Set}_\Delta^+)_{/S}$ are those maps which have the right lifting property with respect to every map which is simultaneously a cofibration and a Cartesian equivalence.*

Proof. It suffices to show that the hypotheses of Proposition A.2.6.13 are satisfied by the class (C) of cofibrations and the class (W).

(1) The class (W) of Cartesian equivalences is perfect (in the sense of Definition A.2.6.10). To prove this, we first observe that the class of marked anodyne maps is generated by the classes of morphisms (1), (2), and (3) of Definition 3.1.1.1 and class (4′) of Corollary 3.1.1.8. By Proposition A.1.2.5, there exists a functor T from $(\mathrm{Set}_\Delta^+)_{/S}$ to itself and a (functorial) factorization

$$X \xrightarrow{i_X} T(X) \xrightarrow{j_X} S^\sharp,$$

where i_X is marked anodyne (and therefore a Cartesian equivalence) and j_X has the right lifting property with respect to all marked anodyne maps and therefore corresponds to a Cartesian fibration over S. Moreover, the functor T commutes with filtered colimits. According to Proposition 3.1.3.5, a map $X \to Y$ in $(\mathrm{Set}_\Delta^+)_{/S}$ is a Cartesian equivalence if and only if, for each vertex $s \in S$, the induced map $T(X)_s \to T(Y)_s$ is a categorical equivalence. It follows from Corollary A.2.6.12 that (W) is a perfect class of morphisms.

(2) The class of weak equivalences is stable under pushouts by cofibrations. Suppose we are given a pushout diagram

$$
\begin{array}{ccc}
X & \xrightarrow{\ p\ } & Y \\
\downarrow{\scriptstyle i} & & \downarrow \\
X' & \xrightarrow{\ p'\ } & Y'
\end{array}
$$

where i is a cofibration and p is a Cartesian equivalence. We wish to show that p' is also a Cartesian equivalence. In other words, we must show that for any Cartesian fibration $Z \to S$, the associated map $\mathrm{Map}_S^{\sharp}(Y', Z^{\natural}) \to \mathrm{Map}_S^{\sharp}(X', Z^{\natural})$ is a homotopy equivalence. Consider the pullback diagram

$$
\begin{array}{ccc}
\mathrm{Map}_S^{\sharp}(Y', Z^{\natural}) & \longrightarrow & \mathrm{Map}_S^{\sharp}(X', Z^{\natural}) \\
\downarrow & & \downarrow \\
\mathrm{Map}_S^{\sharp}(Y, Z^{\natural}) & \longrightarrow & \mathrm{Map}_S^{\sharp}(X, Z^{\natural}).
\end{array}
$$

Since p is a Cartesian equivalence, the bottom horizontal arrow is a homotopy equivalence. According to Lemma 3.1.3.6, the right vertical arrow is a Kan fibration; it follows that the diagram is homotopy Cartesian, so that the top horizontal arrow is an equivalence as well.

(3) A map $p : X \to Y$ in $(\mathrm{Set}_{\Delta}^+)_{/S}$ which has the right lifting property with respect to every map in (C) belongs to (W). Unwinding the definition, we see that p is a trivial fibration of simplicial sets and that an edge e of X is marked if and only if $p(e)$ is a marked edge of Y. It follows that p has a section s with $s \circ p$ fiberwise homotopic to id_X. From this, we deduce easily that p is a Cartesian equivalence.

$\square$

Warning 3.1.3.8. Let S be a simplicial set. We must be careful to distinguish between *Cartesian fibrations* of simplicial sets (in the sense of Definition 2.4.2.1) and fibrations with respect to the Cartesian model structure on $(\mathrm{Set}_{\Delta}^+)_{/S}$ (in the sense of Proposition 3.1.3.7). Though distinct, these notions are closely related: for example, the fibrant objects of $(\mathrm{Set}_{\Delta}^+)_{/S}$ are precisely those objects of the form $X^{\natural}$, where $X \to S$ is a Cartesian fibration (Proposition 3.1.4.1).

Remark 3.1.3.9. The definition of the Cartesian model structure on the category $(\mathrm{Set}_{\Delta}^+)_{/S}$ is not self-opposite. Consequently, we can define another model structure on $(\mathrm{Set}_{\Delta}^+)_{/S}$ as follows:

(C) The cofibrations in $(\mathrm{Set}_{\Delta}^+)_{/S}$ are precisely the monomorphisms.

(W) The weak equivalences in $(\mathrm{Set}_{\Delta}^+)_{/S}$ are precisely the *coCartesian equivalences*: that is, those morphisms $f : \overline{X} \to \overline{Y}$ such that the induced map $f^{op} : \overline{X}^{op} \to \overline{Y}^{op}$ is a Cartesian equivalence in $(\mathrm{Set}_{\Delta}^+)_{/S^{op}}$.

(F) The fibrations in $(\mathrm{Set}_{\Delta}^+)_{/S}$ are those morphisms which have the right lifting property with respect to every morphism satisfying both (C) and (W).

We will refer to this model structure on $(\mathrm{Set}_{\Delta}^+)_{/S}$ as the *coCartesian model structure*.

3.1.4 Properties of the Cartesian Model Structure

In this section, we will establish some of the basic properties of Cartesian model structures on $(\mathrm{Set}^+_\Delta)_{/S}$ which was introduced in §3.1.3. In particular, we will show that each $(\mathrm{Set}^+_\Delta)_{/S}$ is a *simplicial* model category and characterize its fibrant objects.

Proposition 3.1.4.1. *An object $X \in (\mathrm{Set}^+_\Delta)_{/S}$ is fibrant (with respect to the Cartesian model structure) if and only if $X \simeq Y^\natural$, where $Y \to S$ is a Cartesian fibration.*

Proof. Suppose first that X is fibrant. The small object argument implies that there exists a marked anodyne map $j : X \to Z^\natural$ for some Cartesian fibration $Z \to S$. Since j is marked anodyne, it is a Cartesian equivalence. Since X is fibrant, it has the extension property with respect to the trivial cofibration j; thus X is a retract of $Z^\natural$. It follows that X is isomorphic to $Y^\natural$, where Y is a retract of Z.

Now suppose that $Y \to S$ is a Cartesian fibration; we claim that $Y^\natural$ has the right lifting property with respect to any trivial cofibration $j : A \to B$ in $(\mathrm{Set}^+_\Delta)_{/S}$. Since j is a Cartesian equivalence, the map $\eta : \mathrm{Map}^\sharp_S(B, Y^\natural) \to \mathrm{Map}^\sharp_S(A, Y^\natural)$ is a homotopy equivalence of Kan complexes. Hence, for any map $f : A \to Z^\natural$, there is a map $g : B \to Z^\natural$ such that $g|A$ and f are joined by an edge e of $\mathrm{Map}^\sharp_S(A, Z^\natural)$. Let $M = (A \times (\Delta^1)^\sharp) \coprod_{A \times \{1\}^\sharp} (B \times \{1\}^\sharp) \subseteq B \times (\Delta^1)^\sharp$. We observe that e and g together determine a map $M \to Z^\natural$. Consider the diagram

$$
\begin{array}{ccc}
M & \longrightarrow & Z^\natural \\
\downarrow & \nearrow^{\displaystyle F} & \downarrow \\
B \times (\Delta^1)^\sharp & \longrightarrow & S^\sharp.
\end{array}
$$

The left vertical arrow is marked anodyne by Proposition 3.1.2.3. Consequently, there exists a dotted arrow F as indicated. We note that $F|B \times \{0\}$ is an extension of f to B, as desired. □

We now study the behavior of the Cartesian model structures with respect to products.

Proposition 3.1.4.2. *Let S and T be simplicial sets and let Z be an object of $(\mathrm{Set}^+_\Delta)_{/T}$. Then the functor*

$$(\mathrm{Set}^+_\Delta)_{/S} \to (\mathrm{Set}^+_\Delta)_{/S \times T}$$

$$X \mapsto X \times Z$$

preserves Cartesian equivalences.

Proof. Let $f : X \to Y$ be a Cartesian equivalence in $(\mathrm{Set}^+_\Delta)_{/S}$. We wish to show that $f \times \mathrm{id}_Z$ is a Cartesian equivalence in $(\mathrm{Set}^+_\Delta)_{/S \times T}$. Let $X \to X'$

be a marked anodyne map, where $X' \in (\mathrm{Set}_\Delta^+)_{/S}$ is fibrant. Now choose a marked anodyne map $X' \coprod_X Y \to Y'$, where $Y' \in (\mathrm{Set}_\Delta^+)_{/S}$ is fibrant. Since the product maps $X \times Z \to X' \times Z$ and $Y \times Z \to Y' \times Z$ are also marked anodyne (by Proposition 3.1.2.3), it suffices to show that $X' \times Z \to Y' \times Z$ is a Cartesian equivalence. In other words, we may reduce to the situation where X and Y are fibrant. By Proposition 3.1.3.5, f has a homotopy inverse g; then $g \times \mathrm{id}_Y$ is a homotopy inverse to $f \times \mathrm{id}_Y$. $\square$

Corollary 3.1.4.3. *Let $f : A \to B$ be a cofibration in $(\mathrm{Set}_\Delta^+)_{/S}$ and $f' : A' \to B'$ a cofibration in $(\mathrm{Set}_\Delta^+)_{/T}$. Then the smash product map*

$$(A \times B') \coprod_{A \times B} (A' \times B) \to A' \times B'$$

is a cofibration in $(\mathrm{Set}_\Delta^+)_{/S \times T}$, which is trivial if either f or g is trivial.

Corollary 3.1.4.4. *Let S be a simplicial set and regard $(\mathrm{Set}_\Delta^+)_{/S}$ as a simplicial category with mapping objects given by $\mathrm{Map}_S^\sharp(X, Y)$. Then $(\mathrm{Set}_\Delta^+)_{/S}$ is a simplicial model category.*

Proof. Unwinding the definitions, we are reduced to proving the following: given a cofibration $i : X \to X'$ in $(\mathrm{Set}_\Delta^+)_{/S}$ and a cofibration $j : Y \to Y'$ in Set_Δ, the induced cofibration

$$(X' \times Y^\sharp) \coprod_{X \times Y^\sharp} (X \times Y'^\sharp) \subseteq X' \times Y'^\sharp$$

in $(\mathrm{Set}_\Delta^+)_{/S}$ is trivial if either i is a Cartesian equivalence or j is a weak homotopy equivalence. If i is trivial, this follows immediately from Corollary 3.1.4.3. If j is trivial, the same argument applies provided that we can verify that $Y^\sharp \to Y'^\sharp$ is a Cartesian equivalence in Set_Δ^+. Unwinding the definitions, we must show that for every ∞-category Z, the restriction map

$$\theta : \mathrm{Map}^\sharp(Y'^\sharp, Z^\natural) \to \mathrm{Map}^\sharp(Y^\sharp, Z^\natural)$$

is a homotopy equivalence of Kan complexes. Let K be the largest Kan complex contained in Z, so that θ can be identified with the restriction map

$$\mathrm{Map}_{\mathrm{Set}_\Delta}(Y', K) \to \mathrm{Map}_{\mathrm{Set}_\Delta}(Y, K).$$

Since j is a weak homotopy equivalence, this map is a trivial fibration. $\square$

Remark 3.1.4.5. There is a second simplicial structure on $(\mathrm{Set}_\Delta^+)_{/S}$, where the simplicial mapping spaces are given by $\mathrm{Map}_S^\flat(X, Y)$. This simplicial structure is *not* compatible with the Cartesian model structure: for fixed $X \in (\mathrm{Set}_\Delta^+)_{/S}$ the functor

$$A \mapsto A^\flat \times X$$

does not carry weak homotopy equivalences (in the A-variable) to Cartesian equivalences. It does, however, carry *categorical* equivalences (in A) to Cartesian equivalences, and consequently $(\mathrm{Set}_\Delta^+)_{/S}$ is endowed with the structure of a Set_Δ-enriched model category, where we regard Set_Δ as equipped with the Joyal model structure. This second simplicial structure reflects the fact that $(\mathrm{Set}_\Delta^+)_{/S}$ is really a model for an ∞-bicategory.

Remark 3.1.4.6. Suppose S is a Kan complex. A map $p : X \to S$ is a Cartesian fibration if and only if it is a coCartesian fibration (this follows in general from Proposition 3.3.1.8; if $S = \Delta^0$, the main case of interest for us, it is obvious). Moreover, the class of p-coCartesian edges of X coincides with the class of p-Cartesian edges of X: both may be described as the class of equivalences in X. Consequently, if $A \in (\mathrm{Set}_{\Delta}^{+})_{/S}$, then

$$\mathrm{Map}_{S}^{\flat}(A, X^{\natural}) \simeq \mathrm{Map}_{S^{op}}^{\flat}(A^{op}, (X^{op})^{\natural})^{op},$$

where A^{op} is regarded as a marked simplicial set in the obvious way. It follows that a map $A \to B$ is a Cartesian equivalence in $(\mathrm{Set}_{\Delta}^{+})_{/S}$ if and only if $A^{op} \to B^{op}$ is a Cartesian equivalence in $(\mathrm{Set}_{\Delta}^{+})_{/S^{op}}$. In other words, the Cartesian model structure on $(\mathrm{Set}_{\Delta}^{+})_{/S}$ is self-dual *when S is a Kan complex*. In particular, if $S = \Delta^0$, we deduce that the functor

$$A \mapsto A^{op}$$

determines an autoequivalence of the model category $\mathrm{Set}_{\Delta}^{+} \simeq (\mathrm{Set}_{\Delta}^{+})_{/\Delta^0}$.

3.1.5 Comparison of Model Categories

Let S be a simplicial set. We now have a plethora of model structures on categories of simplicial sets over S:

(0) Let $\mathcal{C}_0$ denote the category $(\mathrm{Set}_{\Delta})_{/S}$ of simplicial sets over S endowed with the *Joyal* model structure defined in §2.2.5: the cofibrations are monomorphisms of simplicial sets, and the weak equivalences are categorical equivalences.

(1) Let $\mathcal{C}_1$ denote the category $(\mathrm{Set}_{\Delta}^{+})_{/S}$ of marked simplicial sets over S endowed with the *marked* model structure of Proposition 3.1.3.7: the cofibrations are maps $(X, \mathcal{E}_X) \to (Y, \mathcal{E}_Y)$ which induce monomorphisms $X \to Y$, and the weak equivalences are the Cartesian equivalences.

(2) Let $\mathcal{C}_2$ denote the category $(\mathrm{Set}_{\Delta}^{+})_{/S}$ of marked simplicial sets over S endowed with the following *localization* of the Cartesian model structure: a map $f : (X, \mathcal{E}_X) \to (Y, \mathcal{E}_Y)$ is a cofibration if the underlying map $X \to Y$ is a monomorphism, and a weak equivalence if $f : X^{\sharp} \to Y^{\sharp}$ is a marked equivalence in $(\mathrm{Set}_{\Delta}^{+})_{/S}$.

(3) Let $\mathcal{C}_3$ denote the category $(\mathrm{Set}_{\Delta})_{/S}$ of simplicial sets over S, which is endowed with the *contravariant* model structure described in §2.1.4: the cofibrations are the monomorphisms, and the weak equivalences are the contravariant equivalences.

(4) Let $\mathcal{C}_4$ denote the category $(\mathrm{Set}_{\Delta})_{/S}$ of simplicial sets over S endowed with the usual homotopy-theoretic model structure: the cofibrations are the monomorphisms of simplicial sets, and the weak equivalences are the weak homotopy equivalences of simplicial sets.

The goal of this section is to study the relationship between these five model categories. We may summarize the situation as follows:

Theorem 3.1.5.1. *There exists a sequence of Quillen adjunctions*

$$\mathcal{C}_0 \xrightarrow{F_0} \mathcal{C}_1 \xrightarrow{F_1} \mathcal{C}_2 \xrightarrow{F_2} \mathcal{C}_3 \xrightarrow{F_3} \mathcal{C}_4$$

$$\mathcal{C}_0 \xleftarrow{G_0} \mathcal{C}_1 \xleftarrow{G_1} \mathcal{C}_2 \xleftarrow{G_2} \mathcal{C}_3 \xleftarrow{G_3} \mathcal{C}_4,$$

which may be described as follows:

(A0) *The functor G_0 is the forgetful functor from $(\mathrm{Set}_{\Delta}^+)/S$ to $(\mathrm{Set}_{\Delta})/S$, which ignores the collection of marked edges. The functor F_0 is the left adjoint to G_0, which is given by $X \mapsto X^{\flat}$. The Quillen adjunction (F_0, G_0) is a Quillen equivalence if S is a Kan complex.*

(A1) *The functors F_1 and G_1 are the identity functors on $(\mathrm{Set}_{\Delta}^+)/S$.*

(A2) *The functor F_2 is the forgetful functor from $(\mathrm{Set}_{\Delta}^+)/S$ to $(\mathrm{Set}_{\Delta})/S$ which ignores the collection of marked edges. The functor G_2 is the right adjoint to F_2, which is given by $X \mapsto X^{\sharp}$. The Quillen adjunction (F_2, G_2) is a Quillen equivalence for every simplicial set S.*

(A3) *The functors F_3 and G_3 are the identity functors on $(\mathrm{Set}_{\Delta}^+)/S$. The Quillen adjunction (F_3, G_3) is a Quillen equivalence whenever S is a Kan complex.*

The rest of this section is devoted to giving a proof of Theorem 3.1.5.1. We will organize our efforts as follows. First, we verify that the model category $\mathcal{C}_2$ is well-defined (the analogous results for the other model structures have already been established). We then consider each of the adjunctions (F_i, G_i) in turn and show that it has the desired properties.

Proposition 3.1.5.2. *Let S be a simplicial set. There exists a left proper combinatorial model structure on the category $(\mathrm{Set}_{\Delta}^+)/S$ which may be described as follows:*

(C) *A map $f : (X, \mathcal{E}_X) \to (Y, \mathcal{E}_Y)$ is a cofibration if and only if the underlying map $X \to Y$ is a monomorphism of simplicial sets.*

(W) *A map $f : (X, \mathcal{E}_X) \to (Y, \mathcal{E}_Y)$ is a weak equivalence if and only if the induced map $X^{\sharp} \to Y^{\sharp}$ is a Cartesian equivalence in $(\mathrm{Set}_{\Delta}^+)/S$.*

(F) *A map $f : (X, \mathcal{E}_X) \to (Y, \mathcal{E}_Y)$ is a fibration if and only if it has the right lifting property with respect to all trivial cofibrations.*

Proof. It suffices to show that the conditions of Proposition A.2.6.13 are satisfied. We check them in turn:

(1) The class (W) of Cartesian equivalences is perfect (in the sense of Definition A.2.6.10). This follows from Corollary A.2.6.12, since the class of Cartesian equivalences is perfect and the functor $(X, \mathcal{E}_X) \to X^{\sharp}$ commutes with filtered colimits.

(2) The class of weak equivalences is stable under pushouts by cofibrations. This follows from the analogous property of the Cartesian model structure because the functor $(X, \mathcal{E}_X) \mapsto X^\sharp$ preserves pushouts.

(3) A map $p : (X, \mathcal{E}_X) \to (Y, \mathcal{E}_Y)$ which has the right lifting property with respect to every cofibration is a weak equivalence. In this case, the underlying map of simplicial sets is a trivial fibration, so the induced map $X^\sharp \to Y^\sharp$ has the right lifting property with respect to all trivial cofibrations and is a Cartesian equivalence (as observed in the proof of Proposition 3.1.3.7).

$\square$

Proposition 3.1.5.3. *Let S be a simplicial set. Consider the adjoint functors*

$$(\mathrm{Set}_\Delta)_{/S} \underset{G_0}{\overset{F_0}{\rightleftarrows}} (\mathrm{Set}_\Delta^+)_{/S}$$

described by the formulas

$$F_0(X) = X^\flat$$

$$G_0(X, \mathcal{E}) = X.$$

The adjoint functors (F_0, G_0) determine a Quillen adjunction between the category $(\mathrm{Set}_\Delta)_{/S}$ (with the Joyal model structure) and the category $(\mathrm{Set}_\Delta^+)_{/S}$ (with the Cartesian model structure). If S is a Kan complex, then (F_0, G_0) is a Quillen equivalence.

Proof. To prove that (F_0, G_0) is a Quillen adjunction, it will suffice to show that F_1 preserves cofibrations and trivial cofibrations. The first claim is obvious. For the second, we must show that if $X \subseteq Y$ is a categorical equivalence of simplicial sets over S, then the induced map $X^\flat \to Y^\flat$ is a Cartesian equivalence in $(\mathrm{Set}_\Delta^+)_{/S}$. For this, it suffices to show that for any Cartesian fibration $p : Z \to S$, the restriction map

$$\mathrm{Map}_S^\flat(Y^\flat, Z^\natural) \to \mathrm{Map}_S^\flat(X^\flat, Z^\natural)$$

is a trivial fibration of simplicial sets. In other words, we must show that for every inclusion $A \subseteq B$ of simplicial sets, it is possible to solve any lifting problem of the form

$$\begin{array}{ccc}
A & \longrightarrow & \mathrm{Map}_S^\flat(Y^\flat, Z^\natural) \\
\downarrow & \nearrow & \downarrow \\
B & \longrightarrow & \mathrm{Map}_S^\flat(X^\flat, Z^\natural).
\end{array}$$

Replacing Y by $Y \times B$ and X by $(X \times B) \coprod_{X \times A} (Y \times A)$, we may suppose that $A = \emptyset$ and $B = *$. Moreover, we may rephrase the lifting problem as the problem of constructing the dotted arrow indicated in the following diagram:

By Proposition 3.3.1.7, p is a categorical fibration, and the lifting problem has a solution by virtue of the assumption that $X \subseteq Y$ is a categorical equivalence.

Now suppose that S is a Kan complex. We want to prove that (F_0, G_0) is a Quillen equivalence. In other words, we must show that for any fibrant object of $(\mathrm{Set}_\Delta^+)_{/S}$ corresponding to a Cartesian fibration $Z \to S$, a map $X \to Z$ in $(\mathrm{Set}_\Delta)_{/S}$ is a categorical equivalence if and only if the associated map $X^\flat \to Z^\natural$ is a Cartesian equivalence.

Suppose first that $X \to Z$ is a categorical equivalence. Then the induced map $X^\flat \to Z^\flat$ is a Cartesian equivalence by the argument given above. It therefore suffices to show that $Z^\flat \to Z^\natural$ is a Cartesian equivalence. Since S is a Kan complex, Z is an ∞-category; let K denote the largest Kan complex contained in Z. The marked edges of $Z^\natural$ are precisely the edges which belong to K, so we have a pushout diagram

$$
\begin{array}{ccc}
K^\flat & \longrightarrow & K^\sharp \\
\downarrow & & \downarrow \\
Z^\flat & \longrightarrow & Z^\natural.
\end{array}
$$

It follows that $Z^\flat \to Z^\natural$ is marked anodyne and therefore a Cartesian equivalence.

Now suppose that $X^\flat \to Z^\natural$ is a Cartesian equivalence. Choose a factorization $X \xrightarrow{f} Y \xrightarrow{g} Z$, where f is a categorical equivalence and g is a categorical fibration. We wish to show that g is a categorical equivalence. Proposition 3.3.1.8 implies that $Z \to S$ is a categorical fibration, so that $X' \to S$ is a categorical fibration. Applying Proposition 3.3.1.8 again, we deduce that $Y \to S$ is a Cartesian fibration. Thus we have a factorization

$$
X^\flat \to Y^\flat \to Y^\natural \to Z^\natural,
$$

where the first two maps are Cartesian equivalences by the arguments given above and the composite map is a Cartesian equivalence. Thus $Y^\natural \to Z^\natural$ is an equivalence between fibrant objects of $(\mathrm{Set}_\Delta^+)_{/S}$ and therefore admits a homotopy inverse. The existence of this homotopy inverse proves that g is a categorical equivalence, as desired. $\square$

Proposition 3.1.5.4. *Let S be a simplicial set and let $F_1 = G_1$ be the identity functor from $(\mathrm{Set}_\Delta^+)_{/S}$ to itself. Then (F_1, G_1) determines a Quillen adjunction between $\mathcal{C}_1$ and $\mathcal{C}_2$.*

Proof. We must show that F_1 preserves cofibrations and trivial cofibrations. The first claim is obvious. For the second, let $B : (\mathrm{Set}_\Delta^+)_{/S} \to (\mathrm{Set}_\Delta^+)_{/S}$ be the functor defined by

$$B(M, \mathcal{E}_M) = M^\sharp.$$

We wish to show that if $X \to Y$ is a Cartesian equivalence in $(\mathrm{Set}_\Delta^+)_{/S}$, then $B(X) \to B(Y)$ is a Cartesian equivalence.

We first observe that if $X \to Y$ is marked anodyne, then the induced map $B(X) \to B(Y)$ is also marked anodyne: by general nonsense, it suffices to check this for the generators described in Definition 3.1.1.1, for which it is obvious. Now return to the case of a general Cartesian equivalence $p : X \to Y$ and choose a diagram

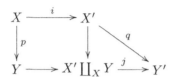

in which X' and Y' are (marked) fibrant and i and j are marked anodyne. It follows that $B(i)$ and $B(j)$ are marked anodyne and therefore Cartesian equivalences. Thus, to prove that $B(p)$ is a Cartesian equivalence, it suffices to show that $B(q)$ is a Cartesian equivalence. But q is a Cartesian equivalence between fibrant objects of $(\mathrm{Set}_\Delta^+)_{/S}$ and therefore has a homotopy inverse. It follows that $B(q)$ also has a homotopy inverse and is therefore a Cartesian equivalence, as desired. □

Remark 3.1.5.5. In the language of model categories, we may summarize Proposition 3.1.5.4 by saying that the model structure of Proposition 3.1.5.2 is a *localization* of the Cartesian model structure on $(\mathrm{Set}_\Delta^+)_{/S}$.

Proposition 3.1.5.6. *Let S be a simplicial set and consider the adjunction*

$$(\mathrm{Set}_\Delta^+)_{/S} \underset{G_2}{\overset{F_2}{\rightleftarrows}} (\mathrm{Set}_\Delta)_{/S}$$

determined by the formulas

$$F_2(X, \mathcal{E}) = X$$

$$G_2(X) = X^\sharp.$$

The adjoint functors (F_2, G_2) determine a Quillen equivalence between $\mathcal{C}_2$ and $\mathcal{C}_3$.

Proof. We first claim that F_2 is conservative: that is, a map $f : (X, \mathcal{E}_X) \to (Y, \mathcal{E}_Y)$ is a weak equivalence in $\mathcal{C}_2$ if and only if the induced map $X \to Y$ is a weak equivalence in $\mathcal{C}_3$. Unwinding the definition, f is a weak equivalence if and only if $X^\sharp \to Y^\sharp$ is a Cartesian equivalence. This holds if and only if, for every Cartesian fibration $Z \to S$, the induced map

$$\phi : \mathrm{Map}_S^\sharp(Y^\sharp, Z^\natural) \to \mathrm{Map}_S^\sharp(X^\sharp, Z^\natural)$$

is a homotopy equivalence. Let $Z^0 \to S$ be the right fibration associated to $Z \to S$ (see Corollary 2.4.2.5). We have natural identifications

$$\mathrm{Map}_S^\sharp(Y^\sharp, Z^\natural) \simeq \mathrm{Map}_S(Y, Z^0) \qquad \mathrm{Map}_S^\sharp(X^\sharp, Z^\natural) \simeq \mathrm{Map}_S(X, Z^0).$$

Consequently, f is a weak equivalence if and only if, for every right fibration $Z^0 \to S$, the associated map

$$\mathrm{Map}_S(Y, Z^0) \to \mathrm{Map}_S(X, Z^0)$$

is a homotopy equivalence. Since $\mathcal{C}_3$ is a simplicial model category for which the fibrant objects are precisely the right fibrations $Z^0 \to S$ (Corollary 2.2.3.12), this is equivalent to the assertion that $X \to Y$ is a weak equivalence in $\mathcal{C}_3$.

To prove that (F_2, G_2) is a Quillen adjunction, it suffices to show that F_2 preserves cofibrations and trivial cofibrations. The first claim is obvious, and the second follows because F_2 preserves all weak equivalences (by the above argument).

To show that (F_2, G_2) is a Quillen equivalence, we must show that the unit and counit

$$LF_2 \circ RG_2 \to \mathrm{id}$$

$$\mathrm{id} \to RG_2 \circ LF_2$$

are weak equivalences. In view of the fact that $F_2 = LF_2$ is conservative, the second assertion follows from the first. To prove the first, it suffices to show that if X is a fibrant object of $\mathcal{C}_3$, then the counit map $(F_2 \circ G_2)(X) \to X$ is a weak equivalence. But this map is an isomorphism. $\square$

Proposition 3.1.5.7. *Let S be a simplicial set and let $F_3 = G_3$ be the identity functor from $(\mathrm{Set}_\Delta)_{/S}$ to itself. Then (F_3, G_3) gives a Quillen adjunction between $\mathcal{C}_3$ and $\mathcal{C}_4$. If S is a Kan complex, then (F_3, G_3) is a Quillen equivalence (in other words, the model structures on $\mathcal{C}_3$ and $\mathcal{C}_4$ coincide).*

Proof. To prove that (F_3, G_3) is a Quillen adjunction, it suffices to prove that F_3 preserves cofibrations and weak equivalences. The first claim is obvious (the cofibrations in $\mathcal{C}_3$ and $\mathcal{C}_4$ are the same). For the second, we note that both $\mathcal{C}_3$ and $\mathcal{C}_4$ are simplicial model categories in which every object is cofibrant. Consequently, a map $f : X \to Y$ is a weak equivalence if and only if, for every fibrant object Z, the associated map $\mathrm{Map}(Y, Z) \to \mathrm{Map}(X, Z)$ is a homotopy equivalence of Kan complexes. Thus, to show that F_3 preserves weak equivalences, it suffices to show that G_3 preserves fibrant objects. A map $p : Z \to S$ is fibrant as an object of $\mathcal{C}_4$ if and only if p is a Kan fibration, and fibrant as an object of $\mathcal{C}_3$ if and only if p is a right fibration (Corollary 2.2.3.12). Since every Kan fibration is a right fibration, it follows that F_3 preserves weak equivalences. If S is a Kan complex, then the converse holds: according to Lemma 2.1.3.4, every right fibration $p : Z \to S$ is a Kan fibration. It follows that G_3 preserves weak equivalences as well, so that the two model structures under consideration coincide. $\square$

3.2 STRAIGHTENING AND UNSTRAIGHTENING

Let $\mathcal{C}$ be a category and let $\chi : \mathcal{C}^{op} \to \mathcal{C}at$ be a functor from $\mathcal{C}$ to the category $\mathcal{C}at$ of small categories. To this data, we can associate (by means of the *Grothendieck construction* discussed in §2.1.1) a new category $\widetilde{\mathcal{C}}$ which may be described as follows:

- The objects of $\widetilde{\mathcal{C}}$ are pairs (C, η), where $C \in \mathcal{C}$ and $\eta \in \chi(C)$.

- Given a pair of objects $(C, \eta), (C', \eta') \in \widetilde{\mathcal{C}}$), a morphism from (C, η) to (C', η') in $\widetilde{\mathcal{C}}$ is a pair (f, α), where $f : C \to C'$ is a morphism in the category $\mathcal{C}$ and $\alpha : \eta \to \chi(f)(\eta')$ is a morphism in the category $\chi(C)$.

- Composition is defined in the obvious way.

This construction establishes an equivalence between $\mathcal{C}at$-valued functors on $\mathcal{C}^{op}$ and categories which are *fibered over* $\mathcal{C}$. (To formulate the equivalence precisely, it is best to view $\mathcal{C}at$ as a *bicategory*, but we will not dwell on this technical point here.)

The goal of this section is to establish an ∞-categorical version of the equivalence described above. We will replace the category $\mathcal{C}$ by a simplicial set S, the category $\mathcal{C}at$ by the ∞-category $\mathcal{C}at_\infty$, and the notion of fibered category with the notion of Cartesian fibration. In this setting, we will obtain an equivalence of ∞-categories, which arises from a Quillen equivalence of simplicial model categories. On one side, we have the category $(\mathrm{Set}_\Delta^+)_{/S}$, equipped with the Cartesian model structure (a simplicial model category whose fibrant objects are precisely the Cartesian fibrations $X \to S$; see §3.1.4). On the other, we have the category of simplicial functors

$$\mathfrak{C}[S]^{op} \to \mathrm{Set}_\Delta^+$$

equipped with the projective model structure (see §A.3.3) whose underlying ∞-category is equivalent to $\mathrm{Fun}(S^{op}, \mathcal{C}at_\infty)$ (Proposition 4.2.4.4). The situation may be summarized as follows:

Theorem 3.2.0.1. *Let S be a simplicial set, $\mathcal{C}$ a simplicial category, and $\phi : \mathfrak{C}[S] \to \mathcal{C}^{op}$ a functor between simplicial categories. Then there exists a pair of adjoint functors*

$$(\mathrm{Set}_\Delta^+)_{/S} \underset{\mathrm{Un}_\phi^+}{\overset{St_\phi^+}{\rightleftarrows}} (\mathrm{Set}_\Delta^+)^{\mathcal{C}}$$

with the following properties:

(1) *The functors $(St_\phi^+, \mathrm{Un}_\phi^+)$ determine a Quillen adjunction between the category $(\mathrm{Set}_\Delta^+)_{/S}$ (with the Cartesian model structure) and the category $(\mathrm{Set}_\Delta^+)^{\mathcal{C}}$ (with the projective model structure).*

(2) *If ϕ is an equivalence of simplicial categories, then $(St_\phi^+, \mathrm{Un}_\phi^+)$ is a Quillen equivalence.*

We will refer to St_ϕ^+ and Un_ϕ^+ as the *straightening* and *unstraightening* functors, respectively. We will construct these functors in §3.2.1 and establish part (1) of Theorem 3.2.0.1. Part (2) is more difficult and requires some preliminary work; we will begin in §3.2.2 by analyzing the structure of Cartesian fibrations $X \to \Delta^n$. We will apply these analyses in §3.2.3 to complete the proof of Theorem 3.2.0.1 when S is a simplex. In §3.2.4, we will deduce the general result by using formal arguments to reduce to the case of a simplex.

In the case where $\mathcal{C}$ is an ordinary category, the straightening and unstraightening procedures of §3.2.1 can be substantially simplified. We will discuss the situation in §3.2.5, where we provide an analogue of Theorem 3.2.0.1 (see Propositions 3.2.5.18 and 3.2.5.21).

3.2.1 The Straightening Functor

Let S be a simplicial set and let $\phi : \mathcal{C}[S] \to \mathcal{C}^{op}$ be a functor between simplicial categories, which we regard as fixed throughout this section. Our objective is to define the *straightening functor* $St_\phi^+ : (Set_\Delta^+)_{/S} \to (Set_\Delta^+)^\mathcal{C}$ and its right adjoint Un_ϕ^+. The intuition is that an object X of $(Set_\Delta^+)_{/S}$ associates ∞-categories to vertices of S in a homotopy coherent fashion, and the functor St_ϕ^+ "straightens" this diagram to obtain an ∞-category valued functor on $\mathcal{C}$. The right adjoint Un_ϕ^+ should be viewed as a forgetful functor which takes a strictly commutative diagram and retains the underlying homotopy coherent diagram.

The functors St_ϕ^+ and Un_ϕ^+ are more elaborate versions of the straightening and unstraightening functors introduced in §2.2.1. We begin by recalling the unmarked version of the construction. For each object $X \in (Set_\Delta)_{/S}$, form a pushout diagram of simplicial categories

$$\begin{array}{ccc} \mathcal{C}[X] & \longrightarrow & \mathcal{C}[X^\triangleright] \\ \downarrow{\scriptstyle\phi} & & \downarrow \\ \mathcal{C}^{op} & \longrightarrow & \mathcal{C}_X^{op}, \end{array}$$

where the left vertical map is given by composing ϕ with the map $\mathcal{C}[X] \to \mathcal{C}[S]$. The functor $St_\phi X : \mathcal{C} \to Set_\Delta$ is defined by the formula

$$(St_\phi X)(C) = Map_{\mathcal{C}_X^{op}}(C, *),$$

where $*$ denotes the cone point of $X^\triangleright$.

We will define St_ϕ^+ by designating certain marked edges on the simplicial sets $(St_\phi X)(C)$ which depend in a natural way on the marked edges of X. In order to describe this dependence, we need to introduce a bit of notation.

Notation 3.2.1.1. Let X be an object of $(Set_\Delta)_{/S}$. Given an n-simplex σ of the simplicial set $Map_{\mathcal{C}^{op}}(C, D)$, we let $\sigma^* : (St_\phi X)(D)_n \to (St_\phi X)(C)_n$ denote the associated map on n-simplices.

Let c be a vertex of X and let $C = \phi(c) \in \mathcal{C}$. We may identify c with a map $c : \Delta^0 \to X$. Then $c \star \mathrm{id}_{\Delta^0} : \Delta^1 \to X^\triangleright$ is an edge of $X^\triangleright$ and so determines a morphism $C \to *$ in $\mathcal{C}_X^{op}$, which we can identify with a vertex $\widetilde{c} \in (St_\phi X)(C)$.

Similarly, suppose that $f : c \to d$ is an edge of X corresponding to a morphism

$$C \overset{F}{\to} D$$

in the simplicial category $\mathcal{C}^{op}$. We may identify f with a map $f : \Delta^1 \to X$. Then $f \star \mathrm{id}_{\Delta^1} : \Delta^2 \to X^\triangleright$ determines a map $\mathfrak{C}[\Delta^2] \to \mathcal{C}_X$, which we may identify with a diagram (not strictly commutative)

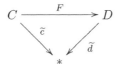

together with an edge

$$\widetilde{f} : \widetilde{c} \to \widetilde{d} \circ F = F^* \widetilde{d}$$

in the simplicial set $\mathrm{Map}_{\mathcal{C}_X^{op}}(C, *) = (St_\phi X)(C)$.

Definition 3.2.1.2. Let S be a simplicial set, $\mathcal{C}$ a simplicial category, and $\phi : \mathfrak{C}[S] \to \mathcal{C}^{op}$ a simplicial functor. Let $(X, \mathcal{E})$ be an object of $(\mathrm{Set}_\Delta^+)_{/S}$. Then

$$St_\phi^+(X, \mathcal{E}) : \mathcal{C} \to \mathrm{Set}_\Delta^+$$

is defined by the formula

$$St_\phi^+(X, \mathcal{E})(C) = ((St_\phi X)(C), \mathcal{E}_\phi(C)),$$

where $\mathcal{E}_\phi(C)$ is the set of all edges of $(St_\phi X)(C)$ having the form

$$G^* \widetilde{f},$$

where $f : d \to e$ is a marked edge of X, giving rise to an edge $\widetilde{f} : \widetilde{d} \to F^* \widetilde{e}$ in $(St_\phi X)(D)$, and G belongs to $\mathrm{Map}_{\mathcal{C}^{op}}(C, D)_1$.

Remark 3.2.1.3. The construction

$$(X, \mathcal{E}) \mapsto St_\phi^+(X, \mathcal{E}) = (St_\phi X, \mathcal{E}_\phi)$$

is obviously functorial in X. Note that we may characterize the subsets $\{\mathcal{E}_\phi(C) \subseteq (St_\phi X)(C)_1\}$ as the smallest collection of sets which contain $\widetilde{f}$ for every $f \in \mathcal{E}$ and depend functorially on C.

The following formal properties of the straightening functor follow immediately from the definition:

Proposition 3.2.1.4. (1) *Let S be a simplicial set, $\mathcal{C}$ a simplicial category, and $\phi : \mathfrak{C}[S] \to \mathcal{C}^{op}$ a simplicial functor; then the associated straightening functor*

$$St_\phi^+ : (\mathrm{Set}_\Delta^+)_{/S} \to (\mathrm{Set}_\Delta^+)^{\mathcal{C}}$$

preserves colimits.

(2) *Let $p : S' \to S$ be a map of simplicial sets, $\mathcal{C}$ a simplicial category, and $\phi : \mathfrak{C}[S] \to \mathcal{C}^{op}$ a simplicial functor, and let $\phi' : \mathfrak{C}[S'] \to \mathcal{C}^{op}$ denote the composition $\phi \circ \mathfrak{C}[p]$. Let $p_! : (\mathrm{Set}_\Delta^+)_{/S'} \to (\mathrm{Set}_\Delta^+)_{/S}$ denote the forgetful functor given by composition with p. There is a natural isomorphism of functors*

$$St_\phi^+ \circ p_! \simeq St_{\phi'}^+$$

from $(\mathrm{Set}_\Delta^+)_{/S'}$ to $(\mathrm{Set}_\Delta^+)^{\mathcal{C}}$.

(3) *Let S be a simplicial set, $\pi : \mathcal{C} \to \mathcal{C}'$ a simplicial functor between simplicial categories, and $\phi : \mathfrak{C}[S] \to \mathcal{C}^{op}$ a simplicial functor. Then there is a natural isomorphism of functors*

$$St_{\pi \circ \phi}^+ \simeq \pi_! \circ St_\phi^+$$

from $(\mathrm{Set}_\Delta^+)_{/S}$ to $(\mathrm{Set}_\Delta^+)^{\mathcal{C}'}$. Here $\pi_! : (\mathrm{Set}_\Delta^+)^{\mathcal{C}} \to (\mathrm{Set}_\Delta^+)^{\mathcal{C}'}$ is the left adjoint to the functor $\pi^* : (\mathrm{Set}_\Delta^+)^{\mathcal{C}'} \to (\mathrm{Set}_\Delta^+)^{\mathcal{C}}$ given by composition with π; see §A.3.3.

Corollary 3.2.1.5. *Let S be a simplicial set, $\mathcal{C}$ a simplicial category, and $\phi : \mathfrak{C}[S] \to \mathcal{C}^{op}$ any simplicial functor. The straightening functor St_ϕ^+ has a right adjoint*

$$\mathrm{Un}_\phi^+ : (\mathrm{Set}_\Delta^+)^{\mathcal{C}} \to (\mathrm{Set}_\Delta^+)_{/S}.$$

Proof. This follows from part (1) of Proposition 3.2.1.4 and the adjoint functor theorem. (Alternatively, one can construct Un_ϕ^+ directly; we leave the details to the reader.) $\square$

Notation 3.2.1.6. Let S be a simplicial set, let $\mathcal{C} = \mathfrak{C}[S]^{op}$, and let $\phi : \mathfrak{C}[S] \to \mathcal{C}^{op}$ be the identity map. In this case, we will denote St_ϕ^+ by St_S^+ and Un_ϕ^+ by Un_S^+.

Our next goal is to show that the straightening and unstraightening functors $(St_\phi^+, \mathrm{Un}_\phi^+)$ give a *Quillen* adjunction between the model categories $(\mathrm{Set}_\Delta^+)_{/S}$ and $(\mathrm{Set}_\Delta^+)^{\mathcal{C}}$. The first step is to show that St_ϕ^+ preserves cofibrations.

Proposition 3.2.1.7. *Let S be a simplicial set, $\mathcal{C}$ a simplicial category, and $\phi : \mathfrak{C}[S] \to \mathcal{C}^{op}$ a simplicial functor. The functor St_ϕ^+ carries cofibrations (with respect to the Cartesian model structure on $(\mathrm{Set}_\Delta^+)_{/S}$) to cofibrations (with respect to the projective model structure on $(\mathrm{Set}_\Delta^+)^{\mathcal{C}})$.*

Proof. Let $j : A \to B$ be a cofibration in $(\mathrm{Set}_\Delta^+)_{/S}$; we wish to show that $St_\phi^+(j)$ is a cofibration. By general nonsense, we may suppose that j is a generating cofibration having either the form $(\partial \Delta^n)^\flat \subseteq (\Delta^n)^\flat$ or the form $(\Delta^1)^\flat \to (\Delta^1)^\sharp$. Using Proposition 3.2.1.4, we may reduce to the case where $S = B$, $\mathcal{C} = \mathfrak{C}[S]$ and ϕ is the identity map. The result now follows from a straightforward computation. $\square$

To complete the proof that (St_ϕ^+, Un_ϕ^+) is a Quillen adjunction, it suffices to show that St_ϕ^+ preserves trivial cofibrations. Since every object of $(Set_\Delta^+)_{/S}$ is cofibrant, this is equivalent to the apparently stronger claim that if $f : X \to Y$ is a Cartesian equivalence in $(Set_\Delta^+)_{/S}$, then $St_\phi^+(f)$ is a weak equivalence in $(Set_\Delta^+)^{\mathcal{C}}$. The main step is to establish this in the case where f is marked anodyne. First, we need a few lemmas.

Lemma 3.2.1.8. *Let $\mathcal{E}$ be the set of all degenerate edges of $\Delta^n \times \Delta^1$ together with the edge $\{n\} \times \Delta^1$. Let $B \subseteq \Delta^n \times \Delta^1$ be the coproduct*

$$(\Delta^n \times \{1\}) \coprod_{\partial \Delta^n \times \{1\}} (\partial \Delta^n \times \Delta^1).$$

Then the map

$$i : (B, \mathcal{E} \cap B_1) \subseteq (\Delta^n \times \Delta^1, \mathcal{E})$$

is marked anodyne.

Proof. We must show that i has the left lifting property with respect to every map $p : X \to S$ satisfying the hypotheses of Proposition 3.1.1.6. This is simply a reformulation of Proposition 2.4.1.8. □

Lemma 3.2.1.9. *Let K be a simplicial set, $K' \subseteq K$ a simplicial subset, and A a set of vertices of K. Let $\mathcal{E}$ denote the set of all degenerate edges of $K \times \Delta^1$ together with the edges $\{a\} \times \Delta^1$, where $a \in A$. Let $B = (K' \times \Delta^1) \coprod_{K' \times \{1\}} (K \times \{1\}) \subseteq K \times \Delta^1$. Suppose that, for every nondegenerate simplex σ of K, either σ belongs to K' or the final vertex of σ belongs to A. Then the inclusion*

$$(B, \mathcal{E} \cap B_1) \subseteq (K \times \Delta^1, \mathcal{E})$$

is marked anodyne.

Proof. Working simplex by simplex, we reduce to Lemma 3.2.1.8. □

Lemma 3.2.1.10. *Let X be a simplicial set, and let $\mathcal{E} \subseteq \mathcal{E}'$ be sets of edges of X containing all degenerate edges. The following conditions are equivalent:*

(1) *The inclusion $(X, \mathcal{E}) \to (X, \mathcal{E}')$ is a trivial cofibration in Set_Δ^+ (with respect to the Cartesian model structure).*

(2) *For every ∞-category $\mathcal{C}$ and every map $f : X \to \mathcal{C}$ which carries each edge of $\mathcal{E}$ to an equivalence in $\mathcal{C}$, f also carries each edge of $\mathcal{E}'$ to an equivalence in $\mathcal{C}$.*

Proof. By definition, (1) holds if and only if for every ∞-category $\mathcal{C}$, the inclusion

$$j : Map^\flat((X, \mathcal{E}'), \mathcal{C}^\natural) \to Map^\flat((X, \mathcal{E}), \mathcal{C}^\natural)$$

is a categorical equivalence. Condition (2) is the assertion that j is an isomorphism. Thus (2) implies (1). Suppose that (1) is satisfied and let $f : X \to \mathcal{C}$

be a vertex of $\mathrm{Map}^\flat((X,\mathcal{E}),\mathcal{C}^\natural)$. By hypothesis, there exists an equivalence $f \simeq f'$, where f' belongs to the image of j. Let $e \in \mathcal{E}'$; then $f'(e)$ is an equivalence in $\mathcal{C}$. Since f and f' are equivalent, $f(e)$ is also an equivalence in $\mathcal{C}$. Consequently, f also belongs to the image of j, and the proof is complete. $\square$

Proposition 3.2.1.11. *Let S be a simplicial set, $\mathcal{C}$ a simplicial category, and $\phi : \mathfrak{C}[S] \to \mathcal{C}^{op}$ a simplicial functor. The functor St_ϕ^+ carries marked anodyne maps in $(\mathrm{Set}_\Delta^+)_{/S}$ (with respect to the Cartesian model structure) to trivial cofibrations in $(\mathrm{Set}_\Delta^+)^{\mathcal{C}}$ (with respect to the projective model structure).*

Proof. Let $f : A \to B$ be a marked anodyne map in $(\mathrm{Set}_\Delta^+)_{/S}$. We wish to prove that $St_\phi^+(f)$ is a trivial cofibration. It will suffice to prove this under the assumption that f is one of the generators for the class of marked anodyne maps given in Definition 3.1.1.1. Using Proposition 3.2.1.4, we may reduce to the case where S is the underlying simplicial set of B, $\mathcal{C} = \mathfrak{C}[S]^{op}$, and ϕ is the identity. There are four cases to consider:

(1) Suppose first that f is among the morphisms listed in (1) of Definition 3.1.1.1; that is, f is an inclusion $(\Lambda_i^n)^\flat \subseteq (\Delta^n)^\flat$, where $0 < i < n$. Let v_k denote the kth vertex of Δ^n, which we may also think of as an object of the simplicial category $\mathcal{C}$. We note that $St_\phi^+(f)$ is an isomorphism when evaluated at v_k for $k \neq 0$. Let K denote the cube $(\Delta^1)^{\{j:0<j\leq n, j\neq i\}}$, let $K' = \partial K$, let A denote the set of all vertices of K corresponding to subsets of $\{j : 0 < j \leq n, j \neq i\}$ which contain an element $> i$, and let $\mathcal{E}$ denote the set of all degenerate edges of $K \times \Delta^1$ together with all edges of the form $\{a\} \times \Delta^1$, where $a \in A$. Finally, let $B = (K \times \{1\}) \coprod_{K' \times \{1\}} (K' \times \Delta^1)$. The morphism $St_\phi^+(f)(v_n)$ is a pushout of $g : (B, \mathcal{E} \cap B_1) \subseteq (K \times \Delta^1, \mathcal{E})$. Since $i > 0$, we may apply Lemma 3.2.1.9 to deduce that g is marked anodyne and therefore a trivial cofibration in Set_Δ^+.

(2) Suppose that f is among the morphisms of part (2) in Definition 3.1.1.1; that is, f is an inclusion

$$(\Lambda_n^n, \mathcal{E} \cap (\Lambda_n^n)_1) \subseteq (\Delta^n, \mathcal{F}),$$

where $\mathcal{F}$ denotes the set of all degenerate edges of Δ^n together with the final edge $\Delta^{\{n-1,n\}}$. If $n > 1$, then one can repeat the argument given above in case (1), except that the set of vertices A needs to be replaced by the set of all vertices of K which correspond to subsets of $\{j : 0 < j < n\}$ which contain $n - 1$. If $n = 1$, then we observe that $St_\phi^+(f)(v_n)$ is isomorphic to the inclusion $\{1\}^\sharp \subseteq (\Delta^1)^\sharp$, which is again a marked anodyne map and therefore a trivial cofibration in Set_Δ^+.

(3) Suppose next that f is the morphism

$$(\Lambda_1^2)^\sharp \coprod_{(\Lambda_1^2)^\flat} (\Delta^2)^\flat \to (\Delta^2)^\sharp$$

specified in (3) of Definition 3.1.1.1. A simple computation shows that $St_\phi^+(f)(v_n)$ is an isomorphism for $n \neq 0$, and $St_\phi^+(f)(v_0)$ may be identified with the inclusion

$$(\Delta^1 \times \Delta^1, \mathcal{E}) \subseteq (\Delta^1 \times \Delta^1)^\sharp,$$

where $\mathcal{E}$ denotes the set of all degenerate edges of $\Delta^1 \times \Delta^1$ together with $\Delta^1 \times \{0\}$, $\Delta^1 \times \{1\}$, and $\{1\} \times \Delta^1$. This inclusion may be obtained as a pushout of

$$(\Lambda_1^2)^\sharp \coprod_{(\Lambda_1^2)^\flat} (\Delta^2)^\flat \to (\Delta^2)^\sharp$$

followed by a pushout of

$$(\Lambda_2^2)^\sharp \coprod_{(\Lambda_2^2)^\flat} (\Delta^2)^\flat \to (\Delta^2)^\sharp.$$

The first of these maps is marked anodyne by definition; the second is marked anodyne by Corollary 3.1.1.7.

(4) Suppose that f is the morphism $K^\flat \to K^\sharp$, where K is a Kan complex, as in (4) of Definition 3.1.1.1. For each vertex v of K, let $St_\phi^+(K^\flat)(v) = (X_v, \mathcal{E}_v)$, so that $St_\phi^+(K^\sharp) = X_v^\sharp$. For each $g \in \mathrm{Map}_{\mathfrak{C}[K]}(v, v')_n$, we let $g^* : X_v \times \Delta^n \to X_{v'}$ denote the induced map. We wish to show that the natural map $(X_v, \mathcal{E}_v) \to X_v^\sharp$ is an equivalence in Set_Δ^+. By Lemma 3.2.1.10, it suffices to show that for every ∞-category Z, if $h : X_v \to Z$ carries each edge belonging to $\mathcal{E}_v$ into an equivalence, then h carries *every* edge of X_v to an equivalence.

We first show that h carries $\widetilde{e}$ to an equivalence for every edge $e : v \to v'$ in K. Let $m_e : \Delta^1 \to \mathrm{Map}_{\mathfrak{C}^{op}}(v, v')$ denote the degenerate edge at the vertex corresponding to e. Since K is a Kan complex, the edge $e : \Delta^1 \to K$ extends to a 2-simplex $\sigma : \Delta^2 \to K$ depicted as follows:

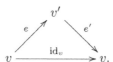

Let $m_{e'} : \Delta^1 \to \mathrm{Map}_\mathfrak{C}(v', v)$ denote the degenerate edge corresponding to e'. The map σ gives rise to a diagram

$$\begin{array}{ccc} \widetilde{v} & \xrightarrow{\ \widetilde{e}\ } & e^*\widetilde{v}' \\ {\scriptstyle \mathrm{id}_{\widetilde{v}}}\downarrow & & \downarrow{\scriptstyle m_e^*\widetilde{e}'} \\ \widetilde{v} & \longrightarrow & e^*(e')^*\widetilde{v} \end{array}$$

in the simplicial set X_v. Since h carries the left vertical arrow and the bottom horizontal arrow into equivalences, it follows that h carries the composition $(m_e^*\widetilde{e}') \circ \widetilde{e}$ to an equivalence in Z; thus $h(\widetilde{e})$ has a left homotopy inverse. A similar argument shows that $h(\tilde{e})$ has a right homotopy inverse, so that $h(\tilde{e})$ is an equivalence.

We observe that every edge of X_v has the form $g^*\tilde{e}$, where g is an edge of $\mathrm{Map}_{\mathbb{C}^{op}}(v, v')$ and $e : v' \to v''$ is an edge of K. We wish to show that $h(g^*\tilde{e})$ is an equivalence in Z. Above, we have shown that this is true if $v = v'$ and g is the identity. We now consider the more general case where g is not necessarily the identity but is a degenerate edge corresponding to some map $v' \to v$ in $\mathbb{C}$. Let h' denote the composition

$$X_{v'} \to X_v \xrightarrow{h} Z.$$

Then $h(g^*\tilde{e}) = h'(\tilde{e})$ is an equivalence in Z by the argument given above.

Now consider the case where $g : \Delta^1 \to \mathrm{Map}_{\mathbb{C}^{op}}(v, v')$ is nondegenerate. In this case, there is a simplicial homotopy $G : \Delta^1 \times \Delta^1 \to \mathrm{Map}_{\mathbb{C}}(v, v')$ with $g = G|\Delta^1 \times \{0\}$ and $g' = G|\Delta^1 \times \{1\}$ a degenerate edge of $\mathrm{Map}_{\mathbb{C}^{op}}(v, v')$ (for example, we can arrange that g' is the constant edge at an endpoint of g). The map G induces a simplicial homotopy $G(e)$ from $g^*\tilde{e}$ to $(g')^*\tilde{e}$. Moreover, the edges $G(e)|\{0\} \times \Delta^1$ and $G(e)|\{1\} \times \Delta^1$ belong to $\mathcal{E}_v$ and are therefore carried by h into equivalences in Z. Since h carries $(g')^*\tilde{e}$ into an equivalence of Z, it carries $g^*\tilde{e}$ into an equivalence of Z, as desired. $\qquad\square$

We now study the behavior of straightening functors with respect to products.

Notation 3.2.1.12. Given two simplicial functors $\mathcal{F} : \mathbb{C} \to \mathrm{Set}_\Delta^+$, $\mathcal{F}' : \mathbb{C}' \to \mathrm{Set}_\Delta^+$, we let $\mathcal{F} \boxtimes \mathcal{F}' : \mathbb{C} \times \mathbb{C}' \to \mathrm{Set}_\Delta^+$ denote the functor described by the formula

$$(\mathcal{F} \boxtimes \mathcal{F}')(C, C') = \mathcal{F}(C) \times \mathcal{F}'(C').$$

Proposition 3.2.1.13. Let S and S' be simplicial sets, $\mathbb{C}$ and $\mathbb{C}'$ simplicial categories, and $\phi : \mathfrak{C}[S] \to \mathbb{C}^{op}$, $\phi' : \mathfrak{C}[S'] \to (\mathbb{C}')^{op}$ simplicial functors; let $\phi \boxtimes \phi'$ denote the induced functor $\mathfrak{C}[S \times S'] \to (\mathbb{C} \times \mathbb{C}')^{op}$. For every $M \in (\mathrm{Set}_\Delta^+)_{/S}$, $M' \in (\mathrm{Set}_\Delta^+)_{/S'}$, the natural map

$$s_{M,M'} : \mathrm{St}_{\phi \boxtimes \phi'}^+(M \times M') \to \mathrm{St}_\phi^+(M) \boxtimes \mathrm{St}_{\phi'}^+(M')$$

is a weak equivalence of functors $\mathbb{C} \times \mathbb{C}' \to \mathrm{Set}_\Delta^+$.

Proof. Since both sides are compatible with the formations of filtered colimits in M, we may suppose that M has only finitely many nondegenerate simplices. We work by induction on the dimension n of M and the number of n-dimensional simplices of M. If $M = \emptyset$, there is nothing to prove. If $n \neq 1$, we may choose a nondegenerate simplex of M having maximal dimension and thereby write $M = N \coprod_{(\partial \Delta^n)^\flat} (\Delta^n)^\flat$. By the inductive hypothesis we may suppose that the result is known for N and $(\partial\Delta^n)^\flat$. The map $s_{M,M'}$ is a pushout of the maps $s_{N,M'}$ and $s_{(\Delta^n)^\flat,M'}$ over $s_{(\partial\Delta^n)^\flat,M'}$. Since Set_Δ^+ is left proper, this pushout is a homotopy pushout; it therefore suffices to prove the result after replacing M by N, $(\partial\Delta^n)^\flat$, or $(\Delta^n)^\flat$. In the first two cases, the inductive hypothesis implies that $s_{M,M'}$ is an equivalence; we are therefore reduced to the case $M = (\Delta^n)^\flat$. If $n = 0$, the result is obvious. If $n > 2$, we set

$$K = \Delta^{\{0,1\}} \coprod_{\{1\}} \Delta^{\{1,2\}} \coprod_{\{2\}} \cdots \coprod_{\{n-1\}} \Delta^{\{n-1,n\}} \subseteq \Delta^n.$$

The inclusion $K \subseteq \Delta^n$ is inner anodyne so that $K^\flat \subseteq M$ is marked anodyne. By Proposition 3.2.1.11, we deduce that $s_{M,M'}$ is an equivalence if and only if $s_{K^\flat,M'}$ is an equivalence, which follows from the inductive hypothesis since K is 1-dimensional.

We may therefore suppose that $n = 1$. Using the above argument, we may reduce to the case where M consists of a single edge, either marked or unmarked. Repeating the above argument with the roles of M and M' interchanged, we may suppose that M' also consists of a single edge. Applying Proposition 3.2.1.4, we may reduce to the case where $S = M$, $S' = M'$, $\mathcal{C} = \mathfrak{C}[S]^{op}$, and $\mathcal{C}' = \mathfrak{C}[S']^{op}$.

Let us denote the vertices of M by x and y, and the unique edge joining them by $e : x \to y$. Similarly, we let x' and y' denote the vertices of M', and $e' : x' \to y'$ the edge which joins them. We note that the map $s_{M,M'}$ induces an isomorphism when evaluated on any object of $\mathcal{C} \times \mathcal{C}'$ except (x, x'). Moreover, the map

$$s_{M,M'}(x, x') : St^+_{\phi\boxtimes\phi'}(M \times M')(x, x') \to St^+_\phi(M)(x) \times St^+_{\phi'}(M')(x')$$

is obtained from $s_{(\Delta^1)^\flat,(\Delta^1)^\flat}$ by successive pushouts along cofibrations of the form $(\Delta^1)^\flat \subseteq (\Delta^1)^\sharp$. Since Set_Δ^+ is left proper, we may reduce to the case where $M = M' = (\Delta^1)^\flat$. The result now follows from a simple explicit computation. $\square$

We now study the situation in which $S = \Delta^0$, $\mathcal{C} = \mathfrak{C}[S]$, and ϕ is the identity map. In this case, St^+_ϕ may be regarded as a functor $T : \mathrm{Set}_\Delta^+ \to \mathrm{Set}_\Delta^+$. The underlying functor of simplicial sets is familiar: we have

$$T(X, \mathcal{E}) = (|X|_{Q^\bullet}, \mathcal{E}'),$$

where Q denotes the cosimplicial object of Set_Δ considered in §2.2.2. In that section, we exhibited a natural map $|X|_{Q^\bullet} \to X$ which we proved to be a weak homotopy equivalence. We now prove a stronger version of that result:

Proposition 3.2.1.14. *For any marked simplicial set $M = (X, \mathcal{E})$, the natural map $|X|_{Q^\bullet} \to X$ induces a Cartesian equivalence*

$$T(M) \to M.$$

Proof. As in the proof of Proposition 3.2.1.13, we may reduce to the case where M consists of a simplex of dimension at most 1 (either marked or unmarked). In these cases, the map $T(M) \to M$ is an isomorphism in Set_Δ^+. □

Corollary 3.2.1.15. *Let S be a simplicial set, $\mathcal{C}$ a simplicial category, $\phi : \mathfrak{C}[S] \to \mathcal{C}^{op}$ a simplicial functor, and $X \in (Set_\Delta^+)_{/S}$ an object. For every $K \in Set_\Delta^+$, there is a natural equivalence*

$$St_\phi^+(M \times K) \to St_\phi^+(M) \boxtimes K$$

of functors from $\mathcal{C}$ to Set_Δ^+.

Proof. Combine the equivalences of Proposition 3.2.1.14 (in the case where $S' = \Delta^0$, $\mathcal{C}' = \mathfrak{C}[S']^{op}$, and ϕ' is the identity) and Proposition 3.2.1.15. □

We can now complete the proof that (St_ϕ^+, Un_ϕ^+) is a Quillen adjunction:

Corollary 3.2.1.16. *Let S be a simplicial set, $\mathcal{C}$ a simplicial category, and $\phi : \mathfrak{C}[S]^{op} \to \mathcal{C}$ a simplicial functor. The straightening functor St_ϕ^+ carries Cartesian equivalences in $(Set_\Delta^+)_{/S}$ to (objectwise) Cartesian equivalences in $(Set_\Delta^+)^{\mathcal{C}}$.*

Proof. Let $f : M \to N$ be a Cartesian equivalence in $(Set_\Delta^+)_{/S}$. Choose a marked anodyne map $M \to M'$, where M' is fibrant; then choose a marked anodyne map $M' \coprod_M N \to N'$, with N' fibrant. Since St_ϕ^+ carries marked anodyne maps to equivalences by Proposition 3.2.1.11, it suffices to prove that the induced map $St_\phi^+(M') \to St_\phi^+(N')$ is an equivalence. In other words, we may replace M by M' and N by N', thereby reducing to the case where M and N are fibrant.

Since f is an Cartesian equivalence of fibrant objects, it has a homotopy inverse g. We claim that $St_\phi^+(g)$ is an inverse to $St_\phi^+(f)$ in the homotopy category of $(Set_\Delta^+)^{\mathcal{C}}$. We will show that $St_\phi^+(f) \circ St_\phi^+(g)$ is homotopic to the identity; applying the same argument with the roles of f and g reversed will then establish the desired result.

Since $f \circ g$ is homotopic to the identity, there is a map $h : N \times K^\sharp \to N$, where K is a contractible Kan complex containing vertices x and y, such that $f \circ g = h|N \times \{x\}$ and $\mathrm{id}_N = h|N \times \{y\}$. The map $St_\phi^+(h)$ factors as

$$St_\phi^+(N \times K^\sharp) \to St_\phi^+(N) \boxtimes K^\sharp \to St_\phi^+(N),$$

where the left map is an equivalence by Corollary 3.2.1.15 and the right map because K is contractible. Since $St_\phi^+(f \circ g)$ and $St_\phi^+(\mathrm{id}_N)$ are both sections of $St_\phi^+(h)$, they represent the same morphism in the homotopy category of $(Set_\Delta^+)^{\mathcal{C}}$. □

3.2.2 Cartesian Fibrations over a Simplex

A map of simplicial sets $p : X \to S$ is a Cartesian fibration if and only if the pullback map $X \times_S \Delta^n \to \Delta^n$ is a Cartesian fibration for each simplex of S. Consequently, we might imagine that Cartesian fibrations $X \to \Delta^n$ are the "primitive building blocks" out of which other Cartesian fibrations are built. The goal of this section is to prove a structure theorem for these building blocks. This result has a number of consequences and will play a vital role in the proof of Theorem 3.2.0.1.

Note that Δ^n is the nerve of the category associated to the linearly ordered set

$$[n] = \{0 < 1 < \cdots < n\}.$$

Since a Cartesian fibration $p : X \to S$ can be thought of as giving a (contravariant) functor from S to ∞-categories, it is natural to expect a close relationship between Cartesian fibrations $X \to \Delta^n$ and composable sequences of maps between ∞-categories

$$A^0 \leftarrow A^1 \leftarrow \cdots \leftarrow A^n.$$

In order to establish this relationship, we need to introduce a few definitions.

Suppose we are given a composable sequence of maps

$$\phi : A^0 \leftarrow A^1 \leftarrow \cdots \leftarrow A^n$$

of simplicial sets. The *mapping simplex* $M(\phi)$ of ϕ is defined as follows. If J is a nonempty finite linearly ordered set with greatest element j, then to specify a map $\Delta^J \to M(\phi)$, one must specify an order-preserving map $f : J \to [n]$ together with a map $\sigma : \Delta^J \to A^{f(j)}$. Given an order-preserving map $p : J \to J'$ of partially ordered sets containing largest elements j and j', there is a natural map $M(\phi)(\Delta^{J'}) \to M(\phi)(\Delta^J)$ which carries (f, σ) to $(f \circ p, e \circ \sigma)$, where $e : A^{f(j')} \to A^{f(p(j))}$ is obtained from ϕ in the obvious way.

Remark 3.2.2.1. The mapping simplex $M(\phi)$ is equipped with a natural map $p : M(\phi) \to \Delta^n$; the fiber of p over the vertex j is isomorphic to the simplicial set A^j.

Remark 3.2.2.2. More generally, let $f : [m] \to [n]$ be an order-preserving map, inducing a map $\Delta^m \to \Delta^n$. Then $M(\phi) \times_{\Delta^n} \Delta^m$ is naturally isomorphic to $M(\phi')$, where the sequence ϕ' is given by

$$A^{f(0)} \leftarrow \cdots \leftarrow A^{f(m)}.$$

Notation 3.2.2.3. Let $\phi : A^0 \leftarrow \cdots \leftarrow A^n$ be a composable sequence of maps of simplicial sets. To give an edge e of $M(\phi)$, one must give a pair of integers $0 \le i \le j \le n$ and an edge $\bar{e} \in A^j$. We will say that e is *marked* if $\bar{e}$ is degenerate; let $\mathcal{E}$ denote the set of all marked edges of $M(\phi)$. Then the pair $(M(\phi), \mathcal{E})$ is a marked simplicial set which we will denote by $M^\natural(\phi)$.

Remark 3.2.2.4. There is a potential ambiguity between the terminology of Definition 3.1.1.9 and that of Notation 3.2.2.3. Suppose that $\phi : A^0 \leftarrow \cdots \leftarrow A^n$ is a composable sequence of maps and that $p : M(\phi) \to \Delta^n$ is a Cartesian fibration. Then $M(\phi)^\natural$ (Definition 3.1.1.9) and $M^\natural(\phi)$ (Notation 3.2.2.3) do not generally coincide as marked simplicial sets. We feel that there is little danger of confusion since it is very rare that p is a Cartesian fibration.

Remark 3.2.2.5. The construction of the mapping simplex is functorial in the sense that a commutative ladder

$$
\begin{array}{ccccc}
\phi : A^0 & \longleftarrow & \cdots & \longleftarrow & A^n \\
\downarrow{\scriptstyle f_0} & & \downarrow & & \downarrow{\scriptstyle f_n} \\
\psi : B^0 & \longleftarrow & \cdots & \longleftarrow & B^n
\end{array}
$$

induces a map $M(f) : M(\phi) \to M(\psi)$. Moreover, if each f_i is a categorical equivalence, then f is a categorical equivalence (this follows by induction on n using the fact that the Joyal model structure is left proper).

Definition 3.2.2.6. Let $p : X \to \Delta^n$ be a Cartesian fibration and let

$$\phi : A^0 \leftarrow \cdots \leftarrow A^n$$

be a composable sequence of maps. A map $q : M(\phi) \to X$ is a *quasi-equivalence* if it has the following properties:

(1) The diagram

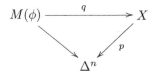

is commutative.

(2) The map q carries marked edges of $M(\phi)$ to p-Cartesian edges of S; in other words, q induces a map $M^\natural(\phi) \to X^\natural$ of marked simplicial sets.

(3) For $0 \leq i \leq n$, the induced map $A^i \to p^{-1}\{i\}$ is a categorical equivalence.

The goal of this section is to prove the following:

Proposition 3.2.2.7. *Let $p : X \to \Delta^n$ be a Cartesian fibration.*

(1) *There exists a composable sequence of maps*

$$\phi : A^0 \leftarrow A^1 \leftarrow \cdots \leftarrow A^n$$

and a quasi-equivalence $q : M(\phi) \to X$.

(2) *Let*

$$\phi : A^0 \leftarrow A^1 \leftarrow \cdots \leftarrow A^n$$

be a composable sequence of maps and let $q : M(\phi) \to X$ be a quasi-equivalence. For any map $T \to \Delta^n$, the induced map

$$M(\phi) \times_{\Delta^n} T \to X \times_{\Delta^n} T$$

is a categorical equivalence.

We first show that, to establish (2) of Proposition 3.2.2.7, it suffices to consider the case where T is a simplex:

Proposition 3.2.2.8. *Suppose we are given a diagram*

$$X \to Y \to Z$$

of simplicial sets. For any map $T \to Z$, we let X_T denote $X \times_Z T$ and Y_T denote $Y \times_Z T$. The following statements are equivalent:

(1) *For any map $T \to Z$, the induced map $X_T \to Y_T$ is a categorical equivalence.*

(2) *For any $n \geq 0$ and any map $\Delta^n \to Z$, the induced map $X_{\Delta^n} \to Y_{\Delta^n}$ is a categorical equivalence.*

Proof. It is clear that (1) implies (2). Let us prove the converse. Since the class of categorical equivalences is stable under filtered colimits, it suffices to consider the case where T has only finitely many nondegenerate simplices. We now work by induction on the dimension of T and the number of nondegenerate simplices contained in T. If T is empty, there is nothing to prove. Otherwise, we may write $T = T' \coprod_{\partial \Delta^n} \Delta^n$. By the inductive hypothesis, the maps

$$X_{T'} \to Y_{T'}$$

$$X_{\partial \Delta^n} \to Y_{\partial \Delta^n}$$

are categorical equivalences, and by assumption, $X_{\Delta^n} \to Y_{\Delta^n}$ is a categorical equivalence as well. We note that

$$X_T = X_{T'} \coprod_{X_{\partial \Delta^n}} X_{\Delta^n}$$

$$Y_T = Y_{T'} \coprod_{Y_{\partial \Delta^n}} Y_{\Delta^n}.$$

Since the Joyal model structure is left proper, these pushouts are homotopy pushouts and therefore categorically equivalent to one another. $\square$

Suppose $p : X \to \Delta^n$ is a Cartesian fibration and $q : M(\phi) \to X$ is a quasi-equivalence. Let $f : \Delta^m \to \Delta^n$ be any map. We note (see Remark 3.2.2.5) that $M(\phi) \times_{\Delta^n} \Delta^m$ may be identified with a mapping simplex $M(\phi')$ and that the induced map

$$M(\phi') \to X \times_{\Delta^n} \Delta^m$$

is again a quasi-equivalence. Consequently, to establish (2) of Proposition 3.2.2.7, it suffices to prove that every quasi-equivalence is a categorical equivalence. First, we need the following lemma.

Lemma 3.2.2.9. *Let*

$$\phi : A^0 \leftarrow \cdots \leftarrow A^n$$

be a composable sequence of maps between simplicial sets, where $n > 0$. Let y be a vertex of A^n and let the edge $e : y' \to y$ be the image of $\Delta^{\{n-1,n\}} \times \{y\}$ under the map $\Delta^n \times A^n \to M(\phi)$. Let x be any vertex of $M(\phi)$ which does not belong to the fiber A^n. Then composition with e induces a weak homotopy equivalence of simplicial sets

$$\mathrm{Map}_{\mathfrak{C}[M(\phi)]}(x, y') \to \mathrm{Map}_{\mathfrak{C}[M(\phi)]}(x, y).$$

Proof. Replacing ϕ by an equivalent diagram if necessary (using Remark 3.2.2.5), we may suppose that the map $A^n \to A^{n-1}$ is a cofibration. Let ϕ' denote the composable subsequence

$$A^0 \leftarrow \cdots \leftarrow A^{n-1}.$$

Let $\mathcal{C} = \mathfrak{C}[M(\phi)]$ and let $\mathcal{C}_- = \mathfrak{C}[M(\phi')] \subseteq \mathcal{C}$. There is a pushout diagram in Cat_Δ

$$
\begin{array}{ccc}
\mathfrak{C}[A^n \times \Delta^{n-1}] & \longrightarrow & \mathfrak{C}[A^n \times \Delta^n] \\
\downarrow & & \downarrow \\
\mathcal{C}_- & \longrightarrow & \mathcal{C}.
\end{array}
$$

This diagram is actually a homotopy pushout since Cat_Δ is a left proper model category and the top horizontal map is a cofibration. Now form the pushout

$$
\begin{array}{ccc}
\mathfrak{C}[A^n \times \Delta^{n-1}] & \longrightarrow & \mathfrak{C}[A^n \times (\Delta^{n-1} \coprod_{\{n-1\}} \Delta^{\{n-1,n\}})] \\
\downarrow & & \downarrow \\
\mathcal{C}_- & \longrightarrow & \mathcal{C}_0.
\end{array}
$$

This diagram is also a homotopy pushout. Since the diagram of simplicial sets

$$
\begin{array}{ccc}
\{n-1\} & \longrightarrow & \Delta^{\{n-1,n\}} \\
\downarrow & & \downarrow \\
\Delta^{n-1} & \longrightarrow & \Delta^n
\end{array}
$$

is homotopy coCartesian (with respect to the Joyal model structure), we deduce that the natural map $\mathcal{C}_0 \to \mathcal{C}$ is an equivalence of simplicial categories. It therefore suffices to prove that composition with e induces a weak homotopy equivalence

$$\mathrm{Map}_{\mathcal{C}_0}(x, y') \to \mathrm{Map}_{\mathcal{C}}(x, y).$$

Form a pushout square

$$
\begin{array}{ccc}
\mathfrak{C}[A^n \times \{n-1, n\}] & \longrightarrow & \mathfrak{C}[A^n] \times \mathfrak{C}[\Delta^{\{n-1,n\}}] \\
\downarrow & & \downarrow \\
\mathcal{C}_0 & \xrightarrow{\quad F \quad} & \mathcal{C}'.
\end{array}
$$

The left vertical map is a cofibration (since $A^n \to A^{n-1}$ is a cofibration of simplicial sets), and the upper horizontal map is an equivalence of simplicial categories (Corollary 2.2.5.6). Invoking the left properness of $\mathcal{Cat}_\Delta$, we conclude that F is an equivalence of simplicial categories. Consequently, it will suffice to prove that $\mathrm{Map}_{\mathcal{C}'}(F(x), F(y')) \to \mathrm{Map}_{\mathcal{C}'}(F(x), F(y))$ is a weak homotopy equivalence. We now observe that this map is an isomorphism of simplicial sets. $\square$

Proposition 3.2.2.10. *Let $p : X \to \Delta^n$ be a Cartesian fibration, let*

$$\phi : A^0 \leftarrow \cdots \leftarrow A^n$$

be a composable sequence of maps of simplicial sets and let $q : M(\phi) \to X$ be a quasi-equivalence. Then q is a categorical equivalence.

Proof. We proceed by induction on n. The result is obvious if $n = 0$, so let us assume that $n > 0$. Let ϕ' denote the composable sequence of maps

$$A^0 \leftarrow A^1 \leftarrow \cdots \leftarrow A^{n-1}$$

which is obtained from ϕ by omitting A^n. Let v denote the final vertex of Δ^n and let $T = \Delta^{\{0,\dots,n-1\}}$ denote the face of Δ^n which is opposite v. Let $X_v = X \times_{\Delta^n} \{v\}$ and $X_T = X \times_{\Delta^n} T$.

We note that $M(\phi) = M(\phi') \coprod_{A^n \times T} (A^n \times \Delta^n)$. We wish to show that the simplicial functor

$$F : \mathcal{C} \simeq \mathfrak{C}[M(\phi)] \simeq \mathfrak{C}[M(\phi')] \coprod_{\mathfrak{C}[A^n \times T]} \mathfrak{C}[A^n \times \Delta^n] \to \mathfrak{C}[X]$$

is an equivalence of simplicial categories. We note that $\mathcal{C}$ decomposes naturally into full subcategories $\mathcal{C}_+ = \mathfrak{C}[A^n \times \{v\}]$ and $\mathcal{C}_- = \mathfrak{C}[M(\phi')]$, having the property that $\mathrm{Map}_{\mathcal{C}}(X, Y) = \emptyset$ if $x \in \mathcal{C}_+$, $y \in \mathcal{C}_-$.

Similarly, $\mathcal{D} = \mathfrak{C}[X]$ decomposes into full subcategories $\mathcal{D}_+ = \mathfrak{C}[X_v]$ and $\mathcal{D}_- = \mathfrak{C}[X_T]$, satisfying $\mathrm{Map}_{\mathcal{D}}(x, y) = \emptyset$ if $x \in \mathcal{D}_+$ and $y \in \mathcal{D}_-$. We observe that F restricts to give an equivalence between $\mathcal{C}_-$ and $\mathcal{D}_-$ by assumption and gives an equivalence between $\mathcal{C}_+$ and $\mathcal{D}_+$ by the inductive hypothesis.

To complete the proof, it will suffice to show that if $x \in \mathcal{C}_-$ and $y \in \mathcal{C}_+$, then F induces a homotopy equivalence

$$\mathrm{Map}_{\mathcal{C}}(x, y) \to \mathrm{Map}_{\mathcal{D}}(F(x), F(y)).$$

We may identify the object $y \in \mathcal{C}_+$ with a vertex of A^n. Let e denote the edge of $M(\phi)$ which is the image of $\{y\} \times \Delta^{\{n-1,n\}}$ under the map $A^n \times \Delta^n \to M(\phi)$. We let $[e] : y' \to y$ denote the corresponding morphism in $\mathcal{C}$. We have a commutative diagram

$$
\begin{array}{ccc}
\mathrm{Map}_{\mathcal{C}_-}(x, y') & \longrightarrow & \mathrm{Map}_{\mathcal{C}}(x, y) \\
\downarrow & & \downarrow \\
\mathrm{Map}_{\mathcal{D}_-}(F(x), F(y')) & \longrightarrow & \mathrm{Map}_{\mathcal{D}}(F(x), F(y)).
\end{array}
$$

Here the left vertical arrow is a weak homotopy equivalence by the inductive hypothesis, and the bottom horizontal arrow (which is given by composition with $[e]$) is a weak homotopy equivalence because $q(e)$ is p-Cartesian. Consequently, to complete the proof, it suffices to show that the top horizontal arrow (given by composition with e) is a weak homotopy equivalence. This follows immediately from Lemma 3.2.2.9. □

To complete the proof of Proposition 3.2.2.7, it now suffices to show that for any Cartesian fibration $p : X \to \Delta^n$, there exists a quasi-equivalence $M(\phi) \to X$. In fact, we will prove something slightly stronger (in order to make our induction work):

Proposition 3.2.2.11. *Let $p : X \to \Delta^n$ be a Cartesian fibration of simplicial sets and A another simplicial set. Suppose we are given a commutative diagram of marked simplicial sets*

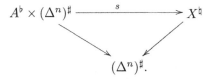

Then there exists a sequence of composable morphisms

$$\phi : A^0 \leftarrow \cdots \leftarrow A^n,$$

a map $A \to A^n$, and an extension

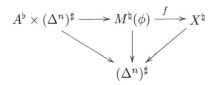

of the previous diagram, such that f is a quasi-equivalence.

Proof. The proof goes by induction on n. We begin by considering the fiber s over the final vertex v of Δ^n. The map $s_v : A \to X_v = X \times_{\Delta^n} \{v\}$ admits a factorization

$$A \xrightarrow{g} A^n \xrightarrow{h} S_v,$$

where g is a cofibration and h is a trivial Kan fibration. The smash product inclusion

$$(\{v\}^\sharp \times (A^n)^\flat) \coprod_{\{v\}^\sharp \times A^\flat} ((\Delta^n)^\sharp \times A^\flat) \subseteq (\Delta^n)^\sharp \times (A^n)^\flat$$

is marked anodyne (Proposition 3.1.2.3). Consequently, we deduce the existence of a dotted arrow f_0 as indicated in the diagram

$$
\begin{array}{ccc}
A^\flat \times (\Delta^n)^\sharp & \longrightarrow & X^\natural \\
\downarrow & \overset{f_0}{\nearrow} & \downarrow \\
(A^n)^\flat \times (\Delta^n)^\sharp & \longrightarrow & (\Delta^n)^\sharp
\end{array}
$$

of marked simplicial sets, where $f_0|(A^n \times \{n\}) = h$.

If $n = 0$, we are now done. If $n > 0$, then we apply the inductive hypothesis to the diagram

$$
\begin{array}{ccc}
(A^n)^\flat \times (\Delta^{n-1})^\sharp & \xrightarrow{\;f_0|A^n \times \Delta^{n-1}\;} & (X \times_{\Delta^n} \Delta^{n-1})^\natural \\
& \searrow \quad \swarrow & \\
& (\Delta^{n-1})^\sharp &
\end{array}
$$

to deduce the existence of a composable sequence of maps

$$\phi' : A^0 \leftarrow \cdots \leftarrow A^{n-1},$$

a map $A^n \to A^{n-1}$, and a commutative diagram

$$
\begin{array}{ccccc}
(A^n)^\flat \times (\Delta^{n-1})^\sharp & \longrightarrow & M^\natural(\phi') & \xrightarrow{\;f'\;} & (X \times_{\Delta^n} \Delta^{n-1})^\natural \\
& \searrow & \downarrow & \swarrow & \\
& & (\Delta^{n-1})^\sharp, & &
\end{array}
$$

where f' is a quasi-equivalence. We now define ϕ to be the result of appending the map $A^n \to A^{n-1}$ to the beginning of ϕ' and $f : M(\phi) \to X$ be the map obtained by amalgamating f_0 and f'. $\qquad\square$

Corollary 3.2.2.12. *Let $p : X \to S$ be a Cartesian fibration of simplicial sets and let $q : Y \to Z$ be a coCartesian fibration. Define new simplicial sets Y' and Z' equipped with maps $Y' \to S$, $Z' \to S$ via the formulas*

$$\mathrm{Hom}_S(K, Y') \simeq \mathrm{Hom}(X \times_S K, Y)$$

$$\mathrm{Hom}_S(K, Z') \simeq \mathrm{Hom}(X \times_S K, Z).$$

Then

(1) *Composition with q determines a coCartesian fibration $q' : Y' \to Z'$.*

(2) *An edge $\Delta^1 \to Y'$ is q'-coCartesian if and only if the induced map $\Delta^1 \times_S X \to Y$ carries p-Cartesian edges to q-coCartesian edges.*

Proof. Let us say that an edge of Y' is *special* if it satisfies the hypothesis of (2). Our first goal is to show that there is a sufficient supply of special edges in Y'. More precisely, we claim that given any edge $e : z \to z'$ in Z' and any vertex $\widetilde{z} \in Y'$ covering z, there exists a special edge $\widetilde{e} : \widetilde{z} \to \widetilde{z}'$ of Y' which covers e.

Suppose that the edge e covers an edge $e_0 : s \to s'$ in S. We can identify $\widetilde{z}$ with a map from X_s to Y. Using Proposition 3.2.2.7, we can choose a morphism $\phi : X'_s \leftarrow X'_{s'}$ and a quasi-equivalence $M(\phi) \to X \times_S \Delta^1$. Composing with $\widetilde{z}$, we obtain a map $X'_s \to Y$. Using Propositions 3.3.1.7 and A.2.3.1, we may reduce to the problem of providing a dotted arrow in the diagram

$$\begin{array}{ccc} X'_s & \longrightarrow & Y \\ \Big\uparrow & \nearrow & \Big\downarrow q \\ M(\phi) & \longrightarrow & Z \end{array}$$

which carries the marked edges of $M^\natural(\phi)$ to q-coCartesian edges of Y. This follows from the fact that $q^{X_s} : Y^{X_s} \to Z^{X_s}$ is a coCartesian fibration and the description of the q^{X_s}-coCartesian edges (Proposition 3.1.2.1).

To complete the proofs of (1) and (2), it will suffice to show that q' is an inner fibration and that every special edge of Y' is q'-coCartesian. For this, we must show that every lifting problem

$$\begin{array}{ccc} \Lambda^n_i & \xrightarrow{\ \sigma_0\ } & Y' \\ \Big\uparrow & \nearrow & \Big\downarrow q' \\ \Delta^n & \longrightarrow & Z' \end{array}$$

has a solution provided that either $0 < i < n$ or $i = 0$, $n \geq 2$, and $\sigma_0 | \Delta^{\{0,1\}}$ is special. We can reformulate this lifting problem using the diagram

$$\begin{array}{ccc} X \times_S \Lambda^n_i & \longrightarrow & Y \\ \Big\uparrow & \nearrow & \Big\downarrow q \\ X \times_S \Delta^n & \longrightarrow & Z. \end{array}$$

Using Proposition 3.2.2.7, we can choose a composable sequence of morphisms

$$\psi : X'_0 \leftarrow \cdots \leftarrow X'_n$$

and a quasi-equivalence $M(\psi) \to X \times_S \Delta^n$. Invoking Propositions 3.3.1.7 and A.2.3.1, we may reduce to the associated mapping problem

$$\begin{array}{ccc} M(\psi) \times_{\Delta^n} \Lambda^n_i & \longrightarrow & Y \\ \Big\downarrow & \nearrow & \Big\downarrow q \\ M(\psi) & \longrightarrow & Z. \end{array}$$

Since $i < n$, this is equivalent to the mapping problem

$$
\begin{array}{ccc}
X'_n \times \Lambda^n_i & \longrightarrow & Y \\
\cap\big\downarrow & & \big\downarrow q \\
X'_n \times \Delta^n & \longrightarrow & Z,
\end{array}
$$

which admits a solution by virtue of Proposition 3.1.2.1. ☐

Corollary 3.2.2.13. *Let $p : X \to S$ be a Cartesian fibration of simplicial sets, and let $q : Y \to S$ be a coCartesian fibration. Define a new simplicial set T equipped with a map $T \to S$ by the formula*

$$
\mathrm{Hom}_S(K, T) \simeq \mathrm{Hom}_S(X \times_S K, Y).
$$

Then:

(1) *The projection $r : T \to S$ is a coCartesian fibration.*

(2) *An edge $\Delta^1 \to Z$ is r-coCartesian if and only if the induced map $\Delta^1 \times_S X \to \Delta^1 \times_S Y$ carries p-Cartesian edges to q-coCartesian edges.*

Proof. Apply Corollary 3.2.2.12 in the case where $Z = S$. ☐

We conclude by noting the following property of quasi-equivalences (which is phrased using the terminology of §3.1.3):

Proposition 3.2.2.14. *Let $S = \Delta^n$, let $p : X \to S$ be a Cartesian fibration, let*

$$
\phi : A^0 \leftarrow \cdots \leftarrow A^n
$$

be a composable sequence of maps, and let $q : M(\phi) \to X$ be a quasi-equivalence. The induced map $M^\natural(\phi) \to X^\natural$ is a Cartesian equivalence in $(\mathrm{Set}^+_\Delta)_{/S}$.

Proof. We must show that for any Cartesian fibration $Y \to S$, the induced map of ∞-categories

$$
\mathrm{Map}^\flat_S(X^\natural, Y^\natural) \to \mathrm{Map}^\flat_S(M^\natural(\phi), Y^\natural)
$$

is a categorical equivalence. Because S is a simplex, the left side may be identified with a full subcategory of Y^X and the right side with a full subcategory of $Y^{M(\phi)}$. Since q is a categorical equivalence, the natural map $Y^X \to Y^{M(\phi)}$ is a categorical equivalence; thus, to complete the proof, it suffices to observe that a map of simplicial sets $f : X \to Y$ is compatible with the projection to S and preserves marked edges if and only if $q \circ f$ has the same properties. ☐

3.2.3 Straightening over a Simplex

Let S be a simplicial set, $\mathcal{C}$ a simplicial category, and $\phi : \mathfrak{C}[S]^{op} \to \mathcal{C}$ a simplicial functor. In §3.2.1, we introduced the straightening and unstraightening functors

$$(\mathrm{Set}_\Delta^+)_{/S} \underset{\mathrm{Un}_\phi^+}{\overset{St_\phi^+}{\rightleftarrows}} (\mathrm{Set}_\Delta^+)^{\mathcal{C}}.$$

In this section, we will prove that $(St_\phi^+, \mathrm{Un}_\phi^+)$ is a Quillen equivalence provided that ϕ is a categorical equivalence and S is a simplex (the case of a general simplicial set S will be treated in §3.2.4).

Our first step is to prove the result in the case where S is a point and ϕ is an isomorphism of simplicial categories. We can identify the functor $St_{\Delta^0}^+$ with the functor $T : \mathrm{Set}_\Delta^+ \to \mathrm{Set}_\Delta^+$ studied in §3.2.1. Consequently, Theorem 3.2.0.1 is an immediate consequence of Proposition 3.2.1.14:

Lemma 3.2.3.1. *The functor $T : \mathrm{Set}_\Delta^+ \to \mathrm{Set}_\Delta^+$ has a right adjoint U, and the pair (T, U) is a Quillen equivalence from Set_Δ^+ to itself.*

Proof. We have already established the existence of the unstraightening functor U in §3.2.1 and proved that (T, U) is a Quillen adjunction. To complete the proof, it suffices to show that the left derived functor of T (which we may identify with T because every object of Set_Δ^+ is cofibrant) is an equivalence from the homotopy category of Set_Δ^+ to itself. But Proposition 3.2.1.14 asserts that T is isomorphic to the identity functor on the homotopy category of Set_Δ^+. $\square$

Let us now return to the case of a general equivalence $\phi : \mathfrak{C}[S] \to \mathcal{C}^{op}$. Since we know that $(St_\phi^+, \mathrm{Un}_\phi^+)$ give a Quillen adjunction between $(\mathrm{Set}_\Delta^+)_{/S}$ and $(\mathrm{Set}_\Delta^+)^{\mathcal{C}}$, it will suffice to prove that the unit and counit

$$u : \mathrm{id} \to R\,\mathrm{Un}_\phi^+ \circ LSt_\phi^+$$

$$v : LSt_\phi^+ \circ R\,\mathrm{Un}_\phi^+ \to \mathrm{id}$$

are weak equivalences. Our first step is to show that $R\,\mathrm{Un}_\phi^+$ detects weak equivalences: this reduces the problem of proving that v is an equivalence to the problem of proving that u is an equivalence.

Lemma 3.2.3.2. *Let S be a simplicial set, $\mathcal{C}$ a simplicial category, and $\phi : \mathfrak{C}[S] \to \mathcal{C}^{op}$ an essentially surjective functor. Let $p : \mathcal{F} \to \mathcal{G}$ be a map between (weakly) fibrant objects of $(\mathrm{Set}_\Delta^+)^{\mathcal{C}}$. Suppose that $\mathrm{Un}_\phi^+(p) : \mathrm{Un}_\phi^+ \mathcal{F} \to \mathrm{Un}_\phi^+ \mathcal{G}$ is a Cartesian equivalence. Then p is an equivalence.*

Proof. Since ϕ is essentially surjective, it suffices to prove that $\mathcal{F}(C) \to \mathcal{F}(D)$ is a Cartesian equivalence for every object $C \in \mathcal{C}$ which lies in the image of ϕ. Let s be a vertex of S with $\psi(s) = C$. Let $i : \{s\} \to S$ denote the inclusion and let $i^* : (\mathrm{Set}_\Delta^+)_{/S} \to \mathrm{Set}_\Delta^+$ denote the functor of passing to the fiber over s:

$$i^* X = X_s = X \times_{S^\sharp} \{s\}^\sharp.$$

Let $i_!$ denote the left adjoint to i^*. Let $\{C\}$ denote the trivial category with one object (and only the identity morphism), and let $j : \{C\} \to \mathcal{C}$ be the simplicial functor corresponding to the inclusion of C as an object of $\mathcal{C}$. According to Proposition 3.2.1.4, we have a natural identification of functors

$$St_\phi^+ \circ i_! \simeq j_! \circ T.$$

Passing to adjoints, we get another identification

$$i^* \circ \mathrm{Un}_\phi^+ \simeq U \circ j^*$$

from $(\mathrm{Set}_\Delta^+)^\mathcal{C}$ to Set_Δ^+. Here U denotes the right adjoint of T.

According to Lemma 3.2.3.1, the functor U detects equivalences between fibrant objects of Set_Δ^+. It therefore suffices to prove that $U(j^* \mathcal{F}) \to U(j^* \mathcal{G})$ is a Cartesian equivalence. Using the identification above, we are reduced to proving that

$$\mathrm{Un}_\phi^+(\mathcal{F})_s \to \mathrm{Un}_\phi^+(\mathcal{G})_s$$

is a Cartesian equivalence. But $\mathrm{Un}_\phi^+(\mathcal{F})$ and $\mathrm{Un}_\phi^+(\mathcal{G})$ are fibrant objects of $(\mathrm{Set}_\Delta^+)_{/S}$ and therefore correspond to Cartesian fibrations over S: the desired result now follows from Proposition 3.1.3.5.   □

We have now reduced the proof of Theorem 3.2.0.1 to the problem of showing that if $\phi : \mathfrak{C}[S] \to \mathcal{C}^{op}$ is an equivalence of simplicial categories, then the unit transformation

$$u : \mathrm{id} \to R\,\mathrm{Un}_\phi^+ \circ St_\phi^+$$

is an isomorphism of functors from the homotopy category $\mathrm{h}(\mathrm{Set}_\Delta^+)_{/S}$ to itself.

Our first step is to analyze the effect of the straightening functor St_ϕ^+ on a mapping simplex. We will need a bit of notation. For any $X \in (\mathrm{Set}_\Delta^+)_{/S}$ and any vertex s of S, we let X_s denote the fiber $X \times_{S^\sharp} \{s\}^\sharp$ and let i^s denote the composite functor

$$\{s\} \hookrightarrow \mathfrak{C}[S] \xrightarrow{\phi} \mathcal{C}^{op}$$

of simplicial categories. According to Proposition 3.2.1.4, there is a natural identification

$$St_\phi^+(X_s) \simeq i_!^s T(X_s)$$

which induces a map

$$\psi_s^X : T(X_s) \to St_\phi^+(X)(s).$$

Lemma 3.2.3.3. *Let*

$$\theta : A^0 \leftarrow \cdots \leftarrow A^n$$

be a composable sequence of maps of simplicial sets and let $M^{\natural}(\theta) \in (\mathrm{Set}^+_{\Delta})_{\Delta^n}$ be its mapping simplex. For each $0 \leq i \leq n$, the map

$$\psi_i^{M^{\natural}(\theta)} : T(A^i)^{\flat} \to St^+_{\Delta^n}(M^{\natural}(\theta))(i)$$

is a Cartesian equivalence in Set^+_{Δ}.

Proof. The proof proceeds by induction on n. We first observe that $\psi_n^{M^{\natural}(\theta)}$ is an isomorphism; we may therefore restrict our attention to $i < n$. Let θ' be the composable sequence

$$A^0 \leftarrow \cdots \leftarrow A^{n-1}$$

and $M^{\natural}(\theta')$ its mapping simplex, which we may regard as an object of either $(\mathrm{Set}^+_{\Delta})_{/\Delta^n}$ or $(\mathrm{Set}^+_{\Delta})_{/\Delta^{n-1}}$.

For $i < n$, we have a commutative diagram

$$
\begin{array}{ccc}
 & St^+_{\Delta^n}(M^{\natural}(\theta'))(i) & \\
 \psi_i^{M^{\natural}(\theta')} \nearrow & & \searrow f_i \\
 T((A^i)^{\flat}) & \xrightarrow{\hspace{3cm}} & St^+_{\Delta^n}(M^{\natural}(\theta))(i).
\end{array}
$$

By Proposition 3.2.1.4, $St^+_{\Delta^n} M^{\natural}(\theta') \simeq j_! St^+_{\Delta^{n-1}} M^{\natural}(\theta')$, where $j : \mathfrak{C}[\Delta^{n-1}] \to \mathfrak{C}[\Delta^n]$ denotes the inclusion. Consequently, the inductive hypothesis implies that the maps

$$T(A^i)^{\flat} \to St^+_{\Delta^{n-1}}(M^{\natural}(\theta'))(i)$$

are Cartesian equivalences for $i < n$. It now suffices to prove that f_i is a Cartesian equivalence for $i < n$.

We observe that there is a (homotopy) pushout diagram

$$
\begin{array}{ccc}
(A^n)^{\flat} \times (\Delta^{n-1})^{\sharp} & \longrightarrow & (A^n)^{\flat} \times (\Delta^n)^{\sharp} \\
\downarrow & & \downarrow \\
M^{\natural}(\theta') & \longrightarrow & M^{\natural}(\theta).
\end{array}
$$

Since $St^+_{\Delta^n}$ is a left Quillen functor, it induces a homotopy pushout diagram

$$
\begin{array}{ccc}
St^+_{\Delta^n}((A^n)^{\flat} \times (\Delta^{n-1})^{\sharp}) & \xrightarrow{g} & St^+_{\Delta^n}((A^n)^{\flat} \times (\Delta^n)^{\sharp}) \\
\downarrow & & \downarrow \\
St^+_{\Delta^n} M^{\natural}(\theta') & \longrightarrow & St^+_{\Delta^n} M^{\natural}(\theta)
\end{array}
$$

in $(\mathrm{Set}^+_{\Delta})^{\mathfrak{C}}$. We are therefore reduced to proving that g induces a Cartesian equivalence after evaluation at any $i < n$.

According to Proposition 3.2.1.13, the vertical maps of the diagram

$$St^+_{\Delta^n}((A^n)^\flat \times (\Delta^{n-1})^\sharp) \longrightarrow St^+_{\Delta^n}((A^n)^\flat \times (\Delta^n)^\sharp)$$

$$\downarrow \qquad\qquad\qquad\qquad\qquad \downarrow$$

$$T(A^n)^\flat \boxtimes St^+_{\Delta^n}(\Delta^{n-1})^\sharp \longrightarrow T(A^n)^\flat \boxtimes St^+_{\Delta^n}(\Delta^n)^\sharp$$

are Cartesian equivalences. To complete the proof we must show that

$$St^+_{\Delta^n}(\Delta^{n-1})^\sharp \to St^+_{\Delta^n}(\Delta^n)^\sharp$$

induces a Cartesian equivalence when evaluated at any $i < n$. Consider the diagram

$$
\begin{array}{ccc}
\{n-1\}^\sharp & \longrightarrow & (\Delta^{n-1})^\sharp \\
\downarrow & & \downarrow \\
(\Delta^{\{n-1,n\}})^\sharp & \longrightarrow & (\Delta^n)^\sharp.
\end{array}
$$

The horizontal arrows are marked anodyne. It therefore suffices to show that

$$St^+_{\Delta^n}\{n-1\}^\sharp \to St^+_{\Delta^n}(\Delta^{\{n-1,n\}})^\sharp$$

induces Cartesian equivalences when evaluated at any $i < n$. This follows from an easy computation.   □

Proposition 3.2.3.4. *Let $n \geq 0$. Then the Quillen adjunction*

$$(\mathsf{Set}^+_\Delta)_{/\Delta^n} \underset{\mathrm{Un}^+_{\Delta^n}}{\overset{St^+_{\Delta^n}}{\rightleftarrows}} (\mathsf{Set}^+_\Delta)^{\mathfrak{C}[\Delta^n]}$$

is a Quillen equivalence.

Proof. As we have argued above, it suffices to show that the unit

$$\mathrm{id} \to R\,\mathrm{Un}^+_\phi \circ St^+_{\Delta^n}$$

is an isomorphism of functors from $h(\mathsf{Set}^+_\Delta)_{\Delta^n}$ to itself. In other words, we must show that given an object $X \in (\mathsf{Set}^+_\Delta)_{/\Delta^n}$ and a weak equivalence

$$St^+_{\Delta^n}X \to \mathcal{F},$$

where $\mathcal{F} \in (\mathsf{Set}^+_\Delta)^{\mathfrak{C}[\Delta^n]}$ is fibrant, the adjoint map

$$j : X \to \mathrm{Un}^+_{\Delta^n}\mathcal{F}$$

is a Cartesian equivalence in $(\mathsf{Set}^+_\Delta)_{/\Delta^n}$.

Choose a fibrant replacement for X: that is, a Cartesian equivalence $X \to Y^\natural$, where $Y \to \Delta^n$ is a Cartesian fibration. According to Proposition 3.2.2.7, there exists a composable sequence of maps

$$\theta : A^0 \leftarrow \cdots \leftarrow A^n$$

and a quasi-equivalence $M^\natural(\theta) \to Y^\natural$. Proposition 3.2.2.14 implies that $M^\natural(\theta) \to Y^\natural$ is a Cartesian equivalence. Thus, X is equivalent to $M^\natural(\theta)$ in the homotopy category of $(\mathrm{Set}_\Delta^+)_{/\Delta^n}$ and we are free to replace X by $M^\natural(\theta)$, thereby reducing to the case where X is a mapping simplex.

We wish to prove that j is a Cartesian equivalence. Since $\mathrm{Un}_{\Delta^n}^+ \mathcal{F}$ is fibrant, Proposition 3.2.2.14 implies that it suffices to show that j is a quasi-equivalence: in other words, we need to show that the induced map of fibers $j_s : X_s \to (\mathrm{Un}_{\Delta^n}^+ \mathcal{F})_s$ is a Cartesian equivalence for each vertex s of Δ^n. As in the proof of Lemma 3.2.3.2, we may identify $(\mathrm{Un}_{\Delta^n}^+ \mathcal{F})_s$ with $U(\mathcal{F}(s))$, where U is the right adjoint to T. By Lemma 3.2.3.1, $X_s \to U(\mathcal{F}(s))$ is a Cartesian equivalence if and only if the adjoint map $T(X_s) \to \mathcal{F}(s)$ is a Cartesian equivalence. This map factors as a composition

$$T(X_s) \to St_{\Delta^n}^+(X)(s) \to \mathcal{F}(s).$$

The map on the left is a Cartesian equivalence by Lemma 3.2.3.3, and the map on the right also a Cartesian equivalence, by virtue of the assumption that $St_{\Delta^n}^+ X \to \mathcal{F}$ is a weak equivalence. □

3.2.4 Straightening in the General Case

Let S be a simplicial set and $\phi : \mathfrak{C}[S] \to \mathcal{C}^{op}$ an equivalence of simplicial categories. Our goal in this section is to complete the proof of Theorem 3.2.0.1 by showing that $(St_\phi^+, \mathrm{Un}_\phi^+)$ is a Quillen equivalence between $(\mathrm{Set}_\Delta^+)_{/S}$ and $(\mathrm{Set}_\Delta^+)^{\mathcal{C}}$. In §3.2.3, we handled the case where S is a simplex (and ϕ an isomorphism) by verifying that the unit map $\mathrm{id} \to R\,\mathrm{Un}_\phi^+ \circ St_\phi^+$ is an isomorphism of functors from $h(\mathrm{Set}_\Delta^+)_{/S}$ to itself.

Here is the idea of the proof. Without loss of generality, we may suppose that ϕ is an isomorphism (since the pair $(\phi_!, \phi^*)$ is a Quillen equivalence between $(\mathrm{Set}_\Delta^+)^{\mathfrak{C}[S]^{op}}$ and $(\mathrm{Set}_\Delta^+)^{\mathcal{C}}$ by Proposition A.3.3.8). We wish to show that Un_ϕ^+ induces an equivalence from the homotopy category of $(\mathrm{Set}_\Delta^+)^{\mathcal{C}}$ to the homotopy category of $(\mathrm{Set}_\Delta^+)_{/S}$. According to Proposition 3.2.3.4, this is true whenever S is a simplex. In the general case, we would like to regard $(\mathrm{Set}_\Delta^+)^{\mathcal{C}}$ and $(\mathrm{Set}_\Delta^+)_{/S}$ as somehow built out of pieces which are associated to simplices and deduce that Un_ϕ^+ is an equivalence because it is an equivalence on each piece. In order to make this argument work, it is necessary to work not just with the homotopy categories of $(\mathrm{Set}_\Delta^+)^{\mathcal{C}}$ and $(\mathrm{Set}_\Delta^+)_{/S}$ but also with the simplicial categories which give rise to them.

We recall that both $(\mathrm{Set}_\Delta^+)^{\mathcal{C}}$ and $(\mathrm{Set}_\Delta^+)_{/S}$ are *simplicial* model categories with respect to the simplicial mapping spaces defined by

$$\mathrm{Hom}_{\mathrm{Set}_\Delta}(K, \mathrm{Map}_{(\mathrm{Set}_\Delta^+)^{\mathcal{C}}}(\mathcal{F}, \mathcal{G})) = \mathrm{Hom}_{(\mathrm{Set}_\Delta^+)^{\mathcal{C}}}(\mathcal{F} \boxtimes K^\sharp, \mathcal{G})$$

$$\mathrm{Hom}_{\mathrm{Set}_\Delta}(K, \mathrm{Map}_{(\mathrm{Set}_\Delta^+)_S}(X, Y)) = \mathrm{Hom}_{(\mathrm{Set}_\Delta^+)_{/S}}(X \times K^\sharp, Y).$$

The functor St_ϕ^+ is not a simplicial functor. However, it is *weakly* compatible with the simplicial structure in the sense that there is a natural map

$$St_\phi^+(X \boxtimes K^\sharp) \to (St_\phi^+ X) \boxtimes K^\sharp$$

for any $X \in (\mathrm{Set}_\Delta^+)_{/S}$, $K \in \mathrm{Set}_\Delta$ (according to Corollary 3.2.1.15, this map is a weak equivalence in $(\mathrm{Set}_\Delta^+)^{\mathcal{C}}$). Passing to adjoints, we get natural maps

$$\mathrm{Map}_{(\mathrm{Set}_\Delta^+)^{\mathcal{C}}}(\mathcal{F}, \mathcal{G}) \to \mathrm{Map}_S^\sharp(\mathrm{Un}_\phi^+ \mathcal{F}, \mathrm{Un}_\phi^+ \mathcal{G}).$$

In other words, Un_ϕ^+ *does* have the structure of a simplicial functor. We now invoke Proposition A.3.1.10 to deduce the following:

Lemma 3.2.4.1. *Let S be a simplicial set, $\mathcal{C}$ a simplicial category, and $\phi : \mathfrak{C}[S] \to \mathcal{C}^{op}$ a simplicial functor. The following are equivalent:*

(1) *The Quillen adjunction $(\mathrm{St}_\phi^+, \mathrm{Un}_\phi^+)$ is a Quillen equivalence.*

(2) *The functor Un_ϕ^+ induces an equivalence of simplicial categories*

$$(\mathrm{Un}_\phi^+)^\circ : ((\mathrm{Set}_\Delta^+)^{\mathcal{C}})^\circ \to ((\mathrm{Set}_\Delta^+)_{/S})^\circ,$$

where $((\mathrm{Set}_\Delta^+)^{\mathcal{C}})^\circ$ denotes the full (simplicial) subcategory of $((\mathrm{Set}_\Delta^+)^{\mathcal{C}})$ consisting of fibrant-cofibrant objects and $((\mathrm{Set}_\Delta^+)_{/S})^\circ$ denotes the full (simplicial) subcategory of $(\mathrm{Set}_\Delta^+)_{/S}$ consisting of fibrant-cofibrant objects.

Consequently, to complete the proof of Theorem 3.2.0.1, it will suffice to show that if ϕ is an equivalence of simplicial categories, then $(\mathrm{Un}_\phi^+)^\circ$ is an equivalence of simplicial categories. The first step is to prove that $(\mathrm{Un}_\phi^+)^\circ$ is fully faithful.

Lemma 3.2.4.2. *Let $S' \subseteq S$ be simplicial sets and let $p : X \to S$, $q : Y \to S$ be Cartesian fibrations. Let $X' = X \times_S S'$ and $Y' = Y \times_S S'$. The restriction map*

$$\mathrm{Map}_S^\sharp(X^\natural, Y^\natural) \to \mathrm{Map}_{S'}^\sharp(X'^\natural, Y'^\natural)$$

is a Kan fibration.

Proof. It suffices to show that the map $Y^\natural \to S$ has the right lifting property with respect to the inclusion

$$(X'^\natural \times B^\sharp) \coprod_{X'^\natural \times A^\sharp} (X^\natural \times A^\sharp) \subseteq X^\natural \times B^\sharp$$

for any anodyne inclusion of simplicial sets $A \subseteq B$.

But this is a smash product of a marked cofibration $X'^\natural \to X^\natural$ (in $(\mathrm{Set}_\Delta^+)_{/S}$) and a trivial marked cofibration $A^\sharp \to B^\sharp$ (in Set_Δ^+) and is therefore a trivial marked cofibration. We conclude by observing that $Y^\natural$ is a fibrant object of $(\mathrm{Set}_\Delta^+)_{/S}$ (Proposition 3.1.4.1). □

Proof of Theorem 3.2.0.1. For each simplicial set S, let $(\mathrm{Set}_\Delta^+)_f^{\mathfrak{C}[S]^{op}}$ denote the category of projectively fibrant objects of $(\mathrm{Set}_\Delta^+)^{\mathfrak{C}[S]^{op}}$ and let W_S be the class of weak equivalences in $(\mathrm{Set}_\Delta^+)_f^{\mathfrak{C}[S]^{op}}$. Let W_S' be the collection of pointwise equivalences in $(\mathrm{Set}_\Delta^+)^\circ_{/S}$. We have a commutative diagram of simplicial categories

$$
\begin{array}{ccc}
((\mathrm{Set}_\Delta^+)^{\mathfrak{C}[S]^{op}})^\circ & \xrightarrow{\;\mathrm{Un}_S^+\;} & (\mathrm{Set}_\Delta^+)^\circ_{/S} \\
\downarrow & & \downarrow{\scriptstyle \psi_S} \\
(\mathrm{Set}_\Delta^+)^{\mathfrak{C}[S]^{op}}_f[W_S^{-1}] & \xrightarrow{\;\phi_S\;} & (\mathrm{Set}_\Delta^+)^\circ_{/S}[W'^{-1}_S]
\end{array}
$$

(see Notation A.3.5.1). In view of Lemma 3.2.4.1, it will suffice to show that the upper horizontal map is an equivalence of simplicial categories. Lemma A.3.6.17 implies that the left vertical map is an equivalence. Using Lemma 2.2.3.6 and Remark A.3.2.14, we deduce that the right vertical map is also an equivalence. It will therefore suffice to show that ϕ_S is an equivalence.

Let $\mathcal{U}$ denote the collection of simplicial sets S for which ϕ_S is an equivalence. We will show that $\mathcal{U}$ satisfies the hypotheses of Lemma 2.2.3.5 and therefore contains every simplicial set S. Conditions (i) and (ii) are obviously satisfied, and condition (iii) follows from Lemma 3.2.4.1 and Proposition 3.2.3.4. We will verify condition (iv); the proof of (v) is similar.

Applying Corollary A.3.6.18, we deduce:

($*$) The functor $S \mapsto (\mathrm{Set}_\Delta^+)^{\mathfrak{C}[S]^{op}}_f[W_S^{-1}]$ carries homotopy colimit diagrams indexed by a partially ordered set to homotopy limit diagrams in $\mathcal{C}at_\Delta$.

Suppose we are given a pushout diagram

$$
\begin{array}{ccc}
X & \longrightarrow & X' \\
\downarrow{\scriptstyle f} & & \downarrow \\
Y & \longrightarrow & Y'
\end{array}
$$

in which $X, X', Y \in \mathcal{U}$, where f is a cofibration. We wish to prove that $Y' \in \mathcal{U}$. We have a commutative diagram

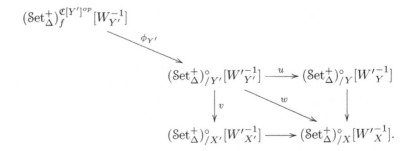

Using ($*$) and Corollary A.3.2.28, we deduce that $\phi_{Y'}$ is an equivalence if and only if, for every pair of objects $x, y \in (\mathrm{Set}_\Delta^+)^\circ_{/Y'}[W'^{-1}_{Y'}]$, the diagram of

simplicial sets

$$\text{Map}_{(\text{Set}_\Delta^+)^\circ_{/Y'}[W'^{-1}_{Y'}]}(x, y) \longrightarrow \text{Map}_{(\text{Set}_\Delta^+)^\circ_{/Y}[W'^{-1}_Y]}(u(x), u(y))$$

$$\downarrow \qquad\qquad\qquad\qquad\qquad\qquad \downarrow$$

$$\text{Map}_{(\text{Set}_\Delta^+)^\circ_{/X'}[W'^{-1}_{X'}]}(v(x), v(y)) \longrightarrow \text{Map}_{(\text{Set}_\Delta^+)^\circ_{/X}[W'^{-1}_X]}(w(x), w(y))$$

is homotopy Cartesian. Since $\psi_{Y'}$ is a weak equivalence of simplicial categories, we may assume without loss of generality that $x = \psi_{Y'}(\overline{x})$ and $y = \psi_{Y'}(\overline{y})$ for some $\overline{x}, \overline{y} \in (\text{Set}_\Delta^+)^\circ_{/Y'}$. It will therefore suffice to prove that the equivalent diagram

$$\text{Map}^\sharp_{Y'}(\overline{x}, \overline{y}) \longrightarrow \text{Map}^\sharp_Y(\overline{u}(\overline{x}), \overline{u}(\overline{y}))$$

$$\downarrow \qquad\qquad\qquad\qquad\qquad \downarrow$$

$$\text{Map}^\sharp_{X'}(\overline{v}(\overline{x}), \overline{v}(\overline{y})) \xrightarrow{\ g\ } \text{Map}^\sharp_X(\overline{w}(\overline{x}), \overline{w}(\overline{y}))$$

is homotopy Cartesian. But this diagram is a pullback square, and the map g is a Kan fibration by Lemma 3.2.4.2. $\qquad\qquad\qquad\qquad\qquad\qquad\qquad\square$

3.2.5 The Relative Nerve

In §3.1.3, we defined the straightening and unstraightening functors, which give rise to a Quillen equivalence of model categories

$$(\text{Set}_\Delta^+)_{/S} \underset{\text{Un}_\phi^+}{\overset{St_\phi^+}{\rightleftarrows}} (\text{Set}_\Delta^+)^{\mathcal{C}}$$

whenever $\phi : \mathfrak{C}[S] \to \mathcal{C}^{op}$ is a weak equivalence of simplicial categories. For many purposes, these constructions are unnecessarily complicated. For example, suppose that $\mathcal{F} : \mathcal{C} \to \text{Set}_\Delta^+$ is a (weakly) fibrant diagram, so that $\text{Un}_\phi^+(\mathcal{F})$ is a fibrant object of $(\text{Set}_\Delta^+)_{/S}$ corresponding to a Cartesian fibration of simplicial sets $X \to S$. For every vertex $s \in S$, the fiber X_s is an ∞-category which is equivalent to $\mathcal{F}(\phi(s))$ but usually not isomorphic to $\mathcal{F}(\phi(s))$. In the special case where $\mathcal{C}$ is an ordinary category and $\phi : \mathfrak{C}[N(\mathcal{C})^{op}] \to \mathcal{C}^{op}$ is the counit map, there is another version of unstraightening construction Un_ϕ^+ which does not share this defect. Our goal in this section is to introduce this simpler construction, which we call the *marked relative nerve* $\mathcal{F} \mapsto N_\mathcal{F}^+(\mathcal{C})$, and to study its basic properties.

Remark 3.2.5.1. To simplify the exposition which follows, the relative nerve functor introduced below will actually be an alternative to the *opposite* of the unstraightening functor

$$\mathcal{F} \mapsto (\text{Un}_\phi^+ \mathcal{F}^{op})^{op},$$

which is a right Quillen functor from the projective model structure on $(\text{Set}_\Delta^+)^{\mathcal{C}}$ to the coCartesian model structure on $(\text{Set}_\Delta^+)_{/N(\mathcal{C})}$.

Definition 3.2.5.2. Let $\mathcal{C}$ be a small category and let $f : \mathcal{C} \to \mathrm{Set}_\Delta$ be a functor. We define a simplicial set $\mathrm{N}_f(\mathcal{C})$, the *nerve of $\mathcal{C}$ relative to f*, as follows. For every nonempty finite linearly ordered set J, a map $\Delta^J \to \mathrm{N}_f(\mathcal{C})$ consists of the following data:

(1) A functor σ from J to $\mathcal{C}$.

(2) For every nonempty subset $J' \subseteq J$ having a maximal element j', a map $\tau(J') : \Delta^{J'} \to \mathcal{F}(\sigma(j'))$.

(3) For nonempty subsets $J'' \subseteq J' \subseteq J$, with maximal elements $j'' \in J''$, $j' \in J'$, the diagram

$$
\begin{array}{ccc}
\Delta^{J''} & \xrightarrow{\ \tau(J'')\ } & f(\sigma(j'')) \\
\downarrow & & \downarrow \\
\Delta^{J'} & \xrightarrow{\ \tau(J')\ } & f(\sigma(j'))
\end{array}
$$

is required to commute.

Remark 3.2.5.3. Let $\mathcal{J}$ denote the linearly ordered set $[n]$, regarded as a category, and let $f : \mathcal{J} \to \mathrm{Set}_\Delta$ correspond to a composable sequence of morphisms $\phi : X_0 \to \cdots \to X_n$. Then $\mathrm{N}_f(\mathcal{J})$ is closely related to the mapping simplex $M^{op}(\phi)$ introduced in §3.2.2. More precisely, there is a canonical map $\mathrm{N}_f(\mathcal{J}) \to M^{op}(\phi)$ compatible with the projection to Δ^n, which induces an isomorphism on each fiber.

Remark 3.2.5.4. The simplicial set $\mathrm{N}_f(\mathcal{C})$ of Definition 3.2.5.2 depends functorially on f. When f takes the constant value Δ^0, there is a canonical isomorphism $\mathrm{N}_f(\mathcal{C}) \simeq \mathrm{N}(\mathcal{C})$. In particular, for *any* functor f, there is a canonical map $\mathrm{N}_f(\mathcal{C}) \to \mathrm{N}(\mathcal{C})$; the fiber of this map over an object $C \in \mathcal{C}$ can be identified with the simplicial set $f(C)$.

Remark 3.2.5.5. Let $\mathcal{C}$ be a small ∞-category. The construction $f \mapsto \mathrm{N}_f(\mathcal{C})$ determines a functor from $(\mathrm{Set}_\Delta)^{\mathcal{C}}$ to $(\mathrm{Set}_\Delta)_{/\mathrm{N}(\mathcal{C})}$. This functor admits a left adjoint, which we will denote by $X \mapsto \mathfrak{F}_X(\mathcal{C})$ (the existence of this functor follows from the adjoint functor theorem). If $X \to \mathrm{N}(\mathcal{C})$ is a left fibration, then $\mathfrak{F}_X(\mathcal{C})$ is a functor $\mathcal{C} \to \mathrm{Set}_\Delta$ which assigns to each $C \in \mathcal{C}$ a simplicial set which is weakly equivalent to the fiber $X_C = X \times_{\mathrm{N}(\mathcal{C})} \{C\}$; this follows from Proposition 3.2.5.18 below.

Example 3.2.5.6. Let $\mathcal{C}$ be a small category and regard $\mathrm{N}(\mathcal{C})$ as an object of $(\mathrm{Set}_\Delta)_{/\mathrm{N}(\mathcal{C})}$ via the identity map. Then $\mathfrak{F}_{\mathrm{N}(\mathcal{C})}(\mathcal{C}) \in (\mathrm{Set}_\Delta)^{\mathcal{C}}$ can be identified with the functor $C \mapsto \mathrm{N}(\mathcal{C}_{/C})$.

Remark 3.2.5.7. Let $g : \mathcal{C} \to \mathcal{D}$ be a functor between small categories and let $f : \mathcal{D} \to \mathrm{Set}_\Delta$ be a diagram. There is a canonical isomorphism of simplicial sets $\mathrm{N}_{f \circ g}(\mathcal{C}) \simeq \mathrm{N}_f(\mathcal{D}) \times_{\mathrm{N}(\mathcal{D})} \mathrm{N}(\mathcal{C})$. In other words, the diagram of categories

$$(\mathfrak{Set}_\Delta)^{\mathcal{D}} \xrightarrow{\ g^* \ } (\mathfrak{Set}_\Delta)^{\mathcal{C}}$$

$$\Big\downarrow \mathrm{N}_\bullet(\mathcal{D}) \qquad\qquad \Big\downarrow \mathrm{N}_\bullet(\mathcal{C})$$

$$(\mathfrak{Set}_\Delta)_{/\,\mathrm{N}(\mathcal{D})} \xrightarrow{\ \mathrm{N}(g)^* \ } (\mathfrak{Set}_\Delta)_{/\,\mathrm{N}(\mathcal{C})}$$

commutes up to canonical isomorphism. Here g^* denotes the functor given by composition with g, and $\mathrm{N}(g)^*$ the functor given by pullback along the map of simplicial sets $\mathrm{N}(g) : \mathrm{N}(\mathcal{C}) \to \mathrm{N}(\mathcal{D})$.

Remark 3.2.5.8. Combining Remarks 3.2.5.5 and 3.2.5.7, we deduce that for any functor $g : \mathcal{C} \to \mathcal{D}$ between small categories, the diagram of left adjoints

$$(\mathfrak{Set}_\Delta)^{\mathcal{D}} \xleftarrow{\ g_! \ } (\mathfrak{Set}_\Delta)^{\mathcal{C}}$$

$$\Big\uparrow \mathfrak{F}_\bullet(\mathcal{D}) \qquad\qquad \Big\uparrow \mathfrak{F}_\bullet(\mathcal{C})$$

$$(\mathfrak{Set}_\Delta)_{/\,\mathrm{N}(\mathcal{D})} \xleftarrow{\qquad} (\mathfrak{Set}_\Delta)_{/\,\mathrm{N}(\mathcal{C})}$$

commutes up to canonical isomorphism; here $g_!$ denotes the functor of left Kan extension along g, and the bottom arrow is the forgetful functor given by composition with $\mathrm{N}(g) : \mathrm{N}(\mathcal{C}) \to \mathrm{N}(\mathcal{D})$.

Notation 3.2.5.9. Let $\mathcal{C}$ be a small category and let $f : \mathcal{C} \to \mathfrak{Set}_\Delta$ be a functor. We let f^{op} denote the functor $\mathcal{C} \to \mathfrak{Set}_\Delta$ described by the formula $f^{op}(C) = f(C)^{op}$. We will use a similar notation in the case where f is a functor from $\mathcal{C}$ to the category $\mathfrak{Set}_\Delta^+$ of marked simplicial sets.

Remark 3.2.5.10. Let $\mathcal{C}$ be a small category, let $S = \mathrm{N}(\mathcal{C})^{op}$, and let $\phi : \mathfrak{C}[S] \to \mathcal{C}^{op}$ be the counit map. For each $X \in (\mathfrak{Set}_\Delta)_{/\,\mathrm{N}(\mathcal{C})}$, there is a canonical map

$$\alpha_{\mathcal{C}}(X) : St_\phi X^{op} \to \mathfrak{F}_X(\mathcal{C})^{op}.$$

The collection of maps $\{\alpha_{\mathcal{C}}(X)\}$ is uniquely determined by the following requirements:

(1) The morphism $\alpha_{\mathcal{C}}(X)$ depends functorially on X. More precisely, suppose we are given a commutative diagram of simplicial sets

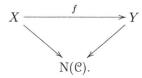

Then the diagram

$$St_\phi X^{op} \xrightarrow{\ \alpha_{\mathcal{C}}(X)\ } \mathfrak{F}_X(\mathcal{C})^{op}$$

$$\Big\downarrow St_\phi f^{op} \qquad\qquad \Big\downarrow \mathfrak{F}_f(\mathcal{C})^{op}$$

$$St_\phi Y^{op} \xrightarrow{\ \alpha_{\mathcal{C}}(Y)\ } \mathfrak{F}_Y(\mathcal{C})^{op}$$

commutes.

(2) The transformation $\alpha_{\mathcal{C}}$ depends functorially on $\mathcal{C}$ in the following sense: for every functor $g : \mathcal{C} \to \mathcal{D}$, if $\phi' : \mathcal{C}[(\mathrm{N}\,\mathcal{D})^{op}] \to \mathcal{D}^{op}$ denotes the counit map and $X \in (\mathrm{Set}_\Delta)_{/\mathrm{N}(\mathcal{C})}$, then the diagram

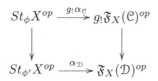

commutes, where the vertical arrows are the isomorphisms provided by Remark 3.2.5.8 and Proposition 2.2.1.1.

(3) Let $\mathcal{C}$ be the category associated to a partially ordered set P and let $X = \mathrm{N}(\mathcal{C})$, regarded as an object of $(\mathrm{Set}_\Delta)_{/\mathrm{N}(\mathcal{C})}$ via the identity map. Then $(St_\phi X^{op}) \in (\mathrm{Set}_\Delta)^{\mathcal{C}}$ can be identified with the functor $p \mapsto \mathrm{N}\,X_p$, where for each $p \in P$ we let X_p denote the collection of nonempty finite chains in P having largest element p. Similarly, Example 3.2.5.6 allows us to identify $\mathfrak{F}_X(\mathcal{C}) \in (\mathrm{Set}_\Delta)^{\mathcal{C}}$ with the functor $p \mapsto \mathrm{N}\{q \in P : q \leq p\}$. The map $\alpha_{\mathcal{C}}(X) : (St_\phi X^{op}) \to \mathfrak{F}_X(\mathcal{C})^{op}$ is induced by the map of partially ordered sets $X_p \to \{q \in P : q \leq p\}$ which carries every chain to its smallest element.

To see that the collection of maps $\{\alpha_{\mathcal{C}}(X)\}_{X \in (\mathrm{Set}_\Delta)_{/\mathrm{N}(\mathcal{C})}}$ is determined by these properties, we first note that because the functors St_ϕ and $\mathfrak{F}_\bullet(\mathcal{C})$ commute with colimits, any natural transformation $\beta_{\mathcal{C}} : St_\phi(\bullet^{op}) \to \mathfrak{F}_\bullet(\mathcal{C})^{op}$ is determined by its values $\beta_{\mathcal{C}}(X) : St_\phi(X^{op}) \to \mathfrak{F}_X(\mathcal{C})^{op}$ in the case where $X = \Delta^n$ is a simplex. In this case, any map $X \to \mathrm{N}\,\mathcal{C}$ factors through the isomorphism $X \simeq \mathrm{N}[n]$, so we can use property (2) to reduce to the case where the category $\mathcal{C}$ is a partially ordered set and the map $X \to \mathrm{N}(\mathcal{C})$ is an isomorphism. The behavior of the natural transformation $\alpha_{\mathcal{C}}$ is then dictated by property (3). This proves the uniqueness of the natural transformations $\alpha_{\mathcal{C}}$; the existence follows by a similar argument.

The following result summarizes some of the basic properties of the relative nerve functor:

Lemma 3.2.5.11. *Let $\mathcal{I}$ be a category and let $\alpha : f \to f'$ be a natural transformation of functors $f, f' : \mathcal{C} \to \mathrm{Set}_\Delta$.*

(1) *Suppose that, for each $I \in \mathcal{C}$, the map $\alpha(I) : f(I) \to f'(I)$ is an inner fibration of simplicial sets. Then the induced map $\mathrm{N}_f(\mathcal{C}) \to \mathrm{N}_{f'}(\mathcal{C})$ is an inner fibration.*

(2) *Suppose that, for each $I \in \mathcal{I}$, the simplicial set $f(I)$ is an ∞-category. Then $\mathrm{N}_f(\mathcal{C})$ is an ∞-category.*

(3) *Suppose that, for each $I \in \mathcal{C}$, the map $\alpha(I) : f(I) \to f'(I)$ is a categorical fibration of ∞-categories. Then the induced map $\mathrm{N}_f(\mathcal{C}) \to \mathrm{N}_{f'}(\mathcal{C})$ is a categorical fibration of ∞-categories.*

Proof. Consider a commutative diagram

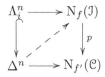

and let I be the image of $\{n\} \subseteq \Delta^n$ under the bottom map. If $0 \le i < n$, then the lifting problem depicted in the diagram above is equivalent to the existence of a dotted arrow in an associated diagram

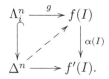

If $\alpha(I)$ is an inner fibration and $0 < i < n$, then we conclude that this lifting problem admits a solution. This proves (1). To prove (2), we apply (1) in the special case where f' is the constant functor taking the value Δ^0. It follows that $N_f(\mathcal{C}) \to N(\mathcal{C})$ is an inner fibration, so that $N_f(\mathcal{C})$ is an ∞-category.

We now prove (3). According to Corollary 2.4.6.5, an inner fibration $\mathcal{D} \to \mathcal{E}$ of ∞-categories is a categorical fibration if and only if the following condition is satisfied:

(∗) For every equivalence $e : E \to E'$ in $\mathcal{E}$ and every object $D \in \mathcal{D}$ lifting E, there exists an equivalence $\overline{e} : D \to D'$ in $\mathcal{D}$ lifting e.

We can identify equivalences in $N_{f'}(\mathcal{C})$ with triples $(g : I \to I', X, e : X' \to Y)$, where g is an isomorphism in $\mathcal{C}$, X is an object of $f'(I)$, X' is the image of X in $f'(I')$, and $e : X' \to Y$ is an equivalence in $f'(I')$. Given a lifting $\overline{X}$ of X to $f(I)$, we can apply the assumption that $\alpha(I')$ is a categorical fibration (and Corollary 2.4.6.5) to lift e to an equivalence $\overline{e} : \overline{X}' \to \overline{Y}$ in $f(I')$. This produces the desired equivalence $(g : I \to I', \overline{X}, \overline{e} : \overline{X}' \to \overline{Y})$ in $N_f(\mathcal{C})$. □

We now introduce a slightly more elaborate version of the relative nerve construction.

Definition 3.2.5.12. Let $\mathcal{C}$ be a small category and $\mathcal{F} : \mathcal{C} \to \mathrm{Set}_\Delta^+$ a functor. We let $N_{\mathcal{F}}^+(\mathcal{C})$ denote the marked simplicial set $(N_f(\mathcal{C}), M)$, where f denotes the composition $\mathcal{C} \xrightarrow{\mathcal{F}} \mathrm{Set}_\Delta^+ \to \mathrm{Set}_\Delta$ and M denotes the collection of all edges $\overline{e}$ of $N_f(\mathcal{C})$ with the following property: if $e : C \to C'$ is the image of $\overline{e}$ in $N(\mathcal{C})$ and σ denotes the edge of $f(C')$ determined by $\overline{e}$, then σ is a marked edge of $\mathcal{F}(C')$. We will refer to $N_{\mathcal{F}}^+(\mathcal{C})$ as the *marked relative nerve* functor.

Remark 3.2.5.13. Let $\mathcal{C}$ be a small category. We will regard the construction $\mathcal{F} \mapsto N_{\mathcal{F}}^+(\mathcal{C})$ as determining a functor from $(\mathrm{Set}_\Delta^+)^{\mathcal{C}}$ to $(\mathrm{Set}_\Delta^+)_{/N(\mathcal{C})}$ (see Remark 3.2.5.4). This functor admits a left adjoint, which we will denote by $\overline{X} \mapsto \mathfrak{F}_{\overline{X}}^+(\mathcal{C})$.

Remark 3.2.5.14. Remark 3.2.5.8 has an evident analogue for the functors $\mathfrak{F}^+$: for any functor $g : \mathcal{C} \to \mathcal{D}$ between small categories, the diagram of left adjoints

$$
\begin{array}{ccc}
(\mathrm{Set}_\Delta^+)^{\mathcal{D}} & \xleftarrow{\;\;g_!\;\;} & (\mathrm{Set}_\Delta^+)^{\mathcal{C}} \\[4pt]
\Big\uparrow {\scriptstyle \mathfrak{F}_\bullet^+(\mathcal{D})} & & \Big\uparrow {\scriptstyle \mathfrak{F}_\bullet^+(\mathcal{C})} \\[4pt]
(\mathrm{Set}_\Delta^+)_{/\,\mathrm{N}(\mathcal{D})} & \longleftarrow & (\mathrm{Set}_\Delta^+)_{/\,\mathcal{C}}
\end{array}
$$

commutes up to canonical isomorphism.

Lemma 3.2.5.15. *Let $\mathcal{C}$ be a small category. Then*

(1) *The functor $X \mapsto \mathfrak{F}_X(\mathcal{C})$ carries cofibrations in $(\mathrm{Set}_\Delta)_{/\,\mathrm{N}(\mathcal{C})}$ to cofibrations in $(\mathrm{Set}_\Delta)^{\mathcal{C}}$ (with respect to the projective model structure).*

(2) *The functor $\overline{X} \mapsto \mathfrak{F}_{\overline{X}}^+(\mathcal{C})$ carries cofibrations (with respect to the co-Cartesian model structure on $(\mathrm{Set}_\Delta^+)_{/\,\mathrm{N}(\mathcal{C})}$) to cofibrations in $(\mathrm{Set}_\Delta^+)^{\mathcal{C}}$ (with respect to the projective model structure).*

Proof. We will give the proof of (2); the proof of (1) is similar. It will suffice to show that the right adjoint functor $\mathrm{N}_\bullet^+(\mathcal{C}) : (\mathrm{Set}_\Delta^+)^{\mathcal{C}} \to \mathrm{Set}_\Delta^+ \, \mathrm{N}(\mathcal{C})$ preserves trivial fibrations. Let $\mathcal{F} \to \mathcal{F}'$ be a trivial fibration in $(\mathrm{Set}_\Delta^+)^{\mathcal{C}}$ with respect to the projective model structure, so that for each $C \in \mathcal{C}$ the induced map $\mathcal{F}(C) \to \mathcal{F}'(C)$ is a trivial fibration of marked simplicial sets. We wish to prove that the induced map $\mathrm{N}_{\mathcal{F}}^+(\mathcal{C}) \to \mathrm{N}_{\mathcal{F}'}^+(\mathcal{C})$ is also a trivial fibration of marked simplicial sets. Let f denote the composition $\mathcal{C} \xrightarrow{\mathcal{F}} \mathrm{Set}_\Delta^+ \to \mathrm{Set}_\Delta$ and let f' be defined likewise. We must verify two things:

(1) Every lifting problem of the form

$$
\begin{array}{ccc}
\partial \Delta^n & \longrightarrow & \mathrm{N}_f(\mathcal{C}) \\[4pt]
\Big\uparrow & & \Big\downarrow \\[4pt]
\Delta^n & \xrightarrow{\;\;u\;\;} & \mathrm{N}_{f'}(\mathcal{C})
\end{array}
$$

admits a solution. Let $C \in \mathcal{C}$ denote the image of the final vertex of Δ^n under the map u. Then it suffices to solve a lifting problem of the form

$$
\begin{array}{ccc}
\partial \Delta^n & \longrightarrow & f(C) \\[4pt]
\Big\uparrow & & \Big\downarrow \\[4pt]
\Delta^n & \longrightarrow & f'(C),
\end{array}
$$

which is possible since the right vertical map is a trivial fibration of simplicial sets.

(2) If $\overline{e}$ is an edge of $N_{\mathcal{F}}^+(\mathcal{C})$ whose image $\overline{e}'$ in $N_{\mathcal{F}'}^+(\mathcal{C})$ is marked, then $\overline{e}$ is itself marked. Let $e : C \to C'$ be the image of $\overline{e}$ in $N(\mathcal{C})$ and let σ denote the edge of $\mathcal{F}(C')$ determined by $\overline{e}$. Since $\overline{e}'$ is a marked edge of $N_{\mathcal{F}'}^+(\mathcal{C})$, the image of σ in $\mathcal{F}'(C')$ is marked. Since the map $\mathcal{F}(C') \to \mathcal{F}'(C')$ is a trivial fibration of marked simplicial sets, we deduce that σ is a marked edge of $\mathcal{F}(C')$, so that $\overline{e}$ is a marked edge of $N_{\mathcal{F}}^+(\mathcal{C})$ as desired.

$\square$

Remark 3.2.5.16. Let $\mathcal{C}$ be a small category, let $S = N(\mathcal{C})^{op}$, and let $\phi : \mathfrak{C}[S] \to \mathcal{C}^{op}$ be the counit map. For every $\overline{X} = (X, M) \in (\mathrm{Set}_\Delta^+)_{/N(\mathcal{C})}$, the morphism $\alpha_{\mathcal{C}}(X) : St_\phi(X^{op}) \to \mathfrak{F}_X(\mathcal{C})^{op}$ of Remark 3.2.5.10 induces a natural transformation $St_\phi^+ \overline{X}^{op} \to \mathfrak{F}_{\overline{X}}^+(\mathcal{C})^{op}$, which we will denote by $\alpha_{\mathcal{C}}^+(\overline{X})$. We will regard the collection of morphisms $\{\alpha_{\mathcal{C}}^+(\overline{X})\}_{\overline{X} \in (\mathrm{Set}_\Delta^+)_{/N(\mathcal{C})}}$ as determining a natural transformation of functors

$$\alpha_{\mathcal{C}} : St_\phi^+(\bullet^{op}) \to \mathfrak{F}_\bullet^+(\mathcal{C})^{op}.$$

Lemma 3.2.5.17. *Let $\mathcal{C}$ be a small category, let $S = N(\mathcal{C})^{op}$, let $\phi : \mathfrak{C}[S] \to \mathcal{C}^{op}$ be the counit map, and let $C \in \mathcal{C}$ be an object. Then*

(1) *For every $X \in (\mathrm{Set}_\Delta)_{/N(\mathcal{C})}$, the map $\alpha_{\mathcal{C}}(X) : St_\phi(X^{op}) \to \mathfrak{F}_X(\mathcal{C})^{op}$ of Remark 3.2.5.10 induces a weak homotopy equivalence of simplicial sets $St_\phi(X^{op})(C) \to \mathfrak{F}_X(\mathcal{C})(C)^{op}$.*

(2) *For every $\overline{X} \in (\mathrm{Set}_\Delta^+)_{/N(\mathcal{C})}$, the map $\alpha_{\mathcal{C}}^+(\overline{X}) : St_\phi^+(\overline{X}^{op}) \to \mathfrak{F}_{\overline{X}}^+(\mathcal{C})^{op}$ of Remark 3.2.5.16 induces a Cartesian equivalence $St_\phi^+(\overline{X}^{op})(C) \to \mathfrak{F}_{\overline{X}}^+(\mathcal{C})(C)^{op}$.*

Proof. We will give the proof of (2); the proof of (1) is similar but easier. Let us say that an object $\overline{X} \in (\mathrm{Set}_\Delta^+)_{/N(\mathcal{C})}$ is *good* if the map $\alpha_{\mathcal{C}}^+(\overline{X})$ is a weak equivalence. We wish to prove that every object $\overline{X} = (X, M) \in (\mathrm{Set}_\Delta^+)_{/N(\mathcal{C})}$ is good. The proof proceeds in several steps.

(A) Since the functors St_ϕ^+ and $\mathfrak{F}_\bullet^+(\mathcal{C})$ both commute with filtered colimits, the collection of good objects of $(\mathrm{Set}_\Delta^+)_{/N(\mathcal{C})}$ is stable under filtered colimits. We may therefore reduce to the case where the simplicial set X has only finitely many nondegenerate simplices.

(B) Suppose we are given a pushout diagram

in the category $(\mathrm{Set}_\Delta^+)_{/N(\mathcal{C})}$. Suppose that either f or g is a cofibration and that the objects $\overline{X}, \overline{X}'$, and $\overline{Y}$ are good. Then $\overline{Y}'$ is good.

This follows from the fact that the functors St_ϕ^+ and $\mathfrak{F}_\bullet^+(\mathcal{C})$ preserve cofibrations (Proposition 3.2.1.7 and Lemma 3.2.5.15) together with the observation that the projective model structure on $(\mathrm{Set}_\Delta^+)^{\mathcal{C}}$ is left proper.

(C) Suppose that $X \simeq \Delta^n$ for $n \le 1$. In this case, the map $\alpha_{\mathcal{C}}^+(\overline{X})$ is an isomorphism (by direct calculation), so that $\overline{X}$ is good.

(D) We now work by induction on the number of nondegenerate marked edges of $\overline{X}$. If this number is nonzero, then there exists a pushout diagram

$$
\begin{array}{ccc}
(\Delta^1)^\flat & \longrightarrow & (\Delta^1)^\sharp \\
\downarrow & & \downarrow \\
\overline{Y} & \longrightarrow & \overline{X},
\end{array}
$$

where $\overline{Y}$ has fewer nondegenerate marked edges than $\overline{X}$, so that $\overline{Y}$ is good by the inductive hypothesis. The marked simplicial sets $(\Delta^1)^\flat$ and $(\Delta^1)^\sharp$ are good by virtue of (C), so that (B) implies that $\overline{X}$ is good. We may therefore reduce to the case where $\overline{X}$ contains no nondegenerate marked edges, so that $\overline{X} \simeq X^\flat$.

(E) We now argue by induction on the dimension n of X and the number of nondegenerate n-simplices of X. If X is empty, there is nothing to prove; otherwise, we have a pushout diagram

$$
\begin{array}{ccc}
\partial\Delta^n & \longrightarrow & \Delta^n \\
\downarrow & & \downarrow \\
Y & \longrightarrow & X.
\end{array}
$$

The inductive hypothesis implies that $(\partial\Delta^n)^\flat$ and $Y^\flat$ are good. Invoking step (B), we can reduce to the case where X is an n-simplex. In view of (C), we may assume that $n \ge 2$.

Let $Z = \Delta^{\{0,1\}} \coprod_{\{1\}} \Delta^{\{1,2\}} \coprod_{\{1\}} \cdots \coprod_{\{n-1\}} \Delta^{\{n-1,n\}}$, so that $Z \subseteq X$ is an inner anodyne inclusion. We have a commutative diagram

$$
\begin{array}{ccc}
St_\phi^+(Z^{op})^\flat & \xrightarrow{\ u\ } & St_\phi^+(X^{op})^\flat \\
\downarrow{\scriptstyle v} & & \downarrow \\
\mathfrak{F}_{Z^\flat}^+(\mathcal{C})^{op} & \xrightarrow{\ w\ } & \mathfrak{F}_{X^\flat}^+(\mathcal{C})^{op}.
\end{array}
$$

The inductive hypothesis implies that v is a weak equivalence, and Proposition 3.2.1.11 implies that u is a weak equivalence. To complete the proof, it will suffice to show that w is a weak equivalence.

(F) The map $X \to N(\mathcal{C})$ factors as a composition

$$\Delta^n \simeq N([n]) \xrightarrow{g} N(\mathcal{C}).$$

Using Remark 3.2.5.14 (together with the fact that the left Kan extension functor $g_!$ preserves weak equivalences between projectively cofibrant objects), we can reduce to the case where $\mathcal{C} = [n]$ and the map $X \to N(\mathcal{C})$ is an isomorphism.

(G) Fix an object $i \in [n]$. A direct computation shows that the map $\mathfrak{F}_{Z^\flat}^+(\mathcal{C})(i) \to \mathfrak{F}_{X^\flat}^+(\mathcal{C})(i)$ can be identified with the inclusion

$$(\Delta^{\{0,1\}} \coprod_{\{1\}} \Delta^{\{1,2\}} \coprod_{\{1\}} \cdots \coprod_{\{i-1\}} \Delta^{\{i-1,i\}})^{op,\flat} \subseteq (\Delta^i)^{op,\flat}.$$

This inclusion is marked anodyne and therefore an equivalence of marked simplicial sets, as desired.

□

Proposition 3.2.5.18. *Let $\mathcal{C}$ be a small category. Then*

(1) *The functors $\mathfrak{F}_\bullet(\mathcal{C})$ and $N_\bullet(\mathcal{C})$ determine a Quillen equivalence between $(\text{Set}_\Delta)_{/N(\mathcal{C})}$ (endowed with the covariant model structure) and $(\text{Set}_\Delta)^{\mathcal{C}}$ (endowed with the projective model structure).*

(2) *The functors $\mathfrak{F}_\bullet^+(\mathcal{C})$ and $N_\bullet^+(\mathcal{C})$ determine a Quillen equivalence between $(\text{Set}_\Delta^+)_{/N(\mathcal{C})}$ (endowed with the coCartesian model structure) and $(\text{Set}_\Delta^+)^{\mathcal{C}}$ (endowed with the projective model structure).*

Proof. We will give the proof of (2); the proof of (1) is similar but easier. We first show that the adjoint pair $(\mathfrak{F}_\bullet^+(\mathcal{C}), N_\bullet^+(\mathcal{C}))$ is a Quillen adjunction. It will suffice to show that the functor $\mathfrak{F}_\bullet^+(\mathcal{C})$ preserves cofibrations and weak equivalences. The case of cofibrations follows from Lemma 3.2.5.15, and the case of weak equivalences from Lemma 3.2.5.17 and Corollary 3.2.1.16. To prove that $(\mathfrak{F}_\bullet^+(\mathcal{C}), N_\bullet^+(\mathcal{C}))$ is a Quillen equivalence, it will suffice to show that the left derived functor $L\mathfrak{F}_\bullet^+(\mathcal{C})$ induces an equivalence from the homotopy category $h(\text{Set}_\Delta^+)_{/N(\mathcal{C})}$ to the homotopy category $h(\text{Set}_\Delta^+)^{\mathcal{C}}$. In view of Lemma 3.2.5.17, it will suffice to prove an analogous result for the straightening functor St_ϕ^+, where ϕ denotes the counit map $\mathfrak{C}[N(\mathcal{C})^{op}] \to \mathcal{C}^{op}$. We now invoke Theorem 3.2.0.1.   □

Corollary 3.2.5.19. *Let $\mathcal{C}$ be a small category and let $\alpha : f \to f'$ be a natural transformation of functors $f, f' : \mathcal{C} \to \text{Set}_\Delta$. Suppose that, for each $C \in \mathcal{C}$, the induced map $f(C) \to f'(C)$ is a Kan fibration. Then the induced map $N_f(\mathcal{C}) \to N_{f'}(\mathcal{C})$ is a covariant fibration in $(\text{Set}_\Delta)_{/N(\mathcal{C})}$. In particular, if each $f(C)$ is Kan complex, then the map $N_f(\mathcal{C}) \to N(\mathcal{C})$ is a left fibration of simplicial sets.*

Corollary 3.2.5.20. *Let* $\mathcal{C}$ *be a small category and* $\mathcal{F} : \mathcal{C} \to \mathrm{Set}_\Delta^+$ *a fibrant object of* $(\mathrm{Set}_\Delta^+)^{\mathcal{C}}$. *Let* $S = \mathrm{N}(\mathcal{C})$ *and let* $\phi : \mathfrak{C}[S^{op}] \to \mathcal{C}^{op}$ *denote the counit map. Then the natural transformation* $\alpha_{\mathcal{C}}^+$ *of Remark 3.2.5.16 induces a weak equivalence* $\mathrm{N}_{\mathcal{F}}(\mathcal{C})^{op} \to (\mathrm{Un}_\phi^+ \, \mathcal{F}^{op})$ *(with respect to the Cartesian model structure on* $(\mathrm{Set}_\Delta^+)_{/S^{op}}$*).*

Proof. It suffices to show that $\alpha_{\mathcal{C}}^+$ induces an isomorphism of right derived functors $R \, \mathrm{N}_\bullet(\mathcal{C})^{op} \to R(\mathrm{Un}_\phi^+ \, \bullet^{op})$, which follows immediately from Lemma 3.2.5.17. □

Proposition 3.2.5.21. *Let* $\mathcal{C}$ *be a category and let* $f : \mathcal{C} \to \mathrm{Set}_\Delta$ *be a functor such that* $f(C)$ *is an* ∞-*category for each* $C \in \mathcal{C}$. *Then*

(1) *The projection* $p : \mathrm{N}_f(\mathcal{C}) \to \mathrm{N}(\mathcal{C})$ *is a coCartesian fibration of simplicial sets.*

(2) *Let* e *be an edge of* $\mathrm{N}_f(\mathcal{C})$ *covering a morphism* $C \to C'$ *in* $\mathcal{C}$. *Then* e *is* p-*coCartesian if and only if the corresponding edge of* $f(C')$ *is an equivalence.*

(3) *The coCartesian fibration* p *is associated to the functor* $\mathrm{N}(f) : \mathrm{N}(\mathcal{C}) \to \mathcal{C}\mathrm{at}_\infty$ *(see* §3.3.2*).*

Proof. Let $\mathcal{F} : \mathcal{C} \to \mathrm{Set}_\Delta^+$ be the functor described by the formula $\mathcal{F}(C) = f(C)^\natural$. Then $\mathcal{F}$ is a projectively fibrant object of $(\mathrm{Set}_\Delta^+)^{\mathcal{C}}$. Invoking Proposition 3.2.5.18, we deduce that $\mathrm{N}_{\mathcal{F}}^+(\mathcal{C})$ is a fibrant object of $(\mathrm{Set}_\Delta^+)_{/\mathrm{N}(\mathcal{C})}$. Invoking Proposition 3.1.4.1, we deduce that the underlying map $p : \mathrm{N}_f(\mathcal{C}) \to \mathrm{N}(\mathcal{C})$ is a coCartesian fibration of simplicial sets and that the p-coCartesian morphisms of $\mathrm{N}_f(\mathcal{C})$ are precisely the marked wedges of $\mathrm{N}_{\mathcal{F}}^+(\mathcal{C})$. This proves (1) and (2). To prove (3), we let $S = \mathrm{N}(\mathcal{C})$ and $\phi : \mathfrak{C}[S]^{op} \to \mathcal{C}^{op}$ be the counit map. By definition, a coCartesian fibration $X \to \mathrm{N}(\mathcal{C})$ is associated to f if and only if it is equivalent to $(\mathrm{Un}_\phi \, f^{op})^{op}$; the desired equivalence is furnished by Corollary 3.2.5.20. □

3.3 APPLICATIONS

The purpose of this section is to survey some applications of technology developed in §3.1 and §3.2. In §3.3.1, we give some applications to the theory of Cartesian fibrations. In §3.3.2, we will introduce the language of *classifying maps* which will allow us to exploit the Quillen equivalence provided by Theorem 3.2.0.1. Finally, in §3.3.3 and §3.3.4, we will use Theorem 3.2.0.1 to give explicit constructions of limits and colimits in the ∞-category $\mathcal{C}\mathrm{at}_\infty$ (and also in the ∞-category $\mathcal{S}$ of spaces).

3.3.1 Structure Theory for Cartesian Fibrations

The purpose of this section is to prove that Cartesian fibrations between simplicial sets enjoy several pleasant properties. For example, every Cartesian fibration is a categorical fibration (Proposition 3.3.1.7), and categorical equivalences are stable under pullbacks by Cartesian fibrations (Proposition 3.3.1.3). These results are fairly easy to prove for Cartesian fibrations $X \to S$ in the case where S is an ∞-category. Theorem 3.2.0.1 provides a method for reducing to this special case:

Proposition 3.3.1.1. *Let $p : S \to T$ be a categorical equivalence of simplicial sets. Then the forgetful functor*

$$p_! : (\mathrm{Set}_\Delta^+)_{/S} \to (\mathrm{Set}_\Delta^+)_{/T}$$

and its right adjoint p^* induce a Quillen equivalence between $(\mathrm{Set}_\Delta^+)_{/S}$ and $(\mathrm{Set}_\Delta^+)_{/T}$.

Proof. Let $\mathcal{C} = \mathfrak{C}[S]^{op}$ and $\mathcal{D} = \mathfrak{C}[T]^{op}$. Consider the following diagram of model categories and left Quillen functors:

$$
\begin{array}{ccc}
(\mathrm{Set}_\Delta^+)_{/S} & \xrightarrow{\ p_!\ } & (\mathrm{Set}_\Delta^+)_{/T} \\
\downarrow{\scriptstyle St_S^+} & & \downarrow{\scriptstyle St_T^+} \\
\mathcal{C} & \xrightarrow{\ \mathfrak{C}[p]_!\ } & \mathcal{D}.
\end{array}
$$

According to Proposition 3.2.1.4, this diagram commutes (up to natural isomorphism). Theorem 3.2.0.1 implies that the vertical arrows are Quillen equivalences. Since p is a categorical equivalence, $\mathfrak{C}[p]$ is an equivalence of simplicial categories, so that $\mathfrak{C}[p]_!$ is a Quillen equivalence (Proposition A.3.3.8). It follows that $(p_!, p^*)$ is a Quillen equivalence as well.   □

Corollary 3.3.1.2. *Let $p : X \to S$ be a Cartesian fibration of simplicial sets and let $S \to T$ be a categorical equivalence. Then there exists a Cartesian fibration $Y \to T$ and an equivalence of X with $S \times_T Y$ (as Cartesian fibrations over X).*

Proof. Proposition 3.3.1.1 implies that the right derived functor Rp^* is essentially surjective.   □

As we explained in Remark 2.2.5.3, the Joyal model structure on Set_Δ is *not* right proper. In other words, it is possible to have a categorical fibration $X \to S$ and a categorical equivalence $T \to S$ such that the induced map $X \times_S T \to X$ is not a categorical equivalence. This poor behavior of categorical fibrations is one of the reasons that they do not play a prominent role in the theory of ∞-categories. Working with a stronger notion of fibration corrects the problem:

Proposition 3.3.1.3. *Let $p : X \to S$ be a Cartesian fibration and let $T \to S$ be a categorical equivalence. Then the induced map $X \times_S T \to X$ is a categorical equivalence.*

Proof. We first suppose that the map $T \to S$ is inner anodyne. By means of a simple argument, we may reduce to the case where $T \to S$ is a middle horn inclusion $\Lambda_i^n \subseteq \Delta^n$, where $0 < i < n$. According to Proposition 3.2.2.7, there exists a sequence of maps

$$\phi : A^0 \leftarrow \cdots \leftarrow A^n$$

and a map $M(\phi) \to X$ which is a categorical equivalence, such that $M(\phi) \times_S T \to X \times_S T$ is also a categorical equivalence. Consequently, it suffices to show that the inclusion $M(\phi) \times_S T \subseteq M(\phi)$ is a categorical equivalence. But this map is a pushout of the inclusion $A^n \times \Lambda_i^n \subseteq A^n \times \Delta^n$, which is inner anodyne.

We now treat the general case. Choose an inner anodyne map $T \to T'$, where T' is an ∞-category. Then choose an inner anodyne map $T' \coprod_T S \to S'$, where S' is also an ∞-category. The map $S \to S'$ is inner anodyne; in particular it is a categorical equivalence, so by Corollary 3.3.1.2 there is a Cartesian fibration $X' \to S'$ and an equivalence $X \to X' \times_{S'} S$ of Cartesian fibrations over S. We have a commutative diagram

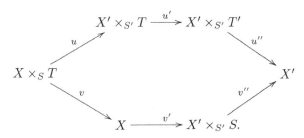

Consequently, to prove that v is a categorical equivalence, it suffices to show that every other arrow in the diagram is a categorical equivalence. The maps u and v' are equivalences of Cartesian fibrations and therefore categorical equivalences. The other three maps correspond to special cases of the assertion we are trying to prove. For the map u'', we have the special case of the map $S' \to T'$, which is an equivalence of ∞-categories: in this case we simply apply Corollary 2.4.4.5. For the maps u' and v'', we need to know that the assertion of the proposition is valid in the special case of the maps $S \to S'$ and $T \to T'$. Since these maps are inner anodyne, the proof is complete. $\square$

Corollary 3.3.1.4. *Let*

be a pullback diagram of simplicial sets, where p' is a Cartesian fibration. Then the diagram is homotopy Cartesian (with respect to the Joyal model structure).

Proof. Choose an inner-anodyne map $S' \to S''$, where S'' is an ∞-category. Using Proposition 3.3.1.1, we may assume without loss of generality that $X' \simeq X'' \times_{S''} S'$, where $X'' \to S''$ is a Cartesian fibration. Now choose a factorization

$$S \xrightarrow{\theta'} T \xrightarrow{\theta''} S'',$$

where θ' is a categorical equivalence and θ'' is a categorical fibration. The diagram

$$T \to S'' \leftarrow X''$$

is fibrant. Consequently, the desired conclusion is equivalent to the assertion that the map $X \to T \times_{S''} X''$ is a categorical equivalence, which follows immediately from Proposition 3.3.1.3. □

We now prove a stronger version of Corollary 2.4.4.4 which does not require that the base S is a ∞-category.

Proposition 3.3.1.5. *Suppose we are given a diagram of simplicial sets*

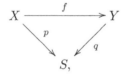

where p and q are Cartesian fibrations and f carries p-Cartesian edges to q-Cartesian edges. The following conditions are equivalent:

(1) *The map f is a categorical equivalence.*

(2) *For each vertex s of S, f induces a categorical equivalence $X_s \to Y_s$.*

(3) *The map $X^\natural \to Y^\natural$ is a Cartesian equivalence in $(\mathrm{Set}_\Delta^+)_{/S}$.*

Proof. The equivalence of (2) and (3) follows from Proposition 3.1.3.5. We next show that (2) implies (1). By virtue of Proposition 3.2.2.8, we may reduce to the case where S is a simplex. Then S is an ∞-category, and the desired result follows from Corollary 2.4.4.4. (Alternatively, we could observe that (2) implies that f has a homotopy inverse.)

To prove that (1) implies (3), we choose an inner anodyne map $j : S \to S'$, where S' is an ∞-category. Let $X^\natural$ denote the object of $(\mathrm{Set}_\Delta^+)_{/S}$ associated to the Cartesian fibration $p : X \to S$ and let $j_! X^\natural$ denote the same marked simplicial set, regarded as an object of $(\mathrm{Set}_\Delta^+)_{/T}$. Choose a marked anodyne map $j_! X^\natural \to X'^\natural$, where $X' \to S'$ is a Cartesian fibration. By Proposition 3.3.1.1, the map $X^\natural \to j^* X'^\natural$ is a Cartesian equivalence, so that $X \to X' \times_{S'} S$ is a categorical equivalence. According to Proposition 3.3.1.3, the map $X' \times_{S'} S \to X'$ is a categorical equivalence; thus the composite map $X \to X'$ is a categorical equivalence.

Similarly, we may choose a marked anodyne map

$$X'^\natural \coprod_{j_! X^\natural} j_! Y^\natural \to Y'^\natural$$

for some Cartesian fibration $Y' \to S'$. Since the Cartesian model structure is left proper, the map $j_! Y^\natural \to Y'^\natural$ is a Cartesian equivalence, so we may argue as above to deduce that $Y \to Y'$ is a categorical equivalence. Now consider the diagram

$$\begin{array}{ccc} X & \xrightarrow{\ f\ } & Y \\ \downarrow & & \downarrow \\ X' & \xrightarrow{\ f'\ } & Y'. \end{array}$$

We have argued that the vertical maps are categorical equivalences. The map f is a categorical equivalence by assumption. It follows that f' is a categorical equivalence. Since S' is an ∞-category, we may apply Corollary 2.4.4.4 to deduce that $X'_s \to Y'_s$ is a categorical equivalence for each object s of S'. It follows that $X'^\natural \to Y'^\natural$ is a Cartesian equivalence in $(\mathrm{Set}_\Delta^+)_{/S}$, so that we have a commutative diagram

$$\begin{array}{ccc} X^\natural & \xrightarrow{\hspace{2em}} & Y^\natural \\ \downarrow & & \downarrow \\ j^* X'^\natural & \xrightarrow{\hspace{2em}} & j^* Y'^\natural \end{array}$$

where the vertical and bottom horizontal arrows are Cartesian equivalences in $(\mathrm{Set}_\Delta^+)_{/S}$. It follows that the top horizontal arrow is a Cartesian equivalence as well, so that (3) is satisfied. $\square$

Corollary 3.3.1.6. *Let*

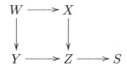

be a diagram of simplicial sets. Suppose that every morphism in this diagram is a right fibration and that the square is a pullback. Then the diagram is homotopy Cartesian with respect to the contravariant model structure on $(\mathrm{Set}_\Delta)_{/S}$.

Proof. Choose a fibrant replacement

$$X' \to Y' \leftarrow Z'$$

for the diagram

$$X \to Y \leftarrow Z$$

in $(\mathrm{Set}_\Delta)_{/S}$ and let $W' = X' \times_{Z'} Y'$. We wish to show that the induced map $i : W \to W'$ is a covariant equivalence in $(\mathrm{Set}_\Delta)_{/S}$. According to Corollary

2.2.3.13, it will suffice to show that, for each vertex s of S, the map of fibers $W_s \to W_s'$ is a homotopy equivalence of Kan complexes. To prove this, we observe that we have a natural transformation of diagrams from

$$\begin{array}{ccc} W_s & \longrightarrow & X_s \\ \downarrow & & \downarrow \\ Y_s & \longrightarrow & Z_s \end{array}$$

to

$$\begin{array}{ccc} W_s' & \longrightarrow & X_s' \\ \downarrow & & \downarrow \\ Y_s' & \longrightarrow & Z_s' \end{array}$$

which induces homotopy equivalences

$$X_s \to X_s' \qquad Y_s \to Y_s' \qquad Z_s \to Z_s'$$

(Corollary 2.2.3.13), where both diagrams are homotopy Cartesian (Proposition 2.1.3.1). □

Proposition 3.3.1.7. *Let $p : X \to S$ be a Cartesian fibration of simplicial sets. Then p is a categorical fibration.*

Proof. Consider a diagram

$$\begin{array}{ccc} A & \longrightarrow & X \\ {\scriptstyle i}\downarrow & {\scriptstyle f}\nearrow & \downarrow{\scriptstyle p} \\ B & \underset{g}{\longrightarrow} & S \end{array}$$

of simplicial sets, where i is an inclusion and a categorical equivalence. We must demonstrate the existence of the indicated dotted arrow. Choose a categorical equivalence $j : S \to T$, where T is an ∞-category. By Corollary 3.3.1.2, there exists a Cartesian fibration $q : Y \to T$ such that $Y \times_T S$ is equivalent to X. Thus there exist maps

$$u : X \to Y \times_T S$$

$$v : Y \times_T S \to X$$

such that $u \circ v$ and $v \circ u$ are homotopic to the identity (over S).

Consider the induced diagram

Since Y is an ∞-category, there exists a dotted arrow f' making the diagram commutative. Let $g' = q \circ f' : B \to T$. We note that $g'|A = (j \circ g)|A$.

Since T is an ∞-category and i is a categorical equivalence, there exists a homotopy $B \times \Delta^1 \to T$ from g' to $j \circ g$ which is fixed on A. Since q is a Cartesian fibration, this homotopy lifts to a homotopy from f' to some map $f'' : B \to Y$, so that we have a commutative diagram

$$
\begin{array}{ccc}
A & \longrightarrow & Y \\
{\scriptstyle i}\downarrow & \nearrow^{f''} & \downarrow{\scriptstyle q} \\
B & \longrightarrow & T.
\end{array}
$$

Consider the composite map

$$f''' : B \xrightarrow{(f'',g)} Y \times_T S \xrightarrow{v} X.$$

Since f' is homotopic to f'' and $v \circ u$ is homotopic to the identity, we conclude that $f'''|A$ is homotopic to f_0 (via a homotopy which is fixed over S). Since p is a Cartesian fibration, we can extend h to a homotopy from f''' to the desired map f. $\square$

In general, the converse to Proposition 3.3.1.7 fails: a categorical fibration of simplicial sets $X \to S$ need not be a Cartesian fibration. This is clear since the property of being a categorical fibration is self-dual, while the condition of being a Cartesian fibration is not. However, in the case where S is a Kan complex, the theory of Cartesian fibrations *is* self-dual, and we have the following result:

Proposition 3.3.1.8. *Let $p : X \to S$ be a map of simplicial sets, where S is a Kan complex. The following assertions are equivalent:*

(1) *The map p is a Cartesian fibration.*

(2) *The map p is a coCartesian fibration.*

(3) *The map p is a categorical fibration.*

Proof. We will prove that (1) is equivalent to (3); the equivalence of (2) and (3) follows from a dual argument. Proposition 3.3.1.7 shows that (1) implies (3) (for this implication, the assumption that S is a Kan complex is not needed).

Now suppose that (3) holds. Then X is an ∞-category. Since every edge of S is an equivalence, the p-Cartesian edges of X are precisely the equivalences in X. It therefore suffices to show that if y is a vertex of X and $\bar{e} : \bar{x} \to p(y)$ is an edge of S, then $\bar{e}$ lifts to an equivalence $e : x \to y$ in S. Since S is a Kan complex, we can find a contractible Kan complex K and a map $\bar{q} : K \to S$ such that $\bar{e}$ is the image of an edge $e' : x' \to y'$ in K. The inclusion $\{y'\} \subseteq K$ is a categorical equivalence; since p is a categorical fibration, we can lift $\bar{q}$ to a map $q : K \to X$ with $q(y') = y$. Then $e = q(e')$ has the desired properties. $\square$

3.3.2 Universal Fibrations

In this section, we will apply Theorem 3.2.0.1 to construct a *universal* Cartesian fibration. Recall that $\mathcal{C}\mathrm{at}_\infty$ is defined to be the nerve of the simplicial category $\mathcal{C}\mathrm{at}_\infty^\Delta = (\mathrm{Set}_\Delta^+)^\circ$ of ∞-categories. In particular, we may regard the inclusion $\mathcal{C}\mathrm{at}_\infty^\Delta \hookrightarrow \mathrm{Set}_\Delta^+$ as a (projectively) fibrant object $\mathcal{F} \in (\mathrm{Set}_\Delta^+)^{\mathcal{C}\mathrm{at}_\infty^\Delta}$. Applying the unstraightening functor $\mathrm{Un}_{\mathcal{C}\mathrm{at}_\infty^{op}}^+$, we obtain a fibrant object of $(\mathrm{Set}_\Delta^+)_{/\mathcal{C}\mathrm{at}_\infty^{op}}$, which we may identify with Cartesian fibration $q : \mathcal{Z} \to \mathcal{C}\mathrm{at}_\infty^{op}$. We will refer to q as the *universal Cartesian fibration*. We observe that the objects of $\mathcal{C}\mathrm{at}_\infty$ can be identified with ∞-categories and that the fiber of q over an ∞-category $\mathcal{C}$ can be identified with $U(\mathcal{C})$, where U is the functor described in Lemma 3.2.3.1. In particular, there is a canonical equivalence of ∞-categories

$$\mathcal{C} \to U(\mathcal{C}) = \mathcal{Z} \times_{\mathcal{C}\mathrm{at}_\infty^{op}} \{\mathcal{C}\}.$$

Thus we may think of q as a Cartesian fibration which associates to each object of $\mathcal{C}\mathrm{at}_\infty$ the associated ∞-category.

Remark 3.3.2.1. The ∞-categories $\mathcal{C}\mathrm{at}_\infty$ and $\mathcal{Z}$ are *large*. However, the universal Cartesian fibration q is small in the sense that for any small simplicial set S and any map $f : S \to \mathcal{C}\mathrm{at}_\infty^{op}$, the fiber product $S \times_{\mathcal{C}\mathrm{at}_\infty^{op}} \mathcal{Z}$ is small. This is because the fiber product can be identified with $\mathrm{Un}_\phi^+(\mathcal{F} \,|\, \mathfrak{C}[S])$, where $\phi : \mathfrak{C}[S] \to \mathrm{Set}_\Delta^+$ is the composition of $\mathfrak{C}[f]$ with the inclusion.

Definition 3.3.2.2. Let $p : X \to S$ be a Cartesian fibration of simplicial sets. We will say that a functor $f : S \to \mathcal{C}\mathrm{at}_\infty^{op}$ *classifies* p if there is an equivalence of Cartesian fibrations $X \to \mathcal{Z} \times_{\mathcal{C}\mathrm{at}_\infty^{op}} S \simeq \mathrm{Un}_S^+ f$.

Dually, if $p : X \to S$ is a coCartesian fibration, then we will say that a functor $f : S \to \mathcal{C}\mathrm{at}_\infty$ *classifies* p if f^{op} classifies the Cartesian fibration $p^{op} : X^{op} \to S^{op}$.

Remark 3.3.2.3. Every Cartesian fibration $X \to S$ between *small* simplicial sets admits a classifying map $\phi : S \to \mathcal{C}\mathrm{at}_\infty^{op}$, which is uniquely determined up to equivalence. This is one expression of the idea that $\mathcal{Z} \to \mathcal{C}\mathrm{at}_\infty^{op}$ is a *universal* Cartesian fibration. However, it is not immediately obvious that this property characterizes $\mathcal{C}\mathrm{at}_\infty$ up to equivalence because $\mathcal{C}\mathrm{at}_\infty$ is not itself small. To remedy the situation, let us consider an arbitrary uncountable regular cardinal κ, and let $\mathcal{C}\mathrm{at}_\infty(\kappa)$ denote the full subcategory of $\mathcal{C}\mathrm{at}_\infty$ spanned by the κ-small ∞-categories. We then deduce the following:

(∗) Let $p : X \to S$ be a Cartesian fibration between small simplicial sets. Then p is classified by a functor $\chi : S \to \mathcal{C}\mathrm{at}_\infty(\kappa)^{op}$ if and only if, for every vertex $s \in S$, the fiber X_s is essentially κ-small. In this case, χ is determined uniquely up to homotopy.

Enlarging the universe and applying (∗) in the case where κ is the supremum of all small cardinals, we deduce the following property:

$(*')$ Let $p : X \to S$ be a Cartesian fibration between simplicial sets which
are not necessarily small. Then p is classified by a functor $\chi : S \to \mathrm{Cat}^{op}_\infty$
if and only if, for every vertex $s \in S$, the fiber X_s is essentially small.
In this case, χ is determined uniquely up to homotopy.

This property evidently determines the ∞-category Cat_∞ (and the Cartesian
fibration $q : \mathcal{Z} \to \mathrm{Cat}^{op}_\infty$) up to equivalence.

Warning 3.3.2.4. The terminology of Definition 3.3.2.2 has the potential
to cause confusion in the case where $p : X \to S$ is both a Cartesian fibration
and a coCartesian fibration. In this case, p is classified both by a functor
$S \to \mathrm{Cat}^{op}_\infty$ (as a Cartesian fibration) and by a functor $S \to \mathrm{Cat}_\infty$ (as a
coCartesian fibration).

The category $\mathcal{K}$an of Kan complexes can be identified with a full (simplicial) subcategory of $\mathrm{Cat}^\Delta_\infty$. Consequently we may identify the ∞-category
$\mathcal{S}$ of spaces with the full simplicial subset of Cat_∞, spanned by the vertices
which represent ∞-groupoids. We let $\mathcal{Z}^0 = \mathcal{Z} \times_{\mathrm{Cat}^{op}_\infty} \mathcal{S}^{op}$ be the restriction
of the universal Cartesian fibration. The fibers of $q^0 : \mathcal{Z}^0 \to \mathcal{S}^{op}$ are Kan
complexes (since they are equivalent to the ∞-categories represented by the
vertices of $\mathcal{S}$). It follows from Proposition 2.4.2.4 that q^0 is a right fibration.
We will refer to q^0 as the *universal right fibration*.

Proposition 2.4.2.4 translates immediately into the following characterization of right fibrations:

Proposition 3.3.2.5. *Let $p : X \to S$ be a Cartesian fibration of simplicial
sets. The following conditions are equivalent:*

(1) *The map p is a right fibration.*

(2) *Every functor $f : S \to \mathrm{Cat}^{op}_\infty$ which classifies p factors through $\mathcal{S}^{op} \subseteq$
Cat^{op}_∞.*

(3) *There exists a functor $f : S \to \mathcal{S}^{op}$ which classifies p.*

Consequently, we may speak of right fibrations $X \to S$ being classified by
functors $S \to \mathcal{S}^{op}$ and left fibrations being classified by functors $S \to \mathcal{S}$.

The ∞-category Δ^0 corresponds to a vertex of Cat_∞ which we will denote
by $*$. The fiber of q over this point may be identified with $U\Delta^0 \simeq \Delta^0$;
consequently, there is a unique vertex $*_{\mathcal{Z}}$ of $\mathcal{Z}$ lying over $*$. We note that
$*$ and $*_{\mathcal{Z}}$ belong to the subcategories $\mathcal{S}$ and $\mathcal{Z}^0$. Moreover, we have the
following:

Proposition 3.3.2.6. *Let $q^0 : \mathcal{Z}^0 \to \mathcal{S}^{op}$ be the universal right fibration.
The vertex $*_{\mathcal{Z}}$ is a final object of the ∞-category $\mathcal{Z}^0$.*

Proof. Let $n > 0$ and let $f_0 : \partial \Delta^n \to \mathcal{Z}^0$ have the property that f_0 carries the final vertex of Δ^n to $*_{\mathcal{Z}}$. We wish to show that there exists an extension

$$
\begin{array}{ccc}
\partial\Delta^n & \xrightarrow{f_0} & \mathcal{Z} \\
\downarrow & \overset{f}{\nearrow} & \\
\Delta^n & &
\end{array}
$$

(in which case the map f automatically factors through $\mathcal{Z}^0$).

Let $\mathcal{D}$ denote the simplicial category containing $\mathcal{S}_{\Delta}^{op}$ as a full subcategory together with one additional object X, with the morphisms given by

$$\mathrm{Map}_{\mathcal{D}}(K, X) = K$$

$$\mathrm{Map}_{\mathcal{D}}(X, X) = *$$

$$\mathrm{Map}_{\mathcal{D}}(X, K) = \emptyset$$

for all $K \in \mathcal{S}_{\Delta}^{op}$. Let $\mathcal{C} = \mathfrak{C}[\Delta^n \star \Delta^0]$ and let $\mathcal{C}_0$ denote the subcategory $\mathcal{C}_0 = \mathfrak{C}[\partial\Delta^n \star \Delta^0] \subseteq \mathcal{C}$. We will denote the objects of $\mathcal{C}$ by $\{v_0, \dots, v_{n+1}\}$. Giving the map f_0 is tantamount to giving a simplicial functor $F_0 : \mathcal{C}_0 \to \mathcal{D}$ with $F_0(v_{n+1}) = X$, and constructing f amounts to giving a simplicial functor $F : \mathcal{C} \to \mathcal{D}$ which extends F_0.

We note that the inclusion $\mathrm{Map}_{\mathcal{C}_0}(v_i, v_j) \to \mathrm{Map}_{\mathcal{C}}(v_i, v_j)$ is an isomorphism unless $i = 0$ and $j \in \{n, n+1\}$. Consequently, to define F, it suffices to find extensions

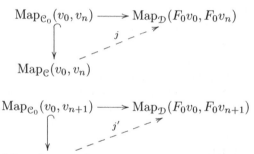

$$
\begin{array}{ccc}
\mathrm{Map}_{\mathcal{C}_0}(v_0, v_n) & \longrightarrow & \mathrm{Map}_{\mathcal{D}}(F_0 v_0, F_0 v_n) \\
\uparrow & \overset{j}{\nearrow} & \\
\mathrm{Map}_{\mathcal{C}}(v_0, v_n) & &
\end{array}
$$

$$
\begin{array}{ccc}
\mathrm{Map}_{\mathcal{C}_0}(v_0, v_{n+1}) & \longrightarrow & \mathrm{Map}_{\mathcal{D}}(F_0 v_0, F_0 v_{n+1}) \\
\uparrow & \overset{j'}{\nearrow} & \\
\mathrm{Map}_{\mathcal{C}}(v_0, v_{n+1}) & &
\end{array}
$$

such that the following diagram commutes:

$$
\begin{array}{ccc}
\mathrm{Map}_{\mathcal{C}}(v_0, v_n) \times \mathrm{Map}_{\mathcal{C}}(v_n, v_{n+1}) & \longrightarrow & \mathrm{Map}_{\mathcal{C}}(v_0, v_{n+1}) \\
\downarrow & & \downarrow \\
\mathrm{Map}_{\mathcal{D}}(F_0 v_0, F_0 v_n) \times \mathrm{Map}_{\mathcal{D}}(F_0 v_n, F_0 v_{n+1}) & \longrightarrow & \mathrm{Map}_{\mathcal{D}}(F_0 v_0, F_0 v_{n+1}).
\end{array}
$$

We note that $\mathrm{Map}_{\mathcal{C}}(v_n, v_{n+1})$ is a point. In view of the assumption that f_0 carries the final vertex of Δ^n to $*_{\mathcal{Z}}$, we see that $\mathrm{Map}_{\mathcal{D}}(F v_n, F v_{n+1})$ is a point. It follows that, for any fixed choice of j', there is a unique choice of j for which the above diagram commutes. It therefore suffices to show that j' exists. Since $\mathrm{Map}_{\mathcal{D}}(F_0 v_0, X)$ is a Kan complex, it will suffice to show that the inclusion $\mathrm{Map}_{\mathcal{C}_0}(v_0, v_{n+1}) \to \mathrm{Map}_{\mathcal{C}}(v_0, v_{n+1})$ is an anodyne map of simplicial sets. In fact, it is isomorphic to the inclusion

$$(\{1\} \times (\Delta^1)^{n-1}) \coprod_{\{1\} \times \partial(\Delta^1)^{n-1}} (\Delta^1 \times \partial(\Delta^1)^{n-1}) \subseteq \Delta^1 \times \Delta^{n-1},$$

which is the smash product of the cofibration $\partial(\Delta^1)^{n-1} \subseteq (\Delta^1)^{n-1}$ and the anodyne inclusion $\{1\} \subseteq \Delta^1$. $\qquad\square$

Corollary 3.3.2.7. *The universal right fibration $q^0 : \mathcal{Z}^0 \to \mathcal{S}^{op}$ is represented by a final object of $\mathcal{S}$.*

Proof. Combine Propositions 3.3.2.6 and 4.4.4.5. $\qquad\square$

Corollary 3.3.2.8. *Let $p : X \to S$ be a left fibration between small simplicial sets. Then there exists a map $S \to \mathcal{S}$ and an equivalence of left fibrations $X \simeq S \times_{\mathcal{S}} \mathcal{S}_{*/}$.*

Proof. Combine Corollary 3.3.2.7 with Remark 3.3.2.3. $\qquad\square$

3.3.3 Limits of ∞-Categories

The ∞-category Cat_∞ can be identified with the simplicial nerve of $(\mathrm{Set}_\Delta^+)^\circ$. It follows from Corollary 4.2.4.8 that Cat_∞ admits (small) limits and colimits, which can be computed in terms of homotopy (co)limits in the model category Set_Δ^+. For many applications, it is convenient to be able to construct limits and colimits while working entirely in the setting of ∞-categories. We will describe the construction of limits in this section; the case of colimits will be discussed in §3.3.4.

Let $p : S^{op} \to \mathrm{Cat}_\infty$ be a diagram in Cat_∞. Then p classifies a Cartesian fibration $q : X \to S$. We will show (Corollary 3.3.3.2 below) that the limit $\varprojlim(p) \in \mathrm{Cat}_\infty$ can be identified with the ∞-category of *Cartesian* sections of q. We begin by proving a more precise assertion:

Proposition 3.3.3.1. *Let K be a simplicial set, $\overline{p} : K^\triangleright \to \mathrm{Cat}_\infty^{op}$ a diagram in the ∞-category of spaces, $\overline{X} \to K^\triangleright$ a Cartesian fibration classified by $\overline{p}$, and $X = \overline{X} \times_{K^\triangleright} K$. The following conditions are equivalent:*

(1) *The diagram $\overline{p}$ is a colimit of $p = \overline{p}|K$.*

(2) *The restriction map*

$$\theta : \mathrm{Map}_{K^\triangleright}^\flat((K^\triangleright)^\sharp, \overline{X}^\natural) \to \mathrm{Map}_K^\flat(K^\sharp, X^\natural)$$

is an equivalence of ∞-categories.

Proof. According to Proposition 4.2.3.14, there exists a small category $\mathcal{C}$ and a cofinal map $f : N(\mathcal{C}) \to K$; let $\overline{\mathcal{C}} = \mathcal{C} \star [0]$ be the category obtained from $\mathcal{C}$ by adjoining a new final object and let $\overline{f} : N(\overline{\mathcal{C}}) \to K^\triangleright$ be the induced map (which is also cofinal). The maps f and $\overline{f}$ are contravariant equivalences in $(\mathcal{S}et_\Delta)_{/K^\triangleright}$ and therefore induce Cartesian equivalences

$$N(\mathcal{C})^\sharp \to K^\sharp \qquad N(\overline{\mathcal{C}})^\sharp \to (K^\triangleright)^\sharp.$$

We have a commutative diagram

$$
\begin{array}{ccc}
\mathrm{Map}^\flat_{K^\triangleright}((K^\triangleright)^\sharp, \overline{X}^\flat) & \xrightarrow{\;\theta\;} & \mathrm{Map}^\flat_{K^\triangleright}(K^\sharp, \overline{X}^\flat) \\
\downarrow & & \downarrow \\
\mathrm{Map}^\flat_{K^\triangleright}(N(\overline{\mathcal{C}})^\sharp, \overline{X}^\flat) & \xrightarrow{\;\theta'\;} & \mathrm{Map}^\flat_{K^\triangleright}(N(\mathcal{C})^\sharp, \overline{X}^\flat).
\end{array}
$$

The vertical arrows are categorical equivalences. Consequently, condition (2) holds for $\overline{p} : K^\triangleright \to \mathcal{C}at^{op}_\infty$ if and only if condition (2) holds for the composition $N(\overline{\mathcal{C}}) \to K^\triangleright \to \mathcal{C}at^{op}_\infty$. We may therefore assume without loss of generality that $K = N(\mathcal{C})$.

Using Corollary 4.2.4.7, we may further suppose that $\overline{p}$ is obtained as the simplicial nerve of a functor $\overline{\mathcal{F}} : \overline{\mathcal{C}}^{op} \to (\mathcal{S}et^+_\Delta)^\circ$. Changing $\overline{\mathcal{F}}$ if necessary, we may suppose that it is a *strongly* fibrant diagram in $\mathcal{S}et^+_\Delta$. Let $\mathcal{F} = \overline{\mathcal{F}}|\,\mathcal{C}^{op}$. Let $\overline{\phi} : \mathfrak{C}[K^\triangleright]^{op} \to \overline{\mathcal{C}}^{op}$ be the counit map and $\phi : \mathfrak{C}[K]^{op} \to \mathcal{C}^{op}$ the restriction of $\overline{\phi}$. We may assume without loss of generality that $\overline{X} = St^+_{\overline{\phi}}\overline{\mathcal{F}}$. We have a (not strictly commutative) diagram of categories and functors

$$
\begin{array}{ccc}
\mathcal{S}et^+_\Delta & \xrightarrow{\;\times K^\sharp\;} & (\mathcal{S}et^+_\Delta)_{/K} \\
{\scriptstyle St^+_*}\downarrow & & \downarrow{\scriptstyle St^+_\phi} \\
\mathcal{S}et^+_\Delta & \xrightarrow{\;\delta\;} & (\mathcal{S}et^+_\Delta)^{\mathcal{C}^{op}},
\end{array}
$$

where δ denotes the diagonal functor. This diagram commutes up to a natural transformation

$$St^+_\phi(K^\sharp \times Z) \to St^+_\phi(K^\sharp) \boxtimes St^+_*(Z) \to \delta(St^+_* Z).$$

Here the first map is a weak equivalence by Proposition 3.2.1.13, and the second map is a weak equivalence because LSt^+_ϕ is an equivalence of categories (Theorem 3.2.0.1) and therefore carries the final object $K^\sharp \in h(\mathcal{S}et^+_\Delta)_{/K}$ to a final object of $h(\mathcal{S}et^+_\Delta)^{\mathcal{C}^{op}}$. We therefore obtain a diagram of *right* derived functors

$$
\begin{array}{ccc}
h\mathcal{S}et^+_\Delta & \xleftarrow{\;\Gamma\;} & h(\mathcal{S}et^+_\Delta)_{/K} \\
{\scriptstyle R\,Un^+_*}\uparrow & & \uparrow{\scriptstyle R\,Un^+_\phi} \\
h\mathcal{S}et^+_\Delta & \longleftarrow & h(\mathcal{S}et^+_\Delta)^{\mathcal{C}^{op}},
\end{array}
$$

which commutes up to natural isomorphism, where we regard $(\mathcal{S}et^+_\Delta)^{\mathcal{C}^{op}}$ as equipped with the *injective* model structure described in §A.3.3. Similarly, we have a commutative diagram

$$
\begin{array}{ccc}
\mathrm{h}\mathcal{S}\mathrm{et}_\Delta^+ & \xleftarrow{\ \Gamma'\ } & \mathrm{h}(\mathcal{S}\mathrm{et}_\Delta^+)_{/K^\triangleright} \\
{\scriptstyle R\,\mathrm{Un}_*^+} \Big\uparrow & & \Big\uparrow {\scriptstyle R\,\mathrm{Un}_{\overline{\phi}}^\pm} \\
\mathrm{h}\mathcal{S}\mathrm{et}_\Delta^+ & \xleftarrow{\hspace{2em}} & \mathrm{h}(\mathcal{S}\mathrm{et}_\Delta^+)^{\overline{\mathcal{C}}^{op}}.
\end{array}
$$

Condition (2) is equivalent to the assertion that the restriction map

$$\Gamma'(\overline{X}^\natural) \to \Gamma(X^\natural)$$

is an isomorphism in $\mathrm{h}\mathcal{S}\mathrm{et}_\Delta^+$. Since the vertical functors in both diagrams are equivalences of categories (Theorem 3.2.0.1), this is equivalent to the assertion that the map

$$\varprojlim \overline{\mathcal{F}} \to \varprojlim \mathcal{F}$$

is a weak equivalence in $\mathcal{S}\mathrm{et}_\Delta^+$. Since $\overline{\mathcal{C}}$ has an initial object v, (2) is equivalent to the assertion that $\overline{\mathcal{F}}$ exhibits $\overline{\mathcal{F}}(v)$ as a homotopy limit of $\mathcal{F}$ in $(\mathcal{S}\mathrm{et}_\Delta^+)^\circ$. Using Theorem 4.2.4.1, we conclude that $(1) \Leftrightarrow (2)$, as desired. $\qquad\square$

It follows from Proposition 3.3.3.1 that limits in $\mathcal{C}\mathrm{at}_\infty$ are computed by forming ∞-categories of Cartesian sections:

Corollary 3.3.3.2. *Let $p : K \to \mathcal{C}\mathrm{at}_\infty^{op}$ be a diagram in the ∞-category $\mathcal{C}\mathrm{at}_\infty$ of spaces and let $X \to K$ be a Cartesian fibration classified by p. There is a natural isomorphism*

$$\varprojlim(p) \simeq \mathrm{Map}_K^\flat(K^\sharp, X^\natural)$$

in the homotopy category $\mathrm{h}\mathcal{C}\mathrm{at}_\infty$.

Proof. Let $\overline{p} : (K^\triangleright)^{op} \to \mathcal{C}\mathrm{at}_\infty^{op}$ be a limit of p and let $X' \to K^\triangleright$ be a Cartesian fibration classified by $\overline{p}$. Without loss of generality, we may suppose $X \simeq X' \times_{K^\triangleright} K$. We have maps

$$\mathrm{Map}_K^\flat(K^\sharp, X^\natural) \leftarrow \mathrm{Map}_{K^\triangleright}^\flat((K^\triangleright)^\sharp, X'^\natural) \to \mathrm{Map}_{K^\triangleright}^\flat(\{v\}^\sharp, X'^\natural),$$

where v denotes the cone point of $K^\triangleright$. Proposition 3.3.3.1 implies that the left map is an equivalence of ∞-categories. Since the inclusion $\{v\}^\sharp \subseteq (K^\triangleright)^\sharp$ is marked anodyne, the map on the right is a trivial fibration. We now conclude by observing that the space $\mathrm{Map}_{K^\triangleright}^\flat(\{v\}^\sharp, X'^\natural) \simeq X' \times_{K^\triangleright} \{v\}$ can be identified with $\overline{p}(v) = \varprojlim(p)$. $\qquad\square$

Using Proposition 3.3.3.1, we can easily deduce an analogous characterization of limits in the ∞-category of spaces.

Corollary 3.3.3.3. *Let K be a simplicial set, $\overline{p} : K^\triangleleft \to \mathcal{S}$ a diagram in the ∞-category of spaces and $X \to K^\triangleleft$ a left fibration classified by $\overline{p}$. The following conditions are equivalent:*

(1) *The diagram $\overline{p}$ is a limit of $p = \overline{p}|K$.*

(2) *The restriction map*

$$\mathrm{Map}_{K^{\triangleleft}}(K^{\triangleleft}, X) \to \mathrm{Map}_{K^{\triangleleft}}(K, X)$$

is a homotopy equivalence of Kan complexes.

Proof. The usual model structure on $\mathcal{S}et_{\Delta}$ is a localization of the Joyal model structure. It follows that the inclusion $\mathcal{K}an \subseteq \mathcal{C}at_{\infty}^{\Delta}$ preserves homotopy limits (of diagrams indexed by categories). Using Theorem 4.2.4.1, Proposition 4.2.3.14, and Corollary 4.2.4.7, we conclude that the inclusion $\mathcal{S} \subseteq \mathcal{C}at_{\infty}$ preserves (small) limits. The desired equivalence now follows immediately from Proposition 3.3.3.1.   □

Corollary 3.3.3.4. *Let $p : K \to \mathcal{S}$ be a diagram in the ∞-category $\mathcal{S}$ of spaces, and let $X \to K$ be a left fibration classified by p. There is a natural isomorphism*

$$\varprojlim(p) \simeq \mathrm{Map}_K(K, X)$$

in the homotopy category $\mathcal{H}$ of spaces.

Proof. Apply Corollary 3.3.3.2.   □

Remark 3.3.3.5. It is also possible to adapt the proof of Proposition 3.3.3.1 to give a direct proof of Corollary 3.3.3.3. We leave the details to the reader.

3.3.4 Colimits of ∞-Categories

In this section, we will address the problem of constructing *colimits* in the ∞-category $\mathcal{C}at_{\infty}$. Let $p : S^{op} \to \mathcal{C}at_{\infty}$ be a diagram classifying a Cartesian fibration $f : X \to S$. In §3.3.3, we saw that $\varprojlim(p)$ can be identified with the ∞-category of Cartesian sections of f. To construct the colimit $\varinjlim(p)$, we need to find an ∞-category which admits a map *from* each fiber X_s. The natural candidate, of course, is X itself. However, because X is generally not an ∞-category, we must take some care to formulate a correct statement.

Lemma 3.3.4.1. *Let*

$$
\begin{array}{ccc}
X' & \longrightarrow & X \\
\downarrow & & \downarrow{\scriptstyle p} \\
S' & \xrightarrow{q} & S
\end{array}
$$

be a pullback diagram of simplicial sets, where p is a Cartesian fibration and q^{op} is cofinal. The induced map $X'^{\natural} \to X^{\natural}$ is a Cartesian equivalence (in $\mathcal{S}et_{\Delta}^{+}$).

Proof. Choose a cofibration $S' \to K$, where K is a contractible Kan complex. The map q factors as a composition

$$S' \xrightarrow{q'} S \times K \xrightarrow{q''} S.$$

It is obvious that the projection $X^\natural \times K^\sharp \to X^\natural$ is a Cartesian equivalence. We may therefore replace S by $S \times K$ and q by q', thereby reducing to the case where q is a cofibration. Proposition 4.1.1.3 now implies that q is left anodyne. It is easy to see that the collection of cofibrations $q : S' \to S$ for which the desired conclusion holds is weakly saturated. We may therefore reduce to the case where q is a horn inclusion $\Lambda_i^n \subseteq \Delta^n$, where $0 \leq i < n$.

We now apply Proposition 3.2.2.7 to choose a sequence of composable maps

$$\phi : A^0 \leftarrow \cdots \leftarrow A^n$$

and a quasi-equivalence $M(\phi) \to X$. We have a commutative diagram of marked simplicial sets:

$$
\begin{array}{ccc}
M^\natural(\phi) \times_{(\Delta^n)^\sharp} (\Lambda_i^n)^\sharp & \longrightarrow & X'^\natural \\
\downarrow{\scriptstyle i} & & \downarrow \\
M^\natural(\phi) & \longrightarrow & X.
\end{array}
$$

Using Proposition 3.2.2.14, we deduce that the horizontal maps are Cartesian equivalences. To complete the proof, it will suffice to show that i is a Cartesian equivalence. We now observe that i is a pushout of the inclusion $i'' : (\Lambda_i^n)^\sharp \times (A^n)^\flat \subseteq (\Delta^n)^\sharp \times (A^n)^\flat$. It will therefore suffice to prove that i'' is a Cartesian equivalence. Using Proposition 3.1.4.2, we are reduced to proving that the inclusion $(\Lambda_i^n)^\sharp \subseteq (\Delta^n)^\sharp$ is a Cartesian equivalence. According to Proposition 3.1.5.7, this is equivalent to the assertion that the horn inclusion $\Lambda_i^n \subseteq \Delta^n$ is a weak homotopy equivalence, which is obvious.   □

Proposition 3.3.4.2. *Let K be a simplicial set, $\overline{p} : K^\triangleleft \to \mathcal{C}at_\infty^{op}$ be a diagram in the ∞-category $\mathcal{C}at_\infty$, $\overline{X} \to K^\triangleleft$ a Cartesian fibration classified by $\overline{p}$, and $X = \overline{X} \times_{K^\triangleleft} K$. The following conditions are equivalent:*

(1) *The diagram $\overline{p}$ is a limit of $p = \overline{p}|K$.*

(2) *The inclusion $X^\natural \subseteq \overline{X}^\natural$ is a Cartesian equivalence in $(\mathrm{Set}_\Delta^+)_{/K^\triangleleft}$.*

(3) *The inclusion $X^\natural \subseteq \overline{X}^\natural$ is a Cartesian equivalence in Set_Δ^+.*

Proof. Using the small object argument, we can construct a factorization

$$X \xrightarrow{i} Y \xrightarrow{j} K^\triangleleft,$$

where j is a Cartesian fibration, i induces a marked anodyne map $X^\natural \to Y^\natural$, and $X \simeq Y \times_{K^\triangleleft} K$. Since i is marked anodyne, we can solve the lifting problem

$$
\begin{array}{ccc}
X^\natural & \longrightarrow & \overline{X}^\natural \\
\downarrow{\scriptstyle i} & \nearrow{\scriptstyle q} & \downarrow \\
Y^\natural & \longrightarrow & (K^\triangleleft)^\sharp.
\end{array}
$$

Since i is a Cartesian equivalence in $(\mathrm{Set}_\Delta^+)_{/K^\triangleleft}$, condition (2) is equivalent to the assertion that q is an equivalence of Cartesian fibrations over $K^\triangleleft$. Since q induces an isomorphism over each vertex of K, this is equivalent to the following assertion:

(2′) The map $q_v : Y_v \to \overline{X}_v$ is an equivalence of ∞-categories, where v denotes the cone point of $K^\triangleleft$.

We have a commutative diagram

$$
\begin{array}{ccc}
Y_v^\natural & \xrightarrow{\ q_v\ } & \overline{X}_v^\natural \\
\uparrow & & \uparrow \\
\\
Y^\natural & \xrightarrow{\ q\ } & \overline{X}^\natural .
\end{array}
$$

Lemma 3.3.4.1 implies that the vertical maps are Cartesian equivalences. It follows that $(2') \Leftrightarrow (3)$, so that $(2) \Leftrightarrow (3)$.

To complete the proof, we will show that $(1) \Leftrightarrow (2)$. According to Proposition 4.2.3.14, there exists a small category $\mathcal{C}$ and a map $p : N(\mathcal{C}) \to K$ such that p^{op} is cofinal. Let $\overline{\mathcal{C}} = [0] \star \mathcal{C}$ be the category obtained by adjoining an initial object to $\mathcal{C}$. Consider the diagram

$$
\begin{array}{ccc}
(X \times_K N(\mathcal{C}))^\natural & \lhook\joinrel\longrightarrow & (\overline{X} \times_{K^\triangleleft} N(\overline{\mathcal{C}}))^\natural \\
\downarrow & & \downarrow \\
\\
X^\natural & \lhook\joinrel\longrightarrow & \overline{X}^\natural .
\end{array}
$$

Lemma 3.3.4.1 implies that the vertical maps are Cartesian equivalences (in Set_Δ^+). It follows that the upper horizontal inclusion is a Cartesian equivalence if and only if the lower horizontal inclusion is a Cartesian equivalence. Consequently, it will suffice to prove the equivalence $(1) \Leftrightarrow (2)$ after replacing K by $N(\mathcal{C})$.

Using Corollary 4.2.4.7, we may further suppose that $\overline{p}$ is the nerve of a functor $\mathcal{F} : \overline{\mathcal{C}} \to (\mathrm{Set}_\Delta^+)^\circ$. Let $\overline{\phi} : \mathfrak{C}[K^\triangleleft] \to \overline{\mathcal{C}}$ be the counit map and let $\phi : \mathfrak{C}[K] \to \mathcal{C}$ be the restriction of $\overline{\phi}$. Without loss of generality, we may suppose that $\overline{X} = \mathrm{Un}_{\overline{\phi}} \mathcal{F}$. We have a commutative diagram of homotopy categories and right derived functors

$$
\begin{array}{ccc}
h(\mathrm{Set}_\Delta^+)^{\overline{\mathcal{C}}} & \xrightarrow{\ G\ } & h(\mathrm{Set}_\Delta^+)^{\mathcal{C}} \\
\Big\downarrow{\scriptstyle R\,\mathrm{Un}_{\overline{\phi}}^+} & & \Big\downarrow{\scriptstyle R\,\mathrm{Un}_\phi^+} \\
h(\mathrm{Set}_\Delta^+)_{/(K^\triangleleft)} & \xrightarrow{\ G'\ } & h(\mathrm{Set}_\Delta^+)_{/K} ,
\end{array}
$$

where G and G' are restriction functors. Let F and F' be the left adjoints to G and G', respectively. According to Theorem 4.2.4.1, assumption (1) is equivalent to the assertion that $\mathcal{F}$ lies in the essential image of F. Since each of the vertical functors is an equivalence of categories (Theorem 3.2.0.1), this is equivalent to the assertion that $\overline{X}$ lies in the essential image of F'. Since F' is fully faithful, this is equivalent to the assertion that the counit map

$$F'G'\overline{X} \to \overline{X}$$

is an isomorphism in $h(\mathrm{Set}_\Delta^+)_{/K^\triangleleft}$, which is clearly a reformulation of (2). $\quad\square$

Corollary 3.3.4.3. *Let* $p : K^{op} \to \mathrm{Cat}_\infty$ *be a diagram classifying a Carte-sian fibration* $X \to K$. *Then there is a natural isomorphism* $\varinjlim(p) \simeq X^\natural$ *in the homotopy category in* $h\mathrm{Cat}_\infty$.

Proof. Let $\overline{p} : (K^{op})^\triangleright \to \mathrm{Cat}_\infty$ be a colimit of p, which classifies a Cartesian fibration $\overline{X} \to K^\triangleleft$. Let v denote the cone point of $K^\triangleleft$, so that $\varinjlim(p) \simeq \overline{X}_v$. We now observe that the inclusions

$$\overline{X}_v^\natural \hookrightarrow \overline{X}^\natural \hookleftarrow X^\natural$$

are both Cartesian equivalences (Lemma 3.3.4.1 and Proposition 3.3.4.2).
$\quad\square$

Warning 3.3.4.4. In the situation of Corollary 3.3.4.3, the marked sim-plicial set $X^\natural$ is usually not a fibrant object of Set_Δ^+ even when K is an ∞-category.

Using exactly the same argument, we can establish a version of Proposition 3.3.4.2 which describes colimits in the ∞-category of spaces:

Proposition 3.3.4.5. *Let* K *be a simplicial set,* $\overline{p} : K^\triangleright \to \mathcal{S}$ *a diagram in the* ∞-*category of spaces,* $\overline{X} \to K^\triangleright$ *a left fibration classified by* $\overline{p}$, *and* $X = \overline{X} \times_{K^\triangleright} K$. *The following conditions are equivalent:*

(1) *The diagram* $\overline{p}$ *is a colimit of* $p = \overline{p}|K$.

(2) *The inclusion* $X \subseteq \overline{X}$ *is a covariant equivalence in* $(\mathrm{Set}_\Delta)_{/K^\triangleright}$.

(3) *The inclusion* $X \subseteq \overline{X}$ *is a weak homotopy equivalence of simplicial sets.*

Proof. Using the small object argument, we can construct a factorization

$$X \xhookrightarrow{i} Y \xrightarrow{j} K^\triangleright,$$

where i is left anodyne, j is a left fibration, and the inclusion $X \subseteq Y \times_{K^\triangleright} K$ is an isomorphism. Choose a dotted arrow q as indicated in the diagram

Since i is a covariant equivalence in $(\mathrm{Set}_\Delta)_{/K^\triangleright}$, condition (2) is equivalent to the assertion that q is an equivalence of left fibrations over $K^\triangleright$. Since q induces an isomorphism over each vertex of K, this is equivalent to the

assertion that $q_v : Y_v \to \overline{X}_v$ is an equivalence, where v denotes the cone point of $K^{\triangleright}$. We have a commutative diagram

$$
\begin{array}{ccc}
Y_v & \xrightarrow{\;q_v\;} & \overline{X}_v \\
\downarrow & & \downarrow \\
Y & \xrightarrow{\;q\;} & \overline{X}.
\end{array}
$$

Proposition 4.1.2.15 implies that the vertical maps are right anodyne and therefore weak homotopy equivalences. Consequently, q_v is a weak homotopy equivalence if and only if q is a weak homotopy equivalence. Since the inclusion $X \subseteq Y$ is a weak homotopy equivalence, this proves that $(2) \Leftrightarrow (3)$.

To complete the proof, we will show that $(1) \Leftrightarrow (2)$. According to Proposition 4.2.3.14, there exists a small category $\mathcal{C}$ and a cofinal map $N(\mathcal{C}) \to K$. Let $\overline{\mathcal{C}} = \mathcal{C} \star [0]$ be the category obtained from $\mathcal{C}$ by adjoining a new final object. Consider the diagram

$$
\begin{array}{ccc}
X \times_K N(\mathcal{C}) & \hookrightarrow & \overline{X} \times_{K^{\triangleright}} N(\overline{\mathcal{C}}) \\
\downarrow & & \downarrow \\
X & \hookrightarrow & \overline{X}.
\end{array}
$$

Proposition 4.1.2.15 implies that $\overline{X} \to K^{\triangleright}$ is smooth, so that the vertical arrows in the above diagram are cofinal. In particular, the vertical arrows are weak homotopy equivalences, so that the upper horizontal inclusion is a weak homotopy equivalence if and only if the lower horizontal inclusion is a weak homotopy equivalence. Consequently, it will suffice to prove the equivalence $(1) \Leftrightarrow (2)$ after replacing K by $N(\mathcal{C})$.

Using Corollary 4.2.4.7, we may further suppose that $\overline{p}$ is obtained as the nerve of a functor $\mathcal{F} : \overline{\mathcal{C}} \to \mathcal{K}an$. Let $\overline{\phi} : \mathfrak{C}[K^{\triangleright}] \to \overline{\mathcal{C}}$ be the counit map and let $\phi : \mathfrak{C}[K] \to \mathcal{C}$ be the restriction of $\overline{\phi}$. Without loss of generality, we may suppose that $\overline{X}^{op} = \mathrm{Un}_{\overline{\phi}} \mathcal{F}$. We have a commutative diagram of homotopy categories and right derived functors

$$
\begin{array}{ccc}
h(\mathcal{S}et_\Delta)^{\overline{\mathcal{C}}} & \xrightarrow{\;G\;} & h(\mathcal{S}et_\Delta)^{\mathcal{C}} \\
\downarrow{\scriptstyle R\,\mathrm{Un}_{\overline{\phi}}} & & \downarrow{\scriptstyle R\,\mathrm{Un}_{\phi}} \\
h(\mathcal{S}et_\Delta)_{/(K^{\triangleright})^{op}} & \xrightarrow{\;G'\;} & h(\mathcal{S}et_\Delta)_{/K},
\end{array}
$$

where G and G' are restriction functors. Let F and F' be the left adjoints to G and G', respectively. According to Theorem 4.2.4.1, assumption (1) is equivalent to the assertion that $\mathcal{F}$ lies in the essential image of F. Since each of the vertical functors is an equivalence of categories (Theorem 2.2.1.2), this is equivalent to the assertion that $\overline{X}^{op}$ lies in the essential image of F'. Since F' is fully faithful, this is equivalent to the assertion that the counit map

$$
F'G'\overline{X}^{op} \to \overline{X}^{op}
$$

is an isomorphism in $h(\mathcal{S}et_\Delta)_{/(K^{\triangleright})^{op}}$, which is clearly equivalent to (2). This shows that $(1) \Leftrightarrow (2)$ and completes the proof. $\qquad\square$

Corollary 3.3.4.6. *Let* $p : K \to \mathcal{S}$ *be a diagram which classifies a left fibration* $\widetilde{K} \to K$ *and let* $X \in \mathcal{S}$ *be a colimit of* p. *Then there is a natural isomorphism*

$$\widetilde{K} \simeq X$$

in the homotopy category $\mathcal{H}$.

Proof. Let $\overline{p} : K^{\triangleright} \to \mathcal{S}$ be a colimit diagram which extends p and let $\widetilde{K}' \to K^{\triangleright}$ be a left fibration classified by $\overline{p}$. Without loss of generality, we may suppose that $\widetilde{K} = \widetilde{K}' \times_{K^{\triangleright}} K$ and $X = \widetilde{K}' \times_{K^{\triangleright}} \{v\}$, where v denotes the cone point of $K^{\triangleright}$. Since the inclusion $\{v\} \subseteq K^{\triangleright}$ is right anodyne and the map $\widetilde{K}' \to K^{\triangleright}$ is a left fibration, Proposition 4.1.2.15 implies that the inclusion $X \subseteq \widetilde{K}'$ is right anodyne and therefore a weak homotopy equivalence. On the other hand, Proposition 3.3.4.5 implies that the inclusion $\widetilde{K} \subseteq \widetilde{K}'$ is a weak homotopy equivalence. The composition

$$X \simeq \widetilde{K}' \simeq \widetilde{K}$$

is the desired isomorphism in $\mathcal{H}$. $\qquad\qquad\qquad\qquad\qquad\qquad\qquad\qquad$ $\square$

Chapter Four

Limits and Colimits

This chapter is devoted to the study of limits and colimits in the setting of ∞-categories. Our goal is to provide tools for proving the existence of limits and colimits, for analyzing them, and for comparing them to the (perhaps more familiar) notion of homotopy limits and colimits in simplicial categories. We will generally confine our remarks to colimits; analogous results for limits can be obtained by passing to the opposite ∞-categories.

We begin in §4.1 by introducing the notion of a *cofinal* map between simplicial sets. If $f : A \to B$ is a cofinal map of simplicial sets, then we can identify colimits of a diagram $p : B \to \mathcal{C}$ with colimits of the induced diagram $p \circ f : A \to \mathcal{C}$. This is a basic maneuver which will appear repeatedly in the later chapters of this book. Consequently, it is important to have a good supply of cofinal maps. This is guaranteed by Theorem 4.1.3.1, which can be regarded as an ∞-categorical generalization of Quillen's Theorem A.

In §4.2, we introduce a battery of additional techniques for analyzing colimits. We will explain how to analyze colimits of complicated diagrams in terms of colimits of simpler diagrams. Using these ideas, we can often reduce questions about the behavior of arbitrary colimits to questions about a few basic constructions, which we will analyze explicitly in §4.4. We will also explain the relationship between the ∞-categorical theory of colimits and the more classical theory of homotopy colimits, which can be studied very effectively using the language of model categories.

The other major topic of this chapter is the theory of *Kan extensions*, which can be viewed as relative versions of limits and colimits. We will study the properties of Kan extensions in §4.3 and prove some fundamental existence theorems which we will need throughout the later chapters of this book.

4.1 COFINALITY

Let $\mathcal{C}$ be an ∞-category and let $p : K \to \mathcal{C}$ be a diagram in $\mathcal{C}$ indexed by a simplicial set K. In §1.2.13, we introduced the definition of a *colimit* $\varinjlim(p)$ for the diagram p. In practice, it is often possible to replace p by a simpler diagram without changing the colimit $\varinjlim(p)$. In this section, we will introduce a general formalism which will allow us to make replacements of this sort: the theory of *cofinal* maps between simplicial sets. We begin in §4.1.1 with a definition of the class of cofinal maps and show (Proposition

4.1.1.8) that, if a map $q : K' \to K$ is cofinal, then there is an equivalence $\varinjlim(p) \simeq \varinjlim(p \circ q)$ (provided that either colimit exists). In §4.1.2, we will reformulate the definition of cofinality using the formalism of contravariant model categories (§2.1.4). We conclude in §4.1.3 by establishing an important recognition criterion for cofinal maps in the special case where K is an ∞-category. This result can be regarded as a refinement of Quillen's Theorem A.

4.1.1 Cofinal Maps

The goal of this section is to introduce the definition of a cofinal map $p : S \to T$ of simplicial sets and study the basic properties of this notion. Our main result is Proposition 4.1.1.8, which characterizes cofinality in terms of the behavior of T-indexed colimits.

Definition 4.1.1.1 (Joyal [44]). Let $p : S \to T$ be a map of simplicial sets. We shall say that p is *cofinal* if, for any right fibration $X \to T$, the induced map of simplicial sets

$$\mathrm{Map}_T(T, X) \to \mathrm{Map}_T(S, X)$$

is a homotopy equivalence.

Remark 4.1.1.2. The simplicial set $\mathrm{Map}_T(S, X)$ parametrizes sections of the right fibration $X \to T$. It may be described as the fiber of the induced map $X^S \to T^S$ over the vertex of T^S corresponding to the map p. Since $X^S \to T^S$ is a right fibration, the fiber $\mathrm{Map}_T(S, X)$ is a Kan complex. Similarly, $\mathrm{Map}_T(T, X)$ is a Kan complex.

We begin by recording a few simple observations about the class of cofinal maps:

Proposition 4.1.1.3.   (1) *Any isomorphism of simplicial sets is cofinal.*

   (2) *Let $f : K \to K'$ and $g : K' \to K''$ be maps of simplicial sets. Suppose that f is cofinal. Then g is cofinal if and only if $g \circ f$ is cofinal.*

   (3) *If $f : K \to K'$ is a cofinal map between simplicial sets, then f is a weak homotopy equivalence.*

   (4) *An inclusion $i : K \subseteq K'$ of simplicial sets is cofinal if and only if it is right anodyne.*

Proof. Assertions (1) and (2) are obvious. We prove (3). Let S be a Kan complex. Since f is cofinal, the composition

$$\mathrm{Map}_{\mathrm{Set}_\Delta}(K', S) = \mathrm{Map}_K(K', S \times K) \to \mathrm{Map}_K(K, S \times K) = \mathrm{Map}_{\mathrm{Set}_\Delta}(K, S)$$

is a homotopy equivalence. Passing to connected components, we deduce that K and K' corepresent the same functor in the homotopy category $\mathcal{H}$ of spaces. It follows that f is a weak homotopy equivalence, as desired.

We now prove (4). Suppose first that i is right anodyne. Let $X \to K'$ be a right fibration. Then the induced map $\mathrm{Hom}_{K'}(K', X) \to \mathrm{Hom}_{K'}(K, X)$ is a trivial fibration and, in particular, a homotopy equivalence.

Conversely, suppose that i is a cofinal inclusion of simplicial sets. We wish to show that i has the left lifting property with respect to any right fibration. In other words, we must show that given any diagram of solid arrows

$$
\begin{array}{ccc}
K & \xrightarrow{\ s\ } & X \\
\downarrow & \nearrow & \downarrow \\
K' & =\!=\!= & K',
\end{array}
$$

for which the right vertical map is a right fibration, there exists a dotted arrow as indicated, rendering the diagram commutative. Since i is cofinal, the map s is homotopic to a map which extends over K'. In other words, there exists a map

$$s' : (K \times \Delta^1) \coprod_{K \times \{1\}} (K' \times \{1\}) \to X$$

compatible with the projection to K', such that $s'|K \times \{0\}$ coincides with s. Since the inclusion

$$(K \times \Delta^1) \coprod_{K \times \{1\}} (K' \times \{1\}) \subseteq K' \times \Delta^1$$

is right anodyne, there exists a map $s'' : K' \times \Delta^1 \to X$ which extends s' and is compatible with the projection to K'. The map $s''|K \times \{0\}$ has the desired properties. $\qquad\square$

Warning 4.1.1.4. The class of cofinal maps does *not* satisfy the two-out-of-three property. If $f : K \to K'$ and $g : K' \to K''$ are such that $g \circ f$ and g are cofinal, then f need not be cofinal.

Our next goal is to establish a characterization of cofinality in terms of the behavior of colimits (Proposition 4.1.1.8). First, we need a lemma.

Lemma 4.1.1.5. *Let $\mathcal{C}$ be an ∞-category and let $p : K \to \mathcal{C}$ and $q : K' \to \mathcal{C}$ be diagrams. Define simplicial sets M and N by the formulas*

$$\mathrm{Hom}(X, M) = \{f : (X \times K) \star K' \to \mathcal{C} : f|(X \times K) = p \circ \pi_K, f|K' = q\}$$

$$\mathrm{Hom}(X, N) = \{g : K \star (X \times K') \to \mathcal{C} : f|K = p, f|(X \times K') = q \circ \pi_{K'}\}.$$

Here π_K and $\pi_{K'}$ denote the projection from a product to the factor indicated by the subscript.

Then M and N are Kan complexes, which are (naturally) homotopy equivalent to one another.

Proof. We define a simplicial set $\mathcal{D}$ as follows. For every finite nonempty linearly ordered set J, to give a map $\Delta^J \to \mathcal{D}$ is to supply the following data:

- A map $\Delta^J \to \Delta^1$ corresponding to a decomposition of J as a disjoint union $J_- \coprod J_+$, where $J_- \subseteq J$ is closed downward and $J_+ \subseteq J$ is closed upward.

- A map $e : (K \times \Delta^{J_-}) \star (K' \times \Delta^{J_+}) \to \mathcal{C}$ such that $e|K \times \Delta^{J_-} = p \circ \pi_K$ and $e|K' \times \Delta^{J_+} = q \circ \pi_{K'}$.

We first claim that $\mathcal{D}$ is an ∞-category. Fix a finite linearly ordered set J as above and let $j \in J$ be neither the largest nor the smallest element of J. Let $f_0 : \Lambda_j^J \to \mathcal{D}$ be any map; we wish to show that there exists a map $f : \Delta^J \to \mathcal{D}$ which extends f_0. We first observe that the induced projection $\Lambda_j^J \to \Delta^1$ extends *uniquely* to Δ^J (since Δ^1 is isomorphic to the nerve of a category). Let $J = J_- \coprod J_+$ be the induced decomposition of J. Without loss of generality, we may suppose that $j \in J_-$. In this case, we may identify f_0 with a map

$$((K \times \Lambda_j^{J_-}) \star (K' \times \Delta^{J_+})) \coprod_{(K \times \Lambda_j^{J_-}) \star (K' \times \partial \Delta^{J_+})} ((K \times \Delta^{J_-}) \star (K' \times \partial \Delta^{J_+})) \to \mathcal{C},$$

and our goal is to find an extension

$$f : (K \times \Delta^{J_-}) \star (K' \times \Delta^{J_+}) \to \mathcal{C}.$$

Since $\mathcal{C}$ is an ∞-category, it will suffice to show that the inclusion

$$(K \times \Lambda_j^{J_-}) \star (K' \times \Delta^{J_+}) \coprod_{(K \times \Lambda_j^{J_-}) \star (K' \times \partial \Delta^{J_+})} (K \times \Delta^{J_-}) \star (K' \times \partial \Delta^{J_+})$$

$$\downarrow$$

$$(K \times \Delta^{J_-}) \star (K' \times \Delta^{J_+})$$

is inner anodyne. According to Lemma 2.1.2.3, it suffices to check that the inclusion $K \times \Lambda_j^{J_-} \subseteq K \times \Delta^{J_-}$ is right anodyne. This follows from Corollary 2.1.2.7 since $\Lambda_j^{J_-} \subseteq \Delta^{J_-}$ is right anodyne.

The ∞-category $\mathcal{D}$ has just two objects, which we will denote by x and y. We observe that $M = \operatorname{Hom}_{\mathcal{D}}^{\mathrm{R}}(x, y)$ and $N = \operatorname{Hom}_{\mathcal{D}}^{\mathrm{L}}(x, y)$. Proposition 1.2.2.3 implies that M and N are Kan complexes. Propositions 2.2.2.7 and 2.2.4.1 imply that each of these Kan complexes is weakly homotopy equivalent to $\operatorname{Map}_{\mathcal{C}[\mathcal{D}]}(x, y)$, so that M and N are homotopy equivalent to one another, as desired. $\qquad\square$

Remark 4.1.1.6. In the situation of Lemma 4.1.1.5, the homotopy equivalence between M and N is furnished by the composition of a chain of weak homotopy equivalences

$$M \leftarrow |M|_{Q^\bullet} \to \operatorname{Hom}_{\mathcal{C}[\mathcal{D}]}(x, y) \leftarrow |N|_{Q^\bullet} \to N,$$

which is functorial in the triple $(\mathcal{C}, p : K \to \mathcal{C}, q : K' \to \mathcal{C})$.

Proposition 4.1.1.7. *Let $v : K' \to K$ be a cofinal map and $p : K \to \mathcal{C}$ a diagram in an ∞-category $\mathcal{C}$. Then the map $\phi : \mathcal{C}_{p/} \to \mathcal{C}_{pv/}$ is an equivalence of left fibrations over $\mathcal{C}$: in other words, it induces a homotopy equivalence of Kan complexes after passing to the fiber over every object x of $\mathcal{C}$.*

Proof. We wish to prove that the map

$$\mathcal{C}_{p/} \times_{\mathcal{C}} \{x\} \to \mathcal{C}_{pv/} \times_{\mathcal{C}} \{x\}$$

is a homotopy equivalence of Kan complexes. Lemma 4.1.1.5 implies that the left hand side is homotopy equivalent to $\operatorname{Map}_{\mathcal{C}}(K, \mathcal{C}_{/x})$. Similarly, the right hand side can be identified with $\operatorname{Map}_{\mathcal{C}}(K', \mathcal{C}_{/x})$. Using the functoriality implicit in the proof of Lemma 4.1.1.5 (see Remark 4.1.1.6), it suffices to show that the restriction map

$$\operatorname{Map}_{\mathcal{C}}(K, \mathcal{C}_{/x}) \to \operatorname{Map}_{\mathcal{C}}(K', \mathcal{C}_{/x})$$

is a homotopy equivalence. Since v is cofinal, this follows immediately from the fact that the projection $\mathcal{C}_{/x} \to \mathcal{C}$ is a right fibration. $\square$

Proposition 4.1.1.8. *Let $v : K' \to K$ be a map of (small) simplicial sets. The following conditions are equivalent:*

(1) *The map v is cofinal.*

(2) *Given any ∞-category $\mathcal{C}$ and any diagram $p : K \to \mathcal{C}$, the induced map $\mathcal{C}_{p/} \to \mathcal{C}_{p'/}$ is an equivalence of ∞-categories, where $p' = p \circ v$.*

(3) *For every ∞-category $\mathcal{C}$ and every diagram $\overline{p} : K^{\triangleright} \to \mathcal{C}$ which is a colimit of $p = \overline{p}|K$, the induced map $\overline{p}' : K'^{\triangleright} \to \mathcal{C}$ is a colimit of $p' = \overline{p}'|K'$.*

Proof. Suppose first that (1) is satisfied. Let $p : K \to \mathcal{C}$ be as in (2). Proposition 4.1.1.7 implies that the induced map $\mathcal{C}_{p/} \to \mathcal{C}_{p'/}$ induces a homotopy equivalence of Kan complexes after passing to the fiber over any object of $\mathcal{C}$. Since both $\mathcal{C}_{p/}$ and $\mathcal{C}_{p'/}$ are left-fibered over $\mathcal{C}$, Corollary 2.4.4.4 implies that $\mathcal{C}_{p/} \to \mathcal{C}_{p'/}$ is a categorical equivalence. This proves that $(1) \Rightarrow (2)$.

Now suppose that (2) is satisfied and let $\overline{p} : K^{\triangleright} \to \mathcal{C}$ be as in (3). Then we may identify $\overline{p}$ with an initial object of the ∞-category $\mathcal{C}_{p/}$. The induced map $\mathcal{C}_{p/} \to \mathcal{C}_{p'/}$ is an equivalence and therefore carries the initial object $\overline{p}$ to an initial object $\overline{p}'$ of $\mathcal{C}_{p'/}$; thus $\overline{p}'$ is a colimit of p'. This proves that $(2) \Rightarrow (3)$.

It remains to prove that $(3) \Rightarrow (1)$. For this, we make use of the theory of classifying right fibrations (§3.3.2). Let $X \to K$ be a right fibration. We wish to show that composition with v induces a homotopy equivalence $\operatorname{Map}_K(K, X) \to \operatorname{Map}_K(K', X)$. It will suffice to prove this result after replacing X by any equivalent right fibration. Let $\mathcal{S}$ denote the ∞-category of spaces. According to Corollary 3.3.2.8, there is a classifying map $p : K \to \mathcal{S}^{op}$ and an equivalence of right fibrations between X and $(\mathcal{S}_{*/})^{op} \times_{\mathcal{S}^{op}} K$, where $*$ denotes a final object of $\mathcal{S}$.

The ∞-category $\mathcal{S}$ admits small limits (Corollary 4.2.4.8). It follows that there exists a map $\overline{p} : K^{\triangleright} \to \mathcal{S}^{op}$ which is a colimit of $p = \overline{p}|K$. Let x denote the image in $\mathcal{S}$ of the cone point of $K^{\triangleright}$. Let $\overline{p}' : K'^{\triangleright} \to \mathcal{S}^{op}$ be the induced map. Then, by hypothesis, $\overline{p}'$ is a colimit of $p' = \overline{p}'|K'$. According to Lemma 4.1.1.5, there is a (natural) chain of weak homotopy equivalences relating $\mathrm{Map}_K(K, X)$ with $(\mathcal{S}^{op})_{p/} \times_{\mathcal{S}^{op}} \{y\}$. Similarly, there is a chain of weak homotopy equivalences connecting $\mathrm{Map}_K(K', X)$ with $(\mathcal{S}^{op})_{p'/} \times_{\mathcal{S}^{op}} \{y\}$. Consequently, we are reduced to proving that the left vertical map in the diagram

$$
\begin{array}{ccccc}
(\mathcal{S}^{op})_{p/} \times_{\mathcal{S}^{op}} \{y\} & \longleftarrow & (\mathcal{S}^{op})_{\overline{p}/} \times_{\mathcal{S}^{op}} \{y\} & \longrightarrow & (\mathcal{S}^{op})_{x/} \times_{\mathcal{S}^{op}} \{y\} \\
\downarrow & & \downarrow & & \downarrow \\
(\mathcal{S}^{op})_{p'/} \times_{\mathcal{S}^{op}} \{y\} & \longleftarrow & (\mathcal{S}^{op})_{\overline{p}'/} \times_{\mathcal{S}^{op}} \{y\} & \longrightarrow & (\mathcal{S}^{op})_{x/} \times_{\mathcal{S}^{op}} \{y\}
\end{array}
$$

is a homotopy equivalence. Since $\overline{p}$ and $\overline{q}$ are colimits of p and q, the left horizontal maps are trivial fibrations. Since the inclusions of the cone points into $K^{\triangleright}$ and $K'^{\triangleright}$ are right anodyne, the right horizontal maps are also trivial fibrations. It therefore suffices to prove that the right vertical map is a homotopy equivalence. But this map is an isomorphism of simplicial sets. $\qquad\square$

Corollary 4.1.1.9. *Let $p : K \to K'$ be a map of simplicial sets and $q : K' \to K''$ a categorical equivalence. Then p is cofinal if and only if $q \circ p$ is cofinal. In particular (taking $p = \mathrm{id}_{S'}$), q itself is cofinal.*

Proof. Let $\mathcal{C}$ be an ∞-category, let $r'' : K'' \to \mathcal{C}$ be a diagram, and set $r' = r'' \circ q$, $r = r' \circ p$. Since q is a categorical equivalence, $\mathcal{C}_{r''/} \to \mathcal{C}_{r'/}$ is a categorical equivalence. It follows that $\mathcal{C}_{r/} \to \mathcal{C}_{r''/}$ is a categorical equivalence if and only if $\mathcal{C}_{r/} \to \mathcal{C}_{r'/}$ is a categorical equivalence. We now apply the characterization (2) of Proposition 4.1.1.8. $\qquad\square$

Corollary 4.1.1.10. *The property of cofinality is homotopy invariant. In other words, if two maps $f, g : K \to K'$ have the same image in the homotopy category of Set_{Δ} obtained by inverting all categorical equivalences, then f is cofinal if and only if g is cofinal.*

Proof. Choose a categorical equivalence $K' \to \mathcal{C}$, where $\mathcal{C}$ is an ∞-category. In view of Corollary 4.1.1.9, we may replace K' by $\mathcal{C}$ and thereby assume that K' is itself an ∞-category. Since f and g are homotopic, there exists a cylinder object S equipped with a trivial fibration $p : S \to K$, a map $q : S \to \mathcal{C}$, and two sections $s, s' : K \to S$ of p, such that $f = q \circ s$, $g = q \circ s'$. Since p is a categorical equivalence, so is every section of p. Consequently, s and s' are cofinal. We now apply Proposition 4.1.1.3 to deduce that f is cofinal if and only if q is cofinal. Similarly, g is cofinal if and only if q is cofinal. $\qquad\square$

Corollary 4.1.1.11. *Let $p : X \to S$ be a map of simplicial sets. The following are equivalent:*

(1) *The map p is a cofinal right fibration.*

(2) *The map p is a trivial fibration.*

Proof. Clearly any trivial fibration is a right fibration. Furthermore, any trivial fibration is a categorical equivalence, hence cofinal by Corollary 4.1.1.9. Thus (2) implies (1). Conversely, suppose that p is a cofinal right fibration. Since p is cofinal, the natural map $\mathrm{Map}_S(S, X) \to \mathrm{Map}_S(X, X)$ is a homotopy equivalence of Kan complexes. In particular, there exists a section $f : S \to X$ of p such that $f \circ p$ is (fiberwise) homotopic to the identity map of X. Consequently, for each vertex s of S, the fiber $X_s = X \times_S \{s\}$ is a contractible Kan complex (since the identity map $X_s \to X_s$ is homotopic to the constant map with value $f(s)$). The dual of Lemma 2.1.3.4 now shows that p is a trivial fibration. $\qquad\qquad\square$

Corollary 4.1.1.12. *A map $X \to Z$ of simplicial sets is cofinal if and only if it admits a factorization*

$$X \xrightarrow{f} Y \xrightarrow{g} Z,$$

where $X \to Y$ is right anodyne and $Y \to Z$ is a trivial fibration.

Proof. The "if" direction is clear: if such a factorization exists, then f is cofinal (since it is right anodyne), g is cofinal (since it is a categorical equivalence), and consequently $g \circ f$ is cofinal (since it is a composition of cofinal maps).

For the "only if" direction, let us suppose that $X \to Z$ is a cofinal map. By the small object argument (Proposition A.1.2.5), there is a factorization

$$X \xrightarrow{f} Y \xrightarrow{g} Z$$

where f is right anodyne and g is a right fibration. The map g is cofinal by Proposition 4.1.1.3 and therefore a trivial fibration by Corollary 4.1.1.11. $\quad\square$

Corollary 4.1.1.13. *Let $p : S \to S'$ be a cofinal map and K any simplicial set. Then the induced map $K \times S \to K \times S'$ is cofinal.*

Proof. Using Corollary 4.1.1.12, we may suppose that p is either right anodyne or a trivial fibration. Then the induced map $K \times S \to K \times S'$ has the same property. $\qquad\qquad\square$

4.1.2 Smoothness and Right Anodyne Maps

In this section, we explain how to characterize the classes of right anodyne and cofinal morphisms in terms of the contravariant model structures studied in §2.1.4. We also introduce a third class of maps between simplicial sets, which we call *smooth*.

We begin with the following characterization of right anodyne maps:

Proposition 4.1.2.1. *Let $i : A \to B$ be a map of simplicial sets. The following conditions are equivalent:*

(1) *The map i is right anodyne.*

(2) *For any map of simplicial sets $j : B \to C$, the map i is a trivial cofibration with respect to the contravariant model structure on $(\mathrm{Set}_\Delta)_{/C}$.*

(3) *The map i is a trivial cofibration with respect to the contravariant model structure on $(\mathrm{Set}_\Delta)_{/B}$.*

Proof. The implication (1) $\Rightarrow$ (2) follows immediately from Proposition 2.1.4.9, and the implication (2) $\Rightarrow$ (3) is obvious. Suppose that (3) holds. To prove (1), it suffices to show that given any diagram

$$
\begin{array}{ccc}
A & \longrightarrow & X \\
{\scriptstyle i}\downarrow & \nearrow^{f} & \downarrow{\scriptstyle p} \\
B & \longrightarrow & Y
\end{array}
$$

such that p is a right fibration, one can supply the dotted arrow f as indicated. Replacing $p : X \to Y$ by the pullback $X \times_Y B \to B$, we may reduce to the case where $Y = B$. Corollary 2.2.3.12 implies that X is a fibrant object of $(\mathrm{Set}_\Delta)_{/B}$ (with respect to contravariant model structure) so that the desired map f can be found. $\qquad\square$

Corollary 4.1.2.2. *Suppose we are given maps $A \xrightarrow{i} B \xrightarrow{j} C$ of simplicial sets. If i and $j \circ i$ are right anodyne and j is a cofibration, then j is right anodyne.*

Proof. By Proposition 4.1.2.1, i and $j \circ i$ are contravariant equivalences in the category $(\mathrm{Set}_\Delta)_{/C}$. It follows that j is a trivial cofibration in $(\mathrm{Set}_\Delta)_{/C}$, so that j is right anodyne (by Proposition 4.1.2.1 again). $\qquad\square$

Corollary 4.1.2.3. *Let*

$$
\begin{array}{ccccc}
A' & \xleftarrow{\;u\;} & A & \longrightarrow & A'' \\
{\scriptstyle f'}\downarrow & & {\scriptstyle f}\downarrow & & \downarrow{\scriptstyle f''} \\
B' & \xleftarrow{\;v\;} & B & \longrightarrow & B''
\end{array}
$$

be a diagram of simplicial sets. Suppose that u and v are monomorphisms and that f, f', and f'' are right anodyne. Then the induced map

$$
A' \coprod_A A'' \to B' \coprod_B B''
$$

is right anodyne.

Proof. According to Proposition 4.1.2.1, each of the maps f, f', and f'' is a contravariant equivalence in the category $(\mathrm{Set}_\Delta)_{/B' \coprod_B B''}$. The assumption on u and v guarantees that $f' \coprod_f f''$ is also a contravariant equivalence in $(\mathrm{Set}_\Delta)_{/B' \coprod_B B''}$, so that $f' \coprod_f f''$ is right anodyne by Proposition 4.1.2.1 again. $\qquad\square$

Corollary 4.1.2.4. *The collection of right anodyne maps of simplicial sets is stable under filtered colimits.*

Proof. Let $f : A \to B$ be a filtered colimit of right anodyne morphisms $f_\alpha : A_\alpha \to B_\alpha$. According to Proposition 4.1.2.1, each f_α is a contravariant equivalence in $(\mathrm{Set}_\Delta)_{/B}$. Since contravariant equivalences are stable under filtered colimits, we conclude that f is a contravariant equivalence in $(\mathrm{Set}_\Delta)_{/B}$, so that f is right anodyne by Proposition 4.1.2.1. $\square$

Proposition 4.1.2.1 has an analogue for cofinal maps:

Proposition 4.1.2.5. *Let $i : A \to B$ be a map of simplicial sets. The following conditions are equivalent:*

(1) *The map i is cofinal.*

(2) *For any map $j : B \to C$, the inclusion i is a contravariant equivalence in $(\mathrm{Set}_\Delta)_{/C}$.*

(3) *The map i is a contravariant equivalence in $(\mathrm{Set}_\Delta)_{/B}$.*

Proof. Suppose (1) is satisfied. By Corollary 4.1.1.12, i admits a factorization as a right anodyne map followed by a trivial fibration. Invoking Proposition 4.1.2.1, we conclude that (2) holds. The implication (2) $\Rightarrow$ (3) is obvious. If (3) holds, then we can choose a factorization

$$A \xrightarrow{i'} A' \xrightarrow{i''} B$$

of i, where i' is right anodyne and i'' is a right fibration. Then i'' is a contravariant fibration (in $\mathrm{Set}_{\Delta/B}$) and a contravariant weak equivalence and is therefore a trivial fibration of simplicial sets. We now apply Corollary 4.1.1.12 to conclude that i is cofinal. $\square$

Corollary 4.1.2.6. *Let $p : X \to S$ be a map of simplicial sets, where S is a Kan complex. Then p is cofinal if and only if it is a weak homotopy equivalence.*

Proof. By Proposition 4.1.2.5, p is cofinal if and only if it is a contravariant equivalence in $(\mathrm{Set}_\Delta)_{/S}$. If S is a Kan complex, then Proposition 3.1.5.7 asserts that the contravariant equivalences are precisely the weak homotopy equivalences. $\square$

Corollary 4.1.2.7. *Suppose we are given a pushout diagram*

$$
\begin{array}{ccc}
A & \xrightarrow{\;g\;} & A' \\
\downarrow{\scriptstyle f} & & \downarrow{\scriptstyle f'} \\
B & \longrightarrow & B'
\end{array}
$$

of simplicial sets. If f is cofinal and either f or g is a cofibration, then f' is cofinal.

Proof. Combine Proposition 4.1.2.5 with the left-properness of the contr-
variant model structure. □

Let $p : X \to Y$ be an arbitrary map of simplicial sets. In §2.1.4, we showed
that p induces a Quillen adjunction $(p_!, p^*)$ between the contravariant model
categories $(\text{Set}_\Delta)_{/X}$ and $(\text{Set}_\Delta)_{/Y}$. The functor p^* itself has a right adjoint,
which we will denote by p_*; it is given by

$$p_*(M) = \text{Map}_Y(X, M).$$

The adjoint functors p^* and p_* are not Quillen adjoints in general. Instead
we have the following result:

Proposition 4.1.2.8. *Let $p : X \to Y$ be a map of simplicial sets. The
following conditions are equivalent:*

(1) *For any right anodyne map $i : A \to B$ in $(\text{Set}_\Delta)_{/Y}$, the induced map
$A \times_Y X \to B \times_Y X$ is right anodyne.*

(2) *For every Cartesian diagram*

$$
\begin{array}{ccc}
X' & \longrightarrow & X \\
\downarrow{\scriptstyle p'} & & \downarrow{\scriptstyle p} \\
Y' & \longrightarrow & Y,
\end{array}
$$

*the functor $p'^* : (\text{Set}_\Delta)_{/Y'} \to (\text{Set}_\Delta)_{/X'}$ preserves contravariant equiv-
alences.*

(3) *For every Cartesian diagram*

$$
\begin{array}{ccc}
X' & \longrightarrow & X \\
\downarrow{\scriptstyle p'} & & \downarrow{\scriptstyle p} \\
Y' & \longrightarrow & Y,
\end{array}
$$

*the adjoint functors (p'^*, p'_*) give rise to a Quillen adjunction between
the contravariant model categories $(\text{Set}_\Delta)_{/Y'}$ and $(\text{Set}_\Delta)_{/X'}$.*

Proof. Suppose that (1) is satisfied; let us prove (2). Since property (1) is
clearly stable under base change, we may suppose that $p' = p$. Let $u : M \to
N$ be a contravariant equivalence in $(\text{Set}_\Delta)_{/Y}$. If M and N are fibrant, then
u is a homotopy equivalence, so that $p^*(u) : p^*M \to p^*N$ is also a homotopy
equivalence. In the general case, we may select a diagram

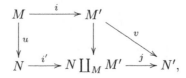

where M' and N' are fibrant and the maps i and j are right anodyne (and therefore i' is also right anodyne). Then $p^*(v)$ is a contravariant equivalence, while the maps $p^*(i)$, $p^*(j)$, and $p^*(i')$ are all right anodyne; by Proposition 4.1.2.1 they are contravariant equivalences as well. It follows that $p^*(u)$ is a contravariant equivalence.

To prove (3), it suffices to show that p'^* preserves cofibrations and trivial cofibrations. The first statement is obvious, and the second follows immediately from (2). Conversely, the existence of a Quillen adjunction (p'^*, p_*) implies that p'^* preserves contravariant equivalences between cofibrant objects. Since every object of $(\mathrm{Set}_\Delta)_{/Y'}$ is cofibrant, we deduce that (3) implies (2).

Now suppose that (2) is satisfied and let $i : A \to B$ be a right anodyne map in $(\mathrm{Set}_\Delta)_{/Y}$ as in (1). Then i is a contravariant equivalence in $(\mathrm{Set}_\Delta)_{/B}$. Let $p' : X \times_Y B \to B$ be base change of p; then (2) implies that the induced map $i' : p'^* A \to p'^* B$ is a contravariant equivalence in $(\mathrm{Set}_\Delta)_{/B \times_Y X}$. By Proposition 4.1.2.1, the map i' is right anodyne. Now we simply note that i' may be identified with the map $A \times_Y X \to B \times_Y X$ in the statement of (1). $\qquad\square$

Definition 4.1.2.9. We will say that a map $p : X \to Y$ of simplicial sets is *smooth* if it satisfies the (equivalent) conditions of Proposition 4.1.2.8.

Remark 4.1.2.10. Let

$$
\begin{array}{ccc}
X' & \xrightarrow{\ f'\ } & X \\
\downarrow & & \downarrow{\scriptstyle p} \\
S' & \xrightarrow{\ f\ } & S
\end{array}
$$

be a pullback diagram of simplicial sets. Suppose that p is smooth and that f is cofinal. Then f' is cofinal: this follows immediately from characterization (2) of Proposition 4.1.2.8 and characterization (3) of Proposition 4.1.2.5.

We next give an alternative characterization of smoothness. Let

$$
\begin{array}{ccc}
X' & \xrightarrow{\ q'\ } & X \\
\downarrow{\scriptstyle p'} & & \downarrow{\scriptstyle p} \\
Y' & \xrightarrow{\ q\ } & Y
\end{array}
$$

be a Cartesian diagram of simplicial sets. Then we obtain an isomorphism $Rp'^* Rq^* \simeq Rq'^* Rp^*$ of right derived functors, which induces a natural transformation

$$
\psi_{p,q} : Lq'_! Rp'^* \to Rp^* Lq_!.
$$

Proposition 4.1.2.11. *Let* $p : X \to Y$ *be a map of simplicial sets. The following conditions are equivalent:*

(1) *The map* p *is smooth.*

(2) *For every Cartesian rectangle*

$$
\begin{array}{ccc}
X'' \xrightarrow{\ q'\ } X' \longrightarrow X \\
\Big\downarrow{\scriptstyle p''} \quad\ \Big\downarrow{\scriptstyle p'} \quad\ \Big\downarrow{\scriptstyle p} \\
Y'' \xrightarrow{\ q\ } Y' \longrightarrow Y,
\end{array}
$$

the natural transformation $\psi_{p',q}$ is an isomorphism of functors from the homotopy category of $(\mathrm{Set}_\Delta)_{/Y''}$ to the homotopy category of $(\mathrm{Set}_\Delta)_{/X'}$ (here we regard all categories as endowed with the contravariant model structure).

Proof. Suppose that (1) is satisfied and consider any Cartesian rectangle as in (2). Since p is smooth, p' and p'' are also smooth. It follows that p'^* and p''^* preserve weak equivalences, so they may be identified with their right derived functors. Similarly, $q_!$ and $q'_!$ preserve weak equivalences, so they may be identified with their left derived functors. Consequently, the natural transformation $\psi_{p',q}$ is simply obtained by passage to the homotopy category from the natural transformation

$$
q'_! p''^* \to p'^* q_!.
$$

But this is an isomorphism of functors before passage to the homotopy categories.

Now suppose that (2) is satisfied. Let $q : Y'' \to Y'$ be a right anodyne map in $(\mathrm{Set}_\Delta)_{/Y}$ and form the Cartesian square as in (2). Let us compute the value of the functors $Lq'_! Rp''^*$ and $Rp'^* Lq_!$ on the object Y'' of $(\mathrm{Set}_\Delta)_{/Y''}$. The composite $Lq'_! Rp''^*$ is easy: because Y'' is fibrant and $X'' = p''^* Y''$ is cofibrant, the result is X'', regarded as an object of $(\mathrm{Set}_\Delta)_{/X'}$. The other composition is slightly trickier: Y'' is cofibrant, but $q_! Y''$ is not fibrant when viewed as an object of $(\mathrm{Set}_\Delta)_{/Y'}$. However, in view of the assumption that q is right anodyne, Proposition 4.1.2.1 ensures that Y' is a fibrant replacement for $q_! Y'$; thus we may identify $Rp'^* Lq_!$ with the object $p'^* Y' = X'$ of $(\mathrm{Set}_\Delta)_{/X'}$. Condition (2) now implies that the natural map $X'' \to X'$ is a contravariant equivalence in $(\mathrm{Set}_\Delta)_{/X'}$. Invoking Proposition 4.1.2.1, we deduce that q' is right anodyne, as desired. $\Box$

Remark 4.1.2.12. The terminology "smooth" is suggested by the analogy of Proposition 4.1.2.11 with the *smooth base change theorem* in the theory of étale cohomology (see, for example, [28]).

Proposition 4.1.2.13. *Suppose we are given a commutative diagram*

$$
\begin{array}{ccc}
X \xrightarrow{\ i\ } X' \\
\Big\downarrow \ {\scriptstyle p}\searrow \ \Big\downarrow{\scriptstyle p'} \\
X'' \xrightarrow{\ p''\ } S
\end{array}
$$

of simplicial sets. Assume that i is a cofibration and that p, p', and p'' are smooth. Then the induced map $X' \coprod_X X'' \to S$ is smooth.

Proof. This follows immediately from Corollary 4.1.2.3 and characterization (1) of Proposition 4.1.2.8. □

Proposition 4.1.2.14. *The collection of smooth maps $p : X \to S$ is stable under filtered colimits in $(\mathrm{Set}_\Delta)_{/S}$.*

Proof. Combine Corollary 4.1.2.4 with characterization (1) of Proposition 4.1.2.8. □

Proposition 4.1.2.15. *Let $p : X \to S$ be a coCartesian fibration of simplicial sets. Then p is smooth.*

Proof. Let $i : B' \to B$ be a right anodyne map in $(\mathrm{Set}_\Delta)_{/S}$; we wish to show that the induced map $B' \times_S X \to B \times_S X$ is right anodyne. By general nonsense, we may reduce ourselves to the case where i is an inclusion $\Lambda_i^n \subseteq \Delta^n$, where $0 < i \leq n$. Making a base change, we may suppose that $S = B$. By Proposition 3.2.2.7, there exists a composable sequence of maps

$$\phi : A^0 \to \cdots \to A^n$$

and a quasi-equivalence $M^{op}(\phi) \to X$. Consider the diagram

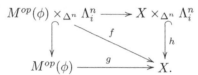

The left vertical map is right anodyne since it is a pushout of the inclusion $A^0 \times \Lambda_i^n \subseteq A^0 \times \Delta^n$. It follows that f is cofinal, being a composition of a right anodyne map and a categorical equivalence. Since g is cofinal (being a categorical equivalence), we deduce from Proposition 4.1.1.3 that h is cofinal. Since h is a monomorphism of simplicial sets, it is right anodyne by Proposition 4.1.1.3. □

Proposition 4.1.2.16. *Let $p : X \to S \times T$ be a bifibration. Then the composite map $\pi_S \circ p : X \to S$ is smooth.*

Proof. For every map $T' \to T$, let $X_{T'} = X \times_T T'$. We note that X is a filtered colimit of $X_{T'}$ as T' ranges over the finite simplicial subsets of T. Using Proposition 4.1.2.14, we can reduce to the case where T is finite. Working by induction on the dimension and the number of nondegenerate simplices of T, we may suppose that $T = T' \coprod_{\partial \Delta^n} \Delta^n$, where the result is known for T' and for $\partial \Delta^n$. Applying Proposition 4.1.2.13, we can reduce to the case $T = \Delta^n$. We now apply Lemma 2.4.7.5 to deduce that p is a coCartesian fibration and therefore smooth (Proposition 4.1.2.15). □

Lemma 4.1.2.17. *Let $\mathcal{C}$ be an ∞-category containing an object C and let $f : X \to Y$ be a covariant equivalence in $(\mathrm{Set}_\Delta)_{/\mathcal{C}}$. The induced map*

$$X \times_{\mathcal{C}} \mathcal{C}^{/C} \to Y \times_{\mathcal{C}} \mathcal{C}^{/C}$$

is also a covariant equivalence in $\mathcal{C}^{/C}$.

Proof. It will suffice to prove that for every object $Z \to \mathcal{C}$ of $(\mathrm{Set}_\Delta)_{/\mathcal{C}}$, the fiber product $Z \times_{\mathcal{C}} \mathcal{C}^{/C}$ is a homotopy product of Z with $\mathcal{C}^{/C}$ in $(\mathrm{Set}_\Delta)_{/\mathcal{C}}$ (with respect to the covariant model structure). Choose a factorization

$$Z \xrightarrow{i} Z' \xrightarrow{j} \mathcal{C},$$

where i is left anodyne and j is a left fibration. According to Corollary 2.2.3.12, we may regard Z' as a fibrant replacement for Z in $(\mathrm{Set}_\Delta)_{/\mathcal{C}}$. It therefore suffices to prove that the map $i' : Z \times_{\mathcal{C}} \mathcal{C}^{/C} \to Z' \times_{\mathcal{C}} \mathcal{C}^{/C}$ is a covariant equivalence. According to Proposition 4.1.2.5, it will suffice to prove that i' is left anodyne. The map i' is a base change of i by the projection $p : \mathcal{C}^{/C} \to \mathcal{C}$; it therefore suffices to prove that p^{op} is smooth. This follows from Proposition 4.1.2.15 since p is a right fibration of simplicial sets.   □

Proposition 4.1.2.18. *Let $\mathcal{C}$ be an ∞-category and*

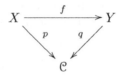

a commutative diagram of simplicial sets. Suppose that p and q are smooth. The following conditions are equivalent:

(1) *The map f is a covariant equivalence in $(\mathrm{Set}_\Delta)_{/\mathcal{C}}$.*

(2) *For each object $C \in \mathcal{C}$, the induced map of fibers $X_C \to Y_C$ is a weak homotopy equivalence.*

Proof. Suppose that (1) is satisfied and let C be an object of $\mathcal{C}$. We have a commutative diagram of simplicial sets

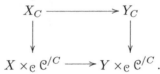

Lemma 4.1.2.17 implies that the bottom horizontal map is a covariant equivalence. The vertical maps are both pullbacks of the right anodyne inclusion $\{C\} \subseteq \mathcal{C}^{/C}$ along smooth maps and are therefore right anodyne. In particular, the vertical arrows and the bottom horizontal arrow are all weak homotopy equivalences; it follows that the map $X_C \to Y_C$ is a weak homotopy equivalence as well.

Now suppose that (2) is satisfied. Choose a commutative diagram

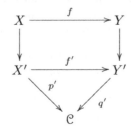

in $(\mathcal{S}et_\Delta)_{/\mathcal{C}}$, where the vertical arrows are left anodyne and the maps p' and q' are left fibrations. Using Proposition 4.1.2.15, we conclude that p' and q' are smooth. Applying (1), we deduce that for each object $C \in \mathcal{C}$, the maps $X_C \to X'_C$ and $Y_C \to Y'_C$ are weak homotopy equivalences. It follows that each fiber $f'_C : X'_C \to Y'_C$ is a homotopy equivalence of Kan complexes, so that f' is an equivalence of left fibrations and therefore a covariant equivalence. Inspecting the above diagram, we deduce that f is also a covariant equivalence, as desired. □

4.1.3 Quillen's Theorem A for ∞-Categories

Suppose that $f : \mathcal{C} \to \mathcal{D}$ is a functor between ∞-categories and that we wish to determine whether or not f is cofinal. According to Proposition 4.1.1.8, the cofinality of f is equivalent to the assertion that for any diagram $p : \mathcal{D} \to \mathcal{E}$, f induces an equivalence

$$\varinjlim(p) \simeq \varinjlim(p \circ f).$$

One can always define a morphism

$$\phi : \varinjlim(p \circ f) \to \varinjlim(p)$$

(provided that both sides are defined); the question is whether or not we can define an inverse $\psi = \phi^{-1}$. Roughly speaking, this involves defining a compatible family of maps $\psi_D : p(D) \to \varinjlim(p \circ f)$ indexed by $D \in \mathcal{D}$. The only reasonable candidate for ψ_D is a composition

$$p(D) \to (p \circ f)(C) \to \varinjlim(p \circ f),$$

where the first map arises from a morphism $D \to f(C)$ in $\mathcal{C}$. Of course, the existence of C is not automatic. Moreover, even if C exists, it is usually not unique. The collection of candidates for C is parametrized by the ∞-category $\mathcal{C}^{D/} = \mathcal{C} \times_\mathcal{D} \mathcal{D}^{D/}$. In order to make the above construction work, we need the ∞-category $\mathcal{C}^{D/}$ to be weakly contractible. More precisely, we will prove the following result:

Theorem 4.1.3.1 (Joyal [44]). *Let $f : \mathcal{C} \to \mathcal{D}$ be a map of simplicial sets, where $\mathcal{D}$ is an ∞-category. The following conditions are equivalent:*

(1) *The functor f is cofinal.*

(2) *For every object $D \in \mathcal{D}$, the simplicial set $\mathcal{C} \times_\mathcal{D} \mathcal{D}_{D/}$ is weakly contractible.*

We first need to establish the following lemma:

Lemma 4.1.3.2. *Let $p : U \to S$ be a Cartesian fibration of simplicial sets. Suppose that for every vertex s of S, the fiber $U_s = p^{-1}\{s\}$ is weakly contractible. Then p is cofinal.*

Proof. Let $q : N \to S$ be a right fibration. For every map of simplicial sets $T \to S$, let $X_T = \text{Map}_S(T, N)$ and $Y_T = \text{Map}_S(T \times_S U, N)$. Our goal is to prove that the natural map $X_S \to Y_S$ is a homotopy equivalence of Kan complexes. We will prove, more generally, that for any map $T \to S$, the map $\phi_T : Y_T \to Z_T$ is a homotopy equivalence. The proof uses induction on the (possibly infinite) dimension of T. Choose a transfinite sequence of simplicial subsets $T(\alpha) \subseteq T$, where each $T(\alpha)$ is obtained from $T(< \alpha) = \bigcup_{\beta < \alpha} T(\beta)$ by adjoining a single nondegenerate simplex of T (if such a simplex exists). We prove that $\phi_{T(\alpha)}$ is a homotopy equivalence by induction on α. Assuming that $\phi_{T(\beta)}$ is a homotopy equivalence for every $\beta < \alpha$, we deduce that $\phi_{T(<\alpha)}$ is the homotopy inverse limit of a tower of equivalences and therefore a homotopy equivalence. If $T(\alpha) = T(< \alpha)$, we are done. Otherwise, we may write $T(\alpha) = T(< \alpha) \coprod_{\partial \Delta^n} \Delta^n$. Then $\phi_{T(\alpha)}$ can be written as a homotopy pullback of $\phi_{T(<\alpha)}$ with ϕ_{Δ^n} over $\phi_{\partial \Delta^n}$. The third map is a homotopy equivalence by the inductive hypothesis. It therefore suffices to prove that ϕ_{Δ^n} is an equivalence. In other words, we may reduce to the case $T = \Delta^n$.

By Proposition 3.2.2.7, there exists a composable sequence of maps

$$\theta : A^0 \leftarrow \cdots \leftarrow A^n$$

and a quasi-equivalence $f : M(\theta) \to X \times_S T$, where $M(\theta)$ denotes the mapping simplex of the sequence θ. Given a map $T' \to T$, we let $Z_{T'} = \text{Map}_S(M(\theta) \times_T T', N)$. Proposition 3.3.1.7 implies that q is a categorical fibration. It follows that for any map $T' \to T$, the categorical equivalence $M(\theta) \times_T T' \to U \times_S T'$ induces another categorical equivalence $\psi_{T'} = Y_{T'} \to Z_{T'}$. Since $Y_{T'}$ and $Z_{T'}$ are Kan complexes, the map $\psi_{T'}$ is a homotopy equivalence. Consequently, to prove that ϕ_T is an equivalence, it suffices to show that the composite map

$$X_T \to Y_T \to Z_T$$

is an equivalence.

Consider the composition

$$u : X_{\Delta^{n-1}} \xrightarrow{u'} Z_{\Delta^{n-1}} \xrightarrow{u''} \text{Map}_S(\Delta^{n-1} \times A^n, N) \xrightarrow{u'''} \text{Map}_S(\{n-1\} \times A^n, N).$$

Using the fact that q is a right fibration and that A^n is weakly contractible, we deduce that u and u''' are homotopy equivalences. The inductive hypothesis implies that u' is a homotopy equivalence. Consequently, u'' is also a homotopy equivalence. The space Z_T fits into a homotopy Cartesian diagram

$$
\begin{array}{ccc}
Z_T & \longrightarrow & Z_{\Delta^{n-1}} \\
\downarrow{\scriptstyle v''} & & \downarrow{\scriptstyle u''} \\
\text{Map}_S(\Delta^n \times A^n, N) & \longrightarrow & \text{Map}_S(\Delta^{n-1} \times A^n, N).
\end{array}
$$

It follows that v'' is a homotopy equivalence. Now consider the composition

$$v : X_{\Delta^n} \xrightarrow{v'} Z_{\Delta^n} \xrightarrow{v''} \text{Map}_S(\Delta^n \times A^n, N) \xrightarrow{v'''} \text{Map}_S(\{n\} \times A^n, N).$$

Again, because q is a right fibration and A^n is weakly contractible, the maps v and v''' are homotopy equivalences. Since v'' is a homotopy equivalence, we deduce that v' is a homotopy equivalence, as desired. $\square$

Proof of Theorem 4.1.3.1. Using the small object argument, we can factor f as a composition

$$\mathcal{C} \xrightarrow{f'} \mathcal{C}' \xrightarrow{f''} \mathcal{D},$$

where f' is a categorical equivalence and f'' is an inner fibration. Then f'' is cofinal if and only if f is cofinal (Corollary 4.1.1.10). For every $D \in \mathcal{D}$, the map $\mathcal{D}_{D/} \to \mathcal{D}$ is a left fibration, so the induced map $\mathcal{C}_{D/} \to \mathcal{C}'_{D/}$ is a categorical equivalence (Proposition 3.3.1.3). Consequently, it will suffice to prove that $(1) \Leftrightarrow (2)$ for the morphism $f'' : \mathcal{C}' \to \mathcal{D}$. In other words, we may assume that the simplicial set $\mathcal{C}$ is an ∞-category.

Suppose first that (1) is satisfied and choose $D \in \mathcal{D}$. The projection $\mathcal{D}_{D/} \to \mathcal{D}$ is a left fibration and therefore smooth (Proposition 4.1.2.15). Applying Remark 4.1.2.10, we deduce that the projection $\mathcal{C} \times_{\mathcal{D}} \mathcal{D}_{D/} \to \mathcal{D}_{D/}$ is cofinal and therefore a weak homotopy equivalence (Proposition 4.1.1.3). Since $\mathcal{D}_{D/}$ has an initial object, it is weakly contractible. Therefore $\mathcal{C} \times_{\mathcal{D}} \mathcal{D}_{D/}$ is weakly contractible, as desired.

We now prove that $(2) \Rightarrow (1)$. Let $\mathcal{M} = \operatorname{Fun}(\Delta^1, \mathcal{D}) \times_{\operatorname{Fun}(\{1\}, \mathcal{D})} \mathcal{C}$. Then the map f factors as a composition

$$\mathcal{C} \xrightarrow{f'} \mathcal{M} \xrightarrow{f''} \mathcal{D},$$

where f' is the obvious map and f'' is given by evaluation at the vertex $\{0\} \subseteq \Delta^1$. Note that there is a natural projection map $\pi : \mathcal{M} \to \mathcal{C}$, that f' is a section of π, and that there is a simplicial homotopy $h : \Delta^1 \times \mathcal{M} \to \mathcal{M}$ from $\operatorname{id}_{\mathcal{M}}$ to $f' \circ \pi$ which is compatible with the projection to $\mathcal{C}$. It follows from Proposition 2.1.2.11 that f' is right anodyne.

Corollary 2.4.7.12 implies that f'' is a Cartesian fibration. The fiber of f'' over an object $D \in \mathcal{D}$ is isomorphic to $\mathcal{C} \times_{\mathcal{D}} \mathcal{D}^{D/}$, which is equivalent to $\mathcal{C} \times_{\mathcal{D}} \mathcal{D}_{D/}$ and therefore weakly contractible (Proposition 4.2.1.5). By assumption, the fibers of f'' are weakly contractible. Lemma 4.1.3.2 asserts that f'' is cofinal. It follows that f, as a composition of cofinal maps, is also cofinal. $\square$

Using Theorem 4.1.3.1, we can easily deduce the following classical result of Quillen:

Corollary 4.1.3.3 (Quillen's Theorem A). *Let $f : \mathcal{C} \to \mathcal{D}$ be a functor between ordinary categories. Suppose that for every object $D \in \mathcal{D}$, the fiber product category $\mathcal{C} \times_{\mathcal{D}} \mathcal{D}_{D/}$ has a weakly contractible nerve. Then f induces a weak homotopy equivalence of simplicial sets $\operatorname{N}(\mathcal{C}) \to \operatorname{N}(\mathcal{D})$.*

Proof. The assumption implies that $\operatorname{N}(f) : \operatorname{N}(\mathcal{C}) \to \operatorname{N}(\mathcal{D})$ satisfies the hypotheses of Theorem 4.1.3.1. It follows that $\operatorname{N}(f)$ is a cofinal map of simplicial sets and therefore a weak homotopy equivalence (Proposition 4.1.1.3). $\square$

4.2 TECHNIQUES FOR COMPUTING COLIMITS

In this section, we will introduce various techniques for computing, analyzing, and manipulating colimits. We begin in §4.2.1 by introducing a variant on the join construction of §1.2. The new join construction is (categorically) equivalent to the version we are already familiar with but has better formal behavior in some contexts. For example, it permits us to define a *parametrized* version of overcategories and undercategories, which we will analyze in §4.2.2.

In §4.2.3, we address the following question: given a diagram $p : K \to \mathcal{C}$ and a decomposition of K into "pieces," how is the colimit $\varinjlim(p)$ related to the colimits of those pieces? For example, if $K = A \cup B$, then it seems reasonable to expect an equation of the form

$$\varinjlim(p) = (\varinjlim p|A) \coprod_{\varinjlim(p|A \cap B)} (\varinjlim p|B).$$

Of course there are many variations on this theme; we will lay out a general framework in §4.2.3 and apply it to specific situations in §4.4.

Although the ∞-categorical theory of colimits is elegant and powerful, it can be be difficult to work with because the colimit $\varinjlim(p)$ of a diagram p is well-defined only up to equivalence. This problem can sometimes be remedied by working in the more rigid theory of model categories, where the notion of an ∞-categorical colimit should be replaced by the notion of a *homotopy colimit* (see §A.3.3). In order to pass smoothly between these two settings, we need to know that the ∞-categorical theory of colimits agrees with the more classical theory of homotopy colimits. A precise statement of this result (Theorem 4.2.4.1) will be formulated and proved in §4.2.4.

4.2.1 Alternative Join and Slice Constructions

In §1.2.8, we introduced the *join* functor $\star$ on simplicial sets. In [44], Joyal introduces a closely related operation $\diamond$ on simplicial sets. This operation is equivalent to $\star$ (Proposition 4.2.1.2) but is technically more convenient in certain contexts. In this section we will review the definition of the operation $\diamond$ and establish some of its basic properties (see also [44] for a discussion).

Definition 4.2.1.1 ([44]). Let X and Y be simplicial sets. The simplicial set $X \diamond Y$ is defined to be pushout

$$X \coprod_{X \times Y \times \{0\}} (X \times Y \times \Delta^1) \coprod_{X \times Y \times \{1\}} Y.$$

We note that since $X \times Y \times (\partial \Delta^1) \to X \times Y \times \Delta^1$ is a monomorphism, the pushout diagram defining $X \diamond Y$ is a homotopy pushout in $\mathcal{S}et_\Delta$ (with respect to the Joyal model structure). Consequently, we deduce that categorical equivalences $X \to X'$, $Y \to Y'$ induce a categorical equivalence $X \diamond Y \to X' \diamond Y'$.

The simplicial set $X \diamond Y$ admits a map $p : X \diamond Y \to \Delta^1$ with $X \simeq p^{-1}\{0\}$ and $Y \simeq p^{-1}\{1\}$. Consequently, there is a unique map $X \diamond Y \to X \star Y$ which is compatible with the projection to Δ^1 and induces the identity maps on X and Y.

Proposition 4.2.1.2. *For any simplicial sets X and Y, the natural map $\phi : X \diamond Y \to X \star Y$ is a categorical equivalence.*

Proof. Since both sides are compatible with the formation of filtered colimits in X, we may suppose that X contains only finitely many nondegenerate simplices. If X is empty, then ϕ is an isomorphism and the result is obvious. Working by induction on the dimension of X and the number of nondegenerate simplices in X, we may write

$$X = X' \coprod_{\partial \Delta^n} \Delta^n,$$

and we may assume that the statement is known for the pairs (X', Y) and $(\partial \Delta^n, Y)$. Since the Joyal model structure on Set_Δ is left proper, we have a map of homotopy pushouts

$$(X' \diamond Y) \coprod_{\partial \Delta^n \diamond Y} (\Delta^n \diamond Y) \to (X' \star Y) \coprod_{\partial \Delta^n \star Y} (\Delta^n \star Y),$$

and we are therefore reduced to proving the assertion in the case where $X = \Delta^n$. The inclusion

$$\Delta^{\{0,1\}} \coprod_{\{1\}} \cdots \coprod_{\{n-1\}} \Delta^{\{n-1,n\}} \subseteq \Delta^n$$

is inner anodyne. Thus if $n > 1$, we can conclude by induction. Thus we may suppose that $X = \Delta^0$ or $X = \Delta^1$. By a similar argument, we may reduce to the case where $Y = \Delta^0$ or $Y = \Delta^1$. The desired result now follows from an explicit calculation. $\square$

Corollary 4.2.1.3. *Let $S \to T$ and $S' \to T'$ be categorical equivalences of simplicial sets. Then the induced map*

$$S \star S' \to T \star T'$$

is a categorical equivalence.

Proof. This follows immediately from Proposition 4.2.1.2 since the operation $\diamond$ has the desired property. $\square$

Corollary 4.2.1.4. *Let X and Y be simplicial sets. Then the natural map*

$$\mathfrak{C}[X \star Y] \to \mathfrak{C}[X] \star \mathfrak{C}[Y]$$

is an equivalence of simplicial categories.

Proof. Using Corollary 4.2.1.3, we may reduce to the case where X and Y are ∞-categories. We note that $\mathfrak{C}[X \star Y]$ is a correspondence from $\mathfrak{C}[X]$ to $\mathfrak{C}[Y]$. To complete the proof, it suffices to show that $\mathrm{Map}_{\mathfrak{C}[X \star Y]}(x, y)$ is weakly contractible for any pair of objects $x \in X$, $y \in Y$. Since $X \star Y$ is an ∞-category, we can apply Theorem 1.1.5.13 to deduce that the mapping space $\mathrm{Map}_{\mathcal{C}[X \star Y]}(x, y)$ is weakly homotopy equivalent to $\mathrm{Hom}^{\mathrm{R}}_{X \star Y}(x, y)$, which consists of a single point.   □

For fixed X, the functor

$$Y \mapsto X \diamond Y$$

$$\mathrm{Set}_\Delta \to (\mathrm{Set}_\Delta)_{X/}$$

preserves all colimits. By the adjoint functor theorem (or by direct construction), this functor has a right adjoint. In other words, for every map of simplicial sets $p : X \to \mathcal{C}$, there exists a simplicial set $\mathcal{C}^{p/}$ with the following universal property: for every simplicial set Y, there is a canonical bijection

$$\mathrm{Hom}_{\mathrm{Set}_\Delta}(Y, \mathcal{C}^{p/}) \simeq \mathrm{Hom}_{(\mathrm{Set}_\Delta)_{X/}}(X \diamond Y, \mathcal{C}).$$

Since the functor $Y \mapsto X \diamond Y$ preserves cofibrations and categorical equivalences, we deduce that the passage from $\mathcal{C}$ to $\mathcal{C}^{p/}$ preserves categorical fibrations and categorical equivalences between ∞-categories. Moreover, Proposition 4.2.1.2 has the following consequence:

Proposition 4.2.1.5. *Let $\mathcal{C}$ be an ∞-category and let $p : X \to \mathcal{C}$ be a diagram. Then the natural map*

$$\mathcal{C}_{p/} \to \mathcal{C}^{p/}$$

is an equivalence of ∞-categories.

According to Definition 1.2.13.4, a colimit for a diagram $p : X \to \mathcal{C}$ is an initial object of the ∞-category $\mathcal{C}_{p/}$. In view of the above remarks, an object of $\mathcal{C}_{p/}$ is a colimit for p if and only if its image in $\mathcal{C}^{p/}$ is an initial object; in other words, we can replace $\mathcal{C}_{p/}$ by $\mathcal{C}^{p/}$ (and $\star$ by $\diamond$) in Definition 1.2.13.4.

By Proposition 2.1.2.1, for any ∞-category $\mathcal{C}$ and any map $p : X \to \mathcal{C}$, the induced map $\mathcal{C}_{p/} \to \mathcal{C}$ is a left fibration. We now show that $\mathcal{C}^{p/}$ has the same property:

Proposition 4.2.1.6. *Suppose we are given a diagram of simplicial sets*

$$K_0 \subseteq K \xrightarrow{p} X \xrightarrow{q} S,$$

where q is a categorical fibration. Let $r = q \circ p : K \to S$, $p_0 = p|K_0$, and $r_0 = r|K_0$. Then the induced map

$$\phi : X^{p/} \to X^{p_0/} \times_{S^{r_0/}} S^{r/}$$

is a left fibration.

Proof. We must show that q has the right lifting property with respect to every left anodyne inclusion $A_0 \subseteq A$. Unwinding the definition, this amounts to proving that q has the right lifting property with respect to the inclusion

$$i : (A_0 \diamond K) \coprod_{A_0 \diamond K_0} (A \diamond K_0) \subseteq A \diamond K.$$

Since q is a categorical fibration, it suffices to show that i is a categorical equivalence. The above pushout is a homotopy pushout, so it will suffice to prove the analogous statement for the weakly equivalent inclusion

$$(A_0 \star K) \coprod_{A_0 \star K_0} (A \star K_0) \subseteq A \star K.$$

But this map is inner anodyne (Lemma 2.1.2.3). $\square$

Corollary 4.2.1.7. *Let $\mathcal{C}$ be an ∞-category and let $p : K \to \mathcal{C}$ be any diagram. For every vertex v of $\mathcal{C}$, the map $\mathcal{C}_{p/} \times_{\mathcal{C}} \{v\} \to \mathcal{C}^{p/} \times_{\mathcal{C}} \{v\}$ is a homotopy equivalence of Kan complexes.*

Proof. The map $\mathcal{C}_{p/} \to \mathcal{C}^{p/}$ is a categorical equivalence of left fibrations over $\mathcal{C}$; now apply Proposition 3.3.1.5. $\square$

Corollary 4.2.1.8. *Let $\mathcal{C}$ be an ∞-category containing vertices x and y. The maps*

$$\mathrm{Hom}^{\mathrm{R}}_{\mathcal{C}}(x, y) \to \mathrm{Hom}_{\mathcal{C}}(x, y) \leftarrow \mathrm{Hom}^{\mathrm{L}}_{\mathcal{C}}(x, y)$$

are homotopy equivalences of Kan complexes (see §1.2.2 for an explanation of this notation).

Proof. Apply Corollary 4.2.1.7 (the dual of Corollary 4.2.1.7) to the case where p is the inclusion $\{x\} \subseteq \mathcal{C}$ (the inclusion $\{y\} \subseteq \mathcal{C}$). $\square$

Remark 4.2.1.9. The above ideas dualize in an evident way; given a map of simplicial sets $p : K \to X$, we can define a simplicial set $X^{/p}$ with the universal mapping property

$$\mathrm{Hom}_{\mathrm{Set}_\Delta}(K', X^{/p}) = \mathrm{Hom}_{(\mathrm{Set}_\Delta)_{K/}}(K' \diamond K, X).$$

4.2.2 Parametrized Colimits

Let $p : K \to \mathcal{C}$ be a diagram in an ∞-category $\mathcal{C}$. The goal of this section is to make precise the idea that the colimit $\varinjlim(p)$ depends functorially on p (provided that $\varinjlim(p)$ exists). We will prove this in a very general context where not only the diagram p but also the simplicial set K is allowed to vary. We begin by introducing a *relative* version of the $\diamond$-operation.

Definition 4.2.2.1. Let S be a simplicial set and let $X, Y \in (\mathrm{Set}_\Delta)_{/S}$. We define

$$X \diamond_S Y = X \coprod_{X \times_S Y \times \{0\}} (X \times_S Y \times \Delta^1) \coprod_{X \times_S Y \times \{1\}} Y \in (\mathrm{Set}_\Delta)_{/S}.$$

We observe that the operation $\diamond_S$ is compatible with base change in the following sense: for any map $T \to S$ of simplicial sets and any objects $X, Y \in (\mathcal{S}\mathrm{et}_\Delta)_{/S}$, there is a natural isomorphism

$$(X_T \diamond_T Y_T) \simeq (X \diamond_S Y)_T,$$

where we let Z_T denote the fiber product $Z \times_S T$. We also note that in the case where S is a point, the operation $\diamond_S$ coincides with the operation $\diamond$ introduced in §4.2.1.

Fix $K \in (\mathcal{S}\mathrm{et}_\Delta)_{/S}$. We note that functor $(\mathcal{S}\mathrm{et}_\Delta)_{/S} \to ((\mathcal{S}\mathrm{et}_\Delta)_{/S})_{K/}$ defined by

$$X \mapsto K \diamond_S S$$

has a right adjoint; this right adjoint associates to a diagram

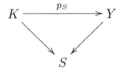

the simplicial set $Y^{p_S/}$, defined by the property that $\mathrm{Hom}_S(X, Y^{p_S/})$ classifies commutative diagrams

$$\begin{array}{ccc} K & \xrightarrow{p_S} & Y \\ \big\uparrow & \nearrow & \big\downarrow \\ K \diamond_S X & \longrightarrow & S. \end{array}$$

The base change properties of the operation $\diamond_S$ imply similar base change properties for the relative slice construction: given a map $p_S : K \to Y$ in $(\mathcal{S}\mathrm{et}_\Delta)_{/S}$ and any map $T \to S$, we have a natural isomorphism

$$Y^{p_S/} \times_S T \simeq (Y \times_S T)^{p_T/},$$

where p_T denotes the induced map $K_T \to Y_T$. In particular, the fiber of $Y^{p_S/}$ over a vertex s of S can be identified with the absolute slice construction $Y_s^{p_s/}$ studied in §4.2.1.

Remark 4.2.2.2. Our notation is somewhat abusive: the simplicial set $Y^{p_S/}$ depends not only on the map $p_S : K \to Y$ but also on the simplicial set S. We will attempt to avoid confusion by always indicating the simplicial set S by a subscript in the notation for the map in question; we will omit this subscript only in the case $S = \Delta^0$, in which case the functor described above coincides with the functor defined in §4.2.1.

Lemma 4.2.2.3. *Let $n > 0$ and let*

$$B = (\Lambda_n^n \times \Delta^1) \coprod_{\Lambda_n^n \times \partial \Delta^1} (\Delta^n \times \partial \Delta^1) \subseteq \Delta^n \times \Delta^1.$$

Suppose we are given a diagram of simplicial sets

$$
\begin{array}{ccc}
A \times B & \xrightarrow{\ f_0\ } & Y \\
\downarrow & \nearrow{\scriptstyle f} & \downarrow{\scriptstyle q} \\
A \times \Delta^n \times \Delta^1 & \longrightarrow & S
\end{array}
$$

in which q is a Cartesian fibration, and that f_0 carries $\{a\} \times \Delta^{\{n-1,n\}} \times \{1\}$ to a q-Cartesian edge of Y for each vertex a of A. Then there exists a morphism f rendering the diagram commutative.

Proof. Invoking Proposition 3.1.2.1, we may replace $q : Y \to S$ by the induced map $Y^A \to S^A$ and thereby reduce to the case where $A = \Delta^0$. We now recall the notation introduced in the proof of Proposition 2.1.2.6: more specifically, the family $\{\sigma_i\}_{0 \leq i \leq n}$ of nondegenerate simplices of $\Delta^n \times \Delta^1$. Let $B(0) = B$ and more generally set $B(n) = B \cup \sigma_n \cup \cdots \cup \sigma_{n+1-i}$ so that we have a filtration

$$
B(0) \subseteq \cdots \subseteq B(n+1) = \Delta^n \times \Delta^1.
$$

A map $f_0 : B(0) \to Y$ has been prescribed for us already; we construct extensions $f_i : B(i) \to Y$ by induction on i. For $i < n$, there is a pushout diagram

$$
\begin{array}{ccc}
\Lambda^{n+1}_{n-i} & \longrightarrow & B(i) \\
\downarrow & & \downarrow \\
\Delta^{n+1} & \longrightarrow & B(i+1).
\end{array}
$$

Thus the extension f_{i+1} can be found by virtue of the assumption that q is an inner fibration. For $i = n$, we obtain instead a pushout diagram

$$
\begin{array}{ccc}
\Lambda^{n+1}_{n+1} & \longrightarrow & B(n) \\
\downarrow & & \downarrow \\
\Delta^{n+1} & \longrightarrow & B(n+1),
\end{array}
$$

and the desired extension can be found by virtue of the assumption that f_0 carries the edge $\Delta^{\{n-1,n\}} \times \{1\}$ to a q-Cartesian edge of Y. $\qquad\square$

Proposition 4.2.2.4. *Suppose we are given a diagram of simplicial sets*

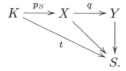

Let $p'_S = q \circ p_S$, and assume that the following conditions are satisfied:

(1) *The map q is a Cartesian fibration.*

(2) *The map t is a coCartesian fibration.*

Then the induced map $r : X^{ps/} \to Y^{ps'/}$ is a Cartesian fibration; moreover, an edge of $X^{ps/}$ is r-Cartesian if and only if its image in X is q-Cartesian.

Proof. We first show that r is an inner fibration. Suppose we are given $0 < i < n$ and a diagram

$$
\begin{array}{ccc}
\Lambda^n_i & \longrightarrow & X^{ps/} \\
\downarrow & \nearrow & \downarrow \\
\Delta^n & \longrightarrow & Y^{ps'/},
\end{array}
$$

we must show that it is possible to provide the dotted arrow. Unwinding the definitions, we see that it suffices to produce the indicated arrow in the diagram

$$
\begin{array}{ccc}
K \diamond_S \Lambda^n_i & \longrightarrow & X \\
\downarrow & \nearrow & \downarrow q \\
K \diamond_S \Delta^n & \longrightarrow & Y.
\end{array}
$$

Since q is a Cartesian fibration, it is a categorical fibration by Proposition 3.3.1.7. Consequently, it suffices to show that the inclusion

$$
K \diamond_S \Lambda^n_i \subseteq K \diamond_S \Delta^n
$$

is a categorical equivalence. In view of the definition of $K \diamond_S M$ as a pushout

$$
K \coprod_{K \times_S M \times \{0\}} (K \times_S M \times \Delta^1) \coprod_{K \times_S M \times \{1\}} M,
$$

it suffices to verify that the inclusions

$$
\Lambda^n_i \subseteq \Delta^n
$$

$$
K \times_S \Lambda^n_i \subseteq K \times_S \Delta^n
$$

are categorical equivalences. The first statement is obvious; the second follows from (the dual of) Proposition 3.3.1.3.

Let us say that an edge of $X^{ps/}$ is *special* if its image in X is q-Cartesian. To complete the proof, it will suffice to show that every special edge of $X^{ps/}$ is r-Cartesian and that there are sufficiently many special edges of $X^{ps/}$. More precisely, consider any $n \geq 1$ and any diagram

$$
\begin{array}{ccc}
\Lambda^n_n & \overset{h}{\longrightarrow} & X^{ps/} \\
\downarrow & \nearrow & \downarrow \\
\Delta^n & \longrightarrow & Y^{ps'/}.
\end{array}
$$

We must verify the following assertions:

- If $n = 1$, then there exists a dotted arrow rendering the diagram commutative, classifying a special edge of $X^{ps/}$.

- If $n > 1$ and $h|\Delta^{\{n-1,n\}}$ classifies a special edge of $X^{ps/}$, then there exists a dotted arrow rendering the diagram commutative.

Unwinding the definitions, we have a diagram

$$\begin{array}{ccc} K \diamond_S \Lambda_n^n & \xrightarrow{f_0} & X \\ \downarrow & \nearrow^{f} & \downarrow q \\ K \diamond_S \Delta^n & \longrightarrow & Y, \end{array}$$

and we wish to prove the existence of the indicated arrow f. As a first step, we consider the restricted diagram

$$\begin{array}{ccc} \Lambda_n^n & \xrightarrow{f_0|\Lambda_n^n} & X \\ \downarrow & \nearrow^{f_1} & \downarrow q \\ \Delta^n & \longrightarrow & Y. \end{array}$$

By assumption, $f_0|\Lambda_n^n$ carries $\Delta^{\{n-1,n\}}$ to a q-Cartesian edge of X (if $n > 1$), so there exists a map f_1 rendering the diagram commutative (and classifying a q-Cartesian edge of X if $n = 1$). It now suffices to produce the dotted arrow in the diagram

$$\begin{array}{ccc} (K \diamond_S \Lambda_n^n) \coprod_{\Lambda_n^n} \Delta^n & \longrightarrow & X \\ \downarrow{\scriptstyle i} & \nearrow^{f} & \downarrow q \\ K \diamond_S \Delta^n & \longrightarrow & Y, \end{array}$$

where the top horizontal arrow is the result of amalgamating f_0 and f_1.

Without loss of generality, we may replace S by Δ^n. By (the dual of) Proposition 3.2.2.7, there exists a composable sequence of maps

$$\phi : A^0 \to \cdots \to A^n$$

and a quasi-equivalence $M^{op}(\phi) \to K$. We have a commutative diagram

$$\begin{array}{ccc} (M^{op}(\phi) \diamond_S \Lambda_n^n) \coprod_{\Lambda_n^n} \Delta^n & \longrightarrow & (K \diamond_S \Lambda_n^n) \coprod_{\Lambda_n^n} \Delta^n \\ \downarrow{\scriptstyle i'} & & \downarrow{\scriptstyle i} \\ M^{op}(\phi) \diamond_S \Delta^n & \longrightarrow & K \diamond_S \Delta^n. \end{array}$$

Since q is a categorical fibration, Proposition A.2.3.1 shows that it suffices to produce a dotted arrow f' in the induced diagram

$$\begin{array}{ccc} (M^{op}(\phi) \diamond_S \Lambda_n^n) \coprod_{\Lambda_n^n} \Delta^n & \longrightarrow & X \\ \downarrow{\scriptstyle i} & \nearrow^{f'} & \downarrow q \\ M^{op}(\phi) \diamond_S \Delta^n & \longrightarrow & Y. \end{array}$$

Let B be as in the statement of Lemma 4.2.2.3; then we have a pushout diagram

$$
\begin{array}{ccc}
A^0 \times B & \longrightarrow & (M^{op}(\phi) \diamond_S \Lambda_n^n) \coprod_{\Lambda_n^n} \Delta^n \\
\Big\downarrow{\scriptstyle i''} & & \Big\downarrow \\
A^0 \times \Delta^n \times \Delta^1 & \longrightarrow & M^{op}(\phi) \diamond_S \Delta^n.
\end{array}
$$

Consequently, it suffices to prove the existence of the map f'' in the diagram

$$
\begin{array}{ccc}
A^0 \times B & \overset{g}{\longrightarrow} & X \\
\Big\downarrow{\scriptstyle i''} \quad {\scriptstyle f''} \nearrow & & \Big\downarrow{\scriptstyle q} \\
A^0 \times \Delta^n \times \Delta^1 & \longrightarrow & Y.
\end{array}
$$

Here the map g carries $\{a\} \times \Delta^{\{n-1,n\}} \times \{1\}$ to a q-Cartesian edge of Y for each vertex a of A^0. The existence of f'' now follows from Lemma 4.2.2.3. $\quad\square$

Remark 4.2.2.5. In most applications of Proposition 4.2.2.4, we will have $Y = S$. In that case, $Y^{p_S/}$ can be identified with S, and the conclusion is that the projection $X^{p_S/} \to S$ is a Cartesian fibration.

Remark 4.2.2.6. The hypothesis on s in Proposition 4.2.2.4 can be weakened: all we need in the proof is the existence of maps $M^{op}(\phi) \to K \times_S \Delta^n$ which are universal categorical equivalences (that is, induce categorical equivalences $M^{op}(\phi) \times_{\Delta^n} T \to K \times_S T$ for any $T \to \Delta^n$). Consequently, Proposition 4.2.2.4 remains valid when $K \simeq S \times K^0$ for *any* simplicial set K^0 (not necessarily an ∞-category). It seems likely that Proposition 4.2.2.4 remains valid whenever s is a smooth map of simplicial sets, but we have not been able to prove this.

We can now express the idea that the colimit of a diagram should depend functorially on the diagram (at least for "smoothly parametrized" families of diagrams):

Proposition 4.2.2.7. *Let $q : Y \to S$ be a Cartesian fibration and let $p_S : K \to Y$ be a diagram. Suppose that the following conditions are satisfied:*

(1) *For each vertex s of S, the restricted diagram $p_s : K_s \to Y_s$ has a colimit in the ∞-category Y_s.*

(2) *The composition $q \circ p_S$ is a coCartesian fibration.*

There exists a map p_S' rendering the diagram

$$
\begin{array}{ccc}
K & \overset{p_S}{\longrightarrow} & Y \\
\Big\downarrow \quad {\scriptstyle p_S'} \nearrow & & \Big\downarrow{\scriptstyle q} \\
K \diamond_S S & \longrightarrow & S
\end{array}
$$

commutative and having the property that for each vertex s of S, the restriction $p'_s : K_s \diamond \{s\} \to Y_s$ is a colimit of p_s. Moreover, the collection of all such maps is parametrized by a contractible Kan complex.

Proof. Apply Proposition 2.4.4.9 to the Cartesian fibration $Y^{ps/}$ and observe that the collection of sections of a trivial fibration constitutes a contractible Kan complex.   □

4.2.3 Decomposition of Diagrams

Let $\mathcal{C}$ be an ∞-category and $p : K \to \mathcal{C}$ a diagram indexed by a simplicial set K. In this section, we will try to analyze the colimit $\varinjlim(p)$ (if it exists) in terms of the colimits $\{\varinjlim(p|K_I)\}$, where $\{K_I\}$ is some family of simplicial subsets of K. In fact, it will be useful to work in slightly more generality: we will allow each K_I to be an arbitrary simplicial set mapping to K (not necessarily via a monomorphism).

Throughout this section, we will fix a simplicial set K, an ordinary category $\mathcal{J}$, and a functor $F : \mathcal{J} \to (\mathrm{Set}_\Delta)_{/K}$. It may be helpful to imagine that $\mathcal{J}$ is a partially ordered set and that F is an order-preserving map from $\mathcal{J}$ to the collection of simplicial subsets of K; this will suffice for many but not all of our applications. We will denote $F(I)$ by K_I and the tautological map $K_I \to K$ by π_I.

Our goal is to show that, under appropriate hypotheses, we can recover the colimit of a diagram $p : K \to \mathcal{C}$ in terms of the colimits of diagrams $p \circ \pi_I : K_I \to \mathcal{C}$. Our first goal is to show that the construction of these colimits is suitably functorial in I. For this, we need an auxiliary construction.

Notation 4.2.3.1. We define a simplicial set K_F as follows. A map $\Delta^n \to K_F$ is determined by the following data:

(i) A map $\Delta^n \to \Delta^1$ corresponding to a decomposition $[n] = \{0, \ldots, i\} \cup \{i+1, \ldots, n\}$.

(ii) A map $e_- : \Delta^{\{0,\ldots,i\}} \to K$.

(iii) A map $e_+ : \Delta^{\{i+1,\ldots,n\}} \to N(\mathcal{J})$ which we may view as a chain of composable morphisms

$$I(i+1) \to \cdots \to I(n)$$

in the category $\mathcal{J}$.

(iv) For each $j \in \{i+1, \ldots, n\}$, a map e_j which fits into a commutative diagram

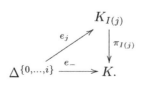

Moreover, for $j \leq k$, we require that e_k is given by the composition

$$\Delta^{\{0,\dots,i\}} \xrightarrow{e_j} K_{I(j)} \to K_{I(k)}.$$

Remark 4.2.3.2. In the case where $i < n$, the maps e_- and $\{e_j\}_{j>i}$ are completely determined by e_{i+1}, which can be arbitrary.

The simplicial set K_F is equipped with a map $K_F \to \Delta^1$. Under this map, the preimage of the vertex $\{0\}$ is $K \subseteq K_F$ and the preimage of the vertex $\{1\}$ is $N(\mathcal{I}) \subseteq K_F$. For $I \in \mathcal{I}$, we will denote the corresponding vertex of $N(\mathcal{I}) \subseteq K_F$ by X_I. We note that, for each $I \in \mathcal{I}$, there is a commutative diagram

$$
\begin{array}{ccc}
K_I & \xrightarrow{\pi_I} & K \\
\downarrow & & \downarrow \\
K_I^{\triangleright} & \xrightarrow{\pi_I'} & K_F,
\end{array}
$$

where π_I' carries the cone point of $K_I^{\triangleright}$ to the vertex X_I of K_F.

Let us now suppose that $p : K \to \mathcal{C}$ is a diagram in an ∞-category $\mathcal{C}$. Our next goal is to prove Proposition 4.2.3.4, which will allow us to extend p to a larger diagram $K_F \to \mathcal{C}$ which carries each vertex X_I to a colimit of $p \circ \pi_I : K_I \to \mathcal{C}$. First, we need a lemma.

Lemma 4.2.3.3. *Let $\mathcal{C}$ be an ∞-category and let $\sigma : \Delta^n \to \mathcal{C}$ be a simplex having the property that $\sigma(0)$ is an initial object of $\mathcal{C}$. Let $\partial\sigma = \sigma|\partial\Delta^n$. The natural map $\mathcal{C}_{\sigma/} \to \mathcal{C}_{\partial\sigma/}$ is a trivial fibration.*

Proof. Unwinding the definition, we are reduced to solving the extension problem depicted in the diagram

$$
\begin{array}{ccc}
(\partial\Delta^n \star \Delta^m) \coprod_{\partial\Delta^n \star \partial\Delta^m} (\Delta^n \star \partial\Delta^m) & \xrightarrow{f_0} & \mathcal{C} \\
\downarrow & \nearrow{f} & \\
\Delta^n \star \Delta^m. & &
\end{array}
$$

We can identify the domain of f_0 with $\partial\Delta^{n+m+1}$. Our hypothesis guarantees that $f_0(0)$ is an initial object of $\mathcal{C}$, which in turn guarantees the existence of f. $\square$

Proposition 4.2.3.4. *Let $p : K \to \mathcal{C}$ be a diagram in an ∞-category $\mathcal{C}$, let $\mathcal{I}$ be an ordinary category, and let $F : \mathcal{I} \to (\mathrm{Set}_\Delta)_{/K}$ be a functor. Suppose that, for each $I \in \mathcal{I}$, the induced diagram $p_I = p \circ \pi_I : K_I \to \mathcal{C}$ has a colimit $q_I : K_I^{\triangleright} \to \mathcal{C}$.*

There exists a map $q : K_F \to \mathcal{C}$ such that $q \circ \pi_I' = q_I$ and $q|K = p$. Furthermore, for any such q, the induced map $\mathcal{C}_{q/} \to \mathcal{C}_{p/}$ is a trivial fibration.

Proof. For each $X \subseteq N(\mathcal{I})$, we let K_X denote the simplicial subset of K_F consisting of all simplices $\sigma \in K_F$ such that $\sigma \cap N(\mathcal{I}) \subseteq X$. We note that $K_\emptyset = K$ and that $K_{N(\mathcal{I})} = K_F$.

Define a transfinite sequence Y_α of simplicial subsets of $N(\mathcal{J})$ as follows. Let $Y_0 = \emptyset$ and let $Y_\lambda = \bigcup_{\gamma < \lambda} Y_\gamma$ when λ is a limit ordinal. Finally, let $Y_{\alpha+1}$ be obtained from Y_α by adjoining a single nondegenerate simplex provided that such a simplex exists. We note that for α sufficiently large, such a simplex will not exist, and we set $Y_\beta = Y_\alpha$ for all $\beta > \alpha$.

We define a sequence of maps $q_\beta : K_{Y_\beta} \to \mathcal{C}$ so that the following conditions are satisfied:

(1) We have $q_0 = p : K_\emptyset = K \to \mathcal{C}$.

(2) If $\alpha < \beta$, then $q_\alpha = q_\beta | K_{Y_\alpha}$.

(3) If $\{X_I\} \subseteq Y_\alpha$, then $q_\alpha \circ \pi_I' = q_I : K_I^\triangleright \to \mathcal{C}$.

Provided that such a sequence can be constructed, we may conclude the proof by setting $q = q_\alpha$ for α sufficiently large.

The construction of q_α goes by induction on α. If $\alpha = 0$, then q_α is determined by condition (1); if α is a (nonzero) limit ordinal, then q_α is determined by condition (2). Suppose that q_α has been constructed; we give a construction of $q_{\alpha+1}$.

There are two cases to consider. Suppose first that $Y_{\alpha+1}$ is obtained from Y_α by adjoining a vertex X_I. In this case, $q_{\alpha+1}$ is uniquely determined by conditions (2) and (3).

Now suppose that $X_{\alpha+1}$ is obtained from X_α by adjoining a nondegenerate simplex σ of positive dimension corresponding to a sequence of composable maps

$$I_0 \to \cdots \to I_n$$

in the category $\mathcal{J}$. We note that the inclusion $K_{Y_\alpha} \subseteq K_{Y_{\alpha+1}}$ is a pushout of the inclusion

$$K_{I_0} \star \partial \sigma \subseteq K_{I_0} \star \sigma.$$

Consequently, constructing the map $q_{\alpha+1}$ is tantamount to finding an extension of a certain map $s_0 : \partial \sigma \to \mathcal{C}_{p_I/}$ to the whole of the simplex σ. By assumption, s_0 carries the initial vertex of σ to an initial object of $\mathcal{C}_{p_I/}$, so that the desired extension s can be found. For use below, we record a further property of our construction: the projection $\mathcal{C}_{q_{\alpha+1}/} \to \mathcal{C}_{q_\alpha/}$ is a pullback of the map $(\mathcal{C}_{p_I/})_{s/} \to (\mathcal{C}_{p_I/})_{s_0/}$, which is a trivial fibration.

We now wish to prove that, for any extension q with the above properties, the induced map $\mathcal{C}_{q/} \to \mathcal{C}_{p/}$ is a trivial fibration. We first observe that the map q can be obtained by the inductive construction given above: namely, we take q_α to be the restriction of q to K_{Y_α}. It will therefore suffice to show that, for every pair of ordinals $\alpha \leq \beta$, the induced map $\mathcal{C}_{q_\beta/} \to \mathcal{C}_{q_\alpha/}$ is a trivial fibration. The proof proceeds by induction on β: the case $\beta = 0$ is clear, and if β is a limit ordinal, we observe that the inverse limit of a transfinite tower of trivial fibrations is itself a trivial fibration. We may therefore suppose that $\beta = \gamma + 1$ is a successor ordinal. Using the factorization

$$\mathcal{C}_{q_\beta/} \to \mathcal{C}_{q_\gamma/} \to \mathcal{C}_{q_\alpha/}$$

and the inductive hypothesis, we are reduced to proving this in the case where β is the successor of α, which was treated above. $\qquad\square$

Let us now suppose that we are given diagrams $p : K \to \mathcal{C}$, $F : \mathcal{I} \to (\mathrm{Set}_\Delta)_{/K}$ as in the statement of Proposition 4.2.3.4 and let $q : K_F \to \mathcal{C}$ be a map which satisfies its conclusions. Since $\mathcal{C}_{q/} \to \mathcal{C}_{p/}$ is a trivial fibration, we may identify colimits of the diagram q with colimits of the diagram p (up to equivalence). Of course, this is not useful in itself since the diagram q is *more* complicated than p. Our objective now is to show that, under the appropriate hypotheses, we may identify the colimits of q with the colimits of $q|\,\mathrm{N}(\mathcal{I})$. First, we need a few lemmas.

Lemma 4.2.3.5 (Joyal [44]). *Let $f : A_0 \subseteq A$ and $g : B_0 \subseteq B$ be inclusions of simplicial sets and suppose that g is a weak homotopy equivalence. Then the induced map*

$$h : (A_0 \star B) \coprod_{A_0 \star B_0} (A \star B_0) \subseteq A \star B$$

is right anodyne.

Proof. Our proof follows the pattern of Lemma 2.1.2.3. The collection of all maps f which satisfy the conclusion (for *any* choice of g) forms a weakly saturated class of morphisms. It will therefore suffice to prove that the h is right anodyne when f is the inclusion $\partial\,\Delta^n \subseteq \Delta^n$. Similarly, the collection of all maps g which satisfy the conclusion (for fixed f) forms a weakly saturated class. We may therefore reduce to the case where g is a horn inclusion $\Lambda_i^m \subseteq \Delta^m$. In this case, we may identify h with the horn inclusion $\Lambda_{i+n+1}^{m+n+1} \subseteq \Delta^{m+n+1}$, which is clearly right anodyne. $\qquad\square$

Lemma 4.2.3.6. *Let $A_0 \subseteq A$ be an inclusion of simplicial sets and let B be weakly contractible. Then the inclusion $A_0 \star B \subseteq A \star B$ is right anodyne.*

Proof. As above, we may suppose that the inclusion $A_0 \subseteq A$ is identified with $\partial\,\Delta^n \subseteq \Delta^n$. If K is a point, then the inclusion $A_0 \times B \subseteq A \times B$ is isomorphic to $\Lambda_{n+1}^{n+1} \subseteq \Delta^{n+1}$, which is clearly right anodyne.

In the general case, B is nonempty, so we may choose a vertex b of B. Since B is weakly contractible, the inclusion $\{b\} \subseteq B$ is a weak homotopy equivalence. We have already shown that $A_0 \star \{b\} \subseteq A \star \{b\}$ is right anodyne. It follows that the pushout inclusion

$$A_0 \star B \subseteq (A \star \{b\}) \coprod_{A_0 \star \{b\}} (A_0 \star B)$$

is right anodyne. To complete the proof, we apply Lemma 4.2.3.5 to deduce that the inclusion

$$(A \star \{b\}) \coprod_{A_0 \star \{b\}} (A_0 \star B) \subseteq A \star B$$

is right anodyne. $\qquad\square$

Notation 4.2.3.7. Let $\sigma \in K_n$ be a simplex of K. We define a category $\mathfrak{I}_\sigma$ as follows. The objects of $\mathfrak{I}_z$ are pairs (I, σ'), where $I \in \mathfrak{I}$, $\sigma' \in (K_I)_n$, and $\pi_I(\sigma') = \sigma$. A morphism from (I', σ') to (I'', σ'') in $\mathfrak{I}_\sigma$ consists of a morphism $\alpha : I' \to I''$ in $\mathfrak{I}$ with the property that $F(\alpha)(\sigma') = \sigma''$. We let $\mathfrak{I}'_\sigma \subseteq \mathfrak{I}_\sigma$ denote the full subcategory consisting of pairs (I, σ'), where σ' is a degenerate simplex in K_I. Note that if σ is nondegenerate, $\mathfrak{I}'_\sigma$ is empty.

Proposition 4.2.3.8. *Let K be a simplicial set, $\mathfrak{I}$ an ordinary category, and $F : \mathfrak{I} \to (\mathrm{Set}_\Delta)_{/K}$ a functor. Suppose that the following conditions are satisfied:*

(1) *For each nondegenerate simplex σ of K, the category $\mathfrak{I}_\sigma$ is acyclic (fthat is, the simplicial set $\mathrm{N}(\mathfrak{I}_\sigma)$ is weakly contractible).*

(2) *For each degenerate simplex σ of K, the inclusion $\mathrm{N}(\mathfrak{I}'_\sigma) \subseteq \mathrm{N}(\mathfrak{I}_\sigma)$ is a weak homotopy equivalence.*

Then the inclusion $\mathrm{N}(\mathfrak{I}) \subseteq K_F$ is right anodyne.

Proof. Consider any family of subsets $\{L_n \subseteq K_n\}$ which is stable under the "face maps" d_i on K (but not necessarily the degeneracy maps s_i, so that the family $\{L_n\}$ does not necessarily have the structure of a simplicial set). We define a simplicial subset $L_F \subseteq K_F$ as follows: a *nondegenerate* simplex $\Delta^n \to K_F$ belongs to L_F if and only if the corresponding (possibly degenerate) simplex $\Delta^{\{0,\dots,i\}} \to K$ belongs to $L_i \subseteq K_i$ (see Notation 4.2.3.1).

We note that if $L = \emptyset$, then $L_F = \mathrm{N}(\mathfrak{I})$. If $L = K$, then $L_F = K_F$ (so that our notation is unambiguous). Consequently, it will suffice to prove that, for any $L \subseteq L'$, the inclusion $L_F \subseteq L'_F$ is right anodyne. By general nonsense, we may reduce to the case where L' is obtained from L by adding a single simplex $\sigma \in K_n$.

We now have two cases to consider. Suppose first that the simplex σ is nondegenerate. In this case, it is not difficult to see that the inclusion $L_F \subseteq L'_F$ is a pushout of $\partial \sigma \star \mathrm{N}(\mathfrak{I}_\sigma) \subseteq \sigma \star \mathrm{N}(\mathfrak{I}_\sigma)$. By hypothesis, $N \mathfrak{I}_z$ is weakly contractible, so that the inclusion $L_F \subseteq L'_F$ is right anodyne by Lemma 4.2.3.6.

In the case where σ is degenerate, we observe that $L_F \subseteq L'_F$ is a pushout of the inclusion

$$(\partial \sigma \star \mathrm{N}(\mathfrak{I}_\sigma)) \coprod_{\partial \sigma \star \mathrm{N}(\mathfrak{I}'_\sigma)} (\sigma \star \mathrm{N}(\mathfrak{I}'_\sigma)) \subseteq \sigma \star \mathrm{N}(\mathfrak{I}_\sigma),$$

which is right anodyne by Lemma 4.2.3.5. $\square$

Remark 4.2.3.9. Suppose that $\mathfrak{I}$ is a partially ordered set and that F is an order-preserving map from $\mathfrak{I}$ to the collection of simplicial subsets of K. In this case, we observe that $\mathfrak{I}'_\sigma = \mathfrak{I}_\sigma$ whenever σ is a degenerate simplex of K and that $\mathfrak{I}_\sigma = \{I \in \mathfrak{I} : \sigma \in K_I\}$ for any σ. Consequently, the conditions of Proposition 4.2.3.8 hold if and only if each of the partially ordered subsets $\mathfrak{I}_\sigma \subseteq \mathfrak{I}$ has a contractible nerve. This holds automatically if $\mathfrak{I}$ is directed and $K = \bigcup_{I \in \mathfrak{I}} K_I$.

Corollary 4.2.3.10. *Let K be a simplicial set, $\mathcal{J}$ a category, and $F : \mathcal{J} \to (\mathrm{Set}_\Delta)_{/K}$ a functor which satisfies the hypotheses of Proposition 4.2.3.8. Let $\mathcal{C}$ be an ∞-category, let $p : K \to \mathcal{C}$ be any diagram, and let $q : K_F \to \mathcal{C}$ be an extension of p which satisfies the conclusions of Proposition 4.2.3.4. The natural maps*

$$\mathcal{C}_{p/} \leftarrow \mathcal{C}_{q/} \to \mathcal{C}_{q|\,\mathrm{N}(\mathcal{J})/}$$

are trivial fibrations. In particular, we may identify colimits of p with colimits of $q|\,\mathrm{N}(\mathcal{J})$.

Proof. This follows immediately from Proposition 4.2.3.8 since the right anodyne inclusion $\mathrm{N}\,\mathcal{J} \subseteq K_F$ is cofinal and therefore induces a trivial fibration $\mathcal{C}_{q/} \to \mathcal{C}_{q|\,\mathrm{N}(\mathcal{J})/}$ by Proposition 4.1.1.8. $\qquad\qquad\square$

We now illustrate the usefulness of Corollary 4.2.3.10 by giving a sample application. First, a bit of terminology. If κ and τ are regular cardinals, we will write $\tau \ll \kappa$ if, for any cardinals $\tau_0 < \tau$, $\kappa_0 < \kappa$, we have $\kappa_0^{\tau_0} < \kappa$ (we refer the reader to Definition 5.4.2.8 and the surrounding discussion for more details concerning this condition).

Corollary 4.2.3.11. *Let $\mathcal{C}$ be an ∞-category and $\tau \ll \kappa$ regular cardinals. Then $\mathcal{C}$ admits κ-small colimits if and only if $\mathcal{C}$ admits τ-small colimits and colimits indexed by (the nerves of) κ-small τ-filtered partially ordered sets.*

Proof. The "only if" direction is obvious. Conversely, let $p : K \to \mathcal{C}$ be any κ-small diagram. Let $\mathcal{J}$ denote the partially ordered set of τ-small simplicial subsets of K. Then $\mathcal{J}$ is directed and $\bigcup_{I \in \mathcal{J}} K_I = K$, so that the hypotheses of Proposition 4.2.3.8 are satisfied. Since each $p_I = p \circ \pi_I$ has a colimit in $\mathcal{C}$, there exists a map $q : K_F \to \mathcal{C}$ satisfying the conclusions of Proposition 4.2.3.4. Because $\mathcal{C}_{q/} \to \mathcal{C}_{p/}$ is an equivalence of ∞-categories, p has a colimit if and only if q has a colimit. By Corollary 4.2.3.10, q has a colimit if and only if $q|\,\mathrm{N}(\mathcal{J})$ has a colimit. It is clear that $\mathcal{J}$ is a τ-filtered partially ordered set. Furthermore, it is κ-small provided that $\tau \ll \kappa$. $\qquad\square$

The following result can be proven by the same argument:

Corollary 4.2.3.12. *Let $f : \mathcal{C} \to \mathcal{C}'$ be a functor between ∞-categories and let $\tau \ll \kappa$ be regular cardinals. Suppose that $\mathcal{C}$ admits κ-small colimits. Then f preserves κ-small colimits if and only if it preserves τ-small colimits and all colimits indexed by (the nerves of) κ-small τ-filtered partially ordered sets.*

We will conclude this section with another application of Proposition 4.2.3.8 in which $\mathcal{J}$ is not a partially ordered set and the maps $\pi_I : K_I \to K$ are not (necessarily) injective. Instead, we take $\mathcal{J}$ to be the *category of simplices* of K. In other words, an object of $I \in \mathcal{J}$ consists of a map $\sigma_I : \Delta^n \to K$, and a morphism from I to I' is given by a commutative diagram

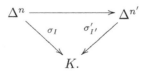

$$\Delta^n \xrightarrow{\hspace{3cm}} \Delta^{n'}$$
$$\sigma_I \searrow \qquad \swarrow \sigma'_{I'}$$
$$K.$$

For each $I \in \mathcal{J}$, we let K_I denote the domain Δ^n of σ_I, and we let $\pi_I = \sigma_I :$ $K_I \to K$.

Lemma 4.2.3.13. *Let K be a simplicial set and let $\mathcal{J}$ denote the category of simplices of K (as defined above). Then there is a retraction $r : K_F \to K$ which fixes $K \subseteq K_F$.*

Proof. Given a map $e : \Delta^n \to K_F$, we will describe the composite map $r \circ e : \Delta^n \to K$. The map e classifies the following data:

(i) A decomposition $[n] = \{0, \ldots, i\} \cup \{i+1, \ldots, n\}$.

(ii) A map $e_- : \Delta^i \to K$.

(iii) A string of morphisms

$$\Delta^{m_{i+1}} \to \cdots \to \Delta^{m_n} \to K.$$

(iv) A compatible family of maps $\{e_j : \Delta^i \to \Delta^{m_j}\}_{j>i}$ having the property that each composition $\Delta^i \xrightarrow{e_j} \Delta^{m_j} \to K$ coincides with e_-.

If $i = n$, we set $r \circ e = e_-$. Otherwise, we let $r \circ e$ denote the composition

$$\Delta^n \xrightarrow{f} \Delta^{m_n} \to K$$

where $f : \Delta^n \to \Delta^{m_n}$ is defined as follows:

- The restriction $f|\Delta^i$ coincides with e_n.

- For $i < j \leq n$, we let $f(j)$ denote the image in Δ^{m_n} of the final vertex of Δ^{m_j}.

$\square$

Proposition 4.2.3.14. *For every simplicial set K, there exists a category $\mathcal{J}$ and a cofinal map $f : \mathrm{N}(\mathcal{J}) \to K$.*

Proof. We take $\mathcal{J}$ to be the category of simplices of K, as defined above, and f to be the composition of the inclusion $\mathrm{N}(\mathcal{J}) \subseteq K_F$ with the retraction r of Lemma 4.2.3.13. To prove that f is cofinal, it suffices to show that the inclusion $\mathrm{N}(\mathcal{J}) \subseteq K_F$ is right anodyne and that the retraction r is cofinal.

To show that $\mathrm{N}(\mathcal{J}) \subseteq K_F$ is right anodyne, it suffices to show that the hypotheses of Proposition 4.2.3.8 are satisfied. Let $\sigma : \Delta^J \to K$ be a simplex of K. We observe that the category $\mathcal{J}_\sigma$ may be described as follows: its objects consist of pairs of maps $(s : \Delta^J \to \Delta^M, t : \Delta^M \to K)$ with $t \circ s = \sigma$. A morphism from (s, t) to (s', t') consists of a map

$$\alpha : \Delta^M \to \Delta^{M'}$$

with $s' = \alpha \circ s$ and $t = t' \circ \alpha$. In particular, we note that $\mathcal{J}_\sigma$ has an initial object $(\mathrm{id}_{\Delta^J}, \sigma)$. It follows that $\mathrm{N}(\mathcal{J}_\sigma)$ is weakly contractible for *any* simplex σ of K. It will therefore suffice to show that $\mathrm{N}(\mathcal{J}'_\sigma)$ is weakly contractible whenever σ is degenerate. Let $\mathcal{I}''_\sigma$ denote the full subcategory of $\mathcal{I}'_\sigma$ spanned by those objects for which the map s is surjective. The inclusion $\mathcal{I}''_\sigma \subseteq \mathcal{I}'_\sigma$ has a right adjoint, so that $N(\mathcal{I}''_\sigma)$ is a deformation retract of $N(\mathcal{I}'_\sigma)$. It will therefore suffice to prove that $N(\mathcal{I}''_\sigma)$ is weakly contractible. For this, we simply observe that $N(\mathcal{I}''_\sigma)$ has a final object $\Delta^J \xrightarrow{s} \Delta^M \xrightarrow{t} K$, characterized by the property that t is a nondegenerate simplex of K.

We now show that r is cofinal. According to Proposition 4.1.1.8, it suffices to show that for any ∞-category $\mathcal{C}$ and any map $p : K \to \mathcal{C}$, the induced map $\mathcal{C}_{q/} \to \mathcal{C}_{p/}$ is a categorical equivalence, where $q = p \circ r$. This follows from Proposition 4.2.3.4. □

Variant 4.2.3.15. Proposition 4.2.3.14 can be strengthened as follows:

(∗) For every simplicial set K, there exists a cofinal map $\phi : K' \to K$, where K' is the nerve of a partially ordered set.

Moreover, the map ϕ can be chosen to depend functorially on K. We will construct ϕ as a composition of four cofinal maps

$$K' = K^{(5)} \xrightarrow{\phi_4} K^{(4)} \xrightarrow{\phi_3} K^{(3)} \xrightarrow{\phi_2} K^{(2)} \xrightarrow{\phi_2} K^{(1)} \xrightarrow{\phi_1} K,$$

which are defined as follows:

(1) The simplicial set $K^{(1)}$ is the nerve $N(\mathcal{I}_1)$, where $\mathcal{I}_1$ denotes the category of simplices of K, and the morphism ϕ_1 is the cofinal map described in Proposition 4.2.3.14.

(2) The simplicial set $K^{(2)}$ is the nerve $N(\mathcal{I}_2)^{op}$, where $\mathcal{I}_2$ denotes the category of simplices of $K^{(1)}$. The map ϕ_2 is induced by the functor $\mathcal{I}_2 \to \mathcal{I}_1$ which carries a chain of morphisms $C_0 \to C_1 \to \cdots \to C_n$ in $\mathcal{I}_1$ to the object C_0. We claim that ϕ_2 is cofinal. To prove this, it will suffice (by virtue of Theorem 4.1.3.1) to prove that for every object $C \in \mathcal{I}_1$, the category $\mathcal{I}_2^{op} \times_{\mathcal{I}_1} (\mathcal{I}_1)_{C/}$ has weakly contractible nerve. Indeed, this is the opposite of the category $\mathcal{J}$ of simplices of $N(\mathcal{I}_1)_{C/}$. Proposition 4.2.3.14 supplies a cofinal map $N(\mathcal{J}) \to N(\mathcal{I}_1)_{C/}$, so that $N(\mathcal{J})$ is weakly homotopy equivalent to $N(\mathcal{I}_1)_{C/}$, which is weakly contractible (since it has an initial object).

(3) Let $\mathcal{I}_3$ denote the subcategory of $\mathcal{I}_2$ consisting of *injective* maps between simplices of $K^{(1)}$, and let $K^{(3)} = N(\mathcal{I}_3)^{op} \subseteq N(\mathcal{I}_2)^{op}$. We have a pullback diagram

$$
\begin{array}{ccc}
N(\mathcal{I}_3)^{op} & \longrightarrow & N(\mathcal{I}_2)^{op} \\
\downarrow & & \downarrow \\
N(\Delta_s)^{op} & \longrightarrow & N(\Delta)^{op}
\end{array}
$$

where lower horizontal map is the cofinal inclusion of Lemma 6.5.3.7. The vertical maps are left fibrations and therefore smooth (Proposition 4.1.2.15), so that the inclusion $\phi_3 : K^{(3)} \subseteq K^{(2)}$ is cofinal.

(4) Let $K^{(4)}$ denote the nerve $N(\mathcal{I}_4)$, where $\mathcal{I}_4$) is the category of simplices of $K^{(3)}$, and let ϕ_4 denote the cofinal map described in Proposition 4.2.3.14.

(5) Let $K^{(5)}$ denote the nerve $N(\mathcal{I}_5) \subseteq N(\mathcal{I}_4)$, where $\mathcal{I}_5$ is the full subcategory spanned by the nondegenerate simplices of $K^{(3)}$. We observe that the category $\mathcal{I}_3$ has the following property: the collection of nonidentity morphisms in $\mathcal{I}_3$ is stable under composition. It follows that every face of a nondegenerate simplex of $K^{(3)}$ is again nondegenerate. Consequently, the inclusion $\mathcal{I}_5 \subseteq \mathcal{I}_4$ admits a left adjoint, so that the inclusion $\phi_5 : N(\mathcal{I}_5) \subseteq N(\mathcal{I}_4)$ is cofinal (this follows easily from Theorem 4.1.3.1). We conclude by observing that the category $\mathcal{I}_5$ is equivalent to a partially ordered set, because if σ is a simplex of $K^{(3)}$, then any face of σ is uniquely determined by the vertices that it contains.

Variant 4.2.3.16. If K is a *finite* simplicial set, then the we can arrange that the simplicial set K' $K' \to K$ appearing in Variant 4.2.3.15 is again finite (though our construction is not functorial in K). First suppose that K satisfies the following condition:

($*$) Every nondegenerate simplex $\sigma : \Delta^n \to K$ is a monomorphism of simplicial sets.

Let $\mathcal{I}$ denote the category of simplices of K, and let $\mathcal{I}_0$ denote the full subcategory spanned by the nondegenerate simplices. Condition ($*$) guarantees that the inclusion $\mathcal{I}_0 \subseteq \mathcal{I}$ admits a left adjoint, so that Theorem 4.1.3.1 implies that the inclusion $N(\mathcal{I}_0) \subseteq N(\mathcal{I})$ is cofinal. Combining this with Proposition 4.2.3.14, we deduce that the map $N(\mathcal{I}_0) \to K$ is cofinal. Moreover, $N(\mathcal{I}_0)$ can be identified with the nerve of the partially ordered set of simplicial subsets $K_0 \subseteq K$ such that K_0 is isomorphic to a simplex. In particular, this partially ordered set is finite.

To handle the general case, it will suffice to establish the following claim:

($*'$) For every finite simplicial set K, there exists a cofinal map $\widetilde{K} \to K$, where $\widetilde{K}$ is a finite simplicial set satisfying ($*$).

The proof proceeds by induction on the number of nondegenerate simplices of K. If K is empty, the result is obvious; otherwise, we have a pushout diagram

$$\begin{array}{ccc} \partial\Delta^n & \longrightarrow & \Delta^n \\ \downarrow & & \downarrow \\ K_0 & \longrightarrow & K. \end{array}$$

The inductive hypothesis guarantees the existence of a map $\widetilde{K}_0 \to K_0$ satisfying ($*$). Now define $\widetilde{K} = (\widetilde{K}_0 \times \Delta^n) \coprod_{\partial\Delta^n} \Delta^n$. It follows from Corollary 4.1.2.7 (and the weak contractibility of Δ^n) that the map $\widetilde{K} \to K$ is cofinal. Moreover, since the map $\partial\Delta^n \to \widetilde{K}_0 \times \Delta^n$ is a monomorphism, we conclude that $\widetilde{K}$ satisfies condition ($*$), as desired.

4.2.4 Homotopy Colimits

Our goal in this section is to compare the ∞-categorical theory of colimits with the more classical theory of homotopy colimits in simplicial categories (see Remark A.3.3.13). Our main result is the following:

Theorem 4.2.4.1. *Let $\mathcal{C}$ and $\mathcal{J}$ be fibrant simplicial categories and $F : \mathcal{J} \to \mathcal{C}$ a simplicial functor. Suppose we are given an object $C \in \mathcal{C}$ and a compatible family of maps $\{\eta_I : F(I) \to C\}_{I \in \mathcal{J}}$. The following conditions are equivalent:*

(1) *The maps η_I exhibit C as a homotopy colimit of the diagram F.*

(2) *Let $f : \mathrm{N}(\mathcal{J}) \to \mathrm{N}(\mathcal{C})$ be the simplicial nerve of F and $\overline{f} : \mathrm{N}(\mathcal{J})^{\triangleright} \to \mathrm{N}(\mathcal{C})$ the extension of f determined by the maps $\{\eta_I\}$. Then $\overline{f}$ is a colimit diagram in $\mathrm{N}(\mathcal{C})$.*

Remark 4.2.4.2. For an analogous result (in a slightly different setting), we refer the reader to [39].

The proof of Theorem 4.2.4.1 will occupy the remainder of this section. We begin with a convenient criterion for detecting colimits in ∞-categories:

Lemma 4.2.4.3. *Let $\mathcal{C}$ be an ∞-category, K a simplicial set, and $\overline{p} : K^{\triangleright} \to \mathcal{C}$ a diagram. The following conditions are equivalent:*

(i) *The diagram $\overline{p}$ is a colimit of $p = \overline{p}|K$.*

(ii) *Let $X \in \mathcal{C}$ denote the image under $\overline{p}$ of the cone point of $K^{\triangleright}$, let $\delta : \mathcal{C} \to \mathrm{Fun}(K, \mathcal{C})$ denote the diagonal embedding, and let $\alpha : p \to \delta(X)$ denote the natural transformation determined by $\overline{p}$. Then, for every object $Y \in \mathcal{C}$, composition with α induces a homotopy equivalence*

$$\phi_Y : \mathrm{Map}_{\mathcal{C}}(X, Y) \to \mathrm{Map}_{\mathrm{Fun}(K, \mathcal{C})}(p, \delta(Y)).$$

Proof. Using Corollary 4.2.1.8, we can identify $\mathrm{Map}_{\mathrm{Fun}(K, \mathcal{C})}(p, \delta(Y))$ with the fiber $\mathcal{C}^{p/} \times_{\mathcal{C}} \{Y\}$ for each object $Y \in \mathcal{C}$. Under this identification, the map ϕ_Y can be identified with the fiber over Y of the composition

$$\mathcal{C}^{X/} \overset{\phi'}{\to} \mathcal{C}^{\overline{p}/} \overset{\phi''}{\to} \mathcal{C}^{p/},$$

where ϕ' is a section to the trivial fibration $\mathcal{C}^{\overline{p}/} \to \mathcal{C}^{X/}$. The map ϕ'' is a left fibration (Proposition 4.2.1.6). Condition (i) is equivalent to the requirement that ϕ'' be a trivial Kan fibration, and condition (ii) is equivalent to the requirement that each of the maps

$$\phi_Y'' : \mathcal{C}^{\overline{p}/} \times_{\mathcal{C}} \{Y\} \to \mathcal{C}^{p/} \times_{\mathcal{C}} \{Y\}$$

is a homotopy equivalence of Kan compexes (which, in view of Lemma 2.1.3.3, is equivalent to the requirement that ϕ_Y'' be a trivial Kan fibration). The equivalence of these two conditions now follows from Lemma 2.1.3.4. $\square$

The key to Theorem 4.2.4.1 is the following result, which compares the construction of diagram categories in the ∞-categorical and simplicial settings:

Proposition 4.2.4.4. *Let S be a small simplicial set, $\mathcal{C}$ a small simplicial category, and $u : \mathfrak{C}[S] \to \mathcal{C}$ an equivalence. Suppose that $\mathbf{A}$ is a combinatorial simplicial model category and let $\mathcal{U}$ be a $\mathcal{C}$-chunk of $\mathbf{A}$ (see Definition A.3.4.9). Then the induced map*

$$\mathrm{N}((\mathcal{U}^{\mathcal{C}})^{\circ}) \to \mathrm{Fun}(S, \mathrm{N}(\mathcal{U}^{\circ}))$$

is a categorical equivalence of simplicial sets.

Remark 4.2.4.5. In the statement of Proposition 4.2.4.4, it makes no difference whether we regard $\mathbf{A}^{\mathcal{C}}$ as endowed with the projective or the injective model structure.

Remark 4.2.4.6. An analogous result was proved by Hirschowitz and Simpson; see [39].

Proof. Choose a regular cardinal κ such that S and $\mathcal{C}$ are κ-small. Using Lemma A.3.4.15, we can write $\mathcal{U}$ as a κ-filtered colimit of small $\mathcal{C}$-chunks $\mathcal{U}'$ contained in $\mathcal{U}$. Since the collection of categorical equivalences is stable under filtered colimits, it will suffice to prove the result after replacing $\mathcal{U}$ by each $\mathcal{U}'$; in other words, we may suppose that $\mathcal{U}$ is small.

According to Theorem 2.2.5.1, we may identify the homotopy category of Set_{Δ} (with respect to the Joyal model structure) with the homotopy category of $\mathcal{C}\mathrm{at}_{\Delta}$. We now observe that because $\mathrm{N}(\mathcal{U}^{\circ})$ is an ∞-category, the simplicial set $\mathrm{Fun}(S, \mathrm{N}(\mathcal{U}^{\circ}))$ can be identified with an exponential $[\mathrm{N}(\mathcal{U}^{\circ})]^{[S]}$ in the homotopy category $h\mathrm{Set}_{\Delta}$. We now conclude by applying Corollary A.3.4.14. $\square$

One consequence of Proposition 4.2.4.4 is that every homotopy coherent diagram in a suitable model category $\mathbf{A}$ can be "straightened," as we indicated in Remark 1.2.6.2.

Corollary 4.2.4.7. *Let $\mathcal{J}$ be a fibrant simplicial category, S a simplicial set, and $p : \mathrm{N}(\mathcal{J}) \to S$ a map. Then it is possible to find the following:*

(1) *A fibrant simplicial category $\mathcal{C}$.*

(2) *A simplicial functor $P : \mathcal{J} \to \mathcal{C}$.*

(3) *A categorical equivalence of simplicial sets $j : S \to \mathrm{N}(\mathcal{C})$.*

(4) *An equivalence between $j \circ p$ and $\mathrm{N}(P)$ as objects of the ∞-category $\mathrm{Fun}(\mathrm{N}(\mathcal{J}), \mathrm{N}(\mathcal{C}))$.*

Proof. Choose an equivalence $i : \mathfrak{C}[S] \to \mathcal{C}_0$, where $\mathcal{C}_0$ is fibrant; let $\mathbf{A}$ denote the model category of simplicial presheaves on $\mathcal{C}_0$ (endowed with the *injective* model structure). Composing i with the Yoneda embedding of $\mathcal{C}_0$, we obtain a fully faithful simplicial functor $\mathfrak{C}[S] \to \mathbf{A}^\circ$, which we may alternatively view as a morphism $j_0 : S \to N(\mathbf{A}^\circ)$.

We now apply Proposition 4.2.4.4 to the case where u is the counit map $\mathfrak{C}[N(\mathcal{I})] \to \mathcal{I}$. We deduce that the natural map

$$N((\mathbf{A}^{\mathcal{I}})^\circ) \to \mathrm{Fun}(N(\mathcal{I}), N(\mathbf{A}^\circ))$$

is an equivalence. From the essential surjectivity, we deduce that $j_0 \circ p$ is equivalent to $N(P_0)$, where $P_0 : \mathcal{I} \to \mathbf{A}^\circ$ is a simplicial functor.

We now take $\mathcal{C}$ to be the essential image of $\mathfrak{C}[S]$ in $\mathbf{A}^\circ$ and note that j_0 and P_0 factor uniquely through maps $j : S \to N(\mathcal{C})$, $P : \mathcal{I} \to \mathcal{C}$ which possess the desired properties. $\qquad\qquad\qquad\qquad\qquad\qquad\qquad\qquad\qquad\qquad\square$

We now return to our main result.

Proof of Theorem 4.2.4.1: Let $\mathbf{A}$ denote the category $\mathrm{Set}_\Delta^{\mathcal{C}}$ endowed with the projective model structure. Let $j : \mathcal{C}^{op} \to \mathbf{A}$ denote the Yoneda embedding and let $\mathcal{U}$ denote the full subcategory of $\mathbf{A}$ spanned by those objects which are weakly equivalent to $j(C)$ for some $C \in \mathcal{C}$, so that j induces an equivalence of simplicial categories $\mathcal{C}^{op} \to \mathcal{U}^\circ$. Choose a trivial injective cofibration $j \circ F \to F'$, where F' is a injectively fibrant object of $\mathbf{A}^{\mathcal{I}^{op}}$. Let $f' : N(\mathcal{I})^{op} \to N(\mathcal{U}^\circ)$ be the nerve of F' and let $C' = j(C)$, so that the maps $\{\eta_I : F(I) \to C\}_{I \in \mathcal{I}}$ induce a natural transformation $\alpha : \delta(C') \to f'$, where $\delta : N(\mathcal{U}^\circ) \to \mathrm{Fun}(N(\mathcal{I})^{op}, N(\mathcal{U}^\circ))$ denotes the diagonal embedding. In view of Lemma 4.2.4.3, condition (1) admits the following reformulation:

(1′) For every object $A \in \mathcal{U}^\circ$, composition with α induces a homotopy equivalence

$$\mathrm{Map}_{N(\mathcal{U}^\circ)}(A, C') \to \mathrm{Map}_{\mathrm{Fun}(N(\mathcal{I})^{op}, N(\mathcal{U}^\circ))}(\delta(A), f').$$

Using Proposition 4.2.4.4, we can reformulate this condition again:

(1″) For every object $A \in \mathcal{U}^\circ$, the canonical map

$$\mathrm{Map}_{\mathbf{A}}(A, C') \to \mathrm{Map}_{\mathbf{A}^{\mathcal{I}^{op}}}(\delta'(A), F')$$

is a homotopy equivalence, where $\delta' : \mathbf{A} \to \mathbf{A}^{\mathcal{I}^{op}}$ denotes the diagonal embedding.

Let $B \in \mathbf{A}$ be a limit of the diagram F', so we have a canonical map $\beta : C' \to B$ between fibrant objects of $\mathbf{A}$. Condition (2) is equivalent to the assertion that β is a weak equivalence in $\mathbf{A}$, while condition (1″) is equivalent to the assertion that composition with β induces a homotopy equivalence

$$\mathrm{Map}_{\mathbf{A}}(A, C') \to \mathrm{Map}_{\mathbf{A}}(A, B)$$

for each $A \in \mathcal{U}^\circ$. The implication (2) $\Rightarrow$ (1″) is clear. Conversely, suppose that (1″) is satisfied. For each $X \in \mathcal{C}$, the object $j(X)$ belongs to $\mathcal{U}^\circ$, so that β induces a homotopy equivalence

$$C'(X) \simeq \mathrm{Map}_{\mathbf{A}}(j(X), C') \to \mathrm{Map}_{\mathbf{A}}(j(X), B) \simeq B(X).$$

It follows that β is a weak equivalence in $\mathbf{A}$, as desired. $\qquad\qquad\square$

Corollary 4.2.4.8. *Let* **A** *be a combinatorial simplicial model category. The associated* ∞-*category* $S = N(\mathbf{A}^\circ)$ *admits (small) limits and colimits.*

Proof. We give the argument for colimits; the case of limits follows by a dual argument. Let $p : K \to S$ be a (small) diagram in S. By Proposition 4.2.3.14, there exists a (small) category $\mathcal{J}$ and a cofinal map $q : N(\mathcal{J}) \to K$. Since q is cofinal, p has a colimit in S if and only if $p \circ q$ has a colimit in S; thus we may reduce to the case where $K = N(\mathcal{J})$.

Using Proposition 4.2.4.4, we may suppose that p is the nerve of a injectively fibrant diagram $p' : \mathcal{J} \to \mathbf{A}^\circ$. Let $\overline{p}' : \mathcal{J} \star \{x\} \to \mathbf{A}^{\mathcal{J}}$ be a limit of p', so that $\overline{p}'$ is a homotopy limit diagram in **A**. Now choose a trivial fibration $\overline{p}'' \to \overline{p}'$ in $\mathbf{A}^{\mathcal{J}}$, where $\overline{p}''$ is cofibrant. The simplicial nerve of $\overline{p}''$ determines a colimit diagram $\overline{f} : N(\mathcal{J})^{\triangleright} \to S$ by Theorem 4.2.4.1. We now observe that $f = \overline{f}| N(\mathcal{J})$ is equivalent to p, so that p also admits a colimit in S. $\square$

4.3 KAN EXTENSIONS

Let $\mathcal{C}$ and $\mathcal{J}$ be ordinary categories. There is an obvious "diagonal" functor $\delta : \mathcal{C} \to \mathcal{C}^{\mathcal{J}}$, which carries an object $C \in \mathcal{C}$ to the constant diagram $\mathcal{J} \to \mathcal{C}$ taking the value C. If $\mathcal{C}$ admits small colimits, then the functor δ has a left adjoint $\mathcal{C}^{\mathcal{J}} \to \mathcal{C}$. This left adjoint admits an explicit description: it carries an arbitrary diagram $f : \mathcal{J} \to \mathcal{C}$ to the colimit $\varinjlim(f)$. Consequently, we can think of the theory of colimits as the study of left adjoints to diagonal functors.

More generally, if one is given a functor $i : \mathcal{J} \to \mathcal{J}'$ between diagram categories, then composition with i induces a functor $i^* : \mathcal{C}^{\mathcal{J}'} \to \mathcal{C}^{\mathcal{J}}$. Assuming that $\mathcal{C}$ has a sufficient supply of colimits, one can construct a left adjoint to i^*. We then refer to this left adjoint as *the left Kan extension along* i.

In this section, we will study the ∞-categorical analogue of the theory of left Kan extensions. In the extreme case where $\mathcal{J}'$ is the one-object category $*$, this theory simply reduces to the theory of colimits introduced in §1.2.13. Our primary interest will be at the opposite extreme, when i is a fully faithful embedding; this is the subject of §4.3.2. We will treat the general case in §4.3.3.

With a view toward later applications, we will treat not only the theory of *absolute* left Kan extensions but also a relative notion which works over a base simplicial set S. The most basic example is the case of a *relative colimit* which we study in §4.3.1.

4.3.1 Relative Colimits

In §1.2.13, we introduced the notions of limit and colimit for a diagram $p : K \to \mathcal{C}$ in an ∞-category $\mathcal{C}$. For many applications, it is convenient to have a *relative* version of these notions, which makes reference not to an ∞-category $\mathcal{C}$ but to an arbitrary inner fibration of simplicial sets.

Definition 4.3.1.1. Let $f : \mathcal{C} \to \mathcal{D}$ be an inner fibration of simplicial sets, let $\overline{p} : K^{\triangleright} \to \mathcal{C}$ be a diagram, and let $p = \overline{p}|K$. We will say that $\overline{p}$ is an f-*colimit* of p if the map

$$\mathcal{C}_{\overline{p}/} \to \mathcal{C}_{p/} \times_{\mathcal{D}_{fp/}} \mathcal{D}_{f\overline{p}/}$$

is a trivial fibration of simplicial sets. In this case, we will also say that $\overline{p}$ is an f-*colimit diagram*.

Remark 4.3.1.2. Let $f : \mathcal{C} \to \mathcal{D}$ and $\overline{p} : K^{\triangleright} \to \mathcal{C}$ be as in Definition 4.3.1.1. Then $\overline{p}$ is an f-colimit of $p = \overline{p}|K$ if and only if the map

$$\phi : \mathcal{C}_{\overline{p}/} \to \mathcal{C}_{p/} \times_{\mathcal{D}_{fp/}} \mathcal{D}_{f\overline{p}/}$$

is a categorical equivalence. The "only if" direction is clear. The converse follows from Proposition 2.1.2.1 (which implies that ϕ is a left fibration), Proposition 3.3.1.7 (which implies that ϕ is a categorical fibration), and the fact that a categorical fibration which is a categorical equivalence is a trivial Kan fibration.

Observe that Proposition 2.1.2.1 also implies that the map

$$\mathcal{D}_{f\overline{p}/} \to \mathcal{D}_{fp/}$$

is a left fibration. Using Propositions 3.3.1.3 and 3.3.1.7, we conclude that the fiber product $\mathcal{C}_{p/} \times_{\mathcal{D}_{fp/}} \mathcal{D}_{f\overline{p}/}$ is also a homotopy fiber product of $\mathcal{C}_{p/}$ with $\mathcal{D}_{f\overline{p}/}$ over $\mathcal{D}_{fp/}$ (with respect to the Joyal model structure on Set_{Δ}). Consequently, we deduce that $\overline{p}$ is an f-colimit diagram if and only if the diagram of simplicial sets

$$
\begin{array}{ccc}
\mathcal{C}_{\overline{p}/} & \longrightarrow & \mathcal{D}_{f\overline{p}/} \\
\downarrow & & \downarrow \\
\mathcal{C}_{p/} & \longrightarrow & \mathcal{D}_{fp/}
\end{array}
$$

is homotopy Cartesian.

Example 4.3.1.3. Let $\mathcal{C}$ be an ∞-category and $f : \mathcal{C} \to *$ the projection of $\mathcal{C}$ to a point. Then a diagram $\overline{p} : K^{\triangleright} \to \mathcal{C}$ is an f-colimit if and only if it is a colimit in the sense of Definition 1.2.13.4.

Example 4.3.1.4. Let $f : \mathcal{C} \to \mathcal{D}$ be an inner fibration of simplicial sets and let $e : \Delta^1 = (\Delta^0)^{\triangleright} \to \mathcal{C}$ be an edge of $\mathcal{C}$. Then e is an f-colimit if and only if it is f-coCartesian.

The following basic stability properties follow immediately from the definition:

Proposition 4.3.1.5. (1) *Let $f : \mathcal{C} \to \mathcal{D}$ be a trivial fibration of simplicial sets. Then every diagram $\overline{p} : K^{\triangleright} \to \mathcal{C}$ is an f-colimit.*

(2) *Let $f : \mathcal{C} \to \mathcal{D}$ and $g : \mathcal{D} \to \mathcal{E}$ be inner fibrations of simplicial sets and let $\overline{p} : K^{\triangleright} \to \mathcal{C}$ be a diagram. Suppose that $f \circ \overline{p}$ is a g-colimit. Then $\overline{p}$ is an f-colimit if and only if $\overline{p}$ is a $(g \circ f)$-colimit.*

(3) *Let $f : \mathcal{C} \to \mathcal{D}$ be an inner fibration of ∞-categories and let $\overline{p}, \overline{q} :$ $K^{\triangleright} \to \mathcal{C}$ be diagrams which are equivalent when viewed as objects of the ∞-category $\mathrm{Fun}(K^{\triangleright}, \mathcal{C})$. Then $\overline{p}$ is an f-colimit if and only if $\overline{q}$ is an f-colimit.*

(4) *Suppose we are given a Cartesian diagram*

$$
\begin{array}{ccc}
\mathcal{C}' & \xrightarrow{g} & \mathcal{C} \\
\downarrow{\scriptstyle f'} & & \downarrow{\scriptstyle f} \\
\mathcal{D}' & \longrightarrow & \mathcal{D}
\end{array}
$$

of simplicial sets, where f (and therefore also f') is an inner fibration. Let $\overline{p} : K^{\triangleright} \to \mathcal{C}'$ be a diagram. If $g \circ \overline{p}$ is an f-colimit, then $\overline{p}$ is an f'-colimit.

Proposition 4.3.1.6. *Suppose we are given a commutative diagram of ∞-categories*

$$
\begin{array}{ccc}
\mathcal{C} & \xrightarrow{f} & \mathcal{C}' \\
\downarrow{\scriptstyle p} & & \downarrow{\scriptstyle p'} \\
\mathcal{D} & \longrightarrow & \mathcal{D}',
\end{array}
$$

where the horizontal arrows are categorical equivalences and the vertical arrows are inner fibrations. Let $\overline{q} : K^{\triangleright} \to \mathcal{C}$ be a diagram and let $q = \overline{q}|K$. Then $\overline{q}$ is a p-colimit of q if and only if $f \circ \overline{q}$ is a p'-colimit of $f \circ q$.

Proof. Consider the diagram

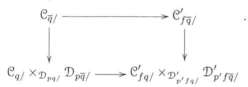

According to Remark 4.3.1.2, it will suffice to show that the left vertical map is a categorical equivalence if and only if the right vertical map is a categorical equivalence. For this, it suffices to show that both of the horizontal maps are categorical equivalences. Proposition 1.2.9.3 implies that the maps $\mathcal{C}_{\overline{q}/} \to \mathcal{C}'_{f\overline{q}/}$, $\mathcal{C}_{q/} \to \mathcal{C}'_{fq/}$, $\mathcal{D}_{p\overline{q}/} \to \mathcal{D}'_{p'f\overline{q}/}$, and $\mathcal{D}_{pq/} \to \mathcal{D}'_{p'fq/}$ are categorical equivalences. It will therefore suffice to show that the diagrams

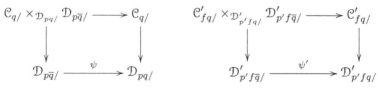

are homotopy Cartesian (with respect to the Joyal model structure). This follows from Proposition 3.3.1.3 because ψ and ψ' are coCartesian fibrations. $\square$

The next pair of results can be regarded as a generalization of Proposition 4.1.1.8. They assert that, when computing relative colimits, we are free to replace any diagram by a cofinal subdiagram.

Proposition 4.3.1.7. *Let $p : \mathcal{C} \to \mathcal{D}$ be an inner fibration of ∞-categories, let $i : A \to B$ be a cofinal map, and let $\overline{q} : B^{\triangleright} \to \mathcal{C}$ be a diagram. Then $\overline{q}$ is a p-colimit if and only if $\overline{q} \circ i^{\triangleright}$ is a p-colimit.*

Proof. Recall (Remark 4.3.1.2) that $\overline{q}$ is a relative colimit diagram if and only if the diagram

$$
\begin{array}{ccc}
\mathcal{C}_{\overline{q}/} & \longrightarrow & \mathcal{C}_{q/} \\
\downarrow & & \downarrow \\
\mathcal{D}_{\overline{q}_0/} & \longrightarrow & \mathcal{D}_{q_0/}
\end{array}
$$

is homotopy Cartesian with respect to the Joyal model structure. Since i and $i^{\triangleright}$ are both cofinal, this is equivalent to the assertion that the diagram

$$
\begin{array}{ccc}
\mathcal{C}_{\overline{q}i^{\triangleright}/} & \longrightarrow & \mathcal{C}_{qi/} \\
\downarrow & & \downarrow \\
\mathcal{D}_{\overline{q}_0 i^{\triangleright}/} & \longrightarrow & \mathcal{D}_{q_0 i/}
\end{array}
$$

is homotopy Cartesian, which (by Remark 4.3.1.2) is equivalent to the assertion that $\overline{q} \circ i^{\triangleright}$ is a relative colimit diagram. $\square$

Proposition 4.3.1.8. *Let $p : \mathcal{C} \to \mathcal{D}$ be a coCartesian fibration of ∞-categories, let $i : A \to B$ be a cofinal map, and let*

$$
\begin{array}{ccc}
B & \overset{q}{\longrightarrow} & \mathcal{C} \\
\downarrow & & \downarrow{\scriptstyle p} \\
B^{\triangleright} & \underset{\overline{q}_0}{\longrightarrow} & \mathcal{D}
\end{array}
$$

be a diagram. Suppose that $q \circ i$ has a relative colimit lifting $\overline{q}_0 \circ i^{\triangleright}$. Then q has a relative colimit lifting $\overline{q}_0$.

Proof. Let $q_0 = \overline{q}_0 | B$. We have a commutative diagram

$$
\begin{array}{ccccc}
\mathcal{C}_{q/} & \overset{f}{\longrightarrow} & \mathcal{C}_{qi/} \times_{\mathcal{D}_{pqi/}} \mathcal{D}_{pq/} & \longrightarrow & \mathcal{C}_{qi/} \\
\downarrow & & \downarrow & & \downarrow \\
\mathcal{D}_{q_0/} & \longrightarrow & \mathcal{D}_{q_0/} & \longrightarrow & \mathcal{D}_{q_0 i/,}
\end{array}
$$

where the horizontal maps are categorical equivalences (this follows from the fact that i is cofinal and Proposition 3.3.1.3). Proposition 2.4.3.2 implies that the vertical maps are coCartesian fibrations and that f preserves co-Cartesian edges. Applying Proposition 3.3.1.5 to f, we deduce that the map

$\phi : \mathcal{C}_{q/} \times_{\mathcal{D}_{q_0/}} \{\overline{q}_0\} \to \mathcal{C}_{qi/} \times_{\mathcal{D}_{q_0i/}} \{\overline{q}_0 i^{\triangleright}\}$ is a categorical equivalence. Since ϕ is essentially surjective, we conclude that there exists an extension $\overline{q} : B^{\triangleright} \to \mathcal{C}$ of q which covers $\overline{q}_0$, such that $\overline{q} \circ i^{\triangleright}$ is a p-colimit diagram. We now apply Proposition 4.3.1.7 to conclude that $\overline{q}$ is itself a p-colimit diagram. $\square$

Let $p : X \to S$ be a coCartesian fibration. The following results will allow us to reduce the theory of p-colimits to the theory of ordinary colimits in the fibers of p.

Proposition 4.3.1.9. *Let* $p : X \to S$ *be an inner fibration of* ∞-*categories,* K *a simplicial set, and* $\overline{h} : \Delta^1 \times K^{\triangleright} \to X$ *a natural transformation from* $\overline{h}_0 = \overline{h}|\{0\} \times K^{\triangleright}$ *to* $\overline{h}_1 = \overline{h}|\{1\} \times K^{\triangleright}$. *Suppose that*

(1) *For every vertex* x *of* $K^{\triangleright}$, *the restriction* $\overline{h}|\Delta^1 \times \{x\}$ *is a* p-*coCartesian edge of* X.

(2) *The composition*

$$\Delta^1 \times \{\infty\} \subseteq \Delta^1 \times K^{\triangleright} \xrightarrow{\overline{h}} X \xrightarrow{p} S$$

is a degenerate edge of S, *where* ∞ *denotes the cone point of* $K^{\triangleright}$.

Then $\overline{h}_0$ *is a* p-*colimit diagram if and only if* $\overline{h}_1$ *is a* p-*colimit diagram.*

Proof. Let $h = \overline{h}|\Delta^1 \times K$, $h_0 = h|\{0\} \times K$, and $h_1 = h|\{1\} \times K$. Consider the diagram

$$
\begin{array}{ccccc}
X_{\overline{h}_0/} & \xleftarrow{\quad \phi \quad} & X_{\overline{h}/} & \xrightarrow{\qquad} & X_{\overline{h}_1/} \\
\downarrow & & \downarrow & & \downarrow \\
X_{h_0/} \times_{S_{ph_0/}} S_{p\overline{h}_0/} & \xleftarrow{\psi} & X_{h/} \times_{S_{ph/}} S_{p\overline{h}/} & \xrightarrow{\quad} & X_{h_1/} \times_{S_{ph_1/}} S_{p\overline{h}_1/}.
\end{array}
$$

According to Remark 4.3.1.2, it will suffice to show that the left vertical map is a categorical equivalence if and only if the right vertical map is a categorical equivalence. For this, it will suffice to show that each of the horizontal arrows is a categorical equivalence. Because the inclusions $\{1\} \times K \subseteq \Delta^1 \times K$ and $\{1\} \times K^{\triangleright} \subseteq \Delta^1 \times K^{\triangleright}$ are right anodyne, the horizontal maps on the right are trivial fibrations. We are therefore reduced to proving that ϕ and ψ are categorical equivalences.

Let $f : x \to y$ denote the edge of X obtained by restricting $\overline{h}$ to the cone point of $K^{\triangleright}$. The map ϕ fits into a commutative diagram

$$
\begin{array}{ccc}
X_{\overline{h}/} & \xrightarrow{\phi} & X_{h_0/} \\
\downarrow & & \downarrow \\
X_{f/} & \xrightarrow{\quad} & X_{x/}.
\end{array}
$$

Since the inclusion of the cone point into $K^{\triangleright}$ is right anodyne, the vertical arrows are trivial fibrations. Moreover, hypotheses (1) and (2) guarantee that

f is an equivalence in X, so that the map $X_{f/} \to X_{x/}$ is a trivial fibration. This proves that ϕ is a categorical equivalence.

The map ψ admits a factorization

$$X_{h/} \times_{S_{ph/}} S_{p\overline{h}/} \xrightarrow{\psi'} X_{h0/} \times_{S_{ph0/}} S_{p\overline{h}/} \xrightarrow{\psi''} X_{h0} \times_{S_{ph0/}} S_{p\overline{h}_0/}.$$

To complete the proof, it will suffice to show that ψ' and ψ'' are trivial fibrations of simplicial sets. We first observe that ψ' is a pullback of the map

$$X_{h/} \to X_{h0/} \times_{S_{ph0/}} S_{ph/},$$

which is a trivial fibration (Proposition 3.1.1.12). The map ψ'' is a pullback of the left fibration $\psi''_0 : S_{p\overline{h}/} \to S_{p\overline{h}_0/}$. It therefore suffices to show that ψ''_0 is a categorical equivalence. To prove this, we consider the diagram

$$
\begin{array}{ccc}
S_{p\overline{h}/} & \xrightarrow{\psi''_0} & S_{p\overline{h}_0/} \\
\downarrow & & \downarrow \\
S_{p(f)/} & \xrightarrow{\psi''_1} & S_{p(x)/}.
\end{array}
$$

As above, we observe that the vertical arrows are trivial fibrations and that ψ''_1 is a trivial fibration (because the morphism $p(f)$ is an equivalence in S). It follows that ψ''_0 is a categorical equivalence, as desired. $\square$

Proposition 4.3.1.10. *Let $q : X \to S$ be a locally coCartesian fibration of ∞-categories, let s be an object of S, and let $\overline{p} : K^{\triangleright} \to X_s$ be a diagram. The following conditions are equivalent:*

(1) *The map $\overline{p}$ is a q-colimit diagram.*

(2) *For every morphism $e : s \to s'$ in S, the associated functor $e_! : X_s \to X_{s'}$ has the property that $e_! \circ \overline{p}$ is a colimit diagram in the ∞-category $X_{s'}$.*

Proof. Assertion (1) is equivalent to the statement that the map

$$\theta : X_{\overline{p}/} \to X_{p/} \times_{S_{qp/}} S_{q\overline{p}/}$$

is a trivial fibration of simplicial sets. Since θ is a left fibration, it will suffice to show that the fibers of θ are contractible. Consider an arbitrary vertex of $S_{q\overline{p}/}$ corresponding to a morphism $t : K \star \Delta^1 \to S$. Since $K \star \Delta^1$ is categorically equivalent to $(K \star \{0\}) \coprod_{\{0\}} \Delta^1$ and $t|K \star \{0\}$ is constant, we may assume without loss of generality that t factors as a composition

$$K \star \Delta^1 \to \Delta^1 \xrightarrow{e} S.$$

Here $e : s \to s'$ is an edge of S. Pulling back by the map e, we can reduce to the problem of proving the following analogue of (1) in the case where $S = \Delta^1$:

(1′) *The projection $h_0 : X_{\overline{p}/} \times_S \{s'\} \to X_{p/} \times_S \{s'\}$ is a trivial fibration of simplicial sets.*

Choose a coCartesian transformation $\overline{\alpha} : K^{\triangleright} \times \Delta^1 \to X$ from $\overline{p}$ to $\overline{p}'$, which covers the projection

$$K^{\triangleright} \times \Delta^1 \to \Delta^1 \simeq S.$$

Consider the diagram

$$
\begin{array}{ccc}
X_{\overline{p}/} \times_S \{s'\} & \longleftarrow X_{\overline{\alpha}/} \times_S \{s'\} \longrightarrow & X_{\overline{p}'/} \times_S \{s'\} \\
\downarrow h_0 & \downarrow h & \downarrow h_1 \\
X_{p/} \times_S \{s'\} & \longleftarrow X_{\alpha/} \times_S \{s'\} \longrightarrow & X_{p'/} \times_S \{s'\}.
\end{array}
$$

Note that the vertical maps are left fibrations (Proposition 2.1.2.1). Since the inclusion $K^{\triangleright} \times \{1\} \subseteq K^{\triangleright} \times \Delta^1$ is right anodyne, the upper right horizontal map is a trivial fibration. Similarly, the lower right horizontal map is a trivial fibration. Since $\overline{\alpha}$ is a coCartesian transformation, we deduce that the left horizontal maps are also trivial fibrations (Proposition 3.1.1.12). Condition (2) is equivalent to the assertion that h_1 is a trivial fibration (for each edge $e : s \to s'$ of the original simplicial set S). Since h_1 is a left fibration and therefore a categorical fibration (Proposition 3.3.1.7), this is equivalent to the assertion that h_1 is a categorical equivalence. Chasing through the diagram, we deduce that (2) is equivalent to the assertion that h_0 is a categorical equivalence, which (by the same argument) is equivalent to the assertion that h_0 is a trivial fibration. $\square$

Corollary 4.3.1.11. *Let* $p : X \to S$ *be a coCartesian fibration of ∞-categories and let K be a simplicial set. Suppose that*

(1) *For each vertex s of S, the fiber $X_s = X \times_S \{s\}$ admits colimits for all diagrams indexed by K.*

(2) *For each edge $f : s \to s'$, the associated functor $X_s \to X_{s'}$ preserves colimits of K-indexed diagrams.*

Then for every diagram

$$
\begin{array}{ccc}
K & \xrightarrow{q} & X \\
\downarrow & \nearrow{\overline{q}} & \downarrow p \\
K^{\triangleright} & \xrightarrow{f} & S
\end{array}
$$

there exists a map $\overline{q}$ as indicated, which is a p-colimit.

Proof. Consider the map $K \times \Delta^1 \to K^{\triangleright}$ which is the identity on $K \times \{0\}$ and carries $K \times \{1\}$ to the cone point of $K^{\triangleright}$. Let F denote the composition

$$K \times \Delta^1 \to K^{\triangleright} \xrightarrow{f} S$$

and let $Q : K \times \Delta^1 \to X$ be a coCartesian lifting of F to X, so that Q is a natural transformation from q to a map $q' : K \to X_s$, where s is the image under f of the cone point of $K^{\triangleright}$. In view of assumption (1), there exists a map $\overline{q}' : K^{\triangleright} \to X_s$ which is a colimit of q'. Assumption (2) and Proposition 4.3.1.10 guarantee that $\overline{q}'$ is also a p-colimit diagram when regarded as a map from $K^{\triangleright}$ to X.

We have a commutative diagram

$$(K \times \Delta^1) \coprod_{K \times \{1\}} (K^{\triangleright} \times \{1\}) \xrightarrow{\ (Q,\overline{q}')\ } X$$

$$(K \times \Delta^1)^{\triangleright} \xrightarrow{\hspace{4cm}} S.$$

with the left vertical map, diagonal r, and right vertical p.

The left vertical map is an inner fibration, so there exists a morphism r as indicated, rendering the diagram commutative. We now consider the map $K^{\triangleright} \times \Delta^1 \to (K \times \Delta^1)^{\triangleright}$ which is the identity on $K \times \Delta^1$ and carries the other vertices of $K^{\triangleright} \times \Delta^1$ to the cone point of $(K \times \Delta^1)^{\triangleright}$. Let $\overline{Q}$ denote the composition

$$K^{\triangleright} \times \Delta^1 \to (K \times \Delta^1)^{\triangleright} \xrightarrow{r} X$$

and let $\overline{q} = \overline{Q}|K^{\triangleright} \times \{0\}$. Then $\overline{Q}$ can be regarded as a natural transformation $\overline{q} \to \overline{q}'$ of diagrams $K^{\triangleright} \to X$. Since $\overline{q}'$ is a p-colimit diagram, Proposition 4.3.1.9 implies that $\overline{q}$ is a p-colimit diagram as well. □

Proposition 4.3.1.12. *Let $p : X \to S$ be a coCartesian fibration of ∞-categories and let $\overline{q} : K^{\triangleright} \to X$ be a diagram. Assume that the following conditions are satisfied:*

(1) *The map $\overline{q}$ carries each edge of K to a p-coCartesian edge of K.*

(2) *The simplicial set K is weakly contractible.*

Then $\overline{q}$ is a p-colimit diagram if and only if it carries every edge of $K^{\triangleright}$ to a p-coCartesian edge of X.

Proof. Let s denote the image under $p \circ \overline{q}$ of the cone point of $K^{\triangleright}$. Consider the map $K^{\triangleright} \times \Delta^1 \to K^{\triangleright}$ which is the identity on $K^{\triangleright} \times \{0\}$ and collapses $K^{\triangleright} \times \{1\}$ to the cone point of $K^{\triangleright}$. Let h denote the composition

$$K^{\triangleright} \times \Delta^1 \to K^{\triangleright} \xrightarrow{\overline{q}} X \xrightarrow{p} S,$$

which we regard as a natural transformation from $p \circ \overline{q}$ to the constant map with value s. Let $H : \overline{q} \to \overline{q}'$ be a coCartesian transformation from $\overline{q}$ to a diagram $\overline{q}' : K^{\triangleright} \to X_s$. Using Proposition 2.4.1.7, we conclude that $\overline{q}'$ carries each edge of K to a p-coCartesian edge of X, which is therefore an equivalence in X_s.

Let us now suppose that $\overline{q}$ carries *every* edge of $K^{\triangleright}$ to a p-coCartesian edge of X. Arguing as above, we conclude that $\overline{q}'$ carries each edge of $K^{\triangleright}$ to an equivalence in X_s. Let $e : s \to s'$ be an edge of S and $e_! : X_s \to X_{s'}$ an associated functor. The composition

$$K^{\triangleright} \xrightarrow{\overline{q}'} X_s \xrightarrow{e_!} X_{s'}$$

carries each edge of $K^\triangleright$ to an equivalence in X_s, and is therefore a colimit diagram in $X_{s'}$ (Corollary 4.4.4.10). Proposition 4.3.1.10 implies that $\overline{q}'$ is a p-colimit diagram, so that Proposition 4.3.1.9 implies that $\overline{q}$ is a p-colimit diagram as well.

For the converse, let us suppose that $\overline{q}$ is a p-colimit diagram. Applying Proposition 4.3.1.9, we conclude that $\overline{q}'$ is a p-colimit diagram. In particular, $\overline{q}'$ is a colimit diagram in the ∞-category X_s. Applying Corollary 4.4.4.10, we conclude that $\overline{q}'$ carries each edge of $K^\triangleright$ to an equivalence in X_s. Now consider an arbitrary edge $f : x \to y$ of $K^\triangleright$. If f belongs to K, then $\overline{q}(f)$ is p-coCartesian by assumption. Otherwise, we may suppose that y is the cone point of K. The map H gives rise to a diagram

$$
\begin{array}{ccc}
\overline{q}(x) & \xrightarrow{\overline{q}(f)} & \overline{q}(y) \\
\downarrow{\scriptstyle\phi} & & \downarrow{\scriptstyle\phi'} \\
\overline{q}'(x) & \xrightarrow{\overline{q}'(f)} & \overline{q}'(y)
\end{array}
$$

in the ∞-category $X \times_S \Delta^1$. Here $\overline{q}'(f)$ and ϕ' are equivalences in X_s, so that $\overline{q}(f)$ and ϕ are equivalent as morphisms $\Delta^1 \to X \times_S \Delta^1$. Since ϕ is p-coCartesian, we conclude that $\overline{q}(f)$ is p-coCartesian, as desired. $\square$

Lemma 4.3.1.13. *Let $p : \mathcal{C} \to \mathcal{D}$ be an inner fibration of ∞-categories, let $C \in \mathcal{C}$ be an object, and let $D = p(C)$. Then C is a p-initial object of $\mathcal{C}$ if and only if (C, id_D) is an initial object of $\mathcal{C} \times_{\mathcal{D}} \mathcal{D}_{D/}$.*

Proof. We have a commutative diagram

$$
\begin{array}{ccc}
\mathcal{C}_{C/} \times_{\mathcal{D}_{D/}} \mathcal{D}_{\mathrm{id}_D/} & \xrightarrow{\psi} & \mathcal{C}_{C/} \\
\downarrow{\scriptstyle\phi} & & \downarrow{\scriptstyle\phi'} \\
\mathcal{C} \times_{\mathcal{D}} \mathcal{D}_{D/} & =\!=\!=\!= & \mathcal{C} \times_{\mathcal{D}} \mathcal{D}_{D/},
\end{array}
$$

where the vertical arrows are left fibrations and therefore categorical fibrations (Proposition 3.3.1.7). We wish to show that ϕ is a trivial fibration if and only if ϕ' is a trivial fibration. This is equivalent to proving that ϕ is a categorical equivalence if and only if ϕ' is a categorical equivalence. For this, it will suffice to show that ψ is a categorical equivalence. But ψ is a pullback of the trivial fibration $\mathcal{D}_{\mathrm{id}_D/} \to \mathcal{D}_{D/}$ and therefore itself a trivial fibration. $\square$

Proposition 4.3.1.14. *Suppose we are given a diagram of ∞-categories*

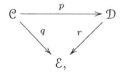

where p and r are inner fibrations, q is a Cartesian fibration, and p carries q-Cartesian morphisms to r-Cartesian morphisms.

Let $C \in \mathcal{C}$ be an object, $D = p(C)$, and $E = q(C)$. Let $\mathcal{C}_E = \mathcal{C} \times_{\mathcal{E}} \{E\}$, $\mathcal{D}_E = \mathcal{D} \times_{\mathcal{E}} \{E\}$, and $p_E : \mathcal{C}_E \to \mathcal{D}_E$ be the induced map. Suppose that C is a p_E-initial object of $\mathcal{C}_E$. Then C is a p-initial object of $\mathcal{C}$.

Proof. Our hypothesis, together with Lemma 4.3.1.13, implies that (C, id_D) is an initial object of
$$\mathcal{C}_E \times_{\mathcal{D}_E} (\mathcal{D}_E)_{D/} \simeq (\mathcal{C} \times_{\mathcal{D}} \mathcal{D}_{D/}) \times_{\mathcal{E}_{E/}} \{\mathrm{id}_E\}.$$
We will prove that the map $\phi : \mathcal{C} \times_{\mathcal{D}} \mathcal{D}_{D/} \to \mathcal{E}_{E/}$ is a Cartesian fibration. Since id_E is an initial object of $\mathcal{E}_{E/}$, Lemma 2.4.4.7 will allow us to conclude that (C, id_D) is an initial object of $\mathcal{C} \times_{\mathcal{D}} \mathcal{D}_{D/}$. We can then conclude the proof by applying Lemma 4.3.1.13 once more.

It remains to prove that ϕ is a Cartesian fibration. Let us say that a morphism of $\mathcal{C} \times_{\mathcal{D}} \mathcal{D}_{D/}$ is *special* if its image in $\mathcal{C}$ is q-Cartesian. Since ϕ is obviously an inner fibration, it will suffice to prove the following assertions:

(1) Given an object X of $\mathcal{C} \times_{\mathcal{D}} \mathcal{D}_{D/}$ and a morphism $\overline{f} : \overline{Y} \to \phi(X)$ in $\mathcal{E}_{E/}$, we can write $\overline{f} = \phi(f)$, where f is a special morphism of $\mathcal{C} \times_{\mathcal{D}} \mathcal{D}_{D/}$.

(2) Every special morphism in $\mathcal{C} \times_{\mathcal{D}} \mathcal{D}_{D/}$ is ϕ-Cartesian.

To prove (1), we first identify X with a pair consisting of an object $C'' \in \mathcal{C}$ and a morphism $D \to p(C'')$ in $\mathcal{D}$, and $\overline{f}$ with a 2-simplex $\overline{\sigma} : \Delta^2 \to \mathcal{E}$ which we depict as a diagram:

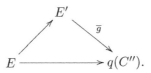

Since q is a Cartesian fibration, the morphism $\overline{g}$ can be written as $q(g)$ for some morphism $g : C' \to C''$ in $\mathcal{C}$. We now have a diagram

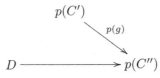

in $\mathcal{D}$. Since p carries q-Cartesian morphisms to r-Cartesian morphisms, we conclude that $p(g)$ is r-Cartesian, so that the above diagram can be completed to a 2-simplex $\sigma : \Delta^2 \to \mathcal{D}$ such that $r(\sigma) = \overline{\sigma}$.

We now prove (2). Suppose that $n \geq 2$ and that we have a commutative diagram

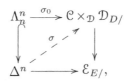

where σ_0 carries the final edge of Λ_n^n to a special morphism of $\mathcal{C} \times_{\mathcal{D}} \mathcal{D}_{D/}$. We wish to prove the existence of the morphism σ indicated in the diagram. We first let τ_0 denote the composite map

$$\Lambda_n^n \xrightarrow{\sigma_0} \mathcal{C} \times_{\mathcal{D}} \mathcal{D}_{D/} \to \mathcal{C}.$$

Consider the diagram

Since $\tau_0(\Delta^{\{n-1,n\}})$ is q-Cartesian, there exists an extension τ as indicated in the diagram. The morphisms τ and σ_0 together determine a map θ_0 which fits into a diagram

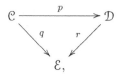

To complete the proof, it suffices to prove the existence of the indicated arrow θ. This follows from the fact that $\theta_0(\Delta^{\{n,n+1\}}) = (p \circ \tau_0)(\Delta^{\{n-1,n\}})$ is an r-Cartesian morphism of $\mathcal{D}$. □

Proposition 4.3.1.14 immediately implies the following slightly stronger statement:

Corollary 4.3.1.15. *Suppose we are given a diagram of ∞-categories*

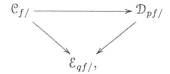

where q and r are Cartesian fibrations, p is an inner fibration, and p carries q-Cartesian morphisms to r-Cartesian morphisms.

Suppose we are given another ∞-category $\mathcal{E}_0$ equipped with a functor $s : \mathcal{E}_0 \to \mathcal{E}$. Set $\mathcal{C}_0 = \mathcal{C} \times_{\mathcal{E}} \mathcal{E}_0$, set $\mathcal{D}_0 = \mathcal{D} \times_{\mathcal{E}} \mathcal{E}_0$, and let $p_0 : \mathcal{C}_0 \to \mathcal{D}_0$ be the functor induced by p. Let $\overline{f}_0 : K^{\triangleright} \to \mathcal{C}_0$ be a diagram, and let $\overline{f}$ denote the composition $K^{\triangleright} \xrightarrow{\overline{f}_0} \mathcal{C}_0 \to \mathcal{C}$. Then $\overline{f}_0$ is a p_0-colimit diagram if and only if $\overline{f}$ is a p-colimit diagram.

Proof. Let $f_0 = \overline{f}_0 | K$ and $f = \overline{f} | K$. Replacing our diagram by

$$\mathcal{C}_{f/} \longrightarrow \mathcal{D}_{pf/}$$
$$\mathcal{E}_{qf/},$$

we can reduce to the case where $K = \emptyset$. Then $\overline{f}_0$ determines an object $C \in \mathcal{C}_0$. Let E denote the image of C in $\mathcal{E}_0$. We have a commutative diagram

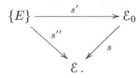

Consequently, to prove Corollary 4.3.1.15 for the map s, it will suffice to prove the analogous assertions for s' and s''; these follow from Proposition 4.3.1.14. □

Corollary 4.3.1.16. *Let* $p : \mathcal{C} \to \mathcal{E}$ *be a Cartesian fibration of* ∞-*categories,* $E \in \mathcal{E}$ *an object, and* $\overline{f} : K^{\triangleright} \to \mathcal{C}_E$ *a diagram. Then* $\overline{f}$ *is a colimit diagram in* $\mathcal{C}_E$ *if and only if it is a* p-*colimit diagram in* $\mathcal{C}$.

Proof. Apply Corollary 4.3.1.15 in the case where $\mathcal{D} = \mathcal{E}$. □

4.3.2 Kan Extensions along Inclusions

In this section, we introduce the theory of *left Kan extensions*. Let $F : \mathcal{C} \to \mathcal{D}$ be a functor between ∞-categories and let $\mathcal{C}^0$ be a full subcategory of $\mathcal{C}$. Roughly speaking, the functor F is a left Kan extension of its restriction $F_0 = F | \mathcal{C}^0$ if the values of F are as "small" as possible given the values of F_0. In order to make this precise, we need to introduce a bit of terminology.

Notation 4.3.2.1. Let $\mathcal{C}$ be an ∞-category and let $\mathcal{C}^0$ be a full subcategory. If $p : K \to \mathcal{C}$ is a diagram, we let $\mathcal{C}^0_{/p}$ denote the fiber product $\mathcal{C}_{/p} \times_{\mathcal{C}} \mathcal{C}^0$. In particular, if C is an object of $\mathcal{C}$, then $\mathcal{C}^0_{/C}$ denotes the full subcategory of $\mathcal{C}_{/C}$ spanned by the morphisms $C' \to C$ where $C' \in \mathcal{C}^0$.

Definition 4.3.2.2. Suppose we are given a commutative diagram of ∞-categories

$$
\begin{array}{ccc}
\mathcal{C}^0 & \xrightarrow{F_0} & \mathcal{D} \\
\cap \Big\downarrow & \nearrow{\scriptstyle F} & \Big\downarrow{\scriptstyle p} \\
\mathcal{C} & \longrightarrow & \mathcal{D}',
\end{array}
$$

where p is an inner fibration and the left vertical map is the inclusion of a full subcategory $\mathcal{C}^0 \subseteq \mathcal{C}$.

We will say that F is a p-*left Kan extension of* F_0 *at* $C \in \mathcal{C}$ if the induced diagram

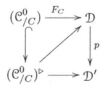

exhibits $F(C)$ as a p-colimit of F_C.

We will say that F is a p-*left Kan extension of F_0* if it is a p-left Kan extension of F_0 at C for every object $C \in \mathcal{C}$.

In the case where $\mathcal{D}' = \Delta^0$, we will omit mention of p and simply say that F is a *left Kan extension of F_0* if the above condition is satisfied.

Remark 4.3.2.3. Consider a diagram

$$
\begin{array}{ccc}
\mathcal{C}^0 & \xrightarrow{F_0} & \mathcal{D} \\
\downarrow & \nearrow{\scriptstyle F} & \downarrow{\scriptstyle p} \\
\mathcal{C} & \longrightarrow & \mathcal{D}'
\end{array}
$$

as in Definition 4.3.2.2. If C is an object of $\mathcal{C}^0$, then the functor $F_C : (\mathcal{C}^0_{/C})^{\triangleright} \to \mathcal{D}$ is automatically a p-colimit. To see this, we observe that $\mathrm{id}_C : C \to C$ is a final object of $\mathcal{C}^0_{/C}$. Consequently, the inclusion $\{\mathrm{id}_C\} \to (\mathcal{C}^0_{/C})$ is cofinal, and we are reduced to proving that $F(\mathrm{id}_C) : \Delta^1 \to \mathcal{D}$ is a colimit of its restriction to $\{0\}$, which is obvious.

Example 4.3.2.4. Consider a diagram

$$
\begin{array}{ccc}
\mathcal{C} & \xrightarrow{q} & \mathcal{D} \\
\downarrow & \nearrow{\scriptstyle \bar{q}} & \downarrow{\scriptstyle p} \\
\mathcal{C}^{\triangleright} & \longrightarrow & \mathcal{D}' .
\end{array}
$$

The map $\bar{q}$ is a p-left Kan extension of q if and only if it is a p-colimit of q. The "only if" direction is clear from the definition, and the converse follows immediately from Remark 4.3.2.3.

We first note a few basic stability properties for the class of left Kan extensions.

Lemma 4.3.2.5. *Consider a commutative diagram of ∞-categories*

$$
\begin{array}{ccc}
\mathcal{C}^0 & \xrightarrow{F_0} & \mathcal{D} \\
\downarrow & \nearrow{\scriptstyle F} & \downarrow{\scriptstyle p} \\
\mathcal{C} & \longrightarrow & \mathcal{D}'
\end{array}
$$

as in Definition 4.3.2.2. Let C and C' be equivalent objects of $\mathcal{C}$. Then F is a p-left Kan extension of F_0 at C if and only if F is a p-left Kan extension of F_0 at C'.

Proof. Let $f : C \to C'$ be an equivalence, so that the restriction maps

$$
\mathcal{C}_{/C} \leftarrow \mathcal{C}_{/f} \to \mathcal{C}_{/C'}
$$

are trivial fibrations of simplicial sets. Let $\mathcal{C}^0_{/f} = \mathcal{C}^0 \times_{\mathcal{C}} \mathcal{C}_{/f}$, so that we have trivial fibrations

$$
\mathcal{C}^0_{/C} \xleftarrow{g} \mathcal{C}^0_{/f} \xrightarrow{g'} \mathcal{C}^0_{/C'} .
$$

Consider the associated diagram

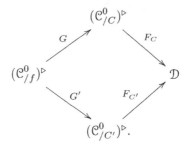

This diagram does not commute, but the functors $F_C \circ G$ and $F_{C'} \circ G'$ are equivalent in the ∞-category $\mathcal{D}^{(\mathcal{C}^0_{/f})^{\triangleright}}$. Consequently, $F_C \circ G$ is a p-colimit diagram if and only if $F_{C'} \circ G'$ is a p-colimit diagram (Proposition 4.3.1.5). Since g and g' are cofinal, we conclude that F_C is a p-colimit diagram if and only if $F_{C'}$ is a p-colimit diagram (Proposition 4.3.1.7). $\square$

Lemma 4.3.2.6. (1) *Let $\mathcal{C}$ be an ∞-category, let $p : \mathcal{D} \to \mathcal{D}'$ be an inner fibration of ∞-categories, and let $F, F' : \mathcal{C} \to \mathcal{D}$ be two functors which are equivalent in $\mathcal{D}^{\mathcal{C}}$. Let $\mathcal{C}^0$ be a full subcategory of $\mathcal{C}$. Then F is a p-left Kan extension of $F| \mathcal{C}^0$ if and only if F' is a p-left Kan extension of $F'| \mathcal{C}^0$.*

(2) *Suppose we are given a commutative diagram of ∞-categories*

$$\begin{array}{ccccccc}
\mathcal{C}^0 & \longrightarrow & \mathcal{C} & \overset{F}{\longrightarrow} & \mathcal{D} & \overset{p}{\longrightarrow} & \mathcal{E} \\
\downarrow{\scriptstyle G_0} & & \downarrow{\scriptstyle G} & & \downarrow & & \downarrow \\
\mathcal{C}'^0 & \longrightarrow & \mathcal{C}' & \underset{F'}{\longrightarrow} & \mathcal{D}' & \underset{p'}{\longrightarrow} & \mathcal{E}',
\end{array}$$

where the left horizontal maps are inclusions of full subcategories, the right horizontal maps are inner fibrations, and the vertical maps are categorical equivalences. Then F is a p-left Kan extension of $F| \mathcal{C}^0$ if and only if F' is a p'-left Kan extension of $F'| \mathcal{C}'^0$.

Proof. Assertion (1) follows immediately from Proposition 4.3.1.5. Let us prove (2). Choose an object $C \in \mathcal{C}$ and consider the diagram

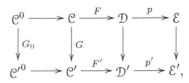

We claim that the upper left horizontal map is a p-colimit diagram if and only if the bottom left horizontal map is a p'-colimit diagram. In view of Proposition 4.3.1.6, it will suffice to show that each of the vertical maps is an equivalence of ∞-categories. For the middle and right vertical maps, this holds by assumption. To prove that the left vertical map is a categorical equivalence, we consider the diagram

$$\begin{array}{ccc} \mathcal{C}^0_{/C} & \longrightarrow & \mathcal{C}'^0_{/G(C)} \\ \downarrow & & \downarrow \\ \mathcal{C}_{/C} & \longrightarrow & \mathcal{C}'_{/G(C)}. \end{array}$$

The bottom horizontal map is a categorical equivalence (Proposition 1.2.9.3), and the vertical maps are inclusions of full subcategories. It follows that the top horizontal map is fully faithful, and its essential image consists of those morphisms $C' \to G(C)$ where C' is equivalent (in $\mathcal{C}'$) to the image of an object of $\mathcal{C}^0$. Since G_0 is essentially surjective, this is the whole of $\mathcal{C}'^0_{/G(C)}$.

It follows that if F' is a p'-left Kan extension of $F'|\mathcal{C}'^0$, then F is a p-left Kan extension of $F|\mathcal{C}^0$. Conversely, if F is a p-left Kan extension of $F|\mathcal{C}^0$, then F' is a p'-left Kan extension of $F'|\mathcal{C}'^0$ at $G(C)$ for every object $C \in \mathcal{C}$. Since G is essentially surjective, Lemma 4.3.2.5 implies that F' is a p'-left Kan extension of $F'|\mathcal{C}'^0$ at every object of $\mathcal{C}'$. This completes the proof of (2). $\qquad\square$

Lemma 4.3.2.7. *Suppose we are given a diagram of ∞-categories*

as in Definition 4.3.2.2, where p is a categorical fibration and F is a left Kan extension of F_0 relative to p. Then the induced map

$$\mathcal{D}_{F/} \to \mathcal{D}'_{pF/} \times_{\mathcal{D}'_{pF_0/}} \mathcal{D}_{F_0/}$$

is a trivial fibration of simplicial sets. In particular, we may identify p-colimits of F with p-colimits of F_0.

Proof. Using Lemma 4.3.2.6, Proposition 2.3.3.9, and Proposition A.2.3.1, we can reduce to the case where $\mathcal{C}$ is minimal. Let us call a simplicial subset $\mathcal{E} \subseteq \mathcal{C}$ *complete* if it has the following property: for any simplex $\sigma : \Delta^n \to \mathcal{C}$, if $\sigma|\Delta^{\{0,\dots,i\}}$ factors through $\mathcal{C}^0$ and $\sigma|\Delta^{\{i+1,\dots,n\}}$ factors through $\mathcal{E}$, then σ factors through $\mathcal{E}$. Note that if $\mathcal{E}$ is complete, then $\mathcal{C}^0 \subseteq \mathcal{E}$. We next define a transfinite sequence of *complete* simplicial subsets of $\mathcal{C}$

$$\mathcal{C}^0 \subseteq \mathcal{C}^1 \subseteq \cdots$$

as follows: if λ is a limit ordinal, we let $\mathcal{C}^\lambda = \bigcup_{\alpha<\lambda} \mathcal{C}^\alpha$. If $\mathcal{C}^\alpha = \mathcal{C}$, then we set $\mathcal{C}^{\alpha+1} = \mathcal{C}$. Otherwise, we choose some simplex $\sigma : \Delta^n \to \mathcal{C}$ which does not belong to $\mathcal{C}^\alpha$, where the dimension n of σ is chosen as small as possible, and let $\mathcal{C}^{\alpha+1}$ be the smallest complete simplicial subset of $\mathcal{C}$ containing $\mathcal{C}^{\alpha}$ and the simplex σ.

Let $F_{\alpha} = F|\,\mathcal{C}^{\alpha}$. We will prove that for every $\beta \leq \alpha$ the projection

$$\phi_{\alpha,\beta} : \mathcal{D}_{F_{\alpha}/} \to \mathcal{D}'_{pF_{\alpha}/} \times_{\mathcal{D}'_{pF_{\beta}/}} \mathcal{D}_{F_{\beta}/}$$

is a trivial fibration of simplicial sets. Taking $\alpha \gg \beta = 0$, we will have $\mathcal{C}^{\alpha} = \mathcal{C}$, and the proof will be complete.

Our proof proceeds by induction on α. If $\alpha = \beta$, then $\phi_{\alpha,\beta}$ is an isomorphism and there is nothing to prove. If $\alpha > \beta$ is a limit ordinal, then the inductive hypothesis implies that $\phi_{\alpha,\beta}$ is the inverse limit of a transfinite tower of trivial fibrations and therefore a trivial fibration. It therefore suffices to prove that if $\phi_{\alpha,\beta}$ is a trivial fibration, then $\phi_{\alpha+1,\beta}$ is a trivial fibration. We observe that $\phi_{\alpha+1,\beta} = \phi'_{\alpha,\beta} \circ \phi_{\alpha+1,\alpha}$, where $\phi'_{\alpha,\beta}$ is a pullback of $\phi_{\alpha,\beta}$ and therefore a trivial fibration by the inductive hypothesis. Consequently, it will suffice to prove that $\phi_{\alpha+1,\alpha}$ is a trivial fibration. The result is obvious if $\mathcal{C}^{\alpha+1} = \mathcal{C}^{\alpha}$, so we may assume without loss of generality that $\mathcal{C}^{\alpha+1}$ is the smallest complete simplicial subset of $\mathcal{C}$ containing $\mathcal{C}^{\alpha}$ together with a simplex $\sigma : \Delta^n \to \mathcal{C}$, where σ does not belong to $\mathcal{C}^{\alpha}$. Since n is chosen to be minimal, we may suppose that σ is nondegenerate and that the boundary of σ already belongs to $\mathcal{C}^{\alpha}$.

Form a pushout diagram

$$
\begin{array}{ccc}
\mathcal{C}^0_{/\sigma} \star \partial\Delta^n & \longrightarrow & \mathcal{C}^{\alpha} \\
\downarrow & & \downarrow \\
\mathcal{C}^0_{/\sigma} \star \Delta^n & \longrightarrow & \mathcal{C}'.
\end{array}
$$

By construction there is an induced map $\mathcal{C}' \to \mathcal{C}$, which is easily shown to be a monomorphism of simplicial sets; we may therefore identify $\mathcal{E}'$ with its image in $\mathcal{C}$. Since $\mathcal{C}$ is minimal, we can apply Proposition 2.3.3.9 to deduce that $\mathcal{C}'$ is complete, so that $\mathcal{C}' = \mathcal{C}^{\alpha+1}$. Let G denote the composition

$$\mathcal{C}^0_{/\sigma} \star \Delta^n \to \mathcal{C} \xrightarrow{F} \mathcal{D}$$

and $G_{\partial} = G|\,\mathcal{C}^0_{/\sigma} \star \partial\Delta^n$. It follows that $\phi_{\alpha+1,\alpha}$ is a pullback of the induced map

$$\psi : \mathcal{D}_{G/} \to \mathcal{D}'_{pG/} \times_{\mathcal{D}'_{pG_{\partial}/}} \mathcal{D}_{G_{\partial}/}.$$

To complete the proof, it will suffice to show that ψ is a trivial fibration of simplicial sets.

Let $G_0 = G|\,\mathcal{C}^0_{/\sigma}$. Let $\mathcal{E} = \mathcal{D}_{G_0/}$, let $\mathcal{E}' = \mathcal{D}'_{p_0 G_0/}$, and let $q : \mathcal{E} \to \mathcal{E}'$ be the induced map. We can identify G with a map $\sigma' : \Delta^n \to \mathcal{E}$. Let $\sigma'_0 = \sigma'|\,\partial\Delta^n$. Then we wish to prove that the map

$$\psi' : \mathcal{E}_{\sigma'/} \to \mathcal{E}'_{q\sigma'/} \times_{\mathcal{E}'_{q\sigma'_0/}} \mathcal{E}_{q\sigma'_0/}$$

is a trivial fibration. Let $C = \sigma(0)$.

The projection $\mathcal{C}^0_{/\sigma} \to \mathcal{C}^0_{/C}$ is a trivial fibration of simplicial sets and therefore cofinal. Since F is a p-left Kan extension of F_0 at C, we conclude that $\sigma'(0)$ is a q-initial object of $\mathcal{E}$.

To prove that ψ is a trivial fibration, it will suffice to prove that ψ has the right lifting property with respect to the inclusion $\partial \Delta^m \subseteq \Delta^m$ for each $m \geq 0$. Unwinding the definitions, this amounts to the existence of a dotted arrow as indicated in the diagram

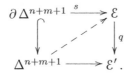

However, the map s carries the initial vertex of Δ^{n+m+1} to a vertex of $\mathcal{E}$ which is q-initial, so that the desired extension can be found. □

Proposition 4.3.2.8. *Let* $F : \mathcal{C} \to \mathcal{D}$ *be a functor between* ∞*-categories,* $p : \mathcal{D} \to \mathcal{D}'$ *a categorical fibration of* ∞*-categories, and* $\mathcal{C}^0 \subseteq \mathcal{C}^1 \subseteq \mathcal{C}$ *full subcategories. Suppose that* $F|\,\mathcal{C}^1$ *is a* p*-left Kan extension of* $F|\,\mathcal{C}^0$. *Then* F *is a* p*-left Kan extension of* $F|\,\mathcal{C}^1$ *if and only if* F *is a* p*-left Kan extension of* $F|\,\mathcal{C}^0$.

Proof. Let C be an object of $\mathcal{C}$; we will show that F is a p-left Kan extension of $F|\,\mathcal{C}^0$ at C if and only if F is a p-left Kan extension of $F|\,\mathcal{C}^1$ at C. Consider the composition

$$F^0_C : (\mathcal{C}^0_{/C})^{\triangleright} \subseteq (\mathcal{C}^1_{/C})^{\triangleright} \xrightarrow{F^1_C} \mathcal{D}.$$

We wish to show that F^0_C is a p-colimit diagram if and only if F^1_C is a p-colimit diagram. According to Lemma 4.3.2.7, it will suffice to show that $F^1_C |\,\mathcal{C}^1_{/C}$ is a left Kan extension of F^0_C. Let $f : C' \to C$ be an object of $\mathcal{C}^1_{/C}$. We wish to show that the composite map

$$(\mathcal{C}^0_{/f})^{\triangleright} \to (\mathcal{C}^0_{/C'})^{\triangleright} \xrightarrow{F^0_{C'}} \mathcal{D}$$

is a p-colimit diagram. Since the projection $\mathcal{C}^0_{/f} \to \mathcal{C}^0_{/C'}$ is cofinal (in fact, a trivial fibration), it will suffice to show that $F^0_{C'}$ is a p-colimit diagram (Proposition 4.3.1.7). This follows from our hypothesis that $F|\,\mathcal{C}^1$ is a p-left Kan extension of $F|\,\mathcal{C}^0$. □

Proposition 4.3.2.9. *Let* $F : \mathcal{C} \times \mathcal{C}' \to \mathcal{D}$ *denote a functor between* ∞*-categories,* $p : \mathcal{D} \to \mathcal{D}'$ *a categorical fibration of* ∞*-categories, and* $\mathcal{C}^0 \subseteq \mathcal{C}$ *a full subcategory. The following conditions are equivalent:*

(1) *The functor* F *is a* p*-left Kan extension of* $F|\,\mathcal{C}^0 \times \mathcal{C}'$.

(2) *For each object* $C' \in \mathcal{C}'$, *the induced functor* $F_{C'} : \mathcal{C} \times \{C'\} \to \mathcal{D}$ *is a* p*-left Kan extension of* $F_{C'}|\,\mathcal{C}^0 \times \{C'\}$.

Proof. It suffices to show that F is a p-left Kan extension of $F|\,\mathcal{C}^0 \times \mathcal{C}'$ at an object $(C, C') \in \mathcal{C} \times \mathcal{C}'$ if and only if $F_{C'}$ is a p-left Kan extension of $F_D|\,\mathcal{C}^0 \times \{D\}$ at C. This follows from the observation that the inclusion $\mathcal{C}^0_{/C} \times \{\mathrm{id}_{C'}\} \subseteq \mathcal{C}^0_{/C} \times \mathcal{C}'_{/C'}$ is cofinal (because $\mathrm{id}_{C'}$ is a final object of $\mathcal{C}'_{/C'}$). $\qquad\square$

Lemma 4.3.2.10. *Let $m \geq 0$, $n \geq 1$ be integers and let*

$$(\partial\,\Delta^m \times \Delta^n) \coprod_{\partial\,\Delta^m \times \partial\,\Delta^n} (\Delta^m \times \partial\,\Delta^n) \xrightarrow{\;f_0\;} X$$

be a diagram of simplicial sets, where p is an inner fibration and $f_0(0,0)$ is a p-initial vertex of X. Then there exists a morphism $f : \Delta^m \times \Delta^n \to X$ rendering the diagram commutative.

Proof. Choose a sequence of simplicial sets

$$(\partial\,\Delta^m \times \Delta^n) \coprod_{\partial\,\Delta^m \times \partial\,\Delta^n} (\Delta^m \times \partial\,\Delta^n) = Y(0) \subseteq \cdots \subseteq Y(k) = \Delta^m \times \Delta^n,$$

where each $Y(i+1)$ is obtained from $Y(i)$ by adjoining a single nondegenerate simplex whose boundary already lies in $Y(i)$. We prove by induction on i that f_0 can be extended to a map f_i such that the diagram

$$
\begin{array}{ccc}
Y(i) & \xrightarrow{\;f_i\;} & X \\
\downarrow & & \downarrow{\scriptstyle p} \\
\Delta^m \times \Delta^n & \longrightarrow & S
\end{array}
$$

is commutative. Having done so, we can then complete the proof by choosing $i = k$.

If $i = 0$, there is nothing to prove. Let us therefore suppose that f_i has been constructed and consider the problem of constructing f_{i+1} which extends f_i. This is equivalent to the lifting problem

$$
\begin{array}{ccc}
\partial\,\Delta^r & \xrightarrow{\;\sigma_0\;} & X \\
\downarrow & \nearrow{\scriptstyle \sigma} & \downarrow{\scriptstyle p} \\
\Delta^r & \longrightarrow & S.
\end{array}
$$

It now suffices to observe that where $r > 0$ and $\sigma_0(0) = f_0(0,0)$ is a p-initial vertex of X (since every simplex of $\Delta^m \times \Delta^n$ which violates one of these conditions already belongs to $Y(0)$). $\qquad\square$

Lemma 4.3.2.11. *Suppose we are given a diagram of simplicial sets*

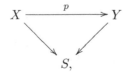

$$
\begin{array}{ccc}
X & \xrightarrow{\;p\;} & Y \\
& \searrow \quad \swarrow & \\
& S, &
\end{array}
$$

where p is an inner fibration. Let K be a simplicial set, let $q_S \in \mathrm{Map}_S(K \times S, X)$, and let $q'_S = p \circ q_S$. Then the induced map

$$p' : X^{q_S/} \to Y^{q'_S/}$$

is an inner fibration (where the above simplicial sets are defined as in §4.2.2).

Proof. Unwinding the definitions, we see that every lifting problem

is equivalent to a lifting problem

$$(A \times (K \diamond \Delta^0)) \coprod_{A \times K} (B \times K) \longrightarrow X$$
$$\downarrow{i'} \qquad\qquad\qquad\qquad \downarrow{p}$$
$$B \times (K \diamond \Delta^0) \longrightarrow Y.$$

We wish to show that this lifting problem has a solution provided that i is inner anodyne. Since p is an inner fibration, it will suffice to prove that i' is inner anodyne, which follows from Corollary 2.3.2.4. $\square$

Lemma 4.3.2.12. *Consider a diagram of ∞-categories*

$$\mathcal{C} \to \mathcal{D}' \xleftarrow{p} \mathcal{D},$$

where p is an categorical fibration. Let $\mathcal{C}^0 \subseteq \mathcal{C}$ be a full subcategory. Suppose we are given $n > 0$ and a commutative diagram

$$\begin{array}{ccc} \partial \Delta^n & \xrightarrow{f_0} & \mathrm{Map}_{\mathcal{D}'}(\mathcal{C}, \mathcal{D}) \\ \downarrow & \nearrow{f} & \downarrow \\ \Delta^n & \xrightarrow{g} & \mathrm{Map}_{\mathcal{D}'}(\mathcal{C}^0, \mathcal{D}) \end{array}$$

with the property that the functor $F : \mathcal{C} \to \mathcal{D}$, determined by evaluating f_0 at the vertex $\{0\} \subseteq \partial \Delta^n$, is a p-left Kan extension of $F|\mathcal{C}^0$. Then there exists a dotted arrow f rendering the diagram commutative.

Proof. The proof uses the same strategy as that of Lemma 4.3.2.7. Using Lemma 4.3.2.6 and Proposition A.2.3.1, we may replace $\mathcal{C}$ by a minimal model and thereby assume that $\mathcal{C}$ is minimal. As in the proof of Lemma 4.3.2.7, let us call a simplicial subset $\mathcal{E} \subseteq \mathcal{C}$ *complete* if it has the following property: for any simplex $\sigma : \Delta^n \to \mathcal{C}$, if $\sigma|\Delta^{\{0,\ldots,i\}}$ factors through $\mathcal{C}^0$ and $\sigma|\Delta^{\{i+1,\ldots,n\}}$ factors through $\mathcal{E}$, then σ factors through $\mathcal{E}$. Let P denote the partially ordered set of pairs $(\mathcal{E}, f_\mathcal{E})$, where $\mathcal{E} \subseteq \mathcal{C}$ is complete and $f_\mathcal{E}$ is a

map rendering commutative the diagram

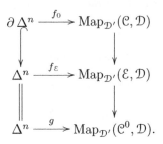

We partially order P as follows: $(\mathcal{E}, f_{\mathcal{E}}) \leq (\mathcal{E}', f_{\mathcal{E}'})$ if $\mathcal{E} \subseteq \mathcal{E}'$ and $f_{\mathcal{E}} = f_{\mathcal{E}'}|\mathcal{E}$. Using Zorn's lemma, we deduce that P has a maximal element $(\mathcal{E}, f_{\mathcal{E}})$. If $\mathcal{E} = \mathcal{C}$, we may take $f = f_{\mathcal{E}}$, and the proof is complete. Otherwise, choose a simplex $\sigma : \Delta^m \to \mathcal{C}$ which does not belong to $\mathcal{E}$, where m is as small as possible. It follows that σ is nondegenerate and that the boundary of σ belongs to $\mathcal{E}$. Form a pushout diagram

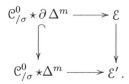

As in the proof of Lemma 4.3.2.7, we may identify $\mathcal{E}'$ with a complete simplicial subset of $\mathcal{C}$, which strictly contains $\mathcal{E}$. Since $(\mathcal{E}, f_{\mathcal{E}})$ is maximal, we conclude that $f_{\mathcal{E}}$ does not extend to $\mathcal{E}'$. Consequently, we deduce that there does not exist a dotted arrow rendering the diagram

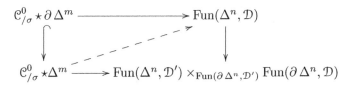

commutative. Let $q : \mathcal{C}^0_{/\sigma} \to \mathrm{Fun}(\Delta^n, \mathcal{D})$ be the restriction of the upper horizontal map and let $q' : \mathcal{C}^0_{/\sigma} \to \mathrm{Fun}(\Delta^n, \mathcal{D}')$, $q_{\partial} : \mathcal{C}^0_{/\sigma} \to \mathrm{Fun}(\partial \Delta^n, \mathcal{D})$, and $q'_{\partial} : \mathcal{C}^0_{/\sigma} \to \mathrm{Fun}(\partial \Delta^n, \mathcal{D}')$ be defined by composition with q. It follows that there exists no solution to the associated lifting problem

$$
\begin{array}{ccc}
\partial \Delta^m & \longrightarrow & \mathrm{Fun}(\Delta^n, \mathcal{D})_{q/} \\
\downarrow & \nearrow & \downarrow \\
\Delta^m & \longrightarrow & \mathrm{Fun}(\Delta^n, \mathcal{D}')_{q'/} \times_{\mathrm{Fun}(\partial \Delta^n, \mathcal{D}')_{q'_{\partial}/}} \mathrm{Fun}(\partial \Delta^n, \mathcal{D})_{q_{\partial}/}.
\end{array}
$$

Applying Proposition A.2.3.1, we deduce also the insolubility of the equivalent lifting problem

$$
\begin{array}{ccc}
\partial \Delta^m & \longrightarrow & \mathrm{Fun}(\Delta^n, \mathcal{D})^{q/} \\
\downarrow & \nearrow & \downarrow \\
\Delta^m & \longrightarrow \mathrm{Fun}(\Delta^n, \mathcal{D}')^{q'/} \times_{\mathrm{Fun}(\partial \Delta^n, \mathcal{D}')^{q'_{\partial}/}} \mathrm{Fun}(\partial \Delta^n, \mathcal{D})^{q_{\partial}/}. &
\end{array}
$$

Let q_{Δ^n} denote the map $\mathcal{C}^0_{/\sigma} \times \Delta^n \to \mathcal{D} \times \Delta^n$ determined by q and let $\mathcal{X} = (\mathcal{D} \times \Delta^n)^{q_{\Delta^n}/}$ be the simplicial set constructed in §4.2.2. Let $q'_{\Delta^n} : \mathcal{C}^0_{/\sigma} \times \Delta^n \to \mathcal{D}' \times \Delta^n$ and $\mathcal{X}' = (\mathcal{D}' \times \Delta^n)^{q'_{\Delta^n}/}$ be defined similarly. We have natural isomorphisms

$$
\mathrm{Fun}(\Delta^n, \mathcal{D})^{q/} \simeq \mathrm{Map}_{\Delta^n}(\Delta^n, \mathcal{X})
$$

$$
\mathrm{Fun}(\partial \Delta^n, \mathcal{D})^{q_{\partial}/} \simeq \mathrm{Map}_{\Delta^n}(\partial \Delta^n, \mathcal{X})
$$

$$
\mathrm{Fun}(\Delta^n, \mathcal{D}')^{q'/} \simeq \mathrm{Map}_{\Delta^n}(\Delta^n, \mathcal{X}')
$$

$$
\mathrm{Fun}(\partial \Delta^n, \mathcal{D}')^{q'_{\partial}/} \simeq \mathrm{Map}_{\Delta^n}(\partial \Delta^n, \mathcal{X}').
$$

These identifications allow us to reformulate our insoluble lifting problem once more:

$$
\begin{array}{ccc}
(\partial \Delta^m \times \Delta^n) \coprod_{\partial \Delta^m \times \partial \Delta^n} (\Delta^m \times \partial \Delta^n) & \xrightarrow{\;g_0\;} & \mathcal{X} \\
\downarrow & \overset{g}{\nearrow} & \downarrow \psi \\
\Delta^m \times \Delta^n & \longrightarrow & \mathcal{X}'.
\end{array}
$$

We have a commutative diagram

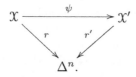

Proposition 4.2.2.4 implies that r and r' are Cartesian fibrations and that ψ carries r-Cartesian edges to r'-Cartesian edges. Lemma 4.3.2.11 implies that ψ is an inner fibration. Let $\psi_0 : \mathcal{X}_{\{0\}} \to \mathcal{X}'_{\{0\}}$ be the diagram induced by taking the fibers over the vertex $\{0\} \subseteq \Delta^n$. We have a commutative diagram

$$
\begin{array}{ccc}
\mathcal{D}_{\mathcal{C}^0_{/\sigma(0)}/} & \longleftarrow \mathcal{D}_{\mathcal{C}^0_{/\sigma}} \longrightarrow & \mathcal{X}_{\{0\}} \\
\downarrow \theta & \downarrow & \downarrow \psi_0 \\
\mathcal{D}'_{\mathcal{C}^0_{/\sigma(0)}/} & \longleftarrow \mathcal{D}'_{\mathcal{C}^0_{/\sigma}/} \longrightarrow & \mathcal{X}'_{\{0\}}
\end{array}
$$

in which the horizontal arrows are categorical equivalences. We can lift $g_0(0,0) \in \mathcal{X}'_{\{0\}}$ to a vertex of $\mathcal{D}_{\mathcal{C}^0_{/\sigma}/}$ whose image in $\mathcal{D}_{\mathcal{C}^0_{/\sigma(0)}/}$ is θ-initial (by virtue of our assumption that F is a p-left Kan extension of $F|\,\mathcal{C}^0$). It follows that $g_0(0,0)$ is ψ_0-initial when regarded as a vertex of $\mathfrak{X}_{\{0\}}$. Applying Proposition 4.3.1.14, we deduce that $g_0(0,0)$ is ψ-initial when regarded as a vertex of $\mathfrak{X}$. Lemma 4.3.2.10 now guarantees the existence of the dotted arrow g, contradicting the maximality of $(\mathcal{E}, f_\mathcal{E})$. $\square$

The following result addresses the existence problem for left Kan extensions:

Lemma 4.3.2.13. *Suppose we are given a diagram of ∞-categories*

$$\begin{array}{ccc} \mathcal{C}^0 & \xrightarrow{\;F_0\;} & \mathcal{D} \\ \cap\,\Big\downarrow & \nearrow{\scriptstyle F} & \Big\downarrow{\scriptstyle p} \\ \mathcal{C} & \xrightarrow{\quad} & \mathcal{D}', \end{array}$$

where p is a categorical fibration and the left vertical arrow is the inclusion of a full subcategory. The following conditions are equivalent:

(1) *There exists a functor $F : \mathcal{C} \to \mathcal{D}$ rendering the diagram commutative, such that F is a p-left Kan extension of F_0.*

(2) *For every object $C \in \mathcal{C}$, the diagram given by the composition*

$$\mathcal{C}^0_{/C} \to \mathcal{C}^0 \xrightarrow{\;F_0\;} \mathcal{D}$$

admits a p-colimit.

Proof. It is clear that (1) implies (2). Let us therefore suppose that (2) is satisfied; we wish to prove that F_0 admits a left Kan extension. We will follow the basic strategy used in the proofs of Lemmas 4.3.2.7 and 4.3.2.12. Using Proposition A.2.3.1 and Lemma 4.3.2.6, we can replace the inclusion $\mathcal{C}^0 \subseteq \mathcal{C}$ by any categorically equivalent inclusion $\mathcal{C}'^0 \subseteq \mathcal{C}'$. Using Proposition 2.3.3.8, we can choose $\mathcal{C}'$ to be a minimal model for $\mathcal{C}$; we thereby reduce to the case where $\mathcal{C}$ is itself a minimal ∞-category.

We will say that a simplicial subset $\mathcal{E} \subseteq \mathcal{C}$ is *complete* if it has the following property: for any simplex $\sigma : \Delta^n \to \mathcal{C}$, if $\sigma|\Delta^{\{0,\dots,i\}}$ factors through $\mathcal{C}^0$ and $\sigma|\Delta^{\{i+1,\dots,n\}}$ factors through $\mathcal{E}$, then σ factors through $\mathcal{E}$. Note that if $\mathcal{E}$ is complete, then $\mathcal{C}^0 \subseteq \mathcal{E}$. Let P be the set of all pairs $(\mathcal{E}, f_\mathcal{E})$ such that $\mathcal{E} \subseteq \mathcal{C}$ is complete, $f_\mathcal{E}$ is a map of simplicial sets which fits into a commutative diagram

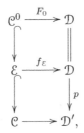

and every object $C \in \mathcal{E}$, the composite map

$$(\mathcal{C}^0_{/C})^{\triangleright} \subseteq (\mathcal{E}_{/C})^{\triangleright} \to \mathcal{E} \xrightarrow{f_{\mathcal{E}}} \mathcal{D}$$

is a p-colimit diagram. We view P as a partially ordered set, with $(\mathcal{E}, f_{\mathcal{E}}) \leq (\mathcal{E}', f_{\mathcal{E}'})$ if $\mathcal{E} \subseteq \mathcal{E}'$ and $f_{\mathcal{E}'}|\mathcal{E} = f_{\mathcal{E}}$. This partially ordered set satisfies the hypotheses of Zorn's lemma and therefore contains a maximal element which we will denote by $(\mathcal{E}, f_{\mathcal{E}})$. If $\mathcal{E} = \mathcal{C}$, then $f_{\mathcal{E}}$ is a p-left Kan extension of F_0, and the proof is complete.

Suppose that $\mathcal{E} \neq \mathcal{C}$. Then there is a simplex $\sigma : \Delta^n \to \mathcal{C}$ which does not factor through $\mathcal{E}$; we choose such a simplex where n is as small as possible. The minimality of n guarantees that σ is nondegenerate, that $\sigma|\partial \Delta^n$ factors through $\mathcal{E}$, and (if $n > 0$) that $\sigma(0) \notin \mathcal{C}^0$. We now form a pushout diagram

$$
\begin{array}{ccc}
\mathcal{C}^0_{/\sigma} \star \partial \Delta^n & \longrightarrow & \mathcal{E} \\
\downarrow & & \downarrow \\
\mathcal{C}^0_{/\sigma} \star \Delta^n & \longrightarrow & \mathcal{E}'.
\end{array}
$$

This diagram induces a map $\mathcal{E}' \to \mathcal{C}$, which is easily shown to be a monomorphism of simplicial sets; we may therefore identify $\mathcal{E}'$ with its image in $\mathcal{C}$. Since $\mathcal{C}$ is minimal, we can apply Proposition 2.3.3.9 to deduce that $\mathcal{E}' \subseteq \mathcal{C}$ is complete. Since $(\mathcal{E}, F_{\mathcal{E}}) \in P$ is maximal, it follows that we cannot extend $F_{\mathcal{E}}$ to a functor $F_{\mathcal{E}'} : \mathcal{E}' \to \mathcal{D}$ such that $(\mathcal{E}', F_{\mathcal{E}'}) \in P$.

Let q denote the composition

$$\mathcal{C}^0_{/\sigma} \to \mathcal{C}^0 \xrightarrow{F_0} \mathcal{D}.$$

The map $f_{\mathcal{E}}$ determines a commutative diagram

$$
\begin{array}{ccc}
\partial \Delta^n & \xrightarrow{g_0} & \mathcal{D}_{q/} \\
\downarrow & \nearrow^{g} & \downarrow{\scriptstyle p'} \\
\Delta^n & \longrightarrow & \mathcal{D}'_{pq/}.
\end{array}
$$

Extending $f_{\mathcal{E}}$ to a map $f_{\mathcal{E}'}$ such that $(\mathcal{E}', f_{\mathcal{E}'}) \in P$ is equivalent to producing a morphism $g : \Delta^n \to \mathcal{D}_{q/}$, rendering the above diagram commutative. which is a p-colimit of q if $n = 0$. In the case where $n = 0$, the existence of such an extension follows from assumption (2). If $n > 0$, let $C = \sigma(0)$; then the projection $\mathcal{C}^0_{/\sigma} \to \mathcal{C}^0_{/C}$ is a trivial fibration of ∞-categories and q factors as a composition

$$\mathcal{C}^0_{/\sigma} \to \mathcal{C}^0_{/C} \xrightarrow{q'} \mathcal{D}.$$

We obtain therefore a commutative diagram

$$
\begin{array}{ccc}
\mathcal{D}_{q/} & \xrightarrow{r} & \mathcal{D}_{q'/} \\
\downarrow{\scriptstyle p'} & & \downarrow{\scriptstyle p''} \\
\mathcal{D}'_{pq/} & \longrightarrow & \mathcal{D}'_{pq'/},
\end{array}
$$

where the horizontal arrows are categorical equivalences. Since $(\mathcal{E}, f_{\mathcal{E}}) \in P$, $(r \circ g_0)(0)$ is a p''-initial vertex of $\mathcal{D}_{q'/}$. Applying Proposition 4.3.1.6, we conclude that $g_0(0)$ is a p'-initial vertex of $\mathcal{D}_{q/}$, which guarantees the existence of the desired extension g. This contradicts the maximality of $(\mathcal{E}, f_{\mathcal{E}})$ and completes the proof. $\qquad\square$

Corollary 4.3.2.14. *Let* $p : \mathcal{D} \to \mathcal{E}$ *be a coCartesian fibration of* ∞-*categories. Suppose that each fiber of* p *admits small colimits and that for every morphism* $E \to E'$ *in* $\mathcal{E}$ *the associated functor* $\mathcal{D}_E \to \mathcal{D}_{E'}$ *preserves small colimits. Let* $\mathcal{C}$ *be a small* ∞-*category and* $\mathcal{C}^0 \subseteq \mathcal{C}$ *a full subcategory. Then every functor* $F_0 : \mathcal{C}^0 \to \mathcal{D}$ *admits a left Kan extension relative to* p.

Proof. This follows easily from Lemma 4.3.2.13 and Corollary 4.3.1.11. $\quad\square$

Combining Lemmas 4.3.2.12 and 4.3.2.13, we deduce the following result:

Proposition 4.3.2.15. *Suppose we are given a diagram of* ∞-*categories*

$$\mathcal{C} \to \mathcal{D}' \xleftarrow{p} \mathcal{D},$$

where p *is a categorical fibration. Let* $\mathcal{C}^0$ *be a full subcategory of* $\mathcal{C}$. *Let* $\mathcal{K} \subseteq \mathrm{Map}_{\mathcal{D}'}(\mathcal{C}, \mathcal{D})$ *be the full subcategory spanned by those functors* $F : \mathcal{C} \to \mathcal{D}$ *which are* p-*left Kan extensions of* $F| \mathcal{C}^0$. *Let* $\mathcal{K}' \subseteq \mathrm{Map}_{\mathcal{D}'}(\mathcal{C}^0, \mathcal{D})$ *be the full subcategory spanned by those functors* $F_0 : \mathcal{C}^0 \to \mathcal{D}$ *with the property that, for each object* $C \in \mathcal{C}$, *the induced diagram* $\mathcal{C}^0_{/C} \to \mathcal{D}$ *has a* p-*colimit. Then the restriction functor* $\mathcal{K} \to \mathcal{K}'$ *is a trivial fibration of simplicial sets.*

Corollary 4.3.2.16. *Suppose we are given a diagram of* ∞-*categories*

$$\mathcal{C} \to \mathcal{D}' \xleftarrow{p} \mathcal{D},$$

where p *is a categorical fibration. Let* $\mathcal{C}^0$ *be a full subcategory of* $\mathcal{C}$. *Suppose further that, for every functor* $F_0 \in \mathrm{Map}_{\mathcal{D}'}(\mathcal{C}^0, \mathcal{D})$, *there exists a functor* $F \in \mathrm{Map}_{\mathcal{D}'}(\mathcal{C}, \mathcal{D})$ *which is a* p-*left Kan extension of* F_0. *Then the restriction map* $i^* : \mathrm{Map}_{\mathcal{D}'}(\mathcal{C}, \mathcal{D}) \to \mathrm{Map}_{\mathcal{D}'}(\mathcal{C}^0, \mathcal{D})$ *admits a section* $i_!$ *whose essential image consists of of precisely those functors* F *which are* p-*left Kan extensions of* $F| \mathcal{C}^0$.

In the situation of Corollary 4.3.2.16, we will refer to $i_!$ as a *left Kan extension functor*. We note that Proposition 4.3.2.15 proves not only the existence of $i_!$ but also its uniqueness up to homotopy (the collection of all such functors is parametrized by a contractible Kan complex). The following characterization of $i_!$ gives a second explanation for its uniqueness:

Proposition 4.3.2.17. *Suppose we are given a diagram of* ∞-*categories*

$$\mathcal{C} \to \mathcal{D}' \xleftarrow{p} \mathcal{D},$$

where p *is a categorical fibration. Let* $i : \mathcal{C}^0 \subseteq \mathcal{C}$ *be the inclusion of a full subcategory and suppose that every functor* $F_0 \in \mathrm{Map}_{\mathcal{D}'}(\mathcal{C}^0, \mathcal{D})$ *admits a* p-*left Kan extension. Then the left Kan extension functor* $i_! : \mathrm{Map}_{\mathcal{D}'}(\mathcal{C}^0, \mathcal{D}) \to \mathrm{Map}_{\mathcal{D}'}(\mathcal{C}, \mathcal{D})$ *is a left adjoint to the restriction functor* $i^* : \mathrm{Map}_{\mathcal{D}'}(\mathcal{C}, \mathcal{D}) \to \mathrm{Map}_{\mathcal{D}'}(\mathcal{C}^0, \mathcal{D})$.

Proof. Since $i^* \circ i_!$ is the identity functor on $\mathrm{Map}_{\mathcal{D}'}(\mathcal{C}^0, \mathcal{D})$, there is an obvious candidate for the unit

$$u : \mathrm{id} \to i^* \circ i_!$$

of the adjunction: namely, the identity. According to Proposition 5.2.2.8, it will suffice to prove that for every $F \in \mathrm{Map}_{\mathcal{D}'}(\mathcal{C}^0, \mathcal{D})$ and $G \in \mathrm{Map}_{\mathcal{D}'}(\mathcal{C}, \mathcal{D})$, the composite map

$$\mathrm{Map}_{\mathrm{Map}_{\mathcal{D}'}(\mathcal{C}, \mathcal{D})}(i_! F, G) \to \mathrm{Map}_{\mathrm{Map}_{\mathcal{D}'}(\mathcal{C}^0, \mathcal{D})}(i^* i_! F, i^* G)$$
$$\xrightarrow{u} \mathrm{Map}_{\mathrm{Map}_{\mathcal{D}'}(\mathcal{C}^0, \mathcal{D})}(F, i^* G)$$

is an isomorphism in the homotopy category $\mathcal{H}$. This morphism in $\mathcal{H}$ is represented by the restriction map

$$\mathrm{Hom}^{\mathrm{R}}_{\mathrm{Map}_{\mathcal{D}'}(\mathcal{C}, \mathcal{D})}(i_! F, G) \to \mathrm{Hom}^{\mathrm{R}}_{\mathrm{Map}_{\mathcal{D}'}(\mathcal{C}^0, \mathcal{D})}(F, i^* G),$$

which is a trivial fibration by Lemma 4.3.2.12. $\square$

Remark 4.3.2.18. Throughout this section we have focused our attention on the theory of (relative) *left* Kan extensions. There is an entirely dual theory of *right* Kan extensions in the ∞-categorical setting, which can be obtained from the theory of left Kan extensions by passing to opposite ∞-categories.

4.3.3 Kan Extensions along General Functors

Our goal in this section is to generalize the theory of Kan extensions to the case where the change of diagram category is not necessarily given by a fully faithful inclusion $\mathcal{C}^0 \subseteq \mathcal{C}$. As in §4.3.2, we will discuss only the theory of *left* Kan extensions; a dual theory of right Kan extensions can be obtained by passing to opposite ∞-categories.

The ideas introduced in this section are relatively elementary extensions of the ideas of §4.3.2. However, we will encounter a new complication. Let $\delta : \mathcal{C} \to \mathcal{C}'$ be a map of diagram ∞-categories, $f : \mathcal{C} \to \mathcal{D}$ a functor, and $\delta_!(f) : \mathcal{C}' \to \mathcal{D}$ its left Kan extension along δ (to be defined below). Then one does not generally expect $\delta^* \delta_!(f)$ to be equivalent to the original functor f. Instead, one has only a unit transformation $f \to \delta^* \delta_!(f)$. To set up the theory, this unit transformation must be taken as part of the data. Consequently, the theory of Kan extensions in general requires more elaborate notation and terminology than the special case treated in §4.3.2. We will compensate for this by considering only the case of *absolute* left Kan extensions. It is straightforward to set up a relative theory as in §4.3.2, but we will not need such a theory in this book.

Definition 4.3.3.1. Let $\delta : K \to K'$ be a map of simplicial sets, let $\mathcal{D}$ be an ∞-category and let $f : K \to \mathcal{D}$ be a diagram. A *left extension of f along δ* consists of a map $f' : K' \to \mathcal{D}$ and a morphism $f \to f' \circ \delta$ in the ∞-category $\mathrm{Fun}(K, \mathcal{D})$.

Equivalently, we may view a left extension of $f : K \to \mathcal{D}$ along $\delta : K \to K'$ as a map $F : M^{op}(\delta) \to \mathcal{D}$ such that $F|K = f$, where $M^{op}(\delta) = M(\delta^{op})^{op} = (K \times \Delta^1) \coprod_{K \times \{1\}} K'$ denotes the mapping cylinder of δ.

Definition 4.3.3.2. Let $\delta : K \to K'$ be a map of simplicial sets, and let $F : M^{op}(\delta) \to \mathcal{D}$ be a diagram in an ∞-category $\mathcal{D}$ (which we view as a left extension of $f = F|K$ along δ). We will say that F is a *left Kan extension of f along δ* if there exists a commutative diagram

$$
\begin{array}{ccccc}
M^{op}(\delta) & \xrightarrow{F''} & \mathcal{K} & \xrightarrow{F'} & \mathcal{D} \\
 & \searrow & \downarrow{\scriptstyle p} & & \\
 & & \Delta^1 & &
\end{array}
$$

where F'' is a categorical equivalence, $\mathcal{K}$ is an ∞-category, $F = F' \circ F''$, and F' is a left Kan extension of $F'|\, \mathcal{K} \times_{\Delta^1} \{0\}$.

Remark 4.3.3.3. In the situation of Definition 4.3.3.2, the map $p : \mathcal{K} \to \Delta^1$ is *automatically* a coCartesian fibration. To prove this, choose a factorization

$$
M(\delta^{op})^\flat \xrightarrow{i} (\mathcal{K}')^\sharp \to (\Delta^1)^\sharp,
$$

where i is marked anodyne and $\mathcal{K}' \to \Delta^1$ is a Cartesian fibration. Then i is a quasi-equivalence, so that Proposition 3.2.2.7 implies that $M(\delta^{op}) \to \mathcal{K}'$ is a categorical equivalence. It follows that $\mathcal{K}$ is equivalent to $(\mathcal{K}')^{op}$ (via an equivalence which respects the projection to Δ^1), so that the projection p is a coCartesian fibration.

The following result asserts that the condition of Definition 4.3.3.2 is essentially independent of the choice of $\mathcal{K}$.

Proposition 4.3.3.4. *Let $\delta : K \to K'$ be a map of simplicial sets and let $F : M^{op}(\delta) \to \mathcal{D}$ be a diagram in an ∞-category $\mathcal{D}$ which is a left Kan extension along δ. Let*

be a diagram where F'' is both a cofibration and a categorical equivalence of simplicial sets. Then $F = F' \circ F''$ for some map $F' : \mathcal{K} \to \mathcal{D}$ which is a left Kan extension of $F'|\, \mathcal{K} \times_{\Delta^1} \{0\}$.

Proof. By hypothesis, there exists a commutative diagram

$$
\begin{array}{ccccc}
M^{op}(\delta) & \xrightarrow{G''} & \mathcal{K}' & \xrightarrow{G'} & \mathcal{D} \\
\downarrow{\scriptstyle F''} & \nearrow{\scriptstyle r} & \downarrow{\scriptstyle q} & & \\
\mathcal{K} & \xrightarrow[p]{} & \Delta^1, & &
\end{array}
$$

where $\mathcal{K}'$ is an ∞-category, $F = G' \circ G''$, G'' is a categorical equivalence, and G' is a left Kan extension of $G'|\mathcal{K}' \times_{\Delta^1}\{0\}$. Since $\mathcal{K}'$ is an ∞-category, there exists a map r as indicated in the diagram such that $G'' = r \circ F''$. We note that r is a categorical equivalence, so that the commutativity of the lower triangle $p = q \circ r$ follows automatically. We now define $F' = G' \circ r$ and note that part (2) of Lemma 4.3.2.6 implies that F' is a left Kan extension of $F'|\mathcal{K} \times_{\Delta^1}\{0\}$. $\square$

We have now introduced two different definitions of left Kan extensions: Definition 4.3.2.2 which applies in the situation of an inclusion $\mathcal{C}^0 \subseteq \mathcal{C}$ of a full subcategory into an ∞-category $\mathcal{C}$, and Definition 4.3.3.2 which applies in the case of a general map $\delta : K \to K'$ of simplicial sets. These two definitions are essentially the same. More precisely, we have the following assertion:

Proposition 4.3.3.5. *Let $\mathcal{C}$ and $\mathcal{D}$ be ∞-categories and let $\delta : \mathcal{C}^0 \to \mathcal{C}$ denote the inclusion of a full subcategory.*

(1) *Let $f : \mathcal{C} \to \mathcal{D}$ be a functor and f_0 its restriction to $\mathcal{C}^0$, so that (f, id_{f_0}) can be viewed as a left extension of f_0 along δ. Then (f, id_{f_0}) is a left Kan extension of f_0 along δ if and only if f is a left Kan extension of f_0.*

(2) *A functor $f_0 : \mathcal{C}^0 \to \mathcal{D}$ has a left Kan extension if and only if it has a left Kan extension along δ.*

Proof. Let $\mathcal{K}$ denote the full subcategory of $\mathcal{C} \times \Delta^1$ spanned by the objects $(C, \{i\})$, where either $C \in \mathcal{C}^0$ or $i = 1$, so that we have inclusions

$$M^{op}(\delta) \subseteq \mathcal{K} \subseteq \mathcal{C} \times \Delta^1.$$

To prove (1), suppose that $f : \mathcal{C} \to \mathcal{D}$ is a left Kan extension of $f_0 = f|\mathcal{C}^0$ and let F denote the composite map

$$\mathcal{K} \subseteq \mathcal{C} \times \Delta^1 \to \mathcal{C} \xrightarrow{f} \mathcal{D}.$$

It follows immediately that F is a left Kan extension of $F|\mathcal{C}^0 \times \{0\}$, so that $F|M^{op}(\delta)$ is a left Kan extension of f_0 along δ.

To prove (2), we observe that the "only if" direction follows from (1); the converse follows from the existence criterion of Lemma 4.3.2.13. $\square$

Suppose that $\delta : K^0 \to K^1$ is a map of simplicial sets, that $\mathcal{D}$ is an ∞-category, and that every diagram $K^0 \to \mathcal{D}$ admits a left Kan extension along δ. Choose a diagram

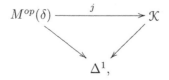

where j is inner anodyne and $\mathcal{K}$ is an ∞-category, which we regard as a correspondence from $\mathcal{K}^0 = \mathcal{K} \times_{\Delta^1} \{0\}$ to $\mathcal{K}^1 = \mathcal{K} \times_{\Delta^1} \{1\}$. Let $\mathcal{C}$ denote the full subcategory of $\operatorname{Fun}(\mathcal{K}, \mathcal{D})$ spanned by those functors $F : \mathcal{K} \to \mathcal{D}$ such that F is a left Kan extension of $F_0 = F | \mathcal{K}^0$. The restriction map $p : \mathcal{C} \to \operatorname{Fun}(K^0, \mathcal{D})$ can be written as a composition of $\mathcal{C} \to \mathcal{D}^{\mathcal{K}^0}$ (a trivial fibration by Proposition 4.3.2.15) and $\operatorname{Fun}(\mathcal{K}^0, \mathcal{D}) \to \operatorname{Fun}(K^0, \mathcal{D})$ (a trivial fibration since $K^0 \to \mathcal{K}^0$ is inner anodyne) and is therefore a trivial fibration. Let $\overline{\delta}_!$ be the composition of a section of p with the restriction map $\mathcal{C} \subseteq \operatorname{Fun}(\mathcal{K}, \mathcal{D}) \to \operatorname{Fun}(M^{op}(\delta), \mathcal{D})$ and let $\delta_!$ denote the composition of $\overline{\delta}_!$ with the restriction map $\operatorname{Fun}(M^{op}(\delta), \mathcal{D}) \to \operatorname{Fun}(K^1, \mathcal{D})$. Then $\overline{\delta}_!$ and $\delta_!$ are well-defined up to equivalence, at least once $\mathcal{K}$ has been fixed (independence of the choice of $\mathcal{K}$ will follow from the characterization given in Proposition 4.3.3.7). We will abuse terminology by referring to *both* $\overline{\delta}_!$ and $\delta_!$ as *left Kan extensions along* δ (it should be clear from the context which of these functors is meant in a given situation). We observe that $\overline{\delta}_!$ assigns to each object $f_0 : K^0 \to \mathcal{D}$ a left Kan extension of f_0 along δ.

Example 4.3.3.6. Let $\mathcal{C}$ and $\mathcal{D}$ be ∞-categories and let $i : \mathcal{C}^0 \to \mathcal{C}$ be the inclusion of a full subcategory. Suppose that $i_! : \operatorname{Fun}(\mathcal{C}^0, \mathcal{D}) \to \operatorname{Fun}(\mathcal{C}, \mathcal{D})$ is a section of i^*, which satisfies the conclusion of Corollary 4.3.2.16. Then $i_!$ is a left Kan extension along i in the sense defined above; this follows easily from Proposition 4.3.3.5.

Left Kan extension functors admit the following characterization:

Proposition 4.3.3.7. *Let $\delta : K^0 \to K^1$ be a map of simplicial sets, let $\mathcal{D}$ be an ∞-category, let $\delta^* : \operatorname{Fun}(K^1, \mathcal{D}) \to \operatorname{Fun}(K^0, \mathcal{D})$ be the restriction functor, and let $\delta_! : \operatorname{Fun}(K^0, \mathcal{D}) \to \operatorname{Fun}(K^1, \mathcal{D})$ be a functor of left Kan extension along δ. Then $\delta_!$ is a left adjoint of δ^*.*

Proof. The map δ can be factored as a composition

$$K^0 \xrightarrow{i} M^{op}(\delta) \xrightarrow{r} K^1$$

where r denotes the natural retraction of $M^{op}(\delta)$ onto K^1. Consequently, $\delta^* = i^* \circ r^*$. Proposition 4.3.2.17 implies that the left Kan extension functor $\overline{\delta}_!$ is a left adjoint to i^*. By Proposition 5.2.2.6, it will suffice to prove that r^* is a right adjoint to the restriction functor $j^* : \operatorname{Fun}(M^{op}(\delta), \mathcal{D}) \to \operatorname{Fun}(K^1, \mathcal{D})$. Using Corollary 2.4.7.12, we deduce that j^* is a coCartesian fibration. Moreover, there is a simplicial homotopy $\operatorname{Fun}(M^{op}(\delta), \mathcal{D}) \times \Delta^1 \to \operatorname{Fun}(M^{op}(\delta), \mathcal{D})$ from the identity to $r^* \circ j^*$, which is a fiberwise homotopy over $\operatorname{Fun}(K^1, \mathcal{D})$. It follows that for every object F of $\operatorname{Fun}(K^1, \mathcal{D})$, $r^* F$ is a final object of the ∞-category $\operatorname{Fun}(M^{op}(\delta), \mathcal{D}) \times_{\operatorname{Fun}(K^1, \mathcal{D})} \{F\}$. Applying Proposition 5.2.4.3, we deduce that r^* is right adjoint to j^*, as desired. $\square$

Let $\delta : K^0 \to K^1$ be a map of simplicial sets and $\mathcal{D}$ an ∞-category for which the left Kan extension $\delta_! : \operatorname{Fun}(K^0, \mathcal{D}) \to \delta_! \operatorname{Fun}(K^1, \mathcal{D})$ is defined. In general, the terminology "Kan extension" is perhaps somewhat unfortunate:

if $F : K^0 \to \mathcal{D}$ is a diagram, then $\delta^* \delta_! F$ need not be equivalent to F. If δ is fully faithful, then the unit map $F \to \delta^* \delta_! F$ is an equivalence: this follows from Proposition 4.3.3.5. We will later need the following more precise assertion:

Proposition 4.3.3.8. *Let $\delta : \mathcal{C}^0 \to \mathcal{C}^1$ and $f_0 : \mathcal{C}^0 \to \mathcal{D}$ be functors between ∞-categories and let $f_1 : \mathcal{C}^1 \to \mathcal{D}$, $\alpha : f_0 \to \delta^* f_1 = f_1 \circ \delta$ be a left Kan extension of f_0 along δ. Let C be an object of $\mathcal{C}^0$ such that, for every $C' \in \mathcal{C}^0$, the functor δ induces an isomorphism*

$$\mathrm{Map}_{\mathcal{C}^0}(C', C) \to \mathrm{Map}_{\mathcal{C}^1}(\delta C', \delta C)$$

in the homotopy category $\mathcal{H}$. Then the morphism $\alpha(C) : f_0(C) \to f_1(\delta C)$ is an equivalence in $\mathcal{D}$.

Proof. Choose a diagram

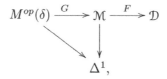

where $\mathcal{M}$ is a correspondence from $\mathcal{C}^0$ to $\mathcal{C}^1$ associated to δ, F is a left Kan extension of $f_0 = F | \mathcal{C}^0$, and $F \circ G$ is the map $M^{op}(\delta) \to \mathcal{D}$ determined by f_0, f_1, and α. Let $u : C \to \delta C$ be the morphism in $\mathcal{M}$ given by the image of $\{C\} \times \Delta^1 \subseteq M^{op}(\delta)$ under G. Then $\alpha(C) = F(u)$, so it will suffice to prove that $F(u)$ is an equivalence. Since F is a left Kan extension of f_0 at δC, the composition

$$(\mathcal{C}^0_{/\delta C})^{\triangleright} \to \mathcal{M} \xrightarrow{F} \mathcal{D}$$

is a colimit diagram. Consequently, it will suffice to prove that $u : C \to \delta C$ is a final object of $\mathcal{C}^0_{/\delta C}$. Consider the diagram

$$\mathcal{C}^0_{/C} \leftarrow \mathcal{C}^0_{/u} \xrightarrow{q} \mathcal{C}^0_{/\delta C}.$$

The ∞-category on the left has a final object id_C, and the map on the left is a trivial fibration of simplicial sets. We deduce that $s^0 u$ is a final object of $\mathcal{C}^0_{/u}$. Since $q(s^0 u) = u \in \mathcal{C}^0_{/\delta C}$, it will suffice to show that q is an equivalence of ∞-categories. We observe that q is a map of right fibrations over $\mathcal{C}^0$. According to Proposition 3.3.1.5, it will suffice to show that, for each object C' in $\mathcal{C}^0$, the map q induces a homotopy equivalence of Kan complexes

$$\mathcal{C}^0_{/u} \times_{\mathcal{C}^0} \{C'\} \to \mathcal{C}^0_{/\delta C} \times_{\mathcal{C}^0} \{C'\}.$$

This map can be identified with the map

$$\mathrm{Map}_{\mathcal{C}^0}(C', C) \to \mathrm{Map}_{\mathcal{M}}(C', \delta C) \simeq \mathrm{Map}_{\mathcal{C}^1}(\delta C', \delta C)$$

in the homotopy category $\mathcal{H}$ and is therefore a homotopy equivalence by assumption. $\qquad\square$

We conclude this section by proving that the construction of left Kan extensions behaves well in families.

Lemma 4.3.3.9. *Suppose we are given a commutative diagram*

$$\mathcal{C}^0 \xrightarrow{\ q\ } \mathcal{C} \xrightarrow{\ F\ } \mathcal{D}$$

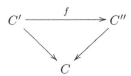

of ∞-categories, where p and q are coCartesian fibrations, i is the inclusion of a full subcategory, and i carries q-coCartesian morphisms of $\mathcal{C}^0$ to p-coCartesian morphisms of $\mathcal{C}$. The following conditions are equivalent:

(1) *The functor F is a left Kan extension of $F|\,\mathcal{C}^0$.*

(2) *For each object $E \in \mathcal{E}$, the induced functor $F_E : \mathcal{C}_E \to \mathcal{D}$ is a left Kan extension of $F_E|\,\mathcal{C}_E^0$.*

Proof. Let C be an object of $\mathcal{C}$ and let $E = p(C)$. Consider the composition

$$(\mathcal{C}_E^0)^{\triangleright}_{/C} \xrightarrow{G^{\triangleright}} (\mathcal{C}_{/C}^0)^{\triangleright} \xrightarrow{F_C} \mathcal{D}.$$

We will show that F_C is a colimit diagram if and only if $F_C \circ G^{\triangleright}$ is a colimit diagram. For this, it suffices to show that the inclusion $G : (\mathcal{C}_E^0)_{/C} \subseteq \mathcal{C}_{/C}^0$ is cofinal. According to Proposition 2.4.3.3, the projection $p' : \mathcal{C}_{/C} \to \mathcal{E}_{/E}$ is a coCartesian fibration, and a morphism

$$C' \xrightarrow{\quad f \quad} C''$$
$$\searrow \qquad \swarrow$$
$$C$$

in $\mathcal{C}_{/C}$ is p'-coCartesian if and only if f is p-coCartesian. It follows that p' restricts to a coCartesian fibration $\mathcal{C}'_{/C} \to \mathcal{E}_{/E}$. We have a pullback diagram of simplicial sets

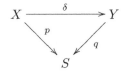

The right vertical map is smooth (Proposition 4.1.2.15) and G_0 is right anodyne, so that G is right anodyne, as desired. □

Proposition 4.3.3.10. *Let*

$$X \xrightarrow{\ \delta\ } Y$$
$$\searrow p \qquad \swarrow q$$
$$S$$

be a commutative diagram of simplicial sets, where p and q are coCartesian fibrations and δ carries p-coCartesian edges to q-coCartesian edges. Let $f_0 : X \to \mathcal{C}$ be a diagram in an ∞-category $\mathcal{C}$ and let $f_1 : Y \to \mathcal{C}$, $\alpha : f_0 \to f_1 \circ \delta$ be a left extension of f_0. The following conditions are equivalent:

(1) *The transformation α exhibits f_1 as a left Kan extension of f_0 along δ.*

(2) *For each vertex $s \in S$, the restriction $\alpha_s : f_0|X_s \to (f_1 \circ \delta)|X_s$ exhibits $f_1|Y_s$ as a left Kan extension of $f_0|X_s$ along $\delta_s : X_s \to Y_s$.*

Proof. Choose an equivalence of simplicial categories $\mathfrak{C}(S) \to \mathcal{E}$, where $\mathcal{E}$ is fibrant, and let $[1]$ denote the linearly ordered set $\{0, 1\}$ regarded as a category. Let ϕ' denote the induced map $\mathfrak{C}(S \times \Delta^1) \to \mathcal{E} \times [1]$. Let M denote the marked simplicial set

$$((X^{op})^\natural \times (\Delta^1)^\sharp) \coprod_{(X^{op})^\natural \times \{0\}} (Y^{op})^\natural.$$

Let $St_\phi^+ : (\mathrm{Set}_\Delta^+)_{(S \times \Delta^1)^{op}} \to (\mathrm{Set}_\Delta^+)^{\mathcal{E} \times [1]}$ denote the straightening functor defined in §3.2.1 and choose a fibrant replacement

$$St_\phi^+ M \to Z$$

in $(\mathrm{Set}_\Delta^+)^{\mathcal{E} \times [1]}$. Let $S' = \mathrm{N}(\mathcal{E})$, so that $S' \times \Delta^1 \simeq \mathrm{N}(\mathcal{E} \times [1])$, and let $\psi : \mathfrak{C}(S' \times \Delta^1) \to \mathcal{E} \times [1]$ be the counit map. Then

$$\mathrm{Un}_\psi^+(Z)$$

is a fibrant object of $(\mathrm{Set}_\Delta^+)_{/(S' \times \Delta^1)^{op}}$, which we may identify with a co-Cartesian fibration of simplicial sets $\mathcal{M} \to S' \times \Delta^1$.

We may regard $\mathcal{M}$ as a correspondence from $\mathcal{M}^0 = \mathcal{M} \times_{\Delta^1} \{0\}$ to $\mathcal{M}^1 = \mathcal{M} \times_{\Delta^1} \{1\}$. By construction, we have a unit map

$$u : M^{op}(\delta) \to \mathcal{M} \times_{S'} S.$$

Theorem 3.2.0.1 implies that the induced maps $u_0 : X \to \mathcal{M}^0 \times_{S'} S$, $u_1 : Y \to \mathcal{M}^1 \times_{S'} S$ are equivalences of coCartesian fibrations. Proposition 3.3.1.3 implies that the maps $\mathcal{M}^0 \times_{S'} S \to \mathcal{M}^0$, $\mathcal{M}^1 \times_{S'} S \to \mathcal{M}^1$ are categorical equivalences.

Let u' denote the composition

$$M^{op}(\delta) \xrightarrow{u} \mathcal{M} \times_{S'} S \to \mathcal{M}$$

and let $u_0' : X \to \mathcal{M}^0$, $u_1' : Y \to \mathcal{M}^1$ be defined similarly. The above argument shows that u_0' and u_1' are categorical equivalences. Consequently, the map u' is a quasi-equivalence of coCartesian fibrations over Δ^1 and therefore a categorical equivalence (Proposition 3.2.2.7). Replacing $\mathcal{M}$ by the product $\mathcal{M} \times K$ if necessary, where K is a contractible Kan complex, we may suppose that u' is a cofibration of simplicial sets. Since $\mathcal{D}$ is an ∞-category, there exists a functor $F : \mathcal{M} \to \mathcal{D}$ as indicated in the diagram below:

$$
\begin{array}{ccc}
M^{op}(\delta) & \xrightarrow{(f_0, f_1, \alpha)} & \mathcal{D} \\
\downarrow & \nearrow{F} & \\
\mathcal{M}. & &
\end{array}
$$

Consequently, we may reformulate condition (1) as follows:

(1′) The functor F is a left Kan extension of $F|\mathcal{M}^0$.

Proposition 3.3.1.5 now implies that, for each vertex s of S, the map $X_s \to \mathcal{M}_s^0$ is a categorical equivalence. Similarly, for each vertex s of S, the inclusion $Y_s \to \mathcal{M}_s^1$ is a categorical equivalence. It follows that the inclusion $M^{op}(\delta)_s \to \mathcal{M}_s$ is a quasi-equivalence and therefore a categorical equivalence (Proposition 3.2.2.7). Consequently, we may reformulate condition (2) as follows:

(2′) For each vertex $s \in S$, the functor $F|\mathcal{M}_s$ is a left Kan extension of $F|\mathcal{M}_s^0$.

Using Lemma 4.3.2.6, it is easy to see that the collection of objects $s \in S'$ such that $F|\mathcal{M}_s$ is a left Kan extension of $F|\mathcal{M}_s^0$ is stable under equivalence. Since the inclusion $S \subseteq S'$ is a categorical equivalence, we conclude that (2′) is equivalent to the following apparently stronger condition:

(2″) For every object $s \in S'$, the functor $F|\mathcal{M}_s$ is a left Kan extension of $F|\mathcal{M}_s^0$.

The equivalence of (1′) and (2″) follows from Lemma 4.3.3.9. $\square$

4.4 EXAMPLES OF COLIMITS

In this section, we will analyze in detail the colimits of some very simple diagrams. Our first three examples are familiar from classical category theory: coproducts (§4.4.1), pushouts (§4.4.2), and coequalizers (§4.4.3).

Our fourth example is slightly more unfamiliar. Let $\mathcal{C}$ be an ordinary category which admits coproducts. Then $\mathcal{C}$ is naturally *tensored* over the category of sets. Namely, for each $C \in \mathcal{C}$ and $S \in \mathcal{S}et$, we can define $C \otimes S$ to be the coproduct of a collection of copies of C indexed by the set S. The object $C \otimes S$ is characterized by the following universal mapping property:

$$\mathrm{Hom}_{\mathcal{C}}(C \otimes S, D) \simeq \mathrm{Hom}_{\mathcal{S}et}(S, \mathrm{Hom}_{\mathcal{C}}(C, D)).$$

In the ∞-categorical setting, it is natural to try to generalize this definition by allowing S to be an object of $\mathcal{S}$. In this case, $C \otimes S$ can again be viewed as a kind of colimit but cannot be written as a *coproduct* unless S is discrete. We will study the situation in §4.4.4.

Our final objective in this section is to study the theory of *retracts* in an ∞-category $\mathcal{C}$. In §4.4.5, we will see that there is a close relationship between retracts in $\mathcal{C}$ and idempotent endomorphisms, just as in classical homotopy theory. Namely, any retract of an object $C \in \mathcal{C}$ determines an idempotent endomorphism of C; conversely, if $\mathcal{C}$ is *idempotent complete*, then every idempotent endomorphism of C determines a retract of C. We will return to this idea in §5.1.4.

4.4.1 Coproducts

In this section, we discuss the simplest type of colimit: namely, coproducts. Let A be a set; we may regard A as a category with

$$\mathrm{Hom}_A(I, J) = \begin{cases} * & \text{if } I = J \\ \emptyset & \text{if } I \neq J. \end{cases}$$

We will also identify A with the (constant) simplicial set which is the nerve of this category. We note that a functor $G : A \to \mathrm{Set}_\Delta$ is injectively fibrant if and only if it takes values in the category $\mathcal{K}$an of Kan complexes. If this condition is satisfied, then the product $\prod_{\alpha \in A} G(\alpha)$ is a homotopy limit for G.

Let $F : A \to \mathcal{C}$ be a functor from A to a fibrant simplicial category; in other words, F specifies a collection $\{X_\alpha\}_{\alpha \in A}$ of objects in $\mathcal{C}$. A homotopy colimit for F will be referred to as a *homotopy coproduct* of the objects $\{X_\alpha\}_{\alpha \in A}$. Unwinding the definition, we see that a homotopy coproduct is an object $X \in \mathcal{C}$ equipped with morphisms $\phi_\alpha : X_\alpha \to X$ such that the induced map

$$\mathrm{Map}_{\mathcal{C}}(X, Y) \to \prod_{\alpha \in A} \mathrm{Map}_{\mathcal{C}}(X_\alpha, Y)$$

is a homotopy equivalence for every object $Y \in \mathcal{C}$. Consequently, we recover the description given in Example 1.2.13.1. As we noted earlier, this characterization can be stated entirely in terms of the enriched homotopy category $h\mathcal{C}$: the maps $\{\phi_\alpha\}$ exhibit X as a homotopy coproduct of the family $\{X_\alpha\}_{\alpha \in A}$ if and only if the induced map

$$\mathrm{Map}_{\mathcal{C}}(X, Y) \to \prod_{\alpha \in A} \mathrm{Map}_{\mathcal{C}}(X_\alpha, Y)$$

is an isomorphism in the homotopy category $\mathcal{H}$ of spaces for each $Y \in \mathcal{C}$.

Now suppose that $\mathcal{C}$ is an ∞-category and let $p : A \to \mathcal{C}$ be a map. As above, we may identify this with a collection of objects $\{X_\alpha\}_{\alpha \in A}$ of $\mathcal{C}$. To specify an object of $\mathcal{C}_{p/}$ is to give an object $X \in \mathcal{C}$ together with morphisms $\phi_\alpha : X_\alpha \to X$ for each $\alpha \in A$. Using Theorem 4.2.4.1, we deduce that X is a colimit of the diagram p if and only if the induced map

$$\mathrm{Map}_{\mathcal{C}}(X, Y) \to \prod_{\alpha \in A} \mathrm{Map}_{\mathcal{C}}(X_\alpha, Y)$$

is an isomorphism in $\mathcal{H}$ for each object $Y \in \mathcal{C}$. In this case, we say that X is a *coproduct* of the family $\{X_\alpha\}_{\alpha \in A}$.

In either setting, we will denote the (homotopy) coproduct of a family of objects $\{X_\alpha\}_{\alpha \in A}$ by

$$\coprod_{\alpha \in A} X_I.$$

It is well-defined up to (essentially unique) equivalence.

Using Corollary 4.2.3.10, we deduce the following:

Proposition 4.4.1.1. *Let $\mathcal{C}$ be an ∞-category and let $\{p_\alpha : K_\alpha \to \mathcal{C}\}_{\alpha \in A}$ be a family of diagrams in $\mathcal{C}$ indexed by a set A. Suppose that each p_α has a colimit X_α in $\mathcal{C}$. Let $K = \coprod K_\alpha$ and let $p : K \to \mathcal{C}$ be the result of amalgamating the maps p_α. Then p has a colimit in $\mathcal{C}$ if and only if the family $\{X_\alpha\}_{\alpha \in A}$ has a coproduct in $\mathcal{C}$; in this case, one may identify colimits of p with coproducts $\coprod_{\alpha \in A} X_\alpha$.*

4.4.2 Pushouts

Let $\mathcal{C}$ be an ∞-category. A *square* in $\mathcal{C}$ is a map $\Delta^1 \times \Delta^1 \to \mathcal{C}$. We will typically denote squares in $\mathcal{C}$ by diagrams

$$
\begin{array}{ccc}
X' & \xrightarrow{\ p'\ } & X \\
{\scriptstyle q'}\downarrow & & \downarrow{\scriptstyle q} \\
Y' & \xrightarrow{\ p\ } & Y,
\end{array}
$$

with the "diagonal" morphism $r : X' \to Y$ and homotopies $r \simeq q \circ p'$, $r \simeq p \circ q'$ being implicit.

We have isomorphisms of simplicial sets

$$
(\Lambda_0^2)^{\triangleright} \simeq \Delta^1 \times \Delta^1 \simeq (\Lambda_2^2)^{\triangleleft}.
$$

Consequently, given a square $\sigma : \Delta^1 \times \Delta^1 \to \mathcal{C}$, it makes sense to ask whether or not σ is a limit or colimit diagram. If σ is a limit diagram, we will also say that σ is a *pullback square* or a *Cartesian square* and we will informally write $X' = X \times_Y Y'$. Dually, if σ is a colimit diagram, we will say that σ is a *pushout square* or a *coCartesian square*, and write $Y = X \coprod_{X'} Y'$.

Now suppose that $\mathcal{C}$ is a (fibrant) simplicial category. By definition, a commutative diagram

$$
\begin{array}{ccc}
X' & \xrightarrow{\ p'\ } & X \\
{\scriptstyle q'}\downarrow & & \downarrow{\scriptstyle q} \\
Y' & \xrightarrow{\ p\ } & Y
\end{array}
$$

is a homotopy pushout square if, for every object $Z \in \mathcal{C}$, the diagram

$$
\begin{array}{ccc}
\mathrm{Map}_{\mathcal{C}}(Y, Z) & \longrightarrow & \mathrm{Map}_{\mathcal{C}}(Y', Z) \\
\downarrow & & \downarrow \\
\mathrm{Map}_{\mathcal{C}}(X, Z) & \longrightarrow & \mathrm{Map}_{\mathcal{C}}(X', Z)
\end{array}
$$

is a homotopy pullback square in $\mathcal{K}$an. Using Theorem 4.2.4.1, we can reduce questions about pushout diagrams in an arbitrary ∞-category to questions about homotopy pullback squares in $\mathcal{K}$an.

The following basic transitivity property for pushout squares will be used repeatedly throughout this book:

Lemma 4.4.2.1. *Let $\mathcal{C}$ be an ∞-category and suppose we are given a map $\sigma : \Delta^2 \times \Delta^1 \to \mathcal{C}$ which we will depict as a diagram*

$$
\begin{array}{ccccc}
X & \longrightarrow & Y & \longrightarrow & Z \\
\downarrow & & \downarrow & & \downarrow \\
X' & \longrightarrow & Y' & \longrightarrow & Z'.
\end{array}
$$

Suppose that the left square is a pushout in $\mathcal{C}$. Then the right square is a pushout if and only if the outer square is a pushout.

Proof. For every subset A of $\{x, y, z, x', y', z'\}$, let $\mathcal{D}(A)$ denote the corresponding full subcategory of $\Delta^2 \times \Delta^1$ and let $\sigma(A)$ denote the restriction of σ to $\mathcal{D}(A)$. We may regard σ as determining an object $\widetilde{\sigma} \in \mathcal{C}_{\sigma(\{y,z,x',y',z'\})/}$. Consider the maps

$$
\mathcal{C}_{\sigma(\{z,x',z'\})/} \xleftarrow{\phi} \mathcal{C}_{\sigma(\{y,z,x',y',z'\})/} \xrightarrow{\psi} \mathcal{C}_{\sigma(\{y,x',y'\})/} \cdot
$$

The map ϕ is the composition of the trivial fibration

$$
\mathcal{C}_{\sigma(\{z,x',y',z'\})/} \to \mathcal{C}_{\sigma(\{z,x',z'\})/}
$$

with a pullback of

$$
\mathcal{C}_{\sigma(\{y,z,y',z'\})/} \to \mathcal{C}_{\sigma(\{z,y',z'\})/},
$$

also a trivial fibration by virtue of our assumption that the square

$$
\begin{array}{ccc}
Y & \longrightarrow & Z \\
\downarrow & & \downarrow \\
Y' & \longrightarrow & Z'
\end{array}
$$

is a pullback in $\mathcal{C}$. The map ψ is a trivial fibration because the inclusion $\mathcal{D}(\{y, x', y'\}) \subseteq \mathcal{D}(\{y, z, x', y', z'\})$ is left anodyne. It follows that $\phi(\widetilde{\sigma})$ is an initial object of $\mathcal{C}_{\sigma(\{z,x',z'\})/}$ if and only if $\psi(\widetilde{\sigma})$ is an initial object of $\mathcal{C}_{\sigma(\{y,x',y'\})/}$, as desired. $\square$

Our next objective is to apply Proposition 4.2.3.8 to show that in many cases complicated colimits may be decomposed as pushouts of simpler colimits. Suppose we are given a pushout diagram of simplicial sets

$$
\begin{array}{ccc}
L' & \xrightarrow{\ i\ } & L \\
\downarrow & & \downarrow \\
K' & \longrightarrow & K
\end{array}
$$

and a diagram $p : K \to \mathcal{C}$, where $\mathcal{C}$ is an ∞-category. Suppose furthermore that $p|K'$, $p|L'$, and $p|L$ admit colimits in $\mathcal{C}$, which we will denote by X, Y, and Z, respectively. If we suppose further that the map i is a cofibration of simplicial sets, then the hypotheses of Proposition 4.2.3.4 are satisfied and we can deduce the following:

Proposition 4.4.2.2. *Let* $\mathcal{C}$ *be an* ∞-*category and let* $p : K \to \mathcal{C}$ *be a map of simplicial sets. Suppose we are given a decomposition* $K = K' \coprod_{L'} L$, *where* $L' \to L$ *is a monomorphism of simplicial sets. Suppose further that* $p|K'$ *has a colimit* $X \in \mathcal{C}$, $p|L'$ *has a colimit* $Y \in \mathcal{C}$, *and* $p|L$ *has a colimit* $Z \in \mathcal{C}$. *Then one may identify colimits for* p *with pushouts* $X \coprod_Y Z$.

Remark 4.4.2.3. The statement of Proposition 4.4.2.2 is slightly vague. Implicit in the discussion is that identifications of X with the colimit of $p|K'$ and Y with the colimit of $p|L'$ induce a morphism $Y \to X$ in $\mathcal{C}$ (and similarly for Y and Z). This morphism is not uniquely determined, but it is determined up to a contractible space of choices: see the proof of Proposition 4.2.3.4.

It follows from Proposition 4.4.2.2 that any finite colimit can be built using initial objects and pushout squares. For example, we have the following:

Corollary 4.4.2.4. *Let* $\mathcal{C}$ *be an* ∞-*category. Then* $\mathcal{C}$ *admits all finite colimits if and only if* $\mathcal{C}$ *admits pushouts and has an initial object.*

Proof. The "only if" direction is clear. For the converse, let us suppose that $\mathcal{C}$ has pushouts and an initial object. Let $p : K \to \mathcal{C}$ be any diagram, where K is a finite simplicial set: that is, K has only finitely many nondegenerate simplices. We will prove that p has a colimit. The proof proceeds by induction: first on the dimension of K, then on the number of simplices of K having the maximal dimension.

If K is empty, then an initial object of $\mathcal{C}$ is a colimit for p. Otherwise, we may fix a nondegenerate simplex of K having the maximal dimension and thereby decompose $K \simeq K_0 \coprod_{\partial \Delta^n} \Delta^n$. By the inductive hypothesis, $p|K_0$ has a colimit X and $p|\partial \Delta^n$ has a colimit Y. The ∞-category Δ^n has a final object, so $p|\Delta^n$ has a colimit Z (which we may take to be $p(v)$, where v is the final vertex of Δ^n). Now we simply apply Proposition 4.4.2.2 to deduce that $X \coprod_Y Z$ is a colimit for p. $\square$

Using the same argument, one can show:

Corollary 4.4.2.5. *Let* $f : \mathcal{C} \to \mathcal{C}'$ *be a functor between* ∞-*categories. Assume that* $\mathcal{C}$ *has all finite colimits. Then* f *preserves all finite colimits if and only if* f *preserves initial objects and pushouts.*

We conclude by showing how *all* colimits can be constructed out of simple ones.

Proposition 4.4.2.6. *Let* $\mathcal{C}$ *be an* ∞-*category which admits pushouts and* κ-*small coproducts. Then* $\mathcal{C}$ *admits colimits for all* κ-*small diagrams.*

Proof. If $\kappa = \omega$, we have already shown this as Corollary 4.4.2.4. Let us therefore suppose that $\kappa > \omega$, and that $\mathcal{C}$ has pushouts and κ-small sums.

Let $p : K \to \mathcal{C}$ be a diagram, where K is κ-small. We first suppose that the dimension n of K is finite: that is, K has no nondegenerate simplices of dimension larger than n. We prove that p has a colimit, working by induction on n.

If $n = 0$, then K consists of a finite disjoint union of fewer than κ vertices. The colimit of p exists by the assumption that $\mathcal{C}$ has κ-small sums.

Now suppose that every diagram indexed by a κ-small simplicial set of dimension n has a colimit. Let $p : K \to \mathcal{C}$ be a diagram, with the dimension of K equal to $n + 1$. Let K^n denote the n-skeleton of K and $K'_{n+1} \subseteq K_{n+1}$ the set of all nondegenerate $(n+1)$-simplices of K, so that there is a pushout diagram of simplicial sets

$$K^n \coprod_{K'_{n+1} \times \partial \Delta^{n+1}} (K'_{n+1} \times \Delta^{n+1}) \simeq K.$$

By Proposition 4.4.2.2, we can construct a colimit of p as a pushout using colimits for $p|K^n$, $p|(K'_{n+1} \times \partial \Delta^{n+1})$, and $p|(K'_{n+1} \times \Delta^{n+1})$. The first two exist by the inductive hypothesis and the last because it is a sum of diagrams which possess colimits.

Now let us suppose that K is not necessarily finite dimensional. In this case, we can filter K by its skeleta $\{K^n\}$. This is a family of simplicial subsets of K indexed by the set $\mathbf{Z}_{\geq 0}$ of nonnegative integers. By what we have shown above, each $p|K^n$ has a colimit x_n in $\mathcal{C}$. Since this family is directed and covers K, Corollary 4.2.3.10 shows that we may identify colimits of p with colimits of a diagram $\mathrm{N}(\mathbf{Z}_{\geq 0}) \to \mathcal{C}$ which we may write informally as

$$x_0 \to x_1 \to \cdots .$$

Let L be the simplicial subset of $\mathrm{N}(\mathbf{Z}_{\geq 0})$ which consists of all vertices together with the edges which join consecutive integers. A simple computation shows that the inclusion $L \subseteq \mathrm{N}(\mathbf{Z}_{\geq 0})$ is a categorical equivalence and therefore cofinal. Consequently, it suffices to construct the colimit of a diagram $L \to \mathcal{C}$. But L is 1-dimensional and is κ-small since $\kappa > \omega$. $\qquad \square$

The same argument also proves the following:

Proposition 4.4.2.7. *Let κ be a regular cardinal and let $f : \mathcal{C} \to \mathcal{D}$ be a functor between ∞-categories, where $\mathcal{C}$ admits κ-small colimits. Then f preserves κ-small colimits if and only if f preserves pushout squares and κ-small coproducts.*

Let $\mathcal{D}$ be an ∞-category containing an object X and suppose that $\mathcal{D}$ admits pushouts. Then $\mathcal{D}_{X/}$ admits pushouts, and these pushouts may be computed in $\mathcal{D}$. In other words, the projection $f : \mathcal{D}_{X/} \to \mathcal{D}$ preserves pushouts. In fact, this is a special case of a very general result; it requires only that f be a left fibration and that the simplicial set Λ_0^2 be weakly contractible.

Lemma 4.4.2.8. *Let $f : \mathcal{C} \to \mathcal{D}$ be a left fibration of ∞-categories and let K be a weakly contractible simplicial set. Then any map $\overline{p} : K^{\triangleright} \to \mathcal{C}$ is an f-colimit diagram.*

Proof. Let $p = \overline{p}|K$. We must show that the map

$$\phi : \mathcal{C}_{\overline{p}/} \to \mathcal{C}_{p/} \times_{\mathcal{D}_{f \circ p/}} \mathcal{D}_{f \circ \overline{p}/}$$

is a trivial fibration of simplicial sets. In other words, we must show that we can solve any lifting problem of the form

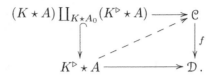

Since f is a left fibration, it will suffice to prove that the left vertical map is left anodyne, which follows immediately from Lemma 4.2.3.5.   □

Proposition 4.4.2.9. *Let* $f : \mathcal{C} \to \mathcal{D}$ *be a left fibration of* ∞-*categories and let* $p : K \to \mathcal{C}$ *be a diagram. Suppose that* K *is weakly contractible. Then*

(1) *Let* $\overline{p} : K^{\triangleright} \to \mathcal{C}$ *be an extension of* p. *Then* $\overline{p}$ *is a colimit of* p *if and only if* $f \circ \overline{p}$ *is a colimit of* $f \circ p$.

(2) *Let* $\overline{q} : K^{\triangleright} \to \mathcal{D}$ *be a colimit of* $f \circ p$. *Then* $\overline{q} = f \circ \overline{p}$, *where* $\overline{p}$ *is an extension* (*automatically a colimit by virtue of* (1)) *of* p.

Proof. To prove (1), fix an extension $\overline{p} : K^{\triangleright} \to \mathcal{C}$. We have a commutative diagram

$$\mathcal{C}_{\overline{p}/} \xrightarrow{\phi} \mathcal{C}_{p/} \times_{\mathcal{D}_{fp/}} \mathcal{D}_{f\overline{p}/} \xrightarrow{\psi'} \mathcal{C}_{p/}$$
$$\downarrow \qquad\qquad\qquad\qquad\qquad \downarrow \theta$$
$$\mathcal{D}_{f\overline{p}/} \xrightarrow{\psi} \mathcal{D}_{fp/}.$$

Lemma 4.4.2.8 implies that ϕ is a trivial Kan fibration. If $f \circ \overline{p}$ is a colimit diagram, the map ψ is a trivial fibration. Since ψ' is a pullback of ψ, we conclude that ψ' is a trivial fibration. It follows that $\psi' \circ \phi$ is a trivial fibration so that $\overline{p}$ is a colimit diagram. This proves the "if" direction of (1).

To prove the converse, let us suppose that $\overline{p}$ is a colimit diagram. The maps ϕ and $\psi' \circ \phi$ are both trivial fibrations. It follows that the fibers of ψ' are contractible. Using Lemma 4.2.3.6, we conclude that the map θ is a trivial fibration, and therefore surjective on vertices. It follows that the fibers of ψ are contractible. Since ψ is a left fibration with contractible fibers, it is a trivial fibration (Lemma 2.1.3.4). Thus $f \circ \overline{p}$ is a colimit diagram, and the proof is complete.

To prove (2), it suffices to show that f has the right lifting property with respect to the inclusion $i : K \subseteq K^{\triangleright}$. Since f is a left fibration, it will suffice to show that i is left anodyne, which follows immediately from Lemma 4.2.3.6.   □

4.4.3 Coequalizers

Let $\mathcal{J}$ denote the category depicted by the diagram

$$X \overset{F}{\underset{G}{\rightrightarrows}} Y.$$

In other words, $\mathcal{J}$ has two objects X and Y satisfying the conditions

$$\operatorname{Hom}_{\mathcal{J}}(X, X) = \operatorname{Hom}_{\mathcal{J}}(Y, Y) = *$$

$$\operatorname{Hom}_{\mathcal{J}}(Y, X) = \emptyset \qquad \operatorname{Hom}_{\mathcal{J}}(X, Y) = \{F, G\}.$$

To give a diagram $p : \mathrm{N}(\mathcal{J}) \to \mathcal{C}$ in an ∞-category $\mathcal{C}$, one must give a pair of morphisms $f = p(F)$, $g = p(G)$ in $\mathcal{C}$ having the same domain $x = p(X)$ and the same codomain $y = p(Y)$. A colimit for the diagram p is said to be a *coequalizer* of f and g.

Applying Corollary 4.2.3.10, we deduce the following:

Proposition 4.4.3.1. *Let K and A be simplicial sets and let $i_0, i_1 : A \to K$ be embeddings having disjoint images in K. Let K' denote the coequalizer of i_0 and i_1: in other words, the simplicial set obtained from K by identifying the image of i_0 with the image of i_1. Let $p : K' \to \mathcal{C}$ be a diagram in an ∞-category S and let $q : K \to \mathcal{C}$ be the composition*

$$K \to K' \overset{p}{\to} S.$$

Suppose that the diagrams $q \circ i_0 = q \circ i_1$ and q possess colimits x and y in S. Then i_0 and i_1 induce maps $j_0, j_1 : x \to y$ (well-defined up to homotopy); colimits for p may be identified with coequalizers of j_0 and j_1.

Like pushouts, coequalizers are a basic construction out of which other colimits can be built. More specifically, we have the following:

Proposition 4.4.3.2. *Let $\mathcal{C}$ be an ∞-category and κ a regular cardinal. Then $\mathcal{C}$ has all κ-small colimits if and only if $\mathcal{C}$ has coequalizers and κ-small coproducts.*

Proof. The "only if" direction is obvious. For the converse, suppose that $\mathcal{C}$ has coequalizers and κ-small coproducts. In view of Proposition 4.4.2.6, it suffices to show that $\mathcal{C}$ has pushouts. Let $p : \Lambda_0^2$ be a pushout diagram in $\mathcal{C}$. We note that Λ_0^2 is the quotient of $\Delta^{\{0,1\}} \coprod \Delta^{\{0,2\}}$ obtained by identifying the initial vertex of $\Delta^{\{0,1\}}$ with the initial vertex of $\Delta^{\{0,2\}}$. In view of Proposition 4.4.3.1, it suffices to show that $p|(\Delta^{\{0,1\}} \coprod \Delta^{\{0,2\}})$ and $p|\{0\}$ possess colimits in $\mathcal{C}$. The second assertion is obvious. Since $\mathcal{C}$ has finite sums, to prove that there exists a colimit for $p|(\Delta^{\{0,1\}} \coprod \Delta^{\{0,2\}})$, it suffices to prove that $p|\Delta^{\{0,1\}}$ and $p|\Delta^{\{0,2\}}$ possess colimits in $\mathcal{C}$. This is immediate because both $\Delta^{\{0,1\}}$ and $\Delta^{\{0,2\}}$ have final objects. $\qquad\square$

Using the same argument, we deduce:

Proposition 4.4.3.3. *Let κ be a regular cardinal and $\mathcal{C}$ be an ∞-category which admits κ-small colimits. A full subcategory $\mathcal{D} \subseteq \mathcal{C}$ is stable under κ-small colimits in $\mathcal{C}$ if and only if $\mathcal{D}$ is stable under coequalizers and under κ-small sums.*

4.4.4 Tensoring with Spaces

Every ordinary category $\mathcal{C}$ can be regarded as a category enriched over $\mathcal{S}$et. Moreover, if $\mathcal{C}$ admits coproducts, then $\mathcal{C}$ can be regarded as *tensored* over $\mathcal{S}$et in an essentially unique way. In the ∞-categorical setting, one has a similar situation: if $\mathcal{C}$ is an ∞-category which admits all small limits, then $\mathcal{C}$ may be regarded as tensored over the ∞-category $\mathcal{S}$ of spaces. To make this idea precise, we would need a good theory of *enriched ∞-categories*, which lies outside the scope of this book. We will instead settle for a slightly ad hoc point of view which nevertheless allows us to construct the relevant tensor products. We begin with a few remarks concerning representable functors in the ∞-categorical setting.

Definition 4.4.4.1. Let $\mathcal{D}$ be a closed monoidal category and let $\mathcal{C}$ be a category enriched over $\mathcal{D}$. We will say that a $\mathcal{D}$-enriched functor $G : \mathcal{C}^{op} \to \mathcal{D}$ is *representable* if there exists an object $C \in \mathcal{C}$ and a map $\eta : 1_{\mathcal{D}} \to G(C)$ such that the induced map

$$\mathrm{Map}_{\mathcal{C}}(X, C) \simeq \mathrm{Map}_{\mathcal{C}}(X, C) \otimes 1_{\mathcal{D}} \to \mathrm{Map}_{\mathcal{C}}(X, C) \otimes G(C) \to G(X)$$

is an isomorphism for every object $X \in \mathcal{C}$. In this case, we will say that (C, η) *represents* the functor F.

Remark 4.4.4.2. In the situation of Definition 4.4.4.1, we will sometimes abuse terminology and simply say that the functor F is *represented* by the object C.

Remark 4.4.4.3. The dual notion of a *corepresentable functor* can be defined in an obvious way.

Definition 4.4.4.4. Let $\mathcal{C}$ be an ∞-category and let $\mathcal{S}$ denote the ∞-category of spaces. We will say that a functor $F : \mathcal{C}^{op} \to \mathcal{S}$ is *representable* if the underlying functor

$$\mathrm{h}F : \mathrm{h}\mathcal{C}^{op} \to \mathrm{h}\mathcal{S} \simeq \mathcal{H}$$

of ($\mathcal{H}$-enriched) homotopy categories is representable. We will say that a pair $C \in \mathcal{C}$, $\eta \in \pi_0 F(C)$ *represents* F if the pair (C, η) represents $\mathrm{h}F$.

Proposition 4.4.4.5. *Let $f : \widetilde{\mathcal{C}} \to \mathcal{C}$ be a right fibration of ∞-categories, let $\widetilde{C}$ be an object of $\widetilde{\mathcal{C}}$, let $C = f(\widetilde{C}) \in \mathcal{C}$, and let $F : \mathcal{C}^{op} \to \mathcal{S}$ be a functor which classifies f (§3.3.2). The following conditions are equivalent:*

(1) *Let $\eta \in \pi_0 F(C) \simeq \pi_0(\widetilde{\mathcal{C}} \times_{\mathcal{C}} \{C\})$ be the connected component containing $\widetilde{C}$. Then the pair (C, η) represents the functor F.*

(2) *The object $\widetilde{C} \in \widetilde{\mathcal{C}}$ is final.*

(3) *The inclusion $\{\widetilde{C}\} \subseteq \widetilde{\mathcal{C}}$ is a contravariant equivalence in $(\mathrm{Set}_{\Delta})_{/\mathcal{C}}$.*

Proof. We have a commutative diagram of right fibrations

$$\begin{array}{ccc} \widetilde{\mathcal{C}}_{/\widetilde{C}} & \xrightarrow{\phi} & \widetilde{\mathcal{C}} \\ \downarrow & & \downarrow \\ \mathcal{C}_{/C} & \longrightarrow & \mathcal{C}. \end{array}$$

Observe that the left vertical map is actually a trivial fibration. Fix an object $D \in \mathcal{C}$. The fiber of the upper horizontal map

$$\phi_D : \widetilde{\mathcal{C}}_{/\widetilde{C}} \times_{\mathcal{C}} \{D\} \to \widetilde{\mathcal{C}} \times_{\mathcal{C}} \{D\}$$

can be identified, in the homotopy category $\mathcal{H}$, with the map $\mathrm{Map}_{\mathcal{C}}(D, C) \to F(C)$. The map ϕ_D is a right fibration of Kan complexes and therefore a Kan fibration. If (1) is satisfied, then ϕ_D is a homotopy equivalence and therefore a trivial fibration. It follows that the fibers of ϕ are contractible. Since ϕ is a right fibration, it is a trivial fibration (Lemma 2.1.3.4). This proves that $\widetilde{C}$ is a final object of $\widetilde{\mathcal{C}}$. Conversely, if (2) is satisfied, then ϕ_D is a trivial Kan fibration and therefore a weak homotopy equivalence. Thus (1) $\Leftrightarrow$ (2).

If (2) is satisfied, then the inclusion $\{\widetilde{C}\} \subseteq \widetilde{\mathcal{C}}$ is right anodyne and therefore a contravariant equivalence by Proposition 4.1.2.1. Thus (2) $\Rightarrow$ (3). Conversely, suppose that (3) is satisfied. The inclusion $\{\mathrm{id}_C\} \subseteq \mathcal{C}_{/C}$ is right anodyne and therefore a contravariant equivalence. It follows that the lifting problem

has a solution. We observe that e is a contravariant equivalence of right fibrations over $\mathcal{C}$ and therefore a categorical equivalence. By construction, e carries a final object of $\mathcal{C}_{/C}$ to $\widetilde{C}$, so that $\widetilde{C}$ is a final object of $\widetilde{\mathcal{C}}$.   □

We will say that a right fibration $\widetilde{\mathcal{C}} \to \mathcal{C}$ is *representable* if $\widetilde{\mathcal{C}}$ has a final object.

Remark 4.4.4.6. Let $\mathcal{C}$ be an ∞-category and let $p : K \to \mathcal{C}$ be a diagram. Then the right fibration $\mathcal{C}_{/p} \to \mathcal{C}$ is representable if and only if p has a limit in $\mathcal{C}$.

Remark 4.4.4.7. All of the above ideas dualize in an evident way, so that we may speak of *corepresentable functors* and *corepresentable left fibrations* in the setting of ∞-categories.

Notation 4.4.4.8. For each diagram $p : K \to \mathcal{C}$ in an ∞-category $\mathcal{C}$, we let $\mathcal{F}_p : h\mathcal{C} \to \mathcal{H}$ denote the $\mathcal{H}$-enriched functor corresponding to the left fibration $\mathcal{C}^{p/} \to \mathcal{C}$.

If $p : * \to \mathcal{C}$ is the inclusion of an object X of $\mathcal{C}$, then we write $\mathcal{F}_X$ for $\mathcal{F}_p$. We note that $\mathcal{F}_X$ is the functor corepresented by X:

$$\mathcal{F}_X(Y) = \operatorname{Map}_{\mathcal{C}}(X, Y).$$

Now suppose that X is an object in an ∞-category $\mathcal{C}$ and let $p : K \to \mathcal{C}$ be a constant map taking the value X. For every object Y of $\mathcal{C}$, we have an isomorphism of simplicial sets $(\mathcal{C}^{p/}) \times_{\mathcal{C}} \{Y\} \simeq (\mathcal{C}^{X/} \times_{\mathcal{C}} \{Y\})^K$. This identification is functorial up to homotopy, so we actually obtain an equivalence

$$\mathcal{F}_p(Y) \simeq \operatorname{Map}_{\mathcal{C}}(X, Y)^{[K]}$$

in the homotopy category $\mathcal{H}$ of spaces, where $[K]$ denotes the simplicial set K regarded as an object of $\mathcal{H}$. Applying Proposition 4.4.4.5, we deduce the following:

Corollary 4.4.4.9. *Let $\mathcal{C}$ be an ∞-category, X an object of $\mathcal{C}$, and K a simplicial set. Let $p : K \to \mathcal{C}$ be the constant map taking the value X. The objects of the fiber $\mathcal{C}^{p/} \times_{\mathcal{C}} \{Y\}$ are classified (up to equivalence) by maps $\psi : [K] \to \operatorname{Map}_{\mathcal{C}}(X, Y)$ in the homotopy category $\mathcal{H}$. Such a map ψ classifies a colimit for p if and only if it induces isomorphisms*

$$\operatorname{Map}_{\mathcal{C}}(Y, Z) \simeq \operatorname{Map}_{\mathcal{C}}(X, Z)^{[K]}$$

in the homotopy category $\mathcal{H}$ for every object Z of $\mathcal{C}$.

In the situation of Corollary 4.4.4.9, we will denote a colimit for p by $X \otimes K$ if such a colimit exists. We note that $X \otimes K$ is well-defined up to (essentially unique) equivalence and that it depends (up to equivalence) only on the weak homotopy type of the simplicial set K.

Corollary 4.4.4.10. *Let $\mathcal{C}$ be an ∞-category, let K be a weakly contractible simplicial set, and let $p : K \to \mathcal{C}$ be a diagram which carries each edge of K to an equivalence in $\mathcal{C}$. Then*

(1) *The diagram p has a colimit in $\mathcal{C}$.*

(2) *An arbitrary extension $\overline{p} : K^{\triangleright} \to \mathcal{C}$ is a colimit for $\mathcal{C}$ if and only if $\overline{p}$ carries each edge of $K^{\triangleright} \to \mathcal{C}$ to an equivalence in $\mathcal{C}$.*

Proof. Let $\mathcal{C}' \subseteq \mathcal{C}$ be the largest Kan complex contained in $\mathcal{C}$. By assumption, p factors through $\mathcal{C}'$. Since K is weakly contractible, we conclude that $p : K \to \mathcal{C}'$ is homotopic to a constant map $p' : K \to \mathcal{C}'$. Replacing p by p' if necessary, we may reduce to the case where p is constant, taking values equal to some fixed object $C \in \mathcal{C}$.

Let $\overline{p} : K^{\triangleright} \to \mathcal{C}$ be the constant map with value C. Using the characterization of colimits in Corollary 4.4.4.9, we deduce that $\overline{p}$ is a colimit diagram in $\mathcal{C}$. This proves (1) and (in view of the uniqueness of colimits up to equivalence) the "only if" direction of (2). To prove the converse, we suppose that $\overline{p}'$ is an arbitrary extension of p which carries each edge of $K^{\triangleright}$ to an equivalence in $\mathcal{C}$. Then $\overline{p}'$ factors through $\mathcal{C}'$. Since $K^{\triangleright}$ is weakly contractible, we conclude as above that $\overline{p}'$ is homotopic to a constant map and is therefore a colimit diagram. $\square$

4.4.5 Retracts and Idempotents

Let $\mathcal{C}$ be a category. An object $Y \in \mathcal{C}$ is said to be a *retract* of an object $X \in \mathcal{C}$ if there is a commutative diagram

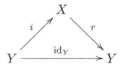

in $\mathcal{C}$. In this case we can identify Y with a subobject of X via the monomorphism i and think of r as a retraction from X onto $Y \subseteq X$. We observe also that the map $i \circ r : X \to X$ is idempotent. Moreover, this idempotent determines Y up to canonical isomorphism: we can recover Y as the equalizer of the pair of maps $(\mathrm{id}_X, i \circ r) : X \to X$ (or, dually, as the coequalizer of the same pair of maps). Consequently, we obtain an injective map from the collection of isomorphism classes of retracts of X to the set of idempotent maps $f : X \to X$. We will say that $\mathcal{C}$ is *idempotent complete* if this correspondence is bijective for every $X \in \mathcal{C}$: that is, if every idempotent map $f : X \to X$ comes from a (uniquely determined) retract of X. If $\mathcal{C}$ admits equalizers (or coequalizers), then $\mathcal{C}$ is idempotent complete.

These ideas can be adapted to the ∞-categorical setting in a straightforward way. If X and Y are objects of an ∞-category $\mathcal{C}$, then we say that Y is a *retract* of X if it is a retract of X in the homotopy category $h\mathcal{C}$. Equivalently, Y is a retract of X if there exists a 2-simplex $\Delta^2 \to \mathcal{C}$ corresponding to a diagram

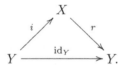

As in the classical case, there is a correspondence between retracts Y of X and idempotent maps $f : X \to X$. However, there are two important differences: first, the notion of an idempotent map needs to be interpreted in an ∞-categorical sense. It is not enough to require that $f = f \circ f$ in the homotopy category $h\mathcal{C}$. This would correspond to the condition that there is a path p joining f to $f \circ f$ in the endomorphism space of X, which would give rise to *two* paths from f to $f \circ f \circ f$. In order to have a hope of recovering Y, we need these paths to be homotopic. This condition does not even make sense unless p is specified; thus we must take p as part of the data of an idempotent map. In other words, in the ∞-categorical setting idempotence is not merely a condition but involves additional data (see Definition 4.4.5.4).

The second important difference between the classical and ∞-categorical theory of retracts is that in the ∞-categorical case one cannot recover a retract Y of X as the limit (or colimit) of a *finite* diagram involving X.

Example 4.4.5.1. Let R be a commutative ring and let $C_\bullet(R)$ be the category of complexes of finite free R-modules, so that an object of $C_\bullet(R)$

is a chain complex

$$\cdots \to M_1 \to M_0 \to M_{-1} \to \cdots$$

such that each M_i is a finite free R-module and $M_i = 0$ for $|i| \gg 0$; morphisms in $C_\bullet(R)$ are given by morphisms of chain complexes. There is a natural simplicial structure on the category $C_\bullet(R)$ for which the mapping spaces are Kan complexes; let $\mathcal{C} = \mathrm{N}(C_\bullet(R))$ be the associated ∞-category. Then $\mathcal{C}$ admits all finite limits and colimits ($\mathcal{C}$ is an example of a *stable* ∞-category; see [50]). However, $\mathcal{C}$ is idempotent complete if and only if every finitely generated projective R-module is stably free.

The purpose of this section is to define the notion of an *idempotent* in an ∞-category $\mathcal{C}$ and to obtain a correspondence between idempotents and retracts in $\mathcal{C}$.

Definition 4.4.5.2. The simplicial set Idem^+ is defined as follows: for every nonempty finite linearly ordered set J, $\mathrm{Hom}_{\mathrm{Set}_\Delta}(\Delta^J, \mathrm{Idem}^+)$ can be identified with the set of pairs $(J_0, \sim)$, where $J_0 \subseteq J$ and $\sim$ is an equivalence relation on J_0 which satisfies the following condition:

(∗) Let $i \leq j \leq k$ be elements of J such that $i, k \in J_0$ and $i \sim k$. Then $j \in J_0$ and $i \sim j \sim k$.

Let Idem denote the simplicial subset of Idem^+ corresponding to those pairs $(J_0, \sim)$ such that $J = J_0$. Let $\mathrm{Ret} \subseteq \mathrm{Idem}^+$ denote the simplicial subset corresponding to those pairs $(J_0, \sim)$ such that the quotient $J_0/\sim$ has at most one element.

Remark 4.4.5.3. The simplicial set Idem has exactly one nondegenerate simplex in each dimension n (corresponding to the equivalence relation $\sim$ on $\{0, 1, \ldots, n\}$ given by $(i \sim j) \Leftrightarrow (i = j)$), and the set of nondegenerate simplices of Idem is stable under passage to faces. In fact, Idem is characterized up to unique isomorphism by these two properties.

Definition 4.4.5.4. Let $\mathcal{C}$ be an ∞-category.

(1) An *idempotent* in $\mathcal{C}$ is a map of simplicial sets $\mathrm{Idem} \to \mathcal{C}$. We will refer to $\mathrm{Fun}(\mathrm{Idem}, \mathcal{C})$ as the ∞-*category of idempotents in* $\mathcal{C}$.

(2) A *weak retraction diagram* in $\mathcal{C}$ is a map of simplicial sets $\mathrm{Ret} \to \mathcal{C}$. We will refer to $\mathrm{Fun}(\mathrm{Ret}, \mathcal{C})$ as the ∞-*category of weak retraction diagrams in* $\mathcal{C}$.

(3) A *strong retraction diagram* in $\mathcal{C}$ is a map of simplicial sets $\mathrm{Idem}^+ \to \mathcal{C}$. We will refer to $\mathrm{Fun}(\mathrm{Idem}^+, \mathcal{C})$ as the ∞-*category of strong retraction diagrams in* $\mathcal{C}$.

We now spell out Definition 4.4.5.4 in more concrete terms. We first observe that Idem^+ has precisely two vertices. Once of these vertices, which we will denote by x, belongs to Idem, and the other, which we will denote by y, does not. The simplicial set Ret can be identified with the quotient of Δ^2 obtained by collapsing $\Delta^{\{0,2\}}$ to the vertex y. A weak retraction diagram $F : \text{Ret} \to \mathcal{C}$ in an ∞-category $\mathcal{C}$ can therefore be identified with a 2-simplex

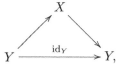

where $X = F(x)$ and $Y = F(y)$. In other words, it is precisely the datum that we need in order to exhibit Y as a retract of X in the homotopy category $h\mathcal{C}$.

To give an idempotent $F : \text{Idem} \to \mathcal{C}$ in $\mathcal{C}$, it suffices to specify the image under F of each nondegenerate simplex of Idem in each dimension $n \geq 0$. Taking $n = 0$, we obtain an object $X = F(x) \in \mathcal{C}$. Taking $n = 1$, we get a morphism $f : X \to X$. Taking $n = 2$, we get a 2-simplex of $\mathcal{C}$ corresponding to a diagram

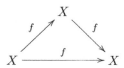

which verifies the equation $f = f \circ f$ in the homotopy category $h\mathcal{C}$. Taking $n > 2$, we get higher-dimensional diagrams which express the idea that f is not only idempotent "up to homotopy," but "up to coherent homotopy."

The simplicial set Idem^+ can be thought of as "interweaving" its simplicial subsets Idem and Ret, so that giving a strong retraction diagram $F : \text{Idem}^+ \to \mathcal{C}$ is equivalent to giving a weak retraction diagram

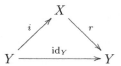

together with a coherently idempotent map $f = i \circ r : X \to X$. Our next result makes precise the sense in which f is "determined" by Y.

Lemma 4.4.5.5. *Let $J \subseteq \{0, \ldots, n\}$ and let $K \subseteq \Delta^n$ be the simplicial subset spanned by the nondegenerate simplices of Δ^n which do not contain Δ^J. Suppose that there exist $0 \leq i < j < k \leq n$ such that $i, k \in J$, $j \notin J$. Then the inclusion $K \subseteq \Delta^n$ is inner anodyne.*

Proof. Let P denote the collection of all subsets $J' \subseteq \{0, \ldots, n\}$ which contain $J \cup \{j\}$. Choose a linear ordering

$$\{J(1) \leq \cdots \leq J(m)\}$$

of P with the property that if $J(i) \subseteq J(j)$, then $i \leq j$. Let

$$K(k) = K \cup \bigcup_{1 \leq i \leq k} \Delta^{J(i)}.$$

Note that there are pushout diagrams

$$
\begin{array}{ccc}
\Lambda_j^{J(i)} & \longrightarrow & \Delta^{J(i)} \\
\downarrow & & \downarrow \\
K(i-1) & \longrightarrow & K(i).
\end{array}
$$

It follows that the inclusions $K(i-1) \subseteq K(i)$ are inner anodyne. Therefore the composite inclusion $K = K(0) \subseteq K(m) = \Delta^n$ is also inner anodyne. □

Proposition 4.4.5.6. *The inclusion* $\mathrm{Ret} \subseteq \mathrm{Idem}^+$ *is an inner anodyne map of simplicial sets.*

Proof. Let $\mathrm{Ret}_m \subseteq \mathrm{Idem}^+$ be the simplicial subset defined so that $(J_0, \sim)$: $\Delta^J \to \mathrm{Idem}^+$ factors through Ret_m if and only if the quotient $J_0/\sim$ has cardinality $\leq m$. We observe that there is a pushout diagram

$$
\begin{array}{ccc}
K & \longrightarrow & \Delta^{2m} \\
\downarrow & & \downarrow \\
\mathrm{Ret}_{m-1} & \longrightarrow & \mathrm{Ret}_m,
\end{array}
$$

where $K \subseteq \Delta^{2m}$ denote the simplicial subset spanned by those faces which do not contain $\Delta^{\{1,3,\ldots,2m-1\}}$. If $m \geq 2$, Lemma 4.4.5.5 implies that the upper horizontal arrow is inner anodyne, so that the inclusion $\mathrm{Ret}_{m-1} \subseteq \mathrm{Ret}_m$ is inner anodyne. The inclusion $\mathrm{Ret} \subseteq \mathrm{Idem}^+$ can be identified with an infinite composition

$$
\mathrm{Ret} = \mathrm{Ret}_1 \subseteq \mathrm{Ret}_2 \subseteq \cdots
$$

of inner anodyne maps and is therefore inner anodyne.   □

Corollary 4.4.5.7. *Let* $\mathcal{C}$ *be an* ∞-*category. Then the restriction map*

$$
\mathrm{Fun}(\mathrm{Idem}^+, \mathcal{C}) \to \mathrm{Fun}(\mathrm{Ret}, \mathcal{C})
$$

from strong retraction diagrams to weak retraction diagrams is a trivial fibration of simplicial sets. In particular, every weak retraction diagram in $\mathcal{C}$ *can be extended to a strong retraction diagram.*

We now study the relationship between strong retraction diagrams and idempotents in an ∞-category $\mathcal{C}$. We will need the following lemma, whose proof is somewhat tedious.

Lemma 4.4.5.8. *The simplicial set* Idem^+ *is an* ∞-*category.*

Proof. Suppose we are given $0 < i < n$ and a map $\Lambda_i^n \to \mathrm{Idem}^+$ corresponding to a compatible family of pairs $\{(J_k, \sim_k)\}_{k \neq i}$, where $J_k \subseteq \{0, \ldots, k-1, k+1, \ldots, n\}$ and $\sim_k$ is an equivalence relation J_k defining an element of $\mathrm{Hom}_{\mathrm{Set}_\Delta}(\Delta^{\{0,\ldots,k-1,k+1,\ldots,n\}}, \mathrm{Idem}^+)$. Let $J = \bigcup J_k$ and define a relation $\sim$ on J as follows: if $a, b \in J$, then $a \sim b$ if and only if either

$$
(\exists k \neq i)[(a, b \in J_k) \wedge (a \sim_k b)]
$$

or

$$(a \neq b \neq i \neq a) \wedge (\exists c \in J_a \cap J_b)[(a \sim_b c) \wedge (b \sim_a c)].$$

We must prove two things: that $(J, \sim) \in \operatorname{Hom}_{\operatorname{Set}_\Delta}(\Delta^n, \operatorname{Idem}^+)$ and that the restriction of $(J, \sim)$ to $\{0, \ldots, k-1, k+1, \ldots, n\}$ coincides with $(J_k, \sim_k)$ for $k \neq i$.

We first check that $\sim$ is an equivalence relation. It is obvious that $\sim$ is reflexive and symmetric. Suppose that $a \sim b$ and that $b \sim c$; we wish to prove that $a \sim c$. There are several cases to consider:

- Suppose that there exists $j \neq i$, $k \neq i$ such that $a, b \in J_j$, $b, c \in J_k$, and $a \sim_j b \sim_k c$. If $a \neq k$, then $a \in J_k$ and $a \sim_k b$, and we conclude that $a \sim c$ by invoking the transitivity of $\sim_k$. Therefore we may suppose that $a = k$. By the same argument, we may suppose that $b = j$; we therefore conclude that $a \sim c$.

- Suppose that there exists $k \neq i$ with $a, b \in J_k$, that $b \neq c \neq i \neq b$, and that there exists $d \in J_b \cap J_c$ with $a \sim_k b \sim_c d \sim_b c$. If $a = b$ or $a = c$ there is nothing to prove; assume therefore that $a \neq b$ and $a \neq c$. Then $a \in J_c$ and $a \sim_c b$, so by transitivity $a \sim_c d$. Similarly, $a \in J_b$ and $a \sim_b d$ so that $a \sim_b c$ by transitivity.

- Suppose that $a \neq b \neq i \neq a$, $b \neq c \neq i \neq b$, and that there exist $d \in J_a \cap J_b$ and $e \in J_b \cap J_c$ such that $a \sim_b d \sim_a b \sim_c e \sim_b c$. It will suffice to prove that $a \sim_b c$. If $c = d$, this is clear; let us therefore assume that $c \neq d$. By transitivity, it suffices to show that $d \sim_b e$. Since $c \neq d$, we have $d \in J_c$ and $d \sim_c b$, so that $d \sim_c e$ by transitivity and therefore $d \sim_b e$.

To complete the proof that $(J, \sim)$ belongs to $\operatorname{Hom}_{\operatorname{Set}_\Delta}(\Delta^n, \operatorname{Idem}^+)$, we must show that if $a < b < c$, $a \in J$, $c \in J$, and $a \sim c$, then $b \in J$ and $a \sim b \sim c$. There are two cases to consider. Suppose first that there exists $k \neq j$ such that $a, c \in J_k$ and $a \sim_k c$. These relations hold for any $k \notin \{i, a, c\}$. If it is possible to choose $k \neq b$, then we conclude that $b \in J_k$ and $a \sim_k b \sim_k c$, as desired. Otherwise, we may suppose that the choices $k = 0$ and $k = n$ are impossible, so that $a = 0$ and $c = n$. Then $a < i < c$, so that $i \in J_b$ and $a \sim_b i \sim_b c$. Without loss of generality, we may suppose $b < i$. Then $a \sim_c i$, so that $b \in J_c$ and $a \sim_c b \sim_c i$, as desired.

We now claim that $(J, \sim) : \Delta^n \to \operatorname{Idem}^+$ is an extension of the original map $\Lambda_i^n \to \operatorname{Idem}^+$. In other words, we claim that for $k \neq i$, $J_k = J \cap \{0, \ldots, k-1, k+1, \ldots, n\}$ and $\sim_k$ is the restriction of $\sim$ to J_k. The first claim is obvious. For the second, let us suppose that $a, b \in J_k$ and $a \sim b$. We wish to prove that $a \sim_k b$. It will suffice to prove that $a \sim_j b$ for any $j \notin \{i, a, b\}$. Since $a \sim b$, either such a j exists or $a \neq b \neq i \neq a$ and there exists $c \in J_a \cap J_b$ such that $a \sim_b c \sim_a b$. If there exists $j \notin \{a, b, c, i\}$, then we conclude that $a \sim_j c \sim_j b$ and hence $a \sim_j b$ by transitivity. Otherwise, we conclude that $c = k \neq i$ and that $0, n \in \{a, b, c\}$. Without loss of generality,

$i < c$; thus $0 \in \{a, b\}$ and we may suppose without loss of generality that $a < i$. Since $a \sim_b c$, we conclude that $i \in J_b$ and $a \sim_b i \sim_b c$. Consequently, $i \in J_a$ and $i \sim_a c \sim_a b$, so that $i \sim_a b$ by transitivity and therefore $i \sim_c b$. We now have $a \sim_c i \sim_c b$, so that $a \sim_c b$, as desired. $\qquad\square$

Remark 4.4.5.9. It is clear that $\mathrm{Idem} \subseteq \mathrm{Idem}^+$ is the full simplicial subset spanned by the vertex x and therefore an ∞-category as well.

According to Corollary 4.4.5.7, every weak retraction diagram

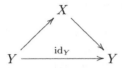

in an ∞-category $\mathcal{C}$ can be extended to a strong retraction diagram $F :$ $\mathrm{Idem}^+ \to \mathcal{C}$, which restricts to give an idempotent in $\mathcal{C}$. Our next goal is to show that F is canonically determined by the restriction $F|\,\mathrm{Idem}$.

Our next result expresses the idea that if an idempotent in $\mathcal{C}$ arises in this manner, then F is essentially unique.

Lemma 4.4.5.10. *The ∞-category* Idem *is weakly contractible.*

Proof. An explicit computation shows that the topological space $|\,\mathrm{Idem}\,|$ is connected, is simply connected, and has vanishing homology in degrees greater than zero. Alternatively, we can deduce this from Proposition 4.4.5.15 below. $\qquad\square$

Lemma 4.4.5.11. *The inclusion* $\mathrm{Idem} \subseteq \mathrm{Idem}^+$ *is a cofinal map of simplicial sets.*

Proof. According to Theorem 4.1.3.1, it will suffice to prove that the simplicial sets $\mathrm{Idem}_{x/}$ and $\mathrm{Idem}_{y/}$ are weakly contractible. The simplicial set $\mathrm{Idem}_{x/}$ is an ∞-category with an initial object and therefore weakly contractible. The projection $\mathrm{Idem}_{y/} \to \mathrm{Idem}$ is an isomorphism, and Idem is weakly contractible by Lemma 4.4.5.10. $\qquad\square$

Proposition 4.4.5.12. *Let $\mathcal{C}$ be an ∞-category and let $F : \mathrm{Idem}^+ \to \mathcal{C}$ be a strong retraction diagram. Then F is a left Kan extension of $F|\,\mathrm{Idem}$.*

Remark 4.4.5.13. Passing to opposite ∞-categories, it follows that a strong retraction diagram $F : \mathrm{Idem}^+ \to \mathcal{C}$ is also a *right* Kan extension of $F|\,\mathrm{Idem}$.

Proof. We must show that the induced map

$$(\mathrm{Idem}_{/y})^{\triangleright} \to (\mathrm{Idem}^+_{/y})^{\triangleright} \xrightarrow{G} \mathrm{Idem}^+ \xrightarrow{F} \mathcal{C}$$

is a colimit diagram. Consider the commutative diagram

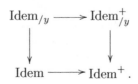

The lower horizontal map is cofinal by Lemma 4.4.5.11, and the vertical maps are isomorphisms: therefore the upper horizontal map is also cofinal. Consequently, it will suffice to prove that $F \circ G$ is a colimit diagram, which is obvious. $\qquad\square$

We will say that an idempotent $F : \mathrm{Idem} \to \mathcal{C}$ in an ∞-category $\mathcal{C}$ is *effective* if it extends to a map $\mathrm{Idem}^+ \to \mathcal{C}$. According to Lemma 4.3.2.13, F is effective if and only if it has a colimit in $\mathcal{C}$. We will say that $\mathcal{C}$ is *idempotent complete* if every idempotent in $\mathcal{C}$ is effective.

Corollary 4.4.5.14. *Let $\mathcal{C}$ be an ∞-category and let $\mathcal{D} \subseteq \mathrm{Fun}(\mathrm{Idem}, \mathcal{C})$ be the full subcategory spanned by the effective idempotents in $\mathcal{C}$. The restriction map $\mathrm{Fun}(\mathrm{Idem}^+, \mathcal{C}) \to \mathcal{D}$ is a trivial fibration. In particular, if $\mathcal{C}$ is idempotent complete, then we have a diagram*

$$\mathrm{Fun}(\mathrm{Ret}, \mathcal{C}) \leftarrow \mathrm{Fun}(\mathrm{Idem}^+, \mathcal{C}) \to \mathrm{Fun}(\mathrm{Idem}, \mathcal{C})$$

of trivial fibrations.

Proof. Combine Proposition 4.4.5.12 with Proposition 4.3.2.15. $\qquad\square$

By definition, an ∞-category $\mathcal{C}$ is idempotent complete if and only if every idempotent $\mathrm{Idem} \to \mathcal{C}$ has a colimit. In particular, if $\mathcal{C}$ admits all small colimits, then it is idempotent complete. As we noted above, this is not necessarily true if $\mathcal{C}$ admits only finite colimits. However, it turns out that filtered colimits do suffice: this assertion is not entirely obvious since the ∞-category Idem itself is not filtered.

Proposition 4.4.5.15. *Let A be a linearly ordered set with no largest element. Then there exists a cofinal map $p : \mathrm{N}(A) \to \mathrm{Idem}$.*

Proof. Let $p : \mathrm{N}(A) \to \mathrm{Idem}$ be the unique map which carries nondegenerate simplices to nondegenerate simplices. Explicitly, this map carries a simplex $\Delta^J \to \mathrm{N}(A)$ corresponding to a map $s : J \to A$ of linearly ordered sets to the equivalence relation $(i \sim j) \Leftrightarrow (s(i) = s(j))$. We claim that p is cofinal. According to Theorem 4.1.3.1, it will suffice to show that the fiber product $\mathrm{N}(A) \times_{\mathrm{Idem}} \mathrm{Idem}_{x/}$ is weakly contractible. We observe that $\mathrm{N}(A) \times_{\mathrm{Idem}} \mathrm{Idem}_x \simeq \mathrm{N}(A')$, where A' denotes the set $A \times \{0, 1\}$ equipped with the partial ordering

$$(\alpha, i) < (\alpha', j) \Leftrightarrow (j = 1) \wedge (\alpha < \alpha').$$

For each $\alpha \in A$, let $A_{<\alpha} = \{\alpha' \in A : \alpha' < \alpha\}$ and let

$$A'_\alpha = \{(\alpha', i) \in A' : (\alpha' < \alpha) \vee ((\alpha', i) = (\alpha, 1))\}.$$

By hypothesis, we can write A as a filtered union $\bigcup_{\alpha \in A} A_{<\alpha}$. It therefore suffices to prove that for each $\alpha \in A$ the map

$$f : \mathrm{N}(A_{<\alpha}) \times_{\mathrm{Idem}} \mathrm{Idem}_{x/} \to \mathrm{N}(A) \times_{\mathrm{Idem}} \mathrm{Idem}_{x/}$$

has a nullhomotopic geometric realization $|f|$. But this map factors through $\mathrm{N}(A'_\alpha)$, and $|\mathrm{N}(A'_\alpha)|$ is contractible because A'_α has a largest element. $\qquad\square$

Corollary 4.4.5.16. *Let κ be a regular cardinal and suppose that $\mathcal{C}$ is an ∞-category which admits κ-filtered colimits. Then $\mathcal{C}$ is idempotent complete.*

Proof. Apply Proposition 4.4.5.15 to the linearly ordered set consisting of all ordinals less than κ (and observe that this linearly ordered set is κ-filtered). $\qquad\square$

Chapter Five

Presentable and Accessible ∞-Categories

Many categories which arise naturally, such as the category $\mathcal{A}$ of abelian groups, are *large*: they have a proper class of objects even when the objects are considered only up to isomorphism. However, though $\mathcal{A}$ itself is large, it is in some sense determined by the much smaller category $\mathcal{A}_0$ of finitely generated abelian groups: $\mathcal{A}$ is naturally equivalent to the category of Ind-objects of $\mathcal{A}_0$. This remark carries more than simply philosophical significance. When properly exploited, it can be used to prove statements such as the following:

Proposition 5.0.0.1. *Let $F : \mathcal{A} \to \mathrm{Set}$ be a contravariant functor from $\mathcal{A}$ to the category of sets. Then F is representable by an object of $\mathcal{A}$ if and only if it carries colimits in $\mathcal{A}$ to limits in Set.*

Proposition 5.0.0.1 is valid not only for the category $\mathcal{A}$ of abelian groups but for any *presentable* category: that is, any category which possesses all (small) colimits and satisfies mild set-theoretic assumptions (such categories are referred to as *locally presentable* in [1]). Our goal in this chapter is to develop an ∞-categorical generalization of the theory of presentable categories and to obtain higher-categorical analogues of Proposition 5.0.0.1 and related results (such as the adjoint functor theorem).

The most basic example of a presentable ∞-category is the ∞-category $\mathcal{S}$ of spaces. More generally, we can define an ∞-category $\mathcal{P}(\mathcal{C})$ of *presheaves* (of spaces) on an arbitrary small ∞-category $\mathcal{C}$. We will study the properties of $\mathcal{P}(\mathcal{C})$ in §5.1; in particular, we will see that there exists a Yoneda embedding $j : \mathcal{C} \to \mathcal{P}(\mathcal{C})$ which is fully faithful, just as in ordinary category theory. Moreover, we give a characterization of $\mathcal{P}(\mathcal{C})$ in terms of $\mathcal{C}$: it is freely generated by the essential image of j under (small) colimits.

For every small ∞-category $\mathcal{C}$, the ∞-category $\mathcal{P}(\mathcal{C})$ is presentable. Conversely, any presentable ∞-category can be obtained as a *localization* of some presheaf ∞-category $\mathcal{P}(\mathcal{C})$ (Proposition 5.5.1.1). To make sense of this statement, we need a theory of localizations of ∞-categories. We will develop such a theory in §5.2 as part of a more general theory of adjoint functors between ∞-categories.

In §5.3, we will introduce, for every small ∞-category $\mathcal{C}$, an ∞-category $\mathrm{Ind}(\mathcal{C})$ of Ind-*objects of* $\mathcal{C}$. Roughly speaking, this is an ∞-category which is obtained from $\mathcal{C}$ by freely adjoining colimits for all *filtered* diagrams. It is characterized up to equivalence by the fact that $\mathrm{Ind}(\mathcal{C})$ contains a full subcategory equivalent to $\mathcal{C}$, which generates $\mathrm{Ind}(\mathcal{C})$ under filtered colimits and consists of *compact* objects.

The construction of Ind-categories will be applied in §5.4 to the study of *accessible* ∞-categories. Roughly speaking, an ∞-category $\mathcal{C}$ is *accessible* if it is generated under (sufficiently) filtered colimits by a small subcategory $\mathcal{C}^0 \subseteq \mathcal{C}$. We will prove that the class of accessible ∞-categories is stable under various categorical constructions. Results of this type will play an important technical role later in this book: they generally allow us to dispense with the set-theoretic aspects of an argument (such as cardinality estimation) and to focus instead on the more conceptual aspects.

We will say that an ∞-category $\mathcal{C}$ is *presentable* if $\mathcal{C}$ is accessible and admits (small) colimits. In §5.5, we will describe the theory of presentable ∞-categories in detail. In particular, we will generalize Proposition 5.0.0.1 to the ∞-categorical setting and prove an analogue of the adjoint functor theorem. We will also study localizations of presentable ∞-categories following ideas of Bousfield. The theory of presentable ∞-categories will play a vital role in the study of ∞-topoi, which is the subject of the next chapter.

5.1 ∞-CATEGORIES OF PRESHEAVES

The category of sets plays a central role in classical category theory. The primary reason for this is Yoneda's lemma, which asserts that for any category $\mathcal{C}$, the Yoneda embedding

$$j : \mathcal{C} \to \mathrm{Set}^{\mathcal{C}^{op}}$$

$$C \mapsto \mathrm{Hom}_{\mathcal{C}}(\bullet, C)$$

is fully faithful. Consequently, objects in $\mathcal{C}$ can be thought of as a kind of "generalized sets," and various questions about the category $\mathcal{C}$ can be reduced to questions about the category of sets.

If $\mathcal{C}$ is an ∞-category, then the mapping *sets* of the above discussion should be replaced by mapping *spaces*. Consequently, one should expect the Yoneda embedding to take values in presheaves of *spaces* rather than in presheaves of sets. To formalize this, we introduce the following notation:

Definition 5.1.0.1. Let S be a simplicial set. We let $\mathcal{P}(S)$ denote the simplicial set $\mathrm{Fun}(S^{op}, \mathcal{S})$; here $\mathcal{S}$ denotes the ∞-category of spaces defined in §1.2.16. We will refer to $\mathcal{P}(S)$ as the ∞-*category of presheaves on* S.

Remark 5.1.0.2. More generally, for any ∞-category $\mathcal{C}$, we might refer to $\mathrm{Fun}(S^{op}, \mathcal{C})$ as the ∞-*category of* $\mathcal{C}$-*valued presheaves on* S. Unless otherwise specified, the word "presheaf" will always refer to a $\mathcal{S}$-valued presheaf. This is somewhat nonstandard terminology: one usually understands the term "presheaf" to refer to a presheaf of sets rather than to a presheaf of spaces. The shift in terminology is justified by the fact that the important role of Set in ordinary category theory is taken on by $\mathcal{S}$ in the ∞-categorical setting.

Our goal in this section is to establish the basic properties of $\mathcal{P}(S)$. We begin in §5.1.1 by reviewing two other possible definitions of $\mathcal{P}(S)$: one via the theory of right fibrations over S, another via simplicial presheaves on the category $\mathfrak{C}[S]$. Using the "straightening" results of §2.2.3 and §4.2.4, we will show that all three of these definitions are equivalent.

The presheaf ∞-categories $\mathcal{P}(S)$ are examples of *presentable* ∞-categories (see §5.5). In particular, each $\mathcal{P}(S)$ admits small limits and colimits. We will give a proof of this assertion in §5.1.2 by reducing to the case where S is a point.

The main question regarding the ∞-category $\mathcal{P}(S)$ is how it relates to the original simplicial set S. In §5.1.3, we will construct a map $j : S \to \mathcal{P}(S)$, which is an ∞-categorical analogue of the usual Yoneda embedding. Just as in classical category theory, the Yoneda embedding is fully faithful. In particular, we note that any ∞-category $\mathcal{C}$ can be embedded in a larger ∞-category which admits limits and colimits; this observation allows us to construct an *idempotent completion* of $\mathcal{C}$, which we will study in §5.1.4.

In §5.1.5, we will characterize the ∞-category $\mathcal{P}(S)$ in terms of the Yoneda embedding $j : S \to \mathcal{P}(S)$. Roughly speaking, we will show that $\mathcal{P}(S)$ is freely generated by S under colimits (Theorem 5.1.5.6). In particular, if $\mathcal{C}$ is a category which admits colimits, then any diagram $f : S \to \mathcal{C}$ extends uniquely (up to homotopy) to a functor $F : \mathcal{P}(S) \to \mathcal{C}$. In §5.1.6, we will give a criterion for determining whether or not F is an equivalence.

5.1.1 Other Models for $\mathcal{P}(S)$

Let S be a simplicial set. We have defined the ∞-category $\mathcal{P}(S)$ of presheaves on S to be the mapping space $\mathrm{Fun}(S^{op}, \mathcal{S})$. However, there are several equivalent models which would serve equally well; we discuss two of them in this section.

Let $\mathcal{P}'_\Delta(S)$ denote the full subcategory of $(\mathrm{Set}_\Delta)_{/S}$ spanned by the right fibrations $X \to S$. We define $\mathcal{P}'(S)$ to be the simplicial nerve $\mathrm{N}(\mathcal{P}'_\Delta(S))$. Because $\mathcal{P}'_\Delta(S)$ is a fibrant simplicial category, $\mathcal{P}'(S)$ is an ∞-category. We will see in a moment that $\mathcal{P}'(S)$ is (naturally) equivalent to $\mathcal{P}(S)$. In order to do this, we need to introduce a third model.

Let $\phi : \mathfrak{C}[S]^{op} \to \mathcal{C}$ be an equivalence of simplicial categories. Let $\mathrm{Set}_\Delta^{\mathcal{C}}$ denote the category of simplicial functors $\mathcal{C} \to \mathrm{Set}_\Delta$ (which we may view as simplicial presheaves on $\mathcal{C}^{op}$). We regard $\mathrm{Set}_\Delta^{\mathcal{C}}$ as being endowed with the *projective* model structure defined in §A.3.3. With respect to this structure, $\mathrm{Set}_\Delta^{\mathcal{C}}$ is a simplicial model category; we let $\mathcal{P}''_\Delta(\phi) = (\mathrm{Set}_\Delta^{\mathcal{C}})^\circ$ denote the full simplicial subcategory consisting of fibrant-cofibrant objects, and we define $\mathcal{P}''(\phi)$ to be the simplicial nerve $\mathrm{N}(\mathcal{P}''_\Delta(\phi))$.

We are now ready to describe the relationship between these different models:

Proposition 5.1.1.1. *Let S be a simplicial set and let $\phi : \mathfrak{C}[S]^{op} \to \mathcal{C}$ be an equivalence of simplicial categories. Then there are (canonical) equivalences* *of ∞-categories*

$$\mathcal{P}(S) \xleftarrow{f} \mathcal{P}''(\phi) \xrightarrow{g} \mathcal{P}'(S).$$

Proof. The map f was constructed in Proposition 4.2.4.4; it therefore suffices to give a construction of g. We will regard the category $(\mathcal{S}et_\Delta)_{/S}$ of simplicial sets over S as endowed with the contravariant model structure defined in §2.1.4. This model structure is simplicial (Proposition 2.1.4.8) and the fibrant objects are precisely the right fibrations over S (Corollary 2.2.3.12). Thus, we may identify $\mathcal{P}'_\Delta(S)$ with the simplicial category $(\mathcal{S}et_\Delta)^\circ_{/S}$ of fibrant-cofibrant objects of $(\mathcal{S}et_\Delta)_{/S}$.

According to Theorem 2.2.1.2, the straightening and unstraightening functors (St_ϕ, Un_ϕ) determine a Quillen equivalence between the model categories $(\mathcal{S}et_\Delta)^{\mathcal{C}}$ and $(\mathcal{S}et_\Delta)_{/S}$. Moreover, for any $X \in (\mathcal{S}et_\Delta)_{/S}$ and any simplicial set K, there is a natural chain of equivalences

$$St_\phi(X \times K) \to (St_\phi X) \otimes |K|_{Q^\bullet} \to (St_\phi X) \otimes K.$$

(The fact that the first map is an equivalence follows easily from Proposition 3.2.1.13.) It follows from Proposition A.3.1.10 that Un_ϕ is endowed with the structure of a simplicial functor and induces an equivalence of simplicial categories

$$(\mathcal{S}et_\Delta^{\mathcal{C}})^\circ \to (\mathcal{S}et_\Delta)^\circ_{/S}.$$

We obtain the desired equivalence g by passing to the simplicial nerve. □

5.1.2 Colimits in ∞-Categories of Functors

Let S be an arbitrary simplicial set. Our goal in this section is to prove that the ∞-category $\mathcal{P}(S)$ of presheaves on S admits small limits and colimits. There are (at least) three approaches to proving this:

(1) According to Proposition 5.1.1.1, we may identify $\mathcal{P}(S)$ with the ∞-category underlying the simplicial model category $\mathcal{S}et_\Delta^{\mathcal{C}[S]^{op}}$. We can then deduce the existence of limits and colimits in $\mathcal{P}(S)$ by invoking Corollary 4.2.4.8.

(2) Since the ∞-category $\mathcal{S}$ classifies left fibrations, the ∞-category $\mathcal{P}(S)$ classifies left fibrations over S^{op}: in other words, homotopy classes of maps $K \to \mathcal{P}(S)$ can be identified with equivalence classes of left fibrations $X \to K \times S^{op}$. It is possible to generalize Proposition 3.3.4.5 and Corollary 3.3.3.3 to describe limits and colimits in $\mathcal{P}(S)$ entirely in the language of left fibrations. The existence problem can then be solved by exhibiting explicit constructions of left fibrations.

(3) Applying either (1) or (2) in the case where S is a point, we can deduce that the ∞-category $\mathcal{S} \simeq \mathcal{P}(*)$ admits limits and colimits. We can then attempt to deduce the same result for $\mathcal{P}(S) = \text{Fun}(S^{op}, \mathcal{S})$ using a general result about (co)limits in functor categories (Proposition 5.1.2.2).

Although approach (1) is probably the quickest, we will adopt approach (3) because it gives additional information: our proof will show that limits and colimits in $\mathcal{P}(S)$ can be computed pointwise. The same proof will also apply to ∞-categories of $\mathcal{C}$-valued presheaves in the case where $\mathcal{C}$ is not necessarily the ∞-category $\mathcal{S}$ of spaces.

Lemma 5.1.2.1. *Let* $q : Y \to S$ *be a Cartesian fibration of simplicial sets and let* $\mathcal{C} = \mathrm{Map}_S(S, Y)$ *denote the* ∞*-category of sections of* q. *Let* $p : S \to Y$ *be an object of* $\mathcal{C}$ *having the property that* $p(s)$ *is an initial object of the fiber* Y_s *for each vertex* s *of* S. *Then* p *is an initial object of* $\mathcal{C}$.

Proof. By Proposition 4.2.2.4, the map $Y^{ps/} \to S$ is a Cartesian fibration. By hypothesis, for each vertex s of S, the map $Y^{ps/} \times_S \{s\} \to Y_s$ is a trivial fibration. It follows that the projection $Y^{ps/} \to Y$ is an equivalence of Cartesian fibrations over S and therefore a categorical equivalence; taking sections over S, we obtain another categorical equivalence

$$\mathrm{Map}_S(S, Y^{ps/}) \to \mathrm{Map}_S(Y, S).$$

But this map is just the left fibration $j : \mathcal{C}^{p/} \to \mathcal{C}$; it follows that j is a categorical equivalence. Applying Propostion 3.3.1.5 to the diagram

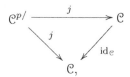

we deduce that j induces categorical equivalences $\mathcal{C}^{p/} \times_{\mathcal{C}} \{t\} \to \{t\}$ for each vertex t of Q. Thus the fibers of j are contractible Kan complexes, so that j is a trivial fibration (by Lemma 2.1.3.4) and p is an initial object of $\mathcal{C}$, as desired. □

Proposition 5.1.2.2. *Let* K *be a simplicial set,* $q : X \to S$ *a Cartesian fibration, and* $p : K \to \mathrm{Map}_S(S, X)$ *a diagram. For each vertex* s *of* S, *we let* $p_s : K \to X_s$ *be the induced map. Suppose, furthermore, that each* p_s *has a colimit in the* ∞*-category* X_s. *Then*

(1) *There exists a map* $\overline{p} : K \diamond \Delta^0 \to \mathrm{Map}_S(S, X)$ *which extends* p *and induces a colimit diagram* $\overline{p} : K \diamond \Delta^0 \to X_s$ *for each vertex* $s \in S$.

(2) *An arbitrary extension* $\overline{p} : K \diamond \Delta^0 \to \mathrm{Map}_S(S, X)$ *of* p *is a colimit for* p *if and only if each* $\overline{p}_s : K \diamond \Delta^0 \to X_s$ *is a colimit for* p_s.

Proof. Choose a factorization $K \to K' \to \mathrm{Map}_S(S, X)$ of p, where $K \to K'$ is inner anodyne (and therefore a categorical equivalence) and $K' \to \mathcal{C}^S$ is an inner fibration (so that K' is an ∞-category). The map $K \to K'$ is a categorical equivalence and therefore cofinal. We are free to replace K by K' and may thereby assume that K is an ∞-category.

We apply Proposition 4.2.2.7 to the Cartesian fibration $X \to S$ and the diagram $p_S : K \times S \to X$ determined by the map p. We deduce that there exists a map

$$\overline{p}_S : (K \times S) \diamond_S S = (K \diamond \Delta^0) \times S \to X$$

having the property that its restriction to the fiber over each $s \in S$ is a colimit of p_s; this proves (1).

The "if" direction of (2) follows immediately from Lemma 5.1.2.1. The "only if" direction follows from (1) and the fact that colimits, when they exist, are unique up to equivalence. $\square$

Corollary 5.1.2.3. *Let K and S be simplicial sets and let $\mathcal{C}$ be an ∞-category which admits K-indexed colimits. Then*

(1) *The ∞-category $\mathrm{Fun}(S, \mathcal{C})$ admits K-indexed colimits.*

(2) *A map $K^{\triangleright} \to \mathrm{Fun}(S, \mathcal{C})$ is a colimit diagram if and only if, for each vertex $s \in S$, the induced map $K^{\triangleright} \to \mathcal{C}$ is a colimit diagram.*

Proof. Apply Proposition 5.1.2.2 to the projection $\mathcal{C} \times S \to S$. $\square$

Corollary 5.1.2.4. *Let S be a simplicial set. The ∞-category $\mathcal{P}(S)$ of presheaves on S admits all small limits and colimits.*

5.1.3 Yoneda's Lemma

In this section, we will construct the ∞-categorical analogue of the Yoneda embedding and prove that it is fully faithful. We begin with a somewhat naive approach based on the formalism of simplicial categories. We note that an analogue of Yoneda's Lemma is valid in enriched category theory (with the usual proof). Namely, suppose that $\mathcal{C}$ is a category enriched over another category $\mathcal{E}$. Then there is an enriched Yoneda embedding

$$i : \mathcal{C} \to \mathcal{E}^{\mathcal{C}^{op}}$$

$$X \mapsto \mathrm{Map}_{\mathcal{C}}(\bullet, X).$$

Consequently, for any simplicial category $\mathcal{C}$, one obtains a fully faithful embedding i of $\mathcal{C}$ into the simplicial category $\mathrm{Map}_{\mathcal{C}at_{\Delta}}(\mathcal{C}^{op}, \mathrm{Set}_{\Delta})$ of simplicial functors from $\mathcal{C}^{op}$ into Set_{Δ}. In fact, i is fully faithful in the strong sense that it induces *isomorphisms* of simplicial sets

$$\mathrm{Map}_{\mathcal{C}}(X, Y) \to \mathrm{Map}_{\mathrm{Set}_{\Delta}^{\mathcal{C}^{op}}}(i(X), i(Y))$$

rather than merely weak homotopy equivalences. Unfortunately, this assertion does not necessarily have any ∞-categorical content because the simplicial category $\mathrm{Set}_{\Delta}^{\mathcal{C}^{op}}$ does not generally represent the correct ∞-category of functors from $\mathcal{C}^{op}$ to Set_{Δ}.

Let us describe an analogous construction in the setting of ∞-categories. Let K be a simplicial set and set $\mathcal{C} = \mathfrak{C}[K]$. Then $\mathcal{C}$ is a simplicial category, so

$$(X, Y) \mapsto \operatorname{Sing} |\operatorname{Hom}_{\mathcal{C}}(X, Y)|$$

determines a simplicial functor from $\mathcal{C}^{op} \times \mathcal{C}$ to the category $\mathcal{K}an$. The functor $\mathfrak{C}$ does not commute with products, but there exists a natural map $\mathfrak{C}[K^{op} \times K] \to \mathcal{C}^{op} \times \mathcal{C}$. Composing with this map, we obtain a simplicial functor $\mathfrak{C}[K^{op} \times K] \to \mathcal{K}an$. Passing to the adjoint, we get a map of simplicial sets $K^{op} \times K \to \mathcal{S}$, which we can identify with

$$j : K \to \operatorname{Fun}(K^{op}, \mathcal{S}) = \mathcal{P}(K).$$

We shall refer to j (or more generally, to any functor equivalent to j) as the *Yoneda embedding*.

Proposition 5.1.3.1 (∞-Categorical Yoneda Lemma). *Let K be a simplicial set. Then the Yoneda embedding $j : K \to \mathcal{P}(K)$ is fully faithful.*

Proof. Let $\mathcal{C}' = \operatorname{Sing} |\mathfrak{C}[K^{op}]|$ be the "fibrant replacement" for $\mathcal{C} = \mathfrak{C}[K^{op}]$. We endow $\operatorname{Set}_{\Delta}^{\mathcal{C}'}$ with the *projective* model structure described in §A.3.3.

We note that the Yoneda embedding factors as a composition

$$K \xrightarrow{j'} \operatorname{N}((\operatorname{Set}_{\Delta}^{\mathcal{C}'})^{\circ}) \xrightarrow{j''} \operatorname{Fun}(K^{op}, \mathcal{S}),$$

where j'' is the map of Proposition 4.2.4.4 and consequently a categorical equivalence. It therefore suffices to prove that j' is fully faithful. For this, we need only show that the adjoint map

$$J : \mathfrak{C}[K] \to \operatorname{Set}_{\Delta}^{\mathcal{C}'}$$

is a fully faithful functor between simplicial categories. We now observe that J is the composition of an equivalence $\mathfrak{C}[K] \to (\mathcal{C}')^{op}$ with the (simplicially enriched) Yoneda embedding $(\mathcal{C}')^{op} \to \operatorname{Set}_{\Delta}^{\mathcal{C}'}$, which is fully faithful by virtue of the classical (enriched) version of Yoneda's Lemma. $\qquad\square$

We conclude by establishing another pleasant property of the Yoneda embedding:

Proposition 5.1.3.2. *Let $\mathcal{C}$ be a small ∞-category and $j : \mathcal{C} \to \mathcal{P}(\mathcal{C})$ the Yoneda embedding. Then j preserves all small limits which exist in $\mathcal{C}$.*

Proof. Let $p : K \to \mathcal{C}$ be a small diagram having a limit in $\mathcal{C}$. We wish to show that j carries any limit for p to a limit of $j \circ p$. Choose a category $\mathcal{J}$ and a cofinal map $\operatorname{N}(\mathcal{J}^{op}) \to K^{op}$ (the existence of which is guaranteed by Proposition 4.2.3.14). Replacing K by $\operatorname{N}(\mathcal{J})$, we may suppose that K is the nerve of a category. Let $\bar{p} : \operatorname{N}(\mathcal{J})^{\triangleleft} \to \mathcal{C}$ be a limit for p.

We recall the definition of the Yoneda embedding. It involves the choice of an equivalence $\mathfrak{C}[\mathcal{C}] \to \mathcal{D}$, where $\mathcal{D}$ is a fibrant simplicial category. For definiteness, we took $\mathcal{D}$ to be $\operatorname{Sing} |\mathfrak{C}[\mathcal{C}]|$. However, we could just as well choose some other fibrant simplicial category $\mathcal{D}'$ equivalent to $\mathfrak{C}[\mathcal{C}]$ and obtain a "modified" Yoneda embedding $j' : \mathcal{C} \to \mathcal{P}(\mathcal{C})$; it is easy to see that j' and j are equivalent functors, so it suffices to show that j' preserves the limit of p. Using Corollary 4.2.4.7, we may suppose that $\overline{p}$ is obtained from a functor between simplicial categories $\overline{q} : \{x\} \star \mathcal{J} \to \mathcal{D}$ by passing to the nerve. According to Theorem 4.2.4.1, $\overline{q}$ is a homotopy limit of $q = \overline{q}| \mathcal{J}$. Consequently, for each object $Z \in \mathcal{D}$, the induced functor

$$\overline{q}_Z : I \mapsto \operatorname{Hom}_{\mathcal{D}}(Z, \overline{q}(I))$$

is a homotopy limit of $q_Z = \overline{q}_Z| \mathcal{J}$. Taking Z to be the image of an object C of $\mathcal{C}$, we deduce that

$$N(\mathcal{J})^{\triangleleft} \to \mathcal{C} \xrightarrow{j'} \mathcal{P}(\mathcal{C}) \to \mathcal{S}$$

is a limit for its restriction to $N(\mathcal{J})$, where the map on the right is given by evaluation at C. Proposition 5.1.2.2 now implies that $j' \circ \overline{p}$ is a limit for $j' \circ p$, as desired. □

5.1.4 Idempotent Completions

Recall that an ∞-category $\mathcal{C}$ is said to be *idempotent complete* if every functor Idem $\to \mathcal{C}$ admits a colimit in $\mathcal{C}$ (see §4.4.5). If an ∞-category $\mathcal{C}$ is not idempotent complete, then we can attempt to correct the situation by passing to a larger ∞-category.

Definition 5.1.4.1. Let $f : \mathcal{C} \to \mathcal{D}$ be a functor between ∞-categories. We will say that f *exhibits* $\mathcal{D}$ *as an idempotent completion of* $\mathcal{C}$ if $\mathcal{D}$ is idempotent complete, f is fully faithful, and every object of $\mathcal{D}$ is a retract of $f(C)$ for some object $C \in \mathcal{C}$.

Our goal in this section is to show that ∞-category $\mathcal{C}$ has an idempotent completion $\mathcal{D}$ which is unique up to equivalence. The uniqueness is a consequence of Proposition 5.1.4.9 below. The existence question is much easier to address.

Proposition 5.1.4.2. *Let $\mathcal{C}$ be an ∞-category. Then $\mathcal{C}$ admits an idempotent completion.*

Proof. Enlarging the universe if necessary, we may suppose that $\mathcal{C}$ is small. Let $\mathcal{C}'$ denote the full subcategory of $\mathcal{P}(\mathcal{C})$ spanned by those objects which are retracts of objects which belong to the image of the Yoneda embedding $j : \mathcal{C} \to \mathcal{P}(\mathcal{C})$. Then $\mathcal{C}'$ is stable under retracts in $\mathcal{P}(\mathcal{C})$. Since $\mathcal{P}(\mathcal{C})$ admits all small colimits, Corollary 4.4.5.16 implies that $\mathcal{P}(\mathcal{C})$ is idempotent complete. It follows that $\mathcal{C}'$ is idempotent complete. Proposition 5.1.3.1 implies that the Yoneda embedding $j : \mathcal{C} \to \mathcal{C}'$ is fully faithful and therefore exhibits $\mathcal{C}'$ as an idempotent completion of $\mathcal{C}$. □

We now address the question of uniqueness for idempotent completions. First, we need a few preliminary results.

Lemma 5.1.4.3. *Let $\mathcal{C}$ be an ∞-category which is idempotent complete and let $p : K \to \mathcal{C}$ be a diagram. Then $\mathcal{C}_{/p}$ and $\mathcal{C}_{p/}$ are also idempotent complete.*

Proof. By symmetry, it will suffice to prove that $\mathcal{C}_{/p}$ is idempotent complete. Let $q : \mathcal{C}_{/p} \to \mathcal{C}$ be the associated right fibration and let $F : \mathrm{Idem} \to \mathcal{C}_{/p}$ be an idempotent. We will show that F has a limit. Since $\mathcal{C}$ is idempotent complete, $q \circ F$ has a limit $\overline{q \circ F} : \mathrm{Idem}^{\triangleleft} \to \mathcal{C}$. Consider the lifting problem

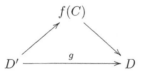

The right vertical map is a right fibration, and the left vertical map is right anodyne (Lemma 4.2.3.6), so that there exists a dotted arrow $\overline{F}$ as indicated. Using Proposition 4.4.2.9, we deduce that $\overline{F}$ is a limit of F. $\qquad\square$

Lemma 5.1.4.4. *Let $f : \mathcal{C} \to \mathcal{D}$ be a functor between ∞-categories which exhibits $\mathcal{D}$ as an idempotent completion of $\mathcal{C}$ and let $p : K \to \mathcal{D}$ be a diagram. Then the induced map $f_{/p} : \mathcal{C} \times_{\mathcal{D}} \mathcal{D}_{/p} \to \mathcal{D}_{/p}$ exhibits $\mathcal{D}_{/p}$ as an idempotent completion of $\mathcal{C} \times_{\mathcal{D}} \mathcal{D}_{/p}$.*

Proof. Lemma 5.1.4.3 asserts that $\mathcal{D}_{/p}$ is idempotent complete. We must show that every object $\overline{D} \in \mathcal{D}_{/p}$ is a retract of $f_{/p}(\overline{C})$ for some $\overline{C} \in \mathcal{C} \times_{\mathcal{D}} \mathcal{D}_{/p}$. Let $q : \mathcal{D}_{/p} \to \mathcal{D}$ be the projection and set $D = q(\overline{D})$. Since f exhibits $\mathcal{D}$ as an idempotent completion of $\mathcal{C}$, there is a diagram

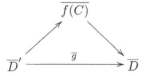

in $\mathcal{D}$, where g is an equivalence. Since q is a right fibration, we can lift this to a diagram

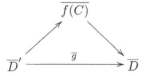

in $\mathcal{D}_{/q}$. Since $\overline{g}$ is q-Cartesian and g is an equivalence, $\overline{g}$ is an equivalence. It follows that $\overline{D}$ is a retract of $\overline{f(C)}$. By construction, $\overline{f(C)} = f_{/p}(\overline{C})$ for an appropriately chosen object $\overline{C} \in \mathcal{C} \times_{\mathcal{D}} \mathcal{D}_{/p}$. $\qquad\square$

Lemma 5.1.4.5. *Let $f : \mathcal{C} \to \mathcal{D}$ be a functor between ∞-categories which exhibits $\mathcal{D}$ as an idempotent completion of $\mathcal{C}$. Suppose that $\mathcal{D}$ has an initial object $\emptyset$. Then $\mathcal{C}$ is weakly contractible as a simplicial set.*

Proof. Without loss of generality, we may suppose that $\mathcal{C}$ is a full subcategory of $\mathcal{D}$ and that f is the inclusion. Since f exhibits $\mathcal{D}$ as an idempotent completion of $\mathcal{C}$, the initial object $\emptyset$ of $\mathcal{D}$ admits a map $f : C \to \emptyset$, where $C \in \mathcal{C}$. The ∞-category $\mathcal{C}_{C/}$ has an initial object and is therefore weakly contractible. Since composition

$$\mathcal{C}_{f/} \to \mathcal{C}_{C/} \to \mathcal{C}$$

is both a weak homotopy equivalence (in fact, a trivial fibration) and weakly nullhomotopic, we conclude that $\mathcal{C}$ is weakly contractible. $\qquad\square$

Lemma 5.1.4.6. *Let $f : \mathcal{C} \to \mathcal{D}$ be a functor between ∞-categories which exhibits $\mathcal{D}$ as an idempotent completion of $\mathcal{C}$. Then f is cofinal.*

Proof. According to Theorem 4.1.3.1, it suffices to prove that for every object $D \in \mathcal{D}$, the simplicial set $\mathcal{C} \times_{\mathcal{D}} \mathcal{D}_{D/}$ is weakly contractible. Lemma 5.1.4.4 asserts that $f_{D/}$ is also an idempotent completion, and Lemma 5.1.4.5 completes the proof. $\qquad\square$

Lemma 5.1.4.7. *Let $F : \mathcal{C} \to \mathcal{D}$ be a functor between ∞-categories and let $\mathcal{C}^0 \subseteq \mathcal{C}$ be a full subcategory such that the inclusion exhibits $\mathcal{C}$ as an idempotent completion of $\mathcal{C}^0$. Then F is a left Kan extension of $F|\,\mathcal{C}^0$.*

Proof. We must show that for every object $C \in \mathcal{C}$, the composite map

$$(\mathcal{C}_{/C}^0)^{\triangleright} \to (\mathcal{C}_{/C})^{\triangleright} \overset{G}{\to} \mathcal{C} \overset{F}{\to} \mathcal{D}$$

is a colimit diagram in $\mathcal{D}$. Lemma 5.1.4.4 guarantees that $\mathcal{C}_{/C}^0 \subseteq \mathcal{C}_{/C}$ is an idempotent completion and therefore cofinal by Lemma 5.1.4.6. Consequently, it suffices to prove that $F \circ G$ is a colimit diagram, which is obvious. $\qquad\square$

Lemma 5.1.4.8. *Let $\mathcal{C}$ and $\mathcal{D}$ be ∞-categories which are idempotent complete and let $\mathcal{C}^0 \subseteq \mathcal{C}$ be a full subcategory such that the inclusion exhibits $\mathcal{C}$ as an idempotent completion of $\mathcal{C}^0$. Then any functor $F_0 : \mathcal{C}^0 \to \mathcal{D}$ has an extension $F : \mathcal{C} \to \mathcal{D}$.*

Proof. We will suppose that the ∞-categories $\mathcal{C}$ and $\mathcal{D}$ are small. Let $\mathcal{P}(\mathcal{D})$ be the ∞-category of presheaves on $\mathcal{D}$ (see §5.1), $j : \mathcal{D} \to \mathcal{P}(\mathcal{D})$ the Yoneda embedding, and $\mathcal{D}'$ the essential image of j. According to Proposition A.2.3.1, it will suffice to prove that $j \circ F_0$ can be extended to a functor $F' : \mathcal{C} \to \mathcal{D}'$. Since $\mathcal{P}(\mathcal{D})$ admits small colimits, we can choose $F' : \mathcal{C} \to \mathcal{P}(\mathcal{D})$ to be a left Kan extension of $j \circ F_0$. Every object of $\mathcal{C}$ is a retract of an object of $\mathcal{C}^0$, so that every object in the essential image of F' is a retract of the Yoneda image of an object of $\mathcal{D}$. Since $\mathcal{D}$ is idempotent complete, it follows that the F' factors through $\mathcal{D}'$. $\qquad\square$

Proposition 5.1.4.9. *Let $f : \mathcal{C} \to \mathcal{D}$ be a functor which exhibits $\mathcal{D}$ as the idempotent completion of $\mathcal{C}$ and let $\mathcal{E}$ be an ∞-category which is idempotent complete. Then composition with f induces an equivalence of ∞-categories $f^* : \mathrm{Fun}(\mathcal{D}, \mathcal{E}) \to \mathrm{Fun}(\mathcal{C}, \mathcal{E})$.*

Proof. Without loss of generality, we may suppose that f is the inclusion of a full subcategory. In this case, we combine Lemma 5.1.4.7, Lemma 5.1.4.8, and Proposition 4.3.2.15 to deduce that f^* is a trivial fibration. □

Remark 5.1.4.10. Let $\mathcal{C}$ be a small ∞-category and let $f : \mathcal{C} \to \mathcal{C}'$ be an idempotent completion of $\mathcal{C}$. The proof of Proposition 5.1.4.2 shows that $\mathcal{C}'$ is equivalent to a full subcategory of $\mathcal{P}(\mathcal{C})$ and therefore locally small (see §5.4.1). Moreover, every object of $h\mathcal{C}'$ is the image of some retraction map in $h\mathcal{C}$; it follows that the set of equivalence classes of objects in $\mathcal{C}'$ is bounded in size. It follows that $\mathcal{C}'$ is essentially small.

5.1.5 The Universal Property of $\mathcal{P}(S)$

Let S be a (small) simplicial set. We have defined $\mathcal{P}(S)$ to be the ∞-category of maps from S^{op} into the ∞-category $\mathcal{S}$ of spaces. Informally, we may view $\mathcal{P}(S)$ as the limit of a diagram in the ∞-bicategory of (large) ∞-categories: namely, the constant diagram carrying S^{op} to $\mathcal{S}$. In more concrete terms, our definition of $\mathcal{P}(S)$ leads immediately to a characacterization of $\mathcal{P}(S)$ by a universal mapping property: for every ∞-category $\mathcal{C}$, there is an equivalence of ∞-categories (in fact, an isomorphism of simplicial sets)

$$\mathrm{Fun}(\mathcal{C}, \mathcal{P}(S)) \simeq \mathrm{Fun}(\mathcal{C} \times S^{op}, \mathcal{S}).$$

The goal of this section is to give a dual characterization of $\mathcal{P}(S)$: it may also be viewed as a *colimit* of copies of $\mathcal{S}$ indexed by S. However, this colimit needs to be understood in an appropriate ∞-bicategory of ∞-categories where the morphisms are given by *colimit-preserving* functors. In other words, we will show that $\mathcal{P}(S)$ is in some sense "freely generated" by S under small colimits (Theorem 5.1.5.6). First, we need to introduce a bit of notation.

Notation 5.1.5.1. Let $\mathcal{C}$ be an ∞-category and S a simplicial set. We will let $\mathrm{Fun}^L(\mathcal{P}(S), \mathcal{C})$ denote the full subcategory of $\mathrm{Fun}(\mathcal{P}(S), \mathcal{C})$ spanned by those functors $\mathcal{P}(S) \to \mathcal{C}$ which preserve small colimits.

The motivation for this notation is as follows: in §5.2.6, we will use the notation $\mathrm{Fun}^L(\mathcal{D}, \mathcal{C})$ to denote the full subcategory of $\mathrm{Fun}(\mathcal{D}, \mathcal{C})$ spanned by those functors which are *left adjoints*. In §5.5.2, we will see that when $\mathcal{D} = \mathcal{P}(S)$ (or more generally, when $\mathcal{D}$ is presentable), then a functor $\mathcal{D} \to \mathcal{C}$ is a left adjoint if and only if it preserves small colimits (see Corollary 5.5.2.9 and Remark 5.5.2.10).

We wish to prove that if $\mathcal{C}$ is an ∞-category which admits small colimits, then any map $S \to \mathcal{C}$ extends in an essentially unique fashion to a colimit-preserving functor $\mathcal{P}(S) \to \mathcal{C}$. To prove this, we need a second characterization of the colimit-preserving functors $f : \mathcal{P}(S) \to \mathcal{C}$: they are precisely those functors which are left Kan extensions of their restriction to the essential image of the Yoneda embedding.

Lemma 5.1.5.2. *Let S be a small simplicial set, let s be a vertex of S, let $e : \mathcal{P}(S) \to \mathcal{S}$ be the map given by evaluation at s, and let $f : \mathcal{C} \to \mathcal{P}(S)$*

be the associated left fibration (see §3.3.2). Then f is corepresentable by the object $j(s) \in \mathcal{P}(S)$, where $j : S \to \mathcal{P}(S)$ denotes the Yoneda embedding.

Proof. Without loss of generality, we may suppose that S is an ∞-category. We make use of the equivalent model $\mathcal{P}'(S)$ of §5.1.1. Observe that the functor $f : \mathcal{P}(S) \to \mathcal{S}$ is equivalent to $f' : \mathcal{P}'(S) \to \mathcal{S}$, where f' is the nerve of the simplicial functor $\mathcal{P}'_{\Delta}(S) \to \mathcal{K}\mathrm{an}$ which associates to each left fibration $Y \to S$ the fiber $Y_s = Y \times_S \{s\}$. Furthermore, under the equivalence of $\mathcal{P}(S)$ with $\mathcal{P}'(S)$, the object $j(s)$ corresponds to a left fibration $X(s) \to S$ which is corepresented by s. Then $X(s)$ contains an initial object x lying over s. The choice of x determines a point $\eta \in \pi_0 f'(X(s))$. According to Proposition 4.4.4.5, to show that $X(s)$ corepresents f', it suffices to show that for every left fibration $X \to S$, the map

$$\mathrm{Map}_S(X(s), Y) \to Y_s,$$

given by evaluation at x, is a homotopy equivalence of Kan complexes. We may rewrite the space on the right hand side as $\mathrm{Map}_S(\{x\}, Y)$. According to Proposition 2.1.4.8, the covariant model structure on $(\mathrm{Set}_{\Delta})_{/S}$ is compatible with the simplicial structure. It therefore suffices to prove that the inclusion $i : \{x\} \subseteq X(s)$ is a covariant equivalence. But this is clear since i is the inclusion of an initial object and therefore left anodyne. $\square$

Lemma 5.1.5.3. *Let S be a small simplicial set and let $j : S \to \mathcal{P}(S)$ denote the Yoneda embedding. Then $\mathrm{id}_{\mathcal{P}(S)}$ is a left Kan extension of j along itself.*

Proof. Let $\mathcal{C} \subseteq \mathcal{P}(S)$ denote the essential image of j. According to Proposition 5.1.3.1, j induces an equivalence $S \to \mathcal{C}$. It therefore suffices to prove that $\mathrm{id}_{\mathcal{P}(S)}$ is a left Kan extension of its restriction to $\mathcal{C}$. Let X be an object of $\mathcal{P}(S)$; we must show that the natural map

$$\phi : \mathcal{C}^{\triangleright}_{/X} \subseteq \mathcal{P}(S)^{\triangleright}_{/X} \to \mathcal{P}(S)$$

is a colimit diagram.

According to Proposition 5.1.2.2, it will suffice to prove that, for each vertex s of S, the map

$$\phi_s : \mathcal{C}^{\triangleright}_{/X} \to \mathcal{S}$$

given by composing ϕ with the evaluation map is a colimit diagram in $\mathcal{S}$. Let $\mathcal{D} \to \mathcal{C}^{\triangleright}_{/X}$ be the pullback of the universal left fibration along ϕ_s and let $\mathcal{D}^0 \subseteq \mathcal{D}$ be the preimage in $\mathcal{D}$ of $\mathcal{C}_{/X} \subseteq \mathcal{C}^{\triangleright}_{/X}$. According to Proposition 3.3.4.5, it will suffice to prove that the inclusion $\mathcal{D}^0 \subseteq \mathcal{D}$ is a weak homotopy equivalence of simplicial sets.

Let $C = j(s)$. Let $\mathcal{E} = \mathcal{C}^{\triangleright}_{/X} \times_{\mathcal{P}(S)} \mathcal{P}(S)_{C/}$, let $\mathcal{E}^0 = \mathcal{C}_{/X} \times_{\mathcal{C}} \mathcal{C}_{C/} \subseteq \mathcal{E}$, and let $\mathcal{E}^1 = \mathcal{C}_{/X} \times_{\mathcal{C}} \{\mathrm{id}_C\} \subseteq \mathcal{E}^0$. Lemma 5.1.5.2 implies that the left fibrations

$$\mathcal{D} \to \mathcal{C}^{\triangleright}_{/X} \leftarrow \mathcal{E}$$

are equivalent. It therefore suffices to show that the inclusion $\mathcal{E}^0 \subseteq \mathcal{E}$ is a weak homotopy equivalence. To prove this, we observe that both $\mathcal{E}$ and $\mathcal{E}^0$

contain $\mathcal{E}^1$ as a deformation retract (that is, there is a retraction $r : \mathcal{E} \to \mathcal{E}^1$ and a homotopy $\mathcal{E} \times \Delta^1 \to \mathcal{E}$ from r to $\mathrm{id}_{\mathcal{E}}$, so that the inclusion $\mathcal{E}^1 \subseteq \mathcal{E}$ is a homotopy equivalence; the situation for $\mathcal{E}^0$ is similar). □

Lemma 5.1.5.4. *Let*

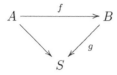

be a diagram of simplicial sets. The following conditions are equivalent:

(1) *The map f is a covariant equivalence in $(\mathrm{Set}_\Delta)_{/S}$.*

(2) *For every diagram $p : S \to \mathcal{C}$ taking values in an ∞-category $\mathcal{C}$ and every limit $\overline{p \circ g} : B^\triangleleft \to \mathcal{C}$ of $p \circ g$, the composition $\overline{p \circ g} \circ f^\triangleleft : A^\triangleleft \to \mathcal{C}$ is a limit diagram.*

(3) *For every diagram $p : S \to \mathcal{S}$ taking values in the ∞-category $\mathcal{S}$ of spaces and every limit $\overline{p \circ g} : B^\triangleleft \to \mathcal{S}$ of $p \circ g$, the composition $\overline{p \circ g} \circ f^\triangleleft : A^\triangleleft \to \mathcal{S}$ is a limit diagram.*

Proof. The equivalence of (1) and (3) follows from Corollary 3.3.3.4 (and the definition of a contravariant equivalence). The implication (2) ⇒ (3) is obvious. We show that (3) ⇒ (2). Let $p : S \to \mathcal{C}$ and $\overline{p \circ g}$ be as in (2). Passing to a larger universe if necessary, we may suppose that $\mathcal{C}$ is small. For each object $C \in \mathcal{C}$, let $j_C : \mathcal{C} \to \mathcal{S}$ denote the composition of the Yoneda embedding $j : \mathcal{C} \to \mathcal{P}(\mathcal{C})$ with the map $\mathcal{P}(\mathcal{C}) \to \mathcal{S}$ given by evaluation at C. Combining Proposition 5.1.3.2 with Proposition 5.1.2.2, we deduce that each $j_C \circ \overline{p \circ g}$ is a limit diagram. Applying (3), we conclude that each $j_C \circ \overline{p \circ g} \circ f^\triangleleft$ is a limit diagram. We now apply Propositions 5.1.3.2 and 5.1.2.2 to conclude that $\overline{p \circ g} \circ f^\triangleleft$ is a limit diagram, as desired. □

Lemma 5.1.5.5. *Let S be a small simplicial set, let $j : S \to \mathcal{P}(S)$ be the Yoneda embedding, and let $\mathcal{C}$ denote the full subcategory of $\mathcal{P}(S)$ spanned by the objects $j(s)$, where s is a vertex of S. Let $\mathcal{D}$ be an arbitrary ∞-category.*

(1) *Let $f : \mathcal{P}(S) \to \mathcal{D}$ be a functor. Then f is a left Kan extension of $f|\,\mathcal{C}$ if and only if f preserves small colimits.*

(2) *Suppose that $\mathcal{D}$ admits small colimits and let $f_0 : \mathcal{C} \to \mathcal{D}$ be an arbitrary functor. There exists an extension $f : \mathcal{P}(S) \to \mathcal{D}$ which is a left Kan extension of $f_0 = f|\,\mathcal{C}$.*

Proof. Assertion (2) follows from Lemma 4.3.2.13 since the ∞-category $\mathcal{C}_{/X}$ is small for each object $X \in \mathcal{P}(S)$. We will prove (1). Suppose first that f preserves small colimits. We must show that for each $X \in \mathcal{P}(S)$, the composition

$$\mathcal{C}_{/X}^{\triangleright} \xrightarrow{\delta} \mathcal{P}(S) \xrightarrow{f} \mathcal{D}$$

is a colimit diagram. Lemma 5.1.5.3 implies that δ is a colimit diagram; if f preserves small colimits, then $f \circ \delta$ is also a colimit diagram.

Now suppose that f is a left Kan extension of $f_0 = f|\,\mathcal{C}$. We wish to prove that f preserves small colimits. Let K be a small simplicial set and let $\overline{p} : K^{\triangleright} \to \mathcal{P}(S)$ be a colimit diagram. We must show that $f \circ \overline{p}$ is also a colimit diagram.

Let

$$\overline{\mathcal{E}} = \mathcal{C} \times_{\mathrm{Fun}(\{0\},\mathcal{P}(S))} \mathrm{Fun}(\Delta^1, \mathcal{P}(S)) \times_{\mathrm{Fun}(\{1\},\mathcal{P}(S))} K^{\triangleright}$$

and let $\mathcal{E} = \overline{\mathcal{E}} \times_{K^{\triangleright}} K \subseteq \overline{\mathcal{E}}$. We have a commutative diagram

where the vertical arrows are coCartesian fibrations (Corollary 2.4.7.12). Let $\overline{\eta} : \overline{\mathcal{E}} \diamond_{K^{\triangleright}} K^{\triangleright} \to \mathcal{P}(S)$ be the natural map and set $\eta = \overline{\eta}|\,\mathcal{E} \diamond_K K$. Proposition 4.3.3.10 implies that $f \circ \eta$ exhibits $f \circ p$ as a left Kan extension of $f \circ (\eta|\,\mathcal{E})$ along $q|\,\mathcal{E}$. Similarly, $f \circ \overline{\eta}$ exhibits $f \circ \overline{p}$ as a left Kan extension of $f \circ (\overline{\eta}|\overline{\mathcal{E}})$. It will therefore suffice to prove that every colimit of $f \circ (\overline{\eta}|\overline{\mathcal{E}})$ is also a colimit of $f \circ (\eta|\,\mathcal{E})$. According to Lemma 5.1.5.4, it suffices to show that the inclusion $\mathcal{E} \subseteq \overline{\mathcal{E}}$ is a contravariant equivalence in $(\mathrm{Set}_{\Delta})_{/\mathcal{C}}$.

Since the map $\overline{\mathcal{E}} \to K^{\triangleright} \times \mathcal{C}$ is a bivariant fibration, we can apply Proposition 4.1.2.16 to deduce that the map $\overline{\mathcal{E}}^{op} \to \mathcal{C}^{op}$ is smooth. Similarly, $\mathcal{E}^{op} \to \mathcal{C}^{op}$ is smooth. According to Proposition 4.1.2.18, the inclusion $\mathcal{E} \subseteq \overline{\mathcal{E}}$ is a contravariant equivalence if and only if, for every object $C \in \mathcal{C}$, the inclusion of fibers $\mathcal{E}_C \subseteq \overline{\mathcal{E}}_C$ is a weak homotopy equivalence. Lemma 5.1.5.2 implies that $\overline{\mathcal{E}}_C \to K^{\triangleright}$ is equivalent to the left fibration given by the pullback of the universal left fibration along the map

$$K^{\triangleright} \xrightarrow{\overline{p}} \mathcal{P}(S) \xrightarrow{s} \mathcal{S}.$$

We now conclude by applying Proposition 3.3.4.5, noting that $\overline{p}$ is a colimit diagram by assumption and that s preserves colimits by Proposition 5.1.2.2. $\qquad\square$

Theorem 5.1.5.6. *Let S be a small simplicial set and let $\mathcal{C}$ be an ∞-category which admits small colimits. Composition with the Yoneda embedding $j : S \to \mathcal{P}(S)$ induces an equivalence of ∞-categories*

$$\mathrm{Fun}^L(\mathcal{P}(S), \mathcal{C}) \to \mathrm{Fun}(S, \mathcal{C}).$$

Proof. Combine Corollary 4.3.2.16 with Lemma 5.1.5.5. $\qquad\square$

Definition 5.1.5.7. Let $\mathcal{C}$ be an ∞-category. A full subcategory $\mathcal{C}' \subseteq \mathcal{C}$ is *stable under colimits* if, for any small diagram $p : K \to \mathcal{C}'$ which has a colimit $\overline{p} : K^{\triangleright} \to \mathcal{C}$ in $\mathcal{C}$, the map $\overline{p}$ factors through $\mathcal{C}'$.

Let $\mathcal{C}$ be an ∞-category which admits all small colimits. Let A be a collection of objects of $\mathcal{C}$. We will say that A *generates* $\mathcal{C}$ *under colimits* if the following condition is satisfied: for any full subcategory $\mathcal{C}' \subseteq \mathcal{C}$ containing every element of A, if $\mathcal{C}'$ is stable under colimits, then $\mathcal{C} = \mathcal{C}'$.

We say that a map $f : S \to \mathcal{C}$ *generates* $\mathcal{C}$ *under colimits* if the image $f(S_0)$ generates $\mathcal{C}$ under colimits.

Corollary 5.1.5.8. *Let S be a small simplicial set. Then the Yoneda embedding $j : S \to \mathcal{P}(S)$ generates $\mathcal{P}(S)$ under small colimits.*

Proof. Let $\mathcal{C}$ be the smallest full subcategory of $\mathcal{P}(S)$ which contains the essential image of j and is stable under small colimits. Applying Theorem 5.1.5.6, we deduce that the diagram $j : S \to \mathcal{C}$ is equivalent to $F \circ j$ for some colimit-preserving functor $F : \mathcal{P}(S) \to \mathcal{C}$. We may regard F as a colimit-preserving functor from $\mathcal{P}(S)$ to itself. Applying Theorem 5.1.5.6 again, we deduce that F is equivalent to the identity functor from $\mathcal{P}(S)$ to itself. It follows that every object of $\mathcal{P}(S)$ is equivalent to an object which lies in $\mathcal{C}$, so that $\mathcal{C} = \mathcal{P}(S)$, as desired. □

5.1.6 Complete Compactness

Let S be a small simplicial set and $f : S \to \mathcal{C}$ a diagram in an ∞-category $\mathcal{C}$. Our goal in this section is to analyze the following question: when is the diagram $f : S \to \mathcal{C}$ equivalent to the Yoneda embedding $j : S \to \mathcal{P}(S)$? An obvious necessary condition is that $\mathcal{C}$ admit small colimits (Corollary 5.1.2.4). Conversely, if $\mathcal{C}$ admits small colimits, then Theorem 5.1.5.6 implies that f is equivalent to $F \circ j$, where $F : \mathcal{P}(S) \to \mathcal{C}$ is a colimit-preserving functor. We are now reduced to the question of deciding whether or not the functor F is an equivalence. There are two obvious necessary conditions for this to be so: f must be fully faithful (Proposition 5.1.3.1), and f must generate $\mathcal{C}$ under colimits (Corollary 5.1.5.8). We will show that the converse holds provided that the essential image of f consists of *completely compact* objects of $\mathcal{C}$ (see Definition 5.1.6.2 below).

We begin by considering an arbitrary simplicial set S and a vertex s of S. Composing the Yoneda embedding $j : S \to \mathcal{P}(S)$ with the evaluation map

$$\mathcal{P}(S) = \mathrm{Fun}(S^{op}, \mathcal{S}) \to \mathrm{Fun}(\{s\}, \mathcal{S}) \simeq \mathcal{S},$$

we obtain a map $j_s : S \to \mathcal{S}$. We will refer to j_s as the *functor corepresented by s*.

Remark 5.1.6.1. The above definition makes sense even when the simplicial set S is not small. However, in this case we need to replace $\mathcal{S}$ (the simplicial nerve of the category of *small* Kan complexes) by the (very large) ∞-category $\widehat{\mathcal{S}}$, where $\widehat{\mathcal{S}}$ is the simplicial nerve of the category of *all* Kan complexes (not necessarily small).

Definition 5.1.6.2. Let $\mathcal{C}$ be an ∞-category which admits small colimits. We will say that an object $C \in \mathcal{C}$ is *completely compact* if the functor $j_C : \mathcal{C} \to \widehat{\mathcal{S}}$ corepresented by C preserves small colimits.

The requirement that an object C of an ∞-category $\mathcal{C}$ be completely compact is *very* restrictive (see Example 5.1.6.9 below). We introduce this notion not because it is a generally useful one but because it is relevant for the purpose of characterizing ∞-categories of presheaves.

Our first goal is to establish that the class of completely compact objects of $\mathcal{C}$ is stable under retracts.

Lemma 5.1.6.3. *Let $\mathcal{C}$ be an ∞-category, K a simplicial set, and $\overline{p}, \overline{q} : K^{\triangleright} \to \mathcal{C}$ a pair of diagrams. Suppose that $\overline{q}$ is a colimit diagram and that $\overline{p}$ is a retract of $\overline{q}$ in the ∞-category $\operatorname{Fun}(K^{\triangleright}, \mathcal{C})$. Then $\overline{p}$ is a colimit diagram.*

Proof. Choose a map $\sigma : \Delta^2 \times K^{\triangleright} \to \mathcal{C}$ such that $\sigma|\{1\} \times K^{\triangleright} = \overline{q}$ and $\sigma|\Delta^{\{0,2\}} \times K^{\triangleright} = \overline{p} \circ \pi_{K^{\triangleright}}$. We have a commutative diagram of simplicial sets:

$$
\begin{array}{ccc}
\mathcal{C}_{\sigma/} & \longrightarrow & \mathcal{C}_{\sigma|\Delta^2 \times K/} \\
\downarrow & & \downarrow \\
\mathcal{C}_{\sigma|\Delta^{\{1,2\}}/ \times K^{\triangleright}} & \overset{f}{\longrightarrow} & \mathcal{C}_{\sigma|\Delta^{\{1,2\}} \times K/} \\
\downarrow & & \downarrow \\
\mathcal{C}_{\sigma|\{2\} \times K^{\triangleright}/} & \overset{f'}{\longrightarrow} & \mathcal{C}_{\sigma|\{2\} \times K/} \; .
\end{array}
$$

We first claim that both vertical compositions are categorical equivalences. We give the argument for the right vertical composition; the other case is similar. We have a factorization

$$
\mathcal{C}_{\sigma|\Delta^2 \times K/} \overset{g'}{\to} \mathcal{C}_{\sigma|\Delta^{\{0,2\}} \times K/} \overset{g''}{\to} \mathcal{C}_{\sigma|\{2\} \times K/},
$$

where g' is a trivial fibration and g'' admits a section s. The map s is also a section of the trivial fibration $\mathcal{C}_{/\sigma|\Delta^{\{0,2\}} \times K} \to \mathcal{C}_{/\sigma|\{0\} \times K}$. Consequently, s and g'' are categorical equivalences. It follows that the map f' is a retract of f in the homotopy category of $\operatorname{Set}_{\Delta}$ (taken with respect to the Joyal model structure).

The map f sits in a commutative diagram

$$
\begin{array}{ccc}
\mathcal{C}_{\sigma|\Delta^{\{1,2\}}/ \times K^{\triangleright}} & \overset{f}{\longrightarrow} & \mathcal{C}_{\sigma|\Delta^{\{1,2\}}/ \times K} \\
\downarrow & & \downarrow \\
\mathcal{C}_{\overline{q}/} & \longrightarrow & \mathcal{C}_{q/},
\end{array}
$$

where the vertical maps and the lower horizontal map are trivial fibrations. It follows that f is a categorical equivalence. Since f' is a retract of f, f' is also a categorical equivalence. Since f' is a left fibration, we deduce that f' is a trivial fibration (Corollary 2.4.4.6), so that $\overline{p}$ is a colimit diagram as desired. $\square$

Lemma 5.1.6.4. *Let $\mathcal{C}$ be an ∞-category which admits small colimits. Let C and D be objects of $\mathcal{C}$. Suppose that C is completely compact and that D*

is a retract of C (that is, there exist maps $f : D \to C$ and $r : C \to D$ with $r \circ f \simeq \mathrm{id}_D$. Then D is completely compact. In particular, if C and D are equivalent, then D is completely compact.

Proof. Let $j : \mathcal{C}^{op} \to \mathcal{S}^{\mathcal{C}}$ denote the Yoneda embedding (for $\mathcal{C}^{op}$). Since D is a retract of C, $j(D)$ is a retract of $j(C)$. Let $\overline{p} : K^{\triangleright} \to \mathcal{C}$ be a diagram. Then $j(D) \circ \overline{p} : K^{\triangleright} \to \mathcal{S}$ is a retract of $j(C) \circ \overline{p} : K^{\triangleright} \to \mathcal{S}$ in the ∞-category $\mathrm{Fun}(K^{\triangleright}, \mathcal{S})$. If $\overline{p}$ is a colimit diagram, then $j(C) \circ \overline{p}$ is a colimit diagram (since C is completely compact). Lemma 5.1.6.3 now implies that $j(D) \circ \overline{p}$ is a colimit diagram as well. $\square$

In order to study the condition of complete compactness in more detail, it is convenient to introduce a slightly more general notion.

Definition 5.1.6.5. Let $\mathcal{C}$ be an ∞-category which admits small colimits and let $\phi : \widetilde{\mathcal{C}} \to \mathcal{C}$ be a left fibration. We will say that ϕ is *completely compact* if it is classified by a functor $\mathcal{C} \to \widehat{\mathcal{S}}$ that preserves small colimits.

Lemma 5.1.6.6. *Let $\mathcal{C}$ be an ∞-category which admits small colimits, let $f : X' \to X$ be a map of Kan complexes, and let*

$$
\begin{array}{ccc}
\mathcal{F}' & \longrightarrow & \mathcal{F} \\
\downarrow & & \downarrow \\
X' \times \mathcal{C} & \xrightarrow{f \times \mathrm{id}_{\mathcal{C}}} & X \times \mathcal{C}
\end{array}
$$

be a diagram of left fibrations over $\mathcal{C}$ which is a homotopy pullback square (with respect to the covariant model structure on $(\mathrm{Set}_\Delta)_{/\mathcal{C}}$). If $\mathcal{F} \to \mathcal{C}$ is completely compact, then $\mathcal{F}' \to \mathcal{C}$ is completely compact.

Proof. Replacing the diagram by an equivalent one if necessary, we may suppose that it is Cartesian and that f is a Kan fibration. Let $\overline{p} : K^{\triangleright} \to \mathcal{C}$ be a colimit diagram and let $F : \mathcal{C} \to \mathcal{S}$ be a functor which classifies the left fibration $\mathcal{F}'$. We wish to show that $F \circ \overline{p}$ is a colimit diagram in $\widehat{\mathcal{S}}$.

We have a pullback diagram

$$
\begin{array}{ccc}
K \times_{\mathcal{C}} \mathcal{F}' & \longrightarrow & K \times_{\mathcal{C}} \mathcal{F} \\
\downarrow{\scriptstyle \psi'} & & \downarrow{\scriptstyle \psi} \\
K^{\triangleright} \times_{\mathcal{C}} \mathcal{F}' & \longrightarrow & K^{\triangleright} \times_{\mathcal{C}} \mathcal{F}
\end{array}
$$

of simplicial sets which is homotopy Cartesian (with respect to the usual model structure on Set_Δ) since the horizontal maps are pullbacks of f. Since $\mathcal{F}$ is completely compact, Proposition 3.3.4.5 implies that the inclusion ψ is a weak homotopy equivalence. It follows that ψ' is also a weak homotopy equivalence. Applying Proposition 3.3.4.5 again, we deduce that $F \circ \overline{p}$ is a colimit diagram, as desired. $\square$

Lemma 5.1.6.7. *Let $\mathcal{C}$ be a presentable ∞-category, let $p : K \to \mathcal{C}$ be a small diagram, and let $X \in \mathcal{C}_{/p}$ be an object whose image in $\mathcal{C}$ is completely compact. Then X is completely compact.*

Proof. Let $\bar{p} : K^\triangleleft \to \mathcal{C}$ be a limit of p carrying the cone point to an object $Z \in \mathcal{C}$. Then we have trivial fibrations

$$\mathcal{C}_{/Z} \leftarrow \mathcal{C}_{/\bar{p}} \to \mathcal{C}_{/p}.$$

Consequently, we may replace the diagram $p : K \to \mathcal{C}$ with the inclusion $\{Z\} \to \mathcal{C}$.

We may identify the object $X \in \mathcal{C}_{/Z}$ with a morphism $f : Y \to Z$ in $\mathcal{C}$. We have a commutative diagram of simplicial sets

$$
\begin{array}{ccccc}
(\mathcal{C}_{/Z})_{f/} & \xrightarrow{\ \theta\ } & (\mathcal{C}_{/Y})_{f/} & \xrightarrow{\ \theta'\ } & (\mathcal{C}_{/Y})^{f/} \\
& \searrow{\scriptstyle\psi} & \downarrow & & \downarrow{\scriptstyle\psi'} \\
& & \mathcal{C}_{/Z} & \xrightarrow{\ \theta'_0\ } & \mathcal{C}^{/Z},
\end{array}
$$

where θ is an isomorphism, the maps θ' and θ'_0 are categorical equivalences (see §4.2.1), and the vertical maps are left fibrations. We wish to prove that ψ is a completely compact left fibration. It will therefore suffice to prove that ψ' is completely compact. We have a (homotopy) pullback diagram

$$
\begin{array}{ccc}
\mathcal{C}_{Y/}^{/f} & \longrightarrow & \mathcal{C}_{Y/}^{\Delta^1} \times_{\mathcal{C}^{\{1\}}} \{Z\} \\
\downarrow & & \downarrow \\
\mathcal{C}^{/Z} & \longrightarrow & (\mathcal{C}_{Y/} \times_{\mathcal{C}} \{Z\}) \times \mathcal{C}^{/Z}
\end{array}
$$

of left fibrations over $\mathcal{C}^{/Z}$. We observe that the left fibrations in the lower part of the diagram are constant. According to Lemma 5.1.6.6, to prove that ψ' is completely compact, it will suffice to prove that the left fibration $\mathcal{C}_{Y/}^{\Delta^1} \times_{\mathcal{C}^{\{1\}}} \{Z\} \xrightarrow{\psi''} \mathcal{C}^{/Z}$ is completely compact. We observe that ψ'' admits a factorization

$$\mathcal{C}_{Y/}^{\Delta^1} \times_{\mathcal{C}^{\{1\}}} \{Z\} \xrightarrow{\ \phi\ } \mathcal{C}_{Y/} \times_{\mathcal{C}^{\{0\}}} \mathcal{C}^{/Z} \xrightarrow{\ \phi'\ } \mathcal{C}^{/Z},$$

where ϕ is a trivial fibration and ϕ' is a pullback of the left fibration $\phi'' : \mathcal{C}_{Y/} \to \mathcal{C}$. Since Y is completely compact, ϕ'' is completely compact. The projection $\mathcal{C}^{/Z} \to \mathcal{C}$ is equivalent to $\mathcal{C}_{/Z} \to \mathcal{C}$ and therefore commutes with colimits by Proposition 1.2.13.8. It follows that ϕ' is completely compact, which completes the proof. $\qquad\square$

Proposition 5.1.6.8. *Let S be a small simplicial set and let $j : S \to \mathcal{P}(S)$ denote the Yoneda embedding. Let C be an object of $\mathcal{P}(S)$. The following conditions are equivalent:*

(1) *The object $C \in \mathcal{P}(S)$ is completely compact.*

(2) *There exists a vertex s of S such that C is a retract of $j(s)$.*

Proof. Suppose first that (1) is satisfied. Let $S_{/C} = S \times_{\mathcal{P}(S)} \mathcal{P}(S)_{/C}$. According to Lemma 5.1.5.3, the natural map

$$S_{/C}^{\triangleright} \xrightarrow{j'} \mathcal{P}(S)_{/C}^{\triangleright} \to \mathcal{P}(S)$$

is a colimit diagram. Let $f : \mathcal{P}(S) \to S$ be the functor corepresented by C. Since C is completely compact. $f(C)$ can be identified with a colimit of the diagram $f|S_{/C}$. The space $f(C)$ is homotopy equivalent to $\mathrm{Map}_{\mathcal{P}(S)}(C,C)$ and therefore contains a point corresponding to id_C. It follows that id_C lies in the image of $\mathrm{Map}_{\mathcal{P}(S)}(C, j'(\widetilde{s})) \to \mathrm{Map}_{\mathcal{P}(S)}(C,C)$ for some vertex $\widetilde{s}$ of $S_{/C}$. The vertex $\widetilde{s}$ classifies a vertex $s \in S$ equipped with a morphism $\alpha : j(s) \to C$. It follows that there is a commutative triangle

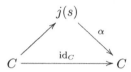

in the ∞-category $\mathcal{P}(S)$, so that C is a retract of $j(s)$.

Now suppose that (2) is satisfied. According to Lemma 5.1.6.4, it suffices to prove that $j(s)$ is completely compact. Using Lemma 5.1.5.2, we may identify the functor $\mathcal{P}(S) \to S$ corepresented by $j(s)$ with the functor given by evaluation at s. Proposition 5.1.2.2 implies that this functor preserves all limits and colimits that exist in $\mathcal{P}(S)$.   □

Example 5.1.6.9. Let $\mathcal{C}$ be the ∞-category S of spaces. Then an object $C \in S$ is completely compact if and only if it is equivalent to $*$, the final object of S.

We now use the theory of completely compact objects to give a characterization of presheaf ∞-categories.

Proposition 5.1.6.10. *Let S be a small simplicial set and $\mathcal{C}$ an ∞-category which admits small colimits. Let $F : \mathcal{P}(S) \to \mathcal{C}$ be a functor which preserves small colimits and let $f = F \circ j$ be its composition with the Yoneda embedding $j : S \to \mathcal{P}(S)$. Suppose further that*

(1) *The functor f is fully faithful.*

(2) *For every vertex s of S, the object $f(s) \in \mathcal{C}$ is completely compact.*

Then F is fully faithful.

Proof. Let C and D be objects of $\mathcal{P}(S)$. We wish to prove that the natural map

$$\eta_{C,D} : \mathrm{Map}_{\mathcal{P}(S)}(C, D) \to \mathrm{Map}_{\mathcal{C}}(F(C), F(D))$$

is an isomorphism in the homotopy category $\mathcal{H}$. Suppose first that C belongs to the essential image of j. Let $G : \mathcal{P}(S) \to S$ be a functor corepresented by C and let $G' : \mathcal{C} \to S$ be a functor corepresented by $F(C)$. Then we have a natural transformation of functors $G \to G' \circ F$. Assumption (2) implies that G' preserves small colimits, so that $G' \circ F$ preserves small colimits. Proposition 5.1.6.8 implies that G preserves small colimits. It follows that the collection of objects $D \in \mathcal{P}(S)$ such that $\eta_{C,D}$ is an equivalence is stable under small colimits. If D belongs to the essential image of j, then assumption (1) implies that $\eta_{C,D}$ is an equivalence. It follows from Lemma 5.1.5.3 that the essential image of j generates $\mathcal{P}(S)$ under small colimits; thus $\eta_{C,D}$ is an isomorphism in $\mathcal{H}$ for every object $D \in \mathcal{P}(S)$.

We now prove the result in general. Fix $D \in \mathcal{P}(S)$. Let $H : \mathcal{P}(S)^{op} \to \mathcal{S}$ be a functor represented by D and let $H' : \mathcal{C}^{op} \to \mathcal{S}$ be a functor represented by FD. Then we have a natural transformation of functors $H \to H' \circ F^{op}$, which we wish to prove is an equivalence. By assumption, F^{op} preserves small limits. Proposition 5.1.3.2 implies that H and H' preserve small limits. It follows that the collection P of objects $C \in \mathcal{P}(S)$ such that $\eta_{C,D}$ is an equivalence is stable under small colimits. The special case above established that P contains the essential image of the Yoneda embedding. We once again invoke Lemma 5.1.5.3 to deduce that every object of $\mathcal{P}(S)$ belongs to P, as desired. $\square$

Corollary 5.1.6.11. *Let $\mathcal{C}$ be an ∞-category which admits small colimits. Let S be a small simplicial set and $F : \mathcal{P}(S) \to \mathcal{C}$ a colimit-preserving functor. Then F is an equivalence if and only if the following conditions are satisfied:*

(1) *The composition $f = F \circ j : S \to \mathcal{C}$ is fully faithful.*

(2) *For every vertex $s \in S$, the object $f(s) \in \mathcal{C}$ is completely compact.*

(3) *The set of objects $\{f(s) : s \in S_0\}$ generates $\mathcal{C}$ under colimits.*

Proof. If (1), (2), and (3) are satisfied, then F is fully faithful (Proposition 5.1.6.10). Since $\mathcal{P}(S)$ admits small colimits and F preserves small colimits, the essential image of F is stable under small colimits. Using (3), we conclude that F is essentially surjective and therefore an equivalence of ∞-categories. For the converse, it suffices to check that $\mathrm{id}_{\mathcal{P}(S)} : \mathcal{P}(S) \to \mathcal{P}(S)$ satisfies (1), (2), and (3). For this, we invoke Propsition 5.1.3.1, Proposition 5.1.6.8, and Lemma 5.1.5.3, respectively. $\square$

Corollary 5.1.6.12. *Let $\mathcal{C}$ be a small ∞-category, let $p : K \to \mathcal{C}$ be a diagram, let $p' : K \to \mathcal{P}(\mathcal{C})$ be the composition of p with the Yoneda embedding $j : \mathcal{C} \to \mathcal{P}(\mathcal{C})$, and let $f : \mathcal{C}_{/p} \to \mathcal{P}(\mathcal{C})_{/p'}$ be the induced map. Let $F : \mathcal{P}(\mathcal{C}_{/p}) \to \mathcal{P}(\mathcal{C})_{/p'}$ be a colimit-preserving functor such that $F \circ j'$ is equivalent to f, where $j' : \mathcal{C}_{/p} \to \mathcal{P}(\mathcal{C}_{/p})$ denotes the Yoneda embedding for $\mathcal{C}_{/p}$ (according to Theorem 5.1.5.6, F exists and is unique up to equivalence). Then F is an equivalence of ∞-categories.*

Proof. We will show that f satisfies conditions (1) through (3) of Corollary 5.1.6.11. The assertion that f is fully faithful follows immediately from the assertion that j is fully faithful (Proposition 5.1.3.1). To prove that the essential image of f consists of completely compact objects, we use Lemma 5.1.6.7 to reduce to proving that the essential image of j consists of completely compact objects of $\mathcal{P}(\mathcal{C})$, which follows from Proposition 5.1.6.8. It remains to prove that $\mathcal{P}(\mathcal{C})_{/p'}$ is generated under colimits by f. Let $\overline{X}$ be an object of $\mathcal{P}(\mathcal{C})_{/p'}$ and X its image in $\mathcal{P}(\mathcal{C})$. Let $\mathcal{D} \subseteq \mathcal{P}(\mathcal{C})$ be the essential image of j and $\overline{\mathcal{D}}$ the inverse image of $\mathcal{D}$ in $\mathcal{P}(\mathcal{C})_{/p'}$, so that $\overline{\mathcal{D}}$ is the essential image of f. Using Lemma 5.1.5.3, we can choose a colimit diagram $\overline{q} : L^{\triangleright} \to \mathcal{P}(\mathcal{C})$ which carries the cone point to X such that $q = \overline{q}|L$ factors through $\mathcal{D}$. Since the inclusion of the cone point into $L^{\triangleright}$ is right anodyne, there exists a map $\overline{q}' : L^{\triangleright} \to \mathcal{P}(\mathcal{C})_{/p'}$ lifting $\overline{q}$, which carries the cone point of $L^{\triangleright}$ to $\overline{X}$. Proposition 1.2.13.8 implies that $\overline{q}'$ is a colimit diagram, so that $\overline{X}$ can be written as the colimit of a diagram $L \to \overline{\mathcal{D}}$. $\square$

5.2 ADJOINT FUNCTORS

Let $\mathcal{C}$ and $\mathcal{D}$ be (ordinary) categories. Two functors

$$\mathcal{C} \underset{G}{\overset{F}{\rightleftarrows}} \mathcal{D}$$

are said to be *adjoint* to one another if there is a functorial bijection

$$\operatorname{Hom}_{\mathcal{D}}(F(C), D) \simeq \operatorname{Hom}_{\mathcal{C}}(C, G(D))$$

defined for $C \in \mathcal{C}$, $D \in \mathcal{D}$. Our goal in this section is to extend the theory of adjoint functors to the ∞-categorical setting.

By definition, a pair of functors F and G (as above) are adjoint if and only if they determine the same correspondence

$$\mathcal{C}^{op} \times \mathcal{D} \to \operatorname{Set}.$$

In §2.3.1, we introduced an ∞-categorical generalization of the notion of a correspondence. In certain cases, a correspondence $\mathcal{M}$ from an ∞-category $\mathcal{C}$ to an ∞-category $\mathcal{D}$ determines a functor $F : \mathcal{C} \to \mathcal{D}$, which we say is a functor *associated* to $\mathcal{M}$. We will study these associated functors in §5.2.1. The notion of a correspondence is self-dual, so it is possible that the correspondence $\mathcal{M}$ also determines an associated functor $G : \mathcal{D} \to \mathcal{C}$. In this case, we will say that F and G are adjoint. We will study the basic properties of adjoint functors in §5.2.2.

One of the most important features of adjoint functors is their behavior with respect to limits and colimits: left adjoints preserve colimits, while right adjoints preserve limits. We will prove an ∞-categorical analogue of this statement in §5.2.3. In certain situations, the *adjoint functor theorem* provides a converse to this statement: see §5.5.2.

The theory of model categories provides a host of examples of adjoint functors between ∞-categories. In §5.2.4, we will show that a simplicial Quillen adjunction between a pair of model categories $(\mathbf{A}, \mathbf{A}')$ determines an adjunction between the associated ∞-categories $(\mathrm{N}(\mathbf{A}^{\circ}), \mathrm{N}(\mathbf{A}'^{\circ}))$. We will also consider some other examples of situations which give rise to adjoint functors.

In §5.2.5, we study the behavior of adjoint functors when restricted to overcategories. Our main result (Proposition 5.2.5.1) can be summarized as follows: suppose that $F : \mathcal{C} \to \mathcal{D}$ is a functor between ∞-categories which admits a right adjoint G. Assume further that the ∞-category $\mathcal{C}$ admits pullbacks. Then for every object C, the induced functor $\mathcal{C}_{/C} \to \mathcal{D}_{/FC}$ admits a right adjoint given by the formula

$$(D \to FC) \mapsto (GD \times_{GFC} C \to C).$$

If a functor $F : \mathcal{C} \to \mathcal{D}$ has a right adjoint G, then G is uniquely determined up to equivalence. In §5.2.6, we will prove a strong version of this statement (which we phrase as an (anti)equivalence of functor ∞-categories).

In §5.2.7, we will restrict the theory of adjoint functors to the special case in which one of the functors is the inclusion of a full subcategory. In this case, we obtain the theory of *localizations* of ∞-categories. This theory will play a central role in our study of presentable ∞-categories (§5.5) and later in the study of ∞-topoi (§6). It is also useful in the study of *factorization systems* on ∞-categories, which we will discuss in §5.2.8.

5.2.1 Correspondences and Associated Functors

Let $p : X \to S$ be a Cartesian fibration of simplicial sets. In §3.3.2, we saw that p is classified by a functor $S^{op} \to \mathcal{C}at_\infty$. In particular, if $S = \Delta^1$, then p determines a diagram

$$G : \mathcal{D} \to \mathcal{C}$$

in the ∞-category $\mathcal{C}at_\infty$, which is well-defined up to equivalence. We can obtain this diagram by applying the straightening functor St_S^+ to the marked simplicial set $X^\natural$ and then taking a fibrant replacement. In general, this construction is rather complicated. However, in the special case where $S = \Delta^1$, it is possible to give a direct construction of G; that is our goal in this section.

Definition 5.2.1.1. Let $p : \mathcal{M} \to \Delta^1$ be a Cartesian fibration and suppose we are given equivalences of ∞-categories $h_0 : \mathcal{C} \to p^{-1}\{0\}$ and $h_1 : \mathcal{D} \to p^{-1}\{1\}$. We will say that a functor $g : \mathcal{D} \to \mathcal{C}$ is *associated to* $\mathcal{M}$ if there is a commutative diagram

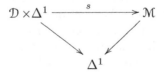

such that $s| \mathcal{D} \times \{1\} = h_1$, $s| \mathcal{D} \times \{0\} = h_0 \circ g$, and $s|\{x\} \times \Delta^1$ is a p-Cartesian edge of $\mathcal{M}$ for every object x of $\mathcal{D}$.

Remark 5.2.1.2. The terminology of Definition 5.2.1.1 is slightly abusive: it would be more accurate to say that g is associated to the triple $(p : \mathcal{M} \to \Delta^1, h_0 : \mathcal{C} \to p^{-1}\{0\}, h_1 : \mathcal{D} \to p^{-1}\{1\})$.

Proposition 5.2.1.3. *Let $\mathcal{C}$ and $\mathcal{D}$ be ∞-categories and let $g : \mathcal{D} \to \mathcal{C}$ be a functor.*

(1) *There exists a diagram*

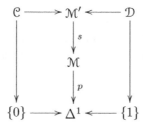

where p is a Cartesian fibration, the associated maps $\mathcal{C} \to p^{-1}\{0\}$ and $\mathcal{D} \to p^{-1}\{1\}$ are isomorphisms, and g is associated to $\mathcal{M}$.

(2) *Suppose we are given a commutative diagram*

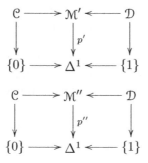

where s is a categorical equivalence, p and $p' = p \circ s$ are Cartesian fibrations, and the maps $\mathcal{C} \to p^{-1}\{0\}$, $\mathcal{D} \to p^{-1}\{1\}$ are categorical equivalences. The functor g is associated to $\mathcal{M}$ if and only if it is associated to $\mathcal{M}'$.

(3) *Suppose we are given diagrams*

$$\mathcal{C} \longrightarrow \mathcal{M}' \longleftarrow \mathcal{D}$$
$$\downarrow \qquad \quad \downarrow {\scriptstyle p'} \qquad \downarrow$$
$$\{0\} \longrightarrow \Delta^1 \longleftarrow \{1\}$$

$$\mathcal{C} \longrightarrow \mathcal{M}'' \longleftarrow \mathcal{D}$$
$$\downarrow \qquad \quad \downarrow {\scriptstyle p''} \qquad \downarrow$$
$$\{0\} \longrightarrow \Delta^1 \longleftarrow \{1\}$$

as above, such that g is associated to both $\mathcal{M}'$ and $\mathcal{M}''$. Then there exists a third such diagram

$$\mathcal{C} \longrightarrow \mathcal{M} \longleftarrow \mathcal{D}$$
$$\downarrow \qquad \quad \downarrow {\scriptstyle p} \qquad \downarrow$$
$$\{0\} \longrightarrow \Delta^1 \longleftarrow \{1\}$$

and a diagram

$$\mathcal{M}' \leftarrow \mathcal{M} \rightarrow \mathcal{M}''$$

of categorical equivalences in $(\mathrm{Set}_\Delta)_{\mathcal{C} \coprod \mathcal{D} / / \Delta^1}$.

Proof. We begin with (1). Let $\mathcal{C}^\natural$ and $\mathcal{D}^\natural$ denote the simplicial sets $\mathcal{C}$ and $\mathcal{D}$ considered as marked simplicial sets, where the marked edges are precisely the equivalences. We set

$$N = (\mathcal{D}^\natural \times (\Delta^1)^\sharp) \coprod_{\mathcal{D}^\natural \times \{0\}^\sharp} \mathcal{C}^\natural.$$

The small object argument implies the existence of a factorization

$$N \rightarrow N(\infty) \rightarrow (\Delta^1)^\sharp,$$

where the left map is marked anodyne and the right map has the right lifting property with respect to all marked anodyne morphisms. We remark that we can obtain $N(\infty)$ as the colimit of a transfinite sequence of simplicial sets $N(\alpha)$, where $N(0) = N$, $N(\alpha)$ is the colimit of the sequence $\{N(\beta)\}_{\beta < \alpha}$ when α is a limit ordinal and each $N(\alpha + 1)$ fits into a pushout diagram

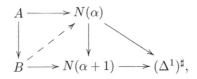

where the left vertical map is one of the generators for the class of marked anodyne maps given in Definition 3.1.1.1. We may furthermore assume that there does *not* exist a dotted arrow as indicated in the diagram. It follows by induction on α that $N(\alpha) \times_{\Delta^1} \{0\} \simeq \mathcal{C}^\natural$ and $N(\alpha) \times_{\Delta^1} \{1\} \simeq \mathcal{D}^\natural$. According to Proposition 3.1.1.6, $N(\infty) \simeq \mathcal{M}^\natural$ for some Cartesian fibration $\mathcal{M} \rightarrow \Delta^1$. It follows immediately that $\mathcal{C} \simeq \mathcal{M}_{\{0\}}$, that $\mathcal{D} \simeq \mathcal{M}_{\{1\}}$, and that g is associated to $\mathcal{M}$.

We now prove (2). The "if" direction is immediate from the definition. Conversely, suppose that g is associated to $\mathcal{M}$. To show that g is associated to $\mathcal{M}'$, we need to produce the dotted arrow indicated in the diagram

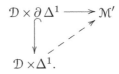

According to Proposition A.2.3.1, we may replace $\mathcal{M}'$ by the equivalent ∞-category $\mathcal{M}$; the desired result then follows from the assumption that g is associated to $\mathcal{M}$.

To prove (3), we take $\mathcal{M}$ to be the correspondence constructed in the course of proving (1). It will suffice to construct an appropriate categorical equivalence $\mathcal{M} \to \mathcal{M}'$; the same argument will construct the desired map $\mathcal{M} \to \mathcal{M}''$. Consider the diagram

(Here we identify N with its underlying simplicial set by forgetting the class of marked edges, and the top horizontal map exhibits g as associated to $\mathcal{M}'$.) In the terminology of §3.2.2, the maps s and s' are both quasi-equivalences. By Proposition 3.2.2.10, they are categorical equivalences. The projection $\mathcal{M}' \to \Delta^1$ is a categorical fibration and s' is a trivial cofibration, which ensures the existence of the arrow s''. The factorization $s = s'' \circ s'$ shows that s'' is a categorical equivalence and completes the proof.   □

Proposition 5.2.1.3 may be informally summarized by saying that every functor $g : \mathcal{D} \to \mathcal{C}$ is associated to some Cartesian fibration $p : \mathcal{M} \to \Delta^1$ and that $\mathcal{M}$ is determined up to equivalence. Conversely, the Cartesian fibration also determines g:

Proposition 5.2.1.4. *Let $p : \mathcal{M} \to \Delta^1$ be a Cartesian fibration and let $h_0 : \mathcal{C} \to p^{-1}\{0\}$ and $h_1 : \mathcal{D} \to p^{-1}\{1\}$ be categorical equivalences. There exists a functor $g : \mathcal{D} \to \mathcal{C}$ associated to $\mathcal{M}$. Any other functor $g' : \mathcal{C} \to \mathcal{D}$ is associated to p if and only if g is equivalent to g' as objects of the ∞-category $\mathcal{C}^{\mathcal{D}}$.*

Proof. Consider the diagram

$$
\begin{array}{ccc}
\mathcal{D}^{\flat} \times \{1\} & \longrightarrow & \mathcal{M}^{\natural} \\
\downarrow & {\scriptstyle s}\nearrow & \downarrow \\
\mathcal{D}^{\flat} \times (\Delta^1)^{\sharp} & \longrightarrow & (\Delta^1)^{\sharp}.
\end{array}
$$

By Proposition 3.1.2.3, the left vertical map is marked anodyne, so the dotted arrow exists. Consider the map $s_0 : s| \mathcal{D} \times \{0\} : \mathcal{D} \to p^{-1}\{0\}$. Since h_0 is a categorical equivalence, there exists a map $g : \mathcal{D} \to \mathcal{C}$ such that the functions $h_0 \circ g$ and s_0 are equivalent. Let $e : \mathcal{D} \times \Delta^1 \to \mathcal{M}$ be an equivalence from $h_0 \circ g$ to s_0. Let $e' : \mathcal{D} \times \Lambda_1^2 \to \mathcal{M}$ be the result of amalgamating e with s. Then we have a commutative diagram of marked simplicial sets

$$
\begin{array}{ccc}
\mathcal{D}^{\flat} \times (\Lambda_1^2)^{\sharp} & \overset{e'}{\longrightarrow} & \mathcal{M}^{\natural} \\
\downarrow & {\scriptstyle e''}\nearrow & \downarrow \\
\mathcal{D}^{\flat} \times (\Delta^2)^{\sharp} & \longrightarrow & (\Delta^1)^{\sharp}.
\end{array}
$$

Because the left vertical map is marked anodyne, there exists a morphism e'' as indicated which renders the diagram commutative. The restriction $e''| \mathcal{D} \times \Delta^{\{0,2\}}$ exhibits g as associated to $\mathcal{M}$.

Now suppose that g' is another functor associated to p. Then there exists a commutative diagram of marked simplicial sets

$$
\begin{array}{ccc}
\mathcal{D}^{\flat} \times \{1\} & \longrightarrow & \mathcal{M}^{\natural} \\
\Big\downarrow & \overset{s'}{\nearrow} & \Big\downarrow \\
\mathcal{D}^{\flat} \times (\Delta^1)^{\sharp} & \longrightarrow & (\Delta^1)^{\sharp},
\end{array}
$$

with $g' = s' | \mathcal{D} \times \{0\}$. Let s'' be the map obtained by amalgamating s and s'. Consider the diagram

$$
\begin{array}{ccc}
\mathcal{D}^{\flat} \times (\Lambda_2^2)^{\sharp} & \overset{s''}{\longrightarrow} & \mathcal{M}^{\natural} \\
\Big\downarrow & \overset{s'''}{\nearrow} & \Big\downarrow \\
\mathcal{D}^{\flat} \times (\Delta^2)^{\sharp} & \longrightarrow & (\Delta^1)^{\sharp}.
\end{array}
$$

Since the left vertical map is marked anodyne, the indicated dotted arrow s'' exists. The restriction $s'' | \mathcal{D} \times \Delta^{\{0,1\}}$ is an equivalence between $h_0 \circ g$ and $h_0 \circ g'$. Since h_0 is a categorical equivalence, g and g' are themselves homotopic.

Conversely, suppose that $f : \mathcal{D} \times \Delta^1 \to \mathcal{C}$ is an equivalence from g' to g. The maps s and $h_0 \circ f$ amalgamate to give a map $f' : \mathcal{D} \times \Lambda_1^2 \to \mathcal{C}$ which fits into a commutative diagram of marked simplicial sets:

$$
\begin{array}{ccc}
\mathcal{D}^{\flat} \times (\Lambda_1^2)^{\sharp} & \overset{f'}{\longrightarrow} & \mathcal{M}^{\natural} \\
\Big\downarrow & \overset{f''}{\nearrow} & \Big\downarrow \\
\mathcal{D}^{\flat} \times (\Delta^2)^{\sharp} & \longrightarrow & (\Delta^1)^{\sharp}.
\end{array}
$$

The left vertical map is marked anodyne, so there exists a dotted arrow f'' as indicated; then the map $f'' | \mathcal{D} \times \Delta^{\{0,2\}}$ exhibits that g' is associated to p. $\square$

Proposition 5.2.1.5. *Let* $p : \mathcal{M} \to \Delta^2$ *be a Cartesian fibration and suppose we are given equivalences of ∞-categories* $\mathcal{C} \to p^{-1}\{0\}$, $\mathcal{D} \to p^{-1}\{1\}$, *and* $\mathcal{E} \to p^{-1}\{2\}$. *Suppose that* $\mathcal{M} \times_{\Delta^2} \Delta^{\{0,1\}}$ *is associated to a functor* $f : \mathcal{D} \to \mathcal{C}$ *and that* $\mathcal{M} \times_{\Delta^2} \Delta^{\{1,2\}}$ *is associated to a functor* $g : \mathcal{E} \to \mathcal{D}$. *Then* $\mathcal{M} \times_{\Delta^2} \Delta^{\{0,2\}}$ *is associated to the composite functor* $f \circ g$.

Proof. Let X be the mapping simplex of the sequence of functors

$$
\mathcal{E} \overset{g}{\to} \mathcal{D} \overset{f}{\to} \mathcal{C}.
$$

Since f and g are associated to restrictions of $\mathcal{M}$, we obtain a commutative diagram

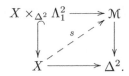

The left vertical inclusion is a pushout of $\mathcal{E} \times \Lambda_1^2 \subseteq \mathcal{E} \times \Delta^2$, which is inner anodyne. Since p is inner anodyne, there exists a dotted arrow s as indicated in the diagram. The restriction $s|X \times_{\Delta^2} \Delta^{\{0,2\}}$ exhibits that the functor $f \circ g$ is associated to the correspondence $\mathcal{M} \times_{\Delta^2} \Delta^{\{0,2\}}$. ☐

Remark 5.2.1.6. Taken together, Propositions 5.2.1.3 and 5.2.1.4 assert that there is a bijective correspondence between equivalence classes of functors $\mathcal{D} \to \mathcal{C}$ and equivalence classes of Cartesian fibrations $p : \mathcal{M} \to \Delta^1$ equipped with equivalences $\mathcal{C} \to p^{-1}\{0\}$, $\mathcal{D} \to p^{-1}\{1\}$.

5.2.2 Adjunctions

In §5.2.1, we established a dictionary that allows us to pass back and forth between functors $g : \mathcal{D} \to \mathcal{C}$ and Cartesian fibrations $p : \mathcal{M} \to \Delta^1$. The dual argument shows that if p is a coCartesian fibration, it also determines a functor $f : \mathcal{C} \to \mathcal{D}$. In this case, we will say that f and g are *adjoint* functors.

Definition 5.2.2.1. Let $\mathcal{C}$ and $\mathcal{D}$ be ∞-categories. An *adjunction* between $\mathcal{C}$ and $\mathcal{D}$ is a map $q : \mathcal{M} \to \Delta^1$ which is both a Cartesian fibration and a coCartesian fibration together with equivalences $\mathcal{C} \to \mathcal{M}_{\{0\}}$ and $\mathcal{D} \to \mathcal{M}_{\{1\}}$.

Let $\mathcal{M}$ be an adjunction between $\mathcal{C}$ and $\mathcal{D}$ and let $f : \mathcal{C} \to \mathcal{D}$ and $g : \mathcal{D} \to \mathcal{C}$ be functors associated to $\mathcal{M}$. In this case, we will say that f is *left adjoint* to g and g is *right adjoint* to f.

Remark 5.2.2.2. Propositions 5.2.1.3 and 5.2.1.4 imply that if a functor $f : \mathcal{C} \to \mathcal{D}$ has a right adjoint $g : \mathcal{D} \to \mathcal{C}$, then g is uniquely determined up to homotopy. In fact, we will later see that g is determined up to a contractible ambiguity.

We now verify a few basic properties of adjunctions:

Lemma 5.2.2.3. *Let $p : X \to S$ be a locally Cartesian fibration of simplicial sets. Let $e : s \to s'$ be an edge of S with the following property:*

$(*)$ *For every 2-simplex*

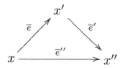

in X such that $p(\bar{e}) = e$, if $\bar{e}$ and $\bar{e}'$ are locally p-Cartesian, then $\bar{e}''$ is locally p-Cartesian.

Let $\bar{e} : x \to y$ be a locally p-coCartesian edge such that $p(\bar{e}) = e$. Then $\bar{e}$ is p-coCartesian.

Proof. We must show that for any $n \geq 2$ and any diagram such that $f|\Delta^{\{0,1\}} = \bar{e}$, there exists a dotted arrow as indicated. Pulling back along the bottom horizontal map, we may reduce to the case $S = \Delta^n$; in particular, X and S are both ∞-categories.

According to (the dual of) Proposition 2.4.4.3, it suffices to show that composition with $\bar{e}$ gives a homotopy Cartesian diagram

$$
\begin{array}{ccc}
\mathrm{Map}_X(y,z) & \longrightarrow & \mathrm{Map}_X(x,z) \\
\downarrow & & \downarrow \\
\mathrm{Map}_S(p(y),p(z)) & \longrightarrow & \mathrm{Map}_S(p(x),p(z)).
\end{array}
$$

There are two cases to consider: if $\mathrm{Map}_S(p(y),p(z)) = \emptyset$, there is nothing to prove. Otherwise, we must show that composition with f induces a homotopy equivalence $\mathrm{Map}_X(y,z) \to \mathrm{Map}_X(x,z)$.

In view of the assumption that $S = \Delta^n$, there is a unique morphism $g_0 : p(y) \to p(z)$. Let $g : y' \to z$ be a locally p-Cartesian edge lifting g_0. We have a commutative diagram

$$
\begin{array}{ccc}
\mathrm{Map}_X(y,y') & \longrightarrow & \mathrm{Map}_X(x,y') \\
\downarrow & & \downarrow \\
\mathrm{Map}_X(y,z) & \longrightarrow & \mathrm{Map}_X(x,z).
\end{array}
$$

Since g is locally p-Cartesian, the left vertical arrow is a homotopy equivalence. Since e is locally p-coCartesian, the top horizontal arrow is a homotopy equivalence. It will therefore suffice to show that the map $\mathrm{Map}_X(x,y') \to \mathrm{Map}_X(x,z)$ is a homotopy equivalence.

Choose a locally p-Cartesian edge $\bar{e}' : x' \to y'$ in X with $p(\bar{e}') = e$, so that we have another commutative diagram

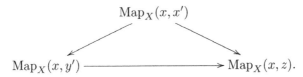

$$
\begin{array}{ccc}
& \mathrm{Map}_X(x,x') & \\
\swarrow & & \searrow \\
\mathrm{Map}_X(x,y') & \longrightarrow & \mathrm{Map}_X(x,z).
\end{array}
$$

Using the two-out-of-three property, we are reduced to proving that both of the diagonal arrows are homotopy equivalences. For the diagonal arrow on the left, this follows from our assumption that $\bar{e}'$ is locally p-Cartesian. For the arrow on the right, it suffices to show that the composition $g \circ \bar{e}'$ is locally p-coCartesian, which follows from assumption $(*)$. $\qquad\square$

Corollary 5.2.2.4. *Let $p : X \to S$ be a Cartesian fibration of simplicial sets. An edge $e : x \to y$ of X is p-coCartesian if and only if it is locally p-coCartesian (see the discussion preceding Proposition 2.4.2.8).*

Corollary 5.2.2.5. *Let $p : X \to S$ be a Cartesian fibration of simplicial sets. The following conditions are equivalent:*

(1) *The map p is a coCartesian fibration.*

(2) *For every edge $f : s \to s'$ of S, the induced functor $f^* : X_{s'} \to X_s$ has a left adjoint.*

Proof. By definition, the functor corresponding to an edge $f : \Delta^1 \to S$ has a left adjoint if and only if the pullback $X \times_S \Delta^1 \to \Delta^1$ is a coCartesian fibration. In other words, condition (2) is equivalent to the assertion that for every edge $f : s \to s'$ and every vertex $\tilde{s}$ of X lifting s, there exists a *locally* p-coCartesian edge $\tilde{f} : \tilde{s} \to \tilde{s}'$ lifting f. Using Corollary 5.2.2.4, we conclude that $\tilde{f}$ is automatically p-coCartesian, so that (2) is equivalent to (1). □

Proposition 5.2.2.6. *Let $f : \mathcal{C} \to \mathcal{D}$ and $f' : \mathcal{D} \to \mathcal{E}$ be functors between ∞-categories. Suppose that f has a right adjoint g and that f' has a right adjoint g'. Then $g \circ g'$ is right adjoint to $f' \circ f$.*

Proof. Let ϕ denote the composable sequence of morphisms

$$\mathcal{C} \xleftarrow{g} \mathcal{D} \xleftarrow{g'} \mathcal{E}.$$

Let $M(\phi)$ denote the mapping simplex and choose a factorization

$$M(\phi) \xrightarrow{s} X \xrightarrow{q} \Delta^2,$$

where s is a quasi-equivalence and $X \to \Delta^2$ is a Cartesian fibration (using Proposition 3.2.2.11). We first show that q is a coCartesian fibration. In other words, we must show that, for every object $\overline{x} \in \mathcal{C}$ and every morphism $e : q(\overline{x}) \to y$, there is a q-Cartesian edge $\overline{e} : \overline{x} \to \overline{y}$ lifting e. This is clear if e is degenerate. If $e = \Delta^{\{0,1\}} \subseteq \Delta^2$, then the existence of a left adjoint to g implies that e has a locally q-coCartesian lift $\overline{e}$. Lemma 5.2.2.3 implies that $\overline{e}$ is q-coCartesian. Similarly, if $e = \Delta^{\{1,2\}}$, then we can find a q-coCartesian lift of e. Finally, if e is the long edge $\Delta^{\{0,2\}}$, then we may write e as a composite $e' \circ e''$; the existence of a q-coCartesian lift of e follows from the existence of q-coCartesian lifts of e' and e''. We now apply Proposition 5.2.1.5 and deduce that the adjunction $X \times_{\Delta^2} \Delta^{\{0,2\}}$ is associated to both $g \circ g'$ and $f' \circ f$. □

In classical category theory, one can spell out the relationship between a pair of adjoint functors $f : \mathcal{C} \to \mathcal{D}$ and $g : \mathcal{D} \to \mathcal{C}$ by specifying a *unit* transformation $\mathrm{id}_{\mathcal{C}} \to g \circ f$ (or, dually, a *counit* $f \circ g \to \mathrm{id}_{\mathcal{D}}$). This concept generalizes to the ∞-categorical setting as follows:

Definition 5.2.2.7. Suppose we are given a pair of functors

$$\mathcal{C} \underset{g}{\overset{f}{\rightleftarrows}} \mathcal{D}$$

between ∞-categories. A *unit transformation* for (f, g) is a morphism $u : \mathrm{id}_{\mathcal{C}} \to g \circ f$ in $\mathrm{Fun}(\mathcal{C}, \mathcal{C})$ with the following property: for every pair of objects $C \in \mathcal{C}$, $D \in \mathcal{D}$, the composition

$$\mathrm{Map}_{\mathcal{D}}(f(C), D) \to \mathrm{Map}_{\mathcal{C}}(g(f(C)), g(D)) \xrightarrow{u(C)} \mathrm{Map}_{\mathcal{C}}(C, g(D))$$

is an isomorphism in the homotopy category $\mathcal{H}$.

Proposition 5.2.2.8. *Let $f : \mathcal{C} \to \mathcal{D}$ and $g : \mathcal{D} \to \mathcal{C}$ be a pair of functors between ∞-categories $\mathcal{C}$ and $\mathcal{D}$. The following conditions are equivalent:*

(1) *The functor f is a left adjoint to g.*

(2) *There exists a unit transformation $u : \mathrm{id}_{\mathcal{C}} \to g \circ f$.*

Proof. Suppose first that (1) is satisfied. Choose an adjunction $p : M \to \Delta^1$ which is associated to f and g; according to (1) of Proposition 5.2.1.3 we may identify $M_{\{0\}}$ with $\mathcal{C}$ and $M_{\{1\}}$ with $\mathcal{D}$. Since f is associated to M, there is a map $F : \mathcal{C} \times \Delta^1 \to M$ such that $F | \mathcal{C} \times \{0\} = \mathrm{id}_{\mathcal{C}}$ and $F | \mathcal{C} \times \{1\} = f$, with each edge $F | \{c\} \times \Delta^1$ p-coCartesian. Similarly, there is a map $G : \mathcal{D} \times \Delta^1 \to M$ with $G | \mathcal{D} \times \{1\} = \mathrm{id}_{\mathcal{D}}$, $G | \mathcal{D} \times \{0\} = g$, and such that $G | \{d\} \times \Delta^1$ is p-Cartesian for each object $d \in \mathcal{D}$. Let $F' : \Lambda_2^2 \times \mathcal{C} \to M$ be such that $F' | \Delta^{\{0,2\}} \times \mathcal{C} = F$ and $F' | \Delta^{\{1,2\}} \times \mathcal{C} = G \circ (f \times \mathrm{id}_{\Delta^1})$. Consider the diagram

$$
\begin{array}{ccc}
\Lambda_2^2 \times \mathcal{C} & \xrightarrow{\ F'\ } & M \\
\big\downarrow & \nearrow^{F''} & \big\downarrow \\
\Delta^2 \times \mathcal{C} & \longrightarrow & \Delta^1.
\end{array}
$$

Using the fact that $F' | \{c\} \times \Delta^{\{1,2\}}$ is p-Cartesian for every object $c \in C$, we deduce the existence of the dotted arrow F''. We now define $u = F' | \mathcal{C} \times \Delta^{\{0,1\}}$. We may regard u as a natural transformation $\mathrm{id}_{\mathcal{C}} \to g \circ f$. We claim that u is a unit transformation. In other words, we must show that for any objects $C \in \mathcal{C}$, $D \in \mathcal{D}$, the composite map

$$
\mathrm{Map}_{\mathcal{D}}(fC, D) \to \mathrm{Map}_{\mathcal{C}}(gfC, gD) \xrightarrow{u} \mathrm{Map}_{\mathcal{C}}(C, gD)
$$

is an isomorphism in the homotopy category $\mathcal{H}$ of spaces. This composite map fits into a commutative diagram

$$
\begin{array}{ccc}
\mathrm{Map}_{\mathcal{D}}(f(C), D) \longrightarrow \mathrm{Map}_{\mathcal{D}}(g(f(C)), g(D)) \longrightarrow \mathrm{Map}_{\mathcal{D}}(C, g(D)) \\
\big\downarrow \qquad\qquad\qquad\qquad\qquad\qquad\qquad\qquad \big\downarrow \\
\mathrm{Map}_M(C, D) \longrightarrow\longrightarrow\longrightarrow\longrightarrow\longrightarrow\longrightarrow \mathrm{Map}_M(C, D).
\end{array}
$$

The left and right vertical arrows in this diagram are given by composition with a p-coCartesian and a p-Cartesian morphism in M, respectively. Proposition 2.4.4.2 implies that these maps are homotopy equivalences.

We now prove that $(2) \Rightarrow (1)$. Choose a correspondence $p : M \to \Delta^1$ from $\mathcal{C}$ to $\mathcal{D}$ which is associated to the functor g via a map $G : \mathcal{D} \times \Delta^1 \to M$ as above. We have natural transformations

$$
\mathrm{id}_{\mathcal{C}} \xrightarrow{\ u\ } g \circ f \xrightarrow{\ G \circ (f \times \mathrm{id}_{\Delta^1})\ } f.
$$

Let $F : \mathcal{C} \times \Delta^1 \to M$ be a composition of these transformations. We will complete the proof by showing that F exhibits M as a correspondence associated to the functor f. It will suffice to show that for each object $C \in \mathcal{C}$,

$F(C) : C \to fC$ is p-coCartesian. According to Proposition 2.4.4.3, it will suffice to show that for each object $D \in \mathcal{D}$, composition with $F(C)$ induces a homotopy equivalence $\mathrm{Map}_{\mathcal{D}}(f(C), D) \to \mathrm{Map}_M(C, D)$. As above, this map fits into a commutative diagram

$$\begin{array}{ccccc} \mathrm{Map}_{\mathcal{D}}(f(C), D) & \longrightarrow & \mathrm{Map}_{\mathcal{D}}(g(f(C)), g(D)) & \longrightarrow & \mathrm{Map}_{\mathcal{D}}(C, g(D)) \\ \downarrow & & & & \downarrow \\ \mathrm{Map}_M(C, D) & & \longrightarrow & & \mathrm{Map}_M(C, D) \end{array}$$

where the upper horizontal composition is an equivalence (since u is a unit transformation) and the right vertical arrow is an equivalence (since it is given by composition with a p-Cartesian morphism). It follows that the left vertical arrow is also a homotopy equivalence, as desired. $\qquad\square$

Proposition 5.2.2.9. *Let* $\mathcal{C}$ *and* $\mathcal{D}$ *be* ∞-*categories and let* $f : \mathcal{C} \to \mathcal{D}$ *and* $g : \mathcal{D} \to \mathcal{C}$ *be adjoint functors. Then* f *and* g *induce adjoint functors* $\mathrm{h}f : \mathrm{h}\mathcal{C} \to \mathrm{h}\mathcal{D}$ *and* $\mathrm{h}g : \mathrm{h}\mathcal{D} \to \mathrm{h}\mathcal{C}$ *between* ($\mathcal{H}$-*enriched*) *homotopy categories.*

Proof. This follows immediately from Proposition 5.2.2.8 because a unit transformation $\mathrm{id}_{\mathcal{C}} \to g \circ f$ induces a unit transformation $\mathrm{id}_{\mathrm{h}\mathcal{C}} \to (\mathrm{h}g) \circ (\mathrm{h}f)$. $\qquad\square$

The converse to Proposition 5.2.2.9 is false. If $f : \mathcal{C} \to \mathcal{D}$ and $g : \mathcal{D} \to \mathcal{C}$ are functors such that $\mathrm{h}f$ and $\mathrm{h}g$ are adjoint to one another, then f and g are not necessarily adjoint. Nevertheless, the existence of adjoints can be tested at the level of (enriched) homotopy categories.

Lemma 5.2.2.10. *Let* $p : \mathcal{M} \to \Delta^1$ *be an inner fibration of simplicial sets giving a correspondence between the* ∞-*categories* $\mathcal{C} = \mathcal{M}_{\{0\}}$ *and* $\mathcal{D} = \mathcal{M}_{\{1\}}$. *Let* c *be an object of* $\mathcal{C}$, d *an object of* $\mathcal{D}$, *and* $f : c \to d$ *a morphism. The following are equivalent:*

(1) *The morphism* f *is* p-*Cartesian.*

(2) *The morphism* f *gives rise to a Cartesian morphism in the enriched homotopy category* $\mathrm{h}\mathcal{M}$; *in other words, composition with* p *induces homotopy equivalences*

$$\mathrm{Map}_{\mathcal{C}}(c', c) \to \mathrm{Map}_{\mathcal{M}}(c', d)$$

for every object $c' \in \mathcal{C}$.

Proof. This follows immediately from Proposition 2.4.4.3. $\qquad\square$

Lemma 5.2.2.11. *Let* $p : \mathcal{M} \to \Delta^1$ *be an inner fibration, so that* $\mathcal{M}$ *can be identified with a correspondence from* $\mathcal{C} = p^{-1}\{0\}$ *to* $\mathcal{D} = p^{-1}\{1\}$. *The following conditions are equivalent:*

(1) *The map* p *is a Cartesian fibration.*

(2) *There exists a $\mathcal{H}$-enriched functor $g : h\mathcal{D} \to h\mathcal{C}$ and a functorial iden-tification*

$$\mathrm{Map}_{\mathcal{M}}(c, d) \simeq \mathrm{Map}_{\mathcal{C}}(c, g(d)).$$

Proof. If p is a Cartesian fibration, then there is a functor $\mathcal{D} \to \mathcal{C}$ asso-ciated to $\mathcal{M}$; we can then take g to be the associated functor on enriched homotopy categories. Conversely, suppose that there exists a functor g as above. We wish to show that p is a Cartesian fibration. In other words, we must show that for every object $d \in \mathcal{D}$, there is an object $c \in \mathcal{C}$ and a p-Cartesian morphism $f : c \to d$. We take $c = g(d)$; in view of the identification $\mathrm{Map}_{\mathcal{M}}(c, d) \simeq \mathrm{Map}_{\mathcal{C}}(c, c)$, there exists a morphism $f : c \to d$ correspond-ing to the identity id_c. Lemma 5.2.2.10 implies that f is p-Cartesian, as desired. $\qquad\square$

Proposition 5.2.2.12. *Let $f : \mathcal{C} \to \mathcal{D}$ be a functor between ∞-categories. Suppose that the induced functor of $\mathcal{H}$-enriched categories $hf : h\mathcal{C} \to h\mathcal{D}$ admits a right adjoint. Then f admits a right adjoint.*

Proof. According to (1) of Proposition 5.2.1.3, there is a coCartesian fibra-tion $p : \mathcal{M} \to \Delta^1$ associated to f. Let hg be the right adjoint of hf. Applying Lemma 5.2.2.11, we deduce that p is a Cartesian fibration. Thus p is an adjunction, so that f has a right adjoint, as desired. $\qquad\square$

5.2.3 Preservation of Limits and Colimits

Let $\mathcal{C}$ and $\mathcal{D}$ be ordinary categories and let $F : \mathcal{C} \to \mathcal{D}$ be a functor. If F has a right adjoint G, then F preserves colimits; we have a chain of natural isomorphisms

$$\mathrm{Hom}_{\mathcal{D}}(F(\varinjlim C_\alpha), D) \simeq \mathrm{Hom}_{\mathcal{C}}(\varinjlim C_\alpha, G(D))$$
$$\simeq \varprojlim \mathrm{Hom}_{\mathcal{C}}(C_\alpha, G(D))$$
$$\simeq \varprojlim \mathrm{Hom}_{\mathcal{D}}(F(C_\alpha), D)$$
$$\simeq \mathrm{Hom}_{\mathcal{D}}(\varinjlim F(C_\alpha), D).$$

In fact, this is in some sense the *defining* feature of left adjoints: under suitable set-theoretic assumptions, the *adjoint functor theorem* asserts that any colimit-preserving functor admits a right adjoint. We will prove an ∞-categorical version of the adjoint functor theorem in §5.5.2. Our goal in this section is to lay the groundwork by showing that left adjoints preserve colimits in the ∞-categorical setting. We will first need to establish several lemmas.

Lemma 5.2.3.1. *Suppose we are given a diagram*

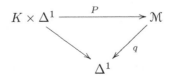

of simplicial sets, where $\mathcal{M}$ *is an* ∞*-category and* $P|\{k\} \times \Delta^1$ *is* q*-coCartesian for every vertex* k *of* K. *Let* $p = P|K \times \{0\}$. *Then the induced map*

$$\psi : \mathcal{M}_{P/} \to \mathcal{M}_{p/}$$

induces a trivial fibration

$$\psi_1 : \mathcal{M}_{P/} \times_{\Delta^1} \{1\} \to \mathcal{M}_{p/} \times_{\Delta^1} \{1\}.$$

Proof. If K is a point, then the assertion reduces immediately to the definition of a coCartesian edge. In the general case, we note that ψ and ψ_1 are both left fibrations between ∞-categories. Consequently, it suffices to show that ψ_1 is a categorical equivalence. In doing so, we are free to replace ψ by the equivalent map $\psi' : \mathcal{M}^{P/} \to \mathcal{M}^{p/}$. To prove that $\psi_1' : \mathcal{M}^{P/} \times_{\Delta^1} \{1\} \to \mathcal{M}^{p/} \times_{\Delta^1} \{1\}$ is a trivial fibration, we must show that for every inclusion $A \subseteq B$ of simplicial sets and any map

$$k_0 : ((K \times \Delta^1) \diamond A) \coprod_{(K \times \{0\}) \diamond A} ((K \times \{0\}) \diamond B) \to \mathcal{M}$$

with $k_0|K \times \Delta^1 = P$ and $k_0(B) \subseteq q^{-1}\{1\}$, there exists an extension of k_0 to a map $k : (K \times \Delta^1) \diamond B \to \mathcal{M}$. Let

$$X = (K \times \Delta^1) \coprod_{K \times \Delta^1 \times B \times \{0\}} (K \times \Delta^1 \times B \times \Delta^1)$$

and let $h : X \to K \diamond B$ be the natural map. Let

$$X' = h^{-1}((K \times \Delta^1) \diamond A) \coprod_{(K \times \{0\}) \diamond A} ((K \times \{0\}) \diamond B) \subseteq X$$

and let $\widetilde{k}_0 : X' \to \mathcal{M}$ be the composition $k_0 \circ h$. It suffices to prove that there exists an extension of $\widetilde{k}_0$ to a map $\widetilde{k} : X \to \mathcal{M}$. Replacing $\mathcal{M}$ by $\mathrm{Map}_{\Delta^1}(K, \mathcal{M})$, we may reduce to the case where K is a point, which we already treated above. $\qquad\square$

Lemma 5.2.3.2. *Let* $q : \mathcal{M} \to \Delta^1$ *be a correspondence between* ∞*-categories* $\mathcal{C} = q^{-1}\{0\}$ *and* $\mathcal{D} = q^{-1}\{1\}$ *and let* $p : K \to \mathcal{C}$ *be a diagram in* $\mathcal{C}$. *Let* $f : c \to d$ *be a* q*-Cartesian morphism in* $\mathcal{M}$ *from* $c \in \mathcal{C}$ *to* $d \in \mathcal{D}$. *Let* $r : \mathcal{M}_{p/} \to \mathcal{M}$ *be the projection and let* $\overline{d}$ *be an object of* $\mathcal{M}_{p/}$ *with* $r(\overline{d}) = d$. *Then*

(1) *There exists a morphism* $\overline{f} : \overline{c} \to \overline{d}$ *in* $\mathcal{M}_{p/}$ *satisfying* $f = r(\overline{f})$.

(2) *Any morphism* $\overline{f} : \overline{c} \to \overline{d}$ *which satisfies* $r(\overline{f}) = f$ *is* r*-Cartesian.*

Proof. We may identify $\overline{d}$ with a map $\overline{d} : K \to \mathcal{M}_{/d}$. Consider the set of pairs (L, s), where $L \subseteq K$ and $s : L \to \mathcal{M}_{/f}$ sits in a commutative diagram

We order these pairs by setting $(L, s) \leq (L', s')$ if $L \subseteq L'$ and $s = s'|L$. By Zorn's lemma, there exists a pair (L, s) which is maximal with respect to this ordering. To prove (1), it suffices to show that $L = K$. Otherwise, we may obtain a larger simplicial subset $L' = L \coprod_{\partial \Delta^n} \Delta^n \subseteq K$ by adjoining a single nondegenerate simplex. By maximality, there is no solution to the associated lifting problem

$$
\begin{array}{ccc}
\partial \Delta^n & \longrightarrow & \mathcal{M}_{/f} \\
\downarrow & \nearrow & \downarrow \\
\Delta^n & \longrightarrow & \mathcal{M}_{/d}
\end{array}
$$

nor to the associated lifting problem

$$
\begin{array}{ccc}
\Lambda^{n+2}_{n+2} & \overset{s}{\longrightarrow} & \mathcal{M} \\
\downarrow & \nearrow & \downarrow q \\
\Delta^{n+2} & \longrightarrow & \Delta^1,
\end{array}
$$

which contradicts the fact that s carries $\Delta^{\{n+1,n+2\}}$ to the q-Cartesian morphism f in $\mathcal{M}$.

Now suppose that $\overline{f}$ is a lift of f. To prove that $\overline{f}$ is r-Cartesian, it suffices to show that for every $m \geq 2$ and every diagram

$$
\begin{array}{ccc}
\Lambda^m_m & \overset{g_0}{\longrightarrow} & \mathcal{M}_{p/} \\
\downarrow & \overset{g}{\nearrow} & \downarrow \\
\Delta^m & \longrightarrow & \mathcal{M}
\end{array}
$$

such that $g_0|\Delta^{\{m-1,m\}} = \widetilde{f}$, there exists a dotted arrow g as indicated, rendering the diagram commutative. We can identify the diagram with a map

$$
t_0 : (K \star \Lambda^m_m) \coprod_{\Lambda^m_m} \Delta^m \to \mathcal{M}.
$$

Consider the set of all pairs (L, t), where $L \subseteq K$ and

$$
t : (K \star \Lambda^m_m) \coprod_{L \star \Lambda^m_m} (L \star \Delta^m) \to \mathcal{M}
$$

is an extension of t_0. As above, we order the set of such pairs by declaring $(L, t) \leq (L', t')$ if $L \subseteq L'$ and $t = t'|L$. Zorn's lemma guarantees the existence of a maximal pair (L, t). If $L = K$, we are done; otherwise let L' be obtained from L by adjoining a single nondegenerate n-simplex of K. By maximality, the map t does not extend to L'; consequently, the associated mapping problem

$$
\begin{array}{ccc}
(\Delta^n \star \Lambda^m_m) \coprod_{\partial \Delta^n \star \Lambda^m_m} (\partial \Delta^n \star \Delta^m) & \longrightarrow & \mathcal{M} \\
\downarrow & \diagup & \downarrow \\
\Delta^n \star \Delta^m & \longrightarrow & \Delta^1
\end{array}
$$

has no solution. But this contradicts the assumption that $r(\widetilde{f}) = f$ is a q-Cartesian edge of $\mathcal{M}$. □

Lemma 5.2.3.3. *Let* $q : \mathcal{M} \to \Delta^1$ *be a correspondence between the* ∞-*categories* $\mathcal{C} = q^{-1}\{0\}$ *and* $\mathcal{D} = q^{-1}\{1\}$. *Let* $f : c \to d$ *be a morphism in* $\mathcal{M}$ *between objects* $c \in \mathcal{C}$ *and* $d \in \mathcal{D}$. *Let* $p : K \to \mathcal{C}$ *be a diagram and consider an associated map*

$$k : \mathcal{M}_{p/} \times_{\mathcal{M}} \{c\} \to \mathcal{M}_{p/} \times_{\mathcal{M}} \{d\}$$

(the map k *is well-defined up to homotopy according to Lemma 2.1.1.4). If* f *is* q-*Cartesian, then* k *is a homotopy equivalence.*

Proof. Let $X = (\mathcal{M}_{p/})^{\Delta^1} \times_{\mathcal{M}^{\Delta^1}} \{f\}$ and consider the diagram

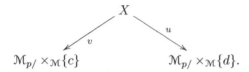

The map u is a homotopy equivalence, and k is defined as the composition of v with a homotopy inverse to u. Consequently, it will suffice to show that v is a trivial fibration. To prove this, we must show that v has the right lifting property with respect to $\partial \Delta^n \subseteq \Delta^n$, which is equivalent to solving a lifting problem

$$
\begin{array}{ccc}
(\partial \Delta^n \times \Delta^1) \coprod_{\partial \Delta^n \times \{1\}} (\Delta^n \times \{1\}) & \longrightarrow & \mathcal{M}_{p/} \\
\downarrow & \nearrow & \downarrow r \\
\Delta^n \times \Delta^1 & \longrightarrow & \mathcal{M}.
\end{array}
$$

If $n = 0$, we invoke (1) of Lemma 5.2.3.2. If $n > 0$, then Proposition 2.4.1.8 implies that it suffices to show that the upper horizontal map carries $\{n\} \times \Delta^1$ to an r-Cartesian edge of $\mathcal{M}_{p/}$, which also follows from assertion (2) of Lemma 5.2.3.2. □

Lemma 5.2.3.4. *Let* $q : \mathcal{M} \to \Delta^1$ *be a Cartesian fibration and let* $\mathcal{C} = q^{-1}\{0\}$. *The inclusion* $\mathcal{C} \subseteq \mathcal{M}$ *preserves all colimits which exist in* $\mathcal{C}$.

Proof. Let $\overline{p} : K^{\triangleright} \to \mathcal{C}$ be a colimit of $p = \overline{p}|K$. We wish to show that $\mathcal{M}_{\overline{p}/} \to \mathcal{M}_{p/}$ is a trivial fibration. Since we have a diagram

$$\mathcal{M}_{\overline{p}/} \to \mathcal{M}_{p/} \to \mathcal{M}$$

of left fibrations, it will suffice to show that the induced map

$$\mathcal{M}_{\overline{p}/} \times_{\mathcal{M}} \{d\} \to \mathcal{M}_{p/} \times_{\mathcal{M}} \{d\}$$

is a homotopy equivalence of Kan complexes for each object d of $\mathcal{M}$. If d belongs to $\mathcal{C}$, this is obvious. In general, we may choose a q-Cartesian morphism $f : c \to d$ in $\mathcal{M}$. Composition with f gives a commutative diagram

$$\begin{array}{ccc} [\mathcal{M}_{\overline{p}/} \times_{\mathcal{M}} \{c\}] & \longrightarrow & [\mathcal{M}_{p/} \times_{\mathcal{M}} \{c\}] \\ \downarrow & & \downarrow \\ [\mathcal{M}_{\overline{p}/} \times_{\mathcal{M}} \{d\}] & \longrightarrow & [\mathcal{M}_{p/} \times_{\mathcal{M}} \{d\}] \end{array}$$

in the homotopy category $\mathcal{H}$ of spaces. The upper horizontal map is a homotopy equivalence since $\overline{p}$ is a colimit of p in $\mathcal{C}$. The vertical maps are homotopy equivalences by Lemma 5.2.3.3. Consequently, the bottom horizontal map is also a homotopy equivalence, as desired. $\qquad\square$

Proposition 5.2.3.5. *Let $f : \mathcal{C} \to \mathcal{D}$ be a functor between ∞-categories which has a right adjoint $g : \mathcal{D} \to \mathcal{C}$. Then f preserves all colimits which exist in $\mathcal{C}$, and g preserves all limits which exist in $\mathcal{D}$.*

Proof. We will show that f preserves colimits; the analogous statement for g follows by a dual argument. Let $\overline{p} : K^{\triangleright} \to \mathcal{C}$ be a colimit for $p = \overline{p}|K$. We must show that $f \circ \overline{p}$ is a colimit of $f \circ p$.

Let $q : \mathcal{M} \to \Delta^1$ be an adjunction between $\mathcal{C} = \mathcal{M}_{\{0\}}$ and $\mathcal{D} = \mathcal{M}_{\{1\}}$ which is associated to f and g. We wish to show that

$$\phi_1 : \mathcal{D}_{f\overline{p}/} \to \mathcal{D}_{fp/}$$

is a trivial fibration. Since ϕ_1 is a left fibration, it suffices to show that ϕ_1 is a categorical equivalence.

Since $\mathcal{M}$ is associated to f, there is a map $F : \mathcal{C} \times \Delta^1 \to \mathcal{M}$ such that $F|\mathcal{C} \times \{0\} = \mathrm{id}_{\mathcal{C}}$, $F|\mathcal{C} \times \{1\} = f$, and $F|\{c\} \times \Delta^1$ a q-coCartesian morphism of $\mathcal{M}$ for every object $c \in \mathcal{C}$. Let $\overline{P} = F \circ (\overline{p} \times \mathrm{id}_{\Delta^1})$ be the induced map $K^{\triangleright} \times \Delta^1 \to \mathcal{M}$ and let $P = \overline{P}|K \times \Delta^1$.

Consider the diagram

$$\begin{array}{ccc} \mathcal{M}_{\overline{p}/} & \xrightarrow{\phi'} & \mathcal{M}_{p/} \\ \overline{v} \uparrow & & \uparrow v \\ \mathcal{M}_{\overline{P}/} & \longrightarrow & \mathcal{M}_{P/} \\ \overline{u} \downarrow & & \downarrow u \\ \mathcal{M}_{f\overline{p}/} & \xrightarrow{\phi} & \mathcal{M}_{fp/} \, . \end{array}$$

We note that every object in this diagram is an ∞-category with a map to Δ^1; moreover, the map ϕ_1 is obtained from ϕ by passage to the fiber over $\{1\} \subseteq \Delta^1$. Consequently, to prove that ϕ_1 is a categorical equivalence, it suffices to verify three things:

(1) The bottom vertical maps u and $\overline{u}$ are trivial fibrations. This follows from the fact that $K \times \{1\} \subseteq K \times \Delta^1$ and $K^{\triangleright} \times \{1\} \subseteq K^{\triangleright} \times \Delta^1$ are right anodyne inclusions (Proposition 2.1.2.5).

(2) The upper vertical maps v and $\bar{v}$ are trivial fibrations when restricted to $\mathcal{D} \subseteq \mathcal{M}$. This follows from Lemma 5.2.3.1 since F carries each $\{c\} \times \Delta^1$ to a q-coCartesian edge of $\mathcal{M}$.

(3) The map ϕ' is a trivial fibration since $\bar{p}$ is a colimit of p in $\mathcal{M}$ according to Lemma 5.2.3.4.

□

Remark 5.2.3.6. Under appropriate set-theoretic hypotheses, one can establish a converse to Proposition 5.2.3.5. See Corollary 5.5.2.9.

5.2.4 Examples of Adjoint Functors

In this section, we describe a few simple criteria for establishing the existence of adjoint functors.

Lemma 5.2.4.1. *Let* $q : \mathcal{M} \to \Delta^1$ *be a coCartesian fibration associated to a functor* $f : \mathcal{C} \to \mathcal{D}$, *where* $\mathcal{C} = q^{-1}\{0\}$ *and* $\mathcal{D} = q^{-1}\{1\}$. *Let* D *be an object of* $\mathcal{D}$. *The following are equivalent:*

(1) *There exists a q-Cartesian morphism* $g : C \to D$ *in* $\mathcal{M}$, *where* $C \in \mathcal{C}$.

(2) *The right fibration* $\mathcal{C} \times_{\mathcal{D}} \mathcal{D}^{/D} \to \mathcal{C}$ *is representable.*

Proof. Let $F : \mathcal{C} \times \Delta^1 \to \mathcal{M}$ be a p-coCartesian natural transformation from $\mathrm{id}_{\mathcal{C}}$ to f. Define a simplicial set X so that for every simplicial set K, $\mathrm{Hom}_{\mathrm{Set}_\Delta}(K, X)$ parametrizes maps $H : K \times \Delta^2 \to \mathcal{M}$ such that $h = H | K \times \{0\}$ factors through $\mathcal{C}$, $H | K \times \Delta^{\{0,1\}} = F \circ (h|(K \times \{0\}) \times \mathrm{id}_{\Delta^1})$, and $H | K \times \{2\}$ is the constant map at the vertex D. We have restriction maps

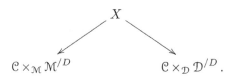

$$\mathcal{C} \times_{\mathcal{M}} \mathcal{M}^{/D} \qquad\qquad \mathcal{C} \times_{\mathcal{D}} \mathcal{D}^{/D}.$$

which are both trivial fibrations (the map on the right because $\mathcal{M}$ is an ∞-category, and the map on the left because F is p-coCartesian). Consequently, (2) is equivalent to the assertion that the ∞-category $\mathcal{C} \times_{\mathcal{M}} \mathcal{M}^{/D}$ has a final object. It now suffices to observe that a final object of $\mathcal{C} \times_{\mathcal{M}} \mathcal{M}^{/D}$ is *precisely* a q-Cartesian morphism $C \to D$, where $C \in \mathcal{C}$.   □

Proposition 5.2.4.2. *Let* $F : \mathcal{C} \to \mathcal{D}$ *be a functor between* ∞*-categories. The following are equivalent:*

(1) *The functor* F *has a right adjoint.*

(2) *For every pullback diagram*

$$
\begin{array}{ccc}
\overline{\mathcal{C}} & \longrightarrow & \overline{\mathcal{D}} \\
{\scriptstyle p'}\downarrow & & \downarrow{\scriptstyle p} \\
\mathcal{C} & \xrightarrow{\ F\ } & \mathcal{D},
\end{array}
$$

if p is a representable right fibration, then p' is also a representable right fibration.

Proof. Let $\mathcal{M}$ be a correspondence from $\mathcal{C}$ to $\mathcal{D}$ associated to F and apply Lemma 5.2.4.1 to each object of D. $\qquad\square$

Proposition 5.2.4.3. *Let $p : \mathcal{C} \to \mathcal{D}$ be a Cartesian fibration of ∞-categories and let $s : \mathcal{D} \to \mathcal{C}$ be a section of p such that $s(D)$ is an initial object of $\mathcal{C}_D = \mathcal{C} \times_{\mathcal{D}} \{D\}$ for every object $D \in \mathcal{D}$. Then s is a left adjoint of p.*

Proof. Let $\mathcal{C}^0 \subseteq \mathcal{C}$ denote the full subcategory of $\mathcal{C}$ spanned by those objects $C \in \mathcal{C}$ such that C is initial in the ∞-category $\mathcal{C}_{p(C)}$. According to Proposition 2.4.4.9, the restriction $p|\,\mathcal{C}^0$ is a trivial fibration from $\mathcal{C}^0$ to $\mathcal{D}$. Consequently, it will suffice to show that the inclusion $\mathcal{C}^0 \subseteq \mathcal{C}$ is left adjoint to the composition $s \circ p : \mathcal{C} \to \mathcal{C}^0$. Let $\mathcal{M} \subseteq \mathcal{C} \times \Delta^1$ be the full subcategory spanned by the vertices $(C, \{i\})$, where $i = 1$ or $C \in \mathcal{C}^0$. Let $q : \mathcal{M} \to \Delta^1$ be the projection. It is clear that q is a coCartesian fibration associated to the inclusion $\mathcal{C}^0 \subseteq \mathcal{C}$. To complete the proof, it will suffice to show that q is also a Cartesian fibration associated to $s \circ p$.

We first show that q is a Cartesian fibration. It will suffice to show that for any object $C \in \mathcal{C}$, there is a q-Cartesian edge $(C', 0) \to (C, 1)$ in $\mathcal{M}$. By assumption, $C' = (s \circ p)(C)$ is an initial object of $\mathcal{C}_{p(C)}$. Consequently, there exists a morphism $f : C' \to C$ in $\mathcal{C}_{p(C)}$; we will show that $f \times \mathrm{id}_{\Delta^1}$ is a q-Cartesian edge of $\mathcal{M}$. To prove this, it suffices to show that for every $n \geq 2$ and every diagram

$$
\begin{array}{ccc}
\Lambda^n_n & \xrightarrow{\ G_0\ } & \mathcal{M} \\
\downarrow & {\scriptstyle G}\nearrow & \downarrow \\
\Delta^n & \longrightarrow & \Delta^1
\end{array}
$$

such that $F_0|\Delta^{\{n-1,n\}} = f \times \mathrm{id}_{\Delta^1}$, there exists a dotted arrow $F : \Delta^n \to \mathcal{M}$ as indicated, rendering the diagram commutative. We may identify G_0 with a map $g_0 : \Lambda^n_n \to \mathcal{C}$. The composite map $p \circ g_0$ carries $\Delta^{\{n-1,n\}}$ to a degenerate edge of $\mathcal{D}$ and therefore admits an extension $\overline{g} : \Delta^n \to \mathcal{D}$. Consider the diagram

$$
\begin{array}{ccc}
\Lambda^n_n & \xrightarrow{\ g_0\ } & \mathcal{C} \\
\downarrow & {\scriptstyle g}\nearrow & \downarrow{\scriptstyle p} \\
\Delta^n & \xrightarrow{\ \overline{g}\ } & \mathcal{D}.
\end{array}
$$

Since g_0 carries the initial vertex v of Δ^n to an initial object of the fiber $\mathcal{C}_{\bar{g}(v)}$, Lemma 2.4.4.8 implies the existence of the indicated map g rendering the diagram commutative. This gives rise to a map $G : \Delta^n \to \mathcal{M}$ with the desired properties and completes the proof that q is a Cartesian fibration.

We now wish to show that $s \circ p$ is associated to q. To prove this, it suffices to prove the existence of a map $H : \mathcal{C} \times \Delta^1 \to \mathcal{C}$ such that $p \circ H = p \circ \pi_{\mathcal{C}}$, $H|\mathcal{C} \times \{1\} = \mathrm{id}_{\mathcal{C}}$, and $H|\mathcal{C} \times \{0\} = s \circ p$. We construct the map H inductively, working simplex by simplex on $\mathcal{C}$. Suppose that we have a nondegenerate simplex $\sigma : \Delta^n \to \mathcal{C}$ and that H has already been defined on $\mathrm{sk}^{n-1} \mathcal{C} \times \Delta^1$. To define $H \circ (\sigma \times \mathrm{id}_{\Delta^1})$, we must solve a lifting problem that may be depicted as follows:

$$
\begin{array}{ccc}
(\partial \Delta^n \times \Delta^1) \coprod_{\partial \Delta^n \times \partial \Delta^1} (\Delta^n \times \partial \Delta^1) & \xrightarrow{\ h_0\ } & \mathcal{C} \\
\downarrow & \nearrow^{h} & \downarrow^{p} \\
\Delta^n \times \Delta^1 & \longrightarrow & \mathcal{D}.
\end{array}
$$

We now consider the filtration

$$
X(n+1) \subseteq X(n) \subseteq \cdots \subseteq X(0) = \Delta^n \times \Delta^1
$$

defined in the proof of Proposition 2.1.2.6. Let $Y(i) = X(i) \coprod_{\partial \Delta^n \times \{0\}} (\Delta^n \times \{1\})$. For $i > 0$, the inclusion $Y(i+1) \subseteq Y(i)$ is a pushout of the inclusion $X(i+1) \subseteq X(i)$ and therefore inner anodyne. Consequently, we may use the assumption that p is an inner fibration to extend h_0 to a map defined on $Y(1)$. The inclusion $Y(1) \subseteq \Delta^n \times \Delta^1$ is a pushout of $\partial \Delta^{n+1} \subseteq \Delta^{n+1}$; we then obtain the desired extension h by applying Lemma 2.4.4.8. □

Proposition 5.2.4.4. *Let $\mathcal{M}$ be a fibrant simplicial category equipped with a functor $p : \mathcal{M} \to \Delta^1$ (here we identify Δ^1 with the two-object category whose nerve is Δ^1), so that we may view $\mathcal{M}$ as a correspondence between the simplicial categories $\mathcal{C} = p^{-1}\{0\}$ and $\mathcal{D} = p^{-1}\{1\}$. The following are equivalent*

(1) *The map p is a Cartesian fibration.*

(2) *For every object $D \in \mathcal{D}$, there exists a morphism $f : C \to D$ in $\mathcal{M}$ which induces homotopy equivalences*

$$
\mathrm{Map}_{\mathcal{C}}(C', C) \to \mathrm{Map}_{\mathcal{M}}(C', D)
$$

for every $C' \in \mathcal{C}$.

Proof. This follows immediately from Proposition 2.4.1.10 since nonempty morphism spaces in Δ^1 are contractible. □

Corollary 5.2.4.5. *Let $\mathcal{C}$ and $\mathcal{D}$ be fibrant simplicial categories and let*

$$
\mathcal{C} \underset{G}{\overset{F}{\rightleftarrows}} \mathcal{D}
$$

be a pair of adjoint functors $F : \mathcal{C} \to \mathcal{D}$ (in the sense of enriched category theory, so that there is a natural isomorphism of simplicial sets

$$\mathrm{Map}_{\mathcal{C}}(F(C), D) \simeq \mathrm{Map}_{\mathcal{D}}(C, G(D))$$

for $C \in \mathcal{C}$, $D \in \mathcal{D}$). Then the induced functors

$$\mathrm{N}(\mathcal{C}) \underset{g}{\overset{f}{\rightleftarrows}} \mathrm{N}(\mathcal{D})$$

are also adjoint to one another.

Proof. Let $\mathcal{M}$ be the correspondence associated to the adjunction (F, G). In other words, $\mathcal{M}$ is a simplicial category containing $\mathcal{C}$ and $\mathcal{D}$ as full (simplicial) subcategories, with

$$\mathrm{Map}_{\mathcal{M}}(C, D) = \mathrm{Map}_{\mathcal{C}}(C, G(D)) = \mathrm{Map}_{\mathcal{D}}(F(C), D)$$

$$\mathrm{Map}_{\mathcal{M}}(D, C) = \emptyset$$

for every pair of objects $C \in \mathcal{C}$, $D \in \mathcal{D}$. Let $M = \mathrm{N}(\mathcal{M})$. Then M is a correspondence between $\mathrm{N}(\mathcal{C})$ and $\mathrm{N}(\mathcal{D})$. By Proposition 5.2.4.4, it is an adjunction. It is easy to see that this adjunction is associated to both f and g. $\qquad\square$

The following variant on the situation of Corollary 5.2.4.5 arises very often in practice:

Proposition 5.2.4.6. *Let $\mathbf{A}$ and $\mathbf{A}'$ be simplicial model categories and let*

$$\mathbf{A} \underset{G}{\overset{F}{\rightleftarrows}} \mathbf{A}'$$

be a (simplicial) Quillen adjunction. Let $\mathcal{M}$ be the simplicial category defined as in the proof of Corollary 5.2.4.5 and let $\mathcal{M}^\circ$ be the full subcategory of $\mathcal{M}$ consisting of those objects which are fibrant-cofibrant (either as objects of $\mathbf{A}$ or as objects of $\mathbf{A}'$). Then $\mathrm{N}(\mathcal{M}^\circ)$ determines an adjunction between $\mathrm{N}(\mathbf{A}^\circ)$ and $\mathrm{N}(\mathbf{A}'^\circ)$.

Proof. We need to show that $\mathrm{N}(\mathcal{M}^\circ) \to \Delta^1$ is both a Cartesian fibration and a coCartesian fibration. We will argue the first point; the second follows from a dual argument. According to Proposition A.2.3.1, it suffices to show that for every fibrant-cofibrant object D of $\mathbf{A}'$, there is a fibrant-cofibrant object C of $\mathbf{A}$ and a morphism $f : C \to D$ in $\mathcal{M}^\circ$ which induces weak homotopy equivalences

$$\mathrm{Map}_{\mathbf{A}}(C', C) \to \mathrm{Map}_{\mathcal{M}}(C', D)$$

for every fibrant-cofibrant object $C' \in \mathbf{A}$. We define C to be a cofibrant replacement for GD: in other words, we choose a cofibrant object C with a trivial fibration $C \to G(D)$ in the model category $\mathbf{A}$. Then $\mathrm{Map}_{\mathbf{A}}(C', C) \to \mathrm{Map}_{\mathcal{M}}(C', D) = \mathrm{Map}_{\mathbf{A}}(C', G(D))$ is a trivial fibration of simplicial sets, whenever C' is a cofibrant object of $\mathbf{A}$. $\qquad\square$

Remark 5.2.4.7. Suppose that $F : \mathbf{A} \to \mathbf{A}'$ and $G : \mathbf{A}' \to \mathbf{A}$ are as in Proposition 5.2.4.6. We may associate to the adjunction $N(M^\circ)$ a pair of adjoint functors $f : N(\mathbf{A}^\circ) \to N(\mathbf{A}'^\circ)$ and $g : N(\mathbf{A}'^\circ) \to N(\mathbf{A}^\circ)$. In this situation, f is often called a (nonabelian) *left derived functor* of F, and g a (nonabelian) *right derived functor* of G. On the level of homotopy categories, f and g reduce to the usual derived functors associated to the Quillen adjunction (see §A.2.5).

5.2.5 Adjoint Functors and Overcategories

Our goal in this section is to prove the following result:

Proposition 5.2.5.1. *Suppose we are given an adjunction of ∞-categories*

$$\mathcal{C} \underset{G}{\overset{F}{\rightleftarrows}} \mathcal{D} .$$

Assume that the ∞-category $\mathcal{C}$ admits pullbacks and let C be an object of $\mathcal{C}$. Then

(1) *The induced functor $f : \mathcal{C}^{/C} \to \mathcal{D}^{/FC}$ admits a right adjoint g.*

(2) *The functor g is equivalent to the composition*

$$\mathcal{D}^{/FC} \xrightarrow{g'} \mathcal{C}^{/GFC} \xrightarrow{g''} \mathcal{C}^{/C} ,$$

   where g' is induced by G and g'' is induced by pullback along the unit map $C \to GFC$.

Proposition 5.2.5.1 is an immediate consequence of the following more general result, which we will prove at the end of this section:

Lemma 5.2.5.2. *Suppose we are given an adjunction between ∞-categories*

$$\mathcal{C} \underset{G}{\overset{F}{\rightleftarrows}} \mathcal{D} .$$

Let K be a simplicial set and suppose we are given a pair of diagrams $p_0 : K \to \mathcal{C}$, $p_1 : K \to \mathcal{D}$ and a natural transformation $h : F \circ p_0 \to p_1$. Assume that $\mathcal{C}$ admits pullbacks and K-indexed limits. Then

(1) *Let $f : \mathcal{C}^{/p_0} \to \mathcal{D}^{/p_1}$ denote the composition*

$$\mathcal{C}^{/p_0} \to \mathcal{D}^{/Fp_0} \overset{\circ \alpha}{\to} \mathcal{D}^{/p_1} .$$

   Then f admits a right adjoint g.

(2) *The functor g is equivalent the composition*

$$\mathcal{D}^{/p_1} \xrightarrow{g'} \mathcal{C}^{/Gp_1} \xrightarrow{g''} \mathcal{C}^{/p_0} .$$

   Here g'' is induced by pullback along the natural transformation $p_0 \to Gp_1$ adjoint to h (see below).

We begin by recalling a bit of notation which will be needed in the proof. Suppose that $q : X \to S$ is an inner fibration of simplicial sets and $p_S : K \to X$ is an arbitrary map, then we have defined a map of simplicial sets $X^{/p_S} \to S$, which is characterized by the following universal property: for every simplicial set Y equipped with a map to S, there is a pullback diagram

$$
\begin{array}{ccc}
\operatorname{Hom}_S(Y, X^{/p_S}) & \longrightarrow & \operatorname{Hom}_S(Y \diamond_S S, X) \\
\downarrow & & \downarrow \\
\{p\} & \longrightarrow & \operatorname{Hom}_S(S, X).
\end{array}
$$

We refer the reader to §4.2.2 for a more detailed discussion.

Lemma 5.2.5.3. *Let $q : \mathcal{M} \to \Delta^1$ be a coCartesian fibration of simplicial sets classifying a functor F from $\mathcal{C} = \mathcal{M} \times_{\Delta^1} \{0\}$ to $\mathcal{D} = \mathcal{M} \times_{\Delta^1} \{1\}$. Let K be a simplicial set and suppose we are given a commutative diagram*

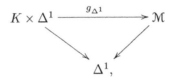

which restricts to give a pair of diagrams

$$
\mathcal{C} \xleftarrow{g_0} K \xrightarrow{g_1} \mathcal{D}.
$$

Then

(1) *The projection $q' : \mathcal{M}^{/g_{\Delta^1}} \to \Delta^1$ is a coCartesian fibration of simplicial sets classifying a functor $F' : \mathcal{C}^{/g_0} \to \mathcal{D}^{/g_1}$. Moreover, an edge of $\mathcal{M}^{/g_{\Delta^1}}$ is q'-coCartesian if and only if its image in $\mathcal{M}$ is q-coCartesian.*

(2) *Suppose that for every vertex k in K, the map g_Δ carries $\{k\} \times \Delta^1$ to a q-coCartesian morphism in $\mathcal{M}$, so that g_{Δ^1} determines an equivalence $g_1 \simeq F \circ g_0$. Then F' is homotopic to the composite functor*

$$
\mathcal{C}^{/g_0} \to \mathcal{D}^{/F g_0} \simeq \mathcal{D}^{/g_1}.
$$

(3) *Suppose that $\mathcal{M} = \mathcal{D} \times \Delta^1$ and that q is the projection onto the second factor, so that we can identify F with the identity functor from $\mathcal{D}$ to itself. Let $\overline{g} : K \times \Delta^1 \to \mathcal{D}$ denote the composition g_{Δ^1} with the projection map $\mathcal{M} \to \mathcal{D}$, so that we can regard $\overline{g}$ as a morphism from g_0 to g_1 in $\operatorname{Fun}(K, \mathcal{D})$. Then the functor $F' : \mathcal{D}^{/g_0} \to \mathcal{D}^{/g_1}$ is induced by composition with $\overline{g}$.*

Proof. Assertion (1) follows immediately from Proposition 4.2.2.4.

We now prove (2). Since F is associated to the correspondence $\mathcal{M}$, there exists a natural transformation $\alpha : \mathcal{C} \times \Delta^1 \to \mathcal{M}$ from $\operatorname{id}_{\mathcal{C}}$ to F, such that for each $C' \in \mathcal{C}$ the induced map $\alpha_C : C' \to FC'$ is q-coCartesian. Without loss of generality, we may assume that g_{Δ^1} is given by the composition

$$
K \times \Delta^1 \xrightarrow{g_0} \mathcal{C} \times \Delta^1 \xrightarrow{\alpha} \mathcal{M}.
$$

In this case, α induces a map $\alpha' : \mathcal{C}^{/g_0} \times \Delta^1 \to \mathcal{M}^{/g_{\Delta^1}}$, which we may identify with a natural transformation from $\mathrm{id}_{\mathcal{C}^{/g_0}}$ to the functor $\mathcal{C}^{/g_0} \to \mathcal{D}^{/Fg_0}$ determined by F. To show that this functor coincides with F', it will suffice to show that α' carries each object of $\mathcal{C}^{/g_0}$ to a q'-coCartesian morphism in $\mathcal{M}^{/g_{\Delta^1}}$. This follows immediately from the description of the q'-coCartesian edges given in assertion (1).

We next prove (3). Consider the diagram

$$\mathcal{D}^{/g_0} \xleftarrow{p} \mathcal{D}^{/\overline{g}} \xrightarrow{p'} \mathcal{D}^{/g_1}.$$

By definition, "composition with $\overline{g}$" refers to a functor from $\mathcal{D}_{/g_0}$ to $\mathcal{D}^{/g_1}$ obtained by composing p' with a section to the trivial fibration p. To prove that this functor is homotopic to F', it will suffice to show that $F' \circ p$ is homotopic to p'. For this, we must produce a map $\beta : \mathcal{D}^{/\overline{g}} \times \Delta^1 \to \mathcal{M}^{/g_{\Delta^1}}$ from p to p', such that β carries each object of $\mathcal{D}^{/\overline{g}}$ to a q'-coCartesian edge of $\mathcal{M}^{/g_{\Delta^1}}$. We observe that $\mathcal{D}^{/\overline{g}} \times \Delta^1$ can be identified with $\mathcal{M}^{/h_{\Delta^1}}$, where $h : \Delta^1 \times \Delta^1 \to \mathcal{M} \simeq \mathcal{D} \times \Delta^1$ is the product of $\overline{g}$ with the identity map. We now take β to be the restriction map $\mathcal{M}^{/h_{\Delta^1}} \to \mathcal{M}^{/g_{\Delta^1}}$ induced by the diagonal inclusion $\Delta^1 \subseteq \Delta^1 \times \Delta^1$. Using (1), we readily deduce that β has the desired properties.

$\square$

We will also need the following counterpart to Proposition 4.2.2.4:

Lemma 5.2.5.4. *Suppose we are given a commutative diagram of simplicial sets*

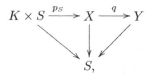

where the left diagonal arrow is projection onto the second factor and q is a Cartesian fibration. Assume further that the following condition holds:

$(*)$ *For every vertex $k \in K$, the map p_S carries each edge of $\{k\} \times S$ to a q-Cartsian edge in X.*

Let $p'_S = q \circ p_S$. Then the map $q' : X^{/p_S} \to Y^{/p'_S}$ is a Cartesian fibration. Moreover, an edge of $X^{/p_S}$ is q'-Cartesian if and only if its image in X is q-Cartesian.

Proof. To give the proof, it is convenient to use the language of marked simplicial sets (see §3.1). Let $X^\natural$ denote the marked simplicial set whose underlying simplicial set is X, where we consider an edge of $X^\natural$ to be marked if it is q-Cartesian. Let $\overline{X}^\natural$ denote the marked simplicial set whose underlying simplicial set is $X^{/p_S}$, where we consider an edge to be marked if and only if its image in X is marked. According to Proposition 3.1.1.6, it will suffice to show that the map $\overline{X}^\natural \to (Y^{/p'_S})^\sharp$ has the right lifting property with respect to every marked anodyne map $i : A \to B$. Let $\overline{A}$ and $\overline{B}$ denote the simplicial sets underlying A and B, respectively. Suppose we are given a diagram of marked simplicial sets

$$
\begin{array}{ccc}
A & \longrightarrow & \overline{X}^{\natural} \\
{\scriptstyle i}\downarrow & \nearrow & \downarrow \\
B & \longrightarrow & (Y/^{p'_{S}})^{\natural}.
\end{array}
$$

We wish to show that there exists a dotted arrow rendering the diagram commutative. We begin by choosing a solution to the associated lifting problem

$$
\begin{array}{ccc}
A & \longrightarrow & X^{\natural} \\
\downarrow & \nearrow & \downarrow \\
B & \longrightarrow & Y^{\natural},
\end{array}
$$

which is possible in view of our assumption that q is a Cartesian fibration. To extend this to a solution to the original problem, it suffices to solve another lifting problem

$$
\begin{array}{ccc}
(\overline{A} \times K \times \Delta^{1}) \coprod_{(\overline{A} \times K \times \partial \Delta^{1})} (\overline{B} \times K \times \partial \Delta^{1}) & \xrightarrow{\ f\ } & X \\
{\scriptstyle j}\downarrow & \qquad\nearrow & \downarrow {\scriptstyle q} \\
\overline{B} \times K \times \Delta^{1} & \longrightarrow & Y.
\end{array}
$$

By construction, the map f induces a map of marked simplicial sets from $B \times K^{\flat} \times \{0\}$ to $X^{\natural}$. Using assumption $(\ast)$, we conclude that f also induces a map of marked simplicial sets from $B \times K^{\flat} \times \{1\}$ to $X^{\natural}$. Using Proposition 3.1.1.6 again (and our assumption that q is a Cartesian fibration), we are reduced to proving that the map j induces a marked anodyne map

$$
(A \times (K \times \Delta^{1})^{\flat}) \coprod_{A \times (K \times \partial \Delta^{1})^{\flat}} (B \times (K \times \partial \Delta^{1})^{\flat}) \to B \times (K \times \Delta^{1})^{\flat}.
$$

Since i is marked anodyne by assumption, this follows immediately from Proposition 3.1.2.3. $\qquad\square$

Lemma 5.2.5.5. *Let $q : \mathcal{M} \to \Delta^{1}$ be a Cartesian fibration of simplicial sets associated to a functor G from $\mathcal{D} = \mathcal{M} \times_{\Delta^{1}} \{1\}$ to $\mathcal{C} = \mathcal{M} \times_{\Delta^{1}} \{0\}$. Suppose we are given a simplicial set K and a commutative diagram*

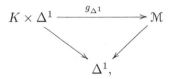

so that $g_{\Delta^{1}}$ restricts to a pair of functors

$$
\mathcal{C} \xleftarrow{\ g_{0}\ } K \xrightarrow{\ g_{1}\ } \mathcal{D}.
$$

Suppose furthermore that, for every vertex k of K, the corresponding morphism $g_{0}(k) \to g_{1}(k)$ is q-Cartesian. Then

(1) *The induced map $q' : \mathcal{M}^{/f_{\Delta^1}} \to \Delta^1$ is a Cartesian fibration. Moreover, an edge of $\mathcal{M}^{/f_{\Delta^1}}$ is q'-Cartesian if and only if its image in $\mathcal{M}$ is q-Cartesian.*

(2) *The associated functor $\mathcal{D}^{/g_1} \to \mathcal{C}^{/g_0}$ is homotopic to the composition of the functor $G' : \mathcal{D}^{/g_1} \to \mathcal{C}^{/Gg_1}$ induced by G and the equivalence $\mathcal{C}^{/Gg_1} \simeq \mathcal{C}^{/g_0}$ determined by the map g_{Δ^1}.*

Proof. Assertion (1) follows immediately from Lemma 5.2.5.4. We will prove (2). Since the functor G is associated to q, there exists a map $\alpha : \mathcal{D} \times \Delta^1 \to \mathcal{M}$ which is a natural transformation from G to $\mathrm{id}_{\mathcal{D}}$, such that for every object $D \in \mathcal{D}$ the induced map $\alpha_D : \{D\} \times \Delta^1 \to \mathcal{M}$ is a q-Cartesian edge of $\mathcal{M}$. Without loss of generality, we may assume that g coincides with the composition

$$K \times \Delta^1 \overset{g_1}{\to} \mathcal{D} \times \Delta^1 \overset{\alpha}{\to} \mathcal{M}.$$

In this case, α induces a map $\alpha' : \mathcal{D}^{/g_1} \times \Delta^1 \to \mathcal{M}^{/f_{\Delta^1}}$, which is a natural transformation from G' to the identity. Using (1), we deduce that α' carries each object of $\mathcal{D}^{/g_1}$ to a q'-Cartesian edge of $\mathcal{M}^{/f_{\Delta^1}}$. It follows that α' exhibits G' as the functor associated to the Cartesian fibration q', as desired.   □

Proof of Lemma 5.2.5.2. Let $q : \mathcal{M} \to \Delta^1$ be a correspondence from $\mathcal{C} = \mathcal{M} \times_{\Delta^1} \{0\}$ to $\mathcal{D} = \mathcal{M} \times_{\Delta^1} \{1\}$, which is associated to the pair of adjoint functors F and G. The natural transformation h determines a map $\alpha : K \times \Delta^1 \to \mathcal{M}$, which is a natural transformation from p_0 to p_1. Using the fact that q is both a Cartesian and a coCartesian fibration, we can form a commutative square σ

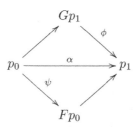

in the ∞-category $\mathrm{Fun}(K, \mathcal{M})$, where the morphism ϕ is q-Cartesian and the morphism ψ is q-coCartesian.

Let $\mathcal{N} = \mathcal{M} \times \Delta^1$. We can identify σ with a map $\sigma_{\Delta^1 \times \Delta^1} : K \times \Delta^1 \times \Delta^1 \to \mathcal{M} \times \Delta^1$. Let $\mathcal{N}' = \mathcal{N}^{/\sigma_{\Delta^1 \times \Delta^1}}$. Proposition 4.2.2.4 implies that the projection $\mathcal{N}' \to \Delta^1 \times \Delta^1$ is a coCartesian fibration associated to some diagram of

∞-categories

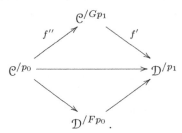

Lemma 5.2.5.3 allows us to identify the functors in the lower triangle, so we see that the horizontal composition is homotopic to the functor f. To complete the proof of (1), it will suffice to show that the functors f' and f'' admit right adjoints. To prove (2), it suffices to show that those right adjoints are given by g' and g'', respectively. The adjointness of f' and g' follows from Lemma 5.2.5.5.

It follows from Lemma 5.2.5.3 that the functor $f'' : \mathcal{C}^{/p_0} \to \mathcal{C}^{/Gp_1}$ is given by composition with the transformation $h' : p_0 \to Gp_1$ which is adjoint to h. The pullback functor g'' is right adjoint to f'' by definition; the only nontrivial point is to establish the existence of g''. Here we must use our hypotheses on the ∞-category $\mathcal{C}$. Let $\overline{p}_0 : K^{\triangleleft} \to \mathcal{C}$ be a limit of p_0 and let $\overline{Gp_1} : K^{\triangleleft} \to \mathcal{C}$ be a limit of Gp_1. Let us identify h' with a map $K \times \Delta^1 \to \mathcal{C}$ and choose an extension $\overline{h}' : K^{\triangleleft} \times \Delta^1 \to \mathcal{C}$ which is a natural transformation from $\overline{p}_0$ to $\overline{Gp_1}$. Let $C \in \mathcal{C}$ denote the image under $\overline{p}_0$ of the cone point of $K^{\triangleleft}$, let $C' \in \mathcal{C}$ denote the image under $\overline{Gp_1}$ of the cone point of $K^{\triangleleft}$, and let $j : C \to C'$ be the morphism induced by $\overline{h}'$. We have a commutative diagram of ∞-categories:

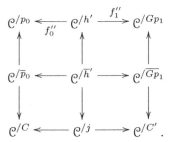

In this diagram, the left horizontal arrows are trivial Kan fibrations, as are all of the vertical arrows. The functor f'' is obtained by composing f''_0 with a section to the trivial Kan fibration f''_1. Utilizing the vertical equivalences, we can identify f'' with the functor $\mathcal{C}^{/C} \to \mathcal{C}^{/C'}$ given by composition with j. But this functor admits a right adjoint in view of our assumption that $\mathcal{C}$ admits pullbacks. $\qquad\square$

5.2.6 Uniqueness of Adjoint Functors

We have seen that if $f : \mathcal{C} \to \mathcal{D}$ is a functor which admits a right adjoint $g : \mathcal{D} \to \mathcal{C}$, then g is uniquely determined up to homotopy. Our next result is a slight refinement of this assertion.

Definition 5.2.6.1. Let $\mathcal{C}$ and $\mathcal{D}$ be ∞-categories. We let $\mathrm{Fun}^{\mathrm{L}}(\mathcal{C}, \mathcal{D}) \subseteq \mathrm{Fun}(\mathcal{C}, \mathcal{D})$ denote the full subcategory of $\mathrm{Fun}(\mathcal{C}, \mathcal{D})$ spanned by those functors $F : \mathcal{C} \to \mathcal{D}$ which are left adjoints. Similarly, we define $\mathrm{Fun}^{\mathrm{R}}(\mathcal{C}, \mathcal{D})$ to be the full subcategory of $\mathrm{Fun}(\mathcal{C}, \mathcal{D})$ spanned by those functors which are right adjoints.

Proposition 5.2.6.2. *Let $\mathcal{C}$ and $\mathcal{D}$ be ∞-categories. Then the ∞-categories $\mathrm{Fun}^{\mathrm{L}}(\mathcal{C}, \mathcal{D})$ and $\mathrm{Fun}^{\mathrm{R}}(\mathcal{D}, \mathcal{C})^{op}$ are (canonically) equivalent to one another.*

Proof. Enlarging the universe if necessary, we may assume without loss of generality that $\mathcal{C}$ and $\mathcal{D}$ are small. Let $j : \mathcal{D} \to \mathcal{P}(\mathcal{D})$ be the Yoneda embedding. Composition with j induces a fully faithful embedding
$$i : \mathrm{Fun}(\mathcal{C}, \mathcal{D}) \to \mathrm{Fun}(\mathcal{C}, \mathcal{P}(\mathcal{D})) \simeq \mathrm{Fun}(\mathcal{C} \times \mathcal{D}^{op}, \mathcal{S}).$$
The essential image of i consists of those functors $G : \mathcal{C} \times \mathcal{D}^{op} \to \mathcal{S}$ with the property that, for each $C \in \mathcal{C}$, the induced functor $G_C : \mathcal{D}^{op} \to \mathcal{S}$ is representable by an object $D \in \mathcal{D}$. The functor i induces a fully faithful embedding
$$i_0 : \mathrm{Fun}^{\mathrm{R}}(\mathcal{C}, \mathcal{D}) \to \mathrm{Fun}(\mathcal{C} \times \mathcal{D}^{op}, \mathcal{S})$$
whose essential image consists of those functors G which belong to the essential image of i and furthermore satisfy the additional condition that for each $D \in \mathcal{D}$, the induced functor $G_D : \mathcal{C} \to \mathcal{S}$ is corepresentable by an object $C \in \mathcal{C}$ (this follows from Proposition 5.2.4.2). Let $\mathcal{E} \subseteq \mathrm{Fun}(\mathcal{C} \times \mathcal{D}^{op}, \mathcal{S})$ be the full subcategory spanned by those functors which satisfy these two conditions, so that the Yoneda embedding induces an equivalence
$$\mathrm{Fun}^{\mathrm{R}}(\mathcal{C}, \mathcal{D}) \to \mathcal{E}.$$
We note that the above conditions are self-dual, so that the same reasoning gives an equivalence of ∞-categories
$$\mathrm{Fun}^{\mathrm{R}}(\mathcal{D}^{op}, \mathcal{C}^{op}) \to \mathcal{E}.$$
We now conclude by observing that there is a natural equivalence of ∞-categories $\mathrm{Fun}^{\mathrm{R}}(\mathcal{D}^{op}, \mathcal{C}^{op}) \simeq \mathrm{Fun}^{\mathrm{L}}(\mathcal{D}, \mathcal{C})^{op}$. $\square$

We will later need a slight refinement of Proposition 5.2.6.2, which exhibits some functoriality in $\mathcal{C}$. We begin with a few preliminary remarks concerning the construction of presheaf ∞-categories.

Let $f : \mathcal{C} \to \mathcal{C}'$ be a functor between small ∞-categories. Then composition with f induces a restriction functor $G : \mathcal{P}(\mathcal{C}') \to \mathcal{P}(\mathcal{C})$. However, there is another slightly less evident functoriality of the construction $\mathcal{C} \mapsto \mathcal{P}(\mathcal{C})$. Namely, according to Theorem 5.1.5.6, there is a colimit-preserving functor $\mathcal{P}(f) : \mathcal{P}(\mathcal{C}) \to \mathcal{P}(\mathcal{C}')$, uniquely determined up to equivalence, such that the diagram

$$
\begin{array}{ccc}
\mathcal{C} & \xrightarrow{\;f\;} & \mathcal{C}' \\
\downarrow & & \downarrow \\
\mathcal{P}(\mathcal{C}) & \xrightarrow{\;\mathcal{P}(f)\;} & \mathcal{P}(\mathcal{C}')
\end{array}
$$

commutes up to homotopy (here the vertical arrows are given by the Yoneda embeddings).

The functor $\mathcal{P}(f)$ has an alternative characterization in the language of adjoint functors:

Proposition 5.2.6.3. *Let* $f : \mathcal{C} \to \mathcal{C}'$ *be a functor between small* ∞-*categories and let* $G : \mathcal{P}(\mathcal{C}') \to \mathcal{P}(\mathcal{C})$ *be the functor given by composition with* f. *Then* G *is right adjoint to* $\mathcal{P}(f)$.

Proof. We first prove that G admits a left adjoint. Let

$$e : \mathcal{P}(\mathcal{C}) \to \mathrm{Fun}(\mathcal{P}(\mathcal{C})^{op}, \widehat{\mathcal{S}})$$

denote the Yoneda embedding. According to Proposition 5.2.4.2, it will suffice to show that for each $M \in \mathcal{P}(\mathcal{C})$, the composite functor $e(M) \circ G$ is corepresentable. Let $\mathcal{D}$ denote the full subcategory of $\mathcal{P}(\mathcal{C})$ spanned by those objects M such that $G \circ e_M$ is corepresentable. Since $\mathcal{P}(\mathcal{C})$ admits small colimits, Proposition 5.1.3.2 implies that the collection of corepresentable functors on $\mathcal{P}(\mathcal{C})$ is stable under small colimits. According to Propositions 5.1.3.2 and 5.1.2.2, the functor $M \mapsto e(M) \circ G$ preserves small colimits. It follows that $\mathcal{D}$ is stable under small colimits in $\mathcal{P}(\mathcal{C})$. Since $\mathcal{P}(\mathcal{C})$ is generated under small colimits by the Yoneda embedding $j_{\mathcal{C}'} : \mathcal{C} \to \mathcal{P}(\mathcal{C})$ (Corollary 5.1.5.8), it will suffice to show that $j_{\mathcal{C}}(C) \in \mathcal{D}$ for each $C \in \mathcal{C}$. According to Lemma 5.1.5.2, $e(j_{\mathcal{C}}(C))$ is equivalent to the functor $\mathcal{P}(\mathcal{C}) \to \widehat{\mathcal{S}}$ given by evaluation at C. Then $e(j_{\mathcal{C}}(C)) \circ G$ is equivalent to the functor given by evaluation at $f(C) \in \mathcal{C}'$, which is corepresentable (Lemma 5.1.5.2 again). We conclude that G has a left adjoint F.

To complete the proof, we must show that F is equivalent to $\mathcal{P}(f)$. To prove this, it will suffice to show that F preserves small colimits and that the diagram

$$
\begin{array}{ccc}
\mathcal{C} & \xrightarrow{\ f\ } & \mathcal{C}' \\
\downarrow & & \downarrow \\
\mathcal{P}(\mathcal{C}) & \xrightarrow{\ F\ } & \mathcal{P}(\mathcal{C}')
\end{array}
$$

commutes up to homotopy. The first point is obvious: since F is a left adjoint, it preserves all colimits which exist in $\mathcal{P}(\mathcal{C})$ (Proposition 5.2.3.5). For the second, choose a counit map $v : F \circ G \to \mathrm{id}_{\mathcal{P}(\mathcal{C}')}$. By construction, the functor f induces a natural transformation $u : j_{\mathcal{C}} \to G \circ j_{\mathcal{C}'} \circ f$. To complete the proof, it will suffice to show that the composition

$$\theta : F \circ j_{\mathcal{C}} \xrightarrow{u} F \circ G \circ j_{\mathcal{C}'} \circ f \xrightarrow{v} j_{\mathcal{C}'} \circ f$$

is an equivalence of functors from $\mathcal{C}$ to $\mathcal{P}(\mathcal{C}')$. Fix objects $C \in \mathcal{C}$, $M \in \mathcal{P}(\mathcal{C}')$.

We have a commutative diagram

$$\begin{array}{ccc}
\mathrm{Map}_{\mathcal{P}(\mathcal{C}')}(j_{\mathcal{C}'}(f(C)), M) & = & \mathrm{Map}_{\mathcal{P}(\mathcal{C}')}(j_{\mathcal{C}'}(f(C)), M) \\
\downarrow & & \downarrow \\
\mathrm{Map}_{\mathcal{P}(\mathcal{C})}(G(j_{\mathcal{C}'}(f(C))), G(M)) & \longrightarrow & \mathrm{Map}_{\mathcal{P}(\mathcal{C}')}(F(G(j_{\mathcal{C}'}(f(C)))), M) \\
\downarrow & & \downarrow \\
\mathrm{Map}_{\mathcal{P}(\mathcal{C})}(j_{\mathcal{C}}(C), G(M)) & \longrightarrow & \mathrm{Map}_{\mathcal{P}(\mathcal{C}')}(F(j_{\mathcal{C}}(C)), M)
\end{array}$$

in the homotopy category $\mathcal{H}$ of spaces, where the vertical arrows are iso-morphisms. Consequently, to prove that the lower horizontal composition is an isomorphism, it suffices to prove that the upper horizontal composition is an isomorphism. Using Lemma 5.1.5.2, we reduce to the assertion that $M(f(C)) \to (G(M))(C)$ is an isomorphism in $\mathcal{H}$, which follows immediately from the definition of G. $\qquad\square$

Remark 5.2.6.4. Suppose we are given a functor $f : \mathcal{D} \to \mathcal{D}'$ which admits a right adjoint g. Let $\mathcal{E} \subseteq \mathrm{Fun}(\mathcal{C} \times \mathcal{D}^{op}, \mathcal{S})$ and $\mathcal{E}' \subseteq \mathrm{Fun}(\mathcal{C} \times (\mathcal{D}')^{op}, \mathcal{S})$ be defined as in the proof of Proposition 5.2.6.2 and consider the diagram

$$\begin{array}{ccccc}
\mathrm{Fun}^R(\mathcal{C}, \mathcal{D}) & \longrightarrow & \mathcal{E} & \longleftarrow & \mathrm{Fun}^L(\mathcal{D}, \mathcal{C})^{op} \\
\downarrow{\scriptstyle \circ g} & & \downarrow & & \downarrow{\scriptstyle \circ f} \\
\mathrm{Fun}^R(\mathcal{C}, \mathcal{D}') & \longrightarrow & \mathcal{E}' & \longleftarrow & \mathrm{Fun}^L(\mathcal{D}', \mathcal{C})^{op}.
\end{array}$$

Here the middle vertical map is given by composition with $\mathrm{id}_{\mathcal{C}} \times f$. The square on the right is manifestly commutative, but the square on the left commutes only up to homotopy. To verify the second point, we observe that the square in question is given by applying the functor $\mathrm{Map}(\mathcal{C}, \bullet)$ to the diagram

$$\begin{array}{ccc}
\mathcal{D} & \longrightarrow & \mathcal{P}(\mathcal{D}) \\
\downarrow{\scriptstyle g} & & \downarrow{\scriptstyle G} \\
\mathcal{D}' & \longrightarrow & \mathcal{P}(\mathcal{D}'),
\end{array}$$

where G is given by composition with f and the horizontal arrows are given by the Yoneda embedding. Let $\mathcal{P}^0(\mathcal{D}) \subseteq \mathcal{P}(\mathcal{D})$ and $\mathcal{P}^0(\mathcal{D}')$ denote the es-sential images of the Yoneda embeddings. Proposition 5.2.4.2 asserts that G carries $\mathcal{P}^0(\mathcal{D}')$ into $\mathcal{P}^0(\mathcal{D})$, so that it will suffice to verify that the diagram

$$\begin{array}{ccc}
\mathcal{D} & \longrightarrow & \mathcal{P}^0(\mathcal{D}) \\
\downarrow{\scriptstyle g} & & \downarrow{\scriptstyle G^0} \\
\mathcal{D}' & \longrightarrow & \mathcal{P}^0(\mathcal{D}')
\end{array}$$

is homotopy commutative. In view of Proposition 5.2.2.6, it will suffice to show that G^0 admits a left adjoint F^0 and that the diagram

$$
\begin{array}{ccc}
\mathcal{D} & \longrightarrow & \mathcal{P}^0(\mathcal{D}) \\
\uparrow & & \uparrow{\scriptstyle F_0} \\
\mathcal{D}' & \longrightarrow & \mathcal{P}^0(\mathcal{D}')
\end{array}
$$

is homotopy commutative. According to Proposition 5.2.6.3, the functor G has a left adjoint $\mathcal{P}(f)$ which fits into a commutative diagram

$$
\begin{array}{ccc}
\mathcal{D} & \longrightarrow & \mathcal{P}(\mathcal{D}) \\
\uparrow{\scriptstyle f} & & \uparrow{\scriptstyle \mathcal{P}(f)} \\
\mathcal{D}' & \longrightarrow & \mathcal{P}(\mathcal{D}').
\end{array}
$$

In particular, $\mathcal{P}(f)$ carries $\mathcal{P}^0(\mathcal{D})$ into $\mathcal{P}^0(\mathcal{D}')$ and therefore restricts to give a left adjoint $F^0 : \mathcal{P}^0(\mathcal{D}) \to \mathcal{P}^0(\mathcal{D}')$ which verifies the desired commutativity.

We conclude this section by establishing the following consequence of Proposition 5.2.6.3:

Corollary 5.2.6.5. *Let $\mathcal{C}$ be a small ∞-category and $\mathcal{D}$ a locally small ∞-category which admits small colimits. Let $F : \mathcal{P}(\mathcal{C}) \to \mathcal{D}$ be a colimit-preserving functor, let $f : \mathcal{C} \to \mathcal{D}$ denote the composition of F with the Yoneda embedding of $\mathcal{C}$, and let $G : \mathcal{D} \to \mathcal{P}(\mathcal{C})$ be the functor given by the composition*

$$
\mathcal{D} \xrightarrow{j'} \mathrm{Fun}(\mathcal{D}^{op}, \mathcal{S}) \xrightarrow{\circ f} \mathcal{P}(\mathcal{C}).
$$

Then G is a right adjoint to F. Moreover, the map

$$
f = F \circ j \to (F \circ (G \circ F)) \circ j = (F \circ G) \circ f
$$

exhibits $F \circ G$ as a left Kan extension of f along itself.

The proof requires a few preliminaries:

Lemma 5.2.6.6. *Suppose we are given a pair of adjoint functors*

$$
\mathcal{C} \underset{g}{\overset{f}{\rightleftarrows}} \mathcal{D}
$$

between ∞-categories. Let $T : \mathcal{C} \to \mathcal{X}$ be any functor. Then $T \circ g : \mathcal{D} \to \mathcal{X}$ is a left Kan extension of T along f.

Proof. Let $p : \mathcal{M} \to \Delta^1$ be a correspondence associated to the pair of adjoint functors f and g. Choose a p-Cartesian homotopy h from r to $\mathrm{id}_{\mathcal{M}}$, where r is a functor from $\mathcal{M}$ to $\mathcal{C}$; thus $r|\mathcal{D}$ is homotopic to g. It will therefore suffice to show that the composition

$$
\overline{T} : \mathcal{M} \xrightarrow{r} \mathcal{C} \xrightarrow{T} \mathcal{X}
$$

is a left Kan extension of $\overline{T}| \, \mathcal{C} \simeq T$. For this, we must show that for each $D \in \mathcal{D}$, the functor $\overline{T}$ exhibits $\overline{T}(D)$ as a colimit of the diagram

$$(\mathcal{C} \times_{\mathcal{M}} \mathcal{M}_{/D}) \to \mathcal{M} \xrightarrow{\ \overline{T}\ } \mathcal{X}.$$

We observe that $\mathcal{C} \times_{\mathcal{M}} \mathcal{M}_{/D}$ has a final object given by any p-Cartesian morphism $e : C \to D$. It therefore suffices to show that $\overline{T}(e)$ is an equivalence in $\mathcal{X}$, which follows immediately from the construction of $\overline{T}$. $\qquad\square$

Lemma 5.2.6.7. *Let $f : \mathcal{C} \to \mathcal{C}'$ be a functor between small ∞-categories and $\mathcal{X}$ an ∞-category which admits small colimits. Let $H : \mathcal{P}(\mathcal{C}) \to \mathcal{X}$ be a functor which preserves small colimits and $h : \mathcal{C} \to \mathcal{X}$ the composition of F with the Yoneda embedding $j_{\mathcal{C}} : \mathcal{C} \to \mathcal{P}(\mathcal{C})$. Then the composition*

$$\mathcal{C}' \xrightarrow{\ j_{\mathcal{C}'}\ } \mathcal{P}(\mathcal{C}') \xrightarrow{\ \circ f\ } \mathcal{P}(\mathcal{C}) \xrightarrow{\ H\ } \mathcal{X}$$

is a left Kan extension of h along f.

Proof. Let $G : \mathcal{P}(\mathcal{C}') \to \mathcal{P}(\mathcal{C})$ be the functor given by composition with f. In view of Proposition 4.3.2.8, it will suffice to show that $H \circ G$ is a left Kan extension of h along $j_{\mathcal{C}} \circ f$.

Theorem 5.1.5.6 implies the existence of a functor $F : \mathcal{P}(\mathcal{C}) \to \mathcal{P}(\mathcal{C}')$ which preserves small colimits, such that $F \circ j_{\mathcal{C}} \simeq j_{\mathcal{C}'} \circ f$. Moreover, Lemma 5.1.5.5 ensures that F is a left Kan extension of f along the fully faithful Yoneda embedding $j_{\mathcal{C}}$. Using Proposition 4.3.2.8 again, we are reduced to proving that $H \circ G$ is a left Kan extension of H along F. This follows immediately from Proposition 5.2.6.3 and Lemma 5.2.6.6. $\qquad\square$

Proof of Corollary 5.2.6.5. The first claim follows from Proposition 5.2.6.3. To prove the second, we may assume without loss of generality that $\mathcal{D}$ is minimal, so that $\mathcal{D}$ is a union of small full subcategories $\{\mathcal{D}_{\alpha}\}$. It will suffice to show that, for each index α such that f factors through $\mathcal{D}_{\alpha}$, the restricted transformation $f \to ((F \circ G)| \, \mathcal{D}_{\alpha}) \circ f$ exhibits $(F \circ G)| \, \mathcal{D}_{\alpha}$ as a left Kan extension of f along the induced map $\mathcal{C} \to \mathcal{D}_{\alpha}$, which follows from Lemma 5.2.6.7. $\qquad\square$

5.2.7 Localization Functors

Suppose we are given an ∞-category $\mathcal{C}$ and a collection S of morphisms of $\mathcal{C}$ which we would like to invert. In other words, we wish to find an ∞-category $S^{-1}\mathcal{C}$ equipped with a functor $\eta : \mathcal{C} \to S^{-1}\mathcal{C}$ which carries each morphism in S to an equivalence and is in some sense universal with respect to these properties. One can give a general construction of $S^{-1}\mathcal{C}$ using the formalism of §3.1.1. Without loss of generality, we may suppose that S contains all the identity morphisms in $\mathcal{C}$. Consequently, the pair $(\mathcal{C}, S)$ may be regarded as a marked simplicial set, and we can choose a marked anodyne map $(\mathcal{C}, S) \to (S^{-1}\mathcal{C}, S')$, where $S^{-1}\mathcal{C}$ is an ∞-category and S' is the collection of all equivalences in $S^{-1}\mathcal{C}$. However, this construction is generally very difficult to analyze, and the properties of $S^{-1}\,\mathcal{C}$ are very difficult to control. For example, it might be the case that $\mathcal{C}$ is locally small and $S^{-1}\,\mathcal{C}$ is not.

Under suitable hypotheses on S (see §5.5.4), there is a drastically simpler approach: we can find the desired ∞-category $S^{-1}\,\mathcal{C}$ *inside* $\mathcal{C}$ as the full subcategory of S-*local* objects of $\mathcal{C}$.

Example 5.2.7.1. Let $\mathcal{C}$ be the (ordinary) category of abelian groups, let p be a prime number, and let S denote the collection of morphisms f whose kernel and cokernel consist entirely of p-power torsion elements. A morphism f lies in S if and only if it induces an isomorphism after inverting the prime number p. In this case, we may identify $S^{-1}\,\mathcal{C}$ with the full subcategory of $\mathcal{C}$ consisting of those abelian groups which are *uniquely p-divisible*. The functor $\mathcal{C} \to S^{-1}\,\mathcal{C}$ is given by

$$M \mapsto M \otimes_{\mathbf{Z}} \mathbf{Z}[\tfrac{1}{p}].$$

In Example 5.2.7.1, the functor $\mathcal{C} \to S^{-1}\,\mathcal{C}$ is actually left adjoint to an inclusion functor. We will take this as our starting point.

Definition 5.2.7.2. A functor $f : \mathcal{C} \to \mathcal{D}$ between ∞-categories is a *localization* if f has a fully faithful right adjoint.

Warning 5.2.7.3. Let $f : \mathcal{C} \to \mathcal{D}$ be a localization functor and let S denote the collection of all morphisms α in $\mathcal{C}$ such that $f(\alpha)$ is an equivalence. Then, for any ∞-category $\mathcal{E}$, composition with f induces a fully faithful functor

$$\operatorname{Fun}(\mathcal{D}, \mathcal{E}) \xrightarrow{\circ f} \operatorname{Fun}(\mathcal{C}, \mathcal{E})$$

whose essential image consists of those functors $p : \mathcal{C} \to \mathcal{E}$ which carry each $\alpha \in S$ to an equivalence in $\mathcal{E}$ (Proposition 5.2.7.12). We may describe the situation more informally by saying that $\mathcal{D}$ is obtained from $\mathcal{C}$ by inverting the morphisms of S.

Some authors use the term "localization" in a more general sense to describe any functor $f : \mathcal{C} \to \mathcal{D}$ in which $\mathcal{D}$ is obtained by inverting some collection S of morphisms in $\mathcal{C}$. Such a morphism f need not be a localization in the sense of Definition 5.2.7.2; however, it is in many cases (see Proposition 5.5.4.15).

If $f : \mathcal{C} \to \mathcal{D}$ is a localization of ∞-categories, then we will also say that $\mathcal{D}$ is a *localization* of $\mathcal{C}$. In this case, a right adjoint $g : \mathcal{D} \to \mathcal{C}$ of f gives an equivalence between $\mathcal{D}$ and a full subcategory of $\mathcal{C}$ (the essential image of g). We let $L : \mathcal{C} \to \mathcal{C}$ denote the composition $g \circ f$. We will abuse terminology by referring to L as a *localization functor* if it arises in this way.

The following result will allow us to recognize localization functors:

Proposition 5.2.7.4. *Let $\mathcal{C}$ be an ∞-category and let $L : \mathcal{C} \to \mathcal{C}$ be a functor with essential image $L\,\mathcal{C} \subseteq \mathcal{C}$. The following conditions are equivalent:*

(1) *There exists a functor $f : \mathcal{C} \to \mathcal{D}$ with a fully faithful right adjoint $g : \mathcal{D} \to \mathcal{C}$ and an equivalence between $g \circ f$ and L.*

(2) *When regarded as a functor from $\mathcal{C}$ to $L\,\mathcal{C}$, L is a left adjoint of the inclusion $L\,\mathcal{C} \subseteq \mathcal{C}$.*

(3) *There exists a natural transformation $\alpha : \mathcal{C} \times \Delta^1 \to \mathcal{C}$ from $\mathrm{id}_{\mathcal{C}}$ to L such that, for every object C of $\mathcal{C}$, the morphisms $L(\alpha(C)), \alpha(LC) : LC \to LLC$ of $\mathcal{C}$ are equivalences.*

Proof. It is obvious that (2) implies (1) (take $\mathcal{D} = L\,\mathcal{C}$, $f = L$, and g to be the inclusion). The converse follows from the observation that, since g is fully faithful, we are free to replace $\mathcal{D}$ by the essential image of g (which is equal to the essential image of L).

We next prove that (2) implies (3). Let $\alpha : \mathrm{id}_{\mathcal{C}} \to L$ be a unit for the adjunction. Then, for each pair of objects $C \in \mathcal{C}$, $D \in L\,\mathcal{C}$, composition with $\alpha(C)$ induces a homotopy equivalence

$$\mathrm{Map}_{\mathcal{C}}(LC, D) \to \mathrm{Map}_{\mathcal{C}}(C, D)$$

and, in particular, a bijection $\mathrm{Hom}_{h\mathcal{C}}(LC, D) \to \mathrm{Hom}_{h\mathcal{C}}(C, D)$. If C belongs to $L\,\mathcal{C}$, then Yoneda's lemma implies that $\alpha(C)$ is an isomorphism in $h\mathcal{C}$. This proves that $\alpha(LC)$ is an equivalence for every $C \in \mathcal{C}$. Since α is a natural transformation, we obtain a diagram

$$
\begin{array}{ccc}
C & \xrightarrow{\ \alpha(C)\ } & LC \\
{\scriptstyle \alpha(C)}\downarrow & & \downarrow{\scriptstyle L\alpha(C)} \\
LC & \xrightarrow{\ \alpha(LC)\ } & LLC.
\end{array}
$$

Since composition with $\alpha(C)$ gives an injective map from $\mathrm{Hom}_{h\mathcal{C}}(LC, LLC)$ to $\mathrm{Hom}_{h\mathcal{C}}(C, LLC)$, we conclude that $\alpha(LC)$ is homotopic to $L\alpha(C)$; in particular, $\alpha(LC)$ is also an equivalence. This proves (3).

Now suppose that (3) is satisfied; we will prove that α is the unit of an adjunction between $\mathcal{C}$ and $L\,\mathcal{C}$. In other words, we must show that for each $C \in \mathcal{C}$ and $D \in \mathcal{C}$, composition with $\alpha(C)$ induces a homotopy equivalence

$$\phi : \mathrm{Map}_{\mathcal{C}}(LC, LD) \to \mathrm{Map}_{\mathcal{C}}(C, LD).$$

By Yoneda's lemma, it will suffice to show that for every Kan complex K, the induced map

$$\mathrm{Hom}_{\mathcal{H}}(K, \mathrm{Map}_{\mathcal{C}}(LC, LD)) \to \mathrm{Hom}_{\mathcal{H}}(K, \mathrm{Map}_{\mathcal{C}}(C, LD))$$

is a bijection of sets, where $\mathcal{H}$ denotes the homotopy category of spaces. Replacing $\mathcal{C}$ by $\mathrm{Fun}(K, \mathcal{C})$, we are reduced to proving the following:

(∗) Suppose that $\alpha : \mathrm{id}_{\mathcal{C}} \to L$ satisfies (3). Then, for every pair of objects $C, D \in \mathcal{C}$, composition with $\alpha(C)$ induces a bijection of sets

$$\phi : \mathrm{Hom}_{h\mathcal{C}}(LC, LD) \to \mathrm{Hom}_{h\mathcal{C}}(C, LD).$$

We first show that ϕ is surjective. Let f be a morphism from C to LD. We then have a commutative diagram

$$
\begin{array}{ccc}
C & \xrightarrow{\ f\ } & LD \\
\downarrow{\scriptstyle \alpha(C)} & & \downarrow{\scriptstyle \alpha(LD)} \\
LC & \xrightarrow{\ Lf\ } & LLD,
\end{array}
$$

so that f is homotopic to the composition $(\alpha(LD)^{-1} \circ Lf) \circ \alpha(C)$; this proves that the homotopy class of f lies in the image of ϕ.

We now show that ϕ is injective. Let $g : LC \to LD$ be an arbitrary morphism. We have a commutative diagram

$$
\begin{array}{ccc}
LC & \xrightarrow{\ g\ } & LD \\
\downarrow{\scriptstyle \alpha(LC)} & & \downarrow{\scriptstyle \alpha(LD)} \\
LLC & \xrightarrow{\ Lg\ } & LLD,
\end{array}
$$

so that g is homotopic to the composition

$$
\alpha(LD)^{-1} \circ Lg \circ \alpha(LC) \simeq \alpha(LD)^{-1} \circ Lg \circ L\alpha(C) \circ (L\alpha(C))^{-1} \circ \alpha(LC)
$$
$$
\simeq \alpha(LD)^{-1} \circ L(g \circ \alpha(C)) \circ (L\alpha(C))^{-1} \circ \alpha(LC).
$$

In particular, g is determined by $g \circ \alpha(C)$ up to homotopy.

$\square$

Remark 5.2.7.5. Let $L : \mathcal{C} \to \mathcal{D}$ be a localization functor and K a simplicial set. Suppose that every diagram $p : K \to \mathcal{C}$ admits a colimit in $\mathcal{C}$. Then the ∞-category $\mathcal{D}$ has the same property. Moreover, we can give an explicit prescription for computing colimits in $\mathcal{D}$. Let $q : K \to \mathcal{D}$ be a diagram and let $p : K \to \mathcal{C}$ be the composition of q with a right adjoint to L. Choose a colimit $\overline{p} : K^{\triangleright} \to \mathcal{C}$. Since L is a left adjoint, $L \circ \overline{p}$ is a colimit diagram in $\mathcal{D}$, and $L \circ p$ is equivalent to the diagram q.

We conclude this section by introducing a few ideas which will allow us to recognize localization functors when they exist.

Definition 5.2.7.6. Let $\mathcal{C}$ be an ∞-category and $\mathcal{C}^0 \subseteq \mathcal{C}$ a full subcategory. We will say that a morphism $f : C \to D$ in $\mathcal{C}$ *exhibits D as a $\mathcal{C}^0$-localization of C* if $D \in \mathcal{C}^0$, and composition with f induces an isomorphism

$$
\mathrm{Map}_{\mathcal{C}^0}(D, E) \to \mathrm{Map}_{\mathcal{C}}(C, E)
$$

in the homotopy category $\mathcal{H}$ for each object $E \in \mathcal{C}^0$.

Remark 5.2.7.7. In the situation of Definition 5.2.7.6, a morphism $f : C \to D$ exhibits D as a localization of C if and only if f is an initial object of the ∞-category $\mathcal{C}^0_{C/} = \mathcal{C}_{C/} \times_{\mathcal{C}} \mathcal{C}^0$. In particular, f is uniquely determined up to equivalence.

Proposition 5.2.7.8. *Let $\mathcal{C}$ be an ∞-category and $\mathcal{C}^0 \subseteq \mathcal{C}$ a full subcategory. The following conditions are equivalent:*

(1) *For every object $C \in \mathcal{C}$, there exists a localization $f : C \to D$ relative to $\mathcal{C}^0$.*

(2) *The inclusion $\mathcal{C}^0 \subseteq \mathcal{C}$ admits a left adjoint.*

Proof. Let $\mathcal{D}$ be the full subcategory of $\mathcal{C} \times \Delta^1$ spanned by objects of the form (C, i), where $C \in \mathcal{C}^0$ if $i = 1$. Then the projection $p : \mathcal{D} \to \Delta^1$ is a correspondence from $\mathcal{C}$ to $\mathcal{C}^0$ which is associated to the inclusion functor $i : \mathcal{C}^0 \subseteq \mathcal{C}$. It follows that i admits a left adjoint if and only if p is a coCartesian fibration. It now suffices to observe that if C is an object of $\mathcal{C}$, then we may identify p-coCartesian edges $f : (C, 0) \to (D, 1)$ of $\mathcal{D}$ with localizations $C \to D$ relative to $\mathcal{C}^0$. $\square$

Remark 5.2.7.9. By analogy with classical category theory, we will say that a full subcategory $\mathcal{C}^0$ of an ∞-category $\mathcal{C}$ is a *reflective subcategory* if the hypotheses of Proposition 5.2.7.8 are satisfied by the inclusion $\mathcal{C}^0 \subseteq \mathcal{C}$.

Example 5.2.7.10. Let $\mathcal{C}$ be an ∞-category which has a final object and let $\mathcal{C}^0$ be the full subcategory of $\mathcal{C}$ spanned by the final objects. Then the inclusion $\mathcal{C}^0 \subseteq \mathcal{C}$ admits a left adjoint.

Corollary 5.2.7.11. *Let $p : \mathcal{C} \to \mathcal{D}$ be a coCartesian fibration between ∞-categories, let $\mathcal{D}^0 \subseteq \mathcal{D}$ be a full subcategory and let $\mathcal{C}^0 = \mathcal{C} \times_{\mathcal{D}} \mathcal{D}^0$. If the inclusion $\mathcal{D}^0 \subseteq \mathcal{D}$ admits a left adjoint, then the inclusion $\mathcal{C}^0 \subseteq \mathcal{C}$ admits a left adjoint.*

Proof. In view of Proposition 5.2.7.8, it will suffice to show that for every object $C \in \mathcal{C}$, there a morphism $f : C \to C_0$ which is a localization of C relative to $\mathcal{C}^0$. Let $D = p(C)$, let $\overline{f} : D \to D_0$ be a localization of D relative to $\mathcal{D}_0$, and let $f : C \to C_0$ be a p-coCartesian morphism in $\mathcal{C}$ lifting $\overline{f}$. We claim that f has the desired property. Choose any object $C' \in \mathcal{C}^0$ and let $D' = p(C') \in \mathcal{D}^0$. We obtain a diagram of spaces

$$
\begin{array}{ccc}
\operatorname{Map}_{\mathcal{C}}(C_0, C') & \xrightarrow{\phi} & \operatorname{Map}_{\mathcal{C}}(C, C') \\
\downarrow & & \downarrow \\
\operatorname{Map}_{\mathcal{D}}(D_0, D') & \xrightarrow{\psi} & \operatorname{Map}_{\mathcal{D}}(D, D')
\end{array}
$$

which commutes up to preferred homotopy. By assumption, the map ψ is a homotopy equivalence. Since f is p-coCartesian, the map ϕ induces a homotopy equivalence after passing to the homotopy fibers over any pair of points $\eta \in \operatorname{Map}_{\mathcal{D}}(D_0, D')$, $\psi(\eta) \in \operatorname{Map}_{\mathcal{D}}(D, D')$. Using the long exact sequence of homotopy groups associated to the vertical fibrations, we conclude that ϕ is a homotopy equivalence, as desired. $\square$

Proposition 5.2.7.12. *Let* $\mathcal{C}$ *be an* ∞-*category and let* $L : \mathcal{C} \to \mathcal{C}$ *be a localization functor with essential image* $L\,\mathcal{C}$. *Let* S *denote the collection of all morphisms* f *in* $\mathcal{C}$ *such that* Lf *is an equivalence. Then, for every* ∞-*category* $\mathcal{D}$, *composition with* f *induces a fully faithful functor*

$$\psi : \mathrm{Fun}(L\,\mathcal{C}, \mathcal{D}) \to \mathrm{Fun}(\mathcal{C}, \mathcal{D}).$$

Moreover, the essential image of ψ *consists of those functors* $F : \mathcal{C} \to \mathcal{D}$ *such that* $F(f)$ *is an equivalence in* $\mathcal{D}$ *for each* $f \in S$.

Proof. Let S_0 be the collection of all morphisms $C \to D$ in $\mathcal{C}$ which exhibit D as an $L\,\mathcal{C}$-localization of C. We first claim that, for any functor $F : \mathcal{C} \to \mathcal{D}$, the following conditions are equivalent:

(a) The functor F is a right Kan extension of $F|L\,\mathcal{C}$.

(b) The functor F carries each morphism in S_0 to an equivalence in $\mathcal{D}$.

(c) The functor F carries each morphism in S to an equivalence in $\mathcal{D}$.

The equivalence of (a) and (b) follows immediately from the definitions (since a morphism $f : C \to D$ exhibits D as an $L\,\mathcal{C}$-localization of C if and only if f is an initial object of $(L\,\mathcal{C}) \times_{\mathcal{C}} \mathcal{C}_{C/}$), and the implication $(c) \Rightarrow (b)$ is obvious. To prove that $(b) \Rightarrow (c)$, let us consider any map $f : C \to D$ which belongs to S. We have a commutative diagram

$$
\begin{array}{ccc}
C & \xrightarrow{\ f\ } & LC \\
\downarrow & & \downarrow{\scriptstyle f} \\
D & \longrightarrow & LD.
\end{array}
$$

Since $f \in S$, the map Lf is an equivalence in $\mathcal{C}$. If F satisfies (b), then F carries each of the horizontal maps to an equivalence in $\mathcal{D}$. It follows from the two-out-of-three property that Ff is an equivalence in $\mathcal{D}$ as well, so that F satisfies (c).

Let $\mathrm{Fun}^0(\mathcal{C}, \mathcal{D})$ denote the full subcategory of $\mathrm{Fun}(\mathcal{C}, \mathcal{D})$ spanned by those functors which satisfy (a), (b), and (c). Using Proposition 4.3.2.15, we deduce that the restriction functor $\phi : \mathrm{Fun}^0(\mathcal{C}, \mathcal{D}) \to \mathrm{Fun}(L\,\mathcal{C}, \mathcal{D})$ is fully faithful. We now observe that ψ is a right homotopy inverse to ϕ. It follows that ϕ is essentially surjective and therefore an equivalence. Being right homotopy inverse to an equivalence, the functor ψ must itself be an equivalence. $\square$

5.2.8 Factorization Systems

Let $f : X \to Z$ be a map of sets. Then f can be written as a composition

$$X \xrightarrow{f'} Y \xrightarrow{f''} Z,$$

where f' is surjective and f'' is injective. This factorization is uniquely determined up to (unique) isomorphism: the set Y can be characterized either as the image of the map f or as the quotient of X by the equivalence relation $R = \{(x, y) \in X^2 : f(x) = f(y)\}$. We can describe the situation formally by saying that the collections of surjective and injective maps form a *factorization system* on the category Set of sets (see Definition 5.2.8.8). In this section, we will describe a theory of factorization systems in the ∞-categorical setting. These ideas are due to Joyal, and we refer the reader to [44] for further details.

Definition 5.2.8.1. Let $f : A \to B$ and $g : X \to Y$ be morphisms in an ∞-category $\mathcal{C}$. We will say that f is *left orthogonal* to g (or that g is *right orthogonal to f*) if the following condition is satisfied:

(∗) For every commutative diagram

$$
\begin{array}{ccc}
A & \longrightarrow & X \\
\downarrow{\scriptstyle f} & & \downarrow{\scriptstyle g} \\
B & \longrightarrow & Y
\end{array}
$$

in $\mathcal{C}$, the mapping space $\mathrm{Map}_{\mathcal{C}_{A//Y}}(B, X)$ is contractible. (Here we abuse notation by identifying B and X with the corresponding objects of $\mathcal{C}_{A//Y}$.)

In this case, we will write $f \perp g$.

Remark 5.2.8.2. More informally, a morphism $f : A \to B$ in an ∞-category $\mathcal{C}$ is left orthogonal to another morphism $g : X \to Y$ if, for every commutative diagram

$$
\begin{array}{ccc}
A & \longrightarrow & X \\
\downarrow{\scriptstyle f} & \nearrow & \downarrow{\scriptstyle g} \\
B & \longrightarrow & Y,
\end{array}
$$

the space of dotted arrows which render the diagram commutative is contractible.

Remark 5.2.8.3. Let $f : A \to B$ and $g : X \to Y$ be morphisms in an ∞-category $\mathcal{C}$. Fix a morphism $A \to Y$, which we can identify with an object $\overline{Y} \in \mathcal{C}_{A/}$. Lifting $g : X \to Y$ to an object of $\widetilde{X} \in \mathcal{C}_{A//Y}$ is equivalent to lifting g to a morphism $\overline{g} : \overline{X} \to \overline{Y}$ in $\mathcal{C}_{A/}$. The map $f : A \to B$ determines an object $\overline{B} \in \mathcal{C}_{A/}$, and lifting f to an object $\widetilde{B} \in \mathcal{C}_{A//Y}$ is equivalent to giving a map $h : \overline{B} \to \overline{Y}$ in $\mathcal{C}_{A/}$. We therefore have a fiber sequence of spaces

$$
\mathrm{Map}_{\mathcal{C}_{A//Y}}(\widetilde{B}, \widetilde{X}) \to \mathrm{Map}_{\mathcal{C}_{A/}}(\overline{B}, \overline{X}) \to \mathrm{Map}_{\mathcal{C}_{A/}}(\overline{B}, \overline{Y}),
$$

where the fiber is taken over the point h. Consequently, condition (∗) of Definition 5.2.8.1 can be reformulated as follows: for every morphism $\overline{g} : \overline{X} \to \overline{Y}$ in $\mathcal{C}_{A/}$ lifting g, composition with $\overline{g}$ induces a homotopy equivalence

$$
\mathrm{Map}_{\mathcal{C}_{A/}}(\overline{B}, \overline{X}) \to \mathrm{Map}_{\mathcal{C}_{A/}}(\overline{B}, \overline{Y}).
$$

Notation 5.2.8.4. Let $\mathcal{C}$ be an ∞-category and let S be a collection of morphisms in $\mathcal{C}$. We let $S^{\perp}$ denote the collection of all morphisms in $\mathcal{C}$ which are right orthogonal to S, and $^{\perp}S$ the collection of all morphisms in $\mathcal{C}$ which are left orthogonal to S.

Remark 5.2.8.5. Let $\mathcal{C}$ be an ordinary category containing a pair of morphisms f and g. If $f \perp g$, then f has the left lifting property with respect to g, and g has the right lifting property with respect to f. It follows that for any collection S of morphisms in $\mathcal{C}$, we have inclusions $S^{\perp} \subseteq S_{\perp}$ and $^{\perp}S \subseteq_{\perp} S$, where the latter classes of morphisms are defined in §A.1.2.

Applying Remark 5.2.8.3 to an ∞-category $\mathcal{C}$ and its opposite, we obtain the following result:

Proposition 5.2.8.6. *Let $\mathcal{C}$ be an ∞-category and S a collection of morphisms in $\mathcal{C}$.*

(1) *The sets of morphisms $S^{\perp}$ and $^{\perp}S$ contain every equivalence in $\mathcal{C}$.*

(2) *The sets of morphisms $S^{\perp}$ and $^{\perp}S$ are closed under the formation of retracts.*

(3) *Suppose we are given a commutative diagram*

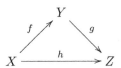

in $\mathcal{C}$, where $g \in S^{\perp}$. Then $f \in S^{\perp}$ if and only if $h \in S^{\perp}$. In particular, $S^{\perp}$ is closed under composition.

(4) *Suppose we are given a commutative diagram*

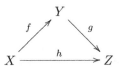

in $\mathcal{C}$, where $f \in {}^{\perp}S$. Then $g \in {}^{\perp}S$ if and only if $h \in {}^{\perp}S$. In particular, $^{\perp}S$ is closed under composition.

(5) *The set of morphisms $S^{\perp}$ is stable under pullbacks: that is, given a pullback diagram*

$$
\begin{array}{ccc}
X' & \longrightarrow & X \\
\downarrow{\scriptstyle g'} & & \downarrow{\scriptstyle g} \\
Y' & \longrightarrow & Y
\end{array}
$$

in $\mathcal{C}$, if g belongs to $S^{\perp}$, then g' belongs to $S^{\perp}$.

(6) *The set of morphisms $^{\perp}S$ is stable under pushouts: that is, given a pushout diagram*

$$
\begin{array}{ccc}
A & \longrightarrow & A' \\
\downarrow f & & \downarrow f' \\
B & \longrightarrow & B',
\end{array}
$$

if f belongs to $^{\perp}S$, then so does f'.

(7) *Let K be a simplicial set such that $\mathcal{C}$ admits K-indexed colimits. Then the full subcategory of $\mathrm{Fun}(\Delta^1, \mathcal{C})$ spanned by the elements of $^{\perp}S$ is closed under K-indexed colimits.*

(8) *Let K be a simplicial set such that $\mathcal{C}$ admits K-indexed limits. Then the full subcategory of $\mathrm{Fun}(\Delta^1, \mathcal{C})$ spanned by the elements of $S^{\perp}$ is closed under K-indexed limits.*

Remark 5.2.8.7. Suppose we are given a pair of adjoint functors

$$
\mathcal{C} \underset{G}{\overset{F}{\rightleftarrows}} \mathcal{D} .
$$

Let f be a morphism in $\mathcal{C}$ and g a morphism in $\mathcal{D}$. Then $f \perp G(g)$ if and only if $F(f) \perp g$.

Definition 5.2.8.8 (Joyal). Let $\mathcal{C}$ be an ∞-category. A *factorization system* on $\mathcal{C}$ is a pair (S_L, S_R), where S_L and S_R are collections of morphisms of $\mathcal{C}$ which satisfy the following axioms:

(1) The collections S_L and S_R are stable under the formation of retracts.

(2) Every morphism in S_L is left orthogonal to every morphism in S_R.

(3) For every morphism $h : X \to Z$ in $\mathcal{C}$, there exists a commutative triangle

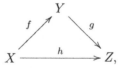

where $f \in S_L$ and $g \in S_R$.

We will call S_L the *left set* of the factorization system and S_R the *right set* of the factorization system.

Example 5.2.8.9. Let $\mathcal{C}$ be an ∞-category. Then $\mathcal{C}$ admits a factorization system (S_L, S_R), where S_L is the collection of all equivalences in $\mathcal{C}$ and S_R consists of all morphisms of $\mathcal{C}$.

Remark 5.2.8.10. Let (S_L, S_R) be a factorization system on an ∞-category $\mathcal{C}$. Then (S_R, S_L) is a factorization system on the opposite ∞-category $\mathcal{C}^{op}$.

Proposition 5.2.8.11. *Let $\mathcal{C}$ be an ∞-category and let (S_L, S_R) be a factorization system on $\mathcal{C}$. Then $S_L = {}^{\perp}S_R$ and $S_R = S_L^{\perp}$.*

Proof. By symmetry, it will suffice to prove the first assertion. The inclusion $S_L \subseteq {}^{\perp}S_R$ follows immediately from the definition. To prove the reverse inclusion, let us suppose that $h : X \to Z$ is a morphism in $\mathcal{C}$ which is left orthogonal to every morphism in S_R. Choose a commutative triangle

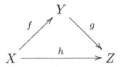

where $f \in S_L$ and $g \in S_R$, and consider the associated diagram

Since $h \perp g$, we can complete this diagram to a 3-simplex of $\mathcal{C}$ as indicated. This 3-simplex exhibits h as a retract of f, so that $h \in S_L$, as desired. $\square$

Remark 5.2.8.12. It follows from Proposition 5.2.8.11 that a factorization system (S_L, S_R) on an ∞-category $\mathcal{C}$ is completely determined by *either* the left set S_L or the right set S_R.

Corollary 5.2.8.13. *Let $\mathcal{C}$ be an ∞-category and let (S_L, S_R) be a factorization system on $\mathcal{C}$. Then the collections of morphisms S_L and S_R contain all equivalences and are stable under composition.*

Proof. Combine Propositions 5.2.8.11 and 5.2.8.6. $\square$

Remark 5.2.8.14. It follows from Corollary 5.2.8.13 that a factorization system (S_L, S_R) on $\mathcal{C}$ determines a pair of subcategories $\mathcal{C}^L, \mathcal{C}^R \subseteq \mathcal{C}$, each containing all the objects of $\mathcal{C}$: the morphisms of $\mathcal{C}^L$ are the elements of S_L, and the morphisms of $\mathcal{C}^R$ are the elements of S_R.

Example 5.2.8.15. Let $p : \mathcal{C} \to \mathcal{D}$ be a coCartesian fibration of ∞-categories. Then there is an associated factorization system (S_L, S_R) on $\mathcal{C}$, where S_L is the class of p-coCartesian morphisms of $\mathcal{C}$ and S_R is the class of morphisms g of $\mathcal{C}$ such that $p(g)$ is an equivalence in $\mathcal{D}$. If $\mathcal{D} \simeq \Delta^0$, this recovers the factorization system of Example 5.2.8.9; if p is an isomorphism, this recovers the opposite of the factorization system of Example 5.2.8.9.

Example 5.2.8.16. Let $\mathcal{X}$ be an ∞-topos and let $n \geq -2$ be an integer. Then there exists a factorization system (S_L, S_R) on $\mathcal{X}$, where S_L denotes the collection of $(n+1)$-connective morphisms of $\mathcal{X}$ and S_R denotes the collection of n-truncated morphisms of $\mathcal{C}$. See §6.5.1.

Let (S_L, S_R) be a factorization system on an ∞-category $\mathcal{C}$, so that any morphism $h : X \to Z$ factors as a composition

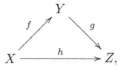

where $f \in S_L$ and $g \in S_R$. For many purposes, it is important to know that this factorization is *canonical*. More precisely, we have the following result:

Proposition 5.2.8.17. *Let $\mathcal{C}$ be an ∞-category and let S_L and S_R be collections of morphisms in $\mathcal{C}$. Suppose that S_L and S_R are stable under equivalence in $\mathrm{Fun}(\Delta^1, \mathcal{C})$ and contain every equivalence in $\mathcal{C}$. The following conditions are equivalent:*

(1) *The pair (S_L, S_R) is a factorization system on $\mathcal{C}$.*

(2) *The restriction map $p : \mathrm{Fun}'(\Delta^2, \mathcal{C}) \to \mathrm{Fun}(\Delta^{\{0,2\}}, \mathcal{C})$ is a trivial Kan fibration. Here $\mathrm{Fun}'(\Delta^2, \mathcal{C})$ denotes the full subcategory of $\mathrm{Fun}(\Delta^2, \mathcal{C})$ spanned by those diagrams*

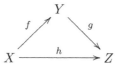

such that $f \in S_L$ and $g \in S_R$.

Corollary 5.2.8.18. *Let $\mathcal{C}$ be an ∞-category equipped with a factorization system (S_L, S_R) and let K be an arbitrary simplicial set. Then the ∞-category $\mathrm{Fun}(K, \mathcal{C})$ admits a factorization system (S_L^K, S_R^K), where S_L^K denotes the collection of all morphisms f in $\mathrm{Fun}(K, \mathcal{C})$ such that $f(v) \in S_L$ for each vertex v of K, and S_R^K is defined likewise.*

The remainder of this section is devoted to the proof of Proposition 5.2.8.17. We begin with a few preliminary results.

Lemma 5.2.8.19. *Let $\mathcal{C}$ be an ∞-category and let (S_L, S_R) be a factorization system on $\mathcal{C}$. Let $\mathcal{D}$ be the full subcategory of $\mathrm{Fun}(\Delta^1, \mathcal{C})$ spanned by the elements of S_R. Then*

(1) *The ∞-category $\mathcal{D}$ is a localization of $\mathrm{Fun}(\Delta^1, \mathcal{C})$; in other words, the inclusion $\mathcal{D} \subseteq \mathrm{Fun}(\Delta^1, \mathcal{C})$ admits a left adjoint.*

(2) *A morphism $\alpha : h \to g$ in $\mathrm{Fun}(\Delta^1, \mathcal{C})$ corresponding to a commutative diagram*

$$
\begin{array}{ccc}
X & \xrightarrow{f} & Y \\
\downarrow{\scriptstyle h} & & \downarrow{\scriptstyle g} \\
Z' & \xrightarrow{e} & Z
\end{array}
$$

exhibits g as a $\mathcal{D}$-localization of h (see Definition 5.2.7.6) if and only if $g \in S_R$, $f \in S_L$, and e is an equivalence.

Proof. We will prove the "if" direction of assertion (2). It follows from the definition of a factorization system that for every object $h \in \mathrm{Fun}(\Delta^1, \mathcal{C})$, there exists a morphism $\alpha : h \to g$ satisfying the condition stated in (2), which therefore exhibits g as a $\mathcal{D}$-localization of h. Invoking Proposition 5.2.7.8, we will deduce (1). Because a $\mathcal{D}$-localization of h is uniquely determined up to equivalence, we will also deduce the "only if" direction of assertion (2).

Suppose we are given a commutative diagram

$$\begin{array}{ccc} X & \xrightarrow{\;f\;} & Y \\ \Big\downarrow{\scriptstyle h} & & \Big\downarrow{\scriptstyle g} \\ Z' & \xrightarrow{\;e\;} & Z, \end{array}$$

where $f \in S_L$, $g \in S_R$, and e is an equivalence, and let $\overline{g} : \overline{Y} \to \overline{Z}$ be another element of S_R. We have a diagram of spaces

$$\begin{array}{ccc} \mathrm{Map}_{\mathrm{Fun}(\Delta^1, \mathcal{C})}(g, \overline{g}) & \xrightarrow{\;\psi\;} & \mathrm{Map}_{\mathrm{Fun}(\Delta^1, \mathcal{C})}(h, \overline{g}) \\ \Big\downarrow & & \Big\downarrow \\ \mathrm{Map}_{\mathcal{C}}(Z, \overline{Z}) & \xrightarrow{\;\psi_0\;} & \mathrm{Map}_{\mathcal{C}}(Z', \overline{Z}) \end{array}$$

which commutes up to canonical homotopy. We wish to prove that ψ is a homotopy equivalence.

Since e is an equivalence in $\mathcal{C}$, the map ψ_0 is a homotopy equivalence. It will therefore suffice to show that ψ induces a homotopy equivalence after passing to the homotopy fibers over any point of $\mathrm{Map}_{\mathcal{C}}(Z, \overline{Z}) \simeq \mathrm{Map}_{\mathcal{C}}(Z', \overline{Z})$. These homotopy fibers can be identified with the homotopy fibers of the vertical arrows in the diagram

$$\begin{array}{ccc} \mathrm{Map}_{\mathcal{C}}(Y, \overline{Y}) & \longrightarrow & \mathrm{Map}_{\mathcal{C}}(X, \overline{Y}) \\ \Big\downarrow & & \Big\downarrow \\ \mathrm{Map}_{\mathcal{C}}(Y, \overline{Z}) & \longrightarrow & \mathrm{Map}_{\mathcal{C}}(X, \overline{Z}). \end{array}$$

It will therefore suffice to show that this diagram (which commutes up to specified homotopy) is a homotopy pullback. Unwinding the definition, this is equivalent to the assertion that f is left orthogonal to $\overline{g}$, which is part of the definition of a factorization system. $\qquad\square$

Lemma 5.2.8.20. *Let K, A, and B be simplicial sets. Then the diagram*

$$\begin{array}{ccc} K \times B & \longrightarrow & K \times (A \star B) \\ \Big\downarrow & & \Big\downarrow \\ B & \longrightarrow & (K \times A) \star B \end{array}$$

is a homotopy pushout square of simplicial sets (with respect to the Joyal model structure).

Proof. We consider the larger diagram

$$
\begin{array}{ccc}
K \times B \longrightarrow K \times (A \diamond B) \longrightarrow K \times (A \star B) \\
\downarrow \qquad\qquad \downarrow \qquad\qquad\qquad \downarrow \\
B \longrightarrow (K \times A) \diamond B \longrightarrow (K \times A) \star B.
\end{array}
$$

The square on the left is a pushout square in which the horizontal maps are monomorphisms of simplicial sets and therefore is a homotopy pushout square (since the Joyal model structure is left proper). The square on the right is a homotopy pushout square since the horizontal arrows are both categorical equivalences (Proposition 4.2.1.2). It follows that the outer rectangle is also a homotopy pushout as desired. □

Notation 5.2.8.21. In the arguments which follow, we let Q denote the simplicial subset of Δ^3 spanned by all simplices which do not contain $\Delta^{\{1,2\}}$. Note that Q is isomorphic to the product $\Delta^1 \times \Delta^1$ as a simplicial set.

Lemma 5.2.8.22. *Let $\mathcal{C}$ be an ∞-category and let $\sigma : Q \to \mathcal{C}$ be a diagram, which we depict as*

$$
\begin{array}{ccc}
A \longrightarrow X \\
\downarrow \qquad \downarrow \\
B \longrightarrow Y.
\end{array}
$$

Then there is a canonical categorical equivalence

$$
\theta : \mathrm{Fun}(\Delta^3, \mathcal{C}) \times_{\mathrm{Fun}(Q,\mathcal{C})} \{\sigma\} \to \mathrm{Map}_{\mathcal{C}_{A//Y}}(B, X).
$$

In particular, $\mathrm{Fun}(\Delta^3, \mathcal{C}) \times_{\mathrm{Fun}(Q,\mathcal{C})} \{\sigma\}$ is a Kan complex.

Proof. We will identify $\mathrm{Map}_{\mathcal{C}_{A//Y}}(B, X)$ with the simplicial set Z defined by the following universal property: for every simplicial set K, we have a pullback diagram of sets

$$
\begin{array}{ccc}
\mathrm{Hom}_{\mathrm{Set}_\Delta}(K, Z) \longrightarrow \mathrm{Hom}_{\mathrm{Set}_\Delta}(\Delta^0 \star (K \times \Delta^1) \star \Delta^0, \mathcal{C}) \\
\downarrow \qquad\qquad\qquad\qquad \downarrow \\
\Delta^0 \longrightarrow \mathrm{Hom}_{\mathrm{Set}_\Delta}(\Delta^0 \star (K \times \partial \Delta^1) \star \Delta^0, \mathcal{C}).
\end{array}
$$

The map θ is then induced by the natural transformation

$$
K \times \Delta^3 \simeq K \times (\Delta^0 \star \Delta^1 \star \Delta^0) \to \Delta^0 \star (K \times \Delta^1) \star \Delta^0.
$$

We wish to prove that θ is a categorical equivalence. Since $\mathcal{C}$ is an ∞-category, it will suffice to show that for every simplicial set K, the bottom square of the diagram

$$
\begin{array}{ccc}
K \times (\Delta^{\{0\}} \coprod \Delta^{\{3\}}) & \longrightarrow & \Delta^{\{0\}} \coprod \Delta^{\{3\}} \\
\downarrow & & \downarrow \\
K \times C & \longrightarrow & \Delta^0 \star (K \times \partial \Delta^1) \star \Delta^0 \\
\downarrow & & \downarrow \\
K \times \Delta^3 & \longrightarrow & \Delta^0 \star (K \times \Delta^1) \star \Delta^0
\end{array}
$$

is a homotopy pushout square (with respect to the Joyal model structure). For this we need only verify that the top and outer squares are homotopy pushout diagrams; this follows from repeated application of Lemma 5.2.8.20. □

Proof of Proposition 5.2.8.17. We first show that (1) $\Rightarrow$ (2). Assume that (S_L, S_R) is a factorization system on $\mathcal{C}$. The restriction map

$$
p : \mathrm{Fun}'(\Delta^2, \mathcal{C}) \to \mathrm{Fun}(\Delta^{\{0,2\}}, \mathcal{C})
$$

is obviously a categorical fibration. It will therefore suffice to show that p is a categorical equivalence.

Let $\mathcal{D}$ be the full subcategory of $\mathrm{Fun}(\Delta^1 \times \Delta^1, \mathcal{C})$ spanned by those diagrams of the form

$$
\begin{array}{ccc}
X & \xrightarrow{\ f\ } & Y \\
\downarrow{\scriptstyle h} & & \downarrow{\scriptstyle g} \\
Z' & \xrightarrow{\ e\ } & Z,
\end{array}
$$

where $f \in S_L$, $g \in S_R$, and e is an equivalence in $\mathcal{C}$. The map p factors as a composition

$$
\mathrm{Fun}'(\Delta^2, \mathcal{C}) \xrightarrow{\ p'\ } \mathcal{D} \xrightarrow{\ p''\ } \mathrm{Fun}(\Delta^1, \mathcal{C}),
$$

where p' carries a diagram

$$
\begin{array}{ccc}
 & Y & \\
{\scriptstyle f}\nearrow & & \searrow{\scriptstyle g} \\
X & \xrightarrow[\ h\ ]{} & Z
\end{array}
$$

to the partially degenerate square

$$
\begin{array}{ccc}
X & \xrightarrow{\ f\ } & Y \\
\downarrow{\scriptstyle h} \ \ {\scriptstyle h}\searrow & & \downarrow{\scriptstyle g} \\
Z & \xrightarrow[\ \mathrm{id}\ ]{} & Z
\end{array}
$$

and p'' is given by restriction to the left vertical edge of the diagram. To complete the proof, it will suffice to show that p' and p'' are categorical equivalences.

We first show that p' is a categorical equivalence. The map p' admits a left inverse q given by composition with an inclusion $\Delta^2 \subseteq \Delta^1 \times \Delta^1$. We note that q is a pullback of the restriction map $q' : \mathrm{Fun}''(\Delta^2, \mathcal{C}) \to \mathrm{Fun}(\Delta^{\{0,2\}}, \mathcal{C})$, where $\mathrm{Fun}''(\Delta^2, \mathcal{C})$ is the full subcategory spanned by diagrams of the form

where e is an equivalence. Since q' is a trivial Kan fibration (Proposition 4.3.2.15), q is a trivial Kan fibration, so that p' is a categorical equivalence, as desired.

We now complete the proof by showing that p'' is a trivial Kan fibration. Let $\mathcal{E}$ denote the full subcategory of $\mathrm{Fun}(\Delta^1, \mathcal{C}) \times \Delta^1$ spanned by those pairs (g, i) where either $i = 0$ or $g \in S_R$. The projection map $r : \mathcal{E} \to \Delta^1$ is a Cartesian fibration associated to the inclusion $\mathrm{Fun}'(\Delta^1, \mathcal{C}) \subseteq \mathrm{Fun}(\Delta^1, \mathcal{C})$, where $\mathrm{Fun}'(\Delta^1, \mathcal{C})$ is the full subcategory spanned by the elements of S_R. Using Lemma 5.2.8.19, we conclude that r is also a coCartesian fibration. Moreover, we can identify

$$\mathcal{D} \subseteq \mathrm{Fun}(\Delta^1 \times \Delta^1, \mathcal{C}) \simeq \mathrm{Map}_{\Delta^1}(\Delta^1, \mathcal{E})$$

with the full subcategory spanned by the coCartesian sections of r. In terms of this identification, p'' is given by evaluation at the initial vertex $\{0\} \subseteq \Delta^1$ and is therefore a trivial Kan fibration, as desired. This completes the proof that $(1) \Rightarrow (2)$.

Now suppose that (2) is satisfied and choose a section s of the trivial Kan fibration p. Let s carry each morphism $f : X \to Z$ to a commutative diagram

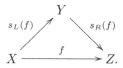

If $s_R(f)$ is an equivalence, then f is equivalent to $s_L(f)$ and therefore belongs to S_L. Conversely, if f belongs to S_L, then the diagram

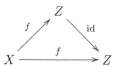

is a preimage of f under p and therefore equivalent to $s(f)$; this implies that $s_L(f)$ is an equivalence. We have proved the following:

($*$) A morphism f of $\mathcal{C}$ belongs to S_L if and only if $s_L(f)$ is an equivalence in $\mathcal{C}$.

It follows immediately from $(*)$ that S_L is stable under the formation of retracts; similarly, S_R is stable under the formation of retracts. To complete the proof, it will suffice to show that $f \perp g$ whenever $f \in S_L$ and $g \in S_R$. Fix a commutative diagram σ

$$
\begin{array}{ccc}
A & \longrightarrow & X \\
\downarrow{\scriptstyle f} & & \downarrow{\scriptstyle g} \\
B & \longrightarrow & Y
\end{array}
$$

in $\mathcal{C}$. In view of Lemma 5.2.8.22, it will suffice to show that the Kan complex $\operatorname{Fun}(\Delta^3, \mathcal{C}) \times_{\operatorname{Fun}(Q,\mathcal{C})} \{\sigma\}$ is contractible.

Let $\mathcal{D}$ denote the full subcategory of $\operatorname{Fun}(\Delta^2 \times \Delta^1, \mathcal{C})$ spanned by those diagrams

$$
\begin{array}{ccc}
C & \longrightarrow & Z \\
\downarrow{\scriptstyle u'} & & \downarrow{\scriptstyle v'} \\
C' & \longrightarrow & Z' \\
\downarrow{\scriptstyle u''} & & \downarrow{\scriptstyle v''} \\
C'' & \longrightarrow & Z''
\end{array}
$$

for which $u' \in S_L$, $v'' \in S_R$, and the maps v' and u'' are equivalences. Let us identify Δ^3 with the full subcategory of $\Delta^2 \times \Delta^1$ spanned by all those vertices except for $(2,0)$ and $(0,1)$. Applying Proposition 4.3.2.15 twice, we deduce that the restriction functor $\operatorname{Fun}(\Delta^2 \times \Delta^1, \mathcal{C}) \to \operatorname{Fun}(\Delta^3, \mathcal{C})$ induces a trivial Kan fibration from $\mathcal{D}$ to the full subcategory $\mathcal{D}' \subseteq \operatorname{Fun}(\Delta^3, \mathcal{C})$ spanned by those diagrams

$$
\begin{array}{ccc}
C & \longrightarrow & Z' \\
\downarrow{\scriptstyle u'} & \nearrow & \downarrow{\scriptstyle v''} \\
C' & \longrightarrow & Z''
\end{array}
$$

such that $u' \in S_L$ and $v'' \in S_R$. It will therefore suffice to show that the fiber $\mathcal{D} \times_{\operatorname{Fun}(Q,\mathcal{C})} \{\sigma\}$ is contractible.

By construction, the restriction functor $\mathcal{D} \to \operatorname{Fun}(Q, \mathcal{C})$ is equivalent to the composition

$$q : \mathcal{D} \subseteq \operatorname{Fun}(\Delta^2 \times \Delta^1, \mathcal{C}) \to \operatorname{Fun}(\Delta^{\{0,2\}} \times \Delta^1, \mathcal{C}).$$

It will therefore suffice to show that $q^{-1}\{\sigma\}$ is a contractible Kan complex. Invoking assumption (2) and $(*)$, we deduce that q induces an equivalence from $\mathcal{D}$ to the full subcategory of $\operatorname{Fun}(\Delta^{\{0,2\}} \times \Delta^1, \mathcal{C})$ spanned by those diagrams

$$
\begin{array}{ccc}
C & \longrightarrow & Z \\
\downarrow{\scriptstyle u} & & \downarrow{\scriptstyle v} \\
C'' & \longrightarrow & Z''
\end{array}
$$

such that $u \in S_L$ and $v \in S_R$. The desired result now follows from our assumption that $f \in S_L$ and $g \in S_R$. $\square$

5.3 ∞-CATEGORIES OF INDUCTIVE LIMITS

Let $\mathcal{C}$ be a category. An Ind-*object* of $\mathcal{C}$ is a diagram $f : \mathcal{I} \to \mathcal{C}$ where $\mathcal{I}$ is a small filtered category. We will informally denote the Ind-object f by

$$[\varinjlim X_i]$$

where $X_i = f(i)$. The collection of all Ind-objects of $\mathcal{C}$ forms a category in which the morphisms are given by the formula

$$\mathrm{Hom}_{\mathrm{Ind}(\mathcal{C})}([\varinjlim X_i], [\varinjlim Y_j]) = \varprojlim \varinjlim \mathrm{Hom}_{\mathcal{C}}(X_i, Y_j).$$

We note that $\mathcal{C}$ may be identified with a full subcategory of $\mathrm{Ind}(\mathcal{C})$ corresponding to diagrams indexed by the one-point category $\mathcal{I} = *$. The idea is that $\mathrm{Ind}(\mathcal{C})$ is obtained from $\mathcal{C}$ by formally adjoining colimits of filtered diagrams. More precisely, $\mathrm{Ind}(\mathcal{C})$ may be described by the following universal property: for any category $\mathcal{D}$ which admits filtered colimits and any functor $F : \mathcal{C} \to \mathcal{D}$, there exists a functor $\widetilde{F} : \mathrm{Ind}(\mathcal{C}) \to \mathcal{D}$ whose restriction to $\mathcal{C}$ is isomorphic to F and which commutes with filtered colimits. Moreover, $\widetilde{F}$ is determined up to (unique) isomorphism.

Example 5.3.0.1. Let $\mathcal{C}$ denote the category of finitely presented groups. Then $\mathrm{Ind}(\mathcal{C})$ is equivalent to the category of groups. (More generally, one could replace "group" by any type of mathematical structure described by algebraic operations which are required to satisfy equational axioms.)

Our objective in this section is to generalize the definition of $\mathrm{Ind}(\mathcal{C})$ to the case where $\mathcal{C}$ is an ∞-category. If we were to work in the setting of simplicial (or topological) categories, we could apply the definition given above directly. However, this leads to a number of problems:

(1) The construction of Ind-categories does not preserve equivalences between simplicial categories.

(2) The obvious generalization of the right hand side in the equation above is given by

$$\varprojlim \varinjlim \mathrm{Map}_{\mathcal{C}}(X_i, Y_j).$$

While the relevant limits and colimits certainly exist in the category of simplicial sets, they are not necessarily the correct objects: one should really replace the limit by a homotopy limit.

(3) In the higher-categorical setting, we should really allow the indexing diagram $\mathcal{I}$ to be a higher category as well. While this does not result in any additional generality (Corollary 5.3.1.16), the restriction to the diagrams indexed by ordinary categories is a technical inconvenience.

Although these difficulties are not insurmountable, it is far more convenient to proceed differently using the theory of ∞-categories. In §5.1, we showed that if $\mathcal{C}$ is a ∞-category, then $\mathcal{P}(\mathcal{C})$ can be interpreted as an ∞-category which is freely generated by $\mathcal{C}$ under colimits. We might therefore hope to find $\mathrm{Ind}(\mathcal{C})$ *inside* $\mathcal{P}(\mathcal{C})$ as a full subcategory. The problem, then, is to characterize this subcategory and to prove that it has the appropriate universal mapping property.

We will begin in §5.3.1 by introducing the definition of a *filtered ∞-category*. Let $\mathcal{C}$ be a small ∞-category. In §5.3.5, we will define $\mathrm{Ind}(\mathcal{C})$ to be the smallest full subcategory of $\mathcal{P}(\mathcal{C})$ which contains all representable presheaves on $\mathcal{C}$ and is stable under filtered colimits. There is also a more direct characterization of which presheaves $F : \mathcal{C} \to \mathcal{S}^{op}$ belong to $\mathrm{Ind}(\mathcal{C})$: they are precisely the *right exact* functors, which we will study in §5.3.2.

In §5.3.5, we will define the Ind-categories $\mathrm{Ind}(\mathcal{C})$ and study their properties. In particular, we will show that morphism spaces in $\mathrm{Ind}(\mathcal{C})$ *are* computed by the naive formula

$$\mathrm{Hom}_{\mathrm{Ind}(\mathcal{C})}([\varinjlim X_i], [\varinjlim Y_j]) = \varprojlim \varinjlim \mathrm{Hom}_{\mathcal{C}}(X_i, Y_j).$$

Unwinding the definitions, this amounts to two conditions:

(1) The (Yoneda) embedding of $j : \mathcal{C} \to \mathrm{Ind}(\mathcal{C})$ is fully faithful (Proposition 5.1.3.1).

(2) For each object $C \in \mathcal{C}$, the corepresentable functor

$$\mathrm{Hom}_{\mathrm{Ind}(\mathcal{C})}(j(C), \bullet)$$

commutes with filtered colimits.

It is useful to translate condition (2) into a definition: an object D of an ∞-category $\mathcal{D}$ is said to be *compact* if the functor $\mathcal{D} \to \mathcal{S}$ corepresented by D commutes with filtered colimits. We will study this compactness condition in §5.3.4.

One of our main results asserts that the ∞-category $\mathrm{Ind}(\mathcal{C})$ is obtained from $\mathcal{C}$ by freely adjoining colimits of filtered diagrams (Proposition 5.3.5.10). In §5.3.6, we will describe a similar construction in the case where the class of filtered diagrams has been replaced by *any* class of diagrams. We will revisit this idea in §5.5.8, where we will study the ∞-category obtained from $\mathcal{C}$ by freely adjoining colimits of *sifted* diagrams.

5.3.1 Filtered ∞-Categories

Recall that a partially ordered set A is *filtered* if every finite subset of A has an upper bound in A. Diagrams indexed by directed partially ordered sets are extremely common in mathematics. For example, if A is the set

$$\mathbf{Z}_{\geq 0} = \{0, 1, \ldots\}$$

of natural numbers, then a diagram indexed by A is a sequence

$$X_0 \to X_1 \to \cdots .$$

The formation of direct limits for such sequences is one of the most basic constructions in mathematics.

In classical category theory, it is convenient to consider not only diagrams indexed by filtered partially ordered sets but also more general diagrams indexed by filtered categories. A category $\mathcal{C}$ is said to be *filtered* if it satisfies the following conditions:

(1) For every finite collection $\{X_i\}$ of objects of $\mathcal{C}$, there exists an object $X \in \mathcal{C}$ equipped with morphisms $\phi_i : X_i \to X$.

(2) Given any two morphisms $f, g : X \to Y$ in $\mathcal{C}$, there exists a morphism $h : Y \to Z$ such that $h \circ f = h \circ g$.

Condition (1) is analogous to the requirement that any finite part of $\mathcal{C}$ admits an "upper bound," while condition (2) guarantees that the upper bound is unique in some asymptotic sense.

If we wish to extend the above definition to the ∞-categorical setting, it is natural to strengthen the second condition.

Definition 5.3.1.1. Let $\mathcal{C}$ be a topological category. We will say that $\mathcal{C}$ is *filtered* if it satisfies the following conditions:

(1′) For every finite set $\{X_i\}$ of objects of $\mathcal{C}$, there exists an object $X \in \mathcal{C}$ and morphisms $\phi_i : X_i \to X$.

(2′) For every pair $X, Y \in \mathcal{C}$ of objects of $\mathcal{C}$, every nonnegative integer $n \geq 0$, and every continuous map $S^n \to \operatorname{Map}_{\mathcal{C}}(X, Y)$, there exists a morphism $Y \to Z$ such that the induced map $S^n \to \operatorname{Map}_{\mathcal{C}}(X, Z)$ is nullhomotopic.

Remark 5.3.1.2. It is easy to see that an ordinary category $\mathcal{C}$ is filtered in the usual sense if and only if it is filtered when regarded as a topological category with discrete mapping spaces. Conversely, if $\mathcal{C}$ is a filtered topological category, then its homotopy category $h\mathcal{C}$ is filtered (when viewed as an ordinary category).

Remark 5.3.1.3. Condition (2′) of Definition 5.3.1.1 is a reasonable analogue of condition (2) in the definition of a filtered category. In the special case $n = 0$, condition (2′) asserts that any pair of morphisms $f, g : X \to Y$ become *homotopic* after composition with some map $Y \to Z$.

Remark 5.3.1.4. Topological spheres S^n need not play any distinguished role in the definition of a filtered topological category. Condition (2′) is equivalent to the following apparently stronger condition:

(2″) For every pair $X, Y \in \mathcal{C}$ of objects of $\mathcal{C}$, every finite cell complex K, and every continuous map $K \to \operatorname{Map}_{\mathcal{C}}(X, Y)$, there exists a morphism $Y \to Z$ such that the induced map $K \to \operatorname{Map}_{\mathcal{C}}(X, Z)$ is nullhomotopic.

Remark 5.3.1.5. The condition that a topological category $\mathcal{C}$ be filtered depends only on the homotopy category $h\mathcal{C}$ (viewed as an $\mathcal{H}$-enriched category). Consequently, if $F : \mathcal{C} \to \mathcal{C}'$ is an equivalence of topological categories, then $\mathcal{C}$ is filtered if and only if $\mathcal{C}'$ is filtered.

Remark 5.3.1.6. Definition 5.3.1.1 has an obvious analogue for (fibrant) simplicial categories: one simply replaces the topological n-sphere S^n by the simplicial n-sphere $\partial \Delta^n$. It is easy to see that a topological category $\mathcal{C}$ is filtered if and only if the simplicial category $\mathrm{Sing}\,\mathcal{C}$ is filtered. Similarly, a (fibrant) simplicial category $\mathcal{D}$ is filtered if and only if the topological category $|\mathcal{D}|$ is filtered.

We now wish to study the analogue of Definition 5.3.1.1 in the setting of ∞-categories. It will be convenient to introduce a slightly more general notion:

Definition 5.3.1.7. Let κ be a regular cardinal and let $\mathcal{C}$ be a ∞-category. We will say that $\mathcal{C}$ is κ-*filtered* if, for every κ-small simplicial set K and every map $f : K \to \mathcal{C}$, there exists a map $\overline{f} : K^{\triangleright} \to \mathcal{C}$ extending f. (In other words, $\mathcal{C}$ is κ-filtered if it has the extension property with respect to the inclusion $K \subseteq K^{\triangleright}$ for every κ-small simplicial set K.)

We will say that $\mathcal{C}$ is *filtered* if it is ω-filtered.

Example 5.3.1.8. Let $\mathcal{C}$ be the nerve of a partially ordered set A. Then $\mathcal{C}$ is κ-filtered if and only if every κ-small subset of A has an upper bound in A.

Remark 5.3.1.9. One may rephrase Definition 5.3.1.7 as follows: an ∞-category $\mathcal{C}$ is κ-filtered if and only if, for every diagram $p : K \to \mathcal{C}$, where K is κ-small, the slice ∞-category $\mathcal{C}_{p/}$ is nonempty. Let $q : \mathcal{C} \to \mathcal{C}'$ be a categorical equivalence of ∞-categories. Proposition 1.2.9.3 asserts that the induced map $\mathcal{C}_{p/} \to \mathcal{C}'_{q \circ p/}$ is a categorical equivalence. Consequently, $\mathcal{C}_{p/}$ is nonempty if and only if $\mathcal{C}'_{q \circ p/}$ is nonempty. It follows that $\mathcal{C}$ is κ-filtered if and only if $\mathcal{C}'$ is κ-filtered.

Remark 5.3.1.10. An ∞-category $\mathcal{C}$ is κ-filtered if and only if, for every κ-small partially ordered set A, $\mathcal{C}$ has the right lifting property with respect to the inclusion $\mathrm{N}(A) \subseteq \mathrm{N}(A)^{\triangleright} \simeq \mathrm{N}(A \cup \{\infty\})$. The "only if" direction is obvious. For the converse, we observe that for every κ-small diagram $p : K \to \mathcal{C}$, the ∞-category $\mathcal{C}_{p/}$ is equivalent to $\mathcal{C}_{q/}$, where q denotes the composition $\mathrm{N}(A) \xrightarrow{p'} K \xrightarrow{p} \mathcal{C}$. Here we have chosen p' to be a cofinal map such that A is a κ-small partially ordered set. (If κ is uncountable, the existence of p' follows from Variant 4.2.3.15; otherwise, we use Variant 4.2.3.16.)

Remark 5.3.1.11. We will say that an arbitrary simplicial set S is κ-*filtered* if there exists a categorical equivalence $j : S \to \mathcal{C}$, where $\mathcal{C}$ is a κ-filtered ∞-category. In view of Remark 5.3.1.9, this condition is independent of the choice of j.

Our next major goal is to prove Proposition 5.3.1.13, which asserts that an ∞-category $\mathcal{C}$ is filtered if and only if the associated topological category $|\mathfrak{C}[\mathcal{C}]|$ is filtered. First, we need a lemma.

Lemma 5.3.1.12. *Let $\mathcal{C}$ be an ∞-category. Then $\mathcal{C}$ is filtered if and only if it has the right extension property with respect to every inclusion $\partial \Delta^n \subseteq \Lambda^{n+1}_{n+1}$, $n \geq 0$.*

Proof. The "only if" direction is clear: we simply take $K = \partial \Delta^n$ in Definition 5.3.1.7. For the converse, let us suppose that the assumption of Definition 5.3.1.7 is satisfied whenever K is the boundary of a simplex; we must then show that it remains satisfied for *any* K which has only finitely many nondegenerate simplices.

We work by induction on the dimension of K and the number of nondegenerate simplices of K. If K is empty, there is nothing to prove (since it is the boundary of a 0-simplex). Otherwise, we may write $K = K' \coprod_{\partial \Delta^n} \Delta^n$, where n is the dimension of K.

Choose a map $p : K \to \mathcal{C}$; we wish to show that p may be extended to a map $\widetilde{p} : K \star \{y\} \to \mathcal{C}$. We first consider the restriction $p|K'$; by the inductive hypothesis, it admits an extension $q : K' \star \{x\} \to \mathcal{C}$. The restriction $q|\partial \Delta^n \star \{x\}$ and the map $p|\Delta^n$ assemble to give a map

$$r : \partial \Delta^{n+1} \simeq (\partial \Delta^n \star \{x\}) \coprod_{\partial \Delta^n} \Delta^n \to \mathcal{C}.$$

By assumption, the map r admits an extension

$$\widetilde{r} : \partial \Delta^{n+1} \star \{y\} \to \mathcal{C}.$$

Let

$$s : (K' \star \{x\}) \coprod_{\partial \Delta^{n+1}} (\partial \Delta^{n+1} \star \{y\})$$

denote the result of amalgamating r with $\widetilde{p}$. We note that the inclusion

$$(K' \star \{x\}) \coprod_{\partial \Delta^n \star \{x\}} (\partial \Delta^{n+1} \star \{y\}) \subseteq (K' \star \{x\} \star \{y\}) \coprod_{\partial \Delta^n \star \{x\} \star \{y\}} (\Delta^n \star \{y\})$$

is a pushout of

$$(K' \star \{x\}) \coprod_{\partial \Delta^n \star \{x\}} (\partial \Delta^n \star \{x\} \star \{y\}) \subseteq K' \star \{x\} \star \{y\}$$

and therefore a categorical equivalence by Lemma 2.4.3.1. It follows that s admits an extension

$$\widetilde{s} : (K' \star \{x\} \star \{y\}) \coprod_{\partial \Delta^n \star \{x\} \star \{y\}} (\Delta^n \star \{y\}) \to \mathcal{C},$$

and we may now define $\widetilde{p} = \widetilde{s}|K \star \{y\}$. $\qquad\square$

Proposition 5.3.1.13. *Let $\mathcal{C}$ be a topological category. Then $\mathcal{C}$ is filtered if and only if the ∞-category $\mathrm{N}(\mathcal{C})$ is filtered.*

Proof. Suppose first that $N(\mathcal{C})$ is filtered. We verify conditions $(1')$ and $(2')$ of Definition 5.3.1.1:

(1') Let $\{X_i\}_{i \in I}$ be a finite collection of objects of $\mathcal{C}$ corresponding to a map $p : I \to N(\mathcal{C})$, where I is regarded as a discrete simplicial set. If $N(\mathcal{C})$ is filtered, then p extends to a map $\widetilde{p} : I \star \{x\} \to N(\mathcal{C})$ corresponding to an object $X = p(x)$ equipped with maps $X_i \to X$ in $\mathcal{C}$.

(2') Let $X, Y \in \mathcal{C}$ be objects, let $n \geq 0$, and let $S^n \to \mathrm{Map}_{\mathcal{C}}(X, Y)$ be a map. We note that this data may be identified with a topological functor $F : |\mathfrak{C}[K]| \to \mathcal{C}$, where K is the simplicial set obtained from $\partial \Delta^{n+2}$ by collapsing the initial face Δ^{n+1} to a point. If $N(\mathcal{C})$ is filtered, then F extends to a functor $\widetilde{F}$ defined on $|\mathfrak{C}[K \star \{z\}]|$; this gives an object $Z = \widetilde{F}(z)$ and a morphism $Y \to Z$ such that the induced map $S^n \to \mathrm{Map}_{\mathcal{C}}(X, Z)$ is nullhomotopic.

For the converse, let us suppose that $\mathcal{C}$ is filtered. We wish to show that $N(\mathcal{C})$ is filtered. By Lemma 5.3.1.12, it will suffice to prove that $N(\mathcal{C})$ has the extension property with respect to the inclusion $\partial \Delta^n \subseteq \Lambda_{n+1}^{n+1}$ for each $n \geq 0$. Equivalently, it suffices to show that $\mathcal{C}$ has the right extension property with respect to the inclusion $|\mathfrak{C}[\partial \Delta^n]| \subseteq |\mathfrak{C}[\Lambda_{n+1}^{n+1}]|$. If $n = 0$, this is simply the assertion that $\mathcal{C}$ is nonempty; if $n = 1$, this is the assertion that for any pair of objects $X, Y \in \mathcal{C}$, there exists an object Z equipped with morphisms $X \to Z$, $Y \to Z$. Both of these conditions follow from part (1) of Definition 5.3.1.1; we may therefore assume that $n > 1$.

Let $\mathcal{A}_0 = |\mathfrak{C}[\partial \Delta^n]|$, $\mathcal{A}_1 = |\mathfrak{C}[\partial \Delta^n \coprod_{\Lambda_n^n} \Lambda_n^n \star \{n+1\}]|$, $\mathcal{A}_2 = |\mathfrak{C}[\Lambda_{n+1}^{n+1}]|$, and $\mathcal{A}_3 = |\mathfrak{C}[\Delta^{n+1}]|$, so that we have inclusions of topological categories

$$\mathcal{A}_0 \subseteq \mathcal{A}_1 \subseteq \mathcal{A}_2 \subseteq \mathcal{A}_3 .$$

We will make use of the description of $\mathcal{A}_3$ given in Remark 1.1.5.2: its objects are integers i satisfying $0 \leq i \leq n + 1$, with $\mathrm{Map}_{\mathcal{A}_3}(i, j)$ given by the cube of all functions $p : \{i, \dots, j\} \to [0, 1]$ satisfying $p(i) = p(j) = 1$ for $i \leq j$ and $\mathrm{Hom}_{\mathcal{A}_3}(i, j) = \emptyset$ for $j < i$. Composition is given by amalgamation of functions.

We note that $\mathcal{A}_1$ and $\mathcal{A}_2$ are subcategories of $\mathcal{A}_3$ having the same objects, whose morphism spaces are may be described as follows:

- $\mathrm{Map}_{\mathcal{A}_1}(i, j) = \mathrm{Map}_{\mathcal{A}_2}(i, j) = \mathrm{Map}_{\mathcal{A}_3}(i, j)$ unless $i = 0$ and $j \in \{n, n+1\}$.

- $\mathrm{Map}_{\mathcal{A}_1}(0, n) = \mathrm{Map}_{\mathcal{A}_2}(0, n)$ is the boundary of the cube
$$\mathrm{Map}_{\mathcal{A}_3}(0, n) \simeq [0, 1]^{n-1}.$$

- $\mathrm{Map}_{\mathcal{A}_1}(0, n+1)$ consists of all functions $p : [n+1] \to [0, 1]$ satisfying $p(0) = p(n+1) = 1$ and $(\exists i)[(1 \leq i \leq n-1) \wedge p(i) \in \{0, 1\}]$.

- $\mathrm{Map}_{\mathcal{A}_2}(0, n+1)$ is the union of $\mathrm{Map}_{\mathcal{A}_1}(0, n+1)$ with the collection of functions $p : \{0, \dots, n+1\} \to [0, 1]$ satisfying $p(0) = p(n) = p(n+1) = 1$.

Finally, we note that $\mathcal{A}_0$ is the full subcategory of $\mathcal{A}_1$ (or $\mathcal{A}_2$) whose set of objects is $\{0, \ldots, n\}$.

We wish to show that any topological functor $F : \mathcal{A}_0 \to \mathcal{C}$ can be extended to a functor $\widetilde{F} : \mathcal{A}_2 \to \mathcal{C}$. Let $X = F(0)$ and let $Y = F(n)$. Then F induces a map $S^{n-1} \simeq \mathrm{Map}_{\mathcal{A}_0}(0, n) \to \mathrm{Map}_{\mathcal{C}}(X, Y)$. Since $\mathcal{C}$ is filtered, there exists a map $\phi : Y \to Z$ such that the induced map $f : S^{n-1} \to \mathrm{Map}_{\mathcal{C}}(X, Z)$ is nullhomotopic.

Now set $\widetilde{F}(n + 1) = Z$; for $p \in \mathrm{Map}_{\mathcal{A}_1}(i, n + 1)$, we set $\widetilde{F}(p) = \phi \circ F(q)$, where $q \in \mathrm{Map}_{\mathcal{A}_1}(i, n)$ is such that $q|\{i, \ldots, n-1\} = p|\{i, \ldots, n-1\}$. Finally, we note that the assumption that f is nullhomotopic allows us to extend $\widetilde{F}$ from $\mathrm{Map}_{\mathcal{A}_1}(0, n + 1)$ to the whole of $\mathrm{Map}_{\mathcal{A}_2}(0, n + 1)$. $\square$

Remark 5.3.1.14. Suppose that $\mathcal{C}$ is a κ-filtered ∞-category and let K be a simplicial set which is categorically equivalent to a κ-small simplicial set. Then $\mathcal{C}$ has the extension property with respect to the inclusion $K \subseteq K^{\triangleright}$. This follows from Proposition A.2.3.1: to test whether or not a map $K \to S$ extends over $K^{\triangleright}$, it suffices to check in the homotopy category of Set_{Δ} (with respect to the Joyal model structure), where we may replace K by an equivalent κ-small simplicial set.

Proposition 5.3.1.15. *Let $\mathcal{C}$ be a ∞-category with a final object. Then $\mathcal{C}$ is κ-filtered for every regular cardinal κ. Conversely, if $\mathcal{C}$ is κ-filtered and there exists a categorical equivalence $K \to \mathcal{C}$, where K is a κ-small simplicial set, then $\mathcal{C}$ has a final object.*

Proof. We remark that $\mathcal{C}$ has a final object if and only if there exists a retraction r of $\mathcal{C}^{\triangleright}$ onto $\mathcal{C}$. If $\mathcal{C}$ is κ-filtered and categorically equivalent to a κ-small simplicial set, then the existence of such a retraction follows from Remark 5.3.1.14. On the other hand, if the retraction r exists, then any map $p : K \to \mathcal{C}$ admits an extension $K^{\triangleright} \to \mathcal{C}$: one merely considers the composition $K^{\triangleright} \to \mathcal{C}^{\triangleright} \xrightarrow{r} \mathcal{C}$. $\square$

A useful observation from classical category theory is that, if we are only interested in using filtered categories to index colimit diagrams, then in fact we do not need the notion of a filtered category at all: we can work instead with diagrams indexed by filtered partially ordered sets. We now prove an ∞-categorical analogue of this statement.

Proposition 5.3.1.16. *Suppose that $\mathcal{C}$ is a κ-filtered ∞-category. Then there exists a κ-filtered partially ordered set A and a cofinal map $\mathrm{N}(A) \to \mathcal{C}$.*

Proof. The proof uses the ideas introduced in §4.2.3 and, in particular, Proposition 4.2.3.8. Let X be a set of cardinality at least κ, and regard X as a category with a unique isomorphism between any pair of objects. We note that $\mathrm{N}(X)$ is a contractible Kan complex; consequently, the projection $\mathcal{C} \times \mathrm{N}(X) \to \mathcal{C}$ is cofinal. Hence, it suffices to produce a cofinal map $\mathrm{N}(A) \to \mathcal{C} \times \mathrm{N}(X)$ with the desired properties.

Let $\{K_\alpha\}_{\alpha \in A}$ be the collection of all simplicial subsets of $K = \mathcal{C} \times \mathrm{N}(X)$ which are κ-small and possess a final vertex. Regard A as a partially ordered by inclusion. We first claim that A is κ-filtered and that $\bigcup_{\alpha \in A} K_\alpha = K$. To prove both of these assertions, it suffices to show that any κ-small simplicial subset $L \subseteq K$ is contained in a κ-small simplicial subset L' which has a final vertex.

Since $\mathcal{C}$ is κ-filtered, the composition

$$L \to \mathcal{C} \times \mathrm{N}(X) \to \mathcal{C}$$

extends to a map $p : L^\triangleright \to \mathcal{C}$. Since X has cardinality at least κ, there exists an element $x \in X$ which is not in the image of $L_0 \to \mathrm{N}(X)_0 = X$. Lift p to a map $\tilde{p} : L^\triangleright \to K$ which extends the inclusion $L \subseteq K \times \mathrm{N}(X)$ and carries the cone point to the element $x \in X = \mathrm{N}(X)_0$. It is easy to see that $\tilde{p}$ is injective, so that we may regard $L^\triangleright$ as a simplicial subset of $K \times \mathrm{N}(X)$. Moreover, it is clearly κ-small and has a final vertex, as desired.

Now regard A as a category and let $F : A \to (\mathcal{S}et_\Delta)_{/K}$ be the functor which carries each $\alpha \in A$ to the simplicial set K_α. For each $\alpha \in A$, choose a final vertex x_α of K_α. Let K_F be defined as in §4.2.3. We claim next that there exists a retraction $r : K_F \to K$ with the property that $r(X_\alpha) = x_\alpha$ for each $I \in \mathcal{I}$.

The construction of r proceeds as in the proof of Proposition 4.2.3.4. Namely, we well-order the finite linearly ordered subsets $B \subseteq A$ and define $r | K'_B$ by induction on B. Moreover, we will select r so that it has the property that if B is nonempty with largest element β, then $r(K'_B) \subseteq K_\beta$.

If B is empty, then $r | K'_B = r | K$ is the identity map. Otherwise, B has a least element α and a largest element β. We are required to construct a map $K_\alpha \star \Delta^B \to K_\beta$ or a map $r_B : \Delta^B \to K_{\mathrm{id} | K_\alpha /}$, where the values of this map on $\partial \Delta^B$ have already been determined. If B is a singleton, we define this map to carry the vertex Δ^B to a final object of $K_{\mathrm{id} | K_\alpha /}$ lying over x_β. Otherwise, we are guaranteed that *some* extension exists by the fact that $r_B | \partial \Delta^B$ carries the final vertex of Δ^B to a final object of $K_{\mathrm{id} | K_\alpha /}$.

Now let $j : \mathrm{N}(A) \to K$ denote the restriction of the retraction of r to $\mathrm{N}(A)$. Using Propositions 4.2.3.4 and 4.2.3.8, we deduce that j is a cofinal map as desired. $\square$

A similar technique can be used to prove the following characterization of κ-filtered ∞-categories:

Proposition 5.3.1.17. *Let S be a simplicial set. The following conditions are equivalent:*

(1) *The simplicial set S is κ-filtered.*

(2) *There exists a diagram of simplicial sets $\{Y_\alpha\}_{\alpha \in \mathcal{I}}$ having colimit Y and a categorical equivalence $S \to Y$, where each Y_α is κ-filtered and the indexing category $\mathcal{I}$ is κ-filtered.*

(3) *There exists a categorical equivalence $S \to \mathcal{C}$, where $\mathcal{C}$ is a κ-filtered union of simplicial subsets $\mathcal{C}_\alpha \subseteq \mathcal{C}$ such that each $\mathcal{C}_\alpha$ is an ∞-category with a final object.*

Proof. Let $T : \mathrm{Set}_\Delta \to \mathrm{Set}_\Delta$ be the fibrant replacement functor given by

$$T(X) = \mathrm{N}(|\,\mathfrak{C}[X]|).$$

There is a natural transformation $j_X : X \to T(X)$ which is a categorical equivalence for every simplicial set X. Moreover, each $T(X)$ is an ∞-category. Furthermore, the functor T preserves inclusions and commutes with filtered colimits.

It is clear that (3) implies (2). Suppose that (2) is satisfied. Replacing the diagram $\{Y_\alpha\}_{\alpha \in \mathcal{I}}$ by $\{T(Y_\alpha)\}_{\alpha \in \mathcal{I}}$ if necessary, we may suppose that each Y_α is an ∞-category. It follows that Y is an ∞-category. If $p : K \to Y$ is a diagram indexed by a κ-small simplicial set, then p factors through a map $p_\alpha : K \to Y_\alpha$ for some $\alpha \in \mathcal{I}$, by virtue of the assumption that $\mathcal{I}$ is κ-filtered. Since Y_α is a κ-filtered ∞-category, we can find an extension $K^\triangleright \to Y_\alpha$ of p_α, hence an extension $K^\triangleright \to Y$ of p.

Now suppose that (1) is satisfied. Replacing S by $T(S)$ if necessary, we may suppose that S is an ∞-category. Choose a set X of cardinality at least κ and let $\mathrm{N}(X)$ be defined as in the proof of Proposition 5.3.1.16. The proof of Proposition 5.3.1.16 shows that we may write $S \times \mathrm{N}(X)$ as a κ-filtered union of simplicial subsets $\{Y_\alpha\}$, where each Y_α has a final vertex. We now take $\mathcal{C} = T(S \times \mathrm{N}(X))$ and let $\mathcal{C}_\alpha = T(Y_\alpha)$: these choices satisfy (3), which completes the proof. □

By definition, an ∞-category $\mathcal{C}$ is κ-filtered if any map $p : K \to \mathcal{C}$ whose domain K is κ-small can be extended over the cone $K^\triangleright$. We now consider the possibility of constructing this extension uniformly in p. First, we need a few lemmas.

Lemma 5.3.1.18. *Let $\mathcal{C}$ be a filtered ∞-category. Then $\mathcal{C}$ is weakly contractible.*

Proof. Since $\mathcal{C}$ is filtered, it is nonempty. Fix an object $C \in \mathcal{C}$. Let $|\mathcal{C}|$ denote the geometric realization of $\mathcal{C}$ as a simplicial set. We identify C with a point of the topological space $|\mathcal{C}|$. By Whitehead's theorem, to show that $\mathcal{C}$ is weakly contractible, it suffices to show that for every $i \geq 0$, the homotopy set $\pi_i(|\mathcal{C}|, C)$ consists of a single point. If not, we can find a finite simplicial subset $K \subseteq \mathcal{C}$ containing C such that the map $f : \pi_i(|K|, C) \to \pi_i(|\mathcal{C}|, C)$ has a nontrivial image. But $\mathcal{C}$ is filtered, so the inclusion $K \subseteq \mathcal{C}$ factors through a map $K^\triangleright \to \mathcal{C}$. It follows that f factors through $\pi_i(|K^\triangleright|, C)$. But this homotopy set is trivial since $K^\triangleright$ is weakly contractible. □

Lemma 5.3.1.19. *Let $\mathcal{C}$ be a κ-filtered ∞-category and let $p : K \to \mathcal{C}$ be a diagram indexed by a κ-small simplicial set K. Then $\mathcal{C}_{p/}$ is κ-filtered.*

Proof. Let K' be a κ-small simplicial set and $p' : K' \to \mathcal{C}_{p/}$ a κ-small diagram. Then we may identify p' with a map $q : K \star K' \to \mathcal{C}$, and we get an isomorphism $(\mathcal{C}_{p/})_{p'/} \simeq \mathcal{C}_{q/}$. Since $K \star K'$ is κ-small, the ∞-category $\mathcal{C}_{q/}$ is nonempty. $\qquad\square$

Proposition 5.3.1.20. *Let $\mathcal{C}$ be an ∞-category and κ a regular cardinal. Then $\mathcal{C}$ is κ-filtered if and only if, for each κ-small simplicial set K, the diagonal map $d : \mathcal{C} \to \mathrm{Fun}(K, \mathcal{C})$ is cofinal.*

Proof. Suppose first that the diagonal map $d : \mathcal{C} \to \mathrm{Fun}(K, \mathcal{C})$ is cofinal for every κ-small simplicial set K. Choose any map $j : K \to \mathcal{C}$; we wish to show that j can be extended to $K^{\triangleright}$. By Proposition A.2.3.1, it suffices to show that j can be extended to the equivalent simplicial set $K \diamond \Delta^0$. In other words, we must produce an object $C \in \mathcal{C}$ and a morphism $j \to d(C)$ in $\mathrm{Fun}(K, \mathcal{C})$. It will suffice to prove that the ∞-category $\mathcal{D} = \mathcal{C} \times_{\mathrm{Fun}(K,\mathcal{C})} \mathrm{Fun}(K, \mathcal{C})_{j/}$ is nonempty. We now invoke Theorem 4.1.3.1 to deduce that $\mathcal{D}$ is weakly contractible.

Now suppose that S is κ-filtered and that K is a κ-small simplicial set. We wish to show that the diagonal map $d : \mathcal{C} \to \mathrm{Fun}(K, \mathcal{C})$ is cofinal. By Theorem 4.1.3.1, it suffices to prove that for every object $X \in \mathrm{Fun}(K, \mathcal{C})$, the ∞-category $\mathrm{Fun}(K, \mathcal{C})^{X/} \times_{\mathrm{Fun}(K,\mathcal{C})} \mathcal{C}$ is weakly contractible. But if we identify X with a map $x : K \to \mathcal{C}$, then we get a natural identification

$$\mathrm{Fun}(K, \mathcal{C})^{X/} \times_{\mathrm{Fun}(K,\mathcal{C})} \mathcal{C} \simeq \mathcal{C}^{x/},$$

which is κ-filtered by Lemma 5.3.1.19 and therefore weakly contractible by Lemma 5.3.1.18. $\qquad\square$

5.3.2 Right Exactness

Let $\mathcal{A}$ and $\mathcal{B}$ be abelian categories. In classical homological algebra, a functor $F : \mathcal{A} \to \mathcal{B}$ is said to be *right exact* if it is additive, and whenever

$$A' \to A \to A'' \to 0$$

is an exact sequence in $\mathcal{A}$, the induced sequence

$$F(A') \to F(A) \to F(A'') \to 0$$

is exact in $\mathcal{B}$.

The notion of right exactness generalizes in a natural way to functors between categories which are not assumed to be abelian. Let $F : \mathcal{A} \to \mathcal{B}$ be a functor between abelian categories as above. Then F is additive if and only if F preserves finite coproducts. Furthermore, an additive functor F is right exact if and only if it preserves coequalizer diagrams. Since every finite colimit can be built out of finite coproducts and coequalizers, right exactness is equivalent to the requirement that F preserve all finite colimits. This condition makes sense whenever the category $\mathcal{A}$ admits finite colimits.

It is possible to generalize even further to the case of a functor between arbitrary categories. To simplify the discussion, let us suppose that $\mathcal{B} =$

Set^{op}. Then we may regard a functor $F : \mathcal{A} \to \mathcal{B}$ as a presheaf of sets on the category $\mathcal{A}$. Using this presheaf, we can define a new category $\mathcal{A}_F$ whose objects are pairs (A, η), where $A \in \mathcal{A}$ and $\eta \in F(A)$, and morphisms from (A, η) to (A', η') are maps $f : A \to A'$ such that $f^*(\eta') = \eta$, where f^* denotes the induced map $F(A') \to F(A)$. If $\mathcal{A}$ admits finite colimits, then the functor F preserves finite colimits if and only if the category $\mathcal{A}_F$ is filtered.

Our goal in this section is to adapt the notion of right exact functors to the ∞-categorical context. We begin with the following:

Definition 5.3.2.1. Let $F : \mathcal{A} \to \mathcal{B}$ be a functor between ∞-categories and κ a regular cardinal. We will say that F is κ-*right exact* if, for any right fibration $\mathcal{B}' \to \mathcal{B}$ where $\mathcal{B}'$ is κ-filtered, the ∞-category $\mathcal{A}' = \mathcal{A} \times_{\mathcal{B}} \mathcal{B}'$ is also κ-filtered. We will say that F is *right exact* if it is ω-right exact.

Remark 5.3.2.2. We also have an dual theory of *left exact* functors.

Remark 5.3.2.3. If $\mathcal{A}$ admits finite colimits, then a functor $F : \mathcal{A} \to \mathcal{B}$ is right exact if and only if F preserves finite colimits (see Proposition 5.3.2.9 below).

We note the following basic stability properties of κ-right exact maps.

Proposition 5.3.2.4. *Let κ be a regular cardinal.*

(1) *If $F : \mathcal{A} \to \mathcal{B}$ and $G : \mathcal{B} \to \mathcal{C}$ are κ-right exact functors between ∞-categories, then $G \circ F : \mathcal{A} \to \mathcal{C}$ is κ-right exact.*

(2) *Any equivalence of ∞-categories is κ-right exact.*

(3) *Let $F : \mathcal{A} \to \mathcal{B}$ be a κ-right exact functor and let $F' : \mathcal{A} \to \mathcal{B}$ be homotopic to F. Then F' is κ-right exact.*

Proof. Property (1) is immediate from the definition. We will establish (2) and (3) as a consequence of the following more general assertion: if $F : \mathcal{A} \to \mathcal{B}$ and $G : \mathcal{B} \to \mathcal{C}$ are functors such that F is a categorical equivalence, then G is κ-right exact if and only if $G \circ F$ is κ-right exact. To prove this, let $\mathcal{C}' \to \mathcal{C}$ be a right fibration. Proposition 3.3.1.3 implies that the induced map

$$\mathcal{A}' = \mathcal{A} \times_{\mathcal{C}} \mathcal{C}' \to \mathcal{B} \times_{\mathcal{C}} \mathcal{C}' = \mathcal{B}'$$

is a categorical equivalence. Thus $\mathcal{A}'$ is κ-filtered if and only if $\mathcal{B}'$ is κ-filtered.

We now deduce (2) by specializing to the case where G is the identity map. To prove (3), we choose a contractible Kan complex K containing a pair of vertices $\{x, y\}$ and a map $g : K \to \mathcal{B}^{\mathcal{A}}$ with $g(x) = F$, $g(y) = F'$. Applying the above argument to the composition

$$\mathcal{A} \simeq \mathcal{A} \times \{x\} \subseteq \mathcal{A} \times K \xrightarrow{G} \mathcal{B},$$

we deduce that G is κ-right exact. Applying the converse to the diagram

$$\mathcal{A} \simeq \mathcal{A} \times \{y\} \subseteq \mathcal{A} \times K \xrightarrow{G} \mathcal{B},$$

we deduce that F' is κ-right exact. $\square$

The next result shows that the κ-right exactness of a functor $F : \mathcal{A} \to \mathcal{B}$ can be tested on a very small collection of right fibrations $\mathcal{B}' \to \mathcal{B}$.

Proposition 5.3.2.5. *Let $F : \mathcal{A} \to \mathcal{B}$ be a functor between ∞-categories and κ a regular cardinal. The following are equivalent:*

(1) *The functor F is κ-right exact.*

(2) *For every object B of $\mathcal{B}$, the ∞-category $\mathcal{A} \times_{\mathcal{B}} \mathcal{B}_{/B}$ is κ-filtered.*

Proof. We observe that for every object $B \in \mathcal{B}$, the ∞-category $\mathcal{B}_{/B}$ is right fibered over $\mathcal{B}$ and is κ-filtered (since it has a final object). Consequently, (1) implies (2). Now suppose that (2) is satisfied. Let $T : (\text{Set}_\Delta)_{/\mathcal{B}} \to (\text{Set}_\Delta)_{/\mathcal{B}}$ denote the composite functor

$$(\text{Set}_\Delta)_{/\mathcal{B}} \overset{St_{\mathcal{B}}}{\to} (\text{Set}_\Delta)^{\mathfrak{C}[\mathcal{B}^{op}]} \overset{\text{Sing}\,|\bullet|}{\to} (\text{Set}_\Delta)^{\mathfrak{C}[\mathcal{B}^{op}]} \overset{\text{Un}_{\mathcal{B}}}{\to} (\text{Set}_\Delta)_{/\mathcal{B}}.$$

We will use the following properties of T:

(i) There is a natural transformation $j_X : X \to T(X)$, where j_X is a contravariant equivalence in $(\text{Set}_\Delta)_{/\mathcal{B}}$ for every $X \in (\text{Set}_\Delta)_{/\mathcal{B}}$.

(ii) For every $X \in (\text{Set}_\Delta)_{/\mathcal{B}}$, the associated map $T(X) \to \mathcal{B}$ is a right fibration.

(iii) The functor T commutes with filtered colimits.

We will say that an object $X \in (\text{Set}_\Delta)_{/\mathcal{B}}$ is *good* if the ∞-category $T(X) \times_{\mathcal{B}} \mathcal{A}$ is κ-filtered. We now make the following observations:

(A) If $X \to Y$ is a contravariant equivalence in $(\text{Set}_\Delta)_{/\mathcal{B}}$, then X is good if and only if Y is good. This follows from the fact that $T(X) \to T(Y)$ is an equivalence of right fibrations, so that the induced map $T(X) \times_{\mathcal{B}} \mathcal{A} \to T(Y) \times_{\mathcal{B}} \mathcal{A}$ is an equivalence of right fibrations and consequently a categorical equivalence of ∞-categories.

(B) If $X \to Y$ is a categorical equivalence in $(\text{Set}_\Delta)_{/\mathcal{B}}$, then X is good if and only if Y is good. This follows from (A) since every categorical equivalence is a contravariant equivalence.

(C) The collection of good objects of $(\text{Set}_\Delta)_{\mathcal{B}}$ is stable under κ-filtered colimits. This follows from the fact that the functor $X \mapsto T(X) \times_{\mathcal{B}} \mathcal{A}$ commutes with κ-filtered colimits (in fact, with all filtered colimits) and Proposition 5.3.1.17.

(D) If $X \in (\text{Set}_\Delta)_{/\mathcal{B}}$ corresponds to a right fibration $X \to \mathcal{B}$, then X is good if and only if $X \times_{\mathcal{B}} \mathcal{A}$ is κ-filtered.

(E) For every object $B \in \mathcal{B}$, the overcategory $\mathcal{B}_{/B}$ is a good object of $(\text{Set}_\Delta)_{/\mathcal{B}}$. In view of (D), this is equivalent to assumption (2).

(F) If X consists of a single vertex x, then X is good. To see this, let $B \in \mathcal{B}$ denote the image of X. The natural map $X \to \mathcal{B}_{/B}$ can be identified with the inclusion of a final vertex; this map is right anodyne and therefore a contravariant equivalence. We now conclude by applying (A) and (E).

(G) If $X \in (\operatorname{Set}_\Delta)_{/\mathcal{B}}$ is an ∞-category with a final object x, then X is good. To prove this, we note that $\{x\}$ is good by (F) and the inclusion $\{x\} \subseteq X$ is right anodyne, hence a contravariant equivalence. We conclude by applying (A).

(H) If $X \in (\operatorname{Set}_\Delta)_{/\mathcal{B}}$ is κ-filtered, then X is good. To prove this, we apply Proposition 5.3.1.17 to deduce the existence of a categorical equivalence $i : X \to \mathcal{C}$, where $\mathcal{C}$ is a κ-filtered union of ∞-categories with final objects. Replacing $\mathcal{C}$ by $\mathcal{C} \times K$ if necessary, where K is a contractible Kan complex, we may suppose that i is a cofibration. Since $\mathcal{B}$ is an ∞-category, the lifting problem

has a solution. Thus we may regard $\mathcal{C}$ as an object of $(\operatorname{Set}_\Delta)_{/\mathcal{B}}$. According to (B), it suffices to show that $\mathcal{C}$ is good. But $\mathcal{C}$ is a κ-filtered colimit of good objects of $(\operatorname{Set}_\Delta)_{\mathcal{B}}$ (by (G)) and is therefore itself good (by (C)).

Now let $\mathcal{B}' \to \mathcal{B}$ be a right fibration, where $\mathcal{B}'$ is κ-filtered. By (H), $\mathcal{B}'$ is a good object of $(\operatorname{Set}_\Delta)_{/\mathcal{B}}$. Applying (D), we deduce that $\mathcal{A}' = \mathcal{B}' \times_\mathcal{B} \mathcal{A}$ is κ-filtered. This proves (1). □

Our next goal is to prove Proposition 5.3.2.9, which gives a very concrete characterization of right exactness under the assumption that there is a sufficient supply of colimits. We first need a few preliminary results.

Lemma 5.3.2.6. *Let $\mathcal{B}' \to \mathcal{B}$ be a Cartesian fibration. Suppose that $\mathcal{B}$ has an initial object B and that $\mathcal{B}'$ is filtered. Then the fiber $\mathcal{B}'_B = \mathcal{B}' \times_\mathcal{B} \{B\}$ is a contractible Kan complex.*

Proof. Since B is an initial object of $\mathcal{B}$, the inclusion $\{B\}^{op} \subseteq \mathcal{B}^{op}$ is cofinal. Proposition 4.1.2.15 implies that the inclusion $(\mathcal{B}'_B)^{op} \subseteq (\mathcal{B}')^{op}$ is also cofinal and therefore a weak homotopy equivalence. It now suffices to prove that $\mathcal{B}'$ is weakly contractible, which follows from Lemma 5.3.1.18. □

Lemma 5.3.2.7. *Let $f : \mathcal{A} \to \mathcal{B}$ be a right exact functor between ∞-categories and let $A \in \mathcal{A}$ be an initial object. Then $f(A)$ is an initial object of $\mathcal{B}$.*

Proof. Let B be an object of $\mathcal{B}$. Proposition 5.3.2.5 implies that $\mathcal{A}' = \mathcal{B}_{/B} \times_{\mathcal{B}} \mathcal{A}$ is filtered. We may identify $\mathrm{Map}_{\mathcal{B}}(f(A), B)$ with the fiber of the right fibration $\mathcal{A}' \to \mathcal{A}$ over the object A. We now apply Lemma 5.3.2.6 to deduce that $\mathrm{Map}_{\mathcal{B}}(f(A), B)$ is contractible. $\qquad\square$

Lemma 5.3.2.8. *Let κ be a regular cardinal, $f : \mathcal{A} \to \mathcal{B}$ a κ-right exact functor between ∞-categories, and $p : K \to \mathcal{A}$ a diagram indexed by a κ-small simplicial set K. The induced map $\mathcal{A}_{p/} \to \mathcal{B}_{fp/}$ is κ-right exact.*

Proof. According to Proposition 5.3.2.5, it suffices to prove that for each object $\overline{B} \in \mathcal{B}_{fop/}$, the ∞-category $\mathcal{A}' = \mathcal{A}_{p/} \times_{\mathcal{B}_{fp/}} (\mathcal{B}_{fp/})_{/\overline{B}}$ is κ-filtered. Let B denote the image of $\overline{B}$ in $\mathcal{B}$ and let $q : K' \to \mathcal{A}'$ be a diagram indexed by a κ-small simplicial set K'; we wish to show that q admits an extension to $K'^{\triangleright}$. We may regard p and q together as defining a diagram $K \star K' \to \mathcal{A} \times_{\mathcal{B}} \mathcal{B}_{/B}$. Since f is κ-filtered, we can extend this to a map $(K \star K')^{\triangleright} \to \mathcal{A} \times_{\mathcal{B}} \mathcal{B}_{/B}$, which can be identified with an extension $\overline{q} : K'^{\triangleright} \to \mathcal{A}'$ of q. $\qquad\square$

Proposition 5.3.2.9. *Let $f : \mathcal{A} \to \mathcal{B}$ be a functor between ∞-categories and let κ be a regular cardinal.*

(1) *If f is κ-right exact, then f preserves all κ-small colimits which exist in $\mathcal{A}$.*

(2) *Conversely, if $\mathcal{A}$ admits κ-small colimits and f preserves κ-small colimits, then f is right exact.*

Proof. Suppose first that f is κ-right exact. Let K be a κ-small simplicial set, and let $\overline{p} : K^{\triangleright} \to \mathcal{A}$ be a colimit of $p = \overline{p}|K$. We wish to show that $f \circ \overline{p}$ is a colimit diagram. Using Lemma 5.3.2.8, we may replace $\mathcal{A}$ by $\mathcal{A}_{p/}$ and $\mathcal{B}$ by $\mathcal{B}_{fp/}$ and thereby reduce to the case $K = \emptyset$. We are then reduced to proving that f preserves initial objects, which follows from Lemma 5.3.2.7.

Now suppose that $\mathcal{A}$ admits κ-small colimits and that f preserves κ-small colimits. We wish to prove that f is κ-right exact. Let B be an object of $\mathcal{B}$ and set $\mathcal{A}' = \mathcal{A} \times_{\mathcal{B}} \mathcal{B}_{/B}$. We wish to prove that $\mathcal{A}'$ is κ-filtered. Let $p' : K \to \mathcal{A}'$ be a diagram indexed by a κ-small simplicial set K; we wish to prove that p' extends to a map $\overline{p}' : K^{\triangleright} \to \mathcal{A}'$. Let $p : K \to \mathcal{A}$ be the composition of p' with the projection $\mathcal{A}' \to \mathcal{A}$ and let $\overline{p} : K^{\triangleright} \to \mathcal{A}$ be a colimit of p. We may identify $f \circ \overline{p}$ and p' with objects of $\mathcal{B}_{fp/}$. Since f preserves κ-small colimits, $f \circ \overline{p}$ is an initial object of $\mathcal{B}_{fp/}$, so that there exists a morphism $\alpha : f \circ \overline{p} \to p'$ in $\mathcal{B}_{fop/}$. The morphism α can be identified with the desired extension $\overline{p}' : K^{\triangleright} \to \mathcal{A}'$. $\qquad\square$

Remark 5.3.2.10. The results of this section all dualize in an evident way: a functor $G : \mathcal{A} \to \mathcal{B}$ is said to be κ-*left exact* if the induced functor $G^{op} : \mathcal{A}^{op} \to \mathcal{B}^{op}$ is κ-right exact. In the case where $\mathcal{A}$ admits κ-small limits, this is equivalent to the requirement that G preserve κ-small limits.

Remark 5.3.2.11. Let $\mathcal{C}$ be an ∞-category, let $F : \mathcal{C} \to \mathcal{S}^{op}$ be a functor, and let $\widetilde{\mathcal{C}} \to \mathcal{C}$ be the associated right fibration (the pullback of the universal right fibration $\mathcal{Q}^0 \to \mathcal{S}^{op}$). If F is κ-right exact, then $\widetilde{\mathcal{C}}$ is κ-filtered (since $\mathcal{Q}^0$ has a final object). If $\mathcal{C}$ admits κ-small colimits, then the converse holds: if $\widetilde{\mathcal{C}}$ is κ-filtered, then F preserves κ-small colimits by Proposition 5.3.5.3 and is therefore κ-right exact by Proposition 5.3.2.5. The converse does not hold in general: it is possible to give an example of right fibration $\widetilde{\mathcal{C}} \to \mathcal{C}$ such that $\widetilde{\mathcal{C}}$ is filtered yet the classifying functor $F : \mathcal{C} \to \mathcal{S}^{op}$ is not right exact.

5.3.3 Filtered Colimits

Filtered categories tend not to be very interesting in themselves. Instead, they are primarily useful for indexing diagrams in other categories. This is because the colimits of filtered diagrams enjoy certain exactness properties not shared by colimits in general. In this section, we will formulate and prove these exactness properties in the ∞-categorical setting. First, we need a few definitions.

Definition 5.3.3.1. Let κ be a regular cardinal. We will say that an ∞-category $\mathcal{C}$ is *κ-closed* if every diagram $p : K \to \mathcal{C}$ indexed by a κ-small simplicial set K admits a colimit $\overline{p} : K^{\triangleright} \to \mathcal{C}$.

In a κ-closed ∞-category, it is possible to construct κ-small colimits functorially. More precisely, suppose that $\mathcal{C}$ is an ∞-category and that K is a simplicial set with the property that every diagram $p : K \to \mathcal{C}$ has a colimit in $\mathcal{C}$. Let $\mathcal{D}$ denote the full subcategory of $\mathrm{Fun}(K^{\triangleright}, \mathcal{C})$ spanned by the colimit diagrams. Proposition 4.3.2.15 implies that the restriction functor $\mathcal{D} \to \mathrm{Fun}(K, \mathcal{C})$ is a trivial fibration. It therefore admits a section s (which is unique up to a contractible ambiguity). Let $e : \mathrm{Fun}(K^{\triangleright}, \mathcal{C}) \to \mathcal{C}$ be the functor given by evaluation at the cone point of $K^{\triangleright}$. We will refer to the composition

$$\mathrm{Fun}(K, \mathcal{C}) \xrightarrow{s} \mathcal{D} \subseteq \mathrm{Fun}(K^{\triangleright}, \mathcal{C}) \xrightarrow{e} \mathcal{C}$$

as a *colimit functor*; it associates to each diagram $p : K \to \mathcal{C}$ a colimit of p in $\mathcal{C}$. We will generally denote colimit functors by $\varinjlim_K : \mathrm{Fun}(K, \mathcal{C}) \to \mathcal{C}$.

Lemma 5.3.3.2. *Let $F \in \mathrm{Fun}(K, \mathcal{S})$ be a corepresentable functor (that is, F lies in the essential image of the Yoneda embedding $K^{op} \to \mathrm{Fun}(K, \mathcal{S})$) and let $X \in \mathcal{S}$ be a colimit of F. Then X is contractible.*

Proof. Without loss of generality, we may suppose that K is an ∞-category. Let $\widetilde{K} \to K$ be a left fibration classified by F. Since F is corepresentable, $\widetilde{K}$ has an initial object and is therefore weakly contractible. Corollary 3.3.4.6 implies that there is an isomorphism $\widetilde{K} \simeq X$ in the homotopy category $\mathcal{H}$, so that X is also contractible. $\square$

Proposition 5.3.3.3. *Let κ be a regular cardinal and let $\mathcal{I}$ be an ∞-category. The following conditions are equivalent:*

(1) *The ∞-category $\mathfrak{I}$ is κ-filtered.*

(2) *The colimit functor $\varinjlim_{\mathfrak{I}} : \mathrm{Fun}(\mathfrak{I}, \mathcal{S}) \to \mathcal{S}$ preserves κ-small limits.*

Proof. Suppose that (1) is satisfied. According to Proposition 5.3.1.16, there exists a κ-filtered partially ordered set A and a cofinal map $i : \mathrm{N}(A) \to \mathcal{S}$. Since i is cofinal, the colimit functor for $\mathfrak{I}$ admits a factorization

$$\mathrm{Fun}(\mathfrak{I}, \mathcal{S}) \xrightarrow{i^*} \mathrm{Fun}(\mathrm{N}(A), \mathcal{S}) \to \mathcal{S}.$$

Proposition 5.1.2.2 implies that i^* preserves limits. We may therefore replace $\mathfrak{I}$ by $\mathrm{N}(A)$ and thereby reduce to the case where $\mathfrak{I}$ is itself the nerve of a κ-filtered partially ordered set A.

We note that the functor $\varinjlim_{\mathfrak{I}} : \mathrm{Fun}(\mathfrak{I}, \mathcal{S}) \to \mathcal{S}$ can be characterized as the left adjoint to the diagonal functor $\delta : \mathcal{S} \to \mathrm{Fun}(\mathfrak{I}, \mathcal{S})$. Let $\mathbf{A}$ denote the category of all functors from A to Set_Δ; we regard $\mathbf{A}$ as a simplicial model category with respect to the *projective* model structure described in §A.3.3. Let $\phi^* : \mathrm{Set}_\Delta \to \mathbf{A}$ denote the diagonal functor which associates to each simplicial set K the constant functor $A \to \mathrm{Set}_\Delta$ with value K, and let $\phi_!$ be a left adjoint of ϕ^*, so that the pair $(\phi^*, \phi_!)$ gives a Quillen adjunction between $\mathbf{A}$ and Set_Δ. Proposition 4.2.4.4 implies that there is an equivalence of ∞-categories $\mathrm{N}(\mathbf{A}^\circ) \to \mathrm{Fun}(\mathfrak{I}, \mathcal{S})$, and δ may be identified with the right derived functor of ϕ^*. Consequently, the functor $\varinjlim_{\mathfrak{I}}$ may be identified with the left derived functor of $\phi_!$. To prove that $\varinjlim_{\mathfrak{I}}$ preserves κ-small limits, it suffices to prove that $\varinjlim_{\mathfrak{I}}$ preserves fiber products and κ-small products. According to Theorem 4.2.4.1, it suffices to prove that $\phi_!$ preserves homotopy fiber products and κ-small homotopy products. For fiber products, this reduces to the classical assertion that if we are given a family of homotopy Cartesian squares

$$
\begin{array}{ccc}
W_\alpha & \longrightarrow & X_\alpha \\
\downarrow & & \downarrow \\
Y_\alpha & \longrightarrow & Z_\alpha
\end{array}
$$

in the category of Kan complexes, indexed by a filtered partially ordered set A, then the colimit square

$$
\begin{array}{ccc}
W & \longrightarrow & X \\
\downarrow & & \downarrow \\
Y & \longrightarrow & Z
\end{array}
$$

is also homotopy Cartesian. The assertion regarding homotopy products is handled similarly.

Now suppose that (2) is satisfied. Let K be a κ-small simplicial set and $p : K \to \mathfrak{I}^{op}$ a diagram; we wish to prove that $\mathfrak{I}^{op}_{/p}$ is nonempty. Suppose otherwise. Let $j : \mathfrak{I}^{op} \to \mathrm{Fun}(\mathfrak{I}, \mathcal{S})$ be the Yoneda embedding, let $q = j \circ p$, let $\overline{q} : K^\triangleleft \to \mathrm{Fun}(\mathfrak{I}, \mathcal{S})$ be a limit of q, and let $X \in \mathrm{Fun}(\mathfrak{I}, \mathcal{S})$ be the image of the cone point of $K^{\triangleleft}$ under $\bar{q}$. Since j is fully faithful and $\mathfrak{I}^{op}_{/p}$ is empty, we have $\mathrm{Map}_{\mathcal{S}^{\mathfrak{I}}}(j(I), X) = \emptyset$ for each $I \in \mathfrak{I}$. Using Lemma 5.1.5.2, we may identify $\mathrm{Map}_{\mathcal{S}^{\mathfrak{I}}}(j(I), X)$ with $X(I)$ in the homotopy category $\mathcal{H}$ of spaces. We therefore conclude that X is an initial object of $\mathrm{Fun}(\mathfrak{I}, \mathcal{S})$. Since the functor $\varinjlim_{\mathfrak{I}} : \mathrm{Fun}(\mathfrak{I}, \mathcal{S}) \to \mathcal{S}$ is a left adjoint, it preserves initial objects. We conclude that $\varinjlim_{\mathfrak{I}} X$ is an initial object of $\mathcal{S}$. On the other hand, if $\varinjlim_{\mathfrak{I}}$ preserves κ-small limits, then $\varinjlim_{\mathfrak{I}} \circ \bar{q}$ exhibits $\varinjlim_{\mathfrak{I}} X$ as the limit of the diagram $\varinjlim_{\mathfrak{I}} \circ q : K \to \mathcal{S}$. For each vertex k in K, Lemmas 5.1.5.2 and 5.3.3.2 imply that $\varinjlim_{\mathfrak{I}} q(k)$ is contractible and therefore a final object of $\mathcal{S}$. It follows that $\varinjlim_{\mathfrak{I}} X$ is also a final object of $\mathcal{S}$. This is a contradiction since the initial object of $\mathcal{S}$ is not final. $\qquad\square$

5.3.4 Compact Objects

Let $\mathcal{C}$ be a category which admits filtered colimits. An object $C \in \mathcal{C}$ is said to be *compact* if the corepresentable functor

$$\mathrm{Hom}_{\mathcal{C}}(C, \bullet)$$

commutes with filtered colimits.

Example 5.3.4.1. Let $\mathcal{C} = \mathrm{Set}$ be the category of sets. An object $C \in \mathcal{C}$ is compact if and only if is finite.

Example 5.3.4.2. Let $\mathcal{C}$ be the category of groups. An object G of $\mathcal{C}$ is compact if and only if it is finitely presented (as a group).

Example 5.3.4.3. Let X be a topological space and let $\mathcal{C}$ be the category of open sets of X (with morphisms given by inclusions). Then an object $U \in \mathcal{C}$ is compact if and only if U is compact when viewed as a topological space: that is, every open cover of U admits a finite subcover.

Remark 5.3.4.4. Because of Example 5.3.4.2, many authors call an object C of a category $\mathcal{C}$ *finitely presented* if $\mathrm{Hom}_{\mathcal{C}}(C, \bullet)$ preserves filtered colimits. Our terminology is motivated instead by Example 5.3.4.3.

Definition 5.3.4.5. Let $\mathcal{C}$ be an ∞-category which admits small κ-filtered colimits. We will say a functor $f : \mathcal{C} \to \mathcal{D}$ is *κ-continuous* if it preserves κ-filtered colimits.

Let $\mathcal{C}$ be an ∞-category containing an object C and let $j_C : \mathcal{C} \to \widehat{\mathcal{S}}$ denote the functor corepresented by C. If $\mathcal{C}$ admits κ-filtered colimits, then we will say that C is *κ-compact* if j_C is κ-continuous. We will say that C is *compact* if it is ω-compact (and $\mathcal{C}$ admits filtered colimits).

Let κ be a regular cardinal and let $\mathcal{C}$ be an ∞-category which admits small κ-filtered colimits. We will say that a left fibration $\widetilde{\mathcal{C}} \to \mathcal{C}$ is *κ-compact* if it is classified by a κ-continuous functor $\mathcal{C} \to \widehat{\mathcal{S}}$.

Notation 5.3.4.6. Let $\mathcal{C}$ be an ∞-category and κ a regular cardinal. We will generally let $\mathcal{C}^{\kappa}$ denote the full subcategory spanned by the κ-compact objects of $\mathcal{C}$.

Lemma 5.3.4.7. *Let* $\mathcal{C}$ *be an* ∞-*category which admits small* κ-*filtered colimits and let* $\mathcal{D} \subseteq \operatorname{Fun}(\mathcal{C}, \widehat{\mathcal{S}})$ *be the full subcategory spanned by the* κ-*continuous functors* $f : \mathcal{C} \to \widehat{\mathcal{S}}$. *Then* $\mathcal{D}$ *is stable under* κ-*small limits in* $\widehat{\mathcal{S}}^{\mathcal{C}}$.

Proof. Let K be a κ-small simplicial set, and let $p : K \to \operatorname{Fun}(\mathcal{C}, \widehat{\mathcal{S}})$ be a diagram which we may identify with a map $p' : \mathcal{C} \to \operatorname{Fun}(K, \widehat{\mathcal{S}})$. Using Proposition 5.1.2.2, we may obtain a limit of the diagram p by composing p' with a limit functor

$$\varprojlim : \operatorname{Fun}(K, \widehat{\mathcal{S}}) \to \widehat{\mathcal{S}}$$

(that is, a right adjoint to the diagonal functor $\widehat{\mathcal{S}} \to \operatorname{Fun}(K, \widehat{\mathcal{S}})$; see §5.3.3). It therefore suffices to show that the functor $\varprojlim$ is κ-continuous. This is simply a reformulation of Proposition 5.3.3.3. $\qquad \square$

The basic properties of κ-compact left fibrations are summarized in the following Lemma:

Lemma 5.3.4.8. *Fix a regular cardinal* κ.

(1) *Let* $\mathcal{C}$ *be an* ∞-*category which admits small* κ-*filtered colimits and let* $C \in \mathcal{C}$ *be an object. Then* C *is* κ-*compact if and only if the left fibration* $\mathcal{C}_{C/} \to \mathcal{C}$ *is* κ-*compact.*

(2) *Let* $f : \mathcal{C} \to \mathcal{D}$ *be a* κ-*continuous functor between* ∞-*categories which admit small* κ-*filtered colimits and let* $\widetilde{\mathcal{D}} \to \mathcal{D}$ *be a* κ-*compact left fibration. Then the associated left fibration* $\mathcal{C} \times_{\mathcal{D}} \widetilde{\mathcal{D}} \to \mathcal{C}$ *is also* κ-*compact.*

(3) *Let* $\mathcal{C}$ *be an* ∞-*category which admits small* κ-*filtered colimits and let* $\mathbf{A} \subseteq (\operatorname{Set}_{\Delta})_{/\mathcal{C}}$ *denote the full subcategory spanned by the* κ-*compact left fibrations over* $\mathcal{C}$. *Then* $\mathbf{A}$ *is stable under* κ-*small homotopy limits (with respect to the covariant model structure on* $(\operatorname{Set}_{\Delta})_{/\mathcal{C}}$*). In particular,* $\mathbf{A}$ *is stable under the formation of homotopy pullbacks,* κ-*small products, and (if* κ *is uncountable) homotopy inverse limits of towers.*

Proof. Assertions (1) and (2) are obvious. To prove (3), let us suppose that $\widetilde{\mathcal{C}}$ is a κ-small homotopy limit of κ-compact left fibrations $\widetilde{\mathcal{C}}_{\alpha} \to \mathcal{C}$. Let $\mathcal{J}$ be a small κ-filtered ∞-category and let $\overline{p} : \mathcal{J}^{\triangleright} \to \mathcal{C}$ be a colimit diagram. We wish to prove that the composition of $\overline{p}$ with the functor $\mathcal{C} \to \widehat{\mathcal{S}}$ classifying $\widetilde{\mathcal{C}}$ is a colimit diagram. Applying Proposition 5.3.1.16, we may reduce to the case where $\mathcal{J}$ is the nerve of a κ-filtered partially ordered set A. According to Theorem 2.2.1.2, it will suffice to show that the collection of homotopy colimit diagrams

$$A \cup \{\infty\} \to \mathcal{K}\mathrm{an}$$

is stable under κ-small homotopy limits in the category $(\operatorname{Set}_{\Delta})^{A \cup \{\infty\}}$, which follows easily from our assumption that A is κ-filtered. $\qquad \square$

Our next goal is to prove a very useful stability result for κ-compact objects (Proposition 5.3.4.13). We first need to establish a few technical lemmas.

Lemma 5.3.4.9. *Let κ be a regular cardinal, let $\mathcal{C}$ be an ∞-category which admits small κ-filtered colimits, and let $f : C \to D$ be a morphism in $\mathcal{C}$. Suppose that C and D are κ-compact objects of $\mathcal{C}$. Then f is a κ-compact object of $\mathrm{Fun}(\Delta^1, \mathcal{C})$.*

Proof. Let $X = \mathrm{Fun}(\Delta^1, \mathcal{C}) \times_{\mathrm{Fun}(\{1\}, \mathcal{C})} \mathcal{C}_{f/}$, $Y = \mathrm{Fun}(\Delta^1, \mathcal{C}_{C/})$ and $Z = \mathrm{Fun}(\Delta^1, \mathcal{C}) \times_{\mathrm{Fun}(\{1\}, \mathcal{C})} \mathcal{C}_{C/}$, so that we have a (homotopy) pullback diagram

$$
\begin{array}{ccc}
\mathrm{Fun}(\Delta^1, \mathcal{C})_{f/} & \longrightarrow & X \\
\downarrow & & \downarrow \\
Y & \longrightarrow & Z
\end{array}
$$

of left fibrations over $\mathrm{Fun}(\Delta^1, \mathcal{C})$. According to Lemma 5.3.4.8, it will suffice to show that X, Y, and Z are κ-compact left fibrations. To show that X is a κ-compact left fibration, it suffices to show that $\mathcal{C}_{f/} \to \mathcal{C}$ is a κ-compact left fibration, which follows since we have a trivial fibration $\mathcal{C}_{f/} \to \mathcal{C}_{D/}$, where D is κ-compact by assumption. Similarly, we have a trivial fibration $Y \to \mathrm{Fun}(\Delta^1, \mathcal{C}) \times_{\mathcal{C}^{(0)}} \mathcal{C}_{C/}$, so that the κ-compactness of C implies that Y is a κ-compact left fibration. Lemma 5.3.4.8 and the compactness of C immediately imply that Z is a κ-compact left fibration, which completes the proof. □

Lemma 5.3.4.10. *Let κ be a regular cardinal and let $\{\mathcal{C}_\alpha\}$ be a κ-small family of ∞-categories having product $\mathcal{C}$. Suppose that each $\mathcal{C}$ admits small κ-filtered colimits. Then*

(1) *The ∞-category $\mathcal{C}$ admits κ-filtered colimits.*

(2) *If $C \in \mathcal{C}$ is an object whose image in each $\mathcal{C}_\alpha$ is κ-compact, then C is κ-compact as an object of $\mathcal{C}$.*

Proof. The first assertion is obvious since colimits in a product can be computed pointwise. For the second, choose an object $C \in \mathcal{C}$ whose images $\{C_\alpha \in \mathcal{C}_\alpha\}$ are κ-compact.

The left fibration $\mathcal{C}_{C/} \to \mathcal{C}$ can be obtained as a κ-small product of the left fibrations $\mathcal{C} \times_{\mathcal{C}_\alpha} (\mathcal{C}_\alpha)_{C_\alpha/} \to \mathcal{C}$. Lemma 5.3.4.8 implies that each factor is κ-compact, so that the product is also κ-compact. □

Lemma 5.3.4.11. *Let S be a simplicial set and suppose we are given a tower*

$$\cdots \to X(1) \xrightarrow{f_1} X(0) \xrightarrow{f_0} S,$$

where each f_i is a left fibration. Then the inverse limit $X(\infty)$ is a homotopy inverse limit of the tower $\{X(i)\}$ with respect to the covariant model structure on $(\mathrm{Set}_\Delta)_{/S}$.

Proof. Construct a ladder

$$
\begin{array}{ccccccc}
\cdots \longrightarrow & X(1) & \xrightarrow{\ f_1\ } & X(0) & \xrightarrow{\ f_0\ } & S \\
& \downarrow & & \downarrow & & \downarrow \\
\cdots \longrightarrow & X'(1) & \xrightarrow{\ f_1'\ } & X'(0) & \xrightarrow{\ f_0'\ } & S
\end{array}
$$

where the vertical maps are covariant equivalences and the tower $\{X'(i)\}$ is fibrant (in the sense that each of the maps f_i' is a covariant fibration). We wish to show that the induced map on inverse limits $X(\infty) \to X'(\infty)$ is a covariant equivalence. Since both $X(\infty)$ and $X'(\infty)$ are left fibered over S, this can be tested by passing to the fibers over each vertex s of S. We may therefore reduce to the case where S is a point, in which case the tower $\{X(i)\}$ is already fibrant (since a left fibration over a Kan complex is a Kan fibration; see Lemma 2.1.3.3). $\qquad\square$

Lemma 5.3.4.12. *Let κ be an uncountable regular cardinal and let*

$$
\cdots \to \mathcal{C}^2 \xrightarrow{\ f_2\ } \mathcal{C}^1 \xrightarrow{\ f_1\ } \mathcal{C}^0
$$

be a tower of ∞-categories. Suppose that each $\mathcal{C}^i$ admits small κ-filtered colimits and that each of the functors f_i is a categorical fibration which preserves κ-filtered colimits. Let $\mathcal{C}$ denote the inverse limit of the tower. Then

(1) *The ∞-category $\mathcal{C}$ admits small κ-filtered colimits, and the projections $p_n : \mathcal{C} \to \mathcal{C}^n$ are κ-continuous.*

(2) *If $C \in \mathcal{C}$ has a κ-compact image in $\mathcal{C}^i$ for each $i \geq 0$, then C is a κ-compact object of $\mathcal{C}$.*

Proof. Let $\overline{q} : K^{\triangleright} \to \mathcal{C}$ be a diagram indexed by an arbitrary simplicial set, let $q = \overline{q}|K$, and set $\overline{q}_n = p_n \circ \overline{q}$, $q_n = p_n \circ q$. Suppose that each $\overline{q}_n$ is a colimit diagram in $\mathcal{C}^n$. Then the map $\mathcal{C}_{\overline{q}/} \to \mathcal{C}_{q/}$ is the inverse limit of a tower of trivial fibrations $\mathcal{C}^n_{\overline{q}_n/} \to \mathcal{C}^n_{q_n/}$ and therefore a trivial fibration.

To complete the proof of (1), it will suffice to show that if K is a κ-filtered ∞-category, then any diagram $q : K \to \mathcal{C}$ can be extended to a map $\overline{q} : K^{\triangleright} \to \mathcal{C}$ with the property described above. To construct $\overline{q}$, it suffices to construct a compatible family $\overline{q}_n : K^{\triangleright} \to \mathcal{C}^n$. We begin by selecting arbitrary colimit diagrams $\overline{q}_n' : K^{\triangleright} \to \mathcal{C}^n$ which extend q_n. We now explain how to adjust these choices to make them compatible with one another using induction on n. Set $\overline{q}_0 = \overline{q}_0'$. Suppose next that $n > 0$. Since f_n preserves κ-filtered colimits, we may identify $\overline{q}_{n-1}$ and $f_n \circ \overline{q}_n'$ with initial objects of $\mathcal{C}^{n-1}_{q_{n-1}/}$. It follows that there exists an equivalence $e : \overline{q}_{n-1} \to f_n \circ \overline{q}_n'$ in $\mathcal{C}^{n-1}_{q_{n-1}/}$. The map f_n induces a categorical fibration $\mathcal{C}^n_{q_n/} \to \mathcal{C}^{n-1}_{q_{n-1}/}$, so that e lifts to an equivalence $\overline{e} : \overline{q}_n \to \overline{q}_n'$ in $\mathcal{C}^n_{q_n/}$. The existence of the equivalence $\overline{e}$ proves that $\overline{q}_n$ is a colimit diagram in $\mathcal{C}^n$, and we have $\overline{q}_{n-1} = f_n \circ \overline{q}_n$ by construction. This proves (1).

Now suppose that $C \in \mathcal{C}$ is as in (2) and let $C^n = p_n(C) \in \mathcal{C}^n$. The left fibration $\mathcal{C}_{/C}$ is the inverse limit of a tower of left fibrations

$$\cdots \to \mathcal{C}^1_{C^1/} \times_{\mathcal{C}^1} \mathcal{C} \to \mathcal{C}^0_{C^0/} \times_{\mathcal{C}^0} \mathcal{C}.$$

Using Lemma 5.3.4.8, we deduce that each term in this tower is a κ-compact left fibration over $\mathcal{C}$. Proposition 2.1.2.1 implies that each map in the tower is a left fibration, so that $\mathcal{C}_{C/}$ is a homotopy inverse limit of a tower of κ-compact left fibrations by Lemma 5.3.4.11. We now apply Lemma 5.3.4.8 again to deduce that $\mathcal{C}_{C/}$ is a κ-compact left fibration, so that $C \in \mathcal{C}$ is κ-compact, as desired. $\square$

Proposition 5.3.4.13. *Let κ be a regular cardinal, let $\mathcal{C}$ be an ∞-category which admits small κ-filtered colimits, and let $f : K \to \mathcal{C}$ be a diagram indexed by a κ-small simplicial set K. Suppose that for each vertex x of K, $f(x) \in \mathcal{C}$ is κ-compact. Then f is a κ-compact object of $\mathrm{Fun}(K, \mathcal{C})$.*

Proof. Let us say that a simplicial set K is *good* if it satisfies the conclusions of the lemma. We wish to prove that all κ-small simplicial sets are good. The proof proceeds in several steps:

(1) Suppose we are given a pushout square

$$\begin{array}{ccc} K' & \longrightarrow & K \\ \downarrow{\scriptstyle i} & & \downarrow \\ L' & \longrightarrow & L, \end{array}$$

where i is a cofibration and the simplicial sets K', K, and L' are good. Then the simplicial set L is also good. To prove this, we observe that the associated diagram of ∞-categories

$$\begin{array}{ccc} \mathrm{Fun}(L, \mathcal{C}) & \longrightarrow & \mathrm{Fun}(L', \mathcal{C}) \\ \downarrow & & \downarrow \\ \mathrm{Fun}(K, \mathcal{C}) & \longrightarrow & \mathrm{Fun}(K', \mathcal{C}) \end{array}$$

is homotopy Cartesian and every arrow in the diagram preserves κ-filtered colimits (by Proposition 5.1.2.2). Now apply Lemma 5.4.5.7.

(2) If $K \to K'$ is a categorical equivalence and K is good, then K' is good: the forgetful functor $\mathrm{Fun}(K', \mathcal{C}) \to \mathrm{Fun}(K, \mathcal{C})$ is an equivalence of ∞-categories and therefore detects κ-compact objects.

(3) Every simplex Δ^n is good. To prove this, we observe that the inclusion

$$\Delta^{\{0,1\}} \coprod_{\{1\}} \cdots \coprod_{\{n-1\}} \Delta^{\{n-1,n\}} \subseteq \Delta^n$$

is a categorical equivalence. Applying (1) and (2), we can reduce to the case $n \leq 1$. If $n = 0$, there is nothing to prove, and if $n = 1$, we apply Lemma 5.3.4.9.

(4) If $\{K_\alpha\}$ is a κ-small collection of good simplicial sets having coproduct K, then K is also good. To prove this, we observe that $\mathrm{Fun}(\mathcal{C}) \simeq \prod_\alpha \mathrm{Fun}(K_\alpha, \mathcal{C})$ and apply Lemma 5.3.4.10.

(5) If K is a κ-small simplicial set of dimension at most n, then K is good. The proof is by induction on n. Let $K^{(n-1)} \subseteq K$ denote the $(n-1)$-skeleton of K, so that we have a pushout diagram

$$
\begin{array}{ccc}
\coprod_{\sigma \in K_n} \partial \Delta^n & \longrightarrow & K^{(n-1)} \\
\downarrow & & \downarrow \\
\coprod_{\sigma \in K_n} \Delta^n & \longrightarrow & K.
\end{array}
$$

The inductive hypothesis implies that $\coprod_{\sigma \in K_n} \partial \Delta^n$ and $K^{(n-1)}$ are good. Applying (3) and (4), we deduce that $\coprod_{\sigma \in K_n} \Delta^n$ is good. We now apply (1) to deduce that K is good.

(6) Every κ-small simplicial set K is good. If $\kappa = \omega$, then this follows immediately from (5) since every κ-small simplicial set is finite-dimensional. If κ is uncountable, then we have an increasing filtration

$$
K^{(0)} \subseteq K^{(1)} \subseteq \cdots
$$

which gives rise to a tower of ∞-categories

$$
\cdots \to \mathrm{Fun}(K^{(1)}, \mathcal{C}) \to \mathrm{Fun}(K^{(0)}, \mathcal{C})
$$

having (homotopy) inverse limit $\mathrm{Fun}(K, \mathcal{C})$. Using Proposition 5.1.2.2, we deduce that the hypotheses of Lemma 5.3.4.12 are satisfied, so that K is good.

$\square$

Corollary 5.3.4.14. *Let κ be a regular cardinal and let $\mathcal{C}$ be an ∞-category which admits small κ-filtered colimits. Suppose that $p : K \to \mathcal{C}$ is a κ-small diagram with the property that for every vertex x of K, $p(x)$ is a κ-compact object of $\mathcal{C}$. Then the left fibration $\mathcal{C}_{p/} \to \mathcal{C}$ is κ-compact.*

Proof. It will suffice to show that the equivalent left fibration $\mathcal{C}^{p/} \to \mathcal{C}$ is κ-compact. Let P be the object of $\mathrm{Fun}(K, \mathcal{C})$ corresponding to p. Then we have an isomorphism of simplicial sets

$$
\mathcal{C}^{p/} \simeq \mathcal{C} \times_{\mathrm{Fun}(K,\mathcal{C})} \mathrm{Fun}(K, \mathcal{C})^{P/}.
$$

Proposition 5.3.4.13 asserts that P is a κ-compact object of $\mathrm{Fun}(K, \mathcal{C})$, so that the left fibration

$$
\mathrm{Fun}(K, \mathcal{C})^{P/} \to \mathrm{Fun}(K, \mathcal{C})
$$

is κ-compact. Proposition 5.1.2.2 guarantees that the diagonal map $\mathcal{C} \to \mathrm{Fun}(K, \mathcal{C})$ preserves κ-filtered colimits, so we can apply part (2) of Lemma 5.3.4.8 to deduce that $\mathcal{C}^{p/} \to \mathcal{C}$ is κ-compact as well. $\square$

Corollary 5.3.4.15. *Let $\mathcal{C}$ be an ∞-category which admits small κ-filtered colimits and let $\mathcal{C}^{\kappa}$ denote the full subcategory of $\mathcal{C}$ spanned by the κ-compact objects. Then $\mathcal{C}^{\kappa}$ is stable under the formation of all κ-small colimits which exist in $\mathcal{C}$.*

Proof. Let K be a κ-small simplicial set and let $\overline{p} : K^{\triangleright} \to \mathcal{C}$ be a colimit diagram. Suppose that, for each vertex x of K, the object $\overline{p}(x) \in \mathcal{C}$ is κ-compact. We wish to show that $C = \overline{p}(\infty) \in \mathcal{C}$ is κ-compact, where ∞ denotes the cone point of $K^{\triangleright}$. Let $p = \overline{p}|K$ and consider the maps

$$\mathcal{C}_{p/} \leftarrow \mathcal{C}_{\overline{p}/} \to \mathcal{C}_{C/} \, .$$

Both are trivial fibrations (the first because $\overline{p}$ is a colimit diagram and the second because the inclusion $\{\infty\} \subseteq K^{\triangleright}$ is right anodyne). Corollary 5.3.4.14 asserts that the left fibration $\mathcal{C}_{p/} \to \mathcal{C}$ is κ-compact. It follows that the equivalent left fibration $\mathcal{C}_{C/}$ is κ-compact, so that C is a κ-compact object of $\mathcal{C}$, as desired. $\qquad\square$

Remark 5.3.4.16. Let κ be a regular cardinal and let $\mathcal{C}$ be an ∞-category which admits κ-filtered colimits. Then the full subcategory $\mathcal{C}^{\kappa} \subseteq \mathcal{C}$ of κ-compact objects is stable under retracts. If $\kappa > \omega$, this follows from Proposition 4.4.5.15 and Corollary 5.3.4.15 (since every retract can be obtained as a κ-small colimit). We give an alternative argument that also works in the most important case $\kappa = \omega$. Let C be κ-compact and let D be a retract of C. Let $j : \mathcal{C}^{op} \to \mathrm{Fun}(\mathcal{C}, \widehat{\mathcal{S}})$ be the Yoneda embedding. Then $j(D) \in \mathrm{Fun}(\mathcal{C}, \widehat{\mathcal{S}})$ is a retract of $j(C)$. Since $j(C)$ preserves κ-filtered colimits, then Lemma 5.1.6.3 implies that $j(D)$ preserves κ-filtered colimits, so that D is κ-compact.

The following result gives a convenient description of the compact objects of an ∞-category of presheaves:

Proposition 5.3.4.17. *Let $\mathcal{C}$ be a small ∞-category, κ a regular cardinal, and $C \in \mathcal{P}(\mathcal{C})$ an object. The following are equivalent:*

(1) *There exists a diagram $p : K \to \mathcal{C}$ indexed by a κ-small simplicial set, such that $j \circ p$ has a colimit D in $\mathcal{P}(\mathcal{C})$ and C is a retract of D.*

(2) *The object C is κ-compact.*

Proof. Proposition 5.1.6.8 asserts that for every object $A \in \mathcal{C}$, $j(A)$ is completely compact and, in particular, κ-compact. According to Corollary 5.3.4.15 and Remark 5.3.4.16, the collection of κ-compact objects of $\mathcal{P}(\mathcal{C})$ is stable under κ-small colimits and retracts. Consequently, $(1) \Rightarrow (2)$.

Now suppose that (2) is satisfied. Let $\mathcal{C}_{/C} = \mathcal{C} \times_{\mathcal{P}(\mathcal{C})} \mathcal{P}(\mathcal{C})_{/C}$. Lemma 5.1.5.3 implies that the composition

$$\overline{p} : \mathcal{C}_{/C}^{\triangleright} \to \mathcal{P}(S)_{/C}^{\triangleright} \to \mathcal{P}(S)$$

is a colimit diagram. As in the proof of Corollary 4.2.3.11, we can write C as the colimit of a κ-filtered diagram $q : \mathcal{J} \to \mathcal{P}(\mathcal{C})$, where each object $q(I)$ is the colimit of $\overline{p}|\mathcal{C}^0$, where $\mathcal{C}^0$ is a κ-small simplicial subset of $\mathcal{C}_{/C}$. Since C is κ-compact, we may argue as in the proof of Proposition 5.1.6.8 to deduce that C is a retract of $q(I)$ for some object $I \in \mathcal{J}$. This proves (1). $\qquad\square$

We close with a result which we will need in §5.5. First, a bit of notation: if $\mathcal{C}$ is a small ∞-category and κ a regular cardinal, we let $\mathcal{P}^\kappa(\mathcal{C})$ denote the full subcategory consisting of κ-compact objects of $\mathcal{P}(\mathcal{C})$.

Proposition 5.3.4.18. *Let $\mathcal{C}$ be a small idempotent complete ∞-category and κ a regular cardinal. The following conditions are equivalent:*

(1) *The ∞-category $\mathcal{C}$ admits κ-small colimits.*

(2) *The Yoneda embedding $j : \mathcal{C} \to \mathcal{P}^\kappa(\mathcal{C})$ has a left adjoint.*

Proof. Suppose that (1) is satisfied. For each object $M \in \mathcal{P}(\mathcal{C})$, let $F_M : \mathcal{P}(\mathcal{C}) \to \widehat{\mathcal{S}}$ denote the associated corepresentable functor. Let $\mathcal{D} \subseteq \mathcal{P}(\mathcal{C})$ denote the full subcategory of $\mathcal{P}(\mathcal{C})$ spanned by those objects M such that $F_M \circ j : \mathcal{C} \to \widehat{\mathcal{S}}$ is corepresentable. According to Proposition 5.1.2.2, composition with j induces a *limit-preserving* functor

$$\operatorname{Fun}(\mathcal{P}(\mathcal{C}), \widehat{\mathcal{S}}) \to \operatorname{Fun}(\mathcal{C}, \widehat{\mathcal{S}}).$$

Applying Proposition 5.1.3.2 to $\mathcal{C}^{op}$, we conclude that the collection of corepresentable functors on $\mathcal{C}$ is stable under retracts and κ-small limits. A second application of Proposition 5.1.3.2 (this time to $\mathcal{P}(\mathcal{C})^{op}$) now shows that $\mathcal{D}$ is stable under retracts and κ-small colimits in $\mathcal{P}(\mathcal{C})$. Since j is fully faithful, $\mathcal{D}$ contains the essential image of j. It follows from Proposition 5.3.4.17 that $\mathcal{D}$ contains $\mathcal{P}^\kappa(\mathcal{C})$. We now apply Proposition 5.2.4.2 to deduce that $j : \mathcal{C} \to \mathcal{P}^\kappa(\mathcal{C})$ admits a left adjoint.

Conversely, suppose that (2) is satisfied. Let L denote a left adjoint to the Yoneda embedding, let $p : K \to \mathcal{C}$ be a κ-small diagram, and let $q = j \circ p$. Using Corollary 5.3.4.15, we deduce that q has a colimit $\overline{q} : K^\triangleright \to \mathcal{P}^\kappa(\mathcal{C})$. Since L is a left adjoint, $L \circ \overline{q}$ is a colimit of $L \circ q$. Since j is fully faithful, the diagram p is equivalent to $L \circ q$, so that p has a colimit as well. $\qquad\square$

5.3.5 Ind-Objects

Let S be a simplicial set. In §5.1.5, we proved that the ∞-category $\mathcal{P}(S)$ is *freely* generated under small colimits by the image of the Yoneda embedding $j : S \to \mathcal{P}(S)$ (Theorem 5.1.5.6). Our goal in this section is to study the analogous construction where we allow only *filtered* colimits.

Definition 5.3.5.1. Let $\mathcal{C}$ be a small ∞-category and let κ be a regular cardinal. We let $\operatorname{Ind}_\kappa(\mathcal{C})$ denote the full subcategory of $\mathcal{P}(\mathcal{C})$ spanned by those functors $f : \mathcal{C}^{op} \to \mathcal{S}$ which classify right fibrations $\widetilde{\mathcal{C}} \to \mathcal{C}$, where the ∞-category $\widetilde{\mathcal{C}}$ is κ-filtered. In the case where $\kappa = \omega$, we will simply write $\operatorname{Ind}(\mathcal{C})$ for $\operatorname{Ind}_\kappa(\mathcal{C})$. We will refer to $\operatorname{Ind}(\mathcal{C})$ as the ∞-category of Ind-*objects* of $\mathcal{C}$.

Remark 5.3.5.2. Let $\mathcal{C}$ be a small ∞-category and κ a regular cardinal. Then the Yoneda embedding $j : \mathcal{C} \to \mathcal{P}(\mathcal{C})$ factors through $\operatorname{Ind}_\kappa(\mathcal{C})$. This follows immediately from Lemma 5.1.5.2 since $j(C)$ classifies the right fibration $\mathcal{C}_{/C} \to \mathcal{C}$. The ∞-category $\mathcal{C}_{/C}$ has a final object and is therefore κ-filtered (Proposition 5.3.1.15).

Proposition 5.3.5.3. *Let $\mathcal{C}$ be a small ∞-category and let κ be a regular cardinal. The full subcategory $\mathrm{Ind}_\kappa(\mathcal{C}) \subseteq \mathcal{P}(\mathcal{C})$ is stable under κ-filtered colimits.*

Proof. Let $\mathcal{P}'_\Delta(\mathcal{C})$ denote the full subcategory of $(\mathrm{Set}_\Delta)_{/\mathcal{C}}$ spanned by the right fibrations $\widetilde{\mathcal{C}} \to \mathcal{C}$. According to Proposition 5.1.1.1, the ∞-category $\mathcal{P}(\mathcal{C})$ is equivalent to the simplicial nerve $\mathrm{N}(\mathcal{P}'_\Delta(\mathcal{C}))$. Let $\mathrm{Ind}'_\kappa(\mathcal{C})$ denote the full subcategory of $\mathcal{P}'_\Delta(\mathcal{C})$ spanned by right fibrations $\widetilde{\mathcal{C}} \to \mathcal{C}$, where $\widetilde{\mathcal{C}}$ is κ-filtered. It will suffice to prove that for any diagram $p : \mathcal{J} \to \mathrm{N}(\mathrm{Ind}'_\Delta(\mathcal{C}))$ indexed by a small κ-filtered ∞-category $\mathcal{J}$, the colimit of p in $\mathrm{N}(\mathcal{P}'_\Delta(\mathcal{C}))$ also belongs to $\mathrm{Ind}'_\kappa(\mathcal{C})$. Using Proposition 5.3.1.16, we may reduce to the case where $\mathcal{J}$ is the nerve of a κ-filtered partially ordered set A. Using Proposition 4.2.4.4, we may further reduce to the case where p is the simplicial nerve of a diagram taking values in the ordinary category $\mathrm{Ind}'_\kappa(\mathcal{C})$. By virtue of Theorem 4.2.4.1, it will suffice to prove that $\mathrm{Ind}'_\kappa(\mathcal{C}) \subseteq \mathcal{P}'_\Delta(\mathcal{C})$ is stable under κ-filtered homotopy colimits. We may identify $\mathcal{P}'_\Delta$ with the collection of fibrant objects of $(\mathrm{Set}_\Delta)_{/\mathcal{C}}$ with respect to the contravariant model structure. Since the class of contravariant equivalences is stable under filtered colimits, any κ-filtered colimit in $(\mathrm{Set}_\Delta)_{/\mathcal{C}}$ is also a homotopy colimit. Consequently, it will suffice to prove that $\mathrm{Ind}'_\kappa(\mathcal{C}) \subseteq \mathcal{P}'_\Delta(\mathcal{C})$ is stable under κ-filtered colimits. This follows immediately from the definition of a κ-filtered ∞-category. $\square$

Corollary 5.3.5.4. *Let $\mathcal{C}$ be a small ∞-category, let κ be a regular cardinal, and let $F : \mathcal{C}^{op} \to \mathcal{S}$ be an object of $\mathcal{P}(\mathcal{C})$. The following conditions are equivalent:*

(1) *There exists a (small) κ-filtered ∞-category $\mathcal{J}$ and a diagram $p : \mathcal{J} \to \mathcal{C}$ such that F is a colimit of the composition $j \circ p : \mathcal{J} \to \mathcal{P}(\mathcal{C})$.*

(2) *The functor F belongs to $\mathrm{Ind}_\kappa(\mathcal{C})$.*

If $\mathcal{C}$ admits κ-small colimits, then (1) and (2) are equivalent to

(3) *The functor F preserves κ-small limits.*

Proof. Lemma 5.1.5.3 implies that F is a colimit of the diagram

$$\mathcal{C}_{/F} \to \mathcal{C} \xrightarrow{j} \mathcal{P}(\mathcal{C}),$$

and Lemma 5.1.5.2 allows us to identify $\mathcal{C}_{/F} = \mathcal{C} \times_{\mathcal{P}(\mathcal{C})} \mathcal{P}(\mathcal{C})_{/F}$ with the right fibration associated to F. Thus $(2) \Rightarrow (1)$. The converse follows from Proposition 5.3.5.3 since every representable functor belongs to $\mathrm{Ind}_\kappa(\mathcal{C})$ (Remark 5.3.5.2).

Now suppose that $\mathcal{C}$ admits κ-small colimits. If (3) is satisfied, then $F^{op} : \mathcal{C} \to \mathcal{S}^{op}$ is κ-right exact by Proposition 5.3.3.3. The right fibration associated to F is the pullback of the universal right fibration by F^{op}. Using Corollary 3.3.2.7, the universal right fibration over $\mathcal{S}^{op}$ is representable by the final object of $\mathcal{S}$. Since F is κ-right exact, the fiber product $(\mathcal{S}^{op})_{/*} \times_{\mathcal{S}^{op}} \mathcal{C}$ is κ-filtered. Thus $(3) \Rightarrow (2)$.

We now complete the proof by showing that $(1) \Rightarrow (3)$. First suppose that F lies in the essential image of the Yoneda embedding $j : \mathcal{C} \to \mathcal{P}(\mathcal{C})$. According to Lemma 5.1.5.2, $j(C)$ is equivalent to the composition of the opposite Yoneda embedding $j' : \mathcal{C}^{op} \to \mathrm{Fun}(\mathcal{C}, \mathcal{S})$ with the evaluation functor $e : \mathrm{Fun}(\mathcal{C}, \mathcal{S}) \to \mathcal{S}$ associated to the object $C \in \mathcal{C}$. Propositions 5.1.3.2 and 5.1.2.2 imply that j' and e preserve κ-small limits, so that $j(C)$ preserves κ-small limits. To conclude the proof, it will suffice to show that the collection of functors $F : \mathcal{C}^{op} \to \mathcal{S}$ which satisfy (3) is stable under κ-filtered colimits: this follows easily from Proposition 5.3.3.3. $\qquad\square$

Proposition 5.3.5.5. *Let $\mathcal{C}$ be a small ∞-category, let κ be a regular cardinal, and let $j : \mathcal{C} \to \mathrm{Ind}_\kappa(\mathcal{C})$ be the Yoneda embedding. For each object $C \in \mathcal{C}$, $j(C)$ is a κ-compact object of $\mathrm{Ind}_\kappa(\mathcal{C})$.*

Proof. The functor $\mathrm{Ind}_\kappa(\mathcal{C}) \to \mathcal{S}$ corepresented by $j(C)$ is equivalent to the composition

$$\mathrm{Ind}_\kappa(\mathcal{C}) \subseteq \mathcal{P}(\mathcal{C}) \to \mathcal{S},$$

where the first map is the canonical inclusion and the second is given by evaluation at C. The second map preserves all colimits (Proposition 5.1.2.2), and the first preserves κ-filtered colimits since $\mathrm{Ind}_\kappa(\mathcal{C})$ is stable under κ-filtered colimits in $\mathcal{P}(\mathcal{C})$ (Proposition 5.3.5.3). $\qquad\square$

Remark 5.3.5.6. Let $\mathcal{C}$ be a small ∞-category and κ a regular cardinal. Suppose that $\mathcal{C}$ is equivalent to an n-category, so that the Yoneda embedding $j : \mathcal{C} \to \mathcal{P}(\mathcal{C})$ factors through $\mathcal{P}_{\leq n-1}(\mathcal{C}) = \mathrm{Fun}(\mathcal{C}^{op}, \tau_{\leq n-1}\mathcal{S})$, where $\tau_{\leq n-1}\mathcal{S}$ denotes the full subcategory of $\mathcal{S}$ spanned by the $(n-1)$-truncated spaces: that is, spaces whose homotopy groups vanish in dimensions n and above. The class of $(n-1)$-truncated spaces is stable under filtered colimits, so that $\mathcal{P}_{\leq n-1}(\mathcal{C})$ is stable under filtered colimits in $\mathcal{P}(\mathcal{C})$. Corollary 5.3.5.4 implies that $\mathrm{Ind}(\mathcal{C}) \subseteq \mathcal{P}_{\leq n-1}(\mathcal{C})$. In particular, $\mathrm{Ind}(\mathcal{C})$ is itself equivalent to an n-category. In particular, if $\mathcal{C}$ is the nerve of an ordinary category $\mathcal{I}$, then $\mathrm{Ind}(\mathcal{C})$ is equivalent to the nerve of an ordinary category $\mathcal{J}$, which is uniquely determined up to equivalence. Moreover, $\mathcal{J}$ admits filtered colimits, and there is a fully faithful embedding $\mathcal{I} \to \mathcal{J}$ which generates $\mathcal{J}$ under filtered colimits and whose essential image consists of compact objects of $\mathcal{J}$. It follows that $\mathcal{J}$ is equivalent to the category of Ind-objects of $\mathcal{I}$ in the sense of ordinary category theory.

According to Corollary 5.3.5.4, we may characterize $\mathrm{Ind}_\kappa(\mathcal{C})$ as the smallest full subcategory of $\mathcal{P}(\mathcal{C})$ which contains the image of the Yoneda embedding $j : \mathcal{C} \to \mathcal{P}(\mathcal{C})$ and is stable under κ-filtered colimits. Our goal is to obtain a more precise characterization of $\mathrm{Ind}_\kappa(\mathcal{C})$: namely, we will show that it is *freely* generated by $\mathcal{C}$ under κ-filtered colimits.

Lemma 5.3.5.7. *Let $\mathcal{D}$ be an ∞-category (not necessarily small). There exists a fully faithful functor $i : \mathcal{D} \to \mathcal{D}'$ with the following properties:*

(1) The ∞-category $\mathcal{D}'$ admits small colimits.

(2) A small diagram $K^\triangleright \to \mathcal{D}$ is a colimit if and only if the composite map $K^\triangleright \to \mathcal{D}'$ is a colimit.

Proof. Let $\mathcal{D}' = \operatorname{Fun}(\mathcal{D}, \widehat{\mathcal{S}})^{op}$ and let i be the opposite of the Yoneda embedding. Then (1) follows from Proposition 5.1.2.2 and (2) from Proposition 5.1.3.2. □

We will need the following analogue of Lemma 5.1.5.5:

Lemma 5.3.5.8. *Let $\mathcal{C}$ be a small ∞-category, κ a regular cardinal, $j : \mathcal{C} \to \operatorname{Ind}_\kappa(\mathcal{C})$ the Yoneda embedding, and $\mathcal{C}' \subseteq \mathcal{C}$ the essential image of j. Let $\mathcal{D}$ be an ∞-category which admits small κ-filtered colimits. Then*

(1) *Every functor $f_0 : \mathcal{C}' \to \mathcal{D}$ admits a left Kan extension $f : \operatorname{Ind}_\kappa(\mathcal{C}) \to \mathcal{D}$.*

(2) *An arbitrary functor $f : \operatorname{Ind}_\kappa(\mathcal{C}) \to \mathcal{D}$ is a left Kan extension of $f | \mathcal{C}'$ if and only if f is κ-continuous.*

Proof. Fix an arbitrary functor $f_0 : \mathcal{C}' \to \mathcal{D}$. Without loss of generality, we may assume that $\mathcal{D}$ is a full subcategory of a larger ∞-category $\mathcal{D}'$, satisfying the conclusions of Lemma 5.3.5.7; in particular, $\mathcal{D}$ is stable under small κ-filtered colimits in $\mathcal{D}'$. We may further assume that $\mathcal{D}$ coincides with its essential image in $\mathcal{D}'$. Lemma 5.1.5.5 guarantees the existence of a functor $F : \mathcal{P}(\mathcal{C}) \to \mathcal{D}'$ which is a left Kan extension of $f_0 = F | \mathcal{C}'$ and such that F preserves small colimits. Since $\operatorname{Ind}_\kappa(\mathcal{C})$ is generated by $\mathcal{C}'$ under κ-filtered colimits (Corollary 5.3.5.4), the restriction $f = F | \operatorname{Ind}_\kappa(\mathcal{C})$ factors through $\mathcal{D}$. It is then clear that $f : \operatorname{Ind}_\kappa(\mathcal{C}) \to \mathcal{D}$ is a left Kan extension of f_0 and that f is κ-continuous. This proves (1) and the "only if" direction of (2) (since left Kan extensions of f_0 are unique up to equivalence).

We now prove the "if" direction of (2). Let $f : \operatorname{Ind}_\kappa(\mathcal{C}) \to \mathcal{D}$ be the functor constructed above and let $f' : \operatorname{Ind}_\kappa(\mathcal{C}) \to \mathcal{D}$ be an arbitrary κ-continuous functor such that $f | \mathcal{C}' = f' | \mathcal{C}'$. We wish to prove that f' is a left Kan extension of $f' | \mathcal{C}'$. Since f is a left Kan extension of $f | \mathcal{C}'$, there exists a natural transformation $\alpha : f \to f'$ which is an equivalence when restricted to $\mathcal{C}'$. Let $\mathcal{E} \subseteq \operatorname{Ind}_\kappa(\mathcal{C})$ be the full subcategory spanned by those objects C for which the morphism $\alpha_C : f(C) \to f'(C)$ is an equivalence in $\mathcal{D}$. By hypothesis, $\mathcal{C}' \subseteq \mathcal{E}$. Since both f and f' are κ-continuous, $\mathcal{E}$ is stable under κ-filtered colimits in $\operatorname{Ind}_\kappa(\mathcal{C})$. We now apply Corollary 5.3.5.4 to conclude that $\mathcal{E} = \operatorname{Ind}_\kappa(\mathcal{C})$. It follows that f' and f are equivalent, so that f' is a left Kan extension of $f' | \mathcal{C}'$, as desired. □

Remark 5.3.5.9. The proof of Lemma 5.3.5.8 is very robust and can be used to establish a number of analogous results. Roughly speaking, given any class S of colimits, one can consider the smallest full subcategory $\mathcal{C}''$ of $\mathcal{P}(\mathcal{C})$ which contains the essential image $\mathcal{C}'$ of the Yoneda embedding and is stable under colimits of type S. Given any functor $f_0 : \mathcal{C}' \to \mathcal{D}$, where $\mathcal{D}$

is an ∞-category which admits colimits of type S, one can show that there exists a functor $f : \mathcal{C}'' \to \mathcal{D}$ which is a left Kan extension of $f_0 = f|\mathcal{C}'$. Moreover, f is characterized by the fact that it preserves colimits of type S. Taking S to be the class of all small colimits, we recover Lemma 5.1.5.5. Taking S to be the class of all small κ-filtered colimits, we recover Lemma 5.3.5.8. Other variations are possible as well: we will exploit this idea further in §5.3.6.

Proposition 5.3.5.10. *Let $\mathcal{C}$ and $\mathcal{D}$ be ∞-categories and let κ be a regular cardinal. Suppose that $\mathcal{C}$ is small and that $\mathcal{D}$ admits small κ-filtered colimits. Then composition with the Yoneda embedding induces an equivalence of ∞-categories*

$$\mathrm{Map}_\kappa(\mathrm{Ind}_\kappa(\mathcal{C}), \mathcal{D}) \to \mathrm{Fun}(\mathcal{C}, \mathcal{D}),$$

where the left hand side denotes the ∞-category of all κ-continuous functors from $\mathrm{Ind}_\kappa(\mathcal{C})$ to $\mathcal{D}$.

Proof. Combine Lemma 5.3.5.8 with Corollary 4.3.2.16. $\qquad\qquad\square$

In other words, if $\mathcal{C}$ is small and $\mathcal{D}$ admits κ-filtered colimits, then any functor $f : \mathcal{C} \to \mathcal{D}$ determines an essentially unique extension $F : \mathrm{Ind}_\kappa(\mathcal{C}) \to \mathcal{D}$ (such that f is equivalent to $F \circ j$). We next give a criterion which will allow us to determine when F is an equivalence.

Proposition 5.3.5.11. *Let $\mathcal{C}$ be a small ∞-category, κ a regular cardinal, and $\mathcal{D}$ an ∞-category which admits κ-filtered colimits. Let $F : \mathrm{Ind}_\kappa(\mathcal{C}) \to \mathcal{D}$ be a κ-continuous functor and $f = F \circ j$ its composition with the Yoneda embedding $j : \mathcal{C} \to \mathrm{Ind}_\kappa(\mathcal{C})$. Then*

(1) *If f is fully faithful and its essential image consists of κ-compact objects of $\mathcal{D}$, then F is fully faithful.*

(2) *The functor F is an equivalence if and only if the following conditions are satisfied:*

    (i) *The functor f is fully faithful.*

    (ii) *The functor f factors through $\mathcal{D}^\kappa$.*

    (iii) *The objects $\{f(C)\}_{C \in \mathcal{C}}$ generate $\mathcal{D}$ under κ-filtered colimits.*

Proof. We first prove (1) using the argument of Proposition 5.1.6.10. Let C and D be objects of $\mathrm{Ind}_\kappa(\mathcal{C})$. We wish to prove that the map

$$\eta_{C,D} : \mathrm{Map}_{\mathcal{P}(\mathcal{C})}(C, D) \to \mathrm{Map}_\mathcal{D}(F(C), F(D))$$

is an isomorphism in the homotopy category $\mathcal{H}$. Suppose first that C belongs to the essential image of j. Let $G : \mathcal{P}(\mathcal{C}) \to \mathcal{S}$ be a functor corepresented by C and let $G' : \mathcal{D} \to \mathcal{S}$ be a functor corepresented by $F(C)$. Then we have a natural transformation of functors $G \to G' \circ F$. Assumption (2) implies that G' preserves small κ-filtered colimits, so that $G' \circ F$ preserves small κ-filtered colimits. Proposition 5.3.5.5 implies that G preserves small κ-filtered colimits. It follows that the collection of objects $D \in \mathrm{Ind}_\kappa(\mathcal{C})$ such that $\eta_{C,D}$ is an equivalence is stable under small κ-filtered colimits. If D belongs to the essential image of j, then the assumption that f is fully faithful implies that $\eta_{C,D}$ is a homotopy equivalence. Since the image of the Yoneda embedding generates $\mathrm{Ind}_\kappa(\mathcal{C})$ under small κ-filtered colimits, we conclude that $\eta_{C,D}$ is a homotopy equivalence for every object $D \in \mathrm{Ind}_\kappa(\mathcal{C})$.

We now drop the assumption that C lies in the essential image of j. Fix $D \in \mathrm{Ind}_\kappa(\mathcal{C})$. Let $H : \mathrm{Ind}_\kappa(\mathcal{C})^{op} \to \mathcal{S}$ be a functor represented by D and let $H' : \mathcal{D}^{op} \to \mathcal{S}$ be a functor represented by FD. Then we have a natural transformation of functors $H \to H' \circ F^{op}$ which we wish to prove is an equivalence. By assumption, F^{op} preserves small κ-filtered limits. Proposition 5.1.3.2 implies that H and H' preserve small limits. It follows that the collection P of objects $C \in \mathcal{P}(S)$ such that $\eta_{C,D}$ is an equivalence is stable under small κ-filtered colimits. The special case above established that P contains the essential image of the Yoneda embedding. Since $\mathrm{Ind}_\kappa(\mathcal{C})$ is generated under small κ-filtered colimits by the image of the Yoneda embedding, we deduce that $\eta_{C,D}$ is an equivalence in general. This completes the proof of (1).

We now prove (2). Suppose first that F is an equivalence. Then (i) follows from Proposition 5.1.3.1, (ii) from Proposition 5.3.5.5, and (iii) from Corollary 5.3.5.4. Conversely, suppose that (i), (ii), and (iii) are satisfied. Using (1), we deduce that F is fully faithful. The essential image of F contains the essential image of f and is stable under small κ-filtered colimits. Therefore F is essentially surjective, so that F is an equivalence as desired. $\qquad\square$

According to Corollary 4.2.3.11, an ∞-category $\mathcal{C}$ admits small colimits if and only if $\mathcal{C}$ admits κ-small colimits and κ-filtered colimits. Using Proposition 5.3.5.11, we can make a much more precise statement:

Proposition 5.3.5.12. *Let $\mathcal{C}$ be a small ∞-category and κ a regular cardinal. The ∞-category $\mathcal{P}^\kappa(\mathcal{C})$ of κ-compact objects of $\mathcal{P}(\mathcal{C})$ is essentially small: that is, there exists a small ∞-category $\mathcal{D}$ and an equivalence $i : \mathcal{D} \to \mathcal{P}^\kappa(\mathcal{C})$. Let $F : \mathrm{Ind}_\kappa(\mathcal{D}) \to \mathcal{P}(\mathcal{C})$ be a κ-continuous functor such that the composition of f with the Yoneda embedding*

$$\mathcal{D} \to \mathrm{Ind}_\kappa(\mathcal{D}) \to \mathcal{P}(\mathcal{C})$$

is equivalent to i (according to Proposition 5.3.5.10, F exists and is unique up to equivalence). Then F is an equivalence of ∞-categories.

Proof. Since $\mathcal{P}(\mathcal{C})$ is locally small, to prove that $\mathcal{P}^\kappa(\mathcal{C})$ is small it will suffice to show that the collection of isomorphism classes of objects in the homotopy category $h\mathcal{P}^\kappa(\mathcal{C})$ is small. For this, we invoke Proposition 5.3.4.17: every κ-compact object X of $\mathcal{P}(\mathcal{C})$ is a retract of some object Y, which is itself the colimit of some composition

$$K \xrightarrow{p} \mathcal{C} \to \mathcal{P}(\mathcal{C}),$$

where K is κ-small. Since there is a bounded collection of possibilities for K and p (up to isomorphism in Set_Δ) and a bounded collection of idempotent maps $Y \rightarrow Y$ in $h\,\mathcal{P}(\mathcal{C})$, there is only a bounded number of possibilities for X.

To prove that F is an equivalence, it will suffice to show that F satisfies conditions (i), (ii), and (iii) of Proposition 5.3.5.11. Conditions (i) and (ii) are obvious. For (iii), we must prove that every object of $X \in \mathcal{P}(\mathcal{C})$ can be obtained as a small κ-filtered colimit of κ-compact objects of $\mathcal{C}$. Using Lemma 5.1.5.3, we can write X as a small colimit taking values in the essential image of $j : \mathcal{C} \rightarrow \mathcal{P}(\mathcal{C})$. The proof of Corollary 4.2.3.11 shows that X can be written as a κ-filtered colimit of a diagram with values in a full subcategory $\mathcal{E} \subseteq \mathcal{P}(\mathcal{C})$, where each object of $\mathcal{E}$ is itself a κ-small colimit of some diagram taking values in the essential image of j. Using Corollary 5.3.4.15, we deduce that $\mathcal{E} \subseteq \mathcal{P}^{\kappa}(\mathcal{C})$, so that X lies in the essential image of F, as desired. $\square$

Note that the construction $\mathcal{C} \mapsto \mathrm{Ind}_{\kappa}(\mathcal{C})$ is functorial in $\mathcal{C}$. Given a functor $f : \mathcal{C} \rightarrow \mathcal{C}'$, Proposition 5.3.5.10 implies that the composition of f with the Yoneda embedding $j_{\mathcal{C}'} : \mathcal{C}' \rightarrow \mathrm{Ind}_{\kappa} \mathcal{C}'$ is equivalent to the composition

$$\mathcal{C} \xrightarrow{j_{\mathcal{C}}} \mathrm{Ind}_{\kappa} \mathcal{C} \xrightarrow{F} \mathrm{Ind}_{\kappa} \mathcal{C}',$$

where F is a κ-continuous functor. The functor F is well-defined up to equivalence (in fact, up to contractible ambiguity). We will denote F by $\mathrm{Ind}_{\kappa} f$ (though this is perhaps a slight abuse of notation since F is uniquely determined only up to equivalence).

Proposition 5.3.5.13. *Let $f : \mathcal{C} \rightarrow \mathcal{C}'$ be a functor between small ∞-categories. The following are equivalent:*

(1) *The functor f is κ-right exact.*

(2) *The map $G : \mathcal{P}(\mathcal{C}') \rightarrow \mathcal{P}(\mathcal{C})$ given by composition with f restricts to a functor $g : \mathrm{Ind}_{\kappa}(\mathcal{C}') \rightarrow \mathrm{Ind}_{\kappa}(\mathcal{C})$.*

(3) *The functor $\mathrm{Ind}_{\kappa} f$ has a right adjoint.*

Moreover, if these conditions are satisfied, then g is a right adjoint to $\mathrm{Ind}_{\kappa} f$.

Proof. The equivalence $(1) \Leftrightarrow (2)$ is just a reformulation of the definition of κ-right exactness. Let $\mathcal{P}(f) : \mathcal{P}(\mathcal{C}) \rightarrow \mathcal{P}(\mathcal{C}')$ be a functor which preserves small colimits such that the diagram of ∞-categories

$$\begin{array}{ccc} \mathcal{C} & \xrightarrow{\;f\;} & \mathcal{C}' \\ \downarrow & & \downarrow \\ \mathcal{P}(\mathcal{C}) & \xrightarrow{\;\mathcal{P}(f)\;} & \mathcal{P}(\mathcal{C}') \end{array}$$

is homotopy commutative. Then we may identify $\mathrm{Ind}_{\kappa}(f)$ with the restriction $\mathcal{P}(f)|\,\mathrm{Ind}_{\kappa}(\mathcal{C})$. Proposition 5.2.6.3 asserts that G is a right adjoint of $\mathcal{P}(f)$. Consequently, if (2) is satisfied, then g is a right adjoint to $\mathrm{Ind}_{\kappa}(f)$. We deduce in particular that $(2) \Rightarrow (3)$. We will complete the proof by showing that (3) implies (2). Suppose that $\operatorname{Ind}_\kappa(f)$ admits a right adjoint $g' : \operatorname{Ind}_\kappa(\mathcal{C}') \to \operatorname{Ind}_\kappa(\mathcal{C})$. Let $X : (\mathcal{C}')^{op} \to \mathcal{S}$ be an object of $\operatorname{Ind}_\kappa(\mathcal{C}')$. Then X^{op} is equivalent to the composition

$$\mathcal{C}' \xrightarrow{j} \operatorname{Ind}_\kappa(\mathcal{C}') \xrightarrow{c_X} \mathcal{S}^{op},$$

where c_X denotes the functor represented by X. Since g' is a left adjoint to $\operatorname{Ind}_\kappa f$, the functor $c_X \circ \operatorname{Ind}_\kappa(f)$ is represented by $g'X$. Consequently, we have a homotopy commutative diagram

$$
\begin{array}{ccccc}
\mathcal{C} & \xrightarrow{j_\mathcal{C}} & \operatorname{Ind}_\kappa(\mathcal{C}) & \xrightarrow{c_{g'X}} & \mathcal{S}^{op} \\
\downarrow{\scriptstyle f} & & \downarrow{\scriptstyle \operatorname{Ind}_\kappa(f)} & & \downarrow \\
\mathcal{C}' & \longrightarrow & \operatorname{Ind}_\kappa(\mathcal{C}') & \xrightarrow{c_X} & \mathcal{S}^{op},
\end{array}
$$

so that $G(X)^{op} = f \circ X^{op} \simeq c_{g'X} \circ j_\mathcal{C}$ and therefore belongs to $\operatorname{Ind}_\kappa(\mathcal{C})$. $\square$

Proposition 5.3.5.14. *Let $\mathcal{C}$ be a small ∞-category and κ a regular cardinal. The Yoneda embedding $j : \mathcal{C} \to \operatorname{Ind}_\kappa(\mathcal{C})$ preserves all κ-small colimits which exist in $\mathcal{C}$.*

Proof. Let K be a κ-small simplicial set and $\overline{p} : K^{\triangleright} \to \mathcal{C}$ a colimit diagram. We wish to show that $j \circ \overline{p} : K^{\triangleright} \to \operatorname{Ind}_\kappa(\mathcal{C})$ is also a colimit diagram. Let $C \in \operatorname{Ind}_\kappa(\mathcal{C})$ be an object and let $F : \operatorname{Ind}_\kappa(\mathcal{C})^{op} \to \widehat{\mathcal{S}}$ be the functor represented by F. According to Proposition 5.1.3.2, it will suffice to show that $F \circ (j \circ \overline{p})^{op}$ is a limit diagram in $\mathcal{S}$. We observe that $F \circ j^{op}$ is equivalent to the object $C \in \operatorname{Ind}_\kappa(\mathcal{C}) \subseteq \operatorname{Fun}(\mathcal{C}^{op}, \mathcal{S})$ and therefore κ-right exact. We now conclude by invoking Proposition 5.3.2.9. $\square$

We conclude this section with a useful result concerning diagrams in ∞-categories of Ind-objects:

Proposition 5.3.5.15. *Let $\mathcal{C}$ be a small ∞-category, κ a regular cardinal, and $j : \mathcal{C} \to \operatorname{Ind}_\kappa(\mathcal{C})$ the Yoneda embedding. Let A be a finite partially ordered set and let $j' : \operatorname{Fun}(\operatorname{N}(A), \mathcal{C}) \to \operatorname{Fun}(\operatorname{N}(A), \operatorname{Ind}_\kappa(\mathcal{C}))$ be the induced map. Then j' induces an equivalence*

$$\operatorname{Ind}_\kappa(\operatorname{Fun}(\operatorname{N}(A), \mathcal{C})) \to \operatorname{Fun}(\operatorname{N}(A), \operatorname{Ind}_\kappa(\mathcal{C})).$$

In other words, every diagram $\operatorname{N}(A) \to \operatorname{Ind}_\kappa(\mathcal{C})$ can be obtained, in an essentially unique way, as a κ-filtered colimit of diagrams $\operatorname{N}(A) \to \mathcal{C}$.

Warning 5.3.5.16. The statement of Proposition 5.3.5.15 fails if we replace $\operatorname{N}(A)$ by an arbitrary finite simplicial set. For example, we may identify the category of abelian groups with the category of Ind-objects of the category of finitely generated abelian groups. If $n > 1$, then the map $q \mapsto \frac{q}{n}$ from the group of rational numbers $\mathbf{Q}$ to itself cannot be obtained as a filtered colimit of endomorphisms of finitely generated abelian groups.

Proof of Proposition 5.3.5.15. According to Proposition 5.3.5.11, it will suffice to prove the following:

(*i*) The functor j' is fully faithful.

(*ii*) The essential image of j' is comprised of of κ-compact objects of $\mathrm{Fun}(\mathrm{N}(A), \mathrm{Ind}_\kappa(\mathcal{C}))$.

(*iii*) The essential image of j' generates $\mathrm{Fun}(\mathrm{N}(A), \mathrm{Ind}_\kappa(\mathcal{C}))$ under small κ-filtered colimits.

Since the Yoneda embedding $j : \mathcal{C} \to \mathrm{Ind}_\kappa(\mathcal{C})$ satisfies the analogues of these conditions, (*i*) is obvious and (*ii*) follows from Proposition 5.3.4.13. To prove (*iii*), we fix an object $F \in \mathrm{Fun}(\mathrm{N}(A), \mathrm{Ind}_\kappa(\mathcal{C}))$. Let $\mathcal{C}'$ denote the essential image of j and form a pullback diagram of simplicial sets

$$\begin{array}{ccc} \mathcal{D} & \longrightarrow & \mathrm{Fun}(\mathrm{N}(A), \mathcal{C}') \\ \downarrow & & \downarrow \\ \mathrm{Fun}(\mathrm{N}(A), \mathrm{Ind}_\kappa(\mathcal{C}))_{/F} & \longrightarrow & \mathrm{Fun}(\mathrm{N}(A), \mathrm{Ind}_\kappa(\mathcal{C})). \end{array}$$

Since $\mathcal{D}$ is essentially small, (*iii*) is a consequence of the following assertions:

(*a*) The ∞-category $\mathcal{D}$ is κ-filtered.

(*b*) The canonical map $\mathcal{D}^{\triangleright} \to \mathrm{Fun}(\mathrm{N}(A), \mathcal{C})$ is a colimit diagram.

To prove (*a*), we need to show that $\mathcal{D}$ has the right lifting property with respect to the inclusion $\mathrm{N}(B) \subseteq \mathrm{N}(B \cup \{\infty\})$ for every κ-small partially ordered set B (Remark 5.3.1.10). Regard $B \cup \{\infty, \infty'\}$ as a partially ordered set with $b < \infty < \infty'$ for each $b \in B$. Unwinding the definitions, we see that (*a*) is equivalent to the following assertion:

(*a'*) Let $\overline{F} : \mathrm{N}(A \times (B \cup \{\infty'\})) \to \mathrm{Ind}_\kappa(\mathcal{C})$ be such that $\overline{F}| \mathrm{N}(A \times \{\infty'\}) = F$ and $\overline{F}'| \mathrm{N}(A \times B)$ factors through $\mathcal{C}'$. Then there exists a map $\overline{F}' :$ $\mathrm{N}(A \times (B \cup \{\infty, \infty'\})) \to \mathrm{Ind}_\kappa(\mathcal{C})$ which extends $\overline{F}$, such that $\overline{F}'| \mathrm{N}(A \times (B \cup \{\infty\}))$ factors through $\mathcal{C}'$.

To find $\overline{F}'$, we write $A = \{a_1, \ldots, a_n\}$, where $a_i \leq a_j$ implies $i \leq j$. We will construct a compatible sequence of maps

$$\overline{F}_k : \mathrm{N}((A \times (B \cup \{\infty'\})) \cup (\{a_1, \ldots, a_k\} \times \{\infty\})) \to \mathcal{C},$$

with $\overline{F}_0 = \overline{F}$ and $\overline{F}_n = \overline{F}'$. For each $a \in A$, we let $A_{\leq a} = \{a' \in A : a' \leq a\}$, and we define $A_{<a}, A_{\geq a}, A_{>a}$ similarly. Supposing that $\overline{F}_{k-1}$ has been constructed, we observe that constructing $\overline{F}_k$ amounts to constructing an object of the ∞-category

$$(\mathcal{C}'_{/F| \mathrm{N}(A_{\geq a_k})})_{\overline{F}_{k-1}| M/},$$

where $M = (A_{\leq a_k} \times B) \cup (A_{<a_k} \times \{\infty\})$. The inclusion $\{a_k\} \subseteq \mathrm{N}(A_{\geq a_k})$ is left anodyne. It will therefore suffice to construct an object in the equivalent

∞-category $(\mathcal{C}'_{/F(a_k)})_{\overline{F}_{k-1}|M/}$. Since M is κ-small, it suffices to show that the ∞-category $\mathcal{C}'_{/F(a_k)}$ is κ-filtered. This is simply a reformulation of the fact that $F(a_k) \in \mathrm{Ind}_\kappa(\mathcal{C})$.

We now prove (b). It will suffice to show that for each $a \in A$, the composition

$$\mathcal{D}^\triangleright \to \mathrm{Fun}(\mathrm{N}(A), \mathrm{Ind}_\kappa(\mathcal{C})) \to \mathrm{Ind}_\kappa(\mathcal{C})$$

is a colimit diagram, where the second map is given by evaluation at a. Let $\mathcal{D}(a) = \mathcal{C}' \times_{\mathrm{Ind}_\kappa(\mathcal{C})} \mathrm{Ind}_\kappa(\mathcal{C})_{/F(a)}$, so that $\mathcal{D}(a)$ is κ-filtered and the associated map $\mathcal{D}(a)^\triangleright \to \mathrm{Ind}_\kappa(\mathcal{C})$ is a colimit diagram. It will therefore suffice to show that the canonical map $\mathcal{D} \to \mathcal{D}(a)$ is cofinal. According to Theorem 4.1.3.1, it will suffice to show that for each object $D \in \mathcal{D}(a)$, the fiber product $\mathcal{E} = \mathcal{D} \times_{\mathcal{D}(a)} \mathcal{D}(a)_{D/}$ is weakly contractible. In view of Lemma 5.3.1.18, it will suffice to show that $\mathcal{E}$ is filtered. This can be established by a minor variation of the argument given above. □

5.3.6 Adjoining Colimits to ∞-Categories

Let $\mathcal{C}$ be a small ∞-category. According to Proposition 5.3.5.10, the ∞-category $\mathrm{Ind}(\mathcal{C})$ enjoys the following properties, which characterize it up to equivalence:

(1) There exists a functor $j : \mathcal{C} \to \mathrm{Ind}(\mathcal{C})$.

(2) The ∞-category $\mathrm{Ind}(\mathcal{C})$ admits small filtered colimits.

(3) Let $\mathcal{D}$ be an ∞-category which admits small filtered colimits and let $\mathrm{Fun}'(\mathrm{Ind}(\mathcal{C}), \mathcal{D})$ be the full subcategory of $\mathrm{Fun}(\mathrm{Ind}(\mathcal{C}), \mathcal{D})$ spanned by those functors which preserve filtered colimits. Then composition with j induces an equivalence $\mathrm{Fun}'(\mathrm{Ind}(\mathcal{C}), \mathcal{D}) \to \mathrm{Fun}(\mathcal{C}, \mathcal{D})$.

We can summarize this characterization as follows: the ∞-category $\mathrm{Ind}(\mathcal{C})$ is obtained from $\mathcal{C}$ by freely adjoining the colimits of all small filtered diagrams. In this section, we will study a generalization of this construction which allows us to freely adjoin to $\mathcal{C}$ the colimits of *any* collection of diagrams.

Notation 5.3.6.1. Let $\mathcal{C}$ and $\mathcal{D}$ be ∞-categories and let $\mathcal{R}$ be a collection of diagrams $\{\overline{p}_\alpha : K_\alpha^\triangleright \to \mathcal{C}\}$. We let $\mathrm{Fun}_\mathcal{R}(\mathcal{C}, \mathcal{D})$ denote the full subcategory of $\mathrm{Fun}(\mathcal{C}, \mathcal{D})$ spanned by those functors which carry each diagram in $\mathcal{R}$ to a colimit diagram in $\mathcal{D}$.

Let $\mathcal{K}$ be a collection of simplicial sets. We will say that an ∞-category $\mathcal{C}$ *admits $\mathcal{K}$-indexed colimits* if it admits K-indexed colimits for each $K \in \mathcal{K}$. If $f : \mathcal{C} \to \mathcal{D}$ is a functor between ∞-categories which admit $\mathcal{K}$-indexed colimits, then we will say that f *preserves $\mathcal{K}$-indexed colimits* if f preserves K-indexed colimits for each $K \in \mathcal{K}$. We let $\mathrm{Fun}_\mathcal{K}(\mathcal{C}, \mathcal{D})$ denote the full subcategory of $\mathrm{Fun}(\mathcal{C}, \mathcal{D})$ spanned by those functors which preserve $\mathcal{K}$-indexed colimits.

Proposition 5.3.6.2. *Let $\mathcal{K}$ be a collection of simplicial sets, $\mathcal{C}$ an ∞-category, and $\mathcal{R} = \{\overline{p}_\alpha : K_\alpha^\triangleright \to \mathcal{C}\}$ a collection of diagrams in $\mathcal{C}$. Assume that each K_α belongs to $\mathcal{K}$. Then there exists a new ∞-category $\mathcal{P}_{\mathcal{R}}^{\mathcal{K}}(\mathcal{C})$ and a map $j : \mathcal{C} \to \mathcal{P}_{\mathcal{R}}^{\mathcal{K}}(\mathcal{C})$ with the following properties:*

(1) *The ∞-category $\mathcal{P}_{\mathcal{R}}^{\mathcal{K}}(\mathcal{C})$ admits $\mathcal{K}$-indexed colimits.*

(2) *For every ∞-category $\mathcal{D}$ which admits $\mathcal{K}$-indexed colimits, composition with j induces an equivalence of ∞-categories*

$$\mathrm{Fun}_{\mathcal{K}}(\mathcal{P}_{\mathcal{R}}^{\mathcal{K}}(\mathcal{C}), \mathcal{D}) \to \mathrm{Fun}_{\mathcal{R}}(\mathcal{C}, \mathcal{D}).$$

Moreover, if every member of $\mathcal{R}$ is already a colimit diagram in $\mathcal{C}$, then we have in addition:

(3) *The functor j is fully faithful.*

Remark 5.3.6.3. In the situation of Proposition 5.3.6.2, assertion (2) (applied in the case $\mathcal{D} = \mathcal{P}_{\mathcal{R}}^{\mathcal{K}}(\mathcal{C})$) guarantees that j carries each diagram in $\mathcal{R}$ to a colimit diagram in $\mathcal{P}_{\mathcal{R}}^{\mathcal{K}}(\mathcal{C})$. We can informally summarize conditions (1) and (2) as follows: the ∞-category $\mathcal{P}_{\mathcal{R}}^{\mathcal{K}}(\mathcal{C})$ is freely generated by $\mathcal{C}$ under $\mathcal{K}$-indexed colimits, subject only to the relation that each diagram in $\mathcal{R}$ determines a colimit diagram in $\mathcal{P}_{\mathcal{R}}^{\mathcal{K}}(\mathcal{C})$. It is clear that this property characterizes $\mathcal{P}_{\mathcal{R}}^{\mathcal{K}}(\mathcal{C})$ (and the map j) up to equivalence.

Example 5.3.6.4. Suppose that $\mathcal{K}$ is the collection of all *small* simplicial sets, that the ∞-category $\mathcal{C}$ is small, and that the set of diagrams $\mathcal{R}$ is empty. Then the Yoneda embedding $j : \mathcal{C} \to \mathcal{P}(\mathcal{C})$ satisfies the conclusions of Proposition 5.3.6.2. This is precisely the assertion of Theorem 5.1.5.6 (save for assertion (3), which follows from Proposition 5.1.3.1). This justifies the notation of Proposition 5.3.6.2; in the general case we can think of $\mathcal{P}_{\mathcal{R}}^{\mathcal{K}}(\mathcal{C})$ as a sort of generalized presheaf category $\mathcal{C}$, and j as an analogue of the Yoneda embedding.

Proof of Proposition 5.3.6.2: We will employ essentially the same argument as in our proof of Proposition 5.3.5.10. First, we may enlarge the universe if necessary to reduce to the case where every element of $\mathcal{K}$ is a small simplicial set, the ∞-category $\mathcal{C}$ is small, and the collection of diagrams $\mathcal{R}$ is small. Let $j_0 : \mathcal{C} \to \mathcal{P}(\mathcal{C})$ denote the Yoneda embedding. For every diagram $\overline{p}_\alpha : K^\triangleright \to \mathcal{C}$, we let p_α denote the restriction $\overline{p}_\alpha | K$, $X_\alpha \in \mathcal{P}(\mathcal{C})$ a colimit for the induced diagram $j \circ p_\alpha : K \to \mathcal{P}(\mathcal{C})$, and $Y_\alpha \in \mathcal{C}$ the image of the cone point under $\overline{p}_\alpha$. The diagram $j_0 \circ \overline{p}_\alpha$ induces a map $s_\alpha : X_\alpha \to j_0(Y_\alpha)$ (well-defined up to homotopy); let $S = \{s_\alpha\}$ be the set of all such morphisms. We let $S^{-1}\mathcal{P}(\mathcal{C}) \subseteq \mathcal{P}(\mathcal{C})$ denote the ∞-category of S-local objects of $\mathcal{P}(\mathcal{C})$ and $L : \mathcal{P}(\mathcal{C}) \to S^{-1}\mathcal{P}(\mathcal{C})$ a left adjoint to the inclusion. We define $\mathcal{P}_{\mathcal{R}}^{\mathcal{K}}(\mathcal{C})$ to be the smallest full subcategory of $S^{-1}\mathcal{P}(\mathcal{C})$ which contains the essential image of the functor $L \circ j_0$ and is closed under $\mathcal{K}$-indexed colimits and let $j = L \circ j_0$ be the induced map. We claim that the map $j : \mathcal{C} \to \mathcal{P}_{\mathcal{R}}^{\mathcal{K}}(\mathcal{C})$ has the desired properties.

Assertion (1) is obvious. We now prove (2). Let $\mathcal{D}$ be an ∞-category which admits $\mathcal{K}$-indexed colimits. In view of Lemma 5.3.5.7, we can assume that there exists a fully faithful inclusion $\mathcal{D} \subseteq \mathcal{D}'$, where $\mathcal{D}'$ admits all small colimits and $\mathcal{D}$ is stable under $\mathcal{K}$-indexed colimits in $\mathcal{D}'$. We have a commutative diagram

$$
\begin{array}{ccc}
\operatorname{Fun}_{\mathcal{K}}(\mathcal{P}_{\mathcal{R}}^{\mathcal{K}}(\mathcal{C}), \mathcal{D}) & \xrightarrow{\phi} & \operatorname{Fun}_{\mathcal{R}}(\mathcal{C}, \mathcal{D}) \\
\downarrow & & \downarrow \\
\operatorname{Fun}_{\mathcal{K}}(\mathcal{P}_{\mathcal{R}}^{\mathcal{K}}(\mathcal{C}), \mathcal{D}') & \xrightarrow{\phi'} & \operatorname{Fun}_{\mathcal{R}}(\mathcal{C}, \mathcal{D}').
\end{array}
$$

We claim that this diagram is homotopy Cartesian. Unwinding the definitions, this is equivalent to the assertion that a functor $f \in \operatorname{Fun}_{\mathcal{K}}(\mathcal{P}_{\mathcal{R}}^{\mathcal{K}}(\mathcal{C}), \mathcal{D}')$ factors through $\mathcal{D}$ if and only if $f \circ j : \mathcal{C} \to \mathcal{D}'$ factors through $\mathcal{D}$. The "only if" direction is obvious. Conversely, if $f \circ j$ factors through $\mathcal{D}$, then $f^{-1} \mathcal{D}$ is a full subcategory of $\mathcal{P}_{\mathcal{R}}^{\mathcal{K}}(\mathcal{C})$ which is stable under $\mathcal{K}$-indexed limits (since f preserves $\mathcal{K}$-indexed limits and $\mathcal{D}$ is stable under $\mathcal{K}$-indexed limits in $\mathcal{D}$) and contains the essential image of j; by minimality, we conclude that $f^{-1} \mathcal{D} = \mathcal{P}_{\mathcal{R}}^{\mathcal{K}}(\mathcal{C})$.

Our goal is to prove that the functor ϕ is an equivalence of ∞-categories. In view of the preceding argument, it will suffice to show that ϕ' is an equivalence of ∞-categories. In other words, we may replace $\mathcal{D}$ by $\mathcal{D}'$ and thereby reduce to the case where $\mathcal{D}'$ admits small colimits.

Let $\mathcal{E} \subseteq \mathcal{P}(\mathcal{C})$ denote the inverse image $L^{-1} \mathcal{P}_{\mathcal{R}}^{\mathcal{K}}(\mathcal{C})$ and let $\overline{S}$ denote the collection of all morphisms α in $\mathcal{E}$ such that $L\alpha$ is an equivalence. Composition with L induces a fully faithful embedding $\operatorname{Fun}(\mathcal{P}_{\mathcal{R}}^{\mathcal{K}}(\mathcal{C}), \mathcal{D}) \to \operatorname{Fun}(\mathcal{E}, \mathcal{D})$ whose essential image consists of those functors $\mathcal{E} \to \mathcal{D}$ which carry every morphism in $\overline{S}$ to an equivalence in $\mathcal{D}$. Furthermore, a functor $f : \mathcal{P}_{\mathcal{R}}^{\mathcal{K}}(\mathcal{C}) \to \mathcal{D}$ preserves $\mathcal{K}$-indexed colimits if and only if the composition $f \circ L : \mathcal{E} \to \mathcal{D}$ preserves $\mathcal{K}$-indexed colimits. The functor ϕ factors as a composition

$$
\operatorname{Fun}_{\mathcal{K}}(\mathcal{P}_{\mathcal{R}}^{\mathcal{K}}(\mathcal{C}), \mathcal{D}) \to \operatorname{Fun}'(\mathcal{E}, \mathcal{D}) \xrightarrow{\psi} \operatorname{Fun}_{\mathcal{R}}(\mathcal{C}, \mathcal{D}),
$$

where $\operatorname{Fun}'(\mathcal{E}, \mathcal{D})$ denotes the full subcategory of $\operatorname{Fun}(\mathcal{E}, \mathcal{D})$ spanned by those functors which carry every morphism in $\overline{S}$ to an equivalence and preserve $\mathcal{K}$-indexed colimits. It will therefore suffice to show that ψ is an equivalence of ∞-categories.

In view of Proposition 4.3.2.15, we need only show that if $F : \mathcal{E} \to \mathcal{D}$ is a functor such that $F \circ j_0$ belongs to $\operatorname{Fun}_{\mathcal{R}}(\mathcal{C}, \mathcal{D})$, then F belongs to $\operatorname{Fun}'(\mathcal{E}, \mathcal{D})$ if and only if F is a left Kan extension of $F | \mathcal{C}'$, where $\mathcal{C}' \subseteq \mathcal{E}$ denotes the essential image of the Yoneda embedding $j_0 : \mathcal{C} \to \mathcal{E}$. We first prove the "if" direction. Let $F_0 = F | \mathcal{C}'$. Since $\mathcal{D}$ admits small colimits, the functor F_0 admits a left Kan extension $\overline{F} : \mathcal{P}(\mathcal{C}) \to \mathcal{D}$; without loss of generality, we may suppose that $F = \overline{F} | \mathcal{E}$. According to Lemma 5.1.5.5, the functor $\overline{F}$ preserves small colimits. Since $\mathcal{E}$ is stable under $\mathcal{K}$-indexed colimits in $\mathcal{P}(\mathcal{C})$, it follows that $\overline{F} | \mathcal{E}$ preserves $\mathcal{K}$-indexed colimits. Furthermore, since $F \circ j_0$ belongs to $\operatorname{Fun}_{\mathcal{R}}(\mathcal{C}, \mathcal{D})$, the functor $\overline{F}$ carries each morphism in S to an equivalence in $\mathcal{D}$. It follows that $\overline{F}$ factors (up to homotopy) through the localization functor L, so that $\overline{F}|\mathcal{E}$ carries each morphism in $\overline{S}$ to an equivalence in $\mathcal{D}$.

For the converse, let us suppose that $F \in \mathrm{Fun}'(\mathcal{E}, \mathcal{D})$; we wish to show that F is a left Kan extension of $F|\mathcal{C}'$. Let F' denote an arbitrary left Kan extension of $F|\mathcal{C}'$, so that the identification $F|\mathcal{C}' = F'|\mathcal{C}'$ induces a natural transformation $\alpha : F' \to F$. We wish to prove that α is an equivalence. Since F' and F both carry each morphism in $\overline{S}$ to an equivalence, we may assume without loss of generality that $F = f \circ L$, $F' = f' \circ L$, and $\alpha = \beta \circ L$, where $\beta : f' \to f$ is a natural transformation of functors from $\mathcal{P}_{\mathcal{R}}^{\mathcal{K}}(\mathcal{C})$ to $\mathcal{D}$. Let $\mathcal{X} \subseteq \mathcal{P}_{\mathcal{R}}^{\mathcal{K}}(\mathcal{C})$ denote the full subcategory spanned by those objects X such that $\beta_X : f'(X) \to f(X)$ is an equivalence. Since both f' and f preserve $\mathcal{K}$-indexed colimits, we conclude that $\mathcal{X}$ is stable under $\mathcal{K}$-indexed colimits in $\mathcal{P}_{\mathcal{R}}^{\mathcal{K}}(\mathcal{C})$. It is clear that $\mathcal{X}$ contains the essential image of the functor $j : \mathcal{C} \to \mathcal{P}_{\mathcal{R}}^{\mathcal{K}}(\mathcal{C})$. It follows by construction that $\mathcal{X} = \mathcal{P}_{\mathcal{R}}^{\mathcal{K}}(\mathcal{C})$, so that β is an equivalence, as desired. This completes the proof of (2).

It remains to prove (3). Suppose that every element of $\mathcal{R}$ is already a colimit diagram in $\mathcal{C}$. We note that the functor j factors as a composition $L \circ j_0$, where the Yoneda embedding $j_0 : \mathcal{C} \to \mathcal{E}$ is already known to be fully faithful (Proposition 5.1.3.1). Since the functor $L|S^{-1}\,\mathcal{P}(\mathcal{C})$ is equivalent to the identity, it will suffice to show that the essential image of j_0 is contained in $S^{-1}\,\mathcal{P}(\mathcal{C})$. In other words, we must show that if $s_\alpha : X_\alpha \to j_0 Y_\alpha$ belongs to S, and $C \in \mathcal{C}$, then the induced map

$$\mathrm{Map}_{\mathcal{P}(\mathcal{C})}(j_0 Y_\alpha, j_0 C) \to \mathrm{Map}_{\mathcal{P}(\mathcal{C})}(X_\alpha, j_0 C)$$

is a homotopy equivalence. Let $\overline{p} : K_\alpha^{\triangleright} \to \mathcal{C}$ be the corresponding diagram (so that $\overline{p}$ carries the cone point of $K_\alpha^{\triangleright}$ to Y_α), let $p = \overline{p}|K_\alpha$, and let $\overline{q} : K_\alpha^{\triangleright} \to \mathcal{P}(\mathcal{C})$ be a colimit diagram extending $q = \overline{q}|K_\alpha = j_0 \circ p$. Consider the diagram

$$\mathcal{P}(\mathcal{C})_{j_0 Y_\alpha /} \xleftarrow{g_0} \mathcal{P}(\mathcal{C})_{j_0 \overline{p}/} \xrightarrow{g_1} \mathcal{P}(\mathcal{C})_{j_0 p/} \xleftarrow{g_2} \mathcal{P}(\mathcal{C})_{\overline{q}/} \xrightarrow{g_3} \mathcal{P}(\mathcal{C})_{X_\alpha /}.$$

The maps g_0 and g_3 are trivial Kan fibrations (since the inclusion of the cone point into $K_\alpha^{\triangleright}$ is cofinal), and the map g_2 is a trivial Kan fibration since $\overline{q}$ is a colimit diagram. Moreover, for every object $Z \in \mathcal{P}(\mathcal{C})$, the above diagram determines the map $\mathrm{Map}_{\mathcal{P}(\mathcal{C})}(j_0 Y_\alpha, Z) \to \mathrm{Map}_{\mathcal{P}(\mathcal{C})}(X_\alpha, Z)$. Consequently, to prove that this map is an equivalence, it suffices to show that g_1 induces a trivial Kan fibration

$$\mathcal{P}(\mathcal{C})_{j_0 \overline{p}/} \times_{\mathcal{P}(\mathcal{C})} \{Z\} \to \mathcal{P}(\mathcal{C})_{j_0 p/} \times_{\mathcal{P}(\mathcal{C})} \{Z\}.$$

Assuming Z belongs to the essential image $\mathcal{C}'$ of the Yoneda embedding j_0, we may reduce to proving that the induced map $\mathcal{C}'_{j_0 \overline{p}/} \to \mathcal{C}'_{j_0 p/}$ is a trivial Kan fibration, which is equivalent to the assertion that $j_0 \circ \overline{p}$ is a colimit diagram in $\mathcal{C}'$. This is clear since $\overline{p}$ is a colimit diagram by assumption and j_0 induces an equivalence of ∞-categories from $\mathcal{C}$ to $\mathcal{C}'$. $\square$

Definition 5.3.6.5. Let $\mathcal{K} \subseteq \mathcal{K}'$ be collections of simplicial sets and let $\mathcal{C}$ be an ∞-category which admits $\mathcal{K}$-indexed limits. We let $\mathcal{P}_{\mathcal{K}}^{\mathcal{K}'}(\mathcal{C})$ denote the

∞-category $\mathcal{P}_{\mathcal{R}}^{\mathcal{K}'}(\mathcal{C})$, where $\mathcal{R}$ is the set of all colimit diagrams $\bar{p} : K^{\triangleright} \to \mathcal{C}$ such that $K \in \mathcal{K}$.

Example 5.3.6.6. Let $\mathcal{K} = \emptyset$ and let $\mathcal{K}'$ denote the class of *all* small simplicial sets. If $\mathcal{C}$ is a small ∞-category, then we have a canonical equivalence $\mathcal{P}_{\mathcal{K}}^{\mathcal{K}'}(\mathcal{C}) \simeq \mathcal{P}(\mathcal{C})$ (Theorem 5.1.5.6).

Example 5.3.6.7. Let $\mathcal{K} = \emptyset$ and let $\mathcal{K}'$ denote the class of all small κ-filtered simplicial sets for some regular cardinal κ. Then for any small ∞-category $\mathcal{C}$, we have a canonical equivalence $\mathcal{P}_{\mathcal{K}}^{\mathcal{K}'}(\mathcal{C}) \simeq \mathrm{Ind}_{\kappa}(\mathcal{C})$ (Proposition 5.3.5.10).

Example 5.3.6.8. Let $\mathcal{K}$ denote the collection of all κ-small simplicial sets for some regular cardinal κ and let $\mathcal{K}'$ be the class of all small simplicial sets. Let $\mathcal{C}$ be a small ∞-category which admits κ-small colimits. Then we have a canonical equivalence $\mathcal{P}_{\mathcal{K}}^{\mathcal{K}'}(\mathcal{C}) \simeq \mathrm{Ind}_{\kappa}(\mathcal{C})$. This follows from Theorem 5.5.1.1 and Proposition 5.5.1.9.

Example 5.3.6.9. Let $\mathcal{K} = \emptyset$ and let $\mathcal{K}' = \{\mathrm{Idem}\}$, where Idem is the simplicial set defined in §4.4.5. Then, for any ∞-category $\mathcal{C}$, $\mathcal{P}_{\mathcal{K}}^{\mathcal{K}'}(\mathcal{C})$ is an idempotent competion of $\mathcal{C}$.

Corollary 5.3.6.10. *Let* $\mathcal{K} \subseteq \mathcal{K}'$ *be classes of simplicial sets. Let* $\widehat{\mathrm{Cat}}_{\infty}$ *denote the ∞-category of (not necessarily small) ∞-categories, let* $\widehat{\mathrm{Cat}}_{\infty}^{\mathcal{K}}$ *denote the subcategory spanned by those ∞-categories which admit $\mathcal{K}$-indexed colimits and those functors which preserve $\mathcal{K}$-indexed colimits, and let* $\widehat{\mathrm{Cat}}_{\infty}^{\mathcal{K}'}$ *be defined likewise. Then the inclusion*

$$\widehat{\mathrm{Cat}}_{\infty}^{\mathcal{K}'} \subseteq \widehat{\mathrm{Cat}}_{\infty}^{\mathcal{K}}$$

admits a left adjoint given by $\mathcal{C} \mapsto \mathcal{P}_{\mathcal{K}}^{\mathcal{K}'}(\mathcal{C})$.

Proof. Combine Proposition 5.3.6.2 with Proposition 5.2.2.12. $\square$

We conclude this section by noting the following transitivity property of the construction $\mathcal{C} \mapsto \mathcal{P}_{\mathcal{R}}^{\mathcal{K}}(\mathcal{C})$:

Proposition 5.3.6.11. *Let* $\mathcal{K} \subseteq \mathcal{K}'$ *be collections of simplicial sets and let* $\mathcal{C}_1, \ldots, \mathcal{C}_n$ *be a sequence of ∞-categories. For $1 \leq i \leq n$, let $\mathcal{R}_i$ be a collection of diagrams $\{\bar{p}_{\alpha} : K_{\alpha}^{\triangleright} \to \mathcal{C}_i\}$, where each K_{α} belongs to $\mathcal{K}$, and let $\mathcal{R}_i'$ denote the collection of all colimit diagrams $\{\bar{q}_{\alpha} : K_{\alpha}^{\triangleright} \to \mathcal{P}_{\mathcal{R}_i}^{\mathcal{K}}(\mathcal{C}_i)\}$ such that $K_{\alpha} \in \mathcal{K}$. Then the canonical map*

$$\mathcal{P}_{\mathcal{R}_1 \boxtimes \cdots \boxtimes \mathcal{R}_n}^{\mathcal{K}'}(\mathcal{C}_1 \times \cdots \times \mathcal{C}_n) \to \mathcal{P}_{\mathcal{R}_1' \boxtimes \cdots \boxtimes \mathcal{R}_n'}^{\mathcal{K}'}(\mathcal{P}_{\mathcal{R}_1}^{\mathcal{K}}(\mathcal{C}_1) \times \cdots \times \mathcal{P}_{\mathcal{R}_n}^{\mathcal{K}}(\mathcal{C}_n))$$

is an equivalence of ∞-categories. Here $\mathcal{R}_1 \boxtimes \cdots \boxtimes \mathcal{R}_n$ denotes the collection of all diagrams of the form

$$K_{\alpha}^{\triangleright} \xrightarrow{\bar{p}_{\alpha}} \mathcal{C}_i \simeq \{C_1\} \times \cdots \times \{C_{i-1}\} \times \mathcal{C}_i \times \cdots \times \{C_n\} \subseteq \mathcal{C}_1 \times \cdots \times \mathcal{C}_n,$$

where $\bar{p}_{\alpha} \in \mathcal{R}_i$ and C_j is an object of $\mathcal{C}_j$ for $j \neq i$, and the collection $\mathcal{R}_1' \boxtimes \cdots \boxtimes \mathcal{R}_n'$ is defined likewise.

Proof. Let $\mathcal{D}$ be an ∞-category which admits $\mathcal{K}'$-indexed colimits. It will suffice to show that the functor

$$\mathrm{Fun}_{\mathcal{K}'}(\mathcal{P}^{\mathcal{K}'}_{\mathcal{R}'_1} \boxtimes \cdots \boxtimes_{\mathcal{R}'_n} (\mathcal{P}^{\mathcal{K}}_{\mathcal{R}_1}(\mathcal{C}_1) \times \cdots \times \mathcal{P}^{\mathcal{K}}_{\mathcal{R}_n}(\mathcal{C}_n)), \mathcal{D})$$

$$\downarrow$$

$$\mathrm{Fun}_{\mathcal{K}'}(\mathcal{P}^{\mathcal{K}'}_{\mathcal{R}_1} \boxtimes \cdots \boxtimes_{\mathcal{R}_n} (\mathcal{C}_1 \times \cdots \times \mathcal{C}_n), \mathcal{D})$$

is an equivalence of ∞-categories. Unwinding the definitions, we are reduced to proving that the functor

$$\mathrm{Fun}_{\mathcal{R}'_1 \boxtimes \cdots \boxtimes \mathcal{R}'_n}(\mathcal{P}^{\mathcal{K}}_{\mathcal{R}_1}(\mathcal{C}_1) \times \cdots \times \mathcal{P}^{\mathcal{K}}_{\mathcal{R}_n}(\mathcal{C}_n), \mathcal{D})$$

$$\downarrow \phi$$

$$\mathrm{Fun}_{\mathcal{R} \boxtimes \cdots \boxtimes \mathcal{R}_n}(\mathcal{C}_1 \times \cdots \times \mathcal{C}_n, \mathcal{D})$$

is an equivalence of ∞-categories. The proof goes by induction on n. If $n = 0$, then both sides are equivalent to $\mathcal{D}$ and there is nothing to prove. If $n > 0$, then set $\mathcal{D}' = \mathrm{Fun}_{\mathcal{R}_n}(\mathcal{C}_n, \mathcal{D})$ and $\mathcal{D}'' = \mathrm{Fun}_{\mathcal{K}}(\mathcal{P}^{\mathcal{K}}_{\mathcal{R}_1}(\mathcal{C}_n), \mathcal{D})$. Proposition 5.3.6.2 implies that the canonical map $\mathcal{D}'' \to \mathcal{D}'$ is an equivalence of ∞-categories. We can identify ϕ with the functor

$$\mathrm{Fun}_{\mathcal{R}'_1 \boxtimes \cdots \boxtimes \mathcal{R}'_{n-1}}(\mathcal{P}^{\mathcal{K}}_{\mathcal{R}_1}(\mathcal{C}) \times \cdots \times \mathcal{P}^{\mathcal{K}}_{\mathcal{R}_{n-1}}(\mathcal{C}), \mathcal{D}'')$$

$$\downarrow$$

$$\mathrm{Fun}_{\mathcal{R}_1 \boxtimes \cdots \boxtimes \mathcal{R}_{n-1}}(\mathcal{C}_1 \times \cdots \times \mathcal{C}_{n-1}, \mathcal{D}').$$

The desired result now follows from the inductive hypothesis. $\square$

5.4 ACCESSIBLE ∞-CATEGORIES

Many of the categories which commonly arise in mathematics can be realized as categories of Ind-objects. For example, the category of sets is equivalent to $\mathrm{Ind}(\mathcal{C})$, where $\mathcal{C}$ is the category of finite sets; the category of rings is equivalent to $\mathrm{Ind}(\mathcal{C})$, where $\mathcal{C}$ is the category of finitely presented rings. The theory of *accessible* categories is an axiomatization of this situation. We refer the reader to [1] for an exposition of the theory of accessible categories. In this section, we will describe an ∞-categorical generalization of the theory of accessible categories.

We will begin in §5.4.1 by introducing the notion of a *locally small* ∞-category. A locally small ∞-category $\mathcal{C}$ need not be small but has small morphism spaces $\mathrm{Map}_{\mathcal{C}}(X, Y)$ for any fixed pair of objects $X, Y \in \mathcal{C}$. This is analogous to the usual set-theoretic conventions taken in category theory: one allows categories which have a proper class of objects but requires that morphisms between any pair of objects form a *set*.

In §5.4.2, we will introduce the definition of an *accessible* ∞-category. An ∞-category $\mathcal{C}$ is accessible if it is locally small and has a good supply of filtered colimits and compact objects. Equivalently, $\mathcal{C}$ is accessible if it is equivalent to $\text{Ind}_\kappa(\mathcal{C}^0)$ for some small ∞-category $\mathcal{C}^0$ and some regular cardinal κ (Proposition 5.4.2.2).

The theory of accessible ∞-categories will play an important technical role throughout the remainder of this book. To understand the usefulness of the hypothesis of accessibility, let us consider the following example. Suppose that $\mathcal{C}$ is an ordinary category, that $F : \mathcal{C} \to \text{Set}$ is a functor, and that we would like to prove that F is representable by an object $C \in \mathcal{C}$. The functor F determines a category $\widetilde{\mathcal{C}} = \{(C, \eta) : C \in \mathcal{C}, \eta \in F(C)\}$, which is fibered over $\mathcal{C}$ in sets. We would like to prove that $\widetilde{\mathcal{C}}$ is equivalent to $\mathcal{C}_{/C}$ for some $C \in \mathcal{C}$. The object C can then be characterized as the colimit of the diagram $p : \widetilde{\mathcal{C}} \to \mathcal{C}$. If $\mathcal{C}$ admits colimits, then we can attempt to construct C by forming the colimit $\varinjlim(p)$.

We now encounter a set-theoretic difficulty. Suppose that we try to ensure the existence of $\varinjlim(p)$ by assuming that $\mathcal{C}$ admits *all* small colimits. In this case, it is not reasonable to expect $\mathcal{C}$ itself to be small. The category $\widetilde{\mathcal{C}}$ is roughly the same size as $\mathcal{C}$ (or larger), so our assumption will not allow us to construct $\varinjlim(p)$. On the other hand, if we assume $\mathcal{C}$ and $\widetilde{\mathcal{C}}$ are small, then it is not reasonable to expect $\mathcal{C}$ to admit colimits of arbitrary small diagrams.

An accessibility hypothesis can be used to circumvent the difficulty described above. An accessible category $\mathcal{C}$ is generally not small but is "controlled" by a small subcategory $\mathcal{C}^0 \subseteq \mathcal{C}$: it therefore enjoys the best features of both the "small" and "large" worlds. More precisely, the fiber product $\widetilde{\mathcal{C}} \times_\mathcal{C} \mathcal{C}^0$ is small enough that we might expect the colimit $\varinjlim(p | \widetilde{\mathcal{C}} \times_\mathcal{C} \mathcal{C}^0)$ to exist on general grounds yet large enough to expect a natural isomorphism

$$\varinjlim(p) \simeq \varinjlim(p | \widetilde{\mathcal{C}} \times_\mathcal{C} \mathcal{C}^0).$$

We refer the reader to §5.5.2 for a detailed account of this argument, which we will use to prove an ∞-categorical version of the adjoint functor theorem.

The discussion above can be summarized as follows: the theory of accessible ∞-categories is a tool which allows us to manipulate large ∞-categories as if they were small without fear of encountering any set-theoretic paradoxes. This theory is quite useful because the condition of accessibility is very robust: the class of accessible ∞-categories is stable under most of the basic constructions of higher category theory. To illustrate this, we will prove the following results:

(1) A small ∞-category $\mathcal{C}$ is accessible if and only if $\mathcal{C}$ is idempotent complete (§5.4.3).

(2) If $\mathcal{C}$ is an accessible ∞-category and K is a small simplicial set, then $\text{Fun}(K, \mathcal{C})$ is accessible (§5.4.4).

(3) If $\mathcal{C}$ is an accessible ∞-category and $p : K \to \mathcal{C}$ is a small diagram, then $\mathcal{C}_{p/}$ and $\mathcal{C}_{/p}$ are accessible (§5.4.5 and §5.4.6).

(4) The collection of accessible ∞-categories is stable under homotopy fiber products (§5.4.6).

We will apply these facts in §5.4.7 to deduce a miscellany of further stability results which will be needed throughout §5.5 and Chapter 6.

5.4.1 Locally Small ∞-Categories

In mathematical practice, it is very common to encounter categories $\mathcal{C}$ for which the collection of all objects is large (too big to form a set), but the collection of morphisms $\operatorname{Hom}_{\mathcal{C}}(X, Y)$ is small for every $X, Y \in \mathcal{C}$. The same situation arises frequently in higher category theory. However, it is a slightly trickier to describe because the formalism of ∞-categories blurs the distinction between objects and morphisms. Nevertheless, there is an adequate notion of "local smallness" in the ∞-categorical setting, which we will describe in this section.

Our first step is to give a characterization of the class of essentially small ∞-categories. We will need the following lemma.

Lemma 5.4.1.1. *Let $\mathcal{C}$ be a simplicial category, n a positive integer, and $f_0 : \partial \Delta^n \to \mathrm{N}(\mathcal{C})$ a map. Let $X = f_0(\{0\})$, $Y = f_0(\{n\})$, and g_0 denote the induced map*

$$\partial (\Delta^1)^{n-1} \to \operatorname{Map}_{\mathcal{C}}(X, Y).$$

Let $f, f' : \Delta^n \to \mathrm{N}(\mathcal{C})$ be extensions of f_0, and let $g, g' : (\Delta^1)^{n-1} \to \operatorname{Map}_{\mathcal{C}}(X, Y)$ be the corresponding extensions of g_0. The following conditions are equivalent:

(1) *The maps f and f' are homotopic relative to $\partial \Delta^n$.*

(2) *The maps g and g' are homotopic relative to $\partial (\Delta^1)^{n-1}$.*

Proof. It is not difficult to show that (1) is equivalent to the assertion that f and f' are left homotopic in the model category $(\operatorname{Set}_\Delta)_{\partial \Delta^n /}$ (with the Joyal model structure) and that (2) is equivalent to the assertion that $\mathfrak{C}[f]$ and $\mathfrak{C}[f']$ are left homotopic in the model category $(\operatorname{Cat}_\Delta)_{\mathfrak{C}[\partial \Delta^n]/}$. We now invoke the Quillen equivalence of Theorem 2.2.5.1 to complete the proof. $\square$

Proposition 5.4.1.2. *Let $\mathcal{C}$ be an ∞-category and κ an uncountable regular cardinal. The following conditions are equivalent:*

(1) *The collection of equivalence classes of objects of $\mathcal{C}$ is κ-small, and for every morphism $f : C \to D$ in $\mathcal{C}$ and every $n \geq 0$, the homotopy set $\pi_i(\operatorname{Hom}_{\mathcal{C}}^{\mathrm{R}}(C, D), f)$ is κ-small.*

(2) *If $\mathcal{C}' \subseteq \mathcal{C}$ is a minimal model for $\mathcal{C}$, then $\mathcal{C}'$ is κ-small.*

(3) *There exists a κ-small ∞-category $\mathcal{C}'$ and an equivalence $\mathcal{C}' \to \mathcal{C}$ of ∞-categories.*

(4) *There exists a κ-small simplicial set K and a categorical equivalence $K \to \mathcal{C}$.*

(5) *The ∞-category $\mathcal{C}$ is κ-compact when regarded as an object of Cat_∞.*

Proof. We begin by proving that $(1) \Rightarrow (2)$. Without loss of generality, we may suppose that $\mathcal{C} = N(\mathcal{D})$, where $\mathcal{D}$ is a topological category. Let $\mathcal{C}' \subseteq \mathcal{C}$ be a minimal model for $\mathcal{C}$. We will prove by induction on $n \geq 0$ that the set $\mathrm{Hom}_{\mathrm{Set}_\Delta}(\Delta^n, \mathcal{C}')$ is κ-small. If $n = 0$, this reduces to the assertion that $\mathcal{C}$ has fewer than κ equivalence classes of objects. Suppose therefore that $n > 0$. By the inductive hypothesis, the set $\mathrm{Hom}_{\mathrm{Set}_\Delta}(\partial \Delta^n, \mathcal{C}')$ is κ-small. Since κ is regular, it will suffice to prove that for each map $f_0 : \partial \Delta^n \to \mathcal{C}'$, the set $S = \{ f \in \mathrm{Hom}_{\mathrm{Set}_\Delta}(\Delta^n, \mathcal{C}') : f | \partial \Delta^n = f_0 \}$ is κ-small. Let $C = f_0(\{0\})$, let $D = f_0(\{n\})$, and let $g_0 : \partial(\Delta^1)^{n-1} \to \mathrm{Map}_{\mathcal{D}}(C, D)$ be the corresponding map. Assumption (1) ensures that there are fewer than κ extensions $g : (\Delta^1)^{n-1} \to \mathrm{Map}_{\mathcal{D}}(C, D)$ modulo homotopy relative to $\partial(\Delta^1)^{n-1}$. Invoking Lemma 5.4.1.1, we deduce that there are fewer than κ maps $f : \Delta^n \to \mathcal{C}$ modulo homotopy relative to $\partial \Delta^n$. Since $\mathcal{C}'$ is minimal, no two distinct elements of S are homotopic in $\mathcal{C}$ relative to $\partial \Delta^n$; therefore S is κ-small, as desired.

It is clear that $(2) \Rightarrow (3) \Rightarrow (4)$. We next show that $(4) \Rightarrow (3)$. Let $K \to \mathcal{C}$ be a categorical equivalence, where K is κ-small. We construct a sequence of inner anodyne inclusions

$$K = K(0) \subseteq K(1) \subseteq \cdots .$$

Supposing that $K(n)$ has been defined, we form a pushout diagram

$$
\begin{array}{ccc}
\coprod \Lambda_i^n & \longrightarrow & \coprod \Delta^n \\
\downarrow & & \downarrow \\
K(n) & \longrightarrow & K(n+1),
\end{array}
$$

where the coproduct is taken over all $0 < i < n$ and all maps $\Lambda_i^n \to K(n)$. It follows by induction on n that each $K(n)$ is κ-small. Since κ is regular and uncountable, the limit $K(\infty) = \bigcup_n K(n)$ is κ-small. The inclusion $K \subseteq K(\infty)$ is inner anodyne; therefore the map $K \to \mathcal{C}$ factors through an equivalence $K(\infty) \to \mathcal{C}$ of ∞-categories; thus (3) is satisfied.

We next show that $(3) \Rightarrow (5)$. Suppose that (3) is satisfied. Without loss of generality, we may replace $\mathcal{C}$ by $\mathcal{C}'$ and thereby suppose that $\mathcal{C}$ is itself κ-small. Let $F : \mathrm{Cat}_\infty \to \mathcal{S}$ denote the functor corepresented by $\mathcal{C}$. According to Lemma 5.1.5.2, we may identify F with the simplicial nerve of the functor $f : \mathrm{Cat}_\infty^\Delta \to \mathcal{K}\mathrm{an}$, which carries an ∞-category $\mathcal{D}$ to the largest Kan complex contained in $\mathcal{D}^{\mathcal{C}}$. Let $\mathcal{J}$ be a κ-filtered ∞-category and $p : \mathcal{J} \to \mathrm{Cat}_\infty$ a diagram. We wish to prove that p has a colimit $\overline{p} : \mathcal{J}^{\triangleright} \to \mathrm{Cat}_\infty$ such that $F \circ \overline{p}$ is a colimit diagram in $\mathcal{S}$. According to Proposition 5.3.1.16, we may suppose that $\mathcal{J}$ is the nerve of a κ-filtered partially ordered set A. Using Proposition 4.2.4.4, we may further reduce to the case where p is the simplicial nerve of a diagram $P : A \to \mathcal{C}at_\infty^\Delta \subseteq \mathcal{S}et_\Delta^+$ taking values in the *ordinary* category of marked simplicial sets. Let $\overline{P}$ be a colimit of P. Since the class of weak equivalences in $\mathcal{S}et_\Delta^+$ is stable under filtered colimits, $\overline{P}$ is a homotopy colimit. Theorem 4.2.4.1 implies that $\overline{p} = \mathrm{N}(\overline{P})$ is a colimit of p. It therefore suffices to show that $F \circ \overline{p} = \mathrm{N}(f \circ \overline{P})$ is a colimit diagram. Using Theorem 4.2.4.1, it suffices to show that $f \circ \overline{P}$ is a homotopy colimit diagram in $\mathcal{S}et_\Delta$. Since the class of weak homotopy equivalences in $\mathcal{S}et_\Delta$ is stable under filtered colimits, it will suffice to prove that $f \circ \overline{P}$ is a colimit diagram in the ordinary category $\mathcal{S}et_\Delta$. It now suffices to observe that f preserves κ-filtered colimits because $\mathcal{C}$ is κ-small.

We now complete the proof by showing that $(5) \Rightarrow (1)$. Let A denote the collection of all κ-small simplicial subsets $K_\alpha \subseteq \mathcal{C}$ and let $A' \subseteq A$ be the subcollection consisting of indices α such that K_α is an ∞-category. It is clear that A is a κ-filtered partially ordered set and that $\mathcal{C} = \bigcup_{\alpha \in A} K_\alpha$. Using the fact that $\kappa > \omega$, it is easy to see that A' is cofinal in A, so that A' is also κ-filtered and $\mathcal{C} = \bigcup_{\alpha \in A'} K_\alpha$. We may therefore regard $\mathcal{C}$ as the colimit of a diagram $P : A' \to \mathcal{S}et_\Delta^+$ in the ordinary category of fibrant objects of $\mathcal{S}et_\Delta^+$. Since A' is filtered, we may also regard $\mathcal{C}$ as a homotopy colimit of P. The above argument shows that $\mathcal{C}^\mathcal{C} = f\, \mathcal{C}$ can be identified with a homotopy colimit of the diagram $f \circ P : A' \to \mathcal{S}et_\Delta$. In particular, the vertex $\mathrm{id}_\mathcal{C} \in \mathcal{C}^\mathcal{C}$ must be homotopic to the image of some map $K_\alpha^\mathcal{C} \to \mathcal{C}^\mathcal{C}$ for some $\alpha \in A'$. It follows that $\mathcal{C}$ is a retract of K_α in the homotopy category $\mathrm{h}\mathcal{C}at_\infty$. Since K_α is κ-small, we easily deduce that K_α satisfies condition (1). Therefore $\mathcal{C}$, being a retract of K_α, satisfies condition (1) as well. $\square$

Definition 5.4.1.3. An ∞-category $\mathcal{C}$ is *essentially κ-small* if it satisfies the equivalent conditions of Proposition 5.4.1.2. We will say that $\mathcal{C}$ is *essentially small* if it is essentially κ-small for some (small) regular cardinal κ.

The following criterion for essential smallness is occasionally useful:

Proposition 5.4.1.4. *Let $p : \mathcal{C} \to \mathcal{D}$ be a Cartesian fibration of ∞-categories and κ an uncountable regular cardinal. Suppose that $\mathcal{D}$ is essentially κ-small and that, for each object $D \in \mathcal{D}$, the fiber $\mathcal{C}_D = \mathcal{C} \times_\mathcal{D} \{D\}$ is essentially κ-small. Then $\mathcal{C}$ is essentially κ-small.*

Proof. We will apply criterion (1) of Proposition 5.4.1.2. Choose a κ-small set of representatives $\{D_\alpha\}$ for the equivalence classes of objects of $\mathcal{D}$. For each α, choose a κ-small set of representatives $\{C_{\alpha,\beta}\}$ for the equivalence classes of objects of $\mathcal{C}_{D_\alpha}$. The collection of all objects $C_{\alpha,\beta}$ is κ-small (since κ is regular) and contains representatives for all equivalence classes of objects of $\mathcal{C}$.

Now suppose that C and C' are objects of $\mathcal{C}$ having images $D, D' \in \mathcal{D}$. Since $\mathcal{D}$ is essentially κ-small, the set $\pi_0 \mathrm{Map}_\mathcal{D}(D, D')$ is κ-small. Let $f : D \to D'$ be a morphism and choose a p-Cartesian morphism $\widetilde{f} : \widetilde{C} \to D'$ covering f. According to Proposition 2.4.4.2, we have a homotopy fiber sequence

$$\mathrm{Map}_{\mathcal{C}_D}(C, \widetilde{C}) \to \mathrm{Map}_\mathcal{C}(C, C') \to \mathrm{Map}_\mathcal{D}(D, D')$$

in the homotopy category $\mathcal{H}$. In particular, we see that $\mathrm{Map}_{\mathcal{C}}(C, C')$ contains fewer than κ connected components lying over $f \in \pi_0 \mathrm{Map}_{\mathcal{D}}(D, D')$ and therefore fewer than κ components in total (since κ is regular). Moreover, the long exact sequence of homotopy groups shows that for every $\overline{f} : C \to C'$ lifting f, the homotopy sets $\pi_i(\mathrm{Hom}^r_{\mathcal{C}}(C, C'), f)$ are κ-small as desired. $\square$

By restricting our attention to *Kan complexes*, we obtain an analogue of Proposition 5.4.1.2 for spaces:

Corollary 5.4.1.5. *Let X be a Kan complex and κ an uncountable regular cardinal. The following conditions are equivalent:*

(1) *For each vertex $x \in X$ and each $n \geq 0$, the homotopy set $\pi_n(X, x)$ is κ-small.*

(2) *If $X' \subseteq X$ is a minimal model for X, then X' is κ-small.*

(3) *There exists a κ-small Kan complex X' and a homotopy equivalence $X' \to X$.*

(4) *There exists a κ-small simplicial set K and a weak homotopy equivalence $K \to X$.*

(5) *The ∞-category $\mathcal{C}$ is κ-compact when regarded as an object of $\mathcal{S}$.*

(6) *The Kan complex X is essentially small (when regarded as an ∞-category).*

Proof. The equivalences $(1) \Leftrightarrow (2) \Leftrightarrow (3) \Leftrightarrow (6)$ follow from Proposition 5.4.1.2. The implication $(3) \Rightarrow (4)$ is obvious. We next prove that $(4) \Rightarrow (5)$. Let $p : K \to \mathcal{S}$ be the constant diagram taking the value $*$, let $\overline{p} : K^{\triangleright} \to \mathcal{S}$ be a colimit of p and let $X' \in \mathcal{S}$ be the image under $\overline{p}$ of the cone point of $K^{\triangleright}$. It follows from Proposition 5.1.6.8 that $*$ is a κ-compact object of $\mathcal{S}$. Corollary 5.3.4.15 implies that X' is a κ-compact object of $\mathcal{S}$. Let $\widetilde{K} \to K^{\triangleright}$ denote the left fibration associated to $\overline{p}$, and let $X'' \subseteq \widetilde{K}$ denote the fiber lying over the cone point of $K^{\triangleright}$. The inclusion of the cone point in $K^{\triangleright}$ is right anodyne. It follows from Proposition 4.1.2.15 that the inclusion $X'' \subseteq \widetilde{K}$ is right anodyne. Since $\overline{p}$ is a colimit diagram, Proposition 3.3.4.5 implies that the inclusion $K \simeq K \times_{K^{\triangleright}} \widetilde{K} \subseteq \widetilde{K}$ is a weak homotopy equivalence. We therefore have a chain of weak homotopy equivalences

$$X \leftarrow K \subseteq \widetilde{K} \leftarrow X'' \leftarrow X',$$

so that X and X' are equivalent objects of $\mathcal{S}$. Since X' is κ-compact, it follows that X is κ-compact.

To complete the proof, we will show that $(5) \Rightarrow (1)$. We employ the argument used in the proof of Proposition 5.4.1.2. Let $F : \mathcal{S} \to \mathcal{S}$ be the functor corepresented by X. Using Lemma 5.1.5.2, we can identify F with the simplicial nerve of the functor $f : \mathcal{K}\mathrm{an} \to \mathcal{K}\mathrm{an}$ given by

$$Y \mapsto Y^X.$$

Let A denote the collection of κ-small simplicial subsets $X_\alpha \subseteq X$ which are Kan complexes. Since κ is uncountable, A is κ-filtered and $X = \bigcup_{\alpha \in A} K_\alpha$. We may regard X as the colimit of a diagram $P : A \to \mathrm{Set}_\Delta$. Since A is filtered, X is also a homotopy colimit of this diagram. Since F preserves κ-filtered colimits, f preserves κ-filtered homotopy colimits; therefore X^X is a homotopy colimit of the diagram $f \circ P$. In particular, the vertex $\mathrm{id}_X \in X^X$ must be homotopic to the image of the map $X_\alpha^X \to X^X$ for some $\alpha \in A$. It follows that X is a retract of X_α in the homotopy category $\mathcal{H}$. Since X_α is κ-small, we can readily verify that X_α satisfies (1). Because X is a retract of X_α, X satisfies (1) as well. $\qquad\square$

Remark 5.4.1.6. When $\kappa = \omega$, the situation is quite a bit more complicated. Suppose that X is a Kan complex representing a compact object of $\mathcal{S}$. Then there exists a simplicial set Y with only finitely many nondegenerate simplices and a map $i : Y \to X$ which realizes X as a *retract* of Y in the homotopy category $\mathcal{H}$ of spaces. However, one cannot generally assume that Y is a Kan complex or that i is a weak homotopy equivalence. The latter can be achieved if X is connected and simply connected, or more generally if a certain K-theoretic invariant of X (the *Wall finiteness obstruction*) vanishes: we refer the reader to [81] for a discussion.

For many applications, it is important to be able to slightly relax the condition that an ∞-category be essentiall small.

Proposition 5.4.1.7. *Let $\mathcal{C}$ be an ∞-category. The following conditions are equivalent:*

(1) *For every pair of objects $X, Y \in \mathcal{C}$, the space $\mathrm{Map}_\mathcal{C}(X, Y)$ is essentially small.*

(2) *For every small collection S of objects of $\mathcal{C}$, the full subcategory of $\mathcal{C}$ spanned by the elements of S is essentially small.*

Proof. This follows immediately from criterion (1) in Propositions 5.4.1.2 and 5.4.1.5. $\qquad\square$

We will say that an ∞-category $\mathcal{C}$ is *locally small* if it satisfies the equivalent conditions of Proposition 5.4.1.7.

Example 5.4.1.8. Let $\mathcal{C}$ and $\mathcal{D}$ be ∞-categories. Suppose that $\mathcal{C}$ is locally small and that $\mathcal{D}$ is essentially small. Then $\mathcal{C}^\mathcal{D}$ is essentially small. To prove this, we may assume without loss of generality that $\mathcal{C}$ and $\mathcal{D}$ are minimal. Let $\{\mathcal{C}_\alpha\}$ denote the collection of all full subcategories of $\mathcal{C}$ spanned by small collections of objects. Since $\mathcal{D}$ is small, every finite collection of functors $\mathcal{D} \to \mathcal{C}$ factors through some small $\mathcal{C}_\alpha \subseteq \mathcal{C}$. It follows that $\mathrm{Fun}(\mathcal{D}, \mathcal{C})$ is the union of small full subcategories $\mathrm{Fun}(\mathcal{D}, \mathcal{C}_\alpha)$ and is therefore locally small. In particular, for every small ∞-category $\mathcal{D}$, the ∞-category $\mathcal{P}(\mathcal{D})$ of presheaves is locally small.

5.4.2 Accessibility

In this section, we will begin our study of the class of accessible ∞-categories.

Definition 5.4.2.1. Let κ be a regular cardinal. An ∞-category $\mathcal{C}$ is κ-*accessible* if there exists a small ∞-category $\mathcal{C}^0$ and an equivalence

$$\mathrm{Ind}_\kappa(\mathcal{C}^0) \to \mathcal{C}.$$

We will say that $\mathcal{C}$ is *accessible* if it is κ-accessible for *some* regular cardinal κ.

The following result gives a few alternative characterizations of the class of accessible ∞-categories.

Proposition 5.4.2.2. *Let $\mathcal{C}$ be an ∞-category and κ a regular cardinal. The following conditions are equivalent:*

(1) *The ∞-category $\mathcal{C}$ is κ-accessible.*

(2) *The ∞-category $\mathcal{C}$ is locally small and admits κ-filtered colimits, the full subcategory $\mathcal{C}^\kappa \subseteq \mathcal{C}$ of κ-compact objects is essentially small, and $\mathcal{C}^\kappa$ generates $\mathcal{C}$ under small, κ-filtered colimits.*

(3) *The ∞-category $\mathcal{C}$ admits small κ-filtered colimits and contains an essentially small full subcategory $\mathcal{C}'' \subseteq \mathcal{C}$ which consists of κ-compact objects and generates $\mathcal{C}$ under small κ-filtered colimits.*

The main obstacle to proving Proposition 5.4.2.2 is in verifying that if $\mathcal{C}_0$ is small, then $\mathrm{Ind}_\kappa(\mathcal{C}_0)$ has only a bounded number of κ-compact objects up to equivalence. It is tempting to guess that any such object must be equivalent to an object of $\mathcal{C}_0$. The following example shows that this is not necessarily the case.

Example 5.4.2.3. Let R be a ring and let $\mathcal{C}_0$ denote the (ordinary) category of finitely generated free R-modules. Then $\mathcal{C} = \mathrm{Ind}(\mathcal{C}_0)$ is equivalent to the category of flat R-modules (by Lazard's theorem; see, for example, the appendix of [47]). The compact objects of $\mathcal{C}$ are precisely the finitely generated projective R-modules, which need not be free.

Nevertheless, the naive guess is not far off, by virtue of the following result:

Lemma 5.4.2.4. *Let $\mathcal{C}$ be a small ∞-category, κ a regular cardinal, and $\mathcal{C}' \subseteq \mathrm{Ind}_\kappa(\mathcal{C})$ the full subcategory of $\mathrm{Ind}_\kappa(\mathcal{C})$ spanned by the κ-compact objects. Then the Yoneda embedding $j : \mathcal{C} \to \mathcal{C}'$ exhibits $\mathcal{C}'$ as an idempotent completion of $\mathcal{C}$. In particular, $\mathcal{C}'$ is essentially small.*

Proof. Corollary 4.4.5.16 implies that $\mathrm{Ind}_\kappa(\mathcal{C})$ is idempotent complete. Since $\mathcal{C}'$ is stable under retracts in $\mathrm{Ind}_\kappa(\mathcal{C})$, $\mathcal{C}'$ is also idempotent complete. Proposition 5.1.3.1 implies that j is fully faithful. It therefore suffices to prove that every object $C' \in \mathcal{C}'$ is a retract of $j(C)$ for some $C \in \mathcal{C}$.

Let $\mathcal{C}_{/C'} = \mathcal{C} \times_{\mathrm{Ind}_\kappa(\mathcal{C})} \mathrm{Ind}_\kappa(\mathcal{C})_{/C'}$. Lemma 5.1.5.3 implies that the diagram

$$\overline{p} : \mathcal{C}^{\triangleright}_{/C'} \to \mathrm{Ind}_\kappa(\mathcal{C})^{\triangleright}_{/C'} \to \mathrm{Ind}_\kappa(\mathcal{C})$$

is a colimit of $p = \overline{p} | \mathcal{C}_{/C'}$. Let $F : \mathrm{Ind}_\kappa(\mathcal{C}) \to \mathcal{S}$ be the functor corepresented by C'; we note that the left fibration associated to F is equivalent to $\mathrm{Ind}_\kappa(\mathcal{C})_{C'/}$. Since F is κ-continuous, Proposition 3.3.4.5 implies that the inclusion

$$\mathcal{C}_{/C'} \times_{\mathrm{Ind}_\kappa(\mathcal{C})} \mathrm{Ind}_\kappa(\mathcal{C})_{C'/} \subseteq \mathcal{C}^{\triangleright}_{/C'} \times_{\mathrm{Ind}_\kappa(\mathcal{C})} \mathrm{Ind}_\kappa(\mathcal{C})_{C'/}$$

is a weak homotopy equivalence. The simplicial set on the right has a canonical vertex, corresponding to the identity map $\mathrm{id}_{C'}$. It follows that there exists a vertex on the left hand side belonging to the same path component. Such a vertex classifies a diagram

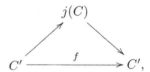

where f is homotopic to the identity, which proves that C' is a retract of $j(C)$ in $\mathrm{Ind}_\kappa(\mathcal{C})$. $\square$

Proof of Proposition 5.4.2.2. Suppose that (1) is satisfied. Without loss of generality, we may suppose that $\mathcal{C} = \mathrm{Ind}_\kappa \mathcal{C}'$, where $\mathcal{C}'$ is small. Since $\mathcal{C}$ is a full subcategory of $\mathcal{P}(\mathcal{C}')$, it is locally small (see Example 5.4.1.8). Proposition 5.3.5.3 implies that $\mathcal{C}$ admits small κ-filtered colimits. Corollary 5.3.5.4 shows that $\mathcal{C}$ is generated under κ-filtered colimits by the essential image of the Yoneda embedding $j : \mathcal{C}' \to \mathcal{C}$, which consists of κ-compact objects by Proposition 5.3.5.5. Lemma 5.4.2.4 implies that the full subcategory of $\mathrm{Ind}_\kappa(\mathcal{C}')$ consisting of compact objects is essentially small. We conclude that $(1) \Rightarrow (2)$.

It is clear that $(2) \Rightarrow (3)$. Suppose that (3) is satisfied. Choose a small ∞-category $\mathcal{C}'$ and an equivalence $i : \mathcal{C}' \to \mathcal{C}''$. Using Proposition 5.3.5.10, we may suppose that i factors as a composition

$$\mathcal{C}' \xrightarrow{j} \mathrm{Ind}_\kappa(\mathcal{C}') \xrightarrow{f} \mathcal{C},$$

where f preserves small κ-filtered colimits. Proposition 5.3.5.11 implies that f is a categorical equivalence. This shows that $(3) \Rightarrow (1)$ and completes the proof. $\square$

Definition 5.4.2.5. If $\mathcal{C}$ is an accessible ∞-category, then a functor $F : \mathcal{C} \to \mathcal{C}'$ is *accessible* if it is κ-continuous for some regular cardinal κ (and therefore for all regular cardinals $\tau \geq \kappa$).

Remark 5.4.2.6. Generally, we will only speak of the accessibility of a functor $F : \mathcal{C} \to \mathcal{C}'$ in the case where both $\mathcal{C}$ and $\mathcal{C}'$ are accessible. However, it is occasionally convenient to use the terminology of Definition 5.4.2.5 in the case where $\mathcal{C}$ is accessible and $\mathcal{C}'$ is not (or $\mathcal{C}'$ is not yet known to be accessible).

Example 5.4.2.7. The ∞-category $\mathcal{S}$ of spaces is accessible. More generally, for any small ∞-category $\mathcal{C}$, the ∞-category $\mathcal{P}(\mathcal{C})$ is accessible: this follows immediately from Proposition 5.3.5.12.

If $\mathcal{C}$ is a κ-accessible ∞-category and $\tau > \kappa$, then $\mathcal{C}$ is not necessarily τ-accessible. Nevertheless, this is true for many values of τ.

Definition 5.4.2.8. Let κ and τ be regular cardinals. We write $\tau \ll \kappa$ if the following condition is satisfied: for every $\tau_0 < \tau$ and every $\kappa_0 < \kappa$, we have $\kappa_0^{\tau_0} < \kappa$.

Note that there exist arbitrarily large regular cardinals κ' with $\kappa' \gg \kappa$: for example, one may take κ' to be the successor of any cardinal having the form τ^{κ}.

Remark 5.4.2.9. Every (infinite) regular cardinal κ satisfies $\omega \ll \kappa$. An uncountable regular cardinal κ satisfies $\kappa \ll \kappa$ if and only if κ is strongly inaccessible.

Lemma 5.4.2.10. *If $\kappa' \gg \kappa$, then any κ'-filtered partially ordered set $\mathcal{I}$ may be written as a union of κ-filtered subsets which are κ'-small. Moreover, the family of all such subsets is κ'-filtered.*

Proof. It will suffice to show that every κ'-small subset $S \subseteq \mathcal{I}$ can be included in a larger κ'-small subset $S' \subseteq \mathcal{I}$, where S' is κ-filtered.

We define a transfinite sequence of subsets $S_\alpha \subseteq \mathcal{I}$ by induction. Let $S_0 = S$. When λ is a limit ordinal, we let $S_\lambda = \bigcup_{\alpha < \lambda} S_\alpha$. Finally, we let $S_{\alpha+1}$ denote a set obtained from S_α by adjoining an upper bound for every κ-small subset of S_α (which exists because $\mathcal{I}$ is κ'-filtered). It follows from the assumption $\kappa' \gg \kappa$ that if S_α is κ'-small, then so is $S_{\alpha+1}$. Since κ' is regular, we deduce easily by induction that $|S_\alpha| < \kappa'$ for all $\alpha < \kappa'$. It is easy to check that the set $S' = S_\kappa$ has the desired properties. $\square$

Proposition 5.4.2.11. *Let $\mathcal{C}$ be a κ-accessible ∞-category. Then $\mathcal{C}$ is κ'-accessible for any $\kappa' \gg \kappa$.*

Proof. Let $\mathcal{C}^\kappa \subseteq \mathcal{C}$ denote the full subcategory consisting of κ-compact objects and let $\mathcal{C}' \subseteq \mathcal{C}$ denote the full subcategory spanned by the colimits of all κ'-small κ-filtered diagrams in $\mathcal{C}^\kappa$. Since $\mathcal{C}$ is locally small and the collection of all equivalence classes of such diagrams is bounded, we conclude that $\mathcal{C}'$ is essentially small. Corollary 5.3.4.15 implies that $\mathcal{C}'$ consists of κ'-compact objects of $\mathcal{C}$. According to Proposition 5.4.2.2, it will suffice to prove that $\mathcal{C}'$ generates $\mathcal{C}$ under small κ'-filtered colimits. Let X be an object of $\mathcal{C}$ and let $p : \mathcal{I} \to \mathcal{C}^\kappa$ be a small κ-filtered diagram with colimit X. Using Proposition 5.3.1.16, we may reduce to the case where $\mathcal{I}$ is the nerve of a κ-filtered partially ordered set A. Lemma 5.4.2.10 implies that A can be written as a κ'-filtered union of κ'-small κ-filtered subsets $\{A_\beta \subseteq A\}_{\beta \in B}$. Using Propositions 4.2.3.4 and 4.2.3.8, we deduce that X can also be obtained as the colimit of a diagram indexed by $N(B)$ which takes values in $\mathcal{C}'$. $\square$

Remark 5.4.2.12. If $\mathcal{C}$ is a κ-accessible ∞-category and $\kappa' > \kappa$, then $\mathcal{C}$ is generally not κ'-accessible. There are counterexamples even in ordinary category theory: see [1].

Remark 5.4.2.13. Let $\mathcal{C}$ be an accessible ∞-category and κ a regular cardinal. Then the full subcategory $\mathcal{C}^{\kappa} \subseteq \mathcal{C}$ consisting of κ-compact objects is essentially small. To prove this, we are free to enlarge κ. Invoking Proposition 5.4.2.11, we can reduce to the case where $\mathcal{C}$ is κ-accessible, in which case the desired result is a consequence of Proposition 5.4.2.2.

Notation 5.4.2.14. If $\mathcal{C}$ and $\mathcal{D}$ are accessible ∞-categories, we will write $\mathrm{Fun}_A(\mathcal{C}, \mathcal{D})$ to denote the full subcategory of $\mathrm{Fun}(\mathcal{C}, \mathcal{D})$ spanned by accessible functors from $\mathcal{C}$ to $\mathcal{D}$.

Remark 5.4.2.15. Accessible ∞-categories are usually not small. However, they are determined by a "small" amount of data: namely, they always have the form $\mathrm{Ind}_{\kappa}(\mathcal{C})$, where $\mathcal{C}$ is a small ∞-category. Similarly, an accessible functor $F : \mathcal{C} \to \mathcal{D}$ between accessible categories is determined by a "small" amount of data in the sense that there always exists a regular cardinal κ such that F is κ-continuous and maps $\mathcal{C}^{\kappa}$ into $\mathcal{D}^{\kappa}$. The restriction $F|\,\mathcal{C}^{\kappa}$ then determines F up to equivalence (Proposition 5.3.5.10). To prove the existence of κ, we first choose a regular cardinal τ such that F is τ-continuous. Enlarging τ if necessary, we may suppose that $\mathcal{C}$ and $\mathcal{D}$ are τ-accessible. The collection of equivalence classes of τ-compact objects of $\mathcal{C}$ is small; consequently, by Remark 5.4.2.13, there exists a (small) regular cardinal τ' such that F carries $\mathcal{C}^{\tau}$ into $\mathcal{D}^{\tau'}$. We may now choose κ to be any regular cardinal such that $\kappa \gg \tau'$.

Definition 5.4.2.16. Let κ be a regular cardinal. We let $\mathrm{Acc}_{\kappa} \subseteq \widehat{\mathrm{Cat}}_{\infty}$ denote the subcategory defined as follows:

(1) The objects of Acc_{κ} are the κ-accessible ∞-categories.

(2) A functor $F : \mathcal{C} \to \mathcal{D}$ between accessible ∞-categories belongs to Acc if and only if F is κ-continuous and preserves κ-compact objects.

Let $\mathrm{Acc} = \bigcup_{\kappa} \mathrm{Acc}_{\kappa}$. We will refer to Acc as the ∞-*category of accessible* ∞-*categories*.

Proposition 5.4.2.17. *Let κ be a regular cardinal and let $\theta : \mathrm{Acc}_{\kappa} \to \widehat{\mathrm{Cat}}_{\infty}$ be the simplicial nerve of the functor which associates to each $\mathcal{C} \in \mathrm{Acc}_{\kappa}$ the full subcategory of $\mathcal{C}$ spanned by the κ-compact objects. Then*

(1) *The functor θ is fully faithful.*

(2) *An ∞-category $\mathcal{C} \in \widehat{\mathrm{Cat}}_{\infty}$ belongs to the essential image of θ if and only if $\mathcal{C}$ is essentially small and idempotent complete.*

Proof. Assertion (1) follows immediately from Proposition 5.3.5.10. If $\mathcal{C} \in \widehat{\mathrm{Cat}}_{\infty}$ belongs to the essential image of θ, then $\mathcal{C}$ is essentially small and idempotent complete (because $\mathcal{C}$ is stable under retracts in an idempotent complete ∞-category). Conversely, suppose that $\mathcal{C}$ is essentially small and idempotent complete and choose a minimal model $\mathcal{C}' \subseteq \mathcal{C}$. Then $\mathrm{Ind}_\kappa(\mathcal{C}')$ is κ-accessible. Moreover, the collection of κ-compact objects of $\mathrm{Ind}_\kappa(\mathcal{C}')$ is an idempotent completion of $\mathcal{C}'$ (Lemma 5.4.2.4) and therefore equivalent to $\mathcal{C}$ (since $\mathcal{C}'$ is already idempotent complete). $\qquad\square$

Let $\mathrm{Cat}_\infty^\vee$ denote the full subcategory of Cat_∞ spanned by the idempotent complete ∞-categories.

Proposition 5.4.2.18. *The inclusion* $\mathrm{Cat}_\infty^\vee \subseteq \mathrm{Cat}_\infty$ *has a left adjoint.*

Proof. Combine Propositions 5.1.4.2, 5.1.4.9, and 5.2.7.8. $\qquad\square$

We will refer to a left adjoint to the inclusion $\mathrm{Cat}_\infty^\vee \subseteq \mathrm{Cat}_\infty$ as the *idempotent completion functor*. Proposition 5.4.2.17 implies that we have fully faithful embeddings $\mathrm{Acc}_\kappa \to \widehat{\mathrm{Cat}_\infty} \hookleftarrow \mathrm{Cat}_\infty^\vee$ with the same essential image. Consequently, there is a (canonical) equivalence of ∞-categories $e : \mathrm{Cat}_\infty^\vee \simeq \mathrm{Acc}_\kappa$ which is well-defined up to homotopy. We let $\mathrm{Ind}_\kappa : \mathrm{Cat}_\infty \to \mathrm{Acc}_\kappa$ denote the composition of e with the idempotent completion functor. In summary:

Proposition 5.4.2.19. *There is a functor* $\mathrm{Ind}_\kappa : \mathrm{Cat}_\infty \to \mathrm{Acc}_\kappa$ *which exhibits* Acc_κ *as a localization of the ∞-category* Cat_∞.

Remark 5.4.2.20. There is a slight danger of confusion with our terminology. The functor $\mathrm{Ind}_\kappa : \mathrm{Cat}_\infty \to \mathrm{Acc}_\kappa$ is well-defined only up to a contractible space of choices. Consequently, if $\mathcal{C}$ is an ∞-category which admits finite colimits, then the image of $\mathcal{C}$ under Ind_κ is well-defined only up to equivalence. Definition 5.3.5.1 produces a canonical representative for this image.

5.4.3 Accessibility and Idempotent Completeness

Let $\mathcal{C}$ be an accessible ∞-category. Then there exists a regular cardinal κ such that $\mathcal{C}$ admits κ-filtered colimits. It follows from Corollary 4.4.5.16 that $\mathcal{C}$ is idempotent complete. Our goal in this section is to prove a converse to this result: if $\mathcal{C}$ is small and idempotent complete, then $\mathcal{C}$ is accessible.

Let $\mathcal{C}$ be a small ∞-category and suppose we want to prove that $\mathcal{C}$ is accessible. The main problem is to show that $\mathcal{C}$ admits κ-filtered colimits provided that κ is sufficiently large. The idea is that if κ is much larger than the size of $\mathcal{C}$, then any κ-filtered diagram $\mathcal{J} \to \mathcal{C}$ is necessarily very "redundant" (Proposition 5.4.3.4). Before making this precise, we will need a few preliminary results.

Lemma 5.4.3.1. *Let* $\kappa < \tau$ *be uncountable regular cardinals, let* A *a* τ-*filtered partially ordered set, and let* $F : A \to \mathfrak{Kan}$ *a diagram of Kan complexes indexed by* A. *Suppose that for each* $\alpha \in A$, *the Kan complex* $F(\alpha)$ *is* *essentially κ-small. For every τ-small subset $A_0 \subseteq A$, there exists a filtered τ-small subset $A_0' \subseteq A$ containing A_0, with the property that the map*

$$\varinjlim_{\alpha \in A_0'} F(\alpha) \to \varinjlim_{\alpha \in A} F(\alpha)$$

is a homotopy equivalence.

Proof. Let $X = \varinjlim_{\alpha \in A} F(\alpha)$. Since F is a filtered diagram, X is also a Kan complex. Let K be a simplicial set with only finitely many nondegenerate simplices. Our first claim is that the set $[K, X]$ of homotopy classes of maps from K into X is κ-small. Suppose we are given a collection $\{g_\beta : K \to X\}$ of pairwise nonhomotopic maps, having cardinality κ. Since A is τ-filtered, we may suppose that there is a fixed index $\alpha \in A$ such that each g_β factors as a composition

$$K \xrightarrow{g_\beta'} F(\alpha) \to X.$$

The maps g_β' are also pairwise nonhomotopic, which contracts our assumption that $F(\alpha)$ is weakly homotopy equivalent to a κ-small simplicial set.

We now define an increasing sequence

$$\alpha_0 \le \alpha_1 \le \cdots$$

of elements of A. Let α_0 be any upper bound for A_0. Assuming that α_i has already been selected, choose a representative for every homotopy class of diagrams

The argument above proves that we can take the set of all such representatives to be κ-small, so that there exists $\alpha_{i+1} \ge \alpha_i$ such that each h_γ factors as a composition

$$\Delta^n \xrightarrow{h_\gamma'} F(\alpha_{i+1}) \to X$$

and the associated diagram

is commutative. We now set $A_0' = A_0 \cup \{\alpha_0, \alpha_1, \dots\}$; it is easy to check that this set has the desired properties. $\qquad \square$

Lemma 5.4.3.2. *Let $\kappa < \tau$ be uncountable regular cardinals, let A be a τ-filtered partially ordered set, and let $\{F_\beta\}_{\beta \in B}$ be a collection of diagrams $A \to \mathrm{Set}_\Delta$ indexed by a τ-small set B. Suppose that for each $\alpha \in A$ and each*

$\beta \in B$, the Kan complex $F_\beta(\alpha)$ is essentially κ-small. Then there exists a filtered τ-small subset $A' \subseteq A$ such that for each $\beta \in B$, the map

$$\varinjlim_{A'} F_\beta(\alpha) \to \varinjlim_A F_\beta(\alpha)$$

is a homotopy equivalence of Kan complexes.

Proof. Without loss of generality, we may suppose that $B = \{\beta : \beta < \beta_0\}$ is a set of ordinals. We will define a sequence of filtered τ-small subsets $A(n) \subseteq A$ by induction on n. For $n = 0$, choose an element $\alpha \in A$ and set $A(0) = \{\alpha\}$. Suppose next that $A(n)$ has been defined. We define a sequence of enlargements $\{A(n)_\beta\}_{\beta \leq \beta_0}$ by induction on β. Let $A(n)_0 = A(n)$, let $A(n)_\lambda = \bigcup_{\beta < \lambda} A(n)_\beta$ when λ is a nonzero limit ordinal, and let $A(n)_{\beta+1}$ be a τ-small filtered subset of A such that the map

$$\varinjlim_{A(n)_{\beta+1}} F_\beta(\alpha) \to \varinjlim_A F_\beta(\alpha)$$

is a weak homotopy equivalence (such a subset exists by virtue of Lemma 5.4.3.1). We now take $A(n+1) = A(n)_{\beta_0}$ and $A' = \bigcup_n A(n)$; it is easy to check that $A' \subseteq A$ has the desired properties. $\square$

Lemma 5.4.3.3. *Let $\kappa < \tau$ be uncountable regular cardinals. Let $\mathcal{C}$ be a τ-small ∞-category with the property that each of the spaces $\mathrm{Map}_{\mathcal{C}}(C, D)$ is essentially κ-small and let $j : \mathcal{C} \to \mathcal{P}(\mathcal{C})$ denote the Yoneda embedding. Let $p : \mathcal{K} \to \mathcal{C}$ be a diagram indexed by a τ-filtered ∞-category $\mathcal{K}$ and let $\overline{p} : \mathcal{K}^\triangleright \to \mathcal{P}(\mathcal{C})$ be a colimit of $j \circ p$. Then there exists a map $i : K \to \mathcal{K}$ such that K is τ-small and the composition $\overline{p} \circ i^\triangleright : K^\triangleright \to \mathcal{K}^\triangleright \to \mathcal{P}(\mathcal{C})$ is a colimit diagram.*

Proof. In view of Proposition 5.3.1.16, we may suppose that $\mathcal{K}$ is the nerve of a τ-filtered partially ordered set A. According to Proposition 5.1.2.2, $\overline{p}$ induces a colimit diagram

$$\overline{p}_C : \mathcal{K}^\triangleright \to \mathcal{P}(\mathcal{C}) \overset{e_C}{\to} \mathcal{S},$$

where e_C denotes the evaluation functor associated to an object $C \in \mathcal{C}$. We will identify $\mathcal{K}^\triangleright$ with the nerve of the partially ordered set $A \cup \{\infty\}$. Proposition 4.2.4.4 implies that we may replace $\overline{p}_C$ with the simplicial nerve of a functor $F_C : A \cup \{\infty\} \to \mathcal{K}$an. Our hypothesis on $\mathcal{C}$ implies that $F_C|A$ takes values in κ-small simplicial sets. Applying Theorem 4.2.4.1, we see that the map $\varinjlim_A F_C(\alpha) \to F_C(\infty)$ is a homotopy equivalence. We now apply Lemma 5.4.3.1 to deduce the existence of a filtered τ-small subset $A' \subseteq A$ such that each of the maps

$$\varinjlim_{A'} F_C(\alpha) \to F_C(\infty)$$

is a homotopy equivalence. Let $K = N(A')$ and let $i : K \to \mathcal{K}$ denote the inclusion. Using Theorem 4.2.4.1 again, we deduce that the composition $e_C \circ \overline{p} \circ i^\triangleright : K^\triangleright \to \mathcal{S}$ is a colimit diagram for each $C \in \mathcal{C}$. Applying Proposition 5.1.2.2, we deduce that $\overline{p} \circ i^\triangleright$ is a colimit diagram, as desired. $\square$

Proposition 5.4.3.4. *Let $\kappa < \tau$ be uncountable regular cardinals. Let $\mathcal{C}$ be an ∞-category which is τ-small, such that the morphism spaces $\operatorname{Map}_{\mathcal{C}}(C, D)$ are essentially κ-small. Let $j : \mathcal{C} \to \mathcal{P}(\mathcal{C})$ denote the Yoneda embedding, let $p : \mathcal{K} \to \mathcal{C}$ be a diagram indexed by a τ-filtered ∞-category $\mathcal{K}$, and let $X \in \mathcal{P}(\mathcal{C})$ be a colimit of $j \circ p : \mathcal{K} \to \mathcal{P}(\mathcal{C})$. Then there exists an object $C \in \mathcal{C}$ such that X is a retract of $j(C)$.*

Proof. Let $i : K \to \mathcal{K}$ be a map satisfying the conclusions of Lemma 5.4.3.3. Since K is τ-small and $\mathcal{K}$ is τ-filtered, there exists an extension $\overline{i} : K^{\triangleright} \to \mathcal{K}$ of i. Let C be the image of the cone point of $K^{\triangleright}$ under $p \circ \overline{i}$ and let $\widetilde{C} \in \mathcal{C}_{p \circ i /}$ be the corresponding lift. Let $\overline{p} : \mathcal{K}^{\triangleright} \to \mathcal{P}(\mathcal{C})$ be a colimit of $j \circ p$ carrying the cone point of $\mathcal{K}^{\triangleright}$ to X. Let $q = j \circ p \circ i : K \to \mathcal{P}(\mathcal{C})$, let $\widetilde{X} \in \mathcal{P}(\mathcal{C})_{q/}$ be the corresponding lift of X, and let $\widetilde{Y} \in \mathcal{P}(\mathcal{C})_{q/}$ be a colimit of q. Since $\widetilde{Y}$ is an initial object of $\mathcal{P}(\mathcal{C})_{q/}$, there is a commutative triangle

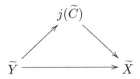

in the ∞-category $\mathcal{P}(\mathcal{C})_{q/}$. Moreover, Lemma 5.4.3.3 asserts that the horizontal map is an equivalence. Thus $\widetilde{X}$ is a retract of $j(\widetilde{C})$ in the homotopy category of $\mathcal{P}(\mathcal{C})_{q/}$, so that X is a retract of $j(C)$ in $\mathcal{P}(\mathcal{C})$. $\square$

Corollary 5.4.3.5. *Let $\kappa < \tau$ be uncountable regular cardinals and let $\mathcal{C}$ be a τ-small ∞-category whose morphism spaces $\operatorname{Map}_{\mathcal{C}}(C, D)$ are essentially κ-small. Then the Yoneda embedding $j : \mathcal{C} \to \operatorname{Ind}_{\tau}(\mathcal{C})$ exhibits $\operatorname{Ind}_{\tau}(\mathcal{C})$ as an idempotent completion of $\mathcal{C}$.*

Proof. Since $\operatorname{Ind}_{\tau}(\mathcal{C})$ admits τ-filtered colimits, it is idempotent complete by Corollary 4.4.5.16. Proposition 5.4.3.4 implies that every object of $\operatorname{Ind}_{\tau}(\mathcal{C})$ is a retract of $j(C)$ for some object $C \in \mathcal{C}$. $\square$

Corollary 5.4.3.6. *A small ∞-category $\mathcal{C}$ is accessible if and only if it is idempotent complete. Moreover, if these conditions are satisfied and $\mathcal{D}$ is an any accessible ∞-category, then every functor $f : \mathcal{C} \to \mathcal{D}$ is accessible.*

Proof. The "only if" direction follows from Corollary 4.4.5.16, and the "if" direction follows from Corollary 5.4.3.5. Now suppose that $\mathcal{C}$ is small and accessible, and let $\mathcal{D}$ be a κ-accessible ∞-category and $f : \mathcal{C} \to \mathcal{D}$ any functor; we wish to prove that f is accessible. By Proposition 5.3.5.10, we may suppose that $f = F \circ j$, where $j : \mathcal{C} \to \operatorname{Ind}_{\kappa}(\mathcal{C})$ is the Yoneda embedding and $F : \operatorname{Ind}_{\kappa}(\mathcal{C}) \to \mathcal{D}$ is a κ-continuous functor and therefore accessible. Enlarging κ if necessary, we may suppose that j is an equivalence of ∞-categories, so that f is accessible as well. $\square$

5.4.4 Accessibility of Functor ∞-Categories

Let $\mathcal{C}$ be an accessible ∞-category and let K be a small simplicial set. Our goal in this section is to prove that $\operatorname{Fun}(K, \mathcal{C})$ is accessible (Proposition 5.4.4.3). In §5.4.7, we will prove a much more general stability result of this kind (Corollary 5.4.7.17), but the proof of that result ultimately rests on the ideas presented here.

Our proof proceeds roughly as follows. If $\mathcal{C}$ is accessible, then $\mathcal{C}$ has many τ-compact objects provided that τ is sufficiently large. Using Proposition 5.3.4.13, we deduce the existence of many τ-compact objects in $\operatorname{Fun}(K, \mathcal{C})$. Our main problem is to show that these objects generate $\operatorname{Fun}(K, \mathcal{C})$ under τ-filtered colimits. To prove this, we will use a rather technical cofinality result (Lemma 5.4.4.2 below). We begin with the following preliminary observation:

Lemma 5.4.4.1. *Let τ be a regular cardinal and let $q : Y \to X$ be a co-Cartesian fibration with the property that for every vertex x of X, the fiber $Y_x = Y \times_X \{x\}$ is τ-filtered. Then q has the right lifting property with respect to $K \subseteq K^\triangleright$ for every τ-small simplicial set K.*

Proof. Using Proposition A.2.3.1, we can reduce to the problem of showing that q has the right lifting property with respect to the inclusion $K \subseteq K \diamond \Delta^0$. In other words, we must show that given any edge $e : C \to D$ in X^K, where D is a constant map, and any vertex $\widetilde{C}$ of Y^K lifting C, there exists an edge $\widetilde{e} : \widetilde{C} \to \widetilde{D}$ lifting $\widetilde{e}$, where $\widetilde{D}$ is a constant map from K to Y. We first choose an arbitrary edge $\widetilde{e}' : \widetilde{C} \to \widetilde{D}'$ lifting e (since the map $q^K : Y^K \to X^K$ is a coCartesian fibration, we can even choose $\widetilde{e}'$ to be q^K-coCartesian, though we will not need this). Suppose that D takes the constant value $x : \Delta^0 \to X$. Since the fiber Y_x is τ-filtered, there exists an edge $\widetilde{e}'' : \widetilde{D}' \to \widetilde{D}$ in Y_x^K, where $\widetilde{D}$ is a constant map from K to Y_x. We now invoke the fact that q^K is an inner fibration to supply the dotted arrow in the diagram

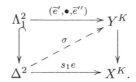

We now define $\widetilde{e} = \sigma | \Delta^{\{0,2\}}$. $\qquad\square$

Lemma 5.4.4.2. *Let $\kappa < \tau$ be regular cardinals. Let $q : Y \to X$ be a map of simplicial sets with the following properties:*

(i) *The simplicial set X is τ-small.*

(ii) *The map q is a coCartesian fibration.*

(iii) *For every vertex $x \in X$, the fiber $Y_x = Y \times_X \{x\}$ is τ-filtered and admits τ-small κ-filtered colimits.*

(iv) *For every edge* $e : x \to y$ *in* X, *the associated functor* $Y_x \to Y_y$ *preserves* τ-*small* κ-*filtered colimits.*

Then

(1) *The* ∞-*category* $\mathcal{C} = \mathrm{Map}_{/X}(X, Y)$ *of sections of* q *is* τ-*filtered.*

(2) *For each vertex* x *of* X, *the evaluation map* $e_x : \mathcal{C} \to Y_x$ *is cofinal.*

Proof. Choose a categorical equivalence $X \to M$, where M is a minimal ∞-category. Since τ is uncountable, Proposition 5.4.1.2 implies that M is τ-small. According to Corollary 3.3.1.2, Y is equivalent to the pullback of a coCartesian fibration $Y' \to M$. We may therefore replace X by M and thereby reduce to the case where X is a minimal ∞-category. For each ordinal α, let $(\alpha) = \{\beta < \alpha\}$.

Let K be a τ-small simplicial set equipped with a map $f : K \to Y$. We define a new object $K'_X \in (\mathrm{Set}_\Delta)_{/X}$ as follows. For every finite nonempty linearly ordered set J, a map $\Delta^J \to K'_X$ is determined by the following data:

- A map $\chi : \Delta^J \to X$.

- A map $\Delta^J \to \Delta^2$ corresponding to a decomposition $J = J_0 \coprod J_1 \coprod J_2$.

- A map $\Delta^{J_0} \to K$.

- An order-preserving map $m : J_1 \to (\kappa)$ having the property that if $m(i) = m(j)$, then $\chi(\Delta^{\{i,j\}})$ is a degenerate edge of X.

We will prove the existence of a dotted arrow F'_X as indicated in the diagram

$$
\begin{array}{ccc}
K & \xrightarrow{\;f\;} & Y \\
\downarrow & \nearrow^{F'_X} & \downarrow{q} \\
\downarrow & \nearrow & \downarrow \\
K'_X & \longrightarrow & X.
\end{array}
$$

Let $K'' \subseteq K'_X$ be the simplicial subset corresponding to simplices as above, where $J_1 = \emptyset$, and let $F'' = F'_X | K''$. Specializing to the case where $K = Z \times X$, Z a τ-small simplicial set, we will deduce that any diagram $Z \to \mathcal{C}$ extends to a map $Z^\triangleright \to \mathcal{C}$ (given by F''), which proves (1). Similarly, by specializing to the case $K = (Z \times X) \coprod_{Z \times \{x\}} (Z^\triangleleft \times \{x\})$, we will deduce that for every object $y \in Y$ with $q(y) = x$, the ∞-category $\mathcal{C} \times_{Y_x} (Y_x)_{y/}$ is τ-filtered and therefore weakly contractible. Applying Theorem 4.1.3.1, we deduce (2).

It remains to construct the map F'_X. There is no harm in enlarging K. We may therefore apply the small object argument to replace K by an ∞-category (which we may also suppose is τ-small since τ is uncountable). We begin by defining, for each $\alpha \leq \kappa$, a simplicial subset $K(\alpha) \subseteq K'_X$. The definition is as follows: we will say that a simplex $\Delta^J \to K'_X$ factors through

$K(\alpha)$ if, in the corresponding decomposition $J = J_0 \coprod J_1 \coprod J_2$, we have $J_2 = \emptyset$ and the map $J_1 \to (\kappa)$ factors through (α). Our first task is to construct $F(\alpha) = F'_X | K(\alpha)$, which we do by induction on α. If $\alpha = 0$, $K(\alpha) = K$ and we set $F(0) = f$. When α is a limit ordinal, we have $K(\alpha) = \bigcup_{\beta < \alpha} K(\beta)$ and we set $F(\alpha) = \bigcup_{\beta < \alpha} F(\beta)$. It therefore suffices to construct $F(\alpha + 1)$, assuming that $F(\alpha)$ has already been constructed. For each vertex x of X, let $\widetilde{x} = (x, \alpha)$ denote the unique vertex of $K(\alpha + 1)$ lying over x which does not belong to $K(\alpha)$. Since X is minimal, Proposition 2.3.3.9 implies that we have a pushout diagram

$$\begin{array}{ccc}
\coprod_x K(\alpha)_{/\widetilde{x}} & \longrightarrow & \coprod_x (K(\alpha)_{/\widetilde{x}})^{\triangleright} \\
\downarrow & & \downarrow \\
K(\alpha) & \longrightarrow & K(\alpha + 1).
\end{array}$$

Therefore, to construct $f_{\alpha+1}$, it suffices to prove that q has the right lifting property with respect to each inclusion $K(\alpha)_{/\widetilde{x}} \subseteq (K(\alpha)_{/\widetilde{x}})^{\triangleright}$, which follows from Lemma 5.4.4.1.

We now define, for each simplicial subset $X' \subseteq X$, a corresponding simplicial subset $K'_{X'} \subseteq K'_X$. The definition is as follows: let $\sigma : \Delta^J \to K'_X$ be a simplex corresponding to a decomposition $J = J_0 \coprod J_1 \coprod J_2$. Then σ factors through $K'_{X'}$ if and only if the induced map $\Delta^{J_2} \to X$ factors through X'. Our next job is to extend the definition of F'_X from $K'_\emptyset = K(\kappa)$ to K'_X by adjoining simplices to X one at a time.

Let $F'_\emptyset = F(\kappa)$ and let x be a vertex of X. We begin by defining a map $F'_{\{x\}} : K'_{\{x\}} \to Y$ which extends $F'_\emptyset$. Since X is minimal, there is a pushout diagram

$$\begin{array}{ccc}
K(\kappa)_{/x} & \longrightarrow & K(\kappa)^{\triangleright}_{/x} \\
\downarrow & & \downarrow \\
K_\emptyset & \longrightarrow & K_{\{x\}}
\end{array}$$

where $K(\kappa)_{/x}$ denotes the fiber product $K(\kappa) \times_X X_{/x}$. Constructing an extension $F'_{\{x\}}$ of $F'_\emptyset$ is therefore equivalent to providing the dotted arrow indicated in the diagram

$$\begin{array}{ccc}
K(\kappa)_{/x} & \xrightarrow{\ p_x\ } & Y \\
\downarrow & \overset{\overline{p}_x}{\nearrow} & \downarrow \\
K(\kappa)^{\triangleright}_{/x} & \longrightarrow & X.
\end{array}$$

We will choose $\overline{p}_x$ to be a relative colimit of p_x over X (see §4.3.1). To prove that such a relative colimit exists, we consider the inclusion $i_x : N(\kappa) \subseteq K(\kappa)_{/x} \times_{X_{/x}} \{\mathrm{id}_x\} \subseteq K(\kappa)_{/x}$. Using Proposition 2.3.3.9, it is not difficult to see that $K(\kappa)_{/x}$ is an ∞-category. For each object $y \in K(\kappa)_{/x}$, the minimality of X implies that $N(\kappa) \times_{\mathcal{K}_{/x}} (\mathcal{K}_{/x})_{y/}$ is isomorphic to $N(\{\alpha : \beta < \alpha < \kappa\})$

for some $\beta < \kappa$ and therefore weakly contractible. Theorem 4.1.3.1 implies that i_x is cofinal. Invoking Proposition 4.3.1.8, it will suffice to prove that $p_x \circ i_x : N(\kappa) \to Y$ admits a relative colimit over X. Using conditions (ii) and (iv) together with Proposition 4.3.1.10, we may reduce to producing a colimit of $p_x \circ i_x$ in the ∞-category Y_x, which is possible by virtue of assumption (iii).

Applying the above argument separately to each vertex of X, we may suppose that $F'_{X^{(0)}}$ has been constructed, where $X^{(0)}$ denotes the 0-skeleton of X. We now consider the collection of all pairs $(X', F'_{X'})$, where X' is a simplicial subset of X containing all vertices of X, and $F'_{X'} : K_{X'} \to Y$ is a map over X whose restriction to K_{X^0} coincides with $F'_{X^{(0)}}$. This collection is partially ordered if we write $(X', F'_{X'}) \leq (X'', F'_{X''})$ to mean that $X' \subseteq X''$ and $F'_{X''}|K_{X'} = F'_{X'}$. The hypotheses of Zorn's lemma are satisfied, so that there exists a maximal such pair $(X', F'_{X'})$. To complete the proof, it suffices to show that $X' = X$. If not, we can choose $X' \subseteq X'' \subseteq X$, where X'' is obtained from X' by adjoining a single nondegenerate simplex $\sigma : \Delta^n \to X$ whose boundary already belongs to X'. Since X' contains $X^{(0)}$, we may suppose that $n > 0$. Let $K(\kappa)_{/\sigma} = K(\kappa) \times_X X_{/\sigma}$ and let $x = \sigma(0)$. Since X is minimal, we have a pushout diagram

$$
\begin{array}{ccc}
K(\kappa)_{/\sigma} \star \partial \Delta^n & \longrightarrow & K(\kappa)_{/\sigma} \star \Delta^n \\
\downarrow & & \downarrow \\
K'_{X'} & \longrightarrow & K'_{X''}.
\end{array}
$$

Let $s : K(\kappa)_{/\sigma} \to Y$ denote the composition of the projection $K(\kappa)_{/\sigma} \to K'_{X'}$ with $F'_{X'}$. We obtain a commutative diagram

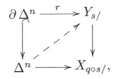

and supplying the indicated dotted arrow is tantamount to giving a map $F'_{X''} : K_{X''} \to Y$ over X which extends $F'_{X'}$. To prove the existence of $F'_{X''}$, it suffices to prove that the map $\bar{s} : K'^{\rhd} \to Y$ associated to $r(0)$ is a q-colimit diagram. We note that $\bar{s}$ is given as a composition

$$
K'^{\rhd} \to K^{\rhd}_{/x} \xrightarrow{\bar{s}'} Y,
$$

where $\bar{s}'$ is a q-colimit diagram by construction. According to Proposition 4.3.1.7, it will suffice to show that the map $K(\kappa)_{/\sigma} \to K(\kappa)_{/x}$ is cofinal. We have a pullback diagram

$$
\begin{array}{ccc}
K(\kappa)_{/\sigma} & \longrightarrow & K(\kappa)_{/x} \\
\downarrow & & \downarrow \\
X_{/\sigma} & \longrightarrow & X_{/x},
\end{array}
$$

where the lower horizontal map is a trivial fibration of simplicial sets. It follows that the upper horizontal map is a trivial fibration and, in particular, cofinal. Consequently, there exists an extension $F_{X''}$ of $F_{X'}$, which contradicts the maximality of $(X', F_{X'})$ and completes the proof.   □

Proposition 5.4.4.3. *Let $\mathcal{C}$ be an accessible ∞-category and let K be a small simplicial set. Then $\mathrm{Fun}(K, \mathcal{C})$ is accessible.*

Proof. Without loss of generality, we may suppose that K is an ∞-category. Choose a regular cardinal κ such that $\mathcal{C}$ admits small κ-filtered colimits and choose a second regular cardinal $\tau > \kappa$ such that $\mathcal{C}$ is also τ-accessible and K is τ-small. We will prove that $\mathrm{Fun}(K, \mathcal{C})$ is τ-accessible. Let $\mathcal{C}' = \mathrm{Fun}(K, \mathcal{C}^\tau) \subseteq \mathrm{Fun}(K, \mathcal{C})$. It is clear that $\mathcal{C}'$ is essentially small. Proposition 5.1.2.2 implies that $\mathrm{Fun}(K, \mathcal{C})$ admits small τ-filtered colimits, and Proposition 5.3.4.13 asserts that $\mathcal{C}'$ consists of τ-compact objects of $\mathrm{Fun}(K, \mathcal{C})$. According to Proposition 5.4.2.2, it will suffice to prove that $\mathcal{C}'$ generates $\mathrm{Fun}(K, \mathcal{C})$ under small τ-filtered colimits.

Without loss of generality, we may suppose that $\mathcal{C} = \mathrm{Ind}_\tau \, \mathcal{D}'$, where $\mathcal{D}'$ is a small ∞-category. Let $\mathcal{D} \subseteq \mathcal{C}$ denote the essential image of the Yoneda embedding. Let $F : K \to \mathcal{C}$ be an arbitrary object of $\mathcal{C}^K$ and let $\mathrm{Fun}(K, \mathcal{D})^{/F} = \mathrm{Fun}(K, \mathcal{D}) \times_{\mathrm{Fun}(K, \mathcal{C})} \mathrm{Fun}(K, \mathcal{C})^{/F}$. Consider the composite diagram

$$\overline{p} : \mathrm{Fun}(K, \mathcal{D})^{/F} \diamond \Delta^0 \to \mathrm{Fun}(K, \mathcal{C})^{/F} \diamond \Delta^0 \to \mathrm{Fun}(K, \mathcal{C}).$$

The ∞-category $\mathrm{Fun}(K, \mathcal{D})^{/F}$ is equivalent to

$$\mathrm{Fun}(K, \mathcal{D}') \times_{\mathrm{Fun}(K, \mathcal{C})} \mathrm{Fun}(K, \mathcal{C})^{/F}$$

and therefore essentially small. To complete the proof, it will suffice to show that $\mathrm{Fun}(K, \mathcal{D})^{/F}$ is τ-filtered and that $\overline{p}$ is a colimit diagram.

We may identify F with a map $f_K : K \to \mathcal{C} \times K$ in $(\mathrm{Set}_\Delta)_{/K}$. According to Proposition 4.2.2.4, we obtain a coCartesian fibration $q : (\mathcal{C} \times K)^{/f_K} \to K$, and the q-coCartesian morphisms are precisely those which project to equivalences in $\mathcal{C}$. Let X denote the full subcategory of $(\mathcal{C} \times K)^{/f_K}$ consisting of those objects whose projection to $\mathcal{C}$ belongs to $\mathcal{D}$. It follows that $q' = q|X : X \to K$ is a coCartesian fibration. We may identify the fiber of q' over a vertex $x \in K$ with $\mathcal{D}^{/F(x)} = \mathcal{D} \times_\mathcal{C} \mathcal{C}^{/F(x)}$. It follows that the fibers of q' are τ-filtered ∞-categories; Lemma 5.4.4.2 now guarantees that $\mathrm{Fun}(K, \mathcal{D})^{/F} \simeq \mathrm{Map}_{/K}(K, X)$ is τ-filtered.

According to Proposition 5.1.2.2, to prove that $\overline{p}$ is a colimit diagram, it will suffice to prove that for every vertex x of K, the composition of $\overline{p}$ with the evaluation map $e_x : \mathrm{Fun}(K, \mathcal{C}) \to \mathcal{C}$ is a colimit diagram. The composition $e_x \circ \overline{p}$ admits a factorization

$$\mathrm{Fun}(K, \mathcal{D})^{/F} \diamond \Delta^0 \to \mathcal{D}^{/F(x)} \diamond \Delta^0 \to \mathcal{C}$$

where $\mathcal{D}^{/F(x)} = \mathcal{D} \times_\mathcal{C} \mathcal{C}^{/F(x)}$ and the second map is a colimit diagram in $\mathcal{C}$ by Lemma 5.1.5.3. It will therefore suffice to prove that the map $g_x : \mathrm{Fun}(K, \mathcal{D})^{/F} \to \mathcal{D}^{/F(x)}$ is cofinal, which follows from Lemma 5.4.4.2.   □

5.4.5 Accessibility of Undercategories

Let $\mathcal{C}$ be an accessible ∞-category and let $p : K \to \mathcal{C}$ be a small diagram. Our goal in this section is to prove that the ∞-category $\mathcal{C}_{p/}$ is accessible (Corollary 5.4.5.16).

Remark 5.4.5.1. The analogous result for the ∞-category $\mathcal{C}_{/p}$ will be proven in §5.4.6 using Propositions 5.4.4.3 and 5.4.6.6. It is possible to use the same argument to give a second proof of Corollary 5.4.5.16; however, we will *need* Corollary 5.4.5.16 in our proof of Proposition 5.4.6.6.

We begin by studying the behavior of colimits with respect to (homotopy) fiber products of ∞-categories.

Lemma 5.4.5.2. *Let*

$$
\begin{array}{ccc}
\mathcal{X}' & \xrightarrow{q'} & \mathcal{X} \\
{\scriptstyle p'}\downarrow & & \downarrow{\scriptstyle p} \\
\mathcal{Y}' & \xrightarrow{q} & \mathcal{Y}
\end{array}
$$

be a diagram of ∞-categories which is homotopy Cartesian (with respect to the Joyal model structure). Suppose that $\mathcal{X}$ and $\mathcal{Y}$ have initial objects and that p and q preserve initial objects. An object $X' \in \mathcal{X}'$ is initial if and only if $p'(X')$ is an initial object of $\mathcal{Y}'$ and $q'(X')$ is an initial object of $\mathcal{X}$. Moreover, there exists an initial object of $\mathcal{X}'$.

Proof. Without loss of generality, we may suppose that p and q are categorical fibrations and that $\mathcal{X}' = \mathcal{X} \times_{\mathcal{Y}} \mathcal{Y}'$. Suppose first that X' is an object of $\mathcal{X}'$ with the property that $X = q'(X')$ and $Y' = p'(X')$ are initial objects of $\mathcal{X}$ and $\mathcal{Y}'$. Then $Y = p(X) = q(Y')$ is an initial object of $\mathcal{Y}$. Let Z be another object of $\mathcal{X}'$. We have a pullback diagram of Kan complexes

$$
\begin{array}{ccc}
\mathrm{Hom}^{\mathrm{R}}_{\mathcal{X}'}(X', Z) & \longrightarrow & \mathrm{Hom}^{\mathrm{R}}_{\mathcal{X}}(X, q'(Z)) \\
\downarrow & & \downarrow \\
\mathrm{Hom}^{\mathrm{R}}_{\mathcal{Y}'}(Y', p'(Z)) & \longrightarrow & \mathrm{Hom}^{\mathrm{R}}_{\mathcal{Y}}(Y, (q \circ p')(Z)).
\end{array}
$$

Since the maps p and q are inner fibrations, Lemma 2.4.4.1 implies that this diagram is homotopy Cartesian (with respect to the usual model structure on Set_Δ). Since X, Y', and Y are initial objects, each one of the Kan complexes $\mathrm{Hom}^{\mathrm{R}}_{\mathcal{X}}(X, q'(Z))$, $\mathrm{Hom}^{\mathrm{R}}_{\mathcal{Y}'}(Y', p'(Z))$, and $\mathrm{Hom}^{\mathrm{R}}_{\mathcal{Y}}(Y, (q \circ p')(Z))$ is contractible. It follows that $\mathrm{Hom}^{\mathrm{R}}_{\mathcal{X}'}(X', Z)$ is contractible as well, so that X' is an initial object of $\mathcal{X}'$.

We now prove that there exists an object $X' \in \mathcal{X}'$ such that $p'(X')$ and $q'(X')$ are initial. The above argument shows that X' is an initial object of $\mathcal{X}'$. Since all initial objects of $\mathcal{X}'$ are equivalent, this will prove that for *any* initial object $X'' \in \mathcal{X}'$, the objects $p'(X'')$ and $q'(X'')$ are initial.

We begin by selecting arbitrary initial objects $X \in \mathfrak{X}$ and $\overline{Y} \in \mathcal{Y}'$. Then $p(X)$ and $q(\overline{Y})$ are both initial objects of $\mathcal{Y}$, so there is an equivalence $e : p(X) \to q(\overline{Y})$. Since q is a categorical fibration, there exists an equivalence $\overline{e} : Y' \to \overline{Y}$ in $\mathcal{Y}$ such that $q(\overline{e}) = e$. It follows that Y' is an initial object of $\mathcal{Y}'$ with $q(Y') = p(X)$, so that the pair (X, Y') can be identified with an object of $\mathfrak{X}'$ which has the desired properties. $\qquad\square$

Lemma 5.4.5.3. *Let $p : \mathfrak{X} \to \mathcal{Y}$ be a categorical fibration of ∞-categories, and let $f : K \to \mathfrak{X}$ be a diagram. Then the induced map $p' : \mathfrak{X}_{f/} \to \mathcal{Y}_{pf/}$ is a categorical fibration.*

Proof. It suffices to show that p' has the right lifting property with respect to every inclusion $A \subseteq B$ which is a categorical equivalence. Unwinding the definitions, it suffices to show that p has the right lifting property with respect to $i : K \star A \subseteq K \star B$. This is immediate since p is a categorical fibration and i is a categorical equivalence. $\qquad\square$

Lemma 5.4.5.4. *Let*

$$
\begin{array}{ccc}
\mathfrak{X}' & \xrightarrow{q'} & \mathfrak{X} \\
\downarrow{\scriptstyle p'} & & \downarrow{\scriptstyle p} \\
\mathcal{Y}' & \xrightarrow{q} & \mathcal{Y}
\end{array}
$$

be a diagram of ∞-categories which is homotopy Cartesian (with respect to the Joyal model structure) and let $f : K \to \mathfrak{X}'$ be a diagram in $\mathfrak{X}'$. Then the induced diagram

$$
\begin{array}{ccc}
\mathfrak{X}'_{f/} & \longrightarrow & \mathfrak{X}_{q'f/} \\
\downarrow & & \downarrow \\
\mathcal{Y}'_{p'f/} & \longrightarrow & \mathcal{Y}_{qp'f/}
\end{array}
$$

is also homotopy Cartesian.

Proof. Without loss of generality, we may suppose that p and q are categorical fibrations and that $\mathfrak{X}' = \mathfrak{X} \times_{\mathcal{Y}} \mathcal{Y}'$. Then $\mathfrak{X}'_{f/} \simeq \mathfrak{X}_{q'f/} \times_{\mathcal{Y}_{qp'f/}} \mathcal{Y}'_{p'f/}$, so the result follows immediately from Lemma 5.4.5.3. $\qquad\square$

Lemma 5.4.5.5. *Let*

$$
\begin{array}{ccc}
\mathfrak{X}' & \xrightarrow{q'} & \mathfrak{X} \\
\downarrow{\scriptstyle p'} & & \downarrow{\scriptstyle p} \\
\mathcal{Y}' & \xrightarrow{q} & \mathcal{Y}
\end{array}
$$

be a diagram of ∞-categories which is homotopy Cartesian (with respect to the Joyal model structure) and let K be a simplicial set. Suppose that $\mathfrak{X}$ and $\mathcal{Y}'$ admit colimits for all diagrams indexed by K and that p and q preserve colimits of diagrams indexed by K. Then

(1) *A diagram* $\overline{f} : K^{\triangleright} \to \mathfrak{X}'$ *is a colimit of* $f = \overline{f}|K$ *if and only if* $p' \circ \overline{f}$ *and* $q' \circ \overline{f}$ *are colimit diagrams. In particular,* p' *and* q' *preserve colimits indexed by* K.

(2) *Every diagram* $f : K \to \mathfrak{X}'$ *has a colimit in* $\mathfrak{X}'$.

Proof. Replacing $\mathfrak{X}'$ by $\mathfrak{X}'_{f/}$, $\mathfrak{X}$ by $\mathfrak{X}_{q'f/}$, $\mathcal{Y}'$ by $\mathcal{Y}'_{p'f/}$, and $\mathcal{Y}$ by $\mathcal{Y}_{qp'f/}$, we may apply Lemma 5.4.5.4 to reduce to the case $K = \emptyset$. Now apply Lemma 5.4.5.2. $\square$

Lemma 5.4.5.6. *Let* $\mathcal{C}$ *be a small filtered category and let* $\mathcal{C}^{\triangleright}$ *be the category obtained by adjoining a (new) final object to* $\mathcal{C}$. *Suppose we are given a homotopy pullback diagram*

$$
\begin{array}{ccc}
F' & \longrightarrow & F \\
\downarrow & & \downarrow{\scriptstyle p} \\
G' & \xrightarrow{\;q\;} & G
\end{array}
$$

in the diagram category $\mathrm{Set}_{\Delta}^{\mathcal{C}^{\triangleright}}$ *(which we endow with the projective model structure). Suppose further that the diagrams* $F, G, G' : \mathcal{C}^{\triangleright} \to \mathrm{Set}_{\Delta}$ *are homotopy colimits. Then* F' *is also a homotopy colimit diagram.*

Proof. Without loss of generality, we may suppose that G is fibrant, that p and q are fibrations, and that $F' = F \times_G G'$. Let $*$ denote the cone point of $\mathcal{C}^{\triangleright}$ and let $F(\infty)$, $G(\infty)$, $F'(\infty)$, and $G'(\infty)$ denote the colimits of the diagrams $F|\mathcal{C}$, $G|\mathcal{C}$, $F'|\mathcal{C}$, and $G'|\mathcal{C}$. Since fibrations in Set_{Δ} are stable under filtered colimits, the pullback diagram

$$
\begin{array}{ccc}
F'(\infty) & \longrightarrow & F(\infty) \\
\downarrow & & \downarrow \\
G'(\infty) & \longrightarrow & G(\infty)
\end{array}
$$

exhibits $F'(\infty)$ as a homotopy fiber product of $F(\infty)$ and $G'(\infty)$ over $G(\infty)$ in Set_{Δ}. Since weak homotopy equivalences are stable under filtered colimits, the natural maps $G(\infty) \to G(*)$, $F'(\infty) \to F'(*)$, and $G'(\infty) \to G'(*)$ are weak homotopy equivalences. Consequently, the diagram

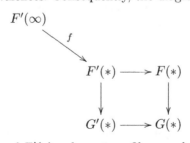

exhibits both $F'(\infty)$ and $F'(*)$ as homotopy fiber products of $F(*)$ and $G'(*)$ over $G(*)$. It follows that f is a weak homotopy equivalence, so that F is a homotopy colimit diagram, as desired. $\square$

Lemma 5.4.5.7. *Let*

$$
\begin{array}{ccc}
\mathfrak{X}' & \xrightarrow{q'} & \mathfrak{X} \\
\downarrow{\scriptstyle p'} & & \downarrow{\scriptstyle p} \\
\mathfrak{Y}' & \xrightarrow{q} & \mathfrak{Y}
\end{array}
$$

be a diagram of ∞-categories which is homotopy Cartesian (with respect to the Joyal model structure) and let κ be a regular cardinal. Suppose that $\mathfrak{X}$ and $\mathfrak{Y}'$ admit small κ-filtered colimits and that p and q preserve small κ-filtered colimits. Then

(1) *The ∞-category $\mathfrak{X}'$ admits small κ-filtered colimits.*

(2) *If X' is an object of $\mathfrak{X}'$ such that $Y' = p'(X')$ and $X = q'(X')$, and $Y = p(X) = q(Y')$ are κ-compact, then X' is a κ-compact object of $\mathfrak{X}'$.*

Proof. Claim (1) follows immediately from Lemma 5.4.5.5. To prove (2), consider a colimit diagram $\overline{f} : \mathfrak{I}^{\triangleright} \to \mathfrak{X}'$. We wish to prove that the composition of $\overline{f}$ with the functor $\mathfrak{X}' \to \widehat{\mathcal{S}}$ corepresented by X' is also a colimit diagram. Using Proposition 5.3.1.16, we may assume without loss of generality that $\mathfrak{I}$ is the nerve of a κ-filtered partially ordered set A. We may further suppose that p and q are categorical fibrations and that $\mathfrak{X}' = \mathfrak{X} \times_{\mathfrak{Y}} \mathfrak{Y}'$. Let $\mathfrak{I}^{\triangleright}_{X'/}$ denote the fiber product $\mathfrak{I}^{\triangleright} \times_{\mathfrak{X}'} \mathfrak{X}_{X'/}$ and define $\mathfrak{I}^{\triangleright}_{X/}$, $\mathfrak{I}^{\triangleright}_{Y'/}$, and $\mathfrak{I}^{\triangleright}_{Y/}$ similarly. We have a pullback diagram

$$
\begin{array}{ccc}
\mathfrak{I}^{\triangleright}_{X'/} & \longrightarrow & \mathfrak{I}^{\triangleright}_{X/} \\
\downarrow & & \downarrow \\
\mathfrak{I}^{\triangleright}_{Y'/} & \longrightarrow & \mathfrak{I}^{\triangleright}_{Y/}
\end{array}
$$

of left fibrations over $\mathfrak{I}^{\triangleright}$. Proposition 2.1.2.1 implies that every arrow in this diagram is a left fibration, so that Corollary 3.3.1.6 implies that $\mathfrak{I}^{\triangleright}_{X'/}$ is a homotopy fiber product of $\mathfrak{I}^{\triangleright}_{X/}$ and $\mathfrak{I}^{\triangleright}_{Y'/}$ over $\mathfrak{I}^{\triangleright}_{Y/}$ in the covariant model category $(\mathcal{S}\mathrm{et}_{\Delta})_{/\mathfrak{I}^{\triangleright}}$. Let $G : (\mathcal{S}\mathrm{et}_{\Delta})^{A\cup\{\infty\}} \to (\mathcal{S}\mathrm{et}_{\Delta})_{\mathfrak{I}^{\triangleright}}$ denote the unstraightening functor of §2.1.4. Since G is the right Quillen functor of a Quillen equivalence, the above diagram is weakly equivalent to the image under G of a homotopy pullback diagram

$$
\begin{array}{ccc}
F_{X'} & \longrightarrow & F_X \\
\downarrow & & \downarrow \\
F_{Y'} & \longrightarrow & F_Y
\end{array}
$$

of (weakly) fibrant objects of $(\mathcal{S}\mathrm{et}_{\Delta})^{A\cup\{\infty\}}$. Moreover, the simplicial nerve of each F_Z can be identified with the composition of $\overline{f}$ with the functor corepresented by Z. According to Theorem 4.2.4.1, it will suffice to show that $F_{X'}$ is a homotopy colimit diagram. We now observe that F_X, $F_{Y'}$, and F_Y are homotopy colimit diagrams (since X, Y', and Y are assumed to be κ-compact) and conclude by applying Lemma 5.4.5.6. $\qquad\square$

In some of the arguments below, it will be important to be able to replace colimits of a diagram $\mathcal{J} \to \mathcal{C}$ by colimits of some composition $\mathcal{I} \xrightarrow{f} \mathcal{J} \to \mathcal{C}$. According to Proposition 4.1.1.8, this maneuver is justified provided that f is cofinal. Unfortunately, the class of cofinal morphisms is not sufficiently robust for our purposes. We will therefore introduce a property somewhat stronger than cofinality which has better stability properties.

Definition 5.4.5.8. Let $f : \mathcal{I} \to \mathcal{J}$ denote a functor between filtered ∞-categories. We will say that f is *weakly cofinal* if, for every object $J \in \mathcal{J}$, there exists an object $I \in \mathcal{I}$ and a morphism $J \to f(I)$ in $\mathcal{J}$. We will say that f is *κ-cofinal* if, for every diagram $p : K \to \mathcal{I}$ where K is κ-small and weakly contractible, the induced functor $\mathcal{I}_{p/} \to \mathcal{J}_{fp/}$ is weakly cofinal.

Example 5.4.5.9. Let $\mathcal{I}$ be a τ-filtered ∞-category and let $p : K \to \mathcal{I}$ be a τ-small diagram. Then the projection $\mathcal{I}_{p/} \to \mathcal{I}$ is τ-cofinal. To prove this, consider a τ-small diagram $K' \to \mathcal{I}_{p/}$, where K' is weakly contractible, corresponding to a map $q : K \star K' \to \mathcal{I}$. According to Lemma 4.2.3.6, the inclusion $K' \subseteq K \star K'$ is right anodyne, so that the map $\mathcal{I}_{q/} \to \mathcal{I}_{q|K'/}$ is a trivial fibration (and therefore weakly cofinal).

Lemma 5.4.5.10. *Let A, B, and C be simplicial sets and suppose that B is weakly contractible. Then the inclusion*

$$(A \star B) \coprod_{B} (B \star C) \subseteq A \star B \star C$$

is a categorical equivalence.

Proof. Let $F(A, B, C) = (A \star B) \coprod_{B} (B \star C)$ and let $G(A, B, C) = A \star B \star C$. We first observe that both F and G preserve filtered colimits and homotopy pushout squares separately in each argument. Using standard arguments (see, for example, the proof of Proposition 2.2.2.7), we can reduce to the case where A and C are simplices.

Let us say that a simplicial set B is *good* if the inclusion $F(A, B, C) \subseteq G(A, B, C)$ is a categorical equivalence. We now make the following observations:

(1) Every simplex is good. Unwinding the definitions, this is equivalent to the assertion that for $0 \le m \le n \le p$, the diagram

$$
\begin{array}{ccc}
\Delta^{\{m,\dots,n\}} & \hookrightarrow & \Delta^{\{0,\dots,n\}} \\
\downarrow & & \downarrow \\
\Delta^{\{m,\dots,p\}} & \hookrightarrow & \Delta^{\{0,\dots,p\}}
\end{array}
$$

is a homotopy pushout square (with respect to the Joyal model structure). For $0 \le i \le j \le p$, set

$$X_{ij} = \Delta^{\{i,i+1\}} \coprod_{\{i\}} \cdots \coprod_{\{j-1\}} \Delta^{\{j-1,j\}} \subseteq \Delta^{\{i,\dots,j\}}$$

(by convention, we agree that $X_{ij} = \{i\}$ if $i = j$). Since each of the inclusions $X_{ij} \subseteq \Delta^{\{i,\dots,j\}}$ is inner anodyne, it will suffice to show that the diagram

$$
\begin{array}{ccc}
X_{mn} & \longrightarrow & X_{0n} \\
\downarrow & & \downarrow \\
X_{mp} & \longrightarrow & X_{0p}
\end{array}
$$

is a homotopy pushout square, which is clear.

(2) Given a pushout diagram of simplicial sets

$$
\begin{array}{ccc}
B & \longrightarrow & B' \\
\uparrow & & \uparrow \\
\downarrow & & \downarrow \\
B'' & \longrightarrow & B'''
\end{array}
$$

in which the vertical arrows are cofibrations, if B, B', and B'' are good, then B''' is good. This follows from the compatibility of the functors F and G with homotopy pushouts in B.

(3) Every horn Λ_i^n is good. This follows by induction on n using (1) and (2).

(4) The collection of good simplicial sets is stable under filtered colimits; this follows from the compatibility of F and G with filtered colimits and the stability of categorical equivalences under filtered colimits.

(5) Every retract of a good simplicial set is good (since the collection of categorical equivalences is stable under the formation of retracts).

(6) If $i : B \to B'$ is an anodyne map of simplicial sets and B is good, then B' is good. This follows by combining observations (1) through (5).

(7) If B is weakly contractible, then B is good. To see this, choose a vertex b of B. The simplicial set $\{b\} \simeq \Delta^0$ is good (by (1)), and the inclusion $\{b\} \subseteq B$ is anodyne. Now apply (6).

$\square$

Lemma 5.4.5.11. *Let κ and τ be regular cardinals, let $f : \mathfrak{I} \to \mathfrak{J}$ be a κ-cofinal functor between τ-filtered ∞-categories, and let $p : K \to \mathfrak{J}$ be a κ-small diagram. Then*

(1) *The ∞-category $\mathfrak{I}_{p/} = \mathfrak{I} \times_{\mathfrak{J}} \mathfrak{J}_{p/}$ is τ-filtered.*

(2) *The induced functor $\mathfrak{I}_{p/} \to \mathfrak{J}_{p/}$ is κ-cofinal.*

Proof. We first prove (1). Let $\widetilde{q} : K' \to \mathcal{I}_{p/}$ be a τ-small diagram classifying a compatible pair of maps $q : K' \to \mathcal{I}$ and $q' : K \star K' \to \mathcal{J}$. Since $\mathcal{I}$ is τ-filtered, we can find an extension $\overline{q} : (K')^{\triangleright} \to \mathcal{I}$ of q. To find a compatible extension of $\widetilde{q}$, it suffices to solve the lifting problem

$$
\begin{array}{ccc}
(K \star K') \coprod_{K'} (K')^{\triangleright} & \longrightarrow & \mathcal{J} \\
\downarrow{\scriptstyle i} & \nearrow & \\
(K \star K')^{\triangleright}, & &
\end{array}
$$

which is possible since i is a categorical equivalence (Lemma 5.4.5.10) and $\mathcal{J}$ is an ∞-category.

To prove (2), we consider a map $\widetilde{q} : K' \to \mathcal{I}_{p/}$ as above, where K is now κ-small and weakly contractible. We have a pullback diagram

$$
\begin{array}{ccc}
(\mathcal{I}_{p/})_{\overline{q}/} & \longrightarrow & \mathcal{I}_{q/} \\
\downarrow & & \downarrow \\
\mathcal{J}_{q'/} & \longrightarrow & \mathcal{J}_{q'|K'/} \, .
\end{array}
$$

Lemma 4.2.3.6 implies that the inclusion $K' \subseteq K \star K'$ is right anodyne, so that the lower horizontal map is a trivial fibration. It follows that the upper horizontal map is also a trivial fibration. Since f is κ-cofinal, the right vertical map is weakly cofinal, so that the left vertical map is weakly cofinal as well. $\qquad\square$

Lemma 5.4.5.12. *Let κ be a regular cardinal and let $f : \mathcal{I} \to \mathcal{J}$ be an κ-cofinal map of filtered ∞-categories. Then f is cofinal.*

Proof. According to Theorem 4.1.3.1, to prove that f is cofinal it suffices to show that for every object $J \in \mathcal{J}$, the fiber product $\mathcal{I}_{J/} = \mathcal{I} \times_{\mathcal{J}} \mathcal{J}_{J/}$ is weakly contractible. Lemma 5.4.5.11 asserts that $\mathcal{I}_{J/}$ is κ-filtered; now apply Lemma 5.3.1.18. $\qquad\square$

Lemma 5.4.5.13. *Let κ be a regular cardinal, let $\mathcal{C}$ be an ∞-category which admits κ-filtered colimits, let $\overline{p} : K^{\triangleright} \to \mathcal{C}^{\tau}$ be a κ-small diagram in the ∞-category of κ-compact objects of $\mathcal{C}$, and let $p = \overline{p}|K$. Then $\overline{p}$ is a κ-compact object of $\mathcal{C}_{p/}$.*

Proof. Let $\overline{p}'$ denote the composition

$$
K \diamond \Delta^0 \to K^{\triangleright} \xrightarrow{\overline{p}} \mathcal{C}^{\kappa};
$$

it will suffice to prove that $\overline{p}'$ is a τ-compact object of $\mathcal{C}^{p/}$. Consider the pullback diagram

$$
\begin{array}{ccc}
\mathcal{C}^{p/} & \longrightarrow & \mathrm{Fun}(K \times \Delta^1, \mathcal{C}) \\
\downarrow & & \downarrow{\scriptstyle f} \\
* & \xrightarrow{\;p\;} & \mathrm{Fun}(K \times \{0\}, \mathcal{C}).
\end{array}
$$

Corollary 2.4.7.12 implies that the f is a Cartesian fibration, so we can apply Proposition 3.3.1.3 to deduce that the diagram is homotopy Cartesian (with respect to the Joyal model structure). Using Proposition 5.1.2.2, we deduce that f preserves κ-filtered colimits and that *any* functor $* \to \mathcal{D}$ preserves filtered colimits (since filtered ∞-categories are weakly contractible; see §4.4.4). Consequently, Lemma 5.4.5.7 implies that $\bar{p}'$ is a κ-compact object of $\mathcal{C}^{p/}$ provided that its images in $*$ and $\operatorname{Fun}(K \times \Delta^1, \mathcal{C})$ are κ-compact. The former condition is obvious, and the latter follows from Proposition 5.3.4.13. $\qquad\square$

Lemma 5.4.5.14. *Let $\mathcal{C}$ be an ∞-category which admits small τ-filtered colimits and let $p : K \to \mathcal{C}$ be a small diagram. Then $\mathcal{C}_{p/}$ admits small τ-filtered colimits.*

Proof. Without loss of generality, we may suppose that K is an ∞-category. Let $\mathcal{J}$ be a τ-filtered ∞-category and let $q_0 : \mathcal{J} \to \mathcal{C}_{p/}$ be a diagram corresponding to a map $q : K \star \mathcal{J} \to \mathcal{C}$. We observe that $K \star \mathcal{J}$ is small and τ-filtered, so that q admits a colimit $\bar{q} : (K \star \mathcal{J})^{\triangleright} \to \mathcal{C}$. The map $\bar{q}$ can also be identified with a colimit of q_0. $\qquad\square$

Proposition 5.4.5.15. *Let $\tau \gg \kappa$ be regular cardinals, let $\mathcal{C}$ be a τ-accessible ∞-category, and let $p : K \to \mathcal{C}^{\tau}$ be a κ-small diagram. Then $\mathcal{C}_{p/}$ is τ-accessible, and an object of $\mathcal{C}_{p/}$ is τ-compact if and only if its image in $\mathcal{C}$ is τ-compact.*

Proof. Let $\mathcal{D} = \mathcal{C}_{p/} \times_{\mathcal{C}} \mathcal{C}^{\tau}$ be the full subcategory of $\mathcal{C}_{p/}$ spanned by those objects whose images in $\mathcal{C}$ are τ-compact. Since $\mathcal{C}_{p/}$ is idempotent complete and the collection of τ-compact objects of $\mathcal{C}$ is stable under the formation of retracts, we conclude that $\mathcal{D}$ is idempotent complete. We also note that $\mathcal{D}$ is essentially small; replacing $\mathcal{C}$ by a minimal model if necessary, we may suppose that $\mathcal{D}$ is actually small. Proposition 5.3.5.10 and Lemma 5.4.5.14 imply that there is an (essentially unique) τ-continuous functor $F : \operatorname{Ind}_{\tau}(\mathcal{D}) \to \mathcal{C}_{p/}$ such that the composition $\mathcal{D} \to \operatorname{Ind}_{\tau}(\mathcal{D}) \xrightarrow{F} \mathcal{C}_{p/}$ is equivalent to the inclusion of $\mathcal{D}$ in $\mathcal{C}_{p/}$. To complete the proof, it will suffice to show that F is an equivalence of ∞-categories. According to Proposition 5.3.5.11, it will suffice to show that $\mathcal{D}$ consists of τ-compact objects of $\mathcal{C}_{p/}$ and generates $\mathcal{C}_{p/}$ under τ-filtered colimits. The first assertion follows from Lemma 5.4.5.13.

To complete the proof, choose an object $\bar{p} : K^{\triangleright} \to \mathcal{C}$ of $\mathcal{C}_{p/}$ and let $C \in \mathcal{C}$ denote the image under $\bar{p}$ of the cone point of $K^{\triangleright}$. Then we may identify $\bar{p}$ with a diagram $\widetilde{p} : K \to \mathcal{C}_{/C}^{\tau}$. Since $\mathcal{C}$ is τ-accessible, the ∞-category $\mathcal{E} = \mathcal{C}_{/C}^{\tau}$ is τ-filtered. It follows that $\mathcal{E}_{\widetilde{p}/}$ is τ-filtered and essentially small; to complete the proof, it will suffice to show that the associated map

$$\mathcal{E}_{\widetilde{p}/}^{\triangleright} \to \mathcal{C}_{p/}$$

is a colimit diagram. Equivalently, we must show that the compositition

$$K \star \mathcal{E}_{\widetilde{p}/}^{\triangleright} \xrightarrow{\theta_0^{\triangleright}} \mathcal{E}^{\triangleright} \xrightarrow{\theta_1} \mathcal{C}$$

is a colimit diagram. Since θ_1 is a colimit diagram, it suffices to prove that θ_0 is cofinal. For this, we consider the composition

$$q : \mathcal{E}_{\widetilde{p}/} \xrightarrow{i} K \star \mathcal{E}_{\widetilde{p}/} \xrightarrow{\theta_0} \mathcal{E} .$$

The ∞-category $\mathcal{E}$ is τ-filtered, so that $\mathcal{E}_{\widetilde{p}/}$ is also τ-filtered and therefore weakly contractible (Lemma 5.3.1.18). It follows that i is right anodyne (Lemma 4.2.3.6) and therefore cofinal. Applying Proposition 4.1.1.3, we conclude that θ_0 is cofinal if and only if q is cofinal. We now observe that that q is τ-cofinal (Example 5.4.5.9) and therefore cofinal (Lemma 5.4.5.12). $\square$

Corollary 5.4.5.16. *Let $\mathcal{C}$ be an accessible ∞-category and let $p : K \to \mathcal{C}$ be a diagram indexed by a small simplicial set K. Then $\mathcal{C}_{p/}$ is accessible.*

Proof. Choose appropriate cardinals $\tau \gg \kappa$ and apply Proposition 5.4.5.15. $\square$

5.4.6 Accessibility of Fiber Products

Our goal in this section is to prove that the class of accessible ∞-categories is stable under (homotopy) fiber products (Proposition 5.4.6.6). The strategy of proof should now be familiar from §5.4.4 and §5.4.5. Suppose we are given a homotopy Cartesian diagram

$$\begin{array}{ccc} \mathcal{X}' & \xrightarrow{q'} & \mathcal{X} \\ {\scriptstyle p'}\downarrow & & \downarrow{\scriptstyle p} \\ \mathcal{Y}' & \xrightarrow{q} & \mathcal{Y} \end{array}$$

of ∞-categories, where $\mathcal{X}$, $\mathcal{Y}'$, and $\mathcal{Y}$ are accessible ∞-categories, and the functors p and q are likewise accessible. If κ is a sufficiently large regular cardinal, then we can use Lemma 5.4.5.7 to produce a good supply of κ-compact objects of $\mathcal{X}'$. Our problem is then to prove that these objects generate $\mathcal{X}'$ under κ-filtered colimits. This requires some rather delicate cofinality arguments.

Lemma 5.4.6.1. *Let $\tau \gg \kappa$ be regular cardinals and let $f : \mathcal{C} \to \mathcal{D}$ be a τ-continuous functor between τ-accessible ∞-categories which carries τ-compact objects of $\mathcal{C}$ to τ-compact objects of $\mathcal{D}$. Let C be an object of $\mathcal{C}$, let $\mathcal{C}^\tau_{/C}$ denote the full subcategory of $\mathcal{C}_{/C}$ spanned by those objects $C' \to C$, where C' is τ-compact, and let $\mathcal{D}^\tau_{/f(C)}$ the full subcategory of $\mathcal{D}_{/f(C)}$ spanned by those objects $D \to f(C)$, where $D \in \mathcal{D}$ is τ-compact. Then f induces a κ-cofinal functor $f' : \mathcal{C}^\tau_{/C} \to \mathcal{D}^\tau_{/f(C)}$.*

Proof. Let $\widetilde{p} : K \to \mathcal{C}^\tau_{/C}$ be a diagram indexed by a τ-small weakly contractible simplicial set K and let $p : K \to \mathcal{C}$ be the underlying map. We need to show that the induced functor $(\mathcal{C}^\tau_{/C})_{\widetilde{p}/} \to (\mathcal{D}^\tau_{/f(C)})_{f'\widetilde{p}/}$ is weakly cofinal. Using Proposition 5.4.5.15, we may replace $\mathcal{C}$ by $\mathcal{C}_{p/}$ and $\mathcal{D}$ by $\mathcal{D}_{fp/}$ and thereby reduce to the problem of showing that f is weakly cofinal. Let $\phi : D \to f(C)$ be an object of $\mathcal{D}^\tau_{/f(C)}$ and let $F_D : \mathcal{D} \to \mathcal{S}$ be the functor corepresented by D. Since D is τ-compact, the functor F_D is τ-continuous, so that $F_D \circ f$ is τ-continuous. Consequently, the space $F_D(f(C))$ can be obtained as a colimit of the τ-filtered diagram

$$p : \mathcal{C}^\tau_{/C} \to \mathcal{D}^\tau_{/f(C)} \to \mathcal{D} \xrightarrow{F_D} \mathcal{S}.$$

In particular, the path component of $F_D(f(C))$ containing ϕ lies in the image of $p(\eta)$ for some $\eta : C' \to C$ as above. It follows that there exists a commutative diagram

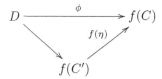

in $\mathcal{D}$, which can be identified with a morphism in $\mathcal{D}^\tau_{/f(C)}$ having the desired properties. $\qquad\square$

Lemma 5.4.6.2. *Let $A = A' \cup \{\infty\}$ be a linearly ordered set containing a largest element ∞ and let $B \subseteq A'$ be a cofinal subset (in other words, for every $\alpha \in A'$, there exists $\beta \in B$ such that $\alpha \leq \beta$). The inclusion*

$$\phi : N(A') \coprod_{N(B)} N(B \cup \{\infty\}) \subseteq N(A)$$

is a categorical equivalence.

Proof. For each $\beta \in B$, let ϕ_β denote the inclusion of

$$N(\{\alpha \in A' : \alpha \leq \beta\}) \coprod_{N(\{\alpha \in B : \alpha \leq \beta\})} N(\{\alpha \in B : \alpha \leq \beta\} \cup \{\infty\})$$

into $N(\{\alpha \in A' : \alpha \leq \beta\} \cup \{\infty\})$. Since B is cofinal in A', ϕ is a filtered colimit of the inclusions ϕ_β. Replacing A' by $\{\alpha \in A' : \alpha \leq \beta\}$ and B by $\{\alpha \in B : \alpha \leq \beta\}$, we may reduce to the case where A' has a largest element (which we will continue to denote by β).

We have a categorical equivalence

$$N(B) \coprod_{\{\beta\}} N(\{\beta, \infty\}) \subseteq N(B \cup \{\infty\}).$$

Consequently, to prove that ϕ is a categorical equivalence, it will suffice to show that the composition

$$N(A') \coprod_{\{\beta\}} N(\{\beta, \infty\}) \subseteq N(A') \coprod_{N(B)} N(B \cup \{\infty\}) \subseteq N(A)$$

is a categorical equivalence, which is clear. $\qquad\square$

Lemma 5.4.6.3. *Let $\tau > \kappa$ be regular cardinals and let*

$$\mathcal{X} \xrightarrow{p} \mathcal{Y} \xleftarrow{p'} \mathcal{X}'$$

be functors between ∞-categories. Assume that the following conditions are satisfied:

(1) *The ∞-categories $\mathfrak{X}$, $\mathfrak{X}'$, and $\mathcal{Y}$ are κ-filtered, and admit τ-small κ-filtered colimits.*

(2) *The functors p and p' preserve τ-small κ-filtered colimits.*

(3) *The functors p and p' are κ-cofinal.*

Then there exist objects $X \in \mathfrak{X}$, $X' \in \mathfrak{X}'$ such that $p(X)$ and $p'(X')$ are equivalent in $\mathcal{Y}$.

Proof. For every ordinal α, we let $[\alpha] = \{\beta : \beta \leq \alpha\}$ and $(\alpha) = \{\beta : \beta < \alpha\}$. Let us say that an ordinal α is *even* if it is of the form $\lambda + n$, where λ is a limit ordinal and n is an even integer; otherwise, we will say that α is *odd*. Let A denote the set of all even ordinals smaller than κ and A' the set of all odd ordinals smaller than κ. We regard A and A' as subsets of the linearly ordered set $A \cup A' = (\kappa)$. We will construct a commutative diagram

$$
\begin{array}{ccccc}
N(A) & \longrightarrow & N(\kappa) & \longleftarrow & N(A') \\
\downarrow q & & \downarrow Q & & \downarrow q' \\
\mathfrak{X} & \xrightarrow{\;p\;} & \mathcal{Y} & \xleftarrow{\;p'\;} & \mathfrak{X}'.
\end{array}
$$

Supposing that this is possible, we choose colimits $X \in \mathfrak{X}$, $X' \in \mathfrak{X}'$, and $Y \in \mathcal{Y}$ for q, q', and Q, respectively. Since the inclusion $N(A) \subseteq N(\kappa)$ is cofinal and p preserves κ-filtered colimits, we conclude that $p(X)$ and Y are equivalent. Similarly, $p'(X')$ and Y are equivalent, so that $p(X)$ and $p'(X')$ are equivalent, as desired.

The construction of q, q', and Q is given by induction. Let $\alpha < \kappa$ and suppose that $q|\,N(\{\beta \in A : \beta < \alpha\})$, $q'|\,N(\{\beta \in A' : \beta < \alpha\})$ and $Q|\,N(\alpha)$ have already been constructed. We will show how to extend the definitions of q, q', and Q to include the ordinal α. We will suppose that α is even; the case where α is odd is similar (but easier).

Suppose first that α is a limit ordinal. In this case, define $q|\,N(\{\beta \in A : \beta \leq \alpha\})$ to be an arbitrary extension of $q|\,N(\{\beta \in A : \beta < \alpha\})$: such an extension exists by virtue of our assumption that $\mathfrak{X}$ is κ-filtered. In order to define $Q|\,N(\alpha)$, it suffices to verify that $\mathcal{Y}$ has the extension property with respect to the inclusion

$$
N(\alpha) \coprod_{N(\{\beta \in A : \beta < \alpha\})} N(\{\beta \in A : \beta \leq \alpha\}) \subseteq N[\alpha].
$$

Since $\mathcal{Y}$ is an ∞-category, this follows immediately from Lemma 5.4.6.2.

We now treat the case where $\alpha = \alpha' + 1$ is a successor ordinal. Let $q_{<\alpha} = q|\{\beta \in A : \beta < \alpha\}$, and regard $Q|\,N(\{\alpha'\} \cup \{\beta \in A : \beta < \alpha\})$ as an object of $\mathcal{Y}_{fq_{<\alpha}/}$. We now observe that $N(\{\beta \in A : \beta < \alpha\})$ is κ-small and weakly contractible. Since p is κ-cofinal, we can construct $q|\{\beta \in A : \beta \leq \alpha\}$ extending $q_{<\alpha}$ and a compatible map $Q|\,N(\{\alpha'\} \cup \{\beta \in A : \beta \leq \alpha\})$. To complete the construction of Q, it suffices to show that $\mathcal{Y}$ has the extension property with respect to the inclusion

$$N(\alpha) \coprod_{N(\{\beta \in A : \beta < \alpha\} \cup \{\alpha'\})} N(\{\beta \in A : \beta \leq \alpha\} \cup \{\alpha'\}) \subseteq N[\alpha].$$

Once again, this follows from Lemma 5.4.6.2.   □

Lemma 5.4.6.4. *Let κ and τ be regular cardinals, let $f : \mathcal{I} \to \mathcal{J}$ be a κ-cofinal functor between τ-filtered ∞-categories, and let $p : K \to \mathcal{I}$ be a diagram indexed by a τ-small simplicial set K. Then the induced functor*

$$\mathcal{I}_{p/} \to \mathcal{J}_{fp/}$$

is κ-cofinal.

Proof. Let K' be a simplicial set which is κ-small and weakly contractible and let $q : K \star K' \to \mathcal{I}$ be a diagram. We have a commutative diagram

$$
\begin{array}{ccc}
\mathcal{I}_{q/} & \longrightarrow & \mathcal{J}_{fq/} \\
\downarrow & & \downarrow \\
\mathcal{I}_{q|K'/} & \longrightarrow & \mathcal{I}_{fq|K'/} .
\end{array}
$$

Lemma 4.2.3.6 implies that $K' \subseteq K \star K'$ is a right anodyne inclusion, so that the vertical maps are trivial fibrations. Since f is κ-cofinal, the lower horizontal map is weakly cofinal; it follows that the upper horizontal map is weakly cofinal as well.   □

Lemma 5.4.6.5. *Let $\tau > \kappa$ be regular cardinals and let*

$$
\begin{array}{ccc}
\mathcal{J}' & \overset{q'}{\longrightarrow} & \mathcal{J} \\
{\scriptstyle p'}\downarrow & & \downarrow{\scriptstyle p} \\
\mathcal{J}' & \overset{q}{\longrightarrow} & \mathcal{J}
\end{array}
$$

be a diagram of ∞-categories which is homotopy Cartesian (with respect to the Joyal model structure). Suppose that $\mathcal{I}$, $\mathcal{J}$, and $\mathcal{J}'$ are τ-filtered ∞-categories which admit τ-small κ-filtered colimits. Suppose further that p and q are κ-cofinal functors which preserve τ-small κ-filtered colimits. Then $\mathcal{J}'$ is τ-filtered, and the functors p' and q' are κ-cofinal.

Proof. Without loss of generality, we may suppose that p and q are categorical fibrations and that $\mathcal{J}' = \mathcal{I} \times_{\mathcal{J}} \mathcal{J}'$. To prove that $\mathcal{J}'$ is τ-filtered, we must show that $\mathcal{J}'_{f/}$ is nonempty for every diagram $f : K \to \mathcal{J}'$ indexed by a τ-small simplicial set K. We have a (homotopy) pullback diagram

$$
\begin{array}{ccc}
\mathcal{J}'_{f/} & \longrightarrow & \mathcal{I}_{q'f/} \\
\downarrow & & \downarrow{\scriptstyle g} \\
\mathcal{J}'_{p'f/} & \overset{h}{\longrightarrow} & \mathcal{J}_{pq'f/} .
\end{array}
$$

Lemma 5.3.1.19 implies that the ∞-categories $\mathcal{I}_{q'f/}$, $\mathcal{I}'_{p'f/}$, and $\mathcal{I}_{pq'f/}$ are τ-filtered, and Lemma 5.4.6.4 implies that g and h are κ-cofinal. We may therefore apply Lemma 5.4.6.3 to deduce that $\mathcal{I}'_{f/}$ is nonempty, as desired.

We now prove that q' is κ-cofinal; the analogous assertion for p' is proven by the same argument. We must show that for every diagram $f : K \to \mathcal{I}'$, where K is κ-small and weakly contractible, the induced map $\mathcal{I}'_{f/} \to \mathcal{I}_{q'f/}$ is weakly cofinal. Replacing $\mathcal{I}'$ by $\mathcal{I}'_{f/}$ as above, we may reduce to the problem of showing that q' itself is weakly cofinal. Let I be an object of $\mathcal{I}$, let $J = p(I) \in \mathcal{J}$ and consider the (homotopy) pullback diagram

$$\begin{array}{ccc} \mathcal{I}'_{I/} & \longrightarrow & \mathcal{I}_{I/} \\ \downarrow & & \downarrow{\scriptstyle u} \\ \mathcal{J}'_{J/} & \xrightarrow{\;v\;} & \mathcal{J}_{J/} \, . \end{array}$$

We wish to show that $\mathcal{I}'_{I/}$ is nonempty. This follows from Lemma 5.4.6.3 because u and v are τ-cofinal (Lemmas 5.4.6.4 and 5.4.5.11, respectively). $\square$

Proposition 5.4.6.6. *Let*

$$\begin{array}{ccc} \mathcal{X}' & \xrightarrow{\;q'\;} & \mathcal{X} \\ \downarrow{\scriptstyle p'} & & \downarrow{\scriptstyle p} \\ \mathcal{Y}' & \xrightarrow{\;q\;} & \mathcal{Y} \end{array}$$

be a diagram of ∞-categories which is homotopy Cartesian (with respect to the Joyal model structure). Suppose further that $\mathcal{X}$, $\mathcal{Y}$, and $\mathcal{Y}'$ are accessible and that both p and q are accessible functors. Then $\mathcal{X}'$ is accessible. Moreover, for any accessible ∞-category $\mathcal{C}$ and any functor $f : \mathcal{C} \to \mathcal{X}$, f is accessible if and only if the compositions $p' \circ f$ and $q' \circ f$ are accessible. In particular (taking $f = \mathrm{id}_{\mathcal{X}}$), the functors p' and q' are accessible.

Proof. Choose a regular cardinal κ such that $\mathcal{X}$, $\mathcal{Y}'$, and $\mathcal{Y}$ are κ-accessible. Enlarging κ if necessary, we may suppose that p and q are κ-continuous. It follows from Lemma 5.4.5.5 that $\mathcal{X}'$ admits small κ-filtered colimits and that for any $\kappa' > \kappa$, a functor $f : \mathcal{C} \to \mathcal{X}$ is κ'-continuous if and only if $p' \circ f$ and $q' \circ f$ are κ'-continuous. This proves the second claim; it now suffices to show that $\mathcal{X}'$ is accessible. For this, we will use characterization (3) of Proposition 5.4.2.2. Without loss of generality, we may suppose that p and q are categorical fibrations and that $\mathcal{X}' = \mathcal{X} \times_{\mathcal{Y}} \mathcal{Y}'$. It then follows easily that $\mathcal{X}'$ is locally small. It will therefore suffice to show that there exists a regular cardinal τ such that $\mathcal{X}'$ is generated by a small collection of τ-compact objects under small τ-filtered colimits.

Since the ∞-categories of κ-compact objects of $\mathcal{X}$ and $\mathcal{Y}'$ are essentially small, there exists $\tau > \kappa$ such that $p| \mathcal{X}^{\kappa} \subseteq \mathcal{Y}^{\tau}$ and $q|\mathcal{Y}'^{\kappa} \subseteq \mathcal{Y}^{\tau}$. Enlarging τ if necessary, we may suppose that $\tau \gg \kappa$. The proof of Proposition 5.4.2.11 shows that every τ-compact object of $\mathcal{X}$ can be written as a τ-small κ-filtered colimit of objects belonging to $\mathfrak{X}^\kappa$. Since p is κ-continuous, it follows that $p(\mathfrak{X}^\tau) \subseteq \mathcal{Y}^\tau$, and similarly $q(\mathcal{Y}'^\tau) \subseteq \mathcal{Y}^\tau$. Let $\mathfrak{X}'' = \mathfrak{X}^\tau \times_{\mathcal{Y}^\tau} \mathcal{Y}'^\tau$. Then $\mathfrak{X}''$ is an essentially small full subcategory of $\mathfrak{X}'$. Lemma 5.4.5.7 implies that $\mathfrak{X}''$ consists of τ-compact objects of $\mathfrak{X}'$. To complete the proof, it will suffice to show that $\mathfrak{X}''$ generates $\mathfrak{X}'$ under small τ-filtered colimits.

Let $X' = (X, Y')$ be an object of $\mathfrak{X}'$ and set $Y = pX = qY'$. We have a (homotopy) pullback diagram

$$
\begin{array}{ccc}
\mathfrak{X}''_{/X'} & \xrightarrow{f'} & \mathfrak{X}^\tau_{/X} \\
\downarrow{\scriptstyle g'} & & \downarrow{\scriptstyle g} \\
\mathcal{Y}'^\tau_{/Y'} & \xrightarrow{f} & \mathcal{Y}^\tau_{/Y}
\end{array}
$$

of essentially small ∞-categories. Lemma 5.4.6.1 asserts that f and g are κ-cofinal. We apply Lemma 5.4.6.5 to conclude that $\mathfrak{X}''_{/X'}$ is τ-filtered and that f' and g' are κ-cofinal. Now consider the diagram

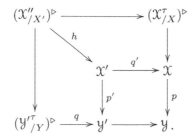

Lemma 5.4.5.12 allows us to conclude that f' and g' are cofinal, so that $p' \circ h$ and $q' \circ h$ are colimit diagrams. Lemma 5.4.5.5 implies that h is a colimit diagram as well, so that X' is the colimit of an essentially small τ-filtered diagram taking values in $\mathfrak{X}''$. □

Corollary 5.4.6.7. *Let $\mathcal{C}$ be an accessible ∞-category and let $p : K \to \mathcal{C}$ be a diagram indexed by a small simplicial set K. Then the ∞-category $\mathcal{C}_{/p}$ is accessible.*

Proof. Since the map $\mathcal{C}_{/p} \to \mathcal{C}^{/p}$ is a categorical equivalence, it will suffice to prove that $\mathcal{C}^{/p}$ is accessible. We have a pullback diagram

$$
\begin{array}{ccc}
\mathcal{C}^{/p} & \longrightarrow & \mathrm{Fun}(K \times \Delta^1, \mathcal{C}) \\
\downarrow & & \downarrow{\scriptstyle p} \\
* & \xrightarrow{q} & \mathrm{Fun}(K \times \{1\}, \mathcal{C})
\end{array}
$$

of ∞-categories. Since p is a coCartesian fibration, Proposition 3.3.1.3 implies that this diagram is homotopy Cartesian. According to Proposition 5.4.4.3, the ∞-categories $\mathcal{C}^{K \times \Delta^1}$ and $\mathcal{C}^{K \times \{1\}}$ are accessible. Using Proposition 5.1.2.2, we conclude that for every regular cardinal κ such that $\mathcal{C}$ admits

κ-filtered colimits, p is κ-continuous; in particular, p is accessible. Corollary 5.4.3.6 implies that $*$ is accessible and that q is an accessible functor. Applying Proposition 5.4.6.6, we deduce that $\mathcal{C}^{/p}$ is accessible. $\qquad\square$

5.4.7 Applications

In §5.4.4 through §5.4.6, we established some of the basic stability properties enjoyed by the class of accessible ∞-categories. In this section, we will reap some of the rewards for our hard work.

Lemma 5.4.7.1. *Let $\{\mathcal{C}_\alpha\}_{\alpha \in A}$ be a family of ∞-categories indexed by a small set A and let $\mathcal{C} = \coprod_{\alpha \in A} \mathcal{C}_\alpha$ be their coproduct. Then $\mathcal{C}$ is an accessible if and only if each $\mathcal{C}_\alpha$ is accessible.*

Proof. This is immediate from the definitions. $\qquad\square$

Lemma 5.4.7.2. *Let $\{\mathcal{C}_\alpha\}_{\alpha \in A}$ be a family of ∞-categories indexed by a small set A and let $\mathcal{C} = \prod_{\alpha \in A} \mathcal{C}_\alpha$ be their product. If each $\mathcal{C}_\alpha$ is accessible, then $\mathcal{C}$ is accessible. Moreover, if $\mathcal{D}$ is an accessible ∞-category, then a functor $\mathcal{D} \to \mathcal{C}$ is accessible if and only if each of the compositions*

$$\mathcal{D} \to \mathcal{C} \to \mathcal{C}_\alpha$$

is accessible.

Proof. Let $\mathcal{D} = \coprod_{\alpha \in A} \mathcal{C}_\alpha$. By Lemma 5.4.7.1, $\mathcal{D}$ is accessible. Let $N(A)$ denote the constant simplicial set with value A. Proposition 5.4.4.3 implies that $\mathrm{Fun}(N(A), \mathcal{D})$ is accessible. We now observe that $\mathrm{Fun}(N(A), \mathcal{D})$ can be written as a disjoint union of $\mathcal{C}$ with another ∞-category; applying Lemma 5.4.7.1 again, we deduce that $\mathcal{C}$ is accessible. The second claim follows immediately from the definitions. $\qquad\square$

Proposition 5.4.7.3. *The ∞-category Acc of accessible ∞-categories admits small limits, and the inclusion $i : \mathrm{Acc} \subseteq \widehat{\mathrm{Cat}}_\infty$ preserves small limits.*

Proof. By Proposition 4.4.2.6, it suffices to prove that Acc admits pullbacks and small products and that i preserves pullbacks and (small) products. Let Acc_Δ be the (simplicial) subcategory of $\widehat{\mathrm{Set}}_\Delta$ defined as follows:

(1) The objects of Acc_Δ are the accessible ∞-categories.

(2) If $\mathcal{C}$ and $\mathcal{D}$ are accessible ∞-categories, then $\mathrm{Map}_{\mathrm{Acc}_\Delta}(\mathcal{C}, \mathcal{D})$ is the subcategory of $\mathrm{Fun}(\mathcal{C}, \mathcal{D})$ whose objects are accessible functors and whose morphisms are *equivalences* of functors.

The ∞-category Acc is isomorphic to the simplicial nerve $N(\mathrm{Acc}_\Delta)$. In view of Theorem 4.2.4.1, it will suffice to prove that the simplicial category Acc_Δ admits homotopy fiber products and (small) homotopy products and that the inclusion $\mathrm{Acc}_\Delta \subseteq (\widehat{\mathrm{Set}_\Delta^+})^\circ$ preserves homotopy fiber products and homotopy products. The case of homotopy fiber products follows from Proposition 5.4.6.6, and the case of (small) homotopy products follows from Lemma 5.4.7.2. $\qquad\square$

If $\mathcal{C}$ is an accessible ∞-category, then $\mathcal{C}$ is the union of full subcategories $\{\mathcal{C}^\tau \subseteq \mathcal{C}\}$, where τ ranges over all (small) regular cardinals. It seems reasonable to expect that if τ is sufficiently large, then the properties of $\mathcal{C}$ are mirrored by properties of $\mathcal{C}^\tau$. The following result provides an illustration of this philosophy:

Proposition 5.4.7.4. *Let $\mathcal{C}$ be a κ-accessible ∞-category and let $\tau \gg \kappa$ be an uncountable regular cardinal such that $\mathcal{C}^\kappa$ is essentially τ-small. Then the full subcategory $\mathcal{C}^\tau \subseteq \mathcal{C}$ is stable under all κ-small limits which exist in $\mathcal{C}$.*

Before giving the proof, we will need to establish a few lemmas.

Lemma 5.4.7.5. *Let $\tau \gg \kappa$ be regular cardinals and assume that τ is uncountable. Let $\mathcal{C}$ be a τ-small ∞-category and let D be an object of $\mathrm{Ind}_\kappa(\mathcal{C})$. The following are equivalent:*

(1) *The object D is τ-compact in $\mathrm{Ind}_\kappa(\mathcal{C})$.*

(2) *For every $C \in \mathcal{C}$, the space $\mathrm{Map}_{\mathrm{Ind}_\kappa(\mathcal{C})}(j(C), D)$ is essentially τ-small, where $j : \mathcal{C} \to \mathrm{Ind}_\kappa(\mathcal{C})$ denotes the Yoneda embedding.*

Proof. Suppose first that (1) is satisfied. Using Lemma 5.1.5.3, we can write D as the colimit of the κ-filtered diagram

$$\mathcal{C}_{/D} = \mathcal{C} \times_{\mathrm{Ind}_\kappa(\mathcal{C})} \mathrm{Ind}_\kappa(\mathcal{C})_{/D} \to \mathrm{Ind}_\kappa(\mathcal{C}).$$

Since $\tau \gg \kappa$, we also write D as a small τ-filtered colimit of objects $\{D_\alpha\}$, where each D_α is the colimit of a τ-small κ-filtered diagram

$$\widetilde{\mathcal{C}} \to \mathcal{C} \to \mathrm{Ind}_\kappa(\mathcal{C}).$$

Since D is τ-compact, we conclude that D is a retract of D_α. Let $F : \mathrm{Ind}_\kappa(\mathcal{C}) \to \mathcal{S}$ denote the functor corepresented by $j(C)$. According to Proposition 5.3.5.5, F is κ-continuous. It follows that $F(D)$ is a retract of $F(D_\alpha)$, which is itself a τ-small colimit of spaces equivalent to

$$\mathrm{Map}_{\mathrm{Ind}_\kappa(\mathcal{C})}(j(C), j(C')) \simeq \mathrm{Map}_{\mathcal{C}}(C, C'),$$

which is essentially τ-small by assumption and therefore a τ-compact object of $\mathcal{S}$. It follows that D is also a τ-compact object of $\mathcal{S}$.

Now assume (2). Once again, we observe that D can be obtained as the colimit of a diagram $\mathcal{C}_{/D} \to \mathrm{Ind}_\kappa(\mathcal{C})$. By assumption, $\mathcal{C}$ is τ-small and the fibers of the right fibration $\mathcal{C}_{/D} \to \mathcal{C}$ are essentially τ-small. Proposition 5.4.1.4 implies that $\mathcal{C}_{/D}$ is essentially τ-small, so that D is a τ-small colimit of κ-compact objects of $\mathrm{Ind}_\kappa(\mathcal{C})$ and therefore τ-compact. $\square$

Lemma 5.4.7.6. *Let $\tau \gg \kappa$ be regular cardinals such that τ is uncountable and let $\mathcal{S}^\tau$ be the full subcategory of $\mathcal{S}$ consisting of essentially τ-small spaces. Then $\mathcal{S}^\tau$ is stable under κ-small limits in $\mathcal{S}$.*

Proof. In view of Proposition 4.4.2.6, it suffices to prove that $\mathcal{S}^\tau$ is stable under pullbacks and κ-small products. Using Theorem 4.2.4.1, it will suffice to show that the full subcategory of $\mathcal{K}an$ spanned by essentially τ-small spaces is stable under κ-small products and homotopy fiber products. This follows immediately from characterization (1) given in Proposition 5.4.1.5. $\qquad\square$

Proof of Proposition 5.4.7.4. Let K be a κ-small simplicial set and let $p :$ $K \to \mathcal{C}^\tau$ be a diagram which admits a limit $X \in \mathcal{C}$. We wish to show that X is τ-compact. According to Lemma 5.4.7.5, it suffices to prove that the space $F(X)$ is essentially τ-small, where $F : \mathcal{C} \to \mathcal{S}$ denotes the functor corepresented by a κ-compact object $C \in \mathcal{C}$. Since F preserves limits, we note that $F(X)$ is a limit of $F \circ p$. Lemma 5.4.7.5 implies that the diagram $F \circ p$ takes values in $\mathcal{S}^\tau \subseteq \mathcal{S}$. We now conclude by applying Lemma 5.4.7.6. $\quad\square$

We note the following useful criterion for establishing that a functor is accessible:

Proposition 5.4.7.7. *Let $G : \mathcal{C} \to \mathcal{C}'$ be a functor between accessible ∞-categories. If G admits a right or a left adjoint, then G is accessible.*

Proof. If G is a left adjoint, then G commutes with all colimits which exist in $\mathcal{C}$. Therefore G is κ-continuous for any cardinal κ having the property that $\mathcal{C}$ is κ-accessible. Let us therefore assume that G is a right adjoint; choose a left adjoint F for G.

Choose a regular cardinal κ such that $\mathcal{C}'$ is κ-accessible. We may suppose without loss of generality that $\mathcal{C}' = \mathrm{Ind}_\kappa \mathcal{D}$, where $\mathcal{D}$ is a small ∞-category. Consider the composite functor

$$\mathcal{D} \xrightarrow{j} \mathrm{Ind}_\kappa(\mathcal{D}) \xrightarrow{F} \mathcal{C}.$$

Since $\mathcal{D}$ is small, there exists a regular cardinal $\tau \gg \kappa$ such that $\mathcal{C}$ is τ-accessible and the essential image of $F \circ j$ consists of τ-compact objects of $\mathcal{C}$. We will show that G is τ-continuous.

Since $\mathrm{Ind}_\kappa(\mathcal{D}) \subseteq \mathcal{P}(\mathcal{D})$ is stable under small τ-filtered colimits, it will suffice to prove that the composition

$$G' : \mathcal{C} \xrightarrow{G} \mathrm{Ind}_\kappa(\mathcal{D}) \to \mathcal{P}(\mathcal{D})$$

is τ-continuous. For each object $D \in \mathcal{D}$, let $G'_D : \mathcal{C} \to \widehat{\mathcal{S}}$ denote the composition of G' with the functor given by evaluation at D. According to Proposition 5.1.2.2, it will suffice to show that each G'_D is τ-continuous. Lemma 5.1.5.2 implies that G'_D is equivalent to the composition of G with the functor $\mathcal{C}' \to \widehat{\mathcal{S}}$ corepresented by $j(D)$. Since F is left adjoint to G, we may identify this with the functor corepresented by $F(j(D))$. Since $F(j(D))$ is τ-compact by construction, this functor is τ-continuous. $\quad\square$

Definition 5.4.7.8. Let $\mathcal{C}$ be an accessible category. A full subcategory $\mathcal{D} \subseteq \mathcal{C}$ is an *accessible subcategory* of $\mathcal{C}$ if $\mathcal{D}$ is accessible, and the inclusion of $\mathcal{D}$ into $\mathcal{C}$ is an accessible functor.

Example 5.4.7.9. Let $\mathcal{C}$ be an accessible ∞-category and K a simplicial set. Suppose that every diagram $K \to \mathcal{C}$ has a limit in $\mathcal{C}$. Let $\mathcal{D} \subseteq \mathrm{Fun}(K^{\triangleleft}, \mathcal{C})$ be the full subcategory spanned by the limit diagrams. Then $\mathcal{D}$ is equivalent to $\mathrm{Fun}(K, \mathcal{C})$ and is therefore accessible (Proposition 5.4.4.3). The inclusion $\mathcal{D} \subseteq \mathrm{Fun}(K^{\triangleleft}, \mathcal{C})$ is a right adjoint and therefore accessible (Proposition 5.4.7.7). Thus $\mathcal{D}$ is an accessible subcategory of $\mathrm{Fun}(K^{\triangleleft}, \mathcal{C})$. Similarly, if every diagram $K \to \mathcal{C}$ has a colimit, then the full subcategory $\mathcal{D}' \subseteq \mathrm{Fun}(K^{\triangleright}, \mathcal{C})$ spanned by the colimit diagrams is an accessible subcategory of $\mathrm{Fun}(K^{\triangleright}, \mathcal{C})$.

Proposition 5.4.7.10. *Let $\mathcal{C}$ be an accessible category and let $\{\mathcal{D}_\alpha \subseteq \mathcal{C}\}_{\alpha \in A}$ be a (small) collection of accessible subcategories of $\mathcal{C}$. Then $\bigcap_{\alpha \in A} \mathcal{D}_\alpha$ is an accessible subcategory of $\mathcal{C}$.*

Proof. We have a homotopy Cartesian diagram

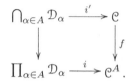

Lemma 5.4.7.2 implies that $\prod_{\alpha \in A} \mathcal{D}_\alpha$ and $\mathcal{C}^A$ are accessible, and it is easy to see that f and i are accessible functors. Applying Proposition 5.4.6.6, we conclude that $\bigcap_{\alpha \in A} \mathcal{D}_\alpha$ is accessible and that i' is an accessible functor, as desired.   □

We conclude this chapter by establishing a generalization of Proposition 5.4.4.3.

Proposition 5.4.7.11. *Let $\mathcal{C}$ be a subcategory of the ∞-category $\widehat{\mathrm{Cat}}_\infty$ of (not necessarily small) ∞-categories satisfying the following conditions:*

(a) *The ∞-category $\mathcal{C}$ admits small limits, and the inclusion $\mathcal{C} \subseteq \widehat{\mathrm{Cat}}_\infty$ preserves small limits.*

(b) *If X belongs to $\mathcal{C}$, then $\mathrm{Fun}(\Delta^1, X)$ belongs to $\mathcal{C}$.*

(c) *If X and Y belong to $\mathcal{C}$, then a functor $X \to \mathrm{Fun}(\Delta^1, Y)$ is a morphism of $\mathcal{C}$ if and only if, for every vertex v of Δ^1, the composite functor $X \to \mathrm{Fun}(\Delta^1, Y) \to \mathrm{Fun}(\{v\}, Y) \simeq Y$ is a morphism of $\mathcal{C}$.*

Let $p : X \to S$ be a map of simplicial sets, where S is small. Assume that

(i) *The map p is a categorical fibration and a locally coCartesian fibration.*

(ii) *For each vertex s in S, the fiber X_s belongs to $\mathcal{C}$.*

(iii) *For each edge $s \to s'$ in S, the associated functor $X_s \to X_{s'}$ is a morphism in $\mathcal{C}$.*

Let $\mathcal{E}$ be a set of edges of S and let Y be the full subcategory of $\mathrm{Map}_S(S, X)$ spanned by those sections $f : S \to X$ of p which satisfy the following condition:

(∗) *For every edge $e : \Delta^1 \to S$ belonging to $\mathcal{E}$, f carries e to a p_e-coCartesian edge of $\Delta^1 \times_S X$, where $p_e : \Delta^1 \times_S X \to \Delta^1$ is the projection.*

Then Y belongs to $\mathcal{C}$. Moreover, if $Z \in \mathcal{C}$, then a functor $Z \to Y$ belongs to $\mathcal{C}$ if and only if, for every vertex s in S, the composite map $Z \to Y \to X_s$ belongs to $\mathcal{C}$.

Remark 5.4.7.12. Hypotheses (i) through (iii) of Proposition 5.4.7.11 are satisfied, in particular, if $p : X \to S$ is a coCartesian fibration classified by a functor $S \to \mathcal{C} \subseteq \widehat{\mathcal{C}\mathrm{at}}_\infty$.

Remark 5.4.7.13. Hypotheses (a), (b), and (c) of Proposition 5.4.7.11 are satisfied for the following subcategories $\mathcal{C} \subseteq \widehat{\mathcal{C}\mathrm{at}}_\infty$:

- Fix a class of simplicial sets $\{K_\alpha\}_{\alpha \in A}$. Then we can take $\mathcal{C}$ to be the subcategory of $\widehat{\mathcal{C}\mathrm{at}}_\infty$ whose objects are ∞-categories which admit K_α-indexed (co)limits for each $\alpha \in A$, and whose morphisms are functors which preserve K_α-indexed (co)limits for each $\alpha \in A$.

- We can take the objects of $\mathcal{C}$ to be accessible ∞-categories and the morphisms in $\mathcal{C}$ to be accessible functors (in view of Propositions 5.4.4.3 and 5.4.7.3).

We will meet some other examples in §5.5.

Remark 5.4.7.14. In the situation of Proposition 5.4.7.11, we can replace "coCartesian" by "Cartesian" everywhere to obtain a dual result. This follows by applying Proposition 5.4.7.11 to the map $X^{op} \to S^{op}$ after replacing $\mathcal{C}$ by its preimage under the "opposition" involution of $\mathrm{h}\widehat{\mathcal{C}\mathrm{at}}_\infty$.

The proof of Proposition 5.4.7.11 makes use of the following observation:

Lemma 5.4.7.15. *Let $p : \mathcal{M} \to \Delta^1$ be a coCartesian fibration classifying a functor $F : \mathcal{C} \to \mathcal{D}$, where $\mathcal{C} = p^{-1}\{0\}$ and $\mathcal{D} = p^{-1}\{1\}$. Let $X = \mathrm{Map}_{\Delta^1}(\Delta^1, \mathcal{M})$ be the ∞-category of sections of p. Then X can be identified with a homotopy limit of the diagram*

$$\mathcal{C} \xrightarrow{F} \mathrm{Fun}(\{0\}, \mathcal{D}) \leftarrow \mathrm{Fun}(\Delta^1, \mathcal{D}).$$

Proof. We first replace the diagram in question by a fibrant one. Let $\mathcal{C}'$ denote the ∞-category of coCartesian sections of p. Then the evaluation map $e : \mathcal{C}' \to \mathcal{C}$ is a trivial fibration of simplicial sets. Moreover, since F is associated to the correspondence $\mathcal{M}$, the map e admits a section s such that the composition

$$\mathcal{C} \xrightarrow{s} \mathcal{C}' \to \mathcal{D}$$

coincides with F. It follows that we have a weak equivalence of diagrams

$$
\begin{array}{ccccc}
\mathcal{C} & \xrightarrow{\ F\ } & \mathrm{Fun}(\{0\}, \mathcal{D}) & \longleftarrow & \mathrm{Fun}(\Delta^1, \mathcal{D}) \\
\downarrow{\scriptstyle s} & & \| & & \| \\
\mathcal{C}' & \xrightarrow{\ F'\ } & \mathrm{Fun}(\{0\}, \mathcal{D}) & \longleftarrow & \mathrm{Fun}(\Delta^1, \mathcal{D}),
\end{array}
$$

where F' is given by evaluation at $\{1\}$ and is therefore a categorical fibration. Let $\mathcal{X}'$ denote the pullback of the lower diagram, which we can identify with the full subcategory of $\mathrm{Map}_{\Delta^1}(\Lambda_1^2, \mathcal{M})$ spanned by those functors which carry the first edge of Λ_1^2 to a coCartesian edge of $\mathcal{M}$.

Regard Δ^2 as an object of $(\mathrm{Set}_\Delta)_{/\Delta^1}$ via the unique retraction $r : \Delta^2 \to \Delta^1$ onto the simplicial subset $\Delta^{\{0,1\}} \subseteq \Delta^{\{0,1,2\}}$. Let $\mathcal{X}''$ denote the full subcategory of $\mathrm{Map}_{\Delta^1}(\Delta^2, \mathcal{M})$ spanned by those maps $\Delta^2 \to \mathcal{M}$ which carry the initial edge of Δ^2 to a p-coCartesian edge of $\mathcal{M}$.

Let T denote the marked simplicial set whose underlying simplicial set is Δ^2, whose sole nondegenerate marked edge is $\Delta^1 \subseteq \Delta^2$, and let $T' = T \times_{(\Delta^2)^\sharp} (\Lambda_1^2)^\sharp$. Since the opposites of the inclusions $T' \subseteq T$, $(\Delta^{\{0,2\}})^\flat \subseteq T$ are marked anodyne, we conclude that the evaluation maps

$$
\mathcal{X} \leftarrow \mathcal{X}'' \to \mathcal{X}'
$$

are trivial fibrations of simplicial sets. It follows that $\mathcal{X}$ and $\mathcal{X}'$ are (canonically) homotopy equivalent, as desired.   □

Remark 5.4.7.16. In the situation of Lemma 5.4.7.15, the full subcategory of $\mathcal{X}$ spanned by the *coCartesian* sections of p is equivalent (via evaluation at $\{0\}$) to $\mathcal{C}$.

Proof of Proposition 5.4.7.11. Let us first suppose that $\mathcal{E} = \emptyset$. Let $\mathrm{sk}^n S$ denote the n-skeleton of S. We observe that $\mathrm{Map}_S(S, X)$ coincides with the (homotopy) inverse limit

$$
\varprojlim \{\mathrm{Map}_S(\mathrm{sk}^n S, X)\}.
$$

In view of assumption (a), it will suffice to prove the result after replacing S by $\mathrm{sk}^n S$. In other words, we may reduce to the case where S is n-dimensional.

We now work by induction on n and observe that there is a homotopy pushout diagram of simplicial sets

$$
\begin{array}{ccc}
S_n \times \partial \Delta^n & \hookrightarrow & S_n \times \Delta^n \\
\downarrow & & \downarrow \\
\mathrm{sk}^{n-1} S & \hookrightarrow & S.
\end{array}
$$

We therefore obtain a homotopy pullback diagram of ∞-categories

$$
\begin{array}{ccc}
\mathrm{Map}_S(S, X) & \longrightarrow & \mathrm{Map}_S(\mathrm{sk}^{n-1} S, X) \\
\downarrow & & \downarrow \\
\mathrm{Map}_S(S_n \times \Delta^n, X) & \longrightarrow & \mathrm{Map}_S(S_n \times \partial \Delta^n, X).
\end{array}
$$

Invoking assumption (a) again, we are reduced to proving the same result after replacing S by $\mathrm{sk}^{n-1}\, S$, $S_n \times \partial \Delta^n$, and $S_n \times \Delta^n$. The first two cases follow from the inductive hypothesis; we may therefore assume that S is a disjoint union of copies of Δ^n. Applying (a) once more, we can reduce to the case $S = \Delta^n$.

If $n = 0$, there is nothing to prove. If $n > 1$, then we have a trivial fibration

$$\mathrm{Map}_S(S, X) \to \mathrm{Map}_S(\Lambda_1^n, X).$$

Since the horn Λ_1^n is of dimension less than n, we may conclude by applying the inductive hypothesis. We are therefore reduced to the case $S = \Delta^1$.

According to Lemma 5.4.7.15, the ∞-category $\mathrm{Map}_{\Delta^1}(\Delta^1, X)$ can be identified with a homotopy limit of the diagram

$$X_{\{0\}} \xrightarrow{F} X_{\{1\}} \leftarrow X_{\{1\}}^{\Delta^1}.$$

In view of (a), it will suffice to prove that all of the ∞-categories and functors in the above diagram belong to $\mathcal{C}$. This follows immediately from (b) and (c).

We now consider the general case where $\mathcal{E}$ is not required to be empty. For each edge $e \in \mathcal{E}$, let $Y(e)$ denote the full subcategory of $\mathrm{Map}_S(S, X)$ spanned by those sections $f : S \to X$ which satisfy the condition $(*)$ for the edge e. We wish to prove:

(1) The intersection $\bigcap_{e \in \mathcal{E}} Y(e)$ belongs to $\mathcal{C}$.

(2) If $Z \in \mathcal{C}$, then a functor $Z \to \bigcap_{e \in \mathcal{E}} Y(e)$ is a morphism of $\mathcal{C}$ if and only if the induced map $Z \to \mathrm{Map}_S(S, X)$ is a morphism of $\mathcal{C}$.

In view of (a), it will suffice to prove the corresponding results when the intersection $\bigcap_{e \in \mathcal{E}} Y(e)$ is replaced by a single subcategory $Y(e) \subseteq \mathrm{Map}_S(S, X)$.

Let $e : s \to s'$ be an edge belonging to $\mathcal{E}$. Lemma 5.4.7.15 implies the existence of a homotopy pullback diagram We now observe that there is a homotopy pullback diagram

$$
\begin{array}{ccc}
Y(e) & \longrightarrow & \mathrm{Map}_S(S, X) \\
\downarrow & & \downarrow \\
\mathrm{Fun}'(\Delta^1, X_{s'}) & \longrightarrow & \mathrm{Fun}(\Delta^1, X_{s'}),
\end{array}
$$

where $\mathrm{Fun}'(\Delta^1, X_{s'}) \simeq X_{s'}$ is the full subcategory of $\mathrm{Fun}(\Delta^1, X_{s'})$ spanned by the equivalences. In view of (a), it suffices to prove the following analogues of (1) and (2):

(1') For each $s' \in S$, the ∞-categories $\mathrm{Fun}'(\Delta^1, X_{s'})$ and $\mathrm{Fun}(\Delta^1, X_{s'})$ belong to $\mathcal{C}$.

(2') Given an object $Z \in \mathcal{C}$, a functor $Z \to \mathrm{Fun}'(\Delta^1, X_{s'})$ is a morphism in $\mathcal{C}$ if and only if the induced map $Z \to \mathrm{Fun}(\Delta^1, X_{s'})$ is a morphism of $\mathcal{C}$.

These assertions follow immediately from (*b*) and (*c*), respectively.   □

Corollary 5.4.7.17. *Let* $p : X \to S$ *be a map of simplicial sets which is a coCartesian fibration (or a Cartesian fibration). Assume that the following conditions are satisfied:*

(1) *The simplicial set* S *is small.*

(2) *For each vertex* s *of* S, *the* ∞-*category* $X_s = X \times_S \{s\}$ *is accessible.*

(3) *For each edge* $e : s \to s'$ *of* S, *the associated functor* $X_s \to X_{s'}$ *(or* $X_{s'} \to X_s$) *is accessible.*

Then $\mathrm{Map}_S(S, X)$ *is an accessible* ∞-*category. Moreover, if* $\mathcal{C}$ *is accessible, then a functor*

$$\mathcal{C} \to \mathrm{Map}_S(S, X)$$

is accessible if and only if, for every vertex s *of* S, *the induced map* $\mathcal{C} \to X_s$ *is accessible.*

5.5 PRESENTABLE ∞-CATEGORIES

Our final object of study in this chapter is the theory of *presentable* ∞-categories.

Definition 5.5.0.1. An ∞-category $\mathcal{C}$ is *presentable* if $\mathcal{C}$ is accessible and admits small colimits.

We will begin in §5.5.1 by giving a number of equivalent reformulations of Definition 5.5.0.1. The main result, Theorem 5.5.1.1, is due to Carlos Simpson: an ∞-category $\mathcal{C}$ is presentable if and only if it arises as an (accessible) localization of an ∞-category of presheaves.

Let $\mathcal{C}$ be an ∞-category and let $F : \mathcal{C} \to \mathcal{S}^{op}$ be a functor. If F is representable by an object of $\mathcal{C}$, then F preserves colimits (Proposition 5.1.3.2). In §5.5.2, we will prove that the converse holds when $\mathcal{C}$ is presentable. This representability criterion has a number of consequences: it implies that $\mathcal{C}$ admits (small) limits (Corollary 5.5.2.4) and leads to an ∞-categorical analogue of the adjoint functor theorem (Corollary 5.5.2.9).

In §5.5.3, we will see that the collection of all presentable ∞-categories can be organized into an ∞-category $\mathcal{P}\mathrm{r}^{\mathrm{L}}$. Moreover, we will explain how to compute limits and colimits in $\mathcal{P}\mathrm{r}^{\mathrm{L}}$. In the course of doing so, we will prove that the class of presentable ∞-categories is stable under most of the basic constructions of higher category theory.

In view of Theorem 5.5.1.1, the theory of localizations plays a central role in the study of presentable ∞-categories. In §5.5.4, we will show that the collection of all (accessible) localizations of a presentable ∞-category $\mathcal{C}$ can be parametrized in a very simple way. Moreover, there is a good supply of localizations of $\mathcal{C}$: given any (small) collection of morphisms S of $\mathcal{C}$, one can construct a corresponding localization functor

$$\mathcal{C} \xrightarrow{L} S^{-1} \mathcal{C} \subseteq \mathcal{C},$$

where $S^{-1}\mathcal{C}$ is the full subcategory of $\mathcal{C}$ spanned by the S-*local* objects. These ideas are due to Bousfield, who works in the setting of model categories; we will give an exposition here in the language of ∞-categories. In §5.5.5, we will employ the same techniques to produce examples of factorization systems on the ∞-category $\mathcal{C}$.

Let $\mathcal{C}$ be an ∞-category and let $C \in \mathcal{C}$ be an object. We will say that $C \in \mathcal{C}$ is *discrete* if, for every $D \in \mathcal{C}$, the nonzero homotopy groups of the mapping space $\mathrm{Map}_\mathcal{C}(D, C)$ vanish. If we let $\tau_{\leq 0}\,\mathcal{C}$ denote the full subcategory of $\mathcal{C}$ spanned by the discrete objects, then $\tau_{\leq 0}\,\mathcal{C}$ is (equivalent to) an ordinary category. If $\mathcal{C}$ is the ∞-category of spaces, then we can identify the discrete objects of $\mathcal{C}$ with the ordinary category of sets. Moreover, the inclusion $\tau_{\leq 0}\,\mathcal{S} \subseteq \mathcal{S}$ has a left adjoint given by

$$X \mapsto \pi_0 X.$$

In §5.5.6, we will show that the preceding remark generalizes to an arbitrary presentable ∞-category $\mathcal{C}$: the discrete objects of $\mathcal{C}$ constitute an (accessible) localization of $\mathcal{C}$. We will also consider a more general condition of k-truncatedness (which specializes to the condition of discreteness when $k = 0$). The truncation functors which we construct will play an important role throughout Chapter 6.

In §5.5.7, we will study the theory of *compactly generated* ∞-categories: ∞-categories which are generated (under colimits) by their compact objects. This class of ∞-categories includes some of the most important examples, such as $\mathcal{S}$ and $\mathcal{C}at_\infty$. In fact, the ∞-category $\mathcal{S}$ satisfies an even stronger condition: it is generated by compact *projective* objects (see Definition 5.5.8.18). The presence of enough compact projective objects in an ∞-category allows us to construct projective resolutions, which gives rise to the theory of nonabelian homological algebra (or "homotopical algebra"). We will review the rudiments of this theory in §5.5.8. Finally, in §5.5.9 we will present the same ideas in a more classical form following Quillen's manuscript [63]. The comparison of these two perspectives is based on a rectification result (Proposition 5.5.9.2) which is of some independent interest.

Remark 5.5.0.2. We refer the reader to [1] for a study of presentability in the setting of ordinary category theory. Note that [1] uses the term "locally presentable category" for what we have chosen to call a *presentable category*.

5.5.1 Presentability

Our main goal in this section is to establish the following characterization of presentable ∞-categories:

Theorem 5.5.1.1 (Simpson [70]). *Let $\mathcal{C}$ be an ∞-category. The following conditions are equivalent:*

(1) *The ∞-category* C *is presentable.*

(2) *The ∞-category* C *is accessible, and for every regular cardinal* κ *the full subcategory* C^κ *admits* κ-*small colimits.*

(3) *There exists a regular cardinal* κ *such* C *is* κ-*accessible and* C^κ *admits* κ-*small colimits.*

(4) *There exists a regular cardinal* κ*, a small ∞-category* D *which admits* κ-*small colimits, and an equivalence* $\mathrm{Ind}_\kappa\, D \to C$.

(5) *There exists a small ∞-category* D *such that* C *is an accessible localization of* $P(D)$.

(6) *The ∞-category* C *is locally small and admits small colimits, and there exists a regular cardinal* κ *and a (small) set* S *of* κ-*compact objects of* C *such that every object of* C *is a colimit of a small diagram taking values in the full subcategory of* C *spanned by* S.

Before giving the proof, we need a few preliminart remarks. We first observe that condition (5) is potentially ambiguous: it is unclear whether the accessibility hypothesis is on C or on the associated localization functor $L : P(D) \to P(D)$. The distinction turns out to be irrelevant by virtue of the following:

Proposition 5.5.1.2. *Let* C *be an accessible ∞-category and let* $L : C \to C$ *be a functor satisfying the equivalent conditions of Proposition 5.2.7.4. The following conditions are equivalent:*

(1) *The essential image* $L\,C$ *of* L *is accessible.*

(2) *There exists a localization* $f : C \to D$*, where* D *is accessible, and an equivalence* $L \simeq g \circ f$.

(3) *The functor* L *is accessible (when regarded as a functor from* C *to itself).*

Proof. Suppose (1) is satisfied. Then we may take $D = L\,C$, $f = L$, and g to be the inclusion $L\,C \subseteq C$; this proves (2). If (2) is satisfied, then Proposition 5.4.7.7 shows that f and g are accessible functors, so their composite $g \circ f \simeq L$ is also accessible; this proves (3). Now suppose that (3) is satisfied. Choose a regular cardinal κ such that C is κ-accessible and L is κ-continuous. The full subcategory C^κ consisting of κ-compact objects of C is essentially small, so there exists a regular cardinal $\tau \gg \kappa$ such that LC is τ-compact for every $C \in C^\kappa$. Let C' denote the full subcategory of C spanned by the colimits of all τ-small κ-filtered diagrams in C^κ and let $L\,C'$ denote the essential image of C' under L. We note that $L\,C'$ is essentially small. Since L is κ-continuous, $L\,C$ is stable under small κ-filtered colimits in C. It follows that any τ-compact object of C which belongs to $L\,C$ is also τ-compact when viewed as an object of $L\,\mathcal{C}$, so that $L\,\mathcal{C}'$ consists of τ-compact objects of $L\,\mathcal{C}$. According to Proposition 5.4.2.2, to complete the proof that $L\,\mathcal{C}$ is accessible, it will suffice to show that $L\,\mathcal{C}'$ generates $L\,\mathcal{C}$ under small τ-filtered colimits.

Let X be an object of $\mathcal{C}$. Then X can be written as a small κ-filtered colimit of objects of $\mathcal{C}^{\kappa}$. The proof of Proposition 5.4.2.11 shows that we can also write X as the colimit of a small τ-filtered diagram in $\mathcal{C}'$. Since L preserves colimits, it follows that LX can be obtained as the colimit of a small τ-filtered diagram in $L\,\mathcal{C}'$. $\square$

The proof of Theorem 5.5.1.1 will require a few easy lemmas:

Lemma 5.5.1.3. *Let* $f : \mathcal{C} \to \mathcal{D}$ *be a functor between small* ∞-*categories which exhibits* $\mathcal{D}$ *as an idempotent completion of* $\mathcal{C}$ *and let* κ *be a regular cardinal. Then* $\mathrm{Ind}_{\kappa}(f) : \mathrm{Ind}_{\kappa}(\mathcal{C}) \to \mathrm{Ind}_{\kappa}(\mathcal{D})$ *is an equivalence of* ∞-*categories.*

Proof. We first apply Proposition 5.3.5.11 to conclude that $\mathrm{Ind}_{\kappa}(f)$ is fully faithful. To prove that $\mathrm{Ind}_{\kappa}(f)$ is an equivalence, we must show that it generates $\mathrm{Ind}_{\kappa}(\mathcal{D})$ under κ-filtered colimits. Since $\mathrm{Ind}_{\kappa}(\mathcal{D})$ is generated under κ-filtered colimits by the essential image of the Yoneda embedding $j_{\mathcal{D}} : \mathcal{D} \to \mathrm{Ind}_{\kappa}(\mathcal{D})$, it will suffice to show that the essential image of $j_{\mathcal{D}}$ is contained in the essential image of $\mathrm{Ind}_{\kappa}(f)$. Let D be an object of $\mathcal{D}$. Then D is a retract of $f(C)$ for some object $C \in \mathcal{C}$. Then $j_{\mathcal{D}}(D)$ is a retract of $(\mathrm{Ind}_{\kappa}(f) \circ j_{\mathcal{C}})(C)$. Since $\mathrm{Ind}_{\kappa}(\mathcal{C})$ is idempotent complete (Corollary 4.4.5.16), we conclude that $j_{\mathcal{D}}(D)$ belongs to the essential image of $\mathrm{Ind}_{\kappa}(f)$. $\square$

Lemma 5.5.1.4. *Let* $F : \mathcal{C} \to \mathcal{D}$ *be a functor between* ∞-*categories which admit small* κ-*filtered colimits and let* G *be a right adjoint to* F. *Suppose that* G *is* κ-*continuous. Then* F *carries* κ-*compact objects of* $\mathcal{C}$ *to* κ-*compact objects of* $\mathcal{D}$.

Proof. Let C be a κ-compact object of $\mathcal{C}$, $e_C : \mathcal{C} \to \widehat{\mathcal{S}}$ the functor corepresented by C, and $e_{F(C)} : \mathcal{D} \to \widehat{\mathcal{S}}$ the functor corepresented by $F(C)$. Since G is a right adjoint to F, we have an equivalence $e_{F(C)} = e_C \circ G$. Since e_C and G are both κ-continuous, e_{FC} is κ-continuous. It follows that $F(C)$ is κ-compact, as desired. $\square$

Proof of Theorem 5.5.1.1. Corollary 5.3.4.15 asserts that the full subcategory $\mathcal{C}^{\kappa}$ is stable under all κ-small colimits which exist in $\mathcal{C}$. This proves that (1) implies (2). The implications (2) $\Rightarrow$ (3) $\Rightarrow$ (4) are obvious. We next prove that (4) implies (5). According to Lemma 5.5.1.3, we may suppose without loss of generality that $\mathcal{D}$ is idempotent complete. Let $\mathcal{P}^{\kappa}(\mathcal{D})$ denote the full subcategory of $\mathcal{P}(\mathcal{D})$ spanned by the κ-compact objects, let $\mathcal{D}'$ be a minimal model for $\mathcal{P}^{\kappa}(\mathcal{D})$, and let g denote the composition

$$\mathcal{D} \xrightarrow{\;j\;} \mathcal{P}^{\kappa}(\mathcal{D}) \to \mathcal{D}',$$

where the second map is a homotopy inverse to the inclusion $\mathcal{D}' \subseteq \mathcal{P}^{\kappa}(\mathcal{D})$. Proposition 5.1.3.1 implies that g is fully faithful, and Proposition 5.3.4.18 implies that g admits a left adjoint f. It follows that $F = \mathrm{Ind}_{\kappa}(f)$ and

$G = \mathrm{Ind}_\kappa(g)$ are adjoint functors, and Proposition 5.3.5.11 implies that G is fully faithful. Moreover, Proposition 5.3.5.12 implies that $\mathrm{Ind}_\kappa \mathcal{D}'$ is equivalent to $\mathcal{P}(\mathcal{D})$, so that $\mathcal{C}$ is equivalent to an accessible localization of $\mathcal{P}(\mathcal{D}')$.

We now prove that (5) implies (6). Let $\mathcal{D}$ be a small ∞-category and $L : \mathcal{P}(\mathcal{D}) \to \mathcal{C}$ an accessible localization. Remark 5.2.7.5 implies that $\mathcal{C}$ admits small colimits and that $\mathcal{C}$ is generated under colimits by the essential image of the composition

$$T : \mathcal{D} \xrightarrow{j} \mathcal{P}(\mathcal{D}) \xrightarrow{L} \mathcal{C}.$$

To complete the proof of (6), it will suffice to show that there exists a regular cardinal κ such that the essential image of T consists of κ-compact objects. Let G denote a left adjoint to L. By assumption, G is an accessible functor so that there exists a regular cardinal κ such that G is κ-continuous. For each object $D \in \mathcal{D}$, the Yoneda image $j(D)$ is a completely compact object of $\mathcal{P}(\mathcal{D})$ and, in particular, κ-compact. Lemma 5.5.1.4 implies that $T(D)$ is a κ-compact object of $\mathcal{C}$.

We now complete the proof by showing that (6) $\Rightarrow$ (1). Assume that there exists a regular cardinal κ and a set S of κ-compact objects of $\mathcal{C}$ such that every object of $\mathcal{C}$ is a colimit of objects in S. Let $\mathcal{C}' \subseteq \mathcal{C}$ be the full subcategory of $\mathcal{C}$ spanned by S and let $\mathcal{C}'' \subseteq \mathcal{C}$ be the full subcategory of $\mathcal{C}$ spanned by the colimits of all κ-small diagrams with values in $\mathcal{C}''$. Since $\mathcal{C}'$ is essentially small, there is only a bounded number of such diagrams up to equivalence, so that $\mathcal{C}''$ is essentially small. Moreover, since every object of $\mathcal{C}$ is a colimit of a small diagram with values in $\mathcal{C}'$, the proof of Corollary 4.2.3.11 shows that every object of $\mathcal{C}$ can also be obtained as the colimit of a small κ-filtered diagram with values in $\mathcal{C}''$. Corollary 5.3.4.15 implies that $\mathcal{C}''$ consists of κ-compact objects of $\mathcal{C}$ (a slightly more refined argument shows that, if $\kappa > \omega$, then $\mathcal{C}''$ consists of *precisely* the κ-compact objects of $\mathcal{C}$). We may therefore apply Proposition 5.4.2.2 to deduce that $\mathcal{C}$ is accessible. $\square$

Remark 5.5.1.5. The characterization of presentable ∞-categories as localizations of presheaf ∞-categories was established by Simpson in [70] (using a somewhat different language). The theory of presentable ∞-categories is essentially equivalent to the theory of *combinatorial* model categories (see §A.3.7 and Proposition A.3.7.6). Since most of the ∞-categories we will meet are presentable, our study could also be phrased in the language of model categories. However, we will try to avoid this language since for many purposes the restriction to presentable ∞-categories seems unnatural and is often technically inconvenient.

Remark 5.5.1.6. Let $\mathcal{C}$ be a presentable ∞-category and let $\mathcal{D}$ be an accessible localization of $\mathcal{C}$. Then $\mathcal{D}$ is presentable: this follows immediately from characterization (5) of Proposition 5.5.1.1.

Remark 5.5.1.7. Let $\mathcal{C}$ be a presentable ∞-category. Since $\mathcal{C}$ admits arbitrary colimits, it is "tensored over spaces," as we explained in §4.4.4. In particular, the homotopy category of $\mathcal{C}$ is naturally tensored over the homotopy category $\mathcal{H}$: for each object C of $\mathcal{C}$ and every simplicial set S, there exists an object $C \otimes S$ of $\mathcal{C}$, well-defined up to equivalence, equipped with isomorphisms

$$\mathrm{Map}_{\mathcal{C}}(C \otimes S, C') \simeq \mathrm{Map}_{\mathcal{C}}(C, C')^{S}$$

in the homotopy category $\mathcal{H}$.

Example 5.5.1.8. The ∞-category $\mathcal{S}$ of spaces is presentable. This follows from characterization (1) of Theorem 5.5.1.1 since $\mathcal{S}$ is accessible (Example 5.4.2.7) and admits (small) colimits by Corollary 4.2.4.8.

According to Theorem 5.5.1.1, if $\mathcal{C}$ is κ-accessible, then $\mathcal{C}$ admits small colimits if and only if the full subcategory $\mathcal{C}^{\kappa} \subseteq \mathcal{C}$ admits κ-small colimits. Roughly speaking, this is because arbitrary colimits in $\mathcal{C}$ can be rewritten in terms of κ-filtered colimits and κ-small colimits of κ-compact objects. Our next result is another variation on this idea; it may also be regarded as an analogue of Theorem 5.5.1.1 (which describes functors rather than ∞-categories):

Proposition 5.5.1.9. *Let $f : \mathcal{C} \to \mathcal{D}$ be a functor between presentable ∞-categories. Suppose that $\mathcal{C}$ is κ-accessible. The following conditions are equivalent:*

(1) *The functor f preserves small colimits.*

(2) *The functor f is κ-continuous, and the restriction $f | \mathcal{C}^{\kappa}$ preserves κ-small colimits.*

Proof. Without loss of generality, we may suppose $\mathcal{C} = \mathrm{Ind}_{\kappa}(\mathcal{C}')$, where $\mathcal{C}'$ is a small idempotent complete ∞-category which admits κ-small colimits. The proof of Theorem 5.5.1.1 shows that the inclusion $\mathrm{Ind}_{\kappa}(\mathcal{C}') \subseteq \mathcal{P}(\mathcal{C}')$ admits a left adjoint L. Let $\alpha : \mathrm{id}_{\mathcal{P}(\mathcal{C}')} \to L$ be a unit for the adjunction and let $f' : \mathcal{C}' \to \mathcal{D}$ denote the composition of f with the Yoneda embedding $j : \mathrm{Ind}_{\kappa}(\mathcal{C}')$. According to Theorem 5.1.5.6, there exists a colimit-preserving functor $F : \mathcal{P}(\mathcal{C}') \to \mathcal{D}$ and an equivalence of f' with $F \circ j$. Proposition 5.3.5.10 implies that f and $F | \mathrm{Ind}_{\kappa}(\mathcal{C})$ are equivalent; we may therefore assume without loss of generality that $f = F | \mathrm{Ind}_{\kappa}(\mathcal{C})$. Let $F' = f \circ L$, so that α induces a natural transformation $\beta : F \to F'$ of functors from $\mathcal{P}(\mathcal{C}')$ to $\mathcal{D}$. We will show that β is an equivalence. Consequently, we deduce that the functor F' is colimit-preserving. It then follows that f is colimit-preserving. To see this, we consider an arbitrary diagram $p : K \to \mathrm{Ind}_{\kappa}(\mathcal{C}')$ and choose a colimit $\overline{p} : K^{\triangleright} \to \mathcal{P}(\mathcal{C}')$. Then $\overline{q} = L \circ \overline{p}$ is a colimit diagram in $\mathrm{Ind}_{\kappa}(\mathcal{C}')$, and $f \circ \overline{q} = F' \circ \overline{p}$ is a colimit diagram in $\mathcal{D}$. Since $q = \overline{q} | K$ is equivalent (via α) to the original diagram p, we conclude that f preserves the colimit of p in $\mathrm{Ind}_{\kappa}(\mathcal{C}')$ as well.

It remains to prove that β is an equivalence of functors. Let $\mathcal{E} \subseteq \mathcal{P}(\mathcal{C}')$ denote the full subcategory spanned by those objects $X \in \mathcal{P}(\mathcal{C}')$ for which

$\beta(X) : F(X) \to F'(X)$ is an equivalence in $\mathcal{D}$. We wish to prove that $\mathcal{E} = \mathcal{P}(\mathcal{C}')$. Since F and F' are both κ-continuous functors, $\mathcal{E}$ is stable under κ-filtered colimits in $\mathcal{P}(\mathcal{C}')$. It will therefore suffice to prove that $\mathcal{E}$ contains $\mathcal{P}^\kappa(\mathcal{C}')$.

It is clear that $\mathcal{E}$ contains $\mathrm{Ind}_\kappa(\mathcal{C}')$; in particular, $\mathcal{E}$ contains the essential image $\mathcal{E}'$ of the Yoneda embedding $j : \mathcal{C}' \to \mathcal{P}(\mathcal{C}')$. According to Proposition 5.3.4.17, every object of $\mathcal{P}^\kappa(\mathcal{C}')$ is a retract of the colimit of a κ-small diagram $p : K \to \mathcal{E}'$. Since $\mathcal{C}'$ is idempotent complete, we may identify $\mathcal{E}'$ with the full subcategory of $\mathrm{Ind}_\kappa(\mathcal{C}')$ consisting of κ-compact objects. In particular, $\mathcal{E}'$ is stable under κ-small colimits and retracts in $\mathrm{Ind}_\kappa(\mathcal{C}')$. It follows that L restricts to a functor $L' : \mathcal{P}^\kappa(\mathcal{C}) \to \mathcal{E}'$ which preserves κ-small colimits.

To complete the proof that $\mathcal{P}^\kappa(\mathcal{C}') \subseteq \mathcal{E}$, it will suffice to prove that $F'|\,\mathcal{P}^\kappa(\mathcal{C})$ preserves κ-small colimits. To see this, we write $F|\,\mathcal{P}^\kappa(\mathcal{C}')$ as a composition

$$\mathcal{P}^\kappa(\mathcal{C}') \overset{L'}{\to} \mathcal{E}' \overset{F|\,\mathcal{E}'}{\to} \mathcal{C},$$

where L' preserves κ-small colimits (as noted above) and $F|\,\mathcal{E}' = f|\,\mathcal{C}^\kappa$ preserves κ-small colimits by assumption.   □

5.5.2 Representable Functors and the Adjoint Functor Theorem

An object F of the ∞-category $\mathcal{P}(\mathcal{C})$ of presheaves on $\mathcal{C}$ is *representable* if it lies in the essential image of the Yoneda embedding $j : \mathcal{C} \to \mathcal{P}(\mathcal{C})$. If $F : \mathcal{C}^{op} \to \mathcal{S}$ is representable, then F preserves limits: this follows from the fact that F is equivalent to the composite map

$$\mathcal{C}^{op} \overset{j}{\to} \mathcal{P}(\mathcal{C}^{op}) \to \mathcal{S},$$

where j denotes the Yoneda embedding for $\mathcal{C}^{op}$ (which is limit-preserving by Proposition 5.1.3.2) and the right map is given by evaluation at C (which is limit-preserving by Proposition 5.1.2.2). If $\mathcal{C}$ is presentable, then the converse holds.

Lemma 5.5.2.1. *Let S be a small simplicial set, let $f : S \to \mathcal{S}$ be an object of $\mathcal{P}(S^{op})$, and let $F : \mathcal{P}(S^{op}) \to \widehat{\mathcal{S}}$ be the functor corepresented by f. Then the composition*

$$S \overset{j}{\to} \mathcal{P}(S^{op}) \overset{F}{\to} \widehat{\mathcal{S}}$$

is equivalent to f.

Proof. According to Corollary 4.2.4.7, we can choose a (small) fibrant simplicial category $\mathcal{C}$ and a categorical equivalence $\phi : S \to \mathrm{N}(\mathcal{C}^{op})$ such that f is equivalent to the composition of ψ^{op} with the nerve of a simplicial functor $f' : \mathcal{C} \to \mathcal{K}\mathrm{an}$. Without loss of generality, we may suppose that $f' \in \mathrm{Set}_\Delta^{\mathcal{C}}$ is projectively cofibrant. Using Proposition 4.2.4.4, we have an equivalence of ∞-categories

$$\psi : \mathrm{N}(\mathrm{Set}_\Delta^{\mathcal{C}})^\circ) \to \mathcal{P}(S).$$

We observe that the composition $F \circ \psi$ can be identified with the simplicial nerve of the functor $G : (\mathrm{Set}_\Delta^{\mathcal{C}})^\circ \to \mathcal{K}\mathrm{an}$ corepresented by f'. The Yoneda embedding factors through ψ via the adjoint of the composition

$$j' : \mathfrak{C}[S] \to \mathcal{C}^{op} \to (\mathrm{Set}_\Delta^{\mathcal{C}})^\circ.$$

It follows that $F \circ j$ can be identified with the adjoint of the composition

$$\mathfrak{C}[S] \xrightarrow{j'} (\mathrm{Set}_\Delta^{\mathcal{C}})^\circ \xrightarrow{G} \mathcal{K}\mathrm{an}.$$

This composition is equal to the functor f', so its simplicial nerve coincides with the original functor f. $\square$

Proposition 5.5.2.2. *Let $\mathcal{C}$ be a presentable ∞-category and let $F : \mathcal{C}^{op} \to \mathcal{S}$ be a functor. The following are equivalent:*

(1) *The functor F is representable by an object $C \in \mathcal{C}$.*

(2) *The functor F preserves small limits.*

Proof. The implication (1) $\Rightarrow$ (2) was proven above (for an arbitrary ∞-category $\mathcal{C}$). For the converse, we first treat the case where $\mathcal{C} = \mathcal{P}(\mathcal{D})$ for some small ∞-category $\mathcal{D}$. Let $f : \mathcal{D}^{op} \to \mathcal{S}$ denote the composition of F with the (opposite) Yoneda embedding $j^{op} : \mathcal{D}^{op} \to \mathcal{P}(\mathcal{D})^{op}$ and let $F' : \mathcal{P}(\mathcal{D})^{op} \to \widehat{\mathcal{S}}$ denote the functor represented by $f \in \mathcal{P}(\mathcal{D})$. We will prove that F and F' are equivalent. We observe that F and F' both preserve small limits; consequently, according to Theorem 5.1.5.6, it will suffice to show that the compositions $f = F \circ j^{op}$ and $f' = F' \circ j^{op}$ are equivalent. This follows immediately from Lemma 5.5.2.1.

We now consider the case where $\mathcal{C}$ is an arbitrary presentable ∞-category. According to Theorem 5.5.1.1, we may suppose that $\mathcal{C}$ is an accessible localization of a presentable ∞-category $\mathcal{C}'$ which has the form $\mathcal{P}(\mathcal{D})$, so that the assertion for $\mathcal{C}'$ has already been established. Let $L : \mathcal{C}' \to \mathcal{C}$ denote the localization functor. The functor $F \circ L^{op} : (\mathcal{C}')^{op} \to \mathcal{S}$ preserves small limits and is therefore representable by an object $C \in \mathcal{C}'$. Let S denote the set of all morphisms ϕ in $\mathcal{C}'$ such that $L(\phi)$ is an equivalence in $\mathcal{C}$. Without loss of generality, we may identify $\mathcal{C}$ with the full subcategory of $\mathcal{C}'$ consisting of S-local objects. By construction, $C \in \mathcal{C}'$ is S-local and therefore belongs to $\mathcal{C}$. It follows that C represents the functor $(F \circ L^{op})|\,\mathcal{C}$, which is equivalent to F. $\square$

The representability criterion provided by Proposition 5.5.2.2 has many consequences, as we now explain.

Lemma 5.5.2.3. *Let X and Y be simplicial sets, let $q : \mathcal{C} \to \mathcal{D}$ be a categorical fibration of ∞-categories, and let $p : X^\triangleright \times Y^\triangleright \to \mathcal{C}$ be a diagram. Suppose that*

(1) *For every vertex x of $X^\triangleright$, the associated map $p_x : Y^\triangleright \to \mathcal{C}$ is a q-colimit diagram.*

(2) *For every vertex y of Y, the associated map $p_y : X^\triangleright \to \mathcal{C}$ is a q-colimit diagram.*

Let ∞ denote the cone point of $Y^\triangleright$. Then the restriction $p_\infty : X^\triangleright \to \mathcal{C}$ is a q-colimit diagram.

Proof. Without loss of generality, we can suppose that X and Y are ∞-categories. Since the inclusion $X \times \{\infty\} \subseteq X \times Y^\triangleright$ is cofinal, it will suffice to show that the restriction $p|(X \times Y^\triangleright)^\triangleright$ is a q-colimit diagram. According to Proposition 4.3.2.9, $p|(X \times Y^\triangleright)$ is a q-left Kan extension of $p|(X \times Y)$. By transitivity, it suffices to show that $p|(X \times Y)^\triangleright$ is a q-colimit diagram. For this, it will suffice to prove the stronger assertion that $p|(X^\triangleright \times Y)^\triangleright$ is a q-left Kan extension of $p|(X \times Y)$. Since Proposition 4.3.2.9 also implies that $p|(X^\triangleright \times Y)$ is a q-left Kan extension of $p|(X \times Y)$, we may again apply transitivity and reduce to the problem of showing that $p|(X^\triangleright \times Y)^\triangleright$ is a q-colimit diagram. Let ∞' denote the cone point of $X^\triangleright$. Since the inclusion $\{\infty'\} \times Y \subseteq X^\triangleright \times Y$ is cofinal, we are reduced to proving that $p_{\infty'} : Y^\triangleright \to \mathcal{C}$ is a q-colimit diagram, which follows from (1). $\qquad\square$

Corollary 5.5.2.4. *A presentable ∞-category $\mathcal{C}$ admits all (small) limits.*

Proof. Let $\widehat{\mathcal{P}}(\mathcal{C}) = \mathrm{Fun}(\mathcal{C}^{op}, \widehat{\mathcal{S}})$, where $\widehat{\mathcal{S}}$ denotes the ∞-category of spaces which are not necessarily small, and let $j : \mathcal{C} \to \widehat{\mathcal{P}}(\mathcal{C})$ be the Yoneda embedding. Since j is fully faithful, it will suffice to show that the essential image of j admits small limits. The ∞-category $\widehat{\mathcal{P}}(\mathcal{C})$ admits all small limits (in fact, even limits which are not necessarily small); it therefore suffices to show that the essential image of j is stable under small limits. This follows immediately from Proposition 5.5.2.2 and Lemma 5.5.2.3. $\qquad\square$

Remark 5.5.2.5. Let A be a (small) partially ordered set. The ∞-category $N(A)$ is presentable if and only if every subset of A has a least upper bound. Corollary 5.5.2.4 can then be regarded as a generalization of the following classical observation: if every subset of A has a least upper bound, then every subset of A has a greatest lower bound (namely, a least upper bound for the collection of all lower bounds).

Remark 5.5.2.6. Now that we know that every presentable ∞-category $\mathcal{C}$ has arbitrary limits, we can apply an argument dual to that of Remark 5.5.1.7 to show that $\mathcal{C}$ is *cotensored over* $\mathcal{S}$. In other words, for any $C \in \mathcal{C}$ and every simplicial set X, there exists an object $C^X \in \mathcal{C}$ (well-defined up to equivalence) together with a collection of natural isomorphisms

$$\mathrm{Map}_{\mathcal{C}}(C', C^X) \simeq \mathrm{Map}_{\mathcal{C}}(C', C)^X$$

in the homotopy category $\mathcal{H}$.

We can now formulate a dual version of Proposition 5.5.2.2, which requires a slightly stronger hypothesis.

Proposition 5.5.2.7. *Let $\mathcal{C}$ be a presentable ∞-category and let $F : \mathcal{C} \to \mathcal{S}$ be a functor. Then F is corepresentable by an object of $\mathcal{C}$ if and only if F is accessible and preserves small limits.*

Proof. The "only if" direction is clear since every object of $\mathcal{C}$ is κ-compact for $\kappa \gg 0$. We will prove the converse. Without loss of generality, we may suppose that $\mathcal{C}$ is minimal (this assumption is a technical convenience which will guarantee that various constructions below stay in the world of small ∞-categories). Let $\widetilde{\mathcal{C}} \to \mathcal{C}$ denote the left fibration represented by F. Choose a regular cardinal κ such that $\mathcal{C}$ is κ-accessible and F is κ-continuous and let $\widetilde{\mathcal{C}}^\kappa$ denote the fiber product $\widetilde{\mathcal{C}} \times_\mathcal{C} \mathcal{C}^\kappa$, where $\mathcal{C}^\kappa \subseteq \mathcal{C}$ denotes the full subcategory spanned by the κ-compact objects of $\mathcal{C}$. The ∞-category $\widetilde{\mathcal{C}}^\kappa$ is small (since $\mathcal{C}$ is assumed minimal). Corollary 5.5.2.4 implies that the diagram $p : \widetilde{\mathcal{C}}^\kappa \to \mathcal{C}$ admits a limit $\overline{p} : (\widetilde{\mathcal{C}}^\kappa)^\triangleleft \to \mathcal{C}$. Since the functor F preserves small limits, Corollary 3.3.3.3 implies that there exists a map $\overline{q} : (\widetilde{\mathcal{C}}^\kappa)^\triangleleft \to \widetilde{\mathcal{C}}$ which extends the inclusion $q : \widetilde{\mathcal{C}}^\kappa \subseteq \widetilde{\mathcal{C}}$ and covers $\overline{p}$. Let $\widetilde{X}_0 \in \widetilde{\mathcal{C}}$ denote the image of the cone point under $\overline{q}$ and X_0 its image in $\mathcal{C}$. Then $\widetilde{X}_0$ determines a connected component of the space $F(X_0)$. Since $\mathcal{C}$ is κ-accessible, we can write X_0 as a κ-filtered colimit of κ-compact objects $\{X_\alpha\}$ of $\mathcal{C}$. Since F is κ-continuous, there exists a κ-compact object $X \in \mathcal{C}$ such that the induced map $F(X) \to F(X_0)$ has nontrivial image in the connected component classified by $\widetilde{X}_0$. It follows that there exists an object $\widetilde{X} \in \widetilde{\mathcal{C}}$ lying over X_α and a morphism $f : \widetilde{X} \to \widetilde{X}_0$ in $\widetilde{\mathcal{C}}$. Since $\widetilde{\mathcal{C}}_{/q} \to \widetilde{\mathcal{C}}$ is a right fibration, we can pull $\overline{q}$ back to obtain a map $\overline{q}' : (\widetilde{\mathcal{C}}^\kappa)^\triangleleft \to \widetilde{\mathcal{C}}$ which extends q and carries the cone point to $\widetilde{X}$. It follows that $\overline{q}'$ factors through $\widetilde{\mathcal{C}}^\kappa$. We have a commutative diagram

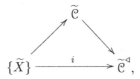

where i denotes the inclusion of the cone point. The map i is left anodyne and therefore a covariant equivalence in $(\mathcal{S}et_\Delta)_{/\mathcal{C}}$. It follows that $\widetilde{\mathcal{C}}^\kappa$ is a retract of $\{X\}$ in the homotopy category of the covariant model category $(\mathcal{S}et_\Delta)_{/\mathcal{C}^\kappa}$. Proposition 5.1.1.1 implies that $F|\,\mathcal{C}^\kappa$ is a retract of the Yoneda image $j(X)$ in $\mathcal{P}(\mathcal{C}^\kappa)$. Since the ∞-category $\mathcal{C}^\kappa$ is idempotent complete and the Yoneda embedding $j : \mathcal{C}^\kappa \to \mathcal{P}(\mathcal{C}^\kappa)$ is fully faithful, we deduce that $F|\,\mathcal{C}^\kappa$ is equivalent to $j(X')$, where $X' \in \mathcal{C}^\kappa$ is a retract of X. Let $F' : \mathcal{C} \to \mathcal{S}$ denote the functor corepresented by X'. We note that $F|\,\mathcal{C}^\kappa$ and $F'|\,\mathcal{C}^\kappa$ are equivalent and that both F and F' are κ-continuous. Since $\mathcal{C}$ is equivalent to $\mathrm{Ind}_\kappa(\mathcal{C}^\kappa)$, Proposition 5.3.5.10 guarantees that F and F' are equivalent, so that F is representable by X'. $\qquad\square$

Remark 5.5.2.8. It is not difficult to adapt our proof of Proposition 5.5.2.7 to obtain an alternative proof of Proposition 5.5.2.2.

From Propositions 5.5.2.2 and 5.5.2.7, we can deduce a version of the adjoint functor theorem:

Corollary 5.5.2.9 (Adjoint Functor Theorem). *Let $F : \mathcal{C} \to \mathcal{D}$ be a functor between presentable ∞-categories.*

(1) *The functor F has a right adjoint if and only if it preserves small colimits.*

(2) *The functor F has a left adjoint if and only if it is accessible and preserves small limits.*

Proof. The "only if" directions follow from Propositions 5.2.3.5 and 5.4.7.7. We now prove the converse direction of (2); the proof of (1) is similar but easier. Suppose that F is accessible and preserves small limits. Let $F' : \mathcal{D} \to \mathcal{S}$ be a corepresentable functor. Then F' is accessible and preserves small limits (Proposition 5.5.2.7). It follows that the composition $F' \circ F : \mathcal{C} \to \mathcal{S}$ is accessible and preserves small limits. Invoking Proposition 5.5.2.7 again, we deduce that $F' \circ F$ is representable. We now apply Proposition 5.2.4.2 to deduce that F has a left adjoint. □

Remark 5.5.2.10. The proof of (1) in Corollary 5.5.2.9 does not require that $\mathcal{D}$ be presentable but only that $\mathcal{D}$ be (essentially) locally small.

5.5.3 Limits and Colimits of Presentable ∞-Categories

In this section, we will introduce and study an ∞-category whose objects are presentable ∞-categories. In fact, we will introduce two such ∞-categories which are (canonically) antiequivalent to one another. The basic observation is the following: given a pair of presentable ∞-categories $\mathcal{C}$ and $\mathcal{D}$, the proper notion of "morphism" between them is a pair of adjoint functors

$$\mathcal{C} \underset{G}{\overset{F}{\rightleftarrows}} \mathcal{D}.$$

Of course, either one of F and G determines the other up to canonical equivalence. We may therefore think of either one as encoding the data of a morphism.

Definition 5.5.3.1. Let $\widehat{\mathcal{C}at}_\infty$ denote the ∞-category of (not necessarily small) ∞-categories. We define subcategories $\mathcal{P}r^R, \mathcal{P}r^L \subseteq \widehat{\mathcal{C}at}_\infty$ as follows:

(1) The objects of both $\mathcal{P}r^R$ and $\mathcal{P}r^L$ are the presentable ∞-categories.

(2) A functor $F : \mathcal{C} \to \mathcal{D}$ between presentable ∞-categories is a morphism in $\mathcal{P}r^L$ if and only if F preserves small colimits.

(3) A functor $G : \mathcal{C} \to \mathcal{D}$ between presentable ∞-categories is a morphism in $\mathcal{P}r^R$ if and only if G is accessible and preserves small limits.

As indicated above, the ∞-categories Pr^{R} and Pr^{L} are antiequivalent to one another. To prove this, it is convenient to introduce the following definition:

Definition 5.5.3.2. A map of simplicial sets $p : X \to S$ is a *presentable fibration* if it is both a Cartesian fibration and a coCartesian fibration and if each fiber $X_s = X \times_S \{s\}$ is a presentable ∞-category.

The following result is simply a reformulation of Corollary 5.5.2.9:

Proposition 5.5.3.3. (1) *Let* $p : X \to S$ *be a Cartesian fibration of simplicial sets classified by a map* $\chi : S^{op} \to \widehat{\mathcal{C}\mathrm{at}}_\infty$. *Then* p *is a presentable fibration if and only if* χ *factors through* $\mathrm{Pr}^{\mathrm{R}} \subseteq \widehat{\mathcal{C}\mathrm{at}}_\infty$.

(2) *Let* $p : X \to S$ *be a coCartesian fibration of simplicial sets classified by a map* $\chi : S \to \widehat{\mathcal{C}\mathrm{at}}_\infty$. *Then* p *is a presentable fibration if and only if* χ *factors through* $\mathrm{Pr}^{\mathrm{L}} \subseteq \widehat{\mathcal{C}\mathrm{at}}_\infty$.

Corollary 5.5.3.4. *For every simplicial set* S, *there is a canonical bijection*

$$[S, \mathrm{Pr}^{\mathrm{L}}] \simeq [S^{op}, \mathrm{Pr}^{\mathrm{R}}],$$

where $[S, \mathcal{C}]$ *denotes the collection of equivalence classes of objects of the* ∞-*category* $\mathrm{Fun}(S, \mathcal{C})$. *In particular, there is a canonical isomorphism* $\mathrm{Pr}^{\mathrm{L}} \simeq (\mathrm{Pr}^{\mathrm{R}})^{op}$ *in the homotopy category of* ∞-*categories.*

Proof. According to Proposition 5.5.3.3, both $[S, \mathrm{Pr}^{\mathrm{L}}]$ and $[S^{op}, \mathrm{Pr}^{\mathrm{R}}]$ can be identified with the collection of equivalence classes of presentable fibrations $X \to S$. $\square$

We now commence our study of the ∞-category Pr^{L} (or equivalently, the antiequivalent ∞-category Pr^{R}). The next few results express the idea that $\mathrm{Pr}^{\mathrm{L}} \subseteq \widehat{\mathcal{C}\mathrm{at}}_\infty$ is stable under a variety of categorical constructions.

Proposition 5.5.3.5. *Let* $\{\mathcal{C}_\alpha\}_{\alpha \in A}$ *be a family of* ∞-*categories indexed by a small set* A *and let* $\mathcal{C} = \prod_{\alpha \in A} \mathcal{C}_\alpha$ *be their product. If each* $\mathcal{C}_\alpha$ *is presentable, then* $\mathcal{C}$ *is presentable.*

Proof. It follows from Lemma 5.4.7.2 that $\mathcal{C}$ is accessible. Let $p : K \to \mathcal{C}$ be a diagram indexed by a small simplicial set K corresponding to a family of diagrams $\{p_\alpha : K \to \mathcal{C}_\alpha\}_{\alpha \in A}$. Since each $\mathcal{C}_\alpha$ is presentable, each p_α has a colimit $\overline{p_\alpha} : K^{\triangleright} \to \mathcal{C}_\alpha$. These colimits determine a map $\overline{p} : K^{\triangleright} \to \mathcal{C}$ which is a colimit of p. $\square$

Proposition 5.5.3.6. *Let* $\mathcal{C}$ *be an presentable* ∞-*category and let* K *be a small simplicial set. Then* $\mathrm{Fun}(K, \mathcal{C})$ *is presentable.*

Proof. According to Proposition 5.4.4.3, $\mathrm{Fun}(K, \mathcal{C})$ is accessible. It follows from Proposition 5.1.2.2 that if $\mathcal{C}$ admits small colimits, then $\mathrm{Fun}(K, \mathcal{C})$ admits small colimits. $\square$

Remark 5.5.3.7. Let S be a (small) simplicial set. It follows from Example 5.4.2.7 and Corollary 5.1.2.4 that $\mathcal{P}(S)$ is a presentable ∞-category. Moreover, Theorem 5.1.5.6 has a natural interpretation in the language of presentable ∞-categories: informally speaking, it asserts that the construction

$$S \mapsto \mathcal{P}(S)$$

is left adjoint to the inclusion functor from presentable ∞-categories to all ∞-categories.

The following is a variant on Proposition 5.5.3.6:

Proposition 5.5.3.8. *Let $\mathcal{C}$ and $\mathcal{D}$ be presentable ∞-categories. The ∞-category $\mathrm{Fun}^L(\mathcal{C}, \mathcal{D})$ is presentable.*

Proof. Since $\mathcal{D}$ admits small colimits, the ∞-category $\mathrm{Fun}(\mathcal{C}, \mathcal{D})$ admits small colimits (Proposition 5.1.2.2). Using Lemma 5.5.2.3, we conclude that

$$\mathrm{Fun}^L(\mathcal{C}, \mathcal{D}) \subseteq \mathrm{Fun}(\mathcal{C}, \mathcal{D})$$

is stable under small colimits. To complete the proof, it will suffice to show that $\mathrm{Fun}^L(\mathcal{C}, \mathcal{D})$ is accessible.

Choose a regular cardinal κ such that $\mathcal{C}$ is κ-accessible and let $\mathcal{C}^\kappa$ be the full subcategory of $\mathcal{C}$ spanned by the κ-compact objects. Propositions 5.5.1.9 and 5.3.5.10 imply that the restriction functor

$$\mathrm{Fun}^L(\mathcal{C}, \mathcal{D}) \to \mathrm{Fun}(\mathcal{C}^\kappa, \mathcal{D})$$

is fully faithful, and its essential image is the full subcategory $\mathcal{E} \subseteq \mathrm{Fun}(\mathcal{C}^\kappa, \mathcal{D})$ spanned by those functors which preserve κ-small colimits.

Since $\mathcal{C}^\kappa$ is essentially small, the ∞-category $\mathrm{Fun}(\mathcal{C}^\kappa, \mathcal{D})$ is accessible (Proposition 5.4.4.3). To complete the proof, we will show that $\mathcal{E}$ is an accessible subcategory of $\mathrm{Fun}(\mathcal{C}^\kappa, \mathcal{D})$. For each κ-small diagram $p : K \to \mathcal{C}^\kappa$, let $\mathcal{E}(p)$ denote the full subcategory of $\mathrm{Fun}(\mathcal{C}^\kappa, \mathcal{D})$ spanned by those functors which preserve the colimit of p. Then $\mathcal{E} = \bigcap_p \mathcal{E}(p)$, where the intersection is taken over a set of representatives for all equivalence classes of κ-small diagrams in $\mathcal{C}^\kappa$. According to Proposition 5.4.7.10, it will suffice to show that each $\mathcal{E}(p)$ is an accessible subcategory of $\mathrm{Fun}(\mathcal{C}^\kappa, \mathcal{D})$. We now observe that there is a (homotopy) pullback diagram of ∞-categories

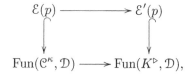

where $\mathcal{E}'$ denotes the full subcategory of $\mathrm{Fun}(K^\triangleright, \mathcal{D})$ spanned by the colimit diagrams. According to Proposition 5.4.4.3, it will suffice to prove that $\mathcal{E}'(p)$ is an accessible subcategory of $\mathrm{Fun}(K^\triangleright, \mathcal{D})$, which follows from Example 5.4.7.9.   □

Remark 5.5.3.9. In the situation of Proposition 5.5.3.8, the presentable ∞-category $\mathrm{Fun}^{\mathrm{L}}(\mathcal{C}, \mathcal{D})$ can be regarded as an *internal mapping object* in $\mathcal{P}\mathrm{r}^{\mathrm{L}}$. For every presentable ∞-category $\mathcal{C}'$, a colimit-preserving functor $\mathcal{C}' \to \mathrm{Fun}^{\mathrm{L}}(\mathcal{C}, \mathcal{D})$ can be identified with a bifunctor $\mathcal{C} \times \mathcal{C}' \to \mathcal{D}$, which is colimit-preserving separately in each variable. There exists a universal recipient for such a bifunctor: a presentable category which we may denote by $\mathcal{C} \otimes \mathcal{C}'$. The operation $\otimes$ endows $\mathcal{P}\mathrm{r}^{\mathrm{L}}$ with the structure of a *symmetric monoidal ∞-category*. Proposition 5.5.3.8 can be interpreted as asserting that this monoidal structure is *closed*.

Proposition 5.5.3.10. *Let $\mathcal{C}$ be an ∞-category and let $p : K \to \mathcal{C}$ be a diagram in $\mathcal{C}$ indexed by a (small) simplicial set K. If $\mathcal{C}$ is presentable, then the ∞-category $\mathcal{C}_{/p}$ is also presentable.*

Proof. According to Corollary 5.4.6.7, $\mathcal{C}_{/p}$ is accessible. The existence of small colimits in $\mathcal{C}_{/p}$ follows from Proposition 1.2.13.8.   □

Proposition 5.5.3.11. *Let $\mathcal{C}$ be an ∞-category and let $p : K \to \mathcal{C}$ be a diagram in $\mathcal{C}$ indexed by a small simplicial set K. If $\mathcal{C}$ is presentable, then the ∞-category $\mathcal{C}_{p/}$ is also presentable.*

Proof. It follows from Corollary 5.4.5.16 that $\mathcal{C}_{p/}$ is accessible. It therefore suffices to prove that every diagram $q : K' \to \mathcal{C}_{p/}$ has a colimit in $\mathcal{C}$. We now observe that $(\mathcal{C}_{p/})_{q/} \simeq \mathcal{C}_{q'/}$, where $q' : K \star K' \to \mathcal{C}$ is the map classified by q. Since $\mathcal{C}$ admits small colimits, $\mathcal{C}_{q'/}$ has an initial object.   □

Proposition 5.5.3.12. *Let*

$$
\begin{array}{ccc}
\mathcal{X}' & \xrightarrow{q'} & \mathcal{X} \\
\downarrow{\scriptstyle p'} & & \downarrow{\scriptstyle p} \\
\mathcal{Y}' & \xrightarrow{q} & \mathcal{Y}
\end{array}
$$

be a diagram of ∞-categories which is homotopy Cartesian (with respect to the Joyal model structure). Suppose further that $\mathcal{X}$, $\mathcal{Y}$, and $\mathcal{Y}'$ are presentable and that p and q are presentable functors. Then $\mathcal{X}'$ is presentable. Moreover, for any presentable ∞-category $\mathcal{C}$ and any functor $f : \mathcal{C} \to \mathcal{X}'$, f is presentable if and only if the compositions $p' \circ f$ and $q' \circ f$ are presentable. In particular (taking $f = \mathrm{id}_{\mathcal{X}}$), p' and q' are presentable functors.

Proof. Proposition 5.4.6.6 implies that $\mathcal{X}'$ is accessible. It therefore suffices to prove that any diagram $f : K \to \mathcal{X}'$ indexed by a small simplicial set K has a colimit in $\mathcal{X}'$. Without loss of generality, we may suppose that p and q are categorical fibrations and that $\mathcal{X}' = \mathcal{X} \times_{\mathcal{Y}} \mathcal{Y}'$. Let X be an initial object of $\mathcal{X}_{q' \circ f/}$ and let Y' be an initial object of $\mathcal{Y}'_{p'f/}$. Since p and q preserve colimits, the images $p(X)$ and $q(Y')$ are initial objects in $\mathcal{Y}_{pq'f/}$ and therefore equivalent to one another. Choose an equivalence $\eta : p(X) \to q(Y')$. Since q is a categorical fibration, η lifts to an equivalence $\overline{\eta} : Y \to Y'$ in $\mathcal{Y}'_{p'f/}$ such that $q(\overline{\eta}) = \eta$. Replacing Y' by Y, we may suppose that $p(X) = q(Y)$, so that the pair (X, Y) may be considered as an object of $\mathfrak{X}'_{f/} = \mathcal{Y}'_{p'f/} \times_{\mathcal{Y}_{pqf/}} \mathfrak{X}_{qf/}$. According to Lemma 5.4.5.2, it is an initial object of $\mathfrak{X}'_{f/}$, so that f has a colimit in $\mathfrak{X}'$. This completes the proof that $\mathfrak{X}'$ is accessible. The second assertion follows immediately from Lemma 5.4.5.5. $\qquad\square$

Proposition 5.5.3.13. *The ∞-category* $\mathcal{P}\mathrm{r}^{\mathrm{L}}$ *admits all small limits, and the inclusion functor* $\mathcal{P}\mathrm{r}^{\mathrm{L}} \subseteq \widehat{\mathcal{C}\mathrm{at}}_\infty$ *preserves all small limits.*

Proof. The proof of Proposition 4.4.2.6 shows that it will suffice to consider the case of pullbacks and small products. The desired result now follows by combining Propositions 5.5.3.12 and 5.5.3.5. $\qquad\square$

Corollary 5.5.3.14. *Let* $p : X \to S$ *be a presentable fibration of simplicial sets, where* S *is small. Then the* ∞*-category* $\mathcal{C}$ *of* coCartesian *sections of* p *is presentable.*

Proof. According to Proposition 5.5.3.3, p is classified by a functor $\chi : S \to \mathcal{P}\mathrm{r}^{\mathrm{L}}$. Using Proposition 5.5.3.13, we deduce that the limit of the composite diagram

$$S \to \mathcal{P}\mathrm{r}^{\mathrm{L}} \to \widehat{\mathcal{C}\mathrm{at}}_\infty$$

is presentable. Corollary 3.3.3.2 allows us to identify this limit with the ∞-category $\mathcal{C}$. $\qquad\square$

Our goal in the remainder of this section is to prove the analogue of Proposition 5.5.3.13 for the ∞-category $\mathcal{P}\mathrm{r}^{\mathrm{R}}$ (which will show that $\mathcal{P}\mathrm{r}^{\mathrm{L}}$ is equipped with all small *colimits* as well as all small limits). The main step is to prove that for every small diagram $S \to \mathcal{P}\mathrm{r}^{\mathrm{R}}$, the limit of the composite functor

$$S \to \mathcal{P}\mathrm{r}^{\mathrm{R}} \to \widehat{\mathcal{C}\mathrm{at}}_\infty$$

is presentable. As in the proof of Corollary 5.5.3.14, this is equivalent to the assertion that for any presentable fibration $p : X \to S$, the ∞-category $\mathcal{C}$ of *Cartesian* sections of p is presentable. To prove this, we will show that the ∞-category $\mathrm{Map}_S(S, X)$ is presentable and that $\mathcal{C}$ is an accessible localization of $\mathrm{Map}_S(S, X)$.

Lemma 5.5.3.15. *Let* $p : \mathcal{M} \to \Delta^1$ *be a Cartesian fibration, let* $\mathcal{C}$ *denote the* ∞*-category of sections of* p, *and let* $e : X \to Y$ *and* $e' : X' \to Y'$ *be objects of* $\mathcal{C}$. *If* e' *is* p-*Cartesian, then the evaluation map* $\mathrm{Map}_{\mathcal{C}}(e, e') \to \mathrm{Map}_{\mathcal{M}}(Y, Y')$ *is a homotopy equivalence.*

Proof. There is a homotopy pullback diagram of simplicial sets whose image in the homotopy category $\mathcal{H}$ is isomorphic to

$$
\begin{array}{ccc}
\mathrm{Map}_{\mathcal{C}}(e, e') & \longrightarrow & \mathrm{Map}_{\mathcal{M}}(Y, Y') \\
\downarrow & & \downarrow \\
\mathrm{Map}_{\mathcal{M}}(X, X') & \longrightarrow & \mathrm{Map}_{\mathcal{M}}(X, Y').
\end{array}
$$

If e' is p-Cartesian, then the lower horizonal map is a homotopy equivalence, so the upper horizonal map is a homotopy equivalence as well. $\square$

Lemma 5.5.3.16. *Let $p : \mathcal{M} \to \Delta^1$ be a Cartesian fibration. Let $\mathcal{C}$ denote the ∞-category of sections of p and $\mathcal{C}' \subseteq \mathcal{C}$ the full subcategory spanned by Cartesian sections of p. Then $\mathcal{C}'$ is a reflective subcategory of $\mathcal{C}$.*

Proof. Let $e : X \to Y$ be an arbitrary section of p and choose a Cartesian section $e' : X' \to Y$ with the same target. Since e' is Cartesian, there exists a diagram

$$
\begin{array}{ccc}
X & \longrightarrow & Y \\
\downarrow & & \downarrow {\scriptstyle \mathrm{id}_Y} \\
X' & \longrightarrow & Y
\end{array}
$$

in $\mathcal{M}$ which we may regard as a morphism ϕ from $e \in \mathcal{C}$ to $e' \in \mathcal{C}'$. In view of Proposition 5.2.7.8, it will suffice to show that ϕ exhibits e' as a $\mathcal{C}'$-localization of $\mathcal{C}$. In other words, we must show that for *any* Cartesian section $e'' : X'' \to Y''$, composition with ϕ induces a homotopy equivalence $\mathrm{Map}_{\mathcal{C}}(e', e'') \to \mathrm{Map}_{\mathcal{C}}(e, e'')$. This follows immediately from Lemma 5.5.3.15. $\square$

Proposition 5.5.3.17. *Let $p : X \to S$ be a presentable fibration, where S is a small simplicial set. Then*

(1) *The ∞-category $\mathcal{C} = \mathrm{Map}_S(S, X)$ of sections of p is presentable.*

(2) *The full subcategory $\mathcal{C}' \subseteq \mathcal{C}$ spanned by Cartesian sections of p is an accessible localization of $\mathcal{C}$.*

Proof. The accessibility of $\mathcal{C}$ follows from Corollary 5.4.7.17. Since p is a Cartesian fibration and the fibers of p admit small colimits, $\mathcal{C}$ admits small colimits by Proposition 5.1.2.2. This proves (1).

For each edge e of S, let $\mathcal{C}(e)$ denote the full subcategory of $\mathcal{C}$ spanned by those maps $S \to X$ which carry e to a p-Cartesian edge of X. By definition, $\mathcal{C}' = \bigcap \mathcal{C}(e)$. According to Lemma 5.5.4.18, it will suffice to show that each $\mathcal{C}(e)$ is an accessible localization of $\mathcal{C}$. Consider the map

$$\theta_e : \mathcal{C} \to \mathrm{Map}_S(\Delta^1, X).$$

Proposition 5.1.2.2 implies that θ_e preserves all limits and colimits. Moreover, $\mathcal{C}(e) = \theta_e^{-1} \mathrm{Map}_S'(\Delta^1, X)$, where $\mathrm{Map}_S'(\Delta^1, X)$ denotes the full subcategory of $\mathrm{Map}_S(\Delta^1, X)$ spanned by p-Cartesian edges. According to Lemma 5.5.4.17, it will suffice to show that $\mathrm{Map}_S'(\Delta^1, X) \subseteq \mathrm{Map}_S(\Delta^1, X)$ is an accessible localization of $\mathrm{Map}_S(\Delta^1, X)$. In other words, we may suppose $S = \Delta^1$. It then follows that evaluation at $\{1\}$ induces a trivial fibration $\mathcal{C}' \to X \times_S \{1\}$, so that $\mathcal{C}'$ is presentable. It therefore suffices to show that $\mathcal{C}'$ is a reflective subcategory of $\mathcal{C}$, which follows from Lemma 5.5.3.16. $\square$

Theorem 5.5.3.18. *The ∞-category $\mathcal{P}r^R$ admits small limits, and the inclusion functor $\mathcal{P}r^R \subseteq \widehat{\mathcal{C}at}_\infty$ preserves small limits.*

Proof. Let $\chi : S^{op} \to \mathcal{P}r^R$ be a small diagram and let $\overline{\chi} : (S^\triangleright)^{op} \to \widehat{\mathcal{C}at}_\infty$ be a limit of χ in $\widehat{\mathcal{C}at}_\infty$. We will show that $\overline{\chi}$ factors through $\mathcal{P}r^R \subseteq \widehat{\mathcal{C}at}_\infty$ and that $\overline{\chi}$ is a limit when regarded as a diagram in $\mathcal{P}r^R$.

We first show that $\overline{\chi}$ carries each vertex to a presentable ∞-category. This is clear with the exception of the cone point of $(S^\triangleright)^{op}$. Let $p : X \to S$ be a presentable fibration classified by χ. According to Corollary 3.3.3.2, we may identify the image of the cone point under $\overline{\chi}$ with the ∞-category $\mathcal{C}$ of Cartesian sections of p. Proposition 5.5.3.17 implies that this ∞-category is presentable.

We next show that $\overline{\chi}$ carries each edge of $(S^\triangleright)^{op}$ to an accessible limit-preserving functor. This is clear for edges which are degenerate or belong to S^{op}. The remaining edges are in bijection with the vertices of s and connect those vertices to the cone point. The corresponding functors can be identified with the composition

$$\mathcal{C} \subseteq \operatorname{Map}_S(S, X) \to X_s,$$

where the second functor is given by evaluation at s. Proposition 5.5.3.17 implies that the inclusion $i : \mathcal{C} \subseteq \operatorname{Map}_S(S, X)$ is accessible and preserves small limits, and Proposition 5.1.2.2 implies that the evaluation map

$$\operatorname{Map}_S(S, X) \to X_s$$

preserves all limits and colimits. This completes the proof that $\overline{\chi}$ factors through $\mathcal{P}r^R$.

We now show that $\overline{\chi}$ is a limit diagram in $\mathcal{P}r^R$. Since $\mathcal{P}r^R$ is a subcategory of $\widehat{\mathcal{C}at}_\infty$ and $\overline{\chi}$ is already a limit diagram in $\widehat{\mathcal{C}at}_\infty$, it will suffice to verify the following assertion:

- If $\mathcal{D}$ is a presentable ∞-category and $F : \mathcal{D} \to \mathcal{C}$ has the property that each of the composite functors

$$\mathcal{D} \xrightarrow{F} \mathcal{C} \overset{i}{\subseteq} \operatorname{Map}_S(S, X) \to X_s$$

  is accessible and limit-preserving, then F is accessible and preseves limits.

Applying Proposition 5.5.3.17, we see that F is accessible and preserves limits if and only if $i \circ F$ is accessible and preserves limits. We now conclude by applying Proposition 5.1.2.2. $\qquad\square$

5.5.4 Local Objects

According to Theorem 5.5.1.1, every presentable ∞-category arises as an (accessible) localization of some presheaf ∞-category $\mathcal{P}(X)$. Consequently, understanding the process of localization is of paramount importance in the study of presentable ∞-categories. In this section, we will classify the accessible localizations of an arbitrary presentable ∞-category $\mathcal{C}$. The basic observation is that a localization functor $L : \mathcal{C} \to \mathcal{C}$ is determined, up to equivalence, by the collection S of all morphisms f such that Lf is an equivalence. Moreover, a collection of morphisms S arises from an accessible localization functor if and only if S is *strongly saturated* (Definition 5.5.4.5) and *of small generation* (Remark 5.5.4.7). Given any small collection of morphisms S in $\mathcal{C}$, there is a smallest strongly saturated collection containing S: this permits us to define a localization $S^{-1}\mathcal{C} \subseteq \mathcal{C}$. The ideas presented in this section go back (at least) to Bousfield; we refer the reader to [12] for a discussion in a more classical setting.

Definition 5.5.4.1. Let $\mathcal{C}$ be an ∞-category and S a collection of morphisms of $\mathcal{C}$. We say that an object Z of $\mathcal{C}$ is *S-local* if, for every morphism $s : X \to Y$ belonging to S, composition with s induces an isomorphism $\mathrm{Map}_{\mathcal{C}}(Y, Z) \to \mathrm{Map}_{\mathcal{C}}(X, Z)$ in the homotopy category $\mathcal{H}$ of spaces.

A morphism $f : X \to Y$ of $\mathcal{C}$ is an *S-equivalence* if, for every S-local object Z, composition with f induces a homotopy equivalence $\mathrm{Map}_{\mathcal{C}}(Y, Z) \to \mathrm{Map}_{\mathcal{C}}(X, Z)$.

The following result provides a dictionary for relating localization functors to classes of morphisms:

Proposition 5.5.4.2. *Let $\mathcal{C}$ be an ∞-category and let $L : \mathcal{C} \to \mathcal{C}$ be a localization functor. Let S denote the collection of all morphisms f in $\mathcal{C}$ such that Lf is an equivalence. Then*

(1) *An object C of $\mathcal{C}$ is S-local if and only if it belongs to $L\mathcal{C}$.*

(2) *Every S-equivalence in $\mathcal{C}$ belongs to S.*

(3) *Suppose that $\mathcal{C}$ is accessible. The following conditions are equivalent:*

   (i) *The ∞-category $L\mathcal{C}$ is accessible.*

   (ii) *The functor $L : \mathcal{C} \to \mathcal{C}$ is accessible.*

   (iii) *There exists a (small) subset $S_0 \subseteq S$ such that every S_0-local object is S-local.*

Proof of (1) and (2). By assumption, L is left adjoint to the inclusion $L\mathcal{C} \subseteq \mathcal{C}$; let $\alpha : \mathrm{id}_{\mathcal{C}} \to L$ be a unit map for the adjunction. We begin by proving (1). Suppose that $X \in L\mathcal{C}$. Let $f : Y \to Z$ belong to S. Then we have a commutative diagram

$$
\begin{array}{ccc}
\mathrm{Map}_{\mathcal{C}}(LZ, X) & \longrightarrow & \mathrm{Map}_{\mathcal{C}}(LY, X) \\
\downarrow & & \downarrow \\
\mathrm{Map}_{\mathcal{C}}(Z, X) & \longrightarrow & \mathrm{Map}_{\mathcal{C}}(Y, X),
\end{array}
$$

in the homotopy category $\mathcal{H}$, where the vertical maps are given by composition with α and are homotopy equivalences by assumption. Since Lf is an equivalence, the top horizontal map is also a homotopy equivalence. It follows that the bottom horizontal map is a homotopy equivalence, so that X is S-local. Conversely, suppose that X is S-local. According to Proposition 5.2.7.4, the map $\alpha(X) : X \to LX$ belongs to S, so that composition with $\alpha(X)$ induces a homotopy equivalence $\mathrm{Map}_{\mathcal{C}}(LX, X) \to \mathrm{Map}_{\mathcal{C}}(X, X)$. In particular, there exists a map $LX \to X$ whose composition with $\alpha(X)$ is homotopic to id_X. Thus X is a retract of LX. Since $\alpha(LX)$ is an equivalence, we conclude that $\alpha(X)$ is an equivalence, so that $X \simeq LX$ and therefore X belongs to the essential image of L, as desired. This proves (1).

Suppose that $f : X \to Y$ is an S-equivalence. We have a commutative diagram

$$
\begin{array}{ccc}
X & \xrightarrow{\ f\ } & Y \\
\downarrow{\scriptstyle \alpha(X)} & & \downarrow{\scriptstyle \alpha(Y)} \\
LX & \xrightarrow{\ Lf\ } & LY
\end{array}
$$

where the vertical maps are S-equivalences (by Proposition 5.2.7.4), so that Lf is also an S-equivalence. Therefore LX and LY corepresent the same functor on the homotopy category $\mathrm{h}L\,\mathcal{C}$. Yoneda's lemma implies that Lf is an equivalence, so that $f \in S$. This proves (2).   □

The proof of (3) is more difficult and will require a few preliminaries.

Lemma 5.5.4.3. *Let $\tau \gg \kappa$ be regular cardinals and suppose that τ is uncountable. Let A be a κ-filtered partially ordered set, $A' \subseteq A$ a τ-small subset, and*

$$\{f_\gamma : X_\gamma \to Y_\gamma\}_{\gamma \in C}$$

a τ-small collection of natural transformations of diagrams in $\mathcal{K}an^A$. Suppose that for each $\alpha \in A$, $\gamma \in C$, the Kan complexes $X_\gamma(\alpha)$ and $Y_\gamma(\alpha)$ are essentially τ-small. Suppose further that, for each $\gamma \in C$, the map of Kan complexes $\varinjlim_A f_\gamma$ is a homotopy equivalence. Then there exists a τ-small κ-filtered subset $A'' \subseteq A$ such that $A' \subseteq A''$, and $\varinjlim_{A''} f_\gamma | A''$ is a homotopy equivalence for each $\gamma \in C$.

Proof. Replacing each f_γ by an equivalent transformation if necessary, we may suppose that for each $\gamma \in C$, $\alpha \in A$, the map $f_\gamma(\alpha)$ is a Kan fibration.

Let $\alpha \in A$, let $\gamma \in C$, and let $\sigma(\alpha, \gamma)$ be a diagram

$$
\begin{array}{ccc}
\partial \Delta^n & \longrightarrow & X_\gamma(\alpha) \\
\downarrow & & \downarrow \\
\Delta^n & \longrightarrow & Y_\gamma(\alpha).
\end{array}
$$

We will say that $\alpha' \geq \alpha$ *trivializes* $\sigma(\alpha, \gamma)$ if the lifting problem depicted in the induced diagram

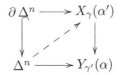

admits a solution. Observe that, if $B \subseteq A$ is filtered, then $\varinjlim_B f_\gamma | B$ is a Kan fibration, which is trivial if and only if for every diagram $\sigma(\alpha, \gamma)$ as above, where $\alpha \in B$, there exists $\alpha' \in B$ such that $\alpha' \geq \alpha$ and α' trivializes $\sigma(\alpha, \gamma)$. In particular, since $\varinjlim_A f_\gamma$ is a homotopy equivalence, every such diagram $\sigma(\alpha, \gamma)$ is trivialized by some $\alpha' \geq \alpha$.

We now define a sequence of τ-small subsets $A(\lambda) \subseteq A$ indexed by ordinals $\lambda \leq \kappa$. Let $A(0) = A'$ and let $A(\lambda) = \bigcup_{\lambda' < \lambda} A(\lambda')$ when λ is a limit ordinal. Supposing that $\lambda < \kappa$ and that $A(\lambda)$ has been defined, we choose a set of representatives $\Sigma = \{\sigma(\alpha, \gamma)\}$ for all homotopy classes of diagrams as above, where $\alpha \in A(\lambda)$ and $\gamma \in C$. Since the Kan complexes $X_\gamma(\alpha)$, $Y_\gamma(\alpha)$ are essentially τ-small, we may choose the set Σ to be τ-small. Each $\sigma \in \Sigma$ is trivialized by some $\alpha'_\sigma \in A$. Let $B = \{\alpha'_\sigma\}_{\sigma \in \Sigma}$; then B is τ-small. Now choose a τ-small κ-filtered subset $A(\lambda + 1) \subseteq A$ containing $A(\lambda) \cup B$ (the existence of $A(\lambda + 1)$ is guaranteed by Lemma 5.4.2.10).

We now define A'' to be $A(\kappa)$; it is easy to see that A'' has the desired properties. $\qquad\qquad\square$

Lemma 5.5.4.4. *Let $\tau \gg \kappa$ be regular cardinals and suppose that τ is uncountable. Let A be a κ-filtered partially ordered set and for every subset $B \subseteq A$, let*

$$\varinjlim_B : \mathrm{Fun}(\mathrm{N}(B), \mathcal{S}) \to \mathcal{S}$$

denote a left adjoint to the diagonal functor. Let $A' \subseteq A$ be a τ-small subset and $\{f_\gamma : X_\gamma \to Y_\gamma\}_{\gamma \in C}$ a τ-small collection of morphisms in the ∞-category $\mathrm{Fun}(\mathrm{N}(A), \mathcal{S}^\tau)$ of diagrams $\mathrm{N}(A) \to \mathcal{S}^\tau$. Suppose that $\varinjlim_A(f_\gamma)$ is an equivalence for each $\gamma \in C$. Then there exists a τ-small κ-filtered subset $A'' \subseteq A$ which contains A' such that each of the morphisms $\varinjlim_{A''}(f_\gamma | \mathrm{N}(A''))$ is an equivalence in $\mathcal{S}$.

Proof. Using Proposition 4.2.4.4, we may assume without loss of generality that each f_γ is the simplicial nerve of a natural transformation of functors from A to $\mathcal{K}$an. According to Theorem 4.2.4.1, we can identify $\varinjlim_B(X_\gamma | \mathrm{N}(B))$ and $\varinjlim_B(Y_\gamma | \mathrm{N}(B))$ with homotopy colimits in $\mathcal{K}$an. If B is filtered, then these homotopy colimits reduce to ordinary colimits (since the class of weak homotopy equivalences in $\mathcal{K}$an is stable under filtered colimits), and we may apply Lemma 5.5.4.3. $\qquad\qquad\square$

Proof of part (3) of Proposition 5.5.4.2. If $L\mathcal{C}$ is accessible, then Proposition 5.4.7.7 implies that both the inclusion $L\mathcal{C} \to \mathcal{C}$ and the functor $L : \mathcal{C} \to$

$L\,\mathcal{C}$ are accessible functors, so that their composition is accessible. Thus $(i) \Rightarrow (ii)$. Suppose next that (ii) is satisfied. Let $\alpha : \mathrm{id}_{\mathcal{C}} \to L$ denote a unit for the adjunction between L and the inclusion $L\,\mathcal{C} \subseteq \mathcal{C}$ and let κ be a regular cardinal such that $\mathcal{C}$ is κ-accessible and L is κ-continuous. Without loss of generality, we may suppose that $\mathcal{C}$ is minimal, so that $\mathcal{C}^{\kappa}$ is a small ∞-category. Let $S_0 = \{\alpha(X) : X \in \mathcal{C}^{\kappa}\}$ and let $Y \in \mathcal{C}$ be S_0-local. We wish to prove that Y is S-local. Let $F_Y : \mathcal{C} \to \mathcal{S}^{op}$ denote the functor represented by Y. Then α induces a natural transformation $F_Y \to F_Y \circ L$. The functors F_Y and $F_Y \circ L$ are both κ-continuous. By assumption, α induces an equivalence of functors $F_Y|\,\mathcal{C}^{\kappa} \to (F_Y \circ L)|\,\mathcal{C}^{\kappa}$ when both sides are restricted to κ-compact objects. Proposition 5.3.5.10 now implies that F_Y and $F_Y \circ L$ are equivalent, so that Y is S-local. This proves (iii).

We complete the proof by showing that (iii) implies (i). Let κ be a regular cardinal such that $\mathcal{C}$ is κ-accessible and S_0 is a set of morphisms between κ-compact objects of $\mathcal{C}$. We claim that $L\,\mathcal{C}$ is stable under κ-filtered colimits in $\mathcal{C}$. To prove this, let $\overline{p} : K^{\triangleright} \to \mathcal{C}$ be a colimit diagram, where K is small and κ-filtered and $p = \overline{p}|K$ factors through $L\,\mathcal{C} \subseteq \mathcal{C}$. Let $s : X \to Y$ be a morphism which belongs to S_0 and let $s' : F_X \to F_Y$ be the corresponding map of corepresentable functors $\mathcal{C} \to \widehat{\mathcal{S}}$. Since X and Y are κ-compact by assumption, both $\overline{p}_X : F_X \circ \overline{p}$ and $\overline{p}_Y : F_Y \circ \overline{p}$ are colimit diagrams in $\widehat{\mathcal{S}}$. The map s' induces a transformation $\overline{p}_X \to \overline{p}_Y$, which is an equivalence when restricted to K and is therefore an equivalence in general. It follows that $\mathrm{Map}_{\mathcal{C}}(Y, \overline{p}(\infty)) \simeq \mathrm{Map}_{\mathcal{C}}(X, \overline{p}(\infty))$, where ∞ denotes the cone point of $K^{\triangleright}$. Thus $\overline{p}(\infty)$ is S_0-local as desired.

Now choose an uncountable regular cardinal $\tau \gg \kappa$ such that $\mathcal{C}^{\kappa}$ is essentially τ-small. According to Proposition 5.4.2.2, to complete the proof that $\mathcal{C}$ is accessible it will suffice to show that $L\,\mathcal{C}$ is generated by τ-compact objects under τ-filtered colimits. Let X be an object of $L\,\mathcal{C}$. Lemma 5.1.5.3 implies that X can be written as the colimit of a small diagram $p : \mathcal{J} \to \mathcal{C}^{\kappa}$, where $\mathcal{J}$ is κ-filtered. Using Proposition 5.3.1.16, we may suppose that $\mathcal{J}$ is the nerve of a κ-filtered partially ordered set A. Let B denote the collection of all κ-filtered τ-small subsets $A_{\beta} \subseteq A$ for which the colimit of $p|\,\mathrm{N}(A_{\beta})$ is S_0-local. Lemma 5.5.4.4 asserts that every τ-small subset of A is contained in A_{β} for some $\beta \in B$. It follows that B is τ-filtered when regarded as partially ordered by inclusion and that $A = \bigcup_{\beta \in B} A_{\beta}$. Using Proposition 4.2.3.8 and Corollary 4.2.3.10, we can obtain X as the colimit of a diagram $q : \mathrm{N}(B) \to \mathcal{C}$, where each $q(\beta)$ is a colimit X_{β} of $p|\,\mathrm{N}(A_{\beta})$. The objects $\{X_{\beta}\}_{\beta \in B}$ are S_0-local and τ-compact by construction.   □

According to Proposition 5.5.4.2, every localization L of an ∞-category $\mathcal{C}$ is determined by the class S of morphisms f such that Lf is an equivalence. This raises the question: which classes of morphisms S arise in this way? To answer this question, we will begin by isolating some of the most obvious properties enjoyed by S.

Definition 5.5.4.5. Let $\mathcal{C}$ be a ∞-category which admits small colimits and let S be a collection of morphisms of $\mathcal{C}$. We will say that S is *strongly* *saturated* if it satisfies the following conditions:

(1) Given a pushout diagram

$$
\begin{array}{ccc}
C & \xrightarrow{f} & D \\
\downarrow & & \downarrow \\
C' & \xrightarrow{f'} & D'
\end{array}
$$

in $\mathcal{C}$, if f belongs to S, then so does f'.

(2) The full subcategory of $\mathrm{Fun}(\Delta^1, \mathcal{C})$ spanned by S is stable under small colimits.

(3) Suppose we are given a 2-simplex of $\mathcal{C}$ corresponding to a diagram

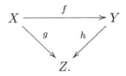

If any two of f, g, and h belong to S, then so does the third.

Remark 5.5.4.6. Let $\mathcal{C}$ be an ∞-category which admits small colimits and let S be a strongly saturated class of morphisms of $\mathcal{C}$. Let $\emptyset$ be an initial object of $\mathcal{C}$. Condition (2) of Definition 5.5.4.5 implies that $\mathrm{id}_\emptyset : \emptyset \to \emptyset$ belongs to S, since it is an initial object of $\mathrm{Fun}(\Delta^1, \mathcal{C})$. Any equivalence in $\mathcal{C}$ is a pushout of $\mathrm{id}_\emptyset$, so condition (1) implies that S contains all equivalences in $\mathcal{C}$. It also follows from condition (1) that if $f : C \to D$ belongs to S and $f' : C \to D$ is homotopic to f, then f' belongs to S (since f' is a pushout of f). Note also that condition (2) implies that S is stable under retracts because any retract of a morphism f can be written as a colimit of copies of f (Proposition 4.4.5.12).

Remark 5.5.4.7. Let $\mathcal{C}$ be an ∞-category which admits colimits. Given any collection $\{S_\alpha\}_{\alpha \in A}$ of strongly saturated classes of morphisms of $\mathcal{C}$, the intersection $S = \bigcap_{\alpha \in A} S_\alpha$ is also strongly saturated. It follows that *any* collection S_0 of morphisms in $\mathcal{C}$ is contained in a smallest strongly saturated class of morphisms S. In this case we will also write $S = \overline{S_0}$; we refer to it as the strongly saturated class of morphisms *generated* by S_0. We will say that S is *of small generation* if $S = \overline{S_0}$, where $S_0 \subseteq S$ is small.

Remark 5.5.4.8. Let $\mathcal{C}$ be an ∞-category which admits small colimits. Let S be a strongly saturated class of morphisms of $\mathcal{C}$. If $f : X \to Y$ lies in S and K is a simplicial set, then the induced map $X \otimes K \to Y \otimes K$ (which is well-defined up to equivalence) lies in S. This follows from the closure of S under colimits. We will use this observation in the proof of Proposition 5.5.4.15 in the case where $K = \partial \Delta^n$ is a (simplicial) sphere.

Example 5.5.4.9. Let $\mathcal{C}$ be an ∞-category which admits small colimits and let S denote the class of all equivalences in $\mathcal{C}$. Then S is strongly saturated; it is clearly the smallest strongly saturated class of morphisms of $\mathcal{C}$.

Remark 5.5.4.10. Let $F : \mathcal{C}' \to \mathcal{C}$ be a functor between ∞-categories. Suppose that $\mathcal{C}$ and $\mathcal{C}'$ admit small colimits and that F preserves small colimits. Let S be a strongly saturated class of morphisms in $\mathcal{C}'$. Then $F^{-1}S$ is a strongly saturated class of morphisms of $\mathcal{C}$. In particular, if we let S denote the collection of all morphisms f of $\mathcal{C}'$ such that $F(f)$ is an equivalence, then S is strongly saturated.

Lemma 5.5.4.11. *Let $\mathcal{C}$ be an ∞-category which admits small colimits, let S_0 be a class of morphisms in $\mathcal{C}$, and let S denote the collection of all S-equivalences. Then S is strongly saturated.*

Proof. For each object $X \in \mathcal{C}$, let $F_X : \mathcal{C} \to \mathcal{S}^{op}$ denote the functor represented by X and let $S(X)$ denote the collection of all morphisms f such that $F_X(f)$ is an equivalence. Since F_X preserves small colimits, Remark 5.5.4.10 implies that $S(X)$ is strongly saturated. We now observe that S is the intersection $\bigcap S(X)$, where X ranges over the class of all S_0-local objects of $\mathcal{C}$.   □

Lemma 5.5.4.12. *Let $\mathcal{C}$ be an ∞-category which admits small colimits, let S be a strongly saturated collection of morphisms of $\mathcal{C}$, and let $C \in \mathcal{C}$ be an object. Let $\mathcal{D} \subseteq \mathcal{C}^{C/}$ be the full subcategory of $\mathcal{C}^{C/}$ spanned by those objects $C \to C'$ which belong to S. Then $\mathcal{D}$ is stable under small colimits in $\mathcal{C}^{C/}$.*

Proof. The proofs of Corollaries 4.2.3.11 and 4.4.2.4 show that it will suffice to prove that $\mathcal{D}$ is stable under filtered colimits and pushouts, and contains the initial objects of $\mathcal{C}^{C/}$. The last condition is equivalent to the requirement that S contain all equivalences, which follows from Remark 5.5.4.6. Now suppose that $\overline{p} : K^{\triangleright} \to \mathcal{C}^{C/}$ is a colimit of $p = \overline{p}|K$, where K is either filtered or equivalent to Λ_0^2, and that $p(K) \subseteq \mathcal{D}$. We can identify $\overline{p}$ with a map $P : K^{\triangleright} \times \Delta^1 \to \mathcal{C}$ such that $P|K^{\triangleright} \times \{0\}$ is the constant map taking the value $C \in \mathcal{C}$. Since K is weakly contractible, $P|K^{\triangleright} \times \{0\}$ is a colimit diagram in $\mathcal{C}$. The map $P|K^{\triangleright} \times \{1\}$ is the image of a colimit diagram under the left fibration $\mathcal{C}^{C/} \to \mathcal{C}$; since K is weakly contractible, Proposition 4.4.2.9 implies that $P|K^{\triangleright} \times \{1\}$ is a colimit diagram. We now apply Proposition 5.1.2.2 to deduce that $P : K^{\triangleright} \to \mathcal{C}^{\Delta^1}$ is a colimit diagram. Since S is stable under colimits in $\mathcal{C}^{\Delta^1}$, we conclude that P carries the cone point of $K^{\triangleright}$ to a morphism belonging to S, as we wished to show.   □

Lemma 5.5.4.13. *Let $\mathcal{C}$ be an ∞-category which admits small filtered colimits, let κ be an uncountable regular cardinal, let A and B be κ-filtered partially ordered sets, and let $p_0 : N(A_0) \to \mathcal{C}^{\kappa}$ and $p_1 : A_1 \to \mathcal{C}^{\kappa}$ be two diagrams which have the same colimit. Let $A_0'' \subseteq A_0$, $A_1'' \subseteq A_1$ be κ-small subsets. Then there exist κ-small filtered subsets $A_0' \subseteq A_0$, $A_1' \subseteq A_1$ such* *that $A_0'' \subseteq A_0'$, $A_1'' \subseteq A_1'$ and the diagrams $p_0 | \mathrm{N}(A_0')$, $p_1 | \mathrm{N}(A_1')$ have the same colimit in $\mathcal{C}$.*

Proof. Let $\bar{p}_0, \bar{p}_1$ be colimits of p_0 and p_1, respectively, which carry the cone points to the same object $C \in \mathcal{C}$. Let $B = A_0'' \cup A_1'' \cup \mathbf{Z}_{\geq 0} \cup \{\infty\}$, which we regard as a partially ordered set so that

$$\mathrm{N}(B) \simeq ((\mathrm{N}(A_0'') \coprod \mathrm{N}(A_1'')) \star \mathrm{N}(\mathbf{Z}_{\geq 0}))^{\triangleright}.$$

We will construct sequences of elements

$$\{a_0 \leq a_2 \leq \ldots\} \subseteq \{a \in A_0 : (\forall a'' \in A_0'')[a'' \leq a]\}$$

$$\{a_1 \leq a_3 \leq \ldots\} \subseteq \{a \in A_1 : (\forall a'' \in A_1'')[a'' \leq a]\}$$

and a diagram $\bar{q} : \mathrm{N}(B) \to \mathcal{C}$ such that

$$\bar{q} | (\mathrm{N}(A_0'') \cup \mathrm{N}\{0, 2, 4, \ldots\})^{\triangleright} = \bar{p}_0 | \mathrm{N}(A_0'' \cup \{a_0, a_2, \ldots\})^{\triangleright}$$

$$\bar{q} | (\mathrm{N}(A_1'') \cup \mathrm{N}\{1, 3, 5, \ldots\})^{\triangleright} = \bar{p}_1 | \mathrm{N}(A_1'' \cup \{a_1, a_3, \ldots\})^{\triangleright}.$$

Supposing that this has been done, we take $A_0' = A_0'' \cup \{a_0, a_2, \ldots\}$, $A_1' = A_1'' \cup \{a_1, a_3, \ldots\}$ and observe that the colimits of $p_0 | \mathrm{N}(A_0')$ and $p_1 | \mathrm{N}(A_1')$ are both equivalent to the colimit of $\bar{q} | \mathrm{N}(\mathbf{Z}_{\geq 0})$.

The construction is by recursion. Let us suppose that the sequence

$$a_0, a_1, \ldots, a_{n-1}$$

and the map $\bar{q}_n = \bar{q} | ((\mathrm{N}(A_0'') \coprod \mathrm{N}(A_1'')) \star (\mathrm{N}\{0, \ldots, n-1\})^{\triangleright}$ have already been constructed (when $n = 0$, we observe that $\bar{q}_0$ is uniquely determined by $\bar{p}_0$ and $\bar{p}_1$). For simplicity we will treat only the case where n is even; the case where n is odd can be handled by a similar argument.

Let $q_n = \bar{q}_n | (\mathrm{N}(A_0'') \coprod \mathrm{N}(A_1'')) \star \mathrm{N}\{0, \ldots, n-1\}$ and $q_n' = \bar{q}_n | \mathrm{N}(A_0'') \star \mathrm{N}\{0, 2, \ldots, n-2\}$. According to Corollary 5.3.4.14, the left fibrations $\mathcal{C}_{q_n/} \to \mathcal{C}$ and $\mathcal{C}_{q_n'/} \to \mathcal{C}$ are κ-compact. Set

$$A_0(n) = \{a \in A_0 : (\forall a'' \in A_0'' \cup \{a_0, \ldots, a_{n-2}\})[a'' \leq a]\}$$

$$X = \mathcal{C}_{q_n/} \times_{\mathcal{C}} \mathrm{N}(A_0(n))^{\triangleright} \qquad X' = \mathcal{C}_{q_n'/} \times_{\mathcal{C}} \mathrm{N}(A_0(n))^{\triangleright}$$

so that X and X' are left fibrations classified by colimit diagrams

$$\mathrm{N}(A_0(n))^{\triangleright} \to \mathcal{S}.$$

Form a pullback diagram

$$\begin{array}{ccc}
Y & \longrightarrow & X \\
\downarrow & & \downarrow \\
\mathrm{N}(A_0(n))^{\triangleright} & \longrightarrow & X',
\end{array}$$

where the left vertical map is a left fibration (by Proposition 2.1.2.1) and the bottom horizontal map is determined by $\bar{p} | \mathrm{N}(A_0'' \cup \{0, \ldots, n-2\}) \star \mathrm{N}(A_0(n))^{\triangleright}$. It follows that the diagram is a homotopy pullback, so that $Y \to \mathrm{N}(A_0(n))^{\triangleright}$

is also a left fibration classified by a colimit diagram $N(A_0(n))^{\triangleright} \to \mathcal{S}$. The map $\bar{q}_n$ determines a vertex v of Y lying over the cone point of $N(A_0(n))^{\triangleright}$. According to Proposition 3.3.4.5, the inclusion $Y \times_{N(A_0(n))^{\triangleright}} N(A_0(n)) \subseteq Y$ is a weak homotopy equivalence of simplicial sets. It follows that there exists an edge $e : v' \to v$ of Y which joins v to some vertex v' lying over an element $a \in A_0(n)$. We now set $a_n = a$ and observe that the edge e corresponds to the desired extension $\bar{q}_{n+1}$ of $\bar{q}_n$. $\qquad\qquad\square$

Lemma 5.5.4.14. *Let $\mathcal{C}$ be a presentable ∞-category, let S be a strongly saturated collection of morphisms in $\mathcal{C}$, and let $\mathcal{D} \subseteq \mathrm{Fun}(\Delta^1, \mathcal{C})$ be the full subcategory spanned by S. The following conditions are equivalent:*

(1) *The ∞-category $\mathcal{D}$ is accessible.*

(2) *The ∞-category $\mathcal{D}$ is presentable.*

(3) *The collection S is of small generation (as a strongly saturated class of morphisms).*

Proof. We observe that $\mathcal{D}$ is stable under small colimits in $\mathrm{Fun}(\Delta^1, \mathcal{C})$ and therefore admits small colimits; thus $(1) \Rightarrow (2)$. To see that (2) implies (3), we choose a small collection S_0 of morphisms in $\mathcal{C}$ which generates $\mathcal{D}$ under colimits; it is then obvious that S_0 generates S as a strongly saturated class of morphisms.

Now suppose that (3) is satisfied. Choose a small collection of morphisms $\{f_\beta : X_\beta \to Y_\beta\}$ which generates S and an uncountable regular cardinal κ such that $\mathcal{C}$ is κ-accessible and each of the objects X_β, Y_β is κ-compact. We will prove that $\mathcal{D}$ is κ-accessible.

It is clear that $\mathcal{D}$ is locally small and admits κ-filtered colimits. Let $\mathcal{D}' \subseteq \mathcal{D}$ be the collection of all morphisms $f : X \to Y$ such that f belongs to S, where both X and Y are κ-compact. Lemma 5.3.4.9 implies that each $f \in \mathcal{D}'$ is a κ-compact object of $\mathrm{Fun}(\Delta^1, \mathcal{C})$ and, in particular, a κ-compact object of $\mathcal{D}$. Assume for simplicity that $\mathcal{C}$ is a minimal ∞-category, so that $\mathcal{D}'$ is small. According to Proposition 5.3.5.10, the inclusion $\mathcal{D}' \subseteq \mathcal{D}$ is equivalent to $j \circ F$, where $j : \mathcal{D}' \to \mathrm{Ind}_\kappa \mathcal{D}'$ is the Yoneda embedding and $F : \mathrm{Ind}_\kappa \mathcal{D}' \to \mathcal{D}$ is κ-continuous. Proposition 5.3.5.11 implies that F is fully faithful; let $\mathcal{D}''$ denote its essential image. To complete the proof, it will suffice to show that $\mathcal{D}'' = \mathcal{D}$. Let $S'' \subseteq S$ denote the collection of objects of $\mathcal{D}''$ (which we may identify with morphisms in $\mathcal{C}$). By construction, S'' contains the collection of morphisms $\{f_\beta\}$ which generates S. Consequently, to prove that $S'' = S$, it will suffice to show that S'' is strongly saturated.

It follows from Proposition 5.5.1.9 that $\mathcal{D}'' \subseteq \mathrm{Fun}(\Delta^1, \mathcal{C})$ is stable under small colimits. We next verify that S'' is stable under pushouts. Let $K = \Lambda_0^2$ and let $\bar{p} : K^{\triangleright} \to \mathcal{C}$ be a colimit of $p = \bar{p}|K$,

$$\begin{array}{ccc} X & \xrightarrow{\ f\ } & Y \\ \downarrow & & \downarrow \\ X' & \xrightarrow{\ f'\ } & Y' \end{array}$$

such that f belongs to S''. The proof of Proposition 5.4.4.3 shows that we can write p as a colimit of a diagram $q : N(A) \to \text{Fun}(\Lambda^2_0, \mathcal{C}^\kappa)$, where A is a κ-filtered partially ordered set. For $\alpha \in A$, we let p_α denote the corresponding diagram, which we may depict as

$$X'_\alpha \leftarrow X_\alpha \xrightarrow{f_\alpha} Y_\alpha.$$

For each $A' \subseteq A$, we let $p_{A'}$ denote a colimit of $q | N(A')$, which we will denote by

$$X'_{A'} \leftarrow X_{A'} \xrightarrow{f_{A'}} Y_{A'}.$$

Let B denote the collection of κ-small filtered subsets $A' \subseteq A$ such that the $f_{A'}$ belongs to S''. Since $f \in S''$, we conclude that f can be obtained as the colimit of a κ-filtered diagram $N(A') \to \mathcal{D}'$. Applying Lemma 5.5.4.13, we deduce that B is κ-filtered and that $A = \bigcup_{A' \in B} A'$. Using Proposition 4.2.3.4 and Corollary 4.2.3.10, we deduce that p is the colimit of a diagram $q' : N(B) \to \text{Fun}(\Lambda^2_0, \mathcal{C})$, where $q'(A') = p_{A'}$. Replacing A by B, we may suppose that each f_α belongs to S'.

Let $\varinjlim : \text{Fun}(\Lambda^2_0, \mathcal{C}) \to \text{Fun}(\Delta^1 \times \Delta^1, \mathcal{C})$ be a colimit functor (that is, a left adjoint to the restriction functor). Lemma 5.5.2.3 implies that we may identify $\overline{p}$ with a colimit of the diagram $\varinjlim \circ q$. Consequently, the morphism f' can be written as a colimit of morphisms f'_α which fit into pushout diagrams

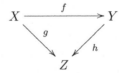

Since $f_\alpha \in S'' \subseteq S$, we conclude that $f'_\alpha \in S$. Since X'_α and Y'_α are κ-compact, we deduce that $f'_\alpha \in S''$. Since $\mathcal{D}''$ is stable under colimits, we deduce that $f' \in S''$, as desired.

We now complete the proof by showing that S'' has the two-out-of-three property, using the same style of argument. Let $\sigma : \Delta^2 \to \mathcal{C}$ be a simplex corresponding to a diagram

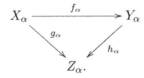

in $\mathcal{C}$. We will show that if $f, g \in S''$, then $h \in S''$: the argument in the other two cases is the same. The proof of Proposition 5.4.4.3 shows that we can write σ as the colimit of a diagram $q : N(A) \to \text{Fun}(\Delta^2, \mathcal{C}^\kappa)$, where A is a κ-filtered partially ordered set. For each $\alpha \in A$, we will denote the corresponding diagram by

$$\begin{array}{ccc} X_\alpha & \xrightarrow{f_\alpha} & Y_\alpha \\ & g_\alpha \searrow & \downarrow h_\alpha \\ & & Z_\alpha. \end{array}$$

Arguing as above, we may assume (possibly after changing A and q) that each f_α belongs to S''. Repeating the same argument, we may suppose that g_α belongs to S''. Since S has the two-out-of-three property, we conclude that each h_α belongs to S. Since X_α and Z_α are κ-compact, we then have $h_\alpha \in S''$. The stability of $\mathcal{D}''$ under colimits now implies that $h \in S''$, as desired.   □

Proposition 5.5.4.15. *Let $\mathcal{C}$ be a presentable ∞-category and let S be a (small) collection of morphisms of $\mathcal{C}$. Let $\overline{S}$ denote the strongly saturated class of morphisms generated by S. Let $\mathcal{C}' \subseteq \mathcal{C}$ denote the full subcategory of $\mathcal{C}$ consisting of S-local objects. Then*

(1) *For each object $C \in \mathcal{C}$, there exists a morphism $s : C \to C'$ such that C' is S-local and s belongs to $\overline{S}$.*

(2) *The ∞-category $\mathcal{C}'$ is presentable.*

(3) *The inclusion $\mathcal{C}' \subseteq \mathcal{C}$ has a left adjoint L.*

(4) *For every morphism f of $\mathcal{C}$, the following are equivalent:*

   (i) *The morphism f is an S-equivalence.*

   (ii) *The morphism f belongs to $\overline{S}$.*

   (iii) *The induced morphism Lf is an equivalence.*

Proof. Assertion (1) is a consequence of Lemma 5.5.5.14, which we will prove in §5.5.5. The equivalence (1) ⇔ (3) follows immediately from Proposition 5.2.7.8. We now prove (4). Lemma 5.5.4.11 implies that the collection of S-equivalences is a strongly saturated class of morphisms containing S; it therefore contains $\overline{S}$, so that $(ii) \Rightarrow (i)$. Now suppose that $f : X \to Y$ is such that Lf is an equivalence and consider the diagram

Our proof of (1) shows that the vertical morphisms belong to $\overline{S}$, and the lower horizontal arrow belongs to $\overline{S}$ by Remark 5.5.4.6. Two applications of the two-out-of-three property now show that $f \in \overline{S}$, so that $(iii) \Rightarrow (ii)$. If f is an S-equivalence, then we may again use the above diagram and the two-out-of-three property to conclude that Lf is an equivalence. It follows that LX and LY corepresent the same functor on the homotopy category $h\,\mathcal{C}'$, so that Yoneda's lemma implies that Lf is an equivalence. Thus $(i) \Rightarrow (iii)$, and the proof of (4) is complete.

It remains to prove (2). Remark 5.2.7.5 implies that $L\,\mathcal{C}$ admits small colimits, so it will suffice to prove that $L\,\mathcal{C}$ is accessible. According to Proposition 5.5.4.2, this follows from the implication $(iii) \Rightarrow (i)$ of assertion (4).   □

Proposition 5.5.4.15 gives a clear picture of the collection of all accessible localizations of a presentable ∞-category $\mathcal{C}$. For any (small) set of morphisms S in $\mathcal{C}$, the full subcategory $S^{-1}\mathcal{C} \subseteq \mathcal{C}$ consisting of S-local objects is a localization of $\mathcal{C}$, and every localization arises in this way. Moreover, the subcategories $S^{-1}\mathcal{C}$ and $T^{-1}\mathcal{C}$ coincide if and only if S and T generate the same strongly saturated class of morphisms. We will also write $S^{-1}\mathcal{C}$ for the class of S-local objects of $\mathcal{C}$ in the case where S is *not* small; however, this is generally only a well-behaved object in the case where there is a (small) subset $S_0 \subseteq S$ which generates the same strongly saturated class of morphisms.

Proposition 5.5.4.16. *Let $f : \mathcal{C} \to \mathcal{D}$ be a presentable functor between presentable ∞-categories and let S be a strongly saturated class of morphisms of $\mathcal{D}$ which is of small generation. Then $f^{-1}S$ is of small generation (as a strongly saturated class of morphisms of $\mathcal{C}$).*

Proof. Replacing $\mathcal{D}$ by $S^{-1}\mathcal{D}$ if necessary, we may suppose that S consists of precisely the equivalences in $\mathcal{D}$. Let $\mathcal{E}_{\mathcal{D}} \subseteq \operatorname{Fun}(\Delta^1, \mathcal{D})$ denote the full subcategory spanned by those morphisms which are equivalences in $\mathcal{D}$ and let $\mathcal{E}_{\mathcal{C}} \subseteq \operatorname{Fun}(\Delta^1, \mathcal{C})$ denote the full subcategory spanned by those morphisms which belong to $f^{-1}S$. We have a homotopy Cartesian diagram of ∞-categories

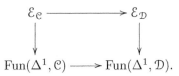

The ∞-category $\mathcal{E}_{\mathcal{D}}$ is equivalent to $\mathcal{D}$ and therefore presentable. The ∞-categories $\operatorname{Fun}(\Delta^1, \mathcal{C})$ and $\operatorname{Fun}(\Delta^1, \mathcal{D})$ are presentable by Proposition 5.5.3.6. It follows from Proposition 5.5.3.12 that $\mathcal{E}_{\mathcal{C}}$ is presentable. In particular, there is a small collection of objects of $\mathcal{E}_{\mathcal{C}}$ which generates $\mathcal{E}_{\mathcal{C}}$ under colimits, as desired. $\square$

Let $\mathcal{C}$ be a presentable ∞-category. We will say that a full subcategory $\mathcal{C}^0 \subseteq \mathcal{C}$ is *strongly reflective* if it is the essential image of an accessible localization functor. Equivalently, $\mathcal{C}^0$ is strongly reflective if it is presentable, it is stable under equivalence in $\mathcal{C}$, and the inclusion $\mathcal{C}^0 \subseteq \mathcal{C}$ admits a left adjoint. According to Proposition 5.5.4.15, $\mathcal{C}^0$ is strongly reflective if and only if there exists a (small) set S of morphisms of $\mathcal{C}$ such that $\mathcal{C}^0$ is the full subcategory of $\mathcal{C}$ spanned by the S-local objects. For later use, we record a few easy stability properties enjoyed by the collection of strongly reflective subcategories of $\mathcal{C}$:

Lemma 5.5.4.17. *Let $f : \mathcal{C} \to \mathcal{D}$ be a colimit-preserving functor between presentable ∞-categories and let $\mathcal{C}^0 \subseteq \mathcal{C}$ be a strongly reflective subcategory. Let f^* denote a right adjoint of f and let $\mathcal{D}^0 \subseteq \mathcal{D}$ be the full subcategory spanned by those objects $D \in \mathcal{D}$ such that $f^*D \in \mathcal{C}^0$. Then $\mathcal{D}^0$ is a strongly reflective subcategory of $\mathcal{D}$.*

Proof. Let S be a (small) set of morphisms of $\mathcal{C}$ such that $\mathcal{C}^0$ is the full subcategory of $\mathcal{C}$ spanned by the S-local objects. Then $\mathcal{D}^0$ is the full subcategory of $\mathcal{D}$ spanned by the $f(S)$-local objects. □

Lemma 5.5.4.18. *Let $\mathcal{C}$ be a presentable ∞-category and let $\{\mathcal{C}_\alpha\}_{\alpha \in A}$ be a family of full subcategories of $\mathcal{C}$ indexed by a (small) set A. Suppose that each $\mathcal{C}_\alpha$ is strongly reflective. Then $\bigcap_{\alpha \in A} \mathcal{C}_\alpha$ is strongly reflective.*

Proof. For each $\alpha \in A$, choose a (small) set $S(\alpha)$ of morphisms of $\mathcal{C}$ such that $\mathcal{C}_\alpha$ is the full subcategory of $\mathcal{C}$ spanned by the $S(\alpha)$-local objects. Then $\bigcap_{\alpha \in A} \mathcal{C}_\alpha$ is the full subcategory of $\mathcal{C}$ spanned by the $\bigcup_{\alpha \in A} S(\alpha)$-local objects. □

Lemma 5.5.4.19. *Let $\mathcal{C}$ be a presentable ∞-category and K a small simplicial set. Let $\mathcal{D}$ denote the full subcategory of $\mathrm{Fun}(K^\triangleleft, \mathcal{C})$ spanned by those diagrams $\overline{p} : K^\triangleleft \to \mathcal{C}$ which are limits of $p = \overline{p}|K$. Then $\mathcal{D}$ is a strongly reflective subcategory of $\mathcal{C}$.*

Proof. The restriction functor $\mathcal{D} \to \mathrm{Fun}(K, \mathcal{C})$ is an equivalence of ∞-categories. This proves that $\mathcal{D}$ is accessible. Let $s : \mathrm{Fun}(K, \mathcal{C}) \to \mathcal{D}$ be a homotopy inverse to the restriction map. Then the composition

$$\mathrm{Fun}(K^\triangleright, \mathcal{C}) \to \mathrm{Fun}(K, \mathcal{C}) \xrightarrow{s} \mathcal{D}$$

is left adjoint to the inclusion. □

We conclude this section by giving a universal property which characterizes the localization $S^{-1}\mathcal{C}$.

Proposition 5.5.4.20. *Let $\mathcal{C}$ be a presentable ∞-category and $\mathcal{D}$ an arbitrary ∞-category. Let S be a (small) set of morphisms of $\mathcal{C}$, and $L : \mathcal{C} \to S^{-1}\mathcal{C} \subseteq \mathcal{C}$ an associated (accessible) localization functor. Composition with L induces a functor*

$$\eta : \mathrm{Fun}^L(S^{-1}\mathcal{C}, \mathcal{D}) \to \mathrm{Fun}^L(\mathcal{C}, \mathcal{D}).$$

The functor η is fully faithful, and the essential image of η consists of those functors $f : \mathcal{C} \to \mathcal{D}$ such that $f(s)$ is an equivalence in $\mathcal{D}$ for each $s \in S$.

Proof. Let $\alpha : \mathrm{id}_\mathcal{C} \to L$ be a unit for the adjunction between L and the inclusion $S^{-1}\mathcal{C} \subseteq \mathcal{C}$. We first observe that every functor $f_0 : S^{-1}\mathcal{C} \to \mathcal{D}$ admits a right Kan extension $f : \mathcal{C} \to \mathcal{D}$. To prove this, we may first replace f_0 by the equivalent diagram $g_0 = f_0 \circ (L|S^{-1}\mathcal{C})$ and define $g = f_0 \circ L$. To prove that g is a right Kan extension of g_0, it suffices to show that for each object $X \in \mathcal{C}$, the diagram

$$\overline{p} : (S^{-1}\mathcal{C})^\triangleleft_{X/} \to \mathcal{C} \xrightarrow{L} S^{-1}\mathcal{C} \xrightarrow{f_0} \mathcal{D}$$

exhibits $f_0(LX)$ as a limit of $p = \overline{p}|(S^{-1}\mathcal{C})_{/X}$. For this, we note that an S-localization map $\alpha(X) : X \to LX$ is an initial object of $(S^{-1}\mathcal{C})_{X/}$ (Remark 5.2.7.7) and that $f_0(L\alpha(X))$ is an equivalence by Proposition 5.2.7.4.

Let $\mathfrak{X}$ denote the full subcategory of $\mathcal{D}^{\mathcal{C}}$ spanned by those functors $f :$ $\mathcal{C} \to \mathcal{D}$ which are right Kan extensions of $f|S^{-1}\,\mathcal{C}$. According to Proposition 4.3.2.15, the restriction map $\mathfrak{X} \to \mathrm{Fun}(S^{-1}\,\mathcal{C}, \mathcal{D})$ is a trivial fibration. Let $\overline{\eta} : \mathrm{Fun}(S^{-1}\,\mathcal{C}, \mathcal{D}) \to \mathrm{Fun}(\mathcal{C}, \mathcal{D})$ be given by composition with L. The above argument shows that $\overline{\eta}$ factors through $\mathfrak{X}$. Moreover, the composition of $\overline{\eta}$ with the restriction map is homotopic to the identity on $\mathrm{Fun}(S^{-1}\,\mathcal{C}, \mathcal{D})$. It follows that $\overline{\eta}$ is an equivalence of ∞-categories.

We have a commutative diagram

$$
\begin{array}{ccc}
\mathrm{Fun}^L(S^{-1}\,\mathcal{C}, \mathcal{D}) & \xrightarrow{\ \eta\ } & \mathrm{Fun}^L(\mathcal{C}, \mathcal{D}) \\
\downarrow & & \downarrow \\
\mathrm{Fun}(S^{-1}\,\mathcal{C}, \mathcal{D}) & \xrightarrow{\ \overline{\eta}\ } & \mathrm{Fun}(\mathcal{C}, \mathcal{D}),
\end{array}
$$

where the vertical maps are inclusions of full subcategories and the lower horizontal map is fully faithful. It follows that η is fully faithful. To complete the proof, we must show that a functor $f : \mathcal{C} \to \mathcal{D}$ belongs to the essential image of η if and only if $f(s)$ is an equivalence for each $s \in S$. The "only if" direction is clear since the functor L carries each element of S to an equivalence in $\mathcal{C}$. Conversely, suppose that f carries each $s \in S$ to an equivalence. The natural transformation α gives a map of functors $\alpha(f) : f \to f \circ L$; we wish to show that $\alpha(f)$ is an equivalence. Equivalently, we wish to show that for each object $X \in \mathcal{C}$, f carries the map $\alpha(X) : X \to LX$ to an equivalence in $\mathcal{D}$. Let S' denote the class of all morphisms ϕ in $\mathcal{C}$ such that $f(\phi)$ is an equivalence in $\mathcal{D}$. By assumption, $S \subseteq S'$. Lemma 5.5.4.11 implies that S' is strongly saturated, so that Proposition 5.5.4.15 asserts that $\alpha(X) \in S'$, as desired. $\qquad\square$

5.5.5 Factorization Systems on Presentable ∞-Categories

Let $\mathcal{C}$ be a presentable ∞-category. In §5.5.4, we saw that it is easy to produce localizations of $\mathcal{C}$: for any small collection of morphisms S in $\mathcal{C}$, the full subcategory $S^{-1}\,\mathcal{C}$ of S-local objects of $\mathcal{C}$ is a presentable localization of $\mathcal{C}$ which depends only on the strongly saturated class of morphisms $\overline{S}$ generated by S. Our goal in this section is to prove a similar result for factorization systems on $\mathcal{C}$. The first step is to introduce the analogue of the notion of "strongly saturated":

Definition 5.5.5.1. Let S be a collection of morphisms in a presentable ∞-category $\mathcal{C}$. We will say that S is *saturated* if the following conditions are satisfied:

(1) The collection S is closed under small colimits in $\mathrm{Fun}(\Delta^1, \mathcal{C})$.

(2) The collection S contains all equivalences and is stable under composition.

(3) The collection S is closed under the formation of pushouts. That is, given a pushout diagram

in $\mathcal{C}$, if f belongs to S, then f' also belongs to S.

Remark 5.5.5.2. Let $\mathcal{C}$ be a presentable ∞-category. Then any intersection of saturated collections of morphisms in $\mathcal{C}$ is again saturated. It follows that for *any* class of morphisms S of $\mathcal{C}$, there exists a smallest saturated collection of morphisms $\overline{S}$ containing S. We will refer to $\overline{S}$ as the *saturated collection of morphisms generated by* S. We will say that a saturated collection of morphisms $\overline{S}$ is *of small generation* if it is generated by some (small) subset $S \subseteq \overline{S}$.

Remark 5.5.5.3. If S is a saturated collection of morphisms of $\mathcal{C}$, then S is closed under retracts.

Remark 5.5.5.4. Let $\mathcal{C}$ be (the nerve of) a presentable category and let S be a saturated class of morphisms in $\mathcal{C}$. Then S is also weakly saturated in the sense of Definition A.1.2.2.

Example 5.5.5.5. Let $\mathcal{C}$ be a presentable ∞-category. Then every strongly saturated class of morphisms in $\mathcal{C}$ is also saturated.

Example 5.5.5.6. Let $\mathcal{C}$ be a presentable ∞-category and let S be any collection of morphisms of $\mathcal{C}$. Then $^\perp S$ is saturated; this follows immediately from Proposition 5.2.8.6. In particular, if (S_L, S_R) is a factorization system on $\mathcal{C}$, then S_L is saturated.

The main result of this section is the following converse to Example 5.5.5.6:

Proposition 5.5.5.7. *Let $\mathcal{C}$ be a presentable ∞-category and let S be a saturated collection of morphisms in $\mathcal{C}$ which is of small generation. Then $(S, S^\perp)$ is a factorization system on $\mathcal{C}$.*

Corollary 5.5.5.8. *Let $\mathcal{C}$ be a presentable ∞-category, let S be a saturated collection of morphisms of $\mathcal{C}$, and suppose that S is of small generation. Let*

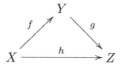

be a commutative diagram in $\mathcal{C}$. If f and h belong to S, then g belongs to S.

Proof. Combine Propositions 5.5.5.7, 5.2.8.11, and 5.2.8.6.   □

In the situation of Proposition 5.5.5.7, we will refer to the elements of $S^\perp$ as *S-local morphisms of* $\mathcal{C}$. Note that an object $X \in \mathcal{C}$ is S-local if and only if a morphism $X \to 1_{\mathcal{C}}$ is S-local, where $1_{\mathcal{C}}$ denotes a final object of $\mathcal{C}$.

The proof of Proposition 5.5.5.7 will be given at the end of this section after we have established a series of technical lemmas.

Lemma 5.5.5.9. *Let* $\mathcal{C}$ *be a presentable* ∞-*category and* S *a saturated collection of morphisms of* $\mathcal{C}$. *The following conditions are equivalent:*

(1) *The collection* S *is of small generation.*

(2) *The full subcategory* $\mathcal{D} \subseteq \operatorname{Fun}(\Delta^1, \mathcal{C})$ *spanned by the elements of* S *is presentable.*

Proof. If $\mathcal{D}$ is presentable, then $\mathcal{D}$ is generated under small colimits by a small set of objects; these objects clearly generate S as a saturated collection of morphisms. This proves that (2) $\Rightarrow$ (1). To prove the reverse implication, choose a small collection of morphisms $\{f_\beta : X_\beta \to Y_\beta\}$ which generates S as a semisaturated class of morphisms and an uncountable regular cardinal κ such that $\mathcal{C}$ is κ-accessible and each of the objects X_β, Y_β is κ-compact. Let $\mathcal{D}' \subseteq \mathcal{D}$ be the collection of all morphisms $f : X \to Y$ such that f belongs to S, where both X and Y are κ-compact. Lemma 5.3.4.9 implies that each $f \in \mathcal{D}'$ is a κ-compact object of $\operatorname{Fun}(\Delta^1, \mathcal{C})$ and, in particular, a κ-compact object of $\mathcal{D}$. Assume for simplicity that $\mathcal{C}$ is a minimal ∞-category, so that $\mathcal{D}'$ is small. According to Proposition 5.3.5.10, the inclusion $\mathcal{D}' \subseteq \mathcal{D}$ is equivalent to $j \circ F$, where $j : \mathcal{D}' \to \operatorname{Ind}_\kappa \mathcal{D}'$ is the Yoneda embedding and $F : \operatorname{Ind}_\kappa \mathcal{D}' \to \mathcal{D}$ is κ-continuous. Proposition 5.3.5.11 implies that F is fully faithful; let $\mathcal{D}''$ denote its essential image. To complete the proof, it will suffice to show that $\mathcal{D}'' = \mathcal{D}$.

Let $S' \subseteq S$ denote the collection of objects of $\mathcal{D}''$ (which we may identify with morphisms in $\mathcal{C}$). By construction, S' contains the collection of morphisms $\{f_\beta\}$ which generates S. Consequently, to prove that $S' = S$, it will suffice to show that S' is saturated.

It follows from Proposition 5.5.1.9 that $\mathcal{D}'' \subseteq \operatorname{Fun}(\Delta^1, \mathcal{C})$ is stable under small colimits. We next verify that S' is stable under pushouts. Let $K = \Lambda_0^2$ and let $\overline{p} : K^\triangleright \to \mathcal{C}$ be a colimit of $p = \overline{p}|K$,

$$
\begin{array}{ccc}
X & \xrightarrow{\ f\ } & Y \\
\downarrow & & \downarrow \\
X' & \xrightarrow{\ f'\ } & Y',
\end{array}
$$

such that f belongs to S'. Using Proposition 5.3.5.15, we can write p as the colimit of a diagram $q : \operatorname{N}(A) \to \operatorname{Fun}(\Lambda_0^2, \mathcal{C}^\kappa)$, where A is a κ-filtered partially ordered set. For $\alpha \in A$, we let p_α denote the corresponding diagram, which we may depict as

$$
X'_\alpha \leftarrow X_\alpha \xrightarrow{f_\alpha} Y_\alpha.
$$

For each $A' \subseteq A$, we let $p_{A'}$ denote a colimit of $q|\,\mathrm{N}(A')$, which we will denote by

$$X'_{A'} \leftarrow X_{A'} \xrightarrow{f_{A'}} Y_{A'}.$$

Let B denote the collection of κ-small filtered subsets $A' \subseteq A$ such that the $f_{A'}$ belongs to S'. Since $f \in S'$, we conclude that f can be obtained as the colimit of a κ-filtered diagram in $\mathcal{D}'$. Applying Lemma 5.5.4.13, we deduce that B is κ-filtered and that $A = \bigcup_{A' \in B} A'$. Using Proposition 4.2.3.4 and Corollary 4.2.3.10, we deduce that p is the colimit of a diagram $q' : \mathrm{N}(B) \to \mathrm{Fun}(\Lambda_0^2, \mathcal{C})$, where $q'(A') = p_{A'}$. Replacing A by B, we may suppose that each f_α belongs to S'.

Let $\varinjlim : \mathrm{Fun}(\Lambda_0^2, \mathcal{C}) \to \mathrm{Fun}(\Delta^1 \times \Delta^1, \mathcal{C})$ be a colimit functor (that is, a left adjoint to the restriction functor). Lemma 5.5.2.3 implies that we may identify $\overline{p}$ with a colimit of the diagram $\varinjlim \circ q$. Consequently, the morphism f' can be written as a colimit of morphisms f'_α which fit into pushout diagrams

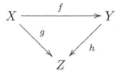

Since $f_\alpha \in S' \subseteq S$, we conclude that $f'_\alpha \in S$. Since X'_α and Y'_α are κ-compact, we deduce that $f'_\alpha \in S'$. Since $\mathcal{D}''$ is stable under colimits, we deduce that $f' \in S'$, as desired.

We now complete the proof by showing that S' is stable under composition. Let $\sigma : \Delta^2 \to \mathcal{C}$ be a simplex corresponding to a diagram

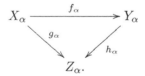

in $\mathcal{C}$. We will show that if $f, g \in S'$, then $h \in S'$. Using Proposition 5.3.5.15, we can write σ as the colimit of a diagram $q : \mathrm{N}(A) \to \mathrm{Fun}(\Delta^2, \mathcal{C}^\kappa)$, where A is a κ-filtered partially ordered set. For each $\alpha \in A$, we will denote the corresponding diagram by

$$X_\alpha \xrightarrow{f_\alpha} Y_\alpha$$
$$\searrow g_\alpha \qquad \swarrow h_\alpha$$
$$Z_\alpha.$$

Arguing as above, we may assume (possibly after changing A and q) that each f_α belongs to S'. Repeating the same argument, we may suppose that g_α belongs to S'. Since S is stable under composition, we conclude that each h_α belongs to S. Since each X_α and each Z_α is κ-compact, we have $h_\alpha \in S'$. The stability of $\mathcal{D}''$ under colimits now implies that $h \in S'$, as desired. $\square$

Lemma 5.5.5.10. *Let $\mathcal{C}$ be a presentable ∞-category and let S be a saturated collection of morphisms in $\mathcal{C}$. For every object $X \in \mathcal{C}$, let S_X denote the collection of all morphisms of $\mathcal{C}_{/X}$ whose image in $\mathcal{C}$ belongs to S. Then each S_X is strongly saturated in $\mathcal{C}_{/X}$. Moreover, if S is of small generation, then each S_X is also of small generation.*

Proof. The first assertion follows immediately from the definitions together with Proposition 1.2.13.8. To prove the second, let $\mathcal{D}$ be the full subcategory of $\mathrm{Fun}(\Delta^1, \mathcal{C})$ spanned by the elements of S, and $\mathcal{D}'$ the full subcategory of $\mathrm{Fun}(\Delta^1, \mathcal{C}_{/X})$ spanned by the elements of S_X. We have a (homotopy) pullback diagram of ∞-categories

$$\begin{array}{ccc} \mathcal{D}' & \longrightarrow & \mathrm{Fun}(\Delta^1, \mathcal{C}_{/X}) \\ \downarrow & & \downarrow{\scriptstyle\psi} \\ \mathcal{D} & \xrightarrow{\ \phi\ } & \mathrm{Fun}(\Delta^1, \mathcal{C}). \end{array}$$

The functors ϕ and ψ preserve small colimits, and the ∞-categories $\mathcal{D}$, $\mathrm{Fun}(\Delta^1, \mathcal{C}_{/X})$, and $\mathrm{Fun}(\Delta^1, \mathcal{C})$ are all presentable (the first in view of Lemma 5.5.5.9). Using Proposition 5.5.3.12, we deduce that $\mathcal{D}'$ is presentable and therefore generated under small colimits by a (small) set of elements of S_X. This proves that S_X is of small generation, as desired. $\qquad\square$

Lemma 5.5.5.11. *Let $\mathcal{C}$ be a presentable ∞-category, S a saturated collection of morphisms of $\mathcal{C}$, and X an object of $\mathcal{C}$. Then the full subcategory $\mathcal{D} \subseteq \mathcal{C}^{X/}$ spanned by the elements which belong to S is closed under small colimits.*

Proof. In view of Corollaries 4.2.3.11 and 4.4.2.4, it will suffice to show that $\mathcal{D}$ is closed under small filtered colimits and pushouts, and contains the initial objects of $\mathcal{C}^{X/}$. The last condition follows from the fact that S contains all equivalences. Now suppose that $\overline{p} : K^{\triangleright} \to \mathcal{C}^{C/}$ is a colimit of $p = \overline{p}|K$, where K is either filtered or equivalent to Λ_0^2, and that $p(K) \subseteq \mathcal{D}$. We can identify $\overline{p}$ with a map $P : K^{\triangleright} \times \Delta^1 \to \mathcal{C}$ such that $P|K^{\triangleright} \times \{0\}$ is the constant map taking the value $C \in \mathcal{C}$. Since K is weakly contractible, $P|K^{\triangleright} \times \{0\}$ is a colimit diagram in $\mathcal{C}$. The map $P|K^{\triangleright} \times \{1\}$ is the image of a colimit diagram under the left fibration $\mathcal{C}^{C/} \to \mathcal{C}$; since K is weakly contractible, Proposition 4.4.2.9 implies that $P|K^{\triangleright} \times \{1\}$ is a colimit diagram. We now apply Proposition 5.1.2.2 to deduce that $P : K^{\triangleright} \to \mathcal{C}^{\Delta^1}$ is a colimit diagram. Since S is stable under colimits in $\mathcal{C}^{\Delta^1}$, we conclude that P carries the cone point of $K^{\triangleright}$ to a morphism belonging to S, as we wished to show. $\qquad\square$

Lemma 5.5.5.12. *Let $\mathcal{C}$ be an ∞-category and let $f : C \to D$, $g : C \to E$ be morphisms in $\mathcal{C}$. Then there is a natural identification of $\mathrm{Map}_{\mathcal{C}_{C/}}(f, g)$ with the homotopy fiber of the map*

$$\mathrm{Map}_{\mathcal{C}}(D, E) \to \mathrm{Map}_{\mathcal{C}}(C, E)$$

induced by composition with f, where the fiber is taken over the point corresponding to g.

Proof. We have a commutative diagram of simplicial sets

$$\begin{array}{ccccc}
\mathcal{C}_{f/} \times_{\mathcal{C}_{C/}} \{g\} & \longrightarrow & \mathcal{C}_{f/} \times_{\mathcal{C}} \{E\} & \longrightarrow & \mathcal{C}_{f/} \\
\downarrow{\scriptstyle\phi} & & \downarrow{\scriptstyle\phi'} & & \downarrow{\scriptstyle\phi''} \\
\{g\} & \longrightarrow & \mathcal{C}_{C/} \times_{\mathcal{C}} \{E\} & \longrightarrow & \mathcal{C}_{C/},
\end{array}$$

where both squares are pullbacks. Proposition 2.1.2.1 asserts that ϕ'' is a left fibration, so that ϕ' and ϕ are left fibrations as well. Since $\mathcal{C}_{C/} \times_{\mathcal{C}} \{E\} = \operatorname{Hom}^{L}_{\mathcal{C}}(C, E)$ is a Kan complex, the map ϕ' is actually a Kan fibration (Lemma 2.1.3.3), so that the square on the left is a homotopy pullback and identifies

$$\mathcal{C}_{f/} \times_{\mathcal{C}_{C/}} \{g\} \simeq \operatorname{Map}_{\mathcal{C}_{C/}}(f, g)$$

with the homotopy fiber of ϕ' over g; we conclude by observing that ϕ' is a model for the map $\operatorname{Map}_{\mathcal{C}}(D, E) \to \operatorname{Map}_{\mathcal{C}}(C, E)$ given by composition with f. $\qquad\square$

Lemma 5.5.5.13. *Let*

$$\begin{array}{ccc}
X & \xrightarrow{\ f\ } & X' \\
\downarrow{\scriptstyle g} & & \downarrow \\
Y & \xrightarrow{\ f'\ } & Y'
\end{array}$$

be a pushout diagram in an ∞-category $\mathcal{C}$. *Then there exists an isomorphism*

$$\operatorname{Map}_{\mathcal{C}_{X/}}(f, g) \simeq \operatorname{Map}_{\mathcal{C}_{Y/}}(f', \operatorname{id}_Y)$$

in the homotopy category $\mathcal{H}$.

Proof. According to Corollary 4.2.4.7, we can assume without loss of generality that $\mathcal{C}$ is the nerve of a fibrant simplicial category $\mathcal{D}$ and that the diagram in question is the nerve of a commutative diagram

$$\begin{array}{ccc}
X & \xrightarrow{\ f\ } & X' \\
\downarrow{\scriptstyle g} & & \downarrow \\
Y & \xrightarrow{\ f'\ } & Y'
\end{array}$$

in $\mathcal{D}$. Theorem 4.2.4.1 implies that this diagram is homotopy coCartesian in $\mathcal{D}$, so that we have a homotopy pullback diagram

$$\begin{array}{ccc}
\operatorname{Map}_{\mathcal{D}}(Y', Y) & \xrightarrow{\ \phi\ } & \operatorname{Map}_{\mathcal{D}}(Y, Y) \\
\downarrow & & \downarrow \\
\operatorname{Map}_{\mathcal{D}}(X', Y) & \xrightarrow{\ \phi'\ } & \operatorname{Map}_{\mathcal{D}}(X, Y)
\end{array}$$

of Kan complexes. Consequently, we obtain an isomorphism in $\mathcal{H}$ between the homotopy fiber of ϕ over id_Y and the homotopy fiber of ϕ' over g. According to Lemma 5.5.5.12, these homotopy fibers may be identified with $\operatorname{Map}_{\mathcal{C}_{Y/}}(f', \operatorname{id}_Y)$ and $\operatorname{Map}_{\mathcal{C}_{X/}}(f, g)$, respectively. $\qquad\square$

Lemma 5.5.5.14. *Let $\mathcal{C}$ be a presentable ∞-category and let S be a saturated collection of morphisms of $\mathcal{C}$ which is of small generation. Then, for every object $X \in \mathcal{C}$, there exists a morphism $f : X \to Y$ in $\mathcal{C}$ such that $f \in S$ and Y is S-local.*

Proof. Let $\mathcal{D}$ be the full subcategory of $\mathrm{Fun}(\Delta^1, \mathcal{C})$ spanned by the elements of S and form a fiber diagram

$$
\begin{array}{ccc}
\mathcal{D}_X & \longrightarrow & \mathcal{D} \\
\downarrow & & \downarrow \\
\{X\} & \longrightarrow & \mathrm{Fun}(\{0\}, \mathcal{C}).
\end{array}
$$

Since S is stable under pushouts, the right vertical map is a coCartesian fibration, so that the above diagram is homotopy Cartesian by Proposition 3.3.1.3. Lemma 5.5.5.9 asserts that $\mathcal{D}$ is accessible, so that $\mathcal{D}_X$ is accessible by Proposition 5.4.6.6. Using Lemma 5.5.5.11, we conclude that $\mathcal{D}_X$ is presentable, so that $\mathcal{D}_X$ has a final object $f : X \to Y$. To complete the proof, it will suffice to show that Y is S-local.

Let $t : A \to B$ be an arbitrary morphism in $\mathcal{C}$ which belongs to S. We wish to show that composition with t induces a homotopy equivalence $\phi : \mathrm{Map}_{\mathcal{C}}(B, Y) \to \mathrm{Map}_{\mathcal{C}}(A, Y)$. Let $g : A \to Y$ be an arbitrary morphism; using Lemma 5.5.5.12, we may identify $\mathrm{Map}_{\mathcal{C}_{A/}}(t, g)$ with the homotopy fiber of ϕ over the base point g of $\mathrm{Map}_{\mathcal{C}}(A, Y)$. We wish to show that this space is contractible. Form a pushout diagram

$$
\begin{array}{ccc}
A & \overset{t}{\longrightarrow} & B \\
{\scriptstyle g}\downarrow & & \downarrow \\
Y & \underset{t'}{\longrightarrow} & Z
\end{array}
$$

in the ∞-category $\mathcal{C}$. Lemma 5.5.5.13 implies the existence of a homotopy equivalence $\mathrm{Map}_{\mathcal{C}_{A/}}(t, g) \simeq \mathrm{Map}_{\mathcal{C}_{Y/}}(t', \mathrm{id}_Y)$. It will therefore suffice to prove that $\mathrm{Map}_{\mathcal{C}_{Y/}}(t', \mathrm{id}_Y)$ is contractible. Since t' is a pushout of t, it belongs to S. Let σ be a 2-simplex of $\mathcal{C}$ classifying a diagram

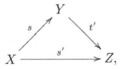

so that s' is a composition of the morphisms s and t' in $\mathcal{C}$ and therefore also belongs to S. Applying Lemma 5.5.5.12 again, we may identify

$$
\mathrm{Map}_{\mathcal{C}_{Y/}}(t', \mathrm{id}_{C'}) \simeq \mathrm{Map}_{\mathcal{C}_s}(\sigma, s_1(s))
$$

with the homotopy fiber of the map $\mathrm{Map}_{\mathcal{C}_{Y/}}(s', s) \to \mathrm{Map}_{\mathcal{C}_{Y/}}(s, s)$ given by composition with σ. By construction, $\mathcal{D}_X$ is a full subcategory of $\mathcal{C}^{X/}$ which contains s and s', and s is a final object of $\mathcal{D}_X$. In view of the equivalence of $\mathcal{C}_{X/}$ with $\mathcal{C}^{X/}$, we conclude that the spaces $\mathrm{Map}_{\mathcal{C}_{X/}}(s', s)$ and $\mathrm{Map}_{\mathcal{C}_{X/}}(s, s)$ are contractible, so that ϕ is a homotopy equivalence, as desired. $\qquad\square$

Proof of Proposition 5.5.5.7. Let $h : X \to Z$ be a morphism in $\mathcal{C}$; we wish to show that h admits a factorization

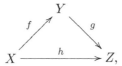

where $f \in S$ and $g \in S^{\perp}$. Using Remark 5.2.8.3, we deduce that a morphism $g : Y \to Z$ belongs to $S^{\perp}$ if and only if it is an S_Z-local object of $\mathcal{C}_{/Z}$, where S_Z is defined as in Lemma 5.5.5.10. The existence of h then follows from Lemma 5.5.5.14. $\qquad\square$

5.5.6 Truncated Objects

Let X be a topological space. The first step in the homotopy-theoretic analysis of the space X is to divide X into path components. The situation can be described as follows: we associate to X a set $\pi_0 X$, which we may view as a discrete topological space. There is a map $f : X \to \pi_0 X$ which collapses each component of X to a point. If X is a sufficiently nice space (for example, a CW complex), then the path components of X are open, so f is continuous. Moreover, f is universal among continuous maps from X into a discrete topological space.

The next step in the analysis of X is to consider its fundamental group $\pi_1 X$, which (provided that X is sufficiently nice) may be studied by means of a universal cover $\widetilde{X}$ of X. However, it is important to realize that neither $\pi_1 X$ nor $\widetilde{X}$ is invariantly associated to X: both require a choice of base point. The situation can be described more canonically as follows: to X we can associate a *fundamental groupoid* $\pi(X)$ and a map ϕ from X to the classifying space $B\pi(X)$. The universal cover $\widetilde{X}$ of X can be identified (up to homotopy equivalence) with the homotopy fibers of the map ϕ. The classifying space $B\pi(X)$ can be regarded as a "quotient" of X obtained by killing all of the higher homotopy groups of X. Like $\pi_0 X$, it can be described by a universal mapping property.

To continue the analysis, we first recall that a space Y is said to be k-*truncated* if the homotopy groups of Y vanish in dimensions larger than k (see Definition 2.3.4.15). Every (sufficiently nice) topological space X admits an essentially unique *Postnikov tower*

$$X \to \cdots \to \tau_{\leq n} X \to \cdots \to \tau_{\leq -1} X,$$

where $\tau_{\leq i} X$ is i-truncated, and is universal (in a suitable homotopy-theoretic sense) among i-truncated spaces which admit a map from X. For example, we can take $\tau_{\leq 0} X = \pi_0 X$, considered as a discrete space, and $\tau_{\leq 1} X = B\pi(X)$. Moreover, we can recover the space X (up to weak homotopy equivalence) by taking the homotopy limit of the tower.

The objective of this section is to construct an analogous theory in the case where X is not a space but instead an object of some (abstract) ∞-category $\mathcal{C}$. We begin by observing that the condition that a space X be

k-truncated can be reformulated in more categorical terms: a Kan complex X is k-truncated if and only if, for every simplicial set S, the mapping space $\mathrm{Map}_{\mathcal{Set}_\Delta}(S, X)$ is k-truncated. This motivates the following:

Definition 5.5.6.1. Let $\mathcal{C}$ be an ∞-category and $k \geq -1$ an integer. We will say that an object C of $\mathcal{C}$ is *k-truncated* if, for every object $D \in \mathcal{C}$, the space $\mathrm{Map}_{\mathcal{C}}(D, C)$ is k-truncated. By convention, we will say that C is *(-2)-truncated* if it is a final object of $\mathcal{C}$. We will say that an object of $\mathcal{C}$ is *discrete* if it is 0-truncated. We will generally let $\tau_{\leq k}\,\mathcal{C}$ denote the full subcategory of $\mathcal{C}$ spanned by the k-truncated objects.

Notation 5.5.6.2. Let $\mathcal{C}$ be an ∞-category. Using Propositions 2.3.4.18 and 2.3.4.5, we conclude that the full subcategory $\tau_{\leq 0}\,\mathcal{C}$ is equivalent to the nerve of its homotopy category. We will denote this homotopy category by $\mathrm{Disc}(\mathcal{C})$ and refer to it as the *category of discrete objects of* $\mathcal{C}$.

Lemma 5.5.6.3. *Let C be an object of an ∞-category $\mathcal{C}$ and let $k \geq -2$. The following conditions are equivalent:*

(1) *The object C is k-truncated.*

(2) *For every $n \geq k + 3$ and every diagram*

$$
\begin{array}{ccc}
\partial\,\Delta^n & \overset{f}{\longrightarrow} & \mathcal{C} \\
\Big\downarrow & \nearrow & \\
\Delta^n & &
\end{array}
$$

for which f carries the final vertex of Δ^n to C, there exists a dotted arrow rendering the diagram commutative.

Proof. Suppose first that (2) is satisfied. Then for every object $D \in \mathcal{D}$, the Kan complex $\mathrm{Hom}_{\mathcal{C}}^{\mathrm{R}}(D, C)$ has the extension property with respect to $\partial\,\Delta^{n-1} \subseteq \Delta^{n-1}$ for all $n \geq k + 3$, and is therefore k-truncated. For the converse, suppose that (1) is satisfied and choose a categorical equivalence $g : \mathcal{C} \to \mathrm{N}\,\mathcal{D}$, where $\mathcal{D}$ is a topological category. According to Proposition A.2.3.1, it will suffice to show that for every $n \geq k + 3$ and every diagram

$$
\begin{array}{ccc}
|\,\mathfrak{C}[\partial\,\Delta^n]\,| & \overset{F}{\longrightarrow} & \mathcal{D} \\
\Big\downarrow & \nearrow & \\
|\,\mathfrak{C}[\Delta^n]\,| & &
\end{array}
$$

having the property that F carries the final object of $|\,\mathfrak{C}[\Delta^n]\,|$ to $g(C)$, there exists a dotted arrow as indicated, rendering the diagram commutative. Let $D \in \mathcal{D}$ denote the image of the initial object of $\partial\,\Delta^n$ under F. Then, constructing the desired extension is equivalent to extending a map $\partial[0, 1]^{n-1} \to \mathrm{Map}_{\mathcal{D}}(D, g(C))$ to a map defined on all of $[0, 1]^{n-1}$, which is possibly by virtue of assumption (1). $\qquad\square$

Remark 5.5.6.4. A Kan complex X is k-truncated if and only if it is k-truncated when regarded as an object in the ∞-category $\mathcal{S}$ of spaces.

Proposition 5.5.6.5. *Let $\mathcal{C}$ be an ∞-category and $k \geq -2$ an integer. The full subcategory $\tau_{\leq k}\,\mathcal{C} \subseteq \mathcal{C}$ of k-truncated objects is stable under all limits which exist in $\mathcal{C}$.*

Proof. Let $j : \mathcal{C} \to \mathcal{P}(\mathcal{C})$ be the Yoneda embedding. By definition, $\tau_{\leq k}\,\mathcal{C}$ is the preimage of $\mathrm{Fun}(\mathcal{C}^{op}, \tau_{\leq k}\,\mathcal{S})$ under j. Since j preserves all limits which exist in $\mathcal{C}$, it will suffice to prove that $\mathrm{Fun}(\mathcal{C}^{op}, \tau_{\leq k}\,\mathcal{S}) \subseteq \mathrm{Fun}(\mathcal{C}^{op}, \mathcal{S})$ is stable under limits. Using Proposition 5.1.2.2, it suffices to prove that the inclusion $i : \tau_{\leq k}\,\mathcal{S} \subseteq \mathcal{S}$ is stable under limits. In other words, we must show that $\tau_{\leq k}\,\mathcal{S}$ admits small limits and that i preserves small limits. According to Propositions 4.4.2.6 and 4.4.2.7, it will suffice to show that $\tau_{\leq k}\,\mathcal{S} \subseteq \mathcal{S}$ is stable under the formation of pullbacks and small products. According to Theorem 4.2.4.1, this is equivalent to the assertion that the full subcategory of $\mathcal{K}$an spanned by the k-truncated Kan complexes is stable under homotopy products and the formation of homotopy pullback squares. Both assertions can be verified easily by computing homotopy groups. $\square$

Remark 5.5.6.6. Let $p : \mathcal{C} \to \mathcal{D}$ be a coCartesian fibration of ∞-categories. Let C and C' be objects of $\mathcal{C}$, let $f : p(C') \to p(C)$ be a morphism in $\mathcal{C}$, and let $\overline{f} : C' \to C''$ be a p-coCartesian morphism lifting f. According to Proposition 2.4.4.2, we may identify $\mathrm{Map}_{\mathcal{C}_{p(C)}}(C'', C)$ with the homotopy fiber of $\mathrm{Map}_{\mathcal{C}}(C', C) \to \mathrm{Map}_{\mathcal{D}}(p(C'), p(C))$ over the base point determined by f. By examining the associated long exact sequences of homotopy groups (as f varies), we conclude that if C is a k-truncated object of the fiber $\mathcal{C}_{p(C)}$ and $p(C)$ is a k-truncated object of $\mathcal{D}$, then C is a k-truncated object of $\mathcal{C}$. This can be considered as a generalization of Lemma 2.4.4.7 (which treats the case $k = -2$).

Remark 5.5.6.7. Let $p : \mathcal{M} \to \Delta^1$ be a coCartesian fibration of simplicial sets, which we regard as a correspondence from the ∞-category $\mathcal{C} = p^{-1}\{0\}$ to $\mathcal{D} = p^{-1}\{1\}$. Suppose that D is a k-truncated object of $\mathcal{D}$. Remark 5.5.6.6 implies that D is a k-truncated object of $\mathcal{M}$. Let $C, C' \in \mathcal{C}$ and let $f : C \to D$ be a p-Cartesian morphism of $\mathcal{M}$. Then composition with f induces a homotopy equivalence $\mathrm{Map}_{\mathcal{C}}(C', C) \to \mathrm{Map}_{\mathcal{M}}(C', D)$; we conclude that C is a k-truncated object of $\mathcal{M}$.

Definition 5.5.6.8. We will say that a map $f : X \to Y$ of Kan complexes is k-*truncated* if the homotopy fibers of f (taken over any base point of Y) are k-truncated. We will say that a morphism $f : C \to D$ in an arbitrary ∞-category $\mathcal{C}$ is k-*truncated* if composition with f induces a k-truncated map $\mathrm{Map}_{\mathcal{C}}(E, C) \to \mathrm{Map}_{\mathcal{C}}(E, D)$ for every object $E \in \mathcal{C}$.

Remark 5.5.6.9. There is an apparent potential for ambiguity in Definition 5.5.6.8 in the case where $\mathcal{C}$ is an ∞-category whose objects are Kan complexes. However, there is no cause for concern: a map $f : X \to Y$ of Kan complexes is k-truncated if and only if it is k-truncated as a morphism in the ∞-category $\mathcal{S}$.

Remark 5.5.6.10. Let $f : C \to D$ and $g : E \to D$ be morphisms in an ∞-category $\mathcal{C}$, and let $\phi : \mathrm{Map}_{\mathcal{C}}(E, C) \to \mathrm{Map}_{\mathcal{C}}(E, D)$ be the map (in the homotopy category $\mathcal{H}$) given by compostion with f. Lemma 5.5.5.12 implies that the homotopy fiber of ϕ over g is homotopy equivalent to $\mathrm{Map}_{\mathcal{C}_{/D}}(f, g)$. Consequently, we deduce that $f : C \to D$ is k-truncated in the sense of Definition 5.5.6.8 if and only if it is k-truncated when viewed as an object of the ∞-category $\mathcal{D}_{/D}$.

Lemma 5.5.6.11. *Let $p : \mathcal{C} \to \mathcal{D}$ be a right fibration of ∞-categories and let $f : X \to Y$ be a morphism in $\mathcal{C}$. Then f is n-truncated if and only if $p(f) : p(X) \to p(Y)$ is n-truncated.*

Proof. The map $\mathcal{C}_{/Y} \to \mathcal{D}_{/p(Y)}$ is a trivial fibration and therefore an equivalence of ∞-categories. $\qquad\square$

Remark 5.5.6.12. A morphism $f : C \to D$ in an ∞-category $\mathcal{C}$ is k-truncated if and only if it is k-truncated when regarded as an object of the ∞-category $\mathcal{C}^{/D}$ (since the natural map $\mathcal{C}_{/D} \to \mathcal{C}^{/D}$ is an equivalence of ∞-categories). We may identify $\mathcal{C}^{/D}$ with $p^{-1}\{D\}$, where p denotes the evaluation map $\mathcal{C}^{\Delta^1} \to \mathcal{C}^{\{1\}}$. Corollary 2.4.7.12 implies that p is a coCartesian fibration. Consequently, Remark 5.5.6.7 translates into the following assertion: if

$$
\begin{array}{ccc}
C' & \xrightarrow{\ f'\ } & D' \\
\downarrow & & \downarrow \\
C & \xrightarrow{\ f\ } & D
\end{array}
$$

is a pullback diagram in $\mathcal{C}$ and f is k-truncated, then f' is k-truncated.

Example 5.5.6.13. A morphism $f : C \to D$ in an ∞-category $\mathcal{C}$ is (-2)-truncated if and only if it is an equivalence.

We will say that a morphism $f : C \to D$ is a *monomorphism* if it is (-1)-truncated; this is equivalent to the assertion that the functor $\mathcal{C}_{/f} \to \mathcal{C}_{/D}$ is fully faithful.

Lemma 5.5.6.14. *Let $\mathcal{C}$ be an ∞-category and $f : X \to Y$ a morphism in $\mathcal{C}$. Suppose that Y is n-truncated. Then X is n-truncated if and only if f is n-truncated.*

Proof. Unwinding the definitions, we reduce immediately to the following statement in classical homotopy theory: given a map $f : X \to Y$ of Kan complexes, where Y is n-truncated, X is n-truncated if and only if the homotopy fibers of f are n-truncated. This can be established easily using the long exact sequence of homotopy groups associated to f. $\qquad\square$

The following lemma gives a recursive characterization of the class of n-truncated morphisms:

Lemma 5.5.6.15. *Let $\mathcal{C}$ be an ∞-category which admits finite limits and let $k \geq -1$ be an integer. A morphism $f : C \to C'$ is k-truncated if and only if the diagonal $\delta : C \to C \times_{C'} C$ (which is well-defined up to homotopy) is $(k-1)$-truncated.*

Proof. For each object $D \in \mathcal{C}$, let $F_D : \mathcal{C} \to \mathcal{S}$ denote the functor corepresented by D. Then each F_D preserves finite limits, and a morphism f in $\mathcal{C}$ is k-truncated if and only if each $F_D(f)$ is a k-truncated morphism in $\mathcal{S}$. We may therefore reduce to the case where $\mathcal{C} = \mathcal{S}$. Without loss of generality, we may suppose that $f : C \to C'$ is a Kan fibration. Then Theorem 4.2.4.1 allows us to identify the fiber product $C \times_{C'} C$ in $\mathcal{S}$ with the same fiber product formed in the ordinary category $\mathcal{K}an$. We now reduce to the following assertion in classical homotopy theory (applied to the fibers of f): if X is a Kan complex, then X is k-truncated if and only if the homotopy fibers of the diagonal map $X \to X \times X$ are $(k-1)$-truncated. This can be proven readily by examining homotopy groups. $\square$

We immediately deduce the following:

Proposition 5.5.6.16. *Let $F : \mathcal{C} \to \mathcal{C}'$ be a left exact functor between ∞-categories which admit finite limits. Then F carries k-truncated objects into k-truncated objects and k-truncated morphisms into k-truncated morphisms.*

Proof. An object C is k-truncated if and only if the morphism $C \to 1$ to the final object is k-truncated. Since F preserves final objects, it suffices to prove the assertion concerning morphisms. Since F commutes with fiber products, Lemma 5.5.6.15 allows us to use induction on k, thereby reducing to the case where $k = -2$. But the (-2)-truncated morphisms are precisely the equivalences, and these are preserved by any functor. $\square$

We now specialize to the case of a *presentable* ∞-category $\mathcal{C}$. In this setting, we can construct an analogue of the Postnikov tower.

Lemma 5.5.6.17. *Let X be a Kan complex and let $k \geq -2$. The following conditions are equivalent:*

(1) *The Kan complex X is k-truncated.*

(2) *The diagonal map $\delta : X \to X^{\partial \Delta^{k+2}}$ is a homotopy equivalence.*

Proof. If $k = -2$, then $X^{\partial \Delta^{k+2}}$ is a point and the assertion is obvious. Assuming $k > -2$, we can choose a vertex v of $\partial \Delta^{k+2}$, which gives rise to an evaluation map $e : X^{\partial \Delta^{k+2}} \to X$. Since $e \circ \delta = \mathrm{id}_X$, (2) is equivalent to the assertion that e is a homotopy equivalence. We observe that e is a Kan fibration. For each x, let $Y_x = X^{\partial \Delta^{k+2}} \times_X \{x\}$ denote the fiber of e over the vertex x. Then Y_x has a canonical base point given by the constant map $\delta(x)$. Moreover, we have a natural isomorphism

$$\pi_i(Y_x, \delta(x)) \simeq \pi_{i+k+1}(X, x).$$

Condition (1) is equivalent to the assertion that $\pi_{i+k+1}(X, x)$ vanishes for all $i \geq 0$ and all $x \in X$. In view of the above isomorphism, this is equivalent to the assertion that each Y_x is contractible, which is true if and only if the Kan fibration e is trivial. $\square$

Proposition 5.5.6.18. *Let $\mathcal{C}$ be a presentable ∞-category, let $k \geq -2$, and let $\tau_{\leq k}\,\mathcal{C}$ denote the full subcategory of $\mathcal{C}$ spanned by the k-truncated objects. Then the inclusion $\tau_{\leq k}\,\mathcal{C} \subseteq \mathcal{C}$ has an accessible left adjoint, which we will denote by $\tau_{\leq k}$.*

Proof. Let $f : \partial \Delta^{k+2} \to \operatorname{Fun}(\mathcal{C}, \mathcal{C})$ denote the constant diagram taking the value $\operatorname{id}_{\mathcal{C}}$. Let $\overline{f} : (\partial \Delta^{k+2})^{\triangleright} \to \operatorname{Fun}(\mathcal{C}, \mathcal{C})$ be a colimit of f and let $F : \mathcal{C} \to \mathcal{C}$ be the image of the cone point under $\overline{f}$. Informally, F is given by the formula

$$C \mapsto C \otimes S^{k+1},$$

where S^{k+1} denotes the $(k+1)$-sphere and we regard $\mathcal{C}$ as tensored over spaces (see Remark 5.5.1.7).

Let $\overline{f}' : (\partial \Delta^{k+2})^{\triangleright} \to \mathcal{C}^{\mathcal{C}}$ be the constant diagram taking the value $\operatorname{id}_{\mathcal{C}}$. It follows that there exists an essentially unique map $\overline{f} \to \overline{f}'$ in $(\mathcal{C}^{\mathcal{C}})_{f/}$, which induces a natural transformation of functors $\alpha : F \to \operatorname{id}_{\mathcal{C}}$. Let $S = \{\alpha(C) : C \in \mathcal{C}\}$. Since F is a colimit of functors which preserve small colimits, F itself preserves small colimits (Lemma 5.5.2.3). Applying Proposition 5.1.2.2, we conclude that $\alpha : \mathcal{C} \to \operatorname{Fun}(\Delta^1, \mathcal{C})$ also preserves small colimits. Consequently, there exists a small subset $S_0 \subseteq S$ which generates S under colimits in $\operatorname{Fun}(\Delta^1, \mathcal{C})$. According to Proposition 5.5.4.15, the collection of S-local objects of $\mathcal{C}$ is an accessible localization of $\mathcal{C}$. It therefore suffices to prove that an object $X \in \mathcal{C}$ is S-local if and only if X is k-truncated.

According to Proposition 5.1.2.2, for each $C \in \mathcal{C}$ we may identify $F(C)$ with the colimit of the constant diagram $\partial \Delta^{k+2} \to \mathcal{C}$ taking the value C. Corollary 4.4.4.9 implies that we have a homotopy equivalence

$$\operatorname{Map}_{\mathcal{C}}(F(C), X) \simeq \operatorname{Map}_{\mathcal{C}}(C, X)^{\partial \Delta^{k+2}}.$$

The map $\alpha(C)$ induces a map

$$\alpha(C)_X : \operatorname{Map}_{\mathcal{C}}(C, X) \to \operatorname{Map}_{\mathcal{C}}(C, X)^{\partial \Delta^{k+2}}$$

which can be identified with the inclusion of $\operatorname{Map}_{\mathcal{C}}(C, X)$ as the space of constant maps from $\partial \Delta^{k+2}$ into $\operatorname{Map}_{\mathcal{C}}(C, X)$. According to Lemma 5.5.6.17, the map $\alpha(C)_X$ is an equivalence if and only if $\operatorname{Map}_{\mathcal{C}}(C, X)$ is k-truncated. Thus X is k-truncated if and only if X is S-local. $\square$

Remark 5.5.6.19. The notation of Proposition 5.5.6.18 is self-consistent in the following sense: the existence of the localization functor $\tau_{\leq k}$ implies that the collection of k-truncated objects of $\mathcal{C}$ may be identified with the essential image of $\tau_{\leq k}$.

Remark 5.5.6.20. If the ∞-category $\mathcal{C}$ is potentially unclear in context, then we will write $\tau_{\leq k}^{\mathcal{C}}$ for the truncation functor in $\mathcal{C}$. Note also that $\tau_{\leq k}^{\mathcal{C}}$ is well-defined up to equivalence (in fact, up to a contractible ambiguity).

Remark 5.5.6.21. It follows from Proposition 5.5.6.18 that if $\mathcal{C}$ is a presentable ∞-category, then the full subcategory $\tau_{\leq k}\,\mathcal{C}$ of k-truncated objects is also presentable. In particular, the ordinary category $\mathrm{Disc}(\mathcal{C})$ of discrete objects of $\mathcal{C}$ is a presentable category in the sense of Definition A.1.1.2.

Recall that, if $\mathcal{C}$ and $\mathcal{D}$ are ∞-categories, then $\mathrm{Fun}^L(\mathcal{C}, \mathcal{D})$ denotes the full subcategory of $\mathrm{Fun}(\mathcal{C}, \mathcal{D})$ spanned by those functors which are left adjoints. The following result gives a characterization of $\tau_{\leq n}\,\mathcal{C}$ by a universal mapping property:

Corollary 5.5.6.22. *Let $\mathcal{C}$ and $\mathcal{D}$ be presentable ∞-categories. Suppose that $\mathcal{D}$ is equivalent to an $(n + 1)$-category. Then composition with $\tau_{\leq n}$ induces an equivalence $s : \mathrm{Fun}^L(\tau_{\leq n}\,\mathcal{C}, \mathcal{D}) \to \mathrm{Fun}^L(\mathcal{C}, \mathcal{D})$.*

Proof. According to Proposition 5.5.4.20, the functor s is fully faithful. A functor $f : \mathcal{C} \to \mathcal{D}$ belongs to the essential image of s if and only if f has a right adjoint g which factors through $\tau_{\leq n}\,\mathcal{C}$. Since g preserves limits, it automatically carries $\mathcal{D} = \tau_{\leq n}\,\mathcal{D}$ into $\tau_{\leq n}\,\mathcal{C}$ (Proposition 5.5.6.16). $\square$

In classical homotopy theory, every space X can be recovered (up to weak homotopy equivalence) as the homotopy inverse limit of its Postnikov tower $\{\tau_{\leq n} X\}_{n \geq 0}$. The analogous statement is not true in an arbitrary presentable ∞-category but often holds in specific examples. We now make a general study of this phenomenon.

Definition 5.5.6.23. Let $\mathbf{Z}_{\geq 0}^{\infty}$ denote the union $\mathbf{Z}_{\geq 0} \cup \{\infty\}$ regarded as a linearly ordered set with largest element ∞. Let $\mathcal{C}$ be a presentable ∞-category. Recall that a *tower* in $\mathcal{C}$ is a functor $\mathrm{N}(\mathbf{Z}_{\geq 0}^{\infty})^{op} \to \mathcal{C}$, which we view as a diagram

$$X_{\infty} \to \cdots \to X_2 \to X_1 \to X_0.$$

A *Postnikov tower* is a tower with the property that for each $n \geq 0$, the map $X_{\infty} \to X_n$ exhibits X_n as an n-truncation of X_{∞}. We define a *pretower* to be a functor from $\mathrm{N}(\mathbf{Z}_{\geq 0})^{op} \to \mathcal{C}$. A *Postnikov pretower* is a pretower

$$\cdots \to X_2 \to X_1 \to X_0$$

which exhibits each X_n as an n-truncation of X_{n+1}. We let $\mathrm{Post}^+(\mathcal{C})$ denote the full subcategory of $\mathrm{Fun}(\mathrm{N}(\mathbf{Z}_{\geq 0}^{\infty})^{op}, \mathcal{C})$ spanned by the Postnikov towers, and $\mathrm{Post}(\mathcal{C})$ the full subcategory of $\mathrm{Fun}(\mathrm{N}(\mathbf{Z}_{\geq 0})^{op}, \mathcal{C})$ spanned by the Postnikov pretowers. We have an evident forgetful functor $\phi : \mathrm{Post}^+(\mathcal{C}) \to \mathrm{Post}(\mathcal{C})$. We will say that *Postnikov towers in $\mathcal{C}$ are convergent* if ϕ is an equivalence of ∞-categories.

Remark 5.5.6.24. Let $\mathcal{C}$ be a presentable ∞-category and let $\mathcal{E}$ denote the full subcategory of $\mathcal{C} \times \mathrm{N}(\mathbf{Z}_{\geq 0}^{\infty})^{op}$ spanned by those pairs (C, n) where $C \in \mathcal{C}$

is n-truncated (by convention, we agree that this condition is always satisfied where $C = \infty$). Then we have a coCartesian fibration $p : \mathcal{E} \to \mathrm{N}(\mathbf{Z}_{\geq 0}^{\infty})^{op}$, which classifies a tower of ∞-categories

$$\mathcal{C} \to \cdots \to \tau_{\leq 2}\,\mathcal{C} \overset{\tau_{\leq 1}}{\to} \tau_{\leq 1}\,\mathcal{C} \overset{\tau_{\leq 0}}{\to} \tau_{\leq 0}\,\mathcal{C}\,.$$

We can identify Postnikov towers with coCartesian sections of p (and Postnikov pretowers with coCartesian sections of the induced fibration

$$\mathcal{E} \times_{\mathrm{N}(\mathbf{Z}_{\geq 0}^{\infty})^{op}} \mathrm{N}(\mathbf{Z}_{\geq 0})^{op} \to \mathrm{N}(\mathbf{Z}_{\geq 0})^{op}).$$

According to Proposition 3.3.3.1, Postnikov towers in $\mathcal{C}$ converge if and only if the tower above exhibits $\mathcal{C}$ as the homotopy limit of the sequence of ∞-categories

$$\cdots \to \tau_{\leq 2}\,\mathcal{C} \to \tau_{\leq 1}\,\mathcal{C} \to \tau_{\leq 0}\,\mathcal{C}\,.$$

Remark 5.5.6.25. Let $\mathcal{C}$ be a presentable ∞-category and assume that Postnikov towers in $\mathcal{C}$ are convergent. Then every Postnikov tower in $\mathcal{C}$ is a limit diagram. Indeed, given objects $X, Y \in \mathcal{C}$, we have natural homotopy equivalences

$$\mathrm{Map}_{\mathcal{C}}(X, Y) \simeq \mathrm{holim}\,\mathrm{Map}_{\mathcal{C}}(\tau_{\leq n}X, \tau_{\leq n}Y) \simeq \mathrm{holim}\,\mathrm{Map}_{\mathcal{C}}(X, \tau_{\leq n}Y),$$

so that Y is the limit of the pretower $\{\tau_{\leq n}Y\}$.

Proposition 5.5.6.26. *Let $\mathcal{C}$ be a presentable ∞-category. Then Postnikov towers in $\mathcal{C}$ are convergent if and only if, for every tower $X : \mathrm{N}(\mathbf{Z}_{\geq 0}^{\infty})^{op} \to \mathcal{C}$, the following conditions are equivalent:*

(1) *The diagram X is a Postnikov tower.*

(2) *The diagram X is a limit in $\mathcal{C}$, and the restriction $X \mid \mathrm{N}(\mathbf{Z}_{\geq 0})^{op}$ is a Postnikov pretower.*

Proof. Let $\mathrm{Post}'(\mathcal{C})$ be the full subcategory of $\mathrm{Fun}(\mathrm{N}(\mathbf{Z}_{\geq 0}^{\infty})^{op}, \mathcal{C})$ spanned by those towers which satisfy condition (2). Using Proposition 4.3.2.15, we deduce that the restriction functor $\mathrm{Post}'(\mathcal{C}) \to \mathrm{Post}(\mathcal{C})$ is a trivial Kan fibration. If conditions (1) and (2) are equivalent, then $\mathrm{Post}'(\mathcal{C}) = \mathrm{Post}^{+}(\mathcal{C})$, so that Postnikov towers in $\mathcal{C}$ are convergent. Conversely, suppose that Postnikov towers in $\mathcal{C}$ are convergent. Using Remark 5.5.6.25, we deduce that $\mathrm{Post}^{+}(\mathcal{C}) \subseteq \mathrm{Post}'(\mathcal{C})$, so we have a commutative diagram

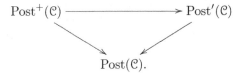

Since both of the vertical arrows are trivial Kan fibrations, we conclude that the inclusion $\mathrm{Post}^{+}(\mathcal{C}) \subseteq \mathrm{Post}'(\mathcal{C})$ is an equivalence, so that $\mathrm{Post}^{+}(\mathcal{C}) = \mathrm{Post}'(\mathcal{C})$. This proves that (1) $\Leftrightarrow$ (2). $\square$

Remark 5.5.6.27. Let $\mathcal{C}$ be a presentable ∞-category. We will say that a tower $X : \mathrm{N}(\mathbf{Z}_{\geq 0}^{\infty})^{op} \to \mathcal{X}$ is *highly connected* if, for every $n \geq 0$, there exists an integer k such that the induced map $\tau_{\leq n} X(\infty) \to \tau_{\leq n} X(k')$ is an equivalence for $k' \geq k$. We will say that a pretower $Y : \mathrm{N}(\mathbf{Z}_{\geq 0}) \to \mathcal{X}$ is *highly connected* if, for every $n \geq 0$, there exists an integer k such that the map $\tau_{\leq n} Y(k'') \to \tau_{\leq n} Y(k')$ is an equivalence for $k'' \geq k' \geq k$. It is clear that every Postnikov (pre)tower is highly connected. Conversely, if X is a highly connected tower and its underlying pretower is highly connected, then X is a Postnikov tower. Indeed, for each $n \geq 0$ we can choose $k \geq n$ such that the map $\tau_{\leq n} X(\infty) \to \tau_{\leq n} X(k)$ is an equivalence. Since X is a Postnikov pretower, this induces an equivalence $\tau_{\leq n} X(\infty) \simeq X(n)$. Consequently, to establish the implication (2) ⇒ (1) in the criterion of Proposition 5.5.6.26, it suffices to verify the following:

(∗) Let $X : \mathrm{N}(\mathbf{Z}_{\geq}^{\infty}) \to \mathcal{C}$ be a tower in $\mathcal{C}$. Assume that X is a limit diagram and that the underlying pretower is highly connected. Then X is highly connected.

In §7.2.1, we will apply this criterion to prove that Postnikov towers are convergent in a large class of ∞-topoi.

We conclude this section with a useful compatibility property between truncation functors in different ∞-categories:

Proposition 5.5.6.28. *Let $\mathcal{C}$ and $\mathcal{D}$ be presentable ∞-categories and let $F : \mathcal{C} \to \mathcal{D}$ be a left exact presentable functor. Then there is an equivalence of functors $F \circ \tau_{\leq k}^{\mathcal{C}} \simeq \tau_{\leq k}^{\mathcal{D}} \circ F$.*

Proof. Since F is left exact, it restricts to a functor from $\tau_{\leq k} \mathcal{C}$ to $\tau_{\leq k} \mathcal{D}$ by Proposition 5.5.6.16. We therefore have a diagram

$$
\begin{array}{ccc}
\mathcal{C} & \xrightarrow{\ F\ } & \mathcal{D} \\
\downarrow{\scriptstyle \tau_{\leq k}^{\mathcal{C}}} & & \downarrow{\scriptstyle \tau_{\leq k}^{\mathcal{D}}} \\
\tau_{\leq k}\,\mathcal{C} & \xrightarrow{\ F\ } & \tau_{\leq k}\,\mathcal{D}
\end{array}
$$

which we wish to prove is commutative up to homotopy. Let G denote a right adjoint to F; then G is left exact and so induces a functor $\tau_{\leq k} \mathcal{D} \to \tau_{\leq k} \mathcal{C}$. Using Proposition 5.2.2.6, we can reduce to proving that the associated diagram of right adjoints

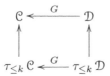

commutes up to homotopy, which is obvious (since the diagram strictly commutes). $\qquad\square$

5.5.7 Compactly Generated ∞-Categories

Definition 5.5.7.1. Let κ be a regular cardinal. We will say that an ∞-category $\mathcal{C}$ is κ-*compactly generated* if it is presentable and κ-accessible. When $\kappa = \omega$, we will simply say that $\mathcal{C}$ is *compactly generated*.

The proof of Theorem 5.5.1.1 shows that an ∞-category $\mathcal{C}$ is κ-compactly generated if and only if there exists a small ∞-category $\mathcal{D}$ which admits κ-small colimits and an equivalence $\mathcal{C} \simeq \mathrm{Ind}_\kappa(\mathcal{D})$. In fact, we can choose $\mathcal{D}$ to be (a minimal model of) the ∞-category of κ-compact objects of $\mathcal{C}$. We would like to assert that this construction establishes an equivalence between two sorts of ∞-categories. In order to make this precise, we need to introduce the appropriate notion of functor between κ-compactly generated ∞-categories.

Proposition 5.5.7.2. *Let κ be a regular cardinal and let $\mathcal{C} \underset{G}{\overset{F}{\rightleftarrows}} \mathcal{D}$ be a pair of adjoint functors, where $\mathcal{C}$ and $\mathcal{D}$ admit small κ-filtered colimits.*

(1) *If G is κ-continuous, then F carries κ-compact objects of $\mathcal{C}$ to κ-compact objects of $\mathcal{D}$.*

(2) *Conversely, if $\mathcal{C}$ is κ-accessible and F preserves κ-compactness, then G is κ-continuous.*

Proof. Suppose first that G is κ-continuous and let $C \in \mathcal{C}$ be a κ-compact object. Let $e : \mathcal{C} \to \widehat{\mathcal{S}}$ be a functor corepresented by C. Then $e \circ G : \mathcal{D} \to \widehat{\mathcal{S}}$ is corepresented by $F(C)$. Since e and G are κ-continuous, so is $e \circ G$; this proves (1).

Conversely, suppose that F preserves κ-compact objects and that $\mathcal{C}$ is κ-accessible. Without loss of generality, we may suppose that there is a small ∞-category $\mathcal{C}'$ such that $\mathcal{C} = \mathrm{Ind}_\kappa(\mathcal{C}') \subseteq \mathcal{P}(\mathcal{C}')$. We wish to prove that G is κ-continuous. Since $\mathrm{Ind}_\kappa(\mathcal{C}')$ is stable under κ-filtered colimits in $\mathcal{P}(\mathcal{C}')$, it will suffice to prove that the composite map

$$\theta : \mathcal{D} \overset{G}{\to} \mathcal{C} \subseteq \mathcal{P}(\mathcal{C}')$$

is κ-continuous. In view of Proposition 5.1.2.2, it will suffice to prove that for every object $C \in \mathcal{C}'$, the composition of θ with evaluation at C is a κ-continuous functor. We conclude by observing that this functor is corepresentable by the image under F of $j(C) \in \mathcal{C}$ (here $j : \mathcal{C}' \to \mathrm{Ind}_\kappa(\mathcal{C})$ denotes the Yoneda embedding). $\qquad\qquad\qquad\qquad\qquad\qquad\square$

Corollary 5.5.7.3. *Let $\mathcal{C}$ be a κ-compactly generated ∞-category and let $L : \mathcal{C} \to \mathcal{C}$ be a localization functor. The following conditions are equivalent:*

(1) *The functor L is κ-continuous.*

(2) *The full subcategory $L\,\mathcal{C} \subseteq \mathcal{C}$ is stable under κ-filtered colimits.*

Suppose that these conditions are satisfied. Then

(3) *The functor L carries κ-compact objects of $\mathcal{C}$ to κ-compact objects of $L\,\mathcal{C}$.*

(4) *The ∞-category $L\,\mathcal{C}$ is κ-compactly generated.*

(5) *An object $D \in L\,\mathcal{C}$ is κ-compact (in $L\,\mathcal{C}$) if and only if there exists a compact object $C \in \mathcal{C}$ such that D is a retract of LC.*

Proof. Suppose that (1) is satisfied. Let $p : K \to L\,\mathcal{C}$ be a κ-filtered diagram. Then the natural transformation $p \to Lp$ is an equivalence. Using (1), we conclude that the induced map $\varinjlim(p) \to L \varinjlim(p)$ is an equivalence, so that $\varinjlim(p) \in L\,\mathcal{C}$. This proves (2).

Conversely, if (2) is satisfied, then the inclusion $L\,\mathcal{C} \subseteq \mathcal{C}$ is κ-continuous, so that $L : \mathcal{C} \to \mathcal{C}$ is a composition of κ-continuous functors

$$\mathcal{C} \xrightarrow{L} L\,\mathcal{C} \to \mathcal{C},$$

which proves (1).

Assume that (1) and (2) are satisfied. Then L is accessible, so that $L\,\mathcal{C}$ is a presentable ∞-category. Assertion (3) follows from Proposition 5.5.7.2. Let $D \in L\,\mathcal{C}$. Since $\mathcal{C}$ is κ-compactly generated, D can be written as the colimit of a κ-filtered diagram $p : K \to \mathcal{C}$ taking values in the κ-compact objects of $\mathcal{C}$. Then $D \simeq LD$ can be written as the colimit of $L \circ p$, which takes values among the κ-compact objects of $L\,\mathcal{C}$. This proves (4). If D is a κ-compact object of $\mathcal{D}$, then we deduce that the identity map $\mathrm{id}_D : D \to D$ factors through $(L \circ p)(k)$ for some vertex $k \in K$, which proves (5). $\qquad\square$

Corollary 5.5.7.4. *Let $\mathcal{C}$ be a κ-compactly generated ∞-category and let $n \geq -2$. Then*

(1) *The full subcategory $\tau_{\leq n}\,\mathcal{C}$ is stable under κ-filtered colimits in $\mathcal{C}$.*

(2) *The truncation functor $\tau_{\leq n} : \mathcal{C} \to \mathcal{C}$ is κ-continuous.*

(3) *The truncation functor $\tau_{\leq n}$ carries compact objects of $\mathcal{C}$ to compact objects of $\mathcal{C}_{\leq n}$.*

(4) *The full subcategory $\tau_{\leq n}\,\mathcal{C}$ is κ-compactly generated.*

(5) *An object $C \in \tau_{\leq n}\,\mathcal{C}$ is compact (in $\tau_{\leq n}\,\mathcal{C}$) if and only if there exists a compact object $C' \in \mathcal{C}$ such that C is a retract of $\tau_{\leq n}C'$.*

Proof. Corollary 5.5.7.3 shows that condition (1) implies (2), (3), (4), and (5). Consequently, it will suffice to prove that (1) is satisfied.

Let C be an object of $\mathcal{C}$. We will show that C is n-truncated if and only if the space $\mathrm{Map}_{\mathcal{C}}(D, C)$ is n-truncated for every κ-compact object $D \in \mathcal{C}$. The "only if" direction is obvious. For the converse, let $F_C : \mathcal{C}^{op} \to \mathcal{S}$ be the functor represented by C and let $\mathcal{C}' \subseteq \mathcal{C}$ be the full subcategory of $\mathcal{C}$ spanned by those objects D such that $F_C(D)$ is n-truncated. Since F_C preserves limits, $\mathcal{C}'$ is stable under colimits in $\mathcal{C}$. If $\mathcal{C}'$ contains every κ-compact object of $\mathcal{C}$, then $\mathcal{C}' = \mathcal{C}$ (since $\mathcal{C}$ is κ-compactly generated).

Now suppose that D is a κ-compact object of $\mathcal{C}$, let $G_D : \mathcal{C} \to \mathcal{S}$ be the functor corepresented by D, and let $\mathcal{C}(D) \subseteq \mathcal{C}$ be the full subcategory of $\mathcal{C}$ spanned by those objects C for which $G_D(C)$ is n-truncated. Then $\tau_{\leq n}\, \mathcal{C} = \bigcap_D \mathcal{C}(D)$. To complete the proof, it will suffice to show that each $\mathcal{C}(D)$ is stable under κ-filtered colimits. Since G_D is κ-continuous, it suffices to observe that $\tau_{\leq n}\, \mathcal{S}$ is stable under κ-filtered colimits in $\mathcal{S}$. $\qquad\square$

Definition 5.5.7.5. If κ is a regular cardinal, we let $\mathrm{Pr}^{\mathrm{R}}_{\kappa}$ denote the full subcategory of $\widehat{\mathcal{Cat}}_{\infty}$ whose objects are κ-compactly generated ∞-categories and whose morphisms are κ-continuous limit-preserving functors.

Proposition 5.5.7.6. *The ∞-category $\mathrm{Pr}^{\mathrm{R}}_{\kappa}$ admits small limits, and the inclusion $\mathrm{Pr}^{\mathrm{R}}_{\kappa} \subseteq \widehat{\mathcal{Cat}}_{\infty}$ preserves small limits.*

Proof. In view of Theorem 5.5.3.18, the only nontrivial point is to verify that if $p : K \to \mathrm{Pr}^{\mathrm{R}}_{\kappa}$ is a diagram of κ-compactly generated ∞-categories $\{\mathcal{C}_\alpha\}$, then the limit $\mathcal{C} = \varprojlim(p)$ in $\widehat{\mathcal{Cat}}_{\infty}$ is κ-compactly generated. In other words, we must show that $\mathcal{C}$ is generated under colimits by its κ-compact objects.

For each vertex α of K, let

$$\mathcal{C}_\alpha \underset{G_\alpha}{\overset{F_\alpha}{\rightleftarrows}} \mathcal{C}$$

denote the corresponding adjunction. Lemma 6.3.3.6 implies that the identity functor $\mathrm{id}_{\mathcal{C}}$ can be obtained as the colimit of a diagram $q : K \to \mathrm{Fun}(\mathcal{C}, \mathcal{C})$, where $q(\alpha) \simeq F_\alpha \circ G_\alpha$. In particular, $\mathcal{C}$ is generated (under small colimits) by the essential images of the functors F_α. Since each $\mathcal{C}_\alpha$ is generated under colimits by κ-compact objects, and the functors F_α preserve colimits and κ-compact objects (Proposition 5.5.7.2), we conclude that $\mathcal{C}$ is generated under colimits by its κ-compact objects, as desired. $\qquad\square$

Notation 5.5.7.7. Let κ be a regular cardinal. We let $\mathrm{Pr}^{\mathrm{L}}_{\kappa}$ denote the full subcategory of $\widehat{\mathcal{Cat}}_{\infty}$ whose objects are κ-compactly generated ∞-categories and whose morphisms are functors which preserve small colimits and κ-compact objects. In view of Proposition 5.5.7.2, the equivalence $\mathrm{Pr}^{\mathrm{L}} \simeq (\mathrm{Pr}^{\mathrm{R}})^{op}$ of Corollary 5.5.3.4 restricts to an equivalence $\mathrm{Pr}^{\mathrm{L}}_{\kappa} \simeq (\mathrm{Pr}^{\mathrm{R}}_{\kappa})^{op}$.

Let $\widehat{\mathcal{Cat}}_{\infty}^{Rex(\kappa)}$ denote the subcategory of $\widehat{\mathcal{Cat}}_{\infty}$ whose objects are (not necessarily small) ∞-categories which admit κ-small colimits and whose morphisms are functors which preserve κ-small colimits, and let $\mathcal{Cat}_{\infty}^{Rex(\kappa)} = \widehat{\mathcal{Cat}}_{\infty}^{Rex(\kappa)} \cap \mathcal{Cat}_{\infty}$.

Proposition 5.5.7.8. *Let κ be a regular cardinal and let*

$$\theta : \mathrm{Pr}^{\mathrm{L}}_{\kappa} \to \widehat{\mathcal{Cat}}_{\infty}^{Rex(\kappa)}$$

be the nerve of the simplicial functor which associates to a κ-compactly generated ∞-category $\mathcal{C}$ the full subcategory $\mathcal{C}^{\kappa} \subseteq \mathcal{C}$ spanned by the κ-compact objects of $\mathcal{C}$. Then

(1) *The functor θ is fully faithful.*

(2) *The essential image of θ consists precisely of those objects of $\widehat{\mathrm{Cat}}_\infty$ which are essentially small and idempotent complete.*

Proof. Combine Propositions 5.4.2.17 and 5.5.1.9. $\qquad\square$

Remark 5.5.7.9. If $\kappa > \omega$, then Corollary 4.4.5.16 shows that the hypothesis of idempotent completeness in (2) is superfluous.

The proof of Proposition 5.4.2.19 yields the following analogue:

Proposition 5.5.7.10. *Let κ be a regular cardinal. The functor $\mathrm{Ind}_\kappa : \mathrm{Cat}_\infty \to \mathrm{Acc}_\kappa$ exhibits $\mathcal{P}r_\kappa^L$ as a localization of $\mathrm{Cat}_\infty^{Rex(\kappa)}$. If $\kappa > \omega$, then Ind_κ induces an equivalence of ∞-categories $\mathrm{Cat}_\infty^{Rex(\kappa)} \to \mathcal{P}r_\kappa^L$.*

Proof. The only additional ingredient needed is the following observation: if $\mathcal{C}$ is an ∞-category which admits κ-small colimits, then the idempotent completion $\mathcal{C}'$ of $\mathcal{C}$ also admits κ-small colimits. To prove this, we observe that $\mathcal{C}'$ can be identified with the collection of κ-compact objects of $\mathrm{Ind}_\kappa(\mathcal{C})$ (Lemma 5.4.2.4). Since $\mathcal{C}$ admits all small colimits (Theorem 5.5.1.1), we conclude that $\mathcal{C}'$ admits κ-small colimits. $\qquad\square$

We conclude this section with a remark about the structure of the ∞-category $\mathrm{Cat}_\infty^{Rex(\kappa)}$.

Proposition 5.5.7.11. *Let κ be a regular cardinal. Then the ∞-category $\mathrm{Cat}_\infty^{Rex(\kappa)}$ admits small κ-filtered colimits and the inclusion $\mathrm{Cat}_\infty^{Rex(\kappa)} \subseteq \mathrm{Cat}_\infty$ preserves small κ-filtered colimits.*

Proof. Let $\mathcal{J}$ be a small κ-filtered ∞-category and let $p : \mathcal{J} \to \mathrm{Cat}_\infty^{Rex(\kappa)}$ be a diagram. Let $\mathcal{C}$ be a colimit of the induced diagram $\mathcal{J} \to \mathrm{Cat}_\infty$. To complete the proof we must prove the following:

(*i*) The ∞-category $\mathcal{C}$ admits κ-small colimits.

(*ii*) For each $I \in \mathcal{J}$, the associated functor $p(I) \to \mathcal{C}$ preserves κ-small colimits.

(*iii*) Let $f : \mathcal{C} \to \mathcal{D}$ be an arbitrary functor. If each of the compositions $p(I) \to \mathcal{C} \to \mathcal{D}$ preserves κ-small colimits, then f preserves κ-small colimits.

Since $\mathcal{J}$ is κ-filtered, any κ-small diagram in $\mathcal{C}$ factors through one of the maps $p(I) \to \mathcal{C}$ (Proposition 5.4.1.2). Thus (*ii*) $\Rightarrow$ (*i*) and (*ii*) $\Rightarrow$ (*iii*). To prove (*ii*), we first use Proposition 5.3.1.16 to reduce to the case where $\mathcal{J} \simeq N(A)$, where A is a κ-filtered partially ordered set. Using Proposition 4.2.4.4, we can reduce to the case where p is the nerve of a functor from $q : A \to \mathrm{Set}_\Delta$. In view of Theorem 4.2.4.1, we can identify $\mathcal{C}$ with a homotopy colimit of q. Since the collection of categorical equivalences is stable under filtered colimits, we can assume that $\mathcal{C}$ is actually the filtered colimit of a family of ∞-categories $\{\mathcal{C}_\alpha\}_{\alpha \in A}$.

Let K be a κ-small simplicial set and let $\overline{g}_\alpha : K^\triangleright \to \mathcal{C}_\alpha$ be a colimit diagram. We wish to show that the induced map $\overline{g} : K^\triangleright \to \mathcal{C}$ is a colimit diagram. Let $g = \overline{g}|K$; we need to show that the map $\theta : \mathcal{C}_{\overline{g}/} \to \mathcal{C}_{g/}$ is a trivial Kan fibration. We observe that θ is a filtered colimit of maps $\theta_\beta : (\mathcal{C}_\beta)_{\overline{g}_\beta/} \to (\mathcal{C}_\beta)_{g_\beta/}$, where β ranges over the set $\{\beta \in A : \beta \geq \alpha\}$. Using the fact that each of the associated maps $\mathcal{C}_\alpha \to \mathcal{C}_\beta$ preserves κ-small colimits, we conclude that each θ_β is a trivial fibration, so that θ is a trivial fibration, as desired. $\qquad\qquad\qquad\qquad\qquad\qquad\qquad\qquad\qquad\qquad\qquad\qquad\qquad\quad\square$

5.5.8 Nonabelian Derived Categories

According to Corollary 4.2.3.11, we can analyze arbitrary colimits in an ∞-category $\mathcal{C}$ in terms of finite colimits and filtered colimits. In particular, suppose that $\mathcal{C}$ admits finite colimits and that we construct a new ∞-category $\mathrm{Ind}(\mathcal{C})$ by formally adjoining filtered colimits to $\mathcal{C}$. Then $\mathrm{Ind}(\mathcal{C})$ admits all small colimits (Theorem 5.5.1.1), and the Yoneda embedding $\mathcal{C} \to \mathrm{Ind}(\mathcal{C})$ preserves finite colimits (Proposition 5.3.5.14). Moreover, we can identify $\mathrm{Ind}(\mathcal{C})$ with the ∞-category of functors $\mathcal{C}^{op} \to \mathcal{S}$ which carry finite colimits in $\mathcal{C}$ to finite limits in $\mathcal{S}$. In this section, we will introduce a variation on the same theme. Instead of assuming $\mathcal{C}$ admits *all* finite colimits, we will assume only that $\mathcal{C}$ admits finite coproducts. We will construct a coproduct-preserving embedding of $\mathcal{C}$ into a larger ∞-category $\mathcal{P}_\Sigma(\mathcal{C})$ which admits all small colimits. Moreover, we can characterize $\mathcal{P}_\Sigma(\mathcal{C})$ as the ∞-category obtained from $\mathcal{C}$ by formally adjoining colimits of *sifted diagrams* (Proposition 5.5.8.15).

Our first goal in this section is to introduce the notion of a *sifted* simplicial set. We begin with a bit of motivation. Let $\mathcal{C}$ denote the (ordinary) category of groups. Then $\mathcal{C}$ admits arbitrary colimits. However, colimits of diagrams in $\mathcal{C}$ can be very difficult to analyze even if the diagram itself is quite simple. For example, the coproduct of a pair of groups G and H is the *amalgamated product* $G \star H$. The group $G \star H$ is typically very complicated even when G and H are not. For example, the amalgamated product $\mathbf{Z}/2\mathbf{Z} \star \mathbf{Z}/3\mathbf{Z}$ is isomorphic to the arithmetic group $\mathrm{PSL}_2(\mathbf{Z})$. In general, $G \star H$ is much larger than the coproduct $G \coprod H$ of the underlying sets of G and H. In other words, the forgetful functor $U : \mathcal{C} \to \mathrm{Set}$ does not preserve coproducts. However, U does preserve *some* colimits: for example, the colimit of a sequence of groups

$$G_0 \to G_1 \to \cdots$$

can be obtained by taking the colimit of the underlying sets and equipping the result with an appropriate group structure.

The forgetful functor U from groups to sets preserves another important type of colimit: namely, the formation of quotients by equivalence relations. If G is a group, then a subgroup $R \subseteq G \times G$ is an equivalence relation on G if and only if there exists a normal subgroup $H \subseteq G$ such that $R = \{(g, g') :$

$g^{-1}g' \in H$}. In this case, the set of R-equivalence classes in G is in bijection with the quotient G/H, which inherits a group structure from G. In other words, the quotient of G by the equivalence relation R can be computed either in the category of groups or in the category of sets; the result is the same.

Each of the examples given above admits a generalization: the colimit of a sequence is a special case of a *filtered colimit*, and the quotient by an equivalence relation is a special case of a *reflexive coequalizer*. The forgetful functor $\mathcal{C} \to \mathrm{Set}$ preserves filtered colimits and reflexive coequalizers; moreover, the same is true if we replace the category of groups by any other category of sets with some sort of finitary algebraic structure (for example, abelian groups or commutative rings). The following definition, which is taken from [66], is an attempt to axiomatize the essence of the situation:

Definition 5.5.8.1 ([66]). A simplicial set K is *sifted* if it satisfies the following conditions:

(1) The simplicial set K is nonempty.

(2) The diagonal map $K \to K \times K$ is cofinal.

Warning 5.5.8.2. In [66], Rosicki uses the term "homotopy sifted" to describe the analogue of Definition 5.5.8.1 for simplicial categories and reserves the term "sifted" for analogous notion in the setting of ordinary categories. There is some danger of confusion with our terminology: if $\mathcal{C}$ is an ordinary category and $\mathrm{N}(\mathcal{C})$ is sifted (in the sense of Definition 5.5.8.1), then $\mathcal{C}$ is sifted in the sense of [66]. However, the converse is false in general.

Example 5.5.8.3. Every filtered ∞-category is sifted (this follows from Proposition 5.3.1.20).

Lemma 5.5.8.4. *The simplicial set* $\mathrm{N}(\mathbf{\Delta})^{op}$ *is sifted.*

Proof. Since $\mathrm{N}(\mathbf{\Delta})^{op}$ is clearly nonempty, it will suffice to show that the diagonal map $\mathrm{N}(\mathbf{\Delta})^{op} \to \mathrm{N}(\mathbf{\Delta})^{op} \times \mathrm{N}(\mathbf{\Delta})^{op}$ is cofinal. According to Theorem 4.1.3.1, this is equivalent to the assertion that for every object $([m], [n]) \in \mathbf{\Delta} \times \mathbf{\Delta}$, the category

$$\mathcal{C} = \mathbf{\Delta}_{/[m]} \times_{\mathbf{\Delta}} \mathbf{\Delta}_{/[n]}$$

has weakly contractible nerve. Let $\mathcal{C}^0$ be the full subcategory of $\mathcal{C}$ spanned by those objects which correspond to *monomorphisms* of partially ordered sets $J \to [m] \times [n]$. The inclusion of $\mathcal{C}^0$ into $\mathcal{C}$ has a left adjoint, so the inclusion $\mathrm{N}(\mathcal{C}^0) \subseteq \mathrm{N}(\mathcal{C})$ is a weak homotopy equivalence. It will therefore suffice to show that $\mathrm{N}(\mathcal{C}^0)$ is weakly contractible. We now observe that $\mathrm{N}(\mathcal{C}^0)$ can be identified with the barycentric subdivision of $\Delta^m \times \Delta^n$ and is therefore weakly homotopy equivalent to $\Delta^m \times \Delta^n$ and so weakly contractible. □

Remark 5.5.8.5. The formation of the geometric realizations of simplicial objects should be thought of as the ∞-categorical analogue of the formation of reflexive coequalizers.

Our next pair of results captures some of the essential features of the theory of sifted simplicial sets:

Proposition 5.5.8.6. *Let K be a sifted simplicial set, let $\mathcal{C}$, $\mathcal{D}$, and $\mathcal{E}$ be ∞-categories which admit K-indexed colimits, and let $f : \mathcal{C} \times \mathcal{D} \to \mathcal{E}$ be a map which preserves K-indexed colimits separately in each variable. Then f preserves K-indexed colimits.*

Proof. Let $p : K \to \mathcal{C}$ and $q : K \to \mathcal{D}$ be diagrams indexed by a small simplicial set K and let $\delta : K \to K \times K$ be the diagonal map. Using the fact that f preserves K-indexed colimits separately in each variable and Lemma 5.5.2.3, we conclude that $\varinjlim(f \circ (p \times q))$ is a colimit for the diagram $f \circ (p \times q) \circ \delta$. Consequently, f preserves K-indexed colimits provided that the diagonal δ is cofinal. We conclude by invoking the assumption that K is sifted. $\qquad\square$

Proposition 5.5.8.7. *Let K be a sifted simplicial set. Then K is weakly contractible.*

Proof. Choose a vertex x in K. According to Whitehead's theorem, it will suffice to show that for each $n \geq 0$, the homotopy set $\pi_n(|K|, x)$ consists of a single element. Let $\delta : K \to K \times K$ be the diagonal map. Since δ is cofinal, Proposition 4.1.1.3 implies that the induced map

$$\pi_n(|K|, x) \to \pi_n(|K \times K|, \delta(x)) \simeq \pi_n(|K|, x) \times \pi_n(|K|, x)$$

is bijective. Since $\pi_n(|K|, x)$ is nonempty, we conclude that it is a singleton. $\qquad\square$

We now return to the problem introduced in the beginning of this section.

Definition 5.5.8.8. Let $\mathcal{C}$ be a small ∞-category which admits finite coproducts. We let $\mathcal{P}_\Sigma(\mathcal{C})$ denote the full subcategory of $\mathcal{P}(\mathcal{C})$ spanned by those functors $\mathcal{C}^{op} \to \mathcal{S}$ which preserve finite products.

Remark 5.5.8.9. The ∞-categories of the form $\mathcal{P}_\Sigma(\mathcal{C})$ have been studied in [66], where they are called *homotopy varieties*. Many of the results proven below can also be found in [66].

Proposition 5.5.8.10. *Let $\mathcal{C}$ be a small ∞-category which admits finite coproducts. Then*

(1) *The ∞-category $\mathcal{P}_\Sigma(\mathcal{C})$ is an accessible localization of $\mathcal{P}(\mathcal{C})$.*

(2) *The Yoneda embedding $j : \mathcal{C} \to \mathcal{P}(\mathcal{C})$ factors through $\mathcal{P}_\Sigma(\mathcal{C})$. Moreover, j carries finite coproducts in $\mathcal{C}$ to finite coproducts in $\mathcal{P}_\Sigma(\mathcal{C})$.*

(3) *Let $\mathcal{D}$ be a presentable ∞-category and let*

$$\mathcal{P}(\mathcal{C}) \underset{G}{\overset{F}{\rightleftarrows}} \mathcal{D}$$

be a pair of adjoint functors. Then G factors through $\mathcal{P}_\Sigma(\mathcal{C})$ if and only if $f = F \circ j : \mathcal{C} \to \mathcal{D}$ preserves finite coproducts.

(4) *The full subcategory $\mathcal{P}_\Sigma(\mathcal{C}) \subseteq \mathcal{P}(\mathcal{C})$ is stable under sifted colimits.*

(5) *Let $L : \mathcal{P}(\mathcal{C}) \to \mathcal{P}_\Sigma(\mathcal{C})$ be a left adjoint to the inclusion. Then L preserves sifted colimits (when regarded as a functor from $\mathcal{P}(\mathcal{C})$ to itself).*

(6) *The ∞-category $\mathcal{P}_\Sigma(\mathcal{C})$ is compactly generated.*

Before giving the proof, we need a preliminary result concerning the interactions between products and sifted colimits.

Lemma 5.5.8.11. *Let K be a sifted simplicial set. Let X be an ∞-category which admits finite products and K-indexed colimits and suppose that the formation of products in X preserves K-indexed colimits separately in each variable. Then the colimit functor $\varinjlim : \mathrm{Fun}(K, X) \to X$ preserves finite products.*

Remark 5.5.8.12. The hypotheses of Lemma 5.5.8.11 are satisfied when X is the ∞-category $\mathcal{S}$ of spaces: see Lemma 6.1.3.14. More generally, Lemma 5.5.8.11 applies whenever the ∞-category X is an ∞-topos (see Definition 6.1.0.2).

Proof. Since the simplicial set K is weakly contractible (Proposition 5.5.8.7), Corollary 4.4.4.9 implies that the functor $\varinjlim$ preserves final objects. To complete the proof, it will suffice to show that the functor $\varinjlim$ preserves pairwise products. Let X and Y be objects of $\mathrm{Fun}(K, X)$. We wish to prove that the canonical map

$$\varinjlim(X \times Y) \to \varinjlim(X) \times \varinjlim(Y)$$

is an equivalence. In other words, we must show that the formation of products commutes with K-indexed colimits, which follows immediately by applying Proposition 5.5.8.6 to the Cartesian product functor $X \times X \to X$. □

Proof of Proposition 5.5.8.10. Assertion (1) is an immediate consequence of Lemmas 5.5.4.17, 5.5.4.18, and 5.5.4.19. To prove (2), it will suffice to show that for every representable functor $e : \mathcal{P}_\Sigma(\mathcal{C})^{op} \to \mathcal{S}$, the composition

$$\mathcal{C}^{op} \xrightarrow{j^{op}} \mathcal{P}_\Sigma(\mathcal{C})^{op} \xrightarrow{e} \mathcal{S}$$

preserves finite products (Proposition 5.1.3.2). This is obvious since the composition can be identified with the object of $\mathcal{P}_\Sigma(\mathcal{C}) \subseteq \mathrm{Fun}(\mathcal{C}^{op}, \mathcal{S})$ representing e.

We next prove (3). We note that f preserves finite coproducts if and only if, for every object $D \in \mathcal{D}$, the composition

$$\mathcal{C}^{op} \xrightarrow{f^{op}} \mathcal{D}^{op} \xrightarrow{e_D} \mathcal{S}$$

preserves finite products, where e_D denotes the functor represented by D. This composition can be identified with $G(D)$, so that f preserves finite coproducts if and only if G factors through $\mathcal{P}_\Sigma(\mathcal{C})$.

Assertion (4) is an immediate consequence of Lemma 5.5.8.11 and Remark 5.5.8.12, and (5) follows formally from (4). To prove (6), we first observe that $\mathcal{P}(\mathcal{C})$ is compactly generated (Proposition 5.3.5.12). Let $\mathcal{E} \subseteq \mathcal{P}(\mathcal{C})$ be the full subcategory spanned by the compact objects and let $L : \mathcal{P}(\mathcal{C}) \to \mathcal{P}_\Sigma(\mathcal{C})$ be a localization functor. Since $\mathcal{E}$ generates $\mathcal{P}(\mathcal{C})$ under filtered colimits, $L(\mathcal{D})$ generates $\mathcal{P}_\Sigma(\mathcal{C})$ under filtered colimits. Consequently, it will suffice to show that for each $E \in \mathcal{E}$, the object $LE \in \mathcal{P}_\Sigma(\mathcal{C})$ is compact. Let $f : \mathcal{P}_\Sigma(\mathcal{C}) \to \mathcal{S}$ be the functor corepresented by LE and let $f' : \mathcal{P}(\mathcal{C}) \to \mathcal{S}$ be the functor corepresented by E. Then the map $E \to LE$ induces an equivalence $f \to f' | \mathcal{P}_\Sigma(\mathcal{C})$. Since f' is continuous and $\mathcal{P}_\Sigma(\mathcal{C})$ is stable under filtered colimits in $\mathcal{P}(\mathcal{C})$, we conclude that f is continuous, so that LE is a compact object of $\mathcal{P}_\Sigma(\mathcal{C})$, as desired. $\qquad\square$

Our next goal is to prove a converse to part (4) of Proposition 5.5.8.10. Namely, we will show that $\mathcal{P}_\Sigma(\mathcal{C})$ is generated by the essential image of the Yoneda embedding under sifted colimits. In fact, we will need to use only special types of sifted colimits: namely, filtered colimits and geometric realizations (Lemma 5.5.8.14). The proof is based on the following technical result:

Lemma 5.5.8.13. *Let $\mathcal{C}$ be a small ∞-category and let X be an object of $\mathcal{P}(\mathcal{C})$. Then there exists a simplicial object $Y_\bullet : \mathrm{N}(\mathbf{\Delta})^{op} \to \mathcal{P}(\mathcal{C})$ with the following properties:*

(1) *The colimit of $Y_\bullet$ is equivalent to X.*

(2) *For each $n \geq 0$, the object $Y_n \in \mathcal{P}(\mathcal{C})$ is equivalent to a small coproduct of objects lying in the image of the Yoneda embedding $j : \mathcal{C} \to \mathcal{P}(\mathcal{C})$.*

We will defer the proof until the end of this section.

Lemma 5.5.8.14. *Let $\mathcal{C}$ be a small ∞-category which admits finite coproducts and let $X \in \mathcal{P}(\mathcal{C})$. The following conditions are equivalent:*

(1) *The object X belongs to $\mathcal{P}_\Sigma(\mathcal{C})$.*

(2) *There exists a simplicial object $U_\bullet : \mathrm{N}(\mathbf{\Delta})^{op} \to \mathrm{Ind}(\mathcal{C})$ whose colimit in $\mathcal{P}(\mathcal{C})$ is X.*

Proof. The full subcategory $\mathcal{P}_\Sigma(\mathcal{C})$ contains the essential image of the Yoneda embedding and is stable under filtered colimits and geometric realizations (Proposition 5.5.8.10); thus (2) $\Rightarrow$ (1). We will prove that (1) $\Rightarrow$ (2).

We first choose a simplicial object $Y_\bullet$ of $\mathcal{P}(\mathcal{C})$ which satisfies the conclusions of Lemma 5.5.8.13. Let L be a left adjoint to the inclusion $\mathcal{P}_\Sigma(\mathcal{C}) \subseteq \mathcal{P}(\mathcal{C})$. Since X is a colimit of $Y_\bullet$, $LX \simeq X$ is a colimit of $LY_\bullet$ (part (5) of Proposition 5.5.8.10). It will therefore suffice to prove that each LY_n belongs to $\mathrm{Ind}(\mathcal{C})$. By hypothesis, each Y_n can be written as a small coproduct $\coprod_{\alpha \in A} j(C_\alpha)$, where $j : \mathcal{C} \to \mathcal{P}(\mathcal{C})$ denotes the Yoneda embedding. Using the results of §4.2.3, we see that Y_n can also be obtained as a filtered colimit of coproducts

$\coprod_{\alpha \in A_0} j(C_\alpha)$, where A_0 ranges over the finite subsets of A. Since L preserves filtered colimits (Proposition 5.5.8.10), it will suffice to show that each of the objects

$$L(\coprod_{\alpha \in A_0} j(C_\alpha))$$

belongs to $\mathrm{Ind}(\mathcal{C})$. We now invoke part (2) of Proposition 5.5.8.10 to identify this object with $j(\coprod_{\alpha \in A_0} C_\alpha)$. $\square$

Proposition 5.5.8.15. *Let $\mathcal{C}$ be a small ∞-category which admits finite coproducts and let $\mathcal{D}$ be an ∞-category which admits filtered colimits and geometric realizations. Let $\mathrm{Fun}_\Sigma(\mathcal{P}_\Sigma(\mathcal{C}), \mathcal{D})$ denote the full subcategory spanned by those functors $\mathcal{P}_\Sigma(\mathcal{C}) \to \mathcal{D}$ which preserve filtered colimits and geometric realizations. Then*

(1) *Composition with the Yoneda embedding $j : \mathcal{C} \to \mathcal{P}_\Sigma(\mathcal{C})$ induces an equivalence of categories*

$$\theta : \mathrm{Fun}_\Sigma(\mathcal{P}_\Sigma(\mathcal{C}), \mathcal{D}) \to \mathrm{Fun}(\mathcal{C}, \mathcal{D}).$$

(2) *Any functor $g \in \mathrm{Fun}_\Sigma(\mathcal{P}_\Sigma(\mathcal{C}), \mathcal{D})$ preserves sifted colimits.*

(3) *Assume that $\mathcal{D}$ admits finite coproducts. A functor $g \in \mathrm{Fun}_\Sigma(\mathcal{P}_\Sigma(\mathcal{C}), \mathcal{D})$ preserves small colimits if and only if $g \circ j$ preserves finite coproducts.*

Proof. Lemma 5.5.8.14 and Proposition 5.5.8.10 imply that $\mathcal{P}_\Sigma(\mathcal{C})$ is the smallest full subcategory of $\mathcal{P}(\mathcal{C})$ which is closed under filtered colimits, is closed under geometric realizations, and contains the essential image of the Yoneda embedding. Consequently, assertion (1) follows from Remark 5.3.5.9 and Proposition 4.3.2.15.

We now prove (2). Let $g \in \mathrm{Fun}_\Sigma(\mathcal{P}_\Sigma(\mathcal{C}), \mathcal{D})$; we wish to show that g preserves sifted colimits. It will suffice to show that for every representable functor $e : \mathcal{D} \to \mathcal{S}^{op}$, the composition $e \circ g$ preserves sifted colimits. In other words, we may replace $\mathcal{D}$ by $\mathcal{S}^{op}$ and thereby reduce to the case where $\mathcal{D}$ itself admits sifted colimits. Let $\mathrm{Fun}'_\Sigma(\mathcal{P}_\Sigma(\mathcal{C}), \mathcal{D})$ denote the full subcategory of $\mathrm{Fun}_\Sigma(\mathcal{P}_\Sigma(\mathcal{C}), \mathcal{D})$ spanned by those functors which preserve sifted colimits. Since $\mathcal{P}_\Sigma(\mathcal{C})$ is also the smallest full subcategory of $\mathcal{P}(\mathcal{C})$ which contains the essential image of the Yoneda embedding and is stable under sifted colimits, Remark 5.3.5.9 implies that θ induces an equivalence

$$\mathrm{Fun}'_\Sigma(\mathcal{P}_\Sigma(\mathcal{C}), \mathcal{D}) \to \mathrm{Fun}(\mathcal{C}, \mathcal{D}).$$

Combining this observation with (1), we deduce that the inclusion

$$\mathrm{Fun}'_\Sigma(\mathcal{P}_\Sigma(\mathcal{C}), \mathcal{D}) \subseteq \mathrm{Fun}_\Sigma(\mathcal{P}_\Sigma(\mathcal{C}), \mathcal{D})$$

is an equivalence of ∞-categories and therefore an equality.

The "only if" direction of (3) is immediate since the Yoneda embedding $j : \mathcal{C} \to \mathcal{P}_\Sigma(\mathcal{C})$ preserves finite coproducts (Proposition 5.5.8.10). To prove the converse, we first apply Lemma 5.3.5.7 to reduce to the case where $\mathcal{D}$ is a full subcategory of an ∞-category $\mathcal{D}'$ with the following properties:

(*i*) The ∞-category $\mathcal{D}'$ admits small colimits.

(*ii*) A small diagram $K^{\triangleright} \to \mathcal{D}$ is a colimit if and only if the induced diagram $K^{\triangleright} \to \mathcal{D}'$ is a colimit.

Let $\mathcal{C}'$ denote the essential image of the Yoneda embedding $j : \mathcal{C} \to \mathcal{P}(\mathcal{C})$. Using Lemma 5.1.5.5, we conclude that there exists a functor $G : \mathcal{P}(\mathcal{C}) \to \mathcal{D}'$ which is a left Kan extension of $G| \mathcal{C}' = g| \mathcal{C}'$ and that G preserves small colimits. Let $G_0 = G| \mathcal{P}_{\Sigma}(\mathcal{C})$. Then G_0 is a left Kan extension of $g| \mathcal{C}'$, so there is a canonical natural transformation $G_0 \to g$. Let $\mathcal{C}''$ denote the full subcategory of $\mathcal{P}_{\Sigma}(\mathcal{C})$ spanned by those objects C for which the map $G_0(C) \to g(C)$ is an equivalence. Then $\mathcal{C}''$ contains $\mathcal{C}'$ and is stable under filtered colimits and geometric realizations and therefore contains all of $\mathcal{P}_{\Sigma}(\mathcal{C})$. We may therefore replace g by G_0 and thereby assume that $G| \mathcal{P}_{\Sigma}(\mathcal{C}) = g$. Since $G \circ j = g \circ j$ preserves finite coproducts, the right adjoint to G factors through $\mathcal{P}_{\Sigma}(\mathcal{C})$ (Proposition 5.5.8.10), so that G is equivalent to the composition

$$\mathcal{P}(\mathcal{C}) \xrightarrow{L} \mathcal{P}_{\Sigma}(\mathcal{C}) \xrightarrow{G'} \mathcal{D}'$$

for some colimit-preserving functor $G' : \mathcal{P}_{\Sigma}(\mathcal{C}) \to \mathcal{D}'$. Restricting to the subcategory $\mathcal{P}_{\Sigma}(\mathcal{C}) \subseteq \mathcal{P}(\mathcal{C})$, we deduce that G' is equivalent to g, so that g preserves small colimits, as desired. $\qquad\square$

Remark 5.5.8.16. Let $\mathcal{C}$ be a small ∞-category which admits finite coproducts. It follows from Proposition 5.5.8.15 that we can identify $\mathcal{P}_{\Sigma}(\mathcal{C})$ with $\mathcal{P}_{\mathcal{K}}^{\mathcal{K}'}(\mathcal{C})$ in each of the following three cases (for an explanation of this notation, we refer the reader to §5.3.6):

(1) The collection $\mathcal{K}$ is empty, and the collection $\mathcal{K}'$ consists of all small filtered simplicial sets together with $N(\Delta)^{op}$.

(2) The collection $\mathcal{K}$ is empty, and the collection $\mathcal{K}'$ consists of all small sifted simplicial sets.

(3) The collection $\mathcal{K}$ consists of all finite discrete simplicial sets, and the collection $\mathcal{K}'$ consists of all small simplicial sets.

Corollary 5.5.8.17. *Let $f : \mathcal{C} \to \mathcal{D}$ be a functor between ∞-categories. Assume that $\mathcal{C}$ admits small colimits. Then f preserves sifted colimits if and only if f preserves filtered colimits and geometric realizations.*

Proof. The "only if" direction is clear. For the converse, suppose that f preserves filtered colimits and geometric realizations. Let $\mathcal{I}$ be a small sifted ∞-category and $\overline{p} : \mathcal{I}^{\triangleright} \to \mathcal{C}$ a colimit diagram; we wish to prove that $f \circ \overline{p}$ is also a colimit diagram. Let $p = \overline{p}| \mathcal{I}$. Let $\mathcal{J} \subseteq \mathcal{P}(\mathcal{I})$ denote a small full subcategory which contains the essential image of the Yoneda embedding $j : \mathcal{I} \to \mathcal{P}(\mathcal{I})$ and is closed under finite coproducts. It follows from Remark 5.3.5.9 that the functor p is homotopic to a composition $q \circ j$, where $q : \mathcal{J} \to \mathcal{C}$

is a functor which preserves finite coproducts. Proposition 5.5.8.15 implies that q is homotopic to a composition

$$\mathcal{J} \xrightarrow{j'} \mathcal{P}_\Sigma(\mathcal{J}) \xrightarrow{q'} \mathcal{C},$$

where j' denotes the Yoneda embedding and q' preserves small colimits. The composition $f \circ q'$ preserves filtered colimits and geometric realization and therefore preserves sifted colimits (Proposition 5.5.8.15).

Let $\overline{p}' : \mathcal{J}^\triangleright \to \mathcal{P}_\Sigma(\mathcal{J})$ be a colimit of the diagram $j' \circ j$. Since q' preserves colimits, the composition $q' \circ \overline{p}'$ is a colimit of $q' \circ j' \circ j \simeq p$ and is therefore equivalent to $\overline{p}$. Consequently, it will suffice to show that $f \circ q' \circ \overline{p}'$ is a colimit diagram. Since $\mathcal{J}$ is sifted, we need only verify that $f \circ q'$ preserves sifted colimits. By Proposition 5.5.8.15, it will suffice to show that $f \circ q'$ preserves filtered colimits and geometric realizations. Since q' preserves all colimits, this follows from our assumption that f preserves filtered colimits and geometric realizations.   □

In the situation of Proposition 5.5.8.15, every functor $f : \mathcal{C} \to \mathcal{D}$ extends (up to homotopy) to a functor $F : \mathcal{P}_\Sigma(\mathcal{C}) \to \mathcal{D}$, which preserves sifted colimits. We will sometimes refer to F as the *left derived functor* of f. In §5.5.9 we will explain the connection of this notion of derived functor with the more classical definition provided by Quillen's theory of homotopical algebra.

Our next goal is to characterize those ∞-categories which have the form $\mathcal{P}_\Sigma(\mathcal{C})$.

Definition 5.5.8.18. Let $\mathcal{C}$ be an ∞-category which admits geometric realizations of simplicial objects. We will say that an object $P \in \mathcal{C}$ is *projective* if the functor $\mathcal{C} \to \mathcal{S}$ corepresented by P commutes with geometric realizations.

Remark 5.5.8.19. Let $\mathcal{C}$ be an ∞-category which admits geometric realizations of simplicial objects. Then the collection of projective objects of $\mathcal{C}$ is stable under all finite coproducts which exist in $\mathcal{C}$. This follows immediately from Lemma 5.5.8.11 and Remark 5.5.8.12.

Remark 5.5.8.20. Let $\mathcal{C}$ be an ∞-category which admits small colimits and let X be an object of $\mathcal{C}$. Then X is compact and projective if and only if X corepresents a functor $\mathcal{C} \to \mathrm{Set}_\Delta$ which preserves sifted colimits. The "only if" direction is obvious, and the converse follows from Corollary 5.5.8.17.

Example 5.5.8.21. Let $\mathcal{A}$ be an abelian category. Then an object $P \in \mathcal{A}$ is projective in the sense of classical homological algebra (that is, the functor $\mathrm{Hom}_\mathcal{A}(P, \bullet)$ is exact) if and only if P corepresents a functor $\mathcal{A} \to \mathrm{Set}$ which commutes with geometric realizations of simplicial objects. This is *not* equivalent to the condition of Definition 5.5.8.18 since the fully faithful embedding $\mathrm{Set} \to \mathcal{S}$ does not preserve geometric realizations. However, it is equivalent to the requirement that P be a projective object (in the sense of Definition 5.5.8.18) in the ∞-category underlying the homotopy theory of simplicial objects of $\mathcal{A}$ (equivalently, the theory of nonpositively graded chain complexes with values in $\mathcal{A}$; we will discuss this example in greater detail in [50]).

Proposition 5.5.8.22. *Let $\mathcal{C}$ be a small ∞-category which admits finite coproducts, $\mathcal{D}$ an ∞-category which admits filtered colimits and geometric realizations, and $F : \mathcal{P}_\Sigma(\mathcal{C}) \to \mathcal{D}$ a left derived functor of $f = F \circ j : \mathcal{C} \to \mathcal{D}$, where $j : \mathcal{C} \to \mathcal{P}_\Sigma(\mathcal{C})$ denotes the Yoneda embedding. Consider the following conditions:*

(i) *The functor f is fully faithful.*

(ii) *The essential image of f consists of compact projective objects of $\mathcal{D}$.*

(iii) *The ∞-category $\mathcal{D}$ is generated by the essential image of f under filtered colimits and geometric realizations.*

If (i) and (ii) are satisfied, then F is fully faithful. Moreover, F is an equivalence if and only if (i), (ii), and (iii) are satisfied.

Proof. If F is an equivalence of ∞-categories, then (i) follows from Proposition 5.1.3.1 and (iii) from Lemma 5.5.8.14. To prove (ii), it suffices to show that for each $C \in \mathcal{C}$, the functor $e : \mathcal{P}_\Sigma(\mathcal{C}) \to \mathcal{S}$ corepresented by C preserves filtered colimits and geometric realizations. We can identify e with the composition

$$\mathcal{P}_\Sigma(\mathcal{C}) \overset{e'}{\subseteq} \mathcal{P}(\mathcal{C}) \overset{e''}{\to} \mathcal{S},$$

where e'' denotes evaluation at C. It now suffices to observe that e' and e'' preserve filtered colimits and geometric realizations (Lemma 5.5.8.14 and Proposition 5.1.2.2).

For the converse, let us suppose that (i) and (ii) are satisfied. We will show that F is fully faithful. First fix an object $C \in \mathcal{C}$ and let $\mathcal{P}'_\sigma(\mathcal{C})$ be the full subcategory of $\mathcal{P}_\Sigma(\mathcal{C})$ spanned by those objects M for which the map

$$\mathrm{Map}_{\mathcal{P}_\Sigma(\mathcal{C})}(j(C), M) \to \mathrm{Map}_{\mathcal{D}}(f(C), F(M))$$

is an equivalence. Condition (i) implies that $\mathcal{P}'_\sigma(\mathcal{C})$ contains the essential image of j, and condition (ii) implies that $\mathcal{P}'_\sigma(\mathcal{C})$ is stable under filtered colimits and geometric realizations. Lemma 5.5.8.14 now implies that $\mathcal{P}'_\Sigma(\mathcal{C}) = \mathcal{P}_\Sigma(\mathcal{C})$.

We now define $\mathcal{P}''_\Sigma(\mathcal{C})$ to be the full subcategory of $\mathcal{P}_\Sigma(\mathcal{C})$ spanned by those objects M such that for all $N \in \mathcal{P}_\Sigma(\mathcal{C})$, the map

$$\mathrm{Map}_{\mathcal{P}_\Sigma(\mathcal{C})}(M, N) \to \mathrm{Map}_{\mathcal{D}}(F(M), F(N))$$

is a homotopy equivalence. The above argument shows that $\mathcal{P}''_\Sigma(\mathcal{C})$ contains the essential image of j. Since F preserves filtered colimits and geometric realizations, $\mathcal{P}''_\Sigma(\mathcal{C})$ is stable under filtered colimits and geometric realizations. Applying Lemma 5.5.8.14, we conclude that $\mathcal{P}''_\Sigma(\mathcal{C}) = \mathcal{P}_\Sigma(\mathcal{C})$; this proves that F is fully faithful.

If F is fully faithful, then the essential image of F contains $f(\mathcal{C})$ and is stable under filtered colimits and geometric realizations. If (iii) is satisfied, it follows that F is an equivalence of ∞-categories. $\qquad\square$

Definition 5.5.8.23. Let $\mathcal{C}$ be an ∞-category which admits small colimits and let S be a collection of objects of $\mathcal{C}$. We will say that S is a *set of compact projective generators for* $\mathcal{C}$ if the following conditions are satisfied:

(1) Each element of S is a compact projective object of $\mathcal{C}$.

(2) The full subcategory of $\mathcal{C}$ spanned by the elements of S is essentially small.

(3) The set S generates $\mathcal{C}$ under small colimits.

We will say that $\mathcal{C}$ is *projectively generated* if there exists a set S of compact projective generators for $\mathcal{C}$.

Example 5.5.8.24. The ∞-category $\mathcal{S}$ of spaces is projectively generated. The compact projective objects of $\mathcal{S}$ are precisely those spaces which are homotopy equivalent to finite sets (endowed with the discrete topology).

Proposition 5.5.8.25. *Let $\mathcal{C}$ be an ∞-category which admits small colimits and let S be a set of compact projective generators for $\mathcal{C}$. Then*

(1) *Let $\mathcal{C}^0 \subseteq \mathcal{C}$ be the full subcategory spanned by finite coproducts of the objects S, let $\mathcal{D} \subseteq \mathcal{C}^0$ be a minimal model for $\mathcal{C}^0$, and let $F : \mathcal{P}_\Sigma(\mathcal{D}) \to \mathcal{C}$ be a left derived functor of the inclusion. Then F is an equivalence of ∞-categories. In particular, $\mathcal{C}$ is a compactly generated presentable ∞-category.*

(2) *Let $C \in \mathcal{C}$ be an object. The following conditions are equivalent:*

   (i) *The object C is compact and projective.*

   (ii) *The functor $e : \mathcal{C} \to \widehat{\mathcal{S}}$ corepresented by C preserves sifted colimits.*

   (iii) *There exists an object $C' \in \mathcal{C}^0$ such that C is a retract of C'.*

Proof. Remark 5.5.8.19 implies that $\mathcal{C}^0$ consists of compact projective objects of $\mathcal{C}$. Assertion (1) now follows immediately from Proposition 5.5.8.22. We now prove (2). The implications $(iii) \Rightarrow (i)$ and $(ii) \Rightarrow (i)$ are obvious. To complete the proof, we will show that $(i) \Rightarrow (iii)$. Using (1), we are free to assume $\mathcal{C} = \mathcal{P}_\Sigma(\mathcal{D})$. Let $C \in \mathcal{C}$ be a compact projective object. Using Lemma 5.5.8.14, we conclude that there exists a simplicial object $X_\bullet$ of $\mathrm{Ind}(\mathcal{D})$ and an equivalence $C \simeq |X_\bullet|$. Since C is projective, we deduce that $\mathrm{Map}_{\mathcal{C}}(C, C)$ is equivalent to the geometric realization of the simplicial space $\mathrm{Map}_{\mathcal{C}}(C, X_\bullet)$. In particular, $\mathrm{id}_C \in \mathrm{Map}_{\mathcal{C}}(C, C)$ is homotopic to the image of some map $f : C \to X_0$. Using our assumption that C is compact, we conclude that f factors as a composition

$$C \xrightarrow{f_0} j(D) \to X_0,$$

where $j : \mathcal{D} \to \mathrm{Ind}(\mathcal{D})$ denotes the Yoneda embedding. It follows that C is a retract of $j(D)$ in $\mathcal{C}$, as desired. □

Remark 5.5.8.26. Let $\mathcal{C}$ be a small ∞-category which admits finite coproducts. Since the truncation functor $\tau_{\leq n} : \mathcal{S} \to \mathcal{S}$ preserves finite products, it induces a map $\tau : \mathcal{P}_\Sigma(\mathcal{C}) \to \mathcal{P}_\Sigma(\mathcal{C})$, which is easily seen to be a localization functor. The essential image of τ consists of those functors $F \in \mathcal{P}_\Sigma(\mathcal{C})$ which take n-truncated values. We claim that these are precisely the n-truncated objects of $\mathcal{P}_\Sigma(\mathcal{C})$. Consequently, we can identify τ with the n-truncation functor on $\mathcal{P}_\Sigma(\mathcal{C})$.

One direction is clear: if $F \in \mathcal{P}_\Sigma(\mathcal{C})$ is n-truncated, then for each $C \in \mathcal{C}$ the space $\mathrm{Map}_{\mathcal{P}_\Sigma(\mathcal{C})}(j(C), F) \simeq F(C)$ must be n-truncated. Conversely, suppose that $F : \mathcal{C}^{op} \to \mathcal{S}$ takes n-truncated values. We wish to prove that the space $\mathrm{Map}_{\mathcal{P}_\Sigma(\mathcal{C})}(F', F)$ is n-truncated for each $F' \in \mathcal{P}_\Sigma(\mathcal{C})$. The collection of all objects F' which satisfy this condition is stable under small colimits in $\mathcal{P}_\Sigma(\mathcal{C})$ and contains the essential image of the Yoneda embedding. It therefore contains the entirety of $\mathcal{P}_\Sigma(\mathcal{C})$, as desired.

We conclude this section by giving the proof of Lemma 5.5.8.13. Our argument uses some concepts and results from Chapter 6 and may be omitted at first reading.

Proof of Lemma 5.5.8.13. For $n \geq 0$, let $\mathbf{\Delta}^{\leq n}$ denote the full subcategory of $\mathbf{\Delta}$ spanned by the objects $\{[k]\}_{k \leq n}$. We will construct a compatible sequence of functors $f_n : \mathrm{N}(\mathbf{\Delta}^{\leq n})^{op} \to \mathcal{P}(\mathcal{C})_{/X}$ with the following properties:

(A) For $n \geq 0$, let L_n denote a colimit of the composite diagram

$$\mathrm{N}(\mathbf{\Delta}^{\leq n-1})^{op} \times_{\mathrm{N}(\mathbf{\Delta})^{op}} \mathrm{N}(\mathbf{\Delta}_{[n]/})^{op} \to \mathrm{N}(\mathbf{\Delta}^{\leq n-1})^{op} \xrightarrow{f_{n-1}} \mathcal{P}(\mathcal{C})_{/X} \to \mathcal{P}(\mathcal{C})$$

(the nth *latching object*). Then there exists an object $Z_n \in \mathcal{P}(\mathcal{C})$, which is a small coproduct of objects in the essential image of the Yoneda embedding $\mathcal{C} \to \mathcal{P}(\mathcal{C})$, and a map $Z_n \to f_n([n])$, which together with the canonical map $L_n \to f_n([n])$ determines an equivalence $L_n \coprod Z_n \simeq f_n([n])$.

(B) For $n \geq 0$, let $\overline{M}_n$ denote the limit of the diagram

$$\mathrm{N}(\mathbf{\Delta}^{\leq n-1})^{op} \times_{\mathrm{N}(\mathbf{\Delta})^{op}} \mathrm{N}(\mathbf{\Delta}_{/[n]})^{op} \to \mathrm{N}(\mathbf{\Delta}^{\leq n-1})^{op} \xrightarrow{f_{n-1}} \mathcal{P}(\mathcal{C})_{/X}$$

(the nth *matching object*) and let M_n denote its image in $\mathcal{P}(\mathcal{C})$. Then the canonical map $f_n([n]) \to M_n$ is an effective epimorphism in $\mathcal{P}(\mathcal{C})$ (see §6.2.3).

The construction of the functors f_n proceeds by induction on n, the case $n < 0$ being trivial. For $n \geq 0$, we invoke Remark A.2.9.16: to extend f_{n-1} to a functor f_n satisfying (A) and (B), it suffices to produce an object Z_n and a morphism $\psi : Z_n \to M_n$ in $\mathcal{P}(\mathcal{C})$, such that the coproduct $L_n \coprod Z_n \to M_n$ is an effective epimorphism. This is satisfied in particular if ψ itself is an effective epimorphism.

The maps f_n together determine a simplicial object $\overline{Y}_\bullet$ of $\mathcal{P}(\mathcal{C})_{/X}$, which we can identify with a simplicial object $Y_\bullet$ in $\mathcal{P}(\mathcal{C})$ equipped with a map θ :

$\varinjlim Y_\bullet \to X$. Assumption (B) guarantees that θ is a hypercovering of X (see §6.5.3), so that the map θ is ∞-connective (Lemma 6.5.3.11). The ∞-topos $\mathcal{P}(\mathcal{C})$ has enough points (given by evaluation at objects of $\mathcal{C}$) and is therefore hypercomplete (Remark 6.5.4.7). It follows that θ is an equivalence. We now complete the proof by observing that for $n \geq 0$, we have an equivalence $Y_n \simeq \coprod_{[n] \to [k]} Z_k$ where the coproduct is taken over all surjective maps of linearly ordered sets $[n] \to [k]$, so that Y_n is itself a small coproduct of objects lying in the essential image of the Yoneda embedding $j : \mathcal{C} \to \mathcal{P}(\mathcal{C})$. □

5.5.9 Quillen's Model for $\mathcal{P}_\Sigma(\mathcal{C})$

Let $\mathcal{C}$ be a small category which admits finite products. Then $N(\mathcal{C})^{op}$ is an ∞-category which admits finite coproducts. In §5.5.8, we studied the ∞-category $\mathcal{P}_\Sigma(N(\mathcal{C})^{op})$, which we can view as the full subcategory of the presheaf ∞-category $\operatorname{Fun}(N(\mathcal{C}), \mathcal{S})$ spanned by those functors which preserve finite products. According to Proposition 4.2.4.4, $\operatorname{Fun}(N(\mathcal{C}), \mathcal{S})$ can be identified with the ∞-category underlying the simplicial model category of diagrams $\operatorname{Set}_\Delta^{\mathcal{C}}$ (which we will endow with the *projective* model structure described in §A.3.2). It follows that every functor $f : N(\mathcal{C}) \to \mathcal{S}$ is equivalent to the (simplicial) nerve of a functor $F : \mathcal{C} \to \mathcal{K}an$. Moreover, f belongs to $\mathcal{P}_\Sigma(N(\mathcal{C})^{op})$ if and only if the functor F is *weakly* product-preserving in the sense that for any finite collection of objects $\{C_i \in \mathcal{C}\}_{1 \leq i \leq n}$, the natural map

$$F(C_1 \times \cdots \times C_n) \to F(C_1) \times \cdots \times F(C_n)$$

is a homotopy equivalence of Kan complexes. Our goal in this section is to prove a refinement of Proposition 4.2.4.4: if f preserves finite products, then it is possible to arrange that F preserves finite products (up to isomorphism rather than up to homotopy equivalence). This result is most naturally phrased as an equivalence between model categories (Proposition 5.5.9.2) and is due to Bergner (see [9]). We begin by recalling the following result of Quillen (for a proof, we refer the reader to [63]):

Proposition 5.5.9.1 (Quillen). *Let $\mathcal{C}$ be a category which admits finite products and let $\mathbf{A}$ denote the category of functors $F : \mathcal{C} \to \operatorname{Set}_\Delta$ which preserve finite products. Then $\mathbf{A}$ admits a simplicial model structure which may be described as follows:*

(W) *A natural transformation $\alpha : F \to F'$ of functors is a weak equivalence in $\mathbf{A}$ if and only if $\alpha(C) : F(C) \to F'(C)$ is a weak homotopy equivalence of simplicial sets for every $C \in \mathcal{C}$.*

(F) *A natural transformation $\alpha : F \to F'$ of functors is a fibration in $\mathbf{A}$ if and only if $\alpha(C) : F(C) \to F'(C)$ is a Kan fibration of simplicial sets for every $C \in \mathcal{C}$.*

Suppose that $\mathcal{C}$ and $\mathbf{A}$ are as in the statement of Proposition 5.5.9.1. Then we may regard $\mathbf{A}$ as a full subcategory of the category $\operatorname{Set}_\Delta^{\mathcal{C}}$ of *all* functors from $\mathcal{C}$ to $\operatorname{Set}_\Delta$, which we regard as endowed with the projective model structure (so that fibrations and weak equivalences are given pointwise). The inclusion $G : \mathbf{A} \subseteq \operatorname{Set}_\Delta^{\mathcal{C}}$ preserves fibrations and trivial fibrations and therefore determines a Quillen adjunction

$$\operatorname{Set}_\Delta^{\mathcal{C}} \underset{G}{\overset{F}{\rightleftarrows}} \mathbf{A}.$$

(A more explicit description of the adjoint functor F will be given below.) Our goal in this section is to prove the following result:

Proposition 5.5.9.2 (Bergner). *Let $\mathcal{C}$ be a small category which admits finite products and let*

$$\operatorname{Set}_\Delta^{\mathcal{C}} \underset{G}{\overset{F}{\rightleftarrows}} \mathbf{A}$$

be as above. Then the right derived functor

$$RG : h\mathbf{A} \to h\operatorname{Set}_\Delta^{\mathcal{C}}$$

is fully faithful, and an object $f \in h\operatorname{Set}_\Delta^{\mathcal{C}}$ belongs to the essential image of RG if and only if f preserves finite products up to weak homotopy equivalence.

Corollary 5.5.9.3. *Let $\mathcal{C}$ be a small category which admits finite products and let $\mathbf{A}$ be as in Proposition 5.5.9.2. Then the natural map $\mathrm{N}(\mathbf{A}^\circ) \to \mathcal{P}_\Sigma(\mathrm{N}(\mathcal{C})^{op})$ is an equivalence of ∞-categories.*

The proof of Proposition 5.5.9.2 is somewhat technical and will occupy the rest of this section. We begin by introducing some preliminaries.

Notation 5.5.9.4. Let $\mathcal{C}$ be a small category. We define a pair of categories $\operatorname{Env}(\mathcal{C}) \subseteq \operatorname{Env}^+(\mathcal{C})$ as follows:

(i) An object of $\operatorname{Env}^+(\mathcal{C})$ is a pair $C = (J, \{C_j\}_{j \in J})$, where J is a finite set and each C_j is an object of $\mathcal{C}$. The object C belongs to $\operatorname{Env}(\mathcal{C})$ if and only if J is nonempty.

(ii) Given objects $C = (J, \{C_j\}_{j \in J})$ and $C' = (J', \{C'_{j'}\}_{j' \in J'})$ of $\operatorname{Env}^+(\mathcal{C})$, a morphism $C \to C'$ consists of the following data:

(a) A map $f : J' \to J$ of finite sets.

(b) For each $j' \in J'$, a morphism $C_{f(j')} \to C'_{j'}$ in the category $\mathcal{C}$.

Such a morphism belongs to $\operatorname{Env}(\mathcal{C})$ if and only if both J and J' are nonempty and f is surjective.

There is a fully faithful embedding functor $\theta : \mathcal{C} \to \operatorname{Env}(\mathcal{C})$ given by $C \mapsto (*, \{C\})$. We can view $\operatorname{Env}^+(\mathcal{C})$ as the category obtained from $\mathcal{C}$ by freely adjoining finite products. In particular, if $\mathcal{C}$ admits finite products, then θ admits a (product-preserving) left inverse $\phi_{\mathcal{C}}^+$ given by the formula $(J, \{C_j\}_{j \in J}) \mapsto \prod_{j \in J} C_j$. We let $\phi_{\mathcal{C}}$ denote the restricton $\phi_{\mathcal{C}}^+ | \operatorname{Env}(\mathcal{C})$.

Given a functor $\mathcal{F} \in \mathrm{Set}_{\Delta}^{\mathcal{C}}$, we let $E^+(\mathcal{F}) \in \mathrm{Set}_{\Delta}^{\mathrm{Env}^+(\mathcal{C})}$ denote the composition

$$\mathrm{Env}^+(\mathcal{C}) \overset{\mathrm{Env}^+(\mathcal{F})}{\to} \mathrm{Env}^+(\mathrm{Set}_{\Delta}) \overset{\phi_{\mathrm{Set}_{\Delta}}^+}{\to} \mathrm{Set}_{\Delta}$$

$$(J, \{C_j\}_{j \in J}) \mapsto \prod f(C_j).$$

We let $E(\mathcal{F})$ denote the restriction $E^+(\mathcal{F})| \mathrm{Env}(\mathcal{C}) \in \mathrm{Set}_{\Delta}^{\mathrm{Env}(\mathcal{C})}$.

If the category $\mathcal{C}$ admits finite products, then we let $L, L^+ : \mathrm{Set}_{\Delta}^{\mathcal{C}} \to \mathrm{Set}_{\Delta}^{\mathcal{C}}$ denote the compositions

$$\mathrm{Set}_{\Delta}^{\mathcal{C}} \overset{E}{\to} \mathrm{Set}_{\Delta}^{\mathrm{Env}(\mathcal{C})} \overset{(\phi_{\mathcal{C}})_!}{\to} \mathrm{Set}_{\Delta}^{\mathcal{C}}$$

$$\mathrm{Set}_{\Delta}^{\mathcal{C}} \overset{E^+}{\to} \mathrm{Set}_{\Delta}^{\mathrm{Env}^+(\mathcal{C})} \overset{(\phi_{\mathcal{C}}^+)_!}{\to} \mathrm{Set}_{\Delta}^{\mathcal{C}},$$

where $(\phi_{\mathcal{C}})_!$ and $(\phi_{\mathcal{C}}^+)_!$ indicate left Kan extension functors. There is a canonical isomorphism $\theta^* \circ E \simeq \mathrm{id}$ which induces a natural transformation $\alpha : \mathrm{id} \to L$. Let $\beta : L \to L^+$ indicate the natural transformation induced by the inclusion $\mathrm{Env}(\mathcal{C}) \subseteq \mathrm{Env}^+(\mathcal{C})$.

Remark 5.5.9.5. Let $\mathcal{C}$ be a small category. The functor $E^+ : \mathrm{Set}_{\Delta}^{\mathcal{C}} \to \mathrm{Set}_{\Delta}^{\mathrm{Env}^+(\mathcal{C})}$ is fully faithful and has a left adjoint given by θ^*.

We begin by constructing the left adjoint which appears in the statement of Proposition 5.5.9.2.

Lemma 5.5.9.6. *Let $\mathcal{C}$ be a simplicial category which admits finite products and let $\mathcal{F} \in \mathrm{Set}_{\Delta}^{\mathcal{C}}$. Then*

(1) *The object $L^+(\mathcal{F}) \in \mathrm{Set}_{\Delta}^{\mathcal{C}}$ is product-preserving.*

(2) *If $\mathcal{F}' \in \mathrm{Set}_{\Delta}^{\mathcal{C}}$ is product-preserving, then composition with $\beta \circ \alpha$ induces an isomorphism of simplicial sets*

$$\mathrm{Map}_{\mathrm{Set}_{\Delta}^{\mathcal{C}}}(L^+(\mathcal{F}), \mathcal{F}') \to \mathrm{Map}_{\mathrm{Set}_{\Delta}^{\mathcal{C}}}(\mathcal{F}, \mathcal{F}').$$

Proof. Suppose we are given a finite collection of objects $\{C_1, \ldots, C_n\}$ in $\mathcal{C}$ and let

$$u : L^+(\mathcal{F})(C_1 \times \cdots \times C_n) \to L^+(\mathcal{F})(C_1) \times \cdots \times L^+(\mathcal{F})(C_n)$$

be the product of the projection maps. We wish to show that u is an isomorphism of simplicial sets. We will give an explicit construction of an inverse to u. For $C \in \mathcal{C}$, we let $\mathrm{Env}^+(\mathcal{C})_{/C}$ denote the fiber product $\mathrm{Env}^+(\mathcal{C}) \times_{\mathcal{D}} \mathcal{C}_{/C}$. For $1 \leq i \leq n$, let $\mathcal{G}_i$ denote the restriction of $E^+(\mathcal{F})$ to $\mathrm{Env}^+(\mathcal{C})_{/C_i}$ and let

$$\mathcal{G} : \prod \mathrm{Env}^+(\mathcal{D})_{/C_i} \to \mathrm{Set}_{\Delta}$$

be the product of the functors $\mathcal{G}_i$. We observe that $L^+(\mathcal{F})(C_i) \simeq \varinjlim(\mathcal{G}_i)$, so that the product $\prod L^+(\mathcal{F})(C_i) \simeq \varinjlim(\mathcal{G})$. We now observe that the formation of products in $\mathcal{E}^+(\mathcal{C})$ gives an identification of $\mathcal{G}$ with the composition

$$\prod \mathrm{Env}^+(\mathcal{C})_{/C_i} \to \mathrm{Env}^+(\mathcal{D})_{/C_1 \times \cdots \times C_n} \overset{E^+(\mathcal{F})}{\to} \mathrm{Set}_{\Delta}.$$

We thereby obtain a morphism

$$v : \varinjlim(\mathcal{G}) \to \varinjlim(E^+(\mathcal{F})| \operatorname{Env}^+(\mathcal{D})_{/C_1 \times \cdots \times C_n} \simeq L^+(\mathcal{F})(C_1 \times \cdots \times C_n).$$

It is not difficult to check that v is an inverse to u.

We observe that (2) is equivalent to the assertion that composition with θ^* induces an isomorphism

$$\operatorname{Map}_{\operatorname{Set}_{\Delta}^{\operatorname{Env}^+(\mathcal{C})}}(E^+(\mathcal{F}), (\phi_{\mathcal{C}}^+)^*(\mathcal{F}')) \to \operatorname{Map}_{\operatorname{Set}_{\Delta}^{\mathcal{C}}}(\mathcal{F}, \mathcal{F}').$$

Because $\mathcal{G}$ is product-preserving, there is a natural isomorphism $(\phi_{\mathcal{C}}^+)^*(\mathcal{F}') \simeq E^+(\mathcal{F}')$. The desired result now follows from Remark 5.5.9.5. $\qquad \square$

It follows that the functor $L^+ : \operatorname{Set}_{\Delta}^{\mathcal{C}} \to \operatorname{Set}_{\Delta}^{\mathcal{C}}$ factors through $\mathbf{A}$ and can be identified with a left adjoint to the inclusion $\mathbf{A} \subseteq \operatorname{Set}_{\Delta}^{\mathcal{C}}$. In order to prove Proposition 5.5.9.2, we need to be able to compute the functor L^+. We will do this in two steps: first, we show that (under mild hypotheses) the natural transformation $L \to L^+$ is a weak equivalence. Second, we will see that the colimit defining L is actually a homotopy colimit and therefore has good properties. More precisely, we have the following pair of lemmas whose proofs will be given at the end of this section.

Lemma 5.5.9.7. *Let $\mathcal{C}$ be a small category which admits finite products and let $\mathcal{F} \in \operatorname{Set}_{\Delta}^{\mathcal{C}}$ be a functor which carries the final object of $\mathcal{C}$ to a contractible Kan complex K. Then the canonical map $\beta : L(\mathcal{F}) \to L^+(\mathcal{F})$ is a weak equivalence in $\operatorname{Set}_{\Delta}^{\mathcal{C}}$.*

Lemma 5.5.9.8. *Let $\mathcal{C}$ be a small simplicial category. If $\mathcal{F}$ is a projectively cofibrant object of $\operatorname{Set}_{\Delta}^{\mathcal{C}}$, then $E(\mathcal{F})$ is a projectively cofibrant object of $\operatorname{Set}_{\Delta}^{\operatorname{Env}(\mathcal{C})}$.*

We are now almost ready to give the proof of Proposition 5.5.9.2. The essential step is contained in the following result:

Lemma 5.5.9.9. *Let $\mathcal{C}$ be a simplicial category that admits finite products and let*

$$\operatorname{Set}_{\Delta}^{\mathcal{C}} \underset{G}{\overset{F}{\rightleftarrows}} \mathbf{A}$$

be as in the statement of Proposition 5.5.9.2. Then

(1) *The functors F and G are Quillen adjoints.*

(2) *If $\mathcal{F} \in \operatorname{Set}_{\Delta}^{\mathcal{C}}$ is projectively cofibrant and weakly product-preserving, then the unit map $\mathcal{F} \to (G \circ F)(\mathcal{F})$ is a weak equivalence.*

Proof. Assertion (1) is obvious since G preserves fibrations and trivial cofibrations. It follows that F preserves weak equivalences between projectively cofibrant objects. Let $K \in \operatorname{Set}_{\Delta}$ denote the image under $\mathcal{F}$ of the final object of $\mathcal{D}$. In proving (2), we are free to replace $\mathcal{F}$ by any weakly equivalent diagram which is also projectively cofibrant. Choosing a fibrant replacement for $\mathcal{F}$, we may suppose that K is a Kan complex. Since $\mathcal{F}$ is weakly product-preserving, K is contractible.

In view of Lemma 5.5.9.6, we can identify the composition $G \circ F$ with L^+ and the unit map with the composition

$$\mathcal{F} \xrightarrow{\alpha} L(\mathcal{F}) \xrightarrow{\beta} L^+(\mathcal{F}).$$

Lemma 5.5.9.7 implies that β is a weak equivalence. Consequently, it will suffice to show that α is a weak equivalence.

We recall the construction of α. Let $\theta : \mathcal{C} \to \mathrm{Env}(\mathcal{C})$ be as in Notation 5.5.9.4, so that there is a canonical isomorphism $\mathcal{F} \simeq \theta^* E(\mathcal{F})$. This isomorphism induces a natural transformation $\overline{\alpha} : \theta_! \mathcal{F} \to E(\mathcal{F})$. The functor α is obtained from $\overline{\alpha}$ by applying the functor $(\phi_{\mathcal{C}})_!$ and identifying $((\phi_{\mathcal{C}})_! \circ \theta_!)(\mathcal{F})$ with $\mathcal{F}$. We observe that $(\phi_{\mathcal{C}})_!$ preserves weak equivalences between projectively cofibrant objects. Since $\theta_!$ preserves projective cofibrations, $\theta_! \mathcal{F}$ is projectively cofibrant. Lemma 5.5.9.8 asserts that $E(\mathcal{F})$ is projectively cofibrant. Consequently, it will suffice to prove that $\overline{\alpha}$ is a weak equivalence in $\mathrm{Set}_{\Delta}^{\mathrm{Env}(\mathcal{C})}$. Unwinding the definitions, this reduces to the condition that $\mathcal{F}$ be weakly compatible with (nonempty) products. $\square$

Proof of Proposition 5.5.9.2. Lemma 5.5.9.9 shows that (F, G) is a Quillen adjunction. To complete the proof, we must show:

(i) The counit transformation $LF \circ RG \to \mathrm{id}$ is an isomorphism of functors from the homotopy category $h\mathbf{A}$ to itself.

(ii) The essential image of $RG : h\mathbf{A} \to h\mathrm{Set}_{\Delta}^{\mathcal{C}}$ consists precisely of those functors which are weakly product-preserving.

We observe that G preserves weak equivalences, so we can identify RG with G. Since G also detects weak equivalences, (i) will follow if we can show that the induced transformation $\theta : G \circ LF \circ G \to G$ is an isomorphism of functors from the homotopy category $h\mathbf{A}$ to itself. This transformation has a right inverse given by composing the unit transformation $\mathrm{id} \to G \circ LF$ with G. Consequently, (i) follows immediately from Lemma 5.5.9.9.

The image of G consists precisely of the product-preserving diagrams $\mathcal{C} \to \mathrm{Set}_{\Delta}$; it follows immediately that every diagram in the essential image of G is weakly product-preserving. Lemma 5.5.9.9 implies the converse: every weakly product-preserving functor belongs to the essential image of G. This proves (ii). $\square$

Remark 5.5.9.10. Proposition 5.5.9.2 can be generalized to the situation where $\mathcal{C}$ is a *simplicial* category which admits finite products. We leave the necessary modifications to the reader.

It remains to prove Lemmas 5.5.9.8 and 5.5.9.7.

Proof of Lemma 5.5.9.8. For every object $C \in \mathcal{C}$ and every simplicial set K, we let $\mathcal{F}_C^K \in \mathrm{Set}_{\Delta}^{\mathcal{C}}$ denote the functor given by the formula $\mathcal{F}_C^K(D) =$

$\mathrm{Map}_{\mathcal{C}}(C, D) \times K$. A cofibration $K \to K'$ induces a projective cofibration $\mathcal{F}_C^K \to \mathcal{F}_C^{K'}$. We will refer to a projective cofibration of this form as a *generating projective cofibration*.

The small object argument implies that if $\mathcal{F} \in \mathrm{Set}_{\Delta}^{\mathcal{C}}$, then there is a transfinite sequence

$$\mathcal{F}_0 \subseteq \mathcal{F}_1 \subseteq \cdots \subseteq \mathcal{F}_{\alpha}$$

with the following properties:

(a) The functor $\mathcal{F}_0 : \mathcal{D} \to \mathrm{Set}_{\Delta}$ is constant, with value $\emptyset$.

(b) If $\lambda \leq \alpha$ is a limit ordinal, then $\mathcal{F}_{\lambda} = \bigcup_{\beta < \lambda} \mathcal{F}_{\beta}$.

(c) For each $\beta < \alpha$, the inclusion $\mathcal{F}_{\beta} \subseteq \mathcal{F}_{\beta+1}$ is a pushout of a generating projective cofibration.

(d) The functor $\mathcal{F}$ is a retract of $\mathcal{F}_{\alpha}$.

The functor $\mathcal{G} \mapsto E(\mathcal{G})$ preserves initial objects, filtered colimits, and retracts. Consequently, to show that $E(\mathcal{F})$ is projectively cofibrant, it will suffice to prove the following assertion:

(∗) Suppose we are given a cofibration $K \to K'$ of simplicial sets and a pushout diagram

$$
\begin{array}{ccc}
\mathcal{F}_C^K & \longrightarrow & \mathcal{G} \\
\downarrow & & \downarrow \\
\mathcal{F}_C^{K'} & \longrightarrow & \mathcal{G}'
\end{array}
$$

in $\mathrm{Set}_{\Delta}^{\mathcal{C}}$. If $E(\mathcal{G})$ is projectively cofibrant, then $E(\mathcal{G}')$ is projectively cofibrant.

To prove this, we will need to analyze the structure of $E(\mathcal{G}')$. Given an object $C' = (J, \{C'_j\}_{j \in J})$ of $\mathrm{Env}(\mathcal{C})$, we have

$$E(\mathcal{G}')(C') = \prod_{j \in J} (\mathcal{G}(C'_j) \coprod_{K \times \mathrm{Map}_{\mathcal{C}}(C, C'_j)} (K' \times \mathrm{Map}_{\mathcal{C}}(C, C'_j))).$$

Let $\sigma : \Delta^n \to E(\mathcal{G}')(C')$ be a simplex and let $J_{\sigma} \subseteq J$ be the collection of all indices j for which the corresponding simplex $\sigma(j) : \Delta^n \to \mathcal{G}'(C'_j)$ does not factor through $\mathcal{G}(C'_j)$. In this case, we can identify $\sigma(j)$ with an n-simplex of K' which does not belong to K. We will say that σ is of *index* $\leq k$ if the set $\{\sigma(j) : j \in J_{\sigma}\}$ has cardinality $\leq k$. Note that σ can be of index smaller than the cardinality of J_{σ} since it is possible for $\sigma(j) = \sigma(j') \in \mathrm{Map}_{\mathrm{Set}_{\Delta}}(\Delta^n, K')$ even if $j \neq j'$.

Let $E(\mathcal{G}')^{(k)}(C')$ be the full simplicial subset of $E(\mathcal{G}')(C')$ spanned by those simplices which are of index $\leq k$. It is easy to see that that $E(\mathcal{G}')^{(k)}(C')$

depends functorially on C', so we can view $E(\mathcal{G}')^{(k)}$ as an object of $\mathrm{Set}_{\Delta}^{\mathrm{Env}(\mathcal{C})}$. We observe that

$$E(\mathcal{G}) \simeq E(\mathcal{G}')^{(0)} \subseteq E(\mathcal{G}')^{(1)} \subseteq \cdots$$

and that the union of this sequence is $E(\mathcal{G}')$. Consequently, it will suffice to prove that each of the inclusions $E(\mathcal{G}')^{(k-1)} \subseteq E(\mathcal{G}')^{(k)}$ is a projective cofibration.

First, we need a bit of notation. Let us say that a simplex of K'^k is *new* if it consists of k distinct simplices of K', none of which belong to K. We will say that a simplex of K'^k is *old* if it is not new. The collection of old simplices of K'^k determines a simplicial subset which we will denote by $K'^{(k)}$. We define a functor $\psi : \mathrm{Env}(\mathcal{C}) \to \mathrm{Env}(\mathcal{C})$ by the formula

$$\psi(J, \{C_j'\}_{j \in J}) = (J \cup \{1, \ldots, k\}, \{C_j'\}_{j \in J} \cup \{C\}_{\{1 \ldots k\}}).$$

Let $\psi^* : \mathrm{Set}_{\Delta}^{\mathrm{Env}(\mathcal{C})} \to \mathrm{Set}_{\Delta}^{\mathrm{Env}(\mathcal{C})}$ be given by composition with ψ and let $\psi_! : \mathrm{Set}_{\Delta}^{\mathrm{Env}(\mathcal{C})} \to \mathrm{Set}_{\Delta}^{\mathrm{Env}(\mathcal{C})}$ be a left adjoint to ψ^* (a functor of left Kan extension). Since ψ^* preserves projective fibrations and weak equivalences, $\psi_!$ preserves projective cofibrations.

Recall that $\mathrm{Set}_{\Delta}^{\mathrm{Env}(\mathcal{C})}$ is tensored over the category of simplicial sets: given an object $\mathcal{M} \in \mathrm{Set}_{\Delta}^{\mathrm{Env}(\mathcal{C})}$ and a simplicial set A, we let $\mathcal{M} \otimes A \in \mathrm{Set}_{\Delta}^{\mathrm{Env}(\mathcal{C})}$ be defined by the formula $(\mathcal{M} \otimes A)(D') = \mathcal{M}(D') \times A$. If $\mathcal{M}$ is projectively cofibrant, then the operation $\mathcal{M} \mapsto \mathcal{M} \otimes A$ preserves cofibrations in A.

There is an obvious map $E(\mathcal{G}) \otimes K'^k \to \psi^* E(\mathcal{G}')^{(k)}$ which restricts to a map $E(\mathcal{G}) \otimes K'^{(k)} \to \psi^* E(\mathcal{G}')^{(k-1)}$. Passing to adjoints, we obtain a commutative diagram

$$
\begin{array}{ccc}
\psi_!(E(\mathcal{G}) \otimes K'^{(k)}) & \longrightarrow & E(\mathcal{G}')^{(k-1)} \\
\downarrow & & \downarrow \\
\psi_!(E(\mathcal{G}) \otimes K'^k) & \longrightarrow & E(\mathcal{G}')^{(k)}.
\end{array}
$$

An easy computation shows that this diagram is coCartesian. Since $E(\mathcal{G})$ is projectively cofibrant, the above remarks imply that the left vertical map is a projective cofibration. It follows that the right vertical map is a projective cofibration as well, which completes the proof. $\square$

The proof of Lemma 5.5.9.7 is somewhat more difficult and will require some preliminaries.

Notation 5.5.9.11. Let $\mathcal{M} : \mathrm{Set} \to \mathrm{Set}$ be the covariant functor which associates to each set S the collection of $\mathcal{M}(S)$ of nonempty finite subsets of S. If K is a simplicial set, we let $\mathcal{M}(K)$ denote the composition of K with $\mathcal{M}$, so that an m-simplex of $\mathcal{M}(K)$ is a finite nonempty collection of m-simplices of K.

Lemma 5.5.9.12. *Let K be a finite simplicial set and let $X \subseteq \mathcal{M}(K^{\triangleright}) \times \Delta^n$ be a simplicial subset with the following properties:*

(i) *The projection* $X \to \Delta^n$ *is surjective.*

(ii) *If* $\tau = (\tau', \tau'') : \Delta^m \to \mathcal{M}(K^\triangleright) \times \Delta^n$ *belongs to* X *and* $\tau' \subseteq \overline{\tau}'$ *as subsets of* $\mathrm{Hom}_{\mathrm{Set}_\Delta}(\Delta^m, K^\triangleright)$, *then* $(\overline{\tau}', \tau'') : \Delta^m \to \mathcal{M}(K^\triangleright) \times \Delta^n$ *belongs to* X.

Then X *is weakly contractible.*

Proof. Let $X' \subseteq X$ be the simplicial subset spanned by those simplices $\tau = (\tau', \tau'') : \Delta^m \to \mathcal{M}(K^\triangleright) \times \Delta^n$ which factor through X and for which $\tau' \subseteq \mathrm{Hom}_{\mathrm{Set}_\Delta}(\Delta^m, K^\triangleright)$ includes the constant simplex at the cone point of $K^\triangleright$. Our first step is to show that X' is a deformation retract of X. More precisely, we will construct a map

$$h : \mathcal{M}(K^\triangleright) \times \Delta^n \times \Delta^1 \to \mathcal{M}(K^\triangleright) \times \Delta^n$$

with the following properties:

(a) The map h carries $X \times \Delta^1$ into X and $X' \times \Delta^1$ into X'.

(b) The restriction $h| \mathcal{M}(K^\triangleright) \times \Delta^n \times \{0\}$ is the identity map.

(c) The restriction $h|X \times \{1\}$ factors through X'.

The map h will be the product of a map $h' : \mathcal{M}(K^\triangleright) \times \Delta^1 \to \mathcal{M}(K^\triangleright)$ and the identity map on Δ^n. To define h', we consider an arbitrary simplex $\tau : \Delta^m \to \mathcal{M}(K^\triangleright) \times \Delta^1$ corresponding to a subset $S \subseteq \mathrm{Hom}_{\mathrm{Set}_\Delta}(\Delta^m, K^\triangleright)$ and a decomposition $[m] = \{0, \ldots, i\} \cup \{i+1, \ldots, m\}$. The subset $h'(\tau) \subseteq \mathrm{Hom}_{\mathrm{Set}_\Delta}(\Delta^m, K^\triangleright)$ is defined as follows: an arbitrary simplex $\sigma : \Delta^m \to K^\triangleright$ belongs to $h'(\tau)$ if there exists $\sigma' \in S$, $i < j \leq n$ such that $\sigma'|\Delta^{\{0, \ldots, j-1\}} = \sigma|\Delta^{\{0, \ldots, j-1\}}$, and $\sigma|\Delta^{\{j, \ldots, m\}}$ is constant at the cone point of $K^\triangleright$. It is easy to check that h' has the desired properties.

It remains to prove that X' is weakly contractible. At this point, it is convenient to work in the setting of *semisimplicial sets*: that is, we will ignore the degeneracy operations. Let X'' be the semisimplicial subset of $\mathcal{M}(K^\triangleright) \times \Delta^n$ spanned by those maps $\tau = (\tau', \tau'') : \Delta^m \to \mathcal{M}(K^\triangleright) \times \Delta^n$ for which $\tau' = \mathrm{Hom}_{\mathrm{Set}_\Delta}(\Delta^m, K^\triangleright)$ (we observe that X'' is not stable under the degeneracy operators on $\mathcal{M}(K^\triangleright) \times \Delta^n$). Assumptions (i) and (ii) guarantee that $X'' \subseteq X'$. Moreover, the projection $X \to \Delta^n$ induces an isomorphism of semisimplicial sets $X'' \to \Delta^n$. Consequently, it will suffice to prove that X'' is a deformation retract of X'.

The proof now proceeds by a variation on our earlier construction. Namely, we will define a map of semisimplicial sets

$$g : \mathcal{M}(K^\triangleright) \times \Delta^n \times \Delta^1 \to \mathcal{M}(K^\triangleright) \times \Delta^n$$

with the following properties:

(a) The map g carries $X' \times \Delta^1$ into X' and $X'' \times \Delta^1$ into X''.

(b) The restriction $g|X' \times \{1\}$ is the identity map.

(c) The restriction $g| \mathcal{M}(K^{\triangleright}) \times \Delta^n \times \{0\}$ factors through X'.

As before, g is the product of a map $g' : \mathcal{M}(K^{\triangleright}) \times \Delta^1 \to \mathcal{M}(K^{\triangleright})$ with the identity map on Δ^n. To define g', we consider an arbitrary simplex $\tau : \Delta^m \to \mathcal{M}(K^{\triangleright}) \times \Delta^1$, corresponding to a subset $S \subseteq \mathrm{Hom}_{\mathrm{Set}_\Delta}(\Delta^m, K^{\triangleright})$ and a decomposition $[m] = \{0, \ldots, i\} \cup \{i + 1, \ldots, m\}$. We let $g'(\tau) \subseteq \mathrm{Hom}_{\mathrm{Set}_\Delta}(\Delta^m, K^{\triangleright}) = S \cup S'$, where S' is the collection of all simplices $\sigma : \Delta^m \to K^{\triangleright}$ such that $\sigma|\Delta^{\{i+1,\ldots,m\}}$ is the constant map at the cone poine of $K^{\triangleright}$. It is readily checked that g' has the desired properties. $\square$

Lemma 5.5.9.13. *Let K be a contractible Kan complex and let*

$$X \subseteq \mathcal{M}(K) \times \Delta^n$$

be a simplicial subset with the following properties:

(i) *The projection $X \to \Delta^n$ is surjective.*

(ii) *If $\tau = (\tau', \tau'') : \Delta^m \to \mathcal{M}(K) \times \Delta^n$ belongs to X and $\tau' \subseteq \overline{\tau}'$ as subsets of $\mathrm{Hom}_{\mathrm{Set}_\Delta}(\Delta^m, K)$, then $(\overline{\tau}', \tau'') : \Delta^m \to \mathcal{M}(K) \times \Delta^n$ belongs to X.*

Then X is weakly contractible.

Proof. It will suffice to show that for every finite simplicial subset $X' \subseteq X$, the inclusion of X' into X is weakly nullhomotopic. Enlarging X' if necessary, we may assume that $X' = (\mathcal{M}(K') \times \Delta^n) \cap X$, where K' is a finite simplicial subset of K. By further enlargement, we may suppose that the map $X' \to \Delta^n$ is surjective. Since K is a contractible Kan complex, the inclusion $K' \subseteq K$ extends to a map $i : K'^{\triangleright} \to K$. Let $\overline{X} \subseteq \mathcal{M}(K'^{\triangleright}) \times \Delta^n$ denote the inverse image of X. Then the inclusion $X' \subseteq X$ factors through $\overline{X}$, and Lemma 5.5.9.12 implies that $\overline{X}$ is weakly contractible. $\square$

Proof of Lemma 5.5.9.7. Fix an object $C \in \mathcal{C}$. The simplicial set $L(\mathcal{F})(C)$ can be described as follows:

(∗) For every $n \geq 1$, every map $f : C_1 \times \cdots \times C_n \to C$ in $\mathcal{C}$, and every collection of simplices $\{\sigma_i : \Delta^k \to \mathcal{F}(C_i)\}$, there is a simplex $f(\{\sigma_i\}) : \Delta^k \to L(\mathcal{F})(C)$.

The simplices $f(\{\sigma_i\})$ satisfy relations which are determined by morphisms in the simplicial category $\mathrm{Env}(\mathcal{C})$.

To every k-simplex $\tau : \Delta^k \to L(\mathcal{F})(D)$ we can associate a nonempty finite subset $S_\tau \subseteq \mathrm{Hom}_{\mathrm{Set}_\Delta}(\Delta^k, K)$. If $\tau = f(\{\sigma_i\})$, we assign the set of images of the simplices σ_i under the canonical maps $\mathcal{F}(C_i) \to \mathcal{F}(1) = K$. It is easy to see that S_τ is independent of the representation $f(\{\sigma_i\})$ chosen for τ and depends functorially on τ. Consequently, we obtain a map of simplicial sets $L(\mathcal{F})(C) \to \mathcal{M}(K)$. Moreover, this map has the following properties:

(i) The product map $\beta' : L(\mathcal{F})(C) \to \mathcal{M}(K) \times L^+(\mathcal{F})(C)$ is a monomorphism of simplicial sets.

(ii) If a k-simplex $\tau = (\tau', \tau'') : \Delta^k \to \mathcal{M}(K) \to L^+(\mathcal{F})(C)$ belongs to the image of β and $\tau' \subseteq \overline{\tau}'$ as finite subsets of $\mathrm{Hom}_{\mathrm{Set}_\Delta}(\Delta^k, K)$, then $(\overline{\tau}', \tau'') : \Delta^k \to \mathcal{M}(K) \times L^+(\mathcal{F})(C)$ belongs to the image of β'.

We wish to show that $\beta : L(\mathcal{F})(C) \to L^+(\mathcal{F})(C)$ is a weak homotopy equivalence. It will suffice to show that for every simplex $\Delta^k \to L^+(\mathcal{F})(C)$, the fiber product $L(\mathcal{F})(C) \times_{L^+(\mathcal{F})(C)} \Delta^k$ is weakly contractible. In view of (i), we can identify this fiber product with a simplicial subset $X \subseteq \Delta^k \times \mathcal{M}(K)$. The surjectivity of β and condition (ii) imply that X satisfies the hypotheses of Lemma 5.5.9.13, so that X is weakly contractible, as desired. $\qquad\square$

The model category $\mathbf{A}$ appearing in Proposition 5.5.9.1 is very well suited to certain calculations, such as the formation of homotopy colimits of simplicial objects. The following result provides a precise formulation of this idea:

Proposition 5.5.9.14. *Let $\mathcal{C}$ be a category which admits finite products, and $\mathbf{A} \subseteq \mathrm{Set}_\Delta^{\mathcal{C}}$, $\mathcal{A} \subseteq \mathrm{Set}^{\mathcal{C}}$ the full subcategories spanned by the product-preserving functors. Let $\mathcal{F} : \Delta^{op} \to \mathbf{A}$ be a simplicial object of $\mathbf{A}$ which we can identify with a bisimplicial object $F : \Delta^{op} \times \Delta^{op} \to \mathcal{A}$. Composition with the diagonal*

$$\Delta^{op} \to \Delta^{op} \times \Delta^{op} \xrightarrow{F} \mathcal{A}$$

gives a simplicial object of $\mathcal{A}$ which we can identify with an object $|\mathcal{F}| \in \mathbf{A}$. Then the homotopy colimit of $\mathcal{F}$ is canonically isomorphic to $|\mathcal{F}|$ in the homotopy category $\mathrm{h}\mathbf{A}$.

The proof requires the following lemma:

Lemma 5.5.9.15. *Let $\mathcal{C}$ be a category which admits finite products and let $\mathbf{A} \subseteq \mathrm{Set}_\Delta^{\mathcal{C}}$ be the full subcategory spanned by the product-preserving functors. For every object $C \in \mathcal{C}$, the evaluation map $\mathbf{A} \to \mathrm{Set}_\Delta$ preserves homotopy colimits of simplicial objects.*

Proof. In view of Corollary 5.5.9.3 and Theorem 4.2.4.1, it will suffice to show that the evaluation functor $\mathcal{P}_\Sigma(\mathrm{N}(\mathcal{C})^{op}) \to \mathrm{Set}_\Delta$ preserves $\mathrm{N}(\Delta)^{op}$-indexed colimits. This follows from Proposition 5.5.8.10 since $\mathrm{N}(\Delta)^{op}$ is sifted (see Lemma 5.5.8.4). $\qquad\square$

Proof of Proposition 5.5.9.14. Since $\mathbf{A}$ is a combinatorial simplicial model category, Corollary A.2.9.30 implies the existence of a canonical map

$$\gamma : \mathrm{hocolim}\, \mathcal{F} \to |\mathcal{F}|$$

in the homotopy category $\mathrm{h}\mathbf{A}$; we wish to prove that γ is an isomorphism. To prove this, it will suffice to show that the induced map

$$\gamma_C : (\mathrm{hocolim}\, \mathcal{F})(C) \to |\mathcal{F}|(C)$$

is an isomorphism in the homotopy category of simplicial sets for each object $C \in \mathcal{C}$. This map fits into a commutative diagram

$$
\begin{array}{ccc}
\mathrm{hocolim}(\mathcal{F}(C)) & \xrightarrow{\gamma'_C} & |\mathcal{F}(C)| \\
\downarrow & & \downarrow \\
\mathrm{hocolim}(\mathcal{F})(C) & \longrightarrow & |\mathcal{F}|(C).
\end{array}
$$

The left vertical map is an isomorphism in the homotopy category of simplicial sets by Lemma 5.5.9.15, the right vertical map is evidently an isomorphism, and the map γ'_C is an isomorphism in the homotopy category by Example A.2.9.31; it follows that γ_C is also an isomorphism, as desired. □

Chapter Six

∞-Topoi

In this chapter, we come to the main subject of this book: the theory of ∞-topoi. Roughly speaking, an ∞-topos is an ∞-category which "looks like" the ∞-category of spaces, just as an ordinary topos is a category which "looks like" the category of sets. As in classical topos theory, there are various ways of making this precise. We will begin in §6.1 by reviewing several possible definitions and proving that they are equivalent to one another.

The main result of §6.1 is Theorem 6.1.0.6, which asserts that an ∞-category $\mathfrak{X}$ is an ∞-topos if and only if $\mathfrak{X}$ arises as an (accessible) left exact localization of an ∞-category of presheaves. In §6.2, we consider the problem of *constructing* left exact localizations. In classical topos theory, there is a bijective correspondence between left exact localizations of a presheaf category $\mathcal{P}(\mathcal{C})$ and Grothendieck topologies on $\mathcal{C}$. In the ∞-category categorical context, one can again use Grothendieck topologies to construct examples of left exact localizations. Unfortunately, not every ∞-topos arises in this way. Nevertheless, the construction of an ∞-category of sheaves $\mathrm{Shv}(\mathcal{C})$ from a Grothendieck topology on $\mathcal{C}$ is an extremely useful construction, which will play an important role throughout Chapter 7.

In order to understand higher topos theory, we will need to consider ∞-topoi not only individually but in relation to one another. In §6.3, we will introduce the notion of a *geometric morphism* of ∞-topoi. The collection of all ∞-topoi and geometric morphisms between them can be organized into an ∞-category $\mathcal{RTop}$. We will study the problem of constructing colimits and (certain) limits in $\mathcal{RTop}$. In the course of doing so, we will show that the class of ∞-topoi is stable under various categorical constructions.

One of our goals in this book is to apply ideas from higher category theory to study more classical mathematical objects such as topological spaces or ordinary topoi. In order to do so, it is convenient to work in a setting where all of these objects can be considered on the same footing. In §6.4, we will introduce the definition of an *n-topos* for all $0 \leq n \leq \infty$. When $n = \infty$, this will reduce to the theory introduced in §6.1. The case $n = 1$ will recover classical topos theory, and the case $n = 0$ is *almost* equivalent to the theory of topological spaces. We will study the theory of n-topoi and introduce constructions which allow us to pass between n-topoi and ∞-topoi. In particular, we associate an ∞-topos $\mathrm{Shv}(X)$ to every topological space X, which will be the primary object of study in Chapter 7.

There are several different ways of thinking about what an ∞-topos $\mathfrak{X}$ is. On the one hand, we can view $\mathfrak{X}$ as a generalized topological space; on the other hand, we can think of $\mathfrak{X}$ as an alternative universe in which we can do homotopy theory. In §6.5, we will reinforce the second point of view by studying the internal homotopy theory of an ∞-topos $\mathfrak{X}$. Just as in classical homotopy theory, one can define homotopy groups, Postnikov towers, Eilenberg MacLane spaces, and so forth. In Chapter 7, we will bring together these two points of view by showing that classical geometric properties of a topological space X are reflected in the internal homotopy of the ∞-topos $\mathrm{Shv}(X)$ of sheaves on X.

There are several papers on higher topos theory in the literature. The papers [74] and [11] both discuss a notion of 2-topos (the second from an elementary point of view). However, the basic model for these 2-topoi is the 2-category of (small) categories rather than the 2-category of (small) groupoids. Jardine ([41]) has exhibited a model structure on the category of simplicial presheaves on a Grothendieck site, and the ∞-category associated to this model category is an ∞-topos in our sense. This construction is generalized from ordinary categories with a Grothendieck topology to simplicial categories with a Grothendieck topology in [78] (and again produces ∞-topoi). However, not every ∞-topos arises in this way: one can construct only ∞-topoi which are *hypercomplete* (called *t-complete* in [78]); we will summarize the situation in Section 6.5.2. Our notion of an ∞-topos is *essentially* equivalent to the notion of a *Segal topos* introduced in [78] and to Charles Rezk's notion of a *model topos*. We note also that the paper [78] has considerable overlap with the ideas discussed here.

6.1 ∞-TOPOI: DEFINITIONS AND CHARACTERIZATIONS

Before we study the ∞-categorical version of topos theory, it seems appropriate to briefly review the classical theory. Recall that a *topos* is a category $\mathcal{C}$ which behaves like the category of sets or (more generally) the category of sheaves of sets on a topological space. There are several (equivalent) ways of making this idea precise. The following result is proved (in a slightly different form) in [2]:

Proposition 6.1.0.1. *Let $\mathcal{C}$ be a category. The following conditions are equivalent:*

(A) *The category $\mathcal{C}$ is (equivalent to) the category of sheaves of sets on some Grothendieck site.*

(B) *The category $\mathcal{C}$ is (equivalent to) a left exact localization of the category of presheaves of sets on some small category $\mathcal{C}_0$.*

(C) *Giraud's axioms are satisfied:*

    (i) *The category $\mathcal{C}$ is presentable (that is, $\mathcal{C}$ has small colimits and a set of small generators).*

(ii) *Colimits in* $\mathcal{C}$ *are universal.*

(iii) *Coproducts in* $\mathcal{C}$ *are disjoint.*

(iv) *Equivalence relations in* $\mathcal{C}$ *are effective.*

Definition 6.1.0.2. A category $\mathcal{C}$ is called a *topos* if it satisfies the equivalent conditions of Proposition 6.1.0.1.

Remark 6.1.0.3. A reader who is unfamiliar with some of the terminology used in the statement of Proposition 6.1.0.1 should not worry: we will review the meaning of each condition in §6.1.1 as we search for ∞-categorical generalizations of axioms (i) through (iv).

Our goal in this section is to introduce the ∞-categorical analogue of Definition 6.1.0.2. Proposition 6.1.0.1 suggests several possible approaches. We begin with the simplest of these:

Definition 6.1.0.4. Let $\mathcal{X}$ be an ∞-category. We will say that $\mathcal{X}$ is an ∞-*topos* if there exists a small ∞-category $\mathcal{C}$ and an accessible left exact localization functor $\mathcal{P}(\mathcal{C}) \to \mathcal{X}$.

Remark 6.1.0.5. Definition 6.1.0.4 involves an accessibility condition which was not mentioned in Proposition 6.1.0.1. This is because every left exact localization of a category of *set*-valued presheaves is automatically accessible (see Proposition 6.4.3.9). We do not know if the corresponding result holds for $\mathcal{S}$-valued presheaves. However, it is true under a suitable hypercompleteness assumption: see [79].

Adopting Definition 6.1.0.4 amounts to selecting an *extrinsic* approach to higher topos theory: the class of ∞-topoi is defined to be the smallest collection of ∞-categories which contains $\mathcal{S}$ and is stable under certain constructions (left exact localizations and the formation of functor categories). The main objective of this section is to give several reformulations of Definition 6.1.0.2 which have a more intrinsic flavor. Our results may be summarized in the following statement (all our our terminology will be explained later in this section):

Theorem 6.1.0.6. *Let* $\mathcal{X}$ *be an* ∞-*category. The following conditions are equivalent:*

(1) *The* ∞-*category* $\mathcal{X}$ *is an* ∞-*topos.*

(2) *The* ∞-*category* $\mathcal{X}$ *is presentable, and for every small simplicial set* K *and every natural transformation* $\overline{\alpha} : \overline{p} \to \overline{q}$ *of diagrams* $\overline{p}, \overline{q} : K^{\triangleright} \to \mathcal{X}$, *the following condition is satisfied:*

  – *If* $\overline{q}$ *is a colimit diagram and* $\alpha = \overline{\alpha}|K$ *is a Cartesian transformation, then* $\overline{p}$ *is a colimit diagram if and only if* $\overline{\alpha}$ *is a Cartesian transformation.*

(3) *The ∞-category X satisfies the following ∞-categorical analogues of Giraud's axioms:*

    (*i*) *The ∞-category X is presentable.*

    (*ii*) *Colimits in X are universal.*

    (*iii*) *Coproducts in X are disjoint.*

    (*iv*) *Every groupoid object of X is effective.*

We will review the meanings of conditions (*i*) through (*iv*) in §6.1.1 and §6.1.2. In §6.1.3, we will give several equivalent formulations of (2) and prove the implications (1) ⇒ (2) ⇒ (3). The implication (3) ⇒ (1) is the most difficult; we will give the proof in §6.1.5 after establishing a crucial technical lemma in §6.1.4. Finally, in §6.1.6 we will establish yet another characterization of ∞-topoi based on the theory of classifying objects.

Remark 6.1.0.7. The characterization of the class of ∞-topoi given in part (2) of Theorem 6.1.0.6 is due to Rezk, as are many of the ideas presented in §6.1.3.

The equivalence (1) ⇔ (3) of Theorem 6.1.0.6 can be viewed as an ∞-categorical analogue of the equivalence (*B*) ⇔ (*C*) in Proposition 6.1.0.1. It is natural to ask if there is also some equivalent of the characterization (*A*). To put the question another way: given a small ∞-category C, does there exist some natural description of the class of all left exact localizations of C? Experience with classical topos theory suggests that we might try to characterize such localizations in terms of *Grothendieck topologies* on C. We will introduce a theory of Grothendieck topologies on ∞-categories in §6.2.2 and show that every Grothendieck topology on C determines a left exact localization of 𝒫(C). However, it turns out that not every ∞-topos arises via this construction. This raises a natural question: is it possible to give an explicit description of *all* left exact localizations of 𝒫(C), perhaps in terms of some more refined theory of Grothendieck topologies? We will give a partial answer to this question in §6.5.

6.1.1 Giraud's Axioms in the ∞-Categorical Setting

Our goal in this section is to formulate higher-categorical analogues of conditions (*i*) through (*iv*) of Proposition 6.1.0.1. We consider each axiom in turn. In each case, our objective is to find an analogous axiom which makes sense in the setting of ∞-categories and is satisfied by the ∞-category 𝒮 of spaces.

(*i*) The category C is presentable.

The generalization to the case where C is a ∞-category is obvious: we should merely require C to be a presentable ∞-category in the sense of Definition 5.5.0.1. According to Example 5.5.1.8, this condition is satisfied when C is the ∞-category of spaces.

(*ii*) Colimits in $\mathcal{C}$ are universal.

Let us first recall the meaning of this condition in classical category theory. If the axiom (*i*) is satisfied, then $\mathcal{C}$ is presentable and therefore admits all (small) limits and colimits. In particular, every diagram

$$X \to S \xleftarrow{f} T$$

has a limit $X_T = X \times_S T$. This construction determines a functor

$$f^* : \mathcal{C}_{/S} \to \mathcal{C}_{/T}$$

$$X \mapsto X_T,$$

which is a right adjoint to the functor given by composition with f.

We say that *colimits in $\mathcal{C}$ are universal* if the functor f^* is preserves colimits for every map $f : T \to S$ in $\mathcal{C}$. (In other words, colimits are universal in $\mathcal{C}$ if any colimit in $\mathcal{C}$ *remains* a colimit in $\mathcal{C}$ after pulling back along a morphism $T \to S$.)

Let us now attempt to make this notion precise in the setting of an arbitrary ∞-category $\mathcal{C}$. Let $\mathcal{O}_{\mathcal{C}} = \mathrm{Fun}(\Delta^1, \mathcal{C})$ and let $p : \mathcal{O}_{\mathcal{C}} \to \mathcal{C}$ be given by evaluation at $\{1\} \subseteq \Delta^1$. Corollary 2.4.7.12 implies that p is a coCartesian fibration.

Lemma 6.1.1.1. *Let $\mathcal{X}$ be an ∞-category and let $p : \mathcal{O}_{\mathcal{X}} \to \mathcal{X}$ be defined as above. Let F be a morphism in $\mathcal{O}_{\mathcal{X}}$ corresponding to a diagram $\sigma : \Delta^1 \times \Delta^1 \simeq (\Lambda_2^2)^{\triangleleft} \to \mathcal{X}$, which we will denote by*

$$
\begin{array}{ccc}
X' & \xrightarrow{f'} & Y' \\
\downarrow & & \downarrow{\scriptstyle g} \\
X & \xrightarrow{f} & Y.
\end{array}
$$

Then F is p-Cartesian if and only if the above diagram is a pullback in $\mathcal{X}$. In particular, p is a Cartesian fibration if and only if the ∞-category $\mathcal{X}$ admits pullbacks.

Proof. For every simplicial set K, let K^+ denote the full simplicial subset of $(K \star \{x\} \star \{y\}) \times \Delta^1$ spanned by all of the vertices except $(x, 0)$ and define a simplicial set $\mathcal{C}$ by setting

$$\mathrm{Fun}(K, \mathcal{C}) = \{m : K^+ \to \mathcal{X} : m|(\{x\} \star \{y\}) \times \{1\} = f, m|\{y\} \times \Delta^1 = g\}.$$

We observe that we have a commutative diagram

$$
\begin{array}{ccc}
\mathcal{C} & \longrightarrow & (\mathcal{O}_{\mathcal{X}})_{/g} \\
\downarrow & & \downarrow \\
\mathcal{X}_{/f} & \longrightarrow & \mathcal{X}_{/Y'}
\end{array}
$$

which induces a map $q : \mathcal{C} \to (\mathcal{O}_{\mathcal{X}})_{/g} \times_{\mathcal{X}_{/Y'}} \mathcal{X}_{/f}$. We first claim that q is a trivial fibration. Unwinding the definitions, we observe that the right lifting property of q with respect to an inclusion $\partial \Delta^n \subseteq \Delta^n$ follows from the extension property of X with respect to Λ_{n+1}^{n+2}, which follows in turn from our assumption that X is an ∞-category.

The inclusion $K^+ \subseteq K \times \Delta^1$ induces a projection $q' : (\mathcal{O}_X)_{/F} \to \mathcal{C}$ which fits into a pullback diagram

$$
\begin{array}{ccc}
(\mathcal{O}_X)_{/F} & \longrightarrow & \mathcal{C} \\
\downarrow & & \downarrow g \\
X_{/\sigma} & \xrightarrow{q''} & X_{/\sigma | \Lambda_2^2} .
\end{array}
$$

It follows that q' is a right fibration and that q' is trivial if σ is a pullback diagram. Conversely, we observe that $(\Lambda_2^2)^\lhd$ is a retract of $(\Delta^0)^+$, so that the map g is surjective on vertices. Consequently, if q' is a trivial fibration, then the fibers of q'' are contractible, so that q'' is a trivial fibration (Lemma 2.1.3.4) and σ is a pullback diagram.

By definition, F is p-Cartesian if and only if the composition

$$ q \circ q' : (\mathcal{O}_X)_{/F} \to (\mathcal{O}_X)_{/g} \times_{X_{/Y'}} X_{/f} $$

is a trivial fibration. Since q is a trivial fibration and q' is a right fibration, this is also equivalent to the assertion that q' is a trivial fibration (Lemma 2.1.3.4). $\qquad\square$

Now suppose that X is an ∞-category which admits pullbacks, so that the projection $p : \mathcal{O}_X \to X$ is both a Cartesian fibration and a coCartesian fibration. Let $f : S \to T$ be a morphism in X. Taking the pullback of p along the corresponding map $\Delta^1 \to X$, we obtain a correspondence from $p^{-1}(S) = X^{/S}$ to $p^{-1}(T) = X^{/T}$ associated to a pair of adjoint functors

$$ X^{/X} \underset{f^*}{\overset{f_!}{\rightleftarrows}} X^{/T} . $$

The functors $f_!$ and f^* are well-defined up to homotopy (in fact, up to a contractible space of choices). We may think of $f_!$ as the functor given by composition with f, and f^* as the functor given by pullback along f (in view of Lemma 6.1.1.1).

We can now formulate the ∞-categorical analogue of (ii):

Definition 6.1.1.2. Let $\mathcal{C}$ be a presentable ∞-category. We will say that *colimits in $\mathcal{C}$ are universal* if, for any morphism $f : T \to S$ in $\mathcal{C}$, the associated pullback functor

$$ f^* : \mathcal{C}^{/S} \to \mathcal{C}^{/T} $$

preserves (small) colimits.

Assume that $\mathcal{C}$ is a presentable ∞-category and let $f : T \to S$ be a morphism in $\mathcal{C}$. By the adjoint functor theorem, $f^* : \mathcal{C}^{/S} \to \mathcal{C}^{/T}$ preserves all colimits if and only if it has a right adjoint f_*. Since the existence of adjoint functors can be tested inside the enriched homotopy category, this gives a convenient criterion which allows us to test whether or not colimits in $\mathcal{C}$ are universal.

Remark 6.1.1.3. Let $\mathfrak{X}$ be an ∞-category. The assumption that colimits in $\mathfrak{X}$ are universal can be viewed as a kind of distributive law. We have the following table of vague analogies:

Higher Category Theory	Algebra
∞-Category	Set
Presentable ∞-category	Abelian group
Colimits	Sums
Limits	Products
$\varinjlim(X_\alpha) \times_S T \simeq \varinjlim(X_\alpha \times_S T)$	$(x+y)z = xz + yz$
∞-Topos	Commutative ring

Definition 6.1.1.2 has a reformulation in the language of classifying functors (§3.3.2):

Proposition 6.1.1.4. *Let $\mathfrak{X}$ be an ∞-category which admits finite limits. The following conditions are equivalent:*

(1) *The ∞-category $\mathfrak{X}$ is presentable, and colimits in $\mathfrak{X}$ are universal.*

(2) *The Cartesian fibration $p : \mathcal{O}_{\mathfrak{X}} \to \mathfrak{X}$ is classified by a functor $\mathfrak{X}^{op} \to \mathfrak{Pr}^{L}$.*

Proof. We can restate condition (2) as follows: each fiber $\mathfrak{X}^{/U}$ of p is presentable, and each of the pullback functors $f^* : \mathfrak{X}^{/V} \to \mathfrak{X}^{/U}$ preserves small colimits. It is clear that $(1) \Rightarrow (2)$ and that (2) implies that colimits in $\mathfrak{X}$ are universal. Since $\mathfrak{X}$ admits finite limits, it has a final object 1; condition (2) implies that $\mathfrak{X} \simeq \mathfrak{X}^{/1}$ is presentable, which proves (1). $\qquad\square$

(iii) Coproducts in $\mathcal{C}$ are disjoint.

If $\mathcal{C}$ is an ∞-category which admits finite coproducts, then we will say that *coproducts in $\mathcal{C}$ are disjoint* if every coCartesian diagram

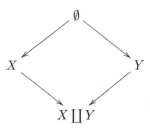

is also Cartesian, provided that $\emptyset$ is an initial object of $\mathcal{C}$. More informally, to say that coproducts are disjoint is to say that the intersection of X and Y inside the union $X \coprod Y$ is empty.

We now come to the most subtle and interesting of Giraud's axioms:

(iv) Every equivalence relation in $\mathcal{C}$ is effective.

Recall that if X is an object in an (ordinary) category $\mathcal{C}$, then an *equivalence relation* R on X is an object of $\mathcal{C}$ equipped with a map $p : R \to X \times X$ such that for any S, the induced map

$$\mathrm{Hom}_{\mathcal{C}}(S, R) \to \mathrm{Hom}_{\mathcal{C}}(S, X) \times \mathrm{Hom}_{\mathcal{C}}(S, X)$$

exhibits $\mathrm{Hom}_{\mathcal{C}}(S, R)$ as an equivalence relation on $\mathrm{Hom}_{\mathcal{C}}(S, X)$.

If $\mathcal{C}$ admits finite limits, then it is easy to construct equivalence relations in $\mathcal{C}$: given any map $X \to Y$ in $\mathcal{C}$, the induced map $X \times_Y X \to X \times X$ is an equivalence relation on X. If the category $\mathcal{C}$ admits finite colimits, then one can attempt to invert this process: given an equivalence relation R on X, one can form the coequalizer of the two projections $R \to X$ to obtain an object which we will denote by X/R. In the category of sets, one can recover R as the fiber product $X \times_{X/R} X$. In general, this need not occur: one always has $R \subseteq X \times_{X/R} X$, but the inclusion may be strict (as subobjects of $X \times X$). If equality holds, then R is said to be an *effective equivalence relation* and the map $X \to X/R$ is said to be an *effective epimorphism*.

Remark 6.1.1.5. Recall that a map $f : X \to Y$ in a category $\mathcal{C}$ is said to be a *categorical epimorphism* if the natural map $\mathrm{Hom}_{\mathcal{C}}(Y, Z) \to \mathrm{Hom}_{\mathcal{C}}(X, Z)$ is *injective* for every object $Z \in \mathcal{C}$, so that we may identify $\mathrm{Hom}_{\mathcal{C}}(Y, Z)$ with a subset of $\mathrm{Hom}_{\mathcal{C}}(X, Z)$. To say that f is an *effective* epimorphism is to say that we can characterize this subset: it is the collection of all maps $g : X \to Z$ such that the diagram

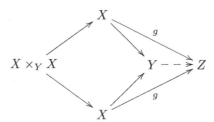

commutes (which is obviously a necessary condition for the indicated dotted arrow to exist).

Using the terminology introduced above, we can neatly summarize some of the fundamental properties of the category of sets:

Fact 6.1.1.6. *In the category of sets, every equivalence relation is effective and the effective epimorphisms are precisely the surjective maps.*

The first assertion of Fact 6.1.1.6 remains valid in any topos. According to the axiomatic point of view, it is one of the defining features of a topos.

If $\mathcal{C}$ is a category with finite limits and colimits in which all equivalence relations are effective, then we obtain a one-to-one correspondence between equivalence relations on an object X and *quotients* of X (that is, isomorphism classes of effective epimorphisms $X \to Y$). This correspondence is extremely useful because it allows us to make elementary descent arguments: one can deduce statements about quotients of X from statements about X and about equivalence relations on X (which live over X). We would like to formulate an ∞-categorical analogue of this condition which will allow us to make similar arguments.

In the ∞-category $\mathcal{S}$ of spaces, the situation is more complicated. The correct notion of surjection of spaces $X \to Y$ is a map which induces a surjection on path components $\pi_0 X \to \pi_0 Y$. However, in this case, the (homotopy) fiber product $R = X \times_Y X$ does not give an equivalence relation on X because the map $R \to X \times X$ is not necessarily injective in any reasonable sense. However, it does retain some of the pleasant features of an equivalence relation: instead of transitivity, we have a *coherently associative* composition law $R \times_X R \to R$ (this is perhaps most familiar in the situation where X is a point: in this case, R can be identified with the based loop space of Y, which is endowed with a multiplication given by concatenation of loops). In §6.1.2, we will make this idea precise and define *groupoid objects* and *effective groupoid objects* in an arbitrary ∞-category. Granting these notions for the moment, we have a natural candidate for the ∞-categorical generalization of condition (iv):

$(iv)'$ Every groupoid object of $\mathcal{C}$ is effective.

6.1.2 Groupoid Objects

Let $\mathcal{C}$ be a category which admits finite limits. A *groupoid object* of $\mathcal{C}$ is a functor F from $\mathcal{C}$ to the category $\mathcal{C}at$ of (small) groupoids, which has the following properties:

(1) There exists an object $X_0 \in \mathcal{C}$ and a (functorial) identification of $\mathrm{Hom}_{\mathcal{C}}(C, X_0)$ with the set of objects in the groupoid $F(C)$ for each $C \in \mathcal{C}$.

(2) There exists an object $X_1 \in \mathcal{C}$ and a (functorial) identification of $\mathrm{Hom}_{\mathcal{C}}(C, X_1)$ with the set of morphisms in groupoid $F(C)$ for each $C \in \mathcal{C}$.

Example 6.1.2.1. Let $\mathcal{C}$ be the category $\mathcal{S}et$ of sets. Then a groupoid object of $\mathcal{C}$ is simply a (small) groupoid.

Giving a groupoid object of a category $\mathcal{C}$ is equivalent to giving a pair of objects $X_0 \in \mathcal{C}$ (the "object classifier") and $X_1 \in \mathcal{C}$ (the "morphism classifier") together with a collection of maps which relate X_0 to X_1 and satisfy appropriate identities, which imitate the usual axiomatics of category theory. These identities can be very efficiently encoded using the formalism of simplicial objects. For every $n \geq 0$, let $[n]$ denote the category associated to the linearly ordered set $\{0, \ldots, n\}$ and consider the functor $F_n : \mathcal{C} \to \mathrm{Set}$ defined so that

$$F_n(C) = \mathrm{Hom}_{\mathcal{C}at}([n], F(C)).$$

By assumption, F_0 and F_1 are representable by objects $X_0, X_1 \in \mathcal{C}$. Since $\mathcal{C}$ is stable under finite limits, it follows that

$$F_n = F_1 \times_{F_0} \cdots \times_{F_0} F_1$$

is representable by an object $X_n = X_1 \times_{X_0} \cdots \times_{X_0} X_1$. The objects X_n can be assembled into a simplicial object $X_\bullet$ of $\mathcal{C}$. We can think of this construction as a generalization of the process which associates to every groupoid $\mathcal{D}$ its nerve $\mathrm{N}(\mathcal{D})$ (a simplicial set). Moreover, as in the classical case, the association $F \mapsto X_\bullet$ is fully faihtful. In other words, we can identify groupoid objects of $\mathcal{C}$ with the corresponding simplicial objects. Of course, not every simplicial object $X_\bullet$ of $\mathcal{C}$ arises via this construction. This is true if and only if certain additional conditions are met: for instance, the diagram

$$
\begin{array}{ccc}
X_2 & \xrightarrow{d_0} & X_1 \\
\downarrow{\scriptstyle d_2} & & \downarrow{\scriptstyle d_1} \\
X_1 & \xrightarrow{d_0} & X_0
\end{array}
$$

must be Cartesian.

The purpose of this section is to generalize the notion of a groupoid object to the setting where $\mathcal{C}$ is an ∞-category. We begin by introducing the class of *simplicial objects* of $\mathcal{C}$; we then define groupoid objects to be simplicial objects which satisfy additional conditions.

Definition 6.1.2.2. Let $\boldsymbol{\Delta}_+$ denote the category of finite (possibly empty) linearly ordered sets. A *simplicial object* of an ∞-category $\mathcal{C}$ is a map of ∞-categories

$$U_\bullet : \mathrm{N}(\boldsymbol{\Delta})^{op} \to \mathcal{C}.$$

An *augmented simplicial object* of $\mathcal{C}$ is a map

$$U_\bullet^+ : \mathrm{N}(\boldsymbol{\Delta}_+)^{op} \to \mathcal{C}.$$

We let $\mathcal{C}_\Delta$ denote the ∞-category $\mathrm{Fun}(\mathrm{N}(\boldsymbol{\Delta})^{op}, \mathcal{C})$; we will refer to $\mathcal{C}_\Delta$ as the ∞-*category of simplicial objects of* $\mathcal{C}$. Similarly, we will refer to the ∞-category $\mathrm{Fun}(\mathrm{N}(\boldsymbol{\Delta}_+)^{op}, \mathcal{C})$ as the ∞-*category of augmented simplicial objects of* $\mathcal{C}$, and we will denote it by $\mathcal{C}_{\Delta_+}$.

If $U_\bullet$ is an (augmented) simplicial object of $\mathcal{C}$ and $n \geq 0$ $(n \geq -1)$, we will write U_n for the object $U([n]) \in \mathcal{C}$.

Remark 6.1.2.3. In the case where $\mathcal{C}$ is the nerve of an ordinary category $\mathcal{D}$, Definition 6.1.2.2 recovers the usual notion of a simplicial object of $\mathcal{D}$. More precisely, the ∞-category $\mathcal{C}_\Delta$ of simplicial objects of $\mathcal{C}$ is naturally isomorphic to the nerve of the category of simplicial objects of $\mathcal{D}$.

Lemma 6.1.2.4. *Let $f : X \to Y$ be a map of simplicial sets satisfying the following conditions:*

(1) *The map f induces a bijection $X_0 \to Y_0$ on vertex sets.*

(2) *The simplicial set Y is a Kan complex.*

(3) *The map f has the right lifting property with respect to every horn inclusion $\Lambda^n_i \subseteq \Delta^n$ for $n \geq 2$.*

(4) *The map f is a weak homotopy equivalence.*

Then f is a trivial Kan fibration.

Proof. In view of condition (4), it suffices to prove that f is a Kan fibration. In other words, we must show that p has the right lifting property with respect to every horn inclusion $\Lambda^n_i \subseteq \Delta^n$. If $n > 1$, this follows from (3). We may therefore reduce to the case where $n = 1$; by symmetry, we may suppose that $i = 0$.

Let $e : y \to y'$ be an edge of Y. Condition (1) implies that there is a (unique) pair of vertices $x, x' \in X_0$ with $y = f(x)$, $y' = f(x')$. Since f is a homotopy equivalence, there is a path p from x to x' in the topological space $|X|$ such that the induced path $|f| \circ p$ in $|Y|$ is homotopic to e via a homotopy which keeps the endpoints fixed. By cellular approximation, we may suppose that this path is contained in the 1-skeleton of $|X|$. Consequently, there is a positive integer k, a sequence of vertices $\{z_0, \ldots, z_k\}$ with $z_0 = x$, $z_k = x'$ such that each adjacent pair (z_i, z_{i+1}) is joined by an edge p_i (running in either direction), such that p is homotopic (relative to its boundary) to the path obtained by concatenating the edges p_i. Using conditions (2) and (3), we note that X has the extension property with respect to the inclusion $\Lambda^n_i \subseteq \Delta^n$ for each $n \geq 2$. It follows that we may assume that p_i runs from z_i to z_{i+1}: if it runs in the opposite direction, then we can extend the map

$$(p_i, s_0 z_i, \bullet) : \Lambda^2_2 \to X$$

to a 2-simplex $\sigma : \Delta^2 \to X$ and then replace p_i by $d_2 \sigma$.

Without loss of generality, we may suppose that $k > 0$ is chosen as small as possible. We claim that $k = 1$. Otherwise, we choose an extension $\tau : \Delta^2 \to X$ of the map

$$(p_1, \bullet, p_0) : \Lambda^2_1 \to X.$$

We can then replace the initial segment

$$z_0 \xrightarrow{p_0} z_1 \xrightarrow{p_1} z_2$$

by the edge $d_1(\tau) : z_0 \to z_2$ and obtain a shorter path from x to x', contradicting the minimality of k.

The edges e and $f(p_0)$ are homotopic in Y relative to their endpoints. Using (3), we see that p_0 is homotopic (relative to its endpoints) to an edge $\overline{e}$ which satisfies $f(\overline{e}) = e$. This completes the proof that f is a Kan fibration. □

Notation 6.1.2.5. Let K be a simplicial set. We let $\Delta_{/K}$ denote the *category of simplices of* K defined in §4.2.3. The objects of $\Delta_{/K}$ are pairs (J, η), where J is an object of Δ and $\eta \in \mathrm{Hom}_{\mathrm{Set}_\Delta}(\Delta^J, K)$. A morphism from (J, η) to (J', η') is a commutative diagram

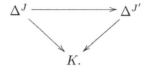

$$K.$$

Equivalently, we can describe $\Delta_{/K}$ as the fiber product $\Delta \times_{\mathrm{Set}_\Delta} (\mathrm{Set}_\Delta)_{/K}$.

If $\mathcal{C}$ is an ∞-category, $U : \mathrm{N}(\Delta)^{op} \to \mathcal{C}$ is a simplicial object of $\mathcal{C}$, and K is a simplicial set, then we let $U[K]$ denote the composite map

$$\mathrm{N}(\Delta_{/K})^{op} \to \mathrm{N}(\Delta)^{op} \to \mathcal{C}.$$

Proposition 6.1.2.6. *Let $\mathcal{C}$ be an ∞-category and $U : \mathrm{N}(\Delta)^{op} \to \mathcal{C}$ a simplicial object of $\mathcal{C}$. The following conditions are equivalent:*

(1) *For every weak homotopy equivalence $f : K \to K'$ of simplicial sets which induces a bijection $K_0 \to K'_0$ on vertices, the induced map $\mathcal{C}_{/U[K']} \to \mathcal{C}_{/U[K]}$ is a categorical equivalence.*

(2) *For every cofibration $f : K \to K'$ of simplicial sets which is a weak homotopy equivalence and bijective on vertices, the induced map*

$$\mathcal{C}_{/U[K']} \to \mathcal{C}_{/U[K]}$$

is a categorical equivalence.

(2′) *For every cofibration $f : K \to K'$ of simplicial sets which is a weak homotopy equivalence and bijective on vertices, the induced map*

$$\mathcal{C}_{/U[K']} \to \mathcal{C}_{/U[K]}$$

is a trivial fibration.

(3) *For every $n \geq 2$ and every $0 \leq i \leq n$, the induced map*

$$\mathcal{C}_{/U[\Delta^n]} \to \mathcal{C}_{/U[\Lambda_i^n]}$$

is a categorical equivalence.

(3′) *For every $n \geq 2$ and every $0 \leq i \leq n$, the induced map*

$$\mathcal{C}_{/U[\Delta^n]} \to \mathcal{C}_{/U[\Lambda_i^n]}$$

is a trivial fibration.

(4) *For every $n \geq 0$ and every partition $[n] = S \cup S'$ such that $S \cap S'$ consists of a single element s, the induced map*

$$\mathcal{C}_{/U[\Delta^n]} \rightarrow \mathcal{C}_{/U[K]}$$

is a categorical equivalence, where $K = \Delta^S \coprod_{\{s\}} \Delta^{S'} \subseteq \Delta^n$.

(4′) *For every $n \geq 0$ and every partition $[n] = S \cup S'$ such that $S \cap S'$ consists of a single element s, the induced map*

$$\mathcal{C}_{/U[\Delta^n]} \rightarrow \mathcal{C}_{/U[K]}$$

is a trivial fibration, where $K = \Delta^S \coprod_{\{s\}} \Delta^{S'} \subseteq \Delta^n$.

(4″) *For every $n \geq 0$ and every partition $[n] = S \cup S'$ such that $S \cap S'$ consists of a single element s, the diagram*

$$
\begin{array}{ccc}
U([n]) & \longrightarrow & U(S) \\
\downarrow & & \downarrow \\
U(S') & \longrightarrow & U(\{s\})
\end{array}
$$

is a pullback square in the ∞-category $\mathcal{C}$.

Proof. The dual of Proposition 2.1.2.1 implies that any monomorphism $K \rightarrow K'$ of simplicial sets induces a right fibration $\mathcal{C}_{/U[K]} \rightarrow \mathcal{C}_{/U[K]}$. By Corollary 2.4.4.6, a right fibration is a trivial fibration if and only if it is a categorical equivalence. This proves that $(2) \Leftrightarrow (2')$, $(3) \Leftrightarrow (3')$, and $(4) \Leftrightarrow (4')$. The implications $(1) \Rightarrow (2) \Rightarrow (3)$ are obvious.

We now prove that (3) implies (1). Let A denote the class of all morphisms $f : K' \rightarrow K$ which induce a categorical equivalence $\mathcal{C}_{/U[K]} \rightarrow \mathcal{C}_{/U[K']}$. Let A' denote the class of all *cofibrations* which have the same property; equivalently, A' is the class of all cofibrations which induce a trivial fibration $\mathcal{C}_{/U[K]} \rightarrow \mathcal{C}_{/U[K']}$. From this characterization it is easy to see that A' is weakly saturated. Let A'' be the weakly saturated class of morphisms generated by the inclusions $\Lambda_i^n \subseteq \Delta^n$ for $n > 1$. If we assume (3), then we have the inclusions $A'' \subseteq A' \subseteq A$.

Let $f : K \rightarrow K'$ be an arbitrary morphism of simplicial sets. By Proposition A.1.2.5, we can choose a map $h' : K' \rightarrow M'$ which belongs to A'', where M' has the extension property with respect to $\Lambda_i^n \subseteq \Delta^n$ for $n > 1$ and is therefore a Kan complex. Applying Proposition A.1.2.5 again, we can construct a commutative diagram

$$
\begin{array}{ccc}
K & \xrightarrow{h} & M \\
\downarrow{f} & & \downarrow{g} \\
K' & \xrightarrow{h'} & M',
\end{array}
$$

where the horizontal maps belong to A'' and g has the right lifting property with respect to every morphism in A''. If f is a weak homotopy equivalence which is bijective on vertices, then g has the same properties, so that g is a trivial fibration by Lemma 6.1.2.4. It follows that g has the right lifting property with respect to the cofibration $g \circ h : K \to M'$, so that $g \circ h$ is a retract of h and therefore belongs to A''. Since $g \circ h = h' \circ f$ and h' belong to $A'' \subseteq A$, it follows that f belongs to A.

It is clear that $(1) \Rightarrow (4)$. We next prove that $(4') \Rightarrow (3)$. We must show that if $n > 1$, then every inclusion $\Lambda_i^n \subseteq \Delta^n$ belongs to the class A defined above. The proof is by induction on n. Replacing i by $n - i$ if necessary, we may suppose that $i < n$. If $(n, i) \neq (2, 0)$, we consider the composition

$$\Delta^{n-1} \coprod_{\{n-1\}} \Delta^{\{n-1,n\}} \xrightarrow{f} \Lambda_i^n \xrightarrow{f'} \Delta^n.$$

Here f belongs to A' by the inductive hypothesis and $f' \circ f$ belongs to A' by virtue of assumption $(4')$; therefore f' also belongs to A. If $n = 2$ and $i = 0$, then we observe that the inclusion $\Lambda_1^2 \subseteq \Delta^2$ is of the form $\Delta^S \coprod_{\{1\}} \Delta^{S'} \subseteq \Delta^2$, where $S = \{0, 1\}$ and $S' = \{0, 2\}$.

To complete the proof, we show that (4) is equivalent to $(4'')$. Fix $n \geq 0$, let $S \cup S' = [n]$ be such that $S \cap S' = \{s\}$, and let $K = \Delta^S \coprod_{\{s\}} \Delta^{S'} \subseteq \Delta^n$. Let $\mathcal{J}'$ denote the full subcategory of $\boldsymbol{\Delta}_{/\Delta^n}$ spanned by the objects $[n]$, S, S', and $\{s\}$. Let $\mathcal{J} \subseteq \mathcal{J}'$ be the full subcategory obtained by omitting the object $[n]$. Let p' denote the composition

$$\mathrm{N}(\mathcal{J}')^{op} \to \mathrm{N}(\boldsymbol{\Delta})^{op} \xrightarrow{U} \mathcal{C}$$

and let $p = p' | \mathrm{N}(\mathcal{J})^{op}$. Consider the diagram

$$\begin{array}{ccc}
\mathcal{C}_{/U[\Delta^n]} & \longrightarrow & \mathcal{C}_{/U[K]} \\
\downarrow{\scriptstyle u} & & \downarrow{\scriptstyle v} \\
\mathcal{C}_{/p'} & \longrightarrow & \mathcal{C}_{/p}.
\end{array}$$

Condition (4) asserts that the upper horizontal map is a categorical equivalence, and condition $(4'')$ asserts that the lower horizontal map is a categorical equivalence. To prove that $(4) \Leftrightarrow (4'')$, it suffices to show that the vertical maps u and v are categorical equivalences.

We have a commutative diagram

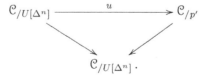

Since Δ^n is a final object of both $\boldsymbol{\Delta}_{/\Delta^n}$ and $\mathcal{J}'$, the unlabelled maps are trivial fibrations. It follows that u is a categorical equivalence.

To prove that v is a categorical equivalence, it suffices to show that the inclusion $g : \mathcal{J} \subseteq \boldsymbol{\Delta}_{/K}$ induces a right anodyne map

$$\mathrm{N}(g) : \mathrm{N}(\mathcal{J}) \subseteq \mathrm{N}(\boldsymbol{\Delta}_{/K})$$

of simplicial sets. We observe that the functor g has a left adjoint f, which associates to each simplex $\sigma : \Delta^m \to K$ the smallest simplex in $\mathcal{I}$ which contains the image of σ. The map $N(g)$ is a section of $N(f)$, and there is a (fiberwise) simplicial homotopy from $\mathrm{id}_{N(\Delta/K)}$ to $N(g) \circ N(f)$. We now invoke Proposition 2.1.2.11 to deduce that $N(g)$ is right anodyne, as desired.   □

Definition 6.1.2.7. Let $\mathcal{C}$ be an ∞-category. We will often denote simplicial objects of $\mathcal{C}$ by $U_\bullet$ and write U_n for $U_\bullet([n]) \in \mathcal{C}$. We will say that a simplicial object $U_\bullet \in \mathcal{C}_\Delta$ is a *groupoid object* of $\mathcal{C}$ if it satisfies the equivalent conditions of Proposition 6.1.2.6. We will let $\mathcal{G}pd(\mathcal{C})$ denote the full subcategory of $\mathcal{C}_\Delta$ spanned by the groupoid objects of $\mathcal{C}$.

Remark 6.1.2.8. It follows from the proof of Proposition 6.1.2.6 that to verify that a simplicial object $X_\bullet \in \mathcal{C}_\Delta$ is a groupoid object, we need only verify condition $(4'')$ in a small class of specific examples, but we will not need this observation.

Proposition 6.1.2.9. *Let $\mathcal{C}$ be a presentable ∞-category. The full subcategory $\mathcal{G}pd(\mathcal{C}) \subseteq \mathcal{C}_\Delta$ is strongly reflective.*

Proof. Let $n \geq 0$ and $[n] = S \cup S'$ be as in statement $(4'')$ of Proposition 6.1.2.6. Let $\mathcal{D}(S, S') \subseteq \mathcal{C}_\Delta$ be the full subcategory consisting of those simplicial objects $U \in \mathcal{C}_\Delta$ for which the associated diagram

$$
\begin{array}{ccc}
U([n]) & \longrightarrow & U(S) \\
\downarrow & & \downarrow \\
U(S') & \longrightarrow & U(\{s\})
\end{array}
$$

is Cartesian. Lemmas 5.5.4.19 and 5.5.4.17 imply that $\mathcal{D}(S, S')$ is a strongly reflective subcategory of $\mathcal{C}_\Delta$. Let $\mathcal{D}$ denote the intersection of all these subcategories taken over all $n \geq 0$ and all such decompositions $[n] = S \cup S'$. Lemma 5.5.4.18 implies that $\mathcal{D} \subseteq \mathcal{C}_\Delta$ is strongly reflective, and Proposition 6.1.2.6 implies that $\mathcal{D} = \mathcal{G}pd(\mathcal{C})$.   □

Our next step is to exhibit a large class of examples of groupoid objects. We first sketch the idea. Suppose that $\mathcal{C}$ is an ∞-category which admits finite limits and let $u : U \to X$ be a morphism in $\mathcal{C}$. Using this data, we can construct a simplicial object $U_\bullet$ of $\mathcal{C}$, where U_n is given by the $(n+1)$-fold fiber power of U over X. In order to describe this construction more precisely, we need to introduce a bit of notation.

Notation 6.1.2.10. Let $\Delta_+^{\leq n}$ denote the full subcategory of Δ_+ spanned by the objects $\{[k]\}_{-1 \leq k \leq n}$.

Proposition 6.1.2.11. *Let $\mathcal{C}$ be an ∞-category and let $U_\bullet^+ : N(\Delta_+)^{op} \to \mathcal{C}$ be an augmented simplicial object of $\mathcal{C}$. The following conditions are equivalent:*

(1) *The augmented simplicial object $U_\bullet^+$ is a right Kan extension of*

$$U_\bullet^+ \mid N(\Delta_+^{\leq 0})^{op}.$$

(2) *The underlying simplicial object $U_\bullet$ is a groupoid object of $\mathcal{C}$, and the diagram $U_\bullet^+ \mid N(\Delta_+^{\leq 1})^{op}$ is a pullback square*

$$
\begin{array}{ccc}
U_1 & \longrightarrow & U_0 \\
\downarrow & & \downarrow \\
U_0 & \longrightarrow & U_{-1}
\end{array}
$$

in the ∞-category $\mathcal{C}$.

Proof. Suppose first that (1) is satisfied. It follows immediately from the definition of right Kan extensions that the diagram

$$
\begin{array}{ccc}
U_1 & \longrightarrow & U_0 \\
\downarrow & & \downarrow \\
U_0 & \longrightarrow & U_{-1}
\end{array}
$$

is a pullback. To prove that $U_\bullet$ is a groupoid, we show that $U_\bullet$ satisfies criterion $(4'')$ of Proposition 6.1.2.6. Let S and S' be sets with union $[n]$ and intersection $S \cap S' = \{s\}$. Let $\mathcal{I}$ be the nerve of the category $(\Delta_+)_{/\Delta^n}$. For each subset $J \subseteq [n]$, let $\mathcal{I}(J)$ denote the full subcategory of $\mathcal{I}$ spanned by the initial object together with the inclusions $\{j\} \to \Delta^n$, $j \in S$. By assumption, $U_\bullet^+$ exhibits $U_\bullet(S)$ as a limit of $U_\bullet^+ \mid N(\mathcal{I}(S))$, $U_\bullet(S')$ as a limit of $U_\bullet^+ \mid N(\mathcal{I}(S'))$, $U_\bullet([n])$ as a limit of $U_\bullet^+ \mid N(\mathcal{I}([n]))$, and $U_\bullet(\{s\})$ as a limit of $U_\bullet^+ \mid N(\mathcal{I}(\{s\}))$. It follows from Corollary 4.2.3.10 that the diagram

$$
\begin{array}{ccc}
U_\bullet([n]) & \longrightarrow & U_\bullet(S) \\
\downarrow & & \downarrow \\
U_\bullet(S') & \longrightarrow & U_\bullet(\{s\})
\end{array}
$$

is a pullback.

We now prove that (2) implies (1). Using the above notation, we must show that for each $n \geq -1$, $U_\bullet^+$ exhibits $U_\bullet^+([n])$ as a limit of $U_\bullet^+ \mid \mathcal{I}([n])$. For $n \leq 0$, this is obvious; for $n = 1$, it is equivalent to the assumption that

$$
\begin{array}{ccc}
U_1 & \longrightarrow & U_0 \\
\downarrow & & \downarrow \\
U_0 & \longrightarrow & U_{-1}
\end{array}
$$

is a pullback diagram. We prove the general case by induction on n. Using the inductive hypothesis, we conclude that $U_\bullet(\Delta^S)$ is a limit of $U_\bullet^+ \mid \mathcal{I}(S)$ for all *proper* subsets $S \subset [n]$. Choose a decomposition $\{0, \ldots, n\} = S \cup S'$, where $S \cap S' = \{s\}$. According to Proposition 4.4.2.2, the desired result is equivalent to the assertion that

$$
\begin{array}{ccc}
U_\bullet([n]) & \longrightarrow & U_\bullet(S) \\
\downarrow & & \downarrow \\
U_\bullet(S') & \longrightarrow & U_\bullet(\{s\})
\end{array}
$$

is a pullback diagram, which follows from our assumption that $U_\bullet$ is a groupoid object of $\mathcal{C}$. $\square$

We will say that an augmented simplicial object $U_\bullet^+$ in an ∞-category $\mathcal{C}$ is a *Čech nerve* if it satisfies the equivalent conditions of Proposition 6.1.2.11. In this case, $U_\bullet^+$ is determined up to equivalence by the map $u : U_0 \to U_{-1}$; we will also say that $U_\bullet^+$ is the *Čech nerve of u*.

Notation 6.1.2.12. Let $U_\bullet$ be a simplicial object in an ∞-category $\mathcal{C}$. We may regard $U_\bullet$ as a diagram in $\mathcal{C}$ indexed by $N(\Delta)^{op}$. We let $|U_\bullet|$: $N(\Delta_+)^{op} \to \mathcal{C}$ denote a colimit for $U_\bullet$ (if such a colimit exists). We will refer to any such colimit as a *geometric realization* of $U_\bullet$.

Remark 6.1.2.13. Note that we are regarding $|U_\bullet|$ as a colimit diagram in $\mathcal{C}$, not as an object of $\mathcal{C}$. We also note that our notation is somewhat abusive since $|U_\bullet|$ is not uniquely determined by $U_\bullet$. However, if a colimit of $U_\bullet$ exists, then it is determined up to contractible ambiguity.

Definition 6.1.2.14. Let $U_\bullet$ be a simplicial object of an ∞-category $\mathcal{C}$. We will say that $U_\bullet$ is an *effective groupoid* if can be extended to a colimit diagram $U_\bullet^+ : N(\Delta_+)^{op} \to \mathcal{C}$ such that $U_\bullet^+$ is a Čech nerve.

Remark 6.1.2.15. It follows easily from characterization (3) of Proposition 6.1.2.11 that any effective groupoid $U_\bullet$ is a groupoid.

We can now state the ∞-categorical counterpart of Fact 6.1.1.6: every groupoid object in $\mathcal{S}$ is effective. This statement is somewhat less trivial than its classical analogue. For example, a groupoid object $U_\bullet$ in $\mathcal{S}$ with $U_0 = *$ can be thought of as a space U_1 equipped with a coherently associative multiplication operation. If $U_\bullet$ is effective, then there exists a fiber diagram

so that U_1 is homotopy equivalent to a loop space. This is a classical result (see, for example, [73]). We will give a somewhat indirect proof in the next section.

6.1.3 ∞-Topoi and Descent

In this section, we will describe an elegant characterization of the notion of an ∞-topos based on the theory of *descent*. We begin by explaining the idea in informal terms. Let $\mathfrak{X}$ be an ∞-category. To each object U of $\mathfrak{X}$ we can associate the overcategory $\mathfrak{X}^{/U}$. If $\mathfrak{X}$ admits finite limits, then this construction gives a contravariant functor from $\mathfrak{X}$ to the ∞-category $\widehat{\mathfrak{Cat}}_\infty$ of (not necessarily small) ∞-categories. If $\mathfrak{X}$ is an ∞-topos, then this functor carries colimits in $\mathfrak{X}$ to limits of ∞-categories. In other words, if an object $X \in \mathfrak{X}$ is obtained as the colimit of some diagram $\{X_\alpha\}$ in $\mathfrak{X}$, then giving a morphism $Y \to X$ is equivalent to giving a suitably compatible diagram of morphisms $\{Y_\alpha \to X_\alpha\}$. Moreover, we will eventually show that this property *characterizes* the class of ∞-topoi. The ideas presented in this section are due to Charles Rezk.

Definition 6.1.3.1. Let $\mathfrak{X}$ be an ∞-category, K a simplicial set, and $p, q : K \to \mathfrak{X}$ two diagrams. We will say that a natural transformation $\alpha : p \to q$ is *Cartesian* if, for each edge $\phi : x \to y$ in K, the associated diagram

$$
\begin{array}{ccc}
p(x) & \xrightarrow{\ p(\phi)\ } & p(y) \\
\downarrow{\scriptstyle \alpha(x)} & & \downarrow{\scriptstyle \alpha(y)} \\
q(x) & \xrightarrow{\ q(\phi)\ } & q(y)
\end{array}
$$

is a pullback in $\mathfrak{X}$.

Lemma 6.1.3.2. *Let $\mathfrak{X}$ be an ∞-category, and let $\alpha : p \to q$ be a natural transformation of diagrams $p, q : K \diamond \Delta^0 \to \mathfrak{X}$. Suppose that, for every vertex x of K, the associated transformation*

$$
p|\{x\} \diamond \Delta^0 \to q|\{x\} \diamond \Delta^0
$$

is Cartesian. Then α is Cartesian.

Proof. Let z be the cone point of $K \diamond \Delta^0$. We note that to every edge $e : x \to y$ in $K \diamond \Delta^0$ we can associate a diagram

The transformation α restricts to a Cartesian transformation on the horizontal edges and the right vertical edge, either by assumption or because they are degenerate. Applying Lemma 4.4.2.1, we deduce first that $\alpha(g)$ is a Cartesian transformation, then that $\alpha(e)$ is a Cartesian transformation. □

The condition that an ∞-category has universal colimits can be formulated in the language of Cartesian transformations:

Lemma 6.1.3.3. *Let $\mathcal{X}$ be a presentable ∞-category. The following conditions are equivalent:*

(1) *Colimits in $\mathcal{X}$ are universal.*

(2) *Let $p, q : (K^{\triangleright} \diamond \Delta^0) \to \mathcal{X}$ be diagrams which carry Δ^0 to vertices $X, Y \in \mathcal{X}$ and let $\alpha : p \to q$ be a Cartesian transformation. If the map $q' : K^{\triangleright} \to \mathcal{X}^{/Y}$ associated to q is a colimit diagram, then the map $p' : K^{\triangleright} \to \mathcal{X}^{/X}$ associated to p is a colimit diagram.*

(3) *Let $p, q : K \star \Delta^1 \to \mathcal{X}$ be diagrams which carry $\{1\}$ to vertices $X, Y \in \mathcal{X}$ and let $\alpha : p \to q$ be a Cartesian transformation. If the map $K^{\triangleright} \to \mathcal{X}_{/Y}$ associated to q is a colimit diagram, then the map $K^{\triangleright} \to \mathcal{X}_{/X}$ associated to p is a colimit diagram.*

(4) *Let $p, q : K \star \Delta^1 \to \mathcal{X}$ be diagrams which carry $\{1\}$ to vertices $X, Y \in \mathcal{X}$ and let $\alpha : p \to q$ be a Cartesian transformation. If $q|K \star \{0\}$ is a colimit diagram, then $p|K \star \{0\}$ is a colimit diagram.*

(5) *Let $\alpha : p \to q$ be a Cartesian transformation of diagrams $K^{\triangleright} \to \mathcal{X}$. If q is a colimit diagram, then p is a colimit diagram.*

Proof. Assume that (1) is satisfied; we will prove (2). The transformation α induces a map $f : X \to Y$. Consider the map

$$\phi : \mathrm{Fun}(K^{\triangleright}, \mathcal{O}_{\mathcal{X}}) \to \mathrm{Fun}(K^{\triangleright}, \mathcal{X})$$

given by evaluation at the final vertex of Δ^1. Let $\delta(f)$ denote the image of f under the diagonal map $\delta : \mathcal{X} \to \mathrm{Fun}(K^{\triangleright}, \mathcal{X})$. Then we may identify α with an edge e of $\mathrm{Fun}(K^{\triangleright}, \mathcal{O}_{\mathcal{X}})$ which covers $\delta(f)$. Since α is Cartesian, we can apply Lemma 6.1.1.1 and Proposition 3.1.2.1 to deduce that e is ϕ-Cartesian. The composition $f^* \circ q'$ is the origin of a ϕ-Cartesian edge $e' : f^* \circ q' \to q'$ of $\mathrm{Fun}(K^{\triangleright}, \mathcal{O}_{\mathcal{X}})$ covering $\delta(f)$, so we conclude that $f^* \circ q'$ and p' are equivalent in $\mathrm{Fun}(K^{\triangleright}, \mathcal{X}^{/X})$. Since q' is a colimit diagram and f^* preserves colimits, $f^* \circ q'$ is a colimit diagram. It follows that p' is a colimit diagram, as desired.

We now prove that (2) $\Rightarrow$ (1). Let $f : X \to Y$ be a morphism in $\mathcal{X}$ and let $q' : K^{\triangleright} \to \mathcal{X}^{/Y}$ be a colimit diagram. Choose a ϕ-Cartesian edge $e' : f^* \circ q' \to q'$ as above, corresponding to a natural transformation $\alpha : p \to q$ of diagrams $p, q : (K^{\triangleright} \diamond \Delta^0) \to \mathcal{X}$. Since e is ϕ-Cartesian, we may invoke Proposition 3.1.2.1 and Lemma 6.1.1.1 to deduce that α restricts to a Cartesian transformation $p|(\{x\} \diamond \Delta^0) \to q|(\{x\} \diamond \Delta^0)$ for every vertex x of $K^{\triangleright}$. It follows from Lemma 6.1.3.2 that α is Cartesian. Invoking (2), we conclude that $f^* \circ q'$ is a colimit diagram, as desired.

The equivalence (2) $\Leftrightarrow$ (3) follows from Proposition 4.2.1.2, and the equivalence (3) $\Leftrightarrow$ (4) follows from Proposition 1.2.13.8. The implication (5) $\Rightarrow$ (4) is obvious. The converse implication (4) $\Rightarrow$ (5) follows from the observation that $K \star \{0\}$ is a retract of $K \star \Delta^1$. $\qquad\square$

Notation 6.1.3.4. Let X be an ∞-category which admits pullbacks and let S be a class of morphisms in X. We will say that S is *stable under pullback* if for any pullback diagram

in X such that f belongs to S, f' also belongs to S. We let $\mathcal{O}_X^S$ denote the full subcategory of $\mathcal{O}_X$ spanned by S, and $\mathcal{O}_X^{(S)}$ the subcategory of $\mathcal{O}_X$ whose objects are elements of S and whose morphisms are pullback diagrams as above. We observe that evaluation at $\{1\} \subseteq \Delta^1$ induces a Cartesian fibration $\mathcal{O}_X^S \to X$, which restricts to a right fibration $\mathcal{O}_X^{(S)} \to X$ (Corollary 2.4.2.5).

Lemma 6.1.3.5. *Let X be a presentable ∞-category and suppose that colimits in X are universal. Let S be a class of morphisms of X which is stable under pullback, K a small simplicial set, and $\overline{q} : K^{\triangleright} \to X$ a colimit diagram. The following conditions are equivalent:*

(1) *The composition $f \circ \overline{q} : K^{\triangleright} \to \widehat{\mathrm{Cat}}_{\infty}^{op}$ is a colimit diagram, where $f : X \to \widehat{\mathrm{Cat}}_{\infty}^{op}$ classifies the Cartesian fibration $\mathcal{O}_X^S \to X$.*

(2) *The composition $f' \circ \overline{q} : K^{\triangleright} \to \widehat{\mathcal{S}}^{op}$ is a colimit diagram, where $f : X \to \widehat{\mathcal{S}}^{op}$ classifies the right fibration $\mathcal{O}_X^{(S)} \to X$.*

(3) *For every natural transformation $\overline{\alpha} : \overline{p} \to \overline{q}$ of colimit diagrams $K^{\triangleright} \to X$, if $\alpha = \overline{\alpha}|K$ is a Cartesian transformation and $\alpha(x) \in S$ for each vertex $x \in K$, then $\overline{\alpha}$ is a Cartesian transformation and $\overline{\alpha}(\infty) \in S$, where ∞ denotes the cone point of $K^{\triangleright}$.*

Proof. Let $\overline{\mathcal{C}} = \mathrm{Fun}(K^{\triangleright}, X)^{/\overline{q}}$ and $\mathcal{C} = \mathrm{Fun}(K, X)^{/q}$. Let $\overline{\mathcal{C}}^0$ denote the full subcategory of $\overline{\mathcal{C}}$ spanned by *Cartesian* natural tranformations $\overline{\alpha} : \overline{p} \to \overline{q}$ with the property that $\overline{\alpha}(x)$ belongs to S for each vertex $x \in K^{\triangleright}$ and let $\mathcal{C}^0$ be defined similarly. Finally, let $\overline{\mathcal{C}}^1$ denote the full subcategory of $\overline{\mathcal{C}}$ spanned by those natural transformations $\overline{\alpha} : \overline{p} \to \overline{q}$ such that $\overline{p}$ is a colimit diagram, $\alpha = \overline{\alpha}|K$ is a Cartesian transformation, and $\overline{\alpha}(x)$ belongs to S for each vertex $x \in K$. Lemma 6.1.3.3 implies that $\overline{\mathcal{C}}^0 \subseteq \overline{\mathcal{C}}^1$.

Let $\mathcal{D}$ denote the full subcategory of $\mathrm{Fun}(K^{\triangleright}, X)$ spanned by the colimit diagrams. Proposition 4.3.2.15 asserts that the restriction map $\mathcal{D} \to \mathrm{Fun}(K, X)$ is a trivial fibration. It follows that the associated map $\mathcal{D}^{/\overline{q}} \to \mathrm{Fun}(K, X)^{/q}$ is also a trivial fibration and therefore restricts to a trivial fibration $\overline{\mathcal{C}}^1 \to \mathcal{C}^0$.

According to Proposition 3.3.3.1, condition (1) is equivalent to the assertion that the projection $\overline{\mathcal{C}}^0 \to \mathcal{C}^0$ is an equivalence of ∞-categories. In view of the above argument, this is equivalent to the assertion that the fully faithful inclusion $\overline{\mathcal{C}}^0 \subseteq \overline{\mathcal{C}}^1$ is essentially surjective. Since $\overline{\mathcal{C}}^0$ is clearly stable under equivalence in $\overline{\mathcal{C}}$, (1) holds if and only if $\overline{\mathcal{C}}^0 = \overline{\mathcal{C}}^1$, which is manifestly equivalent to (3). The proof that (2) $\Leftrightarrow$ (3) is similar but uses Proposition 3.3.3.3 in place of Proposition 3.3.3.1. $\qquad\qquad\qquad\qquad\qquad$ $\square$

Lemma 6.1.3.6. *Let X be a presentable category in which colimits are universal. Let $f : X \to \emptyset$ be a morphism in X, where $\emptyset$ is an initial object of X. Then X is also initial.*

Proof. Observe that $\mathrm{id}_{\emptyset}$ is both an initial object of $X^{/\emptyset}$ (Proposition 1.2.13.8) and a final object of $X^{/\emptyset}$. Let $f^* : X^{/\emptyset} \to X^{/X}$ be a pullback functor. Then f^* preserves limits (since it is a right adjoint) and colimits (since colimits in X are universal). Therefore $f^* \mathrm{id}_{\emptyset}$ is both initial and final in $X^{/X}$. It follows that $\mathrm{id}_X : X \to X$, being another final object of $X^{/X}$, is also initial. Applying Proposition 1.2.13.8, we deduce that X is an initial object of X, as desired. $\qquad\qquad\qquad\qquad\qquad\qquad\qquad\qquad\qquad\qquad$ $\square$

Lemma 6.1.3.7. *Let X be a presentable ∞-category in which colimits are universal and let S be a class of morphisms in X which is stable under pullback. The following conditions are equivalent:*

(1) *The Cartesian fibration $\mathcal{O}_X^S \to X$ is classified by a colimit-preserving functor $X \to \widehat{\mathcal{C}\mathrm{at}}_{\infty}^{op}$.*

(2) *The right fibration $\mathcal{O}_X^{(S)} \to X$ is classified by a colimit-preserving functor $X \to \widehat{\mathcal{S}}^{op}$.*

(3) *The class S is stable under small coproducts, and for every pushout diagram*

$$
\begin{array}{ccc}
f & \xrightarrow{\ \alpha\ } & g \\
{\scriptstyle\beta}\big\downarrow & & \big\downarrow{\scriptstyle\beta'} \\
f' & \xrightarrow{\ \alpha'\ } & g'
\end{array}
$$

in $\mathcal{O}_X$, if α and β are Cartesian transformations and $f, f', g \in S$, then α' and β' are also Cartesian transformations and $g' \in S$.

Proof. The equivalence of (1) and (2) follows easily from Lemma 6.1.3.5. Let $s : X \to \widehat{\mathcal{C}\mathrm{at}}_{\infty}^{op}$ be a functor which classifies $\mathcal{O}_X$. Then (1) is equivalent to the assertion that s preserves small colimits. Supposing that (1) is satisfied, we deduce (3) by applying Lemma 6.1.3.5 in the special cases of sums and coproducts. For the converse, let us suppose that (3) is satisfied. Let $\emptyset$ denote an initial object of X. Since colimits in X are universal, Lemma 6.1.3.6 implies that $X^{/\emptyset}$ is equivalent to a final ∞-category Δ^0. Since the morphism $\mathrm{id}_{\emptyset}$ belongs to S (since S is stable under *empty* coproducts), we conclude that $s(\emptyset)$ is a final ∞-category, so that s preserves initial objects. It follows from Corollary 4.4.2.5 that s preserves finite coproducts. According to Proposition 4.4.2.6, it will suffice to prove that s preserves arbitrary coproducts. To handle the case of infinite coproducts we apply Lemma 6.1.3.5 again: we must show that if $\{f_\alpha\}_{\alpha \in A}$ is a collection of elements of S having a coproduct $f = \coprod_{\alpha \in A} f_\alpha$, then $f \in S$ and each of the maps $f_\alpha \to f$ is a Cartesian transformation. The first condition is true by assumption; for the second we let f' be a coproduct of the family $\{f_\beta\}_{\beta \in A, \beta \neq \alpha}$, so that $f \simeq f' \coprod f_\alpha$ and $f' \in S$. Applying Lemma 6.1.3.5 (and the fact that s preserves finite coproducts), we deduce that $f_\alpha \to f$ is a Cartesian transformation, as desired.   □

Definition 6.1.3.8. Let $\mathcal{X}$ be a presentable ∞-category in which colimits are universal and let S be a class of morphisms in $\mathcal{X}$. We will say that S is *local* if it is stable under pullbacks and satisfies the equivalent conditions of Lemma 6.1.3.7.

Theorem 6.1.3.9. *Let $\mathcal{X}$ be a presentable ∞-category. The following conditions are equivalent:*

(1) *Colimits in $\mathcal{X}$ are universal, and for every pushout diagram*

$$
\begin{array}{ccc}
f & \xrightarrow{\alpha} & g \\
\downarrow{\scriptstyle\beta} & & \downarrow{\scriptstyle\beta'} \\
f' & \xrightarrow{\alpha'} & g'
\end{array}
$$

in $\mathcal{O}_\mathcal{X}$, if α and β are Cartesian transformations, then α' and β' are also Cartesian transformations.

(2) *Colimits in $\mathcal{X}$ are universal, and the class of* all *morphisms in $\mathcal{X}$ is local.*

(3) *The Cartesian fibration $\mathcal{O}_\mathcal{X} \to \mathcal{X}$ is classified by a limit-preserving functor $\mathcal{X}^{op} \to \mathcal{P}\mathrm{r}^L$.*

(4) *Let K be a small simplicial set and $\overline{\alpha} : \overline{p} \to \overline{q}$ a natural transformation of diagrams $\overline{p}, \overline{q} : K^{\triangleright} \to \mathcal{X}$. Suppose that $\overline{q}$ is a colimit diagram, and that $\alpha = \overline{\alpha}|K$ is a Cartesian transformation. Then $\overline{p}$ is a colimit diagram if and only if $\overline{\alpha}$ is a Cartesian transformation.*

Proof. The equivalences (1) ⇔ (2) ⇔ (3) follow from Lemma 6.1.3.7 and Proposition 6.1.1.4. The equivalence (3) ⇔ (4) follows from Lemmas 6.1.3.3 and 6.1.3.5.   □

We now have most of the tools required to establish the implication (1) ⇒ (2) of Theorem 6.1.0.6. In view of Theorem 6.1.3.9, it will suffice to prove the following:

Proposition 6.1.3.10. *Let $\mathcal{X}$ be an ∞-topos. Then*

(1) *Colimits in $\mathcal{X}$ are universal.*

(2) *For every pushout diagram*

$$
\begin{array}{ccc}
f & \xrightarrow{\ \alpha\ } & g \\
\downarrow{\scriptstyle \beta} & & \downarrow{\scriptstyle \beta'} \\
f' & \xrightarrow{\ \alpha'\ } & g'
\end{array}
$$

in $\mathcal{O}_{\mathfrak{X}}$, *if* α *and* β *are Cartesian transformations, then* α' *and* β' *are also Cartesian transformations.*

Remark 6.1.3.11. Once we have established Theorem 6.1.0.6 in its entirety, it will follow from Theorem 6.1.3.9 that the converse of Proposition 6.1.3.10 is also valid: a presentable ∞-category $\mathfrak{X}$ is an ∞-topos if and only if it satisfies conditions (1) and (2) as above. Condition (1) is equivalent to the requirement that for every morphism $f : X \to Y$ in $\mathfrak{X}$, the pullback functor $f^* : \mathfrak{X}_{/Y} \to \mathfrak{X}_{/X}$ has a right adjoint (in the case where Y is a final object of $\mathfrak{X}$, this simply amounts to the requirement that every object $Z \in \mathfrak{X}$ admits an exponential Z^X; in other words, the requirement that $\mathfrak{X}$ be *Cartesian closed*), and condition (2) involves only finite diagrams in the ∞-category $\mathfrak{X}$. One could conceivably obtain a theory of *elementary ∞-topoi* by dropping the requirement that $\mathfrak{X}$ be presentable (or replacing it by weaker conditions which are also finite in nature). We will not pursue this idea further.

Before giving the proof of Proposition 6.1.3.10, we need to establish a few easy lemmas.

Lemma 6.1.3.12. *Let*

$$
\begin{array}{ccc}
\phi & \xrightarrow{\ p\ } & \psi \\
\downarrow{\scriptstyle q} & & \downarrow{\scriptstyle q'} \\
\phi' & \xrightarrow{\ p'\ } & \psi'
\end{array}
$$

be a coCartesian square in the category of arrows of Set_{Δ}. *Suppose that* p *and* q *are homotopy Cartesian and that* q *is a cofibration. Then*

(1) *The maps* p' *and* q' *are homotopy Cartesian.*

(2) *Given any map of arrows* $r : \psi' \to \theta$ *such that* $r \circ p'$ *and* $r \circ q'$ *are homotopy Cartesian, the map* r *is itself homotopy Cartesian.*

Proof. Let $r : \psi' \to \theta$ be as in (2). We must show that r is homotopy Cartesian if and only if $r \circ p'$ and $r \circ q'$ are homotopy Cartesian (taking $r = \mathrm{id}_{\psi'}$, we will deduce (1)). Without loss of generality, we may replace ϕ, ψ, ϕ', and θ with minimal Kan fibrations. We now observe that r, $r \circ p'$, and $r \circ q'$ are homotopy Cartesian if and only if they are Cartesian; the desired result now follows immediately. $\square$

Lemma 6.1.3.13. *Let* $\mathbf{A}$ *be a simplicial model category containing an object* Z *which is both fibrant and cofibrant and let* $\mathbf{A}_{/Z}$ *be endowed with the induced model structure. Then the natural map* $\theta : \mathrm{N}(\mathbf{A}^{\circ}_{/Z}) \to \mathrm{N}(\mathbf{A}^{\circ})_{/Z}$ *is an equivalence of* ∞-categories.

Proof. Let $\phi : Z' \to Z$ be an object of $N(\mathbf{A}^\circ)_{/Z}$. Then we can choose a factorization

$$Z' \xrightarrow{i} Z'' \xrightarrow{\psi} Z,$$

where i is a trivial cofibration and ψ is a fibration, corresponding to a fibrant-cofibrant object of $\mathbf{A}_{/Z}$. The above diagram classifies an equivalence between ϕ and ψ in $N(\mathbf{A}^\circ)_{/Z}$, so that θ is essentially surjective.

Recall that for any simplicial category $\mathcal{C}$ containing a pair of objects X and Y, there is a natural isomorphism of simplicial sets

$$\mathrm{Hom}^{\mathrm{R}}_{N(\mathcal{C})}(X,Y) \simeq \mathrm{Sing}_{Q^\bullet}(\mathrm{Map}_{\mathcal{C}}(X,Y)),$$

where $Q^\bullet$ is the cosimplicial object of Set_Δ introduced in §2.2.2. The same calculation shows that if $\phi : X \to Z$, $\psi : Y \to Z$ are two morphisms in $\mathcal{C}$, then

$$\mathrm{Hom}^{\mathrm{R}}_{N(\mathcal{C})_{/Z}}(\phi,\psi) \simeq \mathrm{Sing}_{Q^\bullet}(P),$$

where P denotes the path space

$$\mathrm{Map}_{\mathcal{C}}(X,Y) \times_{\mathrm{Map}_{\mathcal{C}}(X,Z)^{\{0\}}} \mathrm{Map}_{\mathcal{C}}(X,Z)^{\Delta^1} \times_{\mathrm{Map}_{\mathcal{C}}(X,Z)^{\{1\}}} \{\phi\}.$$

If $\mathcal{C}$ is fibrant, then we may identify M with the homotopy fiber of the map

$$f : \mathrm{Map}_{\mathcal{C}}(X,Y) \xrightarrow{\psi} \mathrm{Map}_{\mathcal{C}}(X,Z)$$

over the vertex ϕ. Consequently, we may identify the natural map

$$\mathrm{Hom}^{\mathrm{R}}_{N(\mathcal{C}_{/Z})}(\phi,\psi) \to \mathrm{Hom}^{\mathrm{R}}_{N(\mathcal{C})_{/Z}}(\phi,\psi)$$

with $\mathrm{Sing}_{Q^\bullet}(\theta)$, where θ denotes the inclusion of the fiber of f into the homotopy fiber of f. Consequently, to show that $\mathrm{Sing}_{Q^\bullet}(\theta)$ is a homotopy equivalence, it suffices to prove that f is a Kan fibration. In the special case where $\mathcal{C} = \mathbf{A}^\circ$ and ψ is a fibration, this follows from the definition of a simplicial model category. $\square$

Lemma 6.1.3.14. *Let $\mathcal{S}$ denote the ∞-category of spaces. Then*

(1) *Colimits in $\mathcal{S}$ are universal.*

(2) *For every pushout diagram*

$$\begin{array}{ccc} f & \xrightarrow{\alpha} & g \\ {\scriptstyle \beta}\downarrow & & \downarrow{\scriptstyle \beta'} \\ f' & \xrightarrow{\alpha'} & g' \end{array}$$

in $\mathcal{O}_{\mathcal{S}}$, if α and β are Cartesian transformations, then α' and β' are also Cartesian transformations.

Proof. We first prove (1). Let $f : X \to Y$ be a morphism in $\mathcal{S}$. Without loss of generality, we may suppose that f is a Kan fibration. We wish to show that the projection

$$\mathcal{S}_{/f} \to \mathcal{S}_{/Y}$$

has a right adjoint which preserves colimits. We obtain a commutative diagram of ∞-categories:

$$
\begin{array}{ccc}
\mathrm{N}((\mathrm{Set}_\Delta)^\circ_{/X}) & \xrightarrow{\;F\;} & \mathrm{N}((\mathrm{Set}_\Delta)^\circ_{/Y}) \\
\downarrow{\scriptstyle\phi} & & \downarrow{\scriptstyle\psi} \\
\mathcal{S}_{/f} & \xrightarrow{\hspace{2cm}} & \mathcal{S}_{/Y} \\
\downarrow{\scriptstyle\phi'} & & \\
\mathcal{S}_{/X}. & &
\end{array}
$$

Lemma 6.1.3.13 asserts that ψ and $\phi' \circ \phi$ are categorical equivalences, and ϕ' is a trivial fibration. It follows that ϕ is also a categorical equivalence. Consequently, it will suffice to show that the functor F has a right adjoint G which preserves colimits.

We observe that F is obtained by restricting the simplicial nerve of the functor $f_! : (\mathrm{Set}_\Delta)_{/X} \to (\mathrm{Set}_\Delta)_{/Y}$ given by composition with f. The functor $f_!$ is a left Quillen functor: it has a right adjoint f^* given by the formula $f^*(Y') = Y' \times_Y X$. According to Proposition 5.2.4.6, F admits a right adjoint G, which is given by restricting the simplicial nerve of the functor f^*. To prove that G preserves colimits, it will suffice to show that G itself admits a right adjoint. Using Proposition 5.2.4.6 again, we are reduced to proving that f^* is a *left* Quillen functor. We observe that f^* admits a right adjoint f_* given by the formula $f_*(X') = \mathrm{Map}_Y(X, X')$. It is clear that f^* preserves cofibrations; it also preserves weak equivalences since f is a fibration and Set_Δ is a *right proper* model category (with its usual model structure).

To prove (2), we first apply Proposition 4.2.4.4 to reduce to the case where the pushout diagram in question arises from a strictly commutative square

$$
\begin{array}{ccc}
f & \xrightarrow{\;\alpha\;} & g \\
\downarrow{\scriptstyle\beta} & & \downarrow{\scriptstyle\beta'} \\
f' & \xrightarrow{\;\alpha'\;} & g'
\end{array}
$$

of morphisms in the category $\mathcal{K}\mathrm{an}$. We now complete the proof by applying Lemma 6.1.3.12 and Theorem 4.2.4.1. $\qquad\qquad\square$

Lemma 6.1.3.15. *Let $\mathcal{X}$ be a presentable ∞-category and let $L : \mathcal{X} \to \mathcal{Y}$ be an accessible left exact localization. If colimits in $\mathcal{X}$ are universal, then colimits in $\mathcal{Y}$ are universal.*

Proof. We will use characterization (5) of Lemma 6.1.3.2. Let G be a right adjoint to L and let $\alpha : \overline{p} \to \overline{q}$ be a Cartesian transformation of diagrams

$K^{\triangleright} \to \mathcal{Y}$. Suppose that $\overline{q}$ is a colimit of $q = \overline{q}|K$. Choose a colimit $\overline{q}'$ of $G \circ q$, so that there exists a morphism $\overline{q}' \to G \circ \overline{q}$ in $\mathcal{X}_{G \circ q/}$ which determines a natural transformation $\beta : \overline{q}' \to \overline{q}$ in $\mathrm{Fun}(K^{\triangleright}, \mathcal{X})$. Form a pullback diagram

$$
\begin{array}{ccc}
\overline{p}' & \xrightarrow{\ \alpha'\ } & \overline{q}' \\
\downarrow & & \downarrow{\scriptstyle \beta} \\
G \circ \overline{p} & \xrightarrow{\ G \circ \alpha\ } & G \circ \overline{q}
\end{array}
$$

in $\mathcal{X}^{K^{\triangleright}}$. Since G is left exact, $G \circ \alpha$ is a Cartesian transformation. It follows that α', being a pullback of $G \circ \alpha$, is also a Cartesian transformation. Since colimits in $\mathcal{X}$ are universal, we conclude that $\overline{p}'$ is a colimit diagram. Since L is left exact, we obtain a pullback diagram

$$
\begin{array}{ccc}
L \circ \overline{p}' & \longrightarrow & L \circ \overline{q}' \\
\downarrow & & \downarrow{\scriptstyle L \circ \beta} \\
L \circ G \circ \overline{p} & \longrightarrow & L \circ G \circ \overline{q}.
\end{array}
$$

Since L preserves colimits, $L \circ \overline{q}'$ and $L \circ \overline{p}'$ are colimit diagrams. The diagram $L \circ G \circ \overline{q}$ is equivalent to q and therefore also a colimit diagram. We deduce that $L \circ \beta$ is an equivalence. Since the diagram is a pullback, the left vertical arrow is an equivalence as well, so that $L \circ G \circ \overline{p}$ is a colimit diagram. We finally conclude that $\overline{p}$ is a colimit diagram, as desired.   □

We are now ready to give the proof of Proposition 6.1.3.10.

Proof of Proposition 6.1.3.10. Let us say that a presentable ∞-category $\mathcal{X}$ is *good* if it satisfies conditions (1) and (2). Lemma 6.1.3.14 asserts that $\mathcal{S}$ is good. Using Proposition 5.1.2.2, it is easy to see that if $\mathcal{X}$ is good, then so is $\mathrm{Fun}(K, \mathcal{X})$ for every small simplicial set K. It follows that every ∞-category $\mathcal{P}(\mathcal{C})$ of presheaves is good. To complete the proof, it will suffice to show that if $\mathcal{X}$ is good and $L : \mathcal{X} \to \mathcal{Y}$ is an accessible left exact localization functor, then $\mathcal{Y}$ is good. Lemma 6.1.3.15 shows that colimits in $\mathcal{Y}$ are universal. Consider a diagram $\sigma : \Lambda^2_0 \to \mathcal{O}_{\mathcal{Y}}$ denoted by

$$
g \xleftarrow{\ \alpha\ } f \xrightarrow{\ \beta\ } h,
$$

where α and β are Cartesian transformations. We wish to show that if $\overline{\sigma}$ is a colimit of σ in $\mathcal{O}_{\mathcal{Y}}$, then $\overline{\sigma}$ carries each edge to a Cartesian transformation. Without loss of generality, we may suppose that $\sigma = L \circ \sigma'$ for some $\sigma' : \Lambda^2_0 \to \mathcal{O}_{\mathcal{X}}$ which is equivalent to $G \circ \sigma$. Since G is left exact, $G(\alpha)$ and $G(\beta)$ are Cartesian transformations. Because $\mathcal{X}$ satisfies (2), there exists a colimit $\overline{\sigma}'$ of σ' which carries each edge to a Cartesian transformation. Then $L \circ \overline{\sigma}'$ is a colimit of σ. Since L is left exact, $L \circ \overline{\sigma}'$ carries each edge to a Cartesian transformation in $\mathcal{O}_{\mathcal{Y}}$.   □

Our final objective in this section is to prove the implication (2) ⇒ (3) of Theorem 6.1.0.6 (Proposition 6.1.3.19 below).

Lemma 6.1.3.16. *Let $\mathfrak{X}$ be an ∞-category and $U_\bullet^+ : N(\mathbf{\Delta}_+)^{op} \to \mathfrak{X}$ an augmented simplicial object of $\mathfrak{X}$. Let $\mathbf{\Delta}_\infty$ denote the category whose objects are finite linearly ordered sets J, where $\mathrm{Hom}_{\mathbf{\Delta}_\infty}(J, J')$ is the collection of all order-preserving maps $J \cup \{\infty\} \to J' \cup \{\infty\}$ which carry ∞ to ∞ (here ∞ is regarded as a maximal element of $J \cup \{\infty\}$ and $J' \cup \{\infty\}$). Suppose that $U_\bullet^+$ extends to a functor $F : N(\mathbf{\Delta}_\infty)^{op} \to \mathfrak{X}$. Then $U_\bullet^+$ is a colimit diagram in $\mathfrak{X}$.*

Proof. Let $\overline{\mathcal{C}}$ denote the category whose objects are triples (J, J_+), where J is a finite linearly ordered set and J_+ is an upward-closed subset of J. We define $\mathrm{Hom}_{\overline{\mathcal{C}}}((J, J_+), (J', J'_+))$ to be the set of all order-preserving maps from J into J' that carry J_+ into J'_+. Observe that we have a functor $\overline{\mathcal{C}} \to \mathbf{\Delta}_\infty$ given by

$$(J, J_+) \mapsto J - J_+.$$

Let F' denote the composite functor

$$N(\overline{\mathcal{C}})^{op} \to N(\mathbf{\Delta}_\infty)^{op} \to \mathfrak{X}.$$

Let $\mathcal{C}$ be the full subcategory of $\overline{\mathcal{C}}$ spanned by those pairs (J, J_+) where $J \neq \emptyset$. Let $\overline{\mathcal{C}}^0$ denote the full subcategory spanned by those pairs (J, J_+) where $J_+ = \emptyset$ and let $\mathcal{C}^0 = \overline{\mathcal{C}}^0 \cap \mathcal{C}$. We observe that $\overline{\mathcal{C}}^0$ can be identified with $\mathbf{\Delta}_+$ and that $\mathcal{C}^0$ can be identified with $\mathbf{\Delta}$ in such a way that $U_\bullet^+$ is identified with $F' | N(\overline{\mathcal{C}}^0)^{op}$.

Our first claim is that the inclusion $N(\mathcal{C}^0)^{op} \subseteq N(\mathcal{C})^{op}$ is cofinal. According to Theorem 4.1.3.1, it will suffice to show that for every object $X = (J, J_+)$ of $\mathcal{C}$, the category $\mathcal{C}^0_{/X}$ has a contractible nerve. This is clear since the relevant category has a final object: namely, the map $(J, \emptyset) \to (J, J_+)$. As a consequence, we conclude that $U_\bullet^+$ is a colimit diagram if and only if F' is a colimit diagram.

We now define $\mathcal{C}^1$ to be the full subcategory of $\mathcal{C}$ spanned by those pairs (J, J_+) such that J_+ is nonempty. We claim that $F' | N(\mathcal{C})^{op}$ is a left Kan extension of $F' | N(\mathcal{C}^1)^{op}$. To prove this, we must show that for every $(J, \emptyset) \in \mathcal{C}^0$, the induced map

$$(N(\mathcal{C}^1_{(J, \emptyset)/})^{op})^{\triangleright} \to N(\mathcal{C})^{op} \to \mathfrak{X}$$

is a colimit diagram. Let $\mathcal{D}$ denote the full subcategory of $\mathcal{C}^1_{(J, \emptyset)/}$ spanned by those morphisms $(J, \emptyset) \to (J', J'_+)$ which induce isomorphisms $J \simeq J' - J'_+$. We claim that the inclusion $N(\mathcal{D})^{op} \subseteq N(\mathcal{C}^1_{(J, \emptyset)/})^{op}$ is cofinal. To prove this, we once again invoke Theorem 4.1.3.1 to reduce to the following assertion: for every morphism $\phi : (J, \emptyset) \to (J'', J''_+)$, if $J''_+ \neq \emptyset$, then the category $\mathcal{D}_{/\phi}$ of all factorizations

$$(J, \emptyset) \to (J', J'_+) \to (J'', J''_+)$$

such that $J'_+ \neq \emptyset$ and such that $J \simeq J' - J'_+$ has a weakly contractible nerve. This is clear since $\mathcal{D}_{/\phi}$ has a final object $(J \coprod J'''_+, J'''_+)$, where $J'''_+ =$

$\{j \in J''_+ : (\forall i \in J)[j \geq \phi(i)]\}$. Consequently, it will suffice to prove that the induced functor

$$N(\mathcal{D}^{op})^{\triangleright} \to \mathfrak{X}$$

is a colimit diagram. This diagram can be identified with the constant diagram

$$N(\mathbf{\Delta}_+)^{op} \to \mathfrak{X}$$

taking the value $U_{\bullet}(J)$ and is a colimit diagram because the category $\mathbf{\Delta}$ has a weakly contractible nerve (Corollary 4.4.4.10).

We now apply Lemma 4.3.2.7, which asserts that F' is a colimit diagram if and only if $F'|(N(\mathcal{C}^1)^{op})^{\triangleright}$ is a colimit diagram. Let $\mathcal{C}^2 \subseteq \mathcal{C}^1$ be the full subcategory spanned by those objects (J, J_+) such that $J = J_+$. We claim that the inclusion $N(\mathcal{C}^2)^{op} \subseteq N(\mathcal{C}^1)^{op}$ is cofinal. According to Theorem 4.1.3.1, it will suffice to show that, for every object $(J, J_+) \in \mathcal{C}^1$, the category $\mathcal{C}^2_{/(J,J_+)}$ has a weakly contractible nerve. This is clear since the map $(J_+, J_+) \to (J, J_+)$ is a final object of the category $\mathcal{C}^{(2)}_{/(J,J_+)}$. Consequently, to prove that $F'|(N(\mathcal{C}^1)^{op})^{\triangleright}$ is a colimit diagram, it will suffice to prove that $F'|(N(\mathcal{C}^2)^{op})^{\triangleright}$ is a colimit diagram. But this diagram can be identified with the constant map $N(\mathbf{\Delta}_+)^{op} \to \mathfrak{X}$ taking the value $U_{\bullet}(\mathbf{\Delta}^{-1})$, which is a colimit diagram because the simplicial set $N(\mathbf{\Delta})^{op}$ is weakly contractible (Corollary 4.4.4.10). □

Lemma 6.1.3.17. *Let $\mathfrak{X}$ be an ∞-category and let $U_{\bullet} : N(\mathbf{\Delta})^{op} \to \mathfrak{X}$ be a simplicial object of $\mathfrak{X}$. Let $U'_{\bullet}$ be the augmented simplicial object given by composing $U_{\bullet}$ with the functor*

$$\mathbf{\Delta}_+ \to \mathbf{\Delta}$$

$$J \to J \coprod \{\infty\}.$$

Then

(1) *The augmented simplicial object $U'_{\bullet}$ is a colimit diagram.*

(2) *If $U_{\bullet}$ is a groupoid object of $\mathfrak{X}$, then the evident natural transformation of simplicial objects $\alpha : U'_{\bullet}|N(\mathbf{\Delta})^{op} \to U_{\bullet}$ is Cartesian.*

Proof. Assertion (1) follows immediately from Lemma 6.1.3.16. To prove (2), let us consider the collection S of all morphisms $f : J \to J'$ in $\mathbf{\Delta}$ such that $\alpha(f)$ is a pullback square

$$
\begin{array}{ccc}
U'_{\bullet}(J') & \longrightarrow & U_{\bullet}(J') \\
\downarrow & & \downarrow \\
U'_{\bullet}(J) & \longrightarrow & U_{\bullet}(J)
\end{array}
$$

in $\mathfrak{X}$. We wish to prove that every morphism of $\mathbf{\Delta}$ belongs to S. Using Lemma 4.4.2.1, we deduce that if $f' \in S$, then $f \in S \Leftrightarrow (f \circ f' \in S)$. Consequently, it will suffice to prove that every inclusion $\{j\} \subseteq J$ belongs to S. Unwinding the definition, this amounts the requirement that the diagram

$$
\begin{array}{ccc}
U_\bullet(J \cup \{\infty\}) & \longrightarrow & U_\bullet(J) \\
\downarrow & & \downarrow \\
U_\bullet(\{j, \infty\}) & \longrightarrow & U_\bullet(\{j\})
\end{array}
$$

be Cartesian, which follows immediately from criterion $(4'')$ of Proposition 6.1.2.6. □

Remark 6.1.3.18. Assertion (2) of Lemma 6.1.3.17 has a converse: if α is a Cartesian transformation, then $U_\bullet$ is a groupoid object of $\mathfrak{X}$. This can be deduced easily by examining the proof of Proposition 6.1.2.6, but we will not have need of it.

Proposition 6.1.3.19. *Let $\mathfrak{X}$ be an ∞-category satisfying the equivalent conditions of Theorem 6.1.3.9. Then $\mathfrak{X}$ satisfies the ∞-categorical Giraud axioms:*

(i) *The ∞-category $\mathfrak{X}$ is presentable.*

(ii) *Colimits in $\mathfrak{X}$ are universal.*

(iii) *Coproducts in $\mathfrak{X}$ are disjoint.*

(iv) *Every groupoid object of $\mathfrak{X}$ is effective.*

Proof. Axioms (i) and (ii) are obvious. To prove (iii), let us consider an arbitrary pair of objects $X, Y \in \mathfrak{X}$ and let $\emptyset$ denote an initial object of $\mathfrak{X}$. Let $f : \emptyset \to X$ be a morphism (unique up to homotopy since $\emptyset$ is initial). We observe that $\mathrm{id}_\emptyset$ is an initial object of $\mathcal{O}_\mathfrak{X}$. Form a pushout diagram

$$
\begin{array}{ccc}
\mathrm{id}_\emptyset & \xrightarrow{\ \alpha\ } & \mathrm{id}_Y \\
{\scriptstyle\beta}\downarrow & & \downarrow{\scriptstyle\beta'} \\
f & \xrightarrow{\ \alpha'\ } & g
\end{array}
$$

in $\mathcal{O}_\mathfrak{X}$. It is clear that α is a Cartesian transformation, and Lemma 6.1.3.6 implies that β is Cartesian as well. Invoking condition (2) of Theorem 6.1.3.9, we deduce that α' is a Cartesian transformation. But α' can be identified with a pushout diagram

$$
\begin{array}{ccc}
\emptyset & \longrightarrow & Y \\
\downarrow & & \downarrow \\
X & \longrightarrow & X \coprod Y.
\end{array}
$$

It remains to prove that every groupoid object in $\mathfrak{X}$ is effective. Let $U_\bullet$ be a groupoid object of $\mathfrak{X}$ and let $\overline{U}_\bullet : \mathrm{N}(\mathbf{\Delta}_+)^{op} \to \mathfrak{X}$ be a colimit of $U_\bullet$. Let $U'_\bullet : \mathrm{N}(\mathbf{\Delta}_+)^{op} \to \mathfrak{X}$ be the result of composing $\overline{U}_\bullet$ with the "shift" functor

$$
\mathbf{\Delta}_+ \to \mathbf{\Delta}_+
$$

$$J \mapsto J \coprod \{\infty\}.$$

(In other words, $U'_\bullet$ is the shifted simplicial object given by $U'_n = U_{n+1}$.)
Lemma 6.1.3.17 implies that $U'_\bullet$ is a colimit diagram in $\mathfrak{X}$. We have a trans-
formation $\overline{\alpha} : U'_\bullet \to \overline{U}_\bullet$. Since $U_\bullet$ is a groupoid, $\alpha = \overline{\alpha}| \, \mathrm{N}(\mathbf{\Delta})^{op}$ is a Cartesian
transformation (Lemma 6.1.3.17 again). Applying (4), we deduce that $\overline{\alpha}$ is
a Cartesian transformation. In particular, we conclude that

$$\begin{array}{ccc} U'_0 & \longrightarrow & U'_{-1} \\ \downarrow & & \downarrow \\ \overline{U}_0 & \longrightarrow & \overline{U}_{-1} \end{array}$$

is a pullback diagram in $\mathfrak{X}$. But this diagram can be identified with

$$\begin{array}{ccc} U_1 & \longrightarrow & U_0 \\ \downarrow & & \downarrow \\ U_0 & \longrightarrow & \overline{U}_{-1}, \end{array}$$

so that $U_\bullet$ is effective by Proposition 6.1.2.11. $\qquad\square$

Corollary 6.1.3.20. *Every groupoid object of $\mathcal{S}$ is effective.*

6.1.4 Free Groupoids

Let $\mathfrak{X}$ be an ∞-category which satisfies the ∞-categorical Giraud axioms (i)
through (iv) of Theorem 6.1.0.6. We wish to prove that $\mathfrak{X}$ is an ∞-topos. It
is clear that any proof will need to make use of the full strength of axioms (i)
through (iv); in particular, we will need to apply (iv) to a class of groupoid
objects of $\mathfrak{X}$ which are not obviously effective. The purpose of this section is
to describe a construction which will yield nontrivial examples of groupoid
objects and to deduce a consequence (Proposition 6.1.4.2) which we will use
in the proof of Theorem 6.1.0.6.

Definition 6.1.4.1. Let $f : \mathfrak{X} \to \mathcal{Y}$ be a functor between ∞-categories
which admit finite limits. Let Z be an object of $\mathfrak{X}$. We will say that f is *left
exact at Z* if, for every pullback square

in $\mathfrak{X}$, the induced square

$$\begin{array}{ccc} f(W) & \longrightarrow & f(Y) \\ \downarrow & & \downarrow \\ f(X) & \longrightarrow & f(Z) \end{array}$$

is a pullback in $\mathcal{Y}$.

We can now state the main result of this section:

Proposition 6.1.4.2. *Let $\mathfrak{X}$ and $\mathfrak{Y}$ be presentable ∞-categories and let $f : \mathfrak{X} \to \mathfrak{Y}$ be a functor which preserves small colimits. Suppose that every groupoid object in either $\mathfrak{X}$ or $\mathfrak{Y}$ is effective. Let*

$$U_1 \rightrightarrows U_0 \xrightarrow{\ s\ } U_{-1}$$

be a coequalizer diagram in $\mathfrak{X}$ and let

$$
\begin{array}{ccc}
X & \longrightarrow & U_0 \\
\downarrow & & \downarrow{\scriptstyle s} \\
U_0 & \xrightarrow{\ s\ } & U_{-1}
\end{array}
$$

be a pullback diagram in $\mathfrak{X}$. Suppose that f is left exact at U_0. Then the associated diagram

$$
\begin{array}{ccc}
f(X) & \longrightarrow & f(U_0) \\
\downarrow & & \downarrow{\scriptstyle s} \\
f(U_0) & \xrightarrow{\ s\ } & f(U_{-1})
\end{array}
$$

is a pullback square in $\mathfrak{Y}$.

Before giving the proof, we must establish some preliminary results.

Lemma 6.1.4.3. *Let $\mathfrak{X}$ and $\mathfrak{Y}$ be ∞-categories which admit finite limits, let $f : \mathfrak{X} \to \mathfrak{Y}$ be a functor, and let $U_\bullet$ be a groupoid object of $\mathfrak{X}$. Suppose that f is left exact at U_0. Then $f \circ U_\bullet$ is a groupoid object of $\mathfrak{Y}$.*

Proof. This follows immediately from characterization $(4'')$ given in Proposition 6.1.2.6. $\qquad\square$

Let $\mathfrak{X}$ be a presentable ∞-category. We define a *simplicial resolution* in $\mathfrak{X}$ to be an augmented simplicial object $U_\bullet^+ : N(\mathbf{\Delta}_+)^{op} \to \mathfrak{X}$ which is a colimit of the underlying simplicial object $U_\bullet = U_\bullet^+ | N(\mathbf{\Delta})^{op}$. We let $\mathcal{R}\mathrm{es}(\mathfrak{X})$ denote the full subcategory of $\mathfrak{X}_{\mathbf{\Delta}_+}$ spanned by the simplicial resolutions. Note that since every simplicial object of $\mathfrak{X}$ has a colimit, the restriction functor $\mathcal{R}\mathrm{es}(\mathfrak{X}) \to \mathfrak{X}_{\mathbf{\Delta}}$ is a trivial fibration and therefore an equivalence of ∞-categories. We will say that a simplicial resolution $U_\bullet^+$ is a *groupoid resolution* if the underlying simplicial object $U_\bullet$ is a groupoid object of $\mathfrak{X}$.

We will say that a map $f : U_\bullet^+ \to V_\bullet^+$ of simplicial resolutions *exhibits $V_\bullet^+$ as the groupoid resolution generated by $U_\bullet^+$* if $V_\bullet^+$ is a groupoid resolution and the induced map

$$\mathrm{Map}_{\mathcal{R}\mathrm{es}(\mathfrak{X})}(V_\bullet^+, W_\bullet^+) \to \mathrm{Map}_{\mathcal{R}\mathrm{es}(\mathfrak{X})}(U_\bullet^+, W_\bullet^+)$$

is a homotopy equivalence for every groupoid resolution $W_\bullet^+ \in \mathcal{R}\mathrm{es}(\mathfrak{X})$.

Remark 6.1.4.4. Let $\mathfrak{X}$ be a presentable ∞-category. Then for every simplicial resolution $U_\bullet^+$ in $\mathfrak{X}$, there is a map $f : U_\bullet^+ \to V_\bullet^+$ which exhibits $V_\bullet^+$ as the groupoid resolution generated by $U_\bullet^+$. In view of the equivalence $\mathrm{Res}(\mathfrak{X}) \to \mathfrak{X}_\Delta$, this is equivalent to the assertion that $\mathrm{Gpd}(\mathfrak{X})$ is a localization of $\mathfrak{X}_\Delta$. This follows from Proposition 6.1.2.6 together with Lemmas 5.5.4.18 and 5.5.4.19.

Lemma 6.1.4.5. *Let $\mathfrak{X}$ be a presentable ∞-category and let $f : U_\bullet^+ \to V_\bullet^+$ be a map of simplicial resolutions which exhibits $V_\bullet^+$ as the groupoid resolution generated by $U_\bullet^+$. Let $W_\bullet^+$ be an augmented simplicial object of $\mathfrak{X}$ such that the underlying simplicial object $W_\bullet \in \mathfrak{X}_\Delta$ is a groupoid. Composition with f induces a homotopy equivalence*

$$\mathrm{Map}_{\mathfrak{X}_{\Delta_+}}(V_\bullet^+, W_\bullet^+) \to \mathrm{Map}_{\mathfrak{X}_{\Delta_+}}(U_\bullet^+, W_\bullet^+).$$

Proof. Let $|W_\bullet|$ be a colimit of $W_\bullet$. Then we have a commutative diagram

$$
\begin{array}{ccc}
\mathrm{Map}_{\mathfrak{X}_{\Delta_+}}(V_\bullet^+, |W_\bullet|) & \longrightarrow & \mathrm{Map}_{\mathfrak{X}_{\Delta_+}}(U_\bullet^+, |W_\bullet|) \\
\downarrow & & \downarrow \\
\mathrm{Map}_{\mathfrak{X}_{\Delta_+}}(V_\bullet^+, W_\bullet^+) & \longrightarrow & \mathrm{Map}_{\mathfrak{X}_{\Delta_+}}(U_\bullet^+, W_\bullet^+)
\end{array}
$$

where the vertical maps are homotopy equivalences (since $U_\bullet^+$ and $V_\bullet^+$ are resolutions) and the upper horizontal map is a homotopy equivalence (since $|W_\bullet|$ is a groupoid resolution). $\square$

Lemma 6.1.4.6. *Let $\mathfrak{X}$ be a presentable ∞-category. Suppose that $f : U_\bullet^+ \to V_\bullet^+$ is a map in $\mathrm{Res}(\mathfrak{X})$ which exhibits $V_\bullet^+$ as the groupoid resolution generated by $U_\bullet^+$. Then f induces equivalences $U_{-1} \to V_{-1}$ and $U_0 \to V_0$.*

Proof. Let $\mathbf{\Delta}_+^{\leq 0}$ be the full subcategory of $\mathbf{\Delta}_+$ spanned by the objects Δ^{-1} and Δ^0. Let $j : \mathbf{\Delta}_+^{\leq 0} \to \mathbf{\Delta}_+$ denote the inclusion functor and let $j^* : \mathfrak{X}_{\Delta_+} \to \mathcal{O}_{\mathfrak{X}}$ be the associated restriction functor. We wish to show that $j^*(f)$ is an equivalence. Equivalently, we show that for every $W \in \mathcal{O}_{\mathfrak{X}}$, composition with $j^*(f)$ induces a homotopy equivalence

$$\mathrm{Map}_{\mathcal{O}_{\mathfrak{X}}}(j^* V_\bullet^+, W) \to \mathrm{Map}_{\mathcal{O}_{\mathfrak{X}}}(j^* U_\bullet^+, W).$$

Let j_* be a right adjoint to j^* (a right Kan extension functor). It will suffice to prove that composition with f induces a homotopy equivalence

$$\mathrm{Map}_{\mathfrak{X}_{\Delta_+}}(V_\bullet^+, j_* W) \to \mathrm{Map}_{\mathfrak{X}_{\Delta_+}}(U_\bullet^+, j_* W).$$

The augmented simplicial object $j_* W$ is a Čech nerve, so that the underlying simplicial object of $j_* W$ is a groupoid by Proposition 6.1.2.11. We now conclude by applying Lemma 6.1.4.5. $\square$

Let $\mathfrak{J}$ denote the subcategory of $\mathbf{\Delta}_+$ spanned by the objects $\emptyset$, $[0]$, and $[1]$, where the morphisms are given by *injective* maps of linearly ordered sets. This category may be depicted as follows:

$$\emptyset \longrightarrow [0] \rightrightarrows [1].$$

We let $\mathcal{J}_0$ denote the full subcategory of $\mathcal{J}$ spanned by the objects $[0]$ and $[1]$. We will say that a diagram $N(\mathcal{J})^{op} \to \mathcal{X}$ is a *coequalizer diagram* if it is a colimit of its restriction to $N(\mathcal{J}_0)^{op} \to \mathcal{X}$.

Let i denote the inclusion $\mathcal{J} \subseteq \mathbf{\Delta}_+$ and let i^* denote the restriction functor $\mathcal{X}_{\mathbf{\Delta}_+} \to \mathrm{Fun}(N(\mathcal{J})^{op}, \mathcal{X})$. If $\mathcal{X}$ is a presentable ∞-category, then i^* has a left adjoint $i_!$ (a left Kan extension).

Lemma 6.1.4.7. *Let $\mathcal{X}$ be a presentable ∞-category. The left Kan extension $i_! : \mathrm{Fun}(N(\mathcal{J})^{op}, \mathcal{X}) \to \mathcal{X}_{\mathbf{\Delta}_+}$ carries coequalizer diagrams to simplicial resolutions.*

Proof. We have a commutative diagram of inclusions of subcategories

$$
\begin{array}{ccc}
\mathcal{J}_0 & \xrightarrow{\ j'\ } & \mathcal{J} \\
\downarrow{\scriptstyle i'} & & \downarrow{\scriptstyle i} \\
\mathbf{\Delta} & \xrightarrow{\ j\ } & \mathbf{\Delta}_+
\end{array}
$$

which gives rise to a homotopy commutative diagram of ∞-categories

$$
\begin{array}{ccc}
\mathrm{Fun}(N(\mathcal{J}_0)^{op}, \mathcal{X}) & \xrightarrow{\ j'_!\ } & \mathrm{Fun}(N(\mathcal{J})^{op}, \mathcal{X}) \\
\downarrow{\scriptstyle i'_!} & & \downarrow{\scriptstyle i_!} \\
\mathcal{X}_{\mathbf{\Delta}} & \xrightarrow{\ j_!\ } & \mathcal{X}_{\mathbf{\Delta}_+}
\end{array}
$$

in which the morphisms are given by left Kan extensions. An object $U \in \mathrm{Fun}(N(\mathcal{J})^{op}\, \mathcal{X})$ is a coequalizer diagram if and only if it lies in the essential image of $j'_!$. In this case, $i_! U$ lies in the essential image of $i_! \circ j'_! \simeq j_! \circ i'_!$, which is contained in the essential image of $j_!$: namely, the resolutions. $\square$

Lemma 6.1.4.8. *Let $\mathcal{X}$ be a presentable ∞-category and suppose we are given a diagram $U : N(\mathcal{J})^{op} \to \mathcal{C}$ which we may depict as*

$$ U_1 \rightrightarrows U_0 \longrightarrow U_{-1}. $$

Let $V_\bullet = i_! U \in \mathcal{X}_{\mathbf{\Delta}_+}$ be a left Kan extension of U along $i : N(\mathcal{J})^{op} \to \mathbf{\Delta}_+^{op}$. Then the augmentation maps $V_0 \to V_{-1}$ and $U_0 \to U_{-1}$ are equivalent in the ∞-category $\mathcal{O}_{\mathcal{X}}$.

Proof. This follows from Proposition 4.3.3.8 since

$$ \mathrm{Hom}_{\mathcal{J}}(\Delta^i, \bullet) \simeq \mathrm{Hom}_{\mathbf{\Delta}_+}(\Delta^i, \bullet) $$

for $i \leq 0$. $\square$

Proof of Proposition 6.1.4.2. Let $U : N(\mathcal{J})^{op} \to \mathcal{C}$ be a coequalizer diagram in $\mathcal{X}$, which we denote by

$$ U_1 \rightrightarrows U_0 \xrightarrow{\ s\ } U_{-1}, $$

and form a pullback square

$$\begin{array}{ccc} X & \longrightarrow & U_0 \\ \downarrow & & \downarrow{\scriptstyle s} \\ U_0 & \xrightarrow{\ s\ } & U_{-1}. \end{array}$$

Let $V_\bullet = i_! U \in \mathfrak{X}_{\Delta_+}$ be a left Kan extension of U. According to Lemma 6.1.4.7, $V_\bullet$ is a simplicial resolution. We may therefore choose a map $V_\bullet \to W_\bullet$ which exhibits $W_\bullet$ as the groupoid resolution generated by $V_\bullet$ (Remark 6.1.4.4). Since every groupoid object in $\mathfrak{X}$ is effective, $W_\bullet$ is a Čech nerve. It follows from the characterization given in Proposition 6.1.2.11 that there is a pullback diagram

$$\begin{array}{ccc} W_1 & \longrightarrow & W_0 \\ \downarrow & & \downarrow \\ W_0 & \longrightarrow & W_{-1} \end{array}$$

in $\mathfrak{X}$. Using Lemma 6.1.4.8 and Lemma 6.1.4.6, we see that this diagram is equivalent to the pullback diagram

$$\begin{array}{ccc} X & \longrightarrow & U_0 \\ \downarrow & & \downarrow{\scriptstyle s} \\ U_0 & \xrightarrow{\ s\ } & U_{-1}. \end{array}$$

It therefore suffices to prove that the induced diagram

$$\begin{array}{ccc} f(W_1) & \longrightarrow & f(W_0) \\ \downarrow & & \downarrow \\ f(W_0) & \longrightarrow & f(W_{-1}) \end{array}$$

is a pullback. We make a slightly stronger claim: the augmented simplicial object $f \circ W_\bullet$ is a Čech nerve. Since every groupoid object in $\mathcal{Y}$ is effective, it will suffice to prove that $f \circ W_\bullet$ is a groupoid resolution. Since f preserves colimits, it is clear that $f \circ W_\bullet$ is a simplicial resolution. It follows from Lemma 6.1.4.3 that the underlying simplicial object of $f \circ W_\bullet$ is a groupoid. $\square$

6.1.5 Giraud's Theorem for ∞-Topoi

In this section, we will complete the proof of Theorem 6.1.0.6 by showing that every ∞-category $\mathfrak{X}$ which satisfies the ∞-categorical Giraud axioms (i) through (iv) arises as a left exact localization of an ∞-category of presheaves. Our strategy is simple: we choose a small category $\mathcal{C}$ equipped with a functor $f : \mathcal{C} \to \mathfrak{X}$. According to Theorem 5.1.5.6, we obtain a colimit-preserving functor $F : \mathcal{P}(\mathcal{C}) \to \mathfrak{X}$ which extends f up to homotopy. We will apply Proposition 6.1.4.2 to show that (under suitable hypotheses) F is a left exact localization functor (Proposition 6.1.5.2).

Lemma 6.1.5.1. *Let* $\mathcal{X}$ *be a presentable* ∞-*category in which colimits are universal and coproducts are disjoint.*

Let $\{\phi_i : Z_i \to Z\}_{i \in I}$ *be a family of morphisms in* $\mathcal{X}$ *which exhibit* Z *as a coproduct of the family of objects* $\{Z_i\}_{i \in I}$. *Let*

$$
\begin{array}{ccc}
W & \xrightarrow{\ \alpha\ } & Z_i \\
\downarrow & & \downarrow{\scriptstyle \phi_i} \\
Z_j & \xrightarrow{\ \phi_j\ } & Z
\end{array}
$$

be a square diagram in $\mathcal{X}$. *Then*

(1) *If* $i \neq j$, *the diagram is a pullback square if and only if* W *is an initial object of* $\mathcal{X}$.

(2) *If* $i = j$, *the diagram is a pullback square if and only if* α *is an equivalence.*

Proof. Let $Z_i^\vee$ be a coproduct for the objects $\{Z_k\}_{k \in I, k \neq i}$ and let $\psi : Z_i^\vee \to Z$ be a morphism such that each of the compositions

$$Z_k \to Z_i^\vee \xrightarrow{\psi} Z$$

is equivalent to Z. Then there is a pushout square

$$
\begin{array}{ccc}
\emptyset & \xrightarrow{\ \beta\ } & Z_i \\
\downarrow & & \downarrow{\scriptstyle \phi_i} \\
Z_i^\vee & \xrightarrow{\ \psi\ } & Z,
\end{array}
$$

where $\emptyset$ denotes an initial object of $\mathcal{X}$. Since coproducts in $\mathcal{X}$ are disjoint, this pushout square is also a pullback.

Let $\phi_i^* : \mathcal{X}^{/Z} \to \mathcal{X}^{/Z_i}$ denote a pullback functor. The above argument shows that $\phi_i^*(\psi)$ is an initial object of $\mathcal{X}^{/Z_i}$. If $j \neq i$, then there is a map of arrows $\phi_j \to \psi$ in $\mathcal{X}^{/Z}$ and therefore a map $\phi_i^*(\phi_j) \to \phi_i^*(\psi)$ in $\mathcal{X}^{/Z_i}$. Consequently, if $\alpha \simeq \phi_i^*(\phi_j)$, then W admits a map to an initial object of $\mathcal{X}$ and is therefore itself initial by Lemma 6.1.3.6. This proves the "only if" direction of (1). The converse follows from the uniqueness of initial objects.

Now suppose that $i = j$. We observe that id_Z is a coproduct of ϕ_i and ψ in the ∞-category $\mathcal{X}^{/Z}$. Since ϕ_i^* preserves coproducts, we deduce that id_{Z_i} is a coproduct of $\phi^*(\phi_j) : X \to Z_i$ and $\beta : \emptyset \to Z_i$ in $\mathcal{X}^{/Z_i}$. Since β is an initial object of $\mathcal{X}^{/Z_i}$, we see that $\phi^*(\phi_j)$ is an equivalence. The natural map $\gamma : \alpha \to \phi_i^*(\phi_i)$ corresponds to a commutative diagram

$$
\begin{array}{ccc}
W & \xrightarrow{\ \alpha\ } & Z_i \\
\downarrow{\scriptstyle \gamma_0} & & \downarrow{\scriptstyle \mathrm{id}_{Z_i}} \\
X & \xrightarrow{\ \phi_i^*(\phi_i)\ } & Z_i
\end{array}
$$

in the ∞-category $\mathcal{X}$. Consequently, α is an equivalence if and only if γ_0 is an equivalence, if and only if γ is an equivalence in $\mathcal{X}^{/Z_i}$. This proves (2). $\quad\square$

Proposition 6.1.5.2. *Let* $\mathcal{C}$ *be a small* ∞-*category which admits finite limits and let* $\mathfrak{X}$ *be an* ∞-*category which satisfies the* ∞-*categorical Giraud axioms* (i) *through* (iv) *of Theorem 6.1.0.6. Let* $F : \mathcal{P}(\mathcal{C}) \to \mathfrak{X}$ *be a colimit-preserving functor. Suppose that the composition* $F \circ j : \mathcal{C} \to \mathfrak{X}$ *is left exact, where* $j : \mathcal{C} \to \mathcal{P}(\mathcal{C})$ *denotes the Yoneda embedding. Then* F *is left exact.*

Proof. According to Corollary 4.4.2.5, to prove that F is left exact, it will suffice to prove that F preserves pullbacks and final objects. Since all final objects are equivalent, to prove that F preserves final objects, it suffices to exhibit a single final object Z of $\mathcal{P}(\mathcal{C})$ such that $FY \in \mathfrak{X}$ is final. Let z be a final object of $\mathcal{C}$ (which exists by virtue of our assumption that $\mathcal{C}$ admits finite limits). Then $Z = j(z)$ is a final object of $\mathcal{P}(\mathcal{C})$ since j preserves limits by Proposition 5.1.3.2. Consequently, $F(Z) = f(z)$ is final since f is left exact.

Let $\alpha : Y \to Z$ be a morphism in $\mathcal{P}(\mathcal{C})$. We will say that α is *good* if for every pullback square

$$\begin{array}{ccc} W & \longrightarrow & Y \\ \downarrow & & \downarrow {\scriptstyle \alpha} \\ X & \longrightarrow & Z \end{array}$$

in $\mathcal{P}(\mathcal{C})$, the induced square

$$\begin{array}{ccc} F(W) & \longrightarrow & F(Y) \\ \downarrow & & \downarrow {\scriptstyle F(\alpha)} \\ F(X) & \xrightarrow{\ \beta\ } & F(Z) \end{array}$$

is a pullback in $\mathfrak{X}$. Note that Lemma 4.4.2.1 implies that the class of good morphisms in $\mathcal{P}(\mathcal{C})$ is stable under composition.

We rephrase this condition that a morphism α be good in terms of the pullback functors $\alpha^* : \mathcal{P}(\mathcal{C})^{/Z} \to \mathcal{P}(\mathcal{C})^{/Y}$, $F(\alpha)^* : \mathfrak{X}^{/F(Z)} \to \mathcal{P}(\mathcal{C})^{/F(Y)}$. Application of the functor F gives a map

$$t : F \circ \alpha^* \to F(\alpha)^* \circ F$$

in the ∞-category of functors from $\mathcal{P}(\mathcal{C})^{/Z}$ to $\mathfrak{X}^{/F(Z)}$, and α is good if and only if t is an equivalence. Note that t is a natural transformation of colimit-preserving functors. Since the image of the Yoneda embedding $j : \mathcal{C} \to \mathcal{P}(\mathcal{C})$ generates $\mathcal{P}(\mathcal{C})$ under colimits, it will suffice to prove that t is an equivalence when evaluated on objects of the form $\beta : j(x) \to Z$, where x is an object of $\mathcal{C}$.

Let us say that an object $Z \in \mathcal{P}(\mathcal{C})$ is *good* if every morphism $\alpha : Y \to Z$ is good. In other words, an object $Z \in \mathcal{P}(\mathcal{C})$ is good if F is left exact at Z in the sense of Definition 6.1.4.1. By repeating the above argument, we deduce that Z is good if and only if every morphism of the form $\alpha : j(y) \to Z$ is good for $y \in \mathcal{C}$.

We next claim that for every object $z \in \mathcal{C}$, the Yoneda image $j(z) \in \mathcal{P}(\mathcal{C})$ is good. In other words, we must show that for every pullback square

$$
\begin{array}{ccc}
W & \longrightarrow & j(y) \\
\downarrow & & \downarrow{\scriptstyle \alpha} \\
j(x) & \xrightarrow{\ \beta\ } & j(z)
\end{array}
$$

in $\mathcal{P}(\mathcal{C})$, the induced square

$$
\begin{array}{ccc}
F(W) & \longrightarrow & f(x) \\
\downarrow & & \downarrow \\
f(y) & \longrightarrow & f(z)
\end{array}
$$

is a pullback in $\mathfrak{X}$. Since the Yoneda embedding is fully faithful, we may suppose that α and β are the Yoneda images of morphisms $x \to z$, $y \to z$. Since j preserves limits, we may reduce to the case where the first diagram is the Yoneda image of a pullback diagram in $\mathcal{C}$. The desired result then follows from the assumption that f is left exact.

To complete the proof that F is left exact, it will suffice to prove that every object of $\mathcal{P}(\mathcal{C})$ is good. Because the Yoneda embedding $j : \mathcal{C} \to \mathcal{P}(\mathcal{C})$ generates $\mathcal{P}(\mathcal{C})$ under colimits, it will suffice to prove that the collection of good objects of $\mathcal{P}(\mathcal{C})$ is stable under colimits. According to Proposition 4.4.3.3, it will suffice to prove that the collection of good objects of $\mathcal{P}(\mathcal{C})$ is stable under coequalizers and small coproducts.

We first consider the case of coproducts. Let $\{Z_i\}_{i \in I}$ be a family of good objects of $\mathcal{P}(\mathcal{C})$ indexed by a (small) set I and let $\{\phi_i : Z_i \to Z\}_{i \in I}$ be a family of morphisms which exhibit Z as a coproduct of the family $\{Z_i\}_{i \in I}$. Suppose we are given a pullback diagram

$$
\begin{array}{ccc}
W & \longrightarrow & j(y) \\
\downarrow & & \downarrow{\scriptstyle \alpha} \\
j(x) & \xrightarrow{\ \beta\ } & Z
\end{array}
$$

in $\mathcal{P}(\mathcal{C})$. According to Proposition 5.1.2.2, evaluation at the object y induces a colimit-preserving functor $\mathcal{P}(\mathcal{C}) \to \mathcal{S}$. Consequently, we have a homotopy equivalence

$$
\operatorname{Map}_{\mathcal{P}(\mathcal{C})}(j(y), Z) \simeq \coprod_{i \in I} \operatorname{Map}_{\mathcal{P}(\mathcal{C})}(j(y), Z_i)
$$

in the homotopy category $\mathcal{H}$. Therefore we may assume that α factors as a composition

$$
j(y) \xrightarrow{\alpha'} Z_i \xrightarrow{\phi_i} Z
$$

for some $i \in I$. By assumption, the morphism α' is good; it therefore suffices to prove that ϕ_i is good. By a similar argument, we can replace β by a map

$\phi_j : Z_j \to Z$ for some $j \in I$. We are now required to show that if

$$
\begin{array}{ccc}
W' & \longrightarrow & Z_i \\
\downarrow & & \downarrow{\scriptstyle \phi_i} \\
Z_j & \xrightarrow{\phi_j} & Z
\end{array}
$$

is a pullback diagram in $\mathcal{P}(\mathcal{C})$, then

$$
\begin{array}{ccc}
F(W') & \longrightarrow & F(Z_i) \\
\downarrow & & \downarrow{\scriptstyle \phi_i} \\
Z_j & \xrightarrow{\phi_j} & Z
\end{array}
$$

is a pullback diagram in $\mathcal{X}$. Since F preserves initial objects, this follows immediately from Lemma 6.1.5.1.

We now complete the proof by showing that the collection of good objects of $\mathcal{P}(\mathcal{C})$ is stable under the formation of coequalizers. Let

$$ Z_1 \rightrightarrows Z_0 \xrightarrow{s} Z_{-1} $$

be a coequalizer diagram in $\mathcal{P}(\mathcal{C})$, and suppose that Z_0 and Z_1 are good. We must show that any pullback diagram

$$
\begin{array}{ccc}
W & \longrightarrow & j(y) \\
\downarrow & & \downarrow{\scriptstyle \alpha} \\
j(x) & \xrightarrow{\beta} & Z_{-1}
\end{array}
$$

remains a pullback diagram after applying the functor F. The functor

$$ \mathcal{P}(\mathcal{C}) \to \mathrm{N}(\mathcal{S}et) $$

$$ T \mapsto \mathrm{Hom}_{\mathrm{h}\mathcal{P}(\mathcal{C})}(j(x), T) $$

can be written as a composition

$$ \mathcal{P}(\mathcal{C}) \to \mathcal{S} \xrightarrow{\pi_0} \mathrm{N}(\mathcal{S}et), $$

where the first functor is given by evaluation at x. Both of these functors commute with colimits. Consequently, we have a coequalizer diagram

$$ \mathrm{Hom}_{\mathrm{h}\mathcal{P}(\mathcal{C})}(j(x), Z_1) \rightrightarrows \mathrm{Hom}_{\mathrm{h}\mathcal{P}(\mathcal{C})}(j(x), Z_0) \longrightarrow \mathrm{Hom}_{\mathrm{h}\mathcal{P}(\mathcal{C})}(j(x), Z_{-1}) $$

in the category of sets. In particular, the map β factors as a composition

$$ j(x) \xrightarrow{\beta'} Z_0 \xrightarrow{s} Z_{-1}. $$

Since we have already assumed that β' is good, we can replace β by the map $s : Z_0 \to Z_{-1}$ in the above diagram. By a similar argument, we can replace $\alpha : Y \to Z_{-1}$ by the map $s : Z_0 \to Z_{-1}$. We now obtain the desired result by applying Proposition 6.1.4.2. $\qquad \square$

We are now ready to complete the proof of Theorem 6.1.0.6:

Proposition 6.1.5.3. *Let $\mathfrak{X}$ be an ∞-category. Suppose that $\mathfrak{X}$ satisfies the ∞-categorical Giraud axioms:*

   (i) The ∞-category $\mathfrak{X}$ is presentable.

   (ii) Colimits in $\mathfrak{X}$ are universal.

   (iii) Coproducts in $\mathfrak{X}$ are disjoint.

   (iv) Every groupoid object of $\mathfrak{X}$ is effective.

Then there exists a small ∞-category $\mathcal{C}$ which admits finite limits and an accessible left exact localization functor $\mathcal{P}(\mathcal{C}) \to \mathfrak{X}$. In particular, $\mathfrak{X}$ is an ∞-topos.

Proof. Let $\mathfrak{X}$ be an ∞-topos. According to Proposition 5.4.7.4, there exists a regular cardinal τ such that $\mathfrak{X}$ is τ-accessible, and the full subcategory $\mathfrak{X}^\tau$ spanned by the τ-compact objects of $\mathfrak{X}$ is stable under finite limits. Let $\mathcal{C}$ be a minimal model for $\mathfrak{X}^\tau$, so that there is an equivalence $\mathrm{Ind}_\tau(\mathcal{C}) \to \mathfrak{X}$. The proof of Theorem 5.5.1.1 shows that the inclusion $\mathrm{Ind}_\tau(\mathcal{C}) \subseteq \mathcal{P}(\mathcal{C})$ has a left adjoint L. The composition of L with the Yoneda embedding $\mathcal{C} \to \mathcal{P}(\mathcal{C})$ can be identified with the Yoneda embedding $\mathcal{C} \to \mathrm{Ind}_\tau(\mathcal{C})$ and therefore preserves all limits which exist in $\mathcal{C}$ (Proposition 5.1.3.2). Applying Proposition 6.1.5.2, we deduce that L is left exact, so that $\mathrm{Ind}_\tau(\mathcal{C})$ is a left exact localization (automatically accessible) of $\mathcal{P}(\mathcal{C})$. Since $\mathfrak{X}$ is equivalent to $\mathrm{Ind}_\tau(\mathcal{C})$, we conclude that $\mathfrak{X}$ is also an accessible left exact localization of $\mathcal{P}(\mathcal{C})$. $\square$

6.1.6 ∞-Topoi and Classifying Objects

Let $\mathfrak{X}$ be an ordinary category and let X be an object of $\mathfrak{X}$. Let $\mathrm{Sub}(X)$ denote the partially ordered collection of *subobjects* of X: an object of $\mathrm{Sub}(X)$ is an equivalence class of monomorphisms $Y \to X$. If $\mathcal{C}$ is accessible, then $\mathrm{Sub}(X)$ is actually a set. If $\mathfrak{X}$ admits finite limits, then $\mathrm{Sub}(X)$ is contravariantly functorial in X: given a subobject $Y \to X$ and any map $X' \to X$, the fiber product $Y' = X' \times_X Y$ is a subobject of X'. A *subobject classifier* is an object Ω of $\mathfrak{X}$ which *represents* the functor Sub. In other words, Ω has a universal subobject $\Omega_0 \subseteq \Omega$ such that every monomorphism $Y \to X$ fits into a *unique* Cartesian diagram

(In this case, Ω_0 is automatically a final object of $\mathcal{C}$.)

Every topos has a subobject classifier. In fact, in the theory of *elementary topoi*, the existence of a subobject classifier is taken as one of the axioms.

Thus the existence of a subobject classifier is one of the defining characteristics of a topos. We would like to discuss the appropriate ∞-categorical generalization of the theory of subobject classifiers. The ideas presented here are due to Charles Rezk.

Definition 6.1.6.1. Let $\mathcal{X}$ be an ∞-category which admits pullbacks and let S be a collection of morphisms of $\mathcal{X}$ which is stable under pullback. We will say that a morphism $f : X \to Y$ *classifies* S if it is a final object of $\mathcal{O}_{\mathcal{X}}^{(S)}$ (see Notation 6.1.3.4). In this situation, we will also say that the object $Y \in \mathcal{X}$ *classifies* S. A *subobject classifier* for $\mathcal{X}$ is an object which classifies the collection of all monomorphisms in $\mathcal{X}$.

Example 6.1.6.2. The ∞-category $\mathcal{S}$ of spaces has a subobject classifier: namely, the discrete space $\{0, 1\}$ with two elements.

The following result provides a necessary and sufficient condition for the existence of a classifying object for S:

Proposition 6.1.6.3. *Let $\mathcal{X}$ be a presentable ∞-category in which colimits are universal and let S be a class of morphisms in $\mathcal{X}$ which is stable under pullbacks. There exists a classifying object for S if and only if the following conditions are satisfied:*

(1) *The class S is local (Definition 6.1.3.8).*

(2) *For every object $X \in \mathcal{X}$, the full subcategory of $\mathcal{X}_{/X}$ spanned by the elements of S is essentially small.*

Proof. Let $s : \mathcal{X}^{op} \to \widehat{\mathcal{S}}$ be a functor which classifies the right fibration $\mathcal{O}_{\mathcal{X}}^{(S)} \to \mathcal{X}$. Then S has a classifying object if and only if s is a representable functor. According to the representability criterion of Proposition 5.5.2.2, this is equivalent to the assertion that s preserves small limits, and the essential image of s consists of essentially small spaces. According to Lemma 6.1.3.7, s preserves small limits if and only if (1) is satisfied. It now suffices to observe that for each $X \in \mathcal{X}$, the space $s(X)$ is essentially small if and only if the full subcategory of $\mathcal{X}_{/X}$ spanned by S is essentially small. □

Using Proposition 6.1.6.3, one can show that every ∞-topos has a subobject classifier. However, in the ∞-categorical context, the emphasis on *subobjects* misses the point. To see why, let us return to considering an ordinary category $\mathcal{X}$ with a subobject classifier Ω. By definition, for every object $X \in \mathcal{X}$, we may identify maps $X \to \Omega$ with subobjects of X: that is, isomorphism classes of maps $Y \to X$ which happen to be monomorphisms. Even better would be an *object classifier*: that is, an object $\widetilde{\Omega}$ such that an element of $\mathrm{Hom}_{\mathcal{X}}(X, \widetilde{\Omega})$ could be identified with an *arbitrary* map $Y \to X$. But this is an unreasonable demand: if $Y \to X$ is not a monomorphism, then there may be automorphisms of Y as an object of $\mathcal{X}_{/X}$. It would be unnatural to ignore these automorphisms. However, it is also not possible to take them into account because $\mathrm{Hom}_{\mathcal{X}}(X, \widetilde{\Omega})$ must be a set rather than a groupoid.

If we allow $\mathfrak{X}$ to be an ∞-category, this objection loses its force. Informally speaking, we can consider the functor which associates to each $X \in \mathfrak{X}$ the maximal ∞-groupoid contained in $\mathfrak{X}_{/X}$ (this is contravariantly functorial in X provided that $\mathfrak{X}$ has finite limits). We might hope that this functor is representable by some $\Omega_\infty \in \mathfrak{X}$, which we would then call an *object classifier*.

Unfortunately, a new problem arises: it is generally unreasonable to ask for the collection of *all* morphisms in $\mathfrak{X}$ to be classified by an object of $\mathfrak{X}$ since this would require each slice $\mathfrak{X}_{/X}$ to be essentially small (Proposition 6.1.6.3). This is essentially a technical difficulty, which we will circumvent by introducing a cardinality bound.

Definition 6.1.6.4. Let $\mathfrak{X}$ be a presentable ∞-category. We will say that a morphism $f : X \to Y$ is *relatively κ-compact* if, for every pullback diagram

$$
\begin{array}{ccc}
X' & \longrightarrow & X \\
\downarrow {\scriptstyle f'} & & \downarrow {\scriptstyle f} \\
Y' & \longrightarrow & Y
\end{array}
$$

such that Y' is κ-compact, X' is also κ-compact.

Lemma 6.1.6.5. *Let $\mathfrak{X}$ be a presentable ∞-category, κ a regular cardinal, $\mathfrak{J}$ a κ-filtered ∞-category, and $\overline{p} : \mathfrak{J}^\triangleright \to \mathfrak{X}$ a colimit diagram. Let $f : X \to Y$ be a morphism in $\mathfrak{X}$, where Y is the image under $\overline{p}$ of the cone point of $\mathfrak{J}^\triangleright$. For each α in $\mathfrak{J}$, let $Y_\alpha = \overline{p}(\alpha)$ and form a pullback diagram*

$$
\begin{array}{ccc}
X_\alpha & \longrightarrow & X \\
\downarrow {\scriptstyle f_\alpha} & & \downarrow {\scriptstyle f} \\
Y_\alpha & \xrightarrow{g_\alpha} & Y.
\end{array}
$$

Suppose that each f_α is relatively κ-compact. Then f is relatively κ-compact.

Proof. Let Z be a κ-compact object of $\mathfrak{X}$ and $g : Z \to Y$ a morphism. Since Z is κ-compact and $\mathfrak{J}$ is κ-filtered, there exists a 2-simplex of $\mathfrak{X}$ corresponding to a diagram

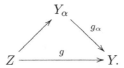

We form a Cartesian rectangle $\Delta^2 \times \Delta^1 \to \mathfrak{X}$, which we will depict as

$$
\begin{array}{ccccc}
Z' & \longrightarrow & X_\alpha & \longrightarrow & X \\
\downarrow {\scriptstyle f'} & & \downarrow {\scriptstyle f_\alpha} & & \downarrow {\scriptstyle f} \\
Z & \longrightarrow & Y_\alpha & \longrightarrow & Y.
\end{array}
$$

Since f' is a pullback of f_α, we conclude that Z' is κ-compact. Lemma 4.4.2.1 implies that f' is also a pullback of f along g, so that f is relatively κ-compact, as desired. $\qquad\square$

Lemma 6.1.6.6. *Let* $\mathfrak{X}$ *be a presentable* ∞*-category in which colimits are universal. Let* $\tau > \kappa$ *be regular cardinals such that* $\mathfrak{X}$ *is* κ*-accessible and the full subcategory* $\mathfrak{X}^\tau$ *consisting of* τ*-compact objects of* $\mathfrak{X}$ *is stable under pullbacks in* $\mathfrak{X}$*. Let* $\alpha : \sigma \to \sigma'$ *be a Cartesian transformation between pushout squares* $\sigma, \sigma' : \Delta^1 \times \Delta^1 \to \mathfrak{X}$*, which we may view as a pushout square*

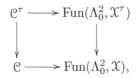

in $\mathrm{Fun}(\Delta^1, \mathfrak{X})$*. Suppose that* f*,* g*, and* f' *are relatively* τ*-compact. Then* g' *is relatively* τ*-compact.*

Proof. Let $\mathcal{C}$ denote the full subcategory of $\mathrm{Fun}(\Delta^1 \times \Delta^1, \mathfrak{X})$ spanned by the pushout squares and let $\mathcal{C}^\tau = \mathcal{C} \cap \mathrm{Fun}(\Delta^1 \times \Delta^1, \mathfrak{X}^\tau)$. Since the class of τ-compact objects of $\mathfrak{X}$ is stable under pushouts (Corollary 5.3.4.15), we have a commutative diagram

$$
\begin{array}{ccc}
\mathcal{C}^\tau & \longrightarrow & \mathrm{Fun}(\Lambda_0^2, \mathfrak{X}^\tau) \\
\downarrow & & \downarrow \\
\mathcal{C} & \longrightarrow & \mathrm{Fun}(\Lambda_0^2, \mathfrak{X}),
\end{array}
$$

where the horizontal arrows are trivial fibrations (Proposition 4.3.2.15). The proof of Proposition 5.4.4.3 shows that every object of $\mathrm{Fun}(\Lambda_0^2, \mathfrak{X})$ can be written as the colimit of a τ-filtered diagram in $\mathrm{Fun}(\Lambda_0^2, \mathfrak{X}^\tau)$. It follows that $\sigma' \in \mathcal{C}$ can be obtained as the colimit of a τ-filtered diagram in $\mathcal{C}^\tau$. Since colimits in $\mathfrak{X}$ are universal, we conclude that the natural transformation α can be obtained as a τ-filtered colimit of natural transformations $\alpha_i : \sigma_i \to \sigma_i'$ in $\mathcal{C}^\tau$. Lemma 5.5.2.3 implies that the inclusion $\mathcal{C} \subseteq \mathrm{Fun}(\Delta^1 \times \Delta^1, \mathfrak{X})$ is colimit-preserving. Consequently, we deduce that g' can be written as a τ-filtered colimit of morphisms $\{g_i'\}$ determined by restricting $\{\alpha_i\}$. According to Lemma 6.1.6.6, it will suffice to prove that each morphism g_i' is relatively τ-compact. In other words, we may replace σ' by σ_i' and thereby reduce to the case where σ' belongs to $\mathcal{C}^\tau$. Since f, g, and f' are relatively τ-compact, we conclude that $\sigma | \Lambda_0^2$ takes values in $\mathfrak{X}^\tau$. Since σ is a pushout diagram, Corollary 5.3.4.15 implies that σ takes values in $\mathfrak{X}^\tau$. Now we observe that g' is a morphism between τ-compact objects of $\mathfrak{X}$ and therefore automatically relatively τ-compact by virtue of our assumption that $\mathfrak{X}^\tau$ is stable under pullbacks in $\mathfrak{X}$. $\square$

Proposition 6.1.6.7. *Let* $\mathfrak{X}$ *be a presentable* ∞*-category in which colimits are universal and let* S *be a local class of morphisms in* $\mathfrak{X}$*. For each regular cardinal* κ*, let* S_κ *denote the collection of all morphisms* f *which belong to* S *and are relatively* κ*-compact. If* κ *is sufficiently large, then* S_κ *has a classifying object.*

Proof. Choose κ' such that $\mathfrak{X}$ is κ'-accessible. The restriction functor $r :$ $\mathrm{Fun}((\Lambda_2^2)^\triangleleft, \mathfrak{X}) \to \mathrm{Fun}(\Lambda_2^2, \mathfrak{X})$ is accessible: in fact, it preserves all colimits (Proposition 5.1.2.2). Let g be a right adjoint to r (a limit functor); Proposition 5.4.7.7 implies that g is also accessible. Choose a regular cardinal $\kappa'' > \kappa'$ such that g is κ''-continuous and choose $\kappa \geq \kappa''$ such that g carries κ''-compact objects of $\mathrm{Fun}(\Lambda_2^2, \mathfrak{X})$ into $\mathrm{Fun}((\Lambda_2^2)^\triangleleft, \mathfrak{X}^\kappa)$. It follows that the class of κ-compact objects of $\mathfrak{X}$ is stable under pullbacks. We will show that S_κ has a classifying object.

We will verify the hypotheses of Proposition 6.1.6.3. First, we must show that S_κ is local. For this, we will verify condition (3) of Lemma 6.1.3.7. We begin by showing that S_κ is stable under small coproducts. Let $\{f_\alpha : X_\alpha \to Y_\alpha\}_{\alpha \in A}$ be a small collection of morphisms belonging to S_κ and let $f : X \to Y$ be a coproduct $\coprod_{\alpha \in A} f_\alpha$ in $\mathrm{Fun}(\Delta^1, \mathfrak{X})$. We wish to show that $f \in S_\kappa$. Since S is local, we conclude that $f \in S$ (using Lemma 6.1.3.7). It therefore suffices to show that f is relatively κ-compact. Suppose we are given a κ-compact object $Z \in \mathfrak{X}$ and a morphism $g : Z \to Y$. Using Proposition 4.2.3.4 and Corollary 4.2.3.10, we conclude that Y can be obtained as a κ-filtered colimit of objects $Y_{A_0} = \coprod_{\alpha \in A_0} Y_\alpha$, where A_0 ranges over the κ-small subsets of A. Since Z is κ-compact, we conclude that there exists a factorization

$$Z \xrightarrow{g'} Y_{A_0} \xrightarrow{g''} Y$$

of g. Form a Cartesian rectangle $\Delta^2 \times \Delta^1 \to \mathfrak{X}$,

$$
\begin{array}{ccc}
Z' & \longrightarrow X_{A_0} & \longrightarrow X \\
\downarrow & \downarrow & \downarrow \\
Z & \longrightarrow Y_{A_0} & \longrightarrow Y.
\end{array}
$$

Since S is local, we can identify X_{A_0} with the coproduct $\coprod_{\alpha \in A_0} X_\alpha$. Since colimits are universal, we conclude that Z' is a coproduct of objects $Z'_\alpha = X_\alpha \times_{Y_\alpha} Z$, where α ranges over A_0. Since each f_α is relatively κ-compact, we conclude that each Z'_α is κ-compact. Thus Z', as a κ-small colimit of κ-compact objects, is also κ-compact (Corollary 5.3.4.15).

We must now show that for every pushout diagram

$$
\begin{array}{ccc}
f & \xrightarrow{\alpha} & g \\
\downarrow{\scriptstyle\beta} & & \downarrow{\scriptstyle\beta'} \\
f' & \xrightarrow{\alpha'} & g'
\end{array}
$$

in $\mathcal{O}_\mathfrak{X}$, if α and β are Cartesian transformations and $f, f', g \in S_\kappa$, then α' and β' are also Cartesian transformations and $g' \in S_\kappa$. The first assertion follows immediately from Lemma 6.1.3.7 (since S is local), and we deduce also that $g' \in S$. It therefore suffices to show that g is relatively κ-compact, which follows from Lemma 6.1.6.6.

It remains to show that, for each $X \in \mathfrak{X}$, the full subcategory of $\mathfrak{X}_{/X}$ spanned by the elements of S is essentially small. Equivalently, we must show that the right fibration $p : \mathcal{O}_{\mathcal{X}}^{(S)} \to \mathcal{X}$ has essentially small fibers. Let $F : \mathcal{X}^{op} \to \widehat{\mathcal{S}}$ classify p. Since S is local, F preserves limits. The full subcategory of $\widehat{\mathcal{S}}$ spanned by the essentially small Kan complexes is stable under small limits, and $\mathcal{X}$ is generated by $\mathcal{X}^{\kappa}$ under small (κ-filtered) colimits. Consequently, it will suffice to show that $F(X)$ is essentially small when X is κ-compact. In other words, we must show that there are only a bounded number of equivalence classes of morphisms $f : Y \to X$ such that $f \in S_{\kappa}$. We now observe that if $f \in S_{\kappa}$, then f is relatively κ-compact, so that Y also belongs to $\mathcal{X}^{\kappa}$. We now conclude by observing that the ∞-category $\mathcal{X}^{\kappa}$ is essentially small.   □

We now give a characterization of ∞-topoi based on the existence of object classifiers.

Theorem 6.1.6.8 (Rezk). *Let $\mathcal{X}$ be a presentable ∞-category. Then $\mathcal{X}$ is an ∞-topos if and only if the following conditions are satisfied:*

(1) *Colimits in $\mathcal{X}$ are universal.*

(2) *For all sufficiently large regular cardinals κ, there exists a classifying object for the class of all relatively κ-compact morphisms in $\mathcal{X}$.*

Proof. Assume that colimits in $\mathcal{X}$ are universal. According to Theorems 6.1.0.6 and 6.1.3.9, $\mathcal{X}$ is an ∞-topos if and only if the class S consisting of all morphisms of $\mathcal{X}$ is local. This clearly implies (2) in view of Proposition 6.1.6.7. Conversely, suppose that (2) is satisfied and let S_{κ} be defined as in the statement of Proposition 6.1.6.7. Proposition 6.1.6.3 ensures that S_{κ} is local for all sufficiently large regular cardinals κ. We note that $S = \bigcup S_{\kappa}$. It follows from Criterion (3) of Lemma 6.1.3.7 that S is also local, so that $\mathcal{X}$ is an ∞-topos.   □

6.2 CONSTRUCTIONS OF ∞-TOPOI

According to Definition 6.1.0.4, an ∞-category $\mathcal{X}$ is an ∞-topos if and only if $\mathcal{X}$ arises as an (accessible) left exact localization of a presheaf ∞-category $\mathcal{P}(\mathcal{C})$. To complete the analogy with classical topos theory, we would like to have some concrete description of the collection of left exact localizations of $\mathcal{P}(\mathcal{C})$. In §6.2.1, we will study left exact localization functors in general and single out a special class which we call *topological* localizations. In §6.2.2, we will study topological localizations of $\mathcal{P}(\mathcal{C})$ and show that they are in bijection with *Grothendieck topologies* on the ∞-category $\mathcal{C}$ by exact analogy with classical topos theory. In particular, given a Grothendieck topology on $\mathcal{C}$, one can define an ∞-topos $\mathrm{Shv}(\mathcal{C}) \subseteq \mathcal{P}(\mathcal{C})$ of *sheaves on* $\mathcal{C}$. In §6.2.3, we will characterize $\mathrm{Shv}(\mathcal{C})$ by a universal mapping property. Unfortunately, not every ∞-topos $\mathcal{X}$ can be obtained as topological localization of an ∞-category of presheaves. Nevertheless, in §6.2.4 we will construct ∞-categories of sheaves which closely approximate $\mathfrak{X}$ using the formalism of *canonical topologies*. These ideas will be applied in §6.4 to obtain a classification theorem for n-topoi.

6.2.1 Left Exact Localizations

Let $\mathfrak{X}$ be an ∞-category. Up to equivalence, a localization $L : \mathfrak{X} \to \mathcal{Y}$ is determined by the collection S of all morphisms $f : X \to Y$ in $\mathfrak{X}$ such that Lf is an equivalence in $\mathcal{Y}$ (Proposition 5.5.4.2). Our first result provides a useful criterion for testing the left exactness of L.

Proposition 6.2.1.1. *Let $L : \mathfrak{X} \to \mathcal{Y}$ be a localization of ∞-categories. Suppose that $\mathfrak{X}$ admits finite limits. The following conditions are equivalent:*

(1) *The functor L is left exact.*

(2) *For every pullback diagram*

$$\begin{array}{ccc} X' & \longrightarrow & X \\ \downarrow{\scriptstyle f'} & & \downarrow{\scriptstyle f} \\ Y' & \longrightarrow & Y \end{array}$$

in $\mathfrak{X}$ such that Lf is an equivalence in $\mathcal{Y}$, Lf' is also an equivalence in $\mathcal{Y}$.

Proof. It is clear that (1) implies (2). Suppose that (2) is satisfied. We wish to show that L is left exact. Let S be the collection of morphisms f in $\mathfrak{X}$ such that Lf is an equivalence. Without loss of generality, we may identify $\mathcal{Y}$ with the full subcategory of $\mathfrak{X}$ spanned by the S-local objects. Since the final object $1 \in \mathfrak{X}$ is obviously S-local, we have $L1 \simeq 1$. Thus it will suffice to show that L commutes with pullbacks. We observe that given any diagram $X \to Y \leftarrow Z$, the pullback $LX \times_{LY} LZ$ is a limit of S-local objects of $\mathfrak{X}$ and therefore S-local. To complete the proof, it will suffice to show that the natural map $f : X \times_Y Z \to LX \times_{LY} LZ$ belongs to S. We can write f as a composition of maps

$$X \times_Y Z \to X \times_{LY} Z \to LX \times_{LY} Z \to LX \times_{LY} LZ.$$

The last two maps are obtained from $X \to LX$ and $Z \to LZ$ by base change. Assumption (2) implies that they belong to S. Thus it will suffice to show that $f' : X \times_Y Z \to X \times_{LY} Z$ belongs to S. This map is a pullback of the diagonal $f'' : Y \to Y \times_{LY} Y$, so it will suffice to prove that $f'' \in S$. Projection to the first factor gives a left homotopy inverse $g : Y \times_{LY} Y \to Y$ of f'', so it suffices to prove that $g \in S$. But g is a base change of the morphism $Y \to LY$. $\qquad\square$

Proposition 6.2.1.2. *Let $\mathfrak{X}$ be a presentable ∞-category in which colimits are universal. Let S be a class of morphisms in $\mathfrak{X}$ and let $\overline{S}$ be the strongly* *saturated class of morphisms generated by S. Suppose that S has the following property: for every pullback diagram*

$$
\begin{array}{ccc}
X' & \longrightarrow & X \\
\downarrow{\scriptstyle f'} & & \downarrow{\scriptstyle f} \\
Y' & \longrightarrow & Y
\end{array}
$$

in $\mathcal{X}$, if $f \in S$, then $f' \in \overline{S}$. Then $\overline{S}$ is stable under pullbacks.

Proof. Let S' be the set of all morphisms f in $\mathcal{X}$ with the property that for any pullback diagram

$$
\begin{array}{ccc}
X' & \longrightarrow & X \\
\downarrow{\scriptstyle f'} & & \downarrow{\scriptstyle f} \\
Y' & \longrightarrow & Y,
\end{array}
$$

the morphism f' belongs to $\overline{S}$. By assumption, $S \subseteq S'$. Using the fact that colimits are universal, we deduce that S' is strongly saturated. Consequently, $\overline{S} \subseteq S'$, as desired.   □

Corollary 6.2.1.3. *Let $\mathcal{X}$ be a presentable ∞-category in which colimits are universal, let S be a (small) set of morphisms in $\mathcal{X}$, and let $\overline{S}$ denote the smallest strongly saturated class of morphisms which contains S and is stable under pullbacks. Then $\overline{S}$ is generated (as a strongly saturated class of morphisms) by a (small) set.*

Proof. Choose a (small) set U of objects of $\mathcal{X}$ which generates $\mathcal{X}$ under colimits. Enlarging U if necessary, we may suppose that U contains the codomain of every morphism belonging to S. Let S' be the set of all morphisms f' which fit into a pullback diagram

$$
\begin{array}{ccc}
X' & \longrightarrow & X \\
\downarrow{\scriptstyle f'} & & \downarrow{\scriptstyle f} \\
Y' & \longrightarrow & Y
\end{array}
$$

where $f \in S$ and $Y' \in U$, and let $\overline{S}'$ denote the strongly saturated class of morphisms generated by S'. To complete the proof it will suffice to show that $\overline{S}' = \overline{S}$. The inclusions $S \subseteq S' \subseteq \overline{S}' \subseteq \overline{S}$ are obvious. To show that $\overline{S} \subseteq \overline{S}'$, it will suffice to show that $\overline{S}'$ is stable under pullbacks. In view of Proposition 6.2.1.2, it will suffice to show that for every pullback diagram

$$
\begin{array}{ccc}
X'' & \longrightarrow & X' \\
\downarrow{\scriptstyle f''} & & \downarrow{\scriptstyle f'} \\
Y'' & \longrightarrow & Y',
\end{array}
$$

such that $f' \in S'$, the morphism f'' belongs to $\overline{S}'$. Using our assumption that colimits in $\mathcal{X}$ are universal and that U generates $\mathcal{X}$ under colimits, we can reduce to the case where $Y'' \in U$. In this case, $f'' \in S'$ by construction.   □

Recall that a morphism $f : Y \to Z$ in an ∞-category $\mathfrak{X}$ is a *monomorphism* if it is a (-1)-truncated object of the ∞-category $\mathfrak{X}_{/Z}$. Equivalently, f is a monomorphism if for every object $X \in \mathfrak{X}$, the induced map

$$\mathrm{Map}_{\mathfrak{X}}(X, Y) \to \mathrm{Map}_{\mathfrak{X}}(X, Z)$$

exhibits $\mathrm{Map}_{\mathfrak{X}}(X, Y) \in \mathcal{H}$ as a summand of $\mathrm{Map}_{\mathfrak{X}}(X, Z)$ in the homotopy category $\mathcal{H}$. If we fix $Z \in \mathfrak{X}$, then the collection of equivalence classes of monomorphisms $Y \to Z$ is partially ordered under inclusion. We will denote this partially ordered collection by $\mathrm{Sub}(Z)$.

Proposition 6.2.1.4. *Let $\mathfrak{X}$ be a presentable ∞-category and let X be an object of $\mathfrak{X}$. Then $\mathrm{Sub}(X)$ is a (small) partially ordered set.*

Proof. By definition, the partially ordered set $\mathrm{Sub}(X)$ is characterized by the existence of an equivalence

$$\tau_{\leq -1} \mathfrak{X}_{/X} \to \mathrm{N}(\mathrm{Sub}(X)).$$

Propositions 5.5.3.10 and 5.5.6.18 imply that $\mathrm{N}(\mathrm{Sub}(X))$ is presentable. Consequently, there exists a small subset $S \subseteq \mathrm{Sub}(X)$ which generates $\mathrm{N}(\mathrm{Sub}(X))$ under colimits. It follows that every element of $\mathrm{Sub}(X)$ can be written as the supremum of a subset of S, so that $\mathrm{Sub}(X)$ is also small. $\square$

Definition 6.2.1.5. Let $\mathfrak{X}$ be a presentable ∞-category and let $\overline{S}$ be a strongly saturated class of morphisms of $\mathfrak{X}$. We will say that $\overline{S}$ is *topological* if the following conditions are satisfied:

(1) There exists $S \subseteq \overline{S}$ consisting of *monomorphisms* such that S generates $\overline{S}$ as a strongly saturated class of morphisms.

(2) Given a pullback diagram

$$\begin{array}{ccc} X' & \longrightarrow & X \\ \downarrow{\scriptstyle f'} & & \downarrow{\scriptstyle f} \\ Y' & \longrightarrow & Y \end{array}$$

in $\mathfrak{X}$ such that f belongs to $\overline{S}$, the morphism f' also belongs to $\overline{S}$.

We will say that a localization $L : \mathfrak{X} \to \mathcal{Y}$ is *topological* if the collection $\overline{S}$ of all morphisms $f : X \to Y$ in $\mathfrak{X}$ such that Lf is an equivalence is topological.

Proposition 6.2.1.6. *Let $\mathfrak{X}$ be a presentable ∞-category in which colimits are universal and let $\overline{S}$ be a strongly saturated class of morphisms of $\mathfrak{X}$ which is topological. Then there exists a (small) subset $S_0 \subseteq \overline{S}$ which consists of monomorphisms and generates $\overline{S}$ as a strongly saturated class of morphisms.*

Proof. For every object $U \in \mathfrak{X}$, let $\mathrm{Sub}'(U) \subseteq \mathrm{Sub}(U)$ denote the collection of equivalence classes of monomorphisms $U' \to U$ which belong to $\overline{S}$. Choose a small collection of objects $\{U_\alpha\}_{\alpha \in A}$ which generates $\mathfrak{X}$ under colimits.

For each $\alpha \in A$ and each element $\widetilde{\alpha} \in \mathrm{Sub}'(U_\alpha)$, choose a representative monomorphism $f_{\widetilde{\alpha}} : V_{\widetilde{\alpha}} \to U_\alpha$ which belongs to $\overline{S}$. Let

$$S_0 = \{ f_{\widetilde{\alpha}} | \alpha \in A, \widetilde{\alpha} \in \mathrm{Sub}'(U_\alpha) \}.$$

It follows from Proposition 6.2.1.4 that S_0 is a (small) set. Let $\overline{S}_0$ denote the strongly saturated class of morphisms generated by S_0. We will show that $\overline{S}_0 = \overline{S}$.

Let $\mathcal{X}^0$ be the full subcategory of $\mathcal{X}$ spanned by objects U with the following property: if $f : V \to U$ is a monomorphism and $f \in \overline{S}$, then $f \in \overline{S}_0$. By construction, for each $\alpha \in A$, $U_\alpha \in \mathcal{X}^0$. Since colimits in $\mathcal{X}$ are universal, it is easy to see that $\mathcal{X}^0$ is stable under colimits in $\mathcal{X}$. It follows that $\mathcal{X}^0 = \mathcal{X}$, so that every monomorphism which belongs to $\overline{S}$ also belongs to $\overline{S}_0$. Since $\overline{S}$ is generated by monomorphisms, we conclude that $\overline{S} = \overline{S}_0$, as desired. □

Corollary 6.2.1.7. *Let $\mathcal{X}$ be a presentable ∞-category. Every topological localization $L : \mathcal{X} \to \mathcal{Y}$ is accessible and left exact.*

6.2.2 Grothendieck Topologies and Sheaves in Higher Category Theory

Every ordinary topos is equivalent to a category of sheaves on some Grothendieck site. This can be deduced from the following pair of statements:

(i) Every topos is equivalent to a left exact localization of some presheaf category $\mathrm{Set}^{\mathcal{C}^{op}}$.

(ii) There is a bijective correspondence between left exact localizations of $\mathrm{Set}^{\mathcal{C}^{op}}$ and Grothendieck topologies on $\mathcal{C}$.

In §6.1, we proved the ∞-categorical analogue of assertion (i). Unfortunately, (ii) is not quite true in the ∞-categorical setting. In this section, we will establish a slightly weaker statement: for every ∞-category $\mathcal{C}$, there is a bijective correspondence between Grothendieck topologies on $\mathcal{C}$ and *topological* localizations of $\mathcal{P}(\mathcal{C})$ (Proposition 6.2.2.9). Our first step is to introduce the ∞-categorical analogue of a Grothendieck site. The following definition is taken from [78]:

Definition 6.2.2.1. Let $\mathcal{C}$ be an ∞-category. A *sieve* on $\mathcal{C}$ is a full subcategory of $\mathcal{C}^{(0)} \subseteq \mathcal{C}$ having the property that if $f : C \to D$ is a morphism in $\mathcal{C}$ and D belongs to $\mathcal{C}^{(0)}$, then C also belongs to $\mathcal{C}^{(0)}$.

Observe that if $f : \mathcal{C} \to \mathcal{D}$ is a functor between ∞-categories and $\mathcal{D}^{(0)} \subseteq \mathcal{D}$ is a sieve on $\mathcal{D}$, then $f^{-1} \mathcal{D}^{(0)} = \mathcal{D}^{(0)} \times_{\mathcal{D}} \mathcal{C}$ is a sieve on $\mathcal{C}$. Moreover, if f is an equivalence, then f^{-1} induces a bijection between sieves on $\mathcal{D}$ and sieves on $\mathcal{C}$.

If $C \in \mathcal{C}$ is an object, then a *sieve on C* is a sieve on the ∞-category $\mathcal{C}_{/C}$. Given a morphism $f : D \to C$ and a sieve $\mathcal{C}^{(0)}_{/C}$ on C, we let $f^* \mathcal{C}^{(0)}_{/C}$ denote the unique sieve on D such that $f^* \mathcal{C}^{(0)}_{/C} \subseteq \mathcal{C}_{/D}$ and $\mathcal{C}^{(0)}_{/C}$ determine the same sieve on $\mathcal{C}_{/f}$.

A *Grothendieck topology* on an ∞-category $\mathcal{C}$ consists of a specification, for each object C of $\mathcal{C}$, of a collection of sieves on C which we will refer to as *covering sieves*. The collections of covering sieves are required to possess the following properties:

(1) If C is an object of $\mathcal{C}$, then the sieve $\mathcal{C}_{/C} \subseteq \mathcal{C}_{/C}$ on C is a covering sieve.

(2) If $f : C \to D$ is a morphism in $\mathcal{C}$ and $\mathcal{C}_{/C}^{(0)}$ is a covering sieve on D, then $f^* \mathcal{C}_{/C}^{(0)}$ is a covering sieve on C.

(3) Let C be an object of $\mathcal{C}$, $\mathcal{C}_{/C}^{(0)}$ a covering sieve on C, and $\mathcal{C}_{/C}^{(1)}$ an arbitrary sieve on C. Suppose that, for each $f : D \to C$ belonging to the sieve $\mathcal{C}_{/C}^{(0)}$, the pullback $f^* \mathcal{C}_{/C}^{(1)}$ is a covering sieve on D. Then $\mathcal{C}_{/C}^{(1)}$ is a covering sieve on C.

Example 6.2.2.2. Any ∞-category $\mathcal{C}$ may be equipped with the *trivial topology* in which a sieve $\mathcal{C}_{/C}^{(0)}$ on an object C of $\mathcal{C}$ is covering if and only if $\mathcal{C}_{/C}^{(0)} = \mathcal{C}_{/C}$.

Remark 6.2.2.3. In the case where $\mathcal{C}$ is (the nerve of) an ordinary category, the definition given above reduces to the usual notion of a Grothendieck topology on $\mathcal{C}$. Even in the general case, a Grothendieck topology on $\mathcal{C}$ is just a Grothendieck topology on the homotopy category $h\mathcal{C}$. This is not completely obvious since for an object C of $\mathcal{C}$, the functor

$$\eta : h(\mathcal{C}_{/C}) \to (h\mathcal{C})_{/C}$$

is usually not an equivalence of categories. A morphism on the left hand side corresponds to a commutative triangle

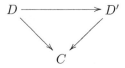

given by a *specified* 2-simplex $\sigma : \Delta^2 \to \mathcal{C}$ (taken modulo homotopy), while on the right hand side one requires only that the above diagram commute up to homotopy: this amounts to requiring the existence of σ, but σ itself is not taken as part of the data.

Although η need not be an equivalence of categories, η^* does induce a bijection from the set of sieves on $(h\mathcal{C})_{/C}$ to the set of sieves on $h(\mathcal{C}_{/C})$: for this, it suffices to observe that η induces surjective maps

$$\mathrm{Hom}_{h(\mathcal{C}_{/C})}(D, D') \to \mathrm{Hom}_{(h\mathcal{C})_{/C}}(D, D')$$

on morphism sets, which is obvious from the description given above.

The main objective of this section is to prove that for any (small) ∞-category $\mathcal{C}$, there is a bijective correspondence between Grothendieck topologies on $\mathcal{C}$ and (equivalence classes of) topological localizations of $\mathcal{P}(\mathcal{C})$. We begin by establishing a correspondence between sieves on $\mathcal{C}$ and (-1)-truncated objects of $\mathcal{P}(\mathcal{C})$. For each object $U \in \mathcal{P}(\mathcal{C})$, let $\mathcal{C}^{(0)}(U) \subseteq \mathcal{C}$ be the full subcategory spanned by those objects $C \in \mathcal{C}$ such that $U(C) \neq \emptyset$. It is easy to see that $\mathcal{C}^{(0)}(U)$ is a sieve on $\mathcal{C}$. Conversely, given a sieve $\mathcal{C}^{(0)} \subseteq \mathcal{C}$, there is a unique map $\mathcal{C} \to \Delta^1$ such that $\mathcal{C}^{(0)}$ is the preimage of $\{0\}$. This construction determines a bijection between sieves on $\mathcal{C}$ and functors $f : \mathcal{C} \to \Delta^1$, and we may identify Δ^1 with the full subcategory of $\mathcal{S}^{op}$ spanned by the objects $\emptyset, \Delta^0 \in \mathcal{K}an$. Since every (-1)-truncated Kan complex is equivalent to either $\emptyset$ or Δ^0, we conclude:

Lemma 6.2.2.4. *For every small ∞-category $\mathcal{C}$, the construction $U \mapsto \mathcal{C}^{(0)}(U)$ determines a bijection between the set of equivalence classes of (-1)-truncated objects of $\mathcal{P}(\mathcal{C})$ and the set of all sieves on $\mathcal{C}$.*

We now introduce a relative version of the above construction. Let $\mathcal{C}$ be a small ∞-category as above and let $j : \mathcal{C} \to \mathcal{P}(\mathcal{C})$ be the Yoneda embedding. Let $C \in \mathcal{C}$ be an object and let $i : U \to j(C)$ be a monomorphism in $\mathcal{P}(\mathcal{C})$. Let $\mathcal{C}_{/C}(U)$ denote the full subcategory of $\mathcal{C}$ spanned by those objects $f : D \to C$ of $\mathcal{C}_{/C}$ such that there exists a commutative triangle

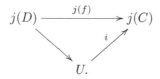

It is easy to see that $\mathcal{C}_{/C}(U)$ is a sieve on C. Moreover, it is clear that if $i : U \to j(C)$ and $i' : U' \to j(C)$ are equivalent subobjects of $j(C)$, then $\mathcal{C}_{/C}(U) = \mathcal{C}_{/C}(U')$.

Proposition 6.2.2.5. *Let $\mathcal{C}$ be a small ∞-category containing an object C and let $j : \mathcal{C} \to \mathcal{P}(\mathcal{C})$ be the Yoneda embedding. The construction described above yields a bijection*

$$(i : U \to j(C)) \mapsto \mathcal{C}_{/C}(U)$$

from $\mathrm{Sub}(j(C))$ to the set of all sieves on C.

Proof. Use Corollary 5.1.6.12 to reduce to Lemma 6.2.2.4.   □

Definition 6.2.2.6. Let $\mathcal{C}$ be a (small) ∞-category equipped with a Grothendieck topology. Let S be the collection of all monomorphisms $U \to j(C)$ which correspond to covering sieves $\mathcal{C}_{/C}^{(0)} \subseteq \mathcal{C}_{/C}$. An object $\mathcal{F} \in \mathcal{P}(\mathcal{C})$ is a *sheaf* if it is S-local. We let $\mathrm{Shv}(\mathcal{C})$ denote the full subcategory of $\mathcal{P}(\mathcal{C})$ spanned by S-local objects.

Lemma 6.2.2.7. *Let $\mathcal{C}$ be a (small) ∞-category equipped with a Grothendieck topology. Then $\mathrm{Shv}(\mathcal{C})$ is a topological localization of $\mathcal{P}(\mathcal{C})$. In particular, $\mathrm{Shv}(\mathcal{C})$ is an ∞-topos.*

Proof. By definition, $\mathrm{Shv}(\mathcal{C}) = S^{-1}\mathcal{P}(\mathcal{C})$, where S is the collection of all monomorphisms $i : U \to j(C)$ which correspond to covering sieves on $C \in \mathcal{C}$. Let $\overline{S}$ be the strongly saturated class of morphisms generated by S; we wish to show that $\overline{S}$ is stable under pullback.

Let S' denote the collection of all morphisms $f : X \to Y$ such that for any pullback diagram $\sigma : \Delta^1 \times \Delta^1 \to \mathcal{P}(\mathcal{C})$ depicted as follows:

$$
\begin{array}{ccc}
X' & \longrightarrow & X \\
\downarrow{\scriptstyle f'} & & \downarrow{\scriptstyle f} \\
Y' & \xrightarrow{g} & Y,
\end{array}
$$

the morphism f' belongs to $\overline{S}$. Since colimits in $\mathcal{P}(\mathcal{C})$ are universal, it is easy to prove that S' is strongly saturated. We wish to prove that $\overline{S} \subseteq S'$. Since $\overline{S}$ is the smallest saturated class containing S, it will suffice to prove that $S \subseteq S'$. We may therefore suppose that $Y = j(C)$ in the diagram above and that $f : X \to j(C)$ is the monomorphism corresponding to a covering sieve $\mathcal{C}^{(0)}_{/C}$ on C.

Since $\mathcal{P}(\mathcal{C})_{/j(C)} \simeq \mathcal{P}(\mathcal{C}_{/C})$ is generated under colimits by the Yoneda embedding, there exists a diagram $p : K \to \mathcal{C}_{/C}$ such that the composite map $j \circ p : K \to \mathcal{P}(\mathcal{C})_{/j(C)}$ has $g : Y' \to j(C)$ as a colimit. Because colimits in $\mathcal{P}(\mathcal{C})$ are universal, we can extend $j \circ p$ to a diagram $P : K \to (\mathcal{P}(\mathcal{C})^{\Delta^1})_{/f}$ which carries each vertex $k \in K$ to a pullback diagram

$$
\begin{array}{ccc}
X_k & \longrightarrow & X \\
\downarrow{\scriptstyle f_k} & & \downarrow \\
j(D_k) & \xrightarrow{j(g_k)} & j(C)
\end{array}
$$

such that σ is a colimit of P. Each f_k is a monomorphism associated to the covering sieve $g_k^* \mathcal{C}^{(0)}_{/C}$ and therefore belongs to $S \subseteq \overline{S}$. It follows that f' is a colimit in $\mathcal{P}(\mathcal{C})^{\Delta^1}$ of morphisms belonging to $\overline{S}$ and thus itself belongs to $\overline{S}$. $\square$

The next lemma ensures us that we can recover a Grothendieck topology on $\mathcal{C}$ from its ∞-category of sheaves $\mathrm{Shv}(\mathcal{C}) \subseteq \mathcal{P}(\mathcal{C})$.

Lemma 6.2.2.8. *Let $\mathcal{C}$ be a (small) ∞-category equipped with a Grothendieck topology and let $L : \mathcal{P}(\mathcal{C}) \to \mathrm{Shv}(\mathcal{C})$ denote a left adjoint to the inclusion. Let $j : \mathcal{C} \to \mathcal{P}(\mathcal{C})$ denote the Yoneda embedding and let $i : U \to j(C)$ be a monomorphism corresponding to a sieve $\mathcal{C}^{(0)}_{/C}$ on C. Then Li is an equivalence if and only if $\mathcal{C}^{(0)}_{/C}$ is a covering sieve.*

Proof. It is clear that if $\mathcal{C}_{/C}^{(0)}$ is a covering sieve, then Li is an equivalence. Conversely, suppose that Li is an equivalence. Then $\tau_{\leq 0}(Li)$ is an equivalence. In view of Proposition 5.5.6.28, we can identify $\tau_{\leq 0}(Li)$ with $L(\tau_{\leq 0}i)$. The morphism $\tau_{\leq 0}i$ can be identified with a monomorphism η : $\mathcal{F} \subseteq \mathrm{Hom}_{h\mathcal{C}}(\bullet, C)$ in the ordinary category of presheaves of sets on $h\mathcal{C}$, where

$$\mathcal{F}(D) = \{f \in \mathrm{Hom}_{h\mathcal{C}}(D, C) : f \in \mathcal{C}_{/C}^{(0)}\}.$$

If η becomes an equivalence after sheafification, then the identity map id_C : $C \to C$ belongs to $\mathcal{F}(C)$ locally; in other words, there exists a collection of morphisms $\{f_\alpha : C_\alpha \to C\}$ which generate a covering sieve on C such that each f_α belongs to $\mathcal{F}(C_\alpha)$ and therefore to $\mathcal{C}_{/C}^{(0)}$. It follows that $\mathcal{C}_{/C}^{(0)}$ contains a covering sieve on C and is therefore itself covering.   □

We may summarize the results of this section as follows:

Proposition 6.2.2.9. *Let $\mathcal{C}$ be a small ∞-category. There is a bijective correspondence between Grothendieck topologies on $\mathcal{C}$ and (equivalence classes of) topological localizations of $\mathcal{P}(\mathcal{C})$.*

Proof. According to Lemma 6.2.2.7, every Grothendieck topology on $\mathcal{C}$ determines a topological localization $\mathrm{Shv}(\mathcal{C}) \subseteq \mathcal{P}(\mathcal{C})$. Lemma 6.2.2.8 shows that two Grothendieck topologies which determine the same ∞-categories of sheaves must coincide. To complete the proof, it will suffice to show that *every* topological localization of $\mathcal{P}(\mathcal{C})$ arises in this way. Let $\overline{S}$ be a strongly saturated collection of morphisms in $\mathcal{P}(\mathcal{C})$ and suppose that $\overline{S}$ is topological. Let $S \subseteq \overline{S}$ be the collection of all monomorphisms $U \to j(C)$ which belong to $\overline{S}$, where $j : \mathcal{C} \to \mathcal{P}(\mathcal{C})$ denotes the Yoneda embedding. Since the objects $\{j(C)\}_{C \in \mathcal{C}}$ generate $\mathcal{P}(\mathcal{C})$ under colimits, and colimits in $\mathcal{P}(\mathcal{C})$ are universal, we conclude that every monomorphism in $\overline{S}$ is a colimit of elements of S. Since $\overline{S}$ is generated by monomorphisms, we conclude that $\overline{S}$ is generated by S.

Let us say that a sieve $\mathcal{C}_{/C}^{(0)} \subseteq \mathcal{C}_{/C}$ on an object $C \in \mathcal{C}$ is *covering* if the corresponding monomorphism $U \to j(C)$ belongs to S. We will show that the collection of covering sieves determines a Grothendieck topology on $\mathcal{C}$. Granting this, we observe that $\overline{S}^{-1} \mathcal{P}(\mathcal{C})$ is the ∞-category $\mathrm{Shv}(\mathcal{C}) \subseteq \mathcal{P}(\mathcal{C})$ of sheaves with respect to this Grothendieck topology, which will complete the proof.

We now verify the axioms (1) through (3) of Definition 6.2.2.1:

(1) Every sieve of the form $\mathcal{C}_{/C} \subseteq \mathcal{C}_{/C}$ is covering since every identity map $\mathrm{id}_{j(C)} : j(C) \to j(C)$ belongs to S.

(2) Let $f : C \to D$ be a morphism in $\mathcal{C}$ and let $\mathcal{C}_{/D}^{(0)} \subseteq \mathcal{C}_{/D}$ be a covering sieve corresponding to a monomorphism $i : U \to j(D)$ which belongs to S. Then $f^* \mathcal{C}_{/C}^{(0)} \subseteq \mathcal{C}_{/C}$ corresponds to a monomorphism $u : U' \to j(C)$

which is a pullback of i along $j(f)$ and therefore belongs to S (since $\overline{S}$ is stable under pullbacks).

(3) Let C be an object of $\mathcal{C}$, $\mathcal{C}^{(0)}_{/C}$ a covering sieve on C corresponding to a monomorphism $i : U \to j(C)$ which belongs to S, and $\mathcal{C}^{(1)}_{/C}$ an arbitrary sieve on C corresponding to a monomorphism $v : U' \to j(C)$. Suppose that, for each $f : D \to C$ belonging to the sieve $\mathcal{C}^{(0)}_{/C}$, the pullback $f^* \, \mathcal{C}^{(1)}_{/C}$ is a covering sieve on D. Since $j' : \mathcal{C}_{/C} \to \mathcal{P}(\mathcal{C})_{/j(C)}$ is a fully faithful embedding which generates $\mathcal{P}(\mathcal{C})_{/j(C)}$ under colimits (see the proof of Corollary 5.1.6.12), we conclude there is a diagram $K \to \mathcal{C}_{/C}$ such that $j' \circ K$ has colimit i'. Since colimits in $\mathcal{P}(\mathcal{C})$ are universal, we conclude that the map $v' : U \times_{j(C)} U' \to U$ is a colimit of morphisms of the form $j(D) \times_{j(C)} U' \to j(D)$, which belong to $\overline{S}$ by assumption. Since $\overline{S}$ is stable under colimits, we conclude that i'' belongs to $\overline{S}$. We now have a pullback diagram

$$
\begin{array}{ccc}
U \times_{j(C)} U' & \xrightarrow{\;v'\;} & U \\
\downarrow{\scriptstyle u'} & & \downarrow{\scriptstyle u} \\
U' & \xrightarrow{\;v\;} & j(C).
\end{array}
$$

By assumption, $u \in S$. Thus $v \circ u' \sim u \circ v' \in \overline{S}$. Since u' is a pullback of u, we conclude that $u' \in \overline{S}$, so that $v \in \overline{S}$. This implies that $\mathcal{C}^{(1)}_{/C} \subseteq \mathcal{C}_{/C}$ is a covering sieve, as we wished to prove.

$\square$

For later use, we record the following characterization of initial objects in ∞-categories of sheaves:

Proposition 6.2.2.10. *Let $\mathcal{C}$ be a small ∞-category equipped with a Grothendieck topology and let $\mathcal{C}' \subseteq \mathcal{C}$ denote the full subcategory spanned by those objects $C \in \mathcal{C}$ such that $\emptyset \subseteq \mathcal{C}_{/C}$ is a covering sieve on C. An object $\mathcal{F} \in \mathrm{Shv}(\mathcal{C})$ is initial if and only if it satisfies the following conditions:*

(1) *If $C \in \mathcal{C}'$, then $\mathcal{F}(C)$ is contractible.*

(2) *If $C \notin \mathcal{C}'$, then $\mathcal{F}(C)$ is empty.*

Proof. Let $L : \mathcal{P}(\mathcal{C}) \to \mathrm{Shv}(\mathcal{C})$ be a left adjoint to the inclusion and let $\emptyset$ be an initial object of $\mathcal{P}(\mathcal{C})$. Then $L\emptyset$ is an initial object of $\mathrm{Shv}(\mathcal{C})$. Since L is left exact, it preserves (-1)-truncated objects, as does the inclusion $\mathrm{Shv}(\mathcal{C}) \subseteq \mathcal{P}(\mathcal{C})$. Thus $L\emptyset$ is (-1)-truncated and corresponds to some sieve $\mathcal{C}^{(0)} \subseteq \mathcal{C}$ (Lemma 6.2.2.4). As we saw in the proof of Lemma 6.2.2.8, a sieve $\mathcal{C}^{(0)}$ classifies an object of $\mathrm{Shv}(\mathcal{C})$ if and only if $\mathcal{C}^{(0)}$ is saturated in the following sense: if $C \in \mathcal{C}$ and the induced sieve $\mathcal{C}^{(0)} \times_{\mathcal{C}} \mathcal{C}_{/C}$ is covering, then $C \in \mathcal{C}^{(0)}$. An initial object of $\mathrm{Shv}(\mathcal{C})$ is an initial object of $\tau_{\leq -1} \mathrm{Shv}(\mathcal{C})$ and must therefore correspond to the *smallest* saturated sieve on $\mathcal{C}$. An easy argument shows that this sieve is $\mathcal{C}'$ and that $\mathcal{F} \in \mathcal{P}(\mathcal{C})$ is a (-1)-truncated object classified by $\mathcal{C}'$ if and only if conditions (1) and (2) are satisfied. $\quad\square$

6.2.3 Effective Epimorphisms

In classical topos theory, the assumption that every equivalence relation is effective leads to a bijective correspondence between equivalence relations on an object X and *effective epimorphisms* $X \to Y$. The purpose of this section is to generalize the notion of an effective epimorphism to the ∞-categorical setting.

Our primary interest is studying the class of effective epimorphisms in an ∞-topos $\mathcal{X}$. However, we will later need to employ the same ideas when $\mathcal{X}$ is an n-topos for $n < \infty$. It is therefore convenient to work in a slightly more general setting.

Definition 6.2.3.1. An ∞-category $\mathcal{X}$ is a *semitopos* if it satisfies the following conditions:

(1) The ∞-category $\mathcal{X}$ is presentable.

(2) Colimits in $\mathcal{X}$ are universal.

(3) For every morphism $f : U \to X$, the underlying groupoid of the Čech nerve $\check{C}(f)$ is effective (see §6.1.2).

Remark 6.2.3.2. Every ∞-topos is a semitopos; this follows immediately from Theorem 6.1.0.6.

Remark 6.2.3.3. If $\mathcal{X}$ is a semitopos, then so is $\mathcal{X}_{/X}$ for every object $X \in \mathcal{X}$.

Proposition 6.2.3.4. *Let $\mathcal{X}$ be a semitopos. Let $p : U \to X$ be a morphism in $\mathcal{X}$, let $U_\bullet$ be the underlying simplicial object of the Čech nerve $\check{C}(p)$, and let $V \in \mathcal{X}$ be a colimit of $U_\bullet$. The induced diagram*

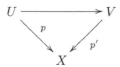

identifies p' with a (-1)-truncation of p in $\mathcal{X}_{/X}$.

Proof. We first show that V is (-1)-truncated. It suffices to show that the diagonal map $V \to V \times_X V$ is an equivalence. We may identify V with $V \times_V V$. Since colimits in $\mathcal{X}$ are universal, it will suffice to prove that for each $m, n \geq 0$, the natural map

$$p_{n.m} : U_m \times_V U_n \to U_m \times_X U_n$$

is an equivalence. We next observe that each $p_{n,m}$ is a pullback of

$$p_{0,0} : U \times_V U \to U \times_X U.$$

Because $U_\bullet$ is an effective groupoid, both sides may be identified with U_1.

To complete the proof, it suffices to show that the natural map

$$\mathrm{Map}_{\mathcal{X}/X}(p', q) \to \mathrm{Map}_{\mathcal{X}/X}(p, q)$$

is an equivalence whenever $q : E \to X$ is a monomorphism. Note that both sides are either empty or contractible. We must show that if $\mathrm{Map}_{\mathcal{X}/X}(p, q)$ is nonempty, then so is $\mathrm{Map}_{\mathcal{X}/X}(p', q)$. We observe that the map $\mathcal{X}_{/q} \to \mathcal{X}_{/X}$ is fully faithful, and that its essential image is a sieve on $\mathcal{X}_{/X}$. If that sieve contains p, then it contains the entire groupoid $U_\bullet$ (viewed as a groupoid in $\mathcal{X}_{/X}$). We conclude that there exists a groupoid object $W_\bullet : \mathrm{N}(\mathbf{\Delta})^{op} \to \mathcal{X}_{/q}$ lifting $U_\bullet$. Let $\widetilde{V} \in \mathcal{X}_{/q}$ be a colimit of $V_\bullet$. According to Proposition 1.2.13.8, the image of $\widetilde{V}$ in $\mathcal{X}_{/X}$ can be identified with the map $p' : V \to X$. The existence of $\widetilde{V}$ proves that $\mathrm{Map}_{\mathcal{X}/X}(p', q)$ is nonempty, as desired. $\qquad\square$

Corollary 6.2.3.5. *Let $\mathcal{X}$ be a semitopos and let $f : U \to X$ be a morphism in $\mathcal{X}$. The following conditions are equivalent:*

(1) *If we regard f as an object of the ∞-category $\mathcal{X}_{/X}$, then $\tau_{\leq -1}(f)$ is a final object of $\mathcal{X}_{/X}$.*

(2) *The Čech nerve $\check{\mathrm{C}}(f)$ is a simplicial resolution of X.*

We will say that a morphism $f : U \to X$ in a semitopos $\mathcal{X}$ is an *effective epimorphism* if it satisfies the equivalent conditions of Corollary 6.2.3.5. There is a one-to-one correspondence between effective epimorphisms and effective groupoids. More precisely, let $\mathcal{R}es_{\mathrm{Eff}}(\mathcal{X})$ denote the full subcategory of the ∞-category $\mathcal{X}_{\Delta_+}$ spanned by those augmented simplicial objects $U_\bullet$ which are both Čech nerves and simplicial resolutions. The restriction functors

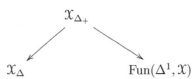

induce equivalences of ∞-categories from $\mathcal{R}es_{\mathrm{Eff}}(\mathcal{X})$ to the full subcategory of $\mathcal{X}_\Delta$ spanned by the effective groupoids, and from $\mathcal{R}es_{\mathrm{Eff}}(\mathcal{X})$ to the full subcategory of $\mathrm{Fun}(\Delta^1, \mathcal{X})$ spanned by the effective epimorphisms.

Remark 6.2.3.6. Let $f_* : \mathcal{X} \to \mathcal{Y}$ be a geometric morphism of ∞-topoi and let $u : U \to Y$ be an effective epimorphism in $\mathcal{Y}$. Then $f^*(u)$ is an effective epimorphism in $\mathcal{X}$. To see this, choose a Čech nerve $U_\bullet$ of u. Since u is an effective epimorphism, $U_\bullet$ is a colimit diagram. The left exactness of f^* implies that $f^* \circ U_\bullet$ is a Čech nerve of $f^*(u)$. Since f^* is a left adjoint, we conclude that $f^* \circ U_\bullet$ is a colimit diagram, so that $f^*(u)$ is an effective epimorphism.

The following result summarizes a few basic properties of effective epimorphisms:

Proposition 6.2.3.7. *Let $\mathfrak{X}$ be a semitopos.*

(1) *Any equivalence $X \to Y$ in $\mathfrak{X}$ is an effective epimorphism.*

(2) *If $f, g : X \to Y$ are homotopic morphisms in $\mathfrak{X}$, then f is an effective epimorphism if and only if g is an effective epimorphism.*

(3) *If $F : \mathcal{Y} \to \mathfrak{X}$ is a left exact presentable functor between semitopoi and $f : U \to X$ is an effective epimorphism in $\mathfrak{X}$, then $F(f)$ is an effective epimorphism in $\mathcal{Y}$.*

Proof. Assertions (1) and (2) are obvious. To prove (3), we observe that f is an effective epimorphism if and only if it can be extended to an augmented simplicial object $U_\bullet$ which is both a simplicial resolution and a Čech nerve. Since F is left exact, it preserves the property of being a Čech nerve; since F preserves colimits, it preserves the property of being a simplicial resolution. $\square$

Remark 6.2.3.8. Let $\mathfrak{X}$ be a semitopos and let $f : X \to T$ be an effective epimorphism in $\mathfrak{X}$. Applying part (3) of Proposition 6.2.3.7 to the geometric morphism $f : \mathfrak{X}_{/S} \to \mathfrak{X}_{/T}$ induced by a morphism $S \to T$ in $\mathfrak{X}$, we deduce that any base change $X \times_T S \to X$ of f is also an effective epimorphism.

In order to verify other basic properties of the class of effective epimorphisms, such as stability under composition, we will need to reformulate the property of surjectivity in terms of subobjects. Let $\mathfrak{X}$ be a presentable ∞-category. For each $X \in \mathfrak{X}$, the ∞-category $\tau_{\leq -1} \mathfrak{X}_{/X}$ of subobjects of X is equivalent to the nerve of a partially ordered set which we will denote by $\mathrm{Sub}(X)$; we may identify $\mathrm{Sub}(X)$ with the set of equivalence classes of monomorphisms $U \to X$. A morphism $f : X \to Y$ in $\mathfrak{X}$ induces a left exact pullback functor $\mathfrak{X}_{/X} \to \mathfrak{X}_{/Y}$. This functor preserves (-1)-truncated objects by Proposition 5.5.6.16 and therefore induces a map $f^* : \mathrm{Sub}(Y) \to \mathrm{Sub}(X)$ of partially ordered sets.

Remark 6.2.3.9. Let $\mathfrak{X}$ be a presentable ∞-category in which colimits are universal. Then any monomorphism $u : U \to \coprod X_\alpha$ can be obtained as a coproduct of maps $u_\alpha : U_\alpha \to X_\alpha$, where each u_α is a pullback of u and therefore also a monomorphism. It follows that the natural map

$$\theta : \mathrm{Sub}(\coprod X_\alpha) \to \prod \mathrm{Sub}(X_\alpha)$$

is a monomorphism of partially ordered sets. If coproducts in $\mathfrak{X}$ are disjoint, then θ is bijective.

Proposition 6.2.3.10. *Let $\mathfrak{X}$ be a semitopos. A morphism $f : U \to X$ in $\mathfrak{X}$ is an effective epimorphism if and only if $f^* : \mathrm{Sub}(X) \to \mathrm{Sub}(U)$ is injective.*

Proof. Suppose first that f^* is injective. Let $U_\bullet$ be the underlying groupoid of a Čech nerve of f, let V be a colimit of $U_\bullet$, let $u : V \to X$ be the corresponding monomorphism, and let $[V]$ denote the corresponding element of $\mathrm{Sub}(X)$. Since f factors through u, we conclude that $f^*[V] = f^*[X] = [U] \in \mathrm{Sub}(U)$. Invoking the injectivity of f^*, we conclude that $[V] = [X]$, so that u is an equivalence.

For the converse, let us suppose that f is an effective epimorphism. Let $[V]$ and $[V']$ be elements of $\mathrm{Sub}(X)$, represented by monomorphisms $u : V \to X$ and $u' : V' \to X$, and suppose that $f^*[V] = f^*[V']$. We wish to prove that $[V] = [V']$. Since f^* is a left exact functor, we have $f^*([V] \cap [V']) = f^*[V \times_X V']$. It will suffice to prove that $[V'] = [V \times_X V']$; the same argument will then establish that $[V] = [V \times_X V']$, and the proof will be complete. In other words, we may assume without loss of generality that $[V] \subseteq [V']$, so that there is a commutative diagram

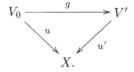

We wish to show that g is an equivalence. The map g induces a natural transformation of augmented simplicial objects

$$\alpha_{\bullet} : u^* \circ \check{\mathrm{C}}(f) \to u'^* \circ \check{\mathrm{C}}(f).$$

We observe that g can be identified with α_{-1}. Since f is an effective epimorphism, $\check{\mathrm{C}}(f)$ is a colimit diagram. Since colimits in $\mathfrak{X}$ are universal, we conclude that α_{-1} is a colimit of $\alpha | \mathrm{N}(\mathbf{\Delta})^{op}$. Consequently, to prove that α_{-1} is an equivalence, it will suffice to prove that α_n is an equivalence for $n \geq 0$. Since each α_n is a pullback of α_0, it will suffice to prove that α_0 is an equivalence. But this is simply a reformulation of the condition that $f^*[V] = f^*[V']$. $\qquad \square$

From this we immediately deduce some corollaries.

Corollary 6.2.3.11. *Let $\mathfrak{X}$ be a semitopos and let $\{f_\alpha : X_\alpha \to Y_\alpha\}$ be a (small) collection of effective epimorphisms in $\mathfrak{X}$. Then the induced map*

$$f : \coprod_\alpha X_\alpha \to \coprod_\alpha Y_\alpha$$

is an effective epimorphism.

Proof. Combine Proposition 6.2.3.10 with Remark 6.2.3.9. $\qquad \square$

Corollary 6.2.3.12. *Let $\mathfrak{X}$ be a semitopos containing a diagram*

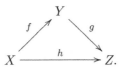

(1) *If f and g are effective epimorphisms, then so is h.*

(2) *If h is an effective epimorphism, then so is g.*

Proof. This follows immediately from Proposition 6.2.3.10 and the observation that we have an equality $f^* \circ g^* = h^*$ of functions $\mathrm{Sub}(Z) \to \mathrm{Sub}(X)$.
$\square$

The theory of effective epimorphisms is a mechanism for proving theorems by descent.

Lemma 6.2.3.13. *Let X be a semitopos, let $\overline{p} : K^{\triangleright} \to X$ be a colimit diagram and let ∞ denote the cone point of $K^{\triangleright}$. Then the associated map*

$$\coprod_{x \in K_0} p(x) \to p(\infty)$$

(which is well-defined up to homotopy) is an effective epimorphism.

Proof. For each vertex x of $K^{\triangleright}$, let $Z_x = p(x)$. If x belongs to K, we will denote the corresponding map $Z_x \to Z_\infty$ by f_x. Let $E'' \subseteq E' \in \mathrm{Sub}(Z_\infty)$ be such that $f_x^* E'' = f_x^* E'$ for each vertex x of K; we wish to show that $E'' = E'$. We can represent E'' and E' by a 2-simplex $\sigma_\infty : \Delta^2 \to X$, which we depict as

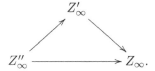

Lift the above diagram to a 2-simplex $\sigma : \Delta^2 \to \mathrm{Fun}(K^{\triangleright}, X)$

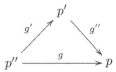

where g, g', and g'' are Cartesian transformations. Our assumption guarantees that the restriction of g' induces an equivalence $p''|K \to p'|K$. Since colimits in X are universal, g' is itself an equivalence, so that $E'' = E'$, as desired. $\square$

Proposition 6.2.3.14. *Let X be an ∞-topos and let S be a collection of morphisms of X which is stable under pullbacks and coproducts. The following conditions are equivalent:*

(1) *The class S is local (Definition 6.1.3.8).*

(2) *Given a pullback diagram*

$$\begin{array}{ccc} X' & \longrightarrow & X \\ \downarrow{\scriptstyle f'} & & \downarrow{\scriptstyle f} \\ Y' & \xrightarrow{g} & Y, \end{array}$$

where g is an effective epimorphism and $f' \in S$, we have $f \in S$.

Proof. We first show that $(1) \Rightarrow (2)$. Let $Y_\bullet : N(\mathbf{\Delta}_+)^{op} \to \mathcal{X}$ be a Čech nerve of the map g and choose a Cartesian transformation $f_\bullet : X_\bullet \to Y_\bullet$ of augmented simplicial objects which extends f. Then we can identify f' with $f_0 : X_0 \to Y_0$. Each f_n is a pullback of f_0, and therefore belongs to S. Applying Lemma 6.1.3.5, we deduce that f belongs to S as well.

Conversely, suppose that (2) is satisfied. We will show that S satisfies criterion (3) of Lemma 6.1.3.7. Let

$$
\begin{array}{ccc}
u & \xrightarrow{\;\alpha\;} & v \\
\downarrow{\scriptstyle\beta} & & \downarrow{\scriptstyle\beta'} \\
u' & \xrightarrow{\;\alpha'\;} & v'
\end{array}
$$

be a pushout diagram in $\mathcal{O}_\mathcal{X}$, where α and β are Cartesian and $u, v, u' \in S$. Since $\mathcal{X}$ is an ∞-topos, we conclude that α' and β' are also Cartesian. To complete the proof, it will suffice to show that $v' \in S$. For this, we observe that there is a pullback diagram

$$
\begin{array}{ccc}
X \coprod X' & \longrightarrow & X'' \\
\downarrow{\scriptstyle v \coprod u'} & & \downarrow{\scriptstyle v'} \\
Y \coprod Y' & \xrightarrow{\;g\;} & Y'',
\end{array}
$$

where g is an effective epimorphism (Lemma 6.2.3.13), and apply hypothesis (2). $\qquad\square$

Proposition 6.2.3.15. *Let $\mathcal{X}$ be a semitopos, and suppose we are given a pullback square*

$$
\begin{array}{ccc}
X' & \xrightarrow{\;g'\;} & X \\
\downarrow{\scriptstyle f'} & & \downarrow{\scriptstyle f} \\
S' & \xrightarrow{\;g\;} & S
\end{array}
$$

in $\mathcal{X}$. If f is an effective epimorphism, then so is f'. The converse holds if g is an effective epimorphism.

Proof. Let $g^* : \mathcal{X}^{/S} \to \mathcal{X}^{/S'}$ be a pullback functor. Without loss of generality, we may suppose that $f' = g^* f$. Let $U_\bullet : N(\mathbf{\Delta}_+)^{op} \to \mathcal{X}$ be a Čech nerve of f. Since g^* is left exact (being a right adjoint), we conclude that $g^* \circ U_\bullet$ is a Čech nerve of f'. If f is an effective epimorphism, then $U_\bullet$ is a colimit diagram. Because colimits in $\mathcal{X}$ are universal, $g^* \circ U_\bullet$ is also a colimit diagram, so that f' is an effective epimorphism.

Conversely, suppose that f' and g are effective epimorphisms. Corollary 6.2.3.12 implies that $g \circ f'$ is an effective epimorphism. The commutativity of the diagram implies that $f \circ g'$ is an effective epimorphism, so that f is an effective epimorphism (Corollary 6.2.3.12 again). $\qquad\square$

Lemma 6.2.3.16. *Let* X *be a semitopos and suppose we are given a pullback square*

$$
\begin{array}{ccc}
X' & \xrightarrow{\ g'\ } & X \\
\downarrow{\scriptstyle f'} & & \downarrow{\scriptstyle f} \\
S' & \xrightarrow{\ g\ } & S
\end{array}
$$

in X. *Suppose that* f' *is an equivalence and that* g *is an effective epimorphism. Then* f *is an equivalence.*

Proof. Let $U_\bullet$ be a Čech nerve of g' and let $V_\bullet$ be a Čech nerve of g. The above diagram induces a transformation $\alpha_\bullet : U_\bullet \to V_\bullet$. The map α_0 can be identified with f' and is therefore an equivalence. For $n \geq 0$, $\alpha_n : U_n \to V_n$ is a pullback of α_0 and is therefore also an equivalence. Since g is an effective epimorphism, $V_\bullet$ is a colimit diagram. Applying Proposition 6.2.3.15, we conclude that g' is also an effective epimorphism, so that $U_\bullet$ is a colimit diagram. It follows that $f = \alpha_{-1}$ is a colimit of equivalences and is therefore an equivalence. □

Proposition 6.2.3.17. *Let* X *be a semitopos and suppose we are given a pullback square*

$$
\begin{array}{ccc}
X' & \longrightarrow & X \\
\downarrow{\scriptstyle f'} & & \downarrow{\scriptstyle f} \\
S' & \xrightarrow{\ g\ } & S
\end{array}
$$

in X. *If* f *is* n-*truncated, then so is* f'. *The converse holds if* g *is an effective epimorphism.*

Proof. Let $g^* : X^{/S} \to X^{/S'}$ be a pullback functor. The first part of (1) asserts that g^* carries n-truncated objects to n-truncated objects. This follows immediately from Proposition 5.5.6.16 since g^* is a right adjoint and therefore left exact. We will prove the converse in a slightly stronger form: if $i : U \to V$ is a morphism in $X^{/S}$ such that $g^*(i)$ is an n-truncated morphism in $X^{/S'}$, then i is n-truncated. The proof is by induction on n. If $n \geq -1$, we can use Lemma 5.5.6.15 to reduce to the problem of showing that the diagonal map $\delta : U \to U \times_V U$ is $(n-1)$-truncated. Since g^* is left exact, we can identify $g^*(\delta)$ with the diagonal map $g^*U \to g^*U \times_{g^*V} g^*U$, which is $(n-1)$-truncated according to Lemma 5.5.6.15; the desired result then follows from the inductive hypothesis. In the case $n = -2$, we have a pullback

diagram

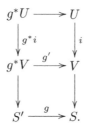

Proposition 6.2.3.15 implies that g' is an effective epimorphism, and g^*i is an equivalence, so that i is also an equivalence by Lemma 6.2.3.16. □

Let $\mathcal{C}$ be a small ∞-category equipped with a Grothendieck topology. Our final goal in this section is to use the language of effective epimorphisms to characterize the ∞-topos $\mathrm{Shv}(\mathcal{C})$ by a universal property.

Lemma 6.2.3.18. *Let $\mathcal{C}$ be a (small) ∞-category containing an object C, let $\{f_\alpha : C_\alpha \to C\}_{\alpha \in A}$ be a collection of morphisms indexed by a set A and let $\mathcal{C}^{(0)}_{/C} \subseteq \mathcal{C}_{/C}$ be the sieve on C that they generate. Let $j : \mathcal{C} \to \mathcal{P}(\mathcal{C})$ denote the Yoneda embedding and $i : U \to j(C)$ a monomorphism corresponding to the sieve $\mathcal{C}^{(0)}_{/C}$. Then i can be identified with a (-1)-truncation of the induced map $\coprod_{\alpha \in A} j(C_\alpha) \to j(C)$ in the ∞-topos $\mathcal{P}(\mathcal{C})_{/C}$.*

Proof. Using Proposition 6.2.2.5, we can identify the equivalence classes of (-1)-truncated object $U \in \mathcal{P}(\mathcal{C})_{/j(C)}$ with sieves $\mathcal{C}^{(0)}_{/C} \subseteq \mathcal{C}_{/C}$. It is not difficult to see that $j(f_\alpha)$ factors through U if and only if $f_\alpha \in \mathcal{C}^{(0)}_{/C}$. Consequently, the (-1)-truncation of $\coprod_{\alpha \in A} j(C_\alpha) \to j(C)$ is associated to the *smallest* sieve on $\mathcal{C}$ which contains each f_α. □

Lemma 6.2.3.19. *Let $\mathcal{X}$ be an ∞-topos, $\mathcal{C}$ a small ∞-category equipped with a Grothendieck topology, and $f_* : \mathcal{X} \to \mathcal{P}(\mathcal{C})$ a functor with a left exact left adjoint $f^* : \mathcal{P}(\mathcal{C}) \to \mathcal{X}$.*
The following conditions are equivalent:

(1) *The functor f_* factors through $\mathrm{Shv}(\mathcal{C}) \subseteq \mathcal{P}(\mathcal{C})$.*

(2) *For every collection of morphisms $\{v_\alpha : C_\alpha \to C\}$ which generate a covering sieve in $\mathcal{C}$, the induced map*

$$\coprod f^*(j(C_\alpha)) \to f^*(j(C))$$

is an effective epimorphism in $\mathcal{X}$, where $j : \mathcal{C} \to \mathcal{P}(\mathcal{C})$ denotes the Yoneda embedding.

Proof. Suppose first that (1) is satisfied and let $\{v_\alpha : C_\alpha \to C\}$ be a collection of morphisms as in the statement of (2). Let $L : \mathcal{P}(\mathcal{C}) \to \mathrm{Shv}(\mathcal{C})$ be a left adjoint to the inclusion. Then we have an equivalence of functors

$f^* \simeq (f^*| \operatorname{Shv}(\mathcal{C})) \circ L$. Applying Remark 6.2.3.6, we are reduced to showing that if

$$u : \coprod j(C_\alpha) \to j(C)$$

is the natural map, then Lu is an effective epimorphism in $\mathcal{P}(\mathcal{C})$. We factor u as a composition

$$\coprod j(C_\alpha) \xrightarrow{u'} U \xrightarrow{u''} j(C),$$

where u' is an effective epimorphism and u'' is a monomorphism. We wish to show that Lu'' is an equivalence. Lemma 6.2.3.18 allows us to identify u'' with the monomorphism associated to the sieve $\mathcal{C}_{/C}^{(0)}$ on C generated by the maps v_α. By assumption, this is a covering sieve, so that Lu'' is an equivalence in $\operatorname{Shv}(\mathcal{C})$ by construction.

Conversely, suppose that (2) is satisfied. Let $C \in \mathcal{C}$ and let $\mathcal{C}_{/C}^{(0)} \subseteq \mathcal{C}_{/C}$ be a covering sieve on C associated to a monomorphism $u'' : U \to j(C)$. We wish to show that $f^* u''$ is an equivalence. According to Lemma 6.2.3.18, we have a factorization

$$\coprod_\alpha j(C_\alpha) \xrightarrow{u'} U \xrightarrow{u''} j(C),$$

where the maps $v_\alpha : C_\alpha \to C$ are chosen to generate the sieve $\mathcal{C}_{/C}^{(0)}$ and u' is an effective epimorphism. Let u be a composition of u' and u''. Then $f^* u'$ is an effective epimorphism (Remark 6.2.3.6), and $f^* u$ is an effective epimorphism by assumption (2). Corollary 6.2.3.12 now shows that $f^* u''$ is an effective epimorphism. Since $f^* u''$ is also a monomorphism, we conclude that $f^* u''$ is an equivalence, as desired. □

Proposition 6.2.3.20. *Let $\mathfrak{X}$ be an ∞-topos and let $\mathcal{C}$ be a small ∞-category equipped with a Grothendieck topology. Let $L : \mathcal{P}(\mathcal{C}) \to \operatorname{Shv}(\mathcal{C})$ denote a left adjoint to the inclusion and $j : \mathcal{C} \to \mathcal{P}(\mathcal{C})$ the Yoneda embedding. Let $\operatorname{Fun}^*(\operatorname{Shv}(\mathcal{C}), \mathfrak{X})$ denote the ∞-category of left exact colimit-preserving functors from $\operatorname{Shv}(\mathcal{C})$ to $\mathfrak{X}$ (Definition 6.3.1.10). The composition*

$$J : \operatorname{Fun}^*(\operatorname{Shv}(\mathcal{C}), \mathfrak{X}) \xrightarrow{L} \operatorname{Fun}^*(\mathcal{P}(\mathcal{C}), \mathfrak{X}) \xrightarrow{j} \operatorname{Fun}(\mathcal{C}, \mathfrak{X})$$

is fully faithful. Suppose furthermore that $\mathcal{C}$ admits finite limits. Then a functor $f : \mathcal{C} \to \mathfrak{X}$ belongs to the essential image of J if and only if the following conditions are satisfied:

(1) *The functor f is left exact.*

(2) *For every collection of morphisms $\{C_\alpha \to C\}_{\alpha \in A}$ which generates a covering sieve on C, the associated morphism*

$$\coprod_{\alpha \in A} f(C_\alpha) \to f(C)$$

is an effective epimorphism in $\mathfrak{X}$.

Proof. If the topology on $\mathcal{C}$ is trivial, then Theorem 5.1.5.6 implies that J is fully faithful, and the description of the essential image of J follows from Proposition 6.1.5.2. In the general case, Proposition 5.5.4.20 implies that composition with L induces a fully faithful embedding

$$J' : \mathrm{Fun}^*(\mathrm{Shv}(\mathcal{C}), \mathcal{X}) \to \mathrm{Fun}^*(\mathcal{P}(\mathcal{C}), \mathcal{X}),$$

so that J is a composition of J' with a fully faithful functor

$$J'' : \mathrm{Fun}^*(\mathcal{P}(\mathcal{C}), \mathcal{X}) \to \mathrm{Fun}(\mathcal{C}, \mathcal{X}).$$

Suppose that $\mathcal{C}$ admits finite limits and that f satisfies (1), so that f is equivalent to $J''(u^*)$ for some left exact colimit-preserving $u^* : \mathcal{P}(\mathcal{C}) \to \mathcal{X}$. The functor u^* is unique up to equivalence, and Lemma 6.2.3.19 ensures that u^* belongs to the essential image of J' if and only if condition (2) is satisfied. $\qquad\square$

Remark 6.2.3.21. It is possible to formulate a generalization of Proposition 6.2.3.20 which describes the essential image of J even when $\mathcal{C}$ does not admit finite limits. The present version will be sufficient for the applications in this book.

6.2.4 Canonical Topologies

Let $\mathcal{X}$ be an ∞-topos. Suppose that we wish to identify $\mathcal{X}$ with an ∞-category of sheaves. The first step is to choose a pair of adjoint functors

$$\mathcal{P}(\mathcal{C}) \underset{G}{\overset{F}{\rightleftarrows}} \mathcal{X}$$

where F is left exact. According to Theorem 5.1.5.6, F is determined up to equivalence by the composition

$$f : \mathcal{C} \overset{j}{\to} \mathcal{P}(\mathcal{C}) \overset{F}{\to} \mathcal{X}.$$

We might then try to choose a topology on $\mathcal{C}$ such that G factors as a composition

$$\mathcal{X} \overset{G'}{\to} \mathrm{Shv}(\mathcal{C}) \subseteq \mathcal{P}(\mathcal{C}).$$

Though it is not always possible to guarantee that G' is an equivalence, we will show that for an appropriately chosen topology (Definition 6.2.4.1), the ∞-topos $\mathrm{Shv}(\mathcal{C})$ is a close approximation to $\mathcal{X}$ (Proposition 6.2.4.6).

Definition 6.2.4.1. Let $\mathcal{X}$ be a semitopos, $\mathcal{C}$ a small ∞-category which admits finite limits, and $f : \mathcal{C} \to \mathcal{X}$ a left exact functor. We will say that a sieve $\mathcal{C}_{/C}^{(0)} \subseteq \mathcal{C}_{/C}$ on an object $C \in \mathcal{C}$ is a *canonical covering relative to* f if there exists a collection of morphisms $\{u_\alpha : C_\alpha \to C\}$ belonging to $\mathcal{C}_{/C}^{(0)}$ such that the induced map $\coprod f(C_\alpha) \to f(C)$ is an effective epimorphism in $\mathcal{X}$.

Our first goal is to verify that the canonical topology is actually a Grothendieck topology on $\mathcal{C}$.

Proposition 6.2.4.2. *Let $f : \mathcal{C} \to \mathfrak{X}$ be as in Definition 6.2.4.1. The collection of canonical coverings relative to f determines a Grothendieck topology on $\mathcal{C}$.*

Proof. Since any identity map $\mathrm{id}_{f(C)} : f(C) \to f(C)$ is an effective epimorphism, it is clear that the sieve $\mathcal{C}_{/C}$ is a canonical covering of C for every $C \in \mathcal{C}$. Suppose that $\mathcal{C}^{(0)}_{/C} \subseteq \mathcal{C}_{/C}$ is a canonical covering of C and that $g : D \to C$ is a morphism in $\mathcal{C}$. We wish to prove that the induced sieve $g^* \, \mathcal{C}^{(0)}_{/C}$ is a canonical covering. Choose a collection of objects $u_\alpha : C_\alpha \to C$ of $\mathcal{C}^{(0)}_{/C}$ such that the induced map $\coprod_\alpha f(C_\alpha) \to f(C)$ is an effective epimorphism and form pullback diagrams

$$
\begin{array}{ccc}
D_\alpha & \xrightarrow{\ v_\alpha\ } & D \\
\downarrow & & \downarrow{\scriptstyle g} \\
C_\alpha & \xrightarrow{\ u_\alpha\ } & C
\end{array}
$$

in $\mathcal{C}$. Using the fact that f is left exact and that colimits in $\mathfrak{X}$ are universal, we conclude that the diagram

$$
\begin{array}{ccc}
\coprod f(D_\alpha) & \longrightarrow & f(D) \\
\downarrow & & \downarrow \\
\coprod f(C_\alpha) & \longrightarrow & f(C)
\end{array}
$$

is a pullback, so that the upper horizontal map is an effective epimorphism by Proposition 6.2.3.15. Since each v_α belongs to $g^* \, \mathcal{C}^{(0)}_{/C}$, it follows that $g^* \, \mathcal{C}^{(0)}_{/C}$ is a canonical covering.

Now suppose that $\mathcal{C}^{(0)}_{/C}$ and $\mathcal{C}^{(1)}_{/C}$ are sieves on $C \in \mathcal{C}$, where $\mathcal{C}^{(0)}_{/C}$ is a canonical covering, and for each $g : D \to C$ in $\mathcal{C}^{(0)}_{/C}$, the covering $g^* \, \mathcal{C}^{(1)}_{/C}$ is a canonical covering of D. Choose a collection of morphisms $g_\alpha : D_\alpha \to C$ belonging to $\mathcal{C}^{(0)}_{/C}$ with the property that $\coprod f(D_\alpha) \to f(C)$ is an effective epimorphism. For each D_α, choose a collection of morphisms $h_{\alpha,\beta} : E_{\alpha,\beta} \to D_\alpha$ belonging to $g_\alpha^* \, \mathcal{C}^{(1)}_{/C}$ such that the map $\coprod_\beta f(E_{\alpha,\beta}) \to f(D_\alpha)$ is an effective epimorphism. Using Corollary 6.2.3.11, we conclude that the map

$$
\coprod_{\alpha,\beta} f(E_{\alpha,\beta}) \to \coprod_\alpha f(D_\alpha)
$$

is an effective epimorphism. Since effective epimorphisms are stable under composition (Corollary 6.2.3.12), we have an effective epimorphism

$$
\coprod_{\alpha,\beta} f(E_{\alpha,\beta}) \to f(C)
$$

induced by the collection of compositions $g_\alpha \circ h_{\alpha,\beta} : E_{\alpha,\beta} \to C$. Each of these compositions belongs to $\mathcal{C}^{(1)}_{/C}$, so that $\mathcal{C}^{(1)}_{/C}$ is a canonical covering of C. $\square$

For later use, we record a few features of the canonical topology:

Lemma 6.2.4.3. *Let $f : \mathcal{C} \to \mathfrak{X}$ be as in Definition 6.2.4.1 and regard $\mathcal{C}$ as endowed with the canonical topology relative to f. Let $j : \mathcal{C} \to \mathcal{P}(\mathcal{C})$ denote the Yoneda embedding and let $L : \mathcal{P}(\mathcal{C}) \to \mathrm{Shv}(\mathcal{C})$ be a left adjoint to the inclusion. Suppose that $C \in \mathcal{C}$ is such that $f(C)$ is an initial object of $\mathfrak{X}$. Then $Lj(C)$ is an initial object of $\mathrm{Shv}(\mathcal{C})$.*

Proof. If $f(C)$ is an initial object of $\mathfrak{X}$, then the empty sieve $\emptyset \subseteq \mathcal{C}_{/C}$ is a covering sieve with respect to the canonical topology. By construction, the associated monomorphism $\emptyset \to j(C)$ becomes an equivalence after applying L, so that $Lj(C)$ is initial in $\mathrm{Shv}(\mathcal{C})$. $\square$

Lemma 6.2.4.4. *Let $f : \mathcal{C} \to \mathfrak{X}$ be as in Definition 6.2.4.1. Suppose that f is fully faithful and that coproducts in $\mathfrak{X}$ are disjoint and let $\{u_\alpha : C_\alpha \to C\}$ be a small collection of morphisms in $\mathcal{C}$ such that the morphisms $f(u_\alpha)$ exhibit $f(C)$ as a coproduct of the family $\{f(C_\alpha)\}$. Let $\mathcal{F} : \mathcal{C}^{op} \to \mathcal{S}$ be a sheaf on $\mathcal{C}$ (with respect to the canonical topology induced by f). Then the morphisms $\{\mathcal{F}(u_\alpha)\}$ exhibit $\mathcal{F}(C)$ as a product of $\{\mathcal{F}(C_\alpha)\}$ in $\mathcal{S}$.*

Proof. We wish to show that the natural map $\mathcal{F}(C) \to \prod \mathcal{F}(C_\alpha)$ is an isomorphism in the homotopy category $\mathcal{H}$. We may identify the left hand side with $\mathrm{Map}_{\mathcal{P}(\mathcal{C})}(j(C), \mathcal{F})$ and the right hand side with $\mathrm{Map}_{\mathcal{P}(\mathcal{C})}(\coprod j(C_\alpha), \mathcal{F})$. Consequently, it will suffice to show that the natural map

$$v : \coprod j(C_\alpha) \to j(C)$$

becomes an equivalence after applying the localization functor $L : \mathcal{P}(\mathcal{C}) \to \mathrm{Shv}(\mathcal{C})$. Choose a factorization of v as a composite

$$\coprod j(C_\alpha) \xrightarrow{v'} U \xrightarrow{v''} j(C),$$

where v' is an effective epimorphism and v'' is a monomorphism. We observe that v'' is the monomorphism associated to the sieve $\mathcal{C}_{/C}^{(0)} \to \mathcal{C}$ generated by the morphisms u_α. This is clearly a covering sieve with respect to the canonical topology, so that Lv'' is an equivalence in $\mathrm{Shv}(\mathcal{C})$. It follows that Lv is equivalent to Lv' and is therefore an effective epimorphism (Remark 6.2.3.6). Form a pullback diagram

$$
\begin{array}{ccc}
V & \xrightarrow{\bar{v}} & \coprod j(C_\beta) \\
\downarrow & & \downarrow{\scriptstyle v} \\
\coprod j(C_\alpha) & \xrightarrow{v} & j(C).
\end{array}
$$

We wish to prove that Lv is an equivalence. According to Lemma 6.2.3.16, it will suffice to show that $L\bar{v}$ is an equivalence. Since colimits in $\mathcal{P}(\mathcal{C})$ are universal, we may identify $\bar{v}$ with a coproduct of morphisms

$$\bar{v}_\beta : V_\beta \to j(C_\beta),$$

where V_β can be written as a coproduct $\coprod_\alpha j(C_\alpha \times_C C_\beta)$. Using Lemma 6.1.5.1, we can identify the summand $j(C_\beta \times_C C_\beta)$ of V_β with $j(C_\beta)$, and the restriction of $\overline{v}_\beta$ to this summand is an equivalence. To complete the proof, it will suffice to show that for every other summand $D_{\alpha,\beta} = j(C_\alpha \times_C C_\beta)$, the localization LD is an initial object of $\mathrm{Shv}(\mathcal{C})$. To prove this, we observe that Lemma 6.1.5.1 implies that $f(C_\alpha \times_C C_\beta)$ is an initial object of $\mathcal{X}$ and apply Lemma 6.2.4.3. $\qquad\square$

Lemma 6.2.4.5. *Let $\mathcal{C}$ be a small ∞-category equipped with a Grothendieck topology and let $u : \mathcal{F}' \to \mathcal{F}$ be a morphism in $\mathrm{Shv}(\mathcal{C})$. Suppose that, for each $C \in \mathcal{C}$ and each $\eta \in \pi_0 \mathcal{F}(C)$, there exists a collection of morphisms $\{C_\alpha \to C\}$ which generates a covering sieve on C and a collection of $\eta_\alpha \in \pi_0 \mathcal{F}'(C_\alpha)$ such that η and η_α have the same image in $\pi_0 \mathcal{F}(C_\alpha)$. Then u is an effective epimorphism.*

Proof. Replacing $\mathcal{F}$ by its image in $\mathcal{F}'$ if necessary, we may suppose that u is a monomorphism. Let $L : \mathcal{P}(\mathcal{C}) \to \mathrm{Shv}(\mathcal{C})$ be a left adjoint to the inclusion and let $\mathcal{D}$ be the full subcategory of $\mathcal{P}(\mathcal{C})$ spanned by those objects $\mathcal{G}$ such that, for every pullback diagram

$$\begin{array}{ccc} \mathcal{G}' & \xrightarrow{u'} & \mathcal{G} \\ \downarrow & & \downarrow \\ \mathcal{F}' & \xrightarrow{u} & \mathcal{F} \end{array}$$

in $\mathcal{P}(\mathcal{C})$, Lu' is an equivalence in $\mathrm{Shv}(\mathcal{C})$. To prove that u is an equivalence, it will suffice to show that the equivalent morphism Lu is an equivalence. For this, it will suffice to prove that $\mathcal{F} \in \mathcal{D}$. We will in fact prove that $\mathcal{D} = \mathcal{P}(\mathcal{C})$. We first observe that since colimits in $\mathcal{P}(\mathcal{C})$ are universal and L commutes with colimits, $\mathcal{D}$ is stable under colimits in $\mathcal{P}(\mathcal{C})$. Since $\mathcal{P}(\mathcal{C})$ is generated under colimits by the image of the Yoneda embedding, it will suffice to prove that $j(C) \in \mathcal{D}$ for each $C \in \mathcal{C}$. Choose a map $j(C) \to \mathcal{F}$, classified up to homotopy by $\eta \in \pi_0 \mathcal{F}(C)$, and form a pullback diagram

$$\begin{array}{ccc} U & \xrightarrow{u'} & j(C) \\ \downarrow & & \downarrow \\ \mathcal{F}' & \xrightarrow{u} & \mathcal{F} \end{array}$$

as above. Then u' is a monomorphism; according to Proposition 6.2.2.5, it is classified by a sieve $\mathcal{C}^{(0)}_{/C}$ on $\mathcal{C}$. Our hypothesis guarantees that $\mathcal{C}^{(0)}_{/C}$ contains a collection of morphisms $\{C_\alpha \to C\}$ which generate a covering sieve, so that $\mathcal{C}^{(0)}_{/C}$ is itself covering. It follows immediately from the construction of $\mathrm{Shv}(\mathcal{C})$ that Lu' is an equivalence. $\qquad\square$

We close with the following result, which implies that any ∞-topos is closely approximated by an ∞-category of sheaves.

Proposition 6.2.4.6. *Let* $\mathfrak{X}$ *be a semitopos,* $\mathfrak{C}$ *a small* ∞-*category which admits finite limits, and*

$$\mathcal{P}(\mathfrak{C}) \underset{G}{\overset{F}{\rightleftarrows}} \mathfrak{X}$$

a pair of adjoint functors. Suppose that the composition

$$f : \mathfrak{C} \xrightarrow{j} \mathcal{P}(\mathfrak{C}) \xrightarrow{F} \mathfrak{X}$$

is left exact and regard $\mathfrak{C}$ *as endowed with the canonical topology relative to* f. *Then:*

(1) *The functor* G *factors through* $\mathrm{Shv}(\mathfrak{C})$.

(2) *Suppose that* f *is fully faithful and generates* $\mathfrak{X}$ *under colimits. Then* G *carries effective epimorphisms in* $\mathfrak{X}$ *to effective epimorphisms in* $\mathrm{Shv}(\mathfrak{C})$.

Proof. In view of the definition of the canonical topology, (1) is equivalent to the following assertion: given a collection of morphisms $\{u_\alpha : C_\alpha \to C\}$ in $\mathfrak{C}$ such that the induced map $u : \coprod_\alpha C_\alpha \to C$ is an effective epimorphism in $\mathfrak{X}$, if $i : U \to j(C)$ is the monomorphism in $\mathcal{P}(\mathfrak{C})$ corresponding to the sieve $\mathfrak{C}_{/C}^{(0)} \subseteq \mathfrak{C}_{/C}$ generated by the collection $\{u_\alpha\}$, then $F(i)$ is an equivalence in $\mathfrak{X}$. Let $u' : \coprod_\alpha j(C_\alpha) \to j(C)$ be the coproduct of the family $\{j(u_\alpha)\}$ and let $V_\bullet : \Delta_+^{op} \to \mathcal{P}(\mathfrak{C})$ be a Čech nerve of u'. Then i can be identified with the induced map from the colimit of $V_\bullet | \mathrm{N}(\Delta)^{op}$ to V_{-1}. Since F preserves colimits, to show that $F(i)$ is an equivalence, it will suffice to show that $F \circ V_\bullet$ is a colimit diagram. Since u is an effective epimorphism, it suffices to observe that $F \circ V_\bullet$ is equivalent to the Čech nerve of u.

We now prove (2). Suppose that $u : Y \to Z$ is an effective epimorphism in $\mathfrak{X}$. We wish to prove that Gu is an effective epimorphism in $\mathrm{Shv}(\mathfrak{C})$. We will show that the criterion of Lemma 6.2.4.5 is satisfied. Choose an object $C \in \mathfrak{C}$ and a point $\eta \in \pi_0 \mathrm{Map}_{\mathcal{P}(\mathfrak{C})}(j(C), GZ) \simeq \pi_0 \mathrm{Map}_{\mathfrak{X}}(f(C), Z)$. Form a pullback diagram

$$
\begin{array}{ccc}
Y' & \xrightarrow{u'} & f(C) \\
\downarrow{\scriptstyle s} & & \downarrow \\
Y & \xrightarrow{u} & Z
\end{array}
$$

so that u' is an effective epimorphism. Since $f(\mathfrak{C})$ generates $\mathfrak{X}$ under colimits, there exists an effective epimorphism $u'' : \coprod_\alpha f(C_\alpha) \to Y$. The composition $u' \circ u''$ is an effective epimorphism and corresponds to a family of maps $w_\alpha : f(C_\alpha) \to f(C)$ in $\mathfrak{X}$. Since f is fully faithful, we may suppose that each $w_\alpha = f v_\alpha$ for some map $v_\alpha : C_\alpha \to C$ in $\mathfrak{C}$. It follows that the collection of maps $\{v_\alpha\}$ generates a covering sieve on C with respect to the canonical topology. Moreover, each of the compositions

$$f(C_\alpha) \to \coprod_\alpha f(C_\alpha) \to Y$$

gives rise to a point $\eta_\alpha \in \pi_0 \operatorname{Map}_{\mathfrak{X}}(f(C_\alpha), Y) \simeq \pi_0 \operatorname{Map}_{\mathcal{P}(\mathcal{C})}(j(C_\alpha), G(Y))$ with the desired properties.   □

6.3 THE ∞-CATEGORY OF ∞-TOPOI

In this section, we will show that the collection of all ∞-topoi can be organized into an ∞-category $\mathcal{R}\mathcal{T}op$. The objects of $\mathcal{R}\mathcal{T}op$ are ∞-topoi, and the morphisms are called *geometric morphisms*; we will give a definition in §6.3.1. In §6.3.2, we will show that $\mathcal{R}\mathcal{T}op$ admits (small) colimits. In §6.3.3, we will show that $\mathcal{R}\mathcal{T}op$ admits (small) *filtered* limits; we will treat the case of general limits in §6.3.4.

Let $\mathfrak{X}$ be an ∞-topos containing an object U. In §6.3.5, we will show that the ∞-category $\mathfrak{X}_{/U}$ is an ∞-topos. Moreover, this ∞-topos is equipped with a canonical geometric morphism $\mathfrak{X}_{/U} \to \mathfrak{X}$. Geometric morphisms which arise via this construction are said to be *étale*. In §6.3.6, we will define a more general notion of *algebraic* morphism between ∞-topoi. We will also prove a structure theorem which implies that every ∞-topos $\mathfrak{X}$ satisfying some mild hypotheses admits an algebraic morphism to an ∞-category of sheaves on a 2-category.

6.3.1 Geometric Morphisms

In classical topos theory, the correct notion of *morphism* between two topoi is that of an adjunction

$$\mathfrak{X} \underset{f_*}{\overset{f^*}{\rightleftarrows}} \mathcal{Y},$$

where the functor f^* is left exact. We will introduce the same ideas in the ∞-categorical setting.

Definition 6.3.1.1. Let $\mathfrak{X}$ and $\mathcal{Y}$ be ∞-topoi. A *geometric morphism* from $\mathfrak{X}$ to $\mathcal{Y}$ is a functor $f_* : \mathfrak{X} \to \mathcal{Y}$ which admits a left exact left adjoint (which we will typically denote by f^*).

Remark 6.3.1.2. Let $f_* : \mathfrak{X} \to \mathcal{Y}$ be a geometric morphism from an ∞-topos $\mathfrak{X}$ to another ∞-topos $\mathcal{Y}$, so that f_* admits a left adjoint f^*. Either of the functors f_* and f^* determines the other up to equivalence (in fact, up to contractible ambiguity). We will often abuse terminology by referring to f^* as a geometric morphism from $\mathfrak{X}$ to $\mathcal{Y}$. We will always indicate in our notation whether the left or the right adjoint is being considered: a superscripted asterisk indicates a left adjoint (pullback functor), and a subscripted asterisk indicates a right adjoint (pushforward functor).

Remark 6.3.1.3. Any equivalence of ∞-topoi is a geometric morphism. If $f_*, g_* : \mathfrak{X} \to \mathcal{Y}$ are homotopic, then f_* is a geometric morphism if and only if g_* is a geometric morphism (because we can identify left adjoints of f_* with left adjoints of g_*).

Remark 6.3.1.4. Let $f_* : \mathfrak{X} \to \mathcal{Y}$ and $g_* : \mathcal{Y} \to \mathcal{Z}$ be geometric morphisms. Then f_* and g_* admit left exact left adjoints, which we will denote by f^* and g^*, respectively. The composite functor $f^* \circ g^*$ is left exact and is a left adjoint to $g_* \circ f_*$ by Proposition 5.2.2.6. We conclude that $g_* \circ f_*$ is a geometric morphism, so the class of geometric morphisms is stable under composition.

Definition 6.3.1.5. Let $\widehat{\mathcal{C}at}_\infty$ denote the ∞-category of (not necessarily small) ∞-categories. We define subcategories $\mathcal{L}\mathcal{T}\mathrm{op}, \mathcal{R}\mathcal{T}\mathrm{op} \subseteq \widehat{\mathcal{C}at}_\infty$ as follows:

(1) The objects of $\mathcal{L}\mathcal{T}\mathrm{op}$ and $\mathcal{R}\mathcal{T}\mathrm{op}$ are the ∞-topoi.

(2) A functor $f^* : \mathfrak{X} \to \mathcal{Y}$ between ∞-topoi belongs to $\mathcal{L}\mathcal{T}\mathrm{op}$ if and only if f^* preserves small colimits and finite limits.

(3) A functor $f_* : \mathfrak{X} \to \mathcal{Y}$ between ∞-topoi belongs to $\mathcal{R}\mathcal{T}\mathrm{op}$ if and only if f_* has a left adjoint which is left exact.

The ∞-categories $\mathcal{L}\mathcal{T}\mathrm{op}$ and $\mathcal{R}\mathcal{T}\mathrm{op}$ are canonically antiequivalent. To prove this, we will use the argument of Corollary 5.5.3.4. First, we need a definition.

Definition 6.3.1.6. A map $p : X \to S$ of simplicial sets is a *topos fibration* if the following conditions are satisfied:

(1) The map p is both a Cartesian fibration and a coCartesian fibration.

(2) For every vertex s of S, the corresponding fiber $X_s = X \times_S \{s\}$ is an ∞-topos.

(3) For every edge $e : s \to s'$ in S, the associated functor $X_s \to X_{s'}$ is left exact.

The following analogue of Proposition 5.5.3.3 follows immediately from the definitions:

Proposition 6.3.1.7. (1) *Let $p : X \to S$ be a Cartesian fibration of simplicial sets classified by a map $\chi : S^{op} \to \widehat{\mathcal{C}at}_\infty$. Then p is a topos fibration if and only if χ factors through $\mathcal{R}\mathcal{T}\mathrm{op} \subseteq \widehat{\mathcal{C}at}_\infty$.*

(2) *Let $p : X \to S$ be a coCartesian fibration of simplicial sets classified by a map $\chi : S \to \widehat{\mathcal{C}at}_\infty$. Then p is a topos fibration if and only if χ factors through $\mathcal{L}\mathcal{T}\mathrm{op} \subseteq \widehat{\mathcal{C}at}_\infty$.*

Corollary 6.3.1.8. *For every simplicial set S, there is a canonical bijection*

$$[S, \mathcal{L}\mathcal{T}\mathrm{op}] \simeq [S^{op}, \mathcal{R}\mathcal{T}\mathrm{op}],$$

where $[K, \mathcal{C}]$ denotes the collection of equivalence classes of objects of the ∞-category $\mathrm{Fun}(K, \mathcal{C})$. In particular, $\mathcal{L}\mathcal{T}\mathrm{op}$ and $\mathcal{R}\mathcal{T}\mathrm{op}^{op}$ are canonically isomorphic in the homotopy category of ∞-categories.

Proof. According to Proposition 6.3.1.7, both $[S, \mathcal{L}\mathcal{T}\text{op}]$ and $[S^{op}, \mathcal{R}\mathcal{T}\text{op}]$ can be identified with the collection of equivalence classes of topos fibrations $X \to S$.   □

The following proposition is a simple reformulation of some of the results of §5.5.6.

Proposition 6.3.1.9. *Let* $f_* : \mathfrak{X} \to \mathfrak{Y}$ *be a geometric morphism between* ∞-*topoi having a left adjoint* f^*. *Then* f^* *and* f_* *carry* k-*truncated objects to* k-*truncated objects and* k-*truncated morphisms to* k-*truncated morphisms, for any integer* $k \geq -2$. *Moreover, there is a (canonical) equivalence of functors* $f^* \tau^{\mathfrak{Y}}_{\leq k} \simeq \tau^{\mathfrak{X}}_{\leq k} f^*$.

Proof. The first assertion follows immediately from Lemma 5.5.6.15 since f_* and f^* are both left exact. The second follows from Proposition 5.5.6.28.   □

Definition 6.3.1.10. Let $\mathfrak{X}$ and $\mathfrak{Y}$ be ∞-topoi. We let $\text{Fun}_*(\mathfrak{X}, \mathfrak{Y})$ denote the full subcategory of $\text{Fun}(\mathfrak{X}, \mathfrak{Y})$ spanned by geometric morphisms $f_* : \mathfrak{X} \to \mathfrak{Y}$, and $\text{Fun}^*(\mathfrak{Y}, \mathfrak{X})$ the full subcategory of $\text{Fun}(\mathfrak{Y}, \mathfrak{X})$ spanned by their left adjoints.

Remark 6.3.1.11. It follows from Proposition 5.2.6.2 that the ∞-categories $\text{Fun}_*(\mathfrak{X}, \mathfrak{Y})$ and $\text{Fun}^*(\mathfrak{Y}, \mathfrak{X})$ are canonically antiequivalent to one another.

Warning 6.3.1.12. If $\mathfrak{X}$ and $\mathfrak{Y}$ are ∞-topoi, then the ∞-category $\text{Fun}_*(\mathfrak{X}, \mathfrak{Y})$ of geometric morphisms from $\mathfrak{X}$ to $\mathfrak{Y}$ is *not* necessarily small or even equivalent to a small ∞-category. This phenomenon is familiar in classical topos theory. For example, there is a classifying topos $\mathcal{A}$ for abelian groups having the property that for *any* topos $\mathfrak{X}$, the category $\mathcal{C}$ of geometric morphisms $\mathfrak{X} \to \mathcal{A}$ is equivalent to the category of abelian group objects of $\mathfrak{X}$. This category is almost never small (for example, when $\mathfrak{X}$ is the topos of sets, $\mathcal{C}$ is equivalent to the category of abelian groups).

In spite of Warning 6.3.1.12, the ∞-category of geometric morphisms between two ∞-topoi can be reasonably controlled:

Proposition 6.3.1.13. *Let* $\mathfrak{X}$ *and* $\mathfrak{Y}$ *be* ∞-*topoi. Then the* ∞-*category* $\text{Fun}^*(\mathfrak{Y}, \mathfrak{X})$ *of geometric morphisms from* $\mathfrak{X}$ *to* $\mathfrak{Y}$ *is accessible.*

Proof. For each regular cardinal κ, let $\mathfrak{Y}^\kappa$ denote the full subcategory of $\mathfrak{Y}$ spanned by κ-compact objects. Choose a regular cardinal κ such that $\mathfrak{Y}$ is κ-accessible and $\mathfrak{Y}^\kappa$ is stable under finite limits in $\mathfrak{Y}$. We may therefore identify $\mathfrak{Y}$ with $\text{Ind}^\kappa(\mathcal{C})$, where $\mathcal{C}$ is a minimal model for $\mathfrak{Y}^\kappa$. According to Proposition 5.3.5.10, composition with the Yoneda embedding $j : \mathcal{C} \to \mathfrak{Y}$ induces an equivalence from the ∞-category of κ-continuous functors $\text{Fun}_\kappa(\mathfrak{Y}, \mathfrak{X})$ to the ∞-category $\text{Fun}(\mathcal{C}, \mathfrak{X})$. We now make the following observations:

(1) A functor $F : \mathfrak{Y} \to \mathfrak{X}$ preserves all small colimits if and only if $F \circ j : \mathcal{C} \to \mathfrak{X}$ preserves κ-small colimits (Proposition 5.5.1.9).

(2) A colimit-preserving functor $F : \mathcal{Y} \to \mathcal{X}$ is left exact if and only if the composition $F \circ j : \mathcal{C} \to \mathcal{X}$ is left exact (Proposition 6.1.5.2).

Invoking Proposition 5.2.6.2, we deduce that the ∞-category $\operatorname{Fun}^*(\mathcal{Y}, \mathcal{X})$ is equivalent to the full subcategory $\mathcal{M} \subseteq \mathcal{X}^{\mathcal{C}}$ consisting of functors which preserve κ-small colimits and finite limits. Proposition 5.4.4.3 implies that $\operatorname{Fun}(\mathcal{C}, \mathcal{X})$ is accessible. For every κ-small (finite) diagram $p : K \to \mathcal{C}$, the full subcategory of $\operatorname{Fun}(\mathcal{C}, \mathcal{X})$ spanned by those functors which preserve colimits (limits) of p is an accessible subcategory of $\operatorname{Fun}(\mathcal{C}, \mathcal{X})$ (Example 5.4.7.9). Up to isomorphism, there are only a bounded number of κ-small (finite) diagrams in $\mathcal{C}$. Consequently, $\mathcal{M}$ is an intersection of a bounded number of accessible subcategories of $\operatorname{Fun}(\mathcal{C}, \mathcal{X})$ and therefore accessible (Proposition 5.4.7.10). $\qquad\square$

6.3.2 Colimits of ∞-Topoi

Our goal in this section is to construct colimits in the ∞-category $\mathcal{R}\mathcal{T}\mathrm{op}$ of ∞-topoi. According to Corollary 6.3.1.8, it will suffice to construct *limits* in the ∞-category $\mathcal{L}\mathcal{T}\mathrm{op}$.

Proposition 6.3.2.1. *Let $\{\mathcal{X}_\alpha\}_{\alpha \in A}$ be a collection of ∞-topoi parametrized by a (small) set A. Then the product $\mathcal{X} = \prod_{\alpha \in A} \mathcal{X}_\alpha$ is an ∞-topos. Moreover, each projection $\pi_\alpha^* : \mathcal{X} \to \mathcal{X}_\alpha$ is left exact and colimit-preserving. The corresponding geometric morphisms exhibit $\mathcal{X}$ as a product of the family $\{\mathcal{X}_\alpha\}_{\alpha \in A}$ in the ∞-category $\mathcal{L}\mathcal{T}\mathrm{op}$.*

Proof. Proposition 5.5.3.5 implies that $\mathcal{X}$ is presentable. It is clear that a diagram $\bar{p} : K^{\triangleright} \to \mathcal{X}$ is a colimit if and only if each composition $\pi_\alpha^* \circ \bar{p} : K^{\triangleright} \to \mathcal{X}_\alpha$ is a colimit diagram in $\mathcal{X}_\alpha$. Similarly, a diagram $\bar{q} : K^{\triangleleft} \to \mathcal{X}$ is a limit if and only if each composition $\pi_\alpha^* \circ \bar{q} : K^{\triangleleft} \to \mathcal{X}_\alpha$ is a limit diagram in $\mathcal{X}_\alpha$. Using criterion (2) of Theorem 6.1.0.6, we deduce that $\mathcal{X}$ is an ∞-topos, and that each π_α^* preserves all limits and colimits that exist in $\mathcal{X}$. Choose a right adjoint $\pi_*^\alpha : \mathcal{X}_\alpha \to \mathcal{X}$ to each π_α^*.

According to Theorem 4.2.4.1, the ∞-category $\mathcal{X}$ is a product of the family $\{\mathcal{X}_\alpha\}_{\alpha \in A}$ in the ∞-category $\widehat{\mathcal{C}\mathrm{at}}_\infty$. Since $\mathcal{L}\mathcal{T}\mathrm{op}$ is a subcategory of $\widehat{\mathcal{C}\mathrm{at}}_\infty$, it will suffice to prove the following assertion:

- For every ∞-topos $\mathcal{Y}$ and every functor $f^* : \mathcal{Y} \to \mathcal{X}$ such that each of the composite functors $\mathcal{Y} \to \mathcal{X}_\alpha$ is left exact and colimit-preserving, f^* is itself left exact and colimit-preserving.

This follows immediately from the fact that limits and colimits are computed pointwise. $\qquad\square$

Proposition 6.3.2.2. *Let*

$$
\begin{array}{ccc}
\mathfrak{X}' & \xrightarrow{q'^*} & \mathfrak{X} \\
\downarrow{\scriptstyle p'^*} & & \downarrow{\scriptstyle p^*} \\
\mathcal{Y}' & \xrightarrow{q^*} & \mathcal{Y}
\end{array}
$$

be a diagram of ∞-categories which is homotopy Cartesian (with respect to the Joyal model structure). Suppose further that $\mathfrak{X}$, $\mathcal{Y}$, and $\mathcal{Y}'$ are ∞-topoi and that p^* and q^* are left exact and colimit-preserving. Then $\mathfrak{X}'$ is an ∞-topos. Moreover, for any ∞-topos $\mathcal{Z}$ and any functor $f^* : \mathcal{Z} \to \mathfrak{X}$, f^* is left exact and colimit-preserving if and only if the compositions $p'^* \circ f^*$ and $q'^* \circ f^*$ are left exact and colimit-preserving. In particular (taking $f^* = \mathrm{id}_{\mathfrak{X}}$), the functors p'^* and q'^* are left exact and colimit-preserving.

Proof. The second claim follows immediately from Lemma 5.4.5.5 and the dual result concerning limits. To prove the first, we observe that $\mathfrak{X}'$ is presentable by Proposition 5.5.3.12. To show that $\mathfrak{X}$ is an ∞-topos, it will suffice to show that it satisfies criterion (2) of Theorem 6.1.0.6. This follows immediately from Lemma 5.4.5.5 since $\mathfrak{X}$ and $\mathcal{Y}'$ satisfy criterion (2) of Theorem 6.1.0.6. $\qquad\square$

Proposition 6.3.2.3. *The ∞-category $\mathcal{L}\mathcal{T}\mathrm{op}$ admits small limits, and the inclusion functor $\mathcal{L}\mathcal{T}\mathrm{op} \subseteq \widehat{\mathrm{Cat}}_\infty$ preserves small limits.*

Proof. According to Proposition 4.4.2.6, it suffices to prove this result for pullbacks and small products. In the case of products, we apply Proposition 6.3.2.1. For pullbacks, we use Proposition 6.3.2.2 and Theorem 4.2.4.1. $\qquad\square$

6.3.3 Filtered Limits of ∞-Topoi

We now consider the problem of computing limits in the ∞-category $\mathcal{R}\mathcal{T}\mathrm{op}$ of ∞-topoi. This is quite a bit more difficult than the analogous problem for colimits because the inclusion functor $i : \mathcal{R}\mathcal{T}\mathrm{op} \subseteq \widehat{\mathrm{Cat}}_\infty$ does not commute with limits in general. However, the inclusion i does commute with *filtered* limits:

Theorem 6.3.3.1. *The ∞-category $\mathcal{R}\mathcal{T}\mathrm{op}$ admits small filtered limits (that is, limits indexed by diagrams $\mathcal{C}^{op} \to \mathcal{R}\mathcal{T}\mathrm{op}$ where $\mathcal{C}$ is a small filtered ∞-category). Moreover, the inclusion $\mathcal{R}\mathcal{T}\mathrm{op} \subseteq \widehat{\mathrm{Cat}}_\infty$ preserves small filtered limits.*

The remainder of this section is devoted to the proof of Theorem 6.3.3.1. Our basic strategy is to mimic the proof of Theorem 5.5.3.18. Our first step is to show that the limit (in $\widehat{\mathrm{Cat}}_\infty$) of a filtered diagram of ∞-topoi is itself an ∞-topos. This is equivalent to a more concrete assertion: if $p : X \to S$ is a topos fibration and S^{op} is a small filtered ∞-category, then the ∞-category $\mathcal{C}$ of Cartesian sections of p is an ∞-topos. We saw in Proposition 5.5.3.17

that $\mathcal{C}$ is an accessible localization of the ∞-category $\mathrm{Map}_S(S,X)$ spanned by *all* sections of p. Our first step will be to show that $\mathrm{Map}_S(S,X)$ is an ∞-topos. For this, the hypothesis that S^{op} is filtered is irrelevant.

Lemma 6.3.3.2. *Let $p : X \to S$ be a topos fibration, where S is a small simplicial set. The ∞-category $\mathrm{Map}_S(S,X)$ of sections of p is an ∞-topos.*

Proof. This is a special case of Proposition 5.4.7.11.

$\square$

Proposition 6.3.3.3. *Let A be a (small) filtered partially ordered set and let $p : X \to \mathrm{N}(A)$ be a topos fibration. Let $\mathcal{C} = \mathrm{Map}_{\mathrm{N}(A)}(\mathrm{N}(A),X)$ be the ∞-category of sections of p and let $\mathcal{C}' \subseteq \mathcal{C}$ be the full subcategory of $\mathcal{C}$ spanned by the Cartesian sections of p. Then $\mathcal{C}'$ is a topological localization of $\mathcal{C}$.*

Proof. Let us say that a subset $A' \subseteq A$ is *dense* if there exists $\alpha \in A$ such that

$$\{\beta \in A : \beta \geq \alpha\} \subseteq A'.$$

For each morphism f in $\mathcal{C}$, let $A(f) \subseteq A$ be the collection of all $\alpha \in A$ such that the image of f in X_α is an equivalence. Let S be the collection of all monomorphisms f in $\mathcal{C}$ such that $A(f)$ is dense. It is clear that S is stable under pullbacks, so that $S^{-1}\mathcal{C}$ is a topological localization of $\mathcal{C}$. To complete the proof, it will suffice to show that $\mathcal{C}' = S^{-1}\mathcal{C}$.

We first claim that each object of $\mathcal{C}'$ is S-local. Let $f : C \to C'$ belong to S and let $D \in \mathcal{C}'$. Choose α_0 such that $A(f)$ contains $A' = \{\beta \in A : \beta \geq \alpha_0\}$ and let R^* denote a right adjoint to the restriction functor

$$R : \mathrm{Map}_{\mathrm{N}(A)}(\mathrm{N}(A),X) \to \mathrm{Map}_{\mathrm{N}(A)}(\mathrm{N}(A'),X).$$

According to Proposition 4.3.2.17, the essential image of R^* consists of those functors $E : \mathrm{N}(A) \to X$ which are p-right Kan extensions of $E|\mathrm{N}(A')$. We claim that D satisfies this condition. In other words, we claim that for each $\alpha \in A$, the map

$$\overline{q} : \mathrm{N}(A'')^{\triangleleft} \to \mathrm{N}(A) \xrightarrow{D} X$$

is a p-limit, where $A'' = \{\beta \in A : \beta \geq \alpha, \beta \geq \alpha_0\}$. Since $\overline{q}$ carries each edge of $\mathrm{N}(A'')^{\triangleleft}$ to a p-Cartesian edge of X, it suffices to verify that the simplicial set $\mathrm{N}(A'')$ is weakly contractible (Proposition 4.3.1.12). This follows immediately from the observation that A'' is a filtered partially ordered set.

We may therefore suppose that $D = R^*\overline{D}$, where $\overline{D} = D|\mathrm{N}(A')$ is a Cartesian section of the induced map $p' : X \times_{\mathrm{N}(A)} \mathrm{N}(A') \to \mathrm{N}(A')$. We wish to prove that composition with f induces a homotopy equivalence

$$\mathrm{Map}_{\mathcal{C}}(C', R^*\overline{D}) \to \mathrm{Map}_{\mathcal{C}}(C, R^*\overline{D}).$$

This follows immediately from the fact that R and R^* are adjoint since $R(f)$ is an equivalence.

We now show that every S-local object of $\mathcal{C}$ belongs to $\mathcal{C}'$. Let $C \in \mathcal{C}$ be a section of p which is S-local. Choose $\alpha \le \beta$ in A and let

$$X_\alpha \underset{G}{\overset{F}{\rightleftarrows}} X_\beta$$

denote the (adjoint) functors associated to the (co)Cartesian fibration $p : X \to N(A)$. The section C gives rise to a pair of objects $C_\alpha \in X_\alpha$, $C_\beta \in X_\beta$, and a morphism $\phi : C_\alpha \to C_\beta$ in the ∞-category X. The map ϕ induces a morphism $u : C_\alpha \to GC_\beta$ in X_α, which is well-defined up to equivalence. We wish to show that ϕ is p-Cartesian, which is equivalent to the assertion that u is an equivalence in X_α. Equivalently, we wish to show that for each object $P \in X_\alpha$, composition with u induces a homotopy equivalence

$$\mathrm{Map}_{X_\alpha}(P, C_\alpha) \to \mathrm{Map}_{X_\alpha}(P, GC_\beta).$$

We may identify P with a diagram

$$
\begin{array}{ccc}
\{\alpha\} & \xrightarrow{\ P\ } & X \\
\downarrow & \overset{D}{\nearrow} & \downarrow \\
N(A) & =\!\!=\!\!= & N(A).
\end{array}
$$

Using Corollary 4.3.2.14, choose an extension D as indicated in the diagram above, so that D is a left Kan extension of $D|\{\alpha\}$ over $N(A)$. Similarly, we have a diagram

$$
\begin{array}{ccc}
\{\beta\} & \xrightarrow{\ F(P)\ } & X \\
\downarrow & \overset{D'}{\nearrow} & \downarrow \\
N(A) & =\!\!=\!\!= & N(A),
\end{array}
$$

and we can choose D' to be a p-left Kan extension of $D'|\{\beta\}$.

Proposition 4.3.2.17 implies that for every object $E \in \mathcal{C}$, the restriction maps

$$\mathrm{Map}_{\mathcal{C}}(D, E) \to \mathrm{Map}_{X_\alpha}(P, E(\alpha))$$

$$\mathrm{Map}_{\mathcal{C}}(D', E) \to \mathrm{Map}_{X_\beta}(F(P), E(\beta))$$

are equivalences. In particular, the equivalence between $D(\beta)$ and $F(P)$ induces a morphism $\theta : D' \to D$.

We have a commutative diagram in the homotopy category $\mathcal{H}$:

$$
\begin{array}{ccc}
\mathrm{Map}_{\mathcal{C}}(D, C) & \xrightarrow{\ \circ\theta\ } & \mathrm{Map}_{\mathcal{C}}(D', C) \\
\downarrow & & \downarrow \\
\mathrm{Map}_{X_\alpha}(P, C_\alpha) \xrightarrow{\circ u} \mathrm{Map}_{X_\alpha}(P, G(C_\beta)) & \longleftarrow & \mathrm{Map}_{X_\beta}(F(P), C_\beta).
\end{array}
$$

The vertical maps are homotopy equivalences, and the the horizontal map on the lower right is a homotopy equivalence because F and G are adjoint.

To complete the proof, it will suffice to show that the upper horizontal map is an equivalence. Since C is S-local, it will suffice to show that $\theta \in S$.

Let $\beta \leq \beta'$ and consider the diagram

$$
\begin{array}{ccc}
D'(\beta) & \xrightarrow{\ w'\ } & D'(\beta') \\
\downarrow{\scriptstyle\theta(\beta)} & & \downarrow{\scriptstyle\theta(\beta')} \\
D(\alpha) \xrightarrow{\ v\ } D(\beta) & \xrightarrow{\ w\ } & D(\beta')
\end{array}
$$

in the ∞-category X. Since D' is a p-left Kan extension of $D'|\{\beta\}$, we conclude that w' is p-coCartesian. Similarly, since D is a p-left Kan extension of $D|\{\alpha\}$, we conclude that v and $w \circ v$ are p-coCartesian. It follows that w is p-coCartesian as well (Proposition 2.4.1.7). Since $\theta(\beta)$ is an equivalence by construction, we conclude that $\theta(\beta')$ is an equivalence. Thus $A(\theta) \subseteq A$ is dense.

It remains only to show that θ is a monomorphism. For this it suffices to show that $\theta(\gamma)$ is a monomorphism in X_γ for each $\gamma \in A$. If $\gamma \geq \beta$, this follows from the above argument. Suppose $\gamma \not\geq \beta$. Since D' is a p-left Kan extension of $D'|\{\beta\}$ over $\mathrm{N}(A)$, we conclude that $D'(\gamma)$ is a p-colimit of the empty diagram and therefore an initial object of X_γ. It follows that any map $D'(\gamma) \to D(\gamma)$ is a monomorphism. $\qquad\square$

Proposition 6.3.3.4. *Let A be a (small) filtered partially ordered set, let $p : X \to \mathrm{N}(A)^{op}$, and let $\mathcal{Y} \subseteq \mathrm{Map}_{\mathrm{N}(A)}(\mathrm{N}(A), X)$ be the full subcategory spanned by the Cartesian sections of p. For each $\alpha \in A$, the evaluation map $\pi_* : \mathcal{Y} \to X_\alpha$ is a geometric morphism of ∞-topoi.*

Proof. Let $A' = \{\beta \in A : \alpha \leq \beta\}$. Using Theorem 4.1.3.1, we conclude that the inclusion $\mathrm{N}(A') \subseteq \mathrm{N}(A)$ is cofinal. Corollary 3.3.3.2 implies that the restriction map

$$
\mathrm{Map}_{\mathrm{N}(A)}(\mathrm{N}(A), X) \to \mathrm{Map}_{\mathrm{N}(A)}(\mathrm{N}(A'), X)
$$

induces an equivalence on the full subcategories spanned by Cartesian sections. Consequently, we are free to replace A by A' and thereby assume that α is a least element of A.

The functor π_* factors as a composition

$$
\mathcal{Y} \xrightarrow{\phi_*} \mathrm{Map}_{\mathrm{N}(A)}(\mathrm{N}(A), X) \xrightarrow{\psi_*} X_\alpha,
$$

where ϕ_* denotes the inclusion functor and ψ_* the evaluation functor. Proposition 6.3.3.3 implies that ϕ_* is a geometric morphism; it therefore suffices to show that ψ_* is a geometric morphism as well.

Let ψ^* be a left adjoint to ψ_* (the existence of ψ^* follows from Proposition 4.3.2.17 as indicated below). We wish to show that ψ^* is left exact. According to Proposition 5.1.2.2, it will suffice to show that the composition

$$
\theta : X_\alpha \xrightarrow{\psi^*} \mathrm{Map}_{\mathrm{N}(A)}(\mathrm{N}(A), X) \xrightarrow{e_\beta} X_\beta
$$

is left exact, where e_β denotes the functor given by evaluation at β.

Let $f : \Delta^1 \to N(A)$ be the edge joining α to β, let $\mathcal{C}$ be the ∞-category of coCartesian sections of p, and let $\mathcal{C}'$ be the ∞-category of coCartesian sections of the induced map $p' : X \times_{N(A)} \Delta^1 \to \Delta^1$. We observe that $\mathcal{C}$ consists precisely of those sections $s : N(A) \to X$ of p which are p-left Kan extensions of $s|\{\alpha\}$. Applying Proposition 4.3.2.15, we conclude that the evaluation map $e_\alpha : \mathcal{C} \to X_\alpha$ is a trivial fibration and that (by Proposition 4.3.2.17) we may identify ψ^* with the composition

$$X_\alpha \xrightarrow{q} \mathcal{C} \subseteq \mathrm{Map}_{N(A)}(N(A), X),$$

where q is a section of $e_\alpha | \mathcal{C}$. Let $q' : X_\alpha \to \mathcal{C}'$ be the composition of q with the restriction map $\mathcal{C} \to \mathcal{C}'$. Then θ can be identified with the composition

$$X_\alpha \xrightarrow{q'} \mathcal{C}' \xrightarrow{e_\beta} X_\beta,$$

which is the functor $X_\alpha \to X_\beta$ associated to $f : \alpha \to \beta$ by the coCartesian fibration p. Since p is a topos fibration, θ is left exact, as desired. □

Let G be a profinite group and let X be a set with a continuous action of G. Then we can recover X as the direct limit of the fixed-point sets X^U, where U ranges over the collection of open subgroups of G. Our next result is an ∞-categorical analogue of this observation.

Lemma 6.3.3.5. *Let $p : X \to S^\triangleright$ be a Cartesian fibration of simplicial sets, which is classified by a colimit diagram $S^\triangleright \to \mathrm{Cat}_\infty^{op}$, and let $\overline{s} : S^\triangleright \to X$ be a Cartesian section of p. Then $\overline{s}$ is a p-colimit diagram.*

Proof. By virtue of Corollary 3.3.1.2, we may suppose that S is an ∞-category. Unwinding the definitions, we must show that the map $X_{\overline{s}/} \to X_{s/}$ induces an equivalence of ∞-categories when restricted to the inverse image of the cone point of $S^\triangleright$. Fix an object $x \in X$ lying over the cone point of $S^\triangleright$. Let $\overline{f} : S^\triangleright \to X$ be the constant map with value x and let $f = \overline{f}|S$. To complete the proof, it will suffice to show that the restriction map

$$\theta : \mathrm{Map}_{\mathrm{Fun}(S^\triangleright, X)}(\overline{s}, \overline{f}) \to \mathrm{Map}_{\mathrm{Fun}(S, X)}(s, f)$$

is a homotopy equivalence. To prove this, we choose a p-Cartesian transformation $\overline{\alpha} : \overline{g} \to \overline{f}$, where $\overline{g} : S^\triangleright \to X$ is a section of p (automatically Cartesian). Let $g = \overline{g}|S$ and let $\alpha : g \to f$ be the associated transformation. Let $\overline{\mathcal{C}}$ be the full subcategory of $\mathrm{Map}_{S^\triangleright}(S^\triangleright, X)$ spanned by the Cartesian sections of p and let $\mathcal{C} \subseteq \mathrm{Map}_{S^\triangleright}(S, X)$ be defined similarly. We have a commutative diagram in the homotopy category $\mathcal{H}$

$$
\begin{array}{ccc}
\mathrm{Map}_{\overline{\mathcal{C}}}(\overline{s}, \overline{g}) & \xrightarrow{\;\theta'\;} & \mathrm{Map}_{\mathcal{C}}(s, g) \\
\downarrow{\overline{\alpha}} & & \downarrow{\alpha} \\
\mathrm{Map}_{\mathrm{Fun}(S^\triangleright, X)}(\overline{s}, \overline{f}) & \xrightarrow{\;\theta\;} & \mathrm{Map}_{\mathrm{Fun}(S, X)}(s, f).
\end{array}
$$

Proposition 2.4.4.2 implies that the vertical maps are homotopy equivalences, and Proposition 3.3.3.1 implies that θ' is a homotopy equivalence (since the restriction map $\overline{\mathcal{C}} \to \mathcal{C}$ is an equivalence of ∞-categories). It follows that θ is a homotopy equivalence as well. □

Lemma 6.3.3.6. *Let $p : X \to S$ be a presentable fibration and let $\mathcal{C}$ be the full subcategory of $\mathrm{Map}_S(S, X)$ spanned by the Cartesian sections of p. For each vertex s of S, let $\psi(s)_* : \mathcal{C} \to X_s$ be the functor given by evaluation at s and let $\psi(s)^*$ be a left adjoint to $\psi(s)_*$. There exists a diagram $\theta : S \to \mathrm{Fun}(\mathcal{C}, \mathcal{C})$ with the following properties:*

(1) *For each vertex s of S, $\theta(s)$ is equivalent to the composition $\psi(s)^* \circ \psi(s)_*$.*

(2) *The identity functor $\mathrm{id}_\mathcal{C}$ is a colimit of θ in the ∞-category of functors $\mathrm{Fun}(\mathcal{C}, \mathcal{C})$.*

Proof. Without loss of generality, we may suppose that p extends to a presentable fibration $\overline{p} : \overline{X} \to S^\triangleright$, which is classified by colimit diagram $S^\triangleright \to \mathcal{P}\mathrm{r}^L$ (and therefore by a colimit diagram $S^\triangleright \to \mathcal{C}\mathrm{at}_\infty^{op}$ by virtue of Theorem 5.5.3.18). Let $\overline{\mathcal{C}}$ be the ∞-category of Cartesian sections of $\overline{p}$, so that we have trivial fibrations

$$\mathcal{C} \leftarrow \overline{\mathcal{C}} \to X_\infty,$$

where $X_\infty = \overline{X} \times_{S^\triangleright} \{\infty\}$ and ∞ denotes the cone point of $S^\triangleright$. For each vertex s of $S^\triangleright$, we let $\overline{\psi}(s)_* : \overline{\mathcal{C}} \to \overline{X}_s$ be the functor given by evaluation at s and $\overline{\psi}(s)^*$ a left adjoint to $\overline{\psi}(s)_*$. To complete the proof, it will suffice to construct a map $\theta' : S \to \mathrm{Fun}(\overline{\mathcal{C}}, X_\infty)$ with the following properties:

(1') *For each vertex s of S, $\theta'(s)$ is equivalent to the composition $\overline{\psi}(\infty)_* \circ \overline{\psi}(s)^* \circ \overline{\psi}(s)_*$.*

(2') *The functor $\overline{\psi}(\infty)_*$ is a colimit of θ'.*

Let $e : \overline{\mathcal{C}} \times S^\triangleright \to \overline{X}$ be the evaluation map. Choose a $\overline{p}$-coCartesian natural transformation $e \to e'$, where e' is a map from $\overline{\mathcal{C}} \times S^\triangleright$ to X_∞. Lemma 6.3.3.5 implies that for each object $X \in \overline{\mathcal{C}}$, the restriction $e|\{X\} \times S^\triangleright$ is a $\overline{p}$-colimit diagram in $\overline{X}$. Applying Proposition 4.3.1.9, we deduce that $e'|\{X\} \times S^\triangleright$ is a colimit diagram in X_∞. According to Proposition 5.1.2.2, e' determines a colimit diagram $S^\triangleright \to \mathrm{Fun}(\overline{\mathcal{C}}, X_\infty)$. Let θ' be the restriction of this diagram to S. Then the colimit of θ' can be identified with $e'|\overline{\mathcal{C}} \times \{\infty\}$, which is equivalent to $e|\overline{\mathcal{C}} \times \{\infty\} = \overline{\psi}(\infty)_*$. This proves (2'). To verify (1'), we observe that $e'|\overline{\mathcal{C}} \times \{s\}$ can be identified with the composition of $\overline{\psi}(s)_* = e|\overline{\mathcal{C}} \times \{s\}$ with the functor $X_s \to X_\infty$ associated to the coCartesian fibration $\overline{p}$, which can in turn be identified with $\overline{\psi}(\infty)_* \circ \overline{\psi}(s)^*$ (both are left adjoints to the pullback functor $X_\infty \to X_s$ associated to $\overline{p}$). $\qquad \square$

Proposition 6.3.3.7. *Let A be a (small) filtered partially ordered set, let $p : X \to \mathrm{N}(A)$, and let $\mathcal{Y} \subseteq \mathrm{Map}_{\mathrm{N}(A)}(\mathrm{N}(A), X)$ be the full subcategory spanned by the Cartesian sections of p. Let $\mathcal{Z}$ be an ∞-topos and let $\pi_* : \mathcal{Z} \to \mathcal{Y}$ be an arbitrary functor. Suppose that, for each $\alpha \in A$, the composition*

$$\mathcal{Z} \xrightarrow{\pi_*} \mathcal{Y} \to X_\alpha$$

is a geometric morphism of ∞-topoi. Then π_* is a geometric morphism of ∞-topoi.

Proof. Let π^* denote a left adjoint to π_*. Since π^* commutes with colimits, Lemma 6.3.3.6 implies that π^* can be written as the colimit of a diagram $q : N(A) \to \mathcal{Z}^{\mathcal{Y}}$ having the property that for each $\alpha \in A$, $q(\alpha)$ is equivalent to $\pi^* \psi(\alpha)^* \psi(\alpha)_*$, where $\psi(\alpha)_*$ denotes the evaluation functor at α and $\psi(\alpha)^*$ is its left adjoint. Each composition $\pi^* \psi(\alpha)^*$ is left adjoint to the geometric morphism $\psi(\alpha)_* \pi_*$ and therefore left exact. It follows that $q(\alpha)$ is left exact. Since filtered colimits in $\mathcal{Z}$ are left exact (Example 7.3.4.7), we conclude that the functor π^* is left exact, as desired. $\square$

Proof of Theorem 6.3.3.1. Let $\mathcal{C}$ be a small filtered ∞-category and let $q : \mathcal{C}^{op} \to \mathcal{RT}op$ be an arbitrary diagram. Choose a limit $\overline{q} : (\mathcal{C}^{\triangleright})^{op} \to \widehat{\mathcal{C}at}_\infty$ of q in the ∞-category $\widehat{\mathcal{C}at}_\infty$. We must show that $\overline{q}$ factors through $\mathcal{RT}op$ and is a limit diagram in $\mathcal{RT}op$.

Using Proposition 5.3.1.16, we may assume without loss of generality that $\mathcal{C}$ is the nerve of a filtered partially ordered set A. Let $p : X \to N(A)^{op}$ be the topos fibration classified by q (Proposition 6.3.1.7). Then the image of the cone point of $(\mathcal{C}^{\triangleright})^{op}$ under $\overline{q}$ is equivalent to the ∞-category $\mathcal{X}$ of Cartesian sections of p (Corollary 3.3.3.2). It follows from Proposition 6.3.3.3 that $\mathcal{X}$ is an ∞-topos. Moreover, Proposition 6.3.3.4 ensures that for each $\alpha \in A$, the evaluation map $\mathcal{X} \to X_\alpha$ is a geometric morphism. This proves that $\overline{q}$ factors through $\mathcal{RT}op$. To complete the proof, we must show that $\overline{q}$ is a limit diagram in $\mathcal{RT}op$. Since $\mathcal{RT}op$ is a subcategory of $\widehat{\mathcal{C}at}_\infty$ and $\overline{q}$ is a limit diagram in $\widehat{\mathcal{C}at}_\infty$, this reduces immediately to the statement of Proposition 6.3.3.7. $\square$

6.3.4 General Limits of ∞-Topoi

Our goal in this section is to construct general limits in the ∞-category $\mathcal{RT}op$. Our strategy is necessarily rather different from that of §6.3.3 because the inclusion $i : \mathcal{RT}op \to \widehat{\mathcal{C}at}$ does not preserve limits in general. In fact, i does not even preserve the final object:

Proposition 6.3.4.1. *Let $\mathcal{X}$ be an ∞-topos. Then $\mathrm{Fun}^*(\mathcal{S}, \mathcal{X})$ is a contractible Kan complex. In particular, $\mathcal{S}$ is a final object in the ∞-category $\mathcal{RT}op$ of ∞-topoi.*

Proof. We observe that $\mathcal{S} \simeq \mathrm{Shv}(\Delta^0)$, where the ∞-category Δ^0 is endowed with the "discrete" topology (so that the empty sieve does not constitute a cover of the unique object). According to Proposition 6.2.3.20, the ∞-category $\mathrm{Fun}^*(\mathcal{S}, \mathcal{X})$ is equivalent to the full subcategory of $\mathcal{X} \simeq \mathrm{Fun}(\Delta^0, \mathcal{X})$ spanned by those objects $X \in \mathcal{X}$ which correspond to left exact functors $\Delta^0 \to \mathcal{X}$. It is clear that these are precisely the final objects of $\mathcal{X}$ and therefore form a contractible Kan complex (Proposition 1.2.12.9). $\square$

To construct limits in general, we first develop some tools for describing ∞-topoi via "generators and relations." This will allow us to reduce the construction of limits in $\mathcal{RJ}$op to the problem of constructing colimits in $\mathcal{C}at_\infty$.

Lemma 6.3.4.2. *Let $\mathcal{C}$ be a small ∞-category, let κ be a regular cardinal, and suppose we are given a (small) collection of κ-small diagrams $\{\overline{f}_\alpha : K_\alpha^{\triangleright} \to \mathcal{C}\}_{\alpha \in A}$. Then there exists a functor $F : \mathcal{C} \to \mathcal{D}$ with the following properties:*

(1) *The ∞-category $\mathcal{D}$ is small and admits κ-small colimits.*

(2) *For each $\alpha \in A$, the induced map $F \circ \overline{f}_\alpha : K_\alpha^{\triangleright} \to \mathcal{C}$ is a colimit diagram.*

(3) *Let $\mathcal{E}$ be an arbitrary ∞-category which admits κ-small colimits. Let $\mathrm{Fun}'(\mathcal{D}, \mathcal{E})$ denote the full subcategory of $\mathrm{Fun}(\mathcal{D}, \mathcal{E})$ spanned by those functors which preserve κ-small colimits. Then composition with F induces a fully faithful embedding*

$$\theta : \mathrm{Fun}'(\mathcal{D}, \mathcal{E}) \to \mathrm{Fun}(\mathcal{C}, \mathcal{E}).$$

The essential image of θ consists of those functors $F' : \mathcal{C} \to \mathcal{E}$ such that each $F' \circ \overline{f}_\alpha$ is a colimit diagram in $\mathcal{E}$.

Proof. Let $j : \mathcal{C} \to \mathcal{P}(\mathcal{C})$ denote the Yoneda embedding. For each $\alpha \in A$, let $f_\alpha = \overline{f}_\alpha | K_\alpha$ and let $C_\alpha \in \mathcal{C}$ denote the image of the cone point under $\overline{f}_\alpha$. Let $D_\alpha \in \mathcal{P}(\mathcal{C})$ denote a colimit of the induced diagram $j \circ f_\alpha$, so that $j \circ \overline{f}_\alpha$ induces a map $s_\alpha = D_\alpha \to j(C_\alpha)$. Let $S = \{s_\alpha\}_{\alpha \in A}$, let $\mathcal{X}$ denote the localization $S^{-1} \mathcal{P}(\mathcal{C})$, and let $L : \mathcal{P}(\mathcal{C}) \to \mathcal{X}$ denote a left adjoint to the inclusion. Let $\mathcal{D}'$ be the smallest full subcategory of $\mathcal{X}$ that contains the essential image of the functor $L \circ j$ and is stable under κ-small colimits, let $\mathcal{D}$ be a minimal model for $\mathcal{D}'$, and let $F : \mathcal{C} \to \mathcal{D}$ denote the composition of $L \circ j$ with a retraction of $\mathcal{D}'$ onto $\mathcal{D}$. It follows immediately from the construction that $\mathcal{D}$ satisfies conditions (1) and (2).

We observe that for each $\alpha \in A$, the domain and codomain of s_α are both κ-compact objects of $\mathcal{P}(\mathcal{C})$. It follows that $\mathcal{X}$ is stable under κ-filtered colimits in $\mathcal{P}(\mathcal{C})$. Corollary 5.5.7.3 implies that L carries κ-compact objects of $\mathcal{P}(\mathcal{C})$ to κ-compact objects of $\mathcal{X}$. Since the collection of κ-compact objects of $\mathcal{X}$ is stable under κ-small colimits, we conclude that $\mathcal{D}'$ consists of κ-compact objects of $\mathcal{X}$. Invoking Proposition 5.3.5.11, we deduce that the inclusion $\mathcal{D} \subseteq \mathcal{X}$ determines an equivalence $\mathrm{Ind}_\kappa(\mathcal{D}) \simeq \mathcal{X}$.

We now prove (3). We observe that there exists a fully faithful embedding $i : \mathcal{E} \to \mathcal{E}'$ which preserves κ-small colimits, where $\mathcal{E}'$ admits arbitrary small colimits (for example, we can take $\mathcal{E}' = \mathrm{Fun}(\mathcal{E}, \widehat{\mathcal{S}})^{op}$ and i to be the Yoneda embedding). Replacing $\mathcal{E}$ by $\mathcal{E}'$ if necessary, we may assume that $\mathcal{E}$ itself admits arbitrary small colimits. We have a homotopy commutative diagram

$$
\begin{array}{ccc}
\mathrm{Fun}^L(\mathcal{X}, \mathcal{E}) & \xrightarrow{\theta'} & \mathrm{Fun}^L(\mathcal{P}(\mathcal{C}), \mathcal{E}) \\
\downarrow & & \downarrow \\
\mathrm{Fun}'(\mathcal{D}, \mathcal{E}) & \xrightarrow{\theta} & \mathrm{Fun}(\mathcal{C}, \mathcal{E}),
\end{array}
$$

where $\mathrm{Fun}^L(\mathcal{Y}, \mathcal{E})$ denotes the full subcategory of $\mathrm{Fun}(\mathcal{Y}, \mathcal{E})$ spanned by those functors which preserve small colimits. Propositions 5.3.5.10 and 5.5.1.9 imply that the left vertical arrow is an equivalence, while Theorem 5.1.5.6 implies that the right vertical arrow is an equivalence. It will therefore suffice to show that θ' is fully faithful and that the essential image of θ' consists of those colimit-preserving functors F' from $\mathcal{P}(\mathcal{C})$ to $\mathcal{E}$ such that $F' \circ j \circ \overline{f}_\alpha$ is a colimit diagram for each $\alpha \in A$. This follows immediately from Proposition 5.5.4.20. □

Definition 6.3.4.3. Let $\mathcal{C}\mathrm{at}^{\mathrm{lex}}_\infty$ denote the subcategory of $\mathcal{C}\mathrm{at}_\infty$ defined as follows:

(1) A small ∞-category $\mathcal{C}$ belongs to $\mathcal{C}\mathrm{at}^{\mathrm{lex}}_\infty$ if and only if $\mathcal{C}$ admits finite limits.

(2) Let $f : \mathcal{C} \to \mathcal{D}$ be a functor between small ∞-categories which admit finite limits. Then f is a morphism in $\mathcal{C}\mathrm{at}^{\mathrm{lex}}_\infty$ if and only if f preserves finite limits.

Lemma 6.3.4.4. *The ∞-category $\mathcal{C}\mathrm{at}^{\mathrm{lex}}_\infty$ admits small colimits.*

Proof. Let $p : \mathcal{J} \to \mathcal{C}\mathrm{at}^{\mathrm{lex}}_\infty$ be a small diagram which carries each vertex $j \in \mathcal{J}$ to an ∞-category $\mathcal{C}_j$. Let $\mathcal{C}$ be a colimit of the diagram p in $\mathcal{C}\mathrm{at}_\infty$, and for each $j \in J$ let $\phi_j : \mathcal{C}_j \to \mathcal{C}$ be the associated functor. Consider the collection of all isomorphism classes of diagrams $\{f : K^\lhd \to \mathcal{C}\}$, where K is a finite simplicial set and the map f admits a factorization

$$K^\lhd \xrightarrow{f_0} \mathcal{C}_j \xrightarrow{\phi_j} \mathcal{C},$$

where f_0 is a limit diagram in $\mathcal{C}_j$. Invoking the dual of Lemma 6.3.4.2, we deduce the existence of a functor $F : \mathcal{C} \to \mathcal{D}$ with the following properties:

(1) The ∞-category $\mathcal{D}$ is small and admits finite limits.

(2) Each of the compositions $F \circ \phi_j$ is left exact.

(3) For every ∞-category $\mathcal{E}$ which admits finite limits, composition with F induces an equivalence from the full subcategory of $\mathrm{Fun}(\mathcal{D}, \mathcal{E})$ spanned by the left exact functors to the full subcategory of $\mathrm{Fun}(\mathcal{C}, \mathcal{E})$ spanned by those functors $F' : \mathcal{C} \to \mathcal{E}$ such that each $F' \circ \phi_j$ is left exact.

It follows that that $\mathcal{D}$ can be identified with a colimit of the diagram p in the ∞-category $\mathcal{C}\mathrm{at}^{\mathrm{lex}}_\infty$. □

Lemma 6.3.4.5. *Let $\mathcal{C}$ be a small ∞-category which admits finite limits and let $f_* : \mathcal{X} \to \mathcal{P}(\mathcal{C})$ be a geometric morphism of ∞-topoi. Then there exists a small ∞-category $\mathcal{D}$ which admits finite limits and a left exact functor $f'' : \mathcal{C} \to \mathcal{D}$ such that f_* is equivalent to the composition*

$$\mathcal{X} \xrightarrow{f'_*} \mathcal{P}(\mathcal{D}) \xrightarrow{f''_*} \mathcal{P}(\mathcal{C}),$$

where f'_* is a fully faithful geometric morphism and f''_* is given by composition with f''.

Proof. Without loss of generality, we may assume that $\mathcal{X}$ is minimal. Let f^* be a left adjoint to f_*. Choose a regular cardinal κ large enough that the composition

$$\mathcal{C} \xrightarrow{j_{\mathcal{C}}} \mathcal{P}(\mathcal{C}) \xrightarrow{f^*} \mathcal{X}$$

carries each object $C \in \mathcal{C}$ to a κ-compact object of $\mathcal{X}$. Enlarging κ if necessary, we may assume that $\mathcal{X}$ is κ-accessible and that the collection of κ-compact objects is stable under finite limits (Proposition 5.4.7.4). Let $\mathcal{D}$ be the collection of κ-compact objects of $\mathcal{X}$. The proof of Proposition 6.1.5.3 shows that the inclusion $\mathcal{D} \subseteq \mathcal{X}$ can be extended to a left exact localization functor $f'^* : \mathcal{P}(\mathcal{D}) \to \mathcal{X}$.

Using Theorem 5.1.5.6, we conclude that the composition $j_{\mathcal{D}} \circ f^* \circ j_{\mathcal{C}} : \mathcal{C} \to \mathcal{P}(\mathcal{D})$ can be extended to a colimit-preserving functor $f''^* : \mathcal{P}(\mathcal{C}) \to \mathcal{P}(\mathcal{D})$ and that $f'^* \circ f''^*$ is homotopic to f^*. Proposition 6.1.5.2 implies that f''^* is left exact. It follows that f'^* and f''^* admit right adjoints f'_* and f''_* with the desired properties. $\qquad\square$

Proposition 6.3.4.6. *The ∞-category $\mathcal{R}\mathfrak{T}op$ of ∞-topoi admits pullbacks.*

Proof. Suppose first that we are given a pullback square

$$\begin{array}{ccc} \mathcal{W} & \xrightarrow{f'_*} & \mathcal{X} \\ {\scriptstyle g'_*}\downarrow & & \downarrow{\scriptstyle g_*} \\ \mathcal{Y} & \xrightarrow{f_*} & \mathcal{Z} \end{array}$$

in the ∞-category of $\mathcal{R}\mathfrak{T}op$. We make the following observations:

(a) Suppose that $\mathcal{Z}$ is a left exact localization of another ∞-topos $\mathcal{Z}'$. Then the induced diagram

$$\begin{array}{ccc} \mathcal{W} & \longrightarrow & \mathcal{X} \\ \downarrow & & \downarrow \\ \mathcal{Y} & \longrightarrow & \mathcal{Z} \end{array}$$

is also a pullback square.

(b) Let $S^{-1}\mathcal{X}$ and $T^{-1}\mathcal{Y}$ be left exact localizations of $\mathcal{X}$ and $\mathcal{Y}$, respectively. Let U be the smallest strongly saturated collection of morphisms in $\mathcal{W}$ which contains f'^*S and g'^*T and is closed under pullbacks. Using Corollary 6.2.1.3, we deduce that U is generated by a (small) set of morphisms in $\mathcal{W}$. It follows that the diagram

$$\begin{array}{ccc} U^{-1}\mathcal{W} & \longrightarrow & S^{-1}\mathcal{X} \\ \downarrow & & \downarrow \\ T^{-1}\mathcal{Y} & \longrightarrow & \mathcal{Z} \end{array}$$

is again a pullback in $\mathcal{R}\mathfrak{T}op$.

Now suppose we are given an arbitrary diagram

$$\mathfrak{X} \xrightarrow{g_*} \mathfrak{Z} \xleftarrow{f_*} \mathfrak{Y}$$

in $\mathcal{R}\mathfrak{Top}$. We wish to prove that there exists a fiber product $\mathfrak{X} \times_{\mathfrak{Z}} \mathfrak{Y}$ in $\mathcal{R}\mathfrak{Top}$. The proof of Proposition 6.1.5.3 implies that there exists a small ∞-category $\mathcal{C}$ which admits finite limits, such that $\mathfrak{Z}$ is a left exact localization of $\mathcal{P}(\mathcal{C})$. Using (a), we can reduce to the case where $\mathfrak{Z} = \mathcal{P}(\mathcal{C})$. Using (b) and Lemma 6.3.4.5, we can reduce to the case where $\mathfrak{X} = \mathcal{P}(\mathcal{D})$ for some small ∞-category $\mathcal{D}$ which admits finite limits, and g_* is induced by composition with a left exact functor $g : \mathcal{C} \to \mathcal{D}$. Similarly, we can assume that f_* is determined by a left exact functor $f : \mathcal{C} \to \mathcal{D}'$. Using Lemma 6.3.4.4, we can form a pushout diagram

$$
\begin{array}{ccc}
\mathcal{E} & \longleftarrow & \mathcal{D} \\
\uparrow & & \uparrow{\scriptstyle g} \\
\mathcal{D}' & \xleftarrow{\;f\;} & \mathcal{C}
\end{array}
$$

in the ∞-category $\mathcal{Cat}_\infty^{\mathrm{lex}}$. Using Proposition 6.1.5.2 and Theorem 5.1.5.6, it is not difficult to see that the induced diagram

$$
\begin{array}{ccc}
\mathcal{P}(\mathcal{E}) & \longrightarrow & \mathcal{P}(\mathcal{D}) \\
\downarrow & & \downarrow{\scriptstyle g_*} \\
\mathcal{P}(\mathcal{D}') & \xrightarrow{\;f_*\;} & \mathcal{P}(\mathcal{C})
\end{array}
$$

is the desired pullback square in $\mathcal{R}\mathfrak{Top}$.   □

Corollary 6.3.4.7. *The ∞-category $\mathcal{R}\mathfrak{Top}$ admits small limits.*

Proof. Using Corollaries 4.2.3.11 and 4.4.2.4, it suffices to show that $\mathcal{R}\mathfrak{Top}$ admits filtered limits, a final object, and pullbacks. The existence of filtered limits follows from Theorem 6.3.3.1, the existence of a final object follows from Proposition 6.3.4.1, and the existence of pullbacks follows from Proposition 6.3.4.6.   □

Remark 6.3.4.8. Our construction of fiber products in $\mathcal{R}\mathfrak{Top}$ is somewhat inexplicit. We will later give a more concrete construction in the case of ordinary products; see §7.3.3.

We conclude this section by proving a companion result to Corollary 6.3.4.7. First, a few general remarks. The ∞-category $\mathcal{R}\mathfrak{Top}$ is most naturally viewed as an ∞-*bicategory* since we can also consider noninvertible natural transformations between geometric morphisms. Correspondingly, we can consider a more general theory of ∞-bicategorical limits in $\mathcal{R}\mathfrak{Top}$. While we do not want to give any precise definitions, we would like to point out that Corollary 6.3.4.7 can be generalized to show that $\mathcal{R}\mathfrak{Top}$ admits all (small) ∞-bicategorical limits. In more concrete terms, this just means that $\mathcal{R}\mathfrak{Top}$ is *cotensored* over $\mathcal{Cat}_\infty$ in the following sense:

Proposition 6.3.4.9. *Let X be an ∞-topos and let $\mathcal{D}$ be a small ∞-category. Then there exists an ∞-topos $\mathrm{Mor}(\mathcal{C}, X)$ and a functor*

$$e : \mathcal{C} \to \mathrm{Fun}_*(\mathrm{Mor}(\mathcal{C}, X), X)$$

with the following universal property:

(∗) *For every ∞-topos $\mathcal{Y}$, composition with e induces an equivalence of ∞-categories*

$$\mathrm{Fun}_*(\mathcal{Y}, \mathrm{Mor}(\mathcal{C}, X)) \to \mathrm{Fun}(\mathcal{C}, \mathrm{Fun}_*(\mathcal{Y}, X)).$$

Proof. We first treat the case where $X = \mathcal{P}(\mathcal{D})$, where $\mathcal{D}$ is a small ∞-category which admits finite limits. Using Lemma 6.3.4.2, we conclude that there exists a functor $e_0 : \mathcal{C}^{op} \times \mathcal{D} \to \mathcal{D}'$ with the following properties:

(1) The ∞-category $\mathcal{D}'$ is small and admits finite limits.

(2) For each object $C \in \mathcal{C}$, the induced functor

$$\mathcal{D} \simeq \{C\} \times \mathcal{D} \subseteq \mathcal{C}^{op} \times \mathcal{D} \overset{e_0}{\to} \mathcal{D}'$$

is left exact.

(3) Let $\mathcal{E}$ be an arbitrary ∞-category which admits finite limits. Then composition with e_0 induces an equivalence from the full subcategory of $\mathrm{Fun}(\mathcal{D}', \mathcal{E})$ spanned by the left exact functors to the full subcategory of $\mathrm{Fun}(\mathcal{C}^{op} \times \mathcal{D}, \mathcal{E})$ spanned by those functors which restrict to left exact functors $\{C\} \times \mathcal{D} \to \mathcal{E}$ for each $C \in \mathcal{C}$.

In this case, we can define $\mathrm{Mor}(\mathcal{C}, X)$ to be $\mathcal{P}(\mathcal{D}')$ and

$$e : \mathcal{C} \to \mathrm{Fun}_*(\mathcal{P}(\mathcal{D}'), \mathcal{P}(\mathcal{D}))$$

to be given by composition with e_0; the universal property (∗) follows immediately from Theorem 5.1.5.6 and Proposition 6.1.5.2.

In the general case, we invoke Proposition 6.1.5.3 to reduce to the case where $X = S^{-1} X'$ is an accessible left exact localizaton of an ∞-topos X', where $X' \simeq \mathcal{P}(\mathcal{D})$ is as above so that we can construct an ∞-topos $\mathrm{Mor}(\mathcal{C}, X')$ and a map $e' : \mathcal{C} \to \mathrm{Fun}_*(\mathrm{Mor}(\mathcal{C}, X'), X')$ satisfying the condition (∗). For each $C \in \mathcal{C}$, let $e'(C)_*$ denote the corresponding geometric morphism from $\mathrm{Mor}(\mathcal{C}, X')$ to X', let $e'(C)^*$ denote a left adjoint to $e'(C)_*$, and let $S(C) = e'(C)^* S$ be the image of S in the collection of morphisms of $\mathrm{Mor}(\mathcal{C}, X')$. Since each $e'(C)^*$ is a colimit-preserving functor, each of the sets $S(C)$ is generated under colimits by a small collection of morphisms in $\mathrm{Mor}(\mathcal{C}, X')$. Let T be the smallest collection of morphisms in $\mathrm{Mor}(\mathcal{C}, X')$ which is strongly saturated, is stable under pullbacks, and contains each of the sets S_C. Using Corollary 6.2.1.3, we conclude that T is generated (as a strongly saturated class of morphisms) by a small collection of morphisms in $\mathrm{Mor}(\mathcal{C}, X')$. It follows that $\mathrm{Mor}(\mathcal{C}, X) = T^{-1} \mathrm{Mor}(\mathcal{C}, X')$ is an ∞-topos. By construction, the map e' restricts to give a functor $e : \mathcal{C} \to \mathrm{Fun}_*(\mathrm{Mor}(\mathcal{C}, X), X)$. Unwinding the definitions, we see that e has the desired properties.   □

Remark 6.3.4.10. Let $\mathfrak{X}$ be an ∞-topos and let $\mathcal{RT}\mathrm{op}^\Delta$ denote the simplicial subcategory of $\widehat{\mathcal{C}\mathrm{at}_\infty}$ corresponding to the subcategory $\mathcal{RT}\mathrm{op} \subseteq \widehat{\mathcal{C}\mathrm{at}_\infty}$, so that $\mathcal{RT}\mathrm{op} \simeq \mathrm{N}(\mathcal{RT}\mathrm{op}^\Delta)$. The construction $\mathcal{Y} \mapsto \mathrm{Fun}_*(\mathfrak{X}, \mathcal{Y})$ determines a simplicial functor from $\mathcal{RT}\mathrm{op}^\Delta$ to $\widehat{\mathcal{C}\mathrm{at}_\infty}^\Delta$, which in turn induces a functor

$$\theta_{\mathfrak{X}} : \mathcal{RT}\mathrm{op} \to \widehat{\mathcal{C}\mathrm{at}_\infty}.$$

We claim that $\theta_{\mathfrak{X}}$ preserves small limits (this translates into the condition that limits in $\mathcal{RT}\mathrm{op}$ really give ∞-*bicategorical limits* in the ∞-bicategory of ∞-topoi).

To prove this, fix an arbitrary ∞-category $\mathcal{C}$ and let $e_{\mathcal{C}} : \widehat{\mathcal{C}\mathrm{at}_\infty} \to \widehat{\mathcal{S}}$ be the functor corepresented by $\mathcal{C}$. It will suffice to show that $e_{\mathcal{C}} \circ \theta_{\mathfrak{X}}$ preserves small limits. The collection of all ∞-categories $\mathcal{C}$ which satisfy this condition is stable under all colimits, so we may assume without loss of generality that $\mathcal{C}$ is small. It now suffices to observe that $e_{\mathcal{C}} \circ \theta_{\mathfrak{X}}$ is equivalent to the functor corepresented by the ∞-topos $\mathrm{Fun}(\mathcal{C}, \mathfrak{X})$.

6.3.5 Étale Morphisms of ∞-Topoi

Let $f : X \to Y$ be a continuous map of topological spaces. We say that f is *étale* (or a *local homeomorphism*) if, for every point $x \in X$, there exist open sets $U \subseteq X$ containing x and $V \subseteq Y$ containing $f(x)$ such that f induces a homeomorphism $U \to V$. Let $\mathcal{F}$ denote the sheaf of sections of f: that is, $\mathcal{F}$ is a sheaf of sets on Y such that for every open set $V \subseteq Y$, $\mathcal{F}(V)$ is the set of all continuous maps $s : V \to X$ such that $f \circ s = \mathrm{id} : V \to Y$. The construction $(f : X \to Y) \mapsto \mathcal{F}$ determines an equivalence of categories, from the category of topological spaces which are étale over Y to the category of sheaves (of sets) on Y. In particular, we can recover the topological space X (up to homeomorphism) from the sheaf of sets $\mathcal{F}$ on Y. For example, we can reconstruct the category $\mathrm{Shv}_{\mathrm{Set}}(X)$ of sheaves on X as the overcategory $\mathrm{Shv}_{\mathrm{Set}}(Y)_{/\mathcal{F}}$.

Our goal in this section is to develop an analogous theory of étale morphisms in the setting of ∞-topoi. Suppose we are given a geometric morphism $f_* : \mathfrak{X} \to \mathcal{Y}$. Under what circumstances should we say that f_* is étale? By analogy with the case of topological spaces, we should expect that an étale morphism determines a "sheaf" on $\mathcal{Y}$: that is, an object U of the ∞-category $\mathcal{Y}$. Moreover, we should then be able to recover the ∞-category $\mathfrak{X}$ as an overcategory $\mathcal{Y}_{/U}$. The following result guarantees that this expectation is somewhat reasonable:

Proposition 6.3.5.1. *Let $\mathfrak{X}$ be an ∞-topos and let U be an object of $\mathfrak{X}$.*

(1) *The ∞-category $\mathfrak{X}_{/U}$ is an ∞-topos.*

(2) *The projection $\pi_! : \mathfrak{X}_{/U} \to \mathfrak{X}$ has a right adjoint π^* which commutes with colimits. Consequently, π^* itself has a right adjoint $\pi_* : \mathfrak{X}_{/U} \to \mathfrak{X}$ which is a geometric morphism of ∞-topoi.*

Proof. The existence of a right adjoint π^* to the projection $\pi_! : \mathfrak{X}_{/U} \to \mathfrak{X}$ follows from the assumption that $\mathfrak{X}$ admits finite limits. Moreover, the assertion that π^* preserves colimits is a special case of the assumption that colimits in $\mathfrak{X}$ are universal. This proves (2).

To prove (1), we will show that $\mathfrak{X}_{/U}$ satisfies criterion (2) of Theorem 6.1.0.6. We first observe that $\mathfrak{X}_{/U}$ is presentable (Proposition 5.5.3.10). Let K be a small simplicial set and let $\overline{\alpha} : \overline{p} \to \overline{q}$ be a natural transformation of diagrams $\overline{p}, \overline{q} : K^{\triangleright} \to \mathfrak{X}_{/U}$. Suppose that $\overline{q}$ is a colimit diagram and that $\alpha = \overline{\alpha}|K$ is a Cartesian transformation. The projection $\pi_!$ preserves all colimits (since it is a left adjoint), so that $\pi_! \circ \overline{q}$ is a colimit diagram in $\mathfrak{X}$. Since $\pi_!$ preserves pullback squares (Proposition 4.4.2.9), $\pi_! \circ \alpha$ is a Cartesian transformation. By assumption, $\mathfrak{X}$ is an ∞-topos, so that Theorem 6.1.0.6 implies that $\pi_! \circ \overline{p}$ is a colimit diagram if and only if $\pi_! \circ \overline{\alpha}$ is a Cartesian transformation. Using Propositions 4.4.2.9 and 1.2.13.8, we conclude that $\overline{p}$ is a colimit diagram if and only if $\overline{\alpha}$ is a Cartesian transformation, as desired. $\qquad\square$

A geometric morphism $f_* : \mathfrak{X} \to \mathcal{Y}$ of ∞-topoi is said to be *étale* if it arises via the construction of Proposition 6.3.5.1; that is, if f admits a factorization

$$\mathfrak{X} \xrightarrow{f'_*} \mathcal{Y}_{/U} \xrightarrow{f''_*} \mathcal{Y},$$

where U is an object of $\mathcal{Y}$, f'_* is a categorical equivalence, and f''_* is a right adjoint to the pullback functor $f''^* : \mathcal{Y} \to \mathcal{Y}_{/U}$. We note that in this case, f^* has a *left adjoint* $f_! = f''_! \circ f'_*$. Consequently, f^* preserves *all* limits, not just finite limits.

Remark 6.3.5.2. Given an étale geometric morphism $f : \mathfrak{X}_{/U} \to \mathfrak{X}$ of ∞-topoi, the description of the pushforward functor f_* is slightly more complicated than that of $f_!$ (which is merely the forgetful functor) or f^* (which is given by taking products with U). Given an object $p : X \to U$ of $\mathfrak{X}_{/U}$, the pushforward $f_* X$ is an object of $\mathfrak{X}$ which represents the functor "sections of p."

The collection of étale geometric morphisms contains all equivalences and is stable under composition. Consequently, we can consider the subcategory $\mathcal{RTop}_{\text{ét}} \subseteq \mathcal{RTop}$ containing all objects of $\mathcal{RTop}$ whose morphisms are precisely the étale geometric morphisms. Our goal in this section is to study the ∞-category $\mathcal{RTop}_{\text{ét}}$. Our main results are the following:

(a) If $\mathfrak{X}$ is an ∞-topos containing an object U, then the associated étale geometric morphism $\pi_* : \mathfrak{X}_{/U} \to \mathfrak{X}$ can be described by a universal property. Namely, $\mathfrak{X}_{/U}$ is universal among ∞-topoi $\mathcal{Y}$ with a geometric morphism $\phi_* : \mathcal{Y} \to \mathfrak{X}$ such that $\phi^* U$ admits a global section (Proposition 6.3.5.5).

(b) There is a simple criterion for testing whether a geometric morphism $\pi_* : \mathfrak{X} \to \mathcal{Y}$ is étale. Namely, π_* is étale if and only if the functor π^*

admits a left adjoint $\pi_!$, the functor $\pi_!$ is conservative, and an appropriate push-pull formula holds in the the ∞-category $\mathcal{Y}$ (Proposition 6.3.5.11).

(c) Given a pair of topological spaces X_0 and X_1 and a homeomorphism $\phi :$ $U_0 \simeq U_1$ between open subsets $U_0 \subseteq X_0$ and $U_1 \subseteq X_1$, we can "glue" X_0 to X_1 along ϕ to obtain a new topological space $X = X_0 \coprod_{U_0} X_1$. Moreover, the topological space X contains open subsets homeomorphic to X_0 and X_1. In the setting of ∞-topoi, it is possible to make much more general "gluing" constructions of the same type. We can formulate this idea more precisely as follows: given any diagram $\{\mathcal{X}_\alpha\}$ in the ∞-category $\mathcal{R}\mathcal{T}\mathrm{op}_{\text{ét}}$ having a colimit $\mathcal{X}$ in $\mathcal{R}\mathcal{T}\mathrm{op}$, each of the associated geometric morphisms $\mathcal{X}_\alpha \to \mathcal{X}$ is étale (Theorem 6.3.5.13). Using this fact, we will show that the ∞-category $\mathcal{R}\mathcal{T}\mathrm{op}_{\text{ét}}$ admits small colimits.

Remark 6.3.5.3. We will say that a geometric morphism of ∞-topoi $f^* :$ $\mathcal{Y} \to \mathcal{X}$ is étale if and only if its right adjoint $f_* : \mathcal{X} \to \mathcal{Y}$ is étale. We let $\mathcal{L}\mathcal{T}\mathrm{op}_{\text{ét}}$ denote the subcategory of $\mathcal{L}\mathcal{T}\mathrm{op}$ spanned by the étale geometric morphisms, so that there is a canonical equivalence $\mathcal{R}\mathcal{T}\mathrm{op}_{\text{ét}} \simeq \mathcal{L}\mathcal{T}\mathrm{op}_{\text{ét}}^{op}$.

Our first step is to obtain a more precise formulation of the universal property described in (a):

Definition 6.3.5.4. Let $f^* : \mathcal{X} \to \mathcal{Y}$ be a geometric morphism of ∞-topoi. Let U be an object of $\mathcal{X}$ and $\alpha : 1_{\mathcal{Y}} \to f^*U$ a morphism in $\mathcal{Y}$, where $1_{\mathcal{Y}}$ denotes a final object of $\mathcal{Y}$. We will say that α *exhibits $\mathcal{Y}$ as a classifying ∞-topos for sections of U* if, for every ∞-topos $\mathcal{Z}$, the diagram

$$
\begin{array}{ccc}
\mathrm{Fun}^*(\mathcal{Y}, \mathcal{Z}) & \xrightarrow{\circ f^*} & \mathrm{Fun}^*(\mathcal{X}, \mathcal{Z}) \\
\downarrow{\scriptstyle \phi} & & \downarrow{\scriptstyle \phi_0} \\
\mathcal{Z}_* & \longrightarrow & \mathcal{Z}
\end{array}
$$

is a homotopy pullback square of ∞-categories. Here $\mathcal{Z}_*$ denotes the ∞-category of pointed objects of $\mathcal{Z}$ (that is, the full subcategory of $\mathrm{Fun}(\Delta^1, \mathcal{Z})$ spanned by morphisms $f : Z \to Z'$, where Z is a final object of $\mathcal{Z}$), and the morphisms ϕ and ϕ_0 are given by evaluation on α and U, respectively.

Let $\mathcal{X}$ be an ∞-topos containing an object U. It follows immediately from the definition that a classifying ∞-topos for sections of U is uniquely determined up to equivalence provided that it exists. For the existence, we have the following result:

Proposition 6.3.5.5. *Let $\mathcal{X}$ be an ∞-topos containing an object U, let $\pi_! : \mathcal{X}_{/U} \to \mathcal{X}$ be the projection map and let $\pi^* : \mathcal{X} \to \mathcal{X}_{/U}$ be a right adjoint to $\pi_!$. Let 1_U denote the identity map from U to itself, regarded as a (final) object of $\mathcal{X}_{/U}$, and let $\alpha : 1_U \to \pi^*U$ be adjoint to the identity map $\pi_! 1_U \simeq U$. Then α exhibits $\mathcal{X}_{/U}$ as a classifying ∞-topos for sections of U.*

Before giving the proof of Proposition 6.3.5.5, we summarize some of the pleasant consequences.

Corollary 6.3.5.6. *Let $\mathfrak{X}$ be an ∞-topos containing an object U, and let $\pi^* : \mathfrak{X} \to \mathfrak{X}_{/U}$ be the corresponding étale geometric morphism. For every ∞-topos $\mathfrak{Z}$, composition with π^* induces a left fibration*

$$\mathrm{Fun}^*(\mathfrak{X}_{/U}, \mathfrak{Z}) \to \mathrm{Fun}^*(\mathfrak{X}, \mathfrak{Z}).$$

Moreover, the fiber over a geometric morphism $\phi^* : \mathfrak{X} \to \mathfrak{Z}$ is homotopy equivalent to the mapping space $\mathrm{Map}_{\mathfrak{Z}}(1_{\mathfrak{Z}}, \phi^* U)$.

Remark 6.3.5.7. Corollary 6.3.5.6 implies, in particular, the existence of homotopy fiber sequences

$$\mathrm{Map}_{\mathfrak{Z}}(1_{\mathfrak{Z}}, \phi^* U) \to \mathrm{Map}_{\mathcal{L}\mathfrak{Top}}(\mathfrak{X}_{/U}, \mathfrak{Z}) \to \mathrm{Map}_{\mathcal{L}\mathfrak{Top}}(\mathfrak{X}, \mathfrak{Z})$$

(where the fiber is taken over a geometric morphism $\phi^* \in \mathrm{Map}_{\mathcal{L}\mathfrak{Top}}(\mathfrak{X}, \mathfrak{Z})$).

Suppose that $\mathfrak{Z} = \mathfrak{X}_{/V}$ and that ϕ^* is a right adjoint to the projection $\mathfrak{X}_{/V} \to \mathfrak{Z}$. We then deduce the existence of a canonical homotopy equivalence

$$\mathrm{Map}_{\mathfrak{X}}(V, U) \simeq \mathrm{Map}_{\mathfrak{Z}}(1_{\mathfrak{Z}}, \phi^* U) \simeq \mathrm{Map}_{\mathcal{L}\mathfrak{Top}_{\mathfrak{X}/}}(\mathfrak{X}_{/U}, \mathfrak{X}_{/V}).$$

Remark 6.3.5.8. It follows from Remark 6.3.5.7 that if $f^* : \mathfrak{X} \to \mathfrak{Y}$ is a geometric morphism of ∞-topoi and $U \in \mathfrak{X}$ is an object, then the induced diagram

$$\begin{array}{ccc} \mathfrak{X} & \longrightarrow & \mathfrak{Y} \\ \downarrow & & \downarrow \\ \mathfrak{X}_{/U} & \longrightarrow & \mathfrak{Y}_{/f^* U} \end{array}$$

is a pushout square in $\mathcal{L}\mathfrak{Top}$.

Corollary 6.3.5.9. *Suppose we are given a commutative diagram*

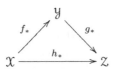

in $\mathcal{L}\mathfrak{Top}^{op}$, where g_* is étale. Then f_* is étale if and only if h_* is étale.

Proof. The "only if" direction is obvious. To prove the converse, let us suppose that g_* and h_* are both étale, so that we have equivalences $\mathfrak{X} \simeq \mathfrak{Z}_{/U}$ and $\mathfrak{U} \simeq \mathfrak{Z}_{/V}$ for some pair of objects $U, V \in \mathfrak{Z}$. Using Remark 6.3.5.7, we deduce that the morphism f_* is determined by a map $U \to V$ in $\mathfrak{Z}$, which we can identify with an object $\overline{V} \in \mathfrak{Y}$ such that $\mathfrak{X} \simeq \mathfrak{Y}_{/\overline{V}}$.   □

Remark 6.3.5.10. Let $\mathfrak{X}$ be an ∞-topos. The projection map

$$p : \mathrm{Fun}(\Delta^1, \mathfrak{X}) \to \mathrm{Fun}(\{1\}, \mathfrak{X}) \simeq \mathfrak{X}$$

is a Cartesian fibration. Moreover, for every morphism $\alpha : U \to V$ in $\mathcal{X}$, the associated functor $\alpha^* : \mathcal{X}^{/V} \to \mathcal{X}^{/U}$ is an étale geometric morphism of ∞-topoi, so that p is classified by a functor $\chi_0 : \mathcal{X}^{op} \to \mathcal{L}\mathcal{T}\mathrm{op}_{\text{ét}}$. The functor χ_0 carries the final object of $\mathcal{X}$ to an ∞-topos equivalent to $\mathcal{X}$ and therefore factors as a composition

$$\mathcal{X}^{op} \xrightarrow{\chi} (\mathcal{L}\mathcal{T}\mathrm{op}_{\text{ét}})_{\mathcal{X}/} \to \mathcal{L}\mathcal{T}\mathrm{op}_{\text{ét}} \, .$$

The argument of Remark 6.3.5.7 shows that χ is fully faithful, and it follows immediately from the definitions that χ is essentially surjective. Corollary 6.3.5.9 allows us to identify $(\mathcal{L}\mathcal{T}\mathrm{op}_{\text{ét}})_{\mathcal{X}/}$ with the full subcategory of $\mathcal{L}\mathcal{T}\mathrm{op}_{\mathcal{X}/}$ spanned by the étale geometric morphisms $f^* : \mathcal{X} \to \mathcal{Y}$. Consequently, we can regard χ as a fully faithful embedding of $\mathcal{X}$ into the ∞-category $(\mathcal{L}\mathcal{T}\mathrm{op}^{op})_{/\mathcal{X}}$ of ∞-topoi over $\mathcal{X}$, whose essential image consists of those ∞-topoi which are étale over $\mathcal{X}$.

Proof of Proposition 6.3.5.5. Let $p : \mathcal{M} \to \Delta^1$ be a correspondence from $\mathcal{X}_{/U} \simeq \mathcal{M} \times_{\Delta^1} \{0\}$ to $\mathcal{X} \simeq \mathcal{M} \times_{\Delta^1} \{1\}$ associated to the adjoint functors

$$\mathcal{X}_{/U} \underset{\pi^*}{\overset{\pi_!}{\rightleftarrows}} \mathcal{X} \, .$$

Let α_0 be a morphism from $1_U \in \mathcal{X}_{/U}$ to $1_{\mathcal{X}} \in \mathcal{X}$ in $\mathcal{M}$ (so that α_0 is determined uniquely up to homotopy). We observe that there is a retraction $r : \mathcal{M} \to \mathcal{X}_{/U}$ which restricts to π^* on $\mathcal{X} \subseteq \mathcal{M}$, and we can identify α with $r(\alpha_0)$.

Let $\mathcal{Z}$ be an arbitrary ∞-topos. Let $\mathcal{C}$ be the full subcategory of $\mathrm{Fun}(\mathcal{M}, \mathcal{Z})$ spanned by those functors $F : \mathcal{M} \to \mathcal{Z}$ with the following properties:

(a) The restriction $F | \mathcal{X}_{/U} : \mathcal{X}_{/U} \to \mathcal{Z}$ preserves small colimits and finite limits.

(b) The functor F is a left Kan extension of $F | \mathcal{X}_{/U}$. In other words, F carries p-Cartesian morphisms in $\mathcal{M}$ to equivalences in $\mathcal{Z}$.

Proposition 4.3.2.15 implies that the restriction map $\mathcal{C} \to \mathrm{Fun}^*(\mathcal{X}_{/U}, \mathcal{Z})$ is a trivial Kan fibration. Moreover, this trivial Kan fibration has a section given by composition with r. It will therefore suffice to show that the diagram

$$
\begin{array}{ccc}
\mathcal{C} & \longrightarrow & \mathrm{Fun}^*(\mathcal{X}, \mathcal{Z}) \\
\downarrow & & \downarrow \\
\mathcal{Z}_* & \longrightarrow & \mathcal{Z}
\end{array}
$$

is a homotopy pullback square. In other words, we wish to show that restriction along α_0 and the inclusion $\mathcal{X} \subseteq \mathcal{M}$ induce a categorical equivalence $\mathcal{C} \to \mathcal{Z}_* \times_{\mathcal{Z}} \mathrm{Fun}^*(\mathcal{X}, \mathcal{Z})$.

We define simplicial subsets $\mathcal{M}'' \subseteq \mathcal{M}' \subseteq \mathcal{M}$ as follows:

(i) Let $\mathcal{M}'' \simeq \mathcal{X} \coprod_{\{1\}} \Delta^1$ be the union of $\mathcal{X}$ with the 1-simplex of $\mathcal{M}$ corresponding to the morphism α_0.

(ii) Let $\mathcal{M}'$ be the full subcategory of $\mathcal{M}$ spanned by $\mathcal{X}$ together with the object 1_U.

We can identify $\mathcal{Z}_* \times_{\mathcal{Z}} \operatorname{Fun}^*(\mathcal{X}, \mathcal{Z})$ with the full subcategory

$$\mathcal{C}'' \subseteq \operatorname{Fun}(\mathcal{M}'', \mathcal{Z})$$

spanned by those functors F satisfying the following conditions:

(a') The restriction $F|\mathcal{X}$ preserves small colimits and finite limits.

(b') The object $F(1_U)$ is final in $\mathcal{Z}$.

Let $\mathcal{C}'$ be the full subcategory of $\operatorname{Fun}(\mathcal{M}', \mathcal{Z})$ spanned by those functors which satisfy (a') and (b'). To complete the proof, it will suffice to show that the restriction maps

$$\mathcal{C} \xrightarrow{u} \mathcal{C}' \xrightarrow{v} \mathcal{C}''$$

are trivial Kan fibrations.

We first show that u is a trivial Kan fibration. In view of Proposition 4.3.2.15, it will suffice to prove the following:

$(*)$ A functor $F : \mathcal{M} \to \mathcal{Z}$ satisfies (a) and (b) if and only if it satisfies (a') and (b') and F is a right Kan extension of $F|\mathcal{M}'$.

To prove the "only if" direction, let us suppose that F satisfies (a) and (b). Without loss of generality, we may suppose $F = F_0 \circ r$, where $F_0 = F|\mathcal{X}_{/U}$. Then $F|\mathcal{X} = F_0 \circ \pi^*$. Since F_0 and π^* both preserve small colimits and finite limits, we deduce (a'). Condition (b') is an immediate consequence of (a). We must show that F is a right Kan extension of $F|\mathcal{X}$. Unwinding the definitions (and applying Corollary 4.1.3.1), we are reduced to showing that for every object $\overline{V} \in \mathcal{X}_{/U}$ corresponding to a morphism $V \to U$ in $\mathcal{X}$, the diagram

$$
\begin{array}{ccc}
F(\overline{V}) & \longrightarrow & F(V) \\
\downarrow & & \downarrow \\
F(1_U) & \longrightarrow & F(U)
\end{array}
$$

is a pullback square in $\mathcal{Z}$. Since $F = F_0 \circ r$ and F_0 preserves finite limits, it suffices to show that the square

$$
\begin{array}{ccc}
\overline{V} & \longrightarrow & \pi^* V \\
\downarrow & & \downarrow \\
1_U & \longrightarrow & \pi^* U
\end{array}
$$

is a pullback square in $\mathcal{X}_{/U}$. In view of Proposition 1.2.13.8, it suffices to observe that the square

$$
\begin{array}{ccc}
V & \longrightarrow & V \times U \\
\downarrow & & \downarrow \\
U & \longrightarrow & U \times U
\end{array}
$$

is a pullback in $\mathfrak{X}$.

Now let us suppose that F is a right Kan extension of $F_1 = F | \mathcal{M}'$ and that F_1 satisfies conditions (a') and (b'). We first claim that F satisfies (b). In other words, we claim that for every object $V \in \mathfrak{X}$, the canonical map $F(\pi^* V) \to F(V)$ is an equivalence. Consider the diagram

$$\begin{array}{ccc}
F(\pi^* V) & \longrightarrow & F(V \times U) & \longrightarrow & F(V) \\
\downarrow & & \downarrow & & \downarrow \\
F(1_U) & \longrightarrow & F(U) & \longrightarrow & F(1_{\mathfrak{X}}).
\end{array}$$

Since F is a right Kan extension of F_1, the left square is a pullback. Since F_1 satisfies (a), the right square is a pullback. Therefore the outer square is a pullback. Condition (b') implies that the lower horizontal composition is an equivalence, so the upper horizontal composition is an equivalence as well.

We now prove that F satisfies (a). Condition (b') implies that the functor $F_0 = F | \mathfrak{X}_{/U}$ preserves final objects. It will therefore suffice to show that F_0 preserves small colimits and pullback squares. Since F is a right Kan extension of F_1, the functor F_0 can be described by the formula

$$V \mapsto F(\pi_! V) \times_{F(U)} F(1_U).$$

It therefore suffices to show that the functors $\pi_!$, $F | \mathfrak{X}$, and $\bullet \times_{F(U)} F(1_U)$ preserve small colimits and pullback squares. For $\pi_!$, this follows from Propositions 1.2.13.8 and 4.4.2.9. For $F | \mathfrak{X}$, we invoke assumption (a'). For the functor $\bullet \times_{F(U)} F(1_U)$, we invoke our assumption that $\mathfrak{Z}$ is an ∞-topos (so that colimits in $\mathfrak{Z}$ are universal). This completes the verification that u is a trivial Kan fibration.

To complete the proof, we must show that the functor v is a trivial Kan fibration. We note that v fits into a pullback diagram

$$\begin{array}{ccc}
\mathcal{C}' & \xrightarrow{\ v\ } & \mathcal{C}'' \\
\downarrow & & \downarrow \\
\mathrm{Fun}(\mathcal{M}', \mathfrak{Z}) & \xrightarrow{\ v'\ } & \mathrm{Fun}(\mathcal{M}'', \mathfrak{Z}).
\end{array}$$

It will therefore suffice to show that v' is a trivial Kan fibration. Since $\mathfrak{Z}$ is an ∞-category, we need only show that the inclusion $\mathcal{M}'' \subseteq \mathcal{M}'$ is a categorical equivalence of simplicial sets. This is a special case of Proposition 3.2.2.7. $\square$

We next establish the recognition principle promised in (b):

Proposition 6.3.5.11. *Let $f^* : \mathfrak{X} \to \mathfrak{Y}$ be a geometric morphism of ∞-topoi. Then f^* is étale if and only if the following conditions are satisfied:*

(1) *The functor f^* admits a left adjoint $f_!$ (in view of Corollary 5.5.2.9, this is equivalent to the assumption that f^* preserves small limits).*

(2) *The functor $f_!$ is conservative. That is, if α is a morphism in $\mathfrak{Y}$ such that $f_! \alpha$ is an equivalence in $\mathfrak{X}$, then α is an equivalence in $\mathfrak{Y}$.*

(3) *For every morphism $X \to Y$ in $\mathcal{X}$, every object $Z \in \mathcal{Y}$, and every morphism $f_! Z \to Y$, the induced diagram*

$$f_!(f^* X \times_{f^* Y} Z) \longrightarrow f_! Z$$

$$\downarrow \qquad\qquad\qquad \downarrow$$

$$X \longrightarrow Y$$

is a pullback square in $\mathcal{X}$.

Remark 6.3.5.12. Condition (3) of Proposition 6.3.5.11 can be regarded as a push-pull formula: it provides a canonical equivalence

$$f_!(f^* X \times_{f^* Y} Z) \simeq X \times_Y f_! Z.$$

In particular, when Y is final in $\mathcal{X}$, we have an equivalence $f_!(f^* X \times Z) \simeq X \times f_! Z$: in other words, the functor $f_!$ is "linear" with respect to the action of $\mathcal{X}$ on $\mathcal{Y}$.

Proof of Proposition 6.3.5.11. Suppose first that f^* is an étale geometric morphism. Without loss of generality, we may suppose that $\mathcal{Y} = \mathcal{X}_{/U}$ and that f^* is right adjoint to the forgetful functor $f_! : \mathcal{X}_{/U} \to \mathcal{X}$. Assertions (1) and (2) are obvious, and assertion (3) follows from the observation that, for every diagram

$$X \to Y \leftarrow Z \to U,$$

the induced map $(X \times U) \times_{Y \times U} Z \to X \times_Y Z$ is an equivalence in $\mathcal{X}$.

For the converse, let us suppose that (1), (2), and (3) are satisfied. We wish to show that f^* is étale. Let $U = f_! 1_{\mathcal{Y}}$. Let F denote the composition $\mathcal{Y} \simeq \mathcal{Y}_{/1_{\mathcal{Y}}} \xrightarrow{f_!} \mathcal{X}_{/U}$. To complete the proof, it will suffice to show that F is an equivalence of ∞-categories. Proposition 5.2.5.1 implies that F admits a right adjoint G given by the formula

$$(X \to U) \mapsto f^* X \times_{f^* U} 1_{\mathcal{Y}}.$$

Assumption (3) guarantees that the counit map $v : FG \to \mathrm{id}_{\mathcal{X}_{/U}}$ is an equivalence. To complete the proof, it suffices to show that for each $Y \in \mathcal{Y}$, the unit map $u_Y : Y \to GFY$ is an equivalence. The map $Fu_Y : FY \to FGFY$ has a left homotopy inverse (given by v_{FY}) which is an equivalence, so that Fu_Y is an equivalence. It follows that $f_! u_Y$ is an equivalence, so that u_Y is an equivalence by virtue of assumption (2). Thus G is a homotopy inverse to F, so that F is an equivalence of ∞-categories, as desired. $\qquad\square$

Our final goal in this section is to prove the following result:

Theorem 6.3.5.13. *The ∞-category $\mathcal{RT}op_{\acute{e}t}$ admits small colimits, and the inclusion $\mathcal{RT}op_{\acute{e}t} \subseteq \mathcal{RT}op$ preserves small colimits.*

The proof of Theorem 6.3.5.13 is rather technical and will occupy our attention for the remainder of this section. However, the analogous result is elementary if we work with ∞-topoi which are assumed to be étale over a fixed base $\mathcal{X}$. In this case, Theorem 6.3.5.13 can be reduced (with the aid of Remark 6.3.5.10) to the following assertion:

Proposition 6.3.5.14. *Let* X *be an* ∞*-topos and let* $\chi : X \to \mathcal{L}\mathfrak{Top}^{op}_{/X}$ *be the functor of Remark 6.3.5.10. Then* χ *preserves small colimits.*

Proof. Combine Propositions 1.2.13.8 and 6.3.2.3 with Theorem 6.1.3.9. □

Proof of Theorem 6.3.5.13. As a first step, we establish the following:

(∗) Suppose we are given a small diagram $p : K \to \mathcal{L}\mathfrak{Top}^{op}$ and a geometric morphism of ∞-topoi $\phi_* : \varinjlim(p) \to \mathcal{Y}$. Suppose further that for each vertex v in K, the induced geometric morphism $\phi(v)_* : p(v) \to \mathcal{Y}$ is étale. Then ϕ_* is étale.

To prove (∗), we note that ϕ_* determines a functor $\overline{p} : K \to \mathcal{L}\mathfrak{Top}^{op}_{/\mathcal{Y}}$ lifting p. Since each $\phi(v)_*$ is étale, Remark 6.3.5.10 implies that $\overline{p}$ factors as a composition

$$K \xrightarrow{q} \mathcal{Y} \xrightarrow{\chi} \mathcal{L}\mathfrak{Top}^{op}_{/\mathcal{Y}} .$$

Let $U \in \mathcal{Y}$ be a colimit of the diagram q. Then Corollary 6.3.5.9 implies that $\varinjlim(p) \simeq \mathcal{Y}_{/U}$, so that ϕ_* is étale, as desired.

We now return to the proof of Theorem 6.3.5.13. Using Proposition 4.4.3.2 and its proof, we can reduce the proof to the following special cases:

(a) The ∞-category $\mathcal{L}\mathfrak{Top}^{op}_{\text{ét}}$ admits small coproducts, and the inclusion $\mathcal{L}\mathfrak{Top}^{op}_{\text{ét}} \subseteq \mathcal{L}\mathfrak{Top}^{op}$ preserves small coproducts.

(b) The ∞-category $\mathcal{L}\mathfrak{Top}^{op}_{\text{ét}}$ admits coequalizers, and the inclusion

$$\mathcal{L}\mathfrak{Top}^{op}_{\text{ét}} \subseteq \mathcal{L}\mathfrak{Top}^{op}$$

preserves coequalizer diagrams.

We first prove (a). In view of (∗), it will suffice to prove the following:

(a′) Let $\{X_\alpha\}$ be a small collection of ∞-topoi and let X be their co-product in $\mathcal{L}\mathfrak{Top}^{op}$ (so that we have an equivalence of ∞-categories $X \simeq \prod_\alpha X_\alpha$). Then each of the associated geometric morphisms $\phi^*_\alpha : X \to X_\alpha$ is étale.

To prove (a′), we may assume without loss of generality that $X = \prod_\alpha X_\alpha$ and that ϕ^*_α is given by projection onto the corresponding factor. The desired result then follows from the criterion of Proposition 6.3.5.11 (more concretely: let let $U \in X$ be an object whose image in X_α is a final object $U_\alpha \in X_\alpha$ and whose image in X_β is an initial object $U_\beta \in X_\beta$ for $\beta \neq \alpha$; then $X_{/U} \simeq \prod_\beta (X_\beta)_{/U_\beta} \simeq X_\alpha$.)

To prove (b), we can again invoke (∗) to reduce to the following assertion:

(b′) Suppose we are given a diagram

$$\mathcal{Y} \rightrightarrows X_0$$

in $\mathcal{L}\mathfrak{Top}^{op}_{\text{ét}}$ having colimit X in $\mathcal{L}\mathfrak{Top}^{op}$. Then the induced geometric morphism $\phi_* : X_0 \to X$ is étale.

To prove (b'), we identify the diagram in question with a functor $p :$ $\mathfrak{I} \to \mathcal{LT}op^{op}$ and $\mathfrak{I}$ with the subcategory of $N(\mathbf{\Delta})^{op}$ spanned by the objects $\{[0], [1]\}$ and injective maps of linearly ordered sets. Let $\mathfrak{X}_\bullet$ be the simplicial object of $\mathcal{LT}op^{op}$ given by left Kan extension along the inclusion $\mathfrak{I} \subseteq N(\mathbf{\Delta})^{op}$, so that each $\mathfrak{X}_n$ is equivalent to a coproduct (in $\mathcal{LT}op^{op}$) of $\mathfrak{X}_0$ with n copies of $\mathcal{Y}$. Using $(*)$ and Corollary 6.3.5.9, we deduce that $\mathfrak{X}_\bullet$ is a simplicial object in $\mathcal{LT}op_{\acute{e}t}^{op}$. Consequently, assertion (b') is an immediate consequence of Lemma 4.3.2.7 and the following:

(b'') Let $\mathfrak{X}_\bullet$ be a simplicial object of $\mathcal{LT}op_{\acute{e}t}^{op}$ and let $\mathfrak{X}$ be its geometric realization in $\mathcal{LT}op^{op}$. Then the induced geometric morphism $\phi_* : \mathfrak{X}_0 \to \mathfrak{X}$ is étale.

The proof of (b'') is based on the following lemma whose proof we defer until the end of this section:

Lemma 6.3.5.15. *Suppose we are given a simplicial object* $\mathfrak{X}_\bullet$ *in* $\mathcal{LT}op_{\acute{e}t}^{op}$. *Then there exists a morphism of simplicial objects* $\mathfrak{X}_\bullet \to \mathfrak{X}_\bullet'$ *of* $\mathcal{LT}op_{\acute{e}t}^{op}$ *with the following properties:*

(1) *The induced map* $\mathfrak{X}_0 \to \mathfrak{X}_0'$ *is an equivalence of* ∞*-topoi.*

(2) *The simplicial object* $\mathfrak{X}_\bullet'$ *is a groupoid object in* $\mathcal{LT}op^{op}$.

(3) *The induced map of geometric realizations (in* $\mathcal{LT}op^{op}$*) is an equivalence* $|\mathfrak{X}_\bullet| \to |\mathfrak{X}_\bullet'|$.

Using Lemma 6.3.5.15, we can reduce the proof of (b'') to the special case where $\mathfrak{X}_\bullet$ is a groupoid object of $\mathcal{LT}op^{op}$. The diagram

$$N(\mathbf{\Delta})^{op} \xrightarrow{\mathfrak{X}_\bullet} \mathcal{LT}op_{\acute{e}t}^{op} \subseteq \widehat{\mathcal{C}at}_\infty^{op}$$

is classified by a Cartesian fibration $q : \mathcal{Z} \to N(\mathbf{\Delta})^{op}$. Here we can identify $\mathfrak{X}_n$ with the fiber $\mathcal{Z}_{[n]} = \mathcal{Z} \times_{N(\mathbf{\Delta})^{op}} \{[n]\}$, and every map of linearly ordered sets $\alpha : [m] \to [n]$ induces a geometric morphism $\alpha^* : \mathcal{Z}_{[m]} \to \mathcal{Z}_{[n]}$. Since the geometric morphism α^* is étale, it admits a left adjoint $\alpha_!$, so that q is also a coCartesian fibration (Corollary 5.2.2.4).

It follows from Propositions 6.3.2.3 and 3.3.3.1 that we can identify $\mathfrak{X}$ with the full subcategory of $\operatorname{Fun}_{N(\mathbf{\Delta})^{op}}(N(\mathbf{\Delta})^{op}, \mathcal{Z})$ spanned by the Cartesian sections of q; under this identification, the pullback functor ϕ^* corresponds to the functor $\mathfrak{X} \to \mathcal{Z}_{[0]} \simeq \mathfrak{X}_0$ given by evaluation at $[0]$.

Let $1_{\mathfrak{X}}$ denote a final object of $\mathfrak{X}$, which we regard as a section of q. Let $T : N(\mathbf{\Delta})^{op} \to N(\mathbf{\Delta})^{op}$ denote the shift functor $[n] \mapsto [n] \star [0]$ and let $\beta_0 : T \to \operatorname{id}_{N(\mathbf{\Delta})^{op}}$ denote the evident natural transformation. Let $\beta : (1_{\mathfrak{X}} \circ T) \to U_\bullet$ be a natural transformation in $\operatorname{Fun}(N(\mathbf{\Delta})^{op}, \mathcal{Z})$ lifting β which is q-coCartesian. Since $\mathfrak{X}_\bullet$ is a groupoid object of $\mathcal{LT}op_{\acute{e}t}^{op}$, we deduce that $U_\bullet$ is a Cartesian section of q, which we can identify with an object of $\mathfrak{X}$.

Let $S = N(\mathbf{\Delta})^{op} \times \Delta^1$, so that β_0 defines a map $S \to N(\mathbf{\Delta})^{op}$. Let $\mathcal{Z}' = \mathcal{Z} \times_{N(\mathbf{\Delta})^{op}} S$ and let $\beta_S = \beta$, regarded as a section of the projection q':

$\mathcal{Z}' \to S$. Let $\mathcal{Z}'' = \mathcal{Z}'^{/\beta_S}$ (see §4.2.2 for an explanation of this notation). Let $q'' : \mathcal{Z}'' \to S$. The fibers of q'' can be described as follows:

- The fiber of q'' over $([n], 0)$ can be identified with $\mathcal{Z}_{[n+1]}^{/1_{\mathcal{Z}_{[n+1]}}} \simeq \mathcal{Z}_{[n+1]}$.

- The fiber of q'' over $([n], 1)$ can be identified with $\mathcal{Z}_{[n]}^{/U_n} \simeq \mathcal{Z}_{[n+1]}$.

Proposition 4.2.2.4 implies that the projection $q'' : \mathcal{Z}'' \to S$ is a coCartesian fibration classified by a map $\chi : S \to \widehat{\mathcal{C}at}_\infty$. The above description shows that χ can be regarded as an equivalence from $\chi^0 = \chi | \mathrm{N}(\mathbf{\Delta})^{op} \times \{0\}$ to $\chi^1 = \chi | \mathrm{N}(\mathbf{\Delta})^{op} \times \{1\}$ in the ∞-category of simplicial objects of $\widehat{\mathcal{C}at}_\infty$. Moreover, the functor χ^0 classifies the pullback of the coCartesian fibration q by the translation map $T : \mathrm{N}(\mathbf{\Delta})^{op} \to \mathrm{N}(\mathbf{\Delta}^{op})$, so that χ^0 and χ^1 factor through $\mathcal{L}\mathcal{T}\mathrm{op}_{\mathrm{\acute{e}t}}^{op}$. Lemma 6.1.3.17 implies that the colimit of χ^0 (hence also of χ^1) in $\mathcal{L}\mathcal{T}\mathrm{op}^{op}$ is canonically equivalent to $\mathcal{X}_0$. On the other hand, Propositions 3.3.3.1 and 6.3.2.3 allow us to identify $\varinjlim(\chi^1)$ with the ∞-category of Cartesian sections of the projection $\mathcal{Z}'' \times_S (\mathrm{N}(\mathbf{\Delta})^{op} \times \{1\}) \to \mathrm{N}(\mathbf{\Delta})^{op}$, which is isomorphic to $\mathcal{X}^{/U_\bullet}$ as a simplicial set. We now complete the proof by observing that the resulting identification $\mathcal{X}_0 \simeq \mathcal{X}^{/U_\bullet}$ is compatible with the projection $\phi_* : \mathcal{X}_0 \to \mathcal{X}$. $\qquad\square$

The remainder of this section is devoted to the proof of Lemma 6.3.5.15. We first need to introduce a bit of notation. We begin with a few remarks about the behavior of ∞-topoi under change of universe.

Notation 6.3.5.16. Let $\mathcal{X}$ be an ∞-topos and $\mathcal{C}$ an arbitrary ∞-category. We let $\mathrm{Shv}_{\mathcal{C}}(\mathcal{X})$ denote the full subcategory of $\mathrm{Fun}(\mathcal{X}^{op}, \mathcal{C})$ spanned by those functors which preserve small limits. We will refer to $\mathrm{Shv}_{\mathcal{C}}(\mathcal{X})$ as the ∞-*category of $\mathcal{C}$-valued sheaves on $\mathcal{X}$*.

Remark 6.3.5.17. Let $\mathcal{X}$ be an ∞-topos. Proposition 5.5.2.2 implies that the Yoneda embedding $\mathcal{X} \to \mathrm{Shv}_{\mathcal{S}}(\mathcal{X})$ is an equivalence; in other words, we can identify $\mathcal{X}$ with the ∞-category of sheaves of (small) spaces on itself. Let $\widehat{\mathcal{S}}$ denote the ∞-category of spaces which belong to some larger universe $\mathcal{U}$. We claim the following:

(a) The ∞-category $\mathrm{Shv}_{\widehat{\mathcal{S}}}(\mathcal{X})$ can be regarded as an ∞-topos in $\mathcal{U}$.

(b) The inclusion $\mathrm{Shv}_{\mathcal{S}}(\mathcal{X}) \subseteq \mathrm{Shv}_{\widehat{\mathcal{S}}}(\mathcal{X})$ preserves small colimits.

To prove (a), let us suppose that $\mathcal{X} = S^{-1}\mathcal{P}(\mathcal{C})$, where $\mathcal{C}$ is a small ∞-category and S is a strongly saturated class of morphisms in $\mathcal{P}(\mathcal{C})$ which is stable under pullbacks and of small generation. Theorem 5.1.5.6 and Proposition 5.5.4.20 allow us to identify $\mathrm{Shv}_{\widehat{\mathcal{S}}}(\mathcal{X})$ with $S^{-1}\widehat{\mathcal{P}}(\mathcal{C})$, where $\widehat{\mathcal{P}}(\mathcal{C})$ denotes the presheaf ∞-category $\mathrm{Fun}(\mathcal{C}^{op}, \widehat{\mathcal{S}})$. Let $\widehat{S}$ denote the strongly saturated class of morphisms of $\widehat{\mathcal{P}}(\mathcal{C})$ generated by S. Then $\widehat{S}$ is of small generation (and therefore of $\mathcal{U}$-small generation); to complete the proof of (a) it will suffice to show that $\widehat{S}$ is stable under pullbacks.

Let $\widehat{\mathcal{P}}(\mathcal{C})^0$ denote the full subcategory of $\widehat{\mathcal{P}}(\mathcal{C})$ spanned by those objects X with the following property:

(∗) Let

$$
\begin{array}{ccc}
Y & \longrightarrow & Y' \\
\downarrow{\scriptstyle f} & & \downarrow{\scriptstyle f'} \\
X & \longrightarrow & X'
\end{array}
$$

be a pullback diagram in $\widehat{\mathcal{P}}(\mathcal{C})$. If $f' \in S$ then $f \in \widehat{S}'$.

Since colimits in $\widehat{\mathcal{P}}(\mathcal{C})$ are universal, the subcategory $\widehat{\mathcal{P}}(\mathcal{C})^0$ is stable under $\mathcal{U}$-small colimits in $\widehat{\mathcal{P}}(\mathcal{C})$. Moreover, since S is stable under pullbacks in $\mathcal{P}(\mathcal{C})$ (and since the inclusion $\mathcal{P}(\mathcal{C}) \subseteq \widehat{\mathcal{P}}(\mathcal{C})$ is fully faithful), the ∞-category $\widehat{\mathcal{P}}(\mathcal{C})^0$ contains $\mathcal{P}(\mathcal{C})$. Since $\widehat{\mathcal{P}}(\mathcal{C})$ is generated (under $\mathcal{U}$-small colimits) by the essential image of the Yoneda embedding $\mathcal{C} \to \mathcal{P}(\mathcal{C})$, we conclude that $\widehat{\mathcal{P}}(\mathcal{C})^0 = \widehat{\mathcal{P}}(\mathcal{C})$.

We now let S' denote the collection of all morphisms in $\widehat{\mathcal{P}}(\mathcal{C})$ such that, for every pullback diagram

$$
\begin{array}{ccc}
Y & \longrightarrow & Y' \\
\downarrow{\scriptstyle f} & & \downarrow{\scriptstyle f'} \\
X & \longrightarrow & X'
\end{array}
$$

in $\mathcal{P}(\mathcal{C})$, if $f' \in S'$ then $f \in \widehat{S}$. The above argument shows that $S \subseteq S'$. Since S' is strongly saturated, we conclude that $\widehat{S} \subseteq S'$, so that $\widehat{S}$ is stable under pullbacks, as desired. This completes the proof of (a).

To prove (b), it will suffice to show that the composite map

$$\mathcal{P}(\mathcal{C}) \to S^{-1}\,\mathcal{P}(\mathcal{C}) \to \widehat{S}^{-1}\widehat{\mathcal{P}}(\mathcal{C})$$

preserves small colimits. We can rewrite this as the composition of a pair of functors

$$\mathcal{P}(\mathcal{C}) \xrightarrow{\,i\,} \widehat{\mathcal{P}}(\mathcal{C}) \xrightarrow{\,L\,} \widehat{S}^{-1}\widehat{\mathcal{P}}(\mathcal{C}).$$

The functor L is left adjoint to the inclusion of $\widehat{S}^{-1}\widehat{\mathcal{P}}(\mathcal{C})$ into $\widehat{\mathcal{P}}(\mathcal{C})$ and therefore preserves all $\mathcal{U}$-small colimits. It therefore suffices to show that the inclusion $i : \mathrm{Fun}(\mathcal{C}^{op}, \mathcal{S}) \to \mathrm{Fun}(\mathcal{C}^{op}, \widehat{\mathcal{S}})$ preserves small colimits. In view of Proposition 5.1.2.2, it will suffice to prove the inclusion $i_0 : \mathcal{S} \to \widehat{\mathcal{S}}$ preserves small colimits. We note that i_0 is an equivalence from $\mathcal{S}$ to the full subcategory $\widehat{\mathcal{S}}^0 \subseteq \widehat{\mathcal{S}}$ spanned by the essentially small spaces. It now suffices to observe that the collection of essentially small spaces is stable under small colimits (this follows from Corollaries 5.4.1.5 and 5.3.4.15).

Remark 6.3.5.18. Let $\mathcal{U}$ be a universe as in Example 6.3.5.17, let $f^* : \mathcal{X} \to \mathcal{Y}$ be a geometric morphism of ∞-topoi, and let $\widehat{f}_* : \mathrm{Shv}_{\widehat{\mathcal{S}}}(\mathcal{X}) \to \mathrm{Shv}_{\widehat{\mathcal{S}}}(\mathcal{Y})$ be given by composition with f^*. Then $\widehat{f}_*$ can be identified with a geometric morphism in the universe $\mathcal{U}$. To prove this, let κ denote the regular cardinal in the universe $\mathcal{U}$ such that small sets (in our original universe) can be identified with κ-small sets in $\mathcal{U}$. It follows from Corollary 5.3.5.4 that we can identify $\mathrm{Shv}_{\widehat{\mathcal{S}}}(\mathcal{X})$ and $\mathrm{Shv}_{\widehat{\mathcal{S}}}(\mathcal{X})$ with $\widehat{\mathrm{Ind}}_\kappa(\mathcal{X})$ and $\widehat{\mathrm{Ind}}_\kappa(\mathcal{Y})$, respectively. Proposition 5.2.6.3 implies that $\widehat{f}_*$ admits a left adjoint $\widehat{f}^*$ which fits into a commutative diagram

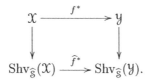

To complete the proof, it will suffice to show that $\widehat{f}^*$ is left exact. Since f^* preserves final objects, the functor $\widehat{f}^*$ preserves final objects as well. It therefore suffices to show that $\widehat{f}^*$ preserves pullback diagrams. Using Proposition 5.3.5.15 and Example 7.3.4.7, we conclude that every pullback diagram in $\mathrm{Shv}_{\widehat{\mathcal{S}}}(\mathcal{X})$ can be obtained as a $\mathcal{U}$-small κ-filtered colimit of pullback diagrams in $\mathcal{X}$. The desired result now follows from the assumption that f^* is left exact and the observation that the class of pullback diagrams in $\mathrm{Shv}_{\widehat{\mathcal{S}}}(\mathcal{Y})$ is stable under $\mathcal{U}$-small filtered colimits (Example 7.3.4.7).

For the remainder of this section, we fix a larger universe $\mathcal{U}$. Let $\widehat{\mathcal{S}}$ denote the ∞-category of $\mathcal{U}$-small spaces.

Notation 6.3.5.19. Let $F : \mathcal{L}\mathcal{T}\mathrm{op} \to \widehat{\mathcal{S}}$ be a functor. For every ∞-topos $\mathcal{X}$, we let $F_\mathcal{X} : \mathcal{X}^{op} \to \widehat{\mathcal{S}}$ denote the composition

$$\mathcal{X}^{op} \simeq \mathcal{L}\mathcal{T}\mathrm{op}_{\mathrm{\acute{e}t}}^{\mathcal{X}/} \to \mathcal{L}\mathcal{T}\mathrm{op} \xrightarrow{F} \widehat{\mathcal{S}}.$$

We will say that $F_\mathcal{X}$ is a *sheaf* if, for every ∞-topos $\mathcal{X}$, the functor $F_\mathcal{X}$ preserves small limits. We let $\widehat{\mathrm{Shv}}(\mathcal{L}\mathcal{T}\mathrm{op}^{op})$ denote the full subcategory of $\mathrm{Fun}(\mathcal{L}\mathcal{T}\mathrm{op}, \widehat{\mathcal{S}})$ spanned by the sheaves.

Example 6.3.5.20. Let $\mathcal{X}$ be an ∞-topos and let $e_\mathcal{X} : \mathcal{L}\mathcal{T}\mathrm{op} \to \widehat{\mathcal{S}}$ be the functor represented by $\mathcal{X}$. Proposition 6.3.5.14 implies that $e_\mathcal{X}$ belongs to $\widehat{\mathrm{Shv}}(\mathcal{L}\mathcal{T}\mathrm{op}^{op})$. We will say that a sheaf $F \in \widehat{\mathrm{Shv}}(\mathcal{L}\mathcal{T}\mathrm{op}^{op})$ is *representable* if $F \simeq e_\mathcal{X}$ for some ∞-topos $\mathcal{X}$.

Lemma 6.3.5.21. *The ∞-category $\widehat{\mathrm{Shv}}(\mathcal{L}\mathcal{T}\mathrm{op}^{op})$ is an ∞-topos in the universe $\mathcal{U}$. Moreover, for every ∞-topos $\mathcal{X}$, the restriction functor $F \mapsto F_\mathcal{X}$ determines a functor $\widehat{\mathrm{Shv}}(\mathcal{L}\mathcal{T}\mathrm{op}^{op}) \to \mathrm{Shv}_{\widehat{\mathcal{S}}}(\mathcal{X})$ which preserves $\mathcal{U}$-small colimits and finite limits.*

Proof. Let $\mathrm{Fun}^{\mathrm{\acute{e}t}}(\Delta^1, \mathcal{L}\mathcal{T}\mathrm{op})$ be the full subcategory $\mathrm{Fun}(\Delta^1, \mathcal{L}\mathcal{T}\mathrm{op})$ spanned by the étale morphisms and let $e : \mathrm{Fun}^{\mathrm{\acute{e}t}}(\Delta^1, \mathcal{L}\mathcal{T}\mathrm{op}) \to \mathcal{L}\mathcal{T}\mathrm{op}$ be given by evaluation at the vertex $\{0\} \in \Delta^1$. Since the collection of étale morphisms in

$\mathcal{L}\mathcal{T}\mathrm{op}$ is stable under pushouts (Remark 6.3.5.8), the map e is a coCartesian fibration.

We define a simplicial set $\mathcal{K}$ equipped with a projection $p : \mathcal{K} \to \mathcal{L}\mathcal{T}\mathrm{op}$ so that the following universal property is satisfied: for every simplicial set K, we have a natural bijection

$$\mathrm{Hom}_{\mathrm{Ind}(\mathcal{G}^{op})}(K, \mathcal{K}) = \mathrm{Hom}_{\mathrm{Set}_\Delta}(K \times_{\mathcal{L}\mathcal{T}\mathrm{op}} \mathrm{Fun}^{\text{ét}}(\Delta^1, \mathcal{L}\mathcal{T}\mathrm{op}), \widehat{\mathcal{S}}).$$

Then $\mathcal{K}$ is an ∞-category whose objects can be identified with pairs $(\mathcal{X}, F_{\mathcal{X}})$, where $\mathcal{X}$ is an ∞-topos and $F_{\mathcal{X}} : \mathcal{L}\mathcal{T}\mathrm{op}^{\mathcal{X}/}_{\text{ét}} \to \widehat{\mathcal{S}}$ is a functor. It follows from Corollary 3.2.2.13 that the projection p is a Cartesian fibration and that a morphism $(\mathcal{X}, F_{\mathcal{X}}) \to (\mathcal{Y}, F_{\mathcal{Y}})$ is p-Cartesian if and only if, for every object $U \in \mathcal{X}$, the canonical map $F_{\mathcal{X}}(\mathcal{X}_{/U}) \to F_{\mathcal{Y}}(\mathcal{Y}_{/f_*U})$ is a homotopy equivalence, where f^* denotes the underlying geometric morphism from $\mathcal{X}$ to $\mathcal{Y}$.

Let $\mathcal{K}_0$ denote the full subcategory of $\mathcal{K}$ spanned by pairs $(\mathcal{X}, F_{\mathcal{X}})$, where the functor $F_{\mathcal{X}}$ preserves small limits. It follows from the above that the Cartesian fibration p restricts to a Cartesian fibration $p_0 : \mathcal{K}_0 \to \mathcal{L}\mathcal{T}\mathrm{op}$ (with the same class of Cartesian morphisms). The fiber of $\mathcal{K}_0$ over an object $\mathcal{X} \in \mathcal{L}\mathcal{T}\mathrm{op}$ can be identified with $\mathrm{Shv}_{\widehat{\mathcal{S}}}(\mathcal{X})$, which is an ∞-topos in the universe $\mathcal{U}$ (Remark 6.3.5.17). Moreover, to every geometric morphism $f^* : \mathcal{X} \to \mathcal{Y}$ in $\mathcal{L}\mathcal{T}\mathrm{op}$, the Cartesian fibration p_0 associates the pushforward functor $\widehat{f}_* : \mathrm{Shv}_{\widehat{\mathcal{S}}}(\mathcal{Y}) \to \mathrm{Shv}_{\widehat{\mathcal{S}}}(\mathcal{X})$ given by composition with f^*. It follows from Remark 6.3.5.18 that $\widehat{f}_*$ admits a left adjoint $\widehat{f}^*$ and that $\widehat{f}^*$ is left exact. We may summarize the situation by saying that p_0 is a topos fibration (see Definition 6.3.1.6); in particular, p_0 is a coCartesian fibration.

Let $\mathcal{Y} = \mathrm{Fun}_{\mathcal{L}\mathcal{T}\mathrm{op}}(\mathcal{L}\mathcal{T}\mathrm{op}, \mathcal{K}_0)$ denote the ∞-category of sections of p_0. Unwinding the definitions, we obtain an identification

$$\mathcal{Y} \simeq \mathrm{Fun}(\mathrm{Fun}^{\text{ét}}(\Delta^1, \mathcal{L}\mathcal{T}\mathrm{op}), \widehat{\mathcal{S}}).$$

Let $\mathcal{L}\mathcal{T}\mathrm{op}'$ denote the essential image of the (fully faithful) diagonal embedding $\mathcal{L}\mathcal{T}\mathrm{op} \to \mathrm{Fun}(\Delta^1, \mathcal{L}\mathcal{T}\mathrm{op})$. Consider the following conditions on a section $s : \mathcal{L}\mathcal{T}\mathrm{op} \to \mathcal{K}_0$ of p_0:

(a) The functor s carries étale morphisms in $\mathcal{L}\mathcal{T}\mathrm{op}$ to p-coCartesian morphisms in $\mathcal{X}$.

(b) Let $S : \mathrm{Fun}^{\text{ét}}(\Delta^1, \mathcal{L}\mathcal{T}\mathrm{op}) \to \widehat{\mathcal{S}}$ be the functor corresponding to s. Then, for every commutative diagram

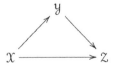

of étale morphisms in $\mathcal{L}\mathcal{T}\mathrm{op}$, the induced map $S(\mathcal{X} \to \mathcal{Z}) \to S(\mathcal{Y} \to \mathcal{Z})$ is an equivalence in $\widehat{\mathcal{S}}$.

(c) For every étale morphism $f^* : \mathcal{X} \to \mathcal{Y}$ in $\mathcal{L}\mathcal{T}\mathrm{op}$, the canonical map $S(\mathrm{id}_{\mathcal{X}}) \to S(f^*)$ is an equivalence in $\widehat{\mathcal{S}}$.

(*d*) The functor S is a left Kan extension of $S | \mathcal{L}\mathcal{T}\mathrm{op}'$.

Unwinding the definitions, we see that (*a*) ⇔ (*b*) ⇒ (*c*) ⇔ (*d*). Moreover, the implication (*c*) ⇒ (*b*) follows by a two-out-of-three argument. Let $\mathcal{Y}'$ denote the full subcategory of $\mathcal{Y}$ spanned by those sections which satisfy the equivalent conditions (*a*) through (*d*); it follows from Proposition 4.3.2.15 that composition with the diagonal embedding $\mathcal{L}\mathcal{T}\mathrm{op} \to \mathrm{Fun}^{\text{ét}}(\Delta^1, \mathcal{L}\mathcal{T}\mathrm{op})$ induces an equivalence $\theta : \mathcal{Y}' \to \widehat{\mathrm{Shv}}(\mathcal{L}\mathcal{T}\mathrm{op}^{op})$. The desired result now follows from Proposition 5.4.7.11. □

To prove Lemma 6.3.5.15, we need a criterion which will allow us to detect étale geometric morphisms of ∞-topoi in terms of the functors that they represent. To formulate this criterion, we introduce a bit of temporary terminology.

Definition 6.3.5.22. Let $\alpha : F \to G$ be a morphism in $\widehat{\mathrm{Shv}}(\mathcal{L}\mathcal{T}\mathrm{op}^{op})$. We will say that α is *universal* if, for every geometric morphism of ∞-topoi $f^* : \mathcal{X} \to \mathcal{Y}$, the induced diagram

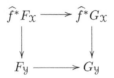

is a pullback square in $\mathrm{Shv}_{\widehat{\mathcal{S}}}(\mathcal{Y})$ (here $\widehat{f}^*$ denotes the geometric morphism described in Remark 6.3.5.18).

Remark 6.3.5.23. The collection of universal morphisms in $\widehat{\mathrm{Shv}}(\mathcal{L}\mathcal{T}\mathrm{op}^{op})$ is stable under pullbacks and composition and contains every equivalence in $\widehat{\mathrm{Shv}}(\mathcal{L}\mathcal{T}\mathrm{op}^{op})$.

Remark 6.3.5.24. Let $p : K \to \widehat{\mathrm{Shv}}(\mathcal{L}\mathcal{T}\mathrm{op}^{op})$ be a small diagram having a colimit F. Assume that for every edge $v \to v'$ of K, the induced map $p(v) \to p(v')$ is universal. Then each of the induced maps $p(v) \to F$ is universal. This follows immediately from Theorem 6.1.3.9.

Lemma 6.3.5.25. *Let* $\alpha : F \to G$ *be a morphism in* $\widehat{\mathrm{Shv}}(\mathcal{L}\mathcal{T}\mathrm{op}^{op})$ *and assume that* G *is representable by an* ∞-*topos* $\mathcal{X}$. *Then* F *is representable by an* ∞-*topos étale over* $\mathcal{X}$ *if the following conditions are satisfied:*

(1) *The morphism* α *is universal (in the sense of Definition 6.3.5.22).*

(2) *For every* ∞-*topos* $\mathcal{Y}$, *the homotopy fibers of the induced map* $F(\mathcal{Y}) \to G(\mathcal{Y})$ *are essentially small.*

Remark 6.3.5.26. In fact, the converse to Lemma 6.3.5.25 is true as well, but we will not need this fact.

Proof. Choose a point $\eta \in G(\mathcal{X})$ which induces an equivalence $e_{\mathcal{X}} \to G$. The map η induces a global section $\overline{\eta}$ of $G_{\mathcal{X}}$ in the ∞-topos $\mathrm{Shv}_{\widehat{\mathcal{S}}}(\mathcal{X})$. Let F_0 denote the fiber product $F_{\mathcal{X}} \times_{G_{\mathcal{X}}} 1_{\mathrm{Shv}_{\widehat{\mathcal{S}}}(\mathcal{X})}$. Assumption (2) implies that the functor $F_0 : \mathcal{X}^{op} \to \widehat{\mathcal{S}}$ takes values which are essentially small. It follows from Proposition 5.5.2.2 that F_0 is representable by an object $U \in \mathcal{X}$. In particular, we have a tautological point $\eta' \in F_0(U)$, which determines a commutative diagram

$$
\begin{array}{ccc}
e_{\mathcal{X}/U} & \longrightarrow & F \\
\downarrow & & \downarrow{\scriptstyle \alpha} \\
e_{\mathcal{X}} & \longrightarrow & G
\end{array}
$$

in $\widehat{\mathrm{Shv}}(\mathcal{L}\mathcal{T}\mathrm{op}^{op})$. To complete the proof, it will suffice to show that the upper horizontal map is an equivalence.

Fix an ∞-topos $\mathcal{Y}$ and let $R : \widehat{\mathrm{Shv}}(\mathcal{L}\mathcal{T}\mathrm{op}^{op}) \to \mathrm{Shv}_{\widehat{\mathcal{S}}}(\mathcal{Y})$ be the restriction map; we will show that the induced map $R(e_{\mathcal{X}/U}) \to R(F)$ is an equivalence in $\mathrm{Shv}_{\widehat{\mathcal{S}}}(\mathcal{Y})$. It will suffice to show that for every $V \in \mathcal{Y}$, the map $R(e_{\mathcal{X}/U}(V)) \to R(F)(V)$ induces a homotopy equivalence after passing to the homotopy fibers over any point $\eta' \in R(G)(V)$. Replacing $\mathcal{Y}$ by $\mathcal{Y}_{/V}$, we may assume that η' is induced by a geometric morphism $f^* : \mathcal{X} \to \mathcal{Y}$ which determines a map $1 \to R(G)$, where 1 denotes the final object of $\mathrm{Shv}_{\widehat{\mathcal{S}}}(\mathcal{Y})$. Let $F' = R(e_{\mathcal{X}/U}) \times_{R(G)} 1$ and $F'' = R(F) \times_{R(G)} 1$; to complete the proof it will suffice to show that the induced map $F' \to F''$ is an equivalence.

We have a commutative diagram

$$
\begin{array}{ccccc}
F'' & \longrightarrow & \widehat{f}^* F_{\mathcal{X}} & \longrightarrow & R(F) \\
\downarrow & & \downarrow & & \downarrow \\
1 & \xrightarrow{\widehat{f}^*\overline{\eta}} & \widehat{f}^* G_{\mathcal{X}} & \longrightarrow & R(G)
\end{array}
$$

in the ∞-category $\mathrm{Shv}_{\widehat{\mathcal{S}}}(\mathcal{Y})$. Here the right square is a pullback since α is universal, and the outer square is a pullback by construction. It follows that the left square is also a pullback, so that

$$
F'' \simeq \widehat{f}^*(1_{\mathcal{X}} \times_{G_{\mathcal{X}}} F_{\mathcal{X}}) \simeq \widehat{f}^* F_0.
$$

We note that $\widehat{f}^* F_0$ can be identified with the functor represented by the object $f^* U \in \mathcal{Y}$, which (by virtue of Remark 6.3.5.7) is equivalent to F', as desired. $\square$

Lemma 6.3.5.27. *Let κ be an uncountable regular cardinal and let $X_\bullet$ be a simplicial object of $\mathcal{S}$ with the following properties:*

(a) *For each $n \geq 0$, the connected components of $X_\bullet$ are essentially κ-small.*

(b) *For every morphism $[m] \to [n]$ in $\mathbf{\Delta}$, the induced map $X_n \to X_m$ has essentially κ-small homotopy fibers.*

Let X be the geometric realization of $X_{\bullet}$. Then the induced map $X_0 \to X$ has essentially κ-small homotopy fibers.

Proof. Replacing X by one of its connected components X' (and each X_n by the inverse image $X' \times_X X_n$), we may suppose that X is connected.

Let $R \subseteq \pi_0 X_0 \times \pi_0 X_0$ denote the image of $\pi_0 X_1$ and let $\sim$ denote the equivalence relation on $\pi_0 X_0$ generated by R. It follows from assumption (b) that for every κ-small subset $A \subseteq \pi_0 X_0$, the intersections $R \cap (A \times \pi_0 X_0)$ and $R \cap (\pi_0 X_0 \times A)$ are again κ-small. Since κ is uncountable, it follows that the every $\sim$-equivalence class is κ-small. Since $(\pi_0 X_0)/\sim$ is isomorphic to $\pi_0 X \simeq *$, we conclude that $\pi_0 X$ is itself κ-small. Combining this with (a), we conclude that X_0 is essentially κ-small. Invoking (b), we deduce that each X_n is essentially κ-small, so that X is essentially κ-small. The desired conclusion now follows from the long exact sequences associated to the fibration sequences $X_0 \times_X \{*\} \to X_0 \to X$. $\qquad\qquad$ □

Lemma 6.3.5.28. *Let $\mathcal{X}$ be an ∞-topos. Then*

(1) *The inclusion $\mathrm{Shv}_{\widehat{\mathcal{S}}}(\mathcal{X}) \subseteq \mathrm{Fun}(\mathcal{X}^{op}, \widehat{\mathcal{S}})$ admits a left exact left adjoint L.*

(2) *Let $F \in \mathrm{Fun}(\mathcal{X}^{op}, \widehat{\mathcal{S}})$ be a functor such that each of the spaces $F(X)$ is essentially small. Then each of the spaces $LF(X)$ is essentially small.*

(3) *Let $\alpha : F \to G$ be a morphism in $\mathrm{Fun}(\mathcal{X}^{op}, \widehat{\mathcal{S}})$ such that, for each $X \in \mathcal{X}$, the homotopy fibers of the induced map $F(X) \to G(X)$ are essentially small. Then for each $X \in \mathcal{X}$, the homotopy fibers of the map $LF(X) \to LG(X)$ are also essentially small.*

Proof. The existence of the left adjoint L follows from Lemma 5.5.4.19. Since $\mathrm{Shv}_{\widehat{\mathcal{S}}}(\mathcal{X})$ contains the essential image of the Yoneda embedding $j : \mathcal{X} \to \mathrm{Fun}(\mathcal{X}^{op}, \widehat{\mathcal{S}})$, we can identify $L \circ j$ with j. Since j is left exact, Proposition 6.1.5.2 implies that L is also left exact. This proves (1).

We now prove (2). Choose a (small) regular cardinal κ such that $\mathcal{X}$ is κ-accessible and let $\mathcal{X}^{\kappa}$ denote the full subcategory of $\mathcal{X}$ spanned by the κ-compact objects. Let T denote the composition

$$\mathrm{Fun}(\mathcal{X}^{op}, \widehat{\mathcal{S}}) \xrightarrow{T'} \mathrm{Fun}((\mathcal{X}^{\kappa})^{op}, \widehat{\mathcal{S}}) \xrightarrow{T''} \mathrm{Fun}(\mathcal{X}^{op}, \widehat{\mathcal{S}}),$$

where T' is the restriction functor and T'' is given by the right Kan extension. We have an evident natural transformation $\mathrm{id} \to T$ which exhibits T as a localization functor on $\mathrm{Fun}(\mathcal{X}^{op}, \widehat{\mathcal{S}})$. Proposition 6.1.3.6 implies that every $\widehat{\mathcal{S}}$-valued sheaf on $\mathcal{X}$ is T-local. It follows that the canonical map $L \to LT$ is an equivalence of functors. In particular, to prove that $LF(X)$ is locally small, we may assume without loss of generality that F is T-local.

Let $\mathrm{Fun}'(\mathcal{X}^{op}, \widehat{\mathcal{S}})$ denote the full subcategory of $\mathrm{Fun}(\mathcal{X}^{op}, \widehat{\mathcal{S}})$ spanned by the T-local functors (in other words, those functors which are right Kan extensions of their restriction to $(\mathcal{X}^{\kappa})^{op}$; by Proposition 4.3.2.15 this ∞-category is equivalent to $\mathrm{Fun}((\mathcal{X}^{\kappa})^{op}, \widehat{\mathcal{S}}))$. We can identify $\mathrm{Fun}'(\mathcal{X}^{op}, \widehat{\mathcal{S}})$ with the ∞-category of $\widehat{S}$-valued sheaves $\mathrm{Shv}_{\widehat{S}}(\mathcal{P}(\mathfrak{X}^{\kappa}))$ on the ∞-topos $\mathcal{P}(\mathfrak{X}^{\kappa})$. Let F' be the image of F under this identification; we observe that the functor $F' : \mathcal{P}(\mathfrak{X}^{\kappa})^{op} \to \widehat{S}$ takes essentially small values. In Remark 6.3.5.17, we saw that this ∞-category contains $\mathrm{Shv}_{\widehat{S}}(\mathfrak{X})$ as a left exact localization and that the localization functor $L' : \mathrm{Shv}_{\widehat{S}}(\mathcal{P}(\mathfrak{X}^{\kappa})) \to \mathrm{Shv}_{\widehat{S}}(\mathfrak{X})$ is equivalent to L when restricted to $\mathrm{Shv}_{S}(\mathcal{P}(\mathfrak{X}^{\kappa})) \simeq \mathcal{P}(\mathfrak{X}^{\kappa})$. Since F' belongs to the essential image of the inclusion $\mathrm{Shv}_{S}(\mathcal{P}(\mathfrak{X}^{\kappa})) \subseteq \mathrm{Shv}_{\widehat{S}}(\mathcal{P}(\mathfrak{X}^{\kappa}))$, the argument given there proves that $L'F'$ belongs to the essential image of the inclusion $\mathrm{Shv}_{S}(\mathfrak{X}) \subseteq \mathrm{Shv}_{\widehat{S}}(\mathfrak{X})$, so that $LF(X)$ is essentially small, as desired.

To prove (3), let us fix a point $\eta \in LG(X)$. We wish to prove the following stronger version of (3):

(3′) For every map $U \to X$ in $\mathfrak{X}$, the homotopy fiber of the induced map $LF \to LG$ is essentially small (here the homotopy fiber is taken over the point determined by η).

Let $\mathfrak{X}^0_{/X}$ denote the full subcategory of $\mathfrak{X}_{/X}$ spanned by those morphisms $U \to X$ for which condition (2′) is satisfied. Since LF and LG belong to $\mathrm{Shv}_{\widehat{S}}(\mathfrak{X})$ (and since the collection of essentially small spaces is stable under small limits), we conclude that $\mathfrak{X}^0_{/X}$ is stable under small colimits in $\mathfrak{X}_{/X}$.

Let $\mathfrak{X}^1_{/X}$ be the largest sieve contained in $\mathfrak{X}^0_{/X}$ (in other words, a morphism $U \to X$ belongs to $\mathfrak{X}^1_{/X}$ if and only if, for every morphism $V \to U$ in $\mathfrak{X}$, the composite map $V \to X$ belongs to $\mathfrak{X}^0_{/X}$). Since colimits in $\mathfrak{X}$ are universal, we conclude that $\mathfrak{X}^1_{/X}$ is stable under small colimits in $\mathfrak{X}_{/X}$. It follows that $\mathfrak{X}^1_{/X} \simeq \mathfrak{X}_{/X_0}$ for some monomorphism $i : X_0 \to X$ in $\mathfrak{X}$. We wish to show that i is an equivalence.

Since L is left exact, we have $L(G \times_{LG} j(X)) \simeq Lj(X) \simeq j(X)$. In particular, the map $G \times_{LG} j(X) \to j(X_0)$ cannot factor through $j(X_0)$ unless i is an equivalence. It will therefore suffice to show that $G \times_{j(X)} j(X_0) \simeq G$. In other words, it will suffice to show that if $U \in \mathfrak{X}_{/X}$ and $\eta' \in G(U)$ is a point such that the images of η and η' lie in the same connected component of $LG(U)$, then $U \in \mathfrak{X}^1_{/X}$. Since the existence of η' is stable under the process of replacing U by some further refinement $V \to U$, it will suffice to show that $U \in \mathfrak{X}^0_{/X}$. Replacing X by U, we obtain the following reformulation of (3′):

(3″) Let $\eta' \in G(X)$. Then the homotopy fiber Z of the induced map $LF(X) \to LG(X)$ (over the point determined by η) is essentially small.

Since L is left exact, we can identify Z with $LF_0(X)$, where $F_0 = F \times_G j(X)$. Since the homotopy fibers of the maps $F(Y) \to G(Y)$ are essentially small, we may assume without loss of generality that $F_0 \in \mathrm{Fun}(\mathfrak{X}^{op}, S)$. Invoking (2), we deduce that the values of LF_0 are essentially small, as desired. $\qquad \square$

Proof of Lemma 6.3.5.15. Let $\mathfrak{X}_\bullet$ be a simplicial object of $\mathcal{L}\mathrm{Top}^{op}_{\text{ét}}$ and let

$F_\bullet$ be its image under the Yoneda embedding $j : \mathcal{L}\mathrm{Top}^{op} \to \widehat{\mathrm{Shv}}(\mathcal{L}\mathrm{Top}^{op})$. Let F be a geometric realization of $|F_\bullet|$. We will prove the following:

(∗) The map $\beta : F_0 \to F$ satisfies conditions (1) and (2) of Lemma 6.3.5.25.

Assuming (∗) for the moment, we will complete the proof of Lemma 6.3.5.15. Let $F'_\bullet$ be a Čech nerve of the induced map $F_0 \to F$ (so that $F'_n \simeq F_0 \times_F F_0 \times \cdots \times_F F_0$; in particular $F'_0 \simeq F_0$). We first claim that each F'_n is representable by an ∞-topos $\mathcal{X}'_n$ and that each inclusion $[0] \hookrightarrow [n]$ induces an étale map of ∞-topos $\mathcal{X}'_n \to \mathcal{X}'_0 \simeq \mathcal{X}_0$. Since $F'_\bullet$ is a groupoid object of $\widehat{\mathrm{Shv}}(\mathcal{L}\mathrm{Top}^{op})$ (and $F'_0 \simeq F_0$ is representable by the ∞-topos $\mathcal{X}_0$), it will suffice to prove this result when $n = 1$. Consider the pullback diagram

$$
\begin{array}{ccc}
F'_1 & \xrightarrow{\ \beta'\ } & F'_0 \\
\downarrow & & \downarrow \\
F_0 & \xrightarrow{\ \beta\ } & F.
\end{array}
$$

It follows from condition (∗) that β' satisfies conditions (1) and (2) of Lemma 6.3.5.25, so that F'_1 is representable by an ∞-topos étale over $\mathcal{X}_0$, as desired.

Since the Yoneda embedding j is fully faithful, we may assume without loss of generality that $F'_\bullet$ is the image under j of a groupoid object $\mathcal{X}'_\bullet$ of $\mathcal{L}\mathrm{Top}^{op}$. Using Corollary 6.3.5.9, we deduce that $\mathcal{X}'_\bullet$ defines a simplicial object of the subcategory $\mathcal{L}\mathrm{Top}^{op}_{\text{ét}}$. The evident natural transformation $F_\bullet \to F'_\bullet$ induces a map of simplicial objects $\alpha : \mathcal{X}_\bullet \to \mathcal{X}'_\bullet$; we claim that α has the desired properties. The only nontrivial point is to verify that the induced map of geometric realizations $|\mathcal{X}_\bullet| \to |\mathcal{X}'_\bullet|$ is an equivalence of ∞-topoi. For this, it suffices to show that for every ∞-topos $\mathcal{Y}$, the upper horizontal map in the diagram

$$
\begin{array}{ccc}
\mathrm{Map}_{\mathcal{L}\mathrm{Top}^{op}}(|\mathcal{X}'_\bullet|, \mathcal{Y}) & \longrightarrow & \mathrm{Map}_{\mathcal{L}\mathrm{Top}^{op}}(|\mathcal{X}_\bullet|, \mathcal{Y}) \\
\downarrow & & \downarrow \\
\varprojlim \mathrm{Map}_{\mathcal{L}\mathrm{Top}^{op}}(\mathcal{X}'_n, \mathcal{Y}) & \longrightarrow & \varprojlim \mathrm{Map}_{\mathcal{L}\mathrm{Top}^{op}}(\mathcal{X}_n, \mathcal{Y})
\end{array}
$$

is a homotopy equivalence. Since the vertical maps are homotopy equivalences, it suffices to show that the lower horizontal map is a homotopy equivalence. Since j is fully faithful, it suffices to show that the lower horizontal map in the analogous diagram

$$
\begin{array}{ccc}
\mathrm{Map}_{\widehat{\mathrm{Shv}}(\mathcal{L}\mathrm{Top}^{op})}(|F'_\bullet|, e_\mathcal{Y}) & \longrightarrow & \mathrm{Map}_{\widehat{\mathrm{Shv}}(\mathcal{L}\mathrm{Top}^{op})}(|F_\bullet|, e_\mathcal{Y}) \\
\downarrow & & \downarrow \\
\varprojlim \mathrm{Map}_{\widehat{\mathrm{Shv}}(\mathcal{L}\mathrm{Top}^{op})}(F'_n, e_\mathcal{Y}) & \longrightarrow & \varprojlim \mathrm{Map}_{\widehat{\mathrm{Shv}}(\mathcal{L}\mathrm{Top}^{op})}(F_n, e_\mathcal{Y})
\end{array}
$$

is a homotopy equivalence. Again, the vertical maps are homotopy equivalences, so we are reduced to showing that the upper horizontal map is a homotopy equivalence. This follows from the fact that we have an equivalence $|F_\bullet| \simeq F \simeq |F'_\bullet|$ in $\widehat{\mathrm{Shv}}(\mathcal{L}\mathcal{T}\mathrm{op}^{op})$ since groupoid objects in $\widehat{\mathrm{Shv}}(\mathcal{L}\mathcal{T}\mathrm{op}^{op})$ are effective (Lemma 6.3.5.21).

It remains to prove (∗). Remark 6.3.5.24 implies that $\beta : F_0 \to F$ is universal. To complete the proof, we must show that for every ∞-topos $\mathcal{Y}$, the homotopy fibers of the induced map $F_0(\mathcal{Y}) \to F(\mathcal{Y})$ are essentially small. Let $R : \widehat{\mathrm{Shv}}(\mathcal{L}\mathcal{T}\mathrm{op}^{op}) \to \mathrm{Shv}_{\widehat{\mathcal{S}}}(\mathcal{Y})$ denote the restriction map, let $G_\bullet$ denote the image of $F_\bullet$ under R, and let $G = |G_\bullet| \simeq R(F)$. Let G' denote the geometric realization of $G_\bullet$ in the larger ∞-category $\mathrm{Fun}(\mathcal{Y}^{op}, \widehat{\mathcal{S}})$ and let $L : \mathrm{Fun}(\mathcal{Y}^{op}, \widehat{\mathcal{S}}) \to \mathrm{Shv}_{\widehat{\mathcal{S}}}(\mathcal{Y})$ be a left adjoint to the inclusion (see Lemma 6.3.5.28). Then we can identify the map $G_0 \to G$ with the image under L of the map $u : G_0 \to G'$. In view of Lemma 6.3.5.28, it will suffice to show that for every object $U \in \mathcal{Y}$, the induced map $G_0(U) \to G'(U)$ has essentially small homotopy fibers.

For each $n \geq 0$ and each object $U \in \mathcal{Y}$, we can identify $G_n(U)$ with the maximal Kan complex contained in $\mathrm{Fun}^*(\mathcal{X}_n, \mathcal{Y}_{/U})$. Since the ∞-category $\mathrm{Fun}^*(\mathcal{X}_n, \mathcal{Y}_{/U})$ is locally small (Proposition 6.3.1.13), we conclude that each connected component of $G_n(U)$ is essentially small. Moreover, for every morphism $[m] \to [n]$ in $\boldsymbol{\Delta}$, the induced map $\beta : G_n(U) \to G_m(U)$ is induced by composition with an étale geometric morphism $g^* : \mathcal{X}_m \to \mathcal{X}_n$, so that the homotopy fibers of β are essentially small by Remark 6.3.5.7. The desired result now follows from Lemma 6.3.5.27. $\square$

6.3.6 Structure Theory for ∞-Topoi

In this section we will analyze the following question: given a geometric morphism $f : \mathcal{X} \to \mathcal{Y}$ of ∞-topoi, when is f an equivalence? Clearly, this is true if and only if the pullback functor f^* is both fully faithful and essentially surjective. It is useful to isolate and study these conditions individually.

Definition 6.3.6.1. Let $f : \mathcal{X} \to \mathcal{Y}$ be a geometric morphism of ∞-topoi. The *image* of f is defined to be the smallest full subcategory of $\mathcal{X}$ which contains $f^* \mathcal{Y}$ and is stable under small colimits and finite limits. We will say that f is *algebraic* if the image of f coincides with $\mathcal{X}$.

Our first goal is to prove that the image of a geometric morphism is itself an ∞-topos.

Proposition 6.3.6.2. *Let $f : \mathcal{X} \to \mathcal{Z}$ be a geometric morphism of ∞-topoi and let $\mathcal{Y}$ be the image of f. Then $\mathcal{Y}$ is an ∞-topos. Moreover, the inclusion $\mathcal{Y} \subseteq \mathcal{X}$ is left exact and colimit-preserving, so we obtain a factorization of f as a composition of geometric morphisms*

$$\mathcal{X} \xrightarrow{g} \mathcal{Y} \xrightarrow{h} \mathcal{Z},$$

where h is algebraic and g^* is fully faithful.

Proof. We will show that $\mathcal{Y}$ satisfies the ∞-categorical versions of Giraud's axioms (see Theorem 6.1.0.6). Axioms (*ii*), (*iii*), and (*iv*) are concerned with the interaction between colimits and finite limits. Since $\mathcal{X}$ satisfies these axioms and $\mathcal{Y} \subseteq \mathcal{X}$ is stable under the relevant constructions, $\mathcal{Y}$ automatically satisfies these axioms as well. The only nontrivial point is to verify (i), which asserts that $\mathcal{Y}$ is presentable.

Choose a small collection of objects $\{Z_\alpha\}$ which generate $\mathcal{Z}$ under colimits. Now choose an uncountable regular cardinal τ with the following properties:

(1) Each $f^*(Z_\alpha)$ is a τ-compact object of $\mathcal{X}$.

(2) The final object $1_{\mathcal{X}}$ is τ-compact.

(3) The limits functor $\mathrm{Fun}(\Lambda_2^2, \mathcal{X}) \to \mathcal{X}$ (a right adjoint to the diagonal functor) is τ-continuous and preserves τ-compact objects.

Let $\mathcal{Y}'$ be the collection of all objects of $\mathcal{Y}$ which are τ-compact when considered as objects of $\mathcal{X}$. Clearly, each object of $\mathcal{Y}'$ is also τ-compact when regarded as an object of $\mathcal{Y}$. Moreover, because $\mathcal{X}$ is accessible, $\mathcal{Y}'$ is essentially small. It will therefore suffice to prove that $\mathcal{Y}'$ generates $\mathcal{Y}$ under colimits.

Choose a minimal model $\mathcal{Y}_0'$ for $\mathcal{Y}$. Since $\mathcal{X}$ is accessible, the full subcategory $\mathcal{X}^\kappa$ spanned by the κ-compact objects is essentially small, so that $\mathcal{Y}_0'$ is small. According to Proposition 5.3.5.10, there exists a τ-continuous functor $F : \mathrm{Ind}_\tau(\mathcal{Y}_0') \to \mathcal{X}$ whose composition with the Yoneda embedding is equivalent to the inclusion $\mathcal{Y}_0' \subseteq \mathcal{X}$. Since $\mathcal{Y}_0'$ admits τ-small colimits, $\mathrm{Ind}_\tau(\mathcal{Y}_0')$ is presentable. Proposition 5.3.5.11 implies that F is fully faithful; let $\mathcal{Y}''$ be its essential image. To complete the proof, it will suffice to show that $\mathcal{Y}'' = \mathcal{Y}$.

Since $\mathcal{Y}$ is stable under colimits in $\mathcal{X}$, we have $\mathcal{Y}'' \subseteq \mathcal{Y}$. According to Proposition 5.5.1.9, F preserves small colimits, so that $\mathcal{Y}''$ is stable under small colimits in $\mathcal{X}$. By construction, $\mathcal{Y}''$ contains each $f^*(Z_\alpha)$. Since f^* preserves colimits, we conclude that $\mathcal{Y}''$ contains $f^* \mathcal{Z}$. By definition $\mathcal{Y}$ is the smallest full subcategory of $\mathcal{X}$ which contains $f^* \mathcal{Z}$ and is stable under small colimits and finite limits. It remains only to show that $\mathcal{Y}''$ is stable under finite limits. Assumption (2) guarantees that $\mathcal{Y}''$ contains the final object of $\mathcal{X}$, so we need only show that $\mathcal{Y}''$ is stable under pullbacks. Consider a diagram $p : \Lambda_2^2 \to \mathcal{Y}''$. The proof of Proposition 5.4.4.3 (applied with $K = \Lambda_2^2$ and $\kappa = \omega$) shows that p can be written as a τ-filtered colimit of diagrams $p_\alpha : \Lambda_2^2 \to \mathcal{Y}''$. Since filtered colimits in $\mathcal{X}$ are left exact (Example 7.3.4.7), we conclude that the limit of p can be obtained as a τ-filtered colimit of limits of the diagrams p_β. In view of assumption (3), each of these limits lies in $\mathcal{Y}'$, so that the limit of p lies in $\mathcal{Y}''$, as desired.   □

Remark 6.3.6.3. The factorization of Proposition 6.3.6.2 is unique up to (canonical) equivalence.

The terminology of Definition 6.3.6.1 is partially justified by the following observations:

Proposition 6.3.6.4. (1) *Every étale geometric morphism between* ∞-*topoi is algebraic.*

(2) *The collection of algebraic geometric morphisms of ∞-topoi is stable under filtered limits (in $\mathcal{R}\mathcal{J}\text{op}$).*

Proof. We first prove (1). Let $\mathcal{X}$ be an ∞-topos, let U be an object of $\mathcal{X}$, let $\pi_! : \mathcal{X}^{/U} \to \mathcal{X}$ be the projection functor and let π^* be a left adjoint to $\pi_!$. Let $f : X \to U$ be an object of $\mathcal{X}^{/U}$, and let $F : f \to \text{id}_U$ be a morphism in $\mathcal{X}^{/U}$ (uniquely determined up to equivalence; for example, we can take F to be the composition of f with a retraction $\Delta^1 \times \Delta^1 \to \Delta^1$). Let $g : F \to \pi^* \pi_! F$ be the unit map for the adjunction between π^* and $\pi_!$. We claim that g is a pullback square in $\mathcal{X}_{/U}$. According to Proposition 4.4.2.9, it will suffice to verify that the image of g under $\pi_!$ is a pullback square in $\mathcal{X}$. But this square can be identified with

$$
\begin{array}{ccc}
X & \longrightarrow & X \times U \\
\downarrow & & \downarrow \\
U & \overset{\delta}{\longrightarrow} & U \times U,
\end{array}
$$

which is easily shown to be Cartesian. It follows that, in $\mathcal{X}_{/U}$, f can be obtained as a fiber product of the final object with objects that lie in the essential image of π^*. It follows that $\pi^* \mathcal{X}$ generates $\mathcal{X}_{/U}$ under finite limits, so that π is algebraic.

To prove (2), we consider a geometric morphism $f : \mathcal{X} \to \mathcal{Y}$ which is a filtered limit of algebraic geometric morphisms $\{f_\alpha : \mathcal{X}_\alpha \to \mathcal{Y}_\alpha\}$ in the ∞-category $\text{Fun}(\Delta^1, \mathcal{R}\mathcal{J}\text{op})$. Let $\mathcal{X}' \subseteq \mathcal{X}$ be a full subcategory which is stable under finite limits, stable under small colimits, and contains $f^* \mathcal{Y}$. We wish to prove that $\mathcal{X}' = \mathcal{X}$. For each α, we have a diagram of ∞-topoi

$$
\begin{array}{ccc}
\mathcal{X} & \overset{f}{\longrightarrow} & \mathcal{Y} \\
\downarrow{\scriptstyle \psi(\alpha)} & & \downarrow \\
\mathcal{X}_\alpha & \overset{f_\alpha}{\longrightarrow} & \mathcal{Y}_\alpha .
\end{array}
$$

Let $\mathcal{X}'_\alpha$ be the preimage of $\mathcal{X}'$ under $\psi(\alpha)^*$. Then $\mathcal{X}'_\alpha \subseteq \mathcal{X}_\alpha$ is stable under finite limits, stable under small colimits, and contains the essential image of f_α^*. Since f_α is algebraic, we conclude that $\mathcal{X}'_\alpha = \mathcal{X}_\alpha$. In other words, $\mathcal{X}'$ contains the essential image of each $\psi(\alpha)^*$. Lemma 6.3.3.6 implies that every object of $\mathcal{X}$ can be realized as a filtered colimit of objects, each of which belongs to the essential image of f_α^* for α appropriately chosen. Since $\mathcal{X}'$ is stable under small colimits, we conclude that $\mathcal{X}' = \mathcal{X}$. It follows that f is algebraic, as desired. $\qquad\square$

Remark 6.3.6.5. It is possible to formulate a converse to Proposition 6.3.6.4. Namely, one can characterize the class of algebraic morphisms as the smallest class of geometric morphisms which contains all étale morphisms and is stable under certain kinds of filtered limits. However, it is necessary to allow limits parametrized not only by filtered ∞-categories but also by filtered *stacks* over ∞-topoi. The precise statement requires ideas which lie outside the scope of this book.

Having achieved a rudimentary understanding of the class of algebraic geometric morphisms, we now turn our attention to the opposite extreme: namely, geometric morphisms $f : \mathcal{X} \to \mathcal{Y}$, where f^* is fully faithful.

Proposition 6.3.6.6. *Let $f : \mathcal{X} \to \mathcal{Y}$ be a geometric morphism of ∞-topoi. Suppose that f^* is fully faithful and essentially surjective on 1-truncated objects. Then f^* is essentially surjective on n-truncated objects for all n.*

The proof uses ideas which will be introduced in §6.5.1 and §7.2.2.

Proof. Without loss of generality, we may identify $\mathcal{Y}$ with the essential image of f^*. We use induction on n. The result is obvious for $n = 1$. Assume that $n > 1$ and let X be an n-truncated object of $\mathcal{X}$. By the inductive hypothesis, $U = \tau_{\leq n-1} X$ belongs to $\mathcal{Y}$. Replacing $\mathcal{X}$ and $\mathcal{Y}$ by $\mathcal{X}_{/U}$ and $\mathcal{Y}_{/U}$, we may suppose that X is n-connective.

We observe that $\pi_n X$ is an abelian group object of the ordinary topos $\mathrm{Disc}(\mathcal{X}_{/X})$. Since X is 2-connective, Proposition 7.2.1.13 implies that the pullback functor $\mathrm{Disc}(\mathcal{X}) \to \mathrm{Disc}(\mathcal{X}_{/X})$ is an equivalence of categories. We may therefore identify $\pi_n X$ with an abelian group object $A \in \mathrm{Disc}(\mathcal{X})$. Since A is discrete, it belongs to $\mathcal{Y}$. It follows that the Eilenberg-MacLane object $K(A, n + 1)$ belongs to $\mathcal{Y}$. Since X is an n-gerb banded by A, Theorem 7.2.2.26 implies the existence of a pullback diagram

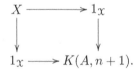

Since $\mathcal{Y}$ is stable under pullbacks in $\mathcal{X}$, we conclude that $X \in \mathcal{Y}$, as desired. $\square$

Corollary 6.3.6.7. *Let $f : \mathcal{X} \to \mathcal{Y}$ be a geometric morphism of ∞-topoi. Suppose that f^* is fully faithful and essentially surjective on 1-truncated objects and that $\mathcal{X}$ is n-localic (see §6.4.5). Then f is an equivalence of ∞-topoi.*

Remark 6.3.6.8. In the situation of Corollary 6.3.6.7, one can eliminate the hypothesis that $\mathcal{X}$ is n-localic in the presence of suitable finite-dimensionality assumptions on $\mathcal{X}$ and $\mathcal{Y}$; see §7.2.1.

Remark 6.3.6.9. Let $\mathcal{X}$ be an n-localic ∞-topos and let $\mathcal{Y}$ be the 2-localic ∞-topos associated to the 2-topos $\tau_{\leq 1} \mathcal{X}$, so that we have a geometric morphism $f : \mathcal{X} \to \mathcal{Y}$. It follows from Corollary 6.3.6.7 that f is algebraic. Roughly speaking, this tells us that there is only a very superficial interaction between the theory of k-categories and "topology," for $k > 2$. On the other hand, this statement fails dramatically if $k = 1$: the relationship between an ordinary topos and its underlying locale is typically very complicated and not algebraic in any reasonable sense. It is natural to ask what happens when $k = 2$. In other words, does Proposition 6.3.6.6 remain valid if f^* is only assumed to be essentially surjective on discrete objects? An affirmative answer would indicate that our theory of ∞-topoi is a relatively modest extension of classical topos theory. A counterexample could be equally interesting if it were to illustrate a nontrivial interaction between higher category theory and geometry.

6.4 N-TOPOI

Roughly speaking, an ordinary topos is a category which resembles the category of sheaves of *sets* on a topological space X. In §6.1, we introduced the definition of an ∞-topos. In the same rough terms, we can think of an ∞-topos as an ∞-category which resembles the ∞-category of sheaves of ∞-groupoids on a topological space X. Phrased in this way, it is natural to guess that these two notions have a common generalization. In §6.4.1, we will introduce the notion of an n-topos for every $0 \leq n \leq \infty$. The idea is that an n-topos should be an n-category which resembles the n-category of sheaves of $(n-1)$-groupoids on a topological space $\mathfrak{X}$. Of course, there are many approaches to making this idea precise. Our main result, Theorem 6.4.1.5, asserts that several candidate definitions are equivalent to one another. The proof of Theorem 6.4.1.5 will occupy our attention for most of this section. In §6.4.3, we study an axiomatization of the class of n-topoi in the spirit of Giraud's theorem, and in §6.4.4 we will give a characterization of n-topoi based on their descent properties. The case of $n = 0$ is somewhat exceptional and merits special treatement. In §6.4.2, we will show that a 0-topos is essentially the same thing as a *locale* (a mild generalization of the notion of a topological space).

   Our main motivation for introducing the definition of an n-topos is that it allows us to study ∞-topoi and topological spaces (or more generally, 0-topoi) in the same setting. In §6.4.5, we will introduce constructions which allow us to pass back and forth between m-topoi and n-topoi for any $0 \leq m \leq n \leq \infty$. We introduce an ∞-category $\mathfrak{Top}_n^{\mathrm{R}}$ of n-topoi for each $n \leq 0$ and show that each $\mathfrak{Top}_n^{\mathrm{R}}$ can be regarded as a *localization* of the ∞-category $\mathcal{R}\mathfrak{Top}$. In other words, the study of n-topoi for $n < \infty$ can be regarded as a special case of the theory of ∞-topoi.

6.4.1 Characterizations of n-Topoi

In this section, we will introduce the definition of n-topos for $0 \leq n < \infty$. In view of Theorem 6.1.0.6, there are several reasonable approaches to the subject. We will begin with an extrinsic approach.

Definition 6.4.1.1. Let $0 \leq n < \infty$. An ∞-category $\mathfrak{X}$ is an *n-topos* if there exists a small ∞-category $\mathcal{C}$ and an (accessible) left exact localization

$$L : \mathcal{P}_{\leq n-1}(\mathcal{C}) \to \mathfrak{X},$$

where $\mathcal{P}_{\leq n-1}(\mathcal{C})$ denotes the full subcategory of $\mathcal{P}(\mathcal{C})$ spanned by the $(n-1)$-truncated objects.

Remark 6.4.1.2. The accessibility condition on the localization functor $L : \mathcal{P}_{\leq n-1}(\mathcal{C}) \to \mathcal{X}$ of Definition 6.4.1.1 is superfluous: we will show that such a left exact localization of $\mathcal{P}_{\leq n-1}(\mathcal{C})$ is automatically accessible (combine Proposition 6.4.3.9 with Corollary 6.2.1.7).

Remark 6.4.1.3. An ∞-category $\mathcal{X}$ is a 1-topos if and only if it is equivalent to the nerve of an ordinary (Grothendieck) topos; this follows immediately from characterization (B) of Proposition 6.1.0.1.

Remark 6.4.1.4. Definition 6.4.1.1 also makes sense in the case $n = -1$ but is not very interesting. Up to equivalence, there is precisely one (-1)-topos: the final ∞-category $*$.

Our main goal is to prove the following result:

Theorem 6.4.1.5. *Let* $\mathcal{X}$ *be a presentable* ∞-*category and let* $0 \leq n < \infty$. *The following conditions are equivalent:*

(1) *There exists a small* n-*category* $\mathcal{C}$ *which admits finite limits, a Grothendieck topology on* $\mathcal{C}$, *and an equivalence of* $\mathcal{X}$ *with the full subcategory of* $\mathrm{Shv}_{\leq n-1}(\mathcal{C}) \subseteq \mathrm{Shv}(\mathcal{C})$ *consisting of* $(n-1)$-*truncated objects of* $\mathrm{Shv}(\mathcal{C})$.

(2) *There exists an* ∞-*topos* $\mathcal{Y}$ *and an equivalence* $\mathcal{X} \to \tau_{\leq n-1} \mathcal{Y}$.

(3) *The* ∞-*category* $\mathcal{X}$ *is an* n-*topos.*

(4) *Colimits in* $\mathcal{X}$ *are universal,* $\mathcal{X}$ *is equivalent to an* n-*category, and the class of* $(n-2)$-*truncated morphisms in* $\mathcal{X}$ *is local (see* §6.1.3*).*

(5) *Colimits in* $\mathcal{X}$ *are universal,* $\mathcal{X}$ *is equivalent to an* n-*category, and for all sufficiently large regular cardinals* κ, *there exists an object of* $\mathcal{X}$ *which classifies* $(n-2)$-*truncated relatively* κ-*compact morphisms in* $\mathcal{X}$.

(6) *The* ∞-*category* $\mathcal{X}$ *satisfies the following* n-*categorical versions of Giraud's axioms:*

   (i) *The* ∞-*category* $\mathcal{X}$ *is equivalent to a presentable* n-*category.*

   (ii) *Colimits in* $\mathcal{X}$ *are universal.*

   (iii) *If* $n > 0$, *then coproducts in* $\mathcal{X}$ *are disjoint.*

   (iv) *Every* n-*efficient (see* §6.4.3*) groupoid object of* $\mathcal{X}$ *is effective.*

Proof. The case $n = 0$ will be analyzed very explicitly in §6.4.2; let us therefore restrict our attention to the case $n > 0$. The implication $(1) \Rightarrow (2)$ is obvious (take $\mathcal{Y} = \mathrm{Shv}(\mathcal{C})$). Suppose that (2) is satisfied. Without loss of generality, we may suppose that $\mathcal{Y}$ is an (accessible) left exact localization of $\mathcal{P}(\mathcal{C})$ for some small ∞-category $\mathcal{C}$. Then $\mathcal{X}$ is a left exact localization of $\mathcal{P}_{\leq n-1}(\mathcal{C})$, which proves (3).

We next prove the converse (3) $\Rightarrow$ (2). We first observe that $\mathcal{P}_{\leq n-1}(\mathcal{C}) = \mathrm{Fun}(\mathcal{C}^{op}, \tau_{\leq n-1}\, \mathcal{S})$. Let $\mathrm{h}_n\,\mathcal{C}$ be the underlying n-category of $\mathcal{C}$, as in Proposition 2.3.4.12. Since $\tau_{\leq n-1}\,\mathcal{S}$ is equivalent to an n-category, we conclude that composition with the projection $\mathcal{C} \to \mathrm{h}_n\,\mathcal{C}$ induces an equivalence

$$\mathcal{P}_{\leq n-1}(\mathrm{h}_n\,\mathcal{C}) \to \mathcal{P}_{\leq n-1}(\mathcal{C}).$$

Consequently, we may assume without loss of generality (replacing $\mathcal{C}$ by $\mathrm{h}_n\,\mathcal{C}$ if necessary) that there is an accessible left exact localization $L : \mathcal{P}_{\leq n-1}(\mathcal{C}) \to \mathcal{X}$, where $\mathcal{C}$ is an n-category. Let S be the collection of all morphisms u in $\mathcal{P}_{\leq n-1}(\mathcal{C})$ such that Lu is an equivalence, so that S is of small generation. Let $\overline{S}$ be the strongly saturated class of morphisms in $\mathcal{P}(\mathcal{C})$ generated by S. We observe that $\tau_{\leq n-1}^{-1}(S)$ is a strongly saturated class of morphisms containing S, so that $\overline{S} \subseteq \tau_{\leq n-1}^{-1}(S)$. It follows that $S^{-1}\mathcal{P}_{\leq n-1}(\mathcal{C})$ is contained in $\mathcal{Y} = \overline{S}^{-1}\mathcal{P}(\mathcal{C})$ and may therefore be identified with the collection of $(n-1)$-truncated objects of $\mathcal{Y}$. To complete the proof, it will suffice to show that $\mathcal{Y}$ is an ∞-topos. For this, it will suffice to show that $\overline{S}$ is stable under pullbacks. Let T be the collection of all morphisms $f : X \to Y$ in $\mathcal{P}(\mathcal{C})$ such that for every pullback diagram

the morphism f' belongs to $\overline{S}$. It is easy to see that T is strongly saturated; we wish to show that $T \subseteq \overline{S}$. It will therefore suffice to prove that $S \subseteq T$. Let us therefore fix $f : X \to Y$ belonging to S and let $\mathcal{D}$ be the full subcategory of $\mathcal{P}(\mathcal{C})$ spanned by those objects Y' such that for *any* pullback diagram

$$
\begin{array}{ccc}
X' & \longrightarrow & X \\
\downarrow{\scriptstyle f'} & & \downarrow{\scriptstyle f} \\
Y' & \longrightarrow & Y,
\end{array}
$$

f' belongs to $\overline{S}$. Since colimits in $\mathcal{P}(\mathcal{C})$ are universal and $\overline{S}$ is stable under colimits, we conclude that $\mathcal{D}$ is stable under colimits in $\mathcal{P}(\mathcal{C})$. Since $\mathcal{P}(\mathcal{C})$ is generated under colimits by the essential image of the Yoneda embedding $j : \mathcal{C} \to \mathcal{P}(\mathcal{C})$, it will suffice to show that $j(C) \in \mathcal{D}$ for each $C \in \mathcal{C}$. We now observe that $\mathcal{P}_{\leq n-1}(\mathcal{C}) \subseteq \mathcal{D}$ (since S is stable under pullbacks in $\mathcal{P}_{\leq n-1}(\mathcal{C})$) and that $j(C) \in \mathcal{P}_{\leq n-1}(\mathcal{C})$ by virtue of our assumption that $\mathcal{C}$ is an n-category.

The implication (2) $\Rightarrow$ (4) will be established in §6.4.4 (Propositions 6.4.4.6 and 6.4.4.7). The proof of Theorem 6.1.6.8 adapts without change to show that (4) $\Leftrightarrow$ (5). The implication (4) $\Rightarrow$ (6) will be proven in §6.4.4 (Proposition 6.4.4.9). Finally, the "difficult" implication (6) $\Rightarrow$ (1) will be proven in §6.4.3 (Proposition 6.4.3.6) using an inductive argument quite similar to the proof of Giraud's original result.   □

Remark 6.4.1.6. Theorem 6.4.1.5 is slightly stronger than its ∞-categorical analogue, Theorem 6.1.0.6: it asserts that every n-topos arises as an n-category of sheaves on some n-category $\mathcal{C}$ equipped with a Grothendieck topology.

Remark 6.4.1.7. Let X be a presentable ∞-category in which colimits are universal. Then there exists a regular cardinal κ such that every monomorphism is relatively κ-compact. In this case, characterization (5) of Theorem 6.4.1.5 recovers a classical description of ordinary topos theory: a category X is a topos if and only if it is presentable, colimits in X are universal, and X has a subobject classifier.

6.4.2 0-Topoi and Locales

Our goal in this section is to prove Theorem 6.4.1.5 in the special case $n = 0$. A byproduct of our proof is a classification result (Corollary 6.4.2.6) which identifies the theory of 0-topoi with the classical theory of *locales* (Definition 6.4.2.3).

We begin by observing that when $n = 0$, a morphism in an ∞-category X is $(n - 2)$-truncated if and only if it is an equivalence. Consequently, any final object of X is an $(n - 2)$-truncated morphism classifier, and the class of $(n - 2)$-truncated morphisms is automatically local (in the sense of Definition 6.1.3.8). Moreover, if X is a 0-category, then every groupoid object in X is equivalent to a constant groupoid and therefore automatically effective. Consequently, characterizations (4) through (6) in Theorem 6.4.1.5 all reduce to the same condition on X and we may restate the desired result as follows:

Theorem 6.4.2.1. *Let X be a presentable 0-category. The following conditions are equivalent:*

(1) *There exists a small 0-category $\mathcal{C}$ which admits finite limits, a Grothendieck topology on $\mathcal{C}$, and an equivalence $X \to \mathrm{Shv}_{\leq -1}(\mathcal{C})$.*

(2) *There exists an ∞-topos $\mathcal{Y}$ and an equivalence $X \to \tau_{\leq -1}\mathcal{Y}$.*

(3) *The ∞-category X is a 0-topos.*

(4) *Colimits in X are universal.*

Before giving a proof of Theorem 6.4.2.1, it is convenient to reformulate condition (4). Recall that any 0-category X is equivalent to $\mathrm{N}(\mathcal{U})$, where $\mathcal{U}$ is a partially ordered set which is well-defined up to canonical isomorphism (see Example 2.3.4.3). The presentability of X is equivalent to the assertion that $\mathcal{U}$ is a *complete lattice*: that is, every subset of $\mathcal{U}$ has a least upper bound in $\mathcal{U}$ (this condition formally implies the existence of greater lower bounds as well).

Remark 6.4.2.2. If $n = 0$, then every presentable n-category is essentially small. This is typically not true for $n > 0$.

We note that the condition that colimits in $\mathcal{X}$ be universal can also be formulated in terms of the partially ordered set $\mathcal{U}$: it is equivalent to the assertion that meets in $\mathcal{U}$ commute with infinite joins in the following sense:

Definition 6.4.2.3. Let $\mathcal{U}$ be a partially ordered set. We will say that $\mathcal{U}$ is a *locale* if the following conditions are satisfied:

(1) Every subset $\{U_\alpha\}$ of elements of $\mathcal{U}$ has a least upper bound $\bigcup_\alpha U_\alpha$ in $\mathcal{U}$.

(2) The formation of least upper bounds commutes with meets in the sense that
$$\bigcup (U_\alpha \cap V) = \left(\bigcup U_\alpha \right) \cap V.$$
(Here $(U \cap V)$ denotes the greatest lower bound of U and V, which exists by virtue of assumption (1).)

Example 6.4.2.4. For every topological space X, the collection $\mathcal{U}(X)$ of open subsets of X forms a locale. Conversely, if $\mathcal{U}$ is a locale, then there is a natural topology on the collection of prime filters of $\mathcal{U}$ which allows us to extract a topological space from $\mathcal{U}$. These two constructions are adjoint to one another, and in good cases they are actually inverse equivalences. More precisely, the adjunction gives rise to an equivalence between the category of *spatial* locales and the category of *sober* topological spaces. In general, a locale can be regarded as a sort of generalized topological space in which one may speak of open sets but one does not generally have a sufficient supply of points. We refer the reader to [42] for details.

We can summarize the above discussion as follows:

Proposition 6.4.2.5. *Let $\mathcal{X}$ be a presentable 0-category. Then colimits in $\mathcal{X}$ are universal if and only if $\mathcal{X}$ is equivalent to $\mathrm{N}(\mathcal{U})$, where $\mathcal{U}$ is a locale.*

We are now ready to give the proof of Theorem 6.4.2.1.

Proof. The implications (1) $\Rightarrow$ (2) $\Rightarrow$ (3) are easy. Suppose that (3) is satisfied, so that $\mathcal{X}$ is a left exact localization of $\mathcal{P}_{\leq -1}(\mathcal{C})$ for some small ∞-category $\mathcal{C}$. Up to equivalence, there are precisely two (-1)-truncated spaces: $\emptyset$ and $*$. Consequently, $\tau_{\leq -1} \mathcal{S}$ is equivalent to the two-object ∞-category Δ^1. It follows that $\mathcal{P}_{\leq -1}(\mathcal{C})$ is equivalent to $\mathrm{Fun}(\mathcal{C}^{op}, \Delta^1)$.

Let $\widetilde{X}$ denote the collection of sieves on $\mathcal{C}$ ordered by inclusion. Then, identifying a functor $f : \mathcal{C} \to \Delta^1$ with the sieve $f^{-1}\{0\} \subseteq \mathcal{C}$, we deduce that $\mathrm{Fun}(\mathcal{C}, \Delta^1)$ is isomorphic to the nerve $\mathrm{N}(\widetilde{X})$.

Without loss of generality, we may identify $\mathcal{X}$ with the essential image of a localization functor $L : \mathrm{N}(\widetilde{X}) \to \mathrm{N}(\widetilde{X})$. The map L may be identified with a map of partially ordered sets from $\widetilde{X}$ to itself. Unwinding the definitions, we find that the condition that L be a left exact localization is equivalent to the following three properties:

(A) The map $L : \widetilde{X} \to \widetilde{X}$ is idempotent.

(B) For each $U \in \widetilde{X}$, $U \subseteq L(U)$.

(C) The map $L : \widetilde{X} \to \widetilde{X}$ preserves finite intersections (since X is a *left exact* localization of $\mathrm{N}(\widetilde{X})$.)

Let $\mathcal{U} = \{U \in \widetilde{X} : LU = U\}$. Then it is easy to see that X is equivalent to the nerve $\mathrm{N}(\mathcal{U})$ and that the partially ordered set X satisfies conditions (1) and (2) of Definition 6.4.2.3. Therefore $\mathcal{U}$ is a locale, so that colimits in $\mathrm{N}(\mathcal{U})$ are universal by Proposition 6.4.2.5. This proves that $(3) \Rightarrow (4)$.

Now suppose that (4) is satisfied. Using Proposition 6.4.2.5, we may suppose without loss of generality that $X = \mathrm{N}(\mathcal{U})$, where $\mathcal{U}$ is a locale. We observe that X is itself small. Let us say that a sieve $\{U_\alpha \to U\}$ on an object $U \in X$ is *covering* if

$$U = \bigcup_\alpha U_\alpha$$

in $\mathcal{U}$. Using the assumption that $\mathcal{U}$ is a locale, it is easy to see that the collection of covering sieves determines a Grothendieck topology on X. The ∞-category $\mathcal{P}_{\leq -1}(X)$ can be identified with the nerve of the partially ordered set of all downward-closed subsets $\mathcal{U}_0 \subseteq \mathcal{U}$. Moreover, an object of $\mathcal{P}_{\leq -1}(X)$ belongs to $\mathrm{Shv}_{\leq -1}(X)$ if and only if the corresponding subset $\mathcal{U}_0 \subseteq \mathcal{U}$ is stable under joins. Every such subset $\mathcal{U}_0 \subseteq \mathcal{U}$ has a largest element $U \in \mathcal{U}$, and we then have an identification $\mathcal{U}_0 = \{V \in \mathcal{U} : V \leq U\}$. It follows that $\mathrm{Shv}_{\leq -1}(X)$ is equivalent to the nerve of the partially ordered set $\mathcal{U}$, which is X. This proves (1) and concludes the argument. □

We may summarize the results of this section as follows:

Corollary 6.4.2.6. *An ∞-category X is a 0-topos if and only if it is equivalent to $\mathrm{N}(\mathcal{U})$, where $\mathcal{U}$ is a locale.*

Remark 6.4.2.7. Coproducts in a 0-topos are typically *not* disjoint.

In classical topos theory, there are functorial constructions for passing back and forth between topoi and locales. Given a locale $\mathcal{U}$ (such as the locale $\mathcal{U}(X)$ of open subsets of a topological space X), one may define a topos X of *sheaves (of sets) on* $\mathcal{U}$. The original locale $\mathcal{U}$ may then be recovered as the partially ordered set of subobjects of the final object of X. In fact, for any topos X, the partially ordered set $\mathcal{U}$ of subobjects of the final object forms a locale. In general, X cannot be recovered as the category of sheaves on $\mathcal{U}$; this is true if and only if X is a *localic* topos: that is, if and only if X is generated under colimits by the collection of subobjects of the final object 1_X. In §6.3, we will discuss a generalization of this picture which will allow us to pass between m-topoi and n-topoi for any $m \leq n$.

6.4.3 Giraud's Axioms for n-Topoi

In §6.1.1, we sketched an axiomatic approach to the theory of ∞-topoi. The axioms we introduced were closely parallel to Giraud's axioms for ordinary topoi with one important difference. If $\mathcal{X}$ is an ∞-topos, then *every* groupoid object of $\mathcal{X}$ is effective. If $\mathcal{X}$ is an ordinary topos, then a groupoid $U_\bullet$ is effective only if the diagram

$$U_1 \rightrightarrows U_0$$

exhibits U_1 as an equivalence relation on U_0. Our first goal in this section is to formulate an analogue of this condition, which will lead us to an axiomatic description of n-topoi for all $0 \leq n \leq \infty$.

Definition 6.4.3.1. Let $\mathcal{X}$ be an ∞-category and $U_\bullet$ a groupoid object of $\mathcal{X}$. We will say that $U_\bullet$ is *n-efficient* if the natural map

$$U_1 \to U_0 \times U_0$$

(which is well-defined up to equivalence) is $(n-2)$-truncated.

Remark 6.4.3.2. By convention, we regard every groupoid object as ∞-efficient.

Example 6.4.3.3. If $\mathcal{C}$ is (the nerve of) an ordinary category, then giving a 1-efficient groupoid object $U_\bullet$ of $\mathcal{C}$ is equivalent to giving an object U_0 of $\mathcal{C}$ and an equivalence relation U_1 on U_0.

Proposition 6.4.3.4. *An ∞-category $\mathcal{X}$ is equivalent to an n-category if and only if every effective groupoid in $\mathcal{X}$ is n-efficient.*

Proof. Suppose first that $\mathcal{C}$ is equivalent to an n-category. Let $U_\bullet$ be an effective groupoid in $\mathcal{X}$. Then $U_\bullet$ has a colimit U_{-1}. The existence of a pullback diagram

$$
\begin{array}{ccc}
U_1 & \longrightarrow & U_0 \\
\downarrow & & \downarrow \\
U_0 & \longrightarrow & U_{-1}
\end{array}
$$

implies that the map $f' : U_1 \to U_0 \times U_0$ is a pullback of the diagonal map $f : U_{-1} \to U_{-1} \times U_{-1}$. We wish to show that f' is $(n-2)$-truncated. By Lemma 5.5.6.12, it suffices to show that f is $(n-2)$-truncated. By Lemma 5.5.6.15, this is equivalent to the assertion that U_{-1} is $(n-1)$-truncated. Since $\mathcal{C}$ is equivalent to an n-category, every object of $\mathcal{C}$ is $(n-1)$-truncated.

Now suppose that every effective groupoid in $\mathcal{X}$ is n-efficient. Let $U \in \mathcal{X}$ be an object; we wish to show that U is $(n-1)$-truncated. The constant simplicial object $U_\bullet$ taking the value U is an effective groupoid and therefore n-efficient. It follows that the diagonal map $U \to U \times U$ is $(n-2)$-truncated. Lemma 5.5.6.15 implies that U is $(n-1)$-truncated, as desired. $\square$

We are now almost ready to supply the "hard" step in the proof of Theorem 6.4.1.5 (namely, the implication $(6) \Rightarrow (1)$). We first need a slightly technical lemma whose proof requires routine cardinality estimates.

Lemma 6.4.3.5. *Let $\mathfrak{X}$ be a presentable ∞-category in which colimits are universal. There exists a regular cardinal τ such that $\mathfrak{X}$ is τ-accessible, and the full subcategory of $\mathfrak{X}^\tau \subseteq \mathfrak{X}$ spanned by the τ-compact objects is stable under the formation of subobjects and finite limits.*

Proof. Choose a regular cardinal κ such that $\mathfrak{X}$ is κ-accessible. We observe that, up to equivalence, there are a bounded number of κ-compact objects of $\mathfrak{X}$ and therefore a bounded number of *subobjects* of κ-compact objects of $\mathfrak{X}$. Now choose an uncountable regular cardinal $\tau \gg \kappa$ such that:

(1) The ∞-category $\mathfrak{X}^\kappa$ is essentially τ-small.

(2) For each $X \in \mathfrak{X}^\kappa$ and each monomorphism $i : U \to X$, U is τ-compact.

It is clear that $\mathfrak{X}$ is τ-accessible, and $\mathfrak{X}^\tau$ is stable under finite limits (in fact, κ-small limits) by Proposition 5.4.7.4. To complete the proof, we must show that $\mathfrak{X}^\tau$ is stable under the formation of subobjects. Let $i : U \to X$ be a monomorphism, where X is τ-compact. Since $\mathfrak{X}$ is κ-accessible, we can write X as the colimit of a κ-filtered diagram $p : \mathcal{J} \to \mathfrak{X}^\kappa$. Since X is τ-compact, it is a retract of the colimit X' of some τ-small subdiagram $p| \mathcal{J}'$. Since τ is uncountable, we can use Proposition 4.4.5.12 to write X as the colimit of a τ-small diagram Idem $\to \mathfrak{X}$ which carries the unique object of Idem to X'. Since colimits in $\mathfrak{X}$ are universal, it follows that U can be written as a τ-small colimit of a diagram Idem $\to \mathfrak{X}$ which takes the value $U' = U \times_X X'$. It will therefore suffice to prove that U' is τ-compact. Invoking the universality of colimits once more, we observe that U' is a τ-small colimit of objects of the form $U'' = U' \times_{X'} p(J)$, where J is an object of $\mathcal{J}'$. We now observe that U'' is a subobject of $p(J) \in \mathfrak{X}^\kappa$ and is therefore τ-compact by assumption (2). It follows that U', being a τ-small colimit of τ-compact objects of $\mathfrak{X}$, is also τ-compact. $\square$

Proposition 6.4.3.6. *Let $0 < n < \infty$ and let $\mathfrak{X}$ be an ∞-category satisfying the following conditions:*

(i) *The ∞-category $\mathfrak{X}$ is presentable.*

(ii) *Colimits in $\mathfrak{X}$ are universal.*

(iii) *Coproducts in $\mathfrak{X}$ are disjoint.*

(iv) *The effective groupoid objects of $\mathfrak{X}$ are precisely the n-efficient groupoids.*

Then there exists a small n-category $\mathcal{C}$ which admits finite limits, a Grothendieck topology on $\mathcal{C}$, and an equivalence $\mathfrak{X} \to \mathrm{Shv}_{\leq n-1}(\mathcal{C})$.

Proof. Without loss of generality, we may suppose that $\mathfrak{X}$ is minimal. Choose a regular cardinal κ such that $\mathfrak{X}$ is κ-accessible, and the full subcategory $\mathcal{C} \subseteq \mathfrak{X}$ spanned by the κ-compact objects of $\mathfrak{X}$ is stable under the formation of subobjects and finite limits (Lemma 6.4.3.5). We endow $\mathcal{C}$ with the canonical topology induced by the inclusion $\mathcal{C} \subseteq \mathfrak{X}$. According to Theorem 5.1.5.6, there is an (essentially unique) colimit-preserving functor $F : \mathcal{P}(\mathcal{C}) \to \mathfrak{X}$ such that $F \circ j$ is equivalent to the inclusion $\mathcal{C} \subseteq \mathfrak{X}$, where $j : \mathcal{C} \to \mathcal{P}(\mathcal{C})$ denotes the Yoneda embedding. The proof of Theorem 5.5.1.1 shows that F has a fully faithful right adjoint $G : \mathfrak{X} \to \mathcal{P}(\mathcal{C})$. We will complete the proof by showing that the essential image of G is precisely $\mathrm{Shv}_{\leq n-1}(\mathcal{C})$.

Since $\mathfrak{X}$ is equivalent to an n-category (Proposition 6.4.3.4) and G is left exact, we conclude that G factors through $\mathcal{P}_{\leq n-1}(\mathcal{C})$. It follows from Proposition 6.2.4.6 that G factors through $\mathrm{Shv}_{\leq n-1}(\mathcal{C})$. Let $\mathfrak{X}' \subseteq \mathrm{Shv}_{\leq n-1}(\mathcal{C})$ denote the essential image of G. To complete the proof, it will suffice to show that $\mathfrak{X}' = \mathrm{Shv}_{\leq n-1}(\mathcal{C})$. Let $\emptyset$ be an initial object of $\mathfrak{X}$. The space $\mathrm{Map}_{\mathfrak{X}}(X, \emptyset)$ is contractible if X is an initial object of $\mathfrak{X}$ and empty otherwise (Lemma 6.1.3.6). It follows from Proposition 6.2.2.10 that $G(\emptyset)$ is an initial object of $\mathrm{Shv}_{\leq n-1}(\mathcal{C})$.

We next claim that $\mathfrak{X}'$ is stable under small coproducts in $\mathrm{Shv}_{\leq n-1}(\mathcal{C})$. It will suffice to show that the map G preserves coproducts. Let $\{U_\alpha\}$ be a small collection of objects of $\mathfrak{X}$ and U their coproduct in $\mathfrak{X}$. According to Lemma 6.1.5.1, we have a pullback diagram

$$
\begin{array}{ccc}
V_{\alpha,\beta} & \xrightarrow{\phi} & U_\alpha \\
{\scriptstyle \phi'}\downarrow & & \downarrow \\
U_\beta & \longrightarrow & U,
\end{array}
$$

where $V_{\alpha,\beta}$ is an initial object of $\mathfrak{X}$ if $\alpha \neq \beta$, while ϕ and ϕ' are equivalences if $\alpha = \beta$. The functor G preserves all limits, so that the diagram

$$
\begin{array}{ccc}
G(V_{\alpha,\beta}) & \longrightarrow & G(U_\alpha) \\
\downarrow & & \downarrow \\
G(U_\beta) & \longrightarrow & G(U)
\end{array}
$$

is a pullback in $\mathrm{Shv}_{\leq n-1}(\mathcal{C})$. Let U' denote a coproduct of the objects $i(U_\alpha)$ in $\mathrm{Shv}_{\leq n-1}(\mathcal{C})$ and let $g : U' \to G(U)$ be the induced map. Since colimits in $\mathfrak{X}$ are universal, we obtain a natural identification of $U' \times_{G(U)} U'$ with the coproduct

$$
\coprod_{\alpha,\beta} (G(U_\alpha) \times_{G(U)} G(U_\beta)) \simeq \coprod_\alpha U_\alpha \simeq U',
$$

where the second equivalence follows from our observation that G preserves initial objects. Applying Lemma 5.5.6.15, we deduce that g is a monomorphism.

To prove that g is an equivalence, it will suffice to show that the map
$$\pi_0 U'(C) \to \pi_0 G(U)(C) = \pi_0 \operatorname{Map}_{\mathfrak{X}}(C, U)$$
is surjective for every object $C \in \mathcal{C}$. Since colimits in $\mathfrak{X}$ are universal, every map $h : C \to U$ can be written as a coproduct of maps $h_\alpha : C_\alpha \to U_\alpha$. Each C_α is a subobject of C (Lemma 6.4.4.8) and therefore belongs to $\mathcal{C}$. Let $h'_\alpha \in \pi_0 U'(C_\alpha)$ denote the homotopy class of the composition $G(C_\alpha) \overset{h_\alpha}{\to} G(U_\alpha) \to U'$. Since the topology on $\mathcal{C}$ is canonical, Lemma 6.2.4.4 implies that $\pi_0 U'(C) \simeq \prod_\alpha \pi_0 U'(C_\alpha)$ contains an element h' which restricts to each h'_α. It is now clear that h is the image of h' under the map $\pi_0 U'(C) \to \pi_0 \operatorname{Map}_{\mathfrak{X}}(C, U)$.

We will prove the following result by induction on k: if there exists a k-truncated morphism $f : X \to Y$, where $Y \in \mathfrak{X}'$ and $X \in \operatorname{Shv}_{\leq n-1}(\mathcal{C})$, then $X \in \mathfrak{X}'$. Taking $k = n - 1$ and Y to be a final object of $\operatorname{Shv}_{\leq n-1}(\mathcal{C})$ (which belongs to $\mathfrak{X}'$ because $\mathcal{C}$ contains a final object), we conclude that every object of $\operatorname{Shv}_{\leq n-1}(\mathcal{C})$ belongs to $\mathfrak{X}'$, which completes the proof.

If $k = -2$, then f is an equivalence so that $X \in \mathfrak{X}'$, as desired. Assume now that $k \geq -1$. Since $\mathfrak{X}'$ contains the essential image of the Yoneda embedding and is stable under coproducts, there exists an effective epimorphism $p : U \to X$ in $\operatorname{Shv}_{\leq n-1}(\mathcal{C})$, where $U \in \mathfrak{X}'$. Let $\overline{U}_\bullet$ be a Čech nerve of p in $\operatorname{Shv}_{\leq n-1}(\mathcal{C})$ and $U_\bullet$ be the associated groupoid object. We claim that $U_\bullet$ is a groupoid object of $\mathfrak{X}'$. Since $\mathfrak{X}'$ is stable under limits in $\operatorname{Shv}_{\leq n-1}(\mathcal{C})$, it suffices to prove that $U_0 = U$ and $U_1 = U \times_X U$ belong to $\mathfrak{X}'$. We now observe that there exists a pullback diagram
$$
\begin{array}{ccc}
U \times_X U & \overset{\delta'}{\longrightarrow} & U \times_Y U \\
\downarrow & & \downarrow \\
X & \overset{\delta}{\longrightarrow} & X \times_Y X.
\end{array}
$$
Since f is k-truncated, δ is $(k-1)$-truncated (Lemma 5.5.6.15), so that δ' is $(k-1)$-truncated. Since $U \times_Y U$ belongs to $\mathfrak{X}'$ (because $\mathfrak{X}'$ is stable under limits), our inductive hypothesis allows us to conclude that $U \times_X U \in \mathfrak{X}'$, as desired.

We observe that $U_\bullet$ is an n-efficient groupoid object of $\mathfrak{X}'$. Invoking assumption (iv), we conclude that $U_\bullet$ is effective in $\mathfrak{X}'$. Let $X' \in \mathfrak{X}'$ be a colimit of $U_\bullet$ in $\mathfrak{X}'$, so that we have a morphism $u : X \to X'$ in $\operatorname{Shv}_{\leq n-1}(\mathcal{C})_{U_\bullet/}$. To complete the proof that $X \in \mathfrak{X}'$, it will suffice to show that u is an equivalence. Since u induces an equivalence
$$U \times_X U \to U \times_{X'} U,$$
it is a monomorphism (Lemma 5.5.6.15). It will therefore suffice to show that u is an effective epimorphism. We have a commutative diagram

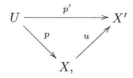

where p is an effective epimorphism; it therefore suffices to show that p' is an effective epimorphism, which follows immediately from Proposition 6.2.4.6. $\square$

Remark 6.4.3.7. Proposition 6.4.3.6 is valid also for $n = 0$ but is almost vacuous: coproducts in a 0-topos $\mathfrak{X}$ are never disjoint unless $\mathfrak{X}$ is trivial (equivalent to the final ∞-category $*$).

Remark 6.4.3.8. In a certain respect, the theory of ∞-topoi is *simpler than* the theory of ordinary topoi: in an ∞-topos, *every* groupoid object is effective; it is not necessary to impose any additional conditions like n-efficiency. The absense of this condition gives the theory of ∞-topoi a slightly different flavor than ordinary topos theory. In an ∞-topos, we are free to form quotients of objects not only by equivalence relations but also by arbitrary groupoid actions. In geometry, this extra flexibility allows the construction of useful objects such as orbifolds and algebraic stacks, which are useful in a variety of mathematical situations.

One can imagine weakening the gluing conditions even further and considering axioms having the form "every category object is effective." This seems like a natural approach to a theory of topos-like (∞, ∞)-categories. However, we will not pursue the matter any further here.

It follows from Proposition 6.4.3.6 (and arguments to be given in §6.4.4) that every left exact localization of a presheaf n-category $\mathcal{P}_{\leq n-1}(\mathcal{C})$ can also be obtained as an n-category of sheaves. According to the next two results, this is no accident: every left exact localization of $\mathcal{P}_{\leq n-1}(\mathcal{C})$ is topological, and the topological localizations of $\mathcal{P}_{\leq n-1}(\mathcal{C})$ correspond precisely to the Grothendieck topologies on $\mathcal{C}$ (provided that $\mathcal{C}$ is an n-category).

Proposition 6.4.3.9. *Let $\mathfrak{X}$ be a presentable n-category, let $0 \leq n < \infty$, and suppose that colimits in $\mathfrak{X}$ are universal. Let $L : \mathfrak{X} \to \mathcal{Y}$ be a left exact localization. Then L is a topological localization.*

Proof. Let S denote the collection of all monomorphisms $f : U \to V$ in $\mathfrak{X}$ such that Lf is an equivalence. Since L is left exact, it is clear that S is stable under pullback. Let $\overline{S}$ be the strongly saturated class of morphisms generated by S. Proposition 6.2.1.2 implies that $\overline{S}$ is stable under pullback and therefore topological. Proposition 6.2.1.6 implies that $\overline{S}$ is generated by a (small) set of morphisms. Let $\mathfrak{X}' \subseteq \mathfrak{X}$ denote the full subcategory spanned by $\overline{S}$-local objects. According to Proposition 5.5.4.15, $\mathfrak{X}'$ is an accessible localization of $\mathfrak{X}$; let L' denote the associated localization functor. Since Lf is an equivalence for each $f \in \overline{S}$, the localization L is equivalent to the composition

$$\mathfrak{X} \xrightarrow{L'} \mathfrak{X}' \xrightarrow{L|\,\mathfrak{X}'} \mathcal{Y} \, .$$

We may therefore replace $\mathfrak{X}$ by $\mathfrak{X}'$ and thereby reduce to the case where S consists precisely of the equivalences in $\mathfrak{X}$; we wish to prove that L is an equivalence.

We now prove the following claim: if $f : X \to Y$ is a k-truncated morphism in $\mathcal{C}$ such that Lf is an equivalence, then f is an equivalence. The proof proceeds by induction on k. If $k = -1$, then f is a monomorphism and so belongs to S; it follows that f is an equivalence. Suppose that $k \geq 0$. Let $\delta : X \to X \times_Y X$ be the diagonal map (which is well-defined up to equivalence). According to Lemma 5.5.6.15, δ is $(k - 1)$-truncated. Since L is left exact, $L(\delta)$ can be identified with a diagonal map $LX \to LX \times_{LY} LX$ which is therefore an equivalence. The inductive hypothesis implies that δ is an equivalence. Applying Lemma 5.5.6.15 again, we deduce that f is a monomorphism, so that $f \in S$ and is therefore an equivalence as noted above.

Since $\mathcal{X}$ is an n-category, every morphism in $\mathcal{X}$ is $(n - 1)$-truncated. We conclude that for *every* morphism f in $\mathcal{X}$, f is an equivalence if and only if Lf is an equivalence. Since L is a localization functor, it must be an equivalence. $\qquad\square$

6.4.4 n-Topoi and Descent

Let $\mathcal{X}$ be an ∞-category which admits finite limits and let $\mathcal{O}_{\mathcal{X}}$ denote the functor ∞-category $\mathrm{Fun}(\Delta^1, \mathcal{X})$ equipped with the Cartesian fibration $e : \mathcal{O}_{\mathcal{X}} \to \mathcal{X}$ (given by evaluation at $\{1\} \subseteq \Delta^1$), as in §6.1.1. Let $F : \mathcal{X}^{op} \to \widehat{\mathcal{C}at}_\infty$ be a functor which classifies e; informally, F associates to each object $U \in \mathcal{X}$ the ∞-category $\mathcal{X}_{/U}$. According to Theorem 6.1.3.9, $\mathcal{X}$ is an ∞-topos if and only if the functor F preserves limits and factors through $\mathcal{P}\mathrm{r}^L \subseteq \widehat{\mathcal{C}at}_\infty$. The assumption that F preserves limits can be viewed as a descent condition: it asserts that if $X \to U$ is a morphism of $\mathcal{X}$ and U is decomposed into "pieces" U_α, then X can be canonically reconstructed from the "pieces" $X \times_U U_\alpha$. The goal of this section is to obtain a similar characterization of the class of n-topoi for $0 \leq n < \infty$.

We begin by considering the case where $\mathcal{X}$ is the (nerve of) the category of sets. In this case, we can think of F as a contravariant functor from sets to categories, which carries a set U to the category $\mathrm{Set}_{/U}$. This functor does not preserve pullbacks: given a pushout square

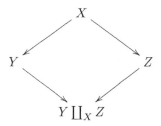

in the category Set, there is an associated functor

$$\theta : \mathrm{Set}_{/Y \amalg_X Z} \to \mathrm{Set}_{/Y} \times_{\mathrm{Set}_{/X}} \mathrm{Set}_{/Z}$$

(here the right hand side indicates a *homotopy* fiber product of categories). The functor θ is generally not an equivalence of categories: for example, θ

fails to be an equivalence if $Y = Z = *$, provided that X has cardinality of at least 2. However, θ is always fully faithful. Moreover, we have the following result:

Fact 6.4.4.1. *The functor θ induces an isomorphism of partially ordered sets*

$$\text{Sub}(Y \coprod_X Z) \to \text{Sub}(Y) \times_{\text{Sub}(X)} \text{Sub}(Z),$$

where $\text{Sub}(M)$ *denotes the partially ordered set of subsets of M.*

In this section, we will show that an appropriate generalization of Fact 6.4.4.1 can be used to characterize the class of n-topoi for all $0 \le n \le \infty$. First, we need to introduce some terminology.

Notation 6.4.4.2. Let $\mathcal{X}$ be an ∞-category which admits pullbacks and let $0 \le n \le \infty$. We let $\mathcal{O}_{\mathcal{X}}^n$ denote the full subcategory of $\mathcal{O}_{\mathcal{X}}$ spanned by morphisms $f : U \to X$ which are $(n-2)$-truncated, and $\mathcal{O}_{\mathcal{X}}^{(n)} \subseteq \mathcal{O}_{\mathcal{X}}^n$ the subcategory whose objects are $(n-2)$-truncated morphisms in $\mathcal{X}$ and whose morphisms are Cartesian transformations (see Notation 6.1.3.4).

Example 6.4.4.3. Let $\mathcal{X}$ be an ∞-category which admits pullbacks. Then $\mathcal{O}_{\mathcal{X}}^0$ is the full subcategory of $\mathcal{O}_{\mathcal{X}}$ spanned by the final objects in each fiber of the morphism $p : \mathcal{O}_{\mathcal{X}} \to \mathcal{X}$. Since p is a coCartesian fibration (Corollary 2.4.7.12), Proposition 2.4.4.9 asserts that the restriction $p|\,\mathcal{O}_{\mathcal{X}}^0$ is a trivial fibration of simplicial sets.

Lemma 6.4.4.4. *Let $\mathcal{X}$ be a presentable ∞-category in which colimits are universal and coproducts are disjoint and let $n \ge -2$. Then the class of n-truncated morphisms in $\mathcal{X}$ is stable under small coproducts.*

Proof. The proof is by induction on n, where the case $n = -2$ is obvious. Suppose that $\{f_\alpha : X_\alpha \to Y_\alpha\}$ is a family of n-truncated morphisms in $\mathcal{X}$ having coproduct $f : X \to Y$. Since colimits in $\mathcal{X}$ are universal, we conclude that $X \times_Y X$ can be written as a coproduct

$$\coprod_{\alpha,\beta} (X_\alpha \times_Y X_\beta) \simeq \coprod_{\alpha,\beta} (X_\alpha \times_{Y_\alpha} (Y_\alpha \times_Y Y_\beta) \times_{Y_\beta} X_\beta).$$

Applying Lemma 6.1.5.1, we can rewrite this coproduct as

$$\coprod_\alpha (X_\alpha \times_{Y_\alpha} X_\alpha).$$

Consequently, the diagonal map $\delta : X \to X \times_Y X$ is a coproduct of diagonal maps $\{\delta_\alpha : X_\alpha \to X_\alpha \times_{Y_\alpha} X_\alpha\}$. Applying Lemma 5.5.6.15, we deduce that each δ_α is $(n-1)$-truncated, so that δ is $(n-1)$-truncated by the inductive hypothesis. We now apply Lemma 5.5.6.15 again to deduce that f is n-truncated, as desired. $\qquad\square$

Combining Lemmas 6.1.3.3, 6.1.3.5, 6.1.3.7, and 6.4.4.4, we deduce the following analogue of Theorem 6.1.3.9.

Theorem 6.4.4.5. *Let X be a presentable ∞-category in which colimits are universal and coproducts are disjoint. The following conditions are equivalent:*

(1) *For every pushout diagram*

$$
\begin{array}{ccc}
f & \xrightarrow{\ \alpha\ } & g \\
\big\downarrow{\scriptstyle\beta} & & \big\downarrow{\scriptstyle\beta'} \\
f' & \xrightarrow{\ \alpha'\ } & g'
\end{array}
$$

in $\mathcal{O}_X^n$, if α and β are Cartesian transformations, then α' and β' are also Cartesian transformations.

(2) *The class of $(n-2)$-truncated morphisms in X is local.*

(3) *The Cartesian fibration $\mathcal{O}_X^n \to X$ is classified by a limit-preserving functor $X^{op} \to \widehat{\mathcal{C}at}_\infty$.*

(4) *The right fibration $\mathcal{O}_X^{(n)} \to X$ is classified by a limit-preserving functor $X^{op} \to \widehat{S}$.*

(5) *Let K be a small simplicial set and $\overline{\alpha} : \overline{p} \to \overline{q}$ a natural transformation of colimit diagrams $\overline{p}, \overline{q} : K^{\triangleright} \to X$. Suppose that $\alpha = \overline{\alpha}|K$ is a Cartesian transformation and that $\alpha(x)$ is $(n-2)$-truncated for every vertex $x \in K$. Then $\overline{\alpha}$ is a Cartesian transformation and $\overline{\alpha}(\infty)$ is $(n-2)$-truncated, where ∞ denotes the cone point of $K^{\triangleright}$.*

Our next goal is to establish the implication $(2) \Rightarrow (4)$ of Theorem 6.4.1.5. We will deduce this from the equivalence $(2) \Leftrightarrow (3)$ (which we have already established) together with Propositions 6.4.4.6 and 6.4.4.7 below.

Proposition 6.4.4.6. *Let X be an n-topos, $0 \leq n \leq \infty$. Then colimits in X are universal.*

Proof. Using Lemma 6.1.3.15, we may reduce to the case $X = \mathcal{P}_{\leq n-1}(\mathcal{C})$ for some small ∞-category $\mathcal{C}$. Using Proposition 5.1.2.2, we may further reduce to the case where $X = \tau_{\leq n-1}\mathcal{S}$.

Let $f : X \to Y$ be a map of $(n-1)$-truncated spaces and let $f^* : \mathcal{S}^{/Y} \to \mathcal{S}^{/X}$ be a pullback functor. Since X is stable under limits in $\mathcal{S}$, f^* restricts to give a functor $X^{/Y} \to X^{/X}$; we wish to prove that this restricted functor commutes with colimits. We observe that $X^{/X}$ and $X^{/Y}$ can be identified with the full subcategories of $\mathcal{S}^{/X}$ and $\mathcal{S}^{/Y}$ spanned by the $(n-1)$-truncated objects, by Lemma 5.5.6.14. Let $\tau_X : \mathcal{S}^{/X} \to X^{/X}$ and $\tau_Y : \mathcal{S}^{/Y} \to X^{/Y}$ denote left adjoints to the inclusions. The functor f^* preserves all colimits (Lemma 6.1.3.14) and all limits (since f^* has a left adjoint). Consequently, Proposition 5.5.6.28 implies that $\tau_X \circ f^* \simeq f^* \circ \tau_Y$.

Let $p : K^{\triangleright} \to X^{/Y}$ be a colimit diagram. We wish to show that $f^* \circ p$ is a colimit diagram. According to Remark 5.2.7.5, we may assume that

$p = \tau_Y \circ p'$ for some colimit diagram $p' : K^{\triangleright} \to \mathcal{S}^{/Y}$. Since colimits in $\mathcal{S}$ are universal (Lemma 6.1.3.14), the composition $f^* \circ p' : K^{\triangleright} \to \mathcal{S}^{/X}$ is a colimit diagram. Since τ_X preserves colimits, we conclude that $\tau_X \circ f^* \circ p' : K^{\triangleright} \to \mathcal{X}^{/X}$ is a colimit diagram, so that $f^* \circ \tau_Y \circ p' = f^* \circ p$ is also a colimit diagram, as desired. $\square$

Proposition 6.4.4.7. *Let $\mathcal{Y}$ be an ∞-topos and let $\mathcal{X} = \tau_{\leq n}\mathcal{Y}$, $0 \leq n \leq \infty$. Then the class of $(n-2)$-truncated morphisms in $\mathcal{X}$ is local.*

Proof. Combining Propositions 6.2.3.17 and 6.2.3.14 with Lemma 6.4.4.4, we conclude that the class of $(n-2)$-truncated morphisms in $\mathcal{Y}$ is local. Consequently, the Cartesian fibration $\mathcal{O}_{\mathcal{Y}}^n \to \mathcal{Y}$ is classified by a colimit-preserving functor $F : \mathcal{Y} \to \widehat{\mathcal{C}at}_{\infty}^{op}$. It follows that $\mathcal{O}_{\mathcal{X}}^{(n)} \to \mathcal{X}$ is classified by $F | \mathcal{X}$. To prove that $F | \mathcal{X}$ is colimit-preserving, it will suffice to show that F is equivalent to $F \circ \tau_{\leq n}$. In other words, we must show that F carries each n-truncation $Y \to \tau_{\leq n}Y$ to an equivalence in $\widehat{\mathcal{C}at}_{\infty}^{op}$. Replacing $\mathcal{Y}$ by $\mathcal{Y}_{/\tau_{\leq n}Y}$, we reduce to Lemma 7.2.1.13. $\square$

We conclude this section by proving the following generalization of Proposition 6.1.3.19, which also establishes the implication (4) $\Rightarrow$ (6) of Theorem 6.4.1.5. We will assume $n > 0$; the case $n = 0$ was analyzed in §6.4.2.

Lemma 6.4.4.8. *Let $\mathcal{X}$ be a presentable ∞-category in which colimits are universal and let $f : \emptyset \to X$ be a morphism in $\mathcal{X}$, where $\emptyset$ is an initial object of $\mathcal{X}$. Then f is a monomorphism.*

Proof. Let Y be an arbitrary object of $\mathcal{X}$. We wish to show that composition with f induces a (-1)-truncated map

$$\mathrm{Map}_{\mathcal{X}}(Y, \emptyset) \to \mathrm{Map}_{\mathcal{X}}(Y, X).$$

If Y is an initial object of $\mathcal{X}$, then both sides are contractible; otherwise the left side is empty (Lemma 6.1.3.6). $\square$

Proposition 6.4.4.9. *Let $1 \leq n \leq \infty$ and let $\mathcal{X}$ be a presentable n-category. Suppose that colimits in $\mathcal{X}$ are universal and that the class of $(n-2)$-truncated morphisms in $\mathcal{X}$ is local. Then $\mathcal{X}$ satisfies the n-categorical Giraud axioms:*

(i) *The ∞-category $\mathcal{X}$ is equivalent to a presentable n-category.*

(ii) *Colimits in $\mathcal{X}$ are universal.*

(iii) *Coproducts in $\mathcal{X}$ are disjoint.*

(iv) *Every n-efficient groupoid object of $\mathcal{X}$ is effective.*

Proof. Axioms (i) and (ii) hold by assumption. To show that coproducts in $\mathcal{X}$ are disjoint, let us consider an arbitrary pair of objects $X, Y \in \mathcal{X}$ and let $\emptyset$ denote an initial object of $\mathcal{X}$. Let $f : \emptyset \to X$ be a morphism (unique up to homotopy since $\emptyset$ is initial). Since colimits in $\mathcal{X}$ are universal, f is a monomorphism (Lemma 6.4.4.8) and therefore belongs to $\mathcal{O}_{\mathcal{X}}^n$ since $n \geq 1$. We observe that $\mathrm{id}_\emptyset$ is an initial object of $\mathcal{O}_{\mathcal{X}}$, so we can form a pushout diagram

$$
\begin{array}{ccc}
\mathrm{id}_\emptyset & \xrightarrow{\ \alpha\ } & \mathrm{id}_Y \\
{\scriptstyle\beta}\downarrow & & \downarrow{\scriptstyle\beta'} \\
f & \xrightarrow{\ \alpha'\ } & g
\end{array}
$$

in $\mathcal{O}_{\mathcal{X}}^n$. It is clear that α is a Cartesian transformation, and Lemma 6.1.3.6 implies that β is Cartesian as well. Invoking Theorem 6.4.4.5, we deduce that α' is a Cartesian transformation. But α' can be identified with a pushout diagram

$$
\begin{array}{ccc}
\emptyset & \longrightarrow & Y \\
\downarrow & & \downarrow \\
X & \longrightarrow & X \coprod Y.
\end{array}
$$

This proves (iii).

Now suppose that $U_\bullet$ is an n-efficient groupoid object of $\mathcal{X}$; we wish to prove that $U_\bullet$ is effective. Let $\overline{U}_\bullet : \mathrm{N}(\Delta_+)^{op} \to \mathcal{X}$ be a colimit of $U_\bullet$. Let $U'_\bullet : \mathrm{N}(\Delta_+)^{op} \to \mathcal{X}$ be the result of composing $\overline{U}_\bullet$ with the shift functor

$$\Delta_+ \to \Delta_+$$
$$J \mapsto J \coprod \{\infty\}.$$

(In other words, $U'_\bullet$ is the shifted simplicial object given by $U'_n = U_{n+1}$.) Lemma 6.1.3.17 asserts that $U'_\bullet$ is a colimit diagram in $\mathcal{X}$. We have a transformation $\overline{\alpha} : U'_\bullet \to \overline{U}_\bullet$. Let $\overline{V}_\bullet$ denote the constant augmented simplicial object of $\mathcal{X}$ taking the value U_0, so that we have a natural transformation $\overline{\beta} : U'_\bullet \to \overline{V}_\bullet$. Let $\overline{W}_\bullet$ denote a product of $\overline{U}_\bullet$ and $\overline{V}_\bullet$ in the ∞-category $\mathcal{X}_{\Delta_+}$ of augmented simplicial objects and let $\overline{\gamma} : U'_\bullet \to \overline{W}_\bullet$ be the induced map. We observe that for each $n \geq 0$, the map $\overline{\gamma}(\Delta^n) : U_{n+1} \to \overline{W}_n$ is a pullback of $U_1 \to U_0 \times U_0$ and therefore $(n-2)$-truncated (since $U_\bullet$ is assumed to be n-efficient). Since $U_\bullet$ is a groupoid, we conclude that $\gamma = \overline{\gamma} | \mathrm{N}(\Delta)^{op}$ is a Cartesian transformation. Invoking Theorem 6.4.4.5, we deduce that $\overline{\gamma}$ is also a Cartesian transformation, so that the diagram

$$
\begin{array}{ccc}
U_1 & \longrightarrow & U_0 \\
\downarrow & & \downarrow \\
\overline{W}_0 & \longrightarrow & U_0 \times \overline{W}_{-1}
\end{array}
$$

is Cartesian. Combining this with the Cartesian diagram

$$
\begin{array}{ccc}
\overline{W}_0 & \longrightarrow & \overline{W}_{-1} \\
\downarrow & & \downarrow \\
U_0 & \longrightarrow & \overline{U}_{-1},
\end{array}
$$

we deduce that U is effective, as desired. $\square$

6.4.5 Localic ∞-Topoi

The standard example of an ordinary topos is the category $\mathrm{Shv}(X; \mathcal{S}et)$ of sheaves (of sets) on a topological space X. Of course, not every topos is of this form: the category $\mathrm{Shv}(X; \mathcal{S}et)$ is generated under colimits by subobjects of its final object (which can be identified with open subsets of X). A topos $\mathfrak{X}$ with this property is said to be *localic* and is determined up to equivalence by the locale $\mathrm{Sub}(1_{\mathfrak{X}})$ which we may view as a 0-topos. The objective of this section is to obtain an ∞-categorical analogue of this picture, which will allow us to relate the theory of n-topoi to that of m-topoi for all $0 \leq m \leq n \leq \infty$.

Definition 6.4.5.1. Let $\mathfrak{X}$ and $\mathfrak{Y}$ be n-topoi for $0 \leq n \leq \infty$. A *geometric morphism* from $\mathfrak{X}$ to $\mathfrak{Y}$ is a functor $f_* : \mathfrak{X} \to \mathfrak{Y}$ which admits a left exact left adjoint (which we will typically denote by f^*).

We let $\mathrm{Fun}_*(\mathfrak{X}, \mathfrak{Y})$ denote the full subcategory of the ∞-category $\mathrm{Fun}(\mathfrak{X}, \mathfrak{Y})$ spanned by the geometric morphisms and let $\mathcal{T}op_n^{\mathrm{R}}$ denote the subcategory of $\widehat{\mathcal{C}at}_\infty$ whose objects are n-topoi and whose morphisms are geometric morphisms.

Remark 6.4.5.2. In the case where $n = 1$, the ∞-category of geometric morphisms $\mathrm{Fun}_*(\mathfrak{X}, \mathfrak{Y})$ between two 1-topoi is equivalent to (the nerve of) the category of geometric morphisms between the ordinary topoi $h\mathfrak{X}$ and $h\mathfrak{Y}$.

Remark 6.4.5.3. In the case where $n = 0$, the ∞-category of geometric morphisms $\mathrm{Fun}_*(\mathfrak{X}, \mathfrak{Y})$ between two 0-topoi is equivalent to the nerve of the partially ordered set of homomorphisms from the underlying locale of $\mathfrak{Y}$ to the underlying locale of $\mathfrak{X}$. (A *homomorphism* between locales is a map of partially ordered sets which preserve finite meets and arbitrary joins.) In the case where $\mathfrak{X}$ and $\mathfrak{Y}$ are associated to (sober) topological spaces X and Y, this is simply the set of continuous maps from X to Y partially ordered by specialization.

If $m \leq n$, then the ∞-categories $\mathcal{T}op_m^{\mathrm{R}}$ and $\mathcal{T}op_n^{\mathrm{R}}$ are related by the following observation:

Proposition 6.4.5.4. *Let $\mathfrak{X}$ be an n-topos and let $0 \leq m \leq n$. Then the full subcategory $\tau_{\leq m-1} \mathfrak{X}$ spanned by the $(m-1)$-truncated objects is an m-topos.*

Proof. If $m = n = \infty$, the result is obvious. Otherwise, it follows immediately from (2) of Theorem 6.4.1.5. $\square$

Lemma 6.4.5.5. *Let $\mathcal{C}$ be a small n-category which admits finite limits and let $\mathfrak{Y}$ be an ∞-topos. Then the restriction map*

$$\mathrm{Fun}_*(\mathfrak{Y}, \mathcal{P}(\mathcal{C})) \to \mathrm{Fun}_*(\tau_{\leq n-1} \mathfrak{Y}, \mathcal{P}_{\leq n-1}(\mathcal{C}))$$

is an equivalence of ∞-categories.

Proof. Let $\mathcal{M} \subseteq \mathrm{Fun}(\mathcal{P}(\mathcal{C}), \mathfrak{Y})$ and $\mathcal{M}' \subseteq \mathrm{Fun}(\mathcal{P}_{\leq n-1}(\mathcal{C}), \tau_{\leq n-1} \mathfrak{Y})$ denote the full subcategories spanned by left exact colimit-preserving functors. In view of Proposition 5.2.6.2, it will suffice to prove that the restriction map $\theta : \mathcal{M} \to \mathcal{M}'$ is an equivalence of ∞-categories.

Let $\mathcal{M}''$ denote the full subcategory of $\operatorname{Fun}(\mathcal{P}(\mathcal{C}), \tau_{\leq n-1} \mathcal{Y})$ spanned by colimit-preserving functors whose restriction to $\mathcal{P}_{\leq n-1}(\mathcal{C})$ is left exact. Corollary 5.5.6.22 implies that the restriction map $\theta' : \mathcal{M}'' \to \mathcal{M}'$ is an equivalence of ∞-categories.

Let $j : \mathcal{C} \to \mathcal{P}_{\leq n-1}(\mathcal{C}) \subseteq \mathcal{P}(\mathcal{C})$ denote the Yoneda embedding. Composition with j yields a commutative diagram

$$
\begin{array}{ccc}
\mathcal{M} & \overset{\theta}{\longrightarrow} & \mathcal{M}' \\
\downarrow{\psi} & & \downarrow{\psi'} \\
\operatorname{Fun}(\mathcal{C}, \tau_{\leq n-1} \mathcal{Y}) & =\!=\!= & \operatorname{Fun}(\mathcal{C}, \tau_{\leq n-1} \mathcal{Y}).
\end{array}
$$

Theorem 5.1.5.6 implies that ψ and $\psi' \circ \theta'$ are fully faithful. Since θ' is an equivalence of ∞-categories, we deduce that ψ' is fully faithful. Thus θ is fully faithful; to complete the proof, we must show that ψ and ψ' have the same essential image. Suppose that $f : \mathcal{C} \to \tau_{\leq n-1} \mathcal{Y}$ belongs to the essential image of ψ'. Without loss of generality, we may suppose that f is a composition

$$
\mathcal{C} \overset{j}{\to} \mathcal{P}_{\leq n-1}(\mathcal{C}) \overset{g^*}{\to} \tau_{\leq n-1} \mathcal{Y}.
$$

As a composition of left exact functors, f is left exact. We may now invoke Proposition 6.1.5.2 to deduce that f belongs to the essential image of ψ. $\qquad \square$

Lemma 6.4.5.6. *Let $\mathcal{C}$ be a small n-category which admits finite limits and is equipped with a Grothendieck topology and let $\mathcal{Y}$ be an ∞-topos. Then the restriction map*

$$
\theta : \operatorname{Fun}_*(\mathcal{Y}, \operatorname{Shv}(\mathcal{C})) \to \operatorname{Fun}_*(\tau_{\leq n-1} \mathcal{Y}, \operatorname{Shv}_{\leq n-1}(\mathcal{C}))
$$

is an equivalence of ∞-categories.

Proof. We have a commutative diagram

$$
\begin{array}{ccc}
\operatorname{Fun}_*(\mathcal{Y}, \operatorname{Shv}(\mathcal{C})) & \overset{\theta}{\longrightarrow} & \operatorname{Fun}_*(\tau_{\leq n-1} \mathcal{Y}, \operatorname{Shv}_{\leq n-1}(\mathcal{C})) \\
\downarrow & & \downarrow \\
\operatorname{Fun}_*(\mathcal{Y}, \mathcal{P}(\mathcal{C})) & \overset{\theta'}{\longrightarrow} & \operatorname{Fun}_*(\tau_{\leq n-1} \mathcal{Y}, \mathcal{P}_{\leq n-1}(\mathcal{C})),
\end{array}
$$

where the vertical arrows are inclusions of full subcategories and θ' is an equivalence of ∞-categories (Lemma 6.4.5.5). To complete the proof, it will suffice to show that if $f_* : \mathcal{Y} \to \mathcal{P}(\mathcal{C})$ is a geometric morphism such that $f_*|\tau_{\leq n-1} \mathcal{Y}$ factors through $\operatorname{Shv}_{\leq n-1}(\mathcal{C})$, then f_* factors through $\operatorname{Shv}(\mathcal{C})$.

Let f^* be a left adjoint to f_* and let $\overline{S}$ denote the collection of all morphisms in $\mathcal{P}(\mathcal{C})$ which localize to equivalences in $\operatorname{Shv}(\mathcal{C})$. We must show that $f^* \overline{S}$ consists of equivalences in $\mathcal{Y}$. Let $S \subseteq \overline{S}$ be the collection of monomorphisms which belong to $\overline{S}$. Since $\operatorname{Shv}(\mathcal{C})$ is a topological localization of $\mathcal{P}(\mathcal{C})$, it will suffice to show that $f^* S$ consists of equivalences in $\mathcal{Y}$. Let $g : X \to Y$

belong to S. Since $\mathcal{P}(\mathcal{C})$ is generated under colimits by the essential image of the Yoneda embedding, we can write Y as a colimit of a diagram $K \to \mathcal{P}_{\leq n-1}(\mathcal{C})$. Since colimits in $\mathcal{P}(\mathcal{C})$ are universal, we obtain a corresponding expression of g as a colimit of morphisms $\{g_\alpha : X_\alpha \to Y_\alpha\}$ which are pullbacks of g, where $Y_\alpha \in \mathcal{P}_{\leq n-1}(\mathcal{C})$. In this case, g_α is again a monomorphism, so that X_α is also $(n-1)$-truncated. Since f^* commutes with colimits, it will suffice to show that each $f^*(g_\alpha)$ is an equivalence. But this follows immediately from our assumption that $f_*|\tau_{\leq n-1}\mathcal{Y}$ factors through $\mathrm{Shv}_{\leq n-1}(\mathcal{Y})$. $\square$

Proposition 6.4.5.7. *Let $0 \leq m \leq n \leq \infty$ and let $\mathcal{Y}$ be an m-topos. There exists an n-topos $\mathcal{X}$ and a categorical equivalence $f_* : \tau_{\leq m-1}\mathcal{X} \to \mathcal{Y}$ with the following universal property: for any n-topos $\mathcal{Z}$, composition with f_* induces an equivalence of ∞-categories*

$$\theta : \mathrm{Fun}_*(\mathcal{Z}, \mathcal{X}) \to \mathrm{Fun}_*(\tau_{\leq m-1}\mathcal{Z}, \mathcal{Y}).$$

Proof. If $m = \infty$, then $n = \infty$, and we may take $\mathcal{X} = \mathcal{Y}$. Otherwise, we may apply Theorem 6.4.1.5 to reduce to the case where $\mathcal{Y} = \mathrm{Shv}_{\leq m-1}(\mathcal{C})$, where $\mathcal{C}$ is a small m-category which admits finite limits and is equipped with a Grothendieck topology. In this case, we let $\mathcal{X} = \mathrm{Shv}_{\leq n-1}(\mathcal{C})$ and define f_* to be the identity. Let $\mathcal{Z}$ be an arbitrary n-topos. According to Theorem 6.4.1.5, we may assume without loss of generality that $\mathcal{Z} = \tau_{\leq n-1}\mathcal{Z}'$, where $\mathcal{Z}'$ is an ∞-topos. We have a commutative diagram

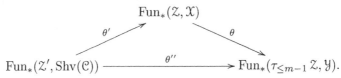

Lemma 6.4.5.6 implies that θ' and θ'' are equivalences of ∞-categories, so that θ is also an equivalence of ∞-categories. $\square$

Definition 6.4.5.8. Let $0 \leq m \leq n \leq \infty$ and let $\mathcal{X}$ be an n-topos. We will say that $\mathcal{X}$ is *m-localic* if, for any n-topos $\mathcal{Y}$, the natural map

$$\mathrm{Fun}_*(\mathcal{Y}, \mathcal{X}) \to \mathrm{Fun}_*(\tau_{\leq m-1}\mathcal{Y}, \tau_{\leq m-1}\mathcal{X})$$

is an equivalence of ∞-categories.

According to Proposition 6.4.5.7, every m-topos $\mathcal{X}$ is equivalent to the subcategory of $(m-1)$-truncated objects in an m-localic n-topos $\mathcal{X}'$, and $\mathcal{X}'$ is determined up to equivalence. More precisely, the truncation functor

$$\mathcal{T}\mathrm{op}_n^{\mathrm{R}} \xrightarrow{\ \tau_{\leq m-1}\ } \mathcal{T}\mathrm{op}_m^{\mathrm{R}}$$

induces an equivalence $\mathcal{C} \to \mathcal{T}\mathrm{op}_m^{\mathrm{R}}$, where $\mathcal{C} \subseteq \mathcal{T}\mathrm{op}_n^{\mathrm{R}}$ is the full subcategory spanned by the m-localic n-topoi. In other words, we may view the ∞-category of m-topoi as a *localization* of the ∞-category of n-topoi. In particular, the theory of m-topoi for $m < \infty$ can be regarded as a special case of the theory of ∞-topoi. For this reason, we will focus our attention on the case $n = \infty$ for most of the remainder of this book.

Proposition 6.4.5.9. *Let* X *be an* n-*localic* ∞-*topos. Then any topological localization of* X *is also* n-*localic.*

Proof. The proof of Proposition 6.4.5.7 shows that X is n-localic if and only if there exists a small n-category $\mathcal{C}$ which admits finite limits, a Grothendieck topology on $\mathcal{C}$, and an equivalence $X \to \mathrm{Shv}(\mathcal{C})$. In other words, X is n-localic if and only if it is equivalent to a topological localization of $\mathcal{P}(\mathcal{C})$, where $\mathcal{C}$ is a small n-category which admits finite limits. It is clear that any topological localization of X has the same property.   □

Let X be an ∞-topos. One should think of the ∞-categories $\tau_{\leq n-1} X$ as "Postnikov sections" of X. The classical 1-truncation $\tau_{\leq 1} X$ of a homotopy type X remembers only the fundamental groupoid of X. It therefore knows all about local systems of sets on X but nothing about fibrations over X with nondiscrete fibers. The relationship between X and $\tau_{\leq 0} X$ is analogous: $\tau_{\leq 0} X$ knows about the sheaves of sets on X but has forgotten about sheaves with nondiscrete stalks.

Remark 6.4.5.10. In view of the above discussion, the notation $\tau_{\leq 0} X$ is unfortunate because the analogous notation for the 1-truncation of a homotopy type X is $\tau_{\leq 1} X$. We caution the reader not to regard $\tau_{\leq 0} X$ as the result of applying an operation $\tau_{\leq 0}$ to X; it instead denotes the essential image of the truncation functor $\tau_{\leq 0} : X \to X$.

6.5 HOMOTOPY THEORY IN AN ∞-TOPOS

In classical homotopy theory, the most important invariants of a (pointed) space X are its homotopy groups $\pi_i(X, x)$. Our first objective in this section is to define analogous invariants in the case where X is an object of an arbitrary ∞-topos X. In this setting, the homotopy groups are not ordinary groups but are instead *sheaves* of groups on the underlying topos $\mathrm{Disc}(X)$. In §6.5.1, we will study these homotopy groups and the closely related theory of n-*connectivity*. The main theme is that the internal homotopy theory of a general ∞-topos X behaves much like the classical case $X = \mathcal{S}$.

One important classical fact which does *not* hold in general for an ∞-topos is Whitehead's theorem. If $f : X \to Y$ is a map of CW complexes, then f is a homotopy equivalence if and only if f induces bijective maps $\pi_i(X, x) \to \pi_i(Y, f(x))$ for any $i \geq 0$ and any base point $x \in X$. If $f : X \to Y$ is a map in an arbitrary ∞-topos X satisfying an analogous condition on (sheaves of) homotopy groups, then we say that f is ∞-*connective*. We will say that an ∞-topos X is *hypercomplete* if every ∞-connective morphism in X is an equivalence. Whitehead's theorem may be interpreted as saying that the ∞-topos $\mathcal{S}$ is hypercomplete. An arbitrary ∞-topos X need not be hypercomplete. We will survey the situation in §6.5.2, where we also give some reformulations of the notion of hypercompleteness and show that every topos X has a *hypercompletion* $X^\wedge$. In §6.5.3, we will show that an ∞-topos

$\mathfrak{X}$ is hypercomplete if and only if $\mathfrak{X}$ satisfies a descent condition with respect to hypercoverings (other versions of this result can be found in [20] and [78]).

Remark 6.5.0.1. The Brown-Joyal-Jardine theory of (pre)sheaves of simplicial sets on a topological space X is a model for the hypercomplete ∞-topos $\mathrm{Shv}(X)^{\wedge}$. In many respects, the ∞-topos $\mathrm{Shv}(X)$ of sheaves of spaces on X is better behaved *before* hypercompletion. We will outline some of the advantages of $\mathrm{Shv}(X)$ in §6.5.4 and in Chapter 7.

6.5.1 Homotopy Groups

Let $\mathfrak{X}$ be an ∞-topos and let X be an object of $\mathfrak{X}$. We will refer to a discrete object of $\mathfrak{X}_{/X}$ as a *sheaf of sets on* X. Since $\mathfrak{X}$ is presentable, it is automatically *cotensored* over spaces as explained in Remark 5.5.2.6. Consequently, for any object X of $\mathfrak{X}$ and any simplicial set K, there exists an object X^K of $\mathfrak{X}$ equipped with natural isomorphisms

$$\mathrm{Map}_{\mathfrak{X}}(Y, X^K) \to \mathrm{Map}_{\mathcal{H}}(K, \mathrm{Map}_{\mathfrak{X}}(Y, X))$$

in the homotopy category $\mathcal{H}$ of spaces.

Definition 6.5.1.1. Let $S^n = \partial \Delta^{n+1} \in \mathcal{H}$ denote the (simplicial) n-sphere and fix a base point $* \in S^n$. Then evaluation at $*$ induces a morphism $s : X^{S^n} \to X$ in $\mathfrak{X}$. We may regard s as an object of $\mathfrak{X}_{/X}$, and we define $\pi_n(X) = \tau_{\leq 0} s \in \mathfrak{X}_{/X}$ to be the associated discrete object of $\mathfrak{X}_{/X}$.

We will generally identify $\pi_n(X)$ with its image in the underlying topos $\mathrm{Disc}(\mathfrak{X}_{/X})$ (where it is well-defined up to canonical isomorphism). The constant map $S^n \to *$ induces a map $X \to X^{S^n}$ which determines a base point of $\pi_n(X)$.

Suppose that K and K' are pointed simplicial sets and let $K \vee K'$ denote the coproduct $K \coprod_* K'$. There is a pullback diagram

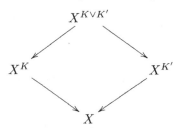

in $\mathfrak{X}$, so that $X^{K \vee K'}$ may be identified with a product of X^K and $X^{K'}$ in the ∞-topos $\mathfrak{X}_{/X}$. We now make the following general observation:

Lemma 6.5.1.2. *Let $\mathfrak{X}$ be an ∞-topos. The truncation functor $\tau_{\leq n} : \mathfrak{X} \to \mathfrak{X}$ preserves finite products.*

Proof. We must show that for any finite collection of objects $\{X_\alpha\}_{\alpha \in A}$ having product X, the induced map

$$\tau_{\leq n} X \to \prod_{\alpha \in A} \tau_{\leq n} X_\alpha$$

is an equivalence. If $\mathcal{X}$ is the ∞-category of spaces, then this follows from Whitehead's theorem: simply compute homotopy groups (and sets) on both sides. If $\mathcal{X} = \mathcal{P}(\mathcal{C})$, then to prove that a map in $\mathcal{X}$ is an equivalence, it suffices to show that it remains an equivalence after evaluation at any object $C \in \mathcal{C}$; thus we may reduce to the case where $\mathcal{X} = \mathcal{S}$ considered above. In the general case, $\mathcal{X}$ is equivalent to the essential image of a left exact localization functor $L : \mathcal{P}(\mathcal{C}) \to \mathcal{P}(\mathcal{C})$ for some small ∞-category $\mathcal{C}$. Without loss of generality, we may identify $\mathcal{X}$ with a full subcategory of $\mathcal{P}(\mathcal{C})$. Then $\mathcal{X} \subseteq \mathcal{X}' = \mathcal{P}(\mathcal{C})$ is stable under limits, so that X may be identified with a product of the family $\{X_\alpha\}_{\alpha \in A}$ in $\mathcal{X}'$. It follows from the case treated above that the natural map

$$\tau_{\leq n}^{\mathcal{X}'} X \to \prod_{\alpha \in A} \tau_{\leq n}^{\mathcal{X}'} X_\alpha$$

is an equivalence. Proposition 5.5.6.28 implies that $L \circ \tau_{\leq n}^{\mathcal{X}'} | \mathcal{X}$ is an n-truncation functor for $\mathcal{X}$. The desired result now follows by applying the functor L to both sides of the above equivalence and invoking the assumption that L is left exact (here we must require the finiteness of A). $\square$

It follows from Lemma 6.5.1.2 that there is a canonical isomorphism

$$\tau_{\leq 0}^{\mathcal{X}/X}(X^{K \vee K'}) \simeq \tau_{\leq 0}^{\mathcal{X}/X}(X^K) \times \tau_{\leq 0}^{\mathcal{X}/X}(X^{K'})$$

in the topos $\mathrm{Disc}(\mathcal{X}_{/X})$. In particular, for $n > 0$, the usual comultiplication $S^n \to S^n \vee S^n$ (a well-defined map in the homotopy category $\mathcal{H}$) induces a multiplication map $\pi_n(X) \times \pi_n(X) \to \pi_n(X)$. As in ordinary homotopy theory, we conclude that $\pi_n(X)$ is a group object of $\mathrm{Disc}(\mathcal{X}_{/X})$ for $n > 0$, which is commutative for $n > 1$.

In order to work effectively with homotopy sets, it is convenient to define the homotopy sets $\pi_n(f)$ of a morphism $f : X \to Y$ to be the homotopy sets of f considered as an object of the ∞-topos $\mathcal{X}_{/Y}$. In view of the equivalences $\mathcal{X}_{/f} \to \mathcal{X}_{/X}$, we may identify $\pi_n(f)$ with an object of $\mathrm{Disc}(\mathcal{X}_{/X})$, which is again a sheaf of groups if $n \geq 1$, and abelian groups if $n \geq 2$. The intuition is that the stalk of these sheaves at a point p of X is the nth homotopy group of the homotopy fiber of f taken with respect to the base point p.

Remark 6.5.1.3. It is useful to have the following recursive definition for homotopy groups. Let $f : X \to Y$ be a morphism in an ∞-topos $\mathcal{X}$. Regarding f as an object of the topos $\mathcal{X}_{/Y}$, we may take its 0th truncation $\tau_{\leq 0}^{\mathcal{X}/Y} f$. This is a discrete object of $\mathcal{X}_{/Y}$, and by definition we have $\pi_0(f) \simeq f^* \tau_{\leq 0}^{\mathcal{X}/Y}(X) \simeq X \times_Y \tau_{\leq 0}^{\mathcal{X}/Y}(f)$. The natural map $X \to \tau_{\leq 0}^{\mathcal{X}/Y}(f)$ gives a global section of $\pi_0(f)$. Note that in this case, $\pi_0(f)$ is the pullback of a discrete object of $\mathcal{X}_{/Y}$: this is because the definition of π_0 does not require a base point.

If $n > 0$, then we have a natural isomorphism $\pi_n(f) \simeq \pi_{n-1}(\delta)$ in the topos $\mathrm{Disc}(\mathcal{X}_{/X})$, where $\delta : X \to X \times_Y X$ is the associated diagonal map.

Remark 6.5.1.4. Let $f : \mathfrak{X} \to \mathfrak{Y}$ be a geometric morphism of ∞-topoi and let $g : Y \to Y'$ be a morphism in $\mathfrak{Y}$. Then there is a canonical isomorphism $f^*(\pi_n(g)) \simeq \pi_n(f^*(g))$ in $\mathrm{Disc}(\mathfrak{X}_{/f_*Y})$. This follows immediately from Proposition 5.5.6.28.

Remark 6.5.1.5. Given a pair of composable morphisms $X \xrightarrow{f} Y \xrightarrow{g} Z$, there is an associated sequence of pointed objects

$$\cdots \to f^*\pi_{n+1}(g) \xrightarrow{\delta_{n+1}} \pi_n(f) \to \pi_n(g \circ f) \to f^*\pi_n(g) \xrightarrow{\delta_n} \pi_{n-1}(f) \to \cdots$$

in the ordinary topos $\mathrm{Disc}(\mathfrak{X}_{/X})$, with the usual exactness properties. To construct the boundary map δ_n, we observe that the n-sphere S^n can be written as a (homotopy) pushout $D^- \coprod_{S^{n-1}} D^+$ of two hemispheres along the equator. By construction, $f^*\pi_n(g)$ can be identified with the 0-truncation of

$$X \times_Y Y^{S^n} \times_{Z^{S^n}} Z \simeq X^{D^-} \times_{Y^{D^-}} Y^{S^n} \times_{Z^{S^n}} Z,$$

which maps by restriction to

$$X^{S^{n-1}} \times_{Y^{S^{n-1}}} Y^{D^+} \simeq X^{S^{n-1}} \times_{Y^{S^{n-1}}} Y.$$

We now observe that the 0-truncation of the latter object is naturally isomorphic to $\pi_{n-1}(f) \in \mathrm{Disc}(\mathfrak{X}_{/X})$.

To prove the exactness of the above sequence in an ∞-topos $\mathfrak{X}$, we first choose an accessible left exact localization $L : \mathcal{P}(\mathcal{C}) \to \mathfrak{X}$. Without loss of generality, we may suppose that the diagram $X \xrightarrow{f} Y \xrightarrow{g} Z$ is the image under L of a diagram in $\mathcal{P}(\mathcal{C})$. Using Remark 6.5.1.4, we conclude that the sequence constructed above is equivalent to the image under L of an analogous sequence in the ∞-topos $\mathcal{P}(\mathcal{C})$. Since L is left exact, it will suffice to prove that this second sequence is exact; in other words, we may reduce to the case $\mathfrak{X} = \mathcal{P}(\mathcal{C})$. Working componentwise, we can reduce further to the case where $\mathfrak{X} = \mathcal{S}$. The desired result now follows from classical homotopy theory. (Special care should be taken regarding the exactness of the above sequence at $\pi_0(f)$: this should really be interpreted in terms of an action of the group $f^*\pi_1(g)$ on $\pi_0(f)$. We leave the details of the construction of this action to the reader.)

Remark 6.5.1.6. If $\mathfrak{X} = \mathcal{S}$ and $\eta : * \to X$ is a pointed space, then $\eta^*\pi_n(X)$ can be identified with the nth homotopy group of X with base point η.

We now study the implications of the vanishing of homotopy groups.

Proposition 6.5.1.7. Let $f : X \to Y$ be an n-truncated morphism in an ∞-topos $\mathfrak{X}$. Then $\pi_k(f) \simeq *$ for all $k > n$. If $n \geq 0$ and $\pi_n(f) \simeq *$, then f is $(n-1)$-truncated.

Proof. The proof goes by induction on n. If $n = -2$, then f is an equivalence and there is nothing to prove. Otherwise, the diagonal map $\delta : X \to X \times_Y X$ is $(n-1)$-truncated (Lemma 5.5.6.15). The inductive hypothesis and Remark

6.5.1.3 allow us to deduce that $\pi_k(f) \simeq \pi_{k-1}(\delta) \simeq *$ whenever $k > n$ and $k > 0$. Similarly, if $n \geq 1$ and $\pi_n(f) \simeq \pi_{n-1}(\delta) \simeq *$, then δ is $(n-2)$-truncated by the inductive hypothesis, so that f is $(n-1)$-truncated (Lemma 5.5.6.15).

The case of small k and n requires special attention: we must show that if f is 0-truncated, then f is (-1)-truncated if and only if $\pi_0(f) \simeq *$. Because f is 0-truncated, we have an equivalence $\tau_{\leq 0}^{X/Y}(f) \simeq f$, so that $\pi_0(f) \simeq X \times_Y X$. To say $\pi_0(f) \simeq *$ is to assert that the diagonal map $\delta : X \to X \times_Y X$ is an equivalence, which is equivalent to the assertion that f is (-1)-truncated (Lemma 5.5.6.15). $\square$

Remark 6.5.1.8. Proposition 6.5.1.7 implies that if f is n-truncated for some $n \gg 0$, then we can test whether or not f is m-truncated for any particular value of m by computing the homotopy groups of f. In contrast to the classical situation, it is not possible to drop the assumption that f is n-truncated for $n \gg 0$.

Lemma 6.5.1.9. *Let X be an object in an ∞-topos $\mathfrak{X}$ and let $p : X \to Y$ be an n-truncation of X. Then p induces isomorphisms $\pi_k(X) \simeq p^*\pi_k(Y)$ for all $k \leq n$.*

Proof. Let $\phi : \mathfrak{X} \to \mathcal{Y}$ be a geometric morphism such that ϕ_* is fully faithful. By Proposition 5.5.6.28 and Remark 6.5.1.4, it will suffice to prove the lemma in the case where $\mathfrak{X} = \mathcal{Y}$. We may therefore assume that $\mathcal{Y}$ is an ∞-category of presheaves. In this case, homotopy groups and truncations are computed pointwise. Thus we may reduce to the case $\mathfrak{X} = \mathcal{S}$, where the conclusion follows from classical homotopy theory. $\square$

Definition 6.5.1.10. Let $f : X \to Y$ be a morphism in an ∞-topos $\mathfrak{X}$ and let $0 \leq n \leq \infty$. We will say that f is *n-connective* if it is an effective epimorphism and $\pi_k(f) = *$ for $0 \leq k < n$. We shall say that the object X is *n-connective* if $f : X \to 1_{\mathfrak{X}}$ is n-connective, where $1_{\mathfrak{X}}$ denotes the final object of $\mathfrak{X}$. By convention, we will say that every morphism f in $\mathfrak{X}$ is (-1)-connective.

Definition 6.5.1.11. Let X be an object of an ∞-topos $\mathfrak{X}$. We will say that X is *connected* if it is 1-connective: that is, if the truncation $\tau_{\leq 0}X$ is a final object in $\mathfrak{X}$.

Proposition 6.5.1.12. *Let X be an object in an ∞-topos $\mathfrak{X}$ and let $n \geq -1$. Then X is n-connective if and only if $\tau_{\leq n-1}X$ is a final object of $\mathfrak{X}$.*

Proof. The case $n = -1$ is trivial. The proof in general proceeds by induction on $n \geq 0$. If $n = 0$, then the conclusion follows from Proposition 6.2.3.4. Suppose $n > 0$. Let $p : X \to \tau_{n-1}X$ be an $(n-1)$-truncation of X. If $\tau_{\leq n-1}X$ is a final object of $\mathfrak{X}$, then

$$\pi_k X \simeq p^*\pi_k(\tau_{n-1}X) \simeq *$$

for $k < n$ by Lemma 6.5.1.9. Since the map $p : X \to \tau_{\leq n-1}X \simeq 1_{\mathfrak{X}}$ is an effective epimorphism (Proposition 7.2.1.14), it follows that X is n-connective.

Conversely, suppose that X is n-connective. Then $p^*\pi_{n-1}(\tau_{\leq n-1}X) \simeq *$. Since p is an effective epimorphism, Lemma 6.2.3.16 implies that

$$\pi_{n-1}(\tau_{\leq n-1}X) \simeq *.$$

Using Proposition 6.5.1.7, we conclude that $\tau_{\leq n-1}X$ is $(n-2)$-truncated, so that $\tau_{\leq n-1}X \simeq \tau_{\leq n-2}X$. Repeating this argument, we reduce to the case where $n = 0$ which was handled above. $\qquad\square$

Corollary 6.5.1.13. *The class of n-connective objects of an ∞-topos $\mathfrak{X}$ is stable under finite products.*

Proof. Combine Proposition 6.5.1.12 with Lemma 6.5.1.2. $\qquad\square$

Let $\mathfrak{X}$ be an ∞-topos and X an object of $\mathfrak{X}$. Since

$$\mathrm{Map}_{\mathfrak{X}}(X, Y) \simeq \mathrm{Map}_{\mathfrak{X}}(\tau_{\leq n}X, Y)$$

whenever Y is n-truncated, we deduce that X is $(n+1)$-connective if and only if the natural map $\mathrm{Map}_{\mathfrak{X}}(1_{\mathfrak{X}}, Y) \to \mathrm{Map}_{\mathfrak{X}}(X, Y)$ is an equivalence for all n-truncated Y. From this, we can immediately deduce the following relative version of Proposition 6.5.1.12:

Corollary 6.5.1.14. *Let $f : X \to X'$ be a morphism in an ∞-topos $\mathfrak{X}$. Then f is $(n+1)$-connective if and only if composition with f induces a homotopy equivalence*

$$\mathrm{Map}_{\mathfrak{X}_{/X'}}(\mathrm{id}_{X'}, Y) \to \mathrm{Map}_{\mathfrak{X}_{/X'}}(f, Y)$$

for every n-truncated object $Y \in \mathfrak{X}_{/X'}$.

Remark 6.5.1.15. Let $L : \mathfrak{X} \to \mathcal{Y}$ be a left exact localization of ∞-topoi and let $f : Y \to Y'$ be an n-connective morphism in $\mathcal{Y}$. Then f is equivalent (in $\mathrm{Fun}(\Delta^1, \mathcal{Y})$) to Lf_0, where f_0 is an n-connective morphism in $\mathfrak{X}$. To see this, we choose a (fully faithful) right adjoint G to L and a factorization

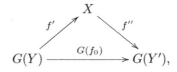

where f' is n-connective and f'' is $(n-1)$-truncated. Then $Lf'' \circ Lf'$ is equivalent to f and is therefore n-connective. It follows that Lf'' is an equivalence, so that Lf' is equivalent to f.

We conclude by noting the following stability properties of the class of n-connective morphisms:

Proposition 6.5.1.16. *Let $\mathfrak{X}$ be an ∞-topos.*

(1) *Let $f : X \to Y$ be a morphism in $\mathfrak{X}$. If f is n-connective, then it is m-connective for all $m \le n$. Conversely, if f is n-connective for all $n < \infty$, then f is ∞-connective.*

(2) *Any equivalence in $\mathfrak{X}$ is ∞-connective.*

(3) *Let $f, g : X \to Y$ be homotopic morphisms in $\mathfrak{X}$. Then f is n-connective if and only if g is n-connective.*

(4) *Let $\pi^* : \mathfrak{X} \to \mathcal{Y}$ be left adjoint to a geometric morphism from $\pi_* : \mathcal{Y} \to \mathfrak{X}$ and let $f : X \to X'$ be an n-connective morphism in $\mathfrak{X}$. Then $\pi^* f$ is an n-connective morphism in $\mathcal{Y}$.*

(5) *Suppose we are given a diagram*

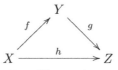

in $\mathfrak{X}$, where f is n-connective. Then g is n-connective if and only if h is n-connective.

(6) *Suppose we are given a pullback diagram*

$$\begin{array}{ccc} X' & \xrightarrow{q'} & X \\ \downarrow{f'} & & \downarrow{f} \\ Y' & \xrightarrow{q} & Y \end{array}$$

in $\mathfrak{X}$. If f is n-connective, then so is f'. The converse holds if q is an effective epimorphism.

Proof. The first three assertions are obvious. Claim (4) follows from Propositions 6.5.1.12 and 5.5.6.28. To prove (5), we first observe that Corollary 6.2.3.12 implies that g is an effective epimorphism if and only if h is an effective epimorphism. According to Remark 6.5.1.5, we have a long exact sequence

$$\cdots \to f^* \pi_{i+1}(g) \to \pi_i(f) \to \pi_i(h) \to f^* \pi_i(g) \to \pi_{i-1}(f) \to \cdots$$

of pointed objects in the topos $\mathrm{Disc}(\mathfrak{X}_{/X})$. It is then clear that if f and g are n-connective, then so is h. Conversely, if f and h are n-connective, then $f^* \pi_i(g) \simeq *$ for $i \le n$. Since f is an effective epimorphism, Lemma 6.2.3.16 implies that $\pi_i(g) \simeq *$ for $i \le n$, so that g is also n-connective.

The first assertion of (6) follows from (4) since a pullback functor $q^* : \mathfrak{X}_{/Y} \to \mathfrak{X}_{/Y'}$ is left adjoint to a geometric morphism. For the converse, let us suppose that q is an effective epimorphism and that f' is n-connective. According to Lemma 6.2.3.15, the maps f and q' are effective epimorphisms. Applying Remark 6.5.1.4, we conclude that there are canonical isomorphisms $q'^* \pi_k(f) \simeq \pi_k(f')$ in the topos $\mathrm{Disc}(\mathfrak{X}_{/X'})$, so that $q'^* \pi_k(f) \simeq *$ for $k < n$. Applying Lemma 6.2.3.16, we conclude that $\pi_k(f) \simeq *$ for $k < n$, so that f is n-connective, as desired. $\square$

Corollary 6.5.1.17. *Let*

$$
\begin{array}{ccc}
X' & \xrightarrow{\;g\;} & X \\
\downarrow{\scriptstyle f'} & & \downarrow{\scriptstyle f} \\
Y' & \longrightarrow & Y
\end{array}
$$

be a pushout diagram in an ∞-topos $\mathfrak{X}$. Suppose that f' is n-connective. Then f is n-connective.

Proof. Choose an accessible left exact localization functor $L : \mathcal{P}(\mathcal{C}) \to \mathfrak{X}$. Using Remark 6.5.1.15, we can assume without loss of generality that $f' = Lf'_0$, where $f'_0 : X'_0 \to Y'_0$ is a morphism in $\mathcal{P}(\mathcal{C})$. Similarly, we may assume $g = Lg_0$ for some morphism $g_0 : X'_0 \to X_0$. Form a pushout diagram

$$
\begin{array}{ccc}
X'_0 & \xrightarrow{\;g_0\;} & X_0 \\
\downarrow{\scriptstyle f'_0} & & \downarrow{\scriptstyle f_0} \\
Y'_0 & \longrightarrow & Y_0
\end{array}
$$

in $\mathcal{P}(\mathcal{C})$. Then the original diagram is equivalent to the image (under L) of the diagram above. In view of Proposition 6.5.1.16, it will suffice to show that f_0 is n-connective. Using Propositions 6.5.1.12 and 5.5.6.28, we see that f_0 is n-connective if and only if its image under the evaluation map $\mathcal{P}(\mathcal{C}) \to \mathcal{S}$ associated to any object $C \in \mathcal{C}$ is n-connective. In other words, we can reduce to the case where $\mathfrak{X} = \mathcal{S}$, and the result now follows from classical homotopy theory. $\square$

We conclude by establishing a few results which will be needed in §7.2:

Proposition 6.5.1.18. *Let $f : X \to Y$ be a morphism in an ∞-topos $\mathfrak{X}$, $\delta : X \to X \times_Y X$ the associated diagonal morphism, and $n \geq 0$. The following conditions are equivalent:*

(1) *The morphism f is n-connective.*

(2) *The diagonal map $\delta : X \to X \times_Y X$ is $(n-1)$-connective and f is an effective epimorphism.*

Proof. The proof is immediate from Definition 6.5.1.10 and Remark 6.5.1.3. $\square$

Proposition 6.5.1.19. *Let $\mathfrak{X}$ be an ∞-topos containing an object X and let $\sigma : \Delta^2 \to \mathfrak{X}$ be a 2-simplex corresponding to a diagram*

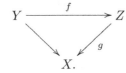

Then f is an n-connective morphism in $\mathfrak{X}$ if and only if σ is an n-connective morphism in $\mathfrak{X}_{/X}$.

Proof. We observe that $\mathcal{X}_{/g} \to \mathcal{X}_{/Z}$ is a trivial fibration, so that an object of $\mathcal{X}_{/g}$ is n-connective if and only if its image in $\mathcal{X}_{/Z}$ is n-connective. □

Proposition 6.5.1.20. *Let $f : X \to Y$ be a morphism in an ∞-topos $\mathcal{X}$, let $s : Y \to X$ be a section of f (so that $f \circ s$ is homotopic to id_Y), and let $n \geq 0$. Then f is n-connective if and only if s is $(n-1)$-connective.*

Proof. We have a 2-simplex $\sigma : \Delta^2 \to \mathcal{X}$ which we may depict as follows:

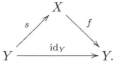

Corollary 6.2.3.12 implies that f is an effective epimorphism; this completes the proof in the case $n = 0$. Suppose that $n > 0$ and that s is $(n-1)$-connective. In particular, s is an effective epimorphism. The long exact sequence of Remark 6.5.1.5 gives an isomorphism $\pi_i(s) \simeq s^*\pi_{i+1}(f)$, so that $s^*\pi_k(f)$ vanishes for $1 \leq k < n$. Applying Lemma 6.2.3.16, we conclude that $\pi_k(f) \simeq *$ for $1 \leq k < n$. Moreover, since s is an effective epimorphism, it induces an effective epimorphism $\pi_0(\mathrm{id}_Y) \to \pi_0(f)$ in the ordinary topos $\mathrm{Disc}(\mathcal{X}_{/Y})$, so that $\pi_0(f) \simeq *$ as well. This proves that f is n-connective.

Conversely, if f is n-connective, then $\pi_i(s) \simeq *$ for $i < n-1$; the only nontrivial point is to verify that s is an effective epimorphism. According to Proposition 6.5.1.19, it will suffice to prove that σ is an effective epimorphism when viewed as a morphism in $\mathcal{X}_{/Y}$. Using Proposition 7.2.1.14, we may reduce to proving that $\sigma' = \tau_{\leq 0}^{\mathcal{X}_{/Y}}(\sigma)$ is an equivalence in $\mathcal{X}_{/Y}$. This is clear since the source and target of σ' are both final objects of $\mathcal{X}_{/Y}$ (by virtue of our assumption that f is 1-connective). □

6.5.2 ∞-Connectedness

Let $\mathcal{C}$ be an ordinary category equipped with a Grothendieck topology and let $\mathbf{A} = \mathrm{Set}_{\Delta}^{\mathcal{C}^{op}}$ be the category of simplicial presheaves on $\mathcal{C}$.

Proposition 6.5.2.1 (Jardine [41]). *There exists a left proper, combinatorial, simplicial model structure on the category $\mathbf{A}$ which admits the following description:*

(C) *A map $f : F_\bullet \to G_\bullet$ of simplicial presheaves on $\mathcal{C}$ is a local cofibration if it is an injective cofibration: that is, if and only if the induced map $F_\bullet(C) \to G_\bullet(C)$ is a cofibration of simplicial sets for each object $C \in \mathcal{C}$.*

(W) *A map $f : F_\bullet \to G_\bullet$ of simplicial presheaves on $\mathcal{C}$ is a local equivalence if and only if, for any object $C \in \mathcal{C}$ and any commutative diagram of topological spaces*

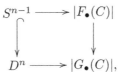

there exists a collection of morphisms $\{C_\alpha \to C\}$ which generates a covering sieve on C, such that in each of the induced diagrams

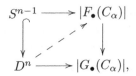

one can produce a dotted arrow so that the upper triangle commutes and the lower triangle commutes up to a homotopy which is fixed on S^{n-1}.

We refer the reader to [41] for a proof (one can also deduce Proposition 6.5.2.1 from Proposition A.2.6.13). We will refer to the model structure of Proposition 6.5.2.1 as the *local* model structure on **A**.

Remark 6.5.2.2. In the case where the topos $\mathfrak{X}$ of sheaves of sets on $\mathcal{C}$ has enough points, there is a simpler description of the class (W) of local equivalences: a map $F \to G$ of simplicial presheaves is a local equivalence if and only if it induces weak homotopy equivalences $F_x \to G_x$ of simplicial sets after passing to the stalk at any point x of $\mathfrak{X}$. We refer the reader to [41] for details.

Let $\mathbf{A}^\circ$ denote the full subcategory of **A** consisting of fibrant-cofibrant objects (with respect to the local model structure) and let $\mathfrak{X} = \mathrm{N}(\mathbf{A}^\circ)$ be the associated ∞-category. We observe that the local model structure on **A** is a localization of the injective model structure on **A**. Consequently, the ∞-category $\mathfrak{X}$ is a localization of the ∞-category associated to the injective model structure on **A**, which (in view of Proposition 5.1.1.1) is equivalent to $\mathcal{P}(\mathrm{N}(\mathcal{C}))$. It is tempting to guess that $\mathfrak{X}$ is equivalent to the left exact localization $\mathrm{Shv}(\mathrm{N}(\mathcal{C}))$ constructed in §6.2.2. This is not true in general; however, as we will explain below, we can always recover $\mathfrak{X}$ as an accessible left exact localization of $\mathrm{Shv}(\mathrm{N}(\mathcal{C}))$. In particular, $\mathfrak{X}$ is itself an ∞-topos.

In general, the difference between $\mathfrak{X}$ and $\mathrm{Shv}(\mathrm{N}(\mathcal{C}))$ is measured by the failure of Whitehead's theorem. Essentially by construction, the equivalences in **A** are those maps which induce isomorphisms on homotopy sheaves. In general, this assumption is not strong enough to guarantee that a morphism in $\mathrm{Shv}(\mathrm{N}(\mathcal{C}))$ is an equivalence. However, this is the only difference: the ∞-category $\mathfrak{X}$ can be obtained from $\mathrm{Shv}(\mathrm{N}(\mathcal{C}))$ by inverting the class of ∞-connective morphisms (Proposition 6.5.2.14). Before proving this, we study the class of ∞-connective morphisms in an arbitrary ∞-topos.

Lemma 6.5.2.3. *Let $p : \mathcal{C} \to \mathcal{D}$ be a Cartesian fibration of ∞-categories, let $\mathcal{C}'$ be a full subcategory of $\mathcal{C}$, and suppose that for every p-Cartesian morphism $f : C \to C'$ in $\mathcal{C}$, if $C' \in \mathcal{C}'$, then $C \in \mathcal{C}'$. Let D be an object of $\mathcal{D}$ and let $f : C \to C'$ be a morphism in the fiber $\mathcal{C}_D = \mathcal{C} \times_{\mathcal{D}} \{D\}$ which exhibits C' as a $\mathcal{C}_D^0$-localization of C (see Definition 5.2.7.6). Then f exhibits C' as a $\mathcal{C}$-localization of $\mathcal{C}$.*

Proof. According to Proposition 2.4.3.3, p induces a Cartesian fibration $\mathcal{C}_{C/} \to \mathcal{D}_{D/}$, which restricts to give a Cartesian fibration $p' : \mathcal{C}'_{C/} \to \mathcal{D}_{D/}$. We observe that f is an object of $\mathcal{C}'_{C/}$ which is an initial object of $(p')^{-1}\{\mathrm{id}_D\}$ (Remark 5.2.7.7), and that id_D is an initial object of $\mathcal{D}_{D/}$. Lemma 2.4.4.7 implies that f is an initial object of $\mathcal{C}'_{C/}$, so that f exhibits C' as $\mathcal{C}'$-localization of C (Remark 5.2.7.7), as desired.   □

Lemma 6.5.2.4. *Let $p : \mathcal{C} \to \mathcal{D}$ be a Cartesian fibration of ∞-categories, let $\mathcal{C}'$ be a full subcategory of $\mathcal{C}$, and suppose that for every p-Cartesian morphism $f : C \to C'$ in $\mathcal{C}$, if $C' \in \mathcal{C}'$, then $C \in \mathcal{C}'$. Suppose that for each object $D \in \mathcal{D}$, the fiber $\mathcal{C}'_D = \mathcal{C}' \times_{\mathcal{D}} \{D\}$ is a reflective subcategory of $\mathcal{C}_D = \mathcal{C} \times_{\mathcal{D}} \{D\}$ (see Remark 5.2.7.9). Then $\mathcal{C}'$ is a reflective subcategory of $\mathcal{C}$.*

Proof. Combine Lemma 6.5.2.3 with Proposition 5.2.7.8.   □

Lemma 6.5.2.5. *Let $\mathcal{X}$ be a presentable ∞-category, let $\mathcal{C}$ be an accessible ∞-category, and let $\alpha : F \to G$ be a natural transformation between accessible functors $F, G : \mathcal{C} \to \mathcal{X}$. Let $\mathcal{C}(n)$ be the full subcategory of $\mathcal{C}$ spanned by those objects C such that $\alpha(C) : F(C) \to G(C)$ is n-truncated. Then $\mathcal{C}(n)$ is an accessible subcategory of $\mathcal{C}$ (see Definition 5.4.7.8).*

Proof. We will work by induction on n. If $n = -2$, then we have a (homotopy) pullback diagram

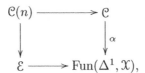

where $\mathcal{E}$ is the full subcategory of $\mathrm{Fun}(\Delta^1, \mathcal{X})$ spanned by equivalences. The inclusion of $\mathcal{E}$ into $\mathrm{Fun}(\Delta^1, \mathcal{X})$ is equivalent to the diagonal map $\mathcal{X} \to \mathrm{Fun}(\Delta^1, \mathcal{X})$ and therefore accessible. Proposition 5.4.6.6 implies that $\mathcal{C}(n)$ is an accessible subcategory of $\mathcal{C}$ as desired.

If $n \geq -1$, we apply the the inductive hypothesis to the diagonal functor $\delta : F \to F \times_G F$ using Lemma 5.5.6.15.   □

Lemma 6.5.2.6. *Let $\mathcal{X}$ be a presentable ∞-category and let $-2 \leq n < \infty$. Let $\mathcal{C}$ be the full subcategory of $\mathrm{Fun}(\Delta^1, \mathcal{X})$ spanned by the n-truncated morphisms. Then $\mathcal{C}$ is a strongly reflective subcategory of $\mathrm{Fun}(\Delta^1, \mathcal{X})$.*

Proof. Applying Lemma 6.5.2.4 to the restriction functor $\mathrm{Fun}(\Delta^1, \mathcal{X}) \to \mathrm{Fun}(\{1\}, \mathcal{X})$, we conclude that $\mathcal{C}$ is a reflective subcategory of $\mathrm{Fun}(\Delta^1, \mathcal{X})$. The accessibility of $\mathcal{C}$ follows from Lemma 6.5.2.5.   □

Lemma 6.5.2.7. *Let $\mathcal{X}$ be an ∞-topos, let $0 \leq n \leq \infty$, and let $\mathcal{D}(n)$ be the full subcategory of $\mathrm{Fun}(\Delta^1, \mathcal{X})$ spanned by the n-connective morphisms of $\mathcal{X}$. Then $\mathcal{D}(n)$ is an accessible subcategory of $\mathcal{X}$ and is stable under colimits in $\mathcal{X}$.*

Proof. Suppose first that $n < \infty$. Let $\mathcal{C}(n) \subseteq \operatorname{Fun}(\Delta^1, \mathfrak{X})$ be the full subcategory spanned by the n-truncated morphisms in $\mathfrak{X}$. According to Lemma 6.5.2.6, the inclusion $\mathcal{C}(n) \subseteq \operatorname{Fun}(\Delta^1, \mathfrak{X})$ has a left adjoint $L : \operatorname{Fun}(\Delta^1, \mathfrak{X}) \to \mathcal{C}(n)$. Moreover, the proof of Lemma 6.5.2.3 shows that f is n-connective if and only if Lf is an equivalence. It is easy to see that the full subcategory $\mathcal{E} \subseteq \mathcal{C}(n)$ spanned by the equivalences is stable under colimits in $\mathcal{C}(n)$, so that $\mathcal{D}(n)$ is stable under colimits in $\operatorname{Fun}(\Delta^1, \mathfrak{X})$. The accessibility of $\mathcal{D}(n)$ follows from the existence of the (homotopy) pullback diagram

$$
\begin{array}{ccc}
\mathcal{D}(n) & \longrightarrow & \operatorname{Fun}(\Delta^1, \mathfrak{X}) \\
\downarrow & & \downarrow {\scriptstyle L} \\
\mathcal{E} & \longrightarrow & \mathcal{C}(n)
\end{array}
$$

and Proposition 5.4.6.6.

If $n = \infty$, we observe that $\mathcal{D}(n) = \cup_{m<\infty} \mathcal{D}(m)$, which is manifestly stable under colimits and is an accessible subcategory of $\mathfrak{X}^{\Delta^1}$ by Proposition 5.4.7.10. $\qquad\square$

Proposition 6.5.2.8. *Let $\mathfrak{X}$ be an ∞-topos and let S denote the collection of ∞-connective morphisms of $\mathfrak{X}$. Then S is strongly saturated and of small generation (see Definition 5.5.4.5).*

Proof. Lemma 6.5.2.7 implies that S is stable under colimits in $\operatorname{Fun}(\Delta^1, \mathfrak{X})$, and Corollary 6.5.1.17 shows that S is stable under pushouts. To prove that S has the two-out-of-three property, we consider a diagram $\sigma : \Delta^2 \to \mathfrak{X}$, which we depict as

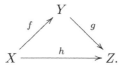

If f is ∞-connective, then Proposition 6.5.1.16 implies that g is ∞-connective if and only if h is ∞-connective. Suppose that g and h are ∞-connective. The long exact sequence

$$
\cdots \to f^* \pi_{n+1}(g) \to \pi_n(f) \to \pi_n(h) \to f^* \pi_n(g) \to \pi_{n-1}(f) \to \cdots
$$

of Remark 6.5.1.5 shows that $\pi_n(f) \simeq *$ for all $n \geq 0$. It will therefore suffice to prove that f is an effective epimorphism. According to Proposition 6.5.1.19, it will suffice to show that σ is an effective epimorphism in $\mathfrak{X}_{/Z}$. According to Proposition 7.2.1.14, it suffices to show that $\tau_{\leq 0}^{\mathfrak{X}_{/Z}}(h)$ and $\tau_{\leq 0}^{\mathfrak{X}_{/Z}}(g)$ are both final objects of $\mathfrak{X}_{/Z}$, which follows from the 0-connectivity of g and h (Proposition 6.5.1.12).

To show that S is of small generation, it suffices (in view of Lemma 5.5.4.14) to show that the full subcategory of $\operatorname{Fun}(\Delta^1, \mathfrak{X})$ spanned by S is accessible. This follows from Lemma 6.5.2.7. $\qquad\square$

Let $\mathfrak{X}$ be an ∞-topos. We will say that an object X of $\mathfrak{X}$ is *hypercomplete* if it is local with respect to the class of ∞-connective morphisms. Let $\mathfrak{X}^\wedge$ denote the full subcategory of $\mathfrak{X}$ spanned by the hypercomplete objects of $\mathfrak{X}$. Combining Propositions 6.5.2.8 and 5.5.4.15, we deduce that $\mathfrak{X}^\wedge$ is an accessible localization of $\mathfrak{X}$. Moreover, since Proposition 6.5.1.16 implies that the class of ∞-connective morphisms is stable under pullback, we deduce from Proposition 6.2.1.1 that $\mathfrak{X}^\wedge$ is a *left exact* localization of $\mathfrak{X}$. It follows that $\mathfrak{X}^\wedge$ is itself an ∞-topos. We will show in a moment that $\mathfrak{X}^\wedge$ can be described by a universal property.

Lemma 6.5.2.9. *Let $\mathfrak{X}$ be an ∞-topos and let $n < \infty$. Then $\tau_{\leq n}\mathfrak{X} \subseteq \mathfrak{X}^\wedge$.*

Proof. Corollary 6.5.1.14 implies that an n-truncated object of $\mathfrak{X}$ is local with respect to every n-connective morphism of $\mathfrak{X}$ and therefore with respect to every ∞-connective morphism of $\mathfrak{X}$. □

Lemma 6.5.2.10. *Let $\mathfrak{X}$ be an ∞-topos, let $L : \mathfrak{X} \to \mathfrak{X}^\wedge$ be a left adjoint to the inclusion, and let $X \in \mathfrak{X}$ be such that LX is an ∞-connective object of $\mathfrak{X}^\wedge$. Then LX is a final object of $\mathfrak{X}^\wedge$.*

Proof. For each $n < \infty$, we have equivalences

$$1_\mathfrak{X} \simeq \tau_{\leq n}^{\mathfrak{X}^\wedge} LX \simeq L\tau_{\leq n}^\mathfrak{X} X \simeq \tau_{\leq n}^\mathfrak{X} X,$$

where the first is because of our hypothesis that LX is ∞-connective, the second is given by Proposition 5.5.6.28, and the third is given by Lemma 6.5.2.9. It follows that X is an ∞-connective object of $\mathfrak{X}$, so that LX is a final object of $\mathfrak{X}^\wedge$ by construction. □

We will say that an ∞-topos $\mathfrak{X}$ is *hypercomplete* if $\mathfrak{X}^\wedge = \mathfrak{X}$; in other words, $\mathfrak{X}$ is hypercomplete if every ∞-connective morphism of $\mathfrak{X}$ is an equivalence, so that Whitehead's theorem holds in $\mathfrak{X}$.

Remark 6.5.2.11. In [78], the authors use the term *t-completeness* to refer to the property that we have called hypercompleteness.

Lemma 6.5.2.12. *Let $\mathfrak{X}$ be an ∞-topos. Then the ∞-topos $\mathfrak{X}^\wedge$ is hypercomplete.*

Proof. Let $f : X \to Y$ be an ∞-connective morphism in $\mathfrak{X}^\wedge$. Applying Lemma 6.5.2.10 to the ∞-topos $(\mathfrak{X}^\wedge)_{/Y} \simeq (\mathfrak{X}_{/Y})^\wedge$, we deduce that f is an equivalence. □

We are now prepared to characterize $\mathfrak{X}^\wedge$ by a universal property:

Proposition 6.5.2.13. *Let $\mathfrak{X}$ and $\mathcal{Y}$ be ∞-topoi. Suppose that $\mathcal{Y}$ is hypercomplete. Then composition with the inclusion $\mathfrak{X}^\wedge \subseteq \mathfrak{X}$ induces an isomorphism*

$$\mathrm{Fun}_*(\mathcal{Y}, \mathfrak{X}^\wedge) \to \mathrm{Fun}_*(\mathcal{Y}, \mathfrak{X}).$$

Proof. Let $f_* : \mathcal{Y} \to \mathcal{X}$ be a geometric morphism; we wish to prove that f_* factors through $\mathcal{X}^\wedge$. Let f^* denote a left adjoint to f_*; it will suffice to show that f^* carries each ∞-connective morphism u of $\mathcal{X}$ to an equivalence in $\mathcal{Y}$. Proposition 6.5.1.16 implies that $f^*(u)$ is ∞-connective, and the hypothesis that $\mathcal{Y}$ is hypercomplete guarantees that u is an equivalence. $\square$

The following result establishes the relationship between our notion of hypercompleteness and the Brown-Joyal-Jardine theory of simplicial presheaves.

Proposition 6.5.2.14. *Let $\mathcal{C}$ be a small category equipped with a Grothendieck topology and let $\mathbf{A}$ denote the category of simplicial presheaves on $\mathcal{C}$ endowed with the local model structure (see Proposition 6.5.2.1). Let $\mathbf{A}^\circ$ denote the full subcategory consisting of fibrant-cofibrant objects and let $\mathcal{A} = \mathrm{N}(\mathbf{A}^\circ)$ be the corresponding ∞-category. Then $\mathcal{A}$ is equivalent to $\mathrm{Shv}(\mathcal{C})^\wedge$; in particular, it is a hypercomplete ∞-topos.*

Proof. Let $\mathcal{P}(\mathcal{C})$ denote the ∞-category $\mathcal{P}(\mathrm{N}(\mathcal{C}))$ of presheaves on $\mathrm{N}(\mathcal{C})$ and let $\mathbf{A}'$ denote the model category of simplicial presheaves on $\mathcal{C}$ endowed with the *injective* model structure of §A.3.3. According to Proposition 4.2.4.4, the simplicial nerve functor induces an equivalence

$$\theta : \mathrm{N}(\mathbf{A}'^\circ) \to \mathcal{P}(\mathcal{C}).$$

We may identify $\mathrm{N}(\mathbf{A}^\circ)$ with the full subcategory of $\mathrm{N}(\mathbf{A}'^\circ)$ spanned by the S-local objects, where S is the class of local equivalences (Proposition A.3.7.3).

We first claim that $\theta | \mathrm{N}(\mathbf{A}^\circ)$ factors through $\mathrm{Shv}(\mathcal{C})$. Consider an object $C \in \mathcal{C}$ and a sieve $\mathcal{C}_{/C}^{(0)} \subseteq \mathcal{C}_{/C}$. Let $\chi_C : \mathcal{C} \to \mathrm{Set}$ be the functor $D \mapsto \mathrm{Hom}_{\mathcal{C}}(D, C)$ represented by C, let $\chi_C^{(0)}$ be the subfunctor of χ_C determined by the sieve $\mathcal{C}_{/C}^{(0)}$, and let $i : \chi_C^{(0)} \to \chi_C$ be the inclusion. We regard χ_C and $\chi_C^{(0)}$ as simplicial presheaves on $\mathcal{C}$ which take values in the full subcategory of Set_Δ spanned by the *constant* simplicial sets. We observe that every simplicial presheaf on $\mathcal{C}$ which is valued in constant simplicial sets is automatically fibrant and every object of $\mathbf{A}'$ is cofibrant. Consequently, we may regard i as a morphism in the ∞-category $\mathrm{N}(\mathbf{A}'^\circ)$. It is easy to see that $\theta(i)$ represents the monomorphism $U \to j(C)$ classified by the sieve $\mathcal{C}_{/C}^{(0)}$. If $\mathcal{C}_{/C}^{(0)}$ is a covering sieve on C, then i is a local equivalence. Consequently, every object $X \in \mathrm{N}(\mathbf{A}^\circ)$ is i-local, so that $\theta(X)$ is $\theta(i)$-local. By construction, $\mathrm{Shv}(\mathcal{C})$ is the full subcategory of $\mathcal{P}(\mathcal{C})$ spanned by those objects which are $\theta(i)$-local for every covering sieve $\mathcal{C}_{/C}^{(0)}$ on every object $C \in \mathcal{C}$. We conclude that $\theta | \mathrm{N}(\mathbf{A}^\circ)$ factors through $\mathrm{Shv}(\mathcal{C})$.

Let $\mathcal{X} = \theta^{-1} \mathrm{Shv}(\mathcal{C})$, so that $\mathrm{N}(\mathbf{A}^\circ)$ can be identified with the collection of S'-local objects of $\mathcal{X}$, where S' is the collection of all morphisms in $\mathcal{X}$ which belong to S. Then θ induces an equivalence $\mathrm{N}(\mathbf{A}^\circ) \to \theta(S')^{-1} \mathrm{Shv}(\mathcal{C})$. We now observe that a morphism f in $\mathcal{X}$ belongs to S' if and only if

$\theta(f)$ is an ∞-connective morphism in $\mathrm{Shv}(\mathcal{C})$ (since the condition of being a local equivalence can be tested on homotopy sheaves). It follows that $\theta(S')^{-1}\mathrm{Shv}(\mathcal{C}) = \mathrm{Shv}(\mathcal{C})^{\wedge}$, as desired. $\square$

Remark 6.5.2.15. In [78], the authors discuss a generalization of Jardine's construction in which the category $\mathcal{C}$ is replaced by a simplicial category. Proposition 6.5.2.14 holds in this more general situation as well.

We conclude this section with a few remarks about localizations of an ∞-topos $\mathcal{X}$. In §6.2.1, we introduced the class of topological localizations of $\mathcal{X}$, which consists of those left exact localizations which can be obtained by inverting monomorphisms in $\mathcal{X}$. The hypercompletion $\mathcal{X}^{\wedge}$ is, in some sense, at the other extreme: it is obtained by inverting the ∞-connective morphisms in $\mathcal{X}$, which are never monomorphisms unless they are already equivalences. In fact, $\mathcal{X}^{\wedge}$ is the *maximal* (left exact) localization of $\mathcal{X}$ which can be obtained without inverting monomorphisms:

Proposition 6.5.2.16. *Let $\mathcal{X}$ and $\mathcal{Y}$ be ∞-topoi and let $f^* : \mathcal{X} \to \mathcal{Y}$ be a left exact colimit-preserving functor. The following conditions are equivalent:*

(1) *For every monomorphism u in $\mathcal{X}$, if f^*u is an equivalence in $\mathcal{Y}$, then u is an equivalence in $\mathcal{X}$.*

(2) *For every morphism $u \in \mathcal{X}$, if f^*u is an equivalence in $\mathcal{Y}$, then f is ∞-connective.*

Proof. Suppose first that (2) is satisfied. If u is a monomorphism and f^*u is an equivalence in $\mathcal{Y}$, then u is ∞-connective. In particular, u is both a monomorphism and an effective epimorphism and therefore an equivalence in $\mathcal{X}$. This proves (1). Conversely, suppose that (1) is satisfied and let $u : X \to Z$ be an arbitrary morphism in $\mathcal{X}$ such that $f^*(u)$ is an equivalence. We will prove by induction on n that u is n-connective.

We first consider the case $n = 0$. Choose a factorization

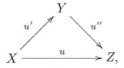

$$X \xrightarrow{\quad u \quad} Z,$$

where u' is an effective epimorphism and u'' is a monomorphism. Since f^*u is an equivalence, Corollary 6.2.3.12 implies that f^*u'' is an effective epimorphism. Since f^*u'' is also a monomorphism (by virtue of our assumption that f is left exact), we conclude that f^*u'' is an equivalence. Applying (1), we deduce that u'' is an equivalence, so that u is an effective epimorphism, as desired.

Now suppose $n > 0$. According to Proposition 6.5.1.18, it will suffice to show that the diagonal map $\delta : X \to X \times_Z X$ is $(n-1)$-connective. By the inductive hypothesis, it will suffice to prove that $f^*(\delta)$ is an equivalence in $\mathcal{Y}$. We conclude by observing that f^* is left exact, so we can identify δ with the diagonal map associated to the equivalence $f^*(u) : f^*X \to f^*Z$. $\square$

Definition 6.5.2.17. Let $\mathcal{X}$ be an ∞-topos and let $\mathcal{Y} \subseteq \mathcal{X}$ be an accessible left exact localization of $\mathcal{X}$. We will say that $\mathcal{Y}$ is an *cotopological* localization of $\mathcal{X}$ if the left adjoint $L : \mathcal{X} \to \mathcal{Y}$ to the inclusion of $\mathcal{Y}$ in $\mathcal{X}$ satisfies the equivalent conditions of Proposition 6.5.2.16.

Remark 6.5.2.18. Let $f^* : \mathcal{X} \to \mathcal{Y}$ be the left adjoint of a geometric morphism between ∞-topoi and suppose that the equivalent conditions of Proposition 6.5.2.16 are satisfied. Let $u : X \to Z$ be a morphism in $\mathcal{X}$ and choose a factorization

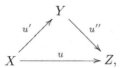

where u' is an effective epimorphism and u'' is a monomorphism. Then u'' is an equivalence if and only if $f^*(u'')$ is an equivalence. Applying Corollary 6.2.3.12, we conclude that u is an effective epimorphism if and only if $f^*(u)$ is an effective epimorphism.

The hypercompletion $\mathcal{X}^\wedge$ of an ∞-topos $\mathcal{X}$ can be characterized as the *maximal* cotopological localization of $\mathcal{X}$ (that is, the cotopological localization which is obtained by inverting as many morphisms as possible). According to our next result, every localization can be obtained by combining topological and cotopological localizations:

Proposition 6.5.2.19. *Let $\mathcal{X}$ be an ∞-topos and let $\mathcal{X}'' \subseteq \mathcal{X}$ be an accessible left exact localization of $\mathcal{X}$. Then there exists a topological localization $\mathcal{X}' \subseteq \mathcal{X}$ such that $\mathcal{X}'' \subseteq \mathcal{X}'$ is a cotopological localization of $\mathcal{X}'$.*

Proof. Let $L : \mathcal{X} \to \mathcal{X}''$ be a left adjoint to the inclusion, let S be the collection of all monomorphisms u in $\mathcal{X}$ such that Lu is an equivalence, and let $\mathcal{X}' = S^{-1}\mathcal{X}$ be the collection of S-local objects of $\mathcal{X}$. Since L is left exact, S is stable under pullbacks and therefore determines a topological localization of $\mathcal{X}$. By construction, we have $\mathcal{X}'' \subseteq \mathcal{X}'$. The restriction $L | \mathcal{X}'$ exhibits $\mathcal{X}''$ as an accessible left exact localization of $\mathcal{X}'$. Let u be a monomorphism in $\mathcal{X}'$ such that Lu is an equivalence. Then u is a monomorphism in $\mathcal{X}$, so that $u \in S$. Since $\mathcal{X}'$ consists of S-local objects, we conclude that u is an equivalence. It follows that $\mathcal{X}''$ is a cotopological localization of $\mathcal{X}'$, as desired. $\square$

Remark 6.5.2.20. It is easy to see that the factorization of Proposition 6.5.2.19 is essentially uniquely determined: more precisely, $\mathcal{X}'$ is unique provided we assume that it is stable under equivalences in $\mathcal{X}$.

Combining Proposition 6.5.2.19 with Remark 7.2.1.16, we see that every ∞-topos $\mathcal{X}$ can be obtained in following way:

(1) Begin with the ∞-category $\mathcal{P}(\mathcal{C})$ of presheaves on some small ∞-category $\mathcal{C}$.

(2) Choose a Grothendieck topology on $\mathcal{C}$: this is equivalent to choosing a left exact localization of the underlying topos $\mathrm{Disc}(\mathcal{P}(\mathcal{C})) = \mathbb{S}\mathrm{et}^{h\mathcal{C}^{op}}$.

(3) Form the associated topological localization $\mathrm{Shv}(\mathcal{C}) \subseteq \mathcal{P}(\mathcal{C})$, which can be described as the pullback

$$\mathcal{P}(\mathcal{C}) \times_{\mathcal{P}(N(h\mathcal{C}))} \mathrm{Shv}(N(h\mathcal{C}))$$

in $\mathcal{R}\mathcal{T}\mathrm{op}$.

(4) Form a cotopological localization of $\mathrm{Shv}(\mathcal{C})$ by inverting some class of ∞-connective morphisms of $\mathrm{Shv}(\mathcal{C})$.

Remark 6.5.2.21. Let $\mathcal{X}$ be an ∞-topos. The collection of all ∞-connective morphisms in $\mathcal{X}$ is saturated. It follows from Proposition 5.5.5.7 that there exists a factorization system (S_L, S_R) on $\mathcal{X}$, where S_L is the collection of all ∞-connective morphisms in $\mathcal{X}$. We will say that a morphism in $\mathcal{X}$ is *hypercomplete* if it belongs to S_R. Unwinding the definitions (and using the fact that a morphism in $\mathcal{X}_{/Y}$ is ∞-connective if and only if its image in $\mathcal{X}$ is ∞-connective), we conclude that a morphism $f : X \to Y$ is hypercomplete if and only if it is hypercomplete when viewed as an object of the ∞-topos $\mathcal{X}_{/Y}$ (see §6.5.2).

Using Proposition 5.2.8.6, we deduce that the collection of hypercomplete morphisms in $\mathcal{X}$ is stable under limits and the formation of pullback squares.

Remark 6.5.2.22. Let $\mathcal{X}$ be an ∞-topos. The condition that a morphism $f : X \to Y$ be hypercomplete is *local*: that is, if $\{Y_\alpha \to Y\}$ is a collection of morphisms which determine an effective epimorphism $\coprod Y_\alpha \to Y$, and each of the induced maps $f_\alpha : X \times_Y Y_\alpha \to Y_\alpha$ is hypercomplete, then f is hypercomplete. To prove this, we set $Y_0 = \coprod_\alpha Y_\alpha$; then $\mathcal{X}_{/Y_0} \simeq \prod_\alpha \mathcal{X}_{/Y_\alpha}$ (since coproducts in $\mathcal{X}$ are disjoint), so it is easy to see that the induced map $f' : X \times_Y Y_0 \to Y_0$ is hypercomplete. Let $Y_\bullet$ be the simplicial object of $\mathcal{X}$ given by the Čech nerve of the effective epimorphism $Y_0 \to Y$. For every map $Z \to Y$, let $Z_\bullet$ be the simplicial object described by the formula $Z_n = Y_n \times_Y Z$ (equivalently, $Z_\bullet$ is the Čech nerve of the effective epimorphism $Z \times_Y Y_0 \to Z$). Using Remark 6.5.2.21, we conclude that each of the maps $X_n \to Y_n$ is hypercomplete.

For every map $A \to Y$, the mapping space $\mathrm{Map}_{\mathcal{X}_{/Y}}(A, X)$ can be obtained as the totalization of a cosimplicial space

$$n \mapsto \mathrm{Map}_{\mathcal{X}_{/Y_n}}(A_n, X_n).$$

If $g : A \to B$ is an ∞-connective morphism in $\mathcal{X}_{/Y}$, then each of the induced maps $A_n \to B_n$ is ∞-connective, so the induced map

$$\mathrm{Map}_{\mathcal{X}_{/Y_n}}(B_n, X_n) \to \mathrm{Map}_{\mathcal{X}_{/Y_n}}(A_n, X_n)$$

is a homotopy equivalence. Passing to the totalization, we obtain a homotopy equivalence $\mathrm{Map}_{\mathcal{X}_{/Y}}(B, X) \to \mathrm{Map}_{\mathcal{X}_{/Y}}(A, X)$. Thus f is hypercomplete, as desired.

6.5.3 Hypercoverings

Let $\mathcal{X}$ be an ∞-topos. In §6.5.2, we defined the *hypercompletion* $\mathcal{X}^\wedge \subseteq \mathcal{X}$ to be the left exact localization of $\mathcal{X}$ obtained by inverting the ∞-connective morphisms. In this section, we will give an alternative description of the hypercomplete objects $X \in \mathcal{X}^\wedge$: they are precisely those objects of $\mathcal{X}$ which satisfy a descent condition with respect to hypercoverings (Theorem 6.5.3.12). We begin by reviewing the definition of a hypercovering.

Let X be a topological space and let $\mathcal{F}$ be a presheaf of sets on X. To construct the sheaf associated to $\mathcal{F}$, it is natural to consider the presheaf $\mathcal{F}^+$ defined by

$$\mathcal{F}^+ = \varinjlim_{\mathcal{U}} \varprojlim_{V \in \mathcal{U}} \mathcal{F}(V).$$

Here the direct limit is taken over all sieves $\mathcal{U}$ which cover U. There is an obvious map $\mathcal{F} \to \mathcal{F}^+$ which is an isomorphism whenever $\mathcal{F}$ is a sheaf. Moreover, $\mathcal{F}^+$ is "closer" to being a sheaf than $\mathcal{F}$ is. More precisely, $\mathcal{F}^+$ is always a separated presheaf: two sections of $\mathcal{F}^+$ which agree locally automatically coincide. If $\mathcal{F}$ is itself a separated presheaf, then $\mathcal{F}^+$ is a sheaf.

For a general presheaf $\mathcal{F}$, we need to apply the above construction twice to construct the associated sheaf $(\mathcal{F}^+)^+$. To understand the problem, let us try to prove that $\mathcal{F}^+$ is a sheaf (to see where the argument breaks down). Suppose we are given an open covering $X = \bigcup U_\alpha$ and a collection of sections $s_\alpha \in \mathcal{F}^+(U_\alpha)$ such that

$$s_\alpha | U_\alpha \cap U_\beta = s_\beta | U_\alpha \cap U_\beta.$$

Refining the covering U_α if necessary, we may assume that each s_α is the image of some section $t_\alpha \in \mathcal{F}(U_\alpha)$. However, the equation

$$t_\alpha | U_\alpha \cap U_\beta = t_\beta | U_\alpha \cap U_\beta$$

holds only locally on $U_\alpha \cap U_\beta$, so the sections t_α do not necessarily determine a global section of $\mathcal{F}^+$. To summarize: the freedom to consider arbitrarily fine open covers $\mathcal{U} = \{U_\alpha\}$ is not enough; we also need to be able to refine the intersections $U_\alpha \cap U_\beta$. This leads very naturally to the notion of a *hypercovering*. Roughly speaking, a hypercovering of X consists of an open covering $\{U_\alpha\}$ of X, an open covering $\{V_{\alpha\beta\gamma}\}$ of each intersection $U_\alpha \cap U_\beta$, and analogous data associated to more complicated intersections (see Definition 6.5.3.2 for a more precise formulation).

In classical sheaf theory, there are two ways to construct the sheaf associated to a presheaf $\mathcal{F}$:

(1) One can apply the construction $\mathcal{F} \mapsto \mathcal{F}^+$ twice.

(2) Using the theory of hypercoverings, one can proceed directly by defining

$$\mathcal{F}^\dagger(U) = \varinjlim_{\mathcal{U}} \varprojlim \mathcal{F}(V),$$

where the direct limit is now taken over arbitrary *hypercoverings* $\mathcal{U}$.

In higher category theory, the difference between these two approaches becomes more prominent. For example, suppose that $\mathcal{F}$ is not a presheaf of sets but a presheaf of *groupoids* on X. In this case, one can construct the associated sheaf of groupoids using either approach. However, in the case of approach (1), it is necessary to apply the construction $\mathcal{F} \mapsto \mathcal{F}^+$ *three* times: the first application guarantees that the automorphism groups of sections of $\mathcal{F}$ are separated presheaves, the second guarantees that they are sheaves, and the third guarantees that $\mathcal{F}$ itself satisfies descent. More generally, if $\mathcal{F}$ is a sheaf of n-truncated spaces, then the sheafification of $\mathcal{F}$ via approach (1) takes place in $(n + 2)$-stages.

When we pass to the case $n = \infty$, the situation becomes more complicated. If $\mathcal{F}$ is a presheaf of spaces on X, then it is not reasonable to expect to obtain a sheaf by applying the construction $\mathcal{F} \mapsto \mathcal{F}^+$ any finite number of times. In fact, it is not obvious that $\mathcal{F}^+$ is any closer than $\mathcal{F}$ to being a sheaf. Nevertheless, this is true: we can construct the sheafification of $\mathcal{F}$ via a *transfinite iteration* of the construction $\mathcal{F} \mapsto \mathcal{F}^+$. More precisely, we define a transfinite sequence of presheaves

$$\mathcal{F}(0) \to \mathcal{F}(1) \to \cdots$$

as follows:

(*i*) Let $\mathcal{F}(0) = \mathcal{F}$.

(*ii*) For every ordinary α, let $\mathcal{F}(\alpha + 1) = \mathcal{F}(\alpha)^+$.

(*iii*) For every limit ordinal λ, let $\mathcal{F}(\lambda) = \varinjlim_\alpha \mathcal{F}(\alpha)$, where α ranges over ordinals less than λ.

One can show that the above construction *converges* in the sense that $\mathcal{F}(\alpha)$ is a sheaf for $\alpha \gg 0$ (and therefore $\mathcal{F}(\alpha) \simeq \mathcal{F}(\beta)$ for $\beta \geq \alpha$). Moreover, $\mathcal{F}(\alpha)$ is universal among sheaves of spaces which admit a map from $\mathcal{F}$.

Alternatively, one can use the construction $\mathcal{F} \mapsto \mathcal{F}^\dagger$ to construct a sheaf of spaces from $\mathcal{F}$ in a single step. The universal property asserted above guarantees the existence of a morphism of sheaves $\theta : \mathcal{F}(\alpha) \to \mathcal{F}^\dagger$. However, the morphism θ is generally *not* an equivalence. Instead, θ realizes $\mathcal{F}^\dagger$ as the *hypercompletion* of $\mathcal{F}(\alpha)$ in the ∞-topos $\mathrm{Shv}(X)$. We will not prove this statement directly but will instead establish a reformulation (Corollary 6.5.3.13) which does not make reference to the sheafification constructions outlined above.

Before we can introduce the definition of a hypercovering, we need to review some simplicial terminology.

Notation 6.5.3.1. For each $n \geq 0$, let $\mathbf{\Delta}^{\leq n}$ denote the full subcategory of $\mathbf{\Delta}$ spanned by the set of objects $\{[0], \dots, [n]\}$. If $\mathcal{X}$ is a presentable ∞-category, the restriction functor

$$\mathrm{sk}_n : \mathcal{X}_\Delta \to \mathrm{Fun}(\mathrm{N}(\mathbf{\Delta}^{\leq n})^{op}, \mathcal{X})$$

has a right adjoint given by right Kan extension along the inclusion functor $\mathrm{N}(\mathbf{\Delta}^{\leq n})^{op} \subseteq \mathrm{N}(\mathbf{\Delta})^{op}$. Let $\mathrm{cosk}_n : \mathcal{X}_\Delta \to \mathcal{X}_\Delta$ be the composition of sk_n with its right adjoint. We will refer to cosk_n as the *n-coskeleton functor*.

Definition 6.5.3.2. Let $\mathfrak{X}$ be an ∞-topos. A simplicial object $U_\bullet \in \mathfrak{X}_\Delta$ is a *hypercovering of $\mathfrak{X}$* if, for each $n \geq 0$, the unit map

$$U_n \to (\mathrm{cosk}_{n-1} U_\bullet)_n$$

is an effective epimorphism. We will say that $U_\bullet$ is an *effective hypercovering of $\mathfrak{X}$* if the colimit of $U_\bullet$ is a final object of $\mathfrak{X}$.

Remark 6.5.3.3. More informally, a simplicial object $U_\bullet \in \mathfrak{X}_\Delta$ is a hypercovering of $\mathfrak{X}$ if each of the associated maps

$$U_0 \to 1_{\mathfrak{X}}$$

$$U_1 \to U_0 \times U_0$$

$$U_2 \to \cdots$$

is an effective epimorphism.

Lemma 6.5.3.4. *Let $\mathfrak{X}$ be an ∞-topos and let $U_\bullet$ be a simplicial object in $\mathfrak{X}$. Let $L : \mathfrak{X} \to \mathfrak{X}^\wedge$ be a left adjoint to the inclusion. The following conditions are equivalent:*

(1) *The simplicial object $U_\bullet$ is a hypercovering of $\mathfrak{X}$.*

(2) *The simplicial object $L \circ U_\bullet$ is a hypercovering of $\mathfrak{X}^\wedge$.*

Proof. Since L is left exact, we can identify $L \circ \mathrm{cosk}_n U_\bullet$ with $\mathrm{cosk}_n(L \circ U_\bullet)$. The desired result now follows from Remark 6.5.2.18. $\square$

Lemma 6.5.3.5. *Let $\mathfrak{X}$ be an ∞-topos and let U be an ∞-connective object of $\mathfrak{X}$. Let $U_\bullet$ be the constant simplicial object with value U. Then $U_\bullet$ is a hypercovering of $\mathfrak{X}$.*

Proof. Using Lemma 6.5.3.4, we can reduce to the case where $\mathfrak{X}$ is hypercomplete. Then $U \simeq 1_{\mathfrak{X}}$, so that $U_\bullet$ is equivalent to the constant functor with value $1_{\mathfrak{X}}$ and is therefore a final object of $\mathfrak{X}_\Delta$. For each $n \geq 0$, the coskeleton functor cosk_{n-1} preserves small limits, so $\mathrm{cosk}_{n-1} U_\bullet$ is also a final object of $U_\bullet$. It follows that the unit map $U_\bullet \to \mathrm{cosk}_{n-1} U_\bullet$ is an equivalence. $\square$

Notation 6.5.3.6. Let $\boldsymbol{\Delta}_s$ be the subcategory of $\boldsymbol{\Delta}$ with the same objects but where the morphisms are given by *injective* order-preserving maps between nonempty linearly ordered sets. If $\mathfrak{X}$ is an ∞-category, we will refer to a diagram $\mathrm{N}(\boldsymbol{\Delta}_s)^{op} \to \mathfrak{X}$ as a *semisimplicial object of $\mathfrak{X}$*.

Lemma 6.5.3.7. *The inclusion $\mathrm{N}(\boldsymbol{\Delta}_s^{op}) \subseteq \mathrm{N}(\boldsymbol{\Delta}^{op})$ is cofinal.*

Proof. According to Theorem 4.1.3.1, it will suffice to prove that for every $n \geq 0$, the category $\mathcal{C} = \boldsymbol{\Delta}_s \times_{\boldsymbol{\Delta}} \boldsymbol{\Delta}_{/[n]}$ has a weakly contractible nerve. To prove this, we let $F : \mathcal{C} \to \mathcal{C}$ be the constant functor taking value given by the inclusion $[0] \subseteq [n]$ and $G : \mathcal{C} \to \mathcal{C}$ be the functor which carries an arbitrary map $[m] \to [n]$ to the induced map $[0] \coprod [m] \to [n]$. We have natural transformations of functors

$$F \to G \leftarrow \mathrm{id}_{\mathcal{C}}.$$

Let X be the topological space $|\mathrm{N}(\mathcal{C})|$. The natural transformations above show that the identity map id_X is homotopic to a constant, so that X is contractible, as desired.   □

Consequently, if $U_\bullet$ is a simplicial object in an ∞-category $\mathcal{X}$ and $U_\bullet^s = U_\bullet | \mathrm{N}(\mathbf{\Delta}_s^{op})$ is the associated semisimplicial object, then we can identify colimits of $U_\bullet$ with colimits of $U_\bullet^s$.

We will say that a simplicial object $U_\bullet$ in an ∞-category $\mathcal{X}$ is *n-coskeletal* if it is a right Kan extension of its restriction to $\mathrm{N}(\mathbf{\Delta}_{\leq n}^{op})$. Similarly, we will say that a semisimplicial object $U_\bullet$ of $\mathcal{X}$ is *n-coskeletal* if it is a right Kan extension of its restriction to $\mathrm{N}(\mathbf{\Delta}_{s,\leq n}^{op})$, where $\mathbf{\Delta}_{s,\leq n} = \mathbf{\Delta}_s \times_{\mathbf{\Delta}} \mathbf{\Delta}_{\leq n}$.

Lemma 6.5.3.8. *Let $\mathcal{X}$ be an ∞-category, let $U_\bullet$ be a simplicial object of $\mathcal{X}$, and let $U_\bullet^s = U_\bullet | \mathrm{N}(\mathbf{\Delta}_s^{op})$ the associated semisimplicial object. Then $U_\bullet$ is n-coskeletal if and only if $U_\bullet^s$ is n-coskeletal.*

Proof. It will suffice to show that, for each $\Delta^m \in \mathbf{\Delta}$, the nerve of the inclusion

$$(\mathbf{\Delta}_s)_{/[m]} \times_{\mathbf{\Delta}_s} \mathbf{\Delta}_{s,\leq n} \subseteq \mathbf{\Delta}_{/[m]} \times_{\mathbf{\Delta}} \mathbf{\Delta}_{\leq n}$$

is cofinal. Let $\theta : [m'] \to [m]$ be an object of $\mathbf{\Delta}_{/[m]} \times_{\mathbf{\Delta}} \mathbf{\Delta}_{\leq n}$. We let $\mathcal{C}$ denote the category of all factorizations

$$[m'] \xrightarrow{\theta'} [m''] \xrightarrow{\theta''} [m]$$

for θ such that θ'' is a monomorphism and $m'' \leq n$. According to Theorem 4.1.3.1, it will suffice to prove that $\mathrm{N}(\mathcal{C})$ is weakly contractible (for every choice of θ). We now simply observe that $\mathcal{C}$ has an initial object (given by the unique factorization where θ' is an epimorphism).   □

Lemma 6.5.3.9 ([20]). *Let $\mathcal{X}$ be an ∞-topos and let $U_\bullet$ be an n-coskeletal hypercovering of $\mathcal{X}$. Then $U_\bullet$ is effective.*

Proof. We will prove this result by induction on n. If $n = 0$, then $U_\bullet$ can be identified with the underlying groupoid of the Čech nerve of the map $\theta : U_0 \to 1_{\mathcal{X}}$, where $1_{\mathcal{X}}$ is a final object of $\mathcal{X}$. Since $U_\bullet$ is a hypercovering, θ is an effective epimorphism, so the Čech nerve of θ is a colimit diagram and the desired result follows. Let us therefore assume that $n > 0$. Let $V_\bullet = \mathrm{cosk}_{n-1} U_\bullet$ and let $f_\bullet : U_\bullet \to V_\bullet$ be the adjunction map. For each $m \geq 0$, the map $f_m : U_m \to V_m$ is a composition of finitely many pullbacks of f_n. Since $U_\bullet$ is a hypercovering, f_n is an effective epimorphism, so each f_m is also an effective epimorphism. We also observe that f_m is an equivalence for $m < n$.

Let $W_+ : \mathrm{N}(\mathbf{\Delta}_+ \times \mathbf{\Delta})^{op} \to \mathcal{X}$ be a Čech nerve of $f_\bullet$ (formed in the ∞-category $\mathcal{X}_{\mathbf{\Delta}}$ of simplicial objects of $\mathcal{X}$). We observe that $W_+ | \mathrm{N}(\{\emptyset\} \times \mathbf{\Delta})^{op}$

can be identified with $V_\bullet$. Since $V_\bullet$ is an $(n-1)$-coskeletal hypercovering of $\mathcal{X}$, the inductive hypothesis implies that any colimit $|V_\bullet|$ is a final object of $\mathcal{X}$. The inclusion $N(\{\emptyset\} \times \boldsymbol{\Delta})^{op} \subseteq N(\boldsymbol{\Delta}_+ \times \boldsymbol{\Delta})^{op}$ is cofinal (being a product of $N(\boldsymbol{\Delta})^{op}$ and the inclusion of a final object into $N(\boldsymbol{\Delta}_+)^{op}$), so we may identify colimits of W_+ with colimits of $V_\bullet$. It follows that any colimit of W_+ is a final object of $\mathcal{X}$. We next observe that each of the augmented simplicial objects $W_+ | N(\boldsymbol{\Delta}_+ \times \{[m]\})^{op}$ is a Čech nerve of f_m and therefore a colimit diagram (since f_m is an effective epimorphism). Applying Lemma 4.3.3.9, we conclude that W_+ is a left Kan extension of the bisimplicial object $W = W_+ | N(\boldsymbol{\Delta} \times \boldsymbol{\Delta})^{op}$. According to Lemma 4.3.2.7, we can identify colimits of W_+ with colimits of W, so any colimit of W is a final object of $\mathcal{X}$.

Let $D_\bullet : N(\boldsymbol{\Delta}^{op}) \to \mathcal{X}$ be the simplicial object of $\mathcal{X}$ obtained by composing W with the diagonal map $\delta : N(\boldsymbol{\Delta}^{op}) \to N(\boldsymbol{\Delta} \times \boldsymbol{\Delta})^{op}$. According to Lemma 5.5.8.4, δ is cofinal. We may therefore identify colimits of W with colimits of $D_\bullet$, so that any colimit $|D_\bullet|$ of $D_\bullet$ is a final object of $\mathcal{X}$.

Let $U_\bullet^s = U_\bullet | N(\boldsymbol{\Delta}_s^{op})$ and let $D_\bullet^s = D_\bullet | N(\boldsymbol{\Delta}_s^{op})$. We will prove that $U_\bullet^s$ is a retract of $D_\bullet^s$ in the ∞-category of semisimplicial objects of $\mathcal{X}$. According to Lemma 6.5.3.7, we can identify colimits of $D_\bullet^s$ with colimits of $D_\bullet$. It will follow that any colimit of $U_\bullet^s$ is a retract of a final object of $\mathcal{X}$ and therefore itself final. Applying Lemma 6.5.3.7 again, we will conclude that any colimit of $U_\bullet$ is a final object of $\mathcal{X}$, and the proof will be complete.

We observe that $D_\bullet^s$ is the result of composing W with the (opposite of the nerve of the) diagonal functor

$$\delta^s : \boldsymbol{\Delta}_s \to \boldsymbol{\Delta} \times \boldsymbol{\Delta}.$$

Similarly, the semisimplicial object $U_\bullet^s$ is obtained from W via the composition

$$\epsilon : \boldsymbol{\Delta}_s \subseteq \boldsymbol{\Delta} \simeq \{[0]\} \times \boldsymbol{\Delta} \subseteq \boldsymbol{\Delta} \times \boldsymbol{\Delta}.$$

There is an obvious natural transformation of functors $\delta^s \to \epsilon$ which yields a map of semisimplicial objects $\theta : U_\bullet^s \to D_\bullet^s$. To complete the proof, it will suffice to show that there exists a map

$$\theta' : D_\bullet^s \to U_\bullet^s$$

such that $\theta' \circ \theta$ is homotopic to the identity on $U_\bullet^s$.

According to Lemma 6.5.3.8, $U_\bullet^s$ is n-coskeletal as a *semisimplicial* object of $\mathcal{X}$. Let $D_{\leq n}^s$ and $U_{\leq n}^s$ denote restrictions of $D_\bullet^s$ and $U_\bullet^s$ to $N(\boldsymbol{\Delta}_{s,\leq n}^{op})$ and $\theta_{\leq n} : U_{\leq n}^s \to D_{\leq n}^s$ be the morphism induced by θ. We have canonical homotopy equivalences

$$\mathrm{Map}_{\mathrm{Fun}(N(\boldsymbol{\Delta}_s^{op}),\mathcal{X})}(D_\bullet^s, U_\bullet^s) \simeq \mathrm{Map}_{\mathrm{Fun}(N(\boldsymbol{\Delta}_{s,\leq n}^{op}),\mathcal{X})}(D_{\leq n}^s, U_{\leq n}^s)$$

$$\mathrm{Map}_{\mathrm{Fun}(N(\boldsymbol{\Delta}_s^{op}),\mathcal{X})}(U_\bullet^s, U_\bullet^s) \simeq \mathrm{Map}_{\mathrm{Fun}(N(\boldsymbol{\Delta}_{s,\leq n}^{op}),\mathcal{X})}(U_{\leq n}^s, U_{\leq n}^s).$$

It will therefore suffice to prove that there exists a map

$$\theta'_{\leq n} : D_{\leq n}^s \to U_{\leq n}^s$$

such that $\theta'_{\leq n} \circ \theta_{\leq n}$ is homotopic to the identity on $U^s_{\leq n}$.

Consider the functors

$$\overline{\delta}^s : \mathbf{\Delta}_{s,\leq n} \to \mathbf{\Delta}_+ \times \mathbf{\Delta}$$

$$\overline{\epsilon} : \mathbf{\Delta}_{s,\leq n} \to \mathbf{\Delta}_+ \times \mathbf{\Delta}$$

defined as follows:

$$\overline{\delta}^s([m]) = \begin{cases} (\emptyset, [m]) & \text{if } m < n \\ ([n], [n]) & \text{if } m = n \end{cases}$$

$$\overline{\epsilon}([m]) = \begin{cases} (\emptyset, [m]) & \text{if } m < n \\ ([0], [n]) & \text{if } m = n. \end{cases}$$

We have a commutative diagram of natural transformations

which gives rise to a diagram

$$\overline{D}^s_{\leq n} \longleftarrow D^s_{\leq n}$$
$$\psi_{\leq n} \uparrow \qquad \theta_{\leq n} \uparrow$$
$$\overline{U}^s_{\leq n} \longleftarrow U^s_{\leq n}$$

in the ∞-category $\mathrm{Fun}(\mathrm{N}(\mathbf{\Delta}^{op}_{s,\leq n}), \mathfrak{X})$. The vertical arrows are equivalences. Consequently, it will suffice to produce a (homotopy) left inverse to $\psi_{\leq n}$.

For $m \geq 0$, let $V^s_{\leq m} = V_\bullet | \mathbf{\Delta}_{s,\leq m}$. We can identify $\overline{D}^s_{\leq n}$ and $\overline{U}^s_{\leq n}$ with objects $X, Y \in \mathfrak{X}_{/V^s_{\leq n-1}}$ and $\psi_{\leq n}$ with a morphism $f : X \to Y$. To complete the proof, it will suffice to produce a left inverse to f in the ∞-category $\mathfrak{X}_{/V^s_{\leq n-1}}$. We observe that because $V_\bullet$ is $(n-1)$-coskeletal, we have a diagram of trivial fibrations

$$\mathfrak{X}_{/V_n} \leftarrow \mathfrak{X}_{/V^s_{\leq n}} \to \mathfrak{X}_{/V^s_{\leq n-1}}.$$

Using this diagram (and the construction of W), we conclude that Y can be identified with a product of $(n+1)$ copies of X in $\mathfrak{X}_{/V^s_{\leq n-1}}$ and that f can be identified with the identity map. The existence of a left homotopy inverse to f is now obvious (choose any of the $(n+1)$-projections from Y onto X). □

Lemma 6.5.3.10. *Let $\mathfrak{X}$ be an ∞-topos and let $f_\bullet : U_\bullet \to V_\bullet$ be a natural transformation between simplicial objects of $\mathfrak{X}$. Suppose that, for each $k \leq n$, the map $f_k : U_k \to V_k$ is an equivalence. Then the induced map $|f_\bullet| : |U_\bullet| \to |V_\bullet|$ of colimits is n-connective.*

Proof. Choose a left exact localization functor $L : \mathcal{P}(\mathcal{C}) \to \mathcal{X}$. Without loss of generality, we may suppose that $f_\bullet = L \circ \overline{f}_\bullet$, where $\overline{f}_\bullet : \overline{U}_\bullet \to \overline{V}_\bullet$ is a transformation between simplicial objects of $\mathcal{P}(\mathcal{C})$, where $\overline{f}_k$ is an equivalence for $k \leq n$. Since L preserves colimits and n-connectivity (Proposition 6.5.1.16), it will suffice to prove that $|f_\bullet|$ is n-connective. Using Propositions 6.5.1.12 and 5.5.6.28, we see that $|f_\bullet|$ is n-connective if and only if, for each object $C \in \mathcal{C}$, the induced morphism in $\mathcal{S}$ is n-connective. In other words, we may assume without loss of generality that $\mathcal{X} = \mathcal{S}$.

According to Proposition 4.2.4.4, we may assume that $f_\bullet$ is obtained by taking the simplicial nerve of a map $f'_\bullet : U'_\bullet \to V'_\bullet$ between simplicial objects in the ordinary category $\mathcal{K}$an. Without loss of generality, we may suppose that $U'_\bullet$ and $V'_\bullet$ are projectively cofibrant (as diagrams in the model category $\mathcal{S}et_\Delta$). According to Theorem 4.2.4.1, it will suffice to prove that the induced map from the (homotopy) colimit of $U'_\bullet$ to the (homotopy) colimit of $V'_\bullet$ has n-connective homotopy fibers, which follows from classical homotopy theory. $\qquad\square$

Lemma 6.5.3.11. *Let $\mathcal{X}$ be an ∞-topos and let $U_\bullet$ be a hypercovering of $\mathcal{X}$. Then the colimit $|U_\bullet|$ is ∞-connective.*

Proof. We will prove that θ is n-connective for every $n \geq 0$. Let $V_\bullet = \operatorname{cosk}_{n+1} U_\bullet$ and let $u : U_\bullet \to V_\bullet$ be the adjunction map. Lemma 6.5.3.10 asserts that the induced map $|U_\bullet| \to |V_\bullet|$ is n-connective, and Lemma 6.5.3.9 asserts that $|V_\bullet|$ is a final object of $\mathcal{X}$. It follows that $|U_\bullet| \in \mathcal{X}$ is n-connective, as desired. $\qquad\square$

The preceding results lead to an easy characterization of the class of hypercomplete ∞-topoi:

Theorem 6.5.3.12. *Let $\mathcal{X}$ be an ∞-topos. The following conditions are equivalent:*

(1) *For every $X \in \mathcal{X}$, every hypercovering $U_\bullet$ of $\mathcal{X}_{/X}$ is effective.*

(2) *The ∞-topos $\mathcal{X}$ is hypercomplete.*

Proof. Suppose that (1) is satisfied. Let $f : U \to X$ be an ∞-connective morphism in $\mathcal{X}$ and let $f_\bullet$ be the constant simplicial object of $\mathcal{X}_{/X}$ with value f. According to Lemma 6.5.3.5, f is a hypercovering of $\mathcal{X}_{/X}$. Invoking (1), we conclude that $f \simeq |f_\bullet|$ is a final object of $\mathcal{X}_{/X}$; in other words, f is an equivalence. This proves that $(1) \Rightarrow (2)$.

Conversely, suppose that $\mathcal{X}$ is hypercomplete. Let $X \in \mathcal{X}$ be an object and $U_\bullet$ a hypercovering of $\mathcal{X}_{/X}$. Then Lemma 6.5.3.11 implies that $|U_\bullet|$ is an ∞-connective object of $\mathcal{X}_{/X}$. Since $\mathcal{X}$ is hypercomplete, we conclude that $|U_\bullet|$ is a final object of $\mathcal{X}_{/X}$, so that $U_\bullet$ is effective. $\qquad\square$

Corollary 6.5.3.13 (Dugger-Hollander-Isaksen [20], Toën-Vezzosi [78]). *Let $\mathcal{X}$ be an ∞-topos. For each $X \in \mathcal{X}$ and each hypercovering $U_\bullet$ of $\mathcal{X}_{/X}$, let $|U_\bullet|$ be the associated morphism of $\mathcal{X}$ (which has target X). Let S denote the* *collection of all such morphisms* $|U_\bullet|$. *Then* $X^\wedge = S^{-1} X$. *In other words, an object of* X *is hypercomplete if and only if it is* S-*local.*

Remark 6.5.3.14. One can generalize Corollary 6.5.3.13 as follows: let $L : X \to Y$ be an arbitrary left exact localization of ∞-topoi and let S be the collection of all morphisms of the form $|U_\bullet|$, where $U_\bullet$ is a simplicial object of $X_{/X}$ such that $L \circ U_\bullet$ is an effective hypercovering of $Y_{/LX}$. Then L induces an equivalence $S^{-1} X \to Y$.

It follows that every ∞-topos can be obtained by starting with an ∞-category of presheaves $\mathcal{P}(\mathcal{C})$, selecting a collection of augmented simplicial objects $U_\bullet^+$, and inverting the corresponding maps $|U_\bullet| \to U_{-1}$. The specification of the desired class of augmented simplicial objects can be viewed as a kind of "generalized topology" on $\mathcal{C}$ in which one specifies not only the covering sieves but also the collection of hypercoverings which are to become effective after localization. It seems plausible that this notion of topology can be described more directly in terms of the ∞-category $\mathcal{C}$, but we will not pursue the matter further.

6.5.4 Descent versus Hyperdescent

Let X be a topological space and let $\mathcal{U}(X)$ denote the category of open subsets of X. The category $\mathcal{U}(X)$ is equipped with a Grothendieck topology in which the covering sieves on U are those sieves $\{U_\alpha \subseteq U\}$ such that $U = \bigcup_\alpha U_\alpha$. We may therefore consider the ∞-topos $\mathrm{Shv}(N(\mathcal{U}(X)))$, which we will call the ∞-topos of *sheaves on* X and denote by $\mathrm{Shv}(X)$. In §6.5.2, we discussed an alternative theory of sheaves on X, which can be obtained either through Jardine's local model structure on the category of simplicial presheaves or by passing to the hypercompletion $\mathrm{Shv}(X)^\wedge$ of $\mathrm{Shv}(X)$. According to Theorem 6.5.3.12, $\mathrm{Shv}(X)^\wedge$ is distinguished from $\mathrm{Shv}(X)$ in that objects of $\mathrm{Shv}(X)^\wedge$ are required to satisfy a descent condition for arbitrary hypercoverings of X, while objects of $\mathrm{Shv}(X)$ are required to satisfy a descent condition only for ordinary coverings.

Warning 6.5.4.1. We will always use the notation $\mathrm{Shv}(X)$ to indicate the ∞-category of $\mathcal{S}$-valued sheaves on X rather than the ordinary category of set-valued sheaves. If we need to indicate the latter, we will denote it by $\mathrm{Shv}_{\mathrm{Set}}(X)$.

The ∞-topos $\mathrm{Shv}(X)^\wedge$ seems to have received more attention than $\mathrm{Shv}(X)$ in the literature (though there is some discussion of $\mathrm{Shv}(X)$ in [20] and [78]). We would like to make the case that for most purposes, $\mathrm{Shv}(X)$ has better properties. A large part of Chapter 7 will be devoted to justifying some of the claims made below.

(1) In §6.4.5, we saw that the construction

$$X \mapsto \mathrm{Shv}(X)$$

could be interpreted as a *right* adjoint to the functor which associates to every ∞-topos Y the underlying locale of subobjects of the final object of $\mathcal{Y}$. In other words, $\mathrm{Shv}(X)$ occupies a universal position among ∞-topoi which are related to the original space X.

(2) Suppose we are given a Cartesian square

$$
\begin{array}{ccc}
X' & \xrightarrow{\ \psi'\ } & X \\
\downarrow{\scriptstyle \pi'} & & \downarrow{\scriptstyle \pi} \\
S' & \xrightarrow{\ \psi\ } & S
\end{array}
$$

in the category of locally compact topological spaces. In classical sheaf theory, there is a *base change* transformation

$$\psi^* \pi_* \to \pi'_* \psi'^*$$

of functors between the derived categories of (left bounded) complexes of (abelian) sheaves on X and on S'. The proper base change theorem asserts that this transformation is an equivalence whenever the map π is proper.

The functors ψ^*, ψ'^*, π_*, and π'_* can be defined on the ∞-topoi $\mathrm{Shv}(X)$, $\mathrm{Shv}(X')$, $\mathrm{Shv}(S)$, and $\mathrm{Shv}(S')$, and on their hypercompletions. Moreover, one has a base change map

$$\psi^* \pi_* \to \pi'^* \psi'^*$$

in this nonabelian situation as well.

It is natural to ask if the base change transformation is an equivalence when π is proper. It turns out that this is *false* if we work with hypercomplete ∞-topoi. Let us sketch a counterexample:

Counterexample 6.5.4.2. Let Q denote the Hilbert cube $[0, 1] \times [0, 1] \times \cdots$. For each i, we let $Q_i \simeq Q$ denote "all but the first i" factors of Q, so that $Q = [0, 1]^i \times Q_i$.

We construct a sheaf of spaces $\mathcal{F}$ on $X = Q \times [0, 1]$ as follows. Begin with the empty stack. Adjoin to it two sections defined over the open sets $[0, 1) \times Q_1 \times [0, 1)$ and $(0, 1] \times Q_1 \times [0, 1)$. These sections both restrict to give sections of $\mathcal{F}$ over the open set $(0, 1) \times Q_1 \times [0, 1)$. We next adjoin paths between these sections defined over the smaller open sets $(0, 1) \times [0, 1) \times Q_2 \times [0, \frac{1}{2})$ and $(0, 1) \times (0, 1] \times Q_2 \times [0, \frac{1}{2})$. These paths are both defined on the smaller open set $(0, 1) \times (0, 1) \times Q_2 \times [0, \frac{1}{2})$, so we next adjoin two homotopies between these paths over the open sets $(0, 1) \times (0, 1) \times [0, 1) \times Q_3 \times [0, \frac{1}{3})$ and $(0, 1) \times (0, 1) \times (0, 1] \times Q_3 \times [0, \frac{1}{3})$. Continuing in this way, we obtain a sheaf $\mathcal{F}$. On the closed subset $Q \times \{0\} \subset X$, the sheaf $\mathcal{F}$ is ∞-connective by construction, and therefore its hypercompletion admits a global section. However, the hypercompletion of $\mathcal{F}$ does not admit a global section in any neighborhood of $Q \times \{0\}$ since such a neighborhood must contain $Q \times [0, \frac{1}{n})$ for $n \gg 0$ and the higher homotopies required for the construction of a section are eventually not globally defined.

However, in the case where π is a proper map, the base change map

$$\psi^* \pi_* \to \pi'_* \psi'^*$$

is an equivalence of functors from $\mathrm{Shv}(X)$ to $\mathrm{Shv}(S')$. One may regard this fact as a nonabelian generalization of the classical proper base change theorem. We refer the reader to §7.3 for a precise statement and proof.

Remark 6.5.4.3. A similar issue arises in classical sheaf theory if one chooses to work with unbounded complexes. In [72], Spaltenstein defines a derived category of unbounded complexes of sheaves on X, where X is a topological space. Spaltenstein's definition forces all quasi-isomorphisms to become invertible, which is analogous to the procedure of obtaining $\mathfrak{X}^\wedge$ from $\mathfrak{X}$ by inverting the ∞-connective morphisms. Spaltenstein's work shows that one can extend the *definitions* of all of the basic objects and functors. However, it turns out that the *theorems* do not all extend: in particular, one does not have the proper base change theorem in Spaltenstein's setting (Counterexample 6.5.4.2 can be adapted to the setting of complexes of abelian sheaves). The problem may be rectified by imposing weaker descent conditions which do not invert all quasi-isomorphisms; we will give a more detailed discussion in [50].

(3) The ∞-topos $\mathrm{Shv}(X)$ often has better finiteness properties than the ∞-topos $\mathrm{Shv}(X)^\wedge$. Recall that a topological space X is *coherent* if the collection of compact open subsets of X is stable under finite intersections and forms a basis for the topology of X.

Proposition 6.5.4.4. *Let X be a coherent topological space. Then the ∞-category $\mathrm{Shv}(X)$ is compactly generated: that is, $\mathrm{Shv}(X)$ is generated under filtered colimits by its compact objects.*

Proof. Let $\mathcal{U}_c(X)$ be the partially ordered set of *compact* open subsets of X, let $\mathcal{P}_c(X) = \mathcal{P}(\mathrm{N}(\mathcal{U}_c(X)))$, and let $\mathrm{Shv}_c(X)$ be the full subcategory of $\mathcal{P}_c(X)$ spanned by those presheaves $\mathcal{F}$ with the following properties:

(1) The object $\mathcal{F}(\emptyset) \in \mathcal{C}$ is final.

(2) For every pair of compact open sets $U, V \subseteq X$, the associated diagram

$$\begin{array}{ccc} \mathcal{F}(U \cap V) & \longrightarrow & \mathcal{F}(U) \\ \downarrow & & \downarrow \\ \mathcal{F}(V) & \longrightarrow & \mathcal{F}(U \cup V) \end{array}$$

is a pullback.

In §7.3.5, we will prove that the restriction functor $\mathrm{Shv}(X) \to \mathrm{Shv}_c(X)$ is an equivalence of ∞-categories (Theorem 7.3.5.2). It will therefore suffice to prove that $\mathrm{Shv}_c(X)$ is compactly generated.

Using Lemmas 5.5.4.19, 5.5.4.17, and 5.5.4.18, we deduce that $\mathrm{Shv}_c(X)$ is an accessible localization of $\mathcal{P}_c(X)$. Let X be a compact object of $\mathcal{P}_c(X)$. We observe that X and LX corepresent the same functor on $\mathrm{Shv}_c(X)$. Proposition 5.3.3.3 implies that the subcategory $\mathrm{Shv}_c(X) \subseteq \mathcal{P}_c(X)$ is stable under filtered colimits in $\mathcal{P}_c(X)$. It follows that LX is a compact object of $\mathrm{Shv}_0(X)$. Since $\mathcal{P}_c(X)$ is generated under filtered colimits by its compact objects (Proposition 5.3.5.12), we conclude that $\mathrm{Shv}_c(X)$ has the same property.   □

It is not possible replace $\mathrm{Shv}(X)$ by $\mathrm{Shv}(X)^\wedge$ in the statement of Proposition 6.5.4.4.

Counterexample 6.5.4.5. Let $S = \{x, y, z\}$ be a topological space consisting of three points, with topology generated by the open subsets $S^+ = \{x, y\} \subset S$ and $S^- = \{x, z\} \subset S$. Let $X = S \times S \times \cdots$ be a product of infinitely many copies of S. Then X is a coherent topological space. We will show that the global sections functor $\Gamma : \mathrm{Shv}(X)^\wedge \to \mathcal{S}$ does not commute with filtered colimits, so that the final object of $\mathrm{Shv}(X)^\wedge$ is not compact. A more elaborate version of the same argument shows that $\mathrm{Shv}(X)^\wedge$ contains no compact objects other than its initial object.

To show that Γ does not commute with filtered colimits, we use a variant on the construction of Counterexample 6.5.4.2. We define a sequence of sheaves

$$\mathcal{F}_0 \to \mathcal{F}_1 \to \cdots$$

as follows. Let $\mathcal{F}_0$ be generated by sections

$$\eta_+^0 \in \mathcal{F}(S^+ \times S \times \cdots)$$

$$\eta_-^0 \in \mathcal{F}(S^- \times S \times \cdots).$$

Let $\mathcal{F}_1$ be the sheaf obtained from $\mathcal{F}_0$ by adjoining paths

$$\eta_+^1 : \Delta^1 \to \mathcal{F}(\{x\} \times S^+ \times S \times \cdots)$$

$$\eta_-^1 : \Delta^1 \to \mathcal{F}(\{x\} \times S^- \times S \times \cdots)$$

from η_+^0 to η_-^0. Similarly, let $\mathcal{F}_2$ be obtained from $\mathcal{F}_1$ by adjoining homotopies

$$\eta_+^2 : (\Delta^1)^2 \to \mathcal{F}(\{x\} \times \{x\} \times S^+ \times S \times \cdots)$$

$$\eta_-^2 : (\Delta^1)^2 \to \mathcal{F}(\{x\} \times \{x\} \times S^- \times S \times \cdots)$$

from η_+^1 to η_-^1. Continuing this procedure, we obtain a sequence of sheaves

$$\mathcal{F}_0 \to \mathcal{F}_1 \to \mathcal{F}_2 \to \cdots$$

whose colimit $\mathcal{F}_\infty \in \mathrm{Shv}(X)^\wedge$ admits a section (since we allow descent with respect to hypercoverings). However, none of the individual sheaves $\mathcal{F}_n$ admits a global section.

Remark 6.5.4.6. The analogue of Proposition 6.5.4.4 fails, in general, if we replace the coherent topological space X by a coherent topos. For example, we cannot take X to be the topos of étale sheaves on an algebraic variety. However, it turns out that the analogue Proposition 6.5.4.4 *is* true for the topos of *Nisnevich* sheaves on an algebraic variety; we refer the reader to [50] for details.

Remark 6.5.4.7. A *point* of an ∞-topos $\mathcal{X}$ is a geometric morphism $p_* : \mathcal{S} \to \mathcal{X}$, where $\mathcal{S}$ denotes the ∞-category of spaces (which is a final object of $\mathcal{RT}\mathrm{op}$ by virtue of Proposition 6.3.4.1). We say that $\mathcal{X}$ has *enough points* if, for every morphism $f : X \to Y$ in $\mathcal{X}$ having the property that $p^*(f)$ is an equivalence for *every* point p of $\mathcal{X}$, f is itself an equivalence in $\mathcal{X}$. If f is ∞-connective, then every stalk $p^*(f)$ is ∞-connective, hence an equivalence by Whitehead's theorem. Consequently, if $\mathcal{X}$ has enough points, then it is hypercomplete.

In classical topos theory, Deligne's version of the Gödel completeness theorem (see [53]) asserts that every coherent topos has enough points. Counterexample 6.5.4.5 shows that there exist coherent topological spaces with $\mathrm{Shv}(X)^\wedge \neq \mathrm{Shv}(X)$, so that $\mathrm{Shv}(X)$ does not necessarily have enough points. Consequently, Deligne's theorem does not hold in the ∞-categorical context.

(4) Let k be a field and let $\mathcal{C}$ denote the category of chain complexes of k-vector spaces. Via the Dold-Kan correspondence, we may regard $\mathcal{C}$ as a simplicial category. We let $\mathrm{Mod}(k) = N(\mathcal{C})$ denote the simplicial nerve. We will refer to $\mathrm{Mod}(k)$ as the ∞-*category of k-modules*; it is a presentable ∞-category which we will discuss at greater length in [50].

Let X be a compact topological space and choose a functorial injective resolution

$$\mathcal{F} \to I^0(\mathcal{F}) \to I^1(\mathcal{F}) \to \cdots$$

on the category of sheaves $\mathcal{F}$ of k-vector spaces on X. For every open subset U on X, we let k_U denote the constant sheaf on U with value k, extended by zero to X. Let $\mathrm{H}^{BM}(U) = \Gamma(X, I^\bullet(k_U))^\vee$, the *dual* of the complex of global sections of the injective resolution $I^\bullet(k_U)$. Then $\mathrm{H}^{BM}(U)$ is a complex of k-vector spaces whose homologies are precisely the Borel-Moore homology of U with coefficients in k (in other words, they are the dual spaces of the compactly supported cohomology groups of U). The assignment

$$U \mapsto \mathrm{H}^{BM}(U)$$

determines a presheaf on X with values in the ∞-category $\mathrm{Mod}(k)$.

In view of the existence of excision exact sequences for Borel-Moore homology, it is natural to suppose that $\mathrm{H}^{BM}(U)$ is actually a *sheaf* on X with values in $\mathrm{Mod}(k)$. This is true provided that the notion of sheaf is suitably interpreted: namely, H^{BM} extends (in an essentially unique fashion) to a colimit-preserving functor

$$\phi : \mathrm{Shv}(X) \to \mathrm{Mod}(k)^{op}.$$

(In other words, the functor $U \mapsto \mathrm{H}^{BM}(U)$ determines a $\mathrm{Mod}(k)$-valued sheaf on X in the sense of Definition 7.3.3.1.) However, the sheaf H^{BM} is not necessarily hypercomplete in the sense that ϕ does not necessarily factor through $\mathrm{Shv}(X)^{\wedge}$.

Counterexample 6.5.4.8. There exists a compact Hausdorff space X and a hypercovering $U_{\bullet}$ of X such that the natural map $\mathrm{H}^{BM}(X) \to \varprojlim \mathrm{H}^{BM}(U_{\bullet})$ is not an equivalence. Let X be the Hilbert cube $Q = [0,1] \times [0,1] \times \cdots$ (more generally, we could take X to be any nonempty Hilbert cube manifold). It is proven in [15] that every point of X has arbitrarily small neighborhoods which are homeomorphic to $Q \times [0,1)$. Consequently, there exists a hypercovering $U_{\bullet}$ of X, where each U_n is a disjoint union of open subsets of X homeomorphic to $Q \times [0,1)$. The Borel-Moore homology of every U_n vanishes; consequently, $\varprojlim \mathrm{H}^{BM}(U_{\bullet})$ is zero. However, the (degree zero) Borel-Moore homology of X itself does not vanish since X is nonempty and compact.

Borel-Moore homology is a very useful tool in the study of a locally compact space X, and its descent properties (in other words, the existence of various Mayer-Vietoris sequences) are very naturally encoded in the statement that H^{BM} is a k-*module in the ∞-topos* $\mathrm{Shv}(X)$ (in other words, a sheaf on X with values in $\mathrm{Mod}(k)$); however, this k-module generally does not lie in $\mathrm{Shv}(X)^{\wedge}$. We see from this example that nonhypercomplete sheaves (with values in $\mathrm{Mod}(k)$ in this case) on X often arise naturally in the study of infinite-dimensional spaces.

(5) Let X be a topological space and $f : \mathrm{Shv}(X) \to \mathrm{Shv}(*) \simeq \mathcal{S}$ be the geometric morphism induced by the projection $X \to *$. Let K be a Kan complex regarded as an object of $\mathcal{S}$. Then $\pi_0 f_* f^* K$ is a natural definition of the sheaf cohomology of X with coefficients in K. If X is paracompact, then the cohomology set defined above is naturally isomorphic to the set $[X, |K|]$ of homotopy classes of maps from X into the geometric realization $|K|$; we will give a proof of this statement in §7.1. The analogous statement fails if we replace $\mathrm{Shv}(X)$ by $\mathrm{Shv}(X)^{\wedge}$.

(6) Let X be a topological space. Combining Remark 6.5.2.2 with Proposition 6.5.2.14, we deduce that $\mathrm{Shv}(X)^\wedge$ has enough points and that $\mathrm{Shv}(X)^\wedge = \mathrm{Shv}(X)$ if and only if $\mathrm{Shv}(X)$ has enough points. The possible failure of Whitehead's theorem in $\mathrm{Shv}(X)$ may be viewed either as a bug or a feature. The existence of enough points for $\mathrm{Shv}(X)$ is extremely convenient; it allows us to reduce many statements about the ∞-topos $\mathrm{Shv}(X)$ to statements about the ∞-topos $\mathcal{S}$ of spaces where we can apply classical homotopy theory. On the other hand, if $\mathrm{Shv}(X)$ does *not* have enough points, then there is the possibility that it detects certain global phenomena which cannot be properly understood by restricting to points. Let us consider an example from geometric topology. A map $f : X \to Y$ of compact metric spaces is called *cell-like* if each fiber $X_y = X \times_Y \{y\}$ has trivial shape (see [18]). This notion has good formal properties provided that we restrict our attention to metric spaces which are *absolute neighborhood retracts*. In the general case, the theory of cell-like maps can be badly behaved: for example, a composition of cell-like maps need not be cell-like.

The language of ∞-topoi provides a convenient formalism for discussing the problem. In §7.3.6, we will introduce the notion of a *cell-like* morphism $p_* : \mathcal{X} \to \mathcal{Y}$ between ∞-topoi. By definition, p_* is cell-like if it is proper and if the unit map $u : \mathcal{F} \to p_* p^* \mathcal{F}$ is an equivalence for each $\mathcal{F} \in \mathcal{Y}$. A cell-like map $p : X \to Y$ of compact metric spaces *need not* give rise to a cell-like morphism $p_* : \mathrm{Shv}(X) \to \mathrm{Shv}(Y)$. The hypothesis that each fiber X_y has trivial shape ensures that the unit $u : \mathcal{F} \to p_* p^* \mathcal{F}$ is an equivalence after passing to stalks at each point $y \in Y$. This implies only that u is ∞-connective, and in general u need not be an equivalence.

Remark 6.5.4.9. It is tempting to try to evade the problem described above by working instead with the hypercomplete ∞-topoi $\mathrm{Shv}(X)^\wedge$ and $\mathrm{Shv}(Y)^\wedge$. In this case, we *can* test whether or not $u : \mathcal{F} \to p_* p^* \mathcal{F}$ is an equivalence by passing to stalks. However, since the proper base change theorem does not hold in the hypercomplete context, the stalk $(p_* p^* \mathcal{F})_y$ is not generally equivalent to the global sections of $p^* \mathcal{F}|X_y$. Thus, we still encounter difficulties if we want to deduce global consequences from information about the individual fibers X_y.

(7) The counterexamples described in this section have one feature in common: the underlying space X is infinite-dimensional. In fact, this is necessary: if the space X is finite-dimensional (in a suitable sense), then the ∞-topos $\mathrm{Shv}(X)$ is hypercomplete (Corollary 7.2.1.12). This finite-dimensionality condition on X is satisfied in many of the situations to which the theory of simplicial presheaves is commonly applied, such as the Nisnevich topology on a scheme of finite Krull dimension.

Chapter Seven

Higher Topos Theory in Topology

In this chapter, we will sketch three applications of the theory of ∞-topoi to the study of classical topology. We begin in §7.1 by showing that if X is a paracompact topological space, then the ∞-topos $\mathrm{Shv}(X)$ of sheaves on X can be interpreted as a homotopy theory of topological spaces Y equipped with a map to X. We will deduce, as an application, that if $p_* : \mathrm{Shv}(X) \to \mathrm{Shv}(*)$ is the geometric morphism induced by the projection $X \to *$, then the composition $p_* p^*$ is equivalent to the functor

$$K \mapsto K^X$$

from (compactly generated) topological spaces to itself.

Our second application is to the dimension theory of topological spaces. There are many different notions of *dimension* for a topological space X, including the notion of covering dimension (when X is paracompact), Krull dimension (when X is Noetherian), and cohomological dimension. We will define the *homotopy dimension* of an ∞-topos $\mathcal{X}$, which specializes to the covering dimension when $\mathcal{X} = \mathrm{Shv}(X)$ for a paracompact space X and is closely related to both cohomological dimension and Krull dimension. We will show that any ∞-topos which is (locally) finite-dimensional is hypercomplete, thereby justifying assertion (7) of §6.5.4. We will conclude by proving a bound on the homotopy dimension of $\mathrm{Shv}(X)$, where X is a *Heyting space* (see §7.2.4 for a definition); this may be regarded as a generalization of Grothendieck's vanishing theorem, which applies to nonabelian cohomology and to (certain) non-Noetherian spaces X.

Our third application is a generalization of the *proper base change theorem*. Suppose we are given a Cartesian diagram

$$\begin{array}{ccc} X' & \xrightarrow{\;p'\;} & X \\ {\scriptstyle q'}\big\downarrow & & \big\downarrow{\scriptstyle q} \\ Y' & \xrightarrow{\;p\;} & Y \end{array}$$

of locally compact topological spaces. There is a natural transformation

$$\eta : p^* q_* \to q'_* p'^*$$

of functors from the derived category of abelian sheaves on X to the derived category of abelian sheaves on Y'. The proper base change theorem asserts that η is an isomorphism whenever q is a proper map. In §7.3, we will generalize this statement to allow nonabelian coefficient systems. To give the proof, we will develop a theory of *proper morphisms* between ∞-topoi, which is of some interest in itself.

7.1 PARACOMPACT SPACES

Let X be a topological space and G an abelian group. There are many different definitions for the cohomology group $\mathrm{H}^n(X;G)$; we will single out three of them for discussion here. First of all, we have the singular cohomology groups $\mathrm{H}^n_{\mathrm{sing}}(X;G)$, which are defined to be cohomology of a chain complex of G-valued singular cochains on X. An alternative is to regard $\mathrm{H}^n(\bullet,G)$ as a representable functor on the homotopy category of topological spaces, so that $\mathrm{H}^n_{\mathrm{rep}}(X;G)$ can be identified with the set of homotopy classes of maps from X into an Eilenberg-MacLane space $K(G,n)$. A third possibility is to use the sheaf cohomology $\mathrm{H}^n_{\mathrm{sheaf}}(X;\underline{G})$ of X with coefficients in the constant sheaf $\underline{G}$ on X.

If X is a sufficiently nice space (for example, a CW complex), then these three definitions give the same result. In general, however, all three give different answers. The singular cohomology of X is defined using continuous maps from Δ^k into X and is useful only when there is a good supply of such maps. Similarly, the cohomology group $\mathrm{H}^n_{\mathrm{rep}}(X;G)$ is defined using continuous maps from X to a simplicial complex and is useful only when there is a good supply of real-valued functions on X. However, the sheaf cohomology of X seems to be a good invariant for arbitrary spaces: it has excellent formal properties and gives sensible answers in situations where the other definitions break down (such as the étale topology of algebraic varieties).

We will take the position that the sheaf cohomology of a space X is the correct answer in all cases. It is then natural to ask for conditions under which the other definitions of cohomology give the same answer. We should expect this to be true for singular cohomology when there are many continuous functions *into* X and for Eilenberg-MacLane cohomology when there are many continuous functions *out of* X. It seems that the latter class of spaces is much larger than the former: it includes, for example, all paracompact spaces, and consequently for paracompact spaces one can show that the sheaf cohomology $\mathrm{H}^n_{\mathrm{sheaf}}(X;G)$ coincides with the Eilenberg-MacLane cohomology $\mathrm{H}^n_{\mathrm{rep}}(X;G)$. Our goal in this section is to prove a generalization of the preceding statement to the setting of nonabelian cohomology (Theorem 7.1.0.1 below; see also Theorem 7.1.4.3 for the case where the coefficient system G it not assumed to be constant).

As we saw in §6.5.4, we can associate to every topological space X an ∞-topos $\mathrm{Shv}(X)$ of sheaves (of spaces) on X. Moreover, given a continuous map $p : X \to Y$ of topological spaces, p^{-1} induces a map from the category of open subsets of Y to the category of open subsets of X. Composition with p^{-1} induces a geometric morphism $p_* : \mathrm{Shv}(X) \to \mathrm{Shv}(Y)$.

Fix a topological space X and let $p : X \to *$ denote the projection from X to a point. Let K be a Kan complex which we may identify with an object of $\mathcal{S} \simeq \mathrm{Shv}(*)$. Then $p^*K \in \mathrm{Shv}(X)$ may be regarded as the constant sheaf on X having value K and $p_*p^*K \in \mathcal{S}$ as the space of global sections of p^*K. Let $|K|$ denote the geometric realization of K (a topological space) and let

$[X, |K|]$ denote the *set* of homotopy classes of maps from X into $|K|$. The main goal of this section is to prove the following:

Theorem 7.1.0.1. *If X is paracompact, then there is a canonical bijection*

$$\phi : [X, |K|] \to \pi_0(p_* p^* K).$$

Remark 7.1.0.2. In fact, the map ϕ exists without the assumption that X is paracompact: the construction in general can be formally reduced to the paracompact case since the universal example $X = |K|$ is paracompact. However, in the case where X is not paracompact, the map ϕ is not necessarily bijective.

Our first step in proving Theorem 7.1.0.1 is to realize the space of maps from X into $|K|$ as a mapping space in an appropriate simplicial category of *spaces over X*. In §7.1.2, we define this category and endow it with a (simplicial) model structure. We may therefore extract an underlying ∞-category $N(\text{Top}^\circ_{/X})$.

Our next goal is to construct an equivalence between $N(\text{Top}^\circ_{/X})$ and the ∞-topos $\text{Shv}(X)$ of sheaves of spaces on X (a very similar comparison result has been obtained by Toën; see [77]). To prove this, we will attempt to realize $N(\text{Top}^\circ_{/X})$ as a localization of a certain ∞-category of presheaves. We will give an explicit description of the relevant localization in §7.1.3 and show that it is equivalent to $N(\text{Top}^\circ_{/X})$ in §7.1.4. In §7.1.5, we will deduce Theorem 7.1.0.1 as a corollary of this more general comparison result. We conclude with §7.1.6, in which we apply our results to obtain a reformulation of classical shape theory in the language of ∞-topoi.

7.1.1 Some Point-Set Topology

Let X be a paracompact topological space. In order to prove Theorem 7.1.0.1, we will need to understand the homotopy theory of presheaves on X. We then encounter the following technical obstacle: an open subset of a paracompact space need not be paracompact. Because we wish to deal only with paracompact spaces, it will be convenient to restrict our attention to presheaves which are defined only with respect to a particular basis $\mathcal{B}$ for X consisting of paracompact open sets. The existence of a well-behaved basis is guaranteed by the following result:

Proposition 7.1.1.1. *Let X be a paracompact topological space and U an open subset of X. The following conditions are equivalent:*

(i) *There exists a continuous function $f : X \to [0, 1]$ such that $U = \{x \in X : f(x) > 0\}$.*

(ii) *There exists a sequence of closed subsets $\{K_n \subseteq X\}_{n \geq 0}$ such that each K_{n+1} contains an open neighborhood of K_n and $U = \bigcup_{n \geq 0} K_n$.*

(iii) *There exists a sequence of closed subsets $\{K_n \subseteq X\}_{n \geq 0}$ such that $U = \bigcup_{n \geq 0} K_n$.*

Let $\mathcal{B}$ denote the collection of all open subsets of X which satisfy these conditions. Then

(1) *The elements of $\mathcal{B}$ form a basis for the topology of X.*

(2) *Each element of $\mathcal{B}$ is paracompact.*

(3) *The collection $\mathcal{B}$ is stable under finite intersections (in particular, $X \in \mathcal{B}$).*

(4) *The empty set $\emptyset$ belongs to $\mathcal{B}$.*

Remark 7.1.1.2. A subset of X which can be written as a countable union of closed subsets of X is called an F_σ-subset of X. Consequently, the basis $\mathcal{B}$ for the topology of X appearing in Proposition 7.1.1.1 can be characterized as the collection of open F_σ-subsets of X.

Remark 7.1.1.3. If the topological space X admits a metric d, then *every* open subset $U \subseteq X$ belongs to the basis $\mathcal{B}$ of Proposition 7.1.1.1. Indeed, we may assume without loss of generality that the diameter of X is at most 1 (adjusting the metric if necessary), in which case the function

$$f(x) = d(x, X - U) = \inf_{y \notin U} d(x, y)$$

satisfies condition (i).

Proof. We first show that (i) and (ii) are equivalent. If (i) is satisfied, then the closed subsets $K_n = \{x \in X : f(x) \geq \frac{1}{n}\}$ satisfy the demands of (ii). Suppose next that (ii) is satisfied. For each $n \geq 0$, let G_n denote the closure of $X - K_{n+1}$, so that $G_n \cap K_n = \emptyset$. It follows that there exists a continuous function $f_n : X \to [0, 1]$ such that that f_n vanishes on G_n and the restriction of f to K_n is the constant function taking the value 1. Then the function $f = \sum_{n > 0} \frac{f_n}{2^n}$ has the property required by (i).

We now prove that $(ii) \Leftrightarrow (iii)$. The implication $(ii) \Rightarrow (iii)$ is obvious. For the converse, suppose that $U = \bigcup_n K_n$, where the K_n are closed subsets of X. We define a new sequence of closed subsets $\{K'_n\}_{n \geq 0}$ by induction as follows. Let $K'_0 = K_0$. Assuming that K'_n has already been defined, let V and W be disjoint open neighborhoods of the closed sets $K'_n \cup K_{n+1}$ and $X - U$, respectively (the existence of such neighborhoods follows from the assumption that X is paracompact; in fact, it would suffice to assume that X is normal) and define K'_{n+1} to be the closure of V. It is then easy to see that the sequence of closed sets $\{K'_n\}_{n \geq 0}$ satisfies the requirements of (ii).

We now verify properties (1) through (4) of the collection of open sets $\mathcal{B}$. Assertions (3) and (4) are obvious. To prove (1), consider an arbitrary point $x \in X$ and an open set U containing x. Then the closed sets $\{x\}$ and $X - U$ are disjoint, so there exists a continuous function $f : X \to [0, 1]$ supported on U such that $f(x) = 1$. Then $U' = \{y \in X : f(y) > 0\}$ is an open neighborhood of x contained in U, and $U' \in \mathcal{B}$.

It remains to prove (2). Let $U \in \mathcal{B}$; we wish to prove that U is paracompact. Write $U = \bigcup_{n \geq 0} K_n$, where each K_n is a closed subset of X containing a neighborhood of K_{n-1} (by convention, we set $K_n = \emptyset$ for $n < 0$). Let $\{U_\alpha\}$ be an open covering of X. Since each K_n is paracompact, we can choose a locally finite covering $\{V_{\alpha,n}\}$ of K_n which refines $\{U_\alpha \cap K_n\}$. Let $V^0_{\alpha,n}$ denote the intersection of $V_{\alpha,n}$ with the interior of K_n and let $W_{\alpha,n} = V^0_{\alpha,n} \cap (X - K_{n-2})$. Then $\{W_{\alpha,n}\}$ is a locally finite open covering of X which refines $\{U_\alpha\}$. $\qquad\square$

Let X be a paracompact topological space and let $\mathcal{B}$ be the basis constructed in Proposition 7.1.1.1. Then $\mathcal{B}$ can be viewed as a category with finite limits and is equipped with a natural Grothendieck topology. To simplify the notation, we will let $\mathrm{Shv}(\mathcal{B})$ denote the ∞-topos $\mathrm{Shv}(\mathrm{N}(\mathcal{B}))$. Note that because $\mathrm{N}(\mathcal{B})$ is the nerve of a partially ordered set, the ∞-topos $\mathrm{Shv}(\mathcal{B})$ is 0-localic. Moreover, the corresponding locale $\mathrm{Sub}(\mathbf{1})$ of subobjects of the final object $\mathbf{1} \in \mathrm{Shv}(\mathcal{B})$ is isomorphic to the lattice of open subsets of X. It follows that the restriction map $\mathrm{Shv}(X) \to \mathrm{Shv}(\mathcal{B})$ is an equivalence of ∞-topoi.

Warning 7.1.1.4. Let X be a topological space and $\mathcal{B}$ a basis of X regarded as a partially ordered set with respect to inclusions. Then $\mathcal{B}$ inherits a Grothendieck topology, and we can define $\mathrm{Shv}(\mathcal{B})$ as above. However, the induced map $\mathrm{Shv}(X) \to \mathrm{Shv}(\mathcal{B})$ is generally not an equivalence of ∞-categories: this requires the assumption that $\mathcal{B}$ is stable under finite intersections. In other words, a sheaf (of spaces) on X generally *cannot* be recovered by knowing its sections on a basis for the topology of X; see Counterexample 6.5.4.8.

7.1.2 Spaces over X

Let X be a topological space with a specified basis $\mathcal{B}$ fixed throughout this section. We wish to study the homotopy theory of *spaces over X*: that is, spaces Y equipped with a map $p : Y \to X$. We should emphasize that we do not wish to assume that the map p is a fibration or that p is equivalent to a fibration in any reasonable sense: we are imagining that p encodes a *sheaf of spaces* on X, and we do not wish to impose any condition of local triviality on this sheaf.

Let Top denote the category of topological spaces and $\mathrm{Top}_{/X}$ the category of topological spaces mapping to X. For each $p : Y \to X$ and every open subset $U \subseteq X$, we define a simplicial set $\mathrm{Sing}_X(Y, U)$ by the formula

$$\mathrm{Sing}_X(Y, U)_n = \mathrm{Hom}_X(U \times |\Delta^n|, Y).$$

Face and degeneracy maps are defined in the obvious way. We note that the simplicial set $\mathrm{Sing}_X(Y, U)$ is *always* a Kan complex. We will simply write $\mathrm{Sing}_X(Y)$ to denote the simplicial presheaf on X given by

$$U \mapsto \mathrm{Sing}_X(Y, U).$$

Proposition 7.1.2.1. *There exists a model structure on the category $\mathrm{Top}_{/X}$ uniquely determined by the following properties:*

(W) *A morphism*

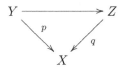

is a weak equivalence if and only if, for every $U \subseteq X$ belonging to $\mathcal{B}$, the induced map $\mathrm{Sing}_X(Z, U)_\bullet \to \mathrm{Sing}_X(Y, U)_\bullet$ is a homotopy equivalence of Kan complexes.

(F) *A morphism*

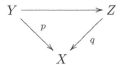

is a fibration if and only if, for every $U \subseteq X$ belonging to $\mathcal{B}$, the induced map $\mathrm{Sing}_X(Z, U)_\bullet \to \mathrm{Sing}_X(Y, U)_\bullet$ is a Kan fibration.

Remark 7.1.2.2. The model structure on $\mathrm{Top}_{/X}$ described in Proposition 7.1.2.1 depends on the chosen basis $\mathcal{B}$ for X and not only on the topological space X itself.

Proof of Proposition 7.1.2.1. The proof uses the theory of *cofibrantly generated* model categories; we give a sketch and refer the reader to [38] for more details. We will say that a morphism $Y \to Z$ in $\mathrm{Top}_{/X}$ is a cofibration if it has the left lifting property with respect to every trivial fibration in $\mathrm{Top}_{/X}$.

We begin by observing that a map $Y \to Z$ in $\mathrm{Top}_{/X}$ is a fibration if and only if it has the right lifting property with respect to every inclusion $U \times \Lambda^n_i \subseteq U \times \Delta^n$, where $0 \leq i \leq n$ and U is in $\mathcal{B}$. Let $\mathcal{I}$ denote the weakly saturated class of morphisms in $\mathrm{Top}_{/X}$ generated by these inclusions. Using the small object argument, one can show that every morphism $Y \to Z$ in $\mathrm{Top}_{/X}$ admits a factorization

$$Y \xrightarrow{f} Y' \xrightarrow{g} Z,$$

where f belongs to $\mathcal{I}$ and g is a fibration. (Although the objects in $\mathrm{Top}_{/X}$ generally are not small, one can still apply the small object argument since they are small *relative* to the class $\mathcal{I}$ of morphisms: see [38].)

Similarly, a map $Y \to Z$ is a trivial fibration if and only if it has the right lifting property with respect to every inclusion $U \times | \partial \Delta^n| \subseteq U \times |\Delta^n|$, where $U \in \mathcal{B}$. Let $\mathcal{J}$ denote the weakly saturated class of morphisms generated by these inclusions: then every morphism $Y \to Z$ admits a factorization

$$Y \xrightarrow{f} Y' \xrightarrow{g} Z,$$

where f belongs to $\mathcal{J}$ and g is a trivial fibration.

The only nontrivial point to verify is that every morphism which belongs to $\mathcal{I}$ is a trivial cofibration; once this is established, the axioms for a model category follow formally. Since it is clear that $\mathcal{I}$ is contained in $\mathcal{J}$ and that $\mathcal{J}$ consists of cofibrations, it suffices to show that every morphism in $\mathcal{I}$ is a weak equivalence. To prove this, let us consider the class $\mathcal{K}$ of all closed immersions $k : Y \to Z$ in $\mathrm{Top}_{/X}$ such that there exist functions $\lambda : Z \to [0, \infty)$ and $h : Z \times [0, \infty) \to Z$ such that $k(Y) = \lambda^{-1}\{0\}$, $h(z, 0) = z$, and $h(z, \lambda(z)) \in k(Y)$. Now we make the following observations:

(1) Every inclusion $U \times |\Lambda_i^n| \subseteq U \times |\Delta^n|$ belongs to $\mathcal{K}$.

(2) The class $\mathcal{K}$ is weakly saturated; consequently, $\mathcal{J} \subseteq \mathcal{K}$.

(3) Every morphism $k : Y \to Z$ which belongs to $\mathcal{K}$ is a homotopy equivalence in $\mathrm{Top}_{/X}$ and is therefore a weak equivalence.

$\square$

The category $\mathrm{Top}_{/X}$ is naturally tensored over simplicial sets if we define $Y \otimes \Delta^n = Y \times |\Delta^n|$ for $Y \in \mathrm{Top}_{/X}$. This induces a simplicial structure on $\mathrm{Top}_{/X}$ which is obviously compatible with the model structure of Proposition 7.1.2.1.

We note that Sing_X is a (simplicial) functor from $\mathrm{Top}_{/X}$ to the category of simplicial presheaves on $\mathcal{B}$ (here we regard $\mathcal{B}$ as a category whose morphisms are given by inclusions of open subsets of X). We regard $\mathrm{Set}_\Delta^{\mathcal{B}^{op}}$ as a simplicial model category via the *projective* model structure described in §A.3.3. By construction, Sing_X preserves fibrations and trivial fibrations. Moreover, the functor Sing_X has a left adjoint

$$F \mapsto |F|_X;$$

we will refer to this left adjoint as *geometric realization* (in the case where X is a point, it coincides with the usual geometric realization functor from Set_Δ to the category of topological spaces). The functor $|F|_X$ is determined by the property that $|F_U|_X \simeq U$ if F_U denotes the presheaf (of sets) represented by U and the requirement that geometric realization commutes with colimits and with tensor products by simplicial sets.

We may summarize the situation as follows:

Proposition 7.1.2.3. *The adjoint functors* $(|\mid_X, \mathrm{Sing}_X)$ *determine a simplicial Quillen adjunction between* $\mathrm{Top}_{/X}$ *(with the model structure of Proposition 7.1.2.1) and* $\mathrm{Set}_\Delta^{\mathcal{B}^{op}}$ *(with the projective model structure).*

7.1.3 The Sheaf Condition

Let X be a topological space and $\mathcal{B}$ a basis for the topology of X which is stable under finite intersections. Let $\mathbf{A}$ denote the category $\mathrm{Set}_\Delta^{\mathcal{B}^{op}}$ of simplicial presheaves on $\mathcal{B}$; we regard $\mathbf{A}$ as a model category with respect to the *projective* model structure defined in §A.3.3. According to Proposition 5.1.1.1, the ∞-category $\mathrm{N}(\mathbf{A}^\circ)$ associated to $\mathbf{A}$ is equivalent to the

∞-category $\mathcal{P}(\mathcal{B}) = \mathcal{P}(\mathrm{N}(\mathcal{B}))$ of presheaves on $\mathcal{B}$. In particular, the homotopy category $\mathrm{h}\mathcal{P}(\mathrm{N}(\mathcal{B}))$ is equivalent to the homotopy category $\mathrm{h}\mathbf{A}$ (the category obtained from $\mathbf{A}$ by formally inverting all weak equivalences of simplicial presheaves). The ∞-category $\mathrm{Shv}(\mathcal{B})$ is a reflective subcategory of $\mathcal{P}(\mathrm{N}(\mathcal{B}))$. Consequently, we may identify the homotopy category $\mathrm{hShv}(\mathcal{B})$ with a reflective subcategory of $\mathrm{h}\mathbf{A}$. We will say that a simplicial presheaf $F : \mathcal{B}^{op} \to \mathrm{Set}_{\Delta}$ is a *sheaf* if it belongs to this reflective subcategory. The purpose of this section is to obtain an explicit criterion which will allow us to test whether or not a given simplicial presheaf $F : \mathcal{B}^{op} \to \mathrm{Set}_{\Delta}$ is a sheaf.

Warning 7.1.3.1. The condition that a simplicial presheaf $F : \mathcal{B}^{op} \to \mathrm{Set}_{\Delta}$ be a sheaf, in the sense defined above, is generally unrelated to the condition that F be a simplicial object in the category of sheaves of sets on X (though these two notions do agree in the special case where the simplicial presheaf F takes values in *constant* simplicial sets).

Let $j : \mathrm{N}(\mathcal{B}) \to \mathcal{P}(\mathcal{B})$ be the Yoneda embedding. By definition, an object $F \in \mathcal{P}(\mathcal{B})$ belongs to $\mathrm{Shv}(\mathcal{B})$ if and only if, for every $U \in \mathcal{B}$ and every monomorphism $i : U^0 \to j(U)$ which corresponds to a *covering* sieve $\mathcal{U}$ on U, the induced map

$$\mathrm{Map}_{\mathcal{P}(\mathcal{B})}(j(U), F) \to \mathrm{Map}_{\mathcal{P}(\mathcal{B})}(U^0, F)$$

is an isomorphism in the homotopy category $\mathcal{H}$. In order to make this condition explicit in terms of simplicial presheaves, we note that $i : U^0 \to j(U)$ can be identified with the inclusion $\chi_{\mathcal{U}} \subseteq \chi_U$ of simplicial presheaves, where

$$\chi_U(V) = \begin{cases} * & \text{if } V \subseteq U \\ \emptyset & \text{otherwise.} \end{cases}$$

$$\chi_{\mathcal{U}}(V) = \begin{cases} * & \text{if } V \in \mathcal{U} \\ \emptyset & \text{otherwise.} \end{cases}$$

However, we encounter a technical issue: in order to extract the correct space of maps $\mathrm{Map}_{\mathcal{P}(\mathcal{B})}(U^0, F)$, we need to select a *projectively cofibrant* model for U^0 in $\mathbf{A}$. In general, the simplicial presheaf $\chi_{\mathcal{U}}$ defined above is not projectively cofibrant. To address this problem, we will construct a new simplicial presheaf, equivalent to $\chi_{\mathcal{U}}$, which has better mapping properties.

Definition 7.1.3.2. Let $\mathcal{U}$ be a linearly ordered set equipped with a map $s : \mathcal{U} \to \mathcal{B}$. We define a simplicial presheaf $N_{\mathcal{U}} : \mathcal{B}^{op} \to \mathrm{Set}_{\Delta}$ as follows: for each $V \in \mathcal{B}$, let $N_{\mathcal{U}}(V)$ be the nerve of the linearly ordered set $\{U \in \mathcal{U} : V \subseteq s(U)\}$. $N_{\mathcal{U}}$ may be viewed as a subobject of the constant presheaf $\underline{\Delta^{\mathcal{U}}}$ taking the value $\mathrm{N}(\mathcal{U}) = \Delta^{\mathcal{U}}$.

Remark 7.1.3.3. The above notation is slightly abusive in that $N_{\mathcal{U}}$ depends not only on $\mathcal{U}$ but also on the map s and on the linear ordering of $\mathcal{U}$. If the map s is injective (as it will be in most applications), we will frequently simply identify $\mathcal{U}$ with its image in $\mathcal{B}$. In practice, $\mathcal{U}$ will usually be a covering sieve on some object $U \in \mathcal{B}$.

Remark 7.1.3.4. The linear ordering of $\mathcal{U}$ is *unrelated* to the partial ordering of $\mathcal{B}$ by inclusion. We will write the former as $\leq$ and the latter as $\subseteq$.

Example 7.1.3.5. Let $\mathcal{U} = \emptyset$. Then $N_{\mathcal{U}} = \emptyset$.

Example 7.1.3.6. Let $\mathcal{U} = \{U\}$ for some $U \in \mathcal{B}$ and let $s : \mathcal{U} \to \mathcal{B}$ be the inclusion. Then $N_{\mathcal{U}} \simeq \chi_U$.

Proposition 7.1.3.7. *Let*

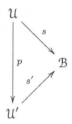

be a commutative diagram, where p is an order-preserving injection between linearly ordered sets. Then the induced map $N_{\mathcal{U}} \to N_{\mathcal{U}'}$ is a projective cofibration of simplicial presheaves.

Proof. Without loss of generality, we may identify $\mathcal{U}$ with a linearly ordered subset of $\mathcal{U}'$ via p. Choose a transfinite sequence of simplicial subsets of $N\,\mathcal{U}'$

$$K_0 \subseteq K_1 \subseteq \cdots,$$

where $K_0 = N\,\mathcal{U}$, $K_\lambda = \bigcup_{\alpha < \lambda} K_\alpha$ if λ is a nonzero limit ordinal, and $K_{\alpha+1}$ is obtained from K_α by adjoining a single nondegenerate simplex (if such a simplex exists). For each ordinal α, let $F_\alpha \subseteq N_{\mathcal{U}'}$ be defined by

$$F_\alpha(V) = N_{\mathcal{U}'}(V) \cap K_\alpha \subseteq \mathrm{N}(\mathcal{U}').$$

Then $F_0 = N_{\mathcal{U}}$ and $F_\lambda = \varinjlim_{\alpha < \lambda} F_\alpha$ when α is a nonzero limit ordinal, and $F_\alpha \simeq N_{\mathcal{U}'}$ for $\alpha \gg 0$. It therefore suffices to show that each map $F_\alpha \to F_{\alpha+1}$ is a projective cofibration. If $K_\alpha = K_{\alpha+1}$, this is clear; otherwise, we may suppose that $K_{\alpha+1}$ is obtained from K_α by adjoining a single nondegenerate simplex $\{U_0 < U_1 < \cdots < U_n\}$ of $\mathrm{N}(\mathcal{U}')$. Let $U = s'(U_0) \cap \cdots \cap s'(U_n) \in \mathcal{B}$.

Then there is a coCartesian square

$$
\begin{array}{ccc}
\chi_U \otimes \partial \Delta^n & \longrightarrow & \chi_U \otimes \Delta^n \\
\downarrow & & \downarrow \\
F_\alpha & \longrightarrow & F_{\alpha+1}.
\end{array}
$$

The desired result now follows since the upper horizontal arrow is clearly a projective cofibration. $\qquad\qquad\square$

Corollary 7.1.3.8. *Let $\mathcal{U}$ be a linearly ordered set and $s : \mathcal{U} \to \mathcal{B}$ a map. Then the simplicial presheaf $N_{\mathcal{U}} \in \mathrm{Set}_\Delta^{\mathcal{B}^{op}}$ is projectively cofibrant.*

Note that $N_{\mathcal{U}}(V)$ is contractible if $V \subseteq s(U)$ for some $U \in \mathcal{U}$ and empty otherwise. Consequently, we deduce:

Corollary 7.1.3.9. *Let $\mathcal{U} \subseteq \mathcal{B}$ be a sieve equipped with a linear ordering. The unique map $N_{\mathcal{U}} \to \chi_{\mathcal{U}}$ is a weak equivalence of simplicial presheaves.*

Notation 7.1.3.10. Let $\mathcal{U}$ be a linearly ordered set equipped with a map $s : \mathcal{U} \to \mathcal{B}$. For any simplicial presheaf $F : \mathcal{B}^{op} \to \mathrm{Set}_{\Delta}$, we let $F(\mathcal{U})$ denote the simplicial set $\mathrm{Map}_{\mathbf{A}}(N_{\mathcal{U}}, F)$.

Remark 7.1.3.11. Let $U \in \mathcal{B}$, let $\mathcal{U} = \{U\}$, and let $s : \mathcal{U} \to \mathcal{B}$ be the inclusion. Then $F(\mathcal{U}) = F(U)$. In general, we can think of $F(\mathcal{U})$ as a *homotopy limit* of $F(V)$ taken over V in the sieve generated by $s : \mathcal{U} \to \mathcal{B}$. To give a vertex of $F(\mathcal{U})$, we must give for each $U \in \mathcal{U}$ a point of $F(sU)$, for every pair of objects $U, V \in \mathcal{U}$ a path between the corresponding points in $F(sU \cap sV)$, and so forth.

Corollary 7.1.3.12. *Let $F : \mathcal{B}^{op} \to \mathcal{K}\mathrm{an}$ be a (projectively fibrant) simplicial presheaf on $\mathcal{B}$. Then F is a sheaf if and only if, for every $U \in \mathcal{B}$ and every sieve $\mathcal{U}$ that covers U, there exists a linearly ordered set $\mathcal{U}_0$ equipped with a map $\mathcal{U}_0 \to \mathcal{U}$, which generates $\mathcal{U}$ as a sieve, such that the induced map $F(U) \to F(\mathcal{U}_0)$ is a weak homotopy equivalence of simplicial sets.*

Lemma 7.1.3.13. *Suppose that $U \subseteq X$ is paracompact and let $\mathcal{U} \subseteq \mathcal{B}$ be a covering of U. Choose a linear ordering of $\mathcal{U}$. Then the natural map $\pi : |N_{\mathcal{U}}|_X \to U$ is a homotopy equivalence in $\mathrm{Top}_{/X}$. (In other words, there exists a section $s : U \to N_{\mathcal{U}}$ of π such that $s \circ \pi$ is fiberwise homotopic to the identity.)*

Proof. Any partition of unity subordinate to the open cover $\mathcal{U}$ gives rise to a section of π. To check that $s \circ \pi$ is fiberwise homotopic to the identity, use a "straight-line" homotopy. $\square$

Proposition 7.1.3.14. *Let X be a topological space and that $\mathcal{B}$ a basis for the topology of X. Assume that $\mathcal{B}$ is stable under finite intersections and that each element of $\mathcal{B}$ is paracompact. For every continuous map of topological spaces $p : Y \to X$, the simplicial presheaf $\mathrm{Sing}_X(Y)$ of sections of p is sheaf.*

Proof. Let $F = \mathrm{Sing}_X(Y)$. We note that F is a projectively fibrant simplicial presheaf on $\mathcal{B}$. By Corollary 7.1.3.12, it suffices to show that for every $U \in \mathcal{B}$, every covering $\mathcal{U}$ of U, and every linear ordering on $\mathcal{U}$, the natural map $F(U) \to F(\mathcal{U})$ is a homotopy equivalence of simplicial sets. In other words, it suffices to show that composition with the projection $\pi : N_{\mathcal{U}} \to U$ induces a homotopy equivalence

$$\mathrm{Map}_{/X}(U, Y) \to \mathrm{Map}_{/X}(|N_{\mathcal{U}}|_X, Y)$$

of simplicial sets. This follows immediately from Lemma 7.1.3.13. $\square$

Remark 7.1.3.15. Under the hypotheses of Proposition 7.1.3.14, the object of $\mathrm{Shv}(X)$ corresponding to the simplicial presheaf $\mathrm{Sing}_X(Y)$ is *not necessarily hypercomplete.*

7.1.4 The Main Result

Suppose that X is a paracompact topological space and $\mathcal{B}$ is the basis for the topology of X described in Proposition 7.1.1.1. Our main goal is to show that the composition of the adjoint functors

$$F \mapsto \mathrm{Sing}_X |F|_X$$

may be identified with a "sheafification" of F, at least in the case where F is a projectively cofibrant simplicial presheaf on $\mathcal{B}$.

In proving this, we have some flexibility regarding the choice of F: it will suffice to treat the question after replacing F by a weakly equivalent simplicial presheaf F' provided that F' is also projectively cofibrant. Our first step is to make a particularly convenient choice for F'.

Lemma 7.1.4.1. *Let $\mathcal{B}$ be a partially ordered set (via $\subseteq$) with a least element $\emptyset$ and let $F : \mathcal{B}^{op} \to \mathrm{Set}_\Delta$ be an arbitrary simplicial presheaf such that $F(\emptyset)$ is weakly contractible.*

There exists a (linearly ordered) set V and a simplicial presheaf $F' : \mathcal{B}^{op} \to \mathrm{Set}_\Delta$ with the following properties:

(1) *There exists a monomorphism $F' \to \underline{\Delta^V}$ from F' to the (constant) simplicial presheaf $\underline{\Delta^V}$ on $\mathcal{B}$ taking the value Δ^V. (Recall that Δ^V denotes the nerve of the linearly ordered set V.)*

(2) *For every finite subset $V_0 \subseteq V$, there exists $U \in \mathcal{B}$ such that $U' \subseteq U$ if and only if $\Delta^{V_0} \subseteq F'(U') \subseteq \Delta^V$.*

(3) *As a simplicial presheaf on $\mathcal{B}$, F' is projectively cofibrant.*

(4) *In the homotopy category of $\mathrm{Set}_\Delta^{\mathcal{B}^{op}}$, F' and F are equivalent to one another.*

Proof. Without loss of generality, we may suppose that F is (weakly) fibrant. We now build a "cellular model" of F. More precisely, we construct the following data:

(A) A transfinite sequence of simplicial sets

$$Y_0 \to Y_1 \to \cdots ,$$

where Y_α is defined for all ordinals less than α_0.

(B) For each $\alpha < \alpha_0$, a subsheaf F_α of the constant presheaf on $\mathcal{B}$ taking the value Y_α.

(C) A compatible family of maps $F_\alpha \to F$, so that we may regard $\{F_\alpha\}$ as a functor from the linearly ordered set $\{\alpha : \alpha < \alpha_0\}$ to $(\mathrm{Set}_\Delta)_{/F}^{\mathcal{B}^{op}}$.

(D) For each $\alpha < \alpha_0$, there exists $U \in \mathcal{B}$, $n \geq 0$, and compatible pushout diagrams

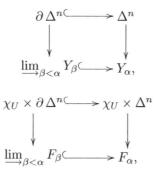

$$
\begin{array}{ccc}
\partial \Delta^n & \longrightarrow & \Delta^n \\
\downarrow & & \downarrow \\
\varinjlim_{\beta<\alpha} Y_\beta & \longrightarrow & Y_\alpha,
\end{array}
$$

$$
\begin{array}{ccc}
\chi_U \times \partial \Delta^n & \longrightarrow & \chi_U \times \Delta^n \\
\downarrow & & \downarrow \\
\varinjlim_{\beta<\alpha} F_\beta & \longrightarrow & F_\alpha,
\end{array}
$$

where

$$
\chi_U(W) = \begin{cases} * & \text{if } W \subseteq U \\ \emptyset & \text{otherwise.} \end{cases}
$$

(E) The canonical map $\varinjlim_{\beta<\alpha_0} F_\beta \to F$ is a weak equivalence in $\mathrm{Set}_\Delta^{\mathcal{B}^{op}}$.

For every simplicial set K, let K'' denote the simplicial set described in Variant 4.2.3.15, so that we have a canonical cofinal map $K'' \to K$ (in particular, a weak homotopy equivalence) and K'' is equivalent to the nerve of a partially ordered set. Let $Y = \varinjlim_{\beta<\alpha_0} Y_\beta$. We note that Y'' can be identified with a simplicial complex: that is, a simplicial subset of Δ^V for some linearly ordered set V. For each α, let F_α'' denote the result of applying the functor $K \mapsto K''$ termwise. Let $F'' = \varinjlim_{\beta<\alpha_0} F_\beta''$. Finally, we define F' by the coCartesian square

$$
\begin{array}{ccc}
F''(\emptyset) \times \chi_\emptyset & \longrightarrow & \Delta^V \times \chi_\emptyset \\
\downarrow & & \downarrow \\
F'' & \longrightarrow & F'.
\end{array}
$$

The simplicial presheaf F' satisfies (1) by construction. Properties (2) and (3) are reasonably clear (in fact, (3) is a formal consequence of (2)). Condition (4) holds for the simplicial presheaf F'' as a consequence of (E). Moreover, the assumption that $F(\emptyset)$ is weakly contractible ensures that (4) remains valid for the pushout F'. $\qquad\square$

Before we can state the next lemma, let us introduce a bit of notation. Let $F : \mathcal{B}^{op} \to \mathrm{Set}_\Delta$ be a simplicial presheaf. Then we let $|F|$ denote the presheaf of topological spaces on $\mathcal{B}$ obtained by composing F with the geometric realization functor; similarly, if G is a presheaf of topological spaces on $\mathcal{B}$, then we let $\mathrm{Sing}\, G$ denote the presheaf of simplicial sets obtained by composing G with the functor Sing. We note that there is a natural transformation

$$
\mathrm{Sing}\, |F| \to \mathrm{Sing}_X |F|_X.
$$

Lemma 7.1.4.2. *Let X be a topological space and let $\mathcal{B}$ be the collection of open F_σ subsets of X (see Proposition 7.1.1.1). Let $F : \mathcal{B}^{op} \to \mathrm{Set}_\Delta$ be a projectively cofibrant simplicial presheaf which is a sheaf (that is, a fibrant model for F satisfies the criterion of Corollary 7.1.3.12). Then the unit map $F \to \mathrm{Sing}_X |F|_X$ is an equivalence.*

Proof. We note that the functor $F \mapsto \mathrm{Sing}\,|F|$ preserves weak equivalences in F and that the functor $F \mapsto \mathrm{Sing}_X |F|_X$ preserves weak equivalences between *projectively cofibrant* presheaves F. Consequently, by Lemma 7.1.4.1, we may suppose without loss of generality that there is a linearly ordered set V and that F is a subsheaf of the constant simplicial presheaf taking the value Δ^V, such that $F(\emptyset) = \Delta^V$.

It will be sufficient to prove that
$$\mathrm{Sing}\,|F| \to \mathrm{Sing}_X |F|_X$$
is an equivalence: in other words, we wish to show that $(\mathrm{Sing}\,|F|)(U) \to (\mathrm{Sing}_X |F|_X)(U)$ is a homotopy equivalence of Kan complexes for every $U \in \mathcal{B}$. Replacing X by U, we can reduce to the problem of showing that
$$p : (\mathrm{Sing}\,|F|)(X) \to (\mathrm{Sing}_X |F|_X)(X)$$
is a homotopy equivalence. It now suffices to show that for every inclusion $K' \subseteq K$ of *finite* simplicial sets (that is, simplicial sets with only finitely many nondegenerate simplices), a commutative diagram

$$
\begin{array}{ccc}
K' \times \{0\} & \longrightarrow & (\mathrm{Sing}\,|F|)(X) \\
\downarrow & & \downarrow \\
K \times \{0\} & \overset{g}{\longrightarrow} & (\mathrm{Sing}_X |F|_X)(X)
\end{array}
$$

can be expanded to a commutative diagram

$$
\begin{array}{ccc}
(K' \times \Delta^1) \coprod_{K' \times \{1\}} (K \times \{1\}) & \longrightarrow & (\mathrm{Sing}\,|F|)(X) \\
\downarrow & & \downarrow \\
K \times \Delta^1 & \longrightarrow & (\mathrm{Sing}_X |F|_X)(X).
\end{array}
$$

(In fact, it suffices to treat the case where $K' \subseteq K$ is the inclusion $\partial \Delta^n \subseteq \Delta^n$; however, this will result in no simplification of the following arguments.)

Now let $\mathcal{B} = \{U_\alpha\}_{\alpha \in A}$, where A is a linearly ordered set. Since F is assumed to be a sheaf on $\mathcal{B}$, the equivalent presheaf $\mathrm{Sing}\,|F|$ is also a sheaf. Consequently, for any covering $\mathcal{U} \subseteq \mathcal{B}$ (and any linear ordering of $\mathcal{U}$), the natural map $(\mathrm{Sing}\,|F|)(X) \to (\mathrm{Sing}\,|F|)(\mathcal{U})$ is an equivalence. Likewise, by Proposition 7.1.3.14, the map $(\mathrm{Sing}_X |F|_X)(X) \to (\mathrm{Sing}_X |F|_X)(\mathcal{U})$ is an equivalence. Consequently, it suffices to find a covering $\mathcal{U} \subseteq \mathcal{B}$ of X and a diagram

$$
\begin{array}{ccc}
(K' \times \Delta^1) \coprod_{K' \times \{1\}} (K \times \{1\}) & \longrightarrow & (\mathrm{Sing}\,|F|)(\mathcal{U}) \\
\downarrow & & \downarrow \\
K \times \Delta^1 & \overset{G}{\longrightarrow} & (\mathrm{Sing}_X |F|_X)(\mathcal{U})
\end{array}
$$

which extends g.

Since K is finite, the map $g : K \to (\mathrm{Sing}_X |F|_X)(X)$ may be identified with a continuous fiber-preserving map $X \times |K| \to |F|_X$, which we will also denote by g. By assumption, F is a subsheaf of the constant presheaf taking the value Δ^V; constantly, we may identify $|F|_X$ with a subspace of $\Delta_X^V = X \otimes \Delta^V$. (We may identify Δ_X^V with the product $X \times |\Delta^V|$ as a set, though it generally has a finer topology.) We may represent a point of Δ_X^V by an ordered pair (x, q), where $x \in X$ and $q : V \to [0, 1]$ has the property that $\{v \in V : q(v) \neq 0\}$ is finite and $\sum_{v \in V} q(v) = 1$. For each $v \in V$, we let $\Delta_{X,v}^V$ denote the *open* subset of Δ_X^V consisting of all pairs (x, g) such that $q(v) > 0$; note that the sets $\{\Delta_{X,v}^V\}_{v \in V}$ form an open cover of Δ_X^V. Consequently, the open sets $\{g^{-1} \Delta_{X,v}^V\}_{v \in V}$ form an open cover of $X \times |K|$. Let x be a point of X. The compactness of $|K|$ implies that there is a finite subset $V_0 \subseteq V$, an open neighborhood U_x of X containing x, and an open covering $\{W_{x,v} : v \in V_0\}$ of $|K|$, such that $g(U_x \times W_{x,v}) \subseteq \Delta_{X,v}^V$. Choose a partition of unity subordinate to the covering $\{W_{x,v}\}$, thereby determining a map $f_x : |K| \to |\Delta^{V_0}|$. The open sets $\{U_x\}$ cover X; since X is paracompact, this covering has a locally finite refinement. Shrinking the U_x if necessary, we may suppose that this refinement is given by $\{U_x\}_{x \in X_0}$ and that each U_x belongs to $\mathcal{B}$. Let $\mathcal{U} = \{U_x\}_{x \in X_0}$ and choose a linear ordering of $\mathcal{U}$.

We now define a new map $g' : K \to (\mathrm{Sing}\,|F|)(\mathcal{U})$. To do so, we must give, for every finite $\mathcal{U}_0 = \{U_{x_0} < \cdots < U_{x_n}\} \subseteq \mathcal{U}$, a map

$$g'_{A_0} : |\Delta^{\{x_0, \ldots, x_n\}}| \times |K| \to |F(U_{x_0} \cap \cdots \cap U_{x_n})| \subseteq |\Delta^V|;$$

moreover, these maps are required to satisfy some obvious compatibilities. Define $g'_{\mathcal{U}_0}$ by the formula

$$g'_{\mathcal{U}_0}\left(\sum \lambda_i x_i, z\right) = \sum \lambda_i f_{x_i}(z).$$

It is clear that $g'_{\mathcal{U}_0}$ is well-defined as a map from $|\Delta^{\{x_0, \ldots, x_n\}}| \times |K|$ to $|\Delta^V|$. We claim that, in fact, this map factors through $|F(U_{x_0} \cap \cdots \cap U_{x_n})|$. Let $z \in |K|$ and consider the set $V' = \{v \in V : (\exists 0 \leq i \leq n)[f_{x_i}(z)(v) \neq 0]\} \subseteq V$. Condition (2) of Lemma 7.1.4.1 ensures that there exists $U \in \mathcal{B}$ such that $\Delta^{V'} \subseteq F(U')$ if and only if $U' \subseteq U$. We note that, for each $y \in U_{x_0} \cap \cdots \cap U_{x_n}$, we have $g(y, z)(v) \neq 0$ for $v \in V'$; it follows that $y \in U$. Consequently, we deduce that $U_{\alpha_0} \cap \cdots \cap U_{\alpha_n} \subseteq U$, so that $\Delta^{V'} \subseteq F(U_{x_0} \cap \cdots \cap U_{x_n})$. It follows that $g'_{\mathcal{U}_0}|\{z\} \times |\Delta^{\mathcal{U}_0}|$ factors through $|F(U_{x_0} \cap \ldots \cap U_{x_n})|$. Since this holds for every $z \in |K|$, it follows that g'_{A_0} is well-defined; evidently these maps are compatible with one another and give the desired map $g' : K \to (\mathrm{Sing}\,|F|)(\mathcal{U})$.

We now observe that the composite maps

$$K \xrightarrow{g'} (\mathrm{Sing}\,|F|)(\mathcal{U}) \to (\mathrm{Sing}_X |F|_X)(\mathcal{U})$$

$$K \xrightarrow{g} (\mathrm{Sing}_X |F|_X)(X) \to (\mathrm{Sing}_X |F|_X)(\mathcal{U})$$

are homotopic via a "straight-line" homotopy

$$G : K \times \Delta^1 \to (\mathrm{Sing}_X |F|_X)(\mathcal{U}),$$

which has the desired properties. $\square$

Now the hard work is done, and we are ready to enjoy the fruits of our labors.

Theorem 7.1.4.3. *Let X be a paracompact topological space and $\mathcal{B}$ the collection of open F_σ subsets of X (see Proposition 7.1.1.1). Then, for any projectively cofibrant $F : \mathcal{B}^{op} \to \mathrm{Set}_\Delta$, the natural map*

$$F \to \mathrm{Sing}_X |F|_X$$

exhibits $\mathrm{Sing}_X |F|_X$ as a sheafification of F.

Proof. Let $\mathrm{hTop}_{/X}$ be the homotopy category of the model category $\mathrm{Top}_{/X}$ (the category obtained by inverting all of the weak equivalences defined in Proposition 7.1.2.1) and $\mathrm{hSet}_\Delta^{\mathcal{B}^{op}}$ the homotopy category of the category of simplicial presheaves on $\mathcal{B}$. It follows from Proposition 7.1.2.3 that the adjoint functors Sing_X and $||_X$ induce adjoint functors

$$\mathrm{hSet}_\Delta^{\mathcal{B}^{op}} \underset{\mathrm{Sing}_X}{\overset{||_X^L}{\rightleftarrows}} \mathrm{hTop}_{/X}.$$

Here $||_X^L$ denotes the left derived functor of the geometric realization (since every object of $\mathrm{Top}_{/X}$ is fibrant, Sing_X may be identified with its right derived functor).

We first claim that for any $Y \in \mathrm{Top}_{/X}$, the counit map $|\mathrm{Sing}_X Y|_X^L \to Y$ is a weak equivalence. To see this, choose a projectively cofibrant model $F \to \mathrm{Sing}_X Y$ for $\mathrm{Sing}_X Y$; we wish to show that the induced map $|F|_X \to Y$ is a weak equivalence. By definition, this is equivalent to the assertion that $\mathrm{Sing}_X |F|_X \to \mathrm{Sing}_X Y$ is a weak equivalence. But we have a commutative triangle

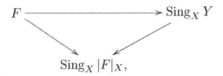

where the left diagonal map is a weak equivalence by Lemma 7.1.4.2 and Proposition 7.1.3.14 and the top horizontal map is a weak equivalence by construction; the desired result now follows from the two-out-of-three property.

It follows that we may identify $\mathrm{hTop}_{/X}$ with a full subcategory $\mathcal{C} \subseteq \mathrm{hSet}_\Delta^{\mathcal{B}^{op}}$. By Proposition 7.1.3.14, the objects of this subcategory are sheaves on $\mathcal{B}$; by Lemma 7.1.4.2, every sheaf on $\mathcal{B}$ is equivalent to $\mathrm{Sing}_X Y$ for an appropriately chosen Y; thus $\mathcal{C}$ consists of precisely the sheaves on $\mathcal{B}$.

The composite functor $F \mapsto \mathrm{Sing}_X |F|_X^L$ may be identified with a localization functor from $\mathrm{hSet}_\Delta^{\mathcal{B}^{op}}$ to the subcategory $\mathcal{C}$. In particular, when F is projectively cofibrant, the unit of the adjunction $F \to \mathrm{Sing}_X |F|_X$ is a localization of F. $\square$

Corollary 7.1.4.4. *Under the hypotheses of Theorem 7.1.4.3, the functor* Sing_X *induces an equivalence of* ∞-*categories*

$$\mathrm{N}(\mathrm{Top}^\circ_{/X}) \to \mathrm{Shv}(X).$$

In particular, the ∞-*category* $\mathrm{N}(\mathrm{Top}^\circ_{/X})$ *is an* ∞-*topos.*

Remark 7.1.4.5. In the language of model categories, we may interpret Corollary 7.1.4.4 as asserting that Sing_X and $\|_X$ furnish a Quillen equivalence between $\mathrm{Top}_{/X}$ (with the model structure of Proposition 7.1.2.1) and $\mathrm{Set}_\Delta^{\mathcal{B}^{op}}$ where the latter is equipped with the following *localization* of the projective model structure:

(1) A map $F \to F'$ in $\mathrm{Set}_\Delta^{\mathcal{B}^{op}}$ is a *cofibration* if it is a projective cofibration (in the sense of Definition A.3.3.1).

(2) A map $F \to F'$ in $\mathrm{Set}_\Delta^{\mathcal{B}^{op}}$ is a *weak equivalence* if it induces an equivalence in the ∞-category $\mathrm{Shv}(X)$.

7.1.5 Base Change

With Corollary 7.1.4.4 in hand, we are *almost* ready to deduce Theorem 7.1.0.1. Suppose we are given a paracompact space X and let $\mathcal{B}$ denote the collection of all open F_σ subsets of X. Let $p : \mathrm{Shv}(X) \to \mathrm{Shv}(*) \simeq \mathcal{S}$ be the geometric morphism induced by the projection $X \to *$.

For any simplicial set K, let F_K denote the constant simplicial presheaf on $\mathcal{B}$ taking the value K. If we endow $\mathrm{Set}_\Delta^{\mathcal{B}^{op}}$ with the localized model structure of Remark 7.1.4.5, then F_K is a model for the sheaf p^*K. Consequently, the space p_*p^*K may be identified up to homotopy with the mapping space

$$\mathrm{Map}_{\mathrm{Shv}(X)}(F_*, F_K)$$

which, by virtue of Corollary 7.1.4.4, is equivalent to

$$\mathrm{Map}_{\mathrm{Top}_{/X}}(X, X \otimes K) = (\mathrm{Sing}_X(X \otimes K))(X).$$

However, at this point a technical wrinkle appears: $X \otimes K$ agrees with $X \times |K|$ as a set, but it is equipped with a finer topology (given by the direct limit of the product topologies $X \times |K_0|$, where $K_0 \subseteq K$ is a finite simplicial subset). In general, we have only an *inclusion* of simplicial presheaves

$$\eta : \mathrm{Sing}_X(X \otimes K) \subseteq \mathrm{Sing}_X(X \times |K|),$$

which need not be an isomorphism. However, we will complete the proof of Theorem 7.1.0.1 by showing that η is an equivalence of simplicial presheaves.

We consider a slightly more general situation. Let $p : X \to Y$ be a continuous map between paracompact spaces and let $\mathcal{B}_X$ and $\mathcal{B}_Y$ denote the collections of open F_σ subsets in X and Y, respectively. Note that the inverse image along p determines a map $q : \mathcal{B}_Y \to \mathcal{B}_X$. Composition with q induces a pushforward functor $q_* : \mathrm{Set}_\Delta^{\mathcal{B}_X^{op}} \to \mathrm{Set}_\Delta^{\mathcal{B}_Y^{op}}$, which has a left adjoint which we will denote by q^*. Similarly, there is a pullback functor $p^* : \mathrm{Top}_{/Y} \to \mathrm{Top}_{/X}$; however, p^* generally does not possess a right adjoint. Consider the square

$$
\begin{array}{ccc}
\mathrm{Set}_{\Delta}^{\mathcal{B}_Y^{op}} & \xrightarrow{\;\|\cdot\|_Y\;} & \mathrm{Top}_{/Y} \\
\Big\downarrow{q^*} & & \Big\downarrow{p^*} \\
\mathrm{Set}_{\Delta}^{\mathcal{B}_X^{op}} & \xrightarrow{\;\|\cdot\|_X\;} & \mathrm{Top}_{/X}\,.
\end{array}
$$

This square is lax commutative in the sense that there exists a natural transformation of functors

$$
\eta_F : |q^* F|_X \to p^* |F|_Y = |F|_Y \times_Y X.
$$

The map η_F is always a bijection of topological spaces but is generally not a homeomorphism. Nevertheless, we have the following:

Proposition 7.1.5.1. *Under the hypotheses above, if $F : \mathcal{B}_Y^{op} \to \mathrm{Set}_{\Delta}$ is a projectively cofibrant simplicial presheaf on Y, then the map $\eta_F : |q^* F|_X \to |F|_Y \times_Y X$ is a weak equivalence in $\mathrm{Top}_{/X}$.*

The proof is based on the following lemma:

Lemma 7.1.5.2. *Let Y be a paracompact topological space and let $\mathcal{B}$ be the collection of open F_σ subsets of Y (see Proposition 7.1.1.1). Let V be a linearly ordered set. Suppose that for every nonempty finite subset $V_0 \subseteq V$, we are given a basic open set $U(V_0) \in \mathcal{B}$ satisfying the following conditions:*

(a) If $V_0 \subseteq V_1$, then $U(V_1) \subseteq U(V_0)$.

(b) The open set $U(\emptyset)$ concides with X.

Let $F : \mathcal{B}^{op} \to \mathrm{Set}_{\Delta}$ be the simplicial presheaf which assigns to each $U \in \mathcal{B}$ the simplicial subset $F(U) \subseteq \Delta^V$ spanned by those nondegenerate simplices σ corresponding to finite subsets $V_0 \subseteq V$ such that $U \subseteq U(V_0)$ (see Lemma 7.1.4.1).

For every object $X \in \mathrm{Top}_{/Y}$, an n-simplex τ of $\mathrm{Map}_{\mathrm{Top}_{/Y}}(Y, |F|_Y)$ determines a map of topological spaces from $X \times |\Delta^n|$ to $|\Delta^V|$, which in turn determines a collection of maps $\phi_v : X \times |\Delta^n| \to [0, 1]$ such that for every $x \in X \times |\Delta^n|$, the sum $\Sigma_{v \in V} \phi_v(x)$ is equal to 1. Let $\mathrm{Map}_{\mathrm{Top}_{/Y}}^0(X, |F|_Y)$ denote the simplicial subset of $\mathrm{Map}_{\mathrm{Top}_{/Y}}(X, |F|_Y)$ spanned by those simplices τ which satisfy the following condition, where $K = \Delta^n$:

(∗) There exists a locally finite collection of open sets $\{U_v \subseteq X \times |K|\}_{v \in V}$ such that each U_v contains the closure of the support of the function ϕ_v and $\bigcap_{v \in V_0} U_v$ is contained in the inverse image of $U(V_0)$ for every finite subset $V_0 \subseteq V$.

If the topological space X is paracompact, then the inclusion

$$
i : \mathrm{Map}_{\mathrm{Top}_{/Y}}^0(X, |F|_Y) \subseteq \mathrm{Map}_{\mathrm{Top}_{/Y}}(X, |F|_Y)
$$

is a homotopy equivalence of Kan complexes.

Proof. Note that for any finite simplicial set K, we can identify

$$\mathrm{Hom}_{\mathrm{Set}_\Delta}(K, \mathrm{Map}^0_{\mathrm{Top}_{/Y}}(X, |F|_Y)$$

with the set of all collections of continuous maps $\{\phi_v : X \times |K| \to [0,1]\}$ satisfying the condition $(*)$. Composition with a retraction of $|\Delta^n|$ onto a horn $|\Lambda^n_i|$ determines a section of the restriction map

$$\mathrm{Hom}_{\mathrm{Set}_\Delta}(\Delta^n, \mathrm{Map}^0_{\mathrm{Top}_{/Y}}(X, |F|_Y)) \to \mathrm{Hom}_{\mathrm{Set}_\Delta}(\Lambda^n_i, \mathrm{Map}^0_{\mathrm{Top}_{/Y}}(X, |F|_Y)),$$

from which it follows that $\mathrm{Map}^0_{\mathrm{Top}_{/Y}}(X, |F|_Y)$ is a Kan complex.

To prove that i is a homotopy equivalence, we argue as in the proof of Lemma 7.1.4.2: it will suffice to show that for every inclusion $K' \subseteq K$ of finite simplicial sets, every commutative diagram

$$
\begin{array}{ccc}
K' \times \{0\} & \longrightarrow & \mathrm{Map}^0_{\mathrm{Top}_{/Y}}(X, |F|_Y) \\
\downarrow & & \downarrow \\
K \times \{0\} & \stackrel{g}{\longrightarrow} & \mathrm{Map}_{\mathrm{Top}_{/Y}}(X, |F|_Y)
\end{array}
$$

can be expanded to a commutative diagram

$$
\begin{array}{ccc}
(K' \times \Delta^1) \coprod_{K' \times \{1\}}(K \times \{1\}) & \longrightarrow & \mathrm{Map}^0_{\mathrm{Top}_{/Y}}(X, |F|_Y) \\
\downarrow & & \downarrow \\
K \times \Delta^1 & \stackrel{G}{\longrightarrow} & \mathrm{Map}_{\mathrm{Top}_{/Y}}(X, |F|_Y).
\end{array}
$$

The map g is classified by a collection of continuous maps $\{g_v : X \times |K| \to [0,1]\}_{v \in V}$ such that $\Sigma_{v \in V} g_v(x) = 1$. Let $\{U_v\}_{v \in V}$ be a collection of open subsets of $X \times |K'|$ satisfying condition $(*)$ for the functions $\{g_v | X \times |K'|\}$. For each $v \in V$, let $W_v = \{x \in X \times |K| : g_v(x) \neq 0\}$. Choose a locally finite open covering $\{U'_v\}_{v \in V}$ of $X \times |K|$ which refines $\{W_v\}$. Let $\{g'_v\}_{v \in V}$ be a partition of unity such that the closure of the support of each g'_v is contained in U'_v. We define maps $\{G_v : X \times |K| \times [0,1] \to [0,1]\}_{v \in V}$ by the formula

$$
G_v(x,t) = \begin{cases} (2t)g'_v(x) + (1 - 2t)g_v(x) & \text{if } t \leq \frac{1}{2} \\ g'_v(x) & \text{if } t \geq \frac{1}{2}. \end{cases}
$$

Then the maps $\{G_v\}$ determine a continuous map $G : X \times |K| \times [0,1] \to |F|_Y$, which we can identify with a map of simplicial sets

$$K \times \Delta^1 \to \mathrm{Map}_{\mathrm{Top}_{/Y}}(X, |F|_Y).$$

The restriction of this map to $(K' \times \Delta^1) \coprod_{K' \times \{1\}}(K \times \{1\})$ factors through $\mathrm{Map}^0_{\mathrm{Top}_{/Y}}(X, |F|_Y)$ since the open subsets $\{(U_v \times [0,1]) \cup (U'_v \times \{1\}\}$ satisfy condition $(*)$. $\square$

Remark 7.1.5.3. In the situation of Lemma 7.1.5.2, suppose we are given a Kan complex

$$\mathrm{Map}^1_{\mathrm{Top}_{/Y}}(X, |F|_Y) \subseteq \mathrm{Map}_{\mathrm{Top}_{/Y}}(X, |F|_Y)$$

which contains $\mathrm{Map}^0_{\mathrm{Top}_{/Y}}(X, |F|_Y)$ and is closed under the formation of "straight-line" homotopies. More precisely, suppose that any map $G : \Delta^n \times \Delta^1 \to \mathrm{Map}_{\mathrm{Top}_{/Y}}(X, |F|_Y)$ factors through $\mathrm{Map}^1_{\mathrm{Top}_{/Y}}(X, |F|_Y)$ provided that it satisfies the properties listed below:

(i) The map G is classified by a collection of continuous functions $\{G_v : X \times |\Delta^n| \times [0,1] \to [0,1]\}_{v \in V}$.

(ii) Each G_v can be described by the formula

$$G_v(x, t) = \begin{cases} (2t)g'_v(x) + (1 - 2t)g_v(x) & \text{if } t \le \frac{1}{2} \\ g'_v(x) & \text{if } t \ge \frac{1}{2}. \end{cases}$$

(iii) The closure of the support of each g'_v is contained in the open set $\{x \in X \times |\Delta^n| : g_v(x) \ne 0\}$.

(iv) The restriction of G to $\Delta^n \times \{0\}$ belongs to $\mathrm{Map}^1_{\mathrm{Top}_{/Y}}(X, |F|_Y)$. By virtue of ($iii$), this implies that $G|\Delta^n \times \{1\}$ factors through the inclusion

$$\mathrm{Map}^0_{\mathrm{Top}_{/Y}}(X, |F|_Y) \subseteq \mathrm{Map}^1_{\mathrm{Top}_{/Y}}(X, |F|_Y).$$

Then the proof of Lemma 7.1.5.2 shows that the inclusions

$$\mathrm{Map}^0_{\mathrm{Top}_{/Y}}(X, |F|_Y) \subseteq \mathrm{Map}^1_{\mathrm{Top}_{/Y}}(X, |F|_Y) \subseteq \mathrm{Map}_{\mathrm{Top}_{/Y}}(X, |F|_Y)$$

are homotopy equivalences.

Proof of Proposition 7.1.5.1. Suppose we are given a weak equivalence $F \to F'$ between projectively cofibrant simplicial presheaves $F, F' : \mathcal{B}_Y^{op} \to \mathrm{Set}_\Delta$. Both q^* and $||_X$ are left Quillen functors and therefore preserve weak equivalences between cofibrant objects; it follows that $|q^*F|_X \to |q^*F'|_X$ is a weak equivalence. Similarly, $|F|_Y \to |F'|_Y$ is a weak equivalence between cofibrant objects of $\mathrm{Top}_{/Y}$. Since *every* object of $\mathrm{Top}_{/Y}$ is fibrant, we conclude that $|F|_Y \to |F'|_Y$ is a homotopy equivalence in $\mathrm{Top}_{/Y}$; thus $|F|_Y \times_Y X \to |F'|_Y \times_Y X$ is a homotopy equivalence in $\mathrm{Top}_{/X}$. Consequently, we deduce that η_F is a weak equivalence if and only if $\eta_{F'}$ is a weak equivalence.

Let F be an arbitrary projectively cofibrant simplicial presheaf; we wish to show that η_F is a weak equivalence. There exists a trivial projective cofibration $F \to F'$, where F' is projectively fibrant. It now suffices to show that $\eta_{F'}$ is a weak equivalence. Replacing F by F', we reduce to the case where F is projectively fibrant.

Let F' be a simplicial presheaf on $\mathcal{B}_Y$ satisfying the conditions of Lemma 7.1.4.1. Then F' and F are equivalent in the homotopy category of simplicial presheaves on $\mathcal{B}_Y$. Since F' is projectively cofibrant and F is projectively fibrant, there exists a weak equivalence $F' \to F$. We may therefore once again reduce to proving that $\eta_{F'}$ is a weak equivalence. Replacing F by F', we may suppose that F satisfies the conditions of Lemma 7.1.4.1, for some linearly ordered set V.

For each $U \in \mathcal{B}_Y$, we have a commutative diagram of Kan complexes

$$
\begin{array}{ccc}
\mathrm{Map}^0_{\mathrm{Top}/X}(U, |q^*F|_X) & \xrightarrow{\phi_0} & \mathrm{Map}^0_{\mathrm{Top}/Y}(U, |F|_Y) \\
\downarrow & & \downarrow \\
\mathrm{Map}_{\mathrm{Top}/X}(U, |q^*F|_X) & \xrightarrow{\phi} & \mathrm{Map}_{\mathrm{Top}/Y}(U, |F|_Y)
\end{array}
$$

where the vertical maps are defined as in Lemma 7.1.5.2. We wish to show that ϕ is a homotopy equivalence. This follows from the observation that ϕ_0 is an isomorphism, and the vertical arrows are homotopy equivalences by Lemma 7.1.5.2.

$\square$

Theorem 7.1.0.1 now follows immediately from Proposition 7.1.5.1 applied in the case where $Y = *$ and F is the constant simplicial presheaf $\mathcal{B}_X \to \mathrm{Set}_\Delta$ taking the value K.

Remark 7.1.5.4. There is another solution to the technical difficulty presented by the fact that the bijection $X \otimes K \to X \times |K|$ is not necessarily a homeomorphism: one can work in a suitable category of compactly generated topological spaces where the base change functor $Z \mapsto X \times_Y Z$ has a right adjoint and therefore automatically commutes with all colimits. This is perhaps a more conceptually satisfying approach; however, it leads to a proof of Theorem 7.1.0.1 only in the special case where the space X is itself compactly generated.

We close this section by describing a few applications of Proposition 7.1.5.1 and its proof to the theory of sheaves (of spaces) on a paracompact topological space X.

Corollary 7.1.5.5. *Let X be a paracompact topological space, Y a closed subset of X, and $i : Y \to X$ the inclusion map. Let $\mathcal{F}$ be an object of $\mathrm{Shv}(X)$ and let η_0 be a global section of $i^* \mathcal{F}$. Then there exists an open subset U of X which contains Y and a section $\eta \in \mathcal{F}(U)$ whose image under the restriction map $\mathcal{F}(U) \to (i^* \mathcal{F})(U \cap Y) = (i^* \mathcal{F})(Y)$ lies in the path component of η_0.*

Proof. Let $\mathcal{B}$ denote the collection of open F_σ subsets of X. Without loss of generality, we may assume that $\mathcal{F}$ is represented by a projectively cofibrant simplicial presheaf $F \subseteq \underline{\Delta}^V$ satisfying the conditions of Lemma 7.1.4.1, where V is a linearly ordered set. Using Proposition 7.1.5.1, Corollary 7.1.4.4, and Lemma 7.1.5.2, it will suffice to prove the following assertion:

(a) Every vertex η_0 of $\mathrm{Map}^0_{\mathrm{Top}/X}(Y, |F|_X)$ can be lifted to a vertex of $\mathrm{Map}^0_{\mathrm{Top}/X}(U, |F|_X)$ for some sufficiently small paracompact neighborhood U of Y.

To prove (a), suppose we are given a vertex of $\eta_0 \in \mathrm{Map}^0_{\mathrm{Top}/X}(Y, |F|_X)$ corresponding to a collection of functions $\{\phi_v : Y \to [0,1]\}$. Since η_0 belongs to $\mathrm{Map}^0_{\mathrm{Top}/X}(Y, |F|_X)$, there exist open sets $U_v \subseteq Y$ satisfying condition $(*)$ of Lemma 7.1.5.2.

For each $y \in Y$, there exists an open neighborhood W_y of y in X for which the set $V(y) = \{v \in V : W_y \cap U_v\}$ is finite. Let $W = \bigcup_{y \in Y} W_y$. Shrinking W if necessary, we may suppose that W is a paracompact open neighborhood of Y in X. Since W is paracompact, there exists a locally finite open covering $\{W'_\alpha\}_\alpha$ of W, so that for each index α there exists a point $y_\alpha \in Y$ such that $W'_\alpha \subseteq W_y$. For $v \in V$, let $U'_v = \bigcup_{v \in V(y_\alpha)} W'_\alpha$. The open sets U'_v form a locally finite open covering of W, and each intersection $U'_v \cap Y$ is an open subset of U_v which contains the closure of the support of ϕ_v.

For each $v \in V$, choose a continuous function $\phi'_v : X \to [0,1]$ such that $\phi'_v|Y = \phi_v$ and the closure of the support of ϕ'_v is contained in U'_v. There exists another open set U''_v whose closure is contained in U'_v which again contains the closure of the support of ϕ'_v. For every finite subset $V_0 \subseteq V$, let K_{V_0} denote the intersection $\bigcap_{v \in V_0} \overline{U}''_v$ and let $K^0_{V_0}$ denote the open subset of K_{V_0} given by the inverse image of the open set $U(V_0) \subseteq X$ (the largest open subset for which Δ^{V_0} belongs to $F(U(V_0)) \subseteq \Delta^V$). Then $\{K_{V_0} - K^0_{V_0}\}$ is a locally finite collection of closed subsets of X, none of which intersects Y. Let $K = \bigcup_{V_0}(K_{V_0} - K^0_{V_0})$; then K is a closed subset of X. Let W' be an open F_σ-subset of W which contains Y and does not intersect K. Replacing X by W', we may assume that $W = X$ and that $K = \emptyset$.

Since the collection of functions $\{\phi'_v\}_{v \in V}$ has locally finite support, the function $\phi' = \Sigma_{v \in V} \phi'_v$ is well-defined and takes the value 1 on Y. The open set $\{x \in X : \phi'(x) > 0\}$ is a paracompact open subset of X (Proposition 7.1.1.1). Shrinking X further, we may suppose that ϕ' is everywhere nonzero on X. Set $\phi''_v = \frac{\phi_v}{\phi}$ for each $v \in V$. Then the functions ϕ''_v determine a vertex $\eta \in \mathrm{Map}_{\mathrm{Top}/X}(X, |F|_X)$. Moreover, the open sets $\{U''_v\}$ satisfy condition $(*)$ appearing in the statement of Lemma 7.1.5.2, so that η belongs to $\mathrm{Map}^0_{\mathrm{Top}/X}(X, |F|_X)$, as desired.   □

Corollary 7.1.5.5 admits the following refinement:

Corollary 7.1.5.6. *Let X be a paracompact topological space, Y a closed subset of X, and $i : Y \to X$ the inclusion map. Let $\mathcal{F}$ be an object of $\mathrm{Shv}(X)$. Then the canonical map*

$$\alpha_{\mathcal{F}} : \varinjlim_{Y \subseteq U} \mathcal{F}(U) \to \varinjlim_{Y \subseteq U} (i^* \mathcal{F})(U \cap Y) \simeq (i^* \mathcal{F})(Y)$$

is a homotopy equivalence. Here the colimit is taken over the filtered partially ordered set of all open subsets of X which contain Y.

Proof. We will prove by induction on $n \geq 0$ that the map $\alpha_{\mathcal{F}}$ is n-connective. The case $n = 0$ follows from Corollary 7.1.5.5. Suppose that $n > 0$. We must show that, for every pair of points $\eta, \eta' \in \varinjlim_{Y \subseteq U} \mathcal{F}(U)$, the induced map of fiber products

$$\alpha'_{\mathcal{F}} : * \times_{\varinjlim_{Y \subseteq U} \mathcal{F}(U)} * \to * \times_{(i_* \mathcal{F})(Y)} *$$

is $(n-1)$-connective. Without loss of generality, we may assume that η and η' arise from sections of $\mathcal{F}$ over some $U \subseteq X$ containing Y. Shrinking U if necessary, we may assume that U is paracompact. Replacing X by U, we may assume that η and η' arise from global sections $f, f' : 1 \to \mathcal{F}$, where 1 denotes the final object of $\mathrm{Shv}(X)$. Let $\mathcal{G} = 1 \times_{\mathcal{F}} 1 \in \mathrm{Shv}(X)$. Using the left exactness of i^* and Proposition 5.3.3.3, we can identify $\alpha'_{\mathcal{F}}$ with $\alpha_{\mathcal{G}}$. We now invoke the inductive hypothesis to deduce that $\alpha_{\mathcal{G}}$ is $(n-1)$-connective, as desired. $\square$

Lemma 7.1.5.7. *Let Y be a paracompact topological and $\mathcal{B}$ the collection of open F_σ subsets of Y (see Proposition 7.1.1.1). Let V be a linearly ordered set and let $F : \mathcal{B}^{op} \to \mathrm{Set}_\Delta$ be as in the statement of Lemma 7.1.5.2. Suppose we are given a paracompact space $X \in \mathrm{Top}_{/Y}$ and a closed subspace $X' \subseteq X$. Then the map*

$$\mathrm{Map}^0_{\mathrm{Top}_{/Y}}(X, |F|_Y) \to \mathrm{Map}^0_{\mathrm{Top}_{/Y}}(X', |F|_Y)$$

is a Kan fibration.

Proof. We must show that every lifting problem of the form

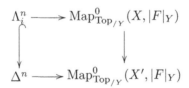

admits a solution. Since the pair $(|\Delta^m|, |\Lambda^m_i|)$ is homeomorphic to $(|\Delta^{m-1}| \times [0,1], |\Delta^{m-1}| \times \{0\})$, we can replace X by $X \times |\Delta^{m-1}|$ and thereby reduce to the case $m = 1$.

Let $Z = (X \times \{0\}) \coprod_{X' \times \{0\}} (X' \times [0,1])$ and let $\eta_0 \in \mathrm{Map}^0_{\mathrm{Top}_{/Y}}(Z, |F|_Y)$; we wish to show that η_0 can be lifted to a point in $\mathrm{Map}^0_{\mathrm{Top}_{/Y}}(X \times [0,1], |F|_Y)$. The proof of Corollary 7.1.5.5 shows that we can lift η_0 to a point $\eta_1 \in \mathrm{Map}^0_{\mathrm{Top}_{/Y}}(U, |F|_Y)$ for some open set $U \subseteq X \times [0,1]$ containing Z.

For each $x \in X$, there exists a real number $\epsilon_x > 0$ and an open neighborhood $V_x \subseteq X$ such that $V_x \times [0, \epsilon_x) \subseteq U$. Since X is paracompact, the open covering $\{V_x\}_{x \in X}$ admits a locally finite refinement $\{W_\alpha\}$, so that for each index α there exists a point $x(\alpha) \in X$ such that $W_\alpha \subseteq V_{x(\alpha)}$. Let $\{\phi_\alpha\}$ be a partition of unity subordinate to the covering W_α and let

$$\psi = \Sigma_\alpha \epsilon_{x(\alpha)} \phi_\alpha.$$

Since the interval $[0, 1]$ is compact, there exists an open neighborhood $V \subseteq X$ containing X' such that $V \times [0, 1] \subseteq U$. Choose a function $\psi' : X \to [0, 1]$ such that $\psi'|(X - V) = \psi|(X - V)$ and $\psi'|X'$ is equal to 1. Set $K = \{(x, t) \in X \times [0, 1] : t \leq \phi(x)\}$, so that $Z \subseteq K \subseteq U$, and let $\eta_2 \in \mathrm{Map}^0_{\mathrm{Top}_{/Y}}(K, |F|_X)$ be the restriction of η_1. Since K is a retract of $X \times [0, 1]$ in the category $\mathrm{Top}_{/Y}$, we can lift η_2 to a point $\eta \in \mathrm{Map}^0_{\mathrm{Top}_{/Y}}(X \times [0, 1], |F|_X)$, as desired. $\qquad\square$

Proposition 7.1.5.8. *Let X be a paracompact topological space. Suppose we are given a sequence of closed subspaces*

$$X_0 \subseteq X_1 \subseteq X_2 \subseteq \cdots \subseteq X$$

with the following properties:

(1) *The union $\bigcup X_i$ coincides with X.*

(2) *A subset $U \subseteq X$ is open if and only if each of the intersections $U \cap X_i$ is an open subset of X_i.*

Then the induced diagram

$$\mathrm{Shv}(X_0) \to \mathrm{Shv}(X_1) \to \cdots \to \mathrm{Shv}(X)$$

exhibits $\mathrm{Shv}(X)$ as the colimit of the sequence $\{\mathrm{Shv}(X_i)\}_{i \geq 0}$ in the ∞-category $\mathcal{RTop}$ of ∞-topoi.

Remark 7.1.5.9. Hypotheses (1) and (2) of Proposition 7.1.5.8 can be summarized by saying that X is the direct limit of the sequence $\{X_i\}$ in the category of topological spaces. It follows from this condition that for any locally compact space Y, the product $X \times Y$ is also the direct limit of the sequence $\{X_i \times Y\}$. To prove this, we observe that for any topological space Z we have bijections

$$\mathrm{Hom}(X \times Y, Z) \simeq \mathrm{Hom}(X, Z^Y) \simeq \varprojlim \mathrm{Hom}(X_i, Z^Y) \simeq \varprojlim \mathrm{Hom}(X_i \times Y, Z),$$

where Z^Y is endowed with the compact-open topology. In particular, we deduce that $X \times \Delta^n$ is the direct limit of the topological spaces $X_i \times \Delta^n$ for each $n \geq 0$.

Proof. For each nonnegative integer n, let $i(n)$ denote the inclusion from X_n to X_{n+1} and $j(n)$ the inclusion of X_n into X. These functors induce geometric morphisms

$$\mathrm{Shv}(X_{n+1}) \underset{i(n)_*}{\overset{i(n)^*}{\rightleftarrows}} \mathrm{Shv}(X_n)$$

$$\mathrm{Shv}(X) \underset{j(n)_*}{\overset{j(n)^*}{\rightleftarrows}} \mathrm{Shv}(X_n).$$

Let $\mathcal{C}$ denote a homotopy inverse limit of the tower of ∞-categories

$$\cdots \longrightarrow \mathrm{Shv}(X_2) \overset{i(1)^*}{\longrightarrow} \mathrm{Shv}(X_1) \overset{i(0)^*}{\longrightarrow} \mathrm{Shv}(X_0).$$

In view of Proposition 6.3.2.3, we can also identify $\mathcal{C}$ with the direct limit of the sequence $\{\mathrm{Shv}(X_i)\}_{i \geq 0}$ in $\mathcal{R}\mathcal{T}\mathrm{op}$. The maps $j(n)$ determine a geometric morphism

$$\mathrm{Shv}(X) \underset{j_*}{\overset{j^*}{\rightleftarrows}} \mathcal{C}.$$

To complete the proof, it will suffice to show that the functor j^* is an equivalence of ∞-categories.

We first show that the unit map $u : \mathrm{id}_{\mathrm{Shv}(X)} \to j_* j^*$ is an equivalence of functors. Let $\mathcal{F} \in \mathrm{Shv}(X)$; we wish to show that the map

$$u_{\mathcal{F}} : \mathcal{F} \to j_* j^* \mathcal{F} \simeq \varprojlim j(n)_* j(n)^* \mathcal{F}$$

is an equivalence in $\mathrm{Shv}(X)$. It will suffice to prove the analogous assertion after evaluating both sides on every open F_σ subset $U \subseteq X$. Replacing X by U, we are reduced to proving that the induced map

$$\alpha_{\mathcal{F}} : \mathcal{F}(X) \to (j_* j^* \mathcal{F})(X) \simeq \varprojlim (j(n)^* \mathcal{F})(X_n)$$

is a homotopy equivalence. Let $\mathcal{B}$ be the collection of all open F_σ subsets of X. Without loss of generality, we may assume that $\mathcal{F}$ is represented by a projectively cofibrant simplicial presheaf $F : \mathcal{B}^{op} \to \mathcal{S}$ satisfying the conditions of Lemma 7.1.4.1. Using Theorem 7.1.4.3, Corollary 7.1.4.4, and Proposition 7.1.5.1, we can identify $\mathcal{F}(X)$ with the Kan complex of sections $K = \mathrm{Map}^0_{\mathrm{Top}/X}(X, \widetilde{X})$ and each $(j(n)^* \mathcal{F})(X_n)$ with the Kan complex of sections $K(n) = \mathrm{Map}_{\mathrm{Top}/X}(X_n, \widetilde{X})$. It will therefore suffice to show that the canonical map $K \to \varprojlim K(n)$ exhibits K as a homotopy inverse limit of the tower $\{K(n)\}$.

It follows from Remark 7.1.5.9 that the map $K \to \varprojlim K(n)$ is an isomorphism of simplicial sets. For each $n \geq 0$, let $K(n)^0 = \mathrm{Map}^0_{\mathrm{Top}/X}(X_n, \widetilde{X}) \subseteq K(n)$ (with notation as in Lemma 7.1.5.2) and let $K^0 = \varprojlim K(n)^0 \subseteq K$. Lemma 7.1.5.2 implies that each inclusion $K(n)^0 \subseteq K(n)$ is a homotopy equivalence. Lemma 7.1.5.7 implies that the restriction maps $K(n+1)^0 \to K(n)^0$ are Kan fibrations. It follows that the inverse limit K^0 of the tower $\{K(n)^0\}$ is a Kan complex and that the map $K^0 \simeq \varprojlim \{K(n)^0\}$ exhibits K^0 as the homotopy inverse limit of $\{K(n)^0\}$. Invoking Remark 7.1.5.3, we deduce that the inclusion $K^0 \subseteq K$ is a homotopy equivalence, so that the equivalent diagram $K \simeq \varprojlim \{K(n)\}$ exhibits K as a homotopy inverse limit of $\{K(n)\}$, as desired.

We now argue that the counit map $v : j^* j_* \to \mathrm{id}$ is an equivalence of functors. Unwinding the definitions, we must prove the following: given a collection of sheaves $\mathcal{F}_n \in \mathrm{Shv}(X_n)$ and equivalences $\mathcal{F}_n \simeq i(n)^* \mathcal{F}_{n+1}$, the canonical map

$$j(n)^* (\varprojlim j(n+k)_* \mathcal{F}_{n+k}) \to \mathcal{F}_n$$

is an equivalence of sheaves on X_n for each $n \geq 0$. It will suffice to show that this map induces a homotopy equivalence after passing to the global sections over every open F_σ subset $U \subseteq X_n$. There exists a function $\phi_0 : X_n \to [0,1]$ such that $U = \{x \in X_n : \phi_0(x) > 0\}$. Choose a map $\phi : X \to [0,1]$ such that $\phi_0 = \phi | X_n$. Replacing X by the paracompact open subset $\{x \in X : \phi(x) > 0\}$, we can reduce to the case where $U = X_n$.

We will prove by induction on k that, for any compatible collection of sheaves $\{\mathcal{F}_n \in \mathrm{Shv}(X_n), \mathcal{F}_n \simeq i(n)^* \mathcal{F}_{n+1}\}$, the map

$$\psi : (j(n)^* \mathcal{F})(X_n) \to \mathcal{F}_n(X_n)$$

is k-connective, where $\mathcal{F} = \varprojlim j(m)_* \mathcal{F}_m$. If $k > 0$, then it will suffice to show that for any pair of points $\eta, \eta' \in (j(n)^* \mathcal{F})(X_n)$, the induced map

$$\psi' : * \times_{(j(n)^* \mathcal{F})(X_n)} * \to * \times_{\mathcal{F}_n(X_n)} *$$

is $(k-1)$-connective. Using Corollary 7.1.5.5, we may assume that η and η' arise from sections $\bar\eta, \bar\eta' \in \mathcal{F}(U)$ for some open neighborhood U of X_n. Shrinking U if necessary, we may assume that U is paracompact. Replacing X by U, we may assume that $\bar\eta$ and $\bar\eta'$ are global sections of $\mathcal{F}$. Since $j(n)^*$ is left exact, we can identify ψ' with the map

$$j(n)^*(* \times_{\mathcal{F}} *)(X_n) \to (* \times_{\mathcal{F}_n} *)(X_n).$$

The $(k-1)$-connectivity of this map now follows from the inductive hypothesis.

It remains to treat the case $k = 0$. Fix an element $\eta_n \in \mathcal{F}_n(X_n)$; we wish to show that η_n lies in the image of $\pi_0 \psi$. For every open set $U \subseteq X$, the composition

$$\pi_0 \mathcal{F}(U) = \pi_0 \varprojlim (j(m)_* \mathcal{F}_m)(U)$$
$$\to \varprojlim (\pi_0 j(m)_* \mathcal{F}_m)(U)$$
$$\simeq \varprojlim \pi_0 \mathcal{F}_m(U \cap X_m)$$

is surjective. Consequently, to prove that η_n lies in the image of $\pi_0 \psi$, it will suffice to show that there exists an open set U containing X_n such that η_n can be lifted to $\varprojlim \pi_0 \mathcal{F}_m(U \cap X_m)$. By virtue of assumption (2), it will suffice to construct a sequence of open F_σ subsets $\{U_m \subseteq X_m\}_{m \geq n}$ and a sequence of compatible sections $\gamma_m \in \pi_0 \mathcal{F}_m(U_m)$ such that $U_n = X_n$ and $\gamma_m = \eta_m$. The construction proceeds by induction on m. Assuming that (U_m, η_m) has already been constructed, we invoke the assumption that U_m is an F_σ to choose a continuous function $f : X_m \to [0,1]$ such that $U_m = \{x \in X_m : f(x) > 0\}$. Let $f' : X_{m+1} \to [0,1]$ be a continuous extension of f and let $V = \{x \in X_{m+1} : f'(x) > 0\}$. Then V is a paracompact open subset of X_{m+1}, and U_m can be identified with a closed subset of V. Applying Corollary 7.1.5.5 to the restriction $\mathcal{F}_{n+1}|V$, we deduce the existence of an open set $U_{m+1} \subseteq V$ such that η_k can be extended to a section $\eta_{k+1} \in \pi_0 \mathcal{F}_{m+1}(U_{m+1})$. Shrinking U_{m+1} if necessary, we may assume that U_{m+1} is itself an F_σ, which completes the induction. $\qquad\square$

Remark 7.1.5.10. Suppose we are given a sequence of closed embeddings of topological spaces

$$X_0 \subseteq X_1 \subseteq X_2 \subseteq \cdots,$$

and let X be the direct limit of the sequence. Suppose further that:

(a) For each $n \geq 0$, the space X_n is paracompact.

(b) For each $n \geq 0$, there exists an open neighborhood Y_n of X_n in X_{n+1} and a retraction r_n of Y_n onto X_n.

Then X is itself paracompact, so that the hypotheses of Proposition 7.1.5.8 are satisfied and $\mathrm{Shv}(X)$ is the direct limit of the sequence of ∞-topoi $\{\mathrm{Shv}(X_n)\}_{n \geq 0}$. To prove this, it will suffice to show that every open covering $\{U_\alpha\}_{\alpha \in A}$ of X admits a refinement $\{V_\beta\}_{\beta \in B}$ which is *countably locally finite*: that is, that there exists a decomposition $B = \bigcup_{n \geq 0} B_n$ such that each of the collections $\{V_\beta\}_{\beta \in B_n}$ is a locally finite collection of open sets, each of which is contained in some U_α (see [59]). To construct this locally finite open covering, we choose for each $n \geq 0$ a locally finite open covering $\{W_\beta\}_{\beta \in B_n}$ of X_n which refines the covering $\{U_\alpha \cap X_n\}_{\alpha \in A}$. For each $\beta \in B_n$, we have $W_\beta \subseteq U_\alpha$ for some $\alpha \in A$. We now define V_β to be the union of a collection of open subsets $\{V_\beta(m) \subseteq X_m\}_{m \geq n}$, which are constructed as follows:

- If $m = n$, we set $V_\beta(m) = W_\beta$.

- Let $m > n$ and let Z_{m-1} be an open neighborhood of X_{m-1} in X_m whose closure is contained in Y_{m-1}. We then set $V_\beta(m) = \{z \in Z_{m-1} : r_{m-1}(z) \in V_\beta(m-1)\} \cap U_\alpha$.

It is clear that each $V_\beta(m)$ is an open subset of X_m contained in U_α and that $V_\beta(m+1) \cap X_m = V_\beta(m)$. Since X is equipped with the direct limit topology, the union $V_\beta = \bigcup_m V_\beta(m)$ is open in X. The only nontrivial point is to verify that the collection $\{V_\beta\}_{\beta \in B_n}$ is locally finite.

Pick a point $x \in X$; we wish to prove the existence of a neighborhood S_x of x such that $\{\beta \in B_n : S \cap V_\beta \neq \emptyset\}$ is finite. Then there exists some $m \geq n$ such that $x \in X_m$; we will construct S_x using induction on m. If $m > n$ and $x \in \overline{Z}_{m-1}$, then let $x' = r_{m-1}(x)$ and set $S_x = S_{x'}$. If $m > n$ and $x \notin \overline{Z}_{m-1}$, or if $m = n$, then we define $S_x = \bigcup_{k \geq m} S_x(k)$, where $S_x(k)$ is an open subset of X_k containing x, defined as follows. If $m > n$, let $S_x(m) = X_m - \overline{Z}_{m-1}$, and if $m = n$, let $S_x(m)$ be an open subset of X_n which intersects only finitely many of the sets $\{W_\beta\}_{\beta \in B_n}$. If $k > m$, we let $S_x(k) = \{z \in Y_{k-1} : r_{k-1}(z) \in S_x(k-1)\}$. It is not difficult to verify that the open set S_x has the desired properties.

7.1.6 Higher Topoi and Shape Theory

If X is a sufficiently nice topological space (for example, an absolute neighborhood retract), then there exists a homotopy equivalence $Y \to X$, where Y is a CW complex. If X is merely assumed to be paracompact, then it is generally not possible to approximate X well by means of a CW complex Y equipped with a map to X. However, in view of Theorem 7.1.4.3, one can still extract a substantial amount of information by considering maps from X to CW complexes. *Shape theory* is an attempt to summarize all of this information in a single invariant called the *shape* of X. In this section, we will sketch a generalization of shape theory to the setting of ∞-topoi.

Definition 7.1.6.1. We let $\mathrm{Pro}(\mathcal{S})$ denote the full subcategory of $\mathrm{Fun}(\mathcal{S}, \mathcal{S})^{op}$ spanned by accessible left exact functors $f : \mathcal{S} \to \mathcal{S}$. We will refer to $\mathrm{Pro}(\mathcal{S})$ as the ∞-category of Pro-*spaces*, or as the ∞-category of *shapes*.

Remark 7.1.6.2. If $\mathcal{C}$ is a small ∞-category which admits finite limits, then any functor $f : \mathcal{C} \to \mathcal{S}$ is accessible and may be viewed as an object of $\mathcal{P}(\mathcal{C}^{op})$. The left exactness of f is then equivalent to the condition that f belongs to $\mathrm{Ind}(\mathcal{C}^{op}) = \mathrm{Pro}(\mathcal{C})^{op}$. Definition 7.1.6.1 constitutes a natural extension of this terminology to a case where $\mathcal{C}$ is not necessarily small; here it is convenient to add a hypothesis of accessibility for technical reasons (which will not play any role in the discussion below).

Definition 7.1.6.3. Let $\mathcal{X}$ be an ∞-topos. According to Proposition 6.3.4.1, there exists a geometric morphism $q_* : \mathcal{X} \to \mathcal{S}$ which is unique up to homotopy. Let q^* be a left adjoint to q_* (also unique up to homotopy). The composition $q_* q^* : \mathcal{S} \to \mathcal{S}$ is an accessible left exact functor, which we will refer to as the *shape of* $\mathcal{X}$ and denote by $Sh(\mathcal{X}) \in \mathrm{Pro}(\mathcal{S})$.

Remark 7.1.6.4. This definition of the shape of an ∞-topos also appears in [78].

Remark 7.1.6.5. Let $p_* : \mathcal{Y} \to \mathcal{X}$ be a geometric morphism of ∞-topoi and let p^* be a left adjoint to p_*. Let $q_* : \mathcal{X} \to \mathcal{S}$ and q^* be as in Definition 7.1.6.3. The unit map $\mathrm{id}_{\mathcal{X}} \to p_* p^*$ induces a transformation

$$q_* q^* \to q_* p_* p^* q^* \simeq (q \circ p)_* (q \circ p)^*,$$

which we may view as a map $Sh(\mathcal{X}) \to Sh(\mathcal{Y})$ in $\mathrm{Pro}(\mathcal{S})$. Via this construction, we may view Sh as a functor from the homotopy category $\mathrm{h}\mathcal{R}\mathcal{J}\mathrm{op}$ of ∞-topoi to the homotopy category $\mathrm{hPro}(\mathcal{S})$. We will say that a geometric morphism $p_* : \mathcal{Y} \to \mathcal{X}$ is a *shape equivalence* if it induces an equivalence $Sh(\mathcal{Y}) \to Sh(\mathcal{X})$ of Pro-spaces.

Remark 7.1.6.6. By construction, the shape of an ∞-topos $\mathcal{X}$ is well-defined up to equivalence in $\mathrm{Pro}(\mathcal{S})$. By refining the above construction, it is possible to construct a shape functor from $\mathcal{R}\mathcal{J}\mathrm{op}$ to the ∞-category $\mathrm{Pro}(\mathcal{S})$ rather than on the level of homotopy.

Remark 7.1.6.7. Our terminology does not quite conform to the usage in classical topology. Recall that if X is a compact metric space, the *shape* of X is defined as a pro-object in the *homotopy* category of spaces. There is a refinement of shape, known as *strong shape*, which takes values in the homotopy category of Pro-spaces. Definition 7.1.6.3 is a generalization of strong shape rather than shape. We refer the reader to [55] for a discussion of classical shape theory.

Proposition 7.1.6.8. *Let* $p : X \to Y$ *be a continuous map of paracompact topological spaces. Then* $p_* : \mathrm{Shv}(X) \to \mathrm{Shv}(Y)$ *is a shape equivalence if and only if, for every Kan complex* K, *the induced map of Kan complexes* $\mathrm{Map}_{\mathrm{Top}}(Y, |K|) \to \mathrm{Map}_{\mathrm{Top}}(X, |K|)$ *is a homotopy equivalence. (Here*

$\mathrm{Map}_{\mathrm{Top}}(Y, |K|)$ denotes the simplicial set whose n-simplices are given by continuous maps $Y \times |\Delta^n| \to |K|$, and $\mathrm{Map}_{\mathrm{Top}}(X, |K|)$ is defined likewise.)

Proof. Corollary 7.1.4.4 and Proposition 7.1.5.1 imply that for any paracompact topological space Z and any Kan complex K, there is a natural isomorphism

$$Sh(\mathrm{Shv}(Z))(K) \simeq \mathrm{Map}_{\mathrm{Top}}(Z, |K|)$$

in the homotopy category $\mathcal{H}$.   □

Example 7.1.6.9. Let X be a scheme, let $\mathfrak{X}$ be the topos of étale sheaves on X, and let $\mathcal{X}$ be the associated 1-localic ∞-topos (see §6.4.5). The shape $Sh(\mathcal{X})$ defined above is closely related to the étale homotopy type introduced by Artin and Mazur (see [3]). There are three important differences:

(1) Artin and Mazur work with pro-objects in the homotopy category $\mathcal{H}$ rather than with actual pro-objects of $\mathcal{S}$. Our definition is closer in spirit to that of Friedlander, who works instead in the homotopy category of pro-objects in Set_Δ (see [30]).

(2) The étale homotopy type of [3] is constructed by considering étale hypercoverings of X; it is therefore more closely related to the shape of the hypercompletion $\mathcal{X}^\wedge$.

(3) Artin and Mazur generally study a certain completion of $Sh(\mathcal{X}^\wedge)$ with respect to the class of truncated spaces, which has the effect of erasing the distinction between $\mathcal{X}$ and $\mathcal{X}^\wedge$ and discarding a bit of (generally irrelevant) information.

Remark 7.1.6.10. Let $*$ denote a topological space consisting of a single point. By definition, $\mathrm{Shv}(*)$ is the full subcategory of $\mathrm{Fun}(\Delta^1, \mathcal{S})$ spanned by those morphisms $f : X \to Y$, where Y is a final object of $\mathcal{S}$. We observe that $\mathrm{Shv}(*)$ is equivalent to the full subcategory spanned by those morphisms f as above, where $Y = \Delta^0 \in \mathcal{S}$, and that this full subcategory is *isomorphic* to $\mathcal{S}$.

Definition 7.1.6.11. We will say that an ∞-topos $\mathcal{X}$ has *trivial shape* if $Sh(\mathcal{X})$ is equivalent to the identity functor $\mathcal{S} \to \mathcal{S}$.

Remark 7.1.6.12. Let $q_* : \mathcal{X} \to \mathcal{S}$ be a geometric morphism. Then the unit map $u : \mathrm{id}_\mathcal{S} \to q_* q^*$ induces a map of Pro-spaces $Sh(X) \to \mathrm{id}_\mathcal{S}$. Since $\mathrm{id}_\mathcal{S}$ is a final object in $\mathrm{Pro}(\mathcal{S})$, we observe that $\mathcal{X}$ has trivial shape if and only if u is an equivalence; in other words, if and only if the pullback functor q^* is fully faithful.

We now sketch another interpretation of shape theory based on the ∞-topoi associated to Pro-spaces. Let $X = \mathcal{S}$, let $\pi : \mathcal{S} \times \mathcal{S} \to \mathcal{S}$ be the projection onto the first factor, let $\delta : \mathcal{S} \to \mathcal{S} \times \mathcal{S}$ denote the diagonal map, and let $\phi : (\mathcal{S} \times \mathcal{S})^{/\delta_\mathcal{S}} \to \mathcal{S}$ be defined as in §4.2.2. Proposition 4.2.2.4 implies that

ϕ is a coCartesian fibration. We may identify the fiber of ϕ over an object $X \in \mathcal{S}$ with the ∞-category $\mathcal{S}^{/X}$. To each morphism $f : X \to Y$ in $\mathcal{S}$, ϕ associates a functor $f_! : \mathcal{S}^{/X} \to \mathcal{S}^{/Y}$ given by composition with f. Since $\mathcal{S}$ admits pullbacks, each $f_!$ admits a right adjoint f^*, so that ϕ is also a Cartesian fibration associated to some functor $\psi : \mathcal{S}^{\mathrm{op}} \to \mathcal{L}\mathcal{T}\mathrm{op}$.

Let $\widehat{X} : \mathcal{S} \to \mathcal{S}$ be a Pro-space. Then $\widehat{X}$ classifies a left fibration $M^{\mathrm{op}} \to \mathcal{S}$, where M is a filtered ∞-category. Let θ denote the composition

$$M^{\mathrm{op}} \to \mathcal{S} \xrightarrow{\psi^{\mathrm{op}}} (\mathcal{L}\mathcal{T}\mathrm{op})^{\mathrm{op}}.$$

Although M is generally not small, the accessibility condition on F guarantees the existence of a cofinal map $M' \to M$, where M' is a small filtered ∞-category. Theorem 6.3.3.1 implies that the diagram θ has a limit, which we will denote by

$$\mathcal{S}_{/\widehat{X}}$$

and refer to as the ∞-*topos of local systems on* $\widehat{X}$.

Remark 7.1.6.13. If $\widehat{X}$ is a Pro-space, then Proposition 6.3.6.4 implies that the associated geometric morphism $\mathcal{S}_{/\widehat{X}} \to \mathcal{S}$ is pro-étale. However, the converse is false in general.

Remark 7.1.6.14. Let G be a profinite group, which we may identify with a Pro-object in the category of finite groups. We let BG denote the corresponding Pro-object of $\mathcal{S}$ obtained by applying the classifying space functor objectwise. Then $\mathcal{S}_{/BG}$ can be identified with the 1-localic ∞-topos associated to the ordinary topos of sets with a continuous G-action. It follows from the construction of filtered limits in $\mathcal{R}\mathcal{T}\mathrm{op}$ (see §6.3.3) that we can describe objects $Y \in \mathcal{S}_{/BG}$ informally as follows: Y associates to each open subgroup $U \subseteq G$ a space Y^U of U-*fixed points* which depends functorially on the finite G-space G/U. Moreover, if U is a normal subgroup of V, then the natural map from Y^V to the (homotopy) fixed-point space $(Y^U)^{V/U}$ should be a homotopy equivalence.

Remark 7.1.6.15. By refining the construction above, it is possible to construct a functor

$$\mathrm{Pro}(\mathcal{S}) \to \mathcal{R}\mathcal{T}\mathrm{op}$$

$$\widehat{X} \mapsto \mathcal{S}_{/\widehat{X}}.$$

This functor has a left adjoint given by

$$\mathcal{X} \mapsto Sh(\mathcal{X}).$$

Warning 7.1.6.16. If $\widehat{X}$ is a Pro-space, then the shape of $\mathcal{S}_{/\widehat{X}}$ is not necessarily equivalent to $\widehat{X}$. In general we have only a counit morphism

$$Sh(\mathcal{S}_{/\widehat{X}}) \to \widehat{X}.$$

7.2 DIMENSION THEORY

In this section, we will discuss the dimension theory of topological spaces from the point of view of higher topos theory. We begin in §7.2.1 by introducing the *homotopy dimension* of an ∞-topos. We will show that the finiteness of the homotopy dimension of an ∞-topos X has pleasant consequences: it implies that every object is the inverse limit of its Postnikov tower and, in particular, that X is hypercomplete.

In §7.2.2, we define the *cohomology groups* of an ∞-topos X. These cohomology groups have a natural interpretation in terms of the classification of higher gerbes on X. Using this interpretation, we will show that the cohomology dimension of an ∞-topos X *almost* coincides with its homotopy dimension.

In §7.2.3, we review the classical theory of *covering dimension* for paracompact topological spaces. Using the results of §7.1, we will show that the covering dimension of a paracompact space X coincides with the homotopy dimension of the ∞-topos $\mathrm{Shv}(X)$.

We conclude in §7.2.4 by introducing a dimension theory for *Heyting spaces*, which generalizes the classical theory of Krull dimension for Noetherian topological spaces. Using this theory, we will prove an upper bound for the homotopy dimension of $\mathrm{Shv}(X)$ for suitable Heyting spaces X. This result can be regarded as a generalization of Grothendieck's vanishing theorem for the cohomology of Noetherian topological spaces.

7.2.1 Homotopy Dimension

Throughout this section, we will use the symbol 1_X to denote the final object of an ∞-topos X.

Definition 7.2.1.1. Let X be an ∞-topos. We shall say that X has *homotopy dimension* $\leq n$ if every n-connective object $U \in X$ admits a global section $1_X \to U$. We say that X has *finite homotopy dimension* if there exists $n \geq 0$ such that X has homotopy dimension $\leq n$.

Example 7.2.1.2. An ∞-topos X is of homotopy dimension ≤ -1 if and only if X is equivalent to the trivial ∞-category $*$ (the ∞-topos of sheaves on the empty space $\emptyset$). The "if" direction is obvious. Conversely, if X has homotopy dimension ≤ -1, then the initial object $\emptyset$ of X admits a global section $1_X \to \emptyset$. For every object $X \in X$, we have a map $X \to 1_X \to \emptyset$, so that X is also initial (Lemma 6.1.3.6). Since the collection of initial objects of X span a contractible Kan complex (Proposition 1.2.12.9), we deduce that X is itself a contractible Kan complex.

Example 7.2.1.3. The ∞-topos S has homotopy dimension 0. More generally, if C is an ∞-category with a final object 1_C, then $\mathcal{P}(C)$ has homotopy dimension ≤ 0. To see this, we first observe that the Yoneda embedding $j : C \to \mathcal{P}(C)$ preserves limits, so that $j(1_C)$ is a final object of $\mathcal{P}(C)$. To prove that $\mathcal{P}(\mathcal{C})$ has homotopy dimension ≤ 0, we need to show that the functor $\mathcal{P}(\mathcal{C}) \rightarrow \mathcal{S}$ corepresented by $j(1_{\mathcal{C}})$ preserves effective epimorphisms. This functor can be identified with evaluation at $1_{\mathcal{C}}$. It therefore preserves all limits and colimits and so carries effective epimorphisms to effective epimorphisms by Proposition 6.2.3.7.

Example 7.2.1.4. Let X be a Kan complex and let $n \geq -1$. The following conditions are equivalent:

(1) The ∞-topos $\mathcal{S}_{/X}$ has homotopy dimension $\leq n$.

(2) The geometric realization $|X|$ is a retract (in the homotopy category $\mathcal{H}$) of a CW complex K of dimension $\leq n$.

To prove that $(2) \Rightarrow (1)$, let us choose an n-connective object of $\mathcal{X}_{/X}$ corresponding to a Kan fibration $p : Y \rightarrow X$ whose homotopy fibers are n-connective. Choose a map $K \rightarrow |X|$ which admits a right homotopy inverse. To prove that p admits a section up to homotopy, it will suffice to show that there exists a dotted arrow

$$
\begin{array}{ccc}
 & & |Y| \\
 & \nearrow & \downarrow p \\
K & \longrightarrow & |X|
\end{array}
$$

in the category of topological spaces, rendering the diagram commutative. The construction of f proceeds simplex by simplex on K, using the n-connectivity of p to solve lifting problems of the form

$$
\begin{array}{ccc}
S^{k-1} & \longrightarrow & |Y| \\
\downarrow & \nearrow & \downarrow p \\
D^k & \longrightarrow & |X|
\end{array}
$$

for $k \leq n$.

To prove that $(1) \Rightarrow (2)$, we choose any n-connective map $q : K \rightarrow |X|$, where K is an n-dimensional CW complex. Condition (1) guarantees that q admits a right homotopy inverse, so that $|X|$ is a retract of K in the homotopy category $\mathcal{H}$.

If $n \neq 2$, then (1) and (2) are equivalent to the following apparently stronger condition:

(3) The geometric realization $|X|$ is homotopy equivalent to a CW complex of dimension $\leq n$.

For a proof, we refer the reader to [81].

Remark 7.2.1.5. If $\mathcal{X}$ is a coproduct (in the ∞-category $\mathcal{R}\mathcal{T}\text{op}$) of ∞-topoi $\mathcal{X}_\alpha$, then $\mathcal{X}$ is of homotopy dimension $\leq n$ if and only if each $\mathcal{X}_\alpha$ is of homotopy dimension $\leq n$.

It is convenient to introduce a relative version of Definition 7.2.1.1.

Definition 7.2.1.6. Let $f : \mathfrak{X} \to \mathcal{Y}$ be a geometric morphism of ∞-topoi. We will say that f is of *homotopy dimension* $\leq n$ if, for every $k \geq n$ and every k-connective morphism $X \to X'$ in $\mathfrak{X}$, the induced map $f_* X \to f_* X'$ is a $(k-n)$-connective morphism in $\mathcal{Y}$ (since f_* is well-defined up to equivalence, this condition is independent of the choice of f_*).

Lemma 7.2.1.7. *Let $\mathfrak{X}$ be an ∞-topos and let $F_* : \mathfrak{X} \to \mathcal{S}$ be a geometric morphism (which is unique up to equivalence). The following are equivalent:*

(1) *The ∞-topos $\mathfrak{X}$ is of homotopy dimension $\leq n$.*

(2) *The geometric morphism F_* is of homotopy dimension $\leq n$.*

Proof. Suppose first that (2) is satisfied and let X be an n-connective object of $\mathfrak{X}$. Then $F_* X$ is a 0-connective object of $\mathcal{S}$: that is, it is a nonempty Kan complex. It therefore has a point $1_{\mathcal{S}} \to F_* X$. By adjointness, we see that there exists a map $1_{\mathfrak{X}} \to X$ in $\mathfrak{X}$, where $1_{\mathfrak{X}} = F^* 1_{\mathcal{S}}$ is a final object of $\mathfrak{X}$ because F^* is left exact. This proves (1).

Now assume (1) and let $s : X \to Y$ be an k-connective morphism in $\mathfrak{X}$; we wish to show that $F_* s$ is $(k-n)$-connective. The proof goes by induction on $k \geq n$. If $k = n$, then we are reduced to proving the surjectivity of the horizontal maps in the diagram

$$\pi_0 \operatorname{Map}_{\mathfrak{X}}(1_{\mathfrak{X}}, X) \longrightarrow \pi_0 \operatorname{Map}_{\mathfrak{X}}(1_{\mathfrak{X}}, Y)$$

$$\pi_0 \operatorname{Map}_{\mathcal{S}}(1_{\mathcal{S}}, F_* X) \longrightarrow \pi_0 \operatorname{Map}_{\mathcal{S}}(1_{\mathcal{S}}, F_* Y)$$

of sets. Let $p : 1_{\mathfrak{X}} \to Y$ be any morphism in $\mathfrak{X}$ and form a pullback diagram

$$
\begin{array}{ccc}
Z & \longrightarrow & X \\
\downarrow{\scriptstyle s'} & & \downarrow{\scriptstyle s} \\
1_{\mathfrak{X}} & \xrightarrow{p} & Y.
\end{array}
$$

The map s' is a pullback of s and therefore n-connective by Proposition 6.5.1.16. Using (1), we deduce the existence of a map $1_{\mathfrak{X}} \to Z$, and a composite map

$$1_{\mathfrak{X}} \to Z \to X$$

is a lifting of p up to homotopy.

We now treat the case where $k > n$. Form a diagram

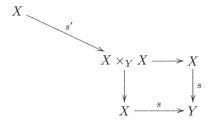

where the square at the bottom right is a pullback in $\mathfrak{X}$. According to Proposition 6.5.1.18, s' is $(k-1)$-connective. Using the inductive hypothesis, we deduce that $F_*(s')$ is $(k-n-1)$-connective. We now invoke Proposition 6.5.1.18 in the ∞-topos $\mathcal{S}$ and deduce that $F_*(s)$ is $(k-n)$-connective, as desired. $\square$

Definition 7.2.1.8. We will say that an ∞-topos $\mathfrak{X}$ is *locally of homotopy dimension $\leq n$* if there exists a collection $\{U_\alpha\}$ of objects of $\mathfrak{X}$ which generate $\mathfrak{X}$ under colimits, such that each $\mathfrak{X}_{/U_\alpha}$ is of homotopy dimension $\leq n$.

Example 7.2.1.9. Let $\mathcal{C}$ be a small ∞-category. Then $\mathcal{P}(\mathcal{C})$ is locally of homotopy dimension ≤ 0. To prove this, we first observe that $\mathcal{P}(\mathcal{C})$ is generated under colimits by the Yoneda embedding $j : \mathcal{C} \to \mathcal{P}(\mathcal{C})$. It therefore suffices to prove that each of the ∞-topoi $\mathcal{P}(\mathcal{C})_{/j(C)}$ has finite homotopy dimension. According to Corollary 5.1.6.12, the ∞-topos $\mathcal{P}(\mathcal{C})_{/j(C)}$ is equivalent to $\mathcal{P}(\mathcal{C}_{/C})$, which is of homotopy dimension 0 (see Example 7.2.1.3).

Our next goal is to prove the following result:

Proposition 7.2.1.10. *Let $\mathfrak{X}$ be an ∞-topos which is locally of homotopy dimension $\leq n$ for some integer n. Then Postnikov towers in $\mathfrak{X}$ are convergent.*

Proof. We will show that $\mathfrak{X}$ satisfies the criterion of Remark 5.5.6.27. Let $X : \mathrm{N}(\mathbf{Z}_{\geq 0}^\infty)^{op} \to \mathfrak{X}$ be a limit tower and assume that the underlying pretower is highly connected. We wish to show that X is highly connected. Choose $m \geq -1$; we wish to show that the map $X(\infty) \to X(k)$ is m-connective for $k \gg 0$. Reindexing the tower if necessary, we may suppose that for every $p \geq q$, the map $X(p) \to X(q)$ is $(m+q)$-connective. We claim that, in this case, we can take $k = 0$. The proof goes by induction on m. If $m > 0$, we can deduce the desired result by applying the inductive hypothesis to the tower

$$X(\infty) \to \cdots \to X(\infty) \times_{X(1)} X(\infty) \to X(\infty) \times_{X(0)} X(\infty).$$

Let us therefore assume that $m = 0$; we wish to show that the map $X(\infty) \to X(0)$ is an effective epimorphism. Since the objects $\{U_\alpha\}$ generate $\mathfrak{X}$ under colimits, there is an effective epimorphism $\phi : U \to X(0)$, where U is a coproduct of objects of the form $\{U_\alpha\}$. Using Remark 7.2.1.5, we deduce that $\mathfrak{X}_{/U}$ has homotopy dimension $\leq n$. Let $F : \mathfrak{X} \to \mathcal{S}$ denote the functor corepresented by U. Then F factors as a composition

$$\mathfrak{X} \xrightarrow{f^*} \mathfrak{X}_{/U} \xrightarrow{\Gamma} \mathcal{S},$$

where f^* is the left adjoint to the geometric morphism $\mathfrak{X}_{/U} \to \mathfrak{X}$ and Γ is the global sections functor. It follows that F carries n-connective morphisms to effective epimorphisms (Lemma 7.2.1.7). The map ϕ determines a point of $F(X(0))$. Since each of the maps $F(X(k+1)) \to F(X(k))$ induces a surjection on connected components, we can lift this point successively to each $F(X(k))$ and thereby obtain a point in $F(X(\infty)) \simeq \operatorname{holim}\{F(X(n))\}$. This point determines a diagram

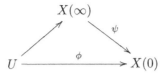

which commutes up to homotopy. Since ϕ is an effective epimorphism, we deduce that the map ψ is an effective epimorphism, as desired. $\qquad\square$

Lemma 7.2.1.11. *Let $\mathcal{X}$ be a presentable ∞-category, let $\operatorname{Fun}(\mathrm{N}(\mathbf{Z}_{\geq 0}^{\infty})^{op}, \mathcal{X})$ be the ∞-category of towers in $\mathcal{X}$, and let $\mathcal{X}_\tau \subseteq \operatorname{Fun}(\mathrm{N}(\mathbf{Z}_{\geq 0}^{\infty})^{op}, \mathcal{X})$ denote the full subcategory spanned by the Postnikov towers. Evaluation at ∞ induces a trivial fibration of simplicial sets $\mathcal{X}_\tau \to \mathcal{X}$. In particular, every object $X(\infty) \in \mathcal{X}$ can be extended to a Postnikov tower*

$$X(\infty) \to \cdots \to X(1) \to X(0).$$

Proof. Let $\mathcal{C}$ be the full subcategory of $\mathcal{X} \times \mathrm{N}(\mathbf{Z}_{\geq 0}^{\infty})^{op}$ spanned by the pairs (X, n), where X is an object of $\mathcal{X}$, $n \in \mathbf{Z}_{\geq 0}^{\infty}$, and X is n-truncated, and let $p : \mathcal{C} \to \mathcal{X}$ denote the natural projection. Since every m-truncated object of $\mathcal{X}$ is also n-truncated for $m \geq n$, it is easy to see that p is a Cartesian fibration. Proposition 5.5.6.18 implies that each of the inclusion functors $\tau_{\leq m} \mathcal{X} \subseteq \tau_{\leq n} \mathcal{X}$ has a left adjoint, so that p is also a coCartesian fibration (Corollary 5.2.2.5). By definition, $\mathcal{X}_\tau$ can be identified with the simplicial set

$$\operatorname{Map}^{\flat}_{\mathrm{N}(\mathbf{Z}_{\geq 0}^{\infty})}(\mathrm{N}(\mathbf{Z}_{\geq 0}^{\infty})^{\sharp}, (\mathcal{C}^{op})^{\natural})^{op},$$

and $\mathcal{X}$ itself can be identified with

$$\operatorname{Map}^{\flat}_{\mathrm{N}(\mathbf{Z}_{\geq 0}^{\infty})}(\{\infty\}^{\sharp}, (\mathcal{C}^{op})^{\natural})^{op}.$$

It now suffices to observe that the inclusion $\{\infty\}^{\sharp} \subseteq \mathrm{N}(\mathbf{Z}_{\geq 0}^{\infty})^{\sharp}$ is marked anodyne. $\qquad\square$

Corollary 7.2.1.12 (Jardine). *Let $\mathcal{X}$ be an ∞-topos which is locally of homotopy dimension $\leq n$ for some integer n. Then $\mathcal{X}$ is hypercomplete.*

Proof. Let $X(\infty)$ be an arbitrary object of $\mathcal{X}$. By Lemma 7.2.1.11, we can find a Postnikov tower

$$X(\infty) \to \cdots \to X(1) \to X(0).$$

Since $X(n)$ is n-truncated, it belongs to $\mathcal{X}^{\wedge}$ by Corollary 6.5.1.14. By Proposition 7.2.1.10, the tower exhibits $X(\infty)$ as a limit of objects of $\mathcal{X}^{\wedge}$, so that $X(\infty)$ belongs to $\mathcal{X}^{\wedge}$ as well since the full subcategory $\mathcal{X}^{\wedge} \subseteq \mathcal{X}$ is stable under limits. $\qquad\square$

Lemma 7.2.1.13. *Let $\mathfrak{X}$ be an ∞-topos, $n \geq 0$, X an $(n+1)$-connective object of $\mathfrak{X}$, and $f^* : \mathfrak{X} \to \mathfrak{X}_{/X}$ a right adjoint to the projection $\mathfrak{X}_{/X} \to \mathfrak{X}$. Then f^* induces a fully faithful functor $\tau_{\leq n} \mathfrak{X} \to \tau_{\leq n} \mathfrak{X}_{/X}$ which restricts to an equivalence from $\tau_{\leq n-1} \mathfrak{X}$ to $\tau_{\leq n-1} \mathfrak{X}_{/X}$.*

Proof. We first prove that f^* is fully faithful when restricted to the ∞-category of n-truncated objects of $\mathfrak{X}$. Let $Y, Z \in \mathfrak{X}$ be objects, where Y is n-truncated. We have a commutative diagram

$$\begin{array}{ccc}
\mathrm{Map}_{\mathfrak{X}_{/X}}(f^*Y, f^*Z) = \mathrm{Map}_{\mathfrak{X}}(X \times Y, Z) \longleftarrow \mathrm{Map}_{\mathfrak{X}}(\tau_{\leq n}(X \times Y), Z) \\
\uparrow \qquad\qquad\qquad\qquad\qquad\qquad\qquad\qquad \uparrow \\
\mathrm{Map}_{\mathfrak{X}}(Y, Z) \longleftarrow\! \mathrm{Map}_{\mathfrak{X}}(\tau_{\leq n}Y, Z)
\end{array}$$

in the homotopy category $\mathcal{H}$, where the horizontal arrows are homotopy equivalences. Consequently, to prove that the left vertical map is a homotopy equivalence, it suffices to show that the projection $\tau_{\leq n}(X \times Y) \to \tau_{\leq n}Y$ is an equivalence. This follows immediately from Lemma 6.5.1.2 and our assumption that X is $(n+1)$-connective.

Now suppose that $\overline{Y}$ is an $(n-1)$-truncated object of $\mathfrak{X}_{/X}$. We wish to show that $\overline{Y}$ lies in the essential image of $f^*|\tau_{\leq n-1} \mathfrak{X}$. Let Y denote the image of $\overline{Y}$ in $\mathfrak{X}$ and let $Y \to Z$ exhibit Z as an $(n-1)$-truncation of Y in $\mathfrak{X}$. To complete the proof, it will suffice to show that the composition

$$u : \overline{Y} \xrightarrow{u'} f^*Y \xrightarrow{u''} f^*Z$$

is an equivalence in $\mathfrak{X}_{/X}$. Since both $\overline{Y}$ and f^*Z are $(n-1)$-truncated, it suffices to prove that u is n-connective. According to Proposition 6.5.1.16, it suffices to prove that u' and u'' are n-connective. Proposition 5.5.6.28 implies that u'' exhibits f^*Z as an $(n-1)$-truncation of f^*Y and is therefore n-connective.

We now complete the proof by showing that u' is n-connective. Let v' denote the image of u' in the ∞-topos $\mathfrak{X}$. According to Proposition 6.5.1.19, it will suffice to show that v' is n-connective. We observe that v' is a section of the projection $q : Y \times X \to Y$. According to Proposition 6.5.1.20, it will suffice to prove that q is $(n+1)$-connective. Since q is a pullback of the projection $X \to 1_{\mathfrak{X}}$, Proposition 6.5.1.16 allows us to conclude the proof (since X is $(n+1)$-connective by assumption). $\qquad\square$

Lemma 7.2.1.13 has some pleasant consequences.

Proposition 7.2.1.14. *Let $\mathfrak{X}$ be an ∞-topos and let $\tau_{\leq 0} : \mathfrak{X} \to \tau_{\leq 0} \mathfrak{X}$ denote a left adjoint to the inclusion. A morphism $\phi : U \to X$ in $\mathfrak{X}$ is an effective epimorphism if and only if $\tau_{\leq 0}(\phi)$ is an effective epimorphism in the ordinary topos $h(\tau_{\leq 0} \mathfrak{X})$.*

Proof. Suppose first that ϕ is an effective epimorphism. Let $U_\bullet : \mathrm{N}\,\boldsymbol{\Delta}_+^{op} \to \mathfrak{X}$ be a Čech nerve of ϕ, so that $U_\bullet$ is a colimit diagram. Since $\tau_{\leq 0}$ is a left adjoint, $\tau_{\leq 0} U_\bullet$ is a colimit diagram in $\tau_{\leq 0} \mathcal{X}$. Using Proposition 6.2.3.10, we deduce easily that $\tau_{\leq 0} \phi$ is an effective epimorphism.

For the converse, choose a factorization of ϕ as a composition

$$U \xrightarrow{\phi'} V \xrightarrow{\phi''} X,$$

where ϕ' is an effective epimorphism and ϕ'' is a monomorphism. Applying Lemma 7.2.1.13 to the ∞-topos $\mathcal{X}_{/\tau_{\leq 0} X}$, we conclude that ϕ'' is the pullback of a monomorphism $i : \overline{V} \to \tau_{\leq 0} X$. Since the effective epimorphism $\tau_{\leq 0}(\phi)$ factors through i, we conclude that i is an equivalence, so that ϕ'' is likewise an equivalence. It follows that ϕ is an effective epimorphism, as desired. $\square$

Proposition 7.2.1.14 can be regarded as a generalization of the following well-known property of the ∞-category of spaces, which can itself be regarded as the ∞-categorical analogue of the second part of Fact 6.1.1.6:

Corollary 7.2.1.15. *Let $f : X \to Y$ be a map of Kan complexes. Then f is an effective epimorphism in the ∞-category $\mathcal{S}$ if and only if the induced map $\pi_0 X \to \pi_0 Y$ is surjective.*

Remark 7.2.1.16. It follows from Proposition 7.2.1.14 that the class of ∞-topoi having the form $\mathrm{Shv}(\mathcal{C})$, where $\mathcal{C}$ is a small ∞-category, is not substantially larger than the class of ordinary topoi. More precisely, every topological localization of $\mathcal{P}(\mathcal{C})$ can be obtained by inverting morphisms between *discrete* objects of $\mathcal{P}(\mathcal{C})$. It follows that there exists a pullback diagram of ∞-topoi

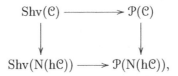

where the ∞-topoi on the bottom line are 1-localic and therefore determined by the ordinary topoi of presheaves of sets on the homotopy category $h\,\mathcal{C}$ and sheaves of sets on $h\,\mathcal{C}$, respectively.

Corollary 7.2.1.17. *Let X be a topological space. Suppose that $\mathrm{Shv}(X)$ is locally of homotopy dimension $\leq n$ for some integer n. Then $\mathrm{Shv}(X)$ has enough points.*

Proof. Note that every point $x \in X$ gives rise to a point $x_* : \mathrm{Shv}(*) \to \mathrm{Shv}(X)$ of the ∞-topos $\mathrm{Shv}(X)$. Let $f : \mathcal{F} \to \mathcal{F}'$ be a morphism in $\mathrm{Shv}(X)$ such that $x^*(f)$ is an equivalence in $\mathcal{S}$ for each $x \in X$. We wish to prove that f is an equivalence. According to Corollary 7.2.1.12, it will suffice to prove that f is ∞-connective. We will prove by induction on n that f is n-connective. If $n > 0$, we simply apply the inductive hypothesis to the diagonal morphism $\delta : \mathcal{F} \to \mathcal{F} \times_{\mathcal{F}'} \mathcal{F}$. We may therefore reduce to the case $n = 0$; we wish to show that f is an effective epimorphism. Since $\mathrm{Shv}(X)$ is generated under colimits by the sheaves χ_U associated to open subsets

$U \subseteq X$, we may assume without loss of generality that $\mathcal{F}' = \chi_U$. We may now invoke Proposition 7.2.1.14 to reduce to the case where $\mathcal{F}$ is an object of $\tau_{\leq 0} \operatorname{Shv}(X)_{/\chi_U}$. This ∞-category is equivalent to the nerve of the category of sheaves of *sets* on U. We are therefore reduced to proving that if $\mathcal{F}$ is a sheaf of sets on U whose stalk $\mathcal{F}_x$ is a singleton at each point $x \in U$, then $\mathcal{F}$ has a global section, which is clear. $\square$

7.2.2 Cohomological Dimension

In classical homotopy theory, one can analyze a space X by means of its Postnikov tower

$$\cdots \to \tau_{\leq n} X \xrightarrow{\phi_n} \tau_{\leq n-1} X \to \cdots .$$

In this diagram, the homotopy fiber F of ϕ_n $(n \geq 1)$ is a space which has only a single nonvanishing homotopy group which appears in dimension n. The space F is determined up to homotopy equivalence by $\pi_n F$: in fact, F is homotopy equivalent to an Eilenberg-MacLane space $K(\pi_n F, n)$ which can be functorially constructed from the group $\pi_n F$. The study of these Eilenberg-MacLane spaces is of central interest because (according to the above analysis) they constitute basic building blocks out of which any arbitrary space can be constructed. Our goal in this section is to generalize the theory of Eilenberg-MacLane spaces to the setting of an arbitrary ∞-topos $\mathcal{X}$.

Definition 7.2.2.1. Let $\mathcal{X}$ be an ∞-category. A *pointed object* is a morphism $X_* : 1 \to X$ in $\mathcal{X}$, where 1 is a final object of $\mathcal{X}$. We let $\mathcal{X}_*$ denote the full subcategory of $\operatorname{Fun}(\Delta^1, \mathcal{X})$ spanned by the pointed objects of $\mathcal{X}$.

A *group object* of $\mathcal{X}$ is a groupoid object $U_\bullet : \mathbf{N} \Delta^{op} \to \mathcal{X}$ for which U_0 is a final object of $\mathcal{X}$. Let $\mathcal{G}rp(\mathcal{X})$ denote the full subcategory of $\mathcal{X}_\Delta$ spanned by the group objects of $\mathcal{X}$.

We will say that a pointed object $1 \to X$ of an ∞-topos $\mathcal{X}$ is an *Eilenberg-MacLane object of degree n* if X is both n-truncated and n-connective. We let $\mathcal{EM}_n(\mathcal{X})$ denote the full subcategory of $\mathcal{X}_*$ spanned by the Eilenberg-MacLane objects of degree n.

Example 7.2.2.2. Let $\mathcal{C}$ be an ordinary category which admits finite limits. A *group object* of $\mathcal{C}$ is an object $X \in \mathcal{C}$ which is equipped with an identity section $1_\mathcal{C} \to X$, an inversion map $X \to X$, and a multiplication $m : X \times X \to X$, which satisfy the usual group axioms. Equivalently, a group object of $\mathcal{C}$ is an object X together with a group structure on each morphism space $\operatorname{Hom}_\mathcal{C}(Y, X)$ which depends functorially on Y. We will denote the category of group objects of $\mathcal{C}$ by $\mathcal{G}rp(\mathcal{C})$. The ∞-category $\mathrm{N}(\mathcal{G}rp(\mathcal{C}))$ is equivalent to the ∞-category of group objects of $\mathrm{N}(\mathcal{C})$ in the sense of Definition 7.2.2.1. Thus the notion of a group object of an ∞-category can be regarded as a generalization of the notion of a group object of an ordinary category.

Remark 7.2.2.3. Let $\mathcal{X}$ be an ∞-topos and $n \geq 0$ an integer. Then the full subcategory of $\operatorname{Fun}(\Delta^1, \mathcal{X})$ consisting of Eilenberg-MacLane objects $p : 1 \to$

X is stable under finite products. This is an immediate consequence of the following observations:

(1) A finite product of n-connective objects of X is n-connective (Corollary 6.5.1.13).

(2) *Any* limit of n-truncated objects of X is n-truncated (since $\tau_{\leq n} X$ is a localization of X).

Proposition 7.2.2.4. *Let X be an ∞-category and let $U_\bullet$ be a simplicial object of X. Then $U_\bullet$ is a group object of X if and only if the following conditions are satisfied:*

(1) *The object U_0 is final in X.*

(2) *For every decomposition $[n] = S \cup S'$, where $S \cap S' = \{s\}$, the maps*

$$U(S) \leftarrow U_n \to U(S')$$

exhibit U_n as a product of $U(S)$ and $U(S)$ in X.

Proof. This follows immediately from characterization $(4'')$ of Proposition 6.1.2.6. □

Corollary 7.2.2.5. *Let X and Y be ∞-categories which admit finite products and let $f : X \to Y$ be a functor which preserves finite products. Then the induced functor $X_\Delta \to Y_\Delta$ carries group objects of X to group objects of Y.*

Corollary 7.2.2.6. *Let X be an ∞-category which admits finite products and let $Y \subseteq X$ be a full subcategory which is stable under finite products. Let $Y_\bullet$ be a simplicial object of Y. Then $Y_\bullet$ is a group object of Y if and only if it is a group object of X.*

Definition 7.2.2.7. Let X be an ∞-category. A *zero object* of X is an object which is both initial and final.

Lemma 7.2.2.8. *Let X be an ∞-category with a final object 1_X. Then the inclusion $i : X^{1_X/} \subseteq X_*$ is an equivalence of ∞-categories.*

Proof. Let K be the full subcategory of X spanned by the final objects and let 1_X be an object of K. Proposition 1.2.12.9 implies that K is a contractible Kan complex, so that the inclusion $\{1_X\} \subseteq K$ is an equivalence of ∞-categories. Corollary 2.4.7.12 implies that the projection $X_* \to K$ is a coCartesian fibration. We now apply Proposition 3.3.1.3 to deduce the desired result. □

Lemma 7.2.2.9. *Let X be an ∞-category with a final object. Then the ∞-category X_* has a zero object. If X already has a zero object, then the forgetful functor $X_* \to X$ is an equivalence of ∞-categories.*

Proof. Let 1_X be a final object of $\mathcal{X}$ and let $U = \mathrm{id}_{1_X} \in \mathcal{X}_*$. We wish to show that U is a zero object of $\mathcal{X}_*$. According to Lemma 7.2.2.8, it will suffice to show that U is a zero object of $\mathcal{X}_{1_X/}$. It is clear that U is initial, and the finality of U follows from Proposition 1.2.13.8.

For the second assertion, let us suppose that 1_X is also an initial object of $\mathcal{X}$. We wish to show that the forgetful functor $\mathcal{X}_* \to \mathcal{X}$ is an equivalence of ∞-categories. Applying Lemma 7.2.2.8, it will suffice to show that the projection $f : \mathcal{X}^{1_X/} \to \mathcal{X}$ is an equivalence of ∞-categories. But f is a trivial fibration of simplicial sets. $\qquad\square$

Lemma 7.2.2.10. *Let $\mathcal{X}$ be an ∞-category and let $f : \mathcal{X}_* \to \mathcal{X}$ be the forgetful functor (which carries a pointed object $1 \to X$ to X). Then f induces an equivalence of ∞-categories*

$$\mathcal{G}\mathrm{rp}(\mathcal{X}_*) \to \mathcal{G}\mathrm{rp}(\mathcal{X}).$$

Proof. The functor f factors as a composition

$$\mathcal{X}_* \subseteq \mathrm{Fun}(\Delta^1, \mathcal{X}) \to \mathcal{X},$$

where the first map is the inclusion of a full subcategory which is stable under limits and the second map preserves all limits (Proposition 5.1.2.2). It follows that f preserves limits so that composition with f induces a functor $F : \mathcal{G}\mathrm{rp}(\mathcal{X}_*) \to \mathcal{G}\mathrm{rp}(\mathcal{X})$ by Corollary 7.2.2.5.

Observe that the 0-simplex Δ^0 is an initial object of $\mathbf{\Delta}^{op}$. Consequently, there exists a functor $T : \Delta^1 \times \mathrm{N}(\mathbf{\Delta})^{op} \to \mathrm{N}(\mathbf{\Delta})^{op}$ which is a natural transformation from the constant functor taking the value Δ^0 to the identity functor. Composition with T induces a functor

$$\mathcal{X}_\Delta \to \mathrm{Fun}(\Delta^1, \mathcal{X})_\Delta.$$

Restricting to group objects, we get a functor $s : \mathcal{G}\mathrm{rp}(\mathcal{X}) \to \mathcal{G}\mathrm{rp}(\mathcal{X}_*)$. It is clear that $F \circ s$ is the identity.

We observe that if $\mathcal{X}$ has a zero object, then f is an equivalence of ∞-categories (Lemma 7.2.2.9). It follows immediately that F is an equivalence of ∞-categories. Since s is a right inverse to F, we conclude that s is an equivalence of ∞-categories as well.

To complete the proof in the general case, it will suffice to show that the composition $s \circ F$ is an equivalence of ∞-categories. To prove this, we set $\mathcal{Y} = \mathcal{X}_*$ and let $F' : \mathcal{G}\mathrm{rp}(\mathcal{Y}_*) \to \mathcal{G}\mathrm{rp}(\mathcal{X}_*)$ and $s' : \mathcal{G}\mathrm{rp}(\mathcal{Y}) \to \mathcal{G}\mathrm{rp}(\mathcal{Y}_*)$ be defined as above. We then have a commutative diagram

$$
\begin{array}{ccc}
\mathcal{G}\mathrm{rp}(\mathcal{Y}) & \xrightarrow{\ F\ } & \mathcal{G}\mathrm{rp}(\mathcal{X}) \\
\downarrow{\scriptstyle s'} & & \downarrow{\scriptstyle s} \\
\mathcal{G}\mathrm{rp}(\mathcal{Y}_*) & \xrightarrow{\ F'\ } & \mathcal{G}\mathrm{rp}(\mathcal{X}_*),
\end{array}
$$

so that $s \circ F = F' \circ s'$. Lemma 7.2.2.9 implies that $\mathcal{Y}$ has a zero object, so that F' and s' are equivalences of ∞-categories. Therefore $F' \circ s' = s \circ F$ is an equivalence of ∞-categories, and the proof is complete. $\qquad\square$

The following proposition guarantees a good supply of Eilenberg-MacLane objects in an ∞-topos $\mathfrak{X}$.

Lemma 7.2.2.11. *Let $\mathfrak{X}$ be an ∞-topos containing a final object $1_{\mathfrak{X}}$ and let $n \geq 1$. Let p denote the composition*

$$\mathrm{Fun}(\Delta^1, \mathfrak{X}) \overset{\check{C}}{\to} \mathfrak{X}_{\Delta_+} \to \mathfrak{X}_{\Delta},$$

which associates to each morphism $U \to X$ the underlying groupoid of its Čech nerve. Then

(1) *Let $\mathfrak{X}'$ denote the full subcategory of $\mathrm{Fun}(\Delta^1, \mathfrak{X})$ consisting of connected pointed objects of $\mathfrak{X}$. Then the restriction of p induces an equivalence of ∞-categories from $\mathfrak{X}'$ to the ∞-category $\mathfrak{Grp}(\mathfrak{X})$.*

(2) *The essential image of $p|\, \mathcal{EM}_n(\mathfrak{X})$ coincides with the essential image of the composition*

$$\mathfrak{Grp}(\mathcal{EM}_{n-1}(\mathfrak{X})) \subseteq \mathfrak{Grp}(\mathfrak{X}_*) \to \mathfrak{Grp}(\mathfrak{X}).$$

Proof. Let $\mathfrak{X}''$ be the full subcategory of $\mathrm{Fun}(\Delta^1, \mathfrak{X})$ spanned by the effective epimorphisms $u : U \to X$. Since $\mathfrak{X}$ is an ∞-topos, p induces an equivalence from $\mathfrak{X}''$ to the ∞-category of groupoid objects of $\mathfrak{X}$. Consequently, to prove (1), it will suffice to show that if $u : 1_{\mathfrak{X}} \to X$ is a morphism in $\mathfrak{X}$ and $1_{\mathfrak{X}}$ is a final object, then u is an effective epimorphism if and only if X is connected. We note that X is connected if and only if the map $\tau_{\leq 0}(u) : \tau_{\leq 0} 1_{\mathfrak{X}} \to \tau_{\leq 0} X$ is an isomorphism in the ordinary topos $\mathrm{Disc}(\mathfrak{X})$. According to Proposition 7.2.1.14, u is an effective epimorphism if and only if $\tau_{\leq 0}(u)$ is an effective epimorphism. We now observe that in any ordinary category $\mathcal{C}$, an effective epimorphism $u' : 1_{\mathcal{C}} \to X'$ whose source is a final object of $\mathcal{C}$ is automatically an isomorphism since the equivalence relation $1_{\mathcal{C}} \times_{X'} 1_{\mathcal{C}} \subseteq 1_{\mathcal{C}} \times 1_{\mathcal{C}}$ consists of the whole of $1_{\mathcal{C}} \times 1_{\mathcal{C}} \simeq 1_{\mathcal{C}}$.

To prove (2), we consider an augmented simplicial object $X_\bullet$ of $\mathfrak{X}$ which is a Čech nerve having the property that X_0 is a final object of $\mathfrak{X}$. We wish to show that the pointed object $X_0 \to X_{-1}$ belongs to $\mathcal{EM}_n(\mathfrak{X})$ if and only if each X_k is $(n-1)$-truncated and $(n-1)$-connective for $k \geq 0$. We conclude by making the following observations:

(a) Since X_k is equivalent to a k-fold product of copies of X_1, the objects X_k are $(n-1)$-truncated $((n-1)$-connective$)$ for all $k \geq 0$ if and only if X_1 is $(n-1)$-truncated $((n-1)$-connective$)$.

(b) We have a pullback diagram

$$
\begin{array}{ccc}
X_1 & \overset{f}{\longrightarrow} & X_0 \\
\downarrow & & \downarrow{\scriptstyle g} \\
X_0 & \overset{g}{\longrightarrow} & X_{-1}.
\end{array}
$$

The object X_1 is $(n-1)$-truncated if and only if f is $(n-1)$-truncated. Since g is an effective epimorphism, f is $(n-1)$-truncated if and only if g is $(n-1)$-truncated (Proposition 6.2.3.17). Using the long exact sequence of Remark 6.5.1.5, we conclude that this is equivalent to the vanishing of $g^*\pi_k X_{-1}$ for $k > n$. Since g is an effective epimorphism, this is equivalent to the vanishing of $\pi_k X_{-1}$ for $k > n$, which is in turn equivalent to the requirement that X_{-1} is n-truncated.

(c) The object X_1 is $(n-1)$-connective if and only if f is $(n-1)$-connective. Arguing as above, we conclude that f is $(n-1)$-connective if and only if g is $(n-1)$-connective (Proposition 6.5.1.16). Using the long exact sequence of Remark 6.5.1.5, this is equivalent to the vanishing of the homotopy sheaf $g^*\pi_k X_{-1}$ for $k < n$. Since g is an effective epimorphism, this is equivalent to the vanishing of $\pi_k X_{-1}$ for $k < n$, which is in turn equivalent to the condition that X_{-1} is $(n-1)$-truncated.

$\square$

Proposition 7.2.2.12. *Let $\mathfrak{X}$ be an ∞-topos, let $n \geq 0$ be a nonnegative integer, and let $\pi_n : \mathfrak{X}_* \to \mathrm{N}(\mathrm{Disc}(\mathfrak{X}))$ denote the associated homotopy group functor.*
Then

(1) *If $n = 0$, then π_n determines an equivalence from the ∞-category $\mathcal{EM}_0(\mathfrak{X})$ to the (nerve of the) category of pointed objects of $\mathrm{Disc}(\mathfrak{X})$.*

(2) *If $n = 1$, then π_n determines an equivalence from the ∞-category $\mathcal{EM}_1(\mathfrak{X})$ to the (nerve of the) category of group objects of $\mathrm{Disc}(\mathfrak{X})$.*

(3) *If $n \geq 2$, then π_n determines an equivalence from the ∞-category $\mathcal{EM}_n(\mathfrak{X})$ to the (nerve of the) category of commutative group objects of $\mathrm{Disc}(\mathfrak{X})$.*

Proof. We use induction on n. The case $n = 0$ follows immediately from the definitions. The case $n = 1$ follows from the case $n = 0$ by combining Lemmas 7.2.2.11 and 7.2.2.10. If $n = 2$, we apply the inductive hypothesis together with Lemma 7.2.2.11 and the observation that if $\mathcal{C}$ is an ordinary category which admits finite products, then $\mathcal{Grp}(\mathcal{Grp}(\mathcal{C}))$ is equivalent to category $\mathcal{Ab}(\mathcal{C})$ of *commutative* group objects of $\mathcal{C}$. The argument in the case $n > 2$ makes use of the inductive hypothesis, Lemma 7.2.2.11, and the observation that $\mathcal{Grp}(\mathcal{Ab}(\mathcal{C}))$ is equivalent to $\mathcal{Ab}(\mathcal{C})$ for any ordinary category $\mathcal{C}$ which admits finite products. $\square$

Fix an ∞-topos $\mathfrak{X}$, a final object $1_{\mathfrak{X}} \in \mathfrak{X}$, and an integer $n \geq 0$. According to Proposition 7.2.2.12, there exists a homotopy inverse to the functor π. We will denote this functor by

$$A \mapsto (p : 1_{\mathfrak{X}} \to K(A, n)),$$

where A is a pointed object of the topos $\mathrm{Disc}(\mathfrak{X})$ if $n = 0$, a group object if $n = 1$, and an abelian group object if $n \geq 2$.

Remark 7.2.2.13. The functor $A \mapsto K(A, n)$ preserves finite products. This is clear since the class of Eilenberg-MacLane objects is stable under finite products (Remark 7.2.2.3) and the homotopy inverse functor π commutes with finite products (since homotopy groups are constructed using pullback and truncation functors, each of which commutes with finite products).

Definition 7.2.2.14. Let $\mathfrak{X}$ be an ∞-topos, $n \geq 0$ an integer, and A an abelian group object of the topos $\mathrm{Disc}(\mathfrak{X})$. We define

$$\mathrm{H}^n(\mathfrak{X}; A) = \pi_0 \, \mathrm{Map}_{\mathfrak{X}}(1_{\mathfrak{X}}, K(A, n));$$

we refer to $\mathrm{H}^n(\mathfrak{X}; A)$ as the *nth cohomology group of $\mathfrak{X}$ with coefficients in A*.

Remark 7.2.2.15. It is clear that we can also make sense of $\mathrm{H}^1(\mathfrak{X}; G)$ when G is a sheaf of nonabelian groups, or $\mathrm{H}^0(\mathfrak{X}; E)$ when E is only a sheaf of (pointed) sets.

Remark 7.2.2.16. It is clear from the definition that $\mathrm{H}^n(\mathfrak{X}; A)$ is functorial in A. Moreover, this functor commutes with finite products by Remark 7.2.2.13 (and the fact that products in $\mathfrak{X}$ are products in the homotopy category $h\mathfrak{X}$). If A is an abelian group, then the multiplication map $A \times A \to A$ induces a (commutative) group structure on $\mathrm{H}^n(\mathfrak{X}; A)$. This justifies our terminology in referring to $\mathrm{H}^n(\mathfrak{X}; A)$ as a cohomology *group*.

Remark 7.2.2.17. Let $\mathcal{C}$ be a small category equipped with a Grothendieck topology and let $\mathfrak{X}$ be the ∞-topos $\mathrm{Shv}(\mathrm{N}\,\mathcal{C})$ of sheaves of *spaces* on $\mathcal{C}$, so that the underlying topos $\mathrm{Disc}(\mathfrak{X})$ is equivalent to the category of sheaves of *sets* on $\mathcal{C}$. Let A be a sheaf of abelian groups on $\mathcal{C}$. Then $\mathrm{H}^n(\mathfrak{X}; A)$ may be identified with the nth cohomology group of $\mathrm{Disc}(\mathfrak{X})$ with coefficients in A in the sense of ordinary sheaf theory. To see this, choose a resolution

$$A \to I^0 \to I^1 \to \cdots \to I^{n-1} \to J$$

of A by abelian group objects of $\mathrm{Disc}(\mathfrak{X})$, where each I^k is injective. The complex

$$I^0 \to \cdots \to J$$

may be identified, via the Dold-Kan correspondence, with a simplicial abelian group object $C_\bullet$ of $\mathrm{Disc}(\mathfrak{X})$. Regard $C_\bullet$ as a presheaf on $\mathcal{C}$ with values in Set_Δ. Then

(1) The induced presheaf $F : \mathrm{N}(\mathcal{C})^{op} \to \mathcal{S}$ belongs to $\mathfrak{X} = \mathrm{Shv}(\mathrm{N}(\mathcal{C})) \subseteq \mathcal{P}(\mathrm{N}(\mathcal{C}))$ (this uses the injectivity of the objects I^k) and is equipped with a canonical base point $p : 1_{\mathfrak{X}} \to F$.

(2) The pointed object $p : 1_X \to F$ is an Eilenberg-MacLane object of X, and there is a canonical identification $A \simeq p^*(\pi_n F)$. We may therefore identify F with $K(A, n)$.

(3) The set of homotopy classes of maps from 1_X to F in X may be identified with the cokernel of the map $\Gamma(\mathrm{Disc}(X); I^{n-1}) \to \Gamma(\mathrm{Disc}(X); J)$, which is also the nth cohomology group of $\mathrm{Disc}(X)$ with coefficients in A in the sense of classical sheaf theory.

For further discussion of this point, we refer the reader to [41].

We are ready to define the cohomological dimension of an ∞-topos.

Definition 7.2.2.18. Let X be an ∞-topos. We will say that X has *cohomological dimension* $\leq n$ if, for any sheaf of abelian groups A on X, the cohomology group $\mathrm{H}^k(X, A)$ vanishes for $k > n$.

Remark 7.2.2.19. For small values of n, some authors prefer to require a stronger vanishing condition which also applies when A is a nonabelian coefficient system. The appropriate definition requires the vanishing of cohomology for coefficient systems which are defined only up to inner automorphisms, as in [31]. With the appropriate modifications, Theorem 7.2.2.29 below remains valid for $n < 2$.

The cohomological dimension of an ∞-topos X is closely related to the homotopy dimension of X. If X has homotopy dimension $\leq n$, then

$$\mathrm{H}^m(X; A) = \pi_0 \mathrm{Map}_X(1_X, K(A, m)) = *$$

for $m > n$ by Lemma 7.2.1.7, so that X is also of cohomological dimension $\leq n$. We will establish a partial converse to this result.

Definition 7.2.2.20. Let X be an ∞-topos. An *n-gerbe* on X is an object $X \in X$ which is n-connective and n-truncated.

Let X be an ∞-topos containing an n-gerbe X and let $f : X_{/X} \to X$ denote the associated geometric morphism. If X is equipped with a base point $p : 1_X \to X$, then X is canonically determined (as a pointed object) by $p^* \pi_n X$ by Proposition 7.2.2.12. We now wish to consider the case in which X is *not* pointed. If $n \geq 2$, then $\pi_n X$ can be regarded as an abelian group object in the topos $\mathrm{Disc}(X_{/X})$. Proposition 7.2.1.13 implies that $\pi_n X \simeq f^* A$, where A is a sheaf of abelian groups on X, which is determined up to canonical isomorphism. (In concrete terms, this boils down to the observation that the 1-connectivity of X allows us to extract higher homotopy groups without specifying a base point on X.) In this situation, we will say that X is *banded by A*.

Remark 7.2.2.21. For $n < 2$, the situation is more complicated. We refer the reader to [31] for a discussion.

Our next goal is to show that the cohomology groups of an ∞-topos $\mathfrak{X}$ can be interpreted as classifying equivalence classes of n-gerbes over $\mathfrak{X}$. Before we can prove this, we need to establish some terminology.

Notation 7.2.2.22. Let $\mathfrak{X}$ be an ∞-topos. We define a category $\mathrm{Band}(\mathfrak{X})$ as follows:

(1) The objects of $\mathrm{Band}(\mathfrak{X})$ are pairs (U, A), where U is an object of $\mathfrak{X}$ and A is an abelian group object of the homotopy category $\mathrm{Disc}(\mathfrak{X}_{/U})$.

(2) Morphisms from (U, A) to (U', A') are given by pairs (η, f), where $\eta \in \pi_0 \mathrm{Map}_{\mathfrak{X}}(U, U')$ and $f : A \to A'$ is a map which induces an isomorphism $A \simeq \eta^* A'$ of abelian group objects. Composition of morphisms is defined in the obvious way.

For $n \geq 2$, let $\mathrm{Gerb}_n(\mathfrak{X})$ denote the subcategory of $\mathrm{Fun}(\Delta^1, \mathfrak{X})$ spanned by those objects $f : X \to S$ which are n-gerbes in $\mathfrak{X}_{/S}$ and those morphisms which correspond to pullback diagrams

$$
\begin{array}{ccc}
X' & \longrightarrow & X \\
\downarrow{\scriptstyle f} & & \downarrow{\scriptstyle f} \\
S' & \longrightarrow & S.
\end{array}
$$

Remark 7.2.2.23. Since the class of morphisms $f : X \to S$ which belong to $\mathfrak{X}^{\Delta^1}$ is stable under pullback, we can apply Corollary 2.4.7.12 (which asserts that $p : \mathrm{Fun}(\Delta^1, \mathfrak{X}) \to \mathrm{Fun}(\{1\}, \mathfrak{X})$ is a Cartesian fibration), Lemma 6.1.1.1 (which characterizes the p-Cartesian morphisms of $\mathrm{Fun}(\Delta^1, \mathfrak{X})$), and Corollary 2.4.2.5 to deduce that the projection $\mathrm{Gerb}_n(\mathfrak{X}) \to \mathfrak{X}$ is a right fibration.

If $f : X \to U$ belongs to $\mathrm{Gerb}_n(\mathfrak{X})$, then there exists an abelian group object A of $\mathrm{Disc}(\mathfrak{X}_{/U})$ such that X is banded by A. The construction

$$(f : X \to U) \mapsto (U, A)$$

determines a functor

$$\chi : \mathrm{Gerb}_n(\mathfrak{X}) \to \mathrm{N}(\mathrm{Band}(\mathfrak{X})).$$

Let A be an abelian group object of $\mathrm{Disc}(\mathfrak{X})$. We let $\mathrm{Band}^A(\mathfrak{X})$ be the category whose objects are triples (X, A_X, ϕ), where $X \in h\mathfrak{X}$, A_X is an abelian group object of $\mathrm{Disc}(\mathfrak{X}_{/X})$, and ϕ is a map $A_X \to A$ which induces an isomorphism $A_X \simeq A \times X$ of abelian group objects of $\mathrm{Disc}(\mathfrak{X}_{/X})$. We have forgetful functors

$$\mathrm{Band}^A(\mathfrak{X}) \xrightarrow{\phi} \mathrm{Band}(\mathfrak{X}) \to h\mathfrak{X},$$

both of which are Grothendieck fibrations and whose composition is an equivalence of categories. We define $\mathrm{Gerb}_n^A(\mathfrak{X})$ by the following pullback diagram:

$$
\begin{array}{ccc}
\mathrm{Gerb}_n^A(\mathfrak{X}) & \longrightarrow & \mathrm{Gerb}_n(\mathfrak{X}) \\
\downarrow & & \downarrow{\scriptstyle \chi} \\
\mathrm{N}(\mathrm{Band}^A(\mathfrak{X})) & \longrightarrow & \mathrm{N}(\mathrm{Band}(\mathfrak{X})).
\end{array}
$$

Note that since ϕ is a Grothendieck fibration, $N\phi$ is a Cartesian fibration (Remark 2.4.2.2), so that the diagram above is homotopy Cartesian (Proposition 3.3.1.3). We will refer to $\mathrm{Gerb}_n^A(\mathcal{X})$ as the *sheaf of gerbes over $\mathcal{X}$ banded by A*.

More informally, an object of $\mathrm{Gerb}_n^A(\mathcal{X})$ is an n-gerbe $f : X \to U$ in $\mathcal{X}_{/U}$ *together with* an isomorphism $\phi_X : \pi_n X \simeq X \times A$ of abelian group objects of $\mathrm{Disc}(\mathcal{X}_{/X})$. Morphisms in Gerb_n^A are given by pullback squares

$$
\begin{array}{ccc}
X' & \xrightarrow{\ f\ } & X \\
\downarrow & & \downarrow \\
U' & \longrightarrow & U
\end{array}
$$

such that the associated diagram of abelian group objects of $\mathrm{Disc}(\mathcal{X}_{/X'})$

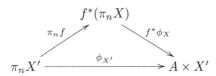

is commutative.

Lemma 7.2.2.24. *Let $\mathcal{X}$ be an ∞-topos, $n \geq 1$, and A an abelian group object in the topos $\mathrm{Disc}(\mathcal{X})$. Let X be an n-gerbe in $\mathcal{X}$ equipped with a fixed isomorphism $\phi : \pi_n X \simeq X \times A$ of abelian group objects of $\mathrm{Disc}(\mathcal{X}_{/X})$, and let $u : 1_{\mathcal{X}} \to K(A, n)$ be an Eilenberg-MacLane object of $\mathcal{X}$ classified by A. Let $\mathrm{Map}_{\mathcal{X}}^{\phi}(K(A, n), X)$ be the summand of $\mathrm{Map}_{\mathcal{X}}(K(A, n), X)$ corresponding to those maps $f : K(A, n) \to X$ for which the composition*

$$A \times K(A, n) \simeq \pi_n K(A, n) \to f^*(\pi_n X) \xrightarrow{f^*\phi} A \times K(A, n)$$

is the identity (in the category of abelian group objects of $\mathcal{X}_{/K(A,n)}$). Then composition with u induces a homotopy equivalence

$$\theta^{\phi} : \mathrm{Map}_{\mathcal{X}}^{\phi}(K(A, n), X) \to \mathrm{Map}_{\mathcal{X}}(1_{\mathcal{X}}, X).$$

Proof. Let $\theta : \mathrm{Map}_{\mathcal{X}}(K(A, n), X) \to \mathrm{Map}_{\mathcal{X}}(1_{\mathcal{X}}, X)$ and let $f : 1_{\mathcal{X}} \to X$ be any map (which we may identify with an Eilenberg-MacLane object of $\mathcal{X}$). The homotopy fiber of θ over the point represented by f can be identified with $\mathrm{Map}_{\mathcal{X}_{1_{\mathcal{X}}/}}(u, f)$. In view of the equivalence between $\mathcal{X}_{1_{\mathcal{X}}/}$ and $\mathcal{X}_*$, we can identify this mapping space with $\mathrm{Map}_{\mathcal{X}_*}(u, f)$. Applying Proposition 7.2.2.12, we deduce that the homotopy fiber of θ is equivalent to the (discrete) set of all endomorphisms $v : A \to A$ (in the category of group objects of $\mathrm{Disc}(\mathcal{X})$). We now observe that the homotopy fiber of θ^{ϕ} over f is a summand of the homotopy fiber of θ over f corresponding to those components for which $v = \mathrm{id}_A$. It follows that the homotopy fibers of θ^{ϕ} are contractible, so that θ^{ϕ} is a homotopy equivalence, as desired. $\square$

Lemma 7.2.2.25. *Let $\mathcal{X}$ be an ∞-topos, $n \geq 1$, and A an abelian group object of $\mathrm{Disc}(\mathcal{X})$. Let $f : K(A,n) \times X \to X$ be a trivial n-gerbe over X banded by A and $g : \widetilde{Y} \to Y$ be any n-gerbe over Y banded by A. Then there is a canonical homotopy equivalence*

$$\mathrm{Map}_{\mathrm{Gerb}_n^A}(f,g) \simeq \mathrm{Map}_{\mathcal{X}}(X, \widetilde{Y}).$$

Proof. Choose a morphism $\alpha : \mathrm{id}_X \to f$ in $\mathcal{X}_{/X}$ corresponding to a diagram

$$
\begin{array}{ccc}
X & \xrightarrow{\ s\ } & X \times K(A,n) \\
\downarrow & & \downarrow{\scriptstyle f} \\
X & \xrightarrow{\ \mathrm{id}_X\ } & X
\end{array}
$$

which exhibits f as an Eilenberg-MacLane object of $\mathcal{X}_{/X}$. We observe that evaluation at $\{0\} \subseteq \Delta^1$ induces a trivial fibration

$$\mathrm{Hom}_{\mathcal{X}^{\Delta^1}}^{\mathrm{L}}(\mathrm{id}_X, g) \to \mathrm{Hom}_{\mathcal{X}}^{\mathrm{L}}(X, \widetilde{Y}).$$

Consequently, we may identify $\mathrm{Map}_{\mathcal{X}}(X, \widetilde{Y})$ with the Kan complex

$$Z = \mathrm{Fun}(\Delta^1, \mathcal{X})_{\mathrm{id}_X \ /} \times_{\mathrm{Fun}(\Delta^1, \mathcal{X})} \{g\}.$$

Similarly, the trivial fibration $\mathrm{Fun}(\Delta^1, \mathcal{X})_{\alpha/} \to \mathrm{Fun}(\Delta^1, \mathcal{X})_{f/}$ allows us to identify $\mathrm{Map}_{\mathrm{Gerb}_n}(f,g)$ with the Kan complex

$$Z' = \mathrm{Fun}(\Delta^1, \mathcal{X})_{\alpha/} \times_{\mathrm{Fun}(\Delta^1, \mathcal{X})} \{g\}$$

and $\mathrm{Map}_{\mathrm{Gerb}_n}(f,g)$ with the summand Z'' of Z' corresponding to those maps which induce the identity isomorphism of $A \times (K(A,n) \times X)$ (in the category of group objects of $\mathrm{Disc}(\mathcal{X}_{/K(A,n) \times X})$). We now observe that evaluation at $\{1\} \subseteq \Delta^1$ gives a commutative diagram

$$
\begin{array}{ccccc}
Z'' & \longrightarrow & Z' & \longrightarrow & Z \\
& \searrow{\scriptstyle \psi''} & \downarrow{\scriptstyle \psi'} & & \downarrow{\scriptstyle \psi} \\
& & \mathcal{X}_{\mathrm{id}_X \ /} \times_{\mathcal{X}} \{Y\} & \longrightarrow & \mathcal{X}_{X/} \times_{\mathcal{X}} \{Y\},
\end{array}
$$

where the vertical maps are Kan fibrations. If we fix a pullback square

$$
\begin{array}{ccc}
\widetilde{X} & \longrightarrow & \widetilde{Y} \\
\downarrow{\scriptstyle g'} & & \downarrow \\
X & \xrightarrow{\ h\ } & Y,
\end{array}
$$

then we can identify the Kan complex $\psi^{-1}\{h\}$ with $\mathrm{Map}_{\mathcal{X}/x}(\mathrm{id}_X, g')$, the Kan complex $\psi'^{-1}\{s^0 h\}$ with $\mathrm{Map}_{\mathcal{X}/x}(X \times K(A,n), g')$, and the Kan complex $\psi'^{-1}\{s^0 h\}$ with the summand of $\mathrm{Map}_{\mathcal{X}/x}(X \times K(A,n), g')$ corresponding to those maps which induce the identity on $A \times (K(A,n) \times X)$ (in the category of group objects of $\mathrm{Disc}(\mathfrak{X}_{/K(A,n)\times X})$). Invoking Lemma 7.2.2.24 in the ∞-topos $\mathfrak{X}^{/X}$, we deduce that the map θ in the diagram

$$
\begin{array}{ccc}
Z'' & \xrightarrow{\ \ \theta\ \ } & Z \\
\downarrow{\scriptstyle \psi''} & & \downarrow{\scriptstyle \psi} \\
\mathfrak{X}_{\mathrm{id}_X\,/}\times_{\mathfrak{X}}\{Y\} & \longrightarrow & \mathfrak{X}_{X/}\times_{\mathfrak{X}}\{Y\}
\end{array}
$$

induces homotopy equivalences from the fibers of ψ'' to the fibers of ψ. Since the lower horizontal map is a trivial fibration of simplicial sets, we conclude that θ is itself a homotopy equivalence, as desired. $\square$

Theorem 7.2.2.26. *Let $\mathfrak{X}$ be an ∞-topos, $n \geq 1$, and A an abelian group object of $\mathrm{Disc}(\mathfrak{X})$. Then*

(1) *The composite map*
$$
\theta : \mathrm{Gerb}_n^A(\mathfrak{X}) \to \mathrm{Gerb}_n(\mathfrak{X}) \subseteq \mathrm{Fun}(\Delta^1, \mathfrak{X}) \to \mathrm{Fun}(\{1\}, \mathfrak{X}) \simeq \mathfrak{X}
$$
is a right fibration.

(2) *The right fibration θ is representable by an Eilenberg-MacLane object $K(A, n+1)$.*

Proof. For each object $X \in \mathfrak{X}$, we let A_X denote the projection $A \times X \to X$ viewed as an abelian group object of $\mathrm{Disc}(\mathfrak{X}_{/X})$. The functor

$$
\phi : \mathrm{Band}^A(\mathfrak{X}) \to \mathrm{Band}(\mathfrak{X})
$$

is a fibration in groupoids, so that $\mathrm{N}\phi$ is a right fibration (Proposition 2.1.1.3). The functor θ admits a factorization

$$
\mathrm{Gerb}_n^A(\mathfrak{X}) \xrightarrow{\theta'} \mathrm{Gerb}_n(\mathfrak{X}) \xrightarrow{\theta''} \mathfrak{X},
$$

where θ'' is a right fibration (Remark 7.2.2.23) and θ' is a pullback of $\mathrm{N}\phi$ and therefore also a right fibration. It follows that θ, being a composition of right fibrations, is a right fibration; this proves (1).

To prove (2), we consider an Eilenberg-MacLane object $u : 1_{\mathfrak{X}} \to K(A, n+1)$. Since $K(A, n+1)$ is $(n+1)$-truncated and $1_{\mathfrak{X}}$ is n-truncated (in fact, (-2)-truncated), Lemma 5.5.6.14 implies that u is n-truncated. The long exact sequence

$$
\cdots \to u^*\pi_{i+1}K(A, n+1) \to \pi_i u \to \pi_i(1_{\mathfrak{X}}) \to i^*\pi_i(K(A, n+1)) \to \cdots
$$

of Remark 6.5.1.5 shows that u is n-connective and provides an isomorphism $\phi : A \simeq \pi_n(u)$ in the category of group objects of $\mathrm{Disc}(\mathfrak{X})$, so that we may view the pair (u, ϕ) as an object of $\mathrm{Gerb}_n^A(\mathfrak{X})$. Since $1_{\mathfrak{X}}$ is a final object of $\mathfrak{X}$, Lemma 7.2.2.25 implies that (u, ϕ) is a final object of $\mathrm{Gerb}_n^A(\mathfrak{X})$, so that the right fibration θ is representable by $\theta(u, \phi) = K(A, n+1)$. $\square$

Corollary 7.2.2.27. *Let $\mathfrak{X}$ be an ∞-topos, let $n \geq 2$, and let A be an abelian group object of $\mathrm{Disc}(\mathfrak{X})$. There is a canonical bijection of $\mathrm{H}^{n+1}(\mathfrak{X}; A)$ with the set of equivalence classes of n-gerbes on $\mathfrak{X}$ banded by A.*

Remark 7.2.2.28. Under the correspondence of Proposition 7.2.2.27, an n-gerbe X on $\mathfrak{X}$ admits a global section $1_{\mathfrak{X}} \to X$ if and only if the associated cohomology class in $\mathrm{H}^{n+1}(\mathfrak{X}; A)$ vanishes.

Theorem 7.2.2.29. *Let $\mathfrak{X}$ be an ∞-topos and $n \geq 2$. Then $\mathfrak{X}$ has cohomological dimension $\leq n$ if and only if it satisfies the following condition: any n-connective truncated object of $\mathfrak{X}$ admits a global section.*

Proof. Suppose that $\mathfrak{X}$ has the property that every n-connective truncated object $X \in \mathfrak{X}$ admits a global section. As in the proof of Lemma 7.2.1.7, we deduce that for any $(n + 1)$-connective truncated object $X \in \mathfrak{X}$, the space of global sections $\mathrm{Map}_{\mathfrak{X}}(1, X)$ is connected. Let $k > n$ and let G be a sheaf of abelian groups on $\mathfrak{X}$. Then $K(G, k)$ is $(n + 1)$-connective, so that $\mathrm{H}^k(\mathfrak{X}, G) = *$. Thus $\mathfrak{X}$ has cohomological dimension $\leq n$.

For the converse, let us assume that $\mathfrak{X}$ has cohomological dimension $\leq n$ and let X denote an n-connective k-truncated object of $\mathfrak{X}$. We will show that X admits a global section by descending induction on k. If $k \leq n-1$, then X is a final object of $\mathfrak{X}$ and there is nothing to prove. In the general case, choose a truncation $X \to \tau_{\leq k-1}X$; we may assume by the inductive hypothesis that $\tau_{\leq k-1}X$ has a global section $s : 1 \to \tau_{\leq k-1}X$. Form a pullback square

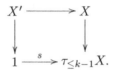

It now suffices to prove that X' has a global section. We note that X' is k-connective, where $k \geq n \geq 2$. It follows that X' is a k-gerbe on $\mathfrak{X}$; suppose it is banded by an abelian group object $A \in \mathrm{Disc}(\mathfrak{X})$. According to Corollary 7.2.2.27, X' is classified up to equivalence by an element in $\mathrm{H}^{k+1}(\mathfrak{X}, A)$, which vanishes by virtue of the fact that $k+1 > n$ and the cohomological dimension of $\mathfrak{X}$ is $\leq n$. Consequently, X' is equivalent to $K(A, k)$ and therefore admits a global section. $\square$

Corollary 7.2.2.30. *Let $\mathfrak{X}$ be an ∞-topos. If $\mathfrak{X}$ has homotopy dimension $\leq n$, then $\mathfrak{X}$ has cohomological dimension $\leq n$. The converse holds provided that $\mathfrak{X}$ has finite homotopy dimension and $n \geq 2$.*

Proof. Only the last claim requires proof. Suppose that $\mathfrak{X}$ has cohomological dimension $\leq n$ and homotopy dimension $\leq k$. We must show that every n-connective object X of $\mathfrak{X}$ has a global section. Choose a truncation $X \to \tau_{\leq k-1}X$. Then $\tau_{\leq k-1}X$ is truncated and n-connective, so it admits a global section by Theorem 7.2.2.29. Form a pullback square

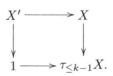

It now suffices to prove that X' has a global section. But X' is k-connective and therefore has a global section by virtue of the assumption that $\mathfrak{X}$ has homotopy dimension $\leq k$. $\qquad\square$

Warning 7.2.2.31. [Weiland] The converse to Corollary 7.2.2.30 is false if we do not assume that $\mathfrak{X}$ has finite homotopy dimension. To see this, we discuss the following example, which we learned from Ben Wieland. Let G denote the group $\mathbf{Z}_p$ of p-adic integers (viewed as a profinite group). Let $\mathcal{C}$ denote the category whose objects are the finite quotients $\{\mathbf{Z}_p/p^n\mathbf{Z}_p\}_{n\geq 0}$ and whose morphisms are given by G-equivariant maps. We regard $\mathcal{C}$ as endowed with a Grothendieck topology in which every nonempty sieve is a covering. The ∞-topos $\mathrm{Shv}(\mathrm{N}\,\mathcal{C})$ is 1-localic, and the underlying ordinary topos $\mathrm{h}\tau_{\leq 0}\,\mathrm{Shv}(\mathrm{N}\,\mathcal{C})$ can be identified with the category BG of continuous G-sets (that is, sets C equipped with an action of G such that the stabilizer of each element $x \in C$ is an open subgroup of G). Since the profinite group G has cohomology dimension 2 (see [69]), we deduce that $\mathfrak{X}$ is of cohomological dimension 2. However, we will show that $\mathfrak{X}$ is not hypercomplete and therefore cannot be of finite homotopy dimension.

Let K be a finite CW complex whose homotopy groups consist entirely of p-torsion (for example, we could take K to be a Moore space $M(\mathbf{Z}/p\mathbf{Z})$) and let $X = \mathrm{Sing}\,K \in \mathcal{S}$. Let $F : \mathrm{N}(\mathcal{C})^{op} \to \mathcal{S}$ denote the constant functor taking the value X. We claim that F belongs to $\mathrm{Shv}(\mathcal{C})$. Unwinding the definitions, we must show that for each $m \leq n$, the diagram F exhibits $F(\mathbf{Z}_p/p^m\mathbf{Z}_p)$ as equivalent to the homotopy invariants for the trivial action of $p^m\mathbf{Z}_p/p^n\mathbf{Z}_p$ on $F(\mathbf{Z}_p/p^n\mathbf{Z}_p)$. In other words, we must show that the diagonal embedding

$$\alpha : X \to \mathrm{Fun}(BH, X)$$

is a homotopy equivalence, where H denotes the quotient group $p^m\mathbf{Z}_p/p^n\mathbf{Z}_p$. Since both sides are p-adically complete, it will suffice to show that α is a p-adic homotopy equivalence, which follows from a suitable version of the Sullivan conjecture (see, for example, [67]).

We define another functor $F' : \mathrm{N}(\mathcal{C})^{op} \to \mathcal{S}$, which is obtained as the simplicial nerve of the functor described by the formula

$$\mathbf{Z}_p/p^n\mathbf{Z}_p \mapsto \mathrm{Sing}(K^{\mathbf{R}/p^n\mathbf{Z}}).$$

For $m \leq n$, the loop space $K^{\mathbf{R}/p^m\mathbf{Z}}$ can be identified with the homotopy fixed points of the (nontrivial) action of $H = p^m\mathbf{Z}_p/p^n\mathbf{Z}_p \simeq p^m\mathbf{Z}/p^n\mathbf{Z}$ on the loop space $K^{\mathbf{R}/p^n\mathbf{Z}}$: this follows from the observation that H acts freely on $\mathbf{R}/p^n\mathbf{Z}$ with quotient $\mathbf{R}/p^m\mathbf{Z}$. Consequently, F' belongs to $\mathrm{Shv}(\mathrm{N}(\mathcal{C}))$.

The inclusion of K into each loop space $K^{\mathbf{R}/p^n\mathbf{Z}}$ induces a morphism $\alpha : F \to F'$ in the ∞-topos $\mathrm{Shv}(\mathrm{N}(\mathcal{C}))$. Using the fact that the homotopy groups of K are p-torsion, we deduce that the morphism α is ∞-connective (this follows from the observation that the map

$$X \simeq \varinjlim F(\mathbf{Z}_p/p^n\mathbf{Z}_p) \to \varinjlim F'(\mathbf{Z}/p^n\mathbf{Z}_P)$$

is a homotopy equivalence). However, the morphism α is not an equivalence in $\mathrm{Shv}(\mathrm{N}(\mathcal{C}))$ unless K is essentially discrete. Consequently, $\mathrm{Shv}(\mathrm{N}(\mathcal{C}))$ is not hypercomplete and therefore cannot be of finite homotopy dimension.

In spite of Warning 7.2.2.31, many situations which guarantee that a topological space (or topos) X is of bounded cohomological dimension also guarantee that the associated ∞-topos is of bounded homotopy dimension. We will see some examples in the next two sections.

7.2.3 Covering Dimension

In this section, we will review the classical theory of covering dimension for paracompact spaces and then show that the covering dimension of a paracompact space X coincides with its homotopy dimension.

Definition 7.2.3.1. A paracompact topological space X has *covering dimension* $\leq n$ if the following condition is satisfied: for any open covering $\{U_\alpha\}$ of X, there exists an open refinement $\{V_\alpha\}$ of X such that each intersection $V_{\alpha_0} \cap \cdots \cap V_{\alpha_{n+1}} = \emptyset$ provided the α_i are pairwise distinct.

Remark 7.2.3.2. When X is paracompact, the condition of Definition 7.2.3.1 is equivalent to the (a priori weaker) requirement that such a refinement exist whenever $\{U_\alpha\}$ is a finite covering of X. This weaker condition gives a good notion whenever X is a normal topological space. Moreover, if X is normal, then the covering dimension of X (by this second definition) coincides with the covering dimension of the Stone-Čech compactification of X. Thus the dimension theory of normal spaces is controlled by the dimension theory of compact Hausdorff spaces.

Remark 7.2.3.3. Suppose that X is a compact Hausdorff space, which is written as a filtered inverse limit of compact Hausdorff spaces $\{X_\alpha\}$, each of which has dimension $\leq n$. Then X has dimension $\leq n$. Conversely, any compact Hausdorff space of dimension $\leq n$ can be written as a filtered inverse limit of finite simplicial complexes having dimension $\leq n$. Thus the dimension theory of compact Hausdorff spaces is controlled by the (completely straightforward) dimension theory of finite simplicial complexes.

Remark 7.2.3.4. There are other approaches to classical dimension theory. For example, a topological space X is said to have *small* (*large*) *inductive dimension* $\leq n$ if every point of X (every closed subset of X) has arbitrarily small open neighborhoods U such that ∂U has small inductive dimension $\leq n - 1$. These notions are well-behaved for separable metric spaces, where they coincide with the covering dimension (and with each other). In general, the covering dimension has better formal properties.

Our goal in this section is to prove that the covering dimension of a paracompact topological space X coincides with the homotopy dimension of $\mathrm{Shv}(X)$. First, we need a technical lemma.

Lemma 7.2.3.5. *Let X be a paracompact space, let $k \geq 0$, and let $\{U_\alpha\}_{\alpha \in A}$ be a covering of X. Suppose that for every $A_0 \subseteq A$ of size $k+1$, we are given a covering $\{V_\beta\}_{\beta \in B(A_0)}$ of the intersection $U_{A_0} = \bigcap_{\alpha \in A_0} U_\alpha$. Then there* *exists a covering* $\{W_\alpha\}_{\alpha \in \widetilde{A}}$ *of* X *and a map* $\pi : \widetilde{A} \to A$ *with the following properties:*

(1) *For* $\widetilde{\alpha} \in \widetilde{A}$ *with* $\pi(\widetilde{\alpha}) = \alpha$, *we have* $W_{\widetilde{\alpha}} \subseteq U_\alpha$.

(2) *Suppose that* $\widetilde{\alpha}_0, \cdots, \widetilde{\alpha}_k$ *is a collection of elements of* $\widetilde{A}$, *with* $\pi(\widetilde{\alpha}_i) = \alpha_i$. *Suppose further that* $A_0 = \{\alpha_0, \ldots, \alpha_k\}$ *has cardinality* $(k+1)$ *(in other words, the* α_i *are all disjoint from one another). Then there exists* $\beta \in B(A_0)$ *such that* $W_{\widetilde{\alpha}_0} \cap \ldots \cap W_{\widetilde{\alpha}_k} \subseteq V_\beta$.

Proof. Since X is paracompact, we can find a locally finite covering $\{U'_\alpha\}_{\alpha \in A}$ of X such that each closure $\overline{U'_\alpha}$ is contained in U_α. Let S denote the set of all subsets $A_0 \subseteq A$ having size $k+1$. For $A_0 \in S$, let $K(A_0) = \bigcap_{\alpha \in A_0} \overline{U_\alpha}$. Now set

$$\widetilde{A} = \{(\alpha, A_0, \beta) : \alpha \in A_0 \in S, \beta \in B(A_0)\} \cup A.$$

For $\widetilde{\alpha} = (\alpha, A_0, \beta) \in \widetilde{A}$, we set $\pi(\widetilde{\alpha}) = \alpha$ and

$$W_{\widetilde{\alpha}} = (U'_\alpha - \bigcup_{\alpha \in A'_0 \in S} K(A'_0)) \cup (V_\beta \cap U'_\alpha).$$

If $\alpha \in A \subseteq \widetilde{A}$, we let $\pi(\alpha) = \alpha$ and $W_\alpha = U'_\alpha - \bigcup_{\alpha \in A_0 \in S} K(A_0)$. The local finiteness of the cover $\{U'_\alpha\}$ ensures that each $W_{\widetilde{\alpha}}$ is an open set. It is now easy to check that the covering $\{W_{\widetilde{\alpha}}\}_{\widetilde{\alpha} \in \widetilde{A}}$ has the desired properties. $\square$

Theorem 7.2.3.6. *Let* X *be a paracompact topological space of covering dimension* $\leq n$. *Then the* ∞-*topos* $\mathrm{Shv}(X)$ *of sheaves on* X *has homotopy dimension* $\leq n$.

Proof. We make use of the results and notations of §7.1. Let $\mathcal{B}$ denote the collection of all open F_σ subsets of X and fix a linear ordering on $\mathcal{B}$. We may identify $\mathrm{Shv}(X)$ with the simplicial nerve of the category of all functors $F : \mathcal{B}^{op} \to \mathcal{K}\mathrm{an}$ which have the property that for any $\mathcal{U} \subseteq \mathcal{B}$ with $U = \bigcup_{V \in \mathcal{U}} V$, the natural map $F(U) \to F(\mathcal{U})$ is a homotopy equivalence.

Suppose that $F : \mathcal{B}^{op} \to \mathrm{Set}_\Delta$ represents an n-connective sheaf; we wish to show that the simplicial set $F(X)$ is nonempty. It suffices to prove that $F(\mathcal{U})$ is nonempty for some covering $\mathcal{U}$ of X; in other words, it suffices to produce a map $N_\mathcal{U} \to F$. The idea is that since X has finite covering dimension, we can choose arbitrarily fine covers $\mathcal{U}$ such that $N_\mathcal{U}$ is n-dimensional (in other words, equal to its n-skeleton).

For every simplicial set K, let $K^{(i)}$ denote the i-*skeleton* of K (the union of all nondegenerate simplices of K of dimension $\leq i$). If $G : \mathcal{B}^{op} \to \mathrm{Set}_\Delta$ is a simplicial presheaf, we let $G^{(i)}$ denote the simplicial presheaf given by the formula

$$G^{(i)}(U) = (G(U))^{(i)}.$$

We will prove the following statement by induction on i, $-1 \leq i \leq n$:

- There exists an open cover $\mathcal{U}_i \subseteq \mathcal{B}$ of X and a map $\eta_i : N_{\mathcal{U}_i}^{(i)} \to F$.

Assume that this statement holds for $i = n$. Passing to a refinement, we may assume that the cover $\mathcal{U}_n$ has the property that no more than $n + 1$ of its members intersect (this is the step where we shall use the assumption on the covering dimension of X). It follows that $N_{\mathcal{U}_n}^{(n)} = N_{\mathcal{U}_n}$, and the proof will be complete.

To begin the induction in the case $i = -1$, we let $\mathcal{U}_{-1} = \{X\}$; the (-1)-skeleton of $N_{\mathcal{U}_{-1}}$ is empty, so that η_{-1} exists (and is unique).

Now suppose that $\mathcal{U}_i = \{U_\alpha\}_{\alpha \in A}$ and η_i have been constructed, $i < n$. Let $A_0 \subseteq A$ have cardinality $i + 2$ and set $U(A_0) = \bigcap_{\alpha \in A_0} U_\alpha$; then A_0 determines an n-simplex of $N_{\mathcal{U}_i}(U(A_0))$, so that η_i restricts to give a map

$$\eta_{i,A_0} : \partial \Delta^{i+1} \to F(U(A_0)).$$

By assumption, F is n-connective; it follows that there is an open covering

$$\{V_\beta\}_{\beta \in B(A_0)}$$

of $U(A_0)$ such that for each V_β there is a commutative diagram

$$
\begin{array}{ccc}
\partial \Delta^{i+1} & \longrightarrow & F(U(A_0)) \\
\uparrow & & \downarrow \\
\Delta^{i+1} & \longrightarrow & F(V_\beta).
\end{array}
$$

We apply Lemma 7.2.3.5 to this data to obtain an new open cover $\mathcal{U}_{i+1} = \{W_{\widetilde{\alpha}}\}_{\widetilde{\alpha} \in \widetilde{A}}$ which refines $\{U_\alpha\}_{\alpha \in A}$. Refining the cover further if necessary, we may assume that each of its members belongs to $\mathcal{B}$. By functoriality, we obtain a map

$$N_{\mathcal{U}_{i+1}}^{(i)} \to F.$$

To complete the proof, it will suffice to extend f to the $(i + 1)$-skeleton of the nerve of $\{W_\alpha\}_{\alpha \in \widetilde{A}}$. Let $\widetilde{A}_0 \subseteq \widetilde{A}$ have cardinality $i + 2$ and let $W(\widetilde{A}_0) = \bigcap_{\widetilde{\alpha} \in \widetilde{A}_0} W_{\widetilde{\alpha}}$; then we must solve a lifting problem

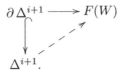

Let $\pi : \widetilde{A} \to A$ denote the map of Lemma 7.2.3.5. If $A_0 = \pi(\widetilde{A}_0)$ has cardinality smaller than $i + 2$, then there is a canonical extension given by applying π and using η_i. Otherwise, Lemma 7.2.3.5 guarantees that $W(\widetilde{A}_0) \subseteq V_\beta$ for some $\beta \in B(A_0)$, so that the desired extension exists by construction. $\square$

Corollary 7.2.3.7. *Let X be a paracompact topological space. The following conditions are equivalent:*

(1) *The covering dimension of X is $\leq n$.*

(2) *The homotopy dimension of* $\mathrm{Shv}(X)$ *is* $\leq n$.

(3) *For every closed subset* $A \subseteq X$, *every* $m \geq n$, *and every continuous map* $f_0 : A \to S^m$, *there exists* $f : X \to S^m$ *extending* f_0.

Proof. The implication $(1) \Rightarrow (2)$ is Theorem 7.2.3.6. The equivalence $(1) \Leftrightarrow (3)$ follows from classical dimension theory (see, for example, [27]). We will complete the proof by showing that $(2) \Rightarrow (3)$. Let A be a closed subset of X, let $m \geq n$, and let $f_0 : A \to S^m$ a continuous map. Let $\mathcal{B}$ be the collection of all open F_σ subsets of X. We define a simplicial presheaf $F : \mathcal{B} \to \mathcal{K}$an, so that an n-simplex of $F(U)$ is a map f rendering the diagram

$$
\begin{array}{ccc}
(U \cap A) \times |\Delta^n| & \longrightarrow & A \\
\downarrow & & \downarrow {\scriptstyle f_0} \\
U \times |\Delta^n| & \xrightarrow{\ f\ } & S^m
\end{array}
$$

commutative. To prove (3), it will suffice to show that $F(X)$ is nonempty. By virtue of the assumption that $\mathrm{Shv}(X)$ has homotopy dimension $\leq n$, it will suffice to show that $\mathcal{F}$ is an n-connective sheaf on X.

We first show that F is a sheaf. Choose a linear ordering on $\mathcal{B}$. We must show that for every open covering $\mathcal{U}$ of $U \in \mathcal{B}$, the natural map $\mathcal{F}(U) \to \mathcal{F}(\mathcal{U})$ is a homotopy equivalence. The proof is similar to that of Proposition 7.1.3.14. Let $\pi : |N_\mathcal{U}|_X \to U$ be the projection; then we may identify $F(\mathcal{U})$ with the simplicial set parametrizing continuous maps $|N_\mathcal{U}|_X \to S^m$, whose restriction to $\pi^{-1}(A)$ is given by f_0. The desired equivalence now follows from the fact that $|N_\mathcal{U}|_X$ is fiberwise homotopy equivalent to U (Lemma 7.1.3.13).

Now we claim that $\mathcal{F}$ is n-connective as an object of $\mathrm{Shv}(X)$. In other words, we must show that for any $U \in \mathcal{B}$, any $k \leq n$, and any map $g : \partial \Delta^k \to F(U)$, there is an open covering $\{U_\alpha\}$ of U and a family of commutative diagrams

$$
\begin{array}{ccc}
\partial \Delta^k & \xrightarrow{\ g\ } & F(U) \\
\uparrow & & \downarrow \\
\Delta^k & \xrightarrow{\ g_\alpha\ } & F(U_\alpha).
\end{array}
$$

We may identify g with a continuous map

$$
g : S^{k-1} \times U \to S^m
$$

such that $g(z, a) = f_0(a)$ for $a \in A$. Choose a point $x \in U$. Consider the map $g|S^{k-1} \times \{x\}$. Since $k - 1 < n \leq m$, this map is nullhomotopic; therefore it admits an extension $g'_x : D^k \times \{x\} \to S^m$. Moreover, if $x \in A$, then we may choose g'_x to be the constant map with value $f_0(x)$. Amalgamating g, g'_x, and f_0, we obtain a continuous map

$$
g'_0 : (S^{k-1} \times U) \cup (D^k \times (A \cup \{x\})) \to S^m.
$$

Since $(S^{k-1} \times U) \cup (D^k \times (A \cup \{x\}))$ is a closed subset of the paracompact space $U \times D^k$ and the sphere S^m is an absolute neighborhood retract, the map g'_0 extends continuously to a map $g'' : W \to S^m$, where W is an open neighborhood of $(S^{k-1} \times U) \cup (D^k \times (A \cup \{x\}))$ in $U \times D^k$. The compactness of D^k implies that W contains $D^k \times U_x$, where $U_x \subseteq U$ is an open neighborhood of x. Shrinking U_x if necessary, we may suppose that U_x belongs to $\mathcal{B}$; these open sets U_x form an open cover of U, with the required extension $\Delta^k \to F(U_x)$ supplied by the map $g'' | D^k \times U_x$. $\qquad\square$

7.2.4 Heyting Dimension

For the purposes of studying paracompact topological spaces, Definition 7.2.3.1 gives a perfectly adequate theory of dimension. However, there are other situations in which Definition 7.2.3.1 is not really appropriate. For example, in algebraic geometry one often considers the Zariski topology on an algebraic variety X. This topology is generally not Hausdorff and is typically of infinite covering dimension. In this setting, there is a better dimension theory: the theory of Krull dimension. In this section, we will introduce a mild generalization of the theory of Krull dimension, which we will call the *Heyting dimension* of a topological space X. We will then study the relationship between the Heyting dimension of X and the homotopy dimension of the associated ∞-topos $\mathrm{Shv}(X)$.

Recall that a topological space X is said to be *Noetherian* if the collection of closed subsets of X satisfies the descending chain condition. A closed subset $K \subseteq X$ is said to be *irreducible* if it cannot be written as a finite union of proper closed subsets of K (in particular, the empty set is *not* irreducible since it can be written as an empty union). The collection of irreducible closed subsets of X forms a well-founded partially ordered set, therefore it has a unique ordinal rank function rk, which may be characterized as follows:

- If K is an irreducible closed subset of X, then $\mathrm{rk}(K)$ is the smallest ordinal which is larger than $\mathrm{rk}(K')$ for all proper irreducible closed subsets $K' \subset K$.

We call $\mathrm{rk}(K)$ the *Krull dimension* of K; the *Krull dimension* of X is the supremum of $\mathrm{rk}(K)$, as K ranges over all irreducible closed subsets of X.

We next introduce a generalization of the Krull dimension to a suitable class of non-Noetherian spaces. We shall say that a topological space X is a *Heyting space* if satisfies the following conditions:

(1) The compact open subsets of X form a basis for the topology of X.

(2) A finite intersection of compact open subsets of X is compact (in particular, X is compact).

(3) If U and V are compact open subsets of X, then the interior of $U \cup (X - V)$ is compact.

Remark 7.2.4.1. Recall that a *Heyting algebra* is a distributive lattice L with the property that for any $x, y \in L$, there exists a maximal element z with the property that $x \wedge z \subseteq y$. It follows immediately from our definition that the lattice of compact open subsets of a Heyting space forms a Heyting algebra. Conversely, given any Heyting algebra, one may form its spectrum, which is a Heyting space. This sets up a duality between the category of *sober* Heyting spaces (Heyting spaces in which every irreducible closed subset has a unique generic point) and the category of Heyting algebras. This duality is a special case of a more general duality between coherent topological spaces and distributive lattices. We refer the reader to [42] for further details.

Remark 7.2.4.2. Suppose that X is a Noetherian topological space. Then X is a Heyting space since every open subset of X is compact.

Remark 7.2.4.3. If X is a Heyting space and $U \subseteq X$ is a compact open subset, then X and $X - U$ are also Heyting spaces. In this case, we say that $X - U$ is a *cocompact* closed subset of X.

We next define the dimension of a Heyting space. The definition is recursive. Let α be an ordinal. A Heyting space X has *Heyting dimension* $\leq \alpha$ if and only if, for any compact open subset $U \subseteq X$, the boundary of U has Heyting dimension $< \alpha$ (we note that the boundary of U is also a Heyting space); a Heyting space has *Heyting dimension* < 0 if and only if it is empty.

Remark 7.2.4.4. A Heyting space has dimension ≤ 0 if and only if it is Hausdorff. The Heyting spaces of dimension ≤ 0 are precisely the compact totally disconnected Hausdorff spaces. In particular, they are also paracompact spaces, and their Heyting dimension coincides with their covering dimension.

Proposition 7.2.4.5. (1) *Let X be a Heyting space of dimension $\leq \alpha$. Then for any compact open subset $U \subseteq X$, both U and $X - U$ have Heyting dimension $\leq \alpha$.*

(2) *Let X be a Heyting space which is a union of finitely many compact open subsets U_α of dimension $\leq \alpha$. Then X has dimension $\leq \alpha$.*

(3) *Let X be a Heyting space which is a union of finitely many cocompact closed subsets K_α of Heyting dimension $\leq \alpha$. Then X has Heyting dimension $\leq \alpha$.*

Proof. All three assertions are proven by induction on α. The first two are easy, so we restrict our attention to (3). Let U be a compact open subset of X having boundary B. Then $U \cap K_\alpha$ is a compact open subset of K_α, so that the boundary B_α of $U \cap K_\alpha$ in K_α has dimension $\leq \alpha$. We see immediately that $B_\alpha \subseteq B \cap K_\alpha$, so that $\bigcup B_\alpha \subseteq B$. Conversely, if $b \notin \bigcup B_\alpha$, then for every β such that $b \in K_\beta$, there exists a neighborhood V_β containing b such that $V_\beta \cap K_\beta \cap U = \emptyset$. Let V be the intersection of the V_β and let $W = V - \bigcup_{b \notin K_\gamma} K_\gamma$. Then by construction, $b \in W$ and $W \cap U = \emptyset$, so that $b \in B$. Consequently, $B = \bigcup B_\alpha$. Each B_α is closed in K_α, thus in X and also in B. The hypothesis implies that B_α has dimension $< \alpha$. Thus the inductive hypothesis guarantees that B has dimension $< \alpha$, as desired. $\square$

Remark 7.2.4.6. It is not necessarily true that a Heyting space which is a union of finitely many *locally closed* subsets of dimension $\leq \alpha$ is also of dimension $\leq \alpha$. For example, a topological space with 2 points and a nondiscrete nontrivial topology has Heyting dimension 1 but is a union of two locally closed subsets of Heyting dimension 0.

Proposition 7.2.4.7. *If X is a Noetherian topological space, then the Krull dimension of X coincides with the Heyting dimension of X.*

Proof. We first prove, by induction on α, that if the Krull dimension of a Noetherian space X is $\leq \alpha$, then the Heyting dimension of X is $\leq \alpha$. Since X is Noetherian, it is a union of finitely many closed irreducible subspaces, each of which automatically has Krull dimension $\leq \alpha$. Using Proposition 7.2.4.5, we may reduce to the case where X is irreducible. Consider any open subset $U \subseteq X$ and let Y be its boundary. We must show that Y has Heyting dimension $\leq \alpha$. Using Proposition 7.2.4.5 again, it suffices to prove this for each irreducible component of Y. Now we simply apply the inductive hypothesis and the definition of the Krull dimension.

For the reverse inequality, we again use induction on α. Assume that X has Heyting dimension $\leq \alpha$. To show that X has Krull dimension $\leq \alpha$, we must show that every irreducible closed subset of X has Krull dimension $\leq \alpha$. Without loss of generality, we may assume that X is irreducible. Now, to show that X has Krull dimension $\leq \alpha$, it will suffice to show that any *proper* closed subset $K \subseteq X$ has Krull dimension $< \alpha$. By the inductive hypothesis, it will suffice to show that K has Heyting dimension $< \alpha$. By the definition of the Heyting dimension, it will suffice to show that K is the boundary of $X - K$. In other words, we must show that $X - K$ is dense in X. This follows immediately from the irreducibility of X. $\square$

We now prepare the way for our vanishing theorem. First, we introduce a modified notion of connectivity:

Definition 7.2.4.8. Let X be a Heyting space and k any integer. Let $\mathcal{F} \in \operatorname{Shv}(V)$ be a sheaf of spaces on a compact open subset $V \subseteq X$. We will say that $\mathcal{F}$ is *strongly k-connective* if the following condition is satisfied: for every compact open subset $U \subseteq V$ and every map $\zeta : \partial \Delta^m \to \mathcal{F}(U)$, there exists a cocompact closed subset $K \subseteq U$ such that $\overline{K} \subseteq X$ has Heyting dimension $< m - k$, an open cover $\{U_\alpha\}$ of $U - K$, and a collection of commutative diagrams

$$
\begin{array}{ccc}
\partial \Delta^m & \xrightarrow{\ \zeta\ } & \mathcal{F}(U) \\
\uparrow & & \downarrow \\
\Delta^m & \xrightarrow{\ \eta_\alpha\ } & \mathcal{F}(U_\alpha).
\end{array}
$$

Remark 7.2.4.9. There is a slight risk of confusion with the terminology of Definition 7.2.4.8. The condition that a sheaf $\mathcal{F}$ on $V \subseteq X$ be strongly k-connective depends not only on V and $\mathcal{F}$ but also on X: this is because the Heyting dimension of a cocompact closed subset $K \subseteq U$ can increase when we take its closure $\overline{K}$ in X.

Remark 7.2.4.10. Strong k-connectivity is an unstable analogue of the connectivity conditions on complexes of sheaves associated to the dual of the standard perversity. For a discussion of perverse sheaves in the abelian context, we refer the reader to [6].

Remark 7.2.4.11. It follows easily from the definition that a strongly k-connective sheaf $\mathcal{F}$ on $V \subseteq X$ is k-connective. Conversely, suppose that X has Heyting dimension $\leq n$ and that $\mathcal{F}$ is k-connective; then $\mathcal{F}$ is strongly $(k - n)$-connective (if $\partial \Delta^m \to \mathcal{F}(U)$ is any map, then we may take $K = U$ for $m > n$ and $K = \emptyset$ for $m \leq n$).

The strong k-connectivity of a sheaf $\mathcal{F}$ is by definition a local property. The key to our vanishing result is that this is equivalent to an apparently stronger *global* property.

Lemma 7.2.4.12. *Let X be a Heyting space, let V be a compact open subset of X, and $\mathcal{F} : \mathcal{U}(V)^{op} \to \mathcal{K}an$ a strongly k-connective sheaf on V. Let $A \subseteq B$ be an inclusion of finite simplicial sets of dimension $\leq m$, let $U \subseteq V$, and let $\zeta : A \to \mathcal{F}(U)$ be a map of simplicial sets.*

There exists a cocompact closed subset $K \subseteq U$ whose closure $\overline{K} \subseteq X$ has Heyting dimension $< m - 1 - k$, an open covering $\{U_\alpha\}$ of $U - K$, and a collection of commutative diagrams

$$
\begin{array}{ccc}
A & \xrightarrow{\;\zeta\;} & \mathcal{F}(U) \\
\downarrow & & \downarrow \\
B & \xrightarrow{\;\eta_\alpha\;} & \mathcal{F}(U_\alpha).
\end{array}
$$

Proof. Induct on the number of simplices of B which do not belong to A, and invoke Definition 7.2.4.8. $\qquad\square$

Lemma 7.2.4.13. *Let X be a Heyting space, V a compact open subset of X, let $\mathcal{F} : \mathcal{U}(V)^{op} \to \mathcal{K}an$ be a sheaf on X, let $\eta : \partial \Delta^m \to \mathcal{F}(V)$ be a map, and form a pullback square*

$$
\begin{array}{ccc}
\mathcal{F}' & \longrightarrow & \mathcal{F}^{\Delta^m} \\
\downarrow & & \downarrow \\
* & \xrightarrow{\;\eta\;} & \mathcal{F}^{\partial \Delta^m}.
\end{array}
$$

If $\mathcal{F}$ is strongly k-connective, then $\mathcal{F}'$ is strongly $(k - m)$-connective.

Proof. Unwinding the definitions, we must show that for every compact $U \subset V$ and every map

$$\zeta : (\partial \Delta^m \times \Delta^n) \coprod_{\partial \Delta^m \times \partial \Delta^n} (\Delta^m \times \partial \Delta^n) \to \mathcal{F}(U)$$

whose restriction $\zeta | \partial \Delta^m \times \Delta^n$ is given by η, there exists a cocompact closed subset $K \subseteq U$ such that $\overline{K} \subseteq X$ has Heyting dimension $< n + m - k$, an open covering $\{U_\alpha\}$ of $U - K$, and a collection of maps

$$\zeta_\alpha : \Delta^m \times \Delta^n \to \mathcal{F}(U_\alpha)$$

which extend ζ. This follows immediately from Lemma 7.2.4.12. $\qquad \square$

Theorem 7.2.4.14. *Let X be a Heyting space of dimension $\leq n$, let $W \subseteq X$ be a compact open set, and let $\mathcal{F} \in \mathrm{Shv}(W)$. The following conditions are equivalent:*

(1) *For any compact open sets $U \subseteq V \subseteq W$ and any commutative diagram*

$$\begin{array}{ccc} \partial \Delta^m & \xrightarrow{\zeta} & \mathcal{F}(V) \\ \uparrow & & \downarrow \\ \Delta^m & \xrightarrow{\eta} & \mathcal{F}(U), \end{array}$$

there exists a cocompact closed subset $K \subseteq V - U$ such that $\overline{K} \subseteq X$ has dimension $< m - k$ and a commutative diagram

$$\begin{array}{ccc} \partial \Delta^m & \xrightarrow{\zeta} & \mathcal{F}(V) \\ \uparrow & & \downarrow \\ \Delta^m & \xrightarrow{\eta'} & \mathcal{F}(V - K) \end{array}$$

such that the composition $\Delta^m \xrightarrow{\eta'} \mathcal{F}(V - K) \to \mathcal{F}(U)$ is homotopic to η relative to $\partial \Delta^m$.

(2) *For any compact open sets $V \subseteq W$ and any map $\zeta : \partial \Delta^m \to \mathcal{F}(V)$, there exists a commutative diagram*

$$\begin{array}{ccc} \partial \Delta^m & \xrightarrow{\zeta} & \mathcal{F}(V) \\ \uparrow & & \downarrow \\ \Delta^m & \xrightarrow{\eta'} & \mathcal{F}(V - K), \end{array}$$

where $K \subseteq V$ is a cocompact closed subset and $\overline{K} \subseteq X$ has dimension $< m - k$.

(3) *The sheaf $\mathcal{F}$ is strongly k-connective.*

Proof. It is clear that (1) implies (2) (take U to be empty) and that (2) implies (3) (by definition). We must show that (3) implies (1). So let $\mathcal{F}$ be a strongly k-connective sheaf on W and

$$
\begin{array}{ccc}
\partial\Delta^m & \xrightarrow{\;\zeta\;} & \mathcal{F}(V) \\
\downarrow & & \downarrow \\
\Delta^m & \xrightarrow{\;\eta\;} & \mathcal{F}(U)
\end{array}
$$

a commutative diagram as above. Without loss of generality, we may replace W by V and $\mathcal{F}$ by $\mathcal{F}|V$.

We may identify $\mathcal{F}$ with a functor from $\mathcal{U}(V)^{op}$ into the category $\mathcal{K}$an of Kan complexes. Form a pullback square

$$
\begin{array}{ccc}
\mathcal{F}' & \longrightarrow & \mathcal{F}^{\Delta^m} \\
\downarrow & & \downarrow \\
* & \xrightarrow{\;\zeta\;} & \mathcal{F}^{\partial\Delta^m}
\end{array}
$$

in $\mathrm{Set}_{\Delta}^{\mathcal{U}(V)^{op}}$. The right vertical map is a projective fibration, so that the diagram is homotopy Cartesian (with respect to the projective model structure). It follows that $\mathcal{F}'$ is also a sheaf on V, which is strongly $(k-m)$-connective by Lemma 7.2.4.13. Replacing $\mathcal{F}$ by $\mathcal{F}'$, we may reduce to the case $m = 0$.

The proof now proceeds by induction on k. For our base case, we take $k = -n - 1$, so that there is no connectivity assumption on the stack $\mathcal{F}$. We are then free to choose $K = V - U$ (it is clear that $\overline{K}$ has dimension $\leq n$).

Now suppose that the theorem is known for strongly $(k-1)$-connective stacks on any compact open subset of X; we must show that for any strongly k-connective $\mathcal{F}$ on V and any $\eta \in \mathcal{F}(U)$, there exists a closed subset $K \subseteq V - U$ such that $\overline{K} \subseteq X$ has Heyting dimension $< -k$ and a point $\eta' \in \mathcal{F}(V-K)$ whose restriction to U lies in the same component of $\mathcal{F}(U)$ as η.

Since $\mathcal{F}$ is strongly k-connective, we deduce that there exists an open cover $\{V_\alpha\}$ of some open subset $V - K_0$, where K_0 has dimension $< -k$ in X, together with points $\psi_\alpha \in \mathcal{F}(V_\alpha)$. Adjoining the open set U and the point η if necessary, we may suppose that $K_0 \cap U = \emptyset$. Replacing V by $V - K_0$, we may reduce to the case $K_0 = \emptyset$.

Since V is compact, we may assume that there exist only finitely many indices α. Proceeding by induction on the number of indices, we may reduce to the case where $V = U \cup V_\alpha$ for some α. Let η' and ψ' denote the images of η and ψ in $U \cap V_\alpha$ and form a pullback diagram

$$
\begin{array}{ccc}
\mathcal{F}' & \longrightarrow & (\mathcal{F}|(U \cap V_\alpha))^{\Delta^1} \\
\downarrow & & \downarrow \\
* & \xrightarrow{(\eta',\psi')} & (\mathcal{F}|(U \cap V_\alpha))^{\partial\Delta^1}.
\end{array}
$$

Again, this diagram is a homotopy pullback, so that $\mathcal{F}'$ is a sheaf on $U \cap V_\alpha$ which is strongly $(k-1)$-connective by Lemma 7.2.4.13. According to the inductive hypothesis, there exists a closed subset $K \subset U \cap V_\alpha$ such that $\overline{K} \subseteq X$ has dimension $< -k + 1$, such that the images of ψ_α and η belong to the same component of $\mathcal{F}((U \cap V_\alpha) - K)$. Since $\overline{K}$ has dimension $< -k + 1$ in X, the boundary ∂K of K has codimension $< -k$ in X. Let $V' = V_\alpha - (V_\alpha \cap \overline{K})$. Since $\mathcal{F}$ is a sheaf, we have a homotopy pullback diagram

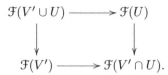

We observe that there is a path joining the images of η and ψ_α in $\mathcal{F}(V' \cap U) = \mathcal{F}((U \cap V_\alpha) - K)$, so that there is a vertex $\widetilde{\eta} \in \mathcal{F}(V' \cup U)$ whose image in $\mathcal{F}(U)$ lies in the same component as η. We now observe that $V' \cup U = V - (V \cap \partial K)$ and that $\overline{V \cap \partial K}$ is contained in ∂K and therefore has Heyting dimension $\leq -k$. □

Corollary 7.2.4.15. *Let $\pi : X \to Y$ be a continuous map between Heyting spaces of finite dimension. Suppose that π has the property that for any cocompact closed subset $K \subseteq X$ of dimension $\leq n$, $\pi(K)$ is contained in a cocompact closed subset of dimension $\leq n$. Then the functor $\pi_* : \mathrm{Shv}(X) \to \mathrm{Shv}(Y)$ carries strongly k-connective sheaves on X to strongly k-connective sheaves on Y.*

Proof. This is clear from the characterization (2) of Theorem 7.2.4.14. □

Corollary 7.2.4.16. *Let X be a Heyting space of finite Heyting dimension and let $\mathcal{F}$ be a strongly k-connective sheaf on X. Then $\mathcal{F}(X)$ is k-connective.*

Proof. Apply Corollary 7.2.4.15 in the case where Y is a point. □

Corollary 7.2.4.17. *Let X be a Heyting space of Heyting dimension $\leq n$ and let $\mathcal{F}$ be an n-connective sheaf on X. Then for any compact open $U \subseteq X$, the map $\pi_0 \mathcal{F}(X) \to \pi_0 \mathcal{F}(U)$ is surjective. In particular, $\mathrm{Shv}(X)$ has homotopy dimension $\leq n$.*

Proof. Suppose first that (1) is satisfied. Let $\mathcal{F}$ be an n-connective sheaf on X. Then $\mathcal{F}$ is strongly 0-connective; by characterization (2) of Theorem 7.2.4.14, we deduce that $\mathcal{F}(X) \to \mathcal{F}(U)$ is surjective. The last claim follows by taking $U = \emptyset$. □

Remark 7.2.4.18. Let X be a Heyting space of Heyting dimension $\leq n$. Then any compact open subset of X also has Heyting dimension $\leq n$. It follows that $\mathrm{Shv}(X)$ is locally of homotopy dimension $\leq n$ and therefore hypercomplete by Corollary 7.2.1.12.

Remark 7.2.4.19. It is not necessarily true that a Heyting space X such that $\mathrm{Shv}(X)$ has homotopy dimension $\leq n$ is itself of Heyting dimension $\leq n$. For example, if X is the Zariski spectrum of a discrete valuation ring (that is, a 2-point space with a nontrivial topology), then X has homotopy dimension zero (see Example 7.2.1.3).

In particular, we obtain Grothendieck's vanishing theorem (see [34] for the original, quite different, proof):

Corollary 7.2.4.20. *Let X be a Noetherian topological space of Krull dimension $\leq n$. Then X has cohomological dimension $\leq n$.*

Proof. Combine Proposition 7.2.4.7 with Corollaries 7.2.4.17 and 7.2.2.30. □

Example 7.2.4.21. Let V be a real algebraic variety (defined over the real numbers, say). Then the lattice of open subsets of V that can be defined by polynomial equations and inequalities is a Heyting algebra, and the spectrum of this Heyting algebra is a Heyting space X having dimension at most equal to the dimension of V. The results of this section therefore apply to X.

More generally, let T be an o-minimal theory (see for example [80]) and let S_n denote the set of complete n-types of T. We endow S_n with the topology generated by the sets $U_\phi = \{p : \phi \in p\}$, where ϕ ranges over formulas with n free variables such that the openness of the set of points satisfying ϕ is provable in T. Then S_n is a Heyting space of Heyting dimension $\leq n$.

Remark 7.2.4.22. The methods of this section can be adapted to slightly more general situations, such as the Nisnevich topology on a Noetherian scheme of finite Krull dimension. It follows that the ∞-topoi associated to such sites have (locally) finite homotopy dimension and are therefore hypercomplete. We will discuss this matter in more detail in [50].

7.3 THE PROPER BASE CHANGE THEOREM

Let

$$
\begin{array}{ccc}
X' & \xrightarrow{q'} & X \\
\downarrow{\scriptstyle p'} & & \downarrow{\scriptstyle p} \\
Y' & \xrightarrow{q} & Y
\end{array}
$$

be a pullback diagram in the category of locally compact Hausdorff spaces. One has a natural isomorphism of pushforward functors

$$q_* p'_* \simeq p_* q'_*$$

from the category of sheaves of sets on Y to the category of sheaves of sets on X'. This isomorphism induces a natural transformation

$$\eta : q^* p_* \to p'_* q'^*.$$

If p (and therefore also p') is a proper map, then η is an isomorphism: this is a simple version of the classical *proper base change theorem*.

The purpose of this section is to generalize the above result, allowing sheaves which take values in the ∞-category $\mathcal{S}$ of spaces rather than in the ordinary category of sets. Our generalization can be viewed as a proper base change theorem for nonabelian cohomology.

We will begin in §7.3.1 by defining the notion of a *proper morphism* of ∞-topoi. Roughly speaking, a geometric morphism $\pi_* : \mathfrak{X} \to \mathfrak{Y}$ of ∞-topoi is proper if and only if it satisfies the conclusion of the proper base change theorem. Using this language, our job is to prove that a proper map of topological spaces $p : X \to Y$ induces a proper morphism $p_* : \mathrm{Shv}(X) \to \mathrm{Shv}(Y)$ of ∞-topoi. We will outline the proof of this result in §7.3.1 by reducing to two special cases: the case where p is a closed embedding and the case where Y is a point. We will treat the first case in §7.3.2, after introducing a general theory of *closed immersions* of ∞-topoi. This allows us to reduce to the case where Y is a point and X is a compact Hausdorff space. Our approach is now in two parts:

(1) In §7.3.3, we will show that we can identify the ∞-category $\mathrm{Shv}(X') = \mathrm{Shv}(X \times Y')$ with an ∞-category of sheaves on X taking values in $\mathrm{Shv}(Y')$.

(2) In §7.3.4, we give an analysis of the category of sheaves on a compact Hausdorff space X taking values in a general ∞-category $\mathcal{C}$. Combining this analysis with (1), we will deduce the desired base change theorem.

The techniques used in §7.3.4 to analyze $\mathrm{Shv}(X)$ can also be applied in the (easier) setting of coherent topological spaces, as we explain in §7.3.5. Finally, we conclude in §7.3.6 by reformulating the classical theory of *cell-like* maps in the language of ∞-topoi.

7.3.1 Proper Maps of ∞-Topoi

In this section, we introduce the notion of a *proper* geometric morphism between ∞-topoi. Here we follow the ideas of [58] and turn the conclusion of the proper base change theorem into a definition. First, we require a bit of terminology.

Suppose we are given a diagram of categories and functors

$$
\begin{array}{ccc}
\mathcal{C}' & \xrightarrow{\;q'_*\;} & \mathcal{D}' \\
{\scriptstyle p'_*}\downarrow & & \downarrow{\scriptstyle p_*} \\
\mathcal{C} & \xrightarrow{\;q_*\;} & \mathcal{D}
\end{array}
$$

which commutes up to a specified isomorphism $\eta : p_* q'_* \to q_* p'_*$. Suppose furthermore that the functors q_* and q'_* admit left adjoints, which we will denote by q^* and q'^*. Consider the composition

$$\gamma : q^* p_* \xrightarrow{u} q^* p_* q'_* q'^* \xrightarrow{\eta} q^* q_* p'_* q'^* \xrightarrow{v} p'_* q'^*,$$

where u denotes a unit for the adjunction (q'^*, q'_*) and v a counit for the adjunction (q^*, q_*). We will refer to γ as the *push-pull* transformation associated to the above diagram.

Definition 7.3.1.1. A diagram of categories

$$
\begin{array}{ccc}
\mathcal{C}' & \xrightarrow{\ q'_* \ } & \mathcal{D}' \\
{\scriptstyle p'_*}\big\downarrow & & \big\downarrow{\scriptstyle p_*} \\
\mathcal{C} & \xrightarrow{\ q_* \ } & \mathcal{D}
\end{array}
$$

which commutes up to a specified isomorphism is *left adjointable* if the functors q_* and q'_* admit left adjoints q^* and q'^* and the associated push-pull transformation

$$
\gamma : q^* p_* \to p'_* q'^*
$$

is an isomorphism of functors.

Definition 7.3.1.2. A diagram of ∞-categories

$$
\begin{array}{ccc}
\mathcal{C}' & \xrightarrow{\ q'_* \ } & \mathcal{D}' \\
{\scriptstyle p'_*}\big\downarrow & & \big\downarrow{\scriptstyle p_*} \\
\mathcal{C} & \xrightarrow{\ q_* \ } & \mathcal{D}
\end{array}
$$

which commutes up to (specified) homotopy is *left adjointable* if the associated diagram of homotopy categories is left adjointable.

Remark 7.3.1.3. Suppose we are given a diagram of simplicial sets

$$
\mathcal{M}' \xrightarrow{P} \mathcal{M} \xrightarrow{f} \Delta^1,
$$

where both f and $f \circ P$ are Cartesian fibrations. Then we may view $\mathcal{M}$ as a correspondence from $\mathcal{D} = f^{-1}\{0\}$ to $\mathcal{C} = f^{-1}\{1\}$ associated to some functor $q_* : \mathcal{C} \to \mathcal{D}$. Similarly, we may view $\mathcal{M}'$ as a correspondence from $\mathcal{D}' = (f \circ P)^{-1}\{0\}$ to $\mathcal{C}' = (f \circ P)^{-1}\{1\}$ associated to some functor $q'_* : \mathcal{C}' \to \mathcal{D}'$. The map P determines functors $p'_* : \mathcal{C}' \to \mathcal{C}$, $q'_* : \mathcal{D}' \to \mathcal{D}$ and (up to homotopy) a natural transformation $\alpha : p_* q'_* \to q_* p'_*$, which is an equivalence if and only if the map P carries $(f \circ P)$-Cartesian edges of $\mathcal{M}'$ to f-Cartesian edges of $\mathcal{M}$. In this case, we obtain a diagram of homotopy categories

$$
\begin{array}{ccc}
h\mathcal{C}' & \xrightarrow{\ q'_* \ } & h\mathcal{D}' \\
{\scriptstyle p'_*}\big\downarrow & & \big\downarrow{\scriptstyle p_*} \\
h\mathcal{C} & \xrightarrow{\ q_* \ } & h\mathcal{D}
\end{array}
$$

which commutes up to canonical isomorphism.

Now suppose that the functors q_* and q'_* admit left adjoints, which we will denote by q^* and q'^*, respectively. Then the maps f and $f \circ P$ are co-Cartesian fibrations. Moreover, the associated push-pull transformation can be described as follows. Choose an object $D' \in \mathcal{D}'$ and a $(f \circ P)$-coCartesian morphism $\phi : D' \to C'$, where $C' \in \mathcal{C}$. Let $D = P(D')$ and choose an f-coCartesian morphism $\psi : D \to C$ in $\mathcal{M}$, where $C \in \mathcal{C}$. Using the fact that ψ is f-coCartesian, we can choose a 2-simplex in $\mathcal{M}$ depicted as follows:

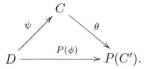

We may then identify C with $q^* p_* D'$, $P(C')$ with $p'_* q'^* D'$, and θ with the value of the push-pull transformation $q^* p_* \to p'_* q'^* D'$ on the object $D' \in \mathcal{D}'$. The morphism θ is an equivalence if and only if $P(\phi)$ is f-coCartesian. Consequently, we deduce that the original diagram

$$
\begin{array}{ccc}
h\mathcal{C}' & \xrightarrow{q'_*} & h\mathcal{D}' \\
\downarrow{\scriptstyle p'_*} & & \downarrow{\scriptstyle p_*} \\
h\mathcal{C} & \xrightarrow{q_*} & h\mathcal{D}
\end{array}
$$

is left adjointable if and only if P carries $(f \circ P)$-coCartesian edges to f-coCartesian edges. We will make use of this criterion in §7.3.4.

Definition 7.3.1.4. Let $p_* : \mathcal{X} \to \mathcal{Y}$ be a geometric morphism of ∞-topoi. We will say that p_* is *proper* if the following condition is satisfied:

$(*)$ For every Cartesian rectangle

$$
\begin{array}{ccccc}
\mathcal{X}'' & \longrightarrow & \mathcal{X}' & \longrightarrow & \mathcal{X} \\
\downarrow & & \downarrow & & \downarrow{\scriptstyle p_*} \\
\mathcal{Y}'' & \longrightarrow & \mathcal{Y}' & \longrightarrow & \mathcal{Y}
\end{array}
$$

of ∞-topoi, the left square is left adjointable.

Remark 7.3.1.5. Let $\mathcal{X}$ be an ∞-topos and let $\mathcal{J}$ be a small ∞-category. The diagonal functor $\delta : \mathcal{X} \to \mathrm{Fun}(\mathcal{J}, \mathcal{X})$ preserves all (small) limits and colimits, by Proposition 5.1.2.2, and therefore admits both a left adjoint $\delta_!$ and a right adjoint δ_*. If $\mathcal{J}$ is filtered, then $\delta_!$ is left exact (Proposition 5.3.3.3). Consequently, we have a diagram of geometric morphisms

$$
\mathcal{X} \xrightarrow{\delta} \mathrm{Fun}(\mathcal{J}, \mathcal{X}) \xrightarrow{\delta_*} \mathcal{X} .
$$

Now suppose that $p_* : \mathcal{X} \to \mathcal{Y}$ is a proper geometric morphism of ∞-topoi. We obtain a rectangle

$$
\begin{array}{ccccc}
\mathcal{X} & \longrightarrow & \mathrm{Fun}(\mathcal{J}, \mathcal{X}) & \longrightarrow & \mathcal{X} \\
\downarrow{\scriptstyle p_*} & & \downarrow{\scriptstyle p_*^{\mathcal{J}}} & & \downarrow \\
\mathcal{Y} & \longrightarrow & \mathrm{Fun}(\mathcal{J}, \mathcal{Y}) & \longrightarrow & \mathcal{Y}
\end{array}
$$

which commutes up to (specified) homotopy. One can show that this is a Cartesian rectangle in $\mathcal{R}\mathcal{T}\mathrm{op}$, so that the square on the left is left adjointable. Unwinding the definitions, we conclude that p_* commutes with filtered colimits. Conversely, if $p_* : \mathfrak{X} \to \mathcal{Y}$ is an arbitrary geometric morphism of ∞-topoi which commutes with colimits indexed by *filtered $\mathcal{Y}$-stacks* (over each object of $\mathcal{Y}$), then p_* is proper. To give a proof (or even a precise formulation) of this statement would require ideas from relative category theory which we will not develop in this book. We refer the reader to [58], where the analogous result is established for proper maps between ordinary topoi.

The following properties of the class of proper morphisms follow immediately from Definition 7.3.1.4:

Proposition 7.3.1.6.   (1) *Every equivalence of ∞-topoi is proper.*

(2) *If p_* and p'_* are equivalent geometric morphisms from an ∞-topos $\mathfrak{X}$ to another ∞-topos $\mathcal{Y}$, then p_* is proper if and only if p'_* is proper.*

(3) *Let*

$$
\begin{array}{ccc}
\mathfrak{X}' & \longrightarrow & \mathfrak{X} \\
\downarrow{\scriptstyle p'_*} & & \downarrow{\scriptstyle p_*} \\
\mathcal{Y}' & \longrightarrow & \mathcal{Y}
\end{array}
$$

be a pullback diagram of ∞-topoi. If p_* is proper, then so is p'_*.

(4) *Let*

$$\mathfrak{X} \xrightarrow{p_*} \mathcal{Y} \xrightarrow{q_*} \mathcal{Z}$$

be proper geometric morphisms between ∞-topoi. Then $q_* \circ p_*$ is a proper geometric morphism.

In order to relate Definition 7.3.1.4 to the classical statement of the proper base change theorem, we need to understand the relationship between products in the category of topological spaces and products in the ∞-category of ∞-topoi. A basic result asserts that these are compatible provided that a certain local compactness condition is met.

Definition 7.3.1.7. Let X be a topological space which is not assumed to be Hausdorff. We say that X is *locally compact* if, for every open set $U \subseteq X$ and every point $x \in U$, there exists a (not necessarily closed) compact set $K \subseteq U$, where K contains an open neighborhood of x.

Example 7.3.1.8. If X is Hausdorff space, then X is locally compact in the sense defined above if and only if X is locally compact in the usual sense.

Example 7.3.1.9. Let X be a topological space for which the compact open subsets of X form a basis for the topology of X. Then X is locally compact.

Remark 7.3.1.10. Local compactness of X is precisely the condition needed for function spaces Y^X, endowed with the compact-open topology, to represent the functor $Z \mapsto \mathrm{Hom}(Z \times X, Y)$.

Proposition 7.3.1.11. *Let X and Y be topological spaces and assume that X is locally compact. The diagram*

$$
\begin{array}{ccc}
\mathrm{Shv}(X \times Y) & \longrightarrow & \mathrm{Shv}(X) \\
\downarrow & & \downarrow \\
\mathrm{Shv}(Y) & \longrightarrow & \mathrm{Shv}(*)
\end{array}
$$

is a pullback square in the ∞-category $\mathcal{R}\mathcal{T}\mathrm{op}$ of ∞-topoi.

Proof. Let $\mathcal{C} \subseteq \mathcal{R}\mathcal{T}\mathrm{op}$ be the full subcategory spanned by the 0-localic ∞-topoi. Since $\mathcal{C}$ is a localization of $\mathcal{R}\mathcal{T}\mathrm{op}$, the inclusion $\mathcal{C} \subseteq \mathcal{R}\mathcal{T}\mathrm{op}$ preserves limits. It therefore suffices to prove that

$$
\begin{array}{ccc}
\mathrm{Shv}(X \times Y) & \longrightarrow & \mathrm{Shv}(X) \\
\downarrow & & \downarrow \\
\mathrm{Shv}(Y) & \longrightarrow & \mathrm{Shv}(*)
\end{array}
$$

gives a pullback diagram in $\mathcal{C}$. Note that $\mathcal{C}^{op}$ is equivalent to the (nerve of the) ordinary category of locales. For each topological space M, let $\mathcal{U}(M)$ denote the locale of open subsets of M. Let

$$
\mathcal{U}(X) \overset{\psi_X}{\to} \mathcal{P} \overset{\psi_Y}{\leftarrow} \mathcal{U}(Y)
$$

be a diagram which exhibits $\mathcal{P}$ as a coproduct of $\mathcal{U}(X)$ and $\mathcal{U}(Y)$ in the category of locales and let $\phi : \mathcal{P} \to \mathcal{U}(X \times Y)$ be the induced map. We wish to prove that ϕ is an isomorphism. This is a standard result in the theory of locales; we will include a proof for completeness.

Given open subsets $U \subseteq X$ and $V \subseteq Y$, let $U \otimes V = (\psi_X U) \cap (\psi_Y V) \in \mathcal{P}$, so that $\phi(U \otimes V) = U \times V \in \mathcal{U}(X \times Y)$. We define a map $\theta : \mathcal{U}(X \times Y) \to \mathcal{P}$ by the formula

$$
\theta(W) = \bigcup_{U \times V \subseteq W} U \otimes V.
$$

Since every open subset of $X \times Y$ can be written as a union of products $U \times V$, where U is an open subset of X and V is an open subset of Y, it is clear that $\phi \circ \theta : \mathcal{U}(X \times Y) \to \mathcal{U}(X \times Y)$ is the identity. To complete the proof, it will suffice to show that $\theta \circ \phi : \mathcal{P} \to \mathcal{P}$ is the identity. Every element of $\mathcal{P}$ can be written as $\bigcup_\alpha U_\alpha \otimes V_\alpha$ for $U_\alpha \subseteq X$ and $V_\alpha \subseteq Y$ appropriately chosen. We therefore wish to show that

$$
\bigcup_{U \times V \subseteq \bigcup_\alpha U_\alpha \otimes V_\alpha} U \times V = \bigcup_\alpha U_\alpha \otimes V_\alpha.
$$

It is clear that the right hand side is contained in the left hand side. The reverse containment is equivalent to the assertion that if $U \times V \subseteq \bigcup_\alpha U_\alpha \times V_\alpha$, then $U \otimes V \subseteq \bigcup_\alpha U_\alpha \otimes V_\alpha$.

We now invoke the local compactness of X. Write $U = \bigcup K_\beta$, where each K_β is a compact subset of U and the interiors $\{K_\beta^\circ\}$ cover U. Then $U \otimes V = \bigcup_\beta K_\beta^\circ \otimes V$; it therefore suffices to prove that $K_\beta^\circ \otimes V \subseteq \bigcup_\alpha U_\alpha \otimes V_\alpha$. Let v be a point of V. Then $K_\beta \times \{v\}$ is a compact subset of $\bigcup_\alpha U_\alpha \times V_\alpha$. Consequently, there exists a finite set of indices $\{\alpha_1, \ldots, \alpha_n\}$ such that $v \in V_{v,\beta} = V_{\alpha_1} \cap \cdots \cap V_{\alpha_n}$ and $K_\beta \subseteq U_{\alpha_1} \cup \cdots \cup U_{\alpha_n}$. It follows that $K_\beta^\circ \otimes V_{v,\beta} \subseteq \bigcup_\alpha U_\alpha \otimes V_\alpha$. Taking a union over all $v \in V$, we deduce the desired result. $\square$

Let us now return to the subject of the proper base change theorem. We have essentially defined a proper morphism of ∞-topoi to be one for which the proper base change theorem holds. The challenge, then, is to produce examples of proper geometric morphisms. The following results will be proven in §7.3.2 and §7.3.4, respectively:

(1) If $p : X \to Y$ is a closed embedding of topological spaces, then $p_* : \mathrm{Shv}(X) \to \mathrm{Shv}(Y)$ is proper.

(2) If X is a compact Hausdorff space, then the global sections functor $\Gamma : \mathrm{Shv}(X) \to \mathrm{Shv}(*)$ is proper.

Granting these statements for the moment, we can deduce the main result of this section. First, we must recall a bit of point-set topology:

Definition 7.3.1.12. A topological space X is said to be *completely regular* if every point of X is closed in X and if for every closed subset $Y \subseteq X$ and every point $x \in X - Y$ there is a continuous function $f : X \to [0, 1]$ such that $f(x) = 0$ and $f|Y$ takes the constant value 1.

Remark 7.3.1.13. A topological space X is completely regular if and only if it is homeomorphic to a subspace of a compact Hausdorff space $\overline{X}$ (see [59]).

Definition 7.3.1.14. A map $p : X \to Y$ of (arbitrary) topological spaces is said to be *proper* if it is universally closed. In other words, p is proper if and only if for every pullback diagram of topological spaces

$$
\begin{array}{ccc}
X' & \longrightarrow & X \\
\downarrow{\scriptstyle p'} & & \downarrow{\scriptstyle p} \\
Y' & \longrightarrow & Y
\end{array}
$$

the map p' is closed.

Remark 7.3.1.15. A map $p : X \to Y$ of topological spaces is proper if and only if it is closed and each of the fibers of p is compact (though not necessarily Hausdorff).

Theorem 7.3.1.16. *Let* $p : X \to Y$ *be a proper map of topological spaces, where* X *is completely regular. Then* $p_* : \mathrm{Shv}(X) \to \mathrm{Shv}(Y)$ *is proper.*

Proof. Let $q : X \to \overline{X}$ be an identification of X with a subspace of a compact Hausdorff space $\overline{X}$. The map p admits a factorization

$$X \overset{q \times p}{\to} \overline{X} \times Y \overset{\pi_Y}{\to} Y.$$

Using Proposition 7.3.1.6, we can reduce to proving that $(q \times p)_*$ and $(\pi_Y)_*$ are proper.

Because q identifies X with a subspace of $\overline{X}$, $q \times p$ identifies X with a subspace over $\overline{X} \times Y$. Moreover, $q \times p$ factors as a composition

$$X \to \overline{X} \times X \to \overline{X} \times Y,$$

where the first map is a closed immersion (since $\overline{X}$ is Hausdorff) and the second map is closed (since p is proper). It follows that $q \times p$ is a closed immersion, so that $(q \times p)_*$ is a proper geometric morphism by Proposition 7.3.2.12.

Proposition 7.3.1.11 implies that the geometric morphism $(\pi_Y)_*$ is a pullback of the global sections functor $\Gamma : \mathrm{Shv}(\overline{X}) \to \mathrm{Shv}(*)$ in the ∞-category $\mathcal{RT}\mathrm{op}$. Using Proposition 7.3.1.6, we may reduce to proving that Γ is proper, which follows from Corollary 7.3.4.11. $\qquad\square$

Remark 7.3.1.17. The converse to Theorem 7.3.1.16 holds as well (and does not require the assumption that X is completely regular): if $p_* : \mathrm{Shv}(X) \to \mathrm{Shv}(Y)$ is a proper geometric morphism, then p is a proper map of topological spaces. This can be proven easily using the characterization of properness described in Remark 7.3.1.5.

Corollary 7.3.1.18 (Nonabelian Proper Base Change Theorem). *Let*

$$
\begin{array}{ccc}
X' & \overset{q'}{\longrightarrow} & X \\
\downarrow{\scriptstyle p'} & & \downarrow{\scriptstyle p} \\
Y' & \overset{q}{\longrightarrow} & Y
\end{array}
$$

be a pullback diagram of locally compact Hausdorff spaces and suppose that p *is proper. Then the associated diagram*

$$
\begin{array}{ccc}
\mathrm{Shv}(X') & \overset{q'_*}{\longrightarrow} & \mathrm{Shv}(X) \\
\downarrow{\scriptstyle p'_*} & & \downarrow{\scriptstyle p_*} \\
\mathrm{Shv}(Y') & \overset{q_*}{\longrightarrow} & \mathrm{Shv}(Y)
\end{array}
$$

is left adjointable.

Proof. In view of Theorem 7.3.1.16, it suffices to show that

$$
\begin{array}{ccc}
\mathrm{Shv}(X') & \overset{q'_*}{\longrightarrow} & \mathrm{Shv}(X) \\
\downarrow{\scriptstyle p'_*} & & \downarrow{\scriptstyle p_*} \\
\mathrm{Shv}(Y') & \overset{q_*}{\longrightarrow} & \mathrm{Shv}(Y)
\end{array}
$$

is a pullback diagram of ∞-topoi. Let $\overline{X}$ denote a compactification of X (for example, the one-point compactification) and consider the larger diagram of ∞-topoi

$$
\begin{array}{ccc}
\mathrm{Shv}(X') & \longrightarrow & \mathrm{Shv}(X) \\
\downarrow & & \downarrow \\
\mathrm{Shv}(\overline{X} \times Y') \longrightarrow \mathrm{Shv}(\overline{X} \times Y) \longrightarrow \mathrm{Shv}(\overline{X}) \\
\downarrow & & \downarrow & & \downarrow \\
\mathrm{Shv}(Y') & \longrightarrow & \mathrm{Shv}(Y) & \longrightarrow & \mathrm{Shv}(*).
\end{array}
$$

The upper square is a (homotopy) pullback by Proposition 7.3.2.12 and Corollary 7.3.2.10. Both the lower right square and the lower rectangle are (homotopy) Cartesian by Proposition 7.3.1.11, so that the lower left square is (homotopy) Cartesian as well. It follows that the vertical rectangle is also (homotopy) Cartesian, as desired. □

Remark 7.3.1.19. The classical proper base change theorem, for sheaves of abelian groups on locally compact topological spaces, is a formal consequence of Corollary 7.3.1.18. We give a brief sketch. The usual formulation of the proper base change theorem (see, for example, [46]) is equivalent to the statement that if

$$
\begin{array}{ccc}
X' & \xrightarrow{q'} & X \\
\downarrow{\scriptstyle p'} & & \downarrow{\scriptstyle p} \\
Y' & \xrightarrow{q} & Y
\end{array}
$$

is a pullback diagram of locally compact topological spaces and p is proper, then the associated diagram

$$
\begin{array}{ccc}
D^-(X') & \xrightarrow{q'_*} & D^-(X) \\
\downarrow{\scriptstyle p'_*} & & \downarrow{\scriptstyle p_*} \\
D^-(Y') & \xrightarrow{q_*} & D^-(Y)
\end{array}
$$

is left adjointable. Here $D^-(Z)$ denotes the (bounded below) derived category of abelian sheaves on a topological space Z.

Let $\mathbf{A}$ denote the category whose objects are chain complexes

$$
\cdots \to A^{-1} \to A^0 \to A^1 \to \cdots
$$

of abelian groups. Then $\mathbf{A}$ admits the structure of a combinatorial model category in which the weak equivalences are given by quasi-isomorphisms. Let $\mathcal{C} = \mathrm{N}(\mathbf{A}^\circ)$ be the underlying ∞-category. For any topological space Z, one can define an ∞-category $\mathrm{Shv}(Z; \mathcal{C})$ of sheaves on Z with values in $\mathcal{C}$; see §7.3.3. The homotopy category $\mathrm{hShv}(Z; \mathcal{C})$ is an unbounded version of the derived category $D^-(Z)$; in particular, it contains $D^-(Z)$ as a full subcategory. Consequently, we obtain a natural generalization of the proper base change theorem where boundedness hypotheses have been removed, which asserts that the diagram

$$
\begin{array}{ccc}
\mathrm{Shv}(X';\mathcal{C}) & \xrightarrow{q'_*} & \mathrm{Shv}(X;\mathcal{C}) \\
\downarrow{\scriptstyle p'_*} & & \downarrow{\scriptstyle p_*} \\
\mathrm{Shv}(Y';\mathcal{C}) & \xrightarrow{q_*} & \mathrm{Shv}(Y;\mathcal{C})
\end{array}
$$

is left adjointable. Using the fact that $\mathcal{C}$ has enough compact objects, one can deduce this statement formally from Corollary 7.3.1.18.

7.3.2 Closed Subtopoi

If X is a topological space and $U \subseteq X$ is an open subset, then we may view the closed complement $X - U \subseteq X$ as a topological space in its own right. Moreover, the inclusion $(X - U) \hookrightarrow X$ is a proper map of topological spaces (that is, a closed map whose fibers are compact). The purpose of this section is to present an analogous construction in the case where X is an ∞-topos.

Lemma 7.3.2.1. *Let X be an ∞-topos and $\emptyset$ an initial object of X. Then $\emptyset$ is (-1)-truncated.*

Proof. Let X be an object of X. The space $\mathrm{Map}_X(X, \emptyset)$ is contractible if X is an initial object of X and empty otherwise (by Lemma 6.1.3.6). In either case, $\mathrm{Map}_X(X, \emptyset)$ is (-1)-truncated. $\square$

Lemma 7.3.2.2. *Let X be an ∞-topos and let $f : \emptyset \to X$ be a morphism in X, where $\emptyset$ is an initial object. Then f is a monomorphism.*

Proof. Apply Lemma 7.3.2.1 to the ∞-topos $X_{/X}$. $\square$

Proposition 7.3.2.3. *Let X be an ∞-topos and let U be an object of X. Let S_U be the smallest strongly saturated class of morphisms of X which is stable under pullbacks and contains a morphism $f : \emptyset \to U$, where $\emptyset$ is an initial object of X. Then S_U is topological (in the sense of Definition 6.2.1.5).*

Proof. For each morphism $g : X \to U$ in $\mathcal{C}$, form a pullback square

$$
\begin{array}{ccc}
\emptyset' & \xrightarrow{f_Y} & Y \\
\downarrow & & \downarrow{\scriptstyle g} \\
\emptyset & \xrightarrow{f} & U.
\end{array}
$$

Let $S = \{f_X\}_{g:X \to U}$ and let $\overline{S}$ be the strongly saturated class of morphisms generated by S. We note that each f_X is a pullback of f and therefore a monomorphism (by Lemma 7.3.2.2). Let S' be the collection of all morphisms $h : V \to W$ with the property that for every pullback diagram

$$
\begin{array}{ccc}
V' & \longrightarrow & V \\
\downarrow{\scriptstyle h'} & & \downarrow{\scriptstyle h} \\
W' & \longrightarrow & W
\end{array}
$$

in $\mathfrak{X}$, the morphism h belongs to $\overline{S}$. Since colimits in $\mathfrak{X}$ are universal, we deduce that S' is strongly saturated, and $S \subseteq S' \subseteq \overline{S}$ by construction. Therefore $S' = \overline{S}$, so that $\overline{S}$ is stable under pullbacks. Since $f \in \overline{S}$, we deduce that $S_U \subseteq \overline{S}$. On the other hand, $S \subseteq S_U$ and S_U is strongly saturated, so $\overline{S} \subseteq S_U$. Therefore $S_U = \overline{S}$. Since S consists of monomorphisms, we conclude that S_U is topological. $\qquad\square$

In the situation of Proposition 7.3.2.3, we will say that a morphism of $\mathfrak{X}$ is an *equivalence away from U* if it belongs to S_U.

Lemma 7.3.2.4. *Let $\mathfrak{X}$ be an ∞-topos containing a pair of objects $U, X \in \mathfrak{X}$ and let S_U denote the class of morphism in $\mathfrak{X}$ which are equivalences away from U. The following are equivalent:*

(1) *The object X is S_U-local.*

(2) *For every map $\widetilde{U} \to U$ in $\mathfrak{X}$, the space $\mathrm{Map}_{\mathfrak{X}}(\widetilde{U}, X)$ is contractible.*

(3) *There exists a morphism $g : U \to X$ such that the diagram*

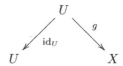

exhibits U as a product of U and X in $\mathfrak{X}$.

Proof. Let S be the collection of all morphisms $f_{\widetilde{U}}$ which come from pullback diagrams

$$
\begin{array}{ccc}
\emptyset' & \xrightarrow{\ f_{\widetilde{U}}\ } & \widetilde{U} \\
\downarrow & & \downarrow \\
\emptyset & \longrightarrow & U,
\end{array}
$$

where $\emptyset$ and therefore $\emptyset'$ also are initial objects of $\mathfrak{X}$. We saw in the proof of Proposition 7.3.2.3 that S generates S_U as a strongly saturated class of morphisms. Therefore X is S_U-local if and only if each $f_{\widetilde{U}}$ induces an isomorphism

$$
\mathrm{Map}_{\mathfrak{X}}(\widetilde{U}, X) \to \mathrm{Map}_{\mathfrak{X}}(\emptyset', X) \simeq *
$$

in the homotopy category $\mathcal{H}$. This proves that $(1) \Leftrightarrow (2)$.

Now suppose that (2) is satisfied. Taking $\widetilde{U} = U$, we deduce that there exists a morphism $g : U \to X$. We will prove that g and id_X exhibit U as a product of U and X. As explained in §4.4.1, this is equivalent to the assertion that for every $Z \in \mathfrak{X}$, the map

$$\mathrm{Map}_{\mathfrak{X}}(Z, U) \to \mathrm{Map}_{\mathfrak{X}}(Z, U) \times \mathrm{Map}_{\mathfrak{X}}(Z, X)$$

is an isomorphism in $\mathcal{H}$. If there are no morphisms from Z to U in $\mathfrak{X}$, then both sides are empty and the result is obvious. Otherwise, we may invoke (2) to deduce that $\mathrm{Map}_{\mathfrak{X}}(Z, X)$ is contractible, and the desired result follows. This completes the proof that $(2) \Rightarrow (3)$.

Suppose now that (3) is satisfied for some morphism $g : U \to X$. For any object $Z \in \mathfrak{X}$, we have a homotopy equivalence

$$\mathrm{Map}_{\mathfrak{X}}(Z, U) \to \mathrm{Map}_{\mathfrak{X}}(Z, U) \times \mathrm{Map}_{\mathfrak{X}}(Z, X).$$

If $\mathrm{Map}_{\mathfrak{X}}(Z, U)$ is nonempty, then we may pass to the fiber over a point of $\mathrm{Map}_{\mathfrak{X}}(Z, U)$ to obtain a homotopy equivalence $* \to \mathrm{Map}_{\mathfrak{X}}(Z, X)$, so that $\mathrm{Map}_{\mathfrak{X}}(Z, X)$ is contractible. This proves (2). $\qquad\square$

If $\mathfrak{X}$ is an ∞-topos and $U \in \mathfrak{X}$, then we will say that an object $X \in \mathfrak{X}$ is *trivial on U* if it satisfies the equivalent conditions of Lemma 7.3.2.4. We let $\mathfrak{X}/U$ denote the full subcategory of $\mathfrak{X}$ spanned by the objects X which are trivial on U. It follows from Proposition 7.3.2.3 that $\mathfrak{X}/U$ is a topological localization of $\mathfrak{X}$ and, in particular, that $\mathfrak{X}/U$ is an ∞-topos. We next show that $\mathfrak{X}/U$ depends only on the support of U.

Lemma 7.3.2.5. *Let $\mathfrak{X}$ be an ∞-topos and let $g : U \to V$ be a morphism in $\mathfrak{X}$. Then $\mathfrak{X}/V \subseteq \mathfrak{X}/U$. Moreover, if g is an effective epimorphism, then $\mathfrak{X}/U = \mathfrak{X}/V$.*

Proof. The first assertion follows immediately from Lemma 7.3.2.4. To prove the second, it will suffice to prove that if g is strongly saturated, then $S_V \subseteq S_U$. Since S_U is strongly saturated and stable under pullbacks, it will suffice to prove that S_U contains a morphism $f : \emptyset \to V$, where $\emptyset$ is an initial object of $\mathfrak{X}$.

Form a pullback diagram $\sigma : \Delta^1 \times \Delta^1 \to \mathfrak{X}$:

$$
\begin{array}{ccc}
\emptyset' & \xrightarrow{f'} & U \\
\downarrow & & \downarrow{\scriptstyle g} \\
\emptyset & \xrightarrow{f} & V.
\end{array}
$$

We may view σ as an effective epimorphism from f' to f in the ∞-topos $\mathfrak{X}^{\Delta^1}$. Let $f_\bullet = \check{C}(\sigma) : \mathbf{\Delta}_+ \to \mathfrak{X}^{\Delta^1}$ be a Čech nerve of $\sigma : f' \to f$. We note that for $n \geq 0$, the map f_n is a pullback of f' and therefore belongs to S_U. Since $f_\bullet$ is a colimit diagram, we deduce that f belongs to S_U, as desired. $\qquad\square$

If $\mathfrak{X}$ is an ∞-topos, we let $\mathrm{Sub}(1_{\mathfrak{X}})$ denote the partially ordered set of equivalence classes of (-1)-truncated objects of $\mathfrak{X}$. We note that this set is independent of the choice of a final object $1_{\mathfrak{X}} \in \mathfrak{X}$ up to canonical isomorphism. Any $U \in \mathrm{Sub}(1_{\mathfrak{X}})$ can be represented by a (-1)-truncated object $\widetilde{U} \in \mathfrak{X}$. We define $\mathfrak{X}/U = \mathfrak{X}/\widetilde{U} \subseteq \mathfrak{X}$. It follows from Lemma 7.3.2.5 that $\mathfrak{X}/U$ is independent of the choice of $\widetilde{U}$ representing U and that for any object $X \in \mathfrak{X}$, we have $\mathfrak{X}/X = \mathfrak{X}/U$, where $U \in \mathrm{Sub}(1_{\mathfrak{X}})$ is the "support" of X (namely, the equivalence class of the truncation $\tau_{-1}X$).

Definition 7.3.2.6. If $\mathfrak{X}$ is an ∞-topos and $U \in \mathrm{Sub}(1_{\mathfrak{X}})$, then we will refer to $\mathfrak{X}/U$ as the *closed subtopos of $\mathfrak{X}$ complementary to U*. More generally, we will say that a geometric morphism $\pi : \mathcal{Y} \to \mathfrak{X}$ is a *closed immersion* if there exists $U \in \mathrm{Sub}(1_{\mathfrak{X}})$ such that π_* induces an equivalence of ∞-categories from $\mathcal{Y}$ to $\mathfrak{X}/U$.

Proposition 7.3.2.7. *Let $\mathfrak{X}$ be an ∞-topos and let $U \in \mathrm{Sub}(1_{\mathfrak{X}})$. Then the closed immersion*

$$\pi : \mathfrak{X}/U \to \mathfrak{X}$$

induces an isomorphism of partially ordered sets from $\mathrm{Sub}(1_{\mathfrak{X}/U})$ to $\{V \in \mathrm{Sub}(1_{\mathfrak{X}}) : U \subseteq V\}$).

Proof. Choose a (-1)-truncated object $\widetilde{U} \in \mathfrak{X}$ representing U. Since π^* is left exact, an object X of $\mathfrak{X}/U$ is (-1)-truncated as an object of $\mathfrak{X}/U$ if and only if it is (-1)-truncated as an object of $\mathfrak{X}$. It therefore suffices to prove that if $\widetilde{V}$ is a (-1)-truncated object of $\mathfrak{X}$ representing an element $V \in \mathrm{Sub}(1_{\mathfrak{X}})$, then $\widetilde{V}$ is S_U-local if and only if $U \subseteq V$. One direction is clear: if $\widetilde{V}$ is S_U-local, then we have an isomorphism

$$\mathrm{Map}_{\mathfrak{X}}(\widetilde{U}, \widetilde{V}) \to \mathrm{Map}_{\mathfrak{X}}(\emptyset, \widetilde{V}) = *$$

in the homotopy category $\mathcal{H}$, so that $U \subseteq V$. The converse follows from characterization (3) given in Lemma 7.3.2.4. $\qquad\square$

Corollary 7.3.2.8. *Let $\mathfrak{X}$ be an ∞-topos and let $U, V \in \mathrm{Sub}(1_{\mathfrak{X}})$. Then $S_U \subseteq S_V$ if and only if $U \subseteq V$.*

Proof. The "if" direction follows from Lemma 7.3.2.5, and the converse from Proposition 7.3.2.7. $\qquad\square$

Corollary 7.3.2.9. *Let $\mathfrak{X}$ be a 0-localic ∞-topos associated to the locale $\mathcal{U}$ and let $U \in \mathcal{U}$. Then $\mathfrak{X}/U$ is a 0-localic ∞-topos associated to the locale $\{V \in \mathcal{U} : U \subseteq V\}$.*

Proof. The ∞-topos $\mathfrak{X}/U$ is a topological localization of a 0-localic ∞-topos and is therefore also 0-localic (Proposition 6.4.5.9). The identification of the underlying locale follows from Proposition 7.3.2.7. $\qquad\square$

Corollary 7.3.2.10. *Let X be a topological space, $U \subseteq X$ an open subset, and $Y = X - U$. The inclusion of Y in X induces a closed immersion of ∞-topoi $\mathrm{Shv}(Y) \to \mathrm{Shv}(X)$ and an equivalence $\mathrm{Shv}(Y) \to \mathrm{Shv}(X)/U$.*

Lemma 7.3.2.11. *Let $\mathcal{X}$ and $\mathcal{Y}$ be ∞-topoi and let $U \in \mathcal{Y}$ be an object. The map*

$$\mathrm{Fun}_*(\mathcal{X}, \mathcal{Y}/U) \to \mathrm{Fun}_*(\mathcal{X}, \mathcal{Y})$$

identifies $\mathrm{Fun}_*(\mathcal{X}, \mathcal{Y}/U)$ with the full subcategory of $\mathrm{Fun}_*(\mathcal{X}, \mathcal{Y})$ spanned by those geometric morphisms $\pi_* : \mathcal{X} \to \mathcal{Y}$ such that π^*U is an initial object of $\mathcal{X}$ (here π^* denotes a left adjoint to π_*).

Proof. Let $\pi_* : \mathcal{X} \to \mathcal{Y}$ be a geometric morphism. Using the adjointness of π_* and π^*, it is easy to see that π_*X is S_U-local if and only if X is $\pi^*(S_U)$-local. In particular, π_* factors through $\mathcal{Y}/U$ if and only if $\pi^*(S_U)$ consists of equivalences in $\mathcal{X}$. Choosing $f \in S_U$ of the form $f : \emptyset \to U$, where $\emptyset$ is an initial object of $\mathcal{X}$, we deduce that π^*f is an equivalence so that $\pi^*U \simeq \pi^*\emptyset$ is an initial object of $\mathcal{X}$. Conversely, suppose that π^*U is an initial object of $\mathcal{X}$. Then π^*f is a morphism between two initial objects of $\mathcal{X}$ and therefore an equivalence. Since π^* is left exact and colimit-preserving, the collection of all morphisms g such that π^*g is an equivalence is strongly saturated, is stable under pullbacks, and contains f; it therefore contains S_U, so that π_* factors through $\mathcal{Y}/U$, as desired. $\square$

Proposition 7.3.2.12. *Let $\pi_* : \mathcal{X} \to \mathcal{Y}$ be a geometric morphism of ∞-topoi and let $\pi^* : \mathrm{Sub}(1_{\mathcal{X}}) \to \mathrm{Sub}(1_{\mathcal{Y}})$ denote the induced map of partially ordered sets. Let $U \in \mathrm{Sub}(1_{\mathcal{X}})$. There is a commutative diagram*

of ∞-topoi and geometric morphisms, where the vertical maps are given by the natural inclusions. This diagram is left adjointable and exhibits $\mathcal{X}/(\pi^*U)$ as a fiber product of $\mathcal{X}$ and $\mathcal{Y}/U$ over $\mathcal{Y}$ in the ∞-category $\mathcal{RT}$op.

Proof. Let π^* denote a left adjoint to π_*. Our first step is to show that the upper horizontal map $\pi_*|(\mathcal{X}/\pi^*U)$ is well-defined. In other words, we must show that if $X \in \mathcal{X}$ is trivial on π^*U, then $\pi_*X \in \mathcal{Y}$ is trivial on U. Suppose that $Y \in \mathcal{Y}$ has support contained in U; we must show that $\mathrm{Map}_{\mathcal{Y}}(Y, \pi_*X)$ is contractible. But this space is homotopy equivalent to $\mathrm{Map}_{\mathcal{X}}(\pi^*Y, X) \simeq *$ since π^*Y has support contained in π^*U and X is trivial on π^*U.

We also note that π^* carries $\mathcal{Y}/U$ into $\mathcal{X}/\pi^*U$. This follows immediately from characterization (3) of Lemma 7.3.2.4 because π^* is left exact. Therefore $\pi^*|\mathcal{Y}/U$ is a left adjoint of $\pi_*|\mathcal{X}/\pi^*U$. From the fact that π^* is left exact, we easily deduce that $\pi^*|\mathcal{Y}/U$ is left exact. It follows that $\pi_*|\mathcal{X}/\pi^*U$ has a left exact left adjoint and is therefore a geometric morphism of ∞-topoi.

Moreover, the diagram

$$\begin{array}{ccc}
\mathcal{X}/\pi^*U & \xleftarrow{\quad \pi^*|\mathcal{Y}/Y \quad} & \mathcal{Y}/U \\
\downarrow & & \downarrow \\
\mathcal{X} & \xleftarrow{\quad \pi^* \quad} & \mathcal{Y}
\end{array}$$

is (strictly) commutative, which proves that the diagram of pushforward functors is left adjointable.

We now claim that the diagram

$$\begin{array}{ccc}
\mathcal{X}/\pi^*U & \xrightarrow{\quad \pi_*|\mathcal{X}/\pi^*U \quad} & \mathcal{Y}/U \\
\downarrow & & \downarrow \\
\mathcal{X} & \xrightarrow{\hspace{3cm}} & \mathcal{Y}
\end{array}$$

is a pullback diagram of ∞-topoi. For every pair of ∞-topoi $\mathcal{A}$ and $\mathcal{B}$, let $[\mathcal{A}, \mathcal{B}]$ denote the largest Kan complex contained in $\mathrm{Fun}_*(\mathcal{A}, \mathcal{B})$. According to Theorem 4.2.4.1, it will suffice to show that for any ∞-topos $\mathcal{Z}$, the associated diagram of Kan complexes

$$\begin{array}{ccc}
[\mathcal{Z}, \mathcal{X}/\pi^*U] & \longrightarrow & [\mathcal{Z}, \mathcal{Y}/U] \\
\downarrow & & \downarrow \\
[\mathcal{Z}, \mathcal{X}] & \longrightarrow & [\mathcal{Z}, \mathcal{Y}]
\end{array}$$

is homotopy Cartesian. Lemma 7.3.2.11 implies that the vertical maps are inclusions of full simplicial subsets. It therefore suffices to show that if $\phi_* : \mathcal{Z} \to \mathcal{Y}$ is a geometric morphism such that $\pi_* \circ \phi_*$ factors through $\mathcal{Y}/U$, then ϕ_* factors through $\mathcal{X}/\pi^*U$. This follows immediately from the characterization given in Lemma 7.3.2.11. $\qquad\qquad\square$

Corollary 7.3.2.13. *Let*

$$\begin{array}{ccc}
\mathcal{X}' & \longrightarrow & \mathcal{X} \\
\downarrow{\scriptstyle p'_*} & & \downarrow{\scriptstyle p_*} \\
\mathcal{Y}' & \longrightarrow & \mathcal{Y}
\end{array}$$

be a pullback diagram in the ∞-category $\mathcal{R}\mathcal{T}\mathrm{op}$ of ∞-topoi. If p_* is a closed immersion, then p'_* is a closed immersion.

7.3.3 Products of ∞-Topoi

In §6.3.4, we showed that the ∞-category $\mathcal{R}\mathcal{T}\mathrm{op}$ of ∞-topoi admits all (small) limits. Unfortunately, the construction of general limits was rather inexplicit. Our goal in this section is to give a very concrete description of the product of two ∞-topoi, at least in a special case.

Definition 7.3.3.1. Let X be a topological space and let $\mathcal{C}$ be an ∞-category. We let $\mathcal{U}(X)$ denote the collection of all open subsets of X partially ordered by inclusion. A *presheaf on X with values in $\mathcal{C}$* is a functor $\mathcal{U}(X)^{op} \to \mathcal{C}$.

Let $\mathcal{F} : \mathcal{U}(X)^{op} \to \mathcal{C}$ be a presheaf on X with values in $\mathcal{C}$. We will say that $\mathcal{F}$ is a *sheaf* with values in $\mathcal{C}$ if, for every $U \subseteq X$ and every covering sieve $\mathcal{U}(X)_{/U}^{(0)} \subseteq \mathcal{U}(X)_{/U}$, the composition

$$N(\mathcal{U}(X)_{/U}^{(0)})^{\triangleright} \subseteq N(\mathcal{U}(X)_{/U})^{\triangleright} \to N(\mathcal{U}(X)) \xrightarrow{\mathcal{F}} \mathcal{C}^{op}$$

is a colimit diagram.

We let $\mathcal{P}(X; \mathcal{C})$ denote the ∞-category $\operatorname{Fun}(\mathcal{U}(X)^{op}, \mathcal{C})$ consisting of all presheaves on X with values in $\mathcal{C}$, and $\operatorname{Shv}(X; \mathcal{C})$ the full subcategory of $\mathcal{P}(X; \mathcal{C})$ spanned by the sheaves on X with values in $\mathcal{C}$.

Remark 7.3.3.2. We can phrase the sheaf condition informally as follows: a $\mathcal{C}$-valued presheaf $\mathcal{F}$ on a topological space X is a sheaf if, for every open subset $U \subseteq X$ and every covering sieve $\{U_\alpha \subseteq U\}$, the natural map $\mathcal{F}(U) \to \varprojlim_\alpha \mathcal{F}(U_\alpha)$ is an equivalence in $\mathcal{C}$.

Remark 7.3.3.3. If X is a topological space, then $\operatorname{Shv}(X) = \operatorname{Shv}(X, \mathcal{S})$, where $\mathcal{S}$ denotes the ∞-category of spaces.

Lemma 7.3.3.4. *Let $\mathcal{C}$, $\mathcal{D}$, and $\mathcal{E}$ be ∞-categories which admit finite limits and let $\mathcal{C}^0 \subseteq \mathcal{C}$ and $\mathcal{D}^0 \subseteq \mathcal{D}$ be the full subcategories of $\mathcal{C}$ and $\mathcal{D}$ consisting of final objects. Let $F : \mathcal{C} \times \mathcal{D} \to \mathcal{E}$ be a functor. The following conditions are equivalent:*

(1) *The functor F preserves finite limits.*

(2) *The functors $F|\,\mathcal{C}^0 \times \mathcal{D}$ and $F|\,\mathcal{C} \times \mathcal{D}^0$ preserve finite limits, and for every pair of morphisms $C \to 1_{\mathcal{C}}$, $D \to 1_{\mathcal{D}}$ where $1_{\mathcal{C}} \in \mathcal{C}$ and $1_{\mathcal{D}} \in \mathcal{D}$ are final objects, the associated diagram*

$$F(1_{\mathcal{C}}, D) \leftarrow F(C, D) \to F(C, 1_{\mathcal{D}})$$

exhibits $F(C, D)$ as a product of $F(1_{\mathcal{C}}, D)$ and $F(C, 1_{\mathcal{D}})$ in $\mathcal{E}$.

(3) *The functors $F|\,\mathcal{C}^0 \times \mathcal{D}$ and $F|\,\mathcal{C} \times \mathcal{D}^0$ preserve finite limits, and F is a right Kan extension of the restriction*

$$F^0 = F|(\mathcal{C} \times \mathcal{D}^0) \coprod_{\mathcal{C}^0 \times \mathcal{D}^0} (\mathcal{C}^0 \times \mathcal{D}).$$

Proof. The implication $(1) \Rightarrow (2)$ is obvious. To see that $(2) \Rightarrow (1)$, we choose final objects $1_{\mathcal{C}} \in \mathcal{C}$, $1_{\mathcal{D}} \in \mathcal{D}$ and natural transformations $\alpha : \operatorname{id}_{\mathcal{C}} \to \underline{1_{\mathcal{C}}}$, $\beta : \operatorname{id}_{\mathcal{D}} \to \underline{1_{\mathcal{D}}}$ (where $\underline{X}$ denotes the constant functor with value X). Let $F_{\mathcal{C}} : \mathcal{C} \to \mathcal{E}$ denote the composition

$$\mathcal{C} \simeq \mathcal{C} \times \{1_{\mathcal{D}}\} \subseteq \mathcal{C} \times \mathcal{D} \xrightarrow{F} \mathcal{E}$$

and define $F_{\mathcal{D}}$ similarly. Then α and β induce natural transformations

$$F_{\mathcal{C}} \circ \pi_{\mathcal{C}} \leftarrow F \rightarrow F_{\mathcal{D}} \circ \pi_{\mathcal{D}}.$$

Assumption (2) implies that the functors $F_{\mathcal{C}}$, $F_{\mathcal{D}}$ preserve finite limits and that the above diagram exhibits F as a product of $F_{\mathcal{C}} \circ \pi_{\mathcal{C}}$ and $F_{\mathcal{D}} \circ \pi_{\mathcal{D}}$ in the ∞-category $\mathcal{E}^{\mathcal{C} \times \mathcal{D}}$. We now apply Lemma 5.5.2.3 to deduce that F preserves finite limits as well.

We now show that (2) $\Leftrightarrow$ (3). Assume that $F|\, \mathcal{C}^0 \times \mathcal{D}$ and $F|\, \mathcal{C} \times \mathcal{D}^0$ preserve finite limits, so that in particular $F|\, \mathcal{C}^0 \times \mathcal{D}^0$ takes values in the full subcategory $\mathcal{E}^0 \subseteq \mathcal{E}$ spanned by the final objects. Fix morphisms $u : C \to 1_{\mathcal{C}}$, $v : D \to 1_{\mathcal{D}}$, where $1_{\mathcal{C}} \in \mathcal{C}$ and $1_{\mathcal{D}} \in \mathcal{D}$ are final obejcts. We will show that the diagram

$$F(1_{\mathcal{C}}, D) \leftarrow F(C, D) \rightarrow F(C, 1_{\mathcal{D}})$$

exhibits $F(C, D)$ as a product of $F(1_{\mathcal{C}}, D)$ and $F(C, 1_{\mathcal{D}})$ if and only if F is a right Kan extension of F^0 at (C, D).

The morphisms u and v determine a map $u \times v : \Delta^1 \times \Delta^1 \to \mathcal{C} \times \mathcal{D}$, which we may identify with a map

$$w : \Lambda_2^2 \rightarrow ((\mathcal{C}^0 \times \mathcal{D}) \coprod_{\mathcal{C}^0 \times \mathcal{D}^0} (\mathcal{C} \times \mathcal{D}^0))_{(C,D)/}.$$

Using Theorem 4.1.3.1, it is easy to see that w^{op} is cofinal. Consequently, F is a right Kan extension of F^0 at (C, D) if and only if the diagram

$$
\begin{array}{ccc}
F(C, D) & \longrightarrow & F(C, 1_{\mathcal{D}}) \\
\downarrow & & \downarrow \\
F(1_{\mathcal{C}}, D) & \longrightarrow & F(1_{\mathcal{C}}, 1_{\mathcal{D}})
\end{array}
$$

is a pullback square. Since $F(1_{\mathcal{C}}, 1_{\mathcal{D}})$ is a final object of $\mathcal{E}$, this is equivalent to assertion (2). $\qquad \square$

Lemma 7.3.3.5. *Let $\mathcal{C}$ and $\mathcal{D}$ be small ∞-categories which admit finite limits, let $1_{\mathcal{C}} \in \mathcal{C}$, $1_{\mathcal{D}} \in \mathcal{D}$ be final objects, and let $\mathcal{X}$ be an ∞-topos. The projections*

$$\mathcal{P}(\mathcal{C} \times \{1_{\mathcal{D}}\}) \xleftarrow{p_*} \mathcal{P}(\mathcal{C} \times \mathcal{D}) \xrightarrow{q_*} \mathcal{P}(\{1_{\mathcal{C}}\} \times \mathcal{D})$$

induce a categorical equivalence

$$\mathrm{Fun}_*(\mathcal{X}, \mathcal{P}(\mathcal{C} \times \mathcal{D})) \rightarrow \mathrm{Fun}_*(\mathcal{X}, \mathcal{P}(\mathcal{C})) \times \mathrm{Fun}_*(\mathcal{X}, \mathcal{P}(\mathcal{D})).$$

In particular, $\mathcal{P}(\mathcal{C} \times \mathcal{D})$ is a product of $\mathcal{P}(\mathcal{C})$ and $\mathcal{P}(\mathcal{D})$ in the ∞-category $\mathcal{RT}op$ of ∞-topoi.

Proof. For every ∞-category $\mathcal{Y}$ which admits finite limits, let $[\mathcal{Y}, \mathcal{X}]$ denote the full subcategory of $\mathrm{Fun}(\mathcal{Y}, \mathcal{X})$ spanned by the left exact functors $\mathcal{Y} \to \mathcal{X}$. If $\mathcal{Y}$ is an ∞-topos, we let $[\mathcal{Y}, \mathcal{X}]_0$ denote the full subcategory of $[\mathcal{Y}, \mathcal{X}]$ spanned by the *colimit-preserving* left exact functors $\mathcal{Y} \to \mathcal{X}$. In view of Proposition

5.2.6.2 and Remark 5.2.6.4, it will suffice to prove that composition with the left adjoints to p_* and q_* induces an equivalence of ∞-categories

$$[\mathcal{P}(\mathcal{C} \times \mathcal{D}), \mathcal{X}]_0 \to [\mathcal{P}(C), \mathcal{X}]_0 \times [\mathcal{P}(\mathcal{D}), \mathcal{X}]_0.$$

Applying Proposition 6.2.3.20, we may reduce to the problem of showing that the map

$$[\mathcal{C} \times \mathcal{D}, \mathcal{X}] \to [\mathcal{C}, \mathcal{X}] \times [\mathcal{D}, \mathcal{X}]$$

is an equivalence of ∞-categories.

Let $\mathcal{C}^0 \subseteq \mathcal{C}$ and $\mathcal{D}^0 \subseteq \mathcal{D}$ denote the full subcategories consisting of final objects of $\mathcal{C}$ and $\mathcal{D}$, respectively. Proposition 1.2.12.9 implies that $\mathcal{C}^0$ and $\mathcal{D}^0$ are contractible. It will therefore suffice to prove that the restriction map

$$\phi : [\mathcal{C} \times \mathcal{D}, \mathcal{X}] \to [\mathcal{C} \times \mathcal{D}^0, \mathcal{X}] \times_{[\mathcal{C}^0 \times \mathcal{D}^0, \mathcal{X}]} [\mathcal{C}^0 \times \mathcal{D}, \mathcal{X}]$$

is a trivial fibration of simplicial sets. This follows immediately from Lemma 7.3.3.4 and Proposition 4.3.2.15. $\qquad \square$

Notation 7.3.3.6. Let $\mathcal{X}$ be an ∞-topos and let $p^* : \mathcal{S} \to \mathcal{X}$ be a geometric morphism (essentially unique in view of Proposition 6.3.4.1). Let $\pi_{\mathcal{X}} : \mathcal{X} \times \mathcal{S} \to \mathcal{X}$ and $\pi_{\mathcal{S}} : \mathcal{X} \times \mathcal{S} \to \mathcal{S}$ denote the projection functors. Let $\otimes$ be a product of $\pi_{\mathcal{X}}$ and $p^* \circ \pi_{\mathcal{S}}$ in the ∞-category of functors from $\mathcal{X} \times \mathcal{S}$ to $\mathcal{X}$. Then $\otimes$ is uniquely defined up to equivalence, and we have natural transformations

$$X \leftarrow X \otimes S \to p^* S$$

which exhibit $X \otimes S$ as product of X and $p^* S$ for all $X \in \mathcal{X}$, $S \in \mathcal{S}$. We observe that $\otimes$ preserves colimits separately in each variable.

If $\mathcal{C}$ is a small ∞-category, we let $\otimes^{\mathcal{C}}$ denote the composition

$$\mathcal{P}(\mathcal{C}; \mathcal{X}) \times \mathcal{P}(\mathcal{C}) \simeq \mathcal{P}(\mathcal{C}; \mathcal{X} \times \mathcal{S}) \xrightarrow{\circ \otimes} \mathcal{P}(\mathcal{C}, \mathcal{X}).$$

We observe that if $F \in \mathcal{P}(\mathcal{C}; \mathcal{X})$ and $G \in \mathcal{P}(\mathcal{C})$, then $F \otimes^{\mathcal{C}} G$ can be identified with a product of F and $p^* \circ G$ in $\mathcal{P}(\mathcal{C}; \mathcal{X})$.

Lemma 7.3.3.7. *Let $\mathcal{C}$ be a small ∞-category and $\mathcal{X}$ an ∞-topos. Let $g : \mathcal{X} \to \mathcal{S}$ a functor corepresented by an object $X \in \mathcal{X}$ and let $G : \mathcal{P}(\mathcal{C}; \mathcal{X}) \to \mathcal{P}(\mathcal{C})$ be the induced functor. Let $\underline{X} \in \mathcal{P}(\mathcal{C}; \mathcal{X})$ denote the constant functor with the value X. Then the functor*

$$F = \underline{X} \otimes^{\mathcal{C}} \mathrm{id}_{\mathcal{P}(\mathcal{C})}$$

is a left adjoint to G.

Proof. Since adjoints and $\otimes^{\mathcal{C}}$ can both be computed pointwise on $\mathcal{C}$, it suffices to treat the case where $\mathcal{C} = \Delta^0$. In this case, we deduce the existence of a left adjoint F' to G using Corollary 5.5.2.9 (the accessibility of G follows from the fact that X is κ-compact for sufficiently large κ since $\mathcal{X}$ is accessible). Now F and F' are both colimit-preserving functors $\mathcal{S} \to \mathcal{X}$. By virtue of Theorem 5.1.5.6, to prove that F and F' are equivalent, it will suffice to show that the objects $F(*), F'(*) \in \mathfrak{X}$ are equivalent. In other words, we must prove that $F'(*) \simeq X$. By adjointness, we have natural isomorphisms

$$\mathrm{Map}_{\mathfrak{X}}(F'(*), Y) \simeq \mathrm{Map}_{\mathcal{H}}(*, G(Y)) \simeq \mathrm{Map}_{\mathfrak{X}}(X, Y)$$

in $\mathcal{H}$ for each $Y \in \mathfrak{X}$, so that $F'(*)$ and X corepresent the same functor on the homotopy category $h\mathfrak{X}$ and are therefore equivalent by Yoneda's lemma. $\square$

Lemma 7.3.3.8. *Let $\mathcal{C}$ be a small ∞-category which admits finite limits and contains a final object $1_{\mathcal{C}}$, let $\mathfrak{X}$ and $\mathcal{Y}$ be ∞-topoi, and let $p_* : \mathfrak{X} \to \mathcal{S}$ be a geometric morphism (essentially unique by virtue of Proposition 6.3.4.1). Then the maps*

$$\mathcal{P}(\mathcal{C}) \xleftarrow{P_*} \mathcal{P}(\mathcal{C}; \mathfrak{X}) \xrightarrow{e_{1_{\mathcal{C}}}} \mathfrak{X}$$

induce equivalences of ∞-categories

$$\mathrm{Fun}_*(\mathcal{Y}, \mathcal{P}(\mathcal{C}; \mathfrak{X})) \to \mathrm{Fun}_*(\mathcal{Y}, \mathfrak{X}) \times \mathrm{Fun}_*(\mathcal{Y}, \mathcal{P}(\mathcal{C})).$$

In particular, $\mathcal{P}(\mathcal{C}; \mathfrak{X})$ is a product of $\mathcal{P}(\mathcal{C})$ and $\mathfrak{X}$ in the ∞-category $\mathcal{RT}op$ of ∞-topoi. Here $e_{1_{\mathcal{C}}}$ denotes the evaluation map at the object $1_{\mathcal{C}} \in \mathcal{C}$, and $P_* : \mathcal{P}(\mathcal{C}; \mathfrak{X}) \to \mathcal{P}(\mathcal{C})$ is given by composition with p_*.

Proof. According to Proposition 6.1.5.3, we may assume without loss of generality that there exists a small ∞-category $\mathcal{D}$ such that $\mathfrak{X}$ is the essential image of an accessible left exact localization functor $L : \mathcal{P}(\mathcal{D}) \to \mathcal{P}(\mathcal{D})$ and that p_* is given by evaluation at a final object $1_{\mathcal{D}} \in \mathcal{D}$. We have a commutative diagram

$$
\begin{array}{ccc}
\mathrm{Fun}_*(\mathcal{Y}, \mathcal{P}(\mathcal{C}; \mathfrak{X})) & \longrightarrow & \mathrm{Fun}_*(\mathcal{Y}, \mathcal{P}(\mathcal{C})) \times \mathrm{Fun}_*(\mathcal{Y}, \mathfrak{X}) \\
\downarrow & & \downarrow \\
\mathrm{Fun}_*(\mathcal{Y}, \mathcal{P}(\mathcal{C} \times \mathcal{D})) & \longrightarrow & \mathrm{Fun}_*(\mathcal{Y}, \mathcal{P}(\mathcal{C})) \times \mathrm{Fun}_*(\mathcal{Y}, \mathcal{P}(\mathcal{D})),
\end{array}
$$

where the vertical arrows are inclusions of full subcategories and the bottom arrow is an equivalence of ∞-categories by Lemma 7.3.3.5. Consequently, it will suffice to show that if $q_* : \mathcal{Y} \to \mathcal{P}(\mathcal{C} \times \mathcal{D})$ is a geometric morphism with the property that the composition

$$r_* : \mathcal{Y} \to \mathcal{P}(\mathcal{C} \times \mathcal{D}) \to \mathcal{P}(\mathcal{D})$$

factors through $\mathfrak{X}$, then q_* factors through $\mathcal{P}(\mathcal{C}; \mathfrak{X})$.

Let $Y \in \mathcal{Y}$ and $C \in \mathcal{C}$; we wish to show that $q_*(Y)(C) \in \mathfrak{X}$. It will suffice to show that if $s : D \to D'$ is a morphism in $\mathcal{P}(\mathcal{D})$ such that $L(s)$ is an equivalence in $\mathfrak{X}$, then $q_*(Y)(C)$ is s-local. Let $F : \mathcal{P}(\mathcal{D}) \to \mathcal{P}(\mathcal{C} \times \mathcal{D})$ be a left adjoint to the functor given by evaluation at C. We have a commutative diagram

$$
\begin{array}{ccc}
\mathrm{Map}_{\mathcal{P}(\mathcal{D})}(D', q_*(Y)(C)) & \longrightarrow & \mathrm{Map}_{\mathcal{Y}}(q^* F(D'), Y) \\
\downarrow & & \downarrow \\
\mathrm{Map}_{\mathcal{P}(\mathcal{D})}(D, q_*(Y)(C)) & \longrightarrow & \mathrm{Map}_{\mathcal{Y}}(q^* F(D), Y),
\end{array}
$$

where the horizontal arrows are homotopy equivalences. Consequently, to prove that the left vertical map is an equivalence, it will suffice to prove that $q^* F(s)$ is an equivalence in $\mathcal{Y}$. According to Lemma 7.3.3.7, the functor F can be identified with a product of a left adjoint r^* to the projection $r_* : \mathcal{P}(\mathcal{C} \times \mathcal{D}) \to \mathcal{P}(\mathcal{D})$ with a constant functor. Since q^* preserves finite products, it will suffice to show that $(q^* \circ r^*)(s)$ is an equivalence in $\mathcal{Y}$. This follows immediately from our assumption that $r_* \circ q_* : \mathcal{Y} \to \mathcal{P}(\mathcal{D})$ factors through $\mathcal{X}$. $\square$

The main result of this section is the following:

Theorem 7.3.3.9. *Let X be a topological space, $\mathcal{X}$ an ∞-topos, and $\pi_* : \mathcal{X} \to \mathcal{S}$ a geometric morphism (which is essentially unique by virtue of Proposition 6.3.4.1). Then $\mathrm{Shv}(X; \mathcal{X})$ is an ∞-topos, and the diagram*

$$\mathcal{X} \xleftarrow{\Gamma} \mathrm{Shv}(X; \mathcal{X}) \xrightarrow{\pi_*} \mathrm{Shv}(X)$$

exhibits $\mathrm{Shv}(X; \mathcal{X})$ as a product of $\mathrm{Shv}(X)$ and $\mathcal{X}$ in the ∞-category $\mathcal{R}\mathcal{J}\mathrm{op}$ of ∞-topoi. Here Γ denotes the global sections functor given by evaluation at $X \in \mathcal{U}(X)$.

Proof. We first show that $\mathrm{Shv}(X; \mathcal{X})$ is an ∞-topos. Let $\mathcal{P}(X; \mathcal{X})$ be the ∞-category $\mathrm{Fun}(\mathrm{N}(\mathcal{U}(X))^{op}, \mathcal{X})$ of $\mathcal{X}$-valued presheaves on $\mathcal{X}$. For each object $Y \in \mathcal{X}$, choose a morphism $e_Y : \emptyset_{\mathcal{X}} \to Y$ in $\mathcal{X}$ whose source is an initial object of $\mathcal{X}$. For each sieve $\mathcal{V}$ on $\mathcal{X}$, let $\chi_{\mathcal{V}}^Y : \mathcal{U}(X)^{op} \to \mathcal{X}$ be the composition

$$\mathcal{U}(X)^{op} \to \Delta^1 \xrightarrow{e_Y} \mathcal{X},$$

so that

$$\chi_{\mathcal{V}}^Y(U) \begin{cases} Y & \text{if } U \in \mathcal{V} \\ \emptyset_{\mathcal{X}} & \text{if } U \notin \mathcal{V}, \end{cases}$$

so that we have a natural map $\chi_{\mathcal{V}}^Y \to \chi_{\mathcal{V}'}^Y$ if $\mathcal{V} \subseteq \mathcal{V}'$. For each open subset $U \subseteq X$, let $\chi_U^Y = \chi_{\mathcal{V}}^Y$, where $\mathcal{V} = \{V \subseteq U\}$. Let S be the set of all morphisms $f_{\mathcal{V}}^Y : \chi_{\mathcal{V}}^Y \to \chi_U^Y$, where $\mathcal{V}$ is a sieve covering U, and let $\overline{S}$ be the strongly saturated class of morphisms generated by $\mathcal{X}$. We first claim that $\overline{S}$ is setwise-generated. To see this, we observe that the passage from Y to $f_{\mathcal{V}}^Y$ is a colimit-preserving functor of Y, so it suffices to consider a set of objects $Y \in \mathcal{X}$ which generates $\mathcal{X}$ under colimits.

We next claim that $\overline{S}$ is topological in the sense of Definition 6.2.1.5. By a standard argument, it will suffice to show that there is a class of objects $F_\alpha \in \mathcal{P}(X; \mathcal{X})$ which generates $\mathcal{P}(X; \mathcal{X})$ under colimits, such that for every pullback diagram

$$\begin{array}{ccc} F_\alpha' & \longrightarrow & \chi_{\mathcal{V}}^Y \\ {\scriptstyle f'}\downarrow & & \downarrow{\scriptstyle f_{\mathcal{V}}^Y} \\ F_\alpha & \longrightarrow & \chi_U^Y, \end{array}$$

the morphism f' belongs to $\overline{S}$. We observe that if $\mathcal{X}$ is a left exact localization of $\mathcal{P}(\mathcal{D})$, then $\mathcal{P}(X;\mathcal{X})$ is a left exact localization of $\mathcal{P}(\mathcal{U}(X) \times \mathcal{D})$ and is therefore generated under colimits by the Yoneda image of $\mathcal{U}(X) \times \mathcal{D}$. In other words, it will suffice to consider F_α of the form $\chi_{U'}^{Y'}$, where $Y' \in \mathcal{X}$ and $U' \subseteq X$. If Y' is an initial object of $\mathcal{X}$, then g is an equivalence and there is nothing to prove. Otherwise, the existence of the lower horizontal map implies that $U' \subseteq U$. Let $\mathcal{V}' = \{V \in \mathcal{V} : V \subseteq U'\}$; then it is easy to see that f' is equivalent to $\chi_{\mathcal{V}'}^{Y'}$ and therefore belongs to $\overline{S}$.

We next claim that $\mathrm{Shv}(X;\mathcal{X})$ consists precisely of the S-local objects of $\mathcal{P}(X;\mathcal{X})$. To see this, let $Y \in \mathcal{X}$ be an arbitrary object and consider the functor $G_Y : \mathcal{X} \to \mathcal{S}$ corepresented by Y. It follows from Proposition 5.1.3.2 that an arbitrary $F \in \mathcal{P}(X;\mathcal{X})$ is a $\mathcal{X}$-valued sheaf on X if and only if, for each $Y \in \mathcal{X}$, the composition $G_Y \circ F \in \mathcal{P}(X)$ belongs to $\mathrm{Shv}(X)$. This is equivalent to the assertion that, for every sieve $\mathcal{V}$ which covers $U \subseteq X$, the presheaf $G_Y \circ F$ is $s_{\mathcal{V}}$-local, where $s_{\mathcal{V}} : \chi_{\mathcal{V}} \to \chi_U$ is the associated monomorphism of presheaves. Let G_Y^* denote a left adjoint to G_Y; then $G_Y \circ F$ is $s_{\mathcal{V}}$-local if and only if F is $G_Y^*(s_{\mathcal{V}})$-local. We now apply Lemma 7.3.3.7 to identify $G_Y^*(s_{\mathcal{V}})$ with $f_{\mathcal{V}}^Y$.

We have an identification $\mathrm{Shv}(X;\mathcal{X}) \simeq \overline{S}^{-1}\mathcal{P}(X;\mathcal{X})$, so that $\mathrm{Shv}(X;\mathcal{X})$ is a topological localization of $\mathcal{P}(X;\mathcal{X})$ and, in particular, an ∞-topos. We now consider an arbitrary ∞-topos $\mathcal{Y}$. We have a commutative diagram

$$\mathrm{Fun}_*(\mathcal{Y}, \mathrm{Shv}(X;\mathcal{X})) \longrightarrow \mathrm{Fun}_*(\mathcal{Y}, \mathrm{Shv}(X)) \times \mathrm{Fun}_*(\mathcal{Y};\mathcal{X})$$

$$\downarrow \qquad\qquad\qquad\qquad\qquad \downarrow$$

$$\mathrm{Fun}_*(\mathcal{Y}, \mathcal{P}(X;\mathcal{X})) \longrightarrow \mathrm{Fun}_*(\mathcal{Y}, \mathcal{P}(X)) \times \mathrm{Fun}_*(\mathcal{Y}, \mathcal{X}),$$

where the vertical arrows are inclusions of full subcategories and the lower horizontal arrow is an equivalence by Lemma 7.3.3.8. To complete the proof, it will suffice to show that the upper horizontal arrow is also an equivalence. In other words, we must show that if $g_* : \mathcal{Y} \to \mathcal{P}(X;\mathcal{X})$ is a geometric morphism with the property that the composition

$$\mathcal{Y} \xrightarrow{g_*} \mathcal{P}(X;\mathcal{X}) \xrightarrow{h_*} \mathcal{P}(X)$$

factors through $\mathrm{Shv}(X)$, then g_* factors through $\mathrm{Shv}(X;\mathcal{X})$. Let g^* and h^* denote left adjoints to g_* and h_*, respectively. It will suffice to show that for every morphism $f_{\mathcal{V}}^Y \in S$, the pullback $g^* f_{\mathcal{V}}^Y$ is an equivalence in $\mathcal{Y}$. We now observe that $f_{\mathcal{V}}^Y$ is a pullback of $f_{\mathcal{V}}^{1_X}$; since g^* is left exact, it will suffice to show that $g^* f_{\mathcal{V}}^{1_X}$ is an equivalence in $\mathcal{Y}$. We have an equivalence $f_{\mathcal{V}}^{1_X} \simeq h^* s_{\mathcal{V}}$, where $s_{\mathcal{V}}$ is the monomorphism in $\mathcal{P}(X)$ associated to the sieve $\mathcal{V}$. The composition $(g^* \circ h^*)(s_{\mathcal{V}})$ is an equivalence because $h_* \circ g_*$ factors through $\mathrm{Shv}(X)$, which consists of $s_{\mathcal{V}}$-local objects of $\mathcal{P}(X)$. $\qquad\square$

7.3.4 Sheaves on Locally Compact Spaces

By definition, a sheaf of sets $\mathcal{F}$ on a topological space X is determined by the sets $\mathcal{F}(U)$ as U ranges over the open subsets of X. If X is a locally compact Hausdorff space, then there is an alternative collection of data which determines X: the values $\mathcal{F}(K)$, where K ranges over the compact subsets of X. Here $\mathcal{F}(K)$ denotes the direct limit $\varinjlim_{K \subseteq U} \mathcal{F}(U)$ taken over all open neighborhoods of K (or equivalently, the collection of global sections of the restriction $\mathcal{F}|K$). The goal of this section is to prove a generalization of this result where the sheaf $\mathcal{F}$ is allowed to take values in a more general ∞-category $\mathcal{C}$.

Definition 7.3.4.1. Let X be a locally compact Hausdorff space. We let $\mathcal{K}(X)$ denote the collection of all compact subsets of X. If $K, K' \subseteq X$, we write $K \Subset K'$ if there exists an open subset $U \subseteq X$ such that $K \subseteq U \subseteq K'$. If $K \in \mathcal{K}(X)$, we let $\mathcal{K}_{K \Subset}(X) = \{K' \in \mathcal{K}(X) : K \Subset K'\}$.

Let $\mathcal{F} : \mathrm{N}(\mathcal{K}(X))^{op} \to \mathcal{C}$ be a presheaf on $\mathrm{N}(\mathcal{K}(X))$ (here $\mathcal{K}(X)$ is viewed as a partially ordered set with respect to inclusion) with values in $\mathcal{C}$. We will say that $\mathcal{F}$ is a $\mathcal{K}$-*sheaf* if the following conditions are satisfied:

(1) The object $\mathcal{F}(\emptyset) \in \mathcal{C}$ is final.

(2) For every pair $K, K' \in \mathcal{K}(X)$, the associated diagram

$$
\begin{array}{ccc}
\mathcal{F}(K \cup K') & \longrightarrow & \mathcal{F}(K) \\
\downarrow & & \downarrow \\
\mathcal{F}(K') & \longrightarrow & \mathcal{F}(K \cap K')
\end{array}
$$

is a pullback square in $\mathcal{C}$.

(3) For each $K \in \mathcal{K}(X)$, the restriction of $\mathcal{F}$ exhibits $\mathcal{F}(K)$ as a colimit of $\mathcal{F}|\mathrm{N}(\mathcal{K}_{K \Subset}(X))^{op}$.

We let $\mathrm{Shv}_{\mathcal{K}}(X; \mathcal{C})$ denote the full subcategory of $\mathrm{Fun}(\mathrm{N}(\mathcal{K}(X))^{op}, \mathcal{C})$ spanned by the $\mathcal{K}$-sheaves. In the case where $\mathcal{C} = \mathcal{S}$, we will write $\mathrm{Shv}_{\mathcal{K}}(X)$ instead of $\mathrm{Shv}_{\mathcal{K}}(X; \mathcal{C})$.

Definition 7.3.4.2. Let $\mathcal{C}$ be a presentable ∞-category. We will say that *filtered colimits in $\mathcal{C}$ are left exact* if the following condition is satisfied: for every small filtered ∞-category $\mathcal{I}$, the colimit functor $\mathrm{Fun}(\mathcal{I}, \mathcal{C}) \to \mathcal{C}$ is left exact.

Example 7.3.4.3. A *Grothendieck abelian category* is an abelian category $\mathcal{A}$ whose nerve $\mathrm{N}(\mathcal{A})$ is a presentable ∞-category with left exact filtered colimits in the sense of Definition 7.3.4.2. We refer the reader to [34] for further discussion.

Example 7.3.4.4. Filtered colimits are left exact in the ∞-category $\mathcal{S}$ of spaces; this follows immediately from Proposition 5.3.3.3. It follows that filtered colimits in $\tau_{\leq n} \mathcal{S}$ are left exact for each $n \geq -2$ since the full subcategory $\tau_{\leq n} \mathcal{S} \subseteq \mathcal{S}$ is stable under filtered colimits and finite limits (in fact, under all limits).

Example 7.3.4.5. Let $\mathcal{C}$ be a presentable ∞-category in which filtered co-limits are left exact and let X be an arbitrary simplicial set. Then filtered colimits are left exact in $\mathrm{Fun}(X, \mathcal{C})$. This follows immediately from Proposition 5.1.2.2, which asserts that the relevant limits and colimits can be computed pointwise.

Example 7.3.4.6. Let $\mathcal{C}$ be a presentable ∞-category in which filtered colimits are left exact and let $\mathcal{D} \subseteq \mathcal{C}$ be the essential image of an (accessible) left exact localization functor L. Then filtered colimits in $\mathcal{D}$ are left exact. To prove this, we consider an arbitrary filtered ∞-category $\mathcal{J}$ and observe that the colimit functor $\varinjlim : \mathrm{Fun}(\mathcal{J}, \mathcal{D}) \to \mathcal{D}$ is equivalent to the composition

$$\mathrm{Fun}(\mathcal{J}, \mathcal{D}) \subseteq \mathrm{Fun}(\mathcal{J}, \mathcal{C}) \to \mathcal{C} \xrightarrow{L} \mathcal{D},$$

where the second arrow is given by the colimit functor $\varinjlim \mathrm{Fun}(\mathcal{J}, \mathcal{C}) \to \mathcal{C}$.

Example 7.3.4.7. Let $\mathcal{X}$ be an n-topos, $0 \leq n \leq \infty$. Then filtered colimits in $\mathcal{X}$ are left exact. This follows immediately from Examples 7.3.4.4, 7.3.4.5, and 7.3.4.6.

Our goal is to prove that if X is a locally compact Hausdorff space and $\mathcal{C}$ is a presentable ∞-category, then the ∞-categories $\mathrm{Shv}(X)$ and $\mathrm{Shv}_{\mathcal{K}}(X)$ are equivalent. As a first step, we prove that a $\mathcal{K}$-sheaf on X is determined locally.

Lemma 7.3.4.8. *Let X be a locally compact Hausdorff space and $\mathcal{C}$ a presentable ∞-category in which filtered colimits are left exact. Let $\mathcal{W}$ be a collection of open subsets of X which covers X and let $\mathcal{K}_{\mathcal{W}}(X) = \{K \in \mathcal{K}(X) : (\exists W \in \mathcal{W})[K \subseteq W]\}$. Suppose that $\mathcal{F} \in \mathrm{Shv}_{\mathcal{K}}(X; \mathcal{C})$. Then $\mathcal{F}$ is a right Kan extension of $\mathcal{F} \,|\, \mathrm{N}(\mathcal{K}_{\mathcal{W}}(X))^{op}$.*

Proof. Let us say that an open covering $\mathcal{W}$ of a locally compact Hausdorff space X is *good* if it satisfies the conclusion of the lemma. Note that $\mathcal{W}$ is a good covering of X if and only if, for every compact subset $K \subseteq X$, the open sets $\{K \cap W : W \in \mathcal{W}\}$ form a good covering of K. We wish to prove that *every* covering $\mathcal{W}$ of a locally compact topological space X is good. By virtue of the preceding remarks, we can reduce to the case where X is compact and thereby assume that $\mathcal{W}$ has a finite subcover.

We will prove, by induction on $n \geq 0$, that if $\mathcal{W}$ is a collection of open subsets of a locally compact Hausdorff space X such that there exist elements $W_1, \cdots, W_n \in \mathcal{W}$ with $W_1 \cup \ldots \cup W_n = X$, then $\mathcal{W}$ is a good covering of X. If $n = 0$, then $X = \emptyset$. In this case, we must prove that $\mathcal{F}(\emptyset)$ is final, which is part of the definition of $\mathcal{K}$-sheaf.

Suppose that $\mathcal{W} \subseteq \mathcal{W}'$ are coverings of X and that for every $W' \in \mathcal{W}'$ the induced covering $\{W \cap W' : W \in \mathcal{W}\}$ is a good covering of W'. It then follows from Proposition 4.3.2.8 that $\mathcal{W}'$ is a good covering of X if and only if $\mathcal{W}$ is a good covering of X.

Now suppose $n > 0$. Let $V = W_2 \cup \cdots \cup W_n$, and let $\mathcal{W}' = \mathcal{W} \cup \{V\}$. Using the above remark and the inductive hypothesis, it will suffice to show that

W' is a good covering of X. Now W' contains a pair of open sets W_1 and V which cover X. We thereby reduce to the case $n = 2$; using the above remark, we can furthermore suppose that $W = \{W_1, W_2\}$.

We now wish to show that for every compact $K \subseteq X$, $\mathcal{F}$ exhibits $\mathcal{F}(K)$ as the limit of $\mathcal{F} \,|\, \mathrm{N}(\mathcal{K}_W(X))^{op}$. Let P be the collection of all pairs $K_1, K_2 \in \mathcal{K}(X)$ such that $K_1 \subseteq W_1$, $K_2 \subseteq W_2$, and $K_1 \cup K_2 = K$. We observe that P is filtered when ordered by inclusion. For $\alpha = (K_1, K_2) \in P$, let $\mathcal{K}_\alpha = \{K' \in \mathcal{K}(X) : (K' \subseteq K_1) \vee (K' \subseteq K_2)\}$. We note that $\mathcal{K}_W(X) = \bigcup_{\alpha \in P} \mathcal{K}_\alpha$. Moreover, Theorem 4.1.3.1 implies that for $\alpha = (K_1, K_2) \in P$, the inclusion $\mathrm{N}\{K_1, K_2, K_1 \cap K_2\} \subseteq \mathrm{N}(\mathcal{K}_\alpha)$ is cofinal. Since $\mathcal{F}$ is a $\mathcal{K}$-sheaf, we deduce that $\mathcal{F}$ exhibits $\mathcal{F}(K)$ as a limit of the diagram $\mathcal{F} \,|\, \mathrm{N}(\mathcal{K}_\alpha)^{op}$ for each $\alpha \in P$. Using Proposition 4.2.3.4, we deduce that $\mathcal{F}(K)$ is a limit of $\mathcal{F} \,|\, \mathrm{N}(\mathcal{K}_W(X))^{op}$ if and only if $\mathcal{F}(K)$ is a limit of the constant diagram $\mathrm{N}(P)^{op} \to \mathcal{S}$ taking the value $\mathcal{F}(K)$. This is clear since P is filtered so that the map $\mathrm{N}(P) \to \Delta^0$ is cofinal by Theorem 4.1.3.1. $\qquad\square$

Theorem 7.3.4.9. *Let X be a locally compact Hausdorff space and let $\mathcal{C}$ be a presentable ∞-category in which filtered colimits are left exact. Let $\mathcal{F} : \mathrm{N}(\mathcal{K}(X) \cup \mathcal{U}(X))^{op} \to \mathcal{C}$ be a presheaf on the partially ordered set $\mathcal{K}(X) \cup \mathcal{U}(X)$. The following conditions are equivalent:*

(1) *The presheaf $\mathcal{F}_{\mathcal{K}} = \mathcal{F} \,|\, \mathrm{N}(\mathcal{K}(X))^{op}$ is a $\mathcal{K}$-sheaf, and $\mathcal{F}$ is a right Kan extension of $\mathcal{F}_{\mathcal{K}}$.*

(2) *The presheaf $\mathcal{F}_{\mathcal{U}} = \mathcal{F} \,|\, \mathrm{N}(\mathcal{U}(X))^{op}$ is a sheaf, and $\mathcal{F}$ is a left Kan extension of $\mathcal{F}_{\mathcal{U}}$.*

Proof. Suppose first that (1) is satisfied. We first prove that $\mathcal{F}$ is a left Kan extension of $\mathcal{F}_{\mathcal{U}}$. Let K be a compact subset of X and let $\mathcal{U}_{K \subseteq}(X) = \{U \in \mathcal{U}(X) : K \subseteq U\}$. Consider the diagram

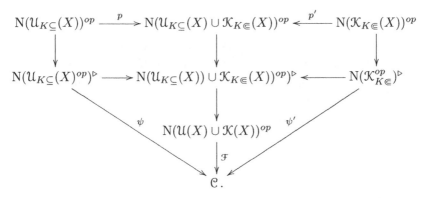

We wish to prove that ψ is a colimit diagram. Since $\mathcal{F}_{\mathcal{K}}$ is a $\mathcal{K}$-sheaf, we deduce that ψ' is a colimit diagram. It therefore suffices to check that p and p' are cofinal. According to Theorem 4.1.3.1, it suffices to show that for every $Y \in \mathcal{U}_{K \subseteq}(X) \cup \mathcal{K}_{K \in}(X)$, the partially ordered sets $\{K' \in \mathcal{K}(X) : K \in K' \subseteq Y\}$ and $\{U \in \mathcal{U}(X) : K \subseteq U \subseteq Y\}$ have contractible nerves. We now observe that both of these partially ordered sets is filtered since they are nonempty and stable under finite unions.

We now show that $\mathcal{F}_\mathcal{U}$ is a sheaf. Let U be an open subset of X and let $\mathcal{W}$ be a sieve which covers U. Let $\mathcal{K}_{\subseteq U}(X) = \{K \in \mathcal{K}(X) : K \subseteq U\}$ and let $\mathcal{K}_\mathcal{W}(X) = \{K \in \mathcal{K}(X) : (\exists W \in \mathcal{W})[K \subseteq W]\}$. We wish to prove that the diagram

$$N(\mathcal{W}^{op})^\triangleleft \to N(\mathcal{U}(X))^{op} \overset{\mathcal{F}_\mathcal{U}}{\to} \mathcal{S}$$

is a limit. Using Theorem 4.1.3.1, we deduce that the inclusion

$$N(\mathcal{W}) \subseteq N(\mathcal{W} \cup \mathcal{K}_\mathcal{W}(X))$$

is cofinal. It therefore suffices to prove that $\mathcal{F}|(\mathcal{W} \cup \mathcal{K}_\mathcal{W}(X) \cup \{U\})^{op}$ is a right Kan extension of $\mathcal{F}|(\mathcal{W} \cup \mathcal{K}_\mathcal{W}(X))^{op}$. Since $\mathcal{F}|(\mathcal{W} \cup \mathcal{K}_\mathcal{W}(X))^{op}$ is a right Kan extension of $\mathcal{F}|\mathcal{K}_\mathcal{W}(X)^{op}$ by assumption, it suffices to prove that $\mathcal{F}|(\mathcal{W} \cup \mathcal{K}_\mathcal{W}(X) \cup \{U\})^{op}$ is a right Kan extension of $\mathcal{F}|\mathcal{K}_\mathcal{W}(X)^{op}$. This is clear at every object distinct from U; it will therefore suffice to prove that $\mathcal{F}|(\mathcal{K}_\mathcal{W}(X) \cup \{U\})^{op}$ is a right Kan extension of $\mathcal{F}|\mathcal{K}_\mathcal{W}(X)^{op}$.

By assumption, the functor $\mathcal{F}|N(\mathcal{K}_{\subseteq U}(X) \cup \{U\})^{op}$ is a right Kan extension of $\mathcal{F}|N(\mathcal{K}_{\subseteq U}(X))^{op}$ and Lemma 7.3.4.8 implies that $\mathcal{F}|N(\mathcal{K}_{\subseteq U}(X))^{op}$ is a right Kan extension of $\mathcal{F}|N(\mathcal{K}_\mathcal{W}(X))^{op}$. Using Proposition 4.3.2.8, we deduce that $\mathcal{F}|N(\mathcal{K}_\mathcal{W}(X) \cup \{U\})^{op}$ is a right Kan extension of $\mathcal{F}|N(\mathcal{K}_\mathcal{W}(X))^{op}$. This shows that $\mathcal{F}_\mathcal{U}$ is a sheaf and completes the proof that $(1) \Rightarrow (2)$.

Now suppose that $\mathcal{F}$ satisfies (2). We first verify that $\mathcal{F}_\mathcal{K}$ is a $\mathcal{K}$-sheaf. The space $\mathcal{F}_\mathcal{K}(\emptyset) = \mathcal{F}_\mathcal{U}(\emptyset)$ is contractible because $\mathcal{F}_\mathcal{U}$ is a sheaf (and because the empty sieve is a covering sieve on $\emptyset \subseteq X$). Suppose next that K and K' are compact subsets of X. We wish to prove that the diagram

$$
\begin{array}{ccc}
\mathcal{F}(K \cup K') & \longrightarrow & \mathcal{F}(K) \\
\downarrow & & \downarrow \\
\mathcal{F}(K') & \longrightarrow & \mathcal{F}(K \cap K')
\end{array}
$$

is a pullback in $\mathcal{S}$. Let us denote this diagram by $\sigma : \Delta^1 \times \Delta^1 \to \mathcal{S}$. Let P be the set of all pairs $U, U' \in \mathcal{U}(X)$ such that $K \subseteq U$ and $K' \subseteq U'$. The functor $\mathcal{F}$ induces a map $\sigma_P : N(P^{op})^\triangleright \to \mathcal{S}^{\Delta^1 \times \Delta^1}$, which carries each pair (U, U') to the diagram

$$
\begin{array}{ccc}
\mathcal{F}(U \cup U') & \longrightarrow & \mathcal{F}(U) \\
\downarrow & & \downarrow \\
\mathcal{F}(U') & \longrightarrow & \mathcal{F}(U \cap U')
\end{array}
$$

and carries the cone point to σ. Since $\mathcal{F}_U$ is a sheaf, each $\sigma_P(U, U')$ is a pullback diagram in $\mathcal{C}$. Since filtered colimits in $\mathcal{C}$ are left exact, it will suffice to show that σ_P is a colimit diagram. By Proposition 5.1.2.2, it suffices to show that each of the four maps

$$N(P^{op})^\triangleright \to \mathcal{S},$$

given by evaluating σ_P at the four vertices of $\Delta^1 \times \Delta^1$, is a colimit diagram. We will treat the case of the final vertex; the other cases are handled in the same way. Let $Q = \{U \in \mathcal{U}(X) : K \cap K' \subseteq U\}$. We are given a map $g : N(P^{op})^{\rhd} \to \mathcal{S}$ which admits a factorization

$$N(P^{op})^{\rhd} \xrightarrow{g''} N(Q^{op})^{\rhd} \xrightarrow{g'} N(\mathcal{U}(X) \cup \mathcal{K}(X))^{op} \xrightarrow{\mathcal{F}} \mathcal{C}.$$

Since $\mathcal{F}$ is a left Kan extension of $\mathcal{F}_{\mathcal{U}}$, the diagram $\mathcal{F} \circ g''$ is a colimit. It therefore suffices to show that g'' induces a cofinal map $N(P)^{op} \to N(Q)^{op}$. Using Theorem 4.1.3.1, it suffices to prove that for every $U'' \in Q$, the partially ordered set $P_{U''} = \{(U, U') \in P : U \cap U' \subseteq U''\}$ has contractible nerve. It now suffices to observe that $P_{U''}^{op}$ is filtered (since $P_{U''}$ is nonempty and stable under intersections).

We next show that for any compact subset $K \subseteq X$, the map

$$N(\mathcal{K}_{K \in}(X)^{op})^{\rhd} \to N(\mathcal{K}(X) \cup \mathcal{U}(X))^{op} \xrightarrow{\mathcal{F}} \mathcal{C}$$

is a colimit diagram. Let $\mathcal{V} = \mathcal{U}(X) \cup \mathcal{K}_{K \in}(X)$ and let $\mathcal{V}' = \mathcal{V} \cup \{K\}$. It follows from Proposition 4.3.2.8 that $\mathcal{F} \mid N(\mathcal{V})^{op}$ and $\mathcal{F} \mid N(\mathcal{V}')^{op}$ are left Kan extensions of $\mathcal{F} \mid N(\mathcal{U}(X))^{op}$, so that $\mathcal{F} \mid N(\mathcal{V}')^{op}$ is a left Kan extension of $\mathcal{F} \mid N(\mathcal{V})^{op}$. Therefore the diagram

$$(N(\mathcal{K}_{K \in}(X) \cup \{U \in \mathcal{U}(X) : K \subseteq U\})^{op})^{\rhd} \to N(\mathcal{K}(X) \cup \mathcal{U}(X))^{op} \xrightarrow{\mathcal{F}} \mathcal{C}$$

is a colimit. It therefore suffices to show that the inclusion

$$N(\mathcal{K}_{K \in}(X))^{op} \subseteq N(\mathcal{K}_{K \in}(X) \cup \{U \in \mathcal{U}(X) : K \subseteq U\})^{op}$$

is cofinal. Using Theorem 4.1.3.1, we are reduced to showing that if

$$Y \in \mathcal{K}_{K \in}(X) \cup \{U \in \mathcal{U}(X) : K \subseteq U\},$$

then the nerve of the partially ordered set $R = \{K' \in \mathcal{K}(X) : K \in K' \subset Y\}$ is weakly contractible. It now suffices to observe that R^{op} is filtered since R is nonempty and stable under intersections. This completes the proof that $\mathcal{F}_{\mathcal{K}}$ is a $\mathcal{K}$-sheaf.

We now show that $\mathcal{F}$ is a right Kan extension of $\mathcal{F}_{\mathcal{K}}$. Let U be an open subset of X and for $V \in \mathcal{U}(X)$ write $V \Subset U$ if the closure $\overline{V}$ is compact and contained in U. Let $\mathcal{U}_{\Subset U}(X) = \{V \in \mathcal{U}(X) : V \Subset U\}$ and consider the diagram

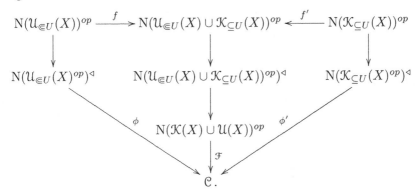

We wish to prove that ϕ' is a limit diagram. Since the sieve $\mathcal{U}_{\in U}(X)$ covers U and $\mathcal{F}_{\mathcal{U}}$ is a sheaf, we conclude that ϕ is a limit diagram. It therefore suffices to prove that f^{op} and $(f')^{op}$ are cofinal maps of simplicial sets. According to Theorem 4.1.3.1, it suffices to prove that if $Y \in \mathcal{K}_{\subseteq U}(X) \cup \mathcal{U}_{\in U}(X)$, then the partially ordered sets $\{V \in \mathcal{U}(X) : Y \subseteq V \in U\}$ and $\{K \in \mathcal{K}(X) : Y \subseteq K \subseteq U\}$ have weakly contractible nerves. We now observe that both of these partially ordered sets are filtered (since they are nonempty and stable under unions). This completes the proof that $\mathcal{F}$ is a right Kan extension of $\mathcal{F}_{\mathcal{K}}$. $\square$

Corollary 7.3.4.10. *Let X be a locally compact topological space and $\mathcal{C}$ a presentable ∞-category in which filtered colimits are left exact. Let*

$$\mathrm{Shv}_{\mathcal{K}\mathcal{U}}(X; \mathcal{C}) \subseteq \mathrm{Fun}(\mathrm{N}(\mathcal{K}(X) \cup \mathcal{U}(X))^{op}, \mathcal{C})$$

be the full subcategory spanned by those presheaves which satisfy the equivalent conditions of Theorem 7.3.4.9. Then the restriction functors

$$\mathrm{Shv}(X; \mathcal{C}) \leftarrow \mathrm{Shv}_{\mathcal{K}\mathcal{U}}(X; \mathcal{C}) \rightarrow \mathrm{Shv}_{\mathcal{K}}(X; \mathcal{C})$$

are equivalences of ∞-categories.

Corollary 7.3.4.11. *Let X be a compact Hausdorff space. Then the global sections functor $\Gamma : \mathrm{Shv}(X) \rightarrow \mathcal{S}$ is a proper morphism of ∞-topoi.*

Proof. The existence of fiber products $\mathrm{Shv}(X) \times_{\mathcal{S}} \mathcal{Y}$ in $\mathcal{RT}\mathrm{op}$ follows from Theorem 7.3.3.9. It will therefore suffice to prove that for any (homotopy) Cartesian rectangle

$$
\begin{array}{ccccc}
\mathcal{X}'' & \longrightarrow & \mathcal{X}' & \longrightarrow & \mathrm{Shv}(X) \\
\downarrow & & \downarrow & & \downarrow \\
\mathcal{Y}'' & \xrightarrow{\;f_*\;} & \mathcal{Y}' & \longrightarrow & \mathcal{S},
\end{array}
$$

the square on the left is left adjointable. Using Theorem 7.3.3.9, we can identify the square on the left with

$$
\begin{array}{ccc}
\mathrm{Shv}(X; \mathcal{Y}'') & \longrightarrow & \mathrm{Shv}(X; \mathcal{Y}') \\
\downarrow & & \downarrow \\
\mathcal{Y}'' & \xrightarrow{\;f_*\;} & \mathcal{Y}',
\end{array}
$$

where the vertical morphisms are given by taking global sections.

Choose a correspondence $\mathcal{M}$ from $\mathcal{Y}'$ to $\mathcal{Y}''$ which is associated to the functor f_*. Since f_* admits a left adjoint f^*, the projection $\mathcal{M} \rightarrow \Delta^1$ is both a Cartesian fibration and a coCartesian fibration. For every simplicial set K, let $\mathcal{M}_K = \mathrm{Fun}(K, \mathcal{M}) \times_{\mathrm{Fun}(K, \Delta^1)} \Delta^1$. Then $\mathcal{M}_K$ determines a correspondence from $\mathrm{Fun}(K, \mathcal{Y}')$ to $\mathrm{Fun}(K, \mathcal{Y}'')$. Using Proposition 3.1.2.1, we conclude that $\mathcal{M}_K \rightarrow \Delta^1$ is both a Cartesian and a coCartesian fibration, and that it is associated to the functors given by composition with f_* and f^*.

Before proceeding further, let us adopt the following convention for the remainder of the proof: given a simplicial set Z with a map $q : Z \rightarrow \Delta^1$, we will say that an edge of Z is *Cartesian* or *coCartesian* if it is q-Cartesian or q-coCartesian, respectively. The map q to which we are referring should be clear from the context.

Let $\mathcal{M}_{\mathcal{U}}$ denote the full subcategory of $\mathcal{M}_{N(\mathcal{U}(X))^{op}}$ whose objects correspond to *sheaves* on X (with values in either $\mathcal{Y}'$ or $\mathcal{Y}''$). Since f_* preserves limits, composition with f_* carries $\mathrm{Shv}(X; \mathcal{Y}'')$ into $\mathrm{Shv}(X; \mathcal{Y}')$. We conclude that the projection $\mathcal{M}_{\mathcal{U}} \to \Delta^1$ is a Cartesian fibration and that the inclusion $\mathcal{M}_{\mathcal{U}} \subseteq \mathcal{M}_{N(\mathcal{U}(X))^{op}}$ preserves Cartesian edges.

Similarly, we define $\mathcal{M}_{\mathcal{K}}$ to be the full subcategory of $\mathcal{M}_{N(\mathcal{K}(X))^{op}}$ whose objects correspond to $\mathcal{K}$-sheaves on X (with values in either $\mathcal{Y}'$ or $\mathcal{Y}''$). Since f^* preserves finite limits and filtered colimits, composition with f^* carries $\mathrm{Shv}_{\mathcal{K}}(X; \mathcal{Y}')$ into $\mathrm{Shv}_{\mathcal{K}}(X; \mathcal{Y}'')$. It follows that the projection $\mathcal{M}_{\mathcal{K}} \to \Delta^1$ is a coCartesian fibration and that the inclusion $\mathcal{M}_{\mathcal{K}} \subseteq \mathcal{M}_{N(\mathcal{U}(X))^{op}}$ preserves coCartesian edges.

Now let $\mathcal{M}'_{\mathcal{K}\mathcal{U}} = \mathcal{M}_{N(\mathcal{K}(X) \cup \mathcal{U}(X))^{op}}$ and let $\mathcal{M}_{\mathcal{K}\mathcal{U}}$ be the full subcategory of $\mathcal{M}'_{\mathcal{K}\mathcal{U}}$ spanned by the objects of $\mathrm{Shv}_{\mathcal{K}\mathcal{U}}(X; \mathcal{Y}')$ and $\mathrm{Shv}_{\mathcal{K}\mathcal{U}}(X; \mathcal{Y}'')$. We have a commutative diagram

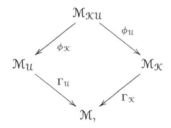

where $\Gamma_{\mathcal{U}}$ and $\Gamma_{\mathcal{K}}$ denote the global sections functors (given by evaluation at $X \in \mathcal{U}(X) \cap \mathcal{K}(X)$). According to Remark 7.3.1.3, to complete the proof it will suffice to show that $\mathcal{M}_{\mathcal{U}} \to \Delta^1$ is a coCartesian fibration and that $\Gamma_{\mathcal{U}}$ preserves both Cartesian and coCartesian edges. It is clear that $\Gamma_{\mathcal{U}}$ preserves Cartesian edges since it is a composition of maps

$$\mathcal{M}_{\mathcal{U}} \subseteq \mathcal{M}_{N(\mathcal{U}(X))^{op}} \to \mathcal{M}$$

which preserve Cartesian edges. Similarly, we already know that $\mathcal{M}_{\mathcal{K}} \to \Delta^1$ is a coCartesian fibration and that $\Gamma_{\mathcal{K}}$ preserves coCartesian edges. To complete the proof, it will therefore suffice to show that $\phi_{\mathcal{U}}$ and $\phi_{\mathcal{K}}$ are equivalences of ∞-categories. We will give the argument for $\phi_{\mathcal{U}}$; the proof in the case of $\phi_{\mathcal{K}}$ is identical and is left to the reader.

According to Corollary 7.3.4.10, the map $\phi_{\mathcal{U}}$ induces equivalences

$$\mathrm{Shv}_{\mathcal{K}\mathcal{U}}(X; \mathcal{Y}') \to \mathrm{Shv}(X; \mathcal{Y}')$$

$$\mathrm{Shv}_{\mathcal{K}\mathcal{U}}(X; \mathcal{Y}'') \to \mathrm{Shv}(X; \mathcal{Y}'')$$

after passing to the fibers over either vertex of Δ^1. We will complete the proof by applying Corollary 2.4.4.4. In order to do so, we must verify that $p : \mathcal{M}_{\mathcal{K}\mathcal{U}} \to \Delta^1$ is a Cartesian fibration and that $\phi_{\mathcal{U}}$ preserves Cartesian edges.

To show that p is a Cartesian fibration, we begin with an arbitrary $\mathcal{F} \in \mathrm{Shv}_{\mathcal{K}\mathcal{U}}(X; \mathcal{Y}'')$. Using Proposition 3.1.2.1, we conclude the existence of a p'-Cartesian morphism $\alpha : \mathcal{F}' \to \mathcal{F}$, where p' denotes the projection $\mathcal{M}'_{\mathcal{K}\mathcal{U}}$ and $\mathcal{F}' = \mathcal{F} \circ p_* \in \mathrm{Fun}(\mathrm{N}(\mathcal{K}(X) \cup \mathcal{U}(X))^{op}, \mathcal{Y}')$. Since p_* preserves limits, we conclude that $\mathcal{F}' \,|\, \mathrm{N}(\mathcal{U}(X))^{op}$ is a sheaf on X with values in $\mathcal{Y}'$; however, $\mathcal{F}'$ is not necessarily a left Kan extension of $\mathcal{F}' \,|\, \mathrm{N}(\mathcal{U}(X))^{op}$. Let $\mathcal{C}$ denote the full subcategory of $\mathrm{Fun}(\mathrm{N}(\mathcal{K}(X) \cup \mathcal{U}(X))^{op}, \mathcal{Y}')$ spanned by those functors $\mathcal{G} : \mathrm{N}(\mathcal{K}(X) \cup \mathcal{U}(X))^{op}$ which are left Kan extensions of $\mathcal{G} \,|\, \mathrm{N}(\mathcal{U}(X))^{op}$, and let s a section of the trivial fibration $\mathcal{C} \to (\mathcal{Y}')^{\mathrm{N}(\mathcal{U}(X))^{op}}$, so that s is a left adjoint to the restriction map $r : \mathcal{M}'_{\mathcal{K}\mathcal{U}} \to (\mathcal{Y}')^{\mathrm{N}(\mathcal{U}(X))^{op}}$. Let $\mathcal{F}'' = (s \circ r)\,\mathcal{F}'$ be a left Kan extension of $\mathcal{F}' \,|\, \mathrm{N}(\mathcal{U}(X))^{op}$. Then $\mathcal{F}''$ is an initial object of the fiber $\mathcal{M}'_{\mathcal{K}\mathcal{U}} \times_{\mathrm{Fun}(\mathrm{N}(\mathcal{U}(X))^{op}, \mathcal{Y}')} \{\mathcal{F}' \,|\, \mathrm{N}(\mathcal{U}(X))^{op}\}$, so that there exists a map $\beta : \mathcal{F}'' \to \mathcal{F}'$ which induces the identity on $\mathcal{F}'' \,|\, \mathrm{N}(\mathcal{U}(X))^{op} = \mathcal{F}' \,|\, \mathrm{N}(\mathcal{U}(X))^{op}$.

Let $\sigma : \Delta^2 \to \mathcal{M}'_{\mathcal{K}\mathcal{U}}$ classify a diagram

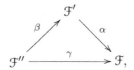

so that γ is a composition of α and β. It is easy to see that $\phi_{\mathcal{U}}(\gamma)$ is a Cartesian edge of $\mathcal{M}_{\mathcal{U}}$ (since it is a composition of a Cartesian edge with an equivalence in $\mathrm{Shv}(X; \mathcal{Y}')$). We claim that γ is p-Cartesian. To prove this, consider the diagram

$$
\begin{array}{ccc}
\mathrm{Shv}_{\mathcal{K}\mathcal{U}}(X; \mathcal{Y}') \times_{\mathcal{M}'_{\mathcal{K}\mathcal{U}}} (\mathcal{M}'_{\mathcal{K}\mathcal{U}})_{/\sigma} & \xrightarrow{\ \eta'\ } & (\mathcal{M}_{\mathcal{K}\mathcal{U}})_{/\gamma} \\
\Big\downarrow{\scriptstyle \theta_0} & & \\
\mathrm{Shv}_{\mathcal{K}\mathcal{U}}(X; \mathcal{Y}')_{/\beta} \times_{\mathrm{Shv}_{\mathcal{K}\mathcal{U}}(X;\mathcal{Y}')_{/\mathcal{F}'}} (\mathcal{M}'_{\mathcal{K}\mathcal{U}})_{/\alpha} & & \Big\downarrow{\scriptstyle \eta} \\
\Big\downarrow{\scriptstyle \theta_1} & & \\
\mathrm{Shv}_{\mathcal{K}\mathcal{U}}(X; \mathcal{Y}') \times_{\mathcal{M}'_{\mathcal{K}\mathcal{U}}} (\mathcal{M}'_{\mathcal{K}\mathcal{U}})_{/\alpha} & \xrightarrow{\ \theta_2\ } & \mathcal{Z},
\end{array}
$$

where $\mathcal{Z}$ denotes the fiber product $\mathrm{Shv}_{\mathcal{K}\mathcal{U}}(X; \mathcal{Y}') \times_{\mathcal{M}_{\mathcal{K}\mathcal{U}}} (\mathcal{M}_{\mathcal{K}\mathcal{U}})_{/\mathcal{F}}$. We wish to show that η is a trivial fibration. Since η is a right fibration, it suffices to show that the fibers of η are contractible. The map η' is a trivial fibration (since the inclusion $\Delta^{\{0,2\}} \subseteq \Delta^2$ is right anodyne), so it will suffice to prove that $\eta \circ \eta'$ is a trivial fibration. In view of the commutativity of the diagram, it will suffice to show that θ_0, θ_1, and θ_2 are trivial fibrations. The triviality of θ_0 follows from the fact that the horn inclusion $\Lambda^2_1 \subseteq \Delta^2$ is right anodyne. The triviality of θ_2 follows from the fact that α is p'-Cartesian. Finally, we observe that θ_1 is a pullback of the map $\theta'_1 : \mathrm{Shv}_{\mathcal{K}\mathcal{U}}(X; \mathcal{Y}')_{/\beta} \to \mathrm{Shv}_{\mathcal{K}\mathcal{U}}(X; \mathcal{Y}')_{/\mathcal{F}'}$. Let $\mathcal{C} = (\mathcal{Y}')^{\mathrm{N}(\mathcal{K}(X) \cup \mathcal{U}(X))^{op}}$. To prove that θ'_1 is a trivial fibration, we must show that for every $\mathcal{G} \in \mathrm{Shv}_{\mathcal{K}\mathcal{U}}$, composition with β induces a homotopy equivalence

$$\mathrm{Map}_{\mathcal{C}}(\mathcal{G}, \mathcal{F}'') \to \mathrm{Map}_{\mathcal{C}}(\mathcal{G}, \mathcal{F}').$$

Without loss of generality, we may suppose that $\mathcal{G} = s(\mathcal{G}')$, where $\mathcal{G}' \in$ $\mathrm{Shv}(X; \mathcal{Y}')$; now we simply invoke the adjointness of s with the restriction functor r and the observation that $r(\beta)$ is an equivalence. $\qquad\square$

Corollary 7.3.4.12. *Let X be a compact Hausdorff space. The global sections functor $\Gamma : \mathrm{Shv}(X) \to \mathcal{S}$ preserves filtered colimits.*

Proof. Applying Theorem 7.3.4.9, we can replace $\mathrm{Shv}(X)$ by $\mathrm{Shv}_{\mathcal{K}}(X)$. Now observe that the full subcategory $\mathrm{Shv}_{\mathcal{K}}(X) \subseteq \mathcal{P}(\mathrm{N}(\mathcal{K}(X))^{op})$ is stable under filtered colimits. We thereby reduce to proving that the evaluation functor $\mathcal{P}(\mathrm{N}(\mathcal{K}(X))^{op}) \to \mathcal{S}$ commutes with filtered colimits, which follows from Proposition 5.1.2.2. Alternatively, one can apply Corollary 7.3.4.10 and Remark 7.3.1.5. $\qquad\square$

Remark 7.3.4.13. One can also deduce Corollary 7.3.4.12 using the geometric model for $\mathrm{Shv}(X)$ introduced in §7.1. Using the characterization of properness in terms of filtered colimits described in Remark 7.3.1.5, one can formally deduce Corollary 7.3.4.11 from Corollary 7.3.4.12. This leads to another proof of the proper base change theorem, which does not make use of Theorem 7.3.4.9 or the other ideas of this section. However, this alternative proof is considerably more difficult than the one described here since it requires a rigorous justification of Remark 7.3.1.5. We also note that Theorem 7.3.4.9 and Corollary 7.3.4.10 are interesting in their own right and could conceivably be applied in other contexts.

7.3.5 Sheaves on Coherent Spaces

Theorem 7.3.4.9 has an analogue in the setting of coherent topological spaces which is somewhat easier to prove. First, we need the analogue of Lemma 7.3.4.8:

Lemma 7.3.5.1. *Let X be a coherent topological space, let $\mathcal{U}_0(X)$ denote the collection of compact open subsets of X, and let $\mathcal{F} : \mathrm{N}(\mathcal{U}_0(X))^{op} \to \mathcal{C}$ be a presheaf taking values in an ∞-category $\mathcal{C}$ having the following properties:*

(1) *The object $\mathcal{F}(\emptyset) \in \mathcal{C}$ is final.*

(2) *For every pair of compact open sets $U, V \subseteq X$, the diagram*

$$
\begin{array}{ccc}
\mathcal{F}(U \cap V) & \longrightarrow & \mathcal{F}(U) \\
\downarrow & & \downarrow \\
\mathcal{F}(V) & \longrightarrow & \mathcal{F}(U \cup V)
\end{array}
$$

is a pullback.

Let $\mathcal{W}$ be a covering of X by compact open subsets and let $\mathcal{U}_1(X) \subseteq \mathcal{U}_0(X)$ be the collection of all compact open subsets of X which are contained in some element of $\mathcal{W}$. Then $\mathcal{F}$ is a right Kan extension of $\mathcal{F} \,|\, \mathrm{N}(\mathcal{U}_1(X))^{op}$.

Proof. The proof is similar to that of Lemma 7.3.4.8 but slightly easier. Let us say that a covering $\mathcal{W}$ of a coherent topological space X by compact open subsets is *good* if it satisfies the conclusions of the lemma. We observe that $\mathcal{W}$ automatically has a finite subcover. We will prove, by induction on $n \geq 0$, that if $\mathcal{W}$ is a collection of open subsets of a locally coherent topological space X such that there exist $W_1, \cdots, W_n \in \mathcal{W}$ with $W_1 \cup \ldots \cup W_n = X$, then $\mathcal{W}$ is a good covering of X. If $n = 0$, then $X = \emptyset$. In this case, we must prove that $\mathcal{F}(\emptyset)$ is final, which is one of our assumptions.

Suppose that $\mathcal{W} \subseteq \mathcal{W}'$ are coverings of X by compact open sets and that for every $W' \in \mathcal{W}'$ the induced covering $\{W \cap W' : W \in \mathcal{W}\}$ is a good covering of W'. It then follows from Proposition 4.3.2.8 that $\mathcal{W}'$ is a good covering of X if and only if $\mathcal{W}$ is a good covering of X.

Now suppose $n > 0$. Let $V = W_2 \cup \cdots \cup W_n$, and let $\mathcal{W}' = \mathcal{W} \cup \{V\}$. Using the above remark and the inductive hypothesis, it will suffice to show that $\mathcal{W}'$ is a good covering of X. Now $\mathcal{W}'$ contains a pair of open sets W_1 and V which cover X. We thereby reduce to the case $n = 2$; using the above remark, we can furthermore suppose that $\mathcal{W} = \{W_1, W_2\}$.

We now wish to show that for every compact $U \subseteq X$, $\mathcal{F}$ exhibits $\mathcal{F}(U)$ as the limit of $\mathcal{F} \,|\, \mathrm{N}(\mathcal{U}_1(X)_{/U})^{op}$. Without loss of generality, we may replace X by U and thereby reduce to the case $U = X$. Let $\mathcal{U}_2(X) = \{W_1, W_2, W_1 \cap W_2\} \subseteq \mathcal{U}_1(X)$. Using Theorem 4.1.3.1, we deduce that the inclusion $\mathrm{N}(\mathcal{U}_2(X)) \subseteq \mathrm{N}(\mathcal{U}_1(X))$ is cofinal. Consequently, it suffices to prove that $\mathcal{F}(X)$ is the limit of the diagram $\mathcal{F} \,|\, \mathrm{N}(\mathcal{U}_2(X))^{op}$. In other words, we must show that the diagram

$$
\begin{array}{ccc}
\mathcal{F}(X) & \longrightarrow & \mathcal{F}(W_1) \\
\downarrow & & \downarrow \\
\mathcal{F}(W_2) & \longrightarrow & \mathcal{F}(W_1 \cap W_2)
\end{array}
$$

is a pullback in $\mathcal{C}$, which is true by assumption. $\qquad\square$

Theorem 7.3.5.2. *Let X be a coherent topological space and let $\mathcal{U}_0(X) \subseteq \mathcal{U}(X)$ denote the collection of compact open subsets of X. Let $\mathcal{C}$ be an ∞-category which admits small limits. The restriction map*

$$\mathrm{Shv}(X; \mathcal{C}) \to \mathrm{Fun}(\mathrm{N}(\mathcal{U}_0(X))^{op}, \mathcal{C})$$

is fully faithful, and its essential image consists of precisely those functors $\mathcal{F}_0 : \mathrm{N}(\mathcal{U}_0(X))^{op} \to \mathcal{C}$ satisfying the following conditions:

(1) *The object $\mathcal{F}_0(\emptyset) \in \mathcal{C}$ is final.*

(2) *For every pair of compact open sets $U, V \subseteq X$, the diagram*

$$
\begin{array}{ccc}
\mathcal{F}_0(U \cap V) & \longrightarrow & \mathcal{F}_0(U) \\
\downarrow & & \downarrow \\
\mathcal{F}_0(V) & \longrightarrow & \mathcal{F}_0(U \cup V)
\end{array}
$$

is a pullback.

Proof. Let $\mathcal{D} \subseteq \mathcal{C}^{N(\mathcal{U}(X))^{op}}$ be the full subcategory spanned by those presheaves $\mathcal{F} : N(\mathcal{U}(X))^{op} \to \mathcal{C}$ which are right Kan extensions of $\mathcal{F}_0 = \mathcal{F} \,|\, N(\mathcal{U}_0(X))^{op}$ and such that $\mathcal{F}_0$ satisfies conditions (1) and (2). According to Proposition 4.3.2.15, it will suffice to show that $\mathcal{D}$ coincides with $\mathrm{Shv}(X; \mathcal{C})$.

Suppose that $\mathcal{F} : N(\mathcal{U}(X))^{op} \to \mathcal{C}$ is a sheaf. We first show that $\mathcal{F}$ is a right Kan extension of $\mathcal{F}_0 = \mathcal{F} \,|\, N(\mathcal{U}_0(X))^{op}$. Let U be an open subset of X, let $\mathcal{U}(X)^{(0)}_{/U}$ denote the collection of compact open subsets of U and let $\mathcal{U}(X)^{(1)}_{/U}$ denote the sieve generated by $\mathcal{U}(X)^{(0)}_{/U}$. Consider the diagram

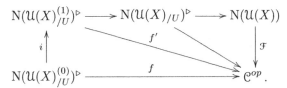

We wish to prove that f is a colimit diagram. Using Theorem 4.1.3.1, we deduce that the inclusion $N(\mathcal{U}(X))^{(0)}_{/U} \subseteq N(\mathcal{U}(X))^{(1)}_{/U}$ is cofinal. It therefore suffices to prove that f' is a colimit diagram. Since $\mathcal{F}$ is a sheaf, it suffices to prove that $\mathcal{U}(X)^{(1)}_{/U}$ is a covering sieve. In other words, we need to prove that U is a union of compact open subsets of X, which follows immediately from our assumption that X is coherent.

We next prove that $\mathcal{F}_0$ satisfies (1) and (2). To prove (1), we simply observe that the empty sieve is a cover of $\emptyset$ and apply the sheaf condition. To prove (2), we may assume without loss of generality that neither U nor V is contained in the other (otherwise the result is obvious). Let $\mathcal{U}(X)^{(0)}_{/U\cup V}$ be the full subcategory spanned by U, V, and $U \cap V$, and let $\mathcal{U}(X)^{(1)}_{/U\cup V}$ be the sieve on $U \cup V$ generated by $\mathcal{U}(X)^{(0)}_{/U\cup V}$. As above, we have a diagram

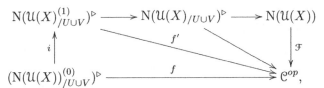

and we wish to show that f is a colimit diagram. Theorem 4.1.3.1 implies that the inclusion $N(\mathcal{U}(X))^{(0)}_{/U\cup V} \subseteq N(\mathcal{U}(X))^{(1)}_{/U\cup V}$ is cofinal. It therefore suffices to prove that f' is a colimit diagram, which follows from the sheaf condition since $\mathcal{U}(X)^{(1)}_{/U\cup V}$ is a covering sieve. This completes the proof that $\mathrm{Shv}(X; \mathcal{C}) \subseteq \mathcal{D}$.

It remains to prove that $\mathcal{D} \subseteq \mathrm{Shv}(X; \mathcal{C})$. In other words, we must show that if $\mathcal{F}$ is a right Kan extension of $\mathcal{F}_0 = \mathcal{F} \,|\, N(\mathcal{U}_0(X))^{op}$ and $\mathcal{F}_0$ satisfies conditions (1) and (2), then $\mathcal{F}$ is a sheaf. Let U be an open subset of X and let $\mathcal{U}(X)^{(0)}_{/U}$ be a sieve which covers U. Let $\mathcal{U}_0(X)_{/U}$ denote the category of compact open subsets of U and $\mathcal{U}_0(X)^{(0)}_{/U}$ the category of compact open subsets of U which belong to the sieve $\mathcal{U}(X)^{(0)}_{/U}$. We wish to prove that $\mathcal{F}(U)$ is a limit of $\mathcal{F} \,|\, \mathrm{N}(\mathcal{U}(X)^{(0)}_{/U})^{op}$. We will in fact prove the slightly stronger assertion that $\mathcal{F} \,|\, \mathrm{N}(\mathcal{U}(X)_{/U})^{op}$ is a right Kan extension of $\mathcal{F} \,|\, \mathrm{N}(\mathcal{U}(X)^{(0)}_{/U})^{op}$.

We have a commutative diagram

$$
\begin{array}{ccc}
\mathcal{U}_0(X)^{(0)}_{/U} & \longrightarrow & \mathcal{U}_0(X)_{/U} \\
\downarrow & & \downarrow \\
\mathcal{U}(X)^{(0)}_{/U} & \longrightarrow & \mathcal{U}(X)_{/U}.
\end{array}
$$

By assumption, $\mathcal{F}$ is a right Kan extension of $\mathcal{F}_0$. It follows that the restriction $\mathcal{F} \,|\, \mathrm{N}(\mathcal{U}(X)^{(0)}_{/U})^{op}$ is a right Kan extension of $\mathcal{F} \,|\, \mathrm{N}(\mathcal{U}_0(X)^{(0)}_{/U})^{op}$ and that $\mathcal{F} \,|\, \mathrm{N}(\mathcal{U}(X)_{/U})^{op}$ is a right Kan extension of $\mathcal{F} \,|\, \mathrm{N}(\mathcal{U}_0(X)_{/U})^{op}$. By the transitivity of Kan extensions (Proposition 4.3.2.8), it will suffice to prove that $\mathcal{F} \,|\, \mathrm{N}(\mathcal{U}_0(X)_{/U})^{op}$ is a right Kan extension of $\mathcal{F} \,|\, \mathrm{N}(\mathcal{U}_0(X)^{(0)}_{/U})^{op}$. This follows immediately from Lemma 7.3.5.1. $\qquad\square$

Corollary 7.3.5.3. *Let X be a coherent topological space. Then the global sections functor $\Gamma : \mathrm{Shv}(X) \to \mathcal{S}$ is a proper map of ∞-topoi.*

Proof. The proof is identical to the proof of Corollary 7.3.4.11 (using Theorem 7.3.5.2 in place of Corollary 7.3.4.10). $\qquad\square$

Corollary 7.3.5.4. *Let X be a coherent topological space. Then the global sections functor*

$$
\Gamma : \mathrm{Shv}(X) \to \mathcal{S}
$$

commutes with filtered colimits.

7.3.6 Cell-Like Maps

Recall that a topological space X is an *absolute neighborhood retract* if X is metrizable and if for any closed immersion $X \hookrightarrow Y$ of X in a metric space Y, there exists an open set $U \subseteq Y$ containing the image of X, such that the inclusion $X \hookrightarrow U$ has a left inverse (in other words, X is a *retract* of U).

Let $p : X \to Y$ be a continuous map between locally compact absolute neighborhood retracts. The map p is said to be *cell-like* if p is proper and each fiber $X_y = X \times_Y \{y\}$ has trivial shape (in the sense of Borsuk; see [55] and §7.1.6). The theory of cell-like maps plays an important role in geometric topology: we refer the reader to [18] for a discussion (and for several equivalent formulations of the condition that a map be cell-like).

The purpose of this section is to describe a class of geometric morphisms between ∞-topoi, which we will call *cell-like* morphisms. We will then compare our theory of cell-like morphisms with the classical theory of cell-like maps. We will also give a "nonclassical" example which arises in the theory of rigid analytic geometry.

Definition 7.3.6.1. Let $p_* : \mathfrak{X} \to \mathcal{Y}$ be a geometric morphism of ∞-topoi. We will say that p_* is *cell-like* if it is proper and if the right adjoint p^* (which is well-defined up to equivalence) is fully faithful.

Warning 7.3.6.2. Many authors refer to a map $p : X \to Y$ of *arbitrary* compact metric spaces as *cell-like* if each fiber $X_y = X \times_Y \{y\}$ has trivial shape. This condition is generally *weaker* than the condition that $p_* : \mathrm{Shv}(X) \to \mathrm{Shv}(Y)$ be cell-like in the sense of Definition 7.3.6.1. However, the two definitions are equivalent provided that X and Y are sufficiently nice (for example, if they are locally compact absolute neighborhood retracts). Our departure from the classical terminology is perhaps justified by the fact that the class of morphisms introduced in Definition 7.3.6.1 has good formal properties: for example, stability under composition.

Remark 7.3.6.3. Let $p_* : \mathfrak{X} \to \mathcal{Y}$ be a cell-like geometric morphism between ∞-topoi. Then the unit map $\mathrm{id}_{\mathcal{Y}} \to p_* p^*$ is an equivalence of functors. It follows immediately that p_* induces an equivalence of shapes $Sh(\mathfrak{X}) \to Sh(\mathcal{Y})$ (see §7.1.6).

Proposition 7.3.6.4. *Let* $p_* : \mathfrak{X} \to \mathcal{Y}$ *be a proper morphism of ∞-topoi. Suppose that* $\mathcal{Y}$ *has enough points. Then* p_* *is cell-like if and only if, for every pullback diagram*

$$\begin{array}{ccc} \mathfrak{X}' & \longrightarrow & \mathfrak{X} \\ \downarrow & & \downarrow {\scriptstyle p_*} \\ \mathcal{S} & \longrightarrow & \mathcal{Y} \end{array}$$

in $\mathcal{R}\mathrm{Top}$, *the ∞-topos* $\mathfrak{X}'$ *has trivial shape.*

Proof. Suppose first that each fiber $\mathfrak{X}'$ has trivial shape. Let $\mathcal{F} \in \mathcal{Y}$. We wish to show that the unit map $u : \mathcal{F} \to p_* p^* \mathcal{F}$ is an equivalence. Since $\mathcal{Y}$ has enough points, it suffices to show that for each point $q_* : \mathcal{S} \to \mathcal{Y}$, the map $q^* u$ is an equivalence in $\mathcal{S}$, where q^* denotes a left adjoint to q_*. Form a pullback diagram of ∞-topoi

$$\begin{array}{ccc} \mathfrak{X}' & \longrightarrow & \mathfrak{X} \\ {\scriptstyle s_*} \downarrow & & \downarrow {\scriptstyle p_*} \\ \mathcal{S} & \xrightarrow{\ q_*\ } & \mathcal{Y} . \end{array}$$

Since p_* is proper, this diagram is left adjointable. Consequently, $q^* u$ can be identified with the unit map

$$K \to s_* s^* K,$$

where $K = q^* \mathcal{F} \in \mathcal{S}$. If $\mathfrak{X}'$ has trivial shape, then this map is an equivalence.

Conversely, if p_* is cell-like, then the above argument shows that for every diagram

$$
\begin{array}{ccc}
\mathcal{X}' & \longrightarrow & \mathcal{X} \\
{\scriptstyle s_*}\downarrow & & \downarrow{\scriptstyle p_*} \\
\mathcal{S} & \xrightarrow{\;q_*\;} & \mathcal{Y}
\end{array}
$$

as above and every $\mathcal{F} \in \mathcal{Y}$, the adjunction map

$$
K \to s_* s^* K
$$

is an equivalence, where $K = q^* \mathcal{F}$. To prove that $\mathcal{X}'$ has trivial shape, it will suffice to show that q^* is essentially surjective. For this, we observe that since $\mathcal{S}$ is a final object in the ∞-category of ∞-topoi, there exists a geometric morphism $r_* : \mathcal{Y} \to \mathcal{S}$ such that $r_* \circ q_*$ is homotopic to $\mathrm{id}_{\mathcal{S}}$. It follows that $q^* \circ r^* \simeq \mathrm{id}_{\mathcal{S}}$. Since $\mathrm{id}_{\mathcal{S}}$ is essentially surjective, we conclude that q^* is essentially surjective. $\qquad\square$

Corollary 7.3.6.5. *Let $p : X \to Y$ be a map of paracompact topological spaces. Assume that p_* is proper and that Y has finite covering dimension. Then $p_* : \mathrm{Shv}(X) \to \mathrm{Shv}(Y)$ is cell-like if and only if each fiber $X_y = X \times_Y \{y\}$ has trivial shape.*

Proof. Combine Proposition 7.3.6.4 with Corollary 7.2.1.17. $\qquad\square$

Proposition 7.3.6.6. *Let $p : X \to Y$ be a proper map of locally compact ANRs. The following conditions are equivalent:*

(1) *The geometric morphism $p_* : \mathrm{Shv}(X) \to \mathrm{Shv}(Y)$ is cell-like.*

(2) *For every open subset $U \subseteq Y$, the restriction map $X \times_Y U \to U$ is a homotopy equivalence.*

(3) *Each fiber $X_y = X \times_Y \{y\}$ has trivial shape.*

Proof. It is easy to see that if p_* is cell-like, then each of the restrictions $p' : X \times_Y U \to U$ induces a cell-like geometric morphism. According to Remark 7.3.6.3, p'_* is a shape equivalence and therefore a homotopy equivalence by Proposition 7.1.6.8. Thus $(1) \Rightarrow (2)$.

We next prove that $(2) \Rightarrow (1)$. Let $\mathcal{F} \in \mathrm{Shv}(Y)$ and let $u : \mathcal{F} \to p_* p^* \mathcal{F}$ be a unit map; we wish to show that u is an equivalence. It will suffice to show that the induced map $\mathcal{F}(U) \to (p_* p^* \mathcal{F})(U)$ is an equivalence in $\mathcal{S}$ for each paracompact open subset $U \subseteq Y$. Replacing Y by u, we may reduce to the problem of showing that the map $\mathcal{F}(Y) \to (p^* \mathcal{F})(X)$ is a homotopy equivalence. According to Corollary 7.1.4.4, we may assume that $\mathcal{F}$ is the simplicial nerve of $\mathrm{Sing}_Y \widetilde{Y}$, where $\widetilde{Y}$ is a fibrant-cofibrant object of $\mathrm{Top}_{/Y}$. According to Proposition 7.1.5.1, we may identify $p^* \mathcal{F}$ with $\mathrm{Sing}_X \widetilde{X}$, where

$\widetilde{X} = X \times_Y \widetilde{Y}$. It therefore suffices to prove that the induced map of simplicial function spaces

$$\mathrm{Map}_Y(Y, \widetilde{Y}) \to \mathrm{Map}_X(X, \widetilde{X}) \simeq \mathrm{Map}_Y(X, \widetilde{Y})$$

is a homotopy equivalence, which follows immediately from (2).

The implication $(1) \Rightarrow (3)$ follows from the proof of Proposition 7.3.6.6, and the implication $(3) \Rightarrow (2)$ is classical (see [37]). $\qquad\square$

Remark 7.3.6.7. It is possible to prove the following generalization of Proposition 7.3.6.6: a proper geometric morphism $p_* : \mathcal{X} \to \mathcal{Y}$ is cell-like if and only if, for each object $U \in \mathcal{Y}$, the associated geometric morphism $\mathcal{X}_{/p^*U} \to \mathcal{Y}_{/U}$ is a shape equivalence (and, in fact, it is necessary to check this only on a collection of objects $U \in \mathcal{Y}$ which generates $\mathcal{Y}$ under colimits).

Remark 7.3.6.8. Another useful property of the class of cell-like morphisms, which we will not prove here, is stability under base change: given a pullback diagram

$$
\begin{array}{ccc}
\mathcal{X}' & \longrightarrow & \mathcal{X} \\
\downarrow{\scriptstyle p'_*} & & \downarrow{\scriptstyle p_*} \\
\mathcal{Y}' & \longrightarrow & \mathcal{Y},
\end{array}
$$

where p_* is cell-like, p'_* is also cell-like.

If $p_* : \mathcal{X} \to \mathcal{Y}$ is a cell-like morphism of ∞-topoi, then many properties of $\mathcal{Y}$ are controlled by the analogous properties of $\mathcal{X}$. For example:

Proposition 7.3.6.9. *Let $p_* : \mathcal{X} \to \mathcal{Y}$ be a cell-like morphism of ∞-topoi. If $\mathcal{X}$ has homotopy dimension $\leq n$, then $\mathcal{Y}$ also has homotopy dimension $\leq n$.*

Proof. Let $1_{\mathcal{Y}}$ be a final object of $\mathcal{Y}$, U an n-connective object of $\mathcal{Y}$, and p^* a left adjoint to p_*. We wish to prove that $\mathrm{Hom}_{h\,\mathcal{Y}}(1_{\mathcal{Y}}, U)$ is nonempty. Since p^* is fully faithful, it will suffice to prove that $\mathrm{Hom}_{h\mathcal{X}}(p^*1_{\mathcal{Y}}, p^*U)$. We now observe that $p^*1_{\mathcal{Y}}$ is a final object of $\mathcal{X}$ (since p is left exact), p^*U is n-connective (Proposition 6.5.1.16), and $\mathcal{X}$ has homotopy dimension $\leq n$, so that $\mathrm{Hom}_{h\mathcal{X}}(p^*1_{\mathcal{Y}}, p^*U)$ is nonempty, as desired. $\qquad\square$

We conclude with a different example of a class of cell-like maps. We will assume in the following discussion that the reader is familiar with the basic ideas of rigid analytic geometry; for an account of this theory we refer the reader to [29]. Let K be a field which is complete with respect to a non-Archimedean absolute value $||_K : K \to \mathbf{R}$. Let A be an affinoid algebra over K: that is, a quotient of an algebra of convergent power series (in several variables) with values in K. Let X be the rigid space associated to A. One can associate to X two different "underlying" topological spaces:

$(ZR1)$ The category $\mathcal{C}$ of rational open subsets of X has a Grothendieck topology given by admissible affine covers. The topos of sheaves of sets on $\mathcal{C}$

is localic, and the underlying locale has enough points: it is therefore isomorphic to the locale of open subsets of a (canonically determined) topological space X_{ZR}, the *Zariski-Riemann* space of X.

$(ZR2)$ In the case where K is a *discretely* valued field with ring of integers R, one may define X_{ZR} to be the inverse limit of the underlying spaces of all formal schemes $\widehat{X} \to \operatorname{Spf} R$ which have generic fiber X.

$(ZR3)$ Concretely, X_{ZR} can be identified with the set of all isomorphism classes of continuous multiplicative seminorms $||_A : A \to M \cup \{\infty\}$, where M is an ordered abelian group containing the value group

$$|K^*|_K \subseteq \mathbf{R}^*$$

and the restriction of $||_A$ to K coincides with $||_K$.

$(B1)$ The category of sheaves of sets on $\mathcal{C}$ contains a full subcategory, consisting of *overconvergent* sheaves. This category is also a localic topos, and the underyling locale is isomorphic to the lattice of open subsets of a (canonically determined) topological space X_B, the *Berkovich space* of X. The category of overconvergent sheaves is a localization of the category of all sheaves on $\mathcal{C}$, and there is an associated map of topological spaces $p : X_{ZR} \to X_B$.

$(B2)$ Concretely, X_B can be identified with the set of all continuous multiplicative seminorms $||_A : A \to \mathbf{R} \cup \{\infty\}$ which extend $||_K$. It is equipped with the topology of pointwise convergence and is a compact Hausdorff space.

The relationship between the Zariski-Riemann space X_{ZR} and the Berkovich space X_B (or more conceptually, the relationship between the category of *all* sheaves on X and the category of *overconvergent* sheaves on X) is neatly summarized by the following result.

Proposition 7.3.6.10. *Let K be a field which is complete with respect to a non-Archimedean absolute value $||_K$, let A be an affinoid algebra over K, let X be the associated rigid space, and let $p : X_{ZR} \to X_B$ be the natural map. Then p induces a cell-like morphism of ∞-topoi $p_* : \operatorname{Shv}(X_{ZR}) \to \operatorname{Shv}(X_B)$.*

Before giving the proof, we need an easy lemma. Recall that a topological space X is *irreducible* if every finite collection of nonempty open subsets of X has nonempty intersections.

Lemma 7.3.6.11. *Let X be an irreducible topological space. Then $\operatorname{Shv}(X)$ has trivial shape.*

Proof. Let $\pi : X \to *$ be the projection from X to a point and let $\pi_* : \operatorname{Shv}(X) \to \operatorname{Shv}(*)$ be the induced geometric morphism. We will construct a left adjoint π^* to π_* such that the unit map $\operatorname{id} \to \pi_* \pi^*$ is an equivalence.

We begin by defining $G : \mathcal{P}(X) \to \mathcal{P}(*)$ to be the functor given by composition with π^{-1}, so that $G|\operatorname{Shv}(X) = \pi_*$. Let

$$i : \mathrm{N}(\mathcal{U}(X))^{op} \to \mathrm{N}(\mathcal{U}(*))^{op}$$

be defined so that

$$i(U) = \begin{cases} \emptyset & \text{if } U = \emptyset \\ \{*\} & \text{if } U \neq \emptyset \end{cases}$$

and let $F : \mathcal{P}(*) \to \mathcal{P}(U)$ be given by composition with i. We observe that F is a left Kan extension functor, so that the identity map

$$\mathrm{id}_{\mathcal{P}(*)} \to G \circ F$$

exhibits F as a left adjoint to G. We will show that $F(\operatorname{Shv}(*)) \subseteq \operatorname{Shv}(X)$. Setting $\pi^* = F|\operatorname{Shv}(*)$, we conclude that the identity map

$$\mathrm{id}_{\operatorname{Shv}(*)} \to \pi_* \pi^*$$

is the unit of an adjunction between π_* and π^*, which will complete the proof.

Let $\mathcal{U} \subseteq \mathcal{U}(X)$ be a sieve which covers the open set $U \subseteq X$. We wish to prove that the diagram

$$p : \mathrm{N}(\mathcal{U}^{op})^{\triangleleft} \to \mathrm{N}(\mathcal{U}(X))^{op} \xrightarrow{i} \mathrm{N}(\mathcal{U}(*))^{op} \xrightarrow{\mathcal{F}} \mathcal{S}$$

is a limit. Let $\mathcal{U}_0 = \{V \in \mathcal{U} : V \neq \emptyset\}$. Since $\mathcal{F}(\emptyset)$ is a final object of $\mathcal{S}$, p is a limit if and only if $p|\mathrm{N}(\mathcal{U}_0^{op})^{\triangleleft}$ is a limit diagram. If $U = \emptyset$, then this follows from the fact that $\mathcal{F}(\emptyset)$ is final in $\mathcal{S}$. If $U \neq \emptyset$, then $p|\mathrm{N}(\mathcal{U}_0^{op})^{\triangleleft}$ is a constant diagram, so it will suffice to prove that the simplicial set $\mathrm{N}(\mathcal{U}_0)^{op}$ is weakly contractible. This follows from the observation that $\mathcal{U}_0^{op}$ is a filtered partially ordered set since $\mathcal{U}_0$ is nonempty and stable under finite intersections (because X is irreducible). $\square$

Proof of Proposition 7.3.6.10. We first show that p_* is a proper map of ∞-topoi. We note that p factors as a composition

$$X_{ZR} \xrightarrow{p'} X_{ZR} \times X_B \xrightarrow{p''} X_B.$$

The map p' is a pullback of the diagonal map $X_B \to X_B \times X_B$. Since X_B is Hausdorff, p' is a closed immersion. It follows that p'_* is a closed immersion of ∞-topoi (Corollary 7.3.2.9) and therefore a proper morphism (Proposition 7.3.2.12). It therefore suffices to prove that p'' is a proper map of ∞-topoi. We note the existence of a commutative diagram

$$\begin{array}{ccc} \operatorname{Shv}(X_{ZR} \times X_B) & \longrightarrow & \operatorname{Shv}(X_{ZR}) \\ \downarrow {\scriptstyle p''_*} & & \downarrow {\scriptstyle g_*} \\ \operatorname{Shv}(X_B) & \longrightarrow & \operatorname{Shv}(*). \end{array}$$

Using Proposition 7.3.1.11, we deduce that this is a homotopy Cartesian diagram of ∞-topoi. It therefore suffices to show that the global sections functor $g_* : \mathrm{Shv}(X_{ZR}) \to \mathrm{Shv}(*)$ is proper, which follows from Corollary 7.3.5.3.

We now observe that the topological space X_B is paracompact and has finite covering dimension ([5], Corollary 3.2.8), so that $\mathrm{Shv}(X_B)$ has enough points (Corollary 7.2.1.17). According to Proposition 7.3.6.4, it suffices to show that for every fiber diagram

$$
\begin{array}{ccc}
\mathfrak{X}' & \longrightarrow & \mathrm{Shv}(X_{ZR}) \\
\downarrow & & \downarrow \\
\mathrm{Shv}(*) & \xrightarrow{\;q_*\;} & \mathrm{Shv}(X_B),
\end{array}
$$

the ∞-topos $\mathfrak{X}'$ has trivial shape. Using Lemma 6.4.5.6, we conclude that q_* is necessarily induced by a homomorphism of locales $\mathcal{U}(X_B) \to \mathcal{U}(*)$, which corresponds to an irreducible closed subset of X_B. Since X_B is Hausdorff, this subset consists of a single (closed) point x. Using Proposition 7.3.2.12 and Corollary 7.3.2.9, we can identify $\mathfrak{X}'$ with the ∞-topos $\mathrm{Shv}(Y)$, where $Y = X_{ZR} \times_{X_B} \{x\}$. We now observe that the topological space Y is coherent and irreducible (it contains a unique "generic" point), so that $\mathrm{Shv}(Y)$ has trivial shape by Lemma 7.3.6.11.   □

Remark 7.3.6.12. Let $p_* : \mathrm{Shv}(X_{ZR}) \to \mathrm{Shv}(X_B)$ be as in Proposition 7.3.6.10. Then p_* has a fully faithful left adjoint p^*. We might say that an object of $\mathrm{Shv}(X_{ZR})$ is *overconvergent* if it belongs to the essential image of p^*; for sheaves of sets, this agrees with the classical terminology.

Remark 7.3.6.13. One can generalize Proposition 7.3.6.10 to rigid spaces which are not affinoid; we leave the details to the reader.

Appendix

This appendix is comprised of three parts. In §A.1, we will review some ideas from classical category theory, such as monoidal structures, enriched categories, and Quillen's small object argument. We give a brief overview of the theory of model categories in §A.2. The main result here is Proposition A.2.6.13, which will allow us to establish the existence of model category structures in a variety of situations with a minimal amount of effort. In §A.3, we will use this result to make a detailed study of the theory of simplicial categories. Our exposition is rather dense; for a more leisurely account of the theory of model categories, we refer the reader to one of the standard texts (such as [40]).

A.1 CATEGORY THEORY

Familiarity with classical category theory is the main prerequisite for reading this book. In this section, we will fix some of the notation that we use when discussing categories and summarize (generally without proofs) some of the concepts employed in the body of the text.

If $\mathcal{C}$ is a category, we let $\mathrm{Ob}(\mathcal{C})$ denote the set of objects of $\mathcal{C}$. We will write $X \in \mathcal{C}$ to mean that X is an object of $\mathcal{C}$. For $X, Y \in \mathcal{C}$, we write $\mathrm{Hom}_{\mathcal{C}}(X, Y)$ for the set of morphisms from X to Y in $\mathcal{C}$. We also write id_X for the identity automorphism of $X \in \mathcal{C}$ (regarded as an element of $\mathrm{Hom}_{\mathcal{C}}(X, X)$).

If Z is an object in a category $\mathcal{C}$, then the *overcategory* $\mathcal{C}_{/Z}$ of *objects over* Z is defined as follows: the objects of $\mathcal{C}_{/Z}$ are diagrams $X \to Z$ in $\mathcal{C}$. A morphism from $f : X \to Z$ to $g : Y \to Z$ is a commutative triangle

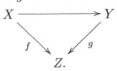

Dually, we have an *undercategory* $\mathcal{C}_{Z/} = ((\mathcal{C}^{op})_{/Z})^{op}$ of *objects under* Z.

If $f : X \to Z$ and $g : Y \to Z$ are objects in $\mathcal{C}_{/Z}$, then we will often write $\mathrm{Hom}_Z(X, Y)$ rather than $\mathrm{Hom}_{\mathcal{C}_{/Z}}(f, g)$.

We let Set denote the category of sets and Cat the category of (small) categories (where the morphisms are given by functors).

If κ is a regular cardinal, we will say that a set S is *κ-small* if it has cardinality less than κ. We will also use this terminology when discussing mathematical objects other than sets, which are built out of sets. For example, we will say that a category $\mathcal{C}$ is *κ-small* if the set of all objects of $\mathcal{C}$ is κ-small and the set of all morphisms in $\mathcal{C}$ is likewise κ-small.

We will need to discuss categories which are not small. In order to minimize the effort spent dealing with set-theoretic complications, we will adopt the usual device of *Grothendieck universes*. We fix a strongly inaccessible cardinal κ and refer to a mathematical object (such as a set or category) as *small* if it is κ-small, and *large* otherwise. It should be emphasized that this is primarily a linguistic device and that none of our results depend in an essential way on the existence of a strongly inaccessible cardinal κ.

Throughout this book, the word "topos" will always mean *Grothendieck topos*. Strictly speaking, a knowledge of classical topos theory is not required to read this book: all of the relevant classical concepts will be introduced (though sometimes in a hurried fashion) in the course of our search for suitable ∞-categorical analogues.

A.1.1 Compactness and Presentability

Let κ be a regular cardinal.

Definition A.1.1.1. A partially ordered set $\mathfrak{I}$ is κ-*filtered* if, for any subset $\mathfrak{I}_0 \subseteq \mathfrak{I}$ having cardinality $< \kappa$, there exists an upper bound for $\mathfrak{I}_0$ in $\mathfrak{I}$.

Let $\mathcal{C}$ be a category which admits (small) colimits and let X be an object of $\mathcal{C}$. Suppose we are given a κ-filtered partially ordered set $\mathfrak{I}$ and a diagram $\{Y_\alpha\}_{\alpha \in \mathfrak{I}}$ in $\mathcal{C}$ indexed by $\mathfrak{I}$. Let Y denote a colimit of this diagram. Then there is an associated map of sets
$$\psi : \varinjlim \operatorname{Hom}_{\mathcal{C}}(X, Y_\alpha) \to \operatorname{Hom}_{\mathcal{C}}(X, Y).$$
We say that X is κ-*compact* if ψ is bijective for *every* κ-filtered partially ordered set $\mathfrak{I}$ and *every* diagram $\{Y_\alpha\}$ indexed by $\mathfrak{I}$. We say that X is *small* if it is κ-compact for some (small) regular cardinal κ. In this case, X is κ-compact for all sufficiently large regular cardinals κ.

Definition A.1.1.2. A category $\mathcal{C}$ is *presentable* if it satisfies the following conditions:

(1) The category $\mathcal{C}$ admits all (small) colimits.

(2) There exists a (small) set S of objects of $\mathcal{C}$ which generates $\mathcal{C}$ under colimits; in other words, every object of $\mathcal{C}$ may be obtained as the colimit of a (small) diagram taking values in S.

(3) Every object in $\mathcal{C}$ is small. (Assuming (2), this is equivalent to the assertion that every object which belongs to S is small.)

(4) For any pair of objects $X, Y \in \mathcal{C}$, the set $\operatorname{Hom}_{\mathcal{C}}(X, Y)$ is small.

Remark A.1.1.3. In §5.5, we describe an ∞-categorical generalization of Definition A.1.1.2.

Remark A.1.1.4. For more details of the theory of presentable categories, we refer the reader to [1]. Note that our terminology differs slightly from that of [1], in which our presentable categories are called *locally presentable* categories.

A.1.2 Lifting Problems and the Small Object Argument

Let $\mathcal{C}$ be a category and let $p : A \to B$ and $q : X \to Y$ be morphisms in $\mathcal{C}$. Recall that p is said to have the *left lifting property* with respect to q, and q the *right lifting property* with respect to p, if given any diagram

there exists a dotted arrow as indicated, rendering the diagram commutative.

Remark A.1.2.1. In the case where Y is a final object of $\mathcal{C}$, we will instead say that X has the *extension property* with respect to $p : A \to B$.

Let S be any collection of morphisms in $\mathcal{C}$. We define $_{\perp}S$ to be the class of all morphisms which have the right lifting property with respect to all morphisms in S, and $S_{\perp}$ to be the class of all morphisms which have the left lifting property with respect to all morphisms in S. We observe that

$$S \subseteq (_{\perp}S)_{\perp}.$$

The class of morphisms $(_{\perp}S)_{\perp}$ enjoys several stability properties which we axiomatize in the following definition.

Definition A.1.2.2. Let $\mathcal{C}$ be a category with all (small) colimits and let S be a class of morphisms of $\mathcal{C}$. We will say that S is *weakly saturated* if it has the following properties:

(1) (Closure under the formation of pushouts) Given a pushout diagram

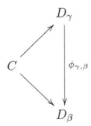

such that f belongs to S, the morphism f' also belongs to S.

(2) (Closure under transfinite composition) Let $C \in \mathcal{C}$ be an object, let α be an ordinal, and let $\{D_\beta\}_{\beta<\alpha}$ be a system of objects of $\mathcal{C}_{C/}$ indexed by α: in other words, for each $\beta < \alpha$, we are supplied with a morphism $C \to D_\beta$, and for each $\gamma \leq \beta < \alpha$ a commutative diagram

satisfying $\phi_{\gamma,\delta} \circ \phi_{\beta,\gamma} = \phi_{\beta,\delta}$. For $\beta \leq \alpha$, we let $D_{<\beta}$ be a colimit of the system $\{D_\gamma\}_{\gamma<\beta}$ taken in the category $\mathcal{C}_{C/}$.

Suppose that, for each $\beta < \alpha$, the natural map $D_{<\beta} \to D_\beta$ belongs to S. Then the induced map $C \to D_{<\alpha}$ belongs to S.

(3) (Closure under the formation of retracts) Given a commutative diagram

$$
\begin{array}{ccccc}
C & \longrightarrow & C' & \longrightarrow & C \\
\downarrow{\scriptstyle f} & & \downarrow{\scriptstyle g} & & \downarrow{\scriptstyle f} \\
D & \longrightarrow & D' & \longrightarrow & D
\end{array}
$$

in which both horizontal compositions are the identity, if g belongs to S, then so does f.

It is worth noting that saturation has the following consequences:

Proposition A.1.2.3. *Let $\mathcal{C}$ be a category which admits all (small) colimits and let S be a weakly saturated class of morphism in $\mathcal{C}$. Then*

(1) *Every isomorphism belongs to S.*

(2) *The class S is stable under composition: if $f : X \to Y$ and $g : Y \to Z$ belong to S, then so does $g \circ f$.*

Proof. Assertion (1) is equivalent to the closure of S under transfinite composition in the special case where $\alpha = 0$; (2) is equivalent to the special case where $\alpha = 2$. $\qquad\square$

Remark A.1.2.4. A reader who is ill at ease with the style of the preceding argument should feel free to take the asserted properties as part of the definition of a weakly saturated class of morphisms.

The intersection of any collection of weakly saturated classes of morphisms is itself weakly saturated. Consequently, for any category $\mathcal{C}$ which admits small colimits, and any collection A of morphisms of $\mathcal{C}$, there exists a *smallest* weakly saturated class of morphisms containing A: we will call this the weakly saturated class of morphisms *generated* by A. We note that $(\perp A)_\perp$ is weakly saturated. Under appropriate set-theoretic assumptions, Quillen's "small object argument" can be used to establish that $(\perp A)_\perp$ is the weakly saturated class generated by A:

Proposition A.1.2.5 (Small Object Argument). *Let $\mathcal{C}$ be a presentable category and $A_0 = \{\phi_i : C_i \to D_i\}_{i \in I}$ a collection of morphisms in $\mathcal{C}$ indexed by a (small) set I. For each $n \geq 0$, let $\mathcal{C}^{[n]}$ denote the category of functors from the linearly ordered set $[n] = \{0, \dots, n\}$ into $\mathcal{C}$. There exists a functor $T : \mathcal{C}^{[1]} \to \mathcal{C}^{[2]}$ with the following properties:*

(1) *The functor T carries a morphism $f : X \to Z$ to a diagram*

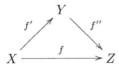

where f' belongs to the weakly saturated class of morphisms generated by A_0 and f'' has the right lifting property with respect to each morphism in A_0.

(2) *If κ is a regular cardinal such that each of the objects C_i, D_i is κ-compact, then T commutes with κ-filtered colimits.*

Proof. Fix a regular cardinal κ as in (2) and fix a morphism $f : X \to Z$ in $\mathcal{C}$. We will give a functorial construction of the desired diagram

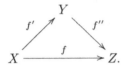

We define a transfinite sequence of objects

$$Y_0 \to Y_1 \to \cdots$$

in $\mathcal{C}_{/Z}$ indexed by ordinals smaller than κ. Let $Y_0 = X$ and let $Y_\lambda = \varinjlim_{\alpha < \lambda} Y_\alpha$ when λ is a nonzero limit ordinal. For $i \in I$, let $F_i : \mathcal{C}_{/Z} \to \mathcal{S}\mathrm{et}$ be the functor

$$(T \to Z) \mapsto \mathrm{Hom}_{\mathcal{C}}(D_i, Z) \times_{\mathrm{Hom}_{\mathcal{C}}(C_i, Z)} \mathrm{Hom}_{\mathcal{C}}(C_i, T).$$

Supposing that Y_α has been defined, we define $Y_{\alpha+1}$ by the following pushout diagram

$$
\begin{array}{ccc}
\coprod_{i \in I, \eta \in F_i(Y_\alpha)} C_i & \longrightarrow & Y_\alpha \\
\downarrow & & \downarrow \\
\coprod_{i \in I, \eta \in F_i(Y_\alpha)} D_i & \longrightarrow & Y_{\alpha+1}.
\end{array}
$$

We conclude by defining Y to be $\varinjlim_{\alpha < \kappa} Y_\alpha$. It is easy to check that this construction has the desired properties. $\square$

Remark A.1.2.6. If $\mathcal{C}$ is enriched, tensored, and cotensored over another presentable monoidal category $\mathbf{S}$ (see §A.1.4), then a similar construction shows that we can choose T to be an $\mathbf{S}$-enriched functor.

Corollary A.1.2.7. *Let $\mathcal{C}$ be a presentable category and let A be a set of morphisms of $\mathcal{C}$. Then $({}_{\perp}A)_{\perp}$ is the smallest weakly saturated class of morphisms containing A.*

Proof. Let $\overline{A}$ be the smallest weakly saturated class of morphisms containing A, so that $\overline{A} \subseteq ({}_\perp A)_\perp$. For the reverse inclusion, let us suppose that $f : X \to Z$ belongs to $({}_\perp A)_\perp$. Proposition A.1.2.5 implies the existence of a factorization

$$X \xrightarrow{f'} Y \xrightarrow{f''} Z,$$

where $f' \in \overline{A}$ and f'' belongs to ${}_\perp A$. It follows that f has the left lifting property with respect to f'', so that f is a retract of f' and therefore belongs to $\overline{A}$. $\square$

Remark A.1.2.8. Let $\mathcal{C}$ be a presentable category, let S be a (small) set of morphisms in $\mathcal{C}$, and suppose that $f : X \to Y$ belongs to the weakly saturated class of morphisms generated by S. The proofs of Proposition A.1.2.5 and Corollary A.1.2.7 show that there exists a transfinite sequence

$$Y_0 \to Y_1 \to \cdots$$

of objects of $\mathcal{C}_{X/}$, indexed by a set of ordinals $\{\beta | \beta < \alpha\}$, with the following properties:

(i) For each $\beta < \alpha$, there is a pushout diagram

$$
\begin{array}{ccc}
C & \xrightarrow{\ g\ } & D \\
\downarrow & & \downarrow \\
\varinjlim_{\gamma < \beta} Y_\gamma & \longrightarrow & Y_\beta,
\end{array}
$$

where the colimit is formed in $\mathcal{C}_{X/}$ and $g \in S$.

(ii) The object Y is a retract of $\varinjlim_{\gamma < \alpha} Y_\gamma$ in the category $\mathcal{C}_{X/}$.

A.1.3 Monoidal Categories

A *monoidal category* is a category $\mathcal{C}$ equipped with a (coherently) associative "product" functor $\otimes : \mathcal{C} \times \mathcal{C} \to \mathcal{C}$ and a unit object $\mathbf{1}$. The associativity is expressed by demanding isomorphisms

$$\eta_{A,B,C} : (A \otimes B) \otimes C \to A \otimes (B \otimes C),$$

and the requirement that $\mathbf{1}$ be unital is expressed by demanding isomorphisms

$$\alpha_A : A \otimes \mathbf{1} \to A$$

$$\beta_A : \mathbf{1} \otimes A \to A.$$

We do not merely require the existence of these isomorphisms: they are part of the structure of a monoidal category. Moreover, these isomorphisms are required to satisfy the following conditions:

- The isomorphism $\eta_{A,B,C}$ depends *functorially* on the triple (A, B, C); in other words, η may be regarded as a natural isomorphism between the functors

$$\mathcal{C} \times \mathcal{C} \times \mathcal{C} \to \mathcal{C}$$

$$(A, B, C) \mapsto (A \otimes B) \otimes C$$

$$(A, B, C) \mapsto A \otimes (B \otimes C).$$

Similarly, α_A and β_A depend functorially on A.

- Given any quadruple (A, B, C, D) of objects of $\mathcal{C}$, the *MacLane pentagon*

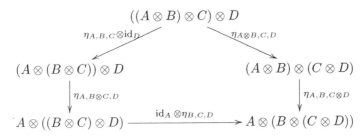

is commutative.

- For any pair (A, B) of objects of $\mathcal{C}$, the triangle

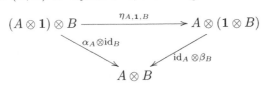

is commutative.

MacLane's coherence theorem asserts that the commutativity of this pair of diagrams implies the commutativity of *all* diagrams that can be written using only the isomorphisms $\eta_{A,B,C}$, α_A, and β_A. More precisely, any monoidal category is equivalent (as a monoidal category) to a *strict* monoidal category: that is, a monoidal category in which $\otimes$ is literally associative, $\mathbf{1}$ is literally a unit with respect to $\otimes$, and the isomorphisms $\eta_{A,B,C}$, α_A, β_A are the identity maps.

Example A.1.3.1. Let $\mathcal{C}$ be a category which admits finite products. Then $\mathcal{C}$ admits the structure of a monoidal category where the operation $\otimes$ is given by Cartesian product

$$A \otimes B \simeq A \times B$$

and the isomorphisms $\eta_{A,B,C}$ are induced from the evident associativity of the Cartesian product. The identity $\mathbf{1}$ is defined to be the final object of $\mathcal{C}$, and the isomorphisms α_A and β_A are determined in the obvious way. We refer to this monoidal structure on $\mathcal{C}$ as the *Cartesian monoidal structure*.

We remark that the Cartesian product $A \times B$ is well-defined only up to (unique) isomorphism (as is the final object $\mathbf{1}$), so that strictly speaking the Cartesian monoidal structure on $\mathcal{C}$ depends on various choices; however, all such choices lead to (canonically) equivalent monoidal categories.

Remark A.1.3.2. Let $(\mathcal{C}, \otimes, \mathbf{1}, \eta, \alpha, \beta)$ be a monoidal category. We will generally abuse notation by simply saying that $\mathcal{C}$ is a monoidal category, that $(\mathcal{C}, \otimes)$ is a monoidal category, or that $\otimes$ is a *monoidal structure* on $\mathcal{C}$; the other structure is implicitly understood to be present as well.

Remark A.1.3.3. Let $\mathcal{C}$ be a category equipped with a monoidal structure $\otimes$. Then we may define a new monoidal structure on $\mathcal{C}$ by setting $A \otimes^{op} B = B \otimes A$. We refer to this monoidal structure $\otimes^{op}$ as the *opposite* of the monoidal structure $\otimes$.

Definition A.1.3.4. A monoidal category $(\mathcal{C}, \otimes)$ is said to be *left-closed* if, for each $A \in \mathcal{C}$, the functor

$$N \mapsto A \otimes N$$

admits a right adjoint

$$Y \mapsto {}^A Y.$$

We say that $(\mathcal{C}, \otimes)$ is *right-closed* if the opposite monoidal structure $(\mathcal{C}, \otimes^{op})$ is left-closed; in other words, if every functor

$$N \mapsto N \otimes A$$

has a right adjoint

$$Y \mapsto Y^A.$$

Finally, we say that $(\mathcal{C}, \otimes)$ is *closed* if it is both right-closed and left-closed.

In the setting of monoidal categories, it is appropriate to consider only those functors which are compatible with the monoidal structures in the following sense:

Definition A.1.3.5. Let $(\mathcal{C}, \otimes)$ and $(\mathcal{D}, \otimes)$ be monoidal categories. A *right-lax monoidal functor* from $\mathcal{C}$ to $\mathcal{D}$ consists of the following data:

- A functor $G : \mathcal{C} \to \mathcal{D}$.

- A natural transformation $\gamma_{A,B} : G(A) \otimes G(B) \to G(A \otimes B)$ rendering commutative the diagram

$$
\begin{array}{ccc}
(G(A) \otimes G(B)) \otimes G(C) & \longrightarrow & G(A) \otimes (G(B) \otimes G(C)) \\
\downarrow {\scriptstyle \gamma_{A,B}} & & \downarrow {\scriptstyle \gamma_{B,C}} \\
G(A \otimes B) \otimes G(C) & & G(A) \otimes G(B \otimes C) \\
\downarrow {\scriptstyle \gamma_{A \otimes B, C}} & & \downarrow {\scriptstyle \gamma_{A, B \otimes C}} \\
G((A \otimes B) \otimes C) & \xrightarrow{\ G(\eta_{A,B,C})\ } & G(A \otimes (B \otimes C)).
\end{array}
$$

- A map $e : \mathbf{1}_{\mathcal{D}} \to G(\mathbf{1}_{\mathcal{C}})$ rendering commutative the diagrams

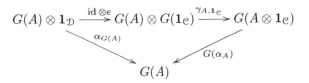

$$G(A) \otimes \mathbf{1}_{\mathcal{D}} \xrightarrow{\mathrm{id} \otimes e} G(A) \otimes G(\mathbf{1}_{\mathcal{C}}) \xrightarrow{\gamma_{A,\mathbf{1}_{\mathcal{C}}}} G(A \otimes \mathbf{1}_{\mathcal{C}})$$
$$\searrow^{\alpha_{G(A)}} \qquad \swarrow_{G(\alpha_A)}$$
$$G(A)$$

$$\mathbf{1}_{\mathcal{D}} \otimes G(B) \xrightarrow{e \otimes \mathrm{id}} G(\mathbf{1}_{\mathcal{C}}) \otimes G(B) \xrightarrow{\gamma_{\mathbf{1}_{\mathcal{C}},A}} G(\mathbf{1}_{\mathcal{C}} \otimes B) \ .$$
$$\searrow^{\beta_{G(B)}} \qquad \swarrow_{G(\alpha_B)}$$
$$G(B)$$

A natural transformation between right-lax monoidal functors is *monoidal* if it commutes with the maps $\gamma_{A,B}$, e.

Dually, a *left-lax monoidal functor* from $\mathcal{C}$ to $\mathcal{D}$ consists of a right-lax monoidal functor from $\mathcal{C}^{op}$ to $\mathcal{D}^{op}$; it is determined by giving a functor $F : \mathcal{C} \to \mathcal{D}$ together with a map $e' : F(\mathbf{1}_{\mathcal{C}}) \to \mathbf{1}_{\mathcal{D}}$ and a natural transformation

$$\gamma'_{A,B} : F(A \otimes B) \to F(A) \otimes F(B)$$

satisfying the appropriate analogues of the conditions listed above.

If F is a right-lax monoidal functor via *isomorphisms*

$$e : \mathbf{1}_{\mathcal{D}} \to F(\mathbf{1}_{\mathcal{C}})$$

$$\gamma_{A,B} : F(A) \otimes F(B) \to F(A \otimes B),$$

then F may be regarded as a left-lax monoidal functor by setting $e' = e^{-1}$, $\gamma'_{A,B} = \gamma_{A,B}^{-1}$. In this case, we simply say that F is a *monoidal* functor.

Remark A.1.3.6. Let

$$\mathcal{C} \underset{G}{\overset{F}{\rightleftarrows}} \mathcal{D}$$

be an adjunction between categories $\mathcal{C}$ and $\mathcal{D}$. Suppose that $\mathcal{C}$ and $\mathcal{D}$ are equipped with monoidal structures. Then endowing G with the structure of a right-lax monoidal functor is equivalent to endowing F with the structure of a left-lax monoidal functor.

Example A.1.3.7. Let $\mathcal{C}$ and $\mathcal{D}$ be categories which admit finite products and let $F : \mathcal{C} \to \mathcal{D}$ be a functor between them. If we regard $\mathcal{C}$ and $\mathcal{D}$ as endowed with the Cartesian monoidal structure, then F acquires the structure of a left-lax monoidal functor in a canonical way via the maps $F(A \times B) \to F(A) \times F(B)$ induced from the functoriality of F. In this case, F is a monoidal functor if and only if it commutes with finite products.

A.1.4 Enriched Category Theory

One frequently encounters categories $\mathcal{D}$ in which the collections of morphisms $\mathrm{Hom}_{\mathcal{D}}(X, Y)$ between two objects $X, Y \in \mathcal{D}$ have additional structure: for example, a topology, a group structure, or the structure of a vector space. These situations may all be efficiently described using the language of *enriched category theory*, which we now introduce.

Let $(\mathcal{C}, \otimes)$ be a monoidal category. A $\mathcal{C}$-*enriched category* $\mathcal{D}$ consists of the following data:

(1) A collection of objects.

(2) For every pair of objects $X, Y \in \mathcal{D}$, a mapping object $\mathrm{Map}_{\mathcal{D}}(X, Y)$ of $\mathcal{C}$.

(3) For every triple of objects $X, Y, Z \in \mathcal{D}$, a composition map
$$\mathrm{Map}_{\mathcal{D}}(Y, Z) \otimes \mathrm{Map}_{\mathcal{D}}(X, Y) \to \mathrm{Map}_{\mathcal{D}}(X, Z).$$

Composition is required to be associative in the sense that for any $W, X, Y, Z \in \mathcal{C}$, the diagram

$$\mathrm{Map}_{\mathcal{D}}(Z, Y) \otimes \mathrm{Map}_{\mathcal{D}}(Y, X) \otimes \mathrm{Map}_{\mathcal{D}}(X, W)$$

$$\mathrm{Map}_{\mathcal{D}}(Z, Y) \otimes \mathrm{Map}_{\mathcal{D}}(Y, W) \qquad \mathrm{Map}_{\mathcal{D}}(Z, X) \otimes \mathrm{Map}_{\mathcal{D}}(X, W)$$

$$\mathrm{Map}_{\mathcal{D}}(Z, W)$$

is commutative.

(4) For every object $X \in \mathcal{D}$, a unit map $\mathbf{1} \to \mathrm{Map}_{\mathcal{D}}(X, X)$ rendering commutative the diagrams

$$\mathbf{1} \otimes \mathrm{Map}_{\mathcal{D}}(Y, X) \longrightarrow \mathrm{Map}_{\mathcal{D}}(X, X) \otimes \mathrm{Map}_{\mathcal{D}}(Y, X)$$

$$\mathrm{Map}_{\mathcal{D}}(Y, X)$$

$$\mathrm{Map}_{\mathcal{D}}(X, Y) \otimes \mathbf{1} \longrightarrow \mathrm{Map}_{\mathcal{D}}(X, Y) \otimes \mathrm{Map}_{\mathcal{D}}(X, X)$$

$$\mathrm{Map}_{\mathcal{D}}(X, Y).$$

Example A.1.4.1. Suppose that $(\mathcal{C}, \otimes)$ is a *right-closed* monoidal category. Then $\mathcal{C}$ is enriched over itself in a natural way if one defines $\mathrm{Map}_{\mathcal{C}}(X, Y) = Y^X$.

Example A.1.4.2. Let $\mathcal{C}$ be the category of sets endowed with the Cartesian monoidal structure. Then a $\mathcal{C}$-enriched category is simply a category in the usual sense.

Remark A.1.4.3. Let $G : \mathcal{C} \to \mathcal{C}'$ be a right-lax monoidal functor between monoidal categories. Suppose that $\mathcal{D}$ is a category enriched over $\mathcal{C}$. We may define a category $G(\mathcal{D})$ enriched over $\mathcal{C}'$ as follows:

(1) The objects of $G(\mathcal{D})$ are the objects of $\mathcal{D}$.

(2) Given objects $X, Y \in \mathcal{D}$, we set
$$\mathrm{Map}_{G(\mathcal{D})}(X, Y) = G(\mathrm{Map}_{\mathcal{D}}(X, Y)).$$

(3) The composition in $G(\mathcal{D})$ is given by the composite map
$$G(\mathrm{Map}_{\mathcal{D}}(Y, Z)) \otimes G(\mathrm{Map}_{\mathcal{D}}(X, Y)) \to G(\mathrm{Map}_{\mathcal{D}}(Y, Z) \otimes \mathrm{Map}_{\mathcal{D}}(X, Y))$$
$$\to G(\mathrm{Map}_{\mathcal{D}}(X, Z)).$$

Here the first map is determined by the right-lax monoidal structure on the functor G, and the second is obtained by applying G to the composition law in the category $\mathcal{D}$.

(4) For every object $X \in \mathcal{D}$, the associated unit $G(\mathcal{D})$ is given by the composition
$$1_{\mathcal{C}'} \to G(1_{\mathcal{C}}) \to G(\mathrm{Map}_{\mathcal{D}}(X, X)).$$

Remark A.1.4.4. If $\mathcal{D}$ and $\mathcal{D}'$ are categories which are enriched over the same monoidal category $\mathcal{C}$, then one can define a category of $\mathcal{C}$-*enriched* functors from $\mathcal{D}$ to $\mathcal{D}'$ in the evident way. Namely, an enriched functor $F : \mathcal{D} \to \mathcal{D}'$ consists of a map from the objects of $\mathcal{D}$ to the objects of $\mathcal{D}'$ and a collection of morphisms
$$\eta_{X,Y} : \mathrm{Map}_{\mathcal{D}}(X, Y) \to \mathrm{Map}_{\mathcal{D}'}(FX, FY)$$

with the following properties:

(*i*) For each object $X \in \mathcal{D}$, the composition
$$1_{\mathcal{C}} \to \mathrm{Map}_{\mathcal{D}}(X, X) \overset{\eta_{X,X}}{\to} \mathrm{Map}_{\mathcal{D}'}(FX, FX)$$
coincides with the unit map for $FX \in \mathcal{D}'$.

(*ii*) For every triple of objects $X, Y, Z \in \mathcal{D}$, the diagram
$$
\begin{array}{ccc}
\mathrm{Map}_{\mathcal{D}}(X, Y) \otimes \mathrm{Map}_{\mathcal{D}}(Y, Z) & \longrightarrow & \mathrm{Map}_{\mathcal{D}}(X, Z) \\
\downarrow & & \downarrow \\
\mathrm{Map}_{\mathcal{D}'}(FX, FY) \otimes \mathrm{Map}_{\mathcal{D}}(FY, FZ) & \longrightarrow & \mathrm{Map}_{\mathcal{D}}(FX, FZ)
\end{array}
$$
is commutative.

If F and F' are enriched functors, an *enriched natural transformation* α from F to F' consists of specifying, for each object $X \in \mathcal{D}$, a morphism $\alpha_X \in \mathrm{Hom}_{\mathcal{D}'}(FX, F'X)$ which renders commutative the diagram

$$
\begin{array}{ccc}
\mathrm{Map}_{\mathcal{D}}(X, Y) & \longrightarrow & \mathrm{Map}_{\mathcal{D}'}(FX, FY) \\
\downarrow & & \downarrow {\alpha_Y} \\
\mathrm{Map}_{\mathcal{D}'}(F'X, F'Y) & \xrightarrow{\alpha_X} & \mathrm{Map}_{\mathcal{D}'}(FX, F'Y).
\end{array}
$$

Suppose that $\mathcal{C}$ is any monoidal category. Consider the functor $\mathcal{C} \to \mathrm{Set}$ given by

$$X \mapsto \mathrm{Hom}_{\mathcal{C}}(\mathbf{1}, X).$$

This is a right-lax monoidal functor from $(\mathcal{C}, \otimes)$ to Set, where the latter is equipped with the Cartesian monoidal structure. By the above remarks, we see that we may equip any $\mathcal{C}$-enriched category $\mathcal{D}$ with the structure of an ordinary category by setting

$$\mathrm{Hom}_{\mathcal{D}}(X, Y) = \mathrm{Hom}_{\mathcal{C}}(\mathbf{1}, \mathrm{Map}_{\mathcal{D}}(X, Y)).$$

We will generally not distinguish notationally between $\mathcal{D}$ as a $\mathcal{C}$-enriched category and this (underlying) category having the same objects. However, to avoid confusion, we use different notations for the morphisms: $\mathrm{Map}_{\mathcal{D}}(X, Y)$ is an object of $\mathcal{C}$, while $\mathrm{Hom}_{\mathcal{D}}(X, Y)$ is a set.

Let $\mathcal{C}$ be a right-closed monoidal category and let $\mathcal{D}$ be a category enriched over $\mathcal{C}$. Fix objects $C \in \mathcal{C}$, $X \in \mathcal{D}$, and consider the functor

$$\mathcal{D} \to \mathcal{C}$$

$$Y \mapsto \mathrm{Map}_{\mathcal{D}}(X, Y)^C.$$

This functor may or may not be *corepresentable* in the sense that there exists an object $Z \in \mathcal{D}$ and an isomorphism of functors

$$\eta : \mathrm{Map}_{\mathcal{D}}(X, \bullet)^C \simeq \mathrm{Map}_{\mathcal{D}}(Z, \bullet).$$

If such an object Z exists, we will denote it by $X \otimes C$. The natural isomorphism η is determined by specifying a single map $\eta(X) : C \to \mathrm{Map}_{\mathcal{D}}(X, X \otimes C)$. By general nonsense, the map $\eta(X)$ determines $X \otimes C$ up to (unique) isomorphism provided that $X \otimes C$ exists. If the object $X \otimes C$ exists for every $C \in \mathcal{C}$, $X \in \mathcal{D}$, then we say that $\mathcal{D}$ is *tensored over* $\mathcal{C}$. In this case, we may regard

$$(X, C) \mapsto X \otimes C$$

as determining a functor $\mathcal{D} \otimes \mathcal{C} \to \mathcal{D}$. Moreover, one has canonical isomorphisms

$$X \otimes (C \otimes D) \simeq (X \otimes C) \otimes D,$$

which express the idea that $\mathcal{D}$ may be regarded as equipped with an "action" of $\mathcal{C}$. Here we imagine $\mathcal{C}$ as a kind of generalized monoid (via its monoidal structure).

Dually, if $\mathcal{C}$ is right-closed, then an object of $\mathcal{D}$ which represents the functor

$$Y \mapsto {}^C \operatorname{Map}_{\mathcal{D}}(Y, X)$$

will be denoted by ${}^C X$; the object ${}^C X$ (if it exists) is determined up to (unique) isomorphism by a map $C \to \operatorname{Map}_{\mathcal{D}}({}^C X, X)$. If this object exists for all $C \in \mathcal{C}$, $X \in \mathcal{D}$, then we say that $\mathcal{D}$ is *cotensored over* $\mathcal{C}$.

Example A.1.4.5. Let $\mathcal{C}$ be a right-closed monoidal category. Then $\mathcal{C}$ may be regarded as enriched over itself in a natural way. It is automatically tensored over itself; it is cotensored over itself if and only if it is left-closed.

A.1.5 Trees

Let $\mathcal{C}$ be a presentable category and S a small collection of morphisms in $\mathcal{C}$. According to Remark A.1.2.8, the smallest weakly saturated class of morphisms $\overline{S}$ containing S can be obtained from S using pushouts, retracts, and transfinite composition. It is natural to ask if the formation of retracts is necessary: that is, does the weakly saturated class of morphisms generated by S coincide with the class of morphisms which generated by transfinite compositions of pushouts of morphisms of S? Our goal for the remainder of this section is to give an affirmative answer, at least after S has been suitably enlarged (Proposition A.1.5.12). This result is of a somewhat technical nature and will be needed only during our discussion of combinatorial model categories in §A.2.6.

We begin by introducing a generalization of the notion of a transfinite chain of morphisms.

Definition A.1.5.1. Let $\mathcal{C}$ be a presentable category and let S be a collection of morphisms in $\mathcal{C}$. An *S-tree* in $\mathcal{C}$ consists of the following data:

(1) An object $X \in \mathcal{C}$ called the *root* of the S-tree.

(2) A partially ordered set A which is *well-founded* (so that every nonempty subset of P has a minimal element).

(3) A diagram $A \to \mathcal{C}_{X/}$, which we will denote by $\alpha \mapsto Y_\alpha$.

(4) For each $\alpha \in A$, a pushout diagram

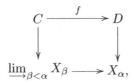

where $f \in S$.

Let κ be a regular cardinal. We will say that an S-tree in $\mathcal{C}$ is κ-*good* if each of the objects C and D appearing above is κ-compact, and if for each $\alpha \in A$, the set $\{\beta \in A : \beta < \alpha\}$ is κ-small.

Notation A.1.5.2. Let $\mathcal{C}$ be a presentable category and S a collection of morphisms in $\mathcal{C}$. We will indicate an S-tree by writing $\{Y_\alpha\}_{\alpha \in A}$. Here the root $X \in \mathcal{C}$ and the relevant pushout diagrams are understood implicitly to be part of the data.

Suppose we are given an S-tree $\{Y_\alpha\}_{\alpha \in A}$ and a subset $B \subseteq A$ which is downward-closed in the following sense: if $\alpha \in B$ and $\beta \leq \alpha$, then $\beta \in B$. Then $\{Y_\alpha\}_{\alpha \in B}$ is an S-tree. We let Y_B denote the colimit $\varinjlim_{\alpha \in B} Y_\alpha$ formed in the category $\mathcal{C}_{X/}$. In particular, we have a canonical isomorphism $Y_\emptyset \simeq X$. If $B = \{\alpha \in A | \alpha \leq \beta\}$, then $Y_B \simeq Y_\alpha$.

Remark A.1.5.3. Let $\mathcal{C}$ be a presentable category, S a collection of morphisms in $\mathcal{C}$, and $\{Y_\alpha\}_{\alpha \in A}$ an S-tree in $\mathcal{C}$ with root X. Given a map $f : X \to X'$, we can form an *associated S-tree* $\{Y_\alpha \coprod_X X'\}_{\alpha \in A}$ having root X'.

Example A.1.5.4. Let $\mathcal{C}$ be a presentable category, S a collection of morphisms in $\mathcal{C}$, and $\{Y_\alpha\}_{\alpha \in A}$ an S-tree in $\mathcal{C}$ with root X. If A is linearly ordered, then we may identify $\{Y_\alpha\}_{\alpha \in A}$ with a (possibly transfinite) sequence of morphisms belonging to S,

$$X \to Y_0 \to Y_1 \to \cdots,$$

as in the statement of (2) in Definition A.1.2.2.

Remark A.1.5.5. Let $\mathcal{C}$ be a presentable category, S a collection of morphisms in $\mathcal{C}$, and $\{Y_\alpha\}_{\alpha \in A}$ an S-tree in $\mathcal{C}$. Let $B \subseteq A$ be downward-closed. For $\alpha \in A - B$, let $B_\alpha = B \cup \{\beta \in A : \beta \leq \alpha\}$ and let $Z_\alpha = Y_{B_\alpha}$. Then $\{Z_\alpha\}_{\alpha \in A-B}$ is an S-tree in $\mathcal{C}$ with root Y_B.

Lemma A.1.5.6. *Let $\mathcal{C}$ be a presentable category and let S be a collection of morphisms in $\mathcal{C}$. Let $\{Y_\alpha\}_{\alpha \in A}$ be an S-tree in $\mathcal{C}$ and let $A'' \subseteq A' \subseteq A$ be subsets which are downward-closed in A. Then the induced map $Y_{A''} \to Y_{A'}$ belongs to the weakly saturated class of morphisms generated by S. In particular, the canonical map $Y_\emptyset \to Y_A$ belongs to the weakly saturated class of morphisms generated by S.*

Proof. Using Remarks A.1.5.5 and A.1.5.3, we can assume without loss of generality that $A'' = \emptyset$ and $A' = A$. Using the assumption that A is well-founded, we can write A as the union of a transfinite sequence (downward-closed) subsets $\{B(\gamma) \subseteq A\}_{\gamma < \beta}$ with the following property:

(∗) For each $\gamma < \beta$, the set $B(\gamma)$ is obtained from $B'(\gamma) = \bigcup_{\gamma' < \gamma} B(\gamma')$ by adjoining a minimal element α_γ of $A - B'(\gamma)$.

For $\gamma < \beta$, let $Z_\gamma = Y_{B(\gamma)}$. We now observe that $Y_A \simeq \varinjlim_{\gamma < \beta} Z_\gamma$ and that for each $\gamma < \beta$ there is a pushout diagram

$$
\begin{array}{ccc}
\varinjlim_{\alpha < \alpha_\gamma} Y_\alpha & \longrightarrow & Y_{\alpha_\gamma} \\
\downarrow & & \downarrow \\
\varinjlim_{\gamma' < \gamma} Z_{\gamma'} & \overset{f}{\longrightarrow} & Z_\gamma,
\end{array}
$$

so that f is the pushout of a morphism belonging to S. $\qquad\square$

Lemma A.1.5.7. *Let $\mathcal{C}$ be a presentable category, let κ be a regular cardinal, and let $S = \{f_s : C_s \to D_s\}$ be a collection of morphisms in $\mathcal{C}$, where each of the objects C_s and D_s is κ-compact. Suppose that $\{Y_\alpha\}_{\alpha \in A}$ is an S-tree in $\mathcal{C}$ indexed by a partially ordered set $(A, \leq)$. Then there exists the following:*

(1) *A new ordering $\preceq$ on A which refines $\leq$ in the following sense: if $\alpha \preceq \beta$, then $\alpha \leq \beta$. Let A' denote the partially ordered set A with this new partial ordering.*

(2) *A κ-good S-tree $\{Y'_\alpha\}_{\alpha \in A'}$ having the same root X as $\{Y_\alpha\}_{\alpha \in A}$.*

(3) *A collection of maps $f_\alpha : Y'_\alpha \to Y_\alpha$ which form a commutative diagram*

$$
\begin{array}{ccc}
Y'_{\alpha'} & \longrightarrow & Y'_\alpha \\
{\scriptstyle f_{\alpha'}}\downarrow & & \downarrow{\scriptstyle f_\alpha} \\
Y_{\alpha'} & \longrightarrow & Y_\alpha
\end{array}
$$

when $\alpha' \preceq \alpha$.

(4) *For every subset $B \subseteq A$ which is downward-closed with respect to $\preceq$, the induced map $f_B : Y'_B \to Y_B$ is an isomorphism.*

Proof. Choose a transfinite sequence of downward-closed subsets $\{A(\gamma) \subseteq A\}_{\gamma \leq \beta}$ so that the following conditions are satisfied:

(i) If $\gamma' \leq \gamma \leq \beta$, then $A(\gamma') \subseteq A(\gamma)$.

(ii) If $\lambda \leq \beta$ is a limit ordinal (possibly zero), then $A(\lambda) = \bigcup_{\gamma < \lambda} A(\gamma)$.

(iii) If $\gamma + 1 \leq \beta$, then $A(\gamma + 1) = A(\gamma) \cup \{\alpha_\gamma\}$, where α_γ is a minimal element of $A - A(\gamma)$.

(iv) The subset $A(\beta)$ coincides with A.

We will construct a compatible family of orderings $A'(\gamma) = (A(\gamma), \preceq)$, S-trees $\{Y'_\alpha\}_{\alpha \in A'(\gamma)}$, and collections of morphisms $\{Y'_\alpha \to Y_\alpha\}_{\alpha \in A(\gamma)}$ by induction on γ, so that the analogues of conditions (1) through (4) are satisfied. If γ is a limit ordinal, there is nothing to do; let us assume therefore that

$\gamma < \beta$ and that the data $(A'(\gamma), \{Y'_\alpha\}_{\alpha \in A'(\gamma)}, \{f_\alpha\}_{\alpha \in A(\gamma)})$ has already been constructed. Let $B = \{\alpha \in A : \alpha < \alpha_\gamma\}$, so that we have a pushout diagram

$$
\begin{array}{ccc}
C & \xrightarrow{\;f\;} & D \\
\downarrow{\scriptstyle i} & & \downarrow \\
Y_B & \longrightarrow & Y_{\alpha},
\end{array}
$$

where $f \in S$. By the inductive hypothesis, we may identify Y_B with Y'_B. Since C is κ-compact, the map i admits a factorization

$$
C \xrightarrow{\;i'\;} Y'_{B'} \xrightarrow{\;i''\;} Y'_B,
$$

where B' is κ-small. Enlarging B' if necessary, we may suppose that B' is downward-closed under $\preceq$. We now extend the partial ordering $\preceq$ to $A'(\gamma + 1) = A'(\gamma) \cup \{\alpha_\gamma\}$ by declaring that $\alpha \leq \alpha_\gamma$ if and only if $\alpha \in B'$. We define Y'_{α_γ} by forming a pushout diagram

$$
\begin{array}{ccc}
C & \xrightarrow{\;f\;} & D \\
\downarrow{\scriptstyle i'} & & \downarrow \\
Y'_{B'} & \longrightarrow & Y'_{\alpha_\gamma},
\end{array}
$$

and we define $f_{\alpha_\gamma} : Y'_{\alpha_\gamma} \to Y_{\alpha_\gamma}$ to be the map induced by i''. It is readily verified that these data satisfy the desired conditions. $\qquad\square$

Lemma A.1.5.8. *Let $\mathcal{C}$ be a presentable category, κ an uncountable regular cardinal, and S a collection of morphisms in $\mathcal{C}$. Let $\{Y_\alpha\}_{\alpha \in A}$ be a κ-good S-tree with root X and let $T_A : Y_A \to Y_A$ be an idempotent endomorphism of Y_A in the category $\mathcal{C}_{X/}$. Let B_0 be an arbitrary κ-small subset of A. Then there exists a κ-small subset $B \subseteq A$ which is downward-closed and contains B_0 and an idempotent endomorphism $T_B : Y_B \to Y_B$ such that the following diagram commutes:*

$$
\begin{array}{ccccc}
X & \longrightarrow & Y_B & \longrightarrow & Y_A \\
\downarrow{\scriptstyle =} & & \downarrow{\scriptstyle T_B} & & \downarrow{\scriptstyle T_A} \\
X & \longrightarrow & Y_B & \longrightarrow & Y_A.
\end{array}
$$

Proof. Enlarging B_0 if necessary, we may assume that B_0 is downward-closed. For every pair of downward-closed subsets $A'' \subseteq A' \subseteq A$, let $i_{A'',A'}$ denote the canonical map from $Y_{A''}$ to $Y_{A'}$. Note that because $\{Y_\alpha\}_{\alpha \in A}$ is a κ-good S-tree, if $A' \subseteq A$ is downward-closed and κ-small, $Y_{A'}$ is κ-compact when viewed as an object of $\mathcal{C}_{X/}$. In particular, Y_{B_0} is a κ-compact object of $\mathcal{C}_{X/}$. It follows that the composition

$$
Y_{B_0} \xrightarrow{\;i_{B_0,A}\;} Y_A \xrightarrow{\;T_A\;} Y_A
$$

can also be factored as a composition

$$Y_{B_0} \overset{T_0}{\to} Y_{B_1} \overset{i_{B_1,A}}{\to} Y_A,$$

where $B_1 \subseteq A$ is downward-closed and κ-small. Enlarging B_1 if necessary, we may suppose that B_1 contains B_0.

We now proceed to define a sequence of κ-small downward-closed subsets

$$B_0 \subseteq B_1 \subseteq B_2 \subseteq \cdots$$

of A and maps $T_i : Y_{B_i} \to Y_{B_{i+1}}$. Suppose that $i > 0$ and that B_i and T_{i-1} have already been constructed. By compactness again, we conclude that the composite map

$$Y_{B_i} \overset{i_{B_i,A}}{\to} Y_A \overset{T_A}{\to} Y_A$$

can be factored as

$$Y_{B_i} \overset{T_i}{\to} Y_{B_{i+1}} \overset{i_{B_{i+1},A}}{\to} Y_A,$$

where B_{i+1} is κ-small. Enlarging B_{i+1} if necessary, we may assume that B_{i+1} contains B_i and that the following diagrams commute:

$$
\begin{array}{ccc}
Y_{B_{i-1}} & \overset{T_{i-1}}{\longrightarrow} & Y_{B_i} \\
{\scriptstyle i_{B_{i-1},B_i}} \downarrow & & \downarrow {\scriptstyle i_{B_i,B_{i+1}}} \\
Y_{B_i} & \overset{T_i}{\longrightarrow} & Y_{B_{i+1}}
\end{array}
\qquad
\begin{array}{ccc}
Y_{B_{i-1}} & \overset{T_{i-1}}{\longrightarrow} & Y_{B_i} \\
{\scriptstyle T_{i-1}} \downarrow & & \downarrow {\scriptstyle T_i} \\
Y_{B_i} & \overset{i_{B_i,B_{i+1}}}{\longrightarrow} & Y_{B_{i+1}}.
\end{array}
$$

Let $B = \bigcup B_i$; then B is κ-small by virtue of our assumption that κ is uncountable. The collection of maps $\{T_i\}$ assemble to a map $T_B : Y_B \to Y_B$ with the desired properties. $\qquad\square$

Lemma A.1.5.9. *Let $\mathcal{C}$ be a presentable category, κ an uncountable regular cardinal, and S a collection of morphisms in $\mathcal{C}$. Let $\{Y_\alpha\}_{\alpha \in A}$ be a κ-good S-tree with root X, let $B \subseteq A$ be downward-closed, and suppose we are given a commutative diagram*

$$
\begin{array}{ccc}
Y_B & \longrightarrow & Y_A \\
{\scriptstyle T_B} \downarrow & & \downarrow {\scriptstyle T_A} \\
Y_B & \longrightarrow & Y_A
\end{array}
$$

in $\mathcal{C}_{X/}$, where T_A and T_B are idempotent. Let $C_0 \subseteq A$ be a κ-small subset. Then there exists a downward-closed κ-small subset $C \subseteq A$ containing C_0 and a pair of idempotent maps

$$T_C : Y_C \to Y_C$$

$$T_{B \cap C} : Y_{B \cap C} \to Y_{B \cap C}$$

such that the following diagram commutes (in $\mathcal{C}_{X/}$):

$$
\begin{array}{ccccccc}
Y_B & \longleftarrow & Y_{B \cap C} & \longrightarrow & Y_C & \longrightarrow & Y_A \\
{\scriptstyle T_B} \downarrow & & {\scriptstyle T_{B \cap C}} \downarrow & & {\scriptstyle T_C} \downarrow & & \downarrow {\scriptstyle T_A} \\
Y_B & \longleftarrow & Y_{B \cap C} & \longrightarrow & Y_C & \longrightarrow & Y_A.
\end{array}
$$

Proof. Enlarging C_0 if necessary, we may suppose that C_0 is downward-closed. We will define sequences of κ-small downward-closed subsets

$$C_0 \subseteq C_1 \subseteq \cdots \subseteq A$$

$$D_1 \subseteq D_2 \subseteq \cdots \subseteq B$$

and idempotent maps $\{T_{C_i} : Y_{C_i} \to Y_{C_i}\}_{i \geq 1}$, $\{T_{D_i} : Y_{D_i} \to Y_{D_i}\}_{i \geq 1}$. Moreover, we will guarantee that the following conditions are satisfied:

(*i*) For each $i > 0$, the set D_i contains the intersection $B \cap C_{i-1}$.

(*ii*) For each $i > 0$, the set C_i contains D_i.

(*iii*) For each $i > 0$, the diagrams

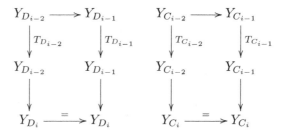

are commutative.

(*iv*) For each $i > 2$, the diagrams

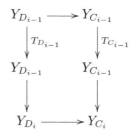

commute.

(*v*) For each $i > 1$, the diagram

$$
\begin{array}{ccc}
Y_{D_{i-1}} & \longrightarrow & Y_{C_{i-1}} \\
\downarrow{\scriptstyle T_{D_{i-1}}} & & \downarrow{\scriptstyle T_{C_{i-1}}} \\
Y_{D_{i-1}} & & Y_{C_{i-1}} \\
\downarrow & & \downarrow \\
Y_{D_i} & \longrightarrow & Y_{C_i}
\end{array}
$$

is commutative.

The construction proceeds by induction on i. Using a compactness argument, we see that conditions (*iv*) and (*v*) are satisfied provided that we choose C_i and D_i to be sufficiently large. The existence of the desired idempotent maps satisfying (*iii*) then follows from Lemma A.1.5.8 applied to the roots $\{Y_\alpha\}_{\alpha \in A}$ and $\{Y_\alpha\}_{\alpha \in B}$. We now take $C = \bigcup C_i$. Conditions (i) and (ii) guarantee that $B \cap C = \bigcup D_i$. Using (iv), it follows that the maps $\{T_{C_i}\}$ and $\{T_{D_i}\}$ glue to give idempotent endomorphisms $T_C : Y_C \to Y_C$, $T_{B \cap C} : Y_{B \cap C} \to Y_{B \cap C}$. Using (iii) and (v), we deduce that all of the desired diagrams are commutative. □

Lemma A.1.5.10. *Let $\mathcal{C}$ be a presentable category, let κ be a regular cardinal, and suppose that $\mathcal{C}$ is κ-accessible: that is, $\mathcal{C}$ is generated under κ-filtered colimits by κ-compact objects (Definition 5.4.2.1). Let $f : C \to D$ be a morphism between κ-compact objects of $\mathcal{C}$, let $g : X \to Y$ be a pushout of f (so that $Y \simeq X \coprod_C D$), and let $g' : X' \to Y'$ be a retract of g in the category of morphisms of $\mathcal{C}$. Then there exists a morphism $f' : C' \to D'$ with the following properties:*

(1) *The objects $C', D' \in \mathcal{C}$ are κ-compact.*

(2) *The morphism g' is a pushout of f'.*

(3) *The morphism f' belongs to the weakly saturated class of morphisms generated by f.*

Proof. Since g' is a retract of g, there exists a commutative diagram

$$
\begin{array}{ccc}
X' \longrightarrow X \longrightarrow X' \\
\downarrow{g'} \qquad \downarrow{g} \qquad \downarrow{g'} \\
Y' \longrightarrow Y \longrightarrow Y'.
\end{array}
$$

Replacing g by the induced map $X' \to X' \coprod_X Y$, we can reduce to the case where $X = X'$ and Y' is a retract of Y in $\mathcal{C}_{X/}$. Then Y' can be identified with the image of some idempotent $i : Y \to Y$.

Since $\mathcal{C}$ is κ-accessible, we can write X as the colimit of a κ-filtered diagram $\{X_\lambda\}$. The object C is κ-compact by assumption. Refining our diagram if necessary, we may assume that it takes values in $\mathcal{C}_{C/}$ and that Y is given as the colimit of the κ-filtered diagram $\{X_\lambda \coprod_C D\}$.

Because D is κ-compact, the composition $D \to Y \xrightarrow{i} Y$ admits a factorization

$$
D \xrightarrow{j} X_\lambda \coprod_C D \to Y.
$$

The κ-compactness of C implies that, after enlarging λ if necessary, we may suppose that the composition $j \circ f$ coincides with the canonical map from C to $X_\lambda \coprod_C D$. Consequently, j and the id_{X_λ} determine a map i' from $Y_\lambda = X_\lambda \coprod_C D$ to itself. Enlarging λ once more, we may suppose that i' is idempotent and that the diagram

$$
\begin{array}{ccc}
Y_\lambda \longrightarrow Y \\
\downarrow{i'} \qquad \downarrow{i} \\
Y_\lambda \longrightarrow Y
\end{array}
$$

is commutative. Let Y'_λ be the image of the idempotent i' and let $f' : X_\lambda \to Y'_\lambda$ be the canonical map. Then f' is a retract of the map $X_\lambda \to Y_\lambda$, which is a pushout of f. This proves (3). The objects X_λ and Y'_λ are κ-compact by construction, so that (1) is satisfied. We now observe that the diagram

$$\begin{array}{ccc} X_\lambda & \longrightarrow & Y'_\lambda \\ \downarrow & & \downarrow \\ X & \longrightarrow & Y' \end{array}$$

is a retract of the pushout diagram

$$\begin{array}{ccc} X_\lambda & \longrightarrow & Y_\lambda \\ \downarrow & & \downarrow \\ X & \longrightarrow & Y \end{array}$$

and therefore itself a pushout diagram. This proves (2) and completes the proof. $\qquad\square$

Lemma A.1.5.11. *Let $\mathcal{C}$ be a presentable category, κ a regular cardinal such that $\mathcal{C}$ is κ-accessible, and $S = \{f_s : C_s \to D_s\}$ a collection of morphisms $\mathcal{C}$ such that each C_s is κ-compact. Let $\{Y_\alpha\}_{\alpha \in A}$ be an S-tree in $\mathcal{C}$ with root X and suppose that A is κ-small. Then there exists a map $X' \to X$, where X is κ-compact, an S-tree $\{Y'_\alpha\}_{\alpha \in A}$ with root X', and an isomorphism of S-trees*

$$\{Y'_\alpha \coprod_{X'} X\}_{\alpha \in A} \simeq \{Y_\alpha\}_{\alpha \in A}$$

(see Remark A.1.5.3).

Proof. Since $\mathcal{C}$ is κ-accessible, we can write X as the colimit of diagram $\{X_i\}_{i \in I}$ indexed by a κ-filtered partially ordered set I, where each X_i is κ-compact. Choose a transfinite sequence of downward-closed subsets $\{A(\gamma) \subseteq A\}_{\gamma \leq \beta}$ so that the following conditions are satisfied:

(i) If $\gamma' \leq \gamma \leq \beta$, then $A(\gamma') \subseteq A(\gamma)$.

(ii) If $\lambda \leq \beta$ is a limit ordinal (possibly zero), then $A(\lambda) = \bigcup_{\gamma < \lambda} A(\gamma)$.

(iii) If $\gamma + 1 \leq \beta$, then $A(\gamma + 1) = A(\gamma) \cup \{\alpha_\gamma\}$, where α_γ is a minimal element of $A - A(\gamma)$.

(iv) The subset $A(\beta)$ coincides with A.

Note that, since A is κ-small, we have $\beta < \kappa$.

We will construct:

(a) A transfinite sequence of elements $\{i_\gamma \in I\}_{\gamma \leq \beta}$ such that $i_\gamma \leq i_{\gamma'}$ for $\gamma \leq \gamma'$.

(b) A sequence of S-trees $\{Y_\alpha^\gamma\}_{\alpha \in A(\gamma)}\}$ having roots X_{i_γ}.

(c) A collection of isomorphisms of S-trees

$$\{Y_\alpha^\gamma \coprod_{X_{i_\gamma}} X_{i_{\gamma'}}\}_{\alpha \in A(\gamma)} \simeq \{Y_\alpha^{\gamma'}\}_{\alpha \in A(\gamma)}$$

$$\{Y_\alpha^\gamma \coprod_{X_{i_\gamma}} X\}_{\alpha \in A(\gamma)} \simeq \{Y_\alpha\}_{\alpha \in A(\gamma)}$$

which are compatible with one another in the obvious sense.

If γ is a limit ordinal (or zero), we simply choose i_γ to be any upper bound for $\{i_{\gamma'}\}_{\gamma' < \gamma}$ in I. The rest of the data is uniquely determined. The existence of such an upper bound is guaranteed by our assumption that I is κ-filtered since $\gamma \leq \beta < \kappa$. Let us therefore suppose that the above data has been constructed for all ordinals $\leq \gamma$, and proceed to define $i_{\gamma+1}$. Let $i = i_\gamma$, let $\alpha = \alpha_\gamma$, and let $B = \{\beta \in A : \beta < \alpha\}$. Then we have canonical isomorphisms

$$Y_B \simeq Y_B^\gamma \coprod_{X_i} X \simeq \varinjlim \{Y_B^\gamma \coprod_{X_i} X_j\}_{j \geq i}$$

and a pushout diagram

$$
\begin{array}{ccc}
C_s & \xrightarrow{f_s} & D_s \\
{\scriptstyle g}\downarrow & & \downarrow \\
Y_B & \longrightarrow & Y_\alpha.
\end{array}
$$

The κ-compactness of C_s implies that g factors as a composition

$$C_s \xrightarrow{g'} Y_B^\gamma \coprod_{X_i} X_j$$

for some $j \geq i$. We now define $i_{\gamma+1} = j$, and $Y_\alpha^{\gamma+1}$ by forming a pushout diagram

$$
\begin{array}{ccc}
C_s & \longrightarrow & D_s \\
{\scriptstyle g_s}\downarrow & & \downarrow \\
Y_B^\gamma \coprod_{X_i} X_j & \longrightarrow & Y_\alpha^{\gamma+1}.
\end{array}
$$

$\square$

Proposition A.1.5.12. *Let $\mathcal{C}$ be a presentable ∞-category, κ a regular cardinal, and $\overline{S}$ a weakly saturated class of morphisms in $\mathcal{C}$. Let $S \subseteq \overline{S}$ be the subset consisting of those morphisms $f : X \to Y$ in $\overline{S}$ such that X and Y are κ-compact. Assume that*

(i) *The regular cardinal κ is uncountable, and $\mathcal{C}$ is κ-accessible.*

(ii) *The set S generates $\overline{S}$ as a weakly saturated class of morphisms.*

Then, for every morphism $f : X \to Y$ belonging to $\overline{S}$, there exists a transfinite sequence of objects $\{Y_\gamma\}_{\gamma < \beta}$ of $\mathcal{C}_{X/}$ with the following properties:

(1) For every ordinal $\gamma < \beta$, the natural map $\varinjlim_{\gamma' < \gamma} Z_{\gamma'} \to Z_\gamma$ is the pushout of a morphism in S.

(2) The colimit $\varinjlim_{\gamma < \beta} Z_\gamma$ is isomorphic to Y (as objects of $\mathcal{C}_{X/}$).

Proof. Remark A.1.2.8 implies the existence of a transfinite sequence of objects

$$Y_0 \to Y_1 \to \cdots$$

in $\mathcal{C}_{X/}$ indexed by a set of ordinals $A = \{\alpha | \alpha < \lambda\}$, satisfying condition (1), such that Y is a *retract* of $\varinjlim_{\alpha < \lambda} Y_\alpha$ in $\mathcal{C}_{X/}$. We may view the sequence $\{Y_\alpha\}_{\alpha \in A}$ as an S-tree in $\mathcal{C}$ having root X. According to Lemma A.1.5.7, we can choose a new S-tree $\{Y'_\alpha\}_{\alpha \in A'}$ which is κ-good, where $Y'_{A'} \simeq Y_A$, so that Y is a retract of $Y'_{A'}$. Choose an idempotent map $T_{A'} : Y'_{A'} \to Y'_{A'}$ in $\mathcal{C}_{X/}$ whose image is isomorphic to Y.

We now define a transfinite sequence

$$B(0) \subseteq B(1) \subseteq B(2) \subseteq \cdots ,$$

indexed by ordinals $\gamma < \beta$, and a compatible system of idempotent maps $T_{B(\gamma)} : Y'_{B_\gamma} \to Y'_{B_\gamma}$. Fix an ordinal γ and suppose that $B(\gamma')$ and $T_{B(\gamma')}$ have been defined for $\gamma' < \gamma$. Let $B'(\gamma) = \bigcup_{\gamma' < \gamma} B(\gamma')$ and let $T_{B'(\gamma)}$ be the result of amalgamating the maps $\{T_{B(\gamma')}\}_{\gamma' < \gamma}$. If $B'(\gamma) = A'$, we set $\beta = \gamma$ and conclude the construction; otherwise, we choose a minimal element $a \in A' - B'(\gamma)$. Applying Lemma A.1.5.9, we deduce the existence of a downward-closed subset $C(\gamma) \subseteq A'$ and a compatible collection of idempotent maps

$$T_{C(\gamma)} : Y'_{C(\gamma)} \to Y'_{C(\gamma)}$$

$$T_{C(\gamma) \cap B'(\gamma)} : Y'_{C(\gamma) \cap B'(\gamma)} \to Y'_{C(\gamma) \cap B'(\gamma)}.$$

We then define $B(\gamma) = B'(\gamma) \cup C(\gamma)$ and define $T_{B(\gamma)}$ to be the result of amalgamating $T_{B'(\gamma)}$ and $T_{C(\gamma)}$.

For every ordinal γ, there is a κ-good S-tree $\{Y''_\alpha\}_{\alpha \in B(\gamma) - B'(\gamma)}$ with root $Y'_{B(\gamma)}$ such that $Y''_{B(\gamma) - B'(\gamma)} \simeq Y'_{B(\gamma)}$ (Remark A.1.5.5). Combining Lemma A.1.5.11 with the observation that $B(\gamma) - B'(\gamma)$ is κ-small, we deduce that the map

$$Y'_{B'(\gamma)} \to Y'_{B(\gamma)}$$

is the pushout of a morphism in S.

For each ordinal $\gamma < \beta$, let Z_γ denote the image of the idempotent map $T_{B(\gamma)}$. Then $\varinjlim_{\gamma < \beta} Z_\gamma \simeq Y$, so that (2) is satisfied. Condition (1) follows from Lemma A.1.5.10. $\qquad\square$

Corollary A.1.5.13. *Under the hypotheses of Proposition A.1.5.12, there exists a κ-good S-tree $\{Y_\alpha\}_{\alpha \in A}$ such that $Y_A \simeq Y$ in $\mathcal{C}_{X/}$.*

Proof. Combine Proposition A.1.5.12 with Lemma A.1.5.7. $\qquad\square$

A.2 MODEL CATEGORIES

One of the oldest and most successful approaches to the study of higher-categorical phenomena is Quillen's theory of model categories. In this book, Quillen's theory will play two (related) roles:

(1) The structures that we use to describe higher categories are naturally organized into model categories. For example, ∞-categories are precisely those simplicial sets which are fibrant with respect to the Joyal model structure (Theorem 2.4.6.1). The theory of model categories provides a convenient framework for phrasing certain results and for comparing different models of higher category theory (see, for example, §2.2.5).

(2) The theory of model categories can itself be regarded as an approach to higher category theory. If $\mathbf{A}$ is a simplicial model category, then the subcategory $\mathbf{A}^\circ \subseteq \mathbf{A}$ of fibrant-cofibrant objects forms a fibrant simplicial category. Proposition 1.1.5.10 implies that the simplicial nerve $N(\mathbf{A}^\circ)$ is an ∞-category. We will refer to $N(\mathbf{A}^\circ)$ as the *underlying ∞-category* of $\mathbf{A}$. Of course, not every ∞-category arises in this way, even up to equivalence: for example, the existence of homotopy limits and homotopy colimits in $\mathbf{A}$ implies the existence of various limits and colimits in $N(\mathbf{A}^\circ)$ (Corollary 4.2.4.8). Nevertheless, we can often use the theory of model categories to prove theorems about general ∞-categories by reducing to the situation of ∞-categories which arise via the above construction (every ∞-category $\mathcal{C}$ admits a fully faithful embedding into $N(\mathbf{A}^\circ)$ for an appropriately chosen simplicial model category $\mathbf{A}$). For example, our proof of the ∞-categorical Yoneda lemma (Proposition 5.1.3.1) uses this strategy.

The purpose of this section is to review the theory of model categories with an eye toward the sort of applications described above. Our exposition is somewhat terse, and we will omit many proofs. For a more detailed account, we refer the reader to [40] (or any other text on the theory of model categories).

A.2.1 The Model Category Axioms

Definition A.2.1.1. A *model category* is a category $\mathcal{C}$ which is equipped with three distinguished classes of morphisms in $\mathcal{C}$, called *cofibrations, fibrations,* and *weak equivalences*, in which the following axioms are satisfied:

(1) The category $\mathcal{C}$ admits (small) limits and colimits.

(2) Given a composable pair of maps $X \xrightarrow{f} Y \xrightarrow{g} Z$, if any two of $g \circ f$, f, and g are weak equivalences, then so is the third.

(3) Suppose $f : X \to Y$ is a retract of $g : X' \to Y'$: that is, suppose there exists a commutative diagram

$$
\begin{array}{ccccc}
X & \xrightarrow{\ i\ } & X' & \xrightarrow{\ r\ } & X \\
\downarrow{\scriptstyle f} & & \downarrow{\scriptstyle g} & & \downarrow{\scriptstyle f} \\
Y & \xrightarrow{\ i'\ } & Y' & \xrightarrow{\ r'\ } & Y,
\end{array}
$$

where $r \circ i = \mathrm{id}_X$ and $r' \circ i' = \mathrm{id}_Y$. Then

   (i) If g is a fibration, so is f.

   (ii) If g is a cofibration, then so is f.

   (iii) If g is a weak equivalence, then so is f.

(4) Given a diagram

$$
\begin{array}{ccc}
A & \longrightarrow & X \\
\downarrow{\scriptstyle i} & \nearrow & \downarrow{\scriptstyle p} \\
B & \longrightarrow & Y,
\end{array}
$$

a dotted arrow can be found rendering the diagram commutative if either

   (i) The map i is a cofibration, and the map p is both a fibration and a weak equivalence.

   (ii) The map i is both a cofibration and a weak equivalence, and the map p is a fibration.

(5) Any map $X \to Z$ in $\mathcal{C}$ admits factorizations

$$
X \xrightarrow{\ f\ } Y \xrightarrow{\ g\ } Z
$$
$$
X \xrightarrow{\ f'\ } Y' \xrightarrow{\ g'\ } Z,
$$

where f is a cofibration, g is a fibration and a weak equivalence, f' is a cofibration and a weak equivalence, and g' is a fibration.

A map f in a model category $\mathcal{C}$ is called a *trivial cofibration* if it is both a cofibration and a weak equivalence; similarly, f is called a *trivial fibration* if it is both a fibration and a weak equivalence. By axiom (1), any model category $\mathcal{C}$ has an initial object $\emptyset$ and a final object $*$. An object $X \in \mathcal{C}$ is said to be *fibrant* if the unique map $X \to *$ is a fibration and *cofibrant* if the unique map $\emptyset \to X$ is a cofibration.

Example A.2.1.2. Let $\mathcal{C}$ be any category which admits small limits and colimits. Then $\mathcal{C}$ can be endowed with the *trivial* model structure:

(W) The weak equivalences in $\mathcal{C}$ are the isomorphisms.

(C) Every morphism in $\mathcal{C}$ is a cofibration.

(F) Every morphism in $\mathcal{C}$ is a fibration.

A.2.2 The Homotopy Category of a Model Category

Let $\mathcal{C}$ be a model category containing an object X. A *cylinder object* for X is an object C together with a diagram $X \coprod X \xrightarrow{i} C \xrightarrow{j} X$ where i is a cofibration, j is a weak equivalence, and the composition $j \circ i$ is the "fold map" $X \coprod X \to X$. Dually, a *path object* for $Y \in \mathcal{C}$ is an object P together with a diagram

$$Y \xrightarrow{q} P \xrightarrow{p} Y \times Y$$

such that q is a weak equivalence, p is a fibration, and $p \circ q$ is the diagonal map $Y \to Y \times Y$. The existence of cylinder and path objects follows from the factorization axiom (5) of Definition A.2.1.1 (factor the fold map $X \coprod X \to X$ as a cofibration followed by a trivial fibration and the diagonal map $Y \to Y \times Y$ as a trivial cofibration followed by a fibration).

Proposition A.2.2.1. *Let $\mathcal{C}$ be a model category. Let X be a cofibrant object of $\mathcal{C}$, Y a fibrant object of $\mathcal{C}$, and $f, g : X \to Y$ two maps. The following conditions are equivalent:*

(1) *For every cylinder object $X \coprod X \xrightarrow{j} C$, there exists a commutative diagram*

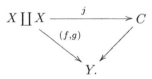

(2) *There exists a cylinder object $X \coprod X \xrightarrow{j} C$ and a commutative diagram*

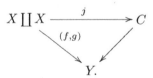

(3) *For every path object $P \xrightarrow{p} Y \times Y$, there exists a commutative diagram*

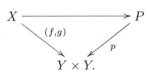

(4) *There exists a path object $P \xrightarrow{p} Y \times Y$ and a commutative diagram*

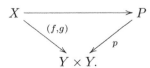

If $\mathcal{C}$ is a model category containing a cofibrant object X and a fibrant object Y, we say two maps $f, g : X \to Y$ are *homotopic* if the hypotheses of Proposition A.2.2.1 are satisfied and write $f \simeq g$. The relation $\simeq$ is an equivalence relation on $\mathrm{Hom}_{\mathcal{C}}(X, Y)$. The *homotopy category* $h\mathcal{C}$ may be defined as follows:

- The objects of $h\mathcal{C}$ are the fibrant-cofibrant objects of $\mathcal{C}$.

- For $X, Y \in h\mathcal{C}$, the set $\mathrm{Hom}_{h\mathcal{C}}(X, Y)$ is the set of $\simeq$-equivalence classes of $\mathrm{Hom}_{\mathcal{C}}(X, Y)$.

Composition is well-defined in $h\mathcal{C}$ by virtue of the fact that if $f \simeq g$, then $f \circ h \simeq g \circ h$ (this is clear from characterization (2) of Proposition A.2.2.1) and $h' \circ f \simeq h' \circ g$ (this is clear from characterization (4) of Proposition A.2.2.1) for any maps h, h' such that the compositions are defined in $\mathcal{C}$.

There is another way of defining $h\mathcal{C}$ (or, at least, a category equivalent to $h\mathcal{C}$): one begins with all of $\mathcal{C}$ and formally adjoins inverses to all weak equivalences. Let $H(\mathcal{C})$ denote the category so obtained. If $X \in \mathcal{C}$ is cofibrant and $Y \in \mathcal{C}$ is fibrant, then homotopic maps $f, g : X \to Y$ have the same image in $H(\mathcal{C})$; consequently, we obtain a functor $h\mathcal{C} \to H(\mathcal{C})$ which can be shown to be an equivalence. We will generally ignore the distinction between these two categories, employing whichever description is more useful for the problem at hand.

Remark A.2.2.2. Since $\mathcal{C}$ is (generally) not a small category, it is not immediately clear that $H(\mathcal{C})$ has small morphism sets; however, this follows from the equivalence between $H(\mathcal{C})$ and $h\mathcal{C}$.

A.2.3 A Lifting Criterion

The following basic principle will be used many times throughout this book:

Proposition A.2.3.1. *Let $\mathcal{C}$ be a model category containing cofibrant objects A and B and a fibrant object X. Suppose we are given a cofibration $i : A \to B$ and any map $f : A \to X$. Suppose moreover that there exists a commutative diagram*

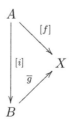

in the homotopy category $h\,\mathcal{C}$. *Then there exists a commutative diagram*

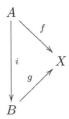

in $\mathcal{C}$, *with* $[g] = \overline{g}$. (*Here we let* $[p]$ *denote the homotopy class in* $h\mathcal{C}$ *of a morphism* p *in* $\mathcal{C}$.)

Proof. Choose a map $g' : B \to X$ representing the homotopy class $\overline{g}$. Choose a cylinder object

$$A \coprod A \to C(A) \to A$$

and a factorization

$$C(A) \coprod_{A \coprod A} (B \coprod B) \to C(B) \to B,$$

where the first map is a cofibration and the second is a trivial fibration. We observe that $C(B)$ is a cylinder object for B.

Since $g' \circ i$ is homotopic to f, there exists a map $h_0 : C(A) \coprod_A B \to X$ with $h_0|B = g'$ and $h_0|A = f$. The inclusion $C(A) \coprod_A B \to C(B)$ is a trivial cofibration, so h_0 extends to a map $h : C(B) \to X$. We may regard h as a homotopy from g' to g, where $g \circ i = f$. $\square$

Proposition A.2.3.1 will often be applied in the following way. Suppose we are given a diagram

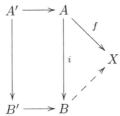

which we would like to extend as indicated by the dotted arrow. If X is fibrant, i is a cofibration between cofibrant objects, and the horizontal arrows are weak equivalences, then it suffices to solve the (frequently easier) problem of constructing the dotted arrow in the diagram

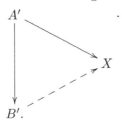

A.2.4 Left Properness and Homotopy Pushout Squares

Definition A.2.4.1. A model category $\mathcal{C}$ is *left proper* if, for any pushout square

$$
\begin{array}{ccc}
A & \xrightarrow{\ i\ } & B \\
\downarrow{\scriptstyle j} & & \downarrow{\scriptstyle j'} \\
A' & \xrightarrow{\ i'\ } & B'
\end{array}
$$

in which i is a cofibration and j is a weak equivalence, the map j' is also a weak equivalence. Dually, $\mathcal{C}$ is *right proper* if, for any pullback square

$$
\begin{array}{ccc}
X' & \xrightarrow{\ p'\ } & Y' \\
\downarrow{\scriptstyle q'} & & \downarrow{\scriptstyle q} \\
X & \xrightarrow{\ p\ } & Y
\end{array}
$$

in which p is a fibration and q is a weak equivalence, the map q' is also a weak equivalence.

In this book, we will deal almost exclusively with left proper model categories. The following provides a useful criterion for establishing left properness.

Proposition A.2.4.2. *Let $\mathcal{C}$ be a model category in which every object is cofibrant. Then $\mathcal{C}$ is left proper.*

Proposition A.2.4.2 is an immediate consequence of the following basic lemma:

Lemma A.2.4.3. *Let $\mathcal{C}$ be a model category containing a pushout diagram*

$$
\begin{array}{ccc}
A & \xrightarrow{\ i\ } & B \\
\downarrow{\scriptstyle j} & & \downarrow{\scriptstyle j'} \\
A' & \xrightarrow{\ i'\ } & B'.
\end{array}
$$

Suppose that A and A' are cofibrant, i is a cofibration, and j is a weak equivalence. Then j' is a weak equivalence.

Proof. We wish to show that j' is an isomorphism in the homotopy category $h\mathcal{C}$. In other words, we need to show that for every fibrant object Z of $\mathcal{C}$, composition with j' induces a bijection $\mathrm{Hom}_{h\mathcal{C}}(B', Z) \to \mathrm{Hom}_{h\mathcal{C}}(B, Z)$.

We first show that composition with j' is surjective on homotopy classes. Suppose we are given a map $f : B \to Z$. Since j is a weak equivalence, the composition $f \circ i$ is homotopic to $g \circ j$ for some $g : A' \to B$. According to Proposition A.2.3.1, there is a map $f' : B \to Z$ such that $f' \circ i = g \circ j$ and such that f' is homotopic to f. The amalgamation of f' and g determines a map $B' \to Z$ which lifts f'.

We now show that j' is injective on homotopy classes. Suppose we are given a pair of maps $s, s' : B' \to Z$. Let P be a path object for Z. If $s \circ j'$ and $s' \circ j'$ are homotopic, then there exists a commutative diagram

$$
\begin{array}{ccc}
B & \xrightarrow{\;\;h\;\;} & P \\
\Big\downarrow{\scriptstyle j'} & & \Big\downarrow \\
B' & \xrightarrow{\;s \times s'\;} & Z \times Z.
\end{array}
$$

We now replace $\mathcal{C}$ by $\mathcal{C}_{/Z \times Z}$ and apply the surjectivity statement above to deduce that there is a map $h' : B' \to P$ such that h is homotopic to $h' \circ j'$. The existence of h' shows that s and s' are homotopic, as desired.   □

Suppose we are given a diagram

$$ A_0 \leftarrow A \to A_1 $$

in a model category $\mathcal{C}$. In general, the pushout $A_0 \coprod_A A_1$ is poorly behaved in the sense that a map of diagrams

$$
\begin{array}{ccc}
A_0 & \longleftarrow A \longrightarrow & A_1 \\
\downarrow & \downarrow & \downarrow \\
B_0 & \longleftarrow B \longrightarrow & B_1
\end{array}
$$

need not induce a weak equivalence $A_0 \coprod_A A_1 \to B_0 \coprod_B B_1$, even if each of the vertical arrows in the diagram is individually a weak equivalence. To correct this difficulty, it is convenient to introduce the left derived functor of "pushout". The *homotopy pushout* of the diagram

$$ A_0 \longleftarrow A \longrightarrow A_1 $$

is defined to be the pushout $A_0' \coprod_{A'} A_1'$, where we have chosen a commutative diagram

$$
\begin{array}{ccc}
A_0' & \xleftarrow{\;j\;} A' \xrightarrow{\;i\;} & A_1' \\
\downarrow & \downarrow & \downarrow \\
A_0 & \longleftarrow A \longrightarrow & A_1
\end{array}
$$

in which the top row is a *cofibrant* diagram in the sense that A' is cofibrant and the maps i and j are both cofibrations. One can show that such a diagram exists and that the pushout $A_0' \coprod_{A'} A_1'$ depends on the choice of diagram only up to weak equivalence. (For a more systematic approach which includes a definition of "cofibrant" for more complicated diagrams, we refer the reader to §A.3.3.)

More generally, we will say that a diagram

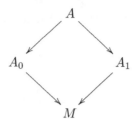

is a *homotopy pushout square* if the composite map

$$A_0' \coprod_{A'} A_1' \to A_0 \coprod_A A_1 \to M$$

is a weak equivalence. In this case we will also say that M is a *homotopy pushout* of A_0 and A_1 over A. One can show that this condition is independent of the choice of a "cofibrant resolution"

$$A_0' \longleftarrow A' \longrightarrow A_1'$$

of the original diagram. In particular, we note that if the diagram

$$A_0 \longleftarrow A \longrightarrow A_1$$

is *already* cofibrant, then the ordinary pushout $A_0 \coprod_A A_1$ is a homotopy pushout. However, the condition that the diagram be cofibrant is quite strong; in good situations we can get away with quite a bit less:

Proposition A.2.4.4. *Let* $\mathcal{C}$ *be a model category and let*

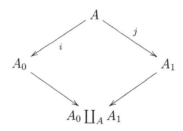

be a pushout square in $\mathcal{C}$. *This diagram is also a homotopy pushout square if either of the following conditions is satisfied:*

(*i*) *The objects* A *and* A_0 *are cofibrant, and* j *is a cofibration.*

(*ii*) *The map* j *is a cofibration, and* $\mathcal{C}$ *is left proper.*

Remark A.2.4.5. The above discussion of homotopy pushouts can be dualized; one obtains the notion of *homotopy pullbacks*, and the analogue of Proposition A.2.4.4 requires either that $\mathcal{C}$ be a *right* proper model category or that the objects in the diagram be fibrant.

A.2.5 Quillen Adjunctions and Quillen Equivalences

Let $\mathcal{C}$ and $\mathcal{D}$ be model categories and suppose we are given a pair of adjoint functors

$$\mathcal{C} \underset{G}{\overset{F}{\rightleftarrows}} \mathcal{D}$$

(here F is the left adjoint and G is the right adjoint). The following conditions are equivalent:

(1) The functor F preserves cofibrations and trivial cofibrations.

(2) The functor G preserves fibrations and trivial fibrations.

(3) The functor F preserves cofibrations, and the functor G preserves fibrations.

(4) The functor F preserves trivial cofibrations, and the functor G preserves trivial fibrations.

If any of these equivalent conditions is satisfied, then we say that the pair (F, G) is a *Quillen adjunction* between $\mathcal{C}$ and $\mathcal{D}$. We also say that F is a *left Quillen functor* and that G is a *right Quillen functor*. In this case, one can show that F preserves weak equivalences between cofibrant objects and G preserves weak equivalences between fibrant objects.

Suppose that $\mathcal{C} \underset{G}{\overset{F}{\rightleftarrows}} \mathcal{D}$ is a Quillen adjunction. We may view the homotopy category $h\mathcal{C}$ as obtained from $\mathcal{C}$ by first passing to the full subcategory consisting of cofibrant objects and then inverting all weak equivalences. Applying a similar procedure with $\mathcal{D}$, we see that because F preserves weak equivalence between cofibrant objects, it induces a functor $h\mathcal{C} \to h\mathcal{D}$; this functor is called the *left derived functor of F* and denoted LF. Similarly, one may define the *right derived functor RG* of G. One can show that LF and RG determine an adjunction between the homotopy categories $h\mathcal{C}$ and $h\mathcal{D}$.

Proposition A.2.5.1. *Let $\mathcal{C}$ and $\mathcal{D}$ be model categories and let*

$$\mathcal{C} \underset{G}{\overset{F}{\rightleftarrows}} \mathcal{D}$$

be a Quillen adjunction. The following are equivalent:

(1) *The left derived functor $LF : h\mathcal{C} \to h\mathcal{D}$ is an equivalence of categories.*

(2) *The right derived functor $RG : h\mathcal{D} \to h\mathcal{C}$ is an equivalence of categories.*

(3) *For every cofibrant object $C \in \mathcal{C}$ and every fibrant object $D \in \mathcal{D}$, a map $C \to G(D)$ is a weak equivalence in $\mathcal{C}$ if and only if the adjoint map $F(C) \to D$ is a weak equivalence in $\mathcal{D}$.*

Proof. Since the derived functors LF and RG are adjoint to one another, it is clear that (1) is equivalent to (2). Moreover, (1) and (2) are equivalent to the assertion that the unit and counit of the adjunction

$$u : \mathrm{id}_{\mathcal{C}} \to RG \circ LF$$

$$v : LF \circ RG \to \mathrm{id}_{\mathcal{D}}$$

are weak equivalences. Let us consider the unit u. Choose a fibrant object C of $\mathcal{C}$. The composite functor $(RG \circ LF)(C)$ is defined to be $G(D)$, where $F(C) \to D$ is a weak equivalence in $\mathcal{D}$ and D is a fibrant object of $\mathcal{D}$. Thus u is a weak equivalence when evaluated on C if and only if for any weak equivalence $F(C) \to D$, the adjoint map $C \to G(D)$ is a weak equivalence. Similarly, the counit v is a weak equivalence if and only if the converse holds. Thus (1) and (2) are equivalent to (3). $\qquad\square$

If the equivalent conditions of Proposition A.2.5.1 are satisfied, then we say that the adjunction (F, G) gives a *Quillen equivalence* between the model categories $\mathcal{C}$ and $\mathcal{D}$.

A.2.6 Combinatorial Model Categories

In this section, we give an overview of Jeff Smith's theory of *combinatorial model categories*. Our main goal is to prove Proposition A.2.6.13, which allows us to construct model structures on a category $\mathcal{C}$ by specifying the class of weak equivalences together with a small amount of additional data.

Definition A.2.6.1 (Smith). Let $\mathbf{A}$ be model category. We say that $\mathbf{A}$ is *combinatorial* if the following conditions are satisfied:

(1) The category $\mathbf{A}$ is presentable.

(2) There exists a set I of *generating cofibrations* such that the collection of all cofibrations in $\mathbf{A}$ is the smallest weakly saturated class of morphisms containing I (see Definition A.1.2.2).

(3) There exists a set J of *generating trivial cofibrations* such that the collection of all trivial cofibrations in $\mathbf{A}$ is the smallest weakly saturated class of morphisms containing J.

If $\mathcal{C}$ is a combinatorial model category, then the model structure on $\mathcal{C}$ is uniquely determined by the generating cofibrations and generating trivial cofibrations. However, in practice these generators might be difficult to find. Our goal in this section is to reformulate Definition A.2.6.1 in a manner which puts more emphasis on the category of weak equivalences in $\mathbf{A}$.

In practice, it is often easier to describe the class of *all* weak equivalences than it is to describe a class of generating trivial cofibrations.

Definition A.2.6.2. Let $\mathcal{C}$ be a presentable category and κ a regular cardinal. We will say that a full subcategory $\mathcal{C}_0 \subseteq \mathcal{C}$ is a *κ-accessible subcategory* of $\mathcal{C}$ if the following conditions are satisfied:

(1) The full subcategory $\mathcal{C}_0 \subseteq \mathcal{C}$ is stable under κ-filtered colimits.

(2) There exists a (small) set of objects of $\mathcal{C}_0$ which generates $\mathcal{C}_0$ under κ-filtered colimits.

We will say that $\mathcal{C}_0 \subseteq \mathcal{C}$ is an *accessible subcategory* if $\mathcal{C}_0$ is a κ-accessible subcategory of $\mathcal{C}$ for some regular cardinal κ.

Condition (2) of Definition A.2.6.2 admits the following reformulation:

Proposition A.2.6.3. *Let κ be a regular cardinal, $\mathcal{C}$ a presentable category, and $\mathcal{C}_0 \subseteq \mathcal{C}$ a full subcategory which is stable under κ-filtered colimits. Then $\mathcal{C}_0$ satisfies condition (2) of Definition A.2.6.2 if and only if the following condition is satisfied for all sufficiently large regular cardinals $\tau \gg \kappa$:*

$(2'_\tau)$ *Let A be a τ-filtered partially ordered set and $\{X_\alpha\}_{\alpha \in A}$ a diagram of τ-compact objects of $\mathcal{C}$ indexed by A. For every κ-filtered subset $B \subseteq A$, we let X_B denote the (κ-filtered) colimit of the diagram $\{X_\alpha\}_{\alpha \in B}$. Suppose that X_A belongs to $\mathcal{C}_0$. Then for every τ-small subset $C \subseteq A$, there exists a τ-small κ-filtered subset $B \subseteq A$ which contains C, such that X_B belongs to $\mathcal{C}_0$.*

First, we need the following preliminary result:

Lemma A.2.6.4. *Let $\tau \gg \kappa$ be regular cardinals such that $\tau > \kappa$, let $\mathcal{D}$ be a presentable ∞-category, and let $\{C_a\}_{a \in A}$ and $\{D_b\}_{b \in B}$ be families of τ-compact objects in $\mathcal{D}$ indexed by τ-filtered partially ordered sets A and B, such that*

$$\varinjlim_{a \in A} C_a \simeq \varinjlim_{b \in B} D_b.$$

Then, for every pair of τ-small subsets $A_0 \subseteq A$, $B_0 \subseteq B$, there exist τ-small κ-filtered subsets $A' \subseteq A$, $B' \subseteq B$ such that $A_0 \subseteq A'$, $B_0 \subseteq B'$, and $\varinjlim_{a \in A'} C_a \simeq \varinjlim_{b \in B'} D_b$.

Proof. Let $\mathcal{A}$ be the partially ordered set of all τ-small κ-filtered subsets of A which contain A_0, let $\mathcal{B}$ be the partially ordered set of all τ-small κ-filtered subsets of B which contain B_0, let $X \in \mathcal{D}$ be the common colimit $\varinjlim_{a \in A} C_a \simeq \varinjlim_{b \in B} D_b$, and let $\mathcal{C}$ be the full subcategory of $\mathcal{D}_{/X}$ spanned by those morphisms $Y \to X$, where Y is a τ-compact object of $\mathcal{D}$. Let $f : \mathcal{A} \to \mathcal{C}$ and $g : \mathcal{B} \to \mathcal{C}$ be the functors described by the formulas

$$f(A') = (\varinjlim_{a \in A'} C_a \to \varinjlim_{a \in A} C_a)$$

$$g(B') = (\varinjlim_{b \in B'} D_b \to \varinjlim_{b \in B} D_b).$$

The desired result now follows by applying Lemma 5.4.6.3 to the associated diagram

$$N(\mathcal{A}) \to N(\mathcal{C}) \leftarrow N(\mathcal{B}).$$

$\square$

Proof of Proposition A.2.6.3. First suppose that $(2'_\tau)$ is satisfied for all sufficiently large $\tau \gg \kappa$. Choose $\tau \gg \kappa$ large enough that $\mathcal{C}$ is generated under colimits by its full subcategory $\mathcal{C}^\tau$ of τ-compact objects and such that $(2'_\tau)$ is satisfied. Let $\mathcal{D} = \mathcal{C}^\tau \cap \mathcal{C}_0$, so that $\mathcal{D}$ is essentially small. We will show that $\mathcal{D}$ generates $\mathcal{C}_0$ under τ-filtered colimits. By assumption, every object $X \in \mathcal{C}$ can be obtained as a τ-filtered colimit of τ-compact objects $\{X_\alpha\}_{\alpha \in A}$. Let A' denote the collection of all τ-small κ-filtered subsets $B \subseteq A$ such that $X_B \in \mathcal{C}_0$. We regard A' as partially ordered via inclusions. Invoking condition $(2'_\tau)$, we deduce that X_A is the colimit of the τ-filtered collection of objects $\{X_{A'}\}_{A' \in B}$. We now observe that each $X_{A'}$ belongs to $\mathcal{D}$.

Now suppose that condition (2) is satisfied, so that $\mathcal{C}_0$ is generated under κ-filtered colimits by a small subcategory $\mathcal{D} \subseteq \mathcal{C}_0$. Choose $\tau \gg \kappa$ large enough that every object of $\mathcal{D}$ is τ-compact. Enlarging τ if necessary, we may suppose that $\tau > \kappa$. We claim that $(2'_\tau)$ is satisfied. To prove this, we consider any system of morphisms $\{X\}_{\alpha \in A}$ satisfying the hypotheses of $(2'_\tau)$. In particular, X_A belongs to $\mathcal{C}_0$, so that X_A may be obtained in some *other* way as a κ-filtered colimit of a system $\{Y_\beta\}_{\beta \in B}$, where each of the objects Y_β belongs to $\mathcal{D}$ and is therefore τ-compact. Let C' denote the family of all τ-small κ-filtered subsets $B_0 \subseteq B$. Replacing B by B' and the family $\{Y_\beta\}_{\beta \in B}$ by $\{Y_{B_0}\}_{B_0 \in B'}$, we may assume that B is τ-filtered.

Let $A_0 \subseteq A$ be a τ-small subset. Applying Lemma A.2.6.4 to the diagram category $\mathcal{C}$, we deduce that $A_0 \subseteq A'$, where A' is a τ-small, κ-filtered subset of A and there is an isomorphism $X_{A'} \simeq Y_{B'}$; here B' is a κ-filtered subset of B, so that $Y_{B'} \in \mathcal{C}_0$ by virtue of our assumption that $\mathcal{C}_0$ is stable under κ-filtered colimits.   □

Corollary A.2.6.5. *Let $f : \mathcal{C} \to \mathcal{D}$ be a functor between presentable categories which preserves κ-filtered colimits and let $\mathcal{D}_0 \subseteq \mathcal{D}$ be a κ-accessible subcategory. Then $f^{-1}\mathcal{D}_0 \subseteq \mathcal{C}$ is a κ-accessible subcategory.*

Corollary A.2.6.6 (Smith). *Let $\mathbf{A}$ be a combinatorial model category, let $\mathbf{A}^{[1]}$ be the category of morphisms in $\mathbf{A}$, let $W \subseteq \mathbf{A}^{[1]}$ be the full subcategory spanned by the weak equivalences, and let $F \subseteq \mathbf{A}^{[1]}$ be the full subcategory spanned by the fibrations. Then F, W, and $F \cap W$ are accessible subcategories of $\mathbf{A}^{[1]}$.*

Proof. For every morphism $i : A \to B$, let $F_i : \mathbf{A}^{[1]} \to \mathrm{Set}^{[1]}$ be the functor which carries a morphism $f : X \to Y$ to the induced map of sets

$$\mathrm{Hom}_{\mathbf{A}}(B, X) \to \mathrm{Hom}_{\mathbf{A}}(B, Y) \times_{\mathrm{Hom}_{\mathbf{A}}(A, Y)} \mathrm{Hom}_{\mathbf{A}}(A, X).$$

We observe that if A and B are κ-compact objects of $\mathbf{A}$, then F_i preserves κ-filtered colimits.

Let $\mathcal{C}_0$ be the full subcategory of $\mathrm{Set}^{[1]}$ spanned by the collection of *surjective* maps between sets. It is easy to see that $\mathcal{C}_0$ is an accessible category of $\mathrm{Set}^{[1]}$. It follows that the full subcategories $R(i) = F_i^{-1}\mathcal{C}_0 \subseteq \mathbf{A}^{[1]}$ are accessible subcategories of $\mathbf{A}^{[1]}$ (Corollary A.2.6.5).

Let I be a set of generating cofibrations for $\mathbf{A}$ and let J be a set of generating trivial cofibrations. Then Proposition 5.4.7.10 implies that the subcategories

$$F = \bigcap_{j \in J} R(j)$$

$$W \cap F = \bigcap_{i \in I} R(i)$$

are accessible subcategories of $\mathbf{A}^{[1]}$.

Applying Proposition A.1.2.5, we deduce that there exists a pair of functors $T', T'' : \mathbf{A}^{[1]} \to \mathbf{A}^{[1]}$, which carry an arbitrary morphism $f : X \to Z$ to a factorization

$$X \overset{T'(f)}{\to} Y \overset{T''(f)}{\to} Z,$$

where $T'(f)$ is a trivial cofibration and $T''(f)$ is a fibration. Moreover, the functor T'' can be chosen to commute with κ-filtered colimits for a sufficiently large regular cardinal κ. We now observe that W is the inverse image of $F \cap W$ under the functor $T'' : \mathbf{A}^{[1]} \to \mathbf{A}^{[1]}$ and is therefore an accessible subcategory of $\mathbf{A}^{[1]}$ by Corollary A.2.6.5. $\qquad\square$

Our next goal is to prove a converse to Corollary A.2.6.6, which will allow us to construct examples of combinatorial model categories. First, we need the following preliminary result.

Lemma A.2.6.7. *Let $\mathbf{A}$ be a presentable category. Suppose W and C are collections of morphisms of $\mathbf{A}$ with the following properties:*

(1) *The collection C is a weakly saturated class of morphisms of $\mathbf{A}$, and there exists a (small) subset $C_0 \subseteq C$ which generates C as a weakly saturated class of morphisms.*

(2) *The intersection $C \cap W$ is a weakly saturated class of morphisms of $\mathbf{A}$.*

(3) *The full subcategory $W \subseteq \mathbf{A}^{[1]}$ is an accessible subcategory of $\mathbf{A}^{[1]}$.*

(4) *The class W has the two-out-of-three property.*

Then $C \cap W$ is generated, as a weakly saturated class of morphisms, by a (small) subset $S \subseteq C \cap W$.

Proof. Let κ be a regular cardinal such that W is κ-accessible. Choose a regular cardinal $\tau \gg \kappa$ such that W satisfies condition $(2'_\tau)$ of Proposition A.2.6.3. Enlarging τ if necessary, we may assume that $\tau > \kappa$ (so that τ is uncountable), that $\mathcal{C}$ is τ-accessible, and that the source and target of every morphism in C_0 is τ-compact. Enlarging C_0 if necessary, we may suppose that C_0 consists of *all* morphisms $f : X \to Y$ in C such that X and Y are τ-compact. Let $S = C_0 \cap W$. We will show that S generates $C \cap W$ as a weakly saturated class of morphisms.

Let $\overline{S}$ be the weakly saturated class of morphisms generated by S and let $f : X \to Y$ be a morphism which belongs to $C \cap W$. We wish to show that $f \in \overline{S}$. Corollary A.1.5.13 implies that there exists a τ-good C_0-tree $\{Y_\alpha\}_{\alpha \in A}$ with root X, such that Y is isomorphic to Y_A as objects of $\mathcal{C}_{X/}$. Let us say that a subset $B \subseteq A$ is *good* if it is downward-closed and the canonical map $i : X \to Y_B$ belongs to W (we note that i automatically belongs to C, by virtue of Lemma A.1.5.6).

We now make the following observations:

(i) Given an increasing transfinite sequence of good subsets $\{A_\gamma\}_{\gamma < \beta}$, the union $\bigcup A_\gamma$ is good. This follows from the assumption that $C \cap W$ is weakly saturated.

(ii) Let $B \subseteq A$ be good and let $B_0 \subseteq B$ be τ-small. Then there exists a τ-small subset $B' \subseteq B$ containing B_0. This follows from our assumption that W satisfies condition $(2'_\tau)$ of Proposition A.2.6.3.

(iii) Suppose that $B, B' \subseteq A$ are such that B, B', and $B \cap B'$ are good. Then $B \cup B'$ is good. To prove this, we consider the pushout diagram

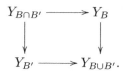

Every morphism in this diagram belongs to C (Lemma A.1.5.6), and the upper horizontal map belongs to W by virtue of assumption (4). Since $C \cap W$ is stable under pushouts, we conclude that the lower vertical map belongs to W. Assumption (4) now implies that the composite map $X \to Y_{B'} \to Y_{B \cup B'}$ belongs to W, as desired.

The next step is to prove the following claim:

(∗) Let A' be a good subset of A and let $B_0 \subseteq A$ be τ-small. Then there exists a τ-small subset $B \subseteq A$ such that $B_0 \subseteq B$, B is good and $B \cap A'$ is good.

To prove (∗), we begin by setting $B'_0 = A' \cap B_0$. We now define sequences of τ-small subsets

$$B_0 \subseteq B_1 \subseteq B_2 \subseteq \cdots$$

$$B'_0 \subseteq B'_1 \subseteq B'_2 \subseteq \cdots$$

as follows. Suppose that B_i and B'_i have been defined. Applying (ii), we choose B_{i+1} to be any τ-small good subset of A which contains $B_i \cup B'_i$. Applying (ii) again, we select B'_{i+1} to be any τ-small good subset of A' which contains $A' \cap B_{i+1}$. Let $B = \bigcup B_i$. It follows from (i) that B and $A' \cap B = \bigcup_i B'_i$ are both good.

We now choose a transfinite sequence of good subsets $\{A(\gamma) \subseteq A\}_{\gamma < \beta}$. Suppose that $A(\gamma')$ has been defined for $\gamma' < \gamma$ and let $A'(\gamma) = \bigcup_{\gamma' < \gamma} A(\gamma')$.

It follows from (i) that $A'(\gamma)$ is good. If $A'(\gamma) = A$, we set $\beta = \gamma$ and conclude the construction. Otherwise, we choose a minimal element $a \in A - A'(\gamma)$. Applying $(*)$, we deduce that there exists a τ-small good subset $B(\gamma) \subseteq A$ containing a, such that $A'(\gamma) \cap B(\gamma)$ is good. Let $A(\gamma) = A'(\gamma) \cup B(\gamma)$. It follows from (iii) that $A(\gamma)$ is good.

We observe that $\{Y_{A(\gamma)}\}_{\gamma < \beta}$ is a transfinite sequence of objects of $\mathcal{C}_{X/}$ having colimit Y. To prove that $f : X \to Y$ belongs to $\overline{S}$, it will suffice to show that for each $\gamma < \beta$, the map $g : Y_{A'(\gamma)} \to Y_{A(\gamma)}$ belongs to $\overline{S}$. Remark A.1.5.5 implies the existence of a C_0-tree $\{Z_\alpha\}_{\alpha \in A(\gamma) - A'(\gamma)}$ with root $Y_{A'(\gamma)}$ and colimit $Y_{A(\gamma)}$. Since $A(\gamma) - A'(\gamma)$ is τ-small, Lemma A.1.5.11 implies the existence of a pushout diagram

$$
\begin{array}{ccc}
M & \longrightarrow & N \\
\downarrow & & \downarrow \\
Y_{A'(\gamma)} & \longrightarrow & Y_{A(\gamma)}
\end{array}
$$

where $g \in C_0$.

Since $\mathcal{C}$ is τ-accessible, we can write $Y_{A'(\gamma)}$ as the colimit of a family of τ-compact objects $\{Z_\lambda\}_{\lambda \in P}$ indexed by a τ-filtered partially ordered set P. Since M is τ-compact, we can assume (reindexing the colimit if necessary) that we have a compatible family of maps $\{M \to Z_\lambda\}$. For each λ, let $g_\lambda : Z_\lambda \to Z_\lambda \coprod_M N$ be the induced map. Then g is the filtered colimit of the family $\{g_\lambda\}_{\lambda \in P}$. Since W satisfies condition $(2'_\tau)$ of Proposition A.2.6.3, we conclude that there exists a τ-small κ-filtered subset $P_0 \subseteq P$, such that $g' = \varinjlim_{\lambda \in P_0} g_\lambda$ belongs to W. We now observe that $g' \in S$ and that g is a pushout of g', so that $g \in \overline{S}$, as desired. $\qquad \square$

Proposition A.2.6.8. *Let* $\mathbf{A}$ *be a presentable category and let* W *and* C *be classes of morphisms in* $\mathbf{A}$ *with the following properties:*

(1) *The collection* C *is a weakly saturated class of morphisms of* $\mathbf{A}$, *and there exists a (small) subset* $C_0 \subseteq C$ *which generates* C *as a weakly saturated class of morphisms.*

(2) *The intersection* $C \cap W$ *is a weakly saturated class of morphisms of* $\mathbf{A}$.

(3) *The full subcategory* $W \subseteq \mathbf{A}^{[1]}$ *is an accessible subcategory of* $\mathbf{A}^{[1]}$.

(4) *The class* W *has the two-out-of-three property.*

(5) *If* f *is a morphism in* $\mathbf{A}$ *which has the right lifting property with respect to each element of* C, *then* $f \in W$.

Then $\mathbf{A}$ *admits a combinatorial model structure, which may be described as follows:*

(C) *The cofibrations in* $\mathbf{A}$ *are the elements of* C.

(W) *The weak equivalences in* $\mathbf{A}$ *are the elements of* W.

(F) *A morphism in* **A** *is a fibration if it has the right lifting property with respect to every morphism in* $C \cap W$.

Proof. The category **A** has all (small) limits and colimits since it is presentable. The two-out-three property for W is among our assumptions, and the stability of W under retracts follows from the accessibility of $W \subseteq \mathbf{A}^{[1]}$ (Corollary 4.4.5.16). The class of cofibrations is stable under retracts by (1), and the class of fibrations is stable under retracts by definition. The classes of fibrations and cofibrations are stable under retracts by definition.

We next establish the factorization axioms. By the small object argument, any morphism $X \to Z$ admits a factorization

$$X \xrightarrow{f} Y \xrightarrow{g} Z,$$

where $f \in C$ and g has the right lifting property with respect to every morphism in C. In particular, g has the right lifting property with respect to every morphism in $C \cap W$, so that g is a fibration; assumption (5) then implies that g is a trivial fibration. Similarly, using Lemma A.2.6.7, we may choose a factorization as above, where $f \in C \cap W$ and g has the right lifting property with respect to $C \cap W$; g is then a fibration by definition.

To complete the proof, it suffices to show that cofibrations have the left lifting property with respect to trivial fibrations, and trivial cofibrations have the left lifting property with respect to fibrations. The second of these statements is clear (it is the definition of a fibration). For the first statement, let us consider an arbitrary trivial fibration $p : X \to Z$. By the small object argument, there exists a factorization of p

$$X \xrightarrow{q} Y \xrightarrow{r} Z,$$

where q is a cofibration and r has the right lifting property with respect to all cofibrations. Then r is a weak equivalence by (3), so that q is a weak equivalence by the two-out-of-three property. Considering the diagram

$$
\begin{array}{ccc}
X & = & X \\
{\scriptstyle q}\downarrow & \nearrow & \downarrow{\scriptstyle p} \\
Y & \xrightarrow{r} & Z,
\end{array}
$$

we deduce the existence of the dotted arrow from the fact that p is a fibration and q is a trivial cofibration. It follows that p is a retract of r, and therefore p also has the right lifting property with respect to all cofibrations. This completes the proof that **A** is a model category. The assertion that **A** is combinatorial follows immediately from (1) and from Lemma A.2.6.7. □

Corollary A.2.6.9. *Let* **A** *be a presentable category equipped with a model structure. Suppose that there exists a (small) set which generates the collection of cofibrations in* **A** *(as a weakly saturated class of morphisms). Then the following are equivalent:*

(1) *The model category* **A** *is combinatorial; in other words, there exists a (small) set which generates the collection of trivial cofibrations in* **A** *(as a weakly saturated class of morphisms).*

(2) *The collection of weak equivalences in* **A** *determines an accessible subcategory of* $\mathbf{A}^{[1]}$.

Proof. The implication (1) $\Rightarrow$ (2) follows from Corollary A.2.6.6, and the reverse implication follows from Proposition A.2.6.8. $\qquad\qquad\square$

Our next goal is to prove a weaker version of Proposition A.2.6.8 which is somewhat easier to apply in practice.

Definition A.2.6.10. Let **A** be a presentable category. A class W of morphisms in $\mathcal{C}$ is *perfect* if it satisfies the following conditions:

(1) Every isomorphism belongs to W.

(2) Given a pair of composable morphisms $X \xrightarrow{f} Y \xrightarrow{g} Z$, if any two of the morphisms f, g, and $g \circ f$ belong to W, then so does the third.

(3) The class W is stable under filtered colimits. More precisely, suppose we are given a family of morphisms $\{f_\alpha : X_\alpha \to Y_\alpha\}$ which is indexed by a filtered partially ordered set. Let X denote a colimit of $\{X_\alpha\}$, Y a colimit of $\{Y_\alpha\}$, and $f : X \to Y$ the induced map. If each f_α belongs to W, then so does f.

(4) There exists a (small) subset $W_0 \subseteq W$ such that every morphism belonging to W can be obtained as a filtered colimit of morphisms belonging to W_0.

Example A.2.6.11. If $\mathcal{C}$ is a presentable category, then the class W consisting of all isomorphisms in $\mathcal{C}$ is perfect.

The following is an immediate consequence of Corollary A.2.6.5:

Corollary A.2.6.12. *Let* $F : \mathcal{C} \to \mathcal{C}'$ *be a functor between presentable categories which preserves filtered colimits and let* $W_{\mathcal{C}'}$ *be a perfect class of morphisms in* $\mathcal{C}'$. *Then* $W_{\mathcal{C}} = F^{-1}W_{\mathcal{C}'}$ *is a perfect class of morphisms in* $\mathcal{C}$.

Proposition A.2.6.13. *Let* **A** *be a presentable category. Suppose we are given a class* W *of morphisms of* $\mathcal{C}$, *which we will call weak equivalences, and a (small) set* C_0 *of morphisms of* $\mathcal{C}$, *which we will call generating cofibrations. Suppose furthermore that the following assumptions are satisfied:*

(1) *The class* W *of weak equivalences is perfect (Definition A.2.6.10).*

(2) *For any diagram*

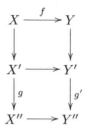

in which both squares are coCartesian, f belongs to C_0, and g belongs to W, the map g' also belongs to W.

(3) *If $g : X \to Y$ is a morphism in $\mathbf{A}$ which has the right lifting property with respect to every morphism in C_0, then g belongs to W.*

Then there exists a left proper combinatorial model structure on $\mathcal{C}$ which may be described as follows:

(C) *A morphism $f : X \to Y$ in $\mathbf{A}$ is a cofibration if it belongs to the weakly saturated class of morphisms generated by C_0.*

(W) *A morphism $f : X \to Y$ in $\mathcal{C}$ is a weak equivalence if it belongs to W.*

(F) *A morphism $f : X \to Y$ in $\mathcal{C}$ is a fibration if it has the right lifting property with respect to every map which is both a cofibration and a weak equivalence.*

Proof. We first show that the class of weak equivalences is stable under pushouts by cofibrations. Let P denote the collection of all morphisms f in $\mathbf{A}$ with the following property: for coCartesian diagram

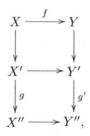

where g belongs to W, the map g' also belongs to W. By assumption, $C_0 \subseteq P$. It is easy to see that P is weakly saturated (using the stability of W under filtered colimits), so that every cofibration belongs to P.

It remains only to show that $\mathbf{A}$ is a model category. In view of Proposition A.2.6.8, it will suffice to show that $C \cap W$ is a weakly saturated class of morphisms. It is clear that $C \cap W$ is stable under retracts. It will therefore suffice to verify the stability of $C \cap W$ under pushouts and transfinite composition. The case of transfinite composition is easy: C is stable under transfinite composition because C is weakly saturated, and W is stable under transfinite composition because it is stable under finite composition and filtered colimits.

It remains to show that $C \cap W$ is stable under pushouts. Suppose we are given a coCartesian diagram

$$
\begin{array}{ccc}
X & \longrightarrow & X'' \\
{\scriptstyle f}\downarrow & & \downarrow{\scriptstyle f''} \\
Y & \longrightarrow & Y''
\end{array}
$$

in which f belongs to $C \cap W$; we wish to show that f'' also belongs to $C \cap W$. Since C is weakly saturated, it will suffice to show that f'' belongs to W. Using the small object argument, we can factor the top horizontal map to produce a coCartesian rectangle

$$
\begin{array}{ccccc}
X & \xrightarrow{\;g\;} & X' & \xrightarrow{\;h\;} & X'' \\
{\scriptstyle f}\downarrow & & {\scriptstyle f'}\downarrow & & \downarrow{\scriptstyle f''} \\
Y & \longrightarrow & Y' & \xrightarrow{\;h'\;} & Y''
\end{array}
$$

in which g is a cofibration and h has the right lifting property with respect to all the morphisms in C_0. Since W is stable under the formation of pushouts by cofibrations, we deduce that f' belongs to W. Moreover, by assumption (3), h belongs to W. Since h' is a pushout of h by the cofibration f', we deduce that h' belongs to W as well. Applying the two-out-of-three property (twice), we deduce that f'' belongs to W. $\square$

Remark A.2.6.14. Let $\mathbf{A}$ be a model category. Then $\mathbf{A}$ arises via the construction of Proposition A.2.6.13 if and only if it is combinatorial and left proper and the collection of weak equivalences in $\mathbf{A}$ is stable under filtered colimits.

A.2.7 Simplicial Sets

The formalism of simplicial sets plays a prominent role throughout this book. In this section, we will review the definition of a simplicial set and establish some notation.

For each $n \geq 0$, we let $[n]$ denote the linearly ordered set $\{0, \ldots, n\}$. We let $\mathbf{\Delta}$ denote the category of *combinatorial simplices*: the objects of $\mathbf{\Delta}$ are the linearly ordered sets $[n]$, and morphisms in $\mathbf{\Delta}$ are given by (nonstrictly) order-preserving maps.

If $\mathcal{C}$ is any category, a *simplicial object* of $\mathcal{C}$ is a functor $\mathbf{\Delta}^{op} \to \mathcal{C}$. Dually, a *cosimplicial object* of $\mathcal{C}$ is a functor $\mathbf{\Delta} \to \mathcal{C}$. A *simplicial set* is a simplicial object in the category of sets. More explicitly, a simplicial set S is determined by the following data:

- A set S_n for each $n \geq 0$ (the value of S on the object $[n] \in \mathbf{\Delta}$).

- A map $p^* : S_n \to S_m$ for each order-preserving map $[m] \to [n]$, the formation of which is compatible with composition (including empty composition, so that $(\mathrm{id}_{[n]})^* = \mathrm{id}_{S_n}$).

Let us recall a bit of standard notation for working with a simplicial set S. For each $0 \leq j \leq n$, the *face map* $d_j : S_n \to S_{n-1}$ is defined to be the pullback p^*, where $p : [n-1] \to [n]$ is given by

$$p(i) = \begin{cases} i & \text{if } i < j \\ i+1 & \text{if } i \geq j. \end{cases}$$

Similarly, the *degeneracy map* $s_j : S_n \to S_{n+1}$ is defined to be the pullback q^*, where $q : [n+1] \to [n]$ is defined by the formula

$$q(i) = \begin{cases} i & \text{if } i \leq j \\ i-1 & \text{if } i > j. \end{cases}$$

Because every order-preserving map from $[n]$ to $[m]$ can be factored as a composition of face and degeneracy maps, the structure of a simplicial set S is completely determined by the sets S_n for $n \geq 0$ together with the face and degeneracy operations defined above. These operations are required to satisfy certain identities, which we will not make explicit here.

Remark A.2.7.1. The category $\mathbf{\Delta}$ is equivalent to the (larger) category of all finite nonempty linearly ordered sets. We will sometimes abuse notation by identifying $\mathbf{\Delta}$ with this larger subcategory and by regarding simplicial sets (or more general simplicial objects) as functors which are defined on all nonempty linearly ordered sets.

Notation A.2.7.2. The category of simplicial sets will be denoted by Set_Δ. If J is a linearly ordered set, we let $\Delta^J \in \mathrm{Set}_\Delta$ denote the representable functor $[n] \mapsto \mathrm{Hom}([n], J)$, where the morphisms are taken in the category of linearly ordered sets. For each $n \geq 0$, we will write Δ^n in place of $\Delta^{[n]}$. We observe that, for any simplicial set S, there is a natural identification of sets $S_n \simeq \mathrm{Hom}_{\mathrm{Set}_\Delta}(\Delta^n, S)$.

Example A.2.7.3. For $0 \leq j \leq n$, we let $\Lambda_j^n \subset \Delta^n$ denote the "jth horn." It is determined by the following property: an element of $(\Lambda_j^n)_m$ is given by an order-preserving map $p : [m] \to [n]$ satisfying the condition that $\{j\} \cup p([m]) \neq [n]$. Geometrically, Λ_j^n corresponds to the subset of an n-simplex Δ^n in which the jth face and the interior have been removed.

More generally, if J is any finite linearly ordered set containing an element j, we let Λ_j^J denote the simplicial subset of Δ^J obtained by removing the interior and the face opposite the vertex j.

The category Set_Δ of simplicial sets has a (combinatorial, left proper, and right proper) model structure, which we will refer to as the *Kan model structure*. It may be described as follows:

- A map of simplicial sets $f : X \to Y$ is a *cofibration* if it is a monomorphism; that is, if the induced map $X_n \to Y_n$ is injective for all $n \geq 0$.

- A map of simplicial sets $f : X \to Y$ is a *fibration* if it is a Kan fibration: that is, if for any diagram

$$
\begin{array}{ccc}
\Lambda_i^n & \longrightarrow & X \\
\downarrow & \nearrow & \downarrow f \\
\Delta^n & \longrightarrow & Y,
\end{array}
$$

it is possible to supply the dotted arrow rendering the diagram commutative.

- A map of simplicial sets $f : X \to Y$ is a *weak equivalence* if the induced map of geometric realizations $|X| \to |Y|$ is a homotopy equivalence of topological spaces.

To prove this, we observe that the class of all cofibrations is generated by the collection of all inclusions $\partial \Delta^n \subseteq \Delta^n$; it is then easy to see that the conditions of Proposition A.2.6.13 are satisfied. The nontrivial point is to verify that the fibrations for the resulting model structure are precisely the Kan fibrations and that Set_Δ is right proper; these facts ultimately rely on a delicate analysis due to Quillen (see [32]).

Remark A.2.7.4. In §2.2.5, we introduce another model structure on Set_Δ, the *Joyal model structure*. This model structure has the same class of cofibrations, but the fibrations and the weak equivalences differ from those defined in this section. To avoid confusion, we will refer to the fibrations and weak equivalences for the usual model structure on simplicial sets as *Kan fibrations* and *weak homotopy equivalences*, respectively.

A.2.8 Diagram Categories and Homotopy (Co)limits

Let $\mathbf{A}$ be a combinatorial model category and $\mathcal{C}$ a small category. We let $\mathrm{Fun}(\mathcal{C}, \mathbf{A})$ denote the category of all functors from $\mathcal{C}$ to $\mathbf{A}$. In this section, we will see that $\mathrm{Fun}(\mathcal{C}, \mathbf{A})$ again admits the structure of a combinatorial model category: in fact, it admits two such structures. Moreover, by considering the functoriality of this construction in the category $\mathcal{C}$, we will obtain the theory of *homotopy limits* and *homotopy colimits*.

Definition A.2.8.1. Let $\mathcal{C}$ be a small category and let $\mathbf{A}$ be a model category. We will say that a natural transformation $\alpha : F \to G$ in $\mathrm{Fun}(\mathcal{C}, \mathbf{A})$ is:

- an *injective cofibration* if the induced map $F(C) \to G(C)$ is a cofibration in $\mathbf{A}$ for each $C \in \mathcal{C}$.

- a *projective fibration* if the induced map $F(C) \to G(C)$ is a fibration in $\mathbf{A}$ for each $C \in \mathcal{C}$.

- a *weak equivalence* if the induced map $F(C) \to G(C)$ is a weak equivalence in $\mathbf{A}$ for each $C \in \mathcal{C}$.

- an *injective fibration* if it has the right lifting property with respect to every morphism β in $\mathrm{Fun}(\mathcal{C}, \mathbf{A})$ which is simultaneously a weak equivalence and a injective cofibration.

- a *projective cofibration* if it has the left lifting property with respect to every morphism β in $\mathrm{Fun}(\mathcal{C}, \mathbf{A})$ which is simultaneously a weak equivalence and a projective fibration.

Proposition A.2.8.2. *Let $\mathbf{A}$ be a combinatorial model category and let $\mathcal{C}$ be a small category. Then there exist two combinatorial model structures on $\mathrm{Fun}(\mathcal{C}, \mathbf{A})$:*

- *The projective model structure determined by the projective cofibrations, weak equivalences, and projective fibrations.*

- *The injective model structure determined by the injective cofibrations, weak equivalences, and injective fibrations.*

The following is the key step in the proof of Proposition A.2.8.2:

Lemma A.2.8.3. *Let $\mathbf{A}$ be a presentable category and let $\mathcal{C}$ be a small category. Let S_0 be a (small) set of morphisms of $\mathbf{A}$ and let $\overline{S}_0$ be the weakly saturated class of morphisms generated by S_0. Let $\widetilde{S}$ be the collection of all morphisms $F \to G$ in $\mathrm{Fun}(\mathcal{C}, \mathbf{A})$ with the following property: for every $C \in \mathcal{C}$, the map $F(C) \to G(C)$ belongs to $\overline{S}_0$. Then there exists a (small) set of morphisms S of $\mathrm{Fun}(\mathcal{C}, \mathbf{A})$ which generates $\widetilde{S}$ as a weakly saturated class of morphisms.*

We prove a generalization of Lemma A.2.8.3 in §A.3.3 (Lemma A.3.3.3).

Proof of Proposition A.2.8.2. We first treat the case of the projective model structure. For each object $C \in \mathcal{C}$ and each $A \in \mathbf{A}$, we define

$$\mathcal{F}_A^C : \mathcal{C} \to \mathbf{A}$$

by the formula

$$\mathcal{F}_A^C(C') = \coprod_{\alpha \in \mathrm{Map}_{\mathcal{C}}(C, C')} A.$$

We note that if $i : A \to A'$ is a (trivial) cofibration in $\mathbf{A}$, then the induced map $\mathcal{F}_A^C \to \mathcal{F}_{A'}^C$ is a strong (trivial) cofibration in $\mathrm{Fun}(\mathcal{C}, \mathbf{A})$.

Let I_0 be a set of generating cofibrations $i : A \to B$ for $\mathbf{A}$ and let I be the set of all induced maps $\mathcal{F}_A^C \to \mathcal{F}_B^C$ (where C ranges over $\mathcal{C}$). Let J_0 be a set of generating trivial cofibrations for $\mathbf{A}$ and define J likewise. It follows immediately from the definitions that a morphism in $\mathrm{Fun}(\mathcal{C}, \mathbf{A})$ is a projective fibration if and only if it has the right lifting property with respect to every morphism in J, and a weak trivial fibration if and only if it has the right lifting property with respect to every morphism in I. Let $\overline{I}$ and $\overline{J}$ be the weakly saturated classes of morphisms of $\mathrm{Fun}(\mathcal{C}, \mathbf{A})$ generated by I and J, respectively. Using the small object argument, we deduce the following:

(i) Every morphism $f : X \to Z$ in $\mathrm{Fun}(\mathcal{C}, \mathbf{A})$ admits a factorization

$$X \xrightarrow{f'} Y \xrightarrow{f''} Z,$$

where $f' \in \overline{I}$ and f'' is a weak trivial fibration.

(ii) Every morphism $f : X \to Z$ in $\mathrm{Fun}(\mathcal{C}, \mathbf{A})$ admits a factorization

$$X \xrightarrow{f'} Y \xrightarrow{f''} Z,$$

where $f' \in \overline{J}$ and f'' is a projective fibration.

(iii) The class $\overline{I}$ coincides with the class of projective cofibrations in $\mathbf{A}$.

Furthermore, since the class of trivial projective cofibrations in $\mathrm{Fun}(\mathcal{C}, \mathbf{A})$ is weakly saturated and contains J, it contains $\overline{J}$. This proves that $\mathrm{Fun}(\mathcal{C}, \mathbf{A})$ satisfies the factorization axioms. The only other nontrivial point to check is that $\mathrm{Fun}(\mathcal{C}, \mathbf{A})$ satisfies the lifting axioms. Consider a diagram

$$
\begin{array}{ccc}
A & \longrightarrow & X \\
{\scriptstyle i}\downarrow & \nearrow & \downarrow {\scriptstyle p} \\
C & \longrightarrow & Y
\end{array}
$$

in $\mathrm{Fun}(\mathcal{C}, \mathbf{A})$, where i is a projective cofibration and p is a projective fibration. We wish to show that there exists a dotted arrow as indicated provided that either i or p is a weak equivalence. If p is a weak equivalence, then this follows immediately from the definition of a injective fibration. Suppose instead that i is a trivial projective cofibration. We wish to show that i has the left lifting property with respect to every projective fibration. It will suffice to show that every trivial injective fibration belongs to $\overline{J}$ (this will also show that J is a set of generating trivial cofibrations for $\mathrm{Fun}(\mathcal{C}, \mathbf{A})$, which will show that the projective model structure on $\mathrm{Fun}(\mathcal{C}, \mathbf{A})$ is combinatorial). Suppose then that i is a trivial weak coibration and choose a factorization

$$A \xrightarrow{i'} B \xrightarrow{i''} C,$$

where $i' \in \overline{J}$ and i'' is a projective fibration. Then i' is a weak equivalence, so that i'' is a weak equivalence by the two-out-of-three property. Consider the diagram

$$
\begin{array}{ccc}
A & \xrightarrow{i'} & B \\
{\scriptstyle i}\downarrow & \nearrow & \downarrow {\scriptstyle i''} \\
C & \xrightarrow{=} & C.
\end{array}
$$

Since i is a cofibration, there exists a dotted arrow as indicated. This proves that i is a retract of i' and therefore belongs to $\overline{J}$, as desired.

We now prove the existence of the injective model structure on $\mathrm{Fun}(\mathcal{C}, \mathbf{A})$. Here it is difficult to proceed directly, so we will instead apply Proposition A.2.6.8. It will suffice to check each of the hypotheses in turn:

(1) The collection of injective cofibrations in $\mathrm{Fun}(\mathcal{C}, \mathbf{A})$ is generated (as a weakly saturated class) by some small set of morphisms. This follows from Lemma A.3.3.3.

(2) The collection of trivial injective cofibrations in $\mathrm{Fun}(\mathcal{C}, \mathbf{A})$ is weakly saturated: this follows immediately from the fact that the class of injective cofibrations in $\mathbf{A}$ is weakly saturated.

(3) The collection of weak equivalences in $\mathrm{Fun}(\mathcal{C}, \mathbf{A})$ is an accessible subcategory of $\mathrm{Fun}(\mathcal{C}, \mathbf{A})^{[1]}$: this follows from the proof of Proposition 5.4.4.3 since the collection of weak equivalences in $\mathbf{A}$ form an accessible subcategory of $\mathbf{A}^{[1]}$.

(4) The collection of weak equivalences in $\mathrm{Fun}(\mathcal{C}, \mathbf{A})$ satisfy the two-out-of-three property: this follows immediately from the fact that the weak equivalences in $\mathbf{A}$ satisfy the two-out-of-three property.

(5) Let $f : X \to Y$ be a morphism in $\mathbf{A}$ which has the right lifting property with respect to every injective cofibration. In particular, f has the right lifting property with respect to each of the morphisms in the class I defined above, so that f is a trivial projective fibration and, in particular, a weak equivalence.

$\square$

Remark A.2.8.4. In the situation of Proposition A.2.8.2, if $\mathbf{A}$ is assumed to be right or left proper, then $\mathrm{Fun}(\mathcal{C}, \mathbf{A})$ is likewise right or left proper (with respect to either the projective or the injective model structures).

Remark A.2.8.5. It follows from the proof of Proposition A.2.8.2 that the class of projective cofibrations is generated (as a weakly saturated class of morphisms) by the maps $j : \mathcal{F}_A^C \to \mathcal{F}_{A'}^C$, where $C \in \mathcal{C}$ and $A \to A'$ is a cofibration in $\mathbf{A}$. We observe that j is an injective cofibration. It follows that every projective cofibration is a injective cofibration; dually, every injective fibration is a projective fibration.

Remark A.2.8.6. The construction of Proposition A.2.8.2 is functorial in the following sense: given a Quillen adjunction of combinatorial model categories $\mathbf{A} \underset{G}{\overset{F}{\rightleftarrows}} \mathbf{B}$ and a small category $\mathcal{C}$, composition with F and G determines a Quillen adjunction

$$\mathrm{Fun}(\mathcal{C}, \mathbf{A}) \underset{G^{\mathcal{C}}}{\overset{F^{\mathcal{C}}}{\rightleftarrows}} \mathrm{Fun}(\mathcal{C}, \mathbf{B})$$

(with respect to either the injective or the projective model structures). Moreover, if (F, G) is a Quillen equivalence, then so is $(F^{\mathcal{C}}, G^{\mathcal{C}})$.

Because the projective and injective model structures on $\mathrm{Fun}(\mathcal{C}, \mathbf{A})$ have the same weak equivalences, the identity functor $\mathrm{id}_{\mathrm{Fun}(\mathcal{C}, \mathbf{A})}$ is a Quillen equivalence between them. However, it is important to distinguish between these two model structures because they have different variance properties, as we now explain.

Let $f : \mathcal{C} \to \mathcal{C}'$ be a functor between small categories. Then composition with f yields a pullback functor $f^* : \mathrm{Fun}(\mathcal{C}', \mathbf{A}) \to \mathrm{Fun}(\mathcal{C}, \mathbf{A})$. Since $\mathbf{A}$ admits small limits and colimits, f^* has a right adjoint, which we will denote by f_*, and a left adjoint, which we shall denote by $f_!$.

Proposition A.2.8.7. *Let $\mathbf{A}$ be a combinatorial model category and let $f : \mathcal{C} \to \mathcal{C}'$ be a functor between small categories. Then*

(1) *The pair $(f_!, f^*)$ determines a Quillen adjunction between the projective model structures on $\mathrm{Fun}(\mathcal{C}, \mathbf{A})$ and $\mathrm{Fun}(\mathcal{C}', \mathbf{A})$.*

(2) *The pair (f^*, f_*) determines a Quillen adjunction between the injective model structures on $\mathrm{Fun}(\mathcal{C}, \mathbf{A})$ and $\mathrm{Fun}(\mathcal{C}', \mathbf{A})$.*

Proof. This follows immediately from the simple observation that f^* preserves weak equivalences, projective fibrations, and weak cofibrations. □

We now review the theory of homotopy limits and colimits in a combinatorial model category $\mathbf{A}$. For simplicity, we will discuss homotopy limits and leave the analogous theory of homotopy colimits to the reader. Let $\mathbf{A}$ be a combinatorial model category and let $f : \mathcal{C} \to \mathcal{C}'$ be a functor betweeen (small) categories. We wish to consider the right derived functor Rf_* of the right Kan extension $f_* : \mathrm{Fun}(\mathcal{C}, \mathbf{A}) \to \mathrm{Fun}(\mathcal{C}', \mathbf{A})$. This derived functor is called the *homotopy right Kan extension* functor. The usual way of defining it involves choosing a fibrant replacement functor $Q : \mathrm{Fun}(\mathcal{C}, \mathbf{A}) \to \mathrm{Fun}(\mathcal{C}, \mathbf{A})$ and setting $Rf_* = f_* \circ Q$. The assumption that $\mathbf{A}$ is combinatorial guarantees that such a fibrant replacement functor exists. However, for our purposes it is more convenient to address the indeterminacy in the definition of Rf_* in another way.

Let $F \in \mathrm{Fun}(\mathcal{C}, \mathbf{A})$, let $G \in \mathrm{Fun}(\mathcal{C}', \mathbf{A})$, and let $\eta : G \to f_*F$ be a map in $\mathrm{Fun}(\mathcal{C}', \mathbf{A})$. We will say that η *exhibits G as the homotopy right Kan extension of F* if, for some weak equivalence $F \to F'$ where F' is injectively fibrant in $\mathrm{Fun}(\mathcal{C}, \mathbf{A})$, the composite map $G \to f_*F \to f_*F'$ is a weak equivalence in $\mathrm{Fun}(\mathcal{C}', \mathbf{A})$. Since f_* preserves weak equivalences between injectively fibrant objects, this condition is independent of the choice of F'.

Remark A.2.8.8. Given an object $F \in \mathrm{Fun}(\mathcal{C}, \mathbf{A})$, it is not necessarily the case that there exists a map $\eta : G \to f_*F$ which exhibits G as a homotopy right Kan extension of F. However, such a map can always be found after replacing F by a weakly equivalent object; for example, if F is injectively fibrant, we may take $G = f_*F$ and η to be the identity map.

Let $[0]$ denote the final object of $\mathcal{C}$at: that is, the category with one object and only the identity morphism. For *any* category $\mathcal{C}$, there is a unique functor $f : \mathcal{C} \to [0]$. If $\mathbf{A}$ is a combinatorial model category, $F : \mathcal{C} \to \mathbf{A}$ is a functor, and $A \in \mathbf{A} \simeq \mathrm{Fun}([0], \mathbf{A})$ is an object, then we will say that a natural transformation $\alpha : f^*A \to F$ *exhibits A as a homotopy limit of F* if it exhibits A as a homotopy right Kan extension of F. Note that we can identify α with a map $A \to \lim_{C \in \mathcal{C}} F(C)$ in the model category $\mathbf{A}$.

The theory of homotopy right Kan extensions in general can be reduced to the theory of homotopy limits in view of the following result:

Proposition A.2.8.9. *Let $\mathbf{A}$ be a combinatorial model category, let $f : \mathcal{C} \to \mathcal{D}$ be a functor between small categories, and let $F : \mathcal{C} \to \mathbf{A}$ and $G : \mathcal{D} \to \mathbf{A}$ be diagrams. A natural transformation $\alpha : f^*G \to F$ exhibits G as a homotopy right Kan extension of F if and only if for each object $D \in \mathcal{D}$, α exhibits $G(D)$ as a homotopy limit of the composite diagram*

$$F_{D/} : \mathcal{C} \times_{\mathcal{D}} \mathcal{D}_{D/} \to \mathcal{C} \xrightarrow{F} \mathbf{A}.$$

To prove Proposition A.2.8.9, we can immediately reduce to the case where F is a injectively fibrant diagram. In this case, α exhibits G as a homotopy right Kan extension of F if and only if it induces a weak homotopy equivalence $G(D) \to \lim F_{D/}$ for each $D \in \mathcal{D}$. It will therefore suffice to prove the following result (in the case $\mathcal{C}' = \mathcal{C} \times_{\mathcal{D}} \mathcal{D}_{D/}$):

Lemma A.2.8.10. *Let $\mathbf{A}$ be a combinatorial model category and let $g : \mathcal{C}' \to \mathcal{C}$ be a functor which exhibits $\mathcal{C}'$ as cofibered in sets over $\mathcal{C}$. Then the pullback functor $g^* : \mathrm{Fun}(\mathcal{C}, \mathbf{A}) \to \mathrm{Fun}(\mathcal{C}', \mathbf{A})$ preserves injective fibrations.*

Proof. It will suffice to show that the left adjoint $g_!$ preserves weak trivial cofibrations. Let $\alpha : F \to F'$ be a map in $\mathrm{Fun}(\mathcal{C}', \mathbf{A})$. We observe that for each object $C \in \mathcal{C}$, the map $(g_!\alpha)(C) : (g_!F)(C) \to (g_!F')(C)$ can be identified with the coproduct of the maps $\{\alpha(C') : F(C') \to F'(C')\}_{C' \in g^{-1}\{C\}}$. If α is a weak trivial cofibration, then each of these maps is a trivial cofibration in $\mathbf{A}$, so that $g_!\alpha$ is again a weak trivial cofibration, as desired. $\square$

Remark A.2.8.11. In the preceding discussion, we considered injective model structures, Rf_*, and homotopy limits. An entirely dual discussion may be carried out with projective model structures and $Lf_!$; one obtains a notion of *homotopy colimit* which is the dual of the notion of homotopy limit.

Example A.2.8.12. Let $\mathbf{A}$ be a combinatorial model category and consider a diagram

$$X' \xleftarrow{f} X \xrightarrow{g} X''.$$

This diagram is projectively cofibrant if and only if the object X is cofibrant and the maps f and g are both cofibrations. Consequently, the definition of homotopy colimits given above recovers, as a special case, the theory of homotopy pushouts presented in §A.2.4.

A.2.9 Reedy Model Structures

Let **A** be a combinatorial model category and $\mathcal{J}$ a small category. In §A.2.8, we saw that the diagram category $\mathrm{Fun}(\mathcal{J}, \mathbf{A})$ can again be regarded as a combinatorial model category via either the projective or the injective model structure of Proposition A.2.8.2. In the special case where $\mathcal{J}$ is a *Reedy category* (see Definition A.2.9.1), it is often useful to consider still another model structure on $\mathrm{Fun}(\mathcal{J}, \mathbf{A})$: the *Reedy model structure*. We will sketch the definition and some of the basic properties of Reedy model categories below; we refer the reader to [38] for a more detailed treatment.

Definition A.2.9.1. A *Reedy category* is a small category $\mathcal{J}$ equipped with a factorization system $\mathcal{J}^L, \mathcal{J}^R \subseteq \mathcal{J}$ satisfying the following conditions:

(1) Every isomorphism in $\mathcal{J}$ is an identity map.

(2) Given a pair of objects $X, Y \in \mathcal{J}$, let us write $X \preceq_0 Y$ if either there exists a morphism $f : X \to Y$ belonging to $\mathcal{J}^R$ or there exists a morphism $g : Y \to X$ belonging to $\mathcal{J}^L$. We will write $X \prec_0 Y$ if $X \preceq_0 Y$ and $X \neq Y$. Then there are no infinite descending chains

$$\cdots \prec_0 X_2 \prec_0 X_1 \prec_0 X_0.$$

Remark A.2.9.2. Let $\mathcal{J}$ be a category equipped with a factorization system $(\mathcal{J}^L, \mathcal{J}^R)$ and let $\preceq_0$ be the relation described in Definition A.2.9.1. This relation is generally not transitive. We will denote its transitive closure by $\preceq$. Then condition (2) of Definition A.2.9.1 guarantees that $\preceq$ is a well-founded partial ordering on the set of objects of $\mathcal{J}$. In other words, every nonempty set S of objects of $\mathcal{J}$ contains a $\preceq$-minimal element.

Remark A.2.9.3. In the situation of Definition A.2.9.1, we will often abuse terminology and simply refer to $\mathcal{J}$ as a Reedy category, implicitly assuming that a factorization system on $\mathcal{J}$ has been specified as well.

Warning A.2.9.4. Condition (1) of Definition A.2.9.1 is not stable under equivalence of categories. Suppose that $\mathcal{J}$ is equivalent to a Reedy category. Then $\mathcal{J}$ can itself be regarded as a Reedy category if and only if every isomorphism class of objects in $\mathcal{J}$ contains a unique representative. (Definition A.2.9.1 can easily be modified so as to be invariant under equivalence, but it is slightly more convenient not to do so.)

Example A.2.9.5. The category $\mathbf{\Delta}$ of combinatorial simplices is a Reedy category with respect to the factorization system $(\mathbf{\Delta}^L, \mathbf{\Delta}^R)$; here a morphism $f : [m] \to [n]$ belongs to $\mathbf{\Delta}^L$ if and only if f is surjective, and to $\mathbf{\Delta}^R$ if and only if f is injective.

Example A.2.9.6. Let $\mathcal{J}$ be a Reedy category with respect to the factorization system $(\mathcal{J}^L, \mathcal{J}^R)$. Then $\mathcal{J}^{op}$ is a Reedy category with respect to the factorization system $((\mathcal{J}^R)^{op}, (\mathcal{J}^L)^{op})$.

Notation A.2.9.7. Let $\mathcal{J}$ be a Reedy category, $\mathcal{C}$ a category which admits small limits and colimits, and $X : \mathcal{J} \to \mathcal{C}$ a functor. For every object $J \in \mathcal{J}$, we define the *latching object* $L_J(X)$ to be the colimit

$$\varinjlim_{J' \in \mathcal{J}^R_{/J}, J' \neq J} X(J').$$

Similarly, we define the *matching object* to be the limit

$$\varprojlim_{J' \in \mathcal{J}^L_{J/}, J' \neq J} X(J').$$

We then have canonical maps $L_J(X) \to X(J) \to M_J(X)$.

Example A.2.9.8. Let $X : \mathbf{\Delta}^{op} \to \mathrm{Set}$ be a simplicial set and regard $\mathbf{\Delta}^{op}$ as a Reedy category using Examples A.2.9.5 and A.2.9.6. For every nonnegative integer n, the latching object $L_{[n]}X$ can be identified with the collection of all degenerate simplices of X. In particular, the map $L_{[n]}(X) \to X([n])$ is always a monomorphism.

More generally, we observe that a map of simplicial sets $f : X \to Y$ is a monomorphism if and only if, for every $n \geq 0$, the map

$$L_{[n]}(Y) \coprod_{L_{[n]}(X)} X([n]) \to Y([n])$$

is a monomorphism of sets. The "if" direction is obvious. For the converse, let us suppose that f is a monomorphism; we must show that if σ is an n-simplex of X such that $f(\sigma)$ is degenerate, then σ is already degenerate. If $f(\sigma)$ is degenerate, then $f(\sigma) = \alpha^* f(\sigma) = f(\alpha^* \sigma)$, where $\alpha : [n] \to [n]$ is a map of linearly ordered sets other than the identity. Since f is a monomorphism, we deduce that $\sigma = \alpha^* \sigma$, so that σ is degenerate, as desired.

Remark A.2.9.9. Let $X : \mathcal{J} \to \mathcal{C}$ be as in Notation A.2.9.7. Then the Jth matching object $M_J(X)$ can be identified with the Jth latching object of the induced functor $X^{op} : \mathcal{J}^{op} \to \mathcal{C}^{op}$.

Remark A.2.9.10. Let $X : \mathcal{J} \to \mathcal{C}$ be as in Notation A.2.9.7. Then the Jth matching object $M_J(X)$ can also be identified with the colimit

$$\varinjlim_{(f:J' \to J) \in S} X(J'),$$

where S is any full subcategory of $\mathcal{J}_{/J}$ with the following properties:

(1) Every morphism $f : J' \to J$ which belongs to $\mathcal{J}^R$ and is not an isomorphism also belongs to S.

(2) If $f : J' \to J$ belongs to S, then $J \not\preceq J'$.

This follows from a cofinality argument since every morphism $f : J' \to J$ in S admits a canonical factorization

$$J' \xrightarrow{f'} J'' \xrightarrow{f''} J,$$

where f' belongs to $\mathcal{J}^L$ and f'' belongs to $\mathcal{J}^R$. Assumption (2) guarantees that the map f'' is not an isomorphism.

Similarly, when convenient, we can replace the limit $\varprojlim_{f: J \to J'} X(J')$ defining the matching object $M_J(X)$ by a limit over a slightly larger category.

Notation A.2.9.11. Let $\mathcal{J}$ be a Reedy category. A *good filtration* of $\mathcal{J}$ is a transfinite sequence

$$\{\mathcal{J}_\beta\}_{\beta < \alpha}$$

of full subcategories of $\mathcal{J}$ with the following properties:

(a) The filtration is exhaustive in the following sense: every object of $\mathcal{J}$ belongs to $\mathcal{J}_\beta$ for sufficiently large $\beta < \alpha$.

(b) For each ordinal $\beta < \alpha$, the category $\mathcal{J}_\beta$ is obtained from the subcategory $\mathcal{J}_{<\beta} = \bigcup_{\gamma < \beta} \mathcal{J}_\gamma$ by adjoining a single new object J_β satisfying the following condition: if $J \in \mathcal{J}$ satisfies $J \prec J_\beta$, then $J \in \mathcal{J}_{<\beta}$.

Remark A.2.9.12. Let $\mathcal{J}$ be a Reedy category. Then there exists a good filtration of $\mathcal{J}$. In fact, the existence of a good filtration is *equivalent* to the second assumption of Definition A.2.9.1.

Remark A.2.9.13. Let $\mathcal{J}$ be a Reedy category with respect to the factorization system $(\mathcal{J}^L, \mathcal{J}^R)$ and let $\{\mathcal{J}_\beta\}_{\beta < \alpha}$ be a good filtration of $\mathcal{J}$. Then each $\mathcal{J}_\beta$ admits a factorization system $(\mathcal{J}^L \cap \mathcal{J}_\beta, \mathcal{J}^R \cap \mathcal{J}_\beta)$. In other words, if $f : I \to K$ is a morphism in $\mathcal{J}_\beta$ which admits a factorization

$$I \xrightarrow{f'} J \xrightarrow{f''} K,$$

where f' belongs to $\mathcal{J}^L$ and f'' belongs to $\mathcal{J}^R$, then the object J also belongs to $\mathcal{J}_\beta$. This is clear: either f'' is an isomorphism, in which case $J = K \in \mathcal{J}_\beta$, or f'' is not an isomorphism, so that $J \prec K$ implies that $J \in \mathcal{J}_{<\beta}$.

The following result summarizes the essential features of a good filtration:

Proposition A.2.9.14. *Let $\mathcal{J}$ be a Reedy category with a good filtration $\{\mathcal{J}_\beta\}_{\beta < \alpha}$ and let $\beta < \alpha$ be an ordinal, so that $\mathcal{J}_\beta$ is obtained from $\mathcal{J}_{<\beta}$ by adjoining a single new object J. Then we have a homotopy pushout square (with respect to the Joyal model structure):*

$$\begin{array}{ccc}
N(\mathcal{J}_{<\beta})_{/J} \star N(\mathcal{J}_{<\beta})_{J/} & \longrightarrow & N(\mathcal{J}_{<\beta}) \\
\big\uparrow & & \big\downarrow \\
N(\mathcal{J}_{<\beta})_{/J} \star \{J\} \star N(\mathcal{J}_{<\beta})_{J/} & \longrightarrow & N(\mathcal{J}_\beta).
\end{array}$$

Corollary A.2.9.15. *Let $\mathcal{J}$ be a Reedy category equipped with a good filtration $\{\mathcal{J}_\beta\}_{\beta < \alpha}$ and let let $\beta < \alpha$ be an ordinal, so that $\mathcal{J}_\beta$ is obtained from $\mathcal{J}_{<\beta}$ by adjoining a single new object J. Let $\mathcal{C}$ be a category which admits small limits and colimits, let $X : \mathcal{J}_{<\beta} \to \mathcal{C}$ be a functor, and let the latching* *and matching objects $L_J(X)$ and $M_J(X)$ be defined as in Notation A.2.9.7 (note that this does not require that the functor X be defined on the object J), so that we have a canonical map $\alpha : L_J(X) \to M_J(X)$. The following data are equivalent:*

(1) *A functor $\overline{X} : \mathfrak{J}_\beta \to \mathcal{C}$ extending X.*

(2) *A commutative diagram*

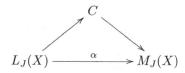

in the category $\mathcal{C}$.

The equivalence carries a functor $\overline{X}$ to the evident diagram with $C = \overline{X}(J)$.

Proof. Using Proposition A.2.9.14, we see that giving an extension $\overline{X} : \mathfrak{J}_\beta \to \mathcal{C}$ of X is equivalent to giving an extension $\overline{Y} : (\mathfrak{J}_{<\beta})_{/J} \star \{J\} \star (\mathfrak{J}_{<\beta})_{J/} \to \mathcal{C}$ of the composite functor

$$Y : (\mathfrak{J}_{<\beta})_{/J} \star (\mathfrak{J}_{<\beta})_{J/} \to \mathfrak{J}_{<\beta} \xrightarrow{X} \mathcal{C}.$$

This, in turn, is equivalent to giving a commutative diagram

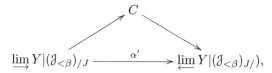

where α' is the map induced by the diagram Y. The equivalence of this with the data (2) follows immediately from Remark A.2.9.10. $\qquad\square$

Remark A.2.9.16. The proof of Corollary A.2.9.15 carries over without essential change to the case where $\mathcal{C}$ is an ∞-category which admits small limits and colimits. In this case, to extend a functor $X : N(\mathfrak{J}_{<\beta}) \to \mathcal{C}$ to a functor $\overline{X}$ defined on the whole of $\mathfrak{J}_\beta$, it will suffice to specify the object

$$\overline{X}(J) \in \mathcal{C}_{X|(\mathfrak{J}_{<\beta})_{/J}/\, X|(\mathfrak{J}_{<\beta})_{J/}} \simeq \mathcal{C}_{L_J(X)/\, /M_J(X)},$$

where the latching and matching objects $L_J(X), M_J(X) \in \mathcal{C}$ are defined in the obvious way.

The proof of Proposition A.2.9.14 will require a few preliminaries.

Lemma A.2.9.17. *Let $\mathfrak{J}$ be a Reedy category equipped with a good filtration $\{\mathfrak{J}_\beta\}_{\beta<\alpha}$. Fix $\beta < \alpha$ and let $\mathfrak{J}_\beta$ be obtained from $\mathfrak{J}_{<\beta}$ by adjoining the object J. Let $f : J \to J$ be a map which is not the identity, let $\mathfrak{I}$ denote the category $(\mathfrak{J}_{J/})_{/f} \simeq (\mathfrak{J}_{/J})_{f/}$ of factorizations of the morphism f, and let $\mathfrak{I}_0 = \mathfrak{I} \times_{\mathfrak{J}} \mathfrak{J}_{<\beta}$. Then the nerve $N\mathfrak{I}_0$ is weakly contractible.*

Proof. Let $\mathfrak{I}_1$ denote the full subcategory of $\mathfrak{I}_0$ spanned by those diagrams

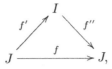

where $I \in \mathcal{J}_{<\beta}$ and f'' is a morphism in $\mathcal{J}^R$. The inclusion $\mathfrak{I}_1 \subseteq \mathfrak{I}_0$ admits
a left adjoint, so that $N\mathfrak{I}_1$ is a deformation retract of $N\mathfrak{I}_0$. It will there-
fore suffice to show that $N\mathfrak{I}_1$ is weakly contractible. Let $\mathfrak{I}_2$ denote the full
subcategory of $\mathfrak{I}_1$ spanned by those diagrams as above where, in addition,
the morphism f' belongs to $\mathcal{J}^L$. Then the inclusion $\mathfrak{I}_2 \subseteq \mathfrak{I}_1$ admits a right
adjoint, so that $N\mathfrak{I}_2$ is a deformation retract of $N\mathfrak{I}_1$. It will therefore suffice
to show that $N\mathfrak{I}_2$ is weakly contractible. This is clear since the category $\mathfrak{I}_2$
consists of a single object (with no nontrivial endomorphisms).   □

Lemma A.2.9.18. *Let $n \geq 1$ and suppose we are given a sequence of weakly
contractible simplicial sets $\{A_i\}_{1 \leq i \leq n}$. Let L denote the iterated join*

$$\{J_0\} \star A_1 \star \{J_1\} \star A_2 \star \cdots \star A_n \star \{J_n\}$$

*and let K denote the simplicial subset of L spanned by those simplices which
do not contain all of the vertices $\{J_i\}_{0 \leq i \leq n}$. Then the inclusion $K \subseteq L$ is a
categorical equivalence of simplicial sets.*

Proof. If $n = 1$, then this follows immediately from Lemma 5.4.5.10. Suppose
that $n > 1$. Let X denote the iterated join $A_1 \star \{J_1\} \star A_2 \star \cdots \star \{J_{n-1}\} \star A_n$.
For every subset $S \subseteq \{1, \ldots, n-1\}$, let $X(S)$ denote the simplicial subset
of X spanned by those simplices which do not contain any vertex J_i for
$i \in S$. Let $X' = \bigcup_{S \neq \emptyset} X(S) \subseteq X(\emptyset) = X$. Then X' is the homotopy colimit
of the diagram of simplicial sets $\{X(S)\}_{S \neq \emptyset}$. Each $X(S)$ is a join of weakly
contractible simplicial sets, and is therefore weakly contractible. Since $n > 1$,
the partially ordered set $\{S \subseteq \{1, \ldots, n-1\} : S \neq \emptyset\}$ has a largest element
and is therefore weakly contractible. It follows that the simplicial set X' is
weakly contractible.

The assertion that the inclusion $K \subseteq L$ is a categorical equivalence is
equivalent to the assertion that the diagram

$$(\{J_0\} \star X') \coprod_{X'} (X' \star \{J_n\}) \longhookrightarrow (\{J_0\} \star X) \coprod_X (X \star \{J_0\})$$

$$\downarrow \qquad\qquad\qquad\qquad\qquad\qquad \downarrow$$

$$\{J_0\} \star X' \star \{J_n\} \longhookrightarrow \{J_0\} \star X \star \{J_n\}$$

is a homotopy pushout square (with respect to the Joyal model structure). To
prove this, it suffices to observe that the vertical maps are both categorical
equivalences (Lemma 5.4.5.10).   □

Proof of Proposition A.2.9.14. Let S denote the collection of all composable
chains of morphisms

$$\overline{f} : J \xrightarrow{f_1} J \xrightarrow{f_2} \cdots \xrightarrow{f_n} J.$$

where $n \geq 1$ and each $f_i \neq \mathrm{id}_J$. For every subset $S' \subseteq S$, let $X(S')$ denote the simplicial subset of $\mathrm{N}(\mathcal{I}_\beta)$ spanned by those simplices σ satisfying the following condition:

$(*)$ For every nondegenerate face τ of σ of positive dimension, if every vertex of τ coincides with J, then τ belongs to S'.

Note that $X(S)$ coincides with $\mathrm{N}(\mathcal{I}_\beta)$, while $X(\emptyset)$ coincides with the pushout

$$(\mathrm{N}(\mathcal{I}_{<\beta})_{/J} \star \{J\} \star \mathrm{N}(\mathcal{I}_{<\beta})_{J/}) \coprod_{\mathrm{N}(\mathcal{I}_{<\beta})_{/J} \star \mathrm{N}(\mathcal{I}_{<\beta})_{J/}} \mathrm{N}(\mathcal{I}_{<\beta}).$$

It will therefore suffice to show that the inclusion $X(\emptyset) \subseteq X(S)$ is a categorical equivalence of simplicial sets.

Choose a well-ordering

$$S = \{\overline{f}_0 < \overline{f}_1 < \overline{f}_2 < \cdots\}$$

with the following property: if $\overline{f}$ has length shorter than $\overline{g}$ (when regarded as a chain of morphisms), then $\overline{f} < \overline{g}$. For every ordinal α, let $S_\alpha = \{\overline{f}_\beta\}_{\beta < \alpha}$. We will prove that for every ordinal α, the inclusion $X(\emptyset) \subseteq X(S_\alpha)$ is a categorical equivalence. The proof proceeds by induction on α. If $\alpha = 0$, there is nothing to prove, and if α is a limit ordinal, then the desired result follows from the inductive hypothesis and the fact that the class of categorical equivalences is stable under filtered colimits. We may therefore assume that $\alpha = \beta + 1$ is a successor ordinal. The inductive hypothesis guarantees that $X(\emptyset) \subseteq X(S_\beta)$ is a categorical equivalence. It will therefore suffice to show that the inclusion $j : X(S_\beta) \subseteq X(S_\alpha)$ is a categorical equivalence. We may also suppose that β is smaller than the order type of S, so that $\overline{f}_\beta$ is well-defined (otherwise, the inclusion j is an isomorphism and the result is obvious).

Let $\overline{f} = \overline{f}_\beta$ be the composable chain of morphisms

$$\overline{f} : J \xrightarrow{f_1} J \xrightarrow{f_2} \cdots \xrightarrow{f_n} J.$$

For $1 \leq i \leq n$, let A_i denote the nerve of the category

$$\mathcal{I}_{<\beta} \times_{\mathcal{I}} (\mathcal{I}_{J/})_{/f_i} \simeq \mathcal{I}_{<\beta} \times_{\mathcal{I}} (\mathcal{I}_{/J})_{f_i/}.$$

Let K denote the simplicial subset of

$$\{J_0\} \star A_1 \star \{J_1\} \star A_2 \star \cdots \star A_n \star \{J_n\}$$

spanned by those simplices which do not contain every vertex J_n. We then have a homotopy pushout diagram

$$\begin{array}{ccc}
\mathrm{N}(\mathcal{I}_{<\beta})_{/J} \star K \star \mathrm{N}(\mathcal{I}_{<\beta})_{J/} & \longrightarrow & X(S_\beta) \\
\Big\uparrow & & \Big\downarrow \\
\mathrm{N}(\mathcal{I}_{<\beta})_{/J} \star \{J_0\} \star A_1 \star \cdots \star A_n \star \{J_n\} \star \mathrm{N}(\mathcal{I}_{<\beta})_{J/} & \longrightarrow & X(S_\alpha).
\end{array}$$

It will therefore suffice to prove that the left vertical map is a categorical equivalence. In view of Corollary 4.2.1.3, it will suffice to show that the inclusion

$$K \subseteq \{J_0\} \star A_1 \star \{J_1\} \star A_2 \star \cdots \star A_n \star \{J_n\}$$

is a categorical equivalence. Since each A_i is weakly contractible (Lemma A.2.9.17), this follows immediately from Lemma A.2.9.18. $\square$

Proposition A.2.9.19. *Let $\mathcal{J}$ be a Reedy category and let $\mathbf{A}$ be a model category. Then there exists a model structure on the category of functors* $\operatorname{Fun}(\mathcal{J}, \mathbf{A})$ *with the following properties:*

(C) *A morphism $X \to Y$ in $\operatorname{Fun}(\mathcal{J}, \mathbf{A})$ is a Reedy cofibration if and only if, for every object $J \in \mathcal{J}$, the induced map $X(J) \coprod_{L_J(X)} L_J(Y) \to Y(J)$ is a cofibration in $\mathbf{A}$.*

(F) *A morphism $X \to Y$ in $\operatorname{Fun}(\mathcal{J}, \mathbf{A})$ is a Reedy fibration if and only if, for every object $J \in \mathcal{J}$, the induced map $X(J) \to Y(J) \times_{M_J(Y)} M_J(X)$ is a fibration in $\mathbf{A}$.*

(W) *A morphism $X \to Y$ in $\operatorname{Fun}(\mathcal{J}, \mathbf{A})$ is a weak equivalence if and only if, for every $J \in \mathcal{J}$, the map $X(J) \to Y(J)$ is a weak equivalence.*

Moreover, a morphism $f : X \to Y$ in $\operatorname{Fun}(\mathcal{J}, \mathbf{A})$ is a trivial cofibration if and only if the following condition is satisfied:

(WC) *For every object $J \in \mathcal{J}$, the map $X(J) \coprod_{L_J(X)} L_J(Y) \to Y(J)$ is a trivial cofibration in $\mathbf{A}$.*

Similarly, f is a fibration if and only if it satisfies the dual condition:

(WF) *For every object $J \in \mathcal{J}$, the map $X(J) \to Y(J) \times_{M_J(Y)} M_J(X)$ is a trivial fibration in $\mathbf{A}$.*

The model structure of Proposition A.2.9.19 is called the *Reedy model structure* on $\operatorname{Fun}(\mathcal{J}, \mathbf{A})$. Note that Proposition A.2.9.19 does not require the model category $\mathbf{A}$ to be combinatorial.

Lemma A.2.9.20. *Let $\mathcal{J}$ be a Reedy category containing an object J, let $\mathbf{A}$ be a model category, and let $f : F \to G$ be a natural transformation in $\operatorname{Fun}(\mathcal{J}, \mathbf{A})$. Let $\mathcal{I} \subseteq \mathcal{J}_{/J}^R$ be a sieve: that is, $\mathcal{I}$ is a full subcategory of $\mathcal{J}_{/J}^R$ with the property that if $I \to I'$ is a morphism in $\mathcal{J}_{/J}^R$ such that $I' \in \mathcal{I}$, then $I \in \mathcal{I}$. Let $\mathcal{I}' \subseteq \mathcal{I}$ be another sieve. Then*

(a) *If the map f satisfies condition (C) of Proposition A.2.9.19 for every object $I \in \mathcal{I}$, then the induced map*

$$\chi_{\mathcal{I}',\mathcal{I}} : \varinjlim(F|\mathcal{I}) \coprod_{\varinjlim(F|\mathcal{I}')} \varinjlim(G|\mathcal{I}') \to \varinjlim(G|\mathcal{I})$$

is a cofibration in $\mathbf{A}$.

(b) *If the map f satisfies condition (WC) of Proposition A.2.9.19 for every object $I \in \mathcal{I}$, then the map $\chi_{\mathcal{I}',\mathcal{I}}$ is a trivial cofibration in* **A**.

Proof. We will prove (a); the proof of (b) is identical. Choose a transfinite sequence of sieves $\{\mathcal{I}_\beta \subseteq \mathcal{I}\}_{\beta < \alpha}$ with the following properties:

(i) The union $\bigcup_{\beta < \alpha} \mathcal{I}_\beta$ coincides with $\mathcal{I}$.

(ii) For each $\beta < \alpha$, the sieve $\mathcal{I}_\beta$ is obtained from $\mathcal{I}_{<\beta} = \mathcal{I}' \cup (\bigcup_{\gamma < \beta} \mathcal{I}_\gamma)$ by adjoining a single new object $J_\beta \in \mathcal{I}_{/J}^R$.

For every triple $\delta \leq \gamma \leq \beta \leq \alpha$, let $\chi_{\delta,\gamma,\beta}$ denote the induced map

$$\varinjlim(F|\mathcal{I}_{<\beta}) \coprod_{\varinjlim(F|\mathcal{I}_{<\delta})} \varinjlim(G|\mathcal{I}_{<\delta}) \to \varinjlim(F|\mathcal{I}_{<\beta}) \coprod_{\varinjlim(F|\mathcal{I}_{<\gamma})} \varinjlim(G|\mathcal{I}_{<\gamma}).$$

We wish to prove that $\chi_{0,\alpha,\alpha}$ is a cofibration. We will prove more generally that $\chi_{\delta,\gamma,\beta}$ is an equivalence for every $\delta \leq \gamma \leq \beta \leq \alpha$. The proof uses induction on γ. If γ is a limit ordinal, then we can write $\chi_{\delta,\gamma,\beta}$ as the transfinite composition of the maps $\{\chi_{\epsilon,\epsilon+1,\beta}\}_{\delta \leq \epsilon < \gamma}$, which are cofibrations by the inductive hypothesis. We may therefore assume that $\gamma = \gamma_0 + 1$ is a successor ordinal. If $\delta = \gamma$, then $\chi_{\delta,\gamma,\beta}$ is an isomorphism; otherwise, we have $\delta \leq \gamma_0$. In this case, we have

$$\chi_{\delta,\gamma,\beta} = \chi_{\gamma_0,\gamma,\beta} \circ \chi_{\delta,\gamma_0,\beta}.$$

Using the inductive hypothesis, we can reduce to the case $\delta = \gamma_0$. The map $\chi_{\gamma_0,\gamma,\beta}$ is a pushout of the map $\chi_{\gamma_0,\gamma,\gamma}$. We are therefore reduced to proving that $\chi_{\gamma_0,\gamma,\gamma}$ is a cofibration. But $\chi_{\gamma_0,\gamma,\gamma}$ is a pushout of the map $L_I(G) \coprod_{L_I(F)} F(I) \to G(I)$ for $I = J_{\gamma_0}$. This map is a cofibration by virtue of our assumption that f satisfies (C). $\qquad\square$

Proof of Proposition A.2.9.19. Let $f : X \to Z$ be a morphism in $\mathrm{Fun}(\mathcal{I}, \mathbf{A})$. We will prove that f admits a factorization

$$X \xrightarrow{f'} Y \xrightarrow{f''} Z,$$

where

(i) The map f'' is a fibration, and f' satisfies (WC).

(ii) The map f' is a cofibration, and f'' satisfies (WF).

By symmetry, it will suffice to consider case (i). Choose a good filtration $\{\mathcal{I}_\beta\}_{\beta < \alpha}$ of $\mathcal{I}$. For $\beta < \alpha$, let $X_\beta = X|\mathcal{I}_\beta$, let $Z_\beta = Z|\mathcal{I}_\beta$, and let $f_\beta : X_\beta \to Z_\beta$ be the restriction of f. We will construct a compatible family of factorizations of f_β as a composition

$$X_\beta \xrightarrow{f'_\beta} Y_\beta \xrightarrow{f''_\beta} Z_\beta.$$

Suppose that $\mathcal{I}_\beta$ is obtained from $\mathcal{I}_{<\beta}$ by adjoining a single new object J. Assuming that (f'_γ, f''_γ) has been constructed for all $\gamma < \beta$, we note that constructing (f'_β, f''_β) is equivalent (by virtue of Corollary A.2.9.15) to giving a commutative diagram

$$
\begin{array}{ccccc}
L_J(X) & \longrightarrow & L_J(Y_{<\beta}) & & \\
\downarrow & & \downarrow & & \\
X(J) & \longrightarrow & Y_\beta(J) & \longrightarrow & Z(J) \\
& & \downarrow & & \downarrow \\
& & M_J(Y_{<\beta}) & \longrightarrow & M_J(Z).
\end{array}
$$

In other words, we must factor a certain map

$$
g : L_J(Y_{<\beta}) \coprod_{L_J(X)} X(J) \to M_J(Y_{<\beta}) \times_{M_J(Z)} Z(J)
$$

as a composition

$$
L_J(Y_{<\beta}) \coprod_{L_J(X)} X(J) \xrightarrow{g'} Y_\beta(J) \xrightarrow{g''} M_J(Y_{<\beta}) \times_{M_J(Z)} Z(J).
$$

Using the fact that $\mathbf{A}$ is a model category, we can choose a factorization where g' is a trivial cofibration and g'' a fibration. It is readily verified that this construction has the desired properties.

We now prove the following:

(i') A morphism $f : X \to Y$ in $\mathrm{Fun}(\mathcal{J}, \mathbf{A})$ satisfies (WC) if and only if f is both a fibration and a weak equivalence.

(ii') A morphism $f : X \to Y$ in $\mathrm{Fun}(\mathcal{J}, \mathbf{A})$ satisfies (WF) if and only if f is both a cofibration and a weak equivalence.

By symmetry, it will suffice to prove (i'). The "only if" direction follows from Lemma A.2.9.20. For the "if" direction, it will suffice to show that for each $\beta < \alpha$, the induced transformation $f_\beta : X_\beta \to Y_\beta$ satisfies (WC) when regarded as a morphism of $\mathrm{Fun}(\mathcal{J}_\beta, \mathbf{A})$. Suppose that $\mathcal{J}_\beta$ is obtained from $\mathcal{J}_{<\beta}$ by adjoining a single new element J. We have a commutative diagram

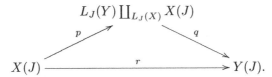

We wish to prove that q is a trivial cofibration in $\mathbf{A}$. Since f is a cofibration in $\mathrm{Fun}(\mathcal{J}, \mathbf{A})$, the map q is a cofibration in $\mathbf{A}$. It will therefore suffice to show that q is a weak equivalence. By the two-out-of-three property, it will suffice to show that p and r are weak equivalences. For r, this follows from our assumption that f is a weak equivalence in $\mathrm{Fun}(\mathcal{J}, \mathbf{A})$. The map p is a pushout of the map of latching objects $L_J(X) \to L_J(Y)$, which is a cofibration in $\mathbf{A}$ by virtue of the inductive hypothesis and Lemma A.2.9.20.

Combining (i) and (i') (and the analogous assertions (ii) and (ii')), we deduce that $\mathrm{Fun}(\mathcal{J}, \mathbf{A})$ satisfies the factorization axioms for a model category. To complete the proof, it will suffice to verify the lifting axioms:

(i'') Every fibration in $\mathrm{Fun}(\mathcal{J}, \mathbf{A})$ has the right lifting property with respect to morphisms in $\mathrm{Fun}(\mathcal{J}, \mathbf{A})$ which satisfy (WC).

(ii'') Every cofibration in $\mathrm{Fun}(\mathcal{J}, \mathbf{A})$ has the left lifting property with respect to morphisms in $\mathrm{Fun}(\mathcal{J}, \mathbf{A})$ which satisfy (WF).

Again, by symmetry it will suffice to prove (i''). Consider a diagram

$$
\begin{array}{ccc}
A & \longrightarrow & X \\
{\scriptstyle f}\downarrow {\scriptstyle\ h} & \nearrow & \downarrow {\scriptstyle g} \\
B & \longrightarrow & Y,
\end{array}
$$

where f satisfies (WC) and g satisfies (F); we wish to prove that there exists a dotted arrow h as indicated, rendering the diagram commutative. To prove this, we will construct a compatible family of natural transformations $\{h_\beta : B|\mathcal{J}_\beta \to X|\mathcal{J}_\beta\}_{\beta < \alpha}$ which render the diagram

$$
\begin{array}{ccc}
A|\mathcal{J}_\beta & \longrightarrow & X|\mathcal{J}_\beta \\
\downarrow {\scriptstyle\ h_\beta} & \nearrow & \downarrow {\scriptstyle g} \\
B|\mathcal{J}_\beta & \longrightarrow & Y|\mathcal{J}_\beta
\end{array}
$$

commutative. Suppose that $\mathcal{J}_\beta$ is obtained from $\mathcal{J}_{<\beta}$ by adjoining a single new object J. Assume that the maps $\{h_\gamma\}_{\gamma < \beta}$ have already been constructed and can be amalgamated into a single natural transformation $h_{<\beta} : B|\mathcal{J}_{<\beta} \to X|\mathcal{J}_{<\beta}$. Using Corollary A.2.9.15, we see that extending $h_{<\beta}$ to a map h_β with the desired properties is equivalent to solving a lifting problem of the kind depicted in the following diagram:

$$
\begin{array}{ccc}
L_J(B) \coprod_{L_J(A)} A(J) & \longrightarrow & X(J) \\
{\scriptstyle f'}\downarrow & \nearrow & \downarrow {\scriptstyle g'} \\
B(J) & \longrightarrow & Y(J) \times_{M_J(Y)} M_J(X).
\end{array}
$$

Since our assumptions guarantee that f' is a trivial cofibration and that g' is a fibration, this lifting problem has a solution, as desired. $\qquad\square$

Example A.2.9.21. Let $\mathbf{A}$ be the category of *bisimplicial* sets, which we will identify with $\mathrm{Fun}(\mathbf{\Delta}^{op}, \mathrm{Set}_\Delta)$ and endow with the Reedy model structure. It follows from Example A.2.9.8 that a morphism $f : X \to Y$ of bisimplicial sets is a Reedy cofibration if and only if it is a monomorphism. Consequently, the Reedy model structure on $\mathbf{A}$ coincides with the injective model structure on $\mathbf{A}$.

Example A.2.9.22. Let $\mathcal{J}$ be a Reedy category with $\mathcal{J}^{L} = \mathcal{J}$ and let $\mathbf{A}$ be a model category. Then the weak equivalences and cofibrations of the Reedy model structure (Proposition A.2.9.19) are the injective cofibrations and the weak equivalences appearing in Definition A.2.8.1. It follows that the Reedy model structure on $\mathrm{Fun}(\mathcal{J}, \mathbf{A})$ coincides with the injective model structure of Proposition A.2.8.2 (in particular, the injective model structure is well-defined in this case even without the assumption that $\mathbf{A}$ is combinatorial). Similarly, if $\mathcal{J}^{R} = \mathcal{J}$, then we can identify the Reedy model structure on $\mathrm{Fun}(\mathcal{J}, \mathbf{A})$ with the projective model structure of Proposition A.2.8.2.

In the general case, we can regard the Reedy model structure on $\mathrm{Fun}(\mathcal{J}, \mathbf{A})$ as a mixture of the projective and injective model structures. More precisely, we have the following:

(i) A natural transformation $F \to G$ in $\mathrm{Fun}(\mathcal{J}, \mathbf{A})$ satisfies condition (C) of Proposition A.2.9.19 if and only if the induced transformation $F|\mathcal{J}^{R} \to G|\mathcal{J}^{R}$ is a projective cofibration in $\mathrm{Fun}(\mathcal{J}^{R}, \mathbf{A})$.

(ii) A natural transformation $F \to G$ in $\mathrm{Fun}(\mathcal{J}, \mathbf{A})$ satisfies condition (F) of Proposition A.2.9.19 if and only if the induced transformation $F|\mathcal{J}^{L} \to G|\mathcal{J}^{L}$ is a injective fibration in $\mathrm{Fun}(\mathcal{J}^{L}, \mathbf{A})$.

(iii) A natural transformation $F \to G$ in $\mathrm{Fun}(\mathcal{J}, \mathbf{A})$ satisfies condition (WC) of Proposition A.2.9.19 if and only if the induced transformation $F|\mathcal{J}^{R} \to G|\mathcal{J}^{R}$ is a trivial projective cofibration in $\mathrm{Fun}(\mathcal{J}^{R}, \mathbf{A})$.

(iv) A natural transformation $F \to G$ in $\mathrm{Fun}(\mathcal{J}, \mathbf{A})$ satisfies condition (WF) of Proposition A.2.9.19 if and only if the induced transformation $F|\mathcal{J}^{L} \to G|\mathcal{J}^{L}$ is a trivial injective fibration in $\mathrm{Fun}(\mathcal{J}^{L}, \mathbf{A})$.

Remark A.2.9.23. Let $\mathcal{J}$ be a Reedy category and $\mathbf{A}$ a combinatorial model category, so that the injective and projective model structures on $\mathrm{Fun}(\mathcal{J}, \mathbf{A})$ are well-defined. The identity functor from $\mathrm{Fun}(\mathcal{J}, \mathbf{A})$ to itself can be regarded as a left Quillen equivalence from the projective model structure to the Reedy model structure and from the Reedy model structure to the injective model structure.

Corollary A.2.9.24. *Let $\mathcal{C}$ be a small category. Suppose that there exists a well-ordering $\leq$ on the collection of objects of $\mathcal{C}$ satisfying the following condition: for every pair of objects $X, Y \in \mathcal{C}$, we have*

$$\mathrm{Hom}_{\mathcal{C}}(X, Y) = \begin{cases} \emptyset & \text{if } X \not> Y \\ \{\mathrm{id}_{X}\} & \text{if } X = Y. \end{cases}$$

Let $\mathbf{A}$ be a model category. Then

(i) *A natural transformation $F \to G$ in $\mathrm{Fun}(\mathcal{C}, \mathbf{A})$ is a (trivial) projective cofibration if and only if, for every object $C \in \mathcal{C}$, the induced map*

$$F(C) \coprod_{\varinjlim_{D \to C, D \neq C} F(D)} \varinjlim_{D \to C, D \neq C} G(D) \to G(C)$$

is a (trivial) cofibration in $\mathbf{A}$.

(ii) *A natural transformation $F \to G$ in $\mathrm{Fun}(\mathcal{C}^{op}, \mathbf{A})$ is a (trivial) injective fibration if and only if, for every object $C \in \mathcal{C}$, the induced map*

$$F(C) \to G(C) \times_{\varprojlim_{D \to C, D \neq C} G(D)} \varprojlim_{D \to C, D \neq C} F(D)$$

is a (trivial) fibration in $\mathbf{A}$.

Proof. Combine Example A.2.9.22 with Proposition A.2.9.19.   □

Corollary A.2.9.25. *Let $\mathbf{A}$ be a model category, let α be an ordinal, and let (α) denote the linearly ordered set $\{\beta < \alpha\}$ regarded as a category. Then*

(1) *Let $F \to F'$ be a natural transformation of diagrams $(\alpha) \to \mathbf{A}$. Suppose that, for each $\beta < \alpha$, the maps*

$$\varinjlim_{\gamma < \beta} F(\gamma) \to F(\beta)$$

$$\varinjlim_{\gamma < \beta} F'(\gamma) \to F'(\beta)$$

are cofibrations, while the map $F(\beta) \to F'(\beta)$ is a weak equivalence. Then the induced map

$$\varinjlim_{\gamma < \alpha} F(\gamma) \to \varinjlim_{\gamma < \alpha} F'(\gamma)$$

is a weak equivalence.

(2) *Let $G \to G'$ be a natural transformation of diagrams $(\alpha)^{op} \to \mathbf{A}$. Suppose that, for each $\beta < \alpha$, the maps*

$$G(\beta) \to \varprojlim_{\gamma < \beta} G(\gamma)$$

$$G'(\beta) \to \varprojlim_{\gamma < \beta} G'(\gamma)$$

are fibrations, while the map $G(\beta) \to G'(\beta)$ is a weak equivalence. Then the induced map

$$\varprojlim_{\gamma < \alpha} G(\gamma) \to \varprojlim_{\gamma < \alpha} G'(\gamma)$$

is a weak equivalence.

Proof. We will prove (1); (2) follows by the same argument. Let $p : (\alpha) \to *$ be the unique map, let $p^* : \mathbf{A} \to \mathbf{A}^{(\alpha)}$ be the diagonal map, and let $p_! : \mathbf{A}^{(\alpha)} \to \mathbf{A}$ be a left adjoint to p^*. Then $p_!$ can be identified with the functor $F \mapsto \varinjlim_{\gamma < \alpha} F(\gamma)$. We observe that $(p_!, p^*)$ is a Quillen adjunction (where $\mathbf{A}^{(\alpha)}$ is endowed with the projective model structure) so that $p_!$ preserves weak equivalence between projectively cofibrant objects. The desired result now follows from Corollary A.2.9.24.   □

Suppose that we are given a bifunctor

$$\otimes : \mathbf{A} \times \mathbf{B} \to \mathbf{C},$$

where $\mathbf{C}$ is a category which admits small limits. For any small category $\mathfrak{J}$, we define the *coend* functor $\int_{\mathfrak{J}} : \mathrm{Fun}(\mathfrak{J}, \mathbf{A}) \times \mathrm{Fun}(\mathfrak{J}^{op}, \mathbf{B}) \to \mathbf{C}$ so that the integral $\int_{\mathfrak{J}}(F, G)$ is defined to be the coequalizer of the diagram

$$\coprod_{J \to J'} F(J) \otimes G(J') \rightrightarrows \coprod_{J} F(J) \otimes G(J) .$$

We then have the following result:

Proposition A.2.9.26. *Let $\otimes : \mathbf{A} \times \mathbf{B} \to \mathbf{C}$ be a left Quillen bifunctor (see Proposition A.3.1.1) and let $\mathfrak{J}$ be a Reedy category. Then the coend functor*

$$\int_{\mathfrak{J}} : \mathrm{Fun}(\mathfrak{J}, \mathbf{A}) \times \mathrm{Fun}(\mathfrak{J}^{op}, \mathbf{B}) \to \mathbf{C}$$

is also a left Quillen bifunctor, where we regard $\mathrm{Fun}(\mathfrak{J}, \mathbf{A})$ *and* $\mathrm{Fun}(\mathfrak{J}^{op}, \mathbf{B})$ *as endowed with the Reedy model structure.*

Proof. Let $f : F \to F'$ be a Reedy cofibration in $\mathrm{Fun}(\mathfrak{J}, \mathbf{A})$ and $g : G \to G'$ a Reedy cofibration in $\mathrm{Fun}(\mathfrak{J}^{op}, \mathbf{B})$. Set $C = \int_{\mathfrak{J}}(F, G') \coprod_{\int_{\mathfrak{J}}(F, G)} \int_{\mathfrak{J}}(F', G) \in \mathbf{C}$ and $C' = \int_{\mathfrak{J}}(F', G')$. We wish to show that the induced map $C \to \int_{\mathfrak{J}}(F', G')$ is a cofibration, which is trivial if either f or g is trivial.
Choose a good filtration $\{\mathfrak{J}_{\beta}\}_{\beta < \alpha}$ of $\mathfrak{J}$. For $\beta \leq \alpha$, we define

$$C_{\beta} = \int_{\mathfrak{J}_{\beta}} (F|\mathfrak{J}_{<\beta}, G'|\mathfrak{J}_{<\beta}) \coprod_{\int \mathfrak{J}_{\beta}(F|\mathfrak{J}_{<\beta}, G|\mathfrak{J}_{<beta})} \int_{\mathfrak{J}_{\beta}} (F'|\mathfrak{J}_{<\beta}, G|\mathfrak{J}_{<\beta})$$

$$C'_{\beta} = \int_{\mathfrak{J}_{\beta}} (F'|\mathfrak{J}_{<\beta}, G'|\mathfrak{J}_{<\beta}).$$

We wish to show that the map

$$C_{\alpha} \simeq C_{\alpha} \coprod_{C_0} C'_0 \to C_{\alpha} \coprod_{C_{\alpha}} C'_{\alpha}$$

is a cofibration (which is trivial if either f or g is trivial). We will prove more generally that for $\delta \leq \gamma \leq \beta \leq \alpha$, the map

$$\eta_{\delta, \gamma, \beta} : C_{\beta} \coprod_{C_{\delta}} C'_{\delta} \to C_{\beta} \coprod_{C_{\gamma}} C'_{\gamma}$$

is a cofibration (trivial if either f or g is trivial). The proof proceeds by induction on γ. If γ is a limit ordinal, then $\eta_{\delta, \gamma, \beta}$ can be obtained as a transfinite composition of the maps $\{\eta_{\epsilon, \epsilon+1, \beta}\}_{\delta \leq \epsilon < \gamma}$, and the result follows from the inductive hypothesis. We may therefore assume that $\gamma = \gamma_0 + 1$ is a successor ordinal. Since $\eta_{\delta, \gamma, \beta} = \eta_{\gamma_0, \gamma, \beta} \circ \eta_{\delta, \gamma_0, \beta}$, we can use the inductive hypothesis to reduce to the case where $\delta = \gamma_0$. Since $\eta_{\delta, \gamma, \beta}$ is a pushout of

$\eta_{\delta,\gamma,\gamma}$, we can also assume that $\beta = \gamma$. In other words, we are reduced to proving that the map

$$h : C_{\gamma_0+1} \coprod_{C_{\gamma_0}} C'_{\gamma_0} \to C'_{\gamma_0}$$

is a cofibration, which is trivial if either f or g is trivial. Let J be the object of $\mathcal{J}_{\gamma_0}$ which does not belong to $\mathcal{J}_{<\gamma_0}$. Form a pushout diagram

$$(F(J) \coprod_{L_J(F)} L_J(F')) \otimes (G(J) \coprod_{L_J(G)} L_J(G'))$$

$$(F(J) \coprod_{L_J(F)} L_J(F')) \otimes G'(J) \qquad (F'(J) \otimes G(J) \coprod_{L_J(G)} L_J(G'))$$

$$X.$$

We have an evident map $h' : X \to F'(J) \otimes G'(J)$ which is a cofibration (trivial if either f or g is trivial) by virtue of our assumptions on f and g (together with the fact that $\otimes$ is a left Quillen bifunctor). We conclude by observing that h is a pushout of h'. □

Remark A.2.9.27. Proposition A.2.9.26 has an analogue for the model structures introduced in Proposition A.2.8.2. That is, suppose that $\mathbf{A}$ and $\mathbf{B}$ are *combinatorial* model categories and let $\mathcal{J}$ be an arbitrary small category. Then any left Quillen bifunctor $\otimes : \mathbf{A} \times \mathbf{B} \to \mathbf{C}$ induces a left Quillen bifunctor

$$\int_{\mathcal{J}} : \mathrm{Fun}(\mathcal{J}, \mathbf{A}) \times \mathrm{Fun}(\mathcal{J}^{op}, \mathbf{B}) \to \mathbf{C},$$

where we regard $\mathrm{Fun}(\mathcal{J}, \mathbf{A})$ as endowed with the projective model structure and $\mathrm{Fun}(\mathcal{J}^{op}, \mathbf{B})$ with the injective model structure. To see this, we must show that for any projective cofibration $f : F \to F'$ in $\mathrm{Fun}(\mathcal{J}, \mathbf{A})$ and any injective cofibration $g : G \to G'$ in $\mathrm{Fun}(\mathcal{J}^{op}, \mathbf{B})$, the induced map

$$h : \int_{\mathcal{J}} (F, G') \coprod_{\int_{\mathcal{J}} (F,G)} \int_{\mathcal{J}} (F', G) \to \int_{\mathcal{J}} (F', G')$$

is a cofibration in $\mathbf{C}$ which is trivial if either f or g is trivial. Without loss of generality, we may suppose that f is a generating projective cofibration of the form $\mathcal{F}_A^J \to \mathcal{F}_{A'}^J$ associated to an object $J \in \mathcal{J}$ and a cofibration $i : A \to A'$ in $\mathbf{A}$, which is trivial if f is trivial (see the proof of Proposition A.2.8.2 for an explanation of this notation). Unwinding the definitions, we can identify h with the map

$$(A \otimes G'(J)) \coprod_{A \otimes G(J)} (A' \otimes G(J)) \to A' \otimes G'(J).$$

Since i is a cofibration in $\mathbf{A}$ and the map $G(J) \to G'(J)$ is a cofibration in $\mathbf{B}$, we deduce that h is a cofibration in $\mathbf{C}$ (since $\otimes$ is a left Quillen bifunctor) which is trivial if either i or h is trivial.

Example A.2.9.28. Let $\mathbf{A}$ be a simplicial model category, so that we have a left Quillen bifunctor

$$\otimes : \mathbf{A} \times \mathrm{Set}_\Delta \to \mathbf{A}.$$

The coend construction determines a left Quillen bifunctor

$$\int_\Delta : \mathrm{Fun}(\Delta, \mathbf{A}) \times \mathrm{Fun}(\Delta^{op}, \mathrm{Set}_\Delta) \to \mathbf{A}.$$

where $\mathrm{Fun}(\Delta, \mathbf{A})$ and $\mathrm{Fun}(\Delta^{op}, \mathrm{Set}_\Delta)$ are both endowed with the Reedy model structure. In particular, if we fix a cosimplicial object $X^\bullet \in \mathrm{Fun}(\Delta, \mathbf{A})$ which is Reedy cofibrant, then forming the coend against $X^\bullet$ determines a left Quillen functor from the category of bisimplicial sets (with the Reedy model structure, which coincides with the injective model structure by Example A.2.9.21) to $\mathbf{A}$.

Example A.2.9.29. Let $\mathbf{A}$ be a simplicial model category, so that we have a left Quillen bifunctor

$$\otimes : \mathbf{A} \times \mathrm{Set}_\Delta \to \mathbf{A},$$

and consider the coend functor

$$\int_{\Delta^{op}} \mathrm{Fun}(\Delta^{op}, \mathbf{A}) \times \mathrm{Fun}(\Delta, \mathrm{Set}_\Delta) \to \mathbf{A}.$$

Let $\Delta^\bullet \in \mathrm{Fun}(\Delta, \mathrm{Set}_\Delta)$ denote the standard simplex (that is, the functor $[n] \mapsto \Delta^n$) and let $\mathbf{1}$ denote the final object of $\mathrm{Fun}(\Delta, \mathrm{Set}_\Delta)$ (that is, the constant functor given by $[n] \mapsto \Delta^0$). The unique map $\Delta^\bullet \to \mathbf{1}$ is a weak equivalence, and $\Delta^\bullet$ is Reedy cofibrant: we may therefore regard $\Delta^\bullet$ as a cofibrant replacement for the constant functor $\mathbf{1}$.

The functor $X_\bullet \mapsto \int_{\Delta^{op}}(X_\bullet, \mathbf{1})$ can be identified with the colimit functor $\mathrm{Fun}(\Delta^{op}, \mathbf{A}) \to \mathbf{A}$. This is a left Quillen functor if $\mathrm{Fun}(\Delta^{op}, \mathbf{A})$ is endowed with the projective model structure but not the Reedy model structure. However, the *geometric realization* functor $X_\bullet \mapsto |X_\bullet| = \int_{\Delta^{op}}(X_\bullet, \Delta^\bullet)$ is a left Quillen functor with respect to the Reedy model structure.

Corollary A.2.9.30. *Let $\mathbf{A}$ be a combinatorial simplicial model category and let $X_\bullet$ be a simplicial object of $\mathbf{A}$. There is a canonical map*

$$\gamma : \mathrm{hocolim}\, X_\bullet \to |X_\bullet|$$

in the homotopy category of $\mathbf{A}$. This map is an equivalence if $X_\bullet$ is Reedy cofibrant.

Proof. Let $\Delta^\bullet$ and $*$ be the cosimplicial objects of Set_Δ described in Example A.2.9.29. Choose a weak equivalence of simplicial objects $X'_\bullet \to X_\bullet$, where $X'_\bullet$ is projectively cofibrant. We then have a diagram

$$\mathrm{hocolim}\, X_\bullet \simeq \varinjlim X'_\bullet \simeq \int_{\Delta^{op}}(X'_\bullet, *) \xleftarrow{\alpha} \int_{\Delta^{op}}(X'_\bullet, \Delta^\bullet) \xrightarrow{\beta} \int_{\Delta^{op}}(X_\bullet, \Delta^\bullet).$$

Since $X'_\bullet$ is projectively cofibrant, Remark A.2.9.27 implies that the coend functor $\int_{\Delta^{op}}(X'_\bullet, \bullet)$ preserves weak equivalences between injectively cofibrant cosimplicial objects of Set_Δ; in particular, α is a weak equivalence in $\mathbf{A}$.

This gives the desired map γ. Proposition A.2.9.26 implies that $\int_{\mathbf{\Delta}^{op}}(\bullet, \Delta^{\bullet})$ preserves weak equivalences between Reedy cofibrant simplicial objects of $\mathbf{A}$, which proves that γ is an isomorphism if $X_{\bullet}$ is Reedy cofibrant. $\qquad\square$

Example A.2.9.31. If $\mathbf{A}$ is the category of simplicial sets, then the map γ of Corollary A.2.9.30 is always an isomorphism; this follows from Example A.2.9.21. In other words, if $X_{\bullet,\bullet}$ is a bisimplicial set, then we can identify the diagonal simplicial set $[n] \mapsto X_{n,n}$ with the homotopy colimit of corresponding diagram $\mathbf{\Delta}^{op} \to \mathrm{Set}_{\Delta}$.

A.3 SIMPLICIAL CATEGORIES

Among the many different models for higher category theory, the theory of simplicial categories is perhaps the most rigid. This can be either a curse or a blessing, depending on the situation. For the most part, we have chosen to use the less rigid theory of ∞-categories (see §1.1.2) throughout this book. However, some arguments are substantially easier to carry out in the setting of simplicial categories. For this reason, we have devoted the final section of this appendix to a review of the theory of simplicial categories.

There exists a model structure on the category $\mathcal{C}at_{\Delta}$ of (small) simplicial categories, which was constructed by Bergner ([7]). In §A.3.2, we will describe an analogous model structure on the category $\mathcal{C}at_{\mathbf{S}}$ of $\mathbf{S}$-enriched categories, where $\mathbf{S}$ is a suitable model category. To formulate this generalization, we will need to employ the language of monoidal model categories, which we review in §A.3.1. Under mild assumptions on $\mathbf{S}$, one can show that an $\mathbf{S}$-enriched category $\mathcal{C}$ is fibrant if and only if each of the mapping objects $\mathrm{Map}_{\mathcal{C}}(X, Y)$ is a fibrant object of $\mathbf{S}$.

In §A.3.3, we will study the category $\mathbf{A}^{\mathcal{C}}$ of diagrams $\mathcal{C} \to \mathbf{A}$, where $\mathcal{C}$ is a small category and $\mathbf{A}$ is a model category, both enriched over some fixed model category $\mathbf{S}$. In the enriched setting we can again endow $\mathbf{A}^{\mathcal{C}}$ with projective and injective model structures, which can be used to define homotopy limits and colimits.

Putting aside set-theoretic technicalities, every $\mathbf{S}$-enriched model category $\mathbf{A}$ gives rise to a fibrant object of $\mathcal{C}at_{\mathbf{S}}$: namely, the full subcategory $\mathbf{A}^{\circ} \subseteq \mathbf{A}$ spanned by the fibrant-cofibrant objects. In §A.3.4, we will introduce a path object for $\mathbf{A}^{\circ}$, which will enable us to perform some calculations in the homotopy category of $\mathcal{C}at_{\mathbf{S}}$.

In §A.3.5, we will consider the problem of constructing homotopy colimits in the category $\mathcal{C}at_{\mathbf{S}}$ of $\mathbf{S}$-enriched categories. Our main result, Theorem A.3.5.15, asserts that the formation of homotopy colimits in $\mathcal{C}at_{\mathbf{S}}$ is compatible with the formation of (tensor) products in $\mathcal{C}at_{\mathbf{S}}$. We will apply this result in §A.3.6 to study the homotopy theory of internal mapping objects in $\mathcal{C}at_{\mathbf{S}}$.

We conclude this section with §A.3.7, where we discuss localizations of (simplicial) model categories.

A.3.1 Enriched and Monoidal Model Categories

Many of the model categories which arise naturally are *enriched* over the category of simplicial sets. Our goal in this section to study enrichments of one model category over another.

Definition A.3.1.1. Let $\mathbf{A}$, $\mathbf{B}$, and $\mathbf{C}$ be model categories. We will say that a functor $F : \mathbf{A} \times \mathbf{B} \to \mathbf{C}$ is a *left Quillen bifunctor* if the following conditions are satisfied:

(a) Let $i : A \to A'$ and $j : B \to B'$ be cofibrations in $\mathbf{A}$ and $\mathbf{B}$, respectively. Then the induced map

$$i \wedge j : F(A', B) \coprod_{F(A,B)} F(A, B') \to F(A', B')$$

is a cofibration in $\mathbf{C}$. Moreover, if either i or j is a trivial cofibration, then $i \wedge j$ is also a trivial cofibration.

(b) The functor F preserves small colimits separately in each variable.

Definition A.3.1.2. A *monoidal model category* is a monoidal category $\mathbf{S}$ equipped with a model structure, which satisfies the following conditions:

(i) The tensor product functor $\otimes : \mathbf{S} \times \mathbf{S} \to \mathbf{S}$ is a left Quillen bifunctor.

(ii) The unit object $\mathbf{1} \in \mathbf{S}$ is cofibrant.

(iii) The monoidal structure on $\mathbf{S}$ is closed.

Remark A.3.1.3. Some authors demand only a weakened form of axiom (ii) in the preceding definition.

Example A.3.1.4. The category of simplicial sets $\mathcal{S}\mathrm{et}_\Delta$ is a monoidal model category with respect to the Cartesian product and the Kan model structure defined in §A.2.7.

Definition A.3.1.5. Let $\mathbf{S}$ be a monoidal model category. A $\mathbf{S}$-*enriched model category* is an $\mathbf{S}$-enriched category $\mathbf{A}$ equipped with a model structure satisfying the following conditions:

(1) The category $\mathbf{A}$ is tensored and cotensored over $\mathbf{S}$.

(2) The tensor product $\otimes : \mathbf{A} \times \mathbf{S} \to \mathbf{A}$ is a left Quillen bifunctor.

In the special case where $\mathbf{S}$ is the category of simplicial sets (regarded as a monoidal model category as in Example A.3.1.4), we will simply refer to $\mathbf{A}$ as a *simplicial model category*.

Remark A.3.1.6. An easy formal argument shows that condition (2) is equivalent to either of the following statements:

(2′) Given any cofibration $i : D \to D'$ in **A** and any fibration $j : X \to Y$ in **A**, the induced map

$$k : \mathrm{Map}_{\mathbf{A}}(D', X) \to \mathrm{Map}_{\mathbf{A}}(D, X) \times_{\mathrm{Map}_{\mathbf{A}}(D,Y)} \mathrm{Map}_{\mathbf{A}}(D', Y)$$

is a fibration in **S**, which is trivial if either i or j is a weak equivalence.

(2″) Given any cofibration $i : C \to C'$ in **S** and any fibration $j : X \to Y$ in **A**, the induced map

$$k : X^{C'} \to X^C \times_{Y^C} Y^{C'}$$

is a fibration in **A**, which is trivial if either i or j is trivial.

The following provides a criterion for detecting simplicial model structures:

Proposition A.3.1.7. *Let $\mathcal{C}$ be a simplicial category that is equipped with a model structure (not assumed to be compatible with the simplicial structure on $\mathcal{C}$). Suppose that every object of $\mathcal{C}$ is cofibrant and that the collection of weak equivalences in $\mathcal{C}$ is stable under filtered colimits. Then $\mathcal{C}$ is a simplicial model category if and only if the following conditions are satisfied:*

(1) *As a simplicial category, $\mathcal{C}$ is both tensored and cotensored over Set_{Δ}.*

(2) *Given a cofibration $i : A \to B$ and a fibration $p : X \to Y$ in $\mathcal{C}$, the induced map of simplicial sets*

$$q : \mathrm{Map}_{\mathcal{C}}(B, X) \to \mathrm{Map}_{\mathcal{C}}(A, X) \times_{\mathrm{Map}_{\mathcal{C}}(A,Y)} \mathrm{Map}_{\mathcal{C}}(B, Y)$$

is a Kan fibration.

(3) *For every $n \geq 0$ and every object C in $\mathcal{C}$, the natural map*

$$C \otimes \Delta^n \to C \otimes \Delta^0 \simeq C$$

is a weak equivalence in $\mathcal{C}$.

Proof. Suppose first that $\mathcal{C}$ is a simplicial model category. It is clear that (1) and (2) are satisfied. To prove (3), we note that the projection $\Delta^n \to \Delta^0$ admits a section $s : \Delta^0 \to \Delta^n$ which is a trivial cofibration. If $\mathcal{C}$ is a simplicial model category, then since C is cofibrant, it follows that $C \otimes \Delta^0 \to C \otimes \Delta^n$ is a trivial cofibration and, in particular, a weak equivalence. Thus the projection $C \otimes \Delta^n \to C \otimes \Delta^0$ is a weak equivalence by the two-out-of-three property.

Now suppose that (1), (2), and (3) are satisfied. We wish to show that $\mathcal{C}$ is a simplicial model category. We first show that the bifunctor

$$(C, K) \mapsto C \otimes K$$

preserves weak equivalences separately in each variable.

Fix the object $C \in \mathcal{C}$ and suppose that $f : K \to K'$ is a weak homotopy equivalence of simplicial sets. Choose a cofibration $K \to K''$, where K'' is a contractible Kan complex. Then we may factor f as a composition

$$K \xrightarrow{f'} K \times K'' \xrightarrow{f''} K.$$

To prove that $\mathrm{id}_C \otimes f$ is a weak equivalence, it suffices to prove that $\mathrm{id}_C \otimes f'$ and $\mathrm{id}_C \otimes f''$ are weak equivalences. Note that the map f'' has a section s which is a trivial cofibration. Thus, to prove that $\mathrm{id}_C \otimes f''$ is a weak equivalence, it suffices to show that $\mathrm{id}_C \otimes s$ is a weak equivalence. In other words, we may reduce to the case where f is itself a trivial cofibration of simplicial sets.

Consider the collection A of all monomorphisms $f : K \to K'$ of simplicial sets having the property that $\mathrm{id}_C \otimes f$ is a weak equivalence in $\mathcal{C}$. It is easy to see that this collection of morphisms is weakly saturated. Thus, to prove that it contains all trivial cofibrations of simplicial sets, it suffices to show that every horn inclusion $\Lambda_i^n \to \Delta^n$ belongs to A. We prove this by induction on $n > 0$. Choose a vertex v belonging to Λ_i^n. We note that the inclusion $\{v\} \to \Lambda_i^n$ is a pushout of horn inclusions in dimensions $< n$; by the inductive hypothesis, this inclusion belongs to A. Thus it suffices to show that $\{v\} \to \Delta^n$ belongs to A, which is equivalent to assumption (3).

Now let us show that for each simplicial set K, the functor

$$C \mapsto C \otimes K$$

preserves weak equivalences. We will prove this by induction on the (possibly infinite) dimension of K. Choose a weak equivalence $g : C \to C'$ in $\mathcal{C}$. Let S denote the collection of all simplicial subsets $L \subseteq K$ such that $g \otimes \mathrm{id}_L$ is a weak equivalence. We regard S as a partially ordered set with respect to inclusions of simplicial subsets. Clearly, $\emptyset \in S$. Since weak equivalences in $\mathcal{C}$ are stable under filtered colimits, the supremum of every chain in S belongs to S. By Zorn's lemma, S has a maximal element L. We wish to show that $L = K$. If not, we may choose some nondegenerate simplex σ of K which does not belong to L. Choose σ of the smallest possible dimension, so that all of the faces of σ belong to L. Thus there is an inclusion $L' = L \coprod_{\partial \sigma} \sigma \subseteq K$. Since $\mathcal{C}$ is left proper, assumption (2) implies that the diagram

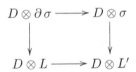

is a homotopy pushout for every object $D \in \mathcal{C}$. We observe that $g \otimes \mathrm{id}_L$ is a weak equivalence by assumption, $g \otimes \mathrm{id}_{\partial \sigma}$ is a weak equivalence by the inductive hypothesis (since $\partial \sigma$ has dimension smaller than the dimension of K), and $g \otimes \mathrm{id}_\sigma$ is a weak equivalence by virtue of assumption (3) and the fact that g is a weak equivalence. It follows that $g \otimes \mathrm{id}_{L'}$ is a weak equivalence, which contradicts the maximality of L. This completes the proof that the bifunctor $\otimes : \mathcal{C} \times \mathrm{Set}_\Delta \to \mathcal{C}$ preserves weak equivalences separately in each variable.

Now suppose we are given a cofibration $i : C \to C'$ in $\mathcal{C}$ and another cofibration $j : S \to S'$ in Set_Δ. We wish to prove that the induced map

$$i \wedge j : (C \otimes S') \coprod_{C \otimes S} (C' \otimes S) \to C' \otimes S'$$

is a cofibration in $\mathcal{C}$, which is trivial if either i or j is trivial. The first point follows immediately from (2). For the triviality, we will assume that i is a weak equivalence (the case where j is a weak equivalence follows using the same argument). Consider the diagram

$$
\begin{array}{ccc}
C \otimes S & \xrightarrow{\;i \otimes \mathrm{id}_S\;} & C' \otimes S \\
\downarrow & & \downarrow \\
C \otimes S' & \xrightarrow{\;\;f\;\;} (C' \otimes S) \coprod_{C \otimes S} (C \otimes S') & \longrightarrow C' \otimes S'.
\end{array}
$$

The arguments above show that $i \otimes \mathrm{id}_S$ and $i \otimes \mathrm{id}_{S'}$ are weak equivalences. The square in the diagram is a homotopy pushout, so Proposition A.2.4.2 implies that f is a weak equivalence as well. Thus $i \wedge j$ is a weak equivalence by the two-out-of-three property. $\square$

If $\mathcal{C}$ is a simplicial model category, then there is automatically a strong relationship between the homotopy theory of the underlying model category and the homotopy theory of the simplicial sets $\mathrm{Map}_{\mathcal{C}}(\bullet, \bullet)$. For example, we have the following:

Remark A.3.1.8. Let $\mathcal{C}$ be a simplicial model category, let X be a cofibrant object of $\mathcal{C}$, and let Y be a fibrant object of $\mathcal{C}$. The simplicial set $K = \mathrm{Map}_{\mathcal{C}}(X, Y)$ is a Kan complex; moreover, there is a canonical bijection

$$
\pi_0 K \simeq \mathrm{Hom}_{\mathrm{h}\mathcal{C}}(X, Y).
$$

We conclude this section by studying a situation which arises in Chapter 3. Let $\mathcal{C}$ and $\mathcal{D}$ be model categories enriched over another model category $\mathbf{S}$, and suppose we are given a Quillen adjunction

$$
\mathcal{C} \underset{G}{\overset{F}{\rightleftarrows}} \mathcal{D}
$$

between the underlying model categories. We wish to study the situation where G (but not F) has the structure of an $\mathbf{S}$-enriched functor. Thus, for every triple of objects $X \in \mathcal{C}$, $Y \in \mathcal{D}$, $S \in \mathbf{S}$, we have a canonical map

$$
\begin{aligned}
\mathrm{Hom}_{\mathcal{C}}(S \otimes X, GY) &\simeq \mathrm{Hom}_{\mathbf{S}}(S, \mathrm{Map}_{\mathcal{C}}(X, GY)) \\
&\to \mathrm{Hom}_{\mathbf{S}}(S, \mathrm{Map}_{\mathcal{D}}(FX, FGY)) \\
&\simeq \mathrm{Hom}_{\mathcal{D}}(S \otimes FX, FGY) \\
&\to \mathrm{Hom}_{\mathcal{D}}(S \otimes FX, Y).
\end{aligned}
$$

Taking $Y = F(S \otimes X)$ and applying this map to the unit of the adjunction between F and G, we obtain a map $S \otimes FX \to F(S \otimes X)$, which we will denote by $\beta_{X,S}$. The collection of maps $\beta_{X,S}$ is simply another way of encoding the data of G as an $\mathbf{S}$-enriched functor. If the maps $\beta_{X,S}$ are isomorphisms, then F is again an $\mathbf{S}$-enriched functor and (F, G) is an adjunction between $\mathbf{S}$-enriched categories. We wish to study an analogous situation where the maps $\beta_{X,S}$ are only assumed to be weak equivalences.

Remark A.3.1.9. Suppose that $\mathbf{S}$ is the category Set_Δ of simplicial sets with its usual model structure. Then the map $\beta_{X,S}$ is automatically a weak equivalence for every cofibrant object $X \in \mathcal{C}$. To prove this, we consider the collection $\mathcal{K}$ of all simplicial sets S such that $\beta_{S,X}$ is an equivalence. It is not difficult to show that $\mathcal{K}$ is closed under weak equivalences, homotopy pushout squares, and coproducts. Since $\Delta^0 \in \mathcal{K}$, we conclude that $\mathcal{K} = \mathrm{Set}_\Delta$.

Proposition A.3.1.10. *Let $\mathcal{C}$ and $\mathcal{D}$ be $\mathbf{S}$-enriched model categories. Let $\mathcal{C} \underset{G}{\overset{F}{\rightleftarrows}} \mathcal{D}$ be a Quillen adjunction between the underlying model categories. Assume that every object of $\mathcal{C}$ is cofibrant, and that the map $\beta_{X,S} : S \otimes F(X) \to F(S \otimes X)$ is a weak equivalence for every pair of cofibrant objects $X \in \mathcal{C}$, $S \in \mathbf{S}$. The following are equivalent:*

(1) *The adjunction (F, G) is a Quillen equivalence.*

(2) *The restriction of G determines a weak equivalence of $\mathbf{S}$-enriched categories $\mathcal{D}^\circ \to \mathcal{C}^\circ$ (see §A.3.2).*

Remark A.3.1.11. Strictly speaking, in §A.3.2, we define only weak equivalences between *small* $\mathbf{S}$-enriched categories; however, the definition extends to large categories in an obvious way.

Proof. Since G preserves fibrant objects and every object of $\mathcal{C}$ is cofibrant, it is clear that G carries $\mathcal{D}^\circ$ into $\mathcal{C}^\circ$. Condition (1) is equivalent to the assertion that for every pair of fibrant-cofibrant objects $C \in \mathcal{C}$, $D \in \mathcal{D}$, a map $g : C \to GD$ is a weak equivalence in $\mathcal{C}$ if and only if the adjoint map $f : FC \to D$ is a weak equivalence in $\mathcal{D}$. Choose a factorization of f' as a composition $FC \overset{f'}{\to} D' \overset{f''}{\to} D$, where f' is a trivial cofibration and f'' is a fibration. By the two-out-of-three property, f is a weak equivalence if and only if f'' is a weak equivalence. We note that g admits an analogous factorization as

$$C \overset{g'}{\to} GD' \overset{g''}{\to} GD.$$

Using (2), we deduce that f'' is an equivalence in $\mathcal{D}^\circ$ if and only if g'' is an equivalence in $\mathcal{C}^\circ$. It will therefore suffice to show that g' is an equivalence in $\mathcal{C}^\circ$. For this, it will suffice to show that C and GD' corepresent the same functor on the homotopy category $\mathrm{h}\mathcal{C}$. Invoking (2) again, it will suffice to show that for every fibrant-cofibrant object $D'' \in \mathcal{D}$, the induced map

$$\mathrm{Hom}_{\mathrm{h}\mathcal{C}}(GD', GD'') \to \mathrm{Hom}_{\mathrm{h}\mathcal{C}}(C, GD'') \simeq \mathrm{Hom}_{\mathrm{h}\mathcal{D}}(FC, D'')$$

is bijective. Using (2), we deduce that the map

$$\mathrm{Hom}_{\mathrm{h}\mathcal{D}}(D', D'') \to \mathrm{Hom}_{\mathrm{h}\mathcal{D}}(GD', GD'')$$

is bijective. The desired result now follows from the fact that f' is a weak equivalence in $\mathcal{D}$.

We now show that $(1) \Rightarrow (2)$. The $\mathbf{S}$-enriched functor $G^\circ : \mathcal{D}^\circ \to \mathcal{C}^\circ$ is essentially surjective since the right derived functor RG is essentially surjective on homotopy categories. It suffices to show that G° is fully faithful: in other words, that for every pair of fibrant-cofibrant objects $X, Y \in \mathcal{D}$, the induced map

$$i : \operatorname{Map}_{\mathcal{D}}(X, Y) \to \operatorname{Map}_{\mathcal{C}}(G(X), G(Y))$$

is a weak equivalence in $\mathbf{S}$.

Since the left derived functor LF is essentially surjective, there exists an object $X' \in \mathcal{C}$ and a weak equivalence $FX' \to X$. We may regard X as a fibrant replacement for FX' in $\mathcal{D}$; it follows that the adjoint map $X' \to GX$ may be identified with the adjunction $X' \to (RG \circ LF)X'$ and is therefore a weak equivalence by (1). Thus we have a diagram

$$
\begin{array}{ccc}
\operatorname{Map}_{\mathcal{D}}(X, Y) & \xrightarrow{\ \ i\ \ } & \operatorname{Map}_{\mathcal{C}}(G(X), G(Y)) \\
\downarrow & & \downarrow \\
\operatorname{Map}_{\mathcal{D}}(F(X'), Y) & \xrightarrow{\ \ i'\ \ } & \operatorname{Map}_{\mathcal{C}}(X', G(Y))
\end{array}
$$

in which the vertical arrows are homotopy equivalences; thus, to show that i is a weak equivalence, it suffices to show that i' is a weak equivalence. For this, it suffices to show that i' induces a bijection from $[S, \operatorname{Map}_{\mathcal{D}}(F(X'), Y)]$ to $[S, \operatorname{Map}_{\mathcal{C}}(X', G(Y))]$ for every cofibrant object $S \in \mathbf{S}$; here $[S, K]$ denotes the set of homotopy classes of maps from S into K in the homotopy category $h\mathbf{S}$. But we may rewrite this map of sets as

$$i'_S : \operatorname{Map}_{h\mathcal{D}}(F(X') \otimes S, Y) \to \operatorname{Map}_{h\mathcal{C}}(X' \otimes S, G(Y)) = \operatorname{Map}_{h\mathcal{D}}(F(X' \otimes S), Y),$$

and it is given by composition with $\beta_{X', S}$. (Here $h\mathcal{C}$ and $h\mathcal{D}$ denote the homotopy categories of $\mathcal{C}$ and $\mathcal{D}$ as *model categories*; these are equivalent to the corresponding homotopy categories of $\mathcal{C}^\circ$ and $\mathcal{D}^\circ$ as $\mathbf{S}$-enriched categories). Since $\beta_{X', S}$ is an isomorphism in the homotopy category $h\mathcal{D}$, the map i'_S is bijective and (2) holds, as desired. $\square$

Corollary A.3.1.12. *Let* $\mathcal{C} \underset{G}{\overset{F}{\rightleftarrows}} \mathcal{D}$ *be a Quillen equivalence between simplicial model categories, where every object of* $\mathcal{C}$ *is cofibrant. Suppose that* G *is a simplicial functor. Then* G *induces an equivalence of* ∞*-categories* $N(\mathcal{D}^\circ) \to N(\mathcal{C}^\circ)$.

A.3.2 The Model Structure on S-Enriched Categories

Throughout this section, we will fix a symmetric monoidal model category $\mathbf{S}$ and study the category of $\mathbf{S}$-enriched categories. The main case of interest to us is that in which $\mathbf{S}$ is the category of simplicial sets (with its usual model structure and the Cartesian monoidal structure). However, the treatment of the general case requires little additional effort, and there are a number of other examples which arise naturally in other contexts:

(i) The category of simplicial sets equipped with the Cartesian monoidal structure and the *Joyal* model structure defined in §2.2.5.

(*ii*) The category of complexes

$$\cdots \to M_n \to M_n \to M_{n-1} \to \cdots,$$

of vector spaces over a field k with its usual model structure (in which weak equivalences are quasi-isomorphisms, fibrations are epimorphisms, and cofibrations are monomorphisms) and monoidal structure given by the formation of tensor products of complexes.

Let $\mathbf{S}$ be an monoidal model category and let $\mathcal{C}at_{\mathbf{S}}$ denote the category of (small) $\mathbf{S}$-enriched categories in which morphisms are given by $\mathbf{S}$-enriched functors. The goal of this section is to describe a model structure on $\mathcal{C}at_{\mathbf{S}}$. We first note that the monoidal structure on $\mathbf{S}$ induces a monoidal structure on its homotopy category $h\mathbf{S}$, which is determined up to (unique) isomorphism by the requirement that there exist a monoidal structure on the functor

$$\mathbf{S} \to h\mathbf{S}$$

given by inverting all weak equivalences. Consequently, we note that any $\mathbf{S}$-enriched category $\mathcal{C}$ gives rise to an $h\mathbf{S}$-enriched category $h\mathcal{C}$ having the same objects as $\mathcal{C}$ and where mapping spaces are given by

$$\mathrm{Map}_{h\mathcal{C}}(X, Y) = [\mathrm{Map}_{\mathcal{C}}(X, Y)].$$

Here we let $[K]$ denote the image in $h\mathbf{S}$ of an object $K \in \mathbf{S}$. We will refer to $h\mathcal{C}$ as the *homotopy category* of $\mathcal{C}$; the passage from $\mathcal{C}$ to $h\mathcal{C}$ is a special case of Remark A.1.4.3.

Definition A.3.2.1. Let $\mathbf{S}$ be an monoidal model category. We say that a functor $F : \mathcal{C} \to \mathcal{C}'$ in $\mathcal{C}at_{\mathbf{S}}$ is a *weak equivalence* if the induced functor $h\mathcal{C} \to h\mathcal{C}'$ is an equivalence of $h\mathbf{S}$-enriched categories. In other words, F is a weak equivalence if and only if:

(1) For every pair of objects $X, Y \in \mathcal{C}$, the induced map

$$\mathrm{Map}_{\mathcal{C}}(X, Y) \to \mathrm{Map}_{\mathcal{C}'}(F(X), F(Y))$$

is a weak equivalence in $\mathbf{S}$.

(2) Every object $Y \in \mathcal{C}'$ is equivalent to $F(X)$ in the homotopy category $h\mathcal{C}'$ for some $X \in \mathcal{C}$.

Remark A.3.2.2. If $\mathbf{S}$ is the category $\mathcal{S}et_{\Delta}$ (endowed with the Kan model structure), then Definition A.3.2.1 reduces to the definition given in §1.1.3.

Remark A.3.2.3. Suppose that the collection of weak equivalences in $\mathbf{S}$ is stable under filtered colimits. Then it is easy to see that the collection of weak equivalences in $\mathcal{C}at_{\mathbf{S}}$ is also stable under filtered colimits. If $\mathbf{S}$ is also a combinatorial model category, then a bit more effort shows that the class of weak equivalences in $\mathcal{C}at_{\mathbf{S}}$ is perfect in the sense of Definition A.2.6.10.

We now introduce a bit of notation for working with **S**-enriched categories. If A is an object of **S**, we will let $[1]_A$ denote the **S**-enriched category having two objects X and Y, with

$$\mathrm{Map}_{[1]_A}(Z, Z') = \begin{cases} \mathbf{1_S} & \text{if } Z = Z' = X \\ \mathbf{1_S} & \text{if } Z = Z' = Y \\ A & \text{if } Z = X, Z' = Y \\ \emptyset & \text{if } Z = Y, Z' = X. \end{cases}$$

Here $\emptyset$ denotes the initial object of **S**, and $\mathbf{1_S}$ denotes the unit object with respect to the monoidal structure on **S**. We will denote $[1]_{\mathbf{1_S}}$ simply by $[1]_\mathbf{S}$. We let $[0]_\mathbf{S}$ denote the **S**-enriched category having only a single object X, with $\mathrm{Map}_*(X, X) = \mathbf{1_S}$.

We let C_0 denote the collection of all morphisms in **S** of the following types:

(*i*) The inclusion $\emptyset \hookrightarrow [0]_\mathbf{S}$.

(*ii*) The induced maps $[1]_S \to [1]_{S'}$, where $S \to S'$ ranges over a set of generators for the weakly saturated class of cofibrations in **S**.

Proposition A.3.2.4. *Let* **S** *be a combinatorial monoidal model category. Assume that every object of* **S** *is cofibrant and that the collection of weak equivalences in* **S** *is stable under filtered colimits. Then there exists a left proper combinatorial model structure on* $\mathcal{C}at_\mathbf{S}$ *characterized by the following conditions:*

(*C*) *The class of cofibrations in* $\mathcal{C}at_\mathbf{S}$ *is the smallest weakly saturated class of morphisms containing the set of morphisms* C_0 *appearing above.*

(*W*) *The weak equivalences in* $\mathcal{C}at_\mathbf{S}$ *are defined as in* §*A.3.2.1.*

Proof. It suffices to verify the hypotheses of Proposition A.2.6.13. Condition (1) follows from Remark A.3.2.3. For condition (3), we must show that any functor $F : \mathcal{C} \to \mathcal{C}'$ having the right lifting property with respect to all morphisms in C_0 is a weak equivalence. Since F has the right lifting property with respect to $i : \emptyset \to [0]_\mathbf{S}$, it is surjective on objects and therefore essentially surjective. The assumption that F has the right lifting property with respect to the remaining morphisms of C_0 guarantees that for every $X, Y \in \mathcal{C}$, the induced map

$$\mathrm{Map}_\mathcal{C}(X, Y) \to \mathrm{Map}_{\mathcal{C}'}(F(X), F(Y))$$

is a trivial fibration in **S** and therefore a weak equivalence.

It remains to verify condition (2): namely, that the class of weak equivalences is stable under pushout by the elements of C_0. We must show that given any pair of functors $F : \mathcal{C} \to \mathcal{D}$, $G : \mathcal{C} \to \mathcal{C}'$ with F a weak equivalence and G a pushout of some morphism in C_0, the induced map $F' : \mathcal{C}' \to \mathcal{D}' = \mathcal{D} \coprod_\mathcal{C} \mathcal{C}'$ is a weak equivalence. There are two cases to consider.

First, suppose that G is a pushout of the generating cofibration $i : \emptyset \to *$. In other words, the category $\mathcal{C}'$ is obtained from $\mathcal{C}$ by adjoining a new object X, which admits no morphisms to or from the objects of $\mathcal{C}$ (and no endomorphisms other than the identity). The category $\mathcal{D}'$ is obtained from $\mathcal{D}$ by adjoining X in the same fashion. It is easy to see that if F is a weak equivalence, then F' is also a weak equivalence.

The other basic case to consider is one in which G is a pushout of one of the generating cofibrations $[1]_S \to [1]_T$, where $S \to T$ is a cofibration in $\mathbf{S}$. Let $H : [1]_S \to \mathcal{C}$ denote the "attaching map," so that H is determined by a pair of objects $x = H(X)$ and $y = H(Y)$ and a map of $h : S \to \mathrm{Map}_{\mathcal{C}}(x, y)$. By definition, $\mathcal{C}'$ is universal with respect to the property that it receives a map from $\mathcal{C}$, and the map h extends to a map $\widetilde{h} : T \to \mathrm{Map}_{\mathcal{C}'}(x, y)$. To carry out the proof, we will give an explicit construction of an $\mathbf{S}$-enriched category $\mathcal{C}'$ which has this universal property.

For the remainder of the proof, we will assume that $\mathbf{S}$ is the category of simplicial sets. This is purely for notational convenience; the same arguments can be employed without change in the general case.

We begin by declaring that the objects of $\mathcal{C}'$ are the objects of $\mathcal{C}$. The definition of the morphisms in $\mathcal{C}'$ is a bit more complicated. Let w and z be objects of $\mathcal{C}$. We define a sequence of simplicial sets $M_{\mathcal{C}'}^k$ as follows:

$$M_{\mathcal{C}}^0 = \mathrm{Map}_{\mathcal{C}}(w, z)$$

$$M_{\mathcal{C}}^1 = \mathrm{Map}_{\mathcal{C}}(y, z) \times T \times \mathrm{Map}_{\mathcal{C}}(w, x)$$

$$M_{\mathcal{C}}^2 = \mathrm{Map}_{\mathcal{C}}(y, z) \times T \times \mathrm{Map}_{\mathcal{C}}(y, x) \times T \times \mathrm{Map}_{\mathcal{C}}(w, x),$$

and so forth. More specifically, for $k \geq 1$, the m-simplices of $M_{\mathcal{C}}^k$ are finite sequences

$$(\sigma_0, \tau_1, \sigma_1, \tau_2, \ldots, \tau_k, \sigma_k),$$

where $\sigma_0 \in \mathrm{Map}_{\mathcal{C}}(y, z)_m$, $\sigma_k \in \mathrm{Map}_{\mathcal{C}}(w, x)_m$, $\sigma_i \in \mathrm{Map}_{\mathcal{C}}(y, x)_m$ for $0 < i < k$, and $\tau_i \in T_m$ for $1 \leq i \leq k$.

We define $\mathrm{Map}_{\mathcal{C}'}(w, z)$ to be the quotient of the disjoint union $\coprod_k M_{\mathcal{C}}^k$ by the equivalence relation which is generated by making the identification

$$(\sigma_0, \tau_1, \ldots, \sigma_k) \simeq (\sigma_0, \tau_1, \ldots, \tau_{j-1}, \sigma_{j-1} \circ h(\tau_j) \circ \sigma_j, \tau_{j+1}, \ldots, \sigma_k)$$

whenever the simplex τ_j belongs to $S_m \subseteq T_m$.

We equip $\mathcal{C}'$ with an associative composition law, which is given on the level of simplices by

$$(\sigma_0, \tau_1, \ldots, \sigma_k) \circ (\sigma_0', \tau_1', \ldots, \sigma_l') = (\sigma_0, \tau_1, \ldots, \tau_k, \sigma_k \circ \sigma_0', \tau_1', \ldots, \sigma_l').$$

It is easy to verify that this composition law is well-defined (that is, compatible with the equivalence relation introduced above) and associative and that the identification $M_{\mathcal{C}}^0 = \mathrm{Map}_{\mathcal{C}}(w, z)$ gives rise to an inclusion of categories $\mathcal{C} \subseteq \mathcal{C}'$. Moreover, the map $h : S \to \mathrm{Map}_{\mathcal{C}}(x, y)$ extends to $\widetilde{h} : T \to \mathrm{Map}_{\mathcal{C}'}(x, y)$ given by the composition

$$T \simeq \{\mathrm{id}_y\} \times T \times \{\mathrm{id}_x\} \subseteq \mathrm{Map}_{\mathcal{C}}(y, y) \times T \times \mathrm{Map}_{\mathcal{C}}(x, x) = M_{\mathcal{C}}^1 \to \mathrm{Map}_{\mathcal{C}'}(x, y).$$

Moreover, it is not difficult to see that $\mathcal{C}'$ has the desired universal property.

We observe that, by construction, the simplicial sets $\text{Map}_{\mathcal{C}'}(w, z)$ come equipped with a natural filtration. Namely, define $\text{Map}_{\mathcal{C}'}(w, z)^k$ to be the image of

$$\coprod_{0 \leq i \leq k} M_{\mathcal{C}}^i$$

in $\text{Map}_{\mathcal{C}'}(w, z)$. Then we have

$$\text{Map}_{\mathcal{C}}(w, z) = \text{Map}_{\mathcal{C}'}(w, z)^0 \subseteq \text{Map}_{\mathcal{C}'}(w, z)^1 \subseteq \cdots$$

and $\bigcup_k \text{Map}_{\mathcal{C}'}(w, z)^k = \text{Map}_{\mathcal{C}'}(w, z)$. Moreover, the inclusion

$$\text{Map}_{\mathcal{C}'}(w, z)^k \subseteq \text{Map}_{\mathcal{C}'}(w, z)^{k+1}$$

is a pushout of the inclusion $N_{\mathcal{C}}^{k+1} \subseteq M_{\mathcal{C}}^{k+1}$, where N^{k+1} is the simplicial subset of $M_{\mathcal{C}}^{k+1}$ whose m-simplices consist of those $(2m + 1)$-tuples $(\sigma_0, \tau_1, \ldots, \sigma_m)$ such that $\tau_i \in S_m$ for at least one value of i.

Let us now return to the problem at hand: namely, we wish to prove that $F' : \mathcal{C}' \to \mathcal{D}'$ is an equivalence. We note that the construction outlined above may also be employed to produce a model for $\mathcal{D}'$ and an analogous filtration on its morphism spaces.

Since $G' : \mathcal{D} \to \mathcal{D}'$ and $F : \mathcal{C} \to \mathcal{D}$ are essentially surjective, we deduce that F' is essentially surjective. Hence it will suffice to show that, for any objects $w, z \in \mathcal{C}'$, the induced map

$$\phi : \text{Map}_{\mathcal{C}'}(w, z) \to \text{Map}_{\mathcal{D}'}(w, z)$$

is a weak homotopy equivalence. For this, it will suffice to show that for each $i \geq 0$, the induced map $\phi_i : \text{Map}_{\mathcal{C}'}(w, z)^i \to \text{Map}_{\mathcal{D}'}(w, z)^i$ is a weak homotopy equivalence; then ϕ, being a filtered colimit of weak homotopy equivalences ϕ_i, will itself be a weak homotopy equivalence.

The proof now proceeds by induction on i. When $i = 0$, ϕ_i is a weak homotopy equivalence by assumption (since F is an equivalence of simplicial categories). For the inductive step, we note that ϕ_{i+1} is obtained as a pushout

$$\text{Map}_{\mathcal{C}'}(w, z)^i \coprod_{N_{\mathcal{C}}^{i+1}} M_{\mathcal{C}}^{i+1} \to \text{Map}_{\mathcal{D}'}(w, z)^i \coprod_{N_{\mathcal{D}}^{i+1}} M_{\mathcal{D}}^{i+1}.$$

Since $\mathbf{S}$ is left proper, both of these pushouts are homotopy pushouts. Consequently, to show that ϕ_{i+1} is a weak equivalence, it suffices to show that ϕ_i is a weak equivalence and that both of the maps

$$N_{\mathcal{C}}^{i+1} \to N_{\mathcal{D}}^{i+1}$$

$$M_{\mathcal{C}}^{i+1} \to M_{\mathcal{D}}^{i+1}$$

are weak equivalences. These statements follow easily from the compatibility of the monoidal structure of $\mathbf{S}$ with the model structure and the assumption that every object of $\mathbf{S}$ is cofibrant. $\qquad\square$

Remark A.3.2.5. It follows from the proof of Proposition A.3.2.4 that if $f : \mathcal{C} \to \mathcal{C}'$ is a cofibration of **S**-enriched categories, then the induced map $\mathrm{Map}_{\mathcal{C}}(X,Y) \to \mathrm{Map}_{\mathcal{C}'}(fX, fY)$ is a cofibration for every pair of objects $X, Y \in \mathcal{C}$.

Remark A.3.2.6. The model structure of Proposition A.3.2.4 enjoys the following functoriality: suppose that $f : \mathbf{S} \to \mathbf{S}'$ is a monoidal left Quillen functor between model categories satisfying the hypotheses of Proposition A.3.2.4, with right adjoint $g : \mathbf{S}' \to \mathbf{S}$. Then f and g induce a Quillen adjunction

$$\mathcal{C}\mathrm{at}_{\mathbf{S}} \underset{G}{\overset{F}{\rightleftarrows}} \mathcal{C}\mathrm{at}_{\mathbf{S}'},$$

where F and G are as in Remark A.1.4.3. Moreover, if (f, g) is a Quillen equivalence, then (F, G) is likewise a Quillen equivalence.

In order for Proposition A.3.2.4 to be useful in practice, we need to understand the fibrations in $\mathcal{C}\mathrm{at}_{\mathbf{S}}$. For this, we first introduce a few definitions.

Definition A.3.2.7. Let $F : \mathcal{C} \to \mathcal{D}$ be a functor between ordinary categories. We will say that F is a *quasi-fibration* if, for every object $X \in \mathcal{C}$ and every isomorphism $f : F(X) \to Y$ in $\mathcal{D}$, there exists an isomorphism $\overline{f} : X \to \overline{Y}$ in $\mathcal{C}$ such that $F(\overline{f}) = f$.

Remark A.3.2.8. The relevance of Definition A.3.2.7 is as follows: the category $\mathcal{C}\mathrm{at}$ admits a model structure in which the weak equivalences are the equivalences of categories and the fibrations are the quasi-fibrations. This is a special case of Theorem A.3.2.24, which we will prove below (namely, the special case where we take $\mathbf{S} = \mathcal{S}\mathrm{et}$ endowed with the trivial model structure of Example A.2.1.2).

Definition A.3.2.9. Let $\mathbf{S}$ be a monoidal model category and let $\mathcal{C}$ be an **S**-enriched category. We will say that a morphism f in $\mathcal{C}$ is an *equivalence* if the homotopy class $[f]$ of f is an isomorphism in $h\mathcal{C}$.

We will say that $\mathcal{C}$ is *locally fibrant* if, for every pair of objects $X, Y \in \mathcal{C}$, the mapping space $\mathrm{Map}_{\mathcal{C}}(X, Y)$ is a fibrant object of $\mathbf{S}$.

We will say that an **S**-enriched functor $F : \mathcal{C} \to \mathcal{C}'$ is a *local fibration* if the following conditions are satisfied:

(i) For every pair of objects $X, Y \in \mathcal{C}$, the induced map $\mathrm{Map}_{\mathcal{C}}(X, Y) \to \mathrm{Map}_{\mathcal{C}'}(FX, FY)$ is a fibration in $\mathbf{S}$.

(ii) The induced map $h\mathcal{C} \to h\mathcal{C}'$ is a quasi-fibration of categories.

Remark A.3.2.10. Let $F : \mathcal{C} \to \mathcal{C}'$ be a functor between **S**-enriched categories which satisfies condition (i) of Definition A.3.2.9. Let $X \in \mathcal{C}$ and $Y \in \mathcal{C}'$ be objects. If $\mathcal{C}'$ is locally fibrant, then every morphism $[f] : F(X) \to Y$ in $h\mathcal{C}'$ can be represented by an equivalence $f : F(X) \to Y$ in $\mathcal{C}'$. Let $\overline{Y}$ be an object of $\mathcal{C}$ such that $F(\overline{Y}) = Y$. Since $\mathbf{1}_{\mathbf{S}}$ is a cofibrant object of $\mathbf{S}$ and the map $\mathrm{Map}_{\mathcal{C}}(X, \overline{Y}) \to \mathrm{Map}_{\mathcal{C}}(F(X), Y)$ is a fibration, Proposition A.2.3.1 implies that $[f]$ can be lifted to an isomorphism $[\overline{f}] : X \to \overline{Y}$ in $h\mathcal{C}$ if and only if f can be lifted to an equivalence $\overline{f} : X \to \overline{Y}$ in $\mathcal{C}$. Consequently, when $h\mathcal{C}$ is locally fibrant, condition (ii) is equivalent to the following analogous assertion:

(ii') For every equivalence $f : F(X) \to Y$ in $\mathcal{C}'$, there exists an equivalence $\overline{f} : X \to \overline{Y}$ in $\mathcal{C}$ such that $F(\overline{f}) = f$.

Notation A.3.2.11. We let $[1]_{\mathbf{S}}^{\sim}$ denote the **S**-enriched category containing a pair of objects X, Y, with

$$\mathrm{Map}_{[1]_{\mathbf{S}}^{\sim}}(Z, Z') = \mathbf{1_S}$$

for all $Z, Z' \in \{X, Y\}$.

Definition A.3.2.12 (Invertibility Hypothesis). Let **S** be a monoidal model category satisfying the hypotheses of Proposition A.3.2.4. We will say that **S** *satisfies the invertibility hypothesis* if the following condition is satisfied:

$(*)$ Let $i : [1]_{\mathbf{S}} \to \mathcal{C}$ be a cofibration of **S**-enriched categories, classifying a morphism f in $\mathcal{C}$ which is invertible in the homotopy category $h\mathcal{C}$, and form a pushout diagram

$$
\begin{array}{ccc}
[1]_{\mathbf{S}} & \xrightarrow{\ i\ } & \mathcal{C} \\
\downarrow & & \downarrow{\scriptstyle j} \\
[1]_{\mathbf{S}}^{\sim} & \longrightarrow & \mathcal{C}\langle f^{-1} \rangle.
\end{array}
$$

Then j is an equivalence of **S**-enriched categories.

In other words, the invertibility hypothesis is the assertion that inverting a morphism f in an **S**-enriched category $\mathcal{C}$ does not change the homotopy type of $\mathcal{C}$ when f is already invertible up to homotopy.

Remark A.3.2.13. Let **S**, f, and $\mathcal{C}$ be as in Definition A.3.2.12 and choose a trivial cofibration $F : \mathcal{C} \to \mathcal{C}'$, where $\mathcal{C}'$ is a fibrant **S**-enriched category. Since $\mathcal{C}\mathrm{at}_{\mathbf{S}}$ is left proper, the induced map $\mathcal{C}\langle f^{-1} \rangle \to \mathcal{C}'\langle F(f)^{-1} \rangle$ is an equivalence of **S**-enriched categories. Consequently, assertion $(*)$ holds for $(\mathcal{C}, f)$ if and only if it holds for $(\mathcal{C}', F(f))$. In other words, to test whether **S** satisfies the invertibility hypothesis, we need only check $(*)$ in the case where $\mathcal{C}$ is fibrant.

Remark A.3.2.14. In Definition A.3.2.12, the condition that i be a cofibration guarantees that the construction $\mathcal{C} \mapsto \mathcal{C}\langle f^{-1} \rangle$ is homotopy invariant. Alternatively, we can guarantee this by choosing a cofibrant replacement for the map $j : [1]_{\mathbf{S}} \to [1]_{\mathbf{S}}^{\sim}$. Namely, choose a factorization for j as a composition

$$[1]_{\mathbf{S}} \xrightarrow{j'} \mathcal{E} \xrightarrow{j''} [1]_{\mathbf{S}}^{\sim},$$

where j'' is a weak equivalence and j' is a cofibration. For every **S**-enriched category containing a morphism f, define $\mathcal{C}[f^{-1}] = \mathcal{C} \coprod_{[1]_{\mathbf{S}}} \mathcal{E}$. Then we have a canonical map $\mathcal{C}[f^{-1}] \to \mathcal{C}\langle f^{-1}\rangle$, which is an equivalence whenever the map $[1]_\mathbf{S} \to \mathcal{C}$ classifying f is a cofibration. Moreover, the construction $\mathcal{C} \mapsto \mathcal{C}[f^{-1}]$ preserves weak equivalences in $\mathcal{C}$. Consequently, we may reformulate the invertibility hypothesis as follows:

($*'$) For every $\mathbf{S}$-enriched category $\mathcal{C}$ containing an equivalence f, the map $\mathcal{C} \to \mathcal{C}[f^{-1}]$ is a weak equivalence of $\mathbf{S}$-enriched categories.

Remark A.3.2.15. Let $\mathcal{C}$ be a fibrant $\mathbf{S}$-enriched category containing an equivalence $f : X \to Y$ and let $\mathcal{C}[f^{-1}]$ be defined as in Remark A.3.2.14. The canonical map $\mathcal{C} \to \mathcal{C}[f^{-1}]$ is a trivial cofibration and therefore admits a section. This section determines a map of $\mathbf{S}$-enriched categories $h : \mathcal{E} \to \mathcal{C}$. We observe that $\mathcal{E}$ is a mapping cylinder for the object $[0]_{\mathcal{C}\mathrm{at}_\mathbf{S}} \in \mathcal{C}\mathrm{at}_\mathbf{S}$, so we can view h as a homotopy between the maps $[0]_{\mathcal{C}\mathrm{at}_\mathbf{S}} \to \mathcal{C}$ classifying the objects X and Y.

More generally, the same argument shows that if $F : \mathcal{C} \to \mathcal{D}$ is a fibration of $\mathbf{S}$-enriched categories and $f : X \to Y$ is an equivalence in $\mathcal{C}$ such that $F(f) = \mathrm{id}_D$ for some object $D \in \mathcal{D}$, then the functors $[0]_\mathbf{S} \to \mathcal{C}$ classifying the objects X and Y are homotopic in the model category $(\mathcal{C}\mathrm{at}_\mathbf{S})_{/\mathcal{D}}$.

Definition A.3.2.16. We will say that a model category $\mathbf{S}$ is *excellent* if it is equipped with a symmetric monoidal structure and satisfies the following conditions:

($A1$) The model category $\mathbf{S}$ is combinatorial.

($A2$) Every monomorphism in $\mathbf{S}$ is a cofibration, and the collection of cofibrations is stable under products.

($A3$) The collection of weak equivalences in $\mathbf{S}$ is stable under filtered colimits.

($A4$) The monoidal structure on $\mathbf{S}$ is compatible with the model structure. In other words, the tensor product functor $\otimes : \mathbf{S} \times \mathbf{S} \to \mathbf{S}$ is a left Quillen bifunctor.

($A5$) The monoidal model category $\mathbf{S}$ satisfies the invertibility hypothesis.

Remark A.3.2.17. Axiom ($A2$) of Definition A.3.2.16 implies that every object of $\mathbf{S}$ is cofibrant. In particular, $\mathbf{S}$ is left proper.

Example A.3.2.18 (Dwyer, Kan). The category of simplicial sets is an excellent model category when endowed with the Kan model structure and the Cartesian product. The only nontrivial point is to show that Set_Δ satisfies the invertibility hypothesis. This is one of the main theorems of [21].

Example A.3.2.19. Let $\mathbf{S}$ be a presentable category equipped with a closed symmetric monoidal structure. Then $\mathbf{S}$ is an excellent model category with respect to the trivial model structure of Example A.2.1.2.

The following lemma guarantees a good supply of examples of excellent model categories:

Lemma A.3.2.20. *Suppose we are given a monoidal left Quillen functor* $T : \mathbf{S} \to \mathbf{S}'$ *between model categories* $\mathbf{S}$ *and* $\mathbf{S}'$ *satisfying axioms* $(A1)$ *through* $(A4)$ *of Definition A.3.2.16. If* $\mathbf{S}$ *satisfies axiom* $(A5)$, *then so does* $\mathbf{S}'$.

Proof. As indicated in Remark A.3.2.6, the functor T determines a Quillen adjunction

$$\mathcal{C}\mathrm{at}_{\mathbf{S}} \underset{G}{\overset{F}{\rightleftarrows}} \mathcal{C}\mathrm{at}_{\mathbf{S}'} \,.$$

Let $\mathcal{C}$ be a $\mathbf{S}'$-enriched category and $i : [1]_{\mathbf{S}'} \to \mathcal{C}$ a cofibration classifying an equivalence f in $\mathcal{C}$. We wish to prove that the map $\mathcal{C} \to \mathcal{C}\langle f^{-1}\rangle$ is an equivalence of $\mathbf{S}'$-enriched categories. In view of Remark A.3.2.13, we may assume that $\mathcal{C}$ is fibrant.

Choose a factorization of the map $[1]_{\mathbf{S}} \to [1]_{\mathbf{S}}^{\sim}$ as a composition

$$[1]_{\mathbf{S}} \overset{j}{\to} \mathcal{E} \overset{j'}{\to} [1]_{\mathbf{S}}^{\sim}$$

as in Remark A.3.2.14, so that we have an analogous factorization

$$[1]_{\mathbf{S}'} \to F(\mathcal{E}) \to [1]_{\mathbf{S}'}^{\sim}$$

in $\mathcal{C}\mathrm{at}_{\mathbf{S}'}$. Using the latter factorization, we can define $\mathcal{C}[f^{-1}]$ as in Remark A.3.2.14; we wish to show that the map $h : \mathcal{C} \to \mathcal{C}[f^{-1}]$ is a trivial cofibration.

Let f_0 be the morphism in $G(\mathcal{C})$ classified by f, and let $G(\mathcal{C})[f_0^{-1}] \in \mathcal{C}\mathrm{at}_{\mathbf{S}}$ be defined as in Remark A.3.2.14. Using the fact that $\mathcal{C}$ is locally fibrant (see Theorem A.3.2.24 below), we conclude that f_0 is an equivalence in $G(\mathcal{C})$. Since $\mathbf{S}$ satisfies the invertibility hypothesis, the map $h_0 : G(\mathcal{C}) \to G(\mathcal{C})[f_0^{-1}]$ is a trivial cofibration. We now conclude by observing that h is a pushout of $F(h_0)$. $\square$

Remark A.3.2.21. Using a similar argument, we can prove a converse to Lemma A.3.2.20 in the case where T is a Quillen equivalence.

Example A.3.2.22. Let $\mathbf{S}$ be the category $\mathcal{S}\mathrm{et}_{\Delta}^{+}$ of marked simplicial sets endowed with the Cartesian model structure defined in §3.1. Then the functor $X \mapsto X^{\sharp}$ is a monoidal left Quillen functor $\mathcal{S}\mathrm{et}_{\Delta} \to \mathbf{S}$. Combining Example A.3.2.18 with Lemma A.3.2.20, we conclude that $\mathbf{S}$ satisfies the invertibility hypothesis, so that $\mathbf{S}$ is an excellent model category (with respect to the Cartesian product).

Example A.3.2.23. Let $\mathbf{S}$ denote the category of simplicial sets, endowed with the Joyal model structure. The functor $X \mapsto X^{\flat}$ determines a monoidal left Quillen equivalence $\mathbf{S} \to \mathcal{S}\mathrm{et}_{\Delta}^{+}$. Using Remark A.3.2.21, we deduce that $\mathbf{S}$ satisfies the invertibility hypothesis, so that $\mathbf{S}$ is an excellent model category (with respect to the Cartesian product).

We are now ready to state our main result:

Theorem A.3.2.24. *Let* **S** *be an excellent model category. Then*

(1) *An* **S***-enriched category* $\mathcal{C}$ *is a fibrant object of* $\mathcal{C}\mathrm{at}_{\mathbf{S}}$ *if and only if it is locally fibrant: that is, if and only if the mapping object* $\mathrm{Map}_{\mathcal{C}}(X, Y) \in$ **S** *is fibrant for every pair of objects* $X, Y \in \mathcal{C}$.

(2) *Let* $F : \mathcal{C} \to \mathcal{D}$ *be an* **S***-enriched functor, where* $\mathcal{D}$ *is a fibrant object of* $\mathcal{C}\mathrm{at}_{\mathbf{S}}$. *Then* F *is a fibration in* $\mathcal{C}\mathrm{at}_{\mathbf{S}}$ *if and only if* F *is a local fibration.*

Remark A.3.2.25. In the case where **S** is the category of simplicial sets (with its usual model structure), Theorem A.3.2.24 is due to Bergner; see [7]. Moreover, Bergner proves a stronger result in this case: assertion (2) holds without the assumption that $\mathcal{D}$ is fibrant.

Before giving the proof of Theorem A.3.2.24, we need to establish some preliminaries. Fix an excellent model category **S**. We observe that $\mathcal{C}\mathrm{at}_{\mathbf{S}}$ is naturally *cotensored* over **S**. That is, for every **S**-enriched category $\mathcal{C}$ and every object $K \in$ **S**, we can define a new **S**-enriched category $\mathcal{C}^K$ as follows:

(*i*) The objects of $\mathcal{C}^K$ are the objects of $\mathcal{C}$.

(*ii*) Given a pair of objects $X, Y \in \mathcal{C}$, we have
$$\mathrm{Map}_{\mathcal{C}^K}(X, Y) = \mathrm{Map}_{\mathcal{C}}(X, Y)^K \in \mathbf{S}.$$

This construction does not endow $\mathcal{C}\mathrm{at}_{\mathbf{S}}$ with the structure of an **S**-enriched category because the construction $\mathcal{D} \mapsto \mathcal{D}^K$ is not compatible with colimits in K. However, we can remedy the situation as follows. Let $\mathcal{C}$ and $\mathcal{D}$ be **S**-enriched categories and let ϕ be a function from the set of objects of $\mathcal{C}$ to the set of objects of $\mathcal{D}$. Then there exists an object $\mathrm{Map}^{\phi}_{\mathcal{C}\mathrm{ats}}(\mathcal{C}, \mathcal{D}) \in \mathbf{S}$ which is characterized by the following universal property: for every $K \in \mathbf{S}$, there is a natural bijection
$$\mathrm{Hom}_{\mathbf{S}}(K, \mathrm{Map}^{u}_{\mathcal{C}\mathrm{ats}}(\mathcal{C}, \mathcal{D})) \simeq \mathrm{Hom}^{\phi}_{\mathcal{C}\mathrm{ats}}(\mathcal{C}, \mathcal{D}^K),$$
where $\mathrm{Hom}^{\phi}_{\mathcal{C}\mathrm{ats}}(\mathcal{C}, \mathcal{D}^K)$ denotes the set of all functors from $\mathcal{C}$ to $\mathcal{D}^K$ which is given on objects by the function ϕ.

Lemma A.3.2.26. *Let* **S** *be an excellent model category. Fix a diagram of* **S***-enriched categories*

$$
\begin{array}{ccc}
\mathcal{C} & \xrightarrow{\;u\;} & \mathcal{C}' \\
{\scriptstyle F}\downarrow & & \downarrow{\scriptstyle F'} \\
\mathcal{D} & \xrightarrow{\;u'\;} & \mathcal{D}'.
\end{array}
$$

Assume that

(*a*) *For every pair of objects* $X, Y \in \mathcal{C}$, *the diagram*

$$
\begin{array}{ccc}
\mathrm{Map}_{\mathcal{C}}(X, Y) & \longrightarrow & \mathrm{Map}_{\mathcal{D}}(FX, FY) \\
\downarrow & & \downarrow \\
\mathrm{Map}_{\mathcal{C}'}(uX, uY) & \longrightarrow & \mathrm{Map}_{\mathcal{D}'}(u'FX, u'FY)
\end{array}
$$

is a homotopy pullback square involving fibrant objects of **S** *and the vertical arrows are fibrations.*

Let $G : \mathcal{A} \to \mathcal{B}$ *be a functor between* **S**-*enriched categories which is a transfinite composition of pushouts of generating cofibrations of the form* $[1]_S \to [1]_{S'}$, *where* $S \to S'$ *is a cofibration in* **S** *and let* ϕ *be a function from the set of objects of* $\mathcal{B}$ *(which is isomorphic to the set of objects of* $\mathcal{A}$*) to* $\mathcal{C}$. *Then the diagram*

$$\mathrm{Map}^{\phi}_{\mathcal{C}\mathrm{ats}}(\mathcal{B},\mathcal{C}) \longrightarrow \mathrm{Map}^{F\phi}_{\mathcal{C}\mathrm{ats}}(\mathcal{B},\mathcal{D}) \times_{\mathrm{Map}^{F\phi}_{\mathcal{C}\mathrm{ats}}(\mathcal{A},\mathcal{D})} \mathrm{Map}^{\phi}_{\mathcal{C}\mathrm{ats}}(\mathcal{A},\mathcal{C})$$

$$\downarrow \qquad\qquad\qquad\qquad\qquad\qquad \downarrow$$

$$\mathrm{Map}^{u\phi}_{\mathcal{C}\mathrm{ats}}(\mathcal{B},\mathcal{C}') \longrightarrow \mathrm{Map}^{u'F\phi}_{\mathcal{C}\mathrm{ats}}(\mathcal{B},\mathcal{D}') \times_{\mathrm{Map}^{u'F\phi}_{\mathcal{C}\mathrm{ats}}(\mathcal{A},\mathcal{D}')} \mathrm{Map}^{u\phi}_{\mathcal{C}\mathrm{ats}}(\mathcal{A},\mathcal{C}')$$

is a homotopy pullback square between fibrant objects of **S**, *and the vertical arrows are fibrations.*

Proof. It is easy to see that the collection of morphisms $G : \mathcal{A} \to \mathcal{B}$ which satisfy the conclusion of the lemma is weakly saturated. It will therefore suffice to show that G contains every morphism of the form $[1]_S \to [1]_{S'}$, where $S \to S'$ is a cofibration in **S**. In this case, ϕ determines a pair of objects $X, Y \in \mathcal{C}$, and we can rewrite the diagram of interest as

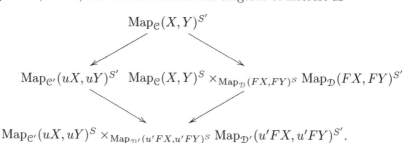

The desired result now follows from (a) since the map $S \to S'$ is a cofibration between cofibrant objects of **S**. □

Proof of Theorem A.3.2.24. Assertion (1) is just a special case of (2) where we take $\mathcal{D}$ to be the final object of $\mathcal{C}\mathrm{ats}$. It will therefore suffice to prove (2).

We first prove the "only if" direction. If F is a fibration, then F has the right lifting property with respect to every trivial cofibration of the form $[1]_S \to [1]_{S'}$, where $S \to S'$ is a trivial cofibration in **S**. It follows that for every pair of objects $X, Y \in \mathcal{C}$, the induced map $\mathrm{Map}_{\mathcal{C}}(X,Y) \to \mathrm{Map}_{\mathcal{D}}(FX,FY)$ is a fibration in **S**. In particular, $\mathcal{C}$ is locally fibrant.

To complete the proof that F is a local fibration, we will show that F satisfies condition (ii') of Remark A.3.2.10. Suppose $X \in \mathcal{C}$ and that $f : FX \to Y$ is an equivalence in $\mathcal{D}$. We wish to show that we can lift f to an equivalence $\overline{f} : X \to \overline{Y}$. Let $\mathcal{E}$ and $\mathcal{D}[f^{-1}]$ be defined as in Remark A.3.2.14. Since **S** satisfies the invertibility hypothesis, the map $h : \mathcal{D} \to \mathcal{D}[f^{-1}]$ is a trivial cofibration. Because we have assumed $\mathcal{D}$ to be fibrant, the map h admits a section. This section determines a map $s : \mathcal{E} \to \mathcal{D}$. We now consider the lifting problem

$$
\begin{array}{ccc}
[0]s & \xrightarrow{\ X\ } & \mathcal{C} \\
\downarrow & \nearrow & \downarrow F \\
\mathcal{E} & \xrightarrow[\ \ s\ \ ]{} & \mathcal{D}.
\end{array}
$$

Since F is a fibration and the left vertical map is a trivial cofibration, there exists a solution as indicated. This solution determines a morphism $\overline{f} : X \to \overline{Y}$ in $\mathcal{C}$ lifting f. Moreover, $\overline{f}$ is the image of a morphism in $\mathcal{E}$. Since every morphism in $\mathcal{E}$ is an equivalence, we deduce that $\overline{f}$ is an equivalence in $\mathcal{C}$.

Let us now suppose that F is a local fibration. We wish to show that F is a fibration. Choose a factorization of F as a composition

$$
\mathcal{C} \xrightarrow{u} \mathcal{C}' \xrightarrow{F'} \mathcal{D},
$$

where u is a weak equivalence and F' is a fibration. We will prove the following:

(∗) Suppose we are given a commutative diagram of **S**-enriched categories

$$
\begin{array}{ccc}
\mathcal{A} & \xrightarrow{\ v\ } & \mathcal{C} \\
\downarrow G & & \downarrow F \\
\mathcal{B} & \xrightarrow[\ v'\ ]{} & \mathcal{D},
\end{array}
$$

where G is a cofibration. If there exists a functor $\alpha : \mathcal{B} \to \mathcal{C}'$ such that $\alpha G = uv$ and $F'\alpha = v'$, then there exists a functor $\beta : \mathcal{B} \to \mathcal{C}$ such that $\beta G = v$ and $F\beta = v'$.

Since the map F' has the right lifting property with respect to all trivial cofibrations, assertion (∗) implies that F also has the right lifting property with respect to all trivial cofibrations, so that F is a fibration as desired.

We now prove (∗). Using the small object argument, we deduce that the functor G is a retract of some functor $G' : \mathcal{A} \to \mathcal{B}'$, where G' is a transfinite composition of morphisms obtained as pushouts of generating cofibrations. It will therefore suffice to prove (∗) after replacing G by G'.

Reordering the transfinite composition if necessary, we may assume that G' factors as a composition

$$
\mathcal{A} \xrightarrow{G'_0} \mathcal{B}'_0 \xrightarrow{G'_1} \mathcal{B}',
$$

where $\mathcal{B}'_0$ is obtained from $\mathcal{A}$ by adjoining a collection of new objects, $\{B_i\}_{i \in I}$, and $\mathcal{B}'$ is obtained from $\mathcal{B}'_0$ by a transfinite sequence of pushouts by generating cofibrations of the form $\mathcal{E}_S \to \mathcal{E}_{S'}$, where $S \to S'$ is a cofibration in **S**. Let $C'_i = \alpha(B_i)$ for each $i \in I$. Since u is an equivalence of

S-enriched categories, there exists a collection of objects $\{C_i\}_{i \in I}$ and equivalences $f_i : uC_i \to C_i'$. Let g_i be the image of f_i in $\mathcal{D}$. Since F is a local fibration, we can lift each g_i to an equivalence $f_i' : C_i \to C_i''$ in $\mathcal{C}$. Since the maps $\mathrm{Map}_{\mathcal{C}'}(uC_i'', C_i') \to \mathrm{Map}_{\mathcal{D}}(FC_i'', F'C_i')$ are fibrations, we can choose morphisms $f_i'' : uC_i'' \to C_i'$ in $\mathcal{C}'$ such that $F'(f_i')$ is the identity for each i, and the diagrams

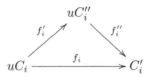

commute up to homotopy. Replacing C_i by C_i'', we may assume that each of the maps f_i projects to the identity in $\mathcal{D}$.

Let $\alpha_0 = \alpha | \mathcal{B}_0'$ and let $\alpha_0' : \mathcal{B}_0' \to \mathcal{C}'$ be defined by the formula

$$\alpha_0'(A) = \begin{cases} \alpha_0(A) & \text{if } A \in \mathcal{A} \\ uC_i & \text{if } A = B_i, i \in I. \end{cases}$$

Remark A.3.2.15 implies that the maps α_0 and α_0' are homotopic in the model category $(\mathcal{C}\mathrm{at}_{\mathbf{S}})_{\mathcal{A}//\mathcal{D}}$. Applying Proposition A.2.3.1, we deduce the existence of a map $\alpha' : \mathcal{B}' \to \mathcal{C}$ which extends α_0 and satisfies $\alpha'G = uv$ and $F'\alpha' = v'$. We may therefore replace α by α', v by α_0', and $\mathcal{A}$ by $\mathcal{B}_0'$ and thereby reduce to the case where the functor $G : \mathcal{A} \to \mathcal{B}$ is a transfinite composition of generating cofibrations of the form $\mathcal{E}_S \to \mathcal{E}_{S'}$, where $S \to S'$ is a cofibration in $\mathbf{S}$.

Let ϕ be the map from the objects of $\mathcal{B}$ to the objects of $\mathcal{C}$ determined by α. Applying Lemma A.3.2.26, we obtain a homotopy pullback diagram

$$\begin{array}{ccc} \mathrm{Map}_{\mathcal{C}\mathrm{at}_{\mathbf{S}}}^{\phi}(\mathcal{B}, \mathcal{C}) & \longrightarrow & \mathrm{Map}_{\mathcal{C}\mathrm{at}_{\mathbf{S}}}^{F\phi}(\mathcal{B}, \mathcal{D}) \times_{\mathrm{Map}_{\mathcal{C}\mathrm{at}_{\mathbf{S}}}^{F\phi}(\mathcal{A}, \mathcal{D})} \mathrm{Map}_{\mathcal{C}\mathrm{at}_{\mathbf{S}}}^{\phi}(\mathcal{A}, \mathcal{C}) \\ \downarrow & & \downarrow \\ \mathrm{Map}^{u\phi}\mathcal{C}\mathrm{at}_{\mathbf{S}}(\mathcal{B}, \mathcal{C}') & \longrightarrow & \mathrm{Map}_{\mathcal{C}\mathrm{at}_{\mathbf{S}}}^{F\phi}(\mathcal{B}, \mathcal{D}) \times_{\mathrm{Map}_{\mathcal{C}\mathrm{at}_{\mathbf{S}}}^{F\phi}(\mathcal{A}, \mathcal{D})} \mathrm{Map}_{\mathcal{C}\mathrm{at}_{\mathbf{S}}}^{u\phi}(\mathcal{A}, \mathcal{C}'). \end{array}$$

in which the horizontal arrows are fibrations. We therefore have a weak equivalence

$$\mathrm{Map}_{\mathcal{C}\mathrm{at}_{\mathbf{S}}}^{\phi}(\mathcal{B}, \mathcal{C}) \to M = \mathrm{Map}_{\mathcal{C}\mathrm{at}_{\mathbf{S}}}^{u\phi}(\mathcal{B}, \mathcal{C}') \times_{\mathrm{Map}_{\mathcal{C}\mathrm{at}_{\mathbf{S}}}^{F\phi}(\mathcal{B}, \mathcal{D})} \mathrm{Map}_{\mathcal{C}\mathrm{at}_{\mathbf{S}}}^{\phi}(\mathcal{A}, \mathcal{C})$$

of fibrations over $N = \mathrm{Map}_{\mathcal{C}\mathrm{at}_{\mathbf{S}}}^{F\phi}(\mathcal{B}, \mathcal{D}) \times_{\mathrm{Map}_{\mathcal{C}\mathrm{at}_{\mathbf{S}}}^{F\phi}(\mathcal{A}, \mathcal{D})} \mathrm{Map}_{\mathcal{C}\mathrm{at}_{\mathbf{S}}}^{u\phi}(\mathcal{A}, \mathcal{C}')$. Moreover, the pair (α, v) determines a map $\mathbf{1}_{\mathbf{S}} \to M$ lifting the map $(v', uv') : \mathbf{1}_{\mathbf{S}} \to N$. Applying Proposition A.2.3.1, we deduce that $(v, uv') : \mathbf{1}_{\mathbf{S}} \to N$ can be lifted to a map $\mathbf{1}_{\mathbf{S}} \to \mathrm{Map}_{\mathcal{C}\mathrm{at}_{\mathbf{S}}}^{\phi}(\mathcal{B}, \mathcal{C})$, which is equivalent to the existence of the desired map β. $\square$

We conclude this section with a few easy results concerning homotopy limits in the model category $\mathcal{C}\mathrm{at}_{\mathbf{S}}$.

Proposition A.3.2.27. *Let* **S** *be an excellent model category,* $\mathcal{J}$ *a small category, and* $\{\mathcal{C}_J\}_{J \in \mathcal{J}}$ *a diagram of* **S***-enriched categories. Suppose we are given a compatible family of functors* $\{f_J : \mathcal{C} \to \mathcal{C}_J\}_{J \in \mathcal{J}}$ *which exhibits* $\mathcal{C}$ *as a homotopy limit of the diagram* $\{\mathcal{C}_J\}_{J \in \mathcal{J}}$ *in* $\mathcal{C}\mathrm{at}_{\mathbf{S}}$*. Then for every pair of objects* $X, Y \in \mathcal{C}$*, the maps* $\{\mathrm{Map}_{\mathcal{C}}(X, Y) \to \mathrm{Map}_{\mathcal{C}_J}(f_J X, f_J Y)\}_{J \in \mathcal{J}}$ *exhibit* $\mathrm{Map}_{\mathcal{C}}(X, Y)$ *as a homotopy limit of the diagram* $\{\mathrm{Map}_{\mathcal{C}_J}(f_J X, f_J Y)\}_{J \in \mathcal{J}}$ *in* **S***.*

Proof. Without loss of generality, we may assume that the diagram $\{\mathcal{C}_J\}_{J \in \mathcal{J}}$ is injectively fibrant and that the maps f_J exhibit $\mathcal{C}$ as a limit of $\{\mathcal{C}_J\}_{J \in \mathcal{J}}$. It follows that $\mathrm{Map}_{\mathcal{C}}(X, Y)$ is a limit of the diagram $\{\mathrm{Map}_{\mathcal{C}_J}(f_J X, f_J Y)\}_{J \in \mathcal{J}}$. It will therefore suffice to show that the diagram $\{\mathrm{Map}_{\mathcal{C}_J}(f_J X, f_J Y)\}_{J \in \mathcal{J}}$ is injectively fibrant. For this, it will suffice to show that $\{\mathrm{Map}_{\mathcal{C}_J}(f_J X, f_J Y)\}_{J \in \mathcal{J}}$ has the right lifting property with respect to every weak trivial cofibration $\alpha : F \to F'$ of diagrams $F, F' : \mathcal{J} \to \mathbf{S}$. Let $G : \mathcal{J} \to \mathcal{C}\mathrm{at}_{\mathbf{S}}$ be defined by the formula $G(J) = [1]_{F(J)}$ and let $G' : \mathcal{J} \to \mathcal{C}\mathrm{at}_{\mathbf{S}}$ be defined likewise. The desired result now follows from the observation that α induces a weak trivial cofibration $G \to G'$ in $\mathrm{Fun}(\mathcal{J}, \mathcal{C}\mathrm{at}_{\mathbf{S}})$. $\qquad\square$

Corollary A.3.2.28. *Let* **S** *be an excellent model category,* $\mathcal{J}$ *a small category, and* $\{\mathcal{C}_J\}_{J \in \mathcal{J}}$ *a diagram of* **S***-enriched categories. Suppose we are given* **S***-enriched functors*

$$\mathcal{D} \xrightarrow{\beta} \mathcal{C} \xrightarrow{\alpha} \lim\{\mathcal{C}_J\}_{J \in \mathcal{J}}$$

such that $\alpha \circ \beta$ *exhibits* $\mathcal{D}$ *as a homotopy limit of the diagram* $\{\mathcal{C}_J\}_{J \in \mathcal{J}}$*. Then the following conditions are equivalent:*

(1) *The functor* α *exhibits* $\mathcal{C}$ *as a homotopy limit of the diagram* $\{\mathcal{C}_J\}_{J \in \mathcal{J}}$*.*

(2) *For every pair of objects* $X, Y \in \mathcal{C}$*, the functor* α *exhibits* $\mathrm{Map}_{\mathcal{C}}(X, Y)$ *as a homotopy limit of the diagram* $\{\mathrm{Map}_{\mathcal{C}_J}(\alpha_J X, \alpha_J Y)\}_{J \in \mathcal{J}}$*.*

Proof. The implication (1) $\Rightarrow$ (2) follows from Proposition A.3.2.27. To prove the converse, we may assume that the diagram $\{\mathcal{C}_J\}_{J \in \mathcal{J}}$ is injectively fibrant. In view of (2), Proposition A.3.2.27 implies that α induces a fully faithful functor between h**S**-enriched homotopy categories. It will therefore suffice to show that α is essentially surjective on homotopy categories, which follows from our assumption that $\alpha \circ \beta$ is a weak equivalence. $\qquad\square$

A.3.3 Model Structures on Diagram Categories

In this section, we consider enriched analogues of the constructions presented in §A.2.8. Namely, suppose that **S** is an excellent model category, **A** a combinatorial **S**-enriched model category, and $\mathcal{C}$ a small **S**-enriched category. Let $\mathbf{A}^{\mathcal{C}}$ denote the category of **S**-enriched functors from $\mathcal{C}$ to **A**. In this section, we will study the associated projective and injective model structures on $\mathbf{A}^{\mathcal{C}}$. The ideas described here will be used in §A.3.4 to construct certain mapping objects in $\mathcal{C}\mathrm{at}_{\mathbf{S}}$.

We begin with the analogue of Definition A.2.8.1.

Definition A.3.3.1. Let $\mathcal{C}$ be a small **S**-category and **A** a combinatorial **S**-enriched model category. A natural transformation $\alpha : F \to G$ in $\mathbf{A}^{\mathcal{C}}$ is:

- an *injective cofibration* if the induced map $F(C) \to G(C)$ is a cofibration in **A** for each $C \in \mathcal{C}$.

- a *projective fibration* if the induced map $F(C) \to G(C)$ is a fibration in **A** for each $C \in \mathcal{C}$.

- a *weak equivalence* if the induced map $F(C) \to G(C)$ is a weak equivalence in **A** for each $C \in \mathcal{C}$.

- an *injective fibration* if it has the right lifting property with respect to every morphism β in $\mathbf{A}^{\mathcal{C}}$ which is simultaneously a weak equivalence and a injective cofibration.

- a *projective cofibration* if it has the left lifting property with respect to every morphism β in $\mathbf{A}^{\mathcal{C}}$ which is simultaneously a weak equivalence and a projective fibration.

Proposition A.3.3.2. *Let* **S** *be an excellent model category, let* **A** *be a combinatorial* **S**-*enriched model category, and let* $\mathcal{C}$ *be a small* **S**-*enriched category. Then there exist two combinatorial model structures on* $\mathbf{A}^{\mathcal{C}}$:

- *The projective model structure determined by the strong cofibrations, weak equivalences, and projective fibrations.*

- *The injective model structure determined by the weak cofibrations, weak equivalences, and injective fibrations.*

The proof of Proposition A.3.3.2 is identical to that of Proposition A.2.8.2, except that it requires the following more general form of Lemma A.2.8.3:

Lemma A.3.3.3. *Let* **A** *be a presentable category which is enriched, tensored, and cotensored over a presentable category* **S**. *Let* S_0 *be a (small) set of morphisms of* **A** *and let* $\overline{S}_0$ *be the weakly saturated class of morphisms generated by* S_0. *Let* $\mathcal{C}$ *be a small* **S**-*enriched category. Let* $\widetilde{S}$ *be the collection of all morphisms* $F \to G$ *in* $\mathbf{A}^{\mathcal{C}}$ *with the following property: for every* $C \in \mathcal{C}$, *the map* $F(C) \to G(C)$ *belongs to* $\overline{S}_0$. *Then there exists a (small) set of morphisms* S *of* $\mathbf{A}^{\mathcal{C}}$ *which generates* $\widetilde{S}$ (*as a weakly saturated class of morphisms*).

We will defer the proof until the end of this section.

Remark A.3.3.4. In the situation of Proposition A.3.3.2, the category $\mathbf{A}^{\mathcal{C}}$ is again enriched, tensored, and cotensored over **S**. The tensor product with an object $K \in \mathbf{S}$ is computed pointwise; in other words, if $\mathcal{F} \in \mathbf{A}^{\mathcal{C}}$, then we have the formula

$$(K \otimes \mathcal{F})(A) = K \otimes \mathcal{F}(A).$$

Using criterion $(2')$ of Remark A.3.1.6, we deduce that $\mathbf{A}^{\mathcal{C}}$ is an **S**-enriched model category with respect to the injective model structure. A dual argument (using criterion $(2'')$ of Remark A.3.1.6) shows that $\mathbf{A}^{\mathcal{C}}$ is also an **S**-enriched model category with respect to the projective model structure.

Remark A.3.3.5. For each object $C \in \mathcal{C}$ and each $A \in \mathbf{A}$, let $\mathcal{F}_A^C \in \mathbf{A}^{\mathcal{C}}$ be the functor given by

$$D \mapsto A \otimes \operatorname{Map}_{\mathcal{C}}(C, D).$$

As in the proof of Proposition A.2.8.2, we learn that the class of projective cofibrations in $\mathbf{A}^{\mathcal{C}}$ is generated by cofibrations of the form $j : \mathcal{F}_A^C \to \mathcal{F}_{A'}^C$, where $A \to A'$ is a cofibration in $\mathbf{A}$. It follows that every projective cofibration is a injective cofibration; dually, every injective fibration is a projective fibration.

As in §A.2.8, the construction $(\mathcal{C}, \mathbf{A}) \mapsto \mathbf{A}^{\mathcal{C}}$ is functorial in both $\mathcal{C}$ and $\mathbf{A}$. We summarize the situation in the following propositions, whose proofs are left to the reader:

Proposition A.3.3.6. *Let* **S** *be an excellent model category,* $\mathcal{C}$ *a small* **S***-enriched model category, and* $\mathbf{A} \underset{G}{\overset{F}{\rightleftarrows}} \mathbf{S}$ *an* **S***-enriched Quillen adjunction between combinatorial* **S***-enriched model categories. The composition with F and G determines another* **S***-enriched Quillen adjunction*

$$\mathbf{A}^{\mathcal{C}} \underset{G^{\mathcal{C}}}{\overset{F^{\mathcal{C}}}{\rightleftarrows}} \mathbf{B}^{\mathcal{C}}$$

with respect to either the projective or the injective model structure. Moreover, if (F, G) is a Quillen equivalence, then $(F^{\mathcal{C}}, G^{\mathcal{C}})$ is also a Quillen equivalence.

Because the projective and injective model structures on $\mathbf{A}^{\mathcal{C}}$ have the same weak equivalences, the identity functor $\operatorname{id}_{\mathbf{A}^{\mathcal{C}}}$ is a Quillen equivalence between them. However, it is important to distinguish between these two model structures because they have different variance properties as we now explain.

Let $f : \mathcal{C} \to \mathcal{C}'$ be an **S**-enriched functor. Then composition with f yields a pullback functor $f^* : \mathbf{A}^{\mathcal{C}'} \to \mathbf{A}^{\mathcal{C}}$. Since $\mathbf{A}$ has all **S**-enriched limits and colimits, f^* has a right adjoint, which we will denote by f_*, and a left adjoint, which we will denote by $f_!$.

Proposition A.3.3.7. *Let* **S** *be an excellent model category,* $\mathbf{A}$ *a combinatorial* **S***-enriched model category, and $f : \mathcal{C} \to \mathcal{C}'$ an* **S***-enriched functor between small* **S***-enriched categories. Let $f^* : \mathbf{A}^{\mathcal{C}'} \to \mathbf{A}^{\mathcal{C}}$ be given by composition with f. Then f^* admits a right adjoint f_* and a left adjoint $f_!$. Moreover:*

(1) *The pair $(f_!, f^*)$ determines a Quillen adjunction between the* projective *model structures on $\mathbf{A}^{\mathcal{C}}$ and $\mathbf{A}^{\mathcal{C}'}$.*

(2) *The pair (f^*, f_*) determines a Quillen adjunction between the* injective *model structures on* $\mathbf{A}^{\mathcal{C}}$ *and* $\mathbf{A}^{\mathcal{C}'}$.

We now study some aspects of the theory which are unique to the enriched context.

Proposition A.3.3.8. *Let* $\mathbf{S}$ *be an excellent model category,* $\mathbf{A}$ *a combinatorial* $\mathbf{S}$-*enriched model category, and* $f : \mathcal{C} \to \mathcal{C}'$ *an equivalence of small* $\mathbf{S}$-*enriched categories. Then*

(1) *The Quillen adjunction* $(f_!, f^*)$ *determines a Quillen equivalence between the projective model structures on* $\mathbf{A}^{\mathcal{C}}$ *and* $\mathbf{A}^{\mathcal{C}'}$.

(2) *The Quillen adjunction* (f^*, f_*) *determines a Quillen equivalence between the injective model structures on* $\mathbf{A}^{\mathcal{C}}$ *and* $\mathbf{A}^{\mathcal{C}'}$.

Proof. We first note that (1) and (2) are equivalent: they are both equivalent to the assertion that f^* induces an equivalence on homotopy categories. It therefore suffices to prove (1). We first prove this under the following additional assumption:

(∗) For every pair of objects $C, D \in \mathcal{C}'$, the map
$$\mathrm{Map}_{\mathcal{C}'}(C, D) \to \mathrm{Map}_{\mathcal{C}}(f(C), f(D))$$
is a cofibration in $\mathbf{S}$.

Let $Lf_! : \mathbf{A}^{\mathcal{C}} \to \mathbf{A}^{\mathcal{C}'}$ denote the left derived functor of $f_!$. We must show that the unit and counit maps
$$h_F : F \mapsto f^* L f_! F$$
$$k_G : L f_! f^* G \to G$$
are isomorphisms for all $F \in \mathrm{h}\mathbf{A}^{\mathcal{C}}$, $G \in \mathrm{h}\mathbf{A}^{\mathcal{C}}$. Since f is essentially surjective on homotopy categories, a natural transformation $K \to K'$ of $\mathbf{S}$-enriched functors $K, K' : \mathcal{C}' \to \mathbf{A}$ is a weak equivalence if and only if $f^* K \to f^* K'$ is a weak equivalence. Consequently, to prove k_G is an isomorphism, it suffices to show that $h_{f^* G}$ is an isomorphism.

Let us say that a map $F \to F'$ in $\mathbf{A}^{\mathcal{C}}$ is *good* if the induced map
$$f^* f_! F \coprod_F F' \to f^* f_! F'$$
is a weak trivial cofibration. To complete the proof, it will suffice to show that every projective cofibration is good. Since the collection of good transformations is weakly saturated, it will suffice to show that each of the generating cofibrations $\mathcal{F}_A^C \to \mathcal{F}_{A'}^C$ is good, where $C \in \mathcal{C}'$ and $j : A \to A'$ is a cofibration in $\mathbf{A}$. Unwinding the definitions, we must show that for each $D \in \mathcal{C}'$ the induced map
$$A' \otimes \mathrm{Map}_{\mathcal{C}'}(C, D)) \coprod_{A \otimes \mathrm{Map}_{\mathcal{C}'}(C, D)} (A \otimes \mathrm{Map}_{\mathcal{C}}(f(C), f(D)))$$
$$\downarrow{\theta}$$
$$A' \otimes \mathrm{Map}_{\mathcal{C}}(f(C), f(D))$$

is a trivial cofibration. This follows from the fact that j is a cofibration and our assumption $(*)$.

We now treat the general case. First, choose a trivial cofibration $g : \mathcal{C} \to \mathcal{C}''$, where $\mathcal{C}''$ is fibrant. Then g satisfies $(*)$, so $g_!$ is a Quillen equivalence. By a two-out-of-three argument, we see that $f_!$ is a Quillen equivalence if and only if $(g \circ f)_!$ is a Quillen equivalence. Replacing $\mathcal{C}$ by $\mathcal{C}''$, we may reduce to the case where $\mathcal{C}$ is itself fibrant.

Choose a cofibration $j : \mathcal{C} \coprod \mathcal{C}' \to \mathcal{D}$, where $\mathcal{D}$ is fibrant and equivalent to the final object of $\mathcal{C}\mathrm{at}_\mathbf{S}$. Then f factors as a composition

$$\mathcal{C}' \xrightarrow{f'} \mathcal{C} \times \mathcal{D} \xrightarrow{f''} \mathcal{C}.$$

Since $\mathcal{C}$ and $\mathcal{D}$ are fibrant, the product $\mathcal{C} \times \mathcal{D}$ is equivalent to $\mathcal{C}$. Moreover, the map f'' admits a section $s : \mathcal{C} \to \mathcal{C} \times \mathcal{D}$. Using another two-out-of-three argument, it will suffice to show that $f'_!$ and $s_!$ are Quillen equivalences. For this, it will suffice to show that f' and s satisfy $(*)$.

We first show that f' satisfies $(*)$. Fix a pair of objects $X, Y \in \mathcal{C}'$. Then f' induces the composite map

$$\mathrm{Map}_{\mathcal{C}'}(X, Y) \xrightarrow{u} \mathrm{Map}_{\mathcal{C}}(fX, fY) \times \mathrm{Map}_{\mathcal{C}'}(X, Y)$$
$$\xrightarrow{u'} \mathrm{Map}_{\mathcal{C}}(fX, fY) \times \mathrm{Map}_{\mathcal{D}}(jX, jY)$$
$$\simeq \mathrm{Map}_{\mathcal{C} \times \mathcal{D}}(f'X, f'Y).$$

The map u is a monomorphism (since it admits a left inverse) and therefore a cofibration in view of axiom $(A2)$ of Definition A.3.2.16. The map u' is a product of cofibrations and therefore also a cofibration (again by axiom $(A2)$).

The proof that s satisfies $(*)$ is similar: for every pair of objects $U, V \in \mathcal{C}$, the map

$$\mathrm{Map}_{\mathcal{C}}(U, V) \to \mathrm{Map}_{\mathcal{C} \times \mathcal{D}}(sU, sV) \simeq \mathrm{Map}_{\mathcal{C}}(U, V) \times \mathrm{Map}_{\mathcal{D}}(jU, jV)$$

is a monomorphism since it admits a left inverse and is therefore a cofibration. $\qquad\square$

In the special case where $f : \mathcal{C} \to \mathcal{C}'$ is a *cofibration* between $\mathbf{S}$-enriched categories, we have some additional functoriality:

Proposition A.3.3.9. *Let $\mathbf{S}$ be an excellent model category and let $f : \mathcal{C} \to \mathcal{C}'$ be a cofibration of small $\mathbf{S}$-enriched categories. Then*

(1) *For every combinatorial $\mathbf{S}$-enriched model category $\mathbf{A}$, the pullback map $f^* : \mathbf{A}^{\mathcal{C}'} \to \mathbf{A}^{\mathcal{C}}$ preserves projective cofibrations.*

(2) *For every projectively cofibrant object $F \in \mathbf{S}^{\mathcal{C}}$, the unit map $F \to f^* f_! F$ is a projective cofibration.*

Lemma A.3.3.10. *Let* **S** *be an excellent model category and suppose we are given a pushout diagram*

$$
\begin{array}{ccc}
[1]_S & \longrightarrow & [1]_{S'} \\
\downarrow{\scriptstyle i} & & \downarrow \\
\mathcal{C} & \xrightarrow{\ f\ } & \mathcal{C}'
\end{array}
$$

of **S**-*enriched categories, where* $j : S \to S'$ *is a cofibration in* **S**. *Let* C *be an object of* $\mathcal{C}$ *and let* $F \in \mathbf{S}^{\mathcal{C}}$ *be the functor given by the formula* $D \mapsto \mathrm{Map}_{\mathcal{C}}(C, D)$. *Then the unit map* $F \to f^* f_! F$ *is a projective cofibration in* $\mathbf{S}^{\mathcal{C}}$.

Proof. The map i determines a pair of objects $X, Y \in \mathcal{C}$ and a map $S \to \mathrm{Map}_{\mathcal{C}}(X, Y)$. The proof of Proposition A.3.2.4 shows that the functor $f^* f_! F$ is the colimit of a sequence

$$ F = F(0) \xrightarrow{h_1} F(1) \xrightarrow{h_2} F(2) \to \cdots, $$

where each h_k is a pushout of a map $\mathcal{F}_A^Y \to \mathcal{F}_{A'}^Y$ induced by a map $t : A \to A'$ in **S**. Moreover, the map t can be identified with the tensor product

$$ \mathrm{id}_{\mathrm{Map}_{\mathcal{C}}(C,X)} \otimes \mathrm{id}_{\mathrm{Map}_{\mathcal{C}}(Y,X)}^{\otimes k-1} \otimes \wedge^k (j), $$

where $\wedge^k(j)$ denotes the kth pushout power of j. It follows that t is a cofibration in **S**, so that each h_k is a projective cofibration in $\mathbf{S}^{\mathcal{C}}$. ☐

Proof of Proposition A.3.3.9. The collection of **S**-enriched functors f which satisfy (1) and (2) is clearly closed under the formation of retracts. We may therefore assume without loss of generality that f is a transfinite composition of pushouts of generating cofibrations (see the discussion preceding Proposition A.3.2.4). Reordering these pushouts if necessary, we can factor f as a composition

$$ \mathcal{C} \xrightarrow{f'} \overline{\mathcal{C}} \xrightarrow{f''} \mathcal{C}', $$

where $\overline{\mathcal{C}}$ is obtained from $\mathcal{C}$ by freely adjoining a collection of new objects and f'' is bijective on objects. Since f' clearly satisfies (1) and (2), it will suffice to prove that f'' satisfies (1) and (2). Replacing f by f'', we may assume that f is bijective on objects.

We now show that $(2) \Rightarrow (1)$. Since the functor f^* preserves colimits, the collection of morphisms α in $\mathbf{A}^{\mathcal{C}'}$ such that f^* is a projective cofibration in $\mathbf{A}^{\mathcal{C}}$ is weakly saturated. It will therefore suffice to show that for every object $X \in \mathcal{C}'$ and every cofibration $A \to A'$ in $\mathbf{A}$, if $\alpha : \mathcal{F}_A^X \to \mathcal{F}_{A'}^X$ denotes the corresponding generating projective cofibration, then $f^*(\alpha)$ is a projective cofibration in **S**.

There is a canonical left Quillen bifunctor

$$ \boxtimes : \mathbf{S}^{\mathcal{C}} \times \mathbf{A} \to \mathbf{A}^{\mathcal{C}} $$

described by the formula $(F \boxtimes A)(C) = F(C) \otimes A$. (Here we regard $\mathbf{S}^{\mathcal{C}}$ as endowed with the projective model structure.) We observe that $f^*(\alpha)$

is the induced map $(f^*F) \boxtimes A \to (f^*F) \boxtimes A'$, where $F \in \mathbf{S}^{\mathcal{C}'}$ is given by $F(C') = \mathrm{Map}_{\mathcal{C}'}(X, C')$. To prove (1), it will suffice to show that f^*F is projectively cofibrant.

Since F is bijective on objects, we can choose an object $X_0 \in \mathcal{C}$ such that $fX_0 = X$. We now observe that $F \simeq f_! F_0$, where $F_0 \in \mathbf{S}^{\mathcal{C}}$ is defined by the formula $F_0(C) = \mathrm{Map}_{\mathcal{C}}(X_0, C)$. If (2) is satisfied, then the unit map $F_0 \to f^*F$ is a projective cofibration in $\mathbf{S}^{\mathcal{C}}$. Since F_0 is projectively cofibrant, we conclude that f^*F is projectively cofibrant as well. This completes the proof that $(2) \Rightarrow (1)$.

To prove (2), let us write f as a transfinite composition of $\mathbf{S}$-enriched functors

$$\mathcal{C} = \mathcal{C}_0 \to \mathcal{C}_1 \to \cdots,$$

each of which is a pushout of a generating cofibration of the form $[1]_S \to [1]_{S'}$, where $S \to S'$ is a cofibration in $\mathbf{S}$. For each $\alpha \leq \beta$, let $f_\alpha^\beta : \mathcal{C}_\alpha \to \mathcal{C}_\beta$ be the corresponding cofibration. We will prove that the following statement holds for every pair of ordinals $\alpha \leq \beta$:

$(2_{\alpha,\beta})$ For every projectively cofibrant object $F \in \mathbf{S}^{\mathcal{C}_\alpha}$, the unit map $u : F \to (f_\alpha^\beta)^*(f_\alpha^\beta)_! F$ is a projective cofibration.

The proof proceeds by induction on β. We observe that u is a transfinite composition of maps of the form

$$u_\gamma : (f_\alpha^\gamma)^*(f_\alpha^\gamma)_! F \to (f_\alpha^\gamma)^*(f_\gamma^{\gamma+1})^*(f_\gamma^{\gamma+1})_! (f_\alpha^\gamma)_! F,$$

where $\gamma < \beta$. It will therefore suffice to show that each u_γ is a projective cofibration. Our inductive hypothesis therefore guarantees that $(2_{\alpha,\gamma})$ holds, so the first part of the proof shows that $(f_\alpha^\gamma)^*$ preserves trivial cofibrations. We are therefore reduced to proving assertion $(2_{\gamma,\gamma+1})$. In other words, to prove (2) in general, it will suffice to treat the case in which f is a pushout of a generating cofibration of the form $[1]_S \to [1]_{S'}$.

We will in fact prove the following stronger version of (2):

(3) For every projective cofibration $\phi : F' \to F$ in $\mathbf{S}^{\mathcal{C}}$, the induced map $\phi' : F \coprod_{F'} f^*f_! F' \to f^*f_! F$ is again a projective cofibration in $\mathbf{S}^{\mathcal{C}}$.

Consider the collection of *all* morphisms $\phi : F' \to F$ in $\mathbf{S}^{\mathcal{C}}$ such that the induced map $\phi' : F \coprod_{F'} f^*f_! F' \to f^*f_! F$ is a projective cofibration. It is easy to see that this collection is weakly saturated. Consequently, to prove (3) it suffices to treat the case where ϕ is a generating projective cofibration of the form $\mathcal{F}_A^C \to \mathcal{F}_{A'}^C$, where $A \to A'$ is a cofibration in $\mathbf{S}$. In this case, we can identify ϕ' with the map

$$(F_C \boxtimes A') \coprod_{F_C \boxtimes A} (f^*f_! F_C \boxtimes A) \to f^*f_! F_C \boxtimes A',$$

where $F_C \in \mathbf{S}^{\mathcal{C}}$ is the functor $D \mapsto \mathrm{Map}_{\mathcal{C}}(C, D)$. Since $\boxtimes$ is a left Quillen bifunctor, it will suffice to show that the unit map $f_C \to f^*f_! F_C$ is a projective cofibration in $\mathbf{S}^{\mathcal{C}}$. This is precisely the content of Lemma A.3.3.10. $\square$

In §A.2.8, we introduced the definitions of homotopy limits and colimits in an arbitrary combinatorial model category $\mathbf{A}$. We now discuss an analogous construction in the case where $\mathbf{A}$ is enriched over an excellent model category $\mathbf{S}$. To simplify the exposition, we will discuss only the case of homotopy limits; the case of homotopy colimits is entirely dual and is left to the reader.

Fix an excellent model category $\mathbf{S}$ and a combinatorial $\mathbf{S}$-enriched model category $\mathbf{A}$. Let $f : \mathcal{C} \to \mathcal{C}'$ be a functor between small $\mathbf{S}$-enriched categories, so that we have an induced Quillen adjunction

$$\mathbf{A}^{\mathcal{C}'} \underset{f_*}{\overset{f^*}{\rightleftarrows}} \mathbf{A}^{\mathcal{C}}.$$

We will refer to the right derived functor Rf_* as the *homotopy right Kan extension* functor. Suppose we are given a pair of functors $F \in \mathbf{A}^{\mathcal{C}}$, $G \in \mathbf{A}^{\mathcal{C}'}$ and a morphism $\eta : G \to f_* F$ in $\mathbf{A}^{\mathcal{C}'}$. We will say that η *exhibits G as the homotopy right Kan extension of F* if, for some weak equivalence $F \to F'$ where F' is injectively fibrant in $\mathbf{A}^{\mathcal{C}}$, the composite map $G \to f_* F \to f_* F'$ is a weak equivalence in $\mathbf{A}^{\mathcal{C}'}$. Since f_* preserves weak equivalences between injectively fibrant objects, this condition is independent of the choice of F'.

Remark A.3.3.11. In §A.2.8, we defined homotopy right Kan extensions in the setting of the diagram categories $\mathrm{Fun}(\mathcal{C}, \mathbf{A})$, where $\mathcal{C}$ is an ordinary category. In fact, this is a special case of the above construction. Namely, there is a unique colimit-preserving monoidal functor $F : \mathrm{Set} \to \mathbf{S}$ given by $F(S) = \coprod_{s \in S} \mathbf{1_S}$. We can therefore define an $\mathbf{S}$-enriched category $\overline{\mathcal{C}}$ whose objects are the objects of $\mathcal{C}$, with $\mathrm{Map}_{\overline{\mathcal{C}}}(X, Y) = F \mathrm{Map}_{\mathcal{C}}(X, Y)$. We now observe that we have an identification $\mathrm{Fun}(\mathcal{C}, \mathbf{A}) \simeq \mathbf{A}^{\overline{\mathcal{C}}}$ which is functorial in both $\mathcal{C}$ and $\mathbf{A}$. This identification is compatible with the definition of the injective model structures on both sides, so that either point of view gives rise to the same theory of homotopy right Kan extensions.

We now discuss a special feature of the enriched theory of homotopy Kan extensions: they can be reduced to the theory of homotopy Kan extensions in the model category $\mathbf{S}$:

Proposition A.3.3.12. *Let $\mathbf{S}$ be an excellent model category, let $\mathbf{A}$ be a combinatorial model category enriched over $\mathbf{S}$, and let $f : \mathcal{C} \to \mathcal{C}'$ be a functor between small $\mathbf{S}$-enriched categories. Suppose given objects $F \in \mathbf{A}^{\mathcal{C}}$, $G \in \mathbf{A}^{\mathcal{C}'}$ and a map $\eta : G \to f_* F$. Assume that F and G are projectively fibrant. The following conditions are equivalent:*

(1) *The map η exhibits G as a homotopy right Kan extension of F.*

(2) *For each cofibrant object $A \in \mathbf{A}$, the induced map*

$$\eta_A : G_A \to f_* F_A$$

exhibits G_A as a homotopy right Kan extension of F_A. Here $F_A \in \mathbf{S}^{\mathcal{C}}$ and $G_A \in \mathbf{S}^{\mathcal{C}'}$ are defined by $F_A(C) = \mathrm{Map}_{\mathbf{A}}(A, F(C)), G_A(C) = \mathrm{Map}_{\mathbf{A}}(A, G(C))$.

(3) *For every fibrant-cofibrant object $A \in \mathbf{A}$, the induced map*

$$\eta_A : G_A \to f_* F_A$$

exhibits G_A as a homotopy right Kan extension of F_A.

Proof. Choose an equivalence $F \to F'$, where F' is injectively fibrant. We note that the induced maps $F_A \to F'_A$ are weak equivalences for any cofibrant $A \in \mathbf{A}$ since $\operatorname{Map}_{\mathbf{A}}(A, \bullet)$ preserves weak equivalences between fibrant objects. Consequently, we may without loss of generality replace F by F' and thereby assume that F is injectively fibrant.

Now suppose that A is any cofibrant object of $\mathbf{A}$; we claim that F_A is injectively fibrant. To show that F_A has the right lifting property with respect to a trivial weak cofibration $H \to H'$ of functors $\mathcal{C} \to \mathbf{S}$, one need only observe that F has the right lifting property with respect to trivial injective cofibration $A \otimes H \to A \otimes H'$ in $\mathbf{A}^{\mathcal{C}}$.

Now we note that (1) is equivalent to the assertion that η is a weak equivalence, (2) is equivalent to the assertion that η_A is a weak equivalence for any cofibrant object A, and (3) is equivalent to the assertion that η_A is a weak equivalence whenever A is fibrant-cofibrant. Because $\operatorname{Map}_{\mathbf{A}}(A, \bullet)$ preserves weak equivalences between fibrant objects, we deduce that $(1) \Rightarrow (2)$. It is clear that $(2) \Rightarrow (3)$. We will complete the proof by showing that $(3) \Rightarrow (1)$. Assume that (3) holds; we must show that $\eta(C') : G(C') \to f_* F(C')$ is an isomorphism in the homotopy category $h\mathbf{A}$ for each $C' \in \mathcal{C}'$. For this, it suffices to show that $G(C')$ and $f_* F(C')$ represent the same $\mathcal{H}$-valued functors on the homotopy category $h\mathbf{A}$, which is precisely the content of (3). □

Remark A.3.3.13. It follows from Proposition A.3.3.12 that we can make sense of homotopy right Kan extensions for diagrams which do not take values in a model category. Let $f : \mathcal{C} \to \mathcal{C}'$ be an **S**-enriched functor as in the discussion above and let $\mathcal{A}$ be an *arbitrary* locally fibrant **S**-enriched category. Suppose we are given objects $F \in \mathcal{A}^{\mathcal{C}}$, $G \in \mathcal{A}^{\mathcal{C}'}$ and $\eta : f^* G \to F$; we say that η *exhibits G as a homotopy right Kan extension of F* if, for each object $A \in \mathcal{A}$, the induced map

$$\eta_A : G_A \to f_* F_A$$

exhibits $G_A \in \mathbf{S}^{\mathcal{C}'}$ as a homotopy right Kan extension of $F_A \in \mathbf{S}^{\mathcal{C}}$.

Suppose that the monoidal structure on **S** is given by the Cartesian product and take $\mathcal{C}'$ to be the final object of $\mathcal{C}at_{\mathbf{S}}$, so that we can identify $\mathcal{A}^{\mathcal{C}'}$ with $\mathcal{A}$. In this case, we can identify G with a single object $B \in \mathcal{A}$ and the map η with a collection of maps $\{B \to F(C)\}_{C \in \mathcal{C}}$. We will say that η *exhibits B as a homotopy limit of F* if it identifies G with a homotopy right Kan extension of F. In other words, η exhibits B as a homotopy limit of F if, for every object $A \in \mathcal{A}$, the induced map

$$\operatorname{Map}_{\mathcal{A}}(A, B) \to \lim\{\operatorname{Map}_{\mathcal{A}}(A, F(C))\}_{C \in \mathcal{C}}$$

exhibits $\mathrm{Map}_A(A, B)$ as a homotopy limit of the diagram

$$\{\mathrm{Map}_A(A, F(C))\}_{C \in \mathcal{C}}$$

in the model category **S**.

We also have a dual notion of *homotopy colimit* in an arbitrary fibrant **S**-enriched category A: a compatible family of maps $\{F(C) \to B\}_{C \in \mathcal{C}}$ *exhibits B as a homotopy colimit of F* if, for every object $A \in A$, the induced maps $\{\mathrm{Map}_A(B, A) \to \mathrm{Map}_A(F(C), A)\}_{C \in \mathcal{C}}$ exhibit $\mathrm{Map}_A(B, A)$ as a homotopy limit of the diagram $\{\mathrm{Map}_A(F(C), A)\}_{C \in \mathcal{C}}$ in **S**.

Remark A.3.3.14. In view of Proposition A.3.3.12, the terminology introduced in Remark A.3.3.13 for general A agrees with the terminology introduced for a combinatorial **S**-enriched model category **A** if we set $A = \mathbf{A}^\circ$. We remark that, in general, the two notions do *not* agree if we take $A = \mathbf{A}$, so that our terminology is potentially ambiguous; however, we feel that there is little danger of confusion.

We conclude this section by giving the proof of Lemma A.3.3.3. Let **A** be a presentable category which is enriched, tensored, and cotensored over a presentable category **S**. Let $\mathcal{C}$ be a small **S**-enriched category and let $\overline{S}_0$ be a weakly saturated class of morphisms of **A** generated by a (small) set S_0. We regard this data as *fixed* for the remainder of this section.

Choose a regular cardinal κ satisfying the following conditions:

(i) The cardinal κ is uncountable.

(ii) The category $\mathcal{C}$ has fewer than κ-objects.

(iii) Let $X, Y \in \mathcal{C}$ and let $K = \mathrm{Map}_{\mathcal{C}}(X, Y)$. Then the functor from **A** to itself given by the formula $A \mapsto A^K$ preserves κ-filtered colimits. This implies, in particular, that the collection of κ-compact objects of **A** is stable with respect to the functors $\bullet \otimes K$.

(iv) The category **A** is κ-accessible. It follows also that $\mathbf{A}^{\mathcal{C}}$ is κ-accessible, and that an object $F \in \mathbf{A}^{\mathcal{C}}$ is κ-compact if and only if each $F(C) \in \mathbf{A}$ is κ-compact. We prove an ∞-category generalization of this statement as Proposition 5.4.4.3. The same proof also works in the setting of ordinary categories.

(v) The source and target of every morphism in S_0 is a κ-compact object of **A**.

Enlarging S_0 if necessary, we may assume that S_0 consists of *all* morphisms in $f \in \overline{S}_0$ such that the source and target of f are κ-compact. Let S be the collection of all injective cofibrations between κ-compact objects of **A** (in view of (iv), we can equally well define S to be the set of morphisms $F \to G$ in $\mathbf{A}^{\mathcal{C}}$ such that each of the induced morphisms $F(C) \to G(C)$ belongs to S_0). Let $\overline{S}$ be the weakly saturated class of morphisms in $\mathbf{A}^{\mathcal{C}}$ generated by S and choose a map $f : F \to G$ in $\mathbf{A}^{\mathcal{C}}$ such that $f(C) \in \overline{S}_0$ for each $C \in \mathcal{C}$. We wish to show that $f \in \overline{S}$. Corollary A.1.5.13 implies that, for each $C \in \mathcal{C}$, there exists a κ-good S_0-tree $\{Y(C)_\alpha\}_{\alpha \in A(C)}$ with root $F(C)$ and colimit $G(C)$.

Let us define a *slice* to be the following data:

(a) For each object $C \in \mathcal{C}$, a downward-closed subset $B(C) \subseteq A(C)$.

(b) For every object $C \in \mathcal{C}$, a morphism

$$\eta_C : \coprod_{C' \in \mathcal{C}} Y(C')_{B(C')} \otimes \mathrm{Map}_{\mathbf{A}}(C', C) \to Y(C)_{B(C)},$$

rendering the following diagrams commutative:

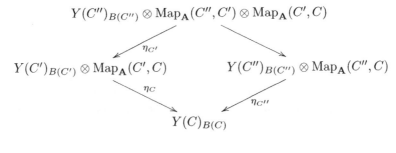

We remark that (b) is precisely the data needed to endow $C \mapsto Y(C)_{B(C)}$ with the structure of an **S**-enriched functor $\mathcal{C} \to \mathbf{A}$ lying between F and G in $\mathbf{A}^{\mathcal{C}}$.

Lemma A.3.3.15. *Suppose we are given a collection of κ-small subsets $\{B_0(C) \subseteq A(C)\}_{C \in \mathcal{C}}$. Then there exists a slice $\{(B(C), \eta_C\}_{C \in \mathcal{C}}$ such that each $B(C)$ is a κ-small subset of $A(C)$ containing $B_0(C)$.*

Proof. Enlarging each $B_0(C)$ if necessary, we may assume that each $B_0(C)$ is downward-closed. Note that because each $\{Y(C)_\alpha\}_{\alpha \in A(C)}$ is a κ-good S_0-tree, if $A' \subseteq A(C)$ is downward-closed and κ-small, $Y(C)_{A'}$ is κ-compact when viewed as an object of $\mathbf{A}_{F(C)/}$. It follows from (iii) that each tensor product $Y(C)_{B_0(C)} \otimes \mathrm{Map}_{\mathbf{A}}(C, C')$ is a κ-compact object of the category $\mathbf{A}_{(F(C) \otimes \mathrm{Map}_{\mathbf{A}}(C,C'))/}$. Consequently, each composition

$$\coprod_{C' \in \mathcal{C}} Y(C')_{B_0(C')} \otimes \mathrm{Map}_{\mathbf{A}}(C', C) \to \coprod_{C' \in \mathcal{C}} G(C') \otimes \mathrm{Map}_{\mathbf{A}}(C', C) \to G(C)$$

admits another factorization

$$\coprod_{C' \in \mathcal{C}} Y(C')_{B_0(C')} \otimes \mathrm{Map}_{\mathbf{A}}(C', C) \overset{\eta_C^1}{\to} Y(C)_{B_1(C)} \to G(C),$$

where $B_1(C)$ is downward-closed and κ-small, and the diagram

$$
\begin{array}{ccc}
\coprod_{C' \in \mathcal{C}} F(C') \otimes \mathrm{Map}_{\mathbf{A}}(C', C) & \longrightarrow & \coprod_{C' \in \mathcal{C}} Y(C')_{B_0(C')} \\
\downarrow & & \downarrow {\scriptstyle \eta_C^1} \\
F(C) & \longrightarrow & Y(C)_{B_1(C)}
\end{array}
$$

commutes. Enlarging $B_1(C)$ if necessary, we may suppose that each $B_1(C)$ contains $B_0(C)$.

We now continue the preceding construction by defining, for each $C \in \mathcal{C}$, a sequence of κ-small downward-closed subsets

$$B_0(C) \subseteq B_1(C) \subseteq B_2(C) \subseteq \cdots$$

of $A(C)$ and maps $\eta_C^i : \coprod_{C' \in \mathcal{C}} Y(C')_{B_{i-1}(C')} \otimes \mathrm{Map}_{\mathbf{A}}(C', C) \to Y(C)_{B_i(C)}$. Suppose that $i > 1$ and that the sets $B_j(C)$ and maps η_C^j have been constructed for $j < i$. Using a compactness argument, we conclude that the composition

$$\coprod_{C' \in \mathcal{C}} Y(C')_{B_{i-1}(C')} \otimes \mathrm{Map}_{\mathbf{A}}(C', C) \to \coprod_{C' \in \mathcal{C}} G(C') \otimes \mathrm{Map}_{\mathbf{A}}(C', C) \to G(C)$$

coincides with

$$\coprod_{C' \in \mathcal{C}} Y(C')_{B_{i-1}(C')} \otimes \mathrm{Map}_{\mathbf{A}}(C', C) \overset{\eta_C^i}{\to} Y(C)_{B_i(C)} \to G(C),$$

where $B_i(C)$ is κ-small and the diagram

$$
\begin{array}{ccc}
\coprod_{C' \in \mathcal{C}} F(C') \otimes \mathrm{Map}_{\mathbf{A}}(C', C) & \longrightarrow & \coprod_{C' \in \mathcal{C}} Y(C')_{B_{i-1}(C')} \otimes \mathrm{Map}_{\mathbf{A}}(C', C) \\
\downarrow & & \downarrow {\scriptstyle \eta_C^i} \\
F(C) & \longrightarrow & Y(C)_{B_i(C)}
\end{array}
$$

commutes. Enlarging $B_i(C)$ if necessary, we may suppose that $B_i(C)$ contains $B_{i-1}(C)$ and that the following diagrams commute as well (for all $C', C'' \in \mathcal{C}$):

$$
\begin{array}{c}
Y(C'')_{B_{i-2}(C'')} \otimes \mathrm{Map}_{\mathbf{A}}(C'', C') \otimes \mathrm{Map}_{\mathbf{A}}(C', C) \\
\swarrow \qquad\qquad \searrow \\
Y(C')_{B_{i-1}(C')} \otimes \mathrm{Map}_{\mathbf{A}}(C', C) \qquad\qquad Y(C'')_{B_{i-1}(C'')} \otimes \mathrm{Map}_{\mathbf{A}}(C'', C) \\
\searrow {\scriptstyle \eta_C^i} \qquad\qquad {\scriptstyle \eta_C^i} \swarrow \\
Y(C)_{B_i(C)}
\end{array}
$$

$$Y(C')_{B_{i-2}(C')} \otimes \mathrm{Map}_{\mathbf{A}}(C', C) \longrightarrow Y(C')_{B_{i-1}(C')} \otimes \mathrm{Map}_{\mathbf{A}}(C', C)$$

$$\downarrow \eta_C^{i-1} \qquad\qquad\qquad\qquad\qquad \downarrow \eta_C^i$$

$$Y(C)_{B_{i-1}(C)} \longrightarrow Y(C)_{B_i(C)}.$$

We now define $B(C) = \bigcup B_i(C)$, and η_C to be the amalgam of the compositions

$$\coprod_{C' \in \mathcal{C}} Y(C')_{B_{i-1}(C')} \otimes \mathrm{Map}_{\mathbf{A}}(C', C) \xrightarrow{\eta_C^i} Y(C)_{B_i}(C) \to Y(C)_{B(C)}.$$

$\square$

We now introduce a bit more terminology. Suppose we are given a pair of slices $M = \{(B(C), \eta_C)\}_{C \in \mathcal{C}}$, $M' = \{(B'(C), \eta'_C)_{C \in \mathcal{C}}\}$. We will say that M is κ-small if each $B(C)$ is κ-small. We say that M' extends M if $B(C) \subseteq B'(C)$ for each $C \in \mathcal{C}$ and each diagram

$$Y(C')_{B(C')} \otimes \mathrm{Map}_{\mathbf{A}}(C', C) \longrightarrow Y(C')_{B'(C')} \otimes \mathrm{Map}_{\mathbf{A}}(C', C)$$

$$\downarrow \eta_C \qquad\qquad\qquad\qquad\qquad \downarrow \eta'_C$$

$$Y(C)_{B(C)} \longrightarrow Y(C)_{B'(C)}$$

is commutative. Equivalently, M' extends M if $B(C) \subseteq B'(C)$ for each $C \in \mathcal{C}$, and the induced maps $Y(C)_{B(C)} \to Y(C)_{B(C')}$ constitute a natural transformation of simplicial functors from $\mathcal{C}$ to $\mathbf{A}$.

Lemma A.3.3.16. *Let $M' = \{(A'(C), \theta_C)\}_{C \in \mathcal{C}}$ be a slice and let $\{B_0(C) \subseteq A(C)\}_{C \in \mathcal{C}}$ be a collection of κ-small subsets of $A(C)$. Then there exists a pair of slices $N = \{(B(C), \eta_C)\}_{C \in \mathcal{C}}$, $N' = \{(B(C) \cap A'(C), \eta'_C)\}$, where $B(C)$ is κ-small and N' is compatible with both N and M'.*

Proof. Let $B'_0(C) = A'(C) \cap B_0(C)$. For every positive integer i, we will construct a pair of slices $N_i = \{(B_i(C), \eta(i)_C)\}$, $N'_i = \{(B'_i(C), \eta'(i)_C)\}$ so that the following conditions are satisfied:

(a) Each $B_i(C)$ is κ-small and contains $B_{i-1}(C)$.

(b) Each $B'_i(C)$ is κ-small, contains $B'_{i-1}(C)$ and the intersection $B_i(C) \cap A'(C)$, and is contained in $A'(C)$.

(c) Each N'_i is compatible with M'.

(d) If $i > 2$ and $C, C' \in \mathcal{C}$, then the diagram

$$Y(C')_{B_{i-2}(C')} \otimes \mathrm{Map}_{\mathbf{A}}(C', C) \longrightarrow Y(C')_{B_{i-1}(C')} \otimes \mathrm{Map}_{\mathbf{A}}(C', C)$$

$$\downarrow \eta(i-2)_C \qquad\qquad\qquad\qquad\qquad \downarrow \eta(i-1)_C$$

$$Y(C)_{B_{i-2}(C)} \qquad\qquad\qquad\qquad Y(C)_{B_{i-1}(C)}$$

$$\downarrow \qquad\qquad\qquad\qquad\qquad\qquad \downarrow$$

$$Y(C)_{B_i(C)} =\!=\!=\!=\!=\!=\!=\!= Y(C)_{B_i(C)}$$

commutes.

(e) If $i > 2$ and $C, C' \in \mathcal{C}$, then the diagram

$$Y(C')_{B'_{i-2}(C')} \otimes \mathrm{Map}_{\mathbf{A}}(C', C) \longrightarrow Y(C')_{B'_{i-1}(C')} \otimes \mathrm{Map}_{\mathbf{A}}(C', C)$$

$$\downarrow{\scriptstyle \eta'(i-2)_C} \qquad\qquad\qquad\qquad \downarrow{\scriptstyle \eta('i-1)_C}$$

$$Y(C)_{B'_{i-2}(C)} \qquad\qquad\qquad\qquad Y(C)_{B'_{i-1}(C)}$$

$$\downarrow \qquad\qquad\qquad\qquad\qquad\qquad\qquad \downarrow$$

$$Y(C)_{B'_i(C)} =\!=\!=\!=\!=\!=\!=\!=\!=\!=\!= Y(C)_{B'_i(C)}$$

commutes.

(f) If $i > 1$ and $C, C' \in \mathcal{C}$, then the diagram

$$Y(C')_{B'_{i-1}(C')} \otimes \mathrm{Map}_{\mathbf{A}}(C', C) \longrightarrow Y(C')_{B_{i-1}(C')} \otimes \mathrm{Map}_{\mathbf{A}}(C', C)$$

$$\downarrow{\scriptstyle \eta'(i-1)_C} \qquad\qquad\qquad\qquad \downarrow{\scriptstyle \eta(i-1)_C}$$

$$Y(C)_{B'_{i-1}(C)} \qquad\qquad\qquad\qquad Y(C)_{B_{i-1}(C)}$$

$$\downarrow \qquad\qquad\qquad\qquad\qquad\qquad\qquad \downarrow$$

$$Y(C)_{B'_i(C)} \longrightarrow Y(C)_{B'_i(C)}$$

commutes.

The construction is by induction on i. The existence of N_i satisfying (a), (d), and (f) follows from Lemma A.3.3.15 (and a compactness argument). Similarly, the existence of N'_i satisfying (b), (c), and (e) follows by applying Lemma A.3.3.15 after replacing $G \in \mathbf{A}^{\mathcal{C}}$ by the functor G' given by $G'(C) = Y(C)_{A'(C)}$, and the S_0-trees $\{Y(C)_\alpha\}_{\alpha \in A(C)}$ by the smaller trees $\{Y(C)_\alpha\}_{\alpha \in A'(C)}$.

We now define $B(C) = \bigcup_i B_i(C)$. It follows from (d) that the $\eta(i)_C$ assemble to a map

$$\eta_C : \coprod_{C' \in \mathcal{C}} Y(C')_{B(C')} \otimes \mathrm{Map}_{\mathbf{A}}(C', C) \to Y(C)_{B(C)}.$$

Taken together, these maps determine a slice $N = \{(B(C), \eta_C)\}$. Similarly, (e) implies that the maps $\eta'(i)_C$ assemble to a slice $N' = \{(B(C) \cap A'(C), \eta'_C)\}$. The compatibility of N and N' follows from (f), while the compatibility of M' and N' follows from (c). $\qquad\square$

We now construct a transfinite sequence of compatible slices $\{M(\gamma) = \{(B(\gamma)(C), \eta(\gamma)_C)\}_{C \in \mathcal{C}}\}_{\gamma < \beta}$. The construction is by recursion. Assume that $M(\gamma')$ has been defined for $\gamma' < \gamma$ and let $M'(\gamma) = \{(B'(\gamma)(C), \eta'(\gamma)_C)\}_{C \in \mathcal{C}}$ denote the slice obtained by amalgamating the family of slices $\{M(\gamma')\}_{\gamma' < \gamma}$.

If $B'(\gamma)(C) = A(C)$ for all $C \in \mathcal{C}$, we set $\beta = \gamma$ and conclude the construction. Otherwise, we choose $C \in \mathcal{C}$ and $a \in A(C) - B'(\gamma)(C)$. According to Lemma A.3.3.16, there exists a pair of slices $N(\gamma) = \{(B''(C), \theta_C)\}_{C \in \mathcal{C}}$, $N'(\gamma) = \{(B''(C) \cap B'(\gamma)(C), \theta'_C)\}_{C \in \mathcal{C}}$ such that $N'(\gamma)$ is compatible with both $N(\gamma)$ and $M'(\gamma)$. We now define $M(\gamma)$ to be the slice obtained by amalgamating $M'(\gamma)$ and $N(\gamma)$.

For $\gamma < \beta$, let $G(\gamma) : \mathcal{C} \to \mathbf{A}$ be the simplicial functor corresponding to the slice $M(\gamma)$. Then we have a transfinite sequence of composable morphisms

$$G(0) \to G(1) \to \cdots$$

in $(\mathbf{A}^{\mathcal{C}})_{F/}$ having colimit $G \simeq \varinjlim_{\gamma < \beta} G(\gamma)$. Since $\overline{S}$ is weakly saturated, to prove that the map $F \to G$ belongs to $\overline{S}$, it will suffice to show that for each $\gamma < \beta$, the map

$$f_\gamma : \varinjlim_{\gamma' < \gamma} G(\gamma') \to G(\gamma)$$

belongs to $\overline{S}$. We observe that for each $C \in \mathcal{C}$, the map $f_\gamma(C)$ can be identified with the map $Y(C)_{B'(\gamma)(C)} \to Y(C)_{B(\gamma)(C)}$. Since $B(\gamma)(C) - B'(\gamma)(C)$ is κ-small, Remark A.1.5.5, Lemma A.1.5.11, and Lemma A.1.5.6 imply that f_γ is the pushout of a morphism belonging to S_0. We now conclude by applying the following result (replacing G by $G(\gamma)$ and F by $\varinjlim_{\gamma' < \gamma} G(\gamma')$):

Lemma A.3.3.17. *Suppose that $f : F \to G$ has the property that, for each $C \in \mathcal{C}$, there exists a pushout diagram*

$$
\begin{array}{ccc}
X_C & \xrightarrow{\ g_C\ } & Y_C \\
\downarrow & & \downarrow \\
F(C) & \xrightarrow{\ f(C)\ } & G(C),
\end{array}
$$

where $g_C \in S_0$. Then f is the pushout of a morphism in S.

Proof. In view of (iv), we can write F as the colimit of a diagram $\{F_\lambda\}_{\lambda \in P}$ indexed by a κ-filtered partially ordered set P, where each F_λ is a κ-compact object of $\mathbf{A}^{\mathcal{C}}$ and is therefore a functor whose values are κ-compact objects of $\mathbf{A}$. Since each $X_C \in \mathbf{A}$ is κ-compact, the map $X_C \to F(C)$ factors through $F_{\lambda(C)}(C)$ for some sufficiently large $\lambda(C) \in P$. Since $\mathcal{C}$ has fewer than κ objects and P is κ-filtered we can choose a single $\lambda \in P$ which works for every object $C \in \mathcal{C}$.

Consider, for each $C \in \mathcal{C}$, the composite map

$$\coprod_{C' \in \mathcal{C}} Y_{C'} \otimes \mathrm{Map}_{\mathbf{A}}(C', C) \to \coprod_{C' \in \mathcal{C}} G(C') \otimes \mathrm{Map}_{\mathbf{A}}(C', C)$$

$$\to G(C)$$

$$\simeq \varinjlim_{\lambda' \in P} F_{\lambda'}(C) \coprod_{X_C} Y_C.$$

Using another compactness argument, we deduce that this map is equivalent to a composition

$$\coprod_{C' \in \mathcal{C}} Y_{C'} \otimes \mathrm{Map}_{\mathbf{A}}(C', C) \to F_{\lambda'(C)}(C) \coprod_{X_C} Y_C.$$

for some sufficiently large $\lambda'(C) \in P$. Once again, because P is κ-filtered we can choose a single $\lambda' \in P$ which works for all C. Enlarging λ and λ', we can assume $\lambda = \lambda'$. Using another compactness argument, we can (after enlarging λ if necessary) assume that each of the diagrams

$$
\begin{array}{ccc}
X_{C'} \otimes \mathrm{Map}_{\mathbf{A}}(C', C) & \longrightarrow & F_\lambda(C) \\
\downarrow & & \downarrow \\
Y_{C'} \otimes \mathrm{Map}_{\mathbf{A}}(C', C) & \longrightarrow & F_\lambda(C) \coprod_{X_C} Y_C
\end{array}
$$

$$
\begin{array}{ccc}
Y_{C''} \otimes \mathrm{Map}_{\mathbf{A}}(C'', C') \otimes \mathrm{Map}_{\mathbf{A}}(C', C) & \longrightarrow & Y_{C''} \otimes \mathrm{Map}_{\mathbf{A}}(C'', C) \\
\downarrow & & \downarrow \\
(F_\lambda(C') \coprod_{X_{C'}} Y_{C'}) \otimes \mathrm{Map}_{\mathbf{A}}(C', C) & \longrightarrow & F_\lambda(C) \coprod_{X_C} Y_C
\end{array}
$$

is commutative. Then the above maps allow us to define an **S**-enriched functor $G_\lambda : \mathcal{C} \to \mathbf{A}$ by the formula $G_\lambda(C) = F_\lambda(C) \coprod_{X_C} Y_C$. We now observe that there is a pushout diagram

$$
\begin{array}{ccc}
F_\lambda & \xrightarrow{f_\lambda} & G_\lambda \\
\downarrow & & \downarrow \\
F & \xrightarrow{f} & G
\end{array}
$$

and that $f_\lambda \in S$.   □

A.3.4 Path Spaces in S-Enriched Categories

Let **S** be a excellent model category. We have seen that there exists a model structure on the category $\mathcal{C}at_{\mathbf{S}}$ of **S**-enriched categories whose fibrant objects are precisely those categories which are enriched over the full subcategory $\mathbf{S}^\circ$ of fibrant objects of **S**.

The theory of model categories provides a plethora of examples: for every **S**-enriched model category **A**, the full subcategory $\mathbf{A}^\circ \subseteq \mathbf{A}$ of fibrant-cofibrant objects is a fibrant object of $\mathcal{C}at_{\mathbf{S}}$. In other words, $\mathbf{A}^\circ$ is suitable to use for computing the homotopy set $[\mathcal{C}, \mathbf{A}^\circ] = \mathrm{Hom}_{\mathrm{h}\mathcal{C}at_{\mathbf{S}}}(\mathcal{C}, \mathbf{A}^\circ)$: if $\mathcal{C}$ is cofibrant, then every map from $\mathcal{C}$ to $\mathbf{A}^\circ$ in the homotopy category of $\mathcal{C}at_{\mathbf{S}}$ is represented by an actual **S**-enriched functor from $\mathcal{C}$ to $\mathbf{A}^\circ$. Moreover, two simplicial functors $F, F' : \mathcal{C} \to \mathbf{A}^\circ$ represent the same morphism in $\mathrm{h}\mathcal{C}at_{\mathbf{S}}$ if and only if they are homotopic to one another. The relation of homotopy can be described in terms of either a cylinder object for $\mathcal{C}$ or a path object for $\mathbf{A}^\circ$. Unfortunately, it is rather difficult to construct a cylinder object for $\mathcal{C}$ explicitly since the cofibrations in $\mathcal{C}at_{\mathbf{S}}$ are difficult to describe directly even when $\mathbf{S} = \mathcal{S}et_\Delta$ (the class of cofibrations of simplicial categories is *not* stable under products, so the usual procedure of constructing mapping cylinders via "product with an interval" cannot be applied). On the other hand, Theorem A.3.2.24 gives a good understanding of the fibrations in $\mathcal{C}at_\mathbf{S}$, which will allow us to give a very explicit construction of a path object for $\mathbf{A}^\circ$.

Let $\mathbf{A}$ be an $\mathbf{S}$-enriched model category. Our goal in this section is to give a direct construction of a path space object for $\mathbf{A}^\circ$ in $\mathcal{C}at_\mathbf{S}$. In other words, we wish to supply a diagram of $\mathbf{S}$-enriched categories

$$\mathbf{A}^\circ \to P(\mathbf{A}) \to \mathbf{A}^\circ \times \mathbf{A}^\circ,$$

where the composite map is the diagonal, the left map is a weak equivalence, and the right map is a fibration. For technical reasons, we will find it convenient to work not with the entire category $\mathbf{A}$ but with some (usually small) subcategory thereof. For this reason, we introduce the following definition:

Definition A.3.4.1. Let $\mathbf{S}$ be an excellent model category and let $\mathbf{A}$ be a combinatorial $\mathbf{S}$-enriched model category. A *chunk of* $\mathbf{A}$ is a full subcategory $\mathcal{U} \subseteq \mathbf{A}$ with the following properties:

(a) Let A be an object of $\mathcal{U}$ and let $\{\phi_i : A \to B_i\}_{i \in I}$ be a finite collection of morphisms in $\mathcal{U}$. Then there exists a factorization

$$A \xrightarrow{p} \overline{A} \xrightarrow{q} \prod_{i \in I} B_i$$

of the product map $\prod_{i \in I} \phi_i$, where p is a trivial cofibration, q is a fibration, and $\overline{A} \in \mathcal{U}$. Moreover, this factorization can be chosen to depend functorially on the collection $\{\phi_i\}$ via an $\mathbf{S}$-enriched functor.

(b) Let A be an object of $\mathcal{U}$ and let $\{\phi_i : B_i \to A\}_{i \in I}$ be a finite collection of morphisms in $\mathcal{U}$. Then there exists a factorization

$$\coprod_{i \in I} B_i \xrightarrow{p} \overline{A} \xrightarrow{q} A$$

of the coproduct map $\coprod_{i \in I} \phi_i$, where p is a cofibration, q is a trivial fibration, and $\overline{A} \in \mathcal{U}$. Moreover, this factorization can be chosen to depend functorially on the collection $\{\phi_i\}$ via an $\mathbf{S}$-enriched functor.

If $\mathcal{U}$ is a chunk of $\mathbf{A}$, we let $\mathcal{U}^\circ$ denote the full subcategory $\mathbf{A}^\circ \cap \mathcal{U} \subseteq \mathcal{U}$ consisting of fibrant-cofibrant objects of $\mathbf{A}$ which belong to $\mathcal{U}$.

We will say that two chunks $\mathcal{U}, \mathcal{U}' \subseteq \mathbf{A}$ are *equivalent* if they have the same essential image in the homotopy category $h\mathbf{A}$.

Remark A.3.4.2. In particular, if $\mathcal{U}$ is a chunk of $\mathbf{A}$, then each object $A \in \mathcal{U}$ admits (functorial) fibrant and cofibrant replacements which also belong to $\mathcal{U}$ (take the set I to be empty in (a) and (b)).

Remark A.3.4.3. If $\mathcal{U} \subseteq \mathcal{U}' \subseteq \mathbf{A}$ are equivalent chunks of $\mathbf{A}$, then the inclusion $\mathcal{U}^\circ \subseteq \mathcal{U}'^\circ$ is a weak equivalence of $\mathbf{S}$-enriched categories.

Example A.3.4.4. Let $\mathbf{S}$ be an excellent model category and let $\mathbf{A}$ be a combinatorial $\mathbf{S}$-enriched model category. Then $\mathbf{A}$ is a chunk of itself; this follows from the small object argument.

Example A.3.4.5. Let $\mathcal{U} \subseteq \mathbf{A}$ be a chunk and let $\{X_\alpha\}$ be a collection of objects in $\mathbf{A}$. Let $\mathcal{V} \subseteq \mathcal{U}$ be the full subcategory spanned by those objects $X \in \mathcal{U}$ such that there exists an isomorphism $[X] \simeq [X_\alpha]$ in the homotopy category $h\mathbf{A}$. Then $\mathcal{V}$ is also a chunk of $\mathbf{A}$.

We will prove a general existence theorem for chunks below (see Lemma A.3.4.15).

Lemma A.3.4.6. *Let $\mathbf{S}$ be an excellent model category and let $\mathcal{C}$ be a small $\mathbf{S}$-enriched category. Then there exists a weak equivalence of $\mathbf{S}$-enriched categories $\mathcal{C} \to \mathcal{U}°$, where $\mathcal{U}$ is a chunk of a combinatorial $\mathbf{S}$-enriched category $\mathbf{A}$.*

Proof. Without loss of generality, we may suppose that $\mathcal{C}$ is fibrant. Let $\mathbf{A} = \mathbf{S}^{\mathcal{C}^{op}}$ endowed with the projective model structure. We can identify $\mathcal{C}$ with a full subcategory of $\mathbf{A}°$ via the Yoneda embedding. Using Lemma A.3.4.15, we can enlarge $\mathcal{C}$ to a chunk in $\mathbf{A}$ having the same image in the homotopy category $h\mathbf{A}$.   □

Notation A.3.4.7. Let $\mathbf{S}$ be an excellent model category, let $\mathbf{A}$ be a combinatorial $\mathbf{S}$-enriched model category, and let $\mathcal{U}$ be a chunk of $\mathbf{A}$. We define a new category $P(\mathcal{U})$ as follows:

(*i*) The objects of $P(\mathcal{U})$ are fibrations $\phi : A \to B \times C$ in $\mathbf{A}$, where $A, B, C \in \mathcal{U}°$ and the composite maps $A \to B$ and $A \to C$ are weak equivalences.

(*ii*) Morphisms in $P(\mathcal{U})$ are given by maps of diagrams

$$
\begin{array}{ccccc}
B & \longleftarrow & A & \longrightarrow & C \\
\downarrow & & \downarrow & & \downarrow \\
B' & \longleftarrow & A' & \longrightarrow & C'.
\end{array}
$$

We let $\pi, \pi' : P(\mathcal{U}) \to \mathcal{U}°$ be the functors described by the formulas

$$\pi(\phi : A \to B \times C) = B \qquad \pi'(\phi : A \to B \times C) = C.$$

We observe that both π and π' have the structure of $\mathbf{S}$-enriched functors. Invoking assumption (*a*) of Proposition A.3.4.1, we deduce the existence of another $\mathbf{S}$-enriched functor $\tau : \mathcal{U}° \to P(\mathcal{U})$, which carries an object $A \in \mathbf{A}°$ to the map q appearing in a functorial factorization

$$A \xrightarrow{p} \overline{A} \xrightarrow{q} A \times A$$

of the diagonal, where p is a trivial cofibration and q is a fibration.

Theorem A.3.4.8. *Let $\mathbf{S}$ be an excellent model category, let $\mathbf{A}$ be a combinatorial $\mathbf{S}$-enriched model category, and let $\mathcal{U}$ be a chunk of $\mathbf{A}$. Then the morphisms $\pi, \pi' : P(\mathcal{U}) \to \mathcal{U}°$ and $\tau : \mathcal{U}° \to P(\mathcal{U})$ furnish $P(\mathcal{U})$ with the structure of a path object for $\mathcal{U}°$ in $\mathcal{C}\mathrm{at}_{\mathbf{S}}$.*

Proof. We first show that $\pi \times \pi'$ is a fibration of **S**-enriched categories. In view of Theorem A.3.2.24, it will suffice to show that $\pi \times \pi'$ is a local fibration. Let $\phi : A \to B \times C$ and $\phi' : A' \to B' \times C'$ be objects of $P(\mathcal{U})$. We must show that the induced map

$$\mathrm{Map}_{P(\mathcal{U})}(\phi, \phi') \to \mathrm{Map}_{\mathbf{A}}(B, B') \times \mathrm{Map}_{\mathbf{A}}(C, C')$$

is a fibration in **S**. This map is a base change of

$$\mathrm{Map}_{\mathbf{A}}(A, A') \to \mathrm{Map}_{\mathbf{A}}(A, B' \times C'),$$

which is a fibration by virtue of the assumption that ϕ' is a fibration (since A is assumed to be cofibrant).

To complete the proof that $\pi \times \pi'$ is a quasi-fibration, we must show that if $\phi : A \to B \times C$ is an object of $P(\mathcal{U})$ and we are given weak equivalences $f : B \to B'$, $g : C \to C'$, then we can lift f and g to an equivalence in $P(\mathcal{U})$. To do so, we factor the composite map $A \to B' \times C'$ as a trivial cofibration $A \to A'$ followed by a fibration $\phi' : A' \to B' \times C'$. Since $\mathcal{U}$ is a chunk of $\mathbf{A}$, we may assume that $A' \in \mathcal{U}$ so that $\phi' \in P(\mathcal{U})$. We have an evident natural transformation $\alpha : \phi \to \phi'$. We will show below that $\pi : P(\mathcal{U}) \to \mathcal{U}^{\circ}$ is an equivalence of **S**-enriched categories; since $\pi(\alpha) = f$ is an isomorphism in $h\mathcal{U}^{\circ}$, we conclude that α is an isomorphism in $hP(\mathcal{U})$, as required.

To complete the proof, we must show that τ is a weak equivalence of **S**-enriched categories. By the two-out-of-three property, it will suffice to show that π is a weak equivalence of **S**-enriched categories. Since τ is a section of π, it is clear that π is essentially surjective. It remains only to prove that π is fully faithful. Let $\phi : A \to B \times C$ and $\phi' : A' \to B' \times C'$ be objects of $P(\mathcal{U})$; we wish to show that the induced map $p : \mathrm{Map}_{P(\mathcal{U})}(\phi, \phi') \to \mathrm{Map}_{\mathbf{A}}(B, B')$ is a weak equivalence in **S**. We have a commutative diagram

$$
\begin{array}{ccc}
\mathrm{Map}_{P(\mathcal{U})}(\phi, \phi') & \longrightarrow & \mathrm{Map}_{\mathbf{A}}(A, A') \\
\downarrow & & \downarrow{\scriptstyle u} \\
\mathrm{Map}_{\mathbf{A}}(B, B') \times \mathrm{Map}_{\mathbf{A}}(C, C') & \longrightarrow & \mathrm{Map}_{\mathbf{A}}(A, B' \times C') \\
\downarrow & & \downarrow \\
\mathrm{Map}_{\mathbf{A}}(B, B') \times \mathrm{Map}_{\mathbf{A}}(A, C') & \longrightarrow & \mathrm{Map}_{\mathbf{A}}(A, B') \times \mathrm{Map}_{\mathbf{A}}(A, C') \\
\downarrow & & \downarrow \\
\mathrm{Map}_{\mathbf{A}}(B, B') & \longrightarrow & \mathrm{Map}_{\mathbf{A}}(A, B').
\end{array}
$$

We note that because the map $A \to B$ is a weak equivalence between cofibrant objects and B' is fibrant, the bottom horizontal map is a weak equivalence in **S**. Consequently, to show that the top horizontal map is a weak equivalence in **S**, it will suffice to show that each square in the diagram is homotopy Cartesian. The bottom square is Cartesian and fibrant, so there is nothing to prove. The middle square is homotopy Cartesian because both of the middle vertical maps are weak equivalences. The upper square is a pullback square between fibrant objects of **S**, and the map u is a fibration; we now complete the proof by invoking Proposition A.2.4.4. □

Fix an excellent model category **S**. The symmetric monoidal structure on **S** induces a symmetric monoidal structure on $\mathcal{C}\text{at}_\mathbf{S}$: if $\mathcal{C}$ and $\mathcal{D}$ are **S**-enriched categories, then we can define a new **S**-enriched category $\mathcal{C}\otimes\mathcal{D}$ as follows:

(i) The objects of $\mathcal{C}\otimes\mathcal{D}$ are pairs (C,D), where $C\in\mathcal{C}$ and $D\in\mathcal{D}$.

(ii) Given a pair of objects $(C,D),(C',D')\in\mathcal{C}\otimes\mathcal{D}$, we have

$$\operatorname{Map}_{\mathcal{C}\otimes\mathcal{D}}((C,D),(C',D')) = \operatorname{Map}_{\mathcal{C}}(C,C')\otimes\operatorname{Map}_{\mathcal{D}}(D,D')\in\mathbf{S}.$$

In the case where the tensor product on **S** is the Cartesian product, this simply reduces to the usual product of **S**-enriched categories.

Note that the operation $\otimes:\mathcal{C}\text{at}_\mathbf{S}\times\mathcal{C}\text{at}_\mathbf{S}\to\mathcal{C}\text{at}_\mathbf{S}$ is *not* a left Quillen bifunctor even when $\mathbf{S}=\mathcal{S}\text{et}_\Delta$: for example, a product of cofibrant simplicial categories is generally not cofibrant. Nevertheless, $\otimes$ behaves much like a left Quillen bifunctor at the level of homotopy categories. For example, the operation $\otimes$ respects weak equivalences in each argument and therefore induces a functor $\otimes:\text{h}\mathcal{C}\text{at}_\mathbf{S}\times\text{h}\mathcal{C}\text{at}_\mathbf{S}\to\text{h}\mathcal{C}\text{at}_\mathbf{S}$, which is characterized by the existence of natural isomorphisms $[\mathcal{C}\otimes\mathcal{D}]\simeq[\mathcal{C}]\otimes[\mathcal{D}]$.

Our goal for the remainder of this section is to show that the monoidal structure $\otimes$ on $\mathcal{C}\text{at}_\mathbf{S}$ is *closed*: that is, there exist internal mapping objects in $\text{h}\mathcal{C}\text{at}_\mathbf{S}$. This is not completely obvious. It is easy to see that the monoidal structure $\otimes$ on $\mathcal{C}\text{at}_\mathbf{S}$ is closed: given a pair of **S**-enriched categories $\mathcal{C}$ and $\mathcal{D}$, the category of **S**-enriched functors $\mathcal{D}^{\mathcal{C}}$ is itself **S**-enriched and possesses the appropriate universal property. However, this is not necessarily the "correct" mapping object in the sense that the homotopy equivalence class $[\mathcal{D}^{\mathcal{C}}]$ does not necessarily coincide with the internal mapping object $[\mathcal{D}]^{[\mathcal{C}]}$ in $\text{h}\mathcal{C}\text{at}_\mathbf{S}$. Roughly speaking, the problem is that $\mathcal{D}^{\mathcal{C}}$ consists of functors which are strictly compatible with composition; the correct mapping object should also incorporate functors which preserve composition only up to (coherent) weak equivalence. However, when $\mathcal{D}$ is the category of fibrant-cofibrant objects of an **S**-enriched *model* category **A**, then we can proceed more directly.

Definition A.3.4.9. Let **S** be an excellent model category, **A** a combinatorial **S**-enriched model category, and $\mathcal{C}$ an **S**-enriched category. We will say that a full subcategory $\mathcal{U}\subseteq\mathbf{A}$ is a $\mathcal{C}$-*chunk of* **A** if it is a chunk of **A** and the subcategory $\mathcal{U}^{\mathcal{C}}$ is a chunk of $\mathbf{A}^{\mathcal{C}}$. Here we regard $\mathbf{A}^{\mathcal{C}}$ as endowed with the *projective* model structure.

Lemma A.3.4.10. *Let* **S** *be an excellent model category,* **A** *a combinatorial* **S**-*enriched model category,* $\mathcal{C}$ *a (small) cofibrant* **S**-*enriched category, and* $\mathcal{U}\subseteq\mathbf{A}$ *a* $\mathcal{C}$-*chunk. Let* $f,f':\mathcal{C}\to\mathcal{U}^\circ$ *be a pair of maps. The following conditions are equivalent:*

(1) *The homotopy classes* $[f]$ *and* $[f']$ *coincide in* $\operatorname{Hom}_{\text{h}\mathcal{C}\text{at}_\mathbf{S}}(\mathcal{C},\mathcal{U}^\circ)$.

(2) *The maps f and f' are weakly equivalent when regarded as objects of $\mathbf{A}^{\mathcal{C}}$.*

Proof. Suppose first that (1) is satisfied. Using Theorem A.3.4.8, we deduce the existence of a homotopy $h : \mathcal{C} \to P(\mathcal{U})$ from $f = \pi \circ h$ to $f' = \pi' \circ h$. The map h determines another simplicial functor $f'' : \mathcal{C} \to \mathcal{U}$ equipped with weak equivalences $f'' \to f$, $f'' \to f'$. This proves that f and f' are isomorphic in the homotopy category of $\mathbf{A}^{\mathcal{C}}$, so that (2) is satisfied.

Now suppose that (2) is satisfied. Since $\mathcal{U}$ is a $\mathcal{C}$-chunk, we can find a projectively cofibrant $f'' : \mathcal{U} \to \mathcal{C}$ equipped with a weak equivalence $\alpha : f'' \to f$. Using (2), we deduce that there is also a weak equivalence $\beta : f'' \to f'$. Again using the assumption that $\mathcal{U}^{\mathcal{C}}$ is a chunk of $\mathbf{A}^{\mathcal{C}}$, we can choose a factorization of $\alpha \times \beta$ as a composition

$$f'' \xrightarrow{u} f''' \xrightarrow{v} f \times f'$$

where u is a trivial projective cofibration, v is a projective fibration, and $f''' \in \mathcal{U}^{\mathcal{C}}$. The map v can be viewed as an object of $\mathcal{P}(\mathcal{U})$, which determines a right homotopy from f to f'. $\square$

Corollary A.3.4.11. *Let* $\mathbf{S}$ *be an excellent model category and let* $f : \mathcal{C} \to \mathcal{C}'$ *be an* $\mathbf{S}$-*enriched functor. Suppose that* f *is fully faithful in the sense that for every pair of objects* $X, Y \in \mathcal{C}$, *the induced map* $\mathrm{Map}_{\mathcal{C}}(X, Y) \to \mathrm{Map}_{\mathcal{C}'}(fX, fY)$ *is a weak equivalence in* $\mathbf{S}$. *Let* $\mathcal{D}$ *be an arbitrary* $\mathbf{S}$-*enriched category. Then*

(1) *Composition with* f *induces an injective map* $\phi : \mathrm{Hom}_{h\mathcal{C}ats}(\mathcal{D}, \mathcal{C}) \to \mathrm{Hom}_{h\mathcal{C}ats}(\mathcal{D}, \mathcal{C}')$.

(2) *The image of* ϕ *consists of those maps* $g : \mathcal{D} \to \mathcal{C}'$ *in* $h\mathcal{C}ats$ *such that the essential image of* $[g]$ *in* $h\mathcal{C}'$ *is contained in the essential image of* $[f]$ *in* $h\mathcal{C}'$.

Proof. Using Lemma A.3.4.6, we may assume without loss of generality that $\mathcal{C}' = \mathcal{U}^{\circ}$, where $\mathcal{U}$ is a chunk of an $\mathbf{S}$-enriched model category. Let $\mathcal{V} \subseteq \mathcal{U}$ be the full subcategory spanned by those objects which are weakly equivalent to an object lying in the image of f. Since f is fully faithful, the induced map $\mathcal{C} \to \mathcal{V}^{\circ}$ is a weak equivalence. We may therefore assume that $\mathcal{C} = \mathcal{V}^{\circ}$.

Without loss of generality, we may suppose that $\mathcal{D}$ is cofibrant. Enlarging $\mathcal{U}$ and $\mathcal{V}$ if necessary (using Lemma A.3.4.15), we may assume that $\mathcal{U}$ and $\mathcal{V}$ are $\mathcal{D}$-chunks. The desired results now follow immediately from Lemma A.3.4.10. $\square$

Let $\pi_0 \mathbf{A}^{\mathcal{C}}$ denote the collection of weak equivalence classes of objects in $\mathbf{A}^{\mathcal{C}}$. Every equivalence class contains a fibrant-cofibrant representative which determines an $\mathbf{S}$-enriched functor $\mathcal{C} \to \mathbf{A}^{\circ}$.

Proposition A.3.4.12. *Let* $\mathbf{S}$ *be an excellent model category,* $\mathbf{A}$ *a combinatorial* $\mathbf{S}$-*enriched model category, and* $\mathcal{C}$ *a (small)* $\mathbf{S}$-*enriched category. Then the map*

$$\phi : \pi_0 \mathbf{A}^{\mathcal{C}} \to \mathrm{Hom}_{h\mathcal{C}ats}(\mathcal{C}, \mathbf{A}^{\circ})$$

described above is bijective.

Proof. In view of Proposition A.3.3.8, we may assume that $\mathcal{C}$ is cofibrant. Lemma A.3.4.10 shows that ϕ is well-defined and injective. We show that ϕ is surjective. Let $[f] \in \mathrm{Hom}_{\mathrm{h}\mathcal{C}\mathrm{ats}}(\mathcal{C}, \mathcal{U}^{\circ})$. Since $\mathcal{C}$ is cofibrant and $\mathbf{A}^{\circ}$ is fibrant in $\mathcal{C}\mathrm{ats}_{\mathbf{S}}$, we can find an $\mathbf{S}$-enriched functor $f : \mathcal{C} \to \mathbf{A}^{\circ}$ representing $[f]$. The simplicial functor f takes values in fibrant-cofibrant objects of $\mathbf{A}$ but is not necessarily fibrant-cofibrant *as an object of* $\mathbf{A}^{\mathcal{C}}$. However, we can choose a weak trivial fibration $f' \to f$, where f' is projectively cofibrant. Consequently, it will suffice to show that a weak equivalence $u : f' \to f$ of $\mathbf{S}$-enriched functors $\mathcal{C} \to \mathbf{A}^{\circ}$ guarantees that $[f] = [f'] \in \mathrm{Hom}_{\mathrm{h}\mathcal{C}\mathrm{ats}}(\mathcal{C}, \mathbf{A}^{\circ})$, which follows from Lemma A.3.4.10. $\qquad\square$

Proposition A.3.4.13. *Let $\mathbf{S}$ be an excellent model category, $\mathbf{A}$ a combinatorial $\mathbf{S}$-enriched model category, and $\mathcal{C}$ a small $\mathbf{S}$-enriched category. Then the evaluation map $e : (\mathbf{A}^{\mathcal{C}})^{\circ} \otimes \mathcal{C} \to \mathbf{A}^{\circ}$ has the following property: for every small $\mathbf{S}$-enriched category $\mathcal{D}$, composition with e induces a bijection*

$$\mathrm{Hom}_{\mathrm{h}\mathcal{C}\mathrm{ats}}(\mathcal{D}, (\mathbf{A}^{\mathcal{C}})^{\circ}) \to \mathrm{Hom}_{\mathrm{h}\mathcal{C}\mathrm{ats}}(\mathcal{C} \otimes \mathcal{D}, \mathbf{A}^{\circ}).$$

Proof. Using Proposition A.3.4.12, we can identify both sides with $\pi_{0} \mathbf{A}^{\mathcal{C} \otimes \mathcal{D}}$. $\qquad\square$

It is not clear that the conclusion of Proposition A.3.4.13 characterizes $(\mathbf{A}^{\mathcal{C}})^{\circ}$ up to equivalence since $(\mathbf{A}^{\mathcal{C}})^{\circ}$ is a *large* $\mathbf{S}$-enriched category, and the proof of the proposition applies only in the case where $\mathcal{D}$ is small. To remedy this defect, we establish a more refined version:

Corollary A.3.4.14. *Let $\mathbf{S}$ be an excellent model category, $\mathbf{A}$ a combinatorial $\mathbf{S}$-enriched model category, and $\mathcal{C}$ a small $\mathbf{S}$-enriched category. Let $\mathcal{U}$ be a $\mathcal{C}$-chunk of $\mathbf{A}$. Then the evaluation map $e : (\mathcal{U}^{\mathcal{C}})^{\circ} \otimes \mathcal{C} \to \mathcal{U}^{\circ}$ has the following property: for every small $\mathbf{S}$-enriched category $\mathcal{D}$, composition with e induces a bijection*

$$\mathrm{Hom}_{\mathrm{h}\mathcal{C}\mathrm{ats}}(\mathcal{D}, (\mathcal{U}^{\mathcal{C}})^{\circ}) \to \mathrm{Hom}_{\mathrm{h}\mathcal{C}\mathrm{ats}}(\mathcal{C} \otimes \mathcal{D}, \mathcal{U}^{\circ}).$$

Proof. Combine Proposition A.3.4.13 with Corollary A.3.4.11. $\qquad\square$

We conclude this section with a technical result which ensures the existence of a good supply of chunks of combinatorial model categories.

Lemma A.3.4.15. *Let $\mathbf{S}$ be an excellent model category, $\mathbf{A}$ a combinatorial $\mathbf{S}$-enriched model category, and $\{\mathcal{C}_{\alpha}\}_{\alpha \in A}$ a (small) collection of (small) cofibrant $\mathbf{S}$-enriched categories. Let $\mathcal{U}$ be a small full subcategory of $\mathbf{A}$. Then there exists a small subcategory $\mathcal{V} \subseteq \mathbf{A}$ containing $\mathcal{U}$, such that $\mathcal{V}$ is a $\mathcal{C}_{\alpha}$-chunk for each $\alpha \in A$. Moreover, we may arrange that $\mathcal{U}$ and $\mathcal{V}$ have the same essential image in the homotopy category $\mathrm{h}\mathbf{A}$.*

Proof. Enlarging A if necessary, we may suppose that the collection $\{\mathcal{C}_\alpha\}_{\alpha \in A}$ includes the unit category $[0]_{\mathbf{S}}$. For each $\alpha \in A$, we can choose $\mathbf{S}$-enriched functors

$$F_\alpha : \mathbf{A}^{\mathcal{C}_\alpha \otimes [1]_{\mathbf{S}}} \to \mathbf{A}^{\mathcal{C}_\alpha \otimes [2]_{\mathbf{S}}} \qquad G_\alpha : \mathbf{A}^{\mathcal{C}_\alpha \otimes [1]_{\mathbf{S}}} \to \mathbf{A}^{\mathcal{C}_\alpha \otimes [2]_{\mathbf{S}}},$$

such that F carries each morphism $u : f \to g$ in $\mathbf{A}^{\mathcal{C}}$ to a factorization

$$f \xrightarrow{u'} f' \xrightarrow{u''} g,$$

where u' is a strong trivial cofibration and u'' is a projective fibration, and G carries u to a factorization

$$f \xrightarrow{v'} g' \xrightarrow{v''} g,$$

where v' is a projective cofibration and v'' is a weak trivial cofibration. For $C \in \mathcal{C}_\alpha$, let F_α^C be the functor $u \mapsto f'(C)$ and define G_α^C likewise.

Choose a regular cardinal κ such that each $\mathcal{C}_\alpha$ is κ-small. We define a sequence of full subcategories $\{\mathcal{U}_\alpha \subseteq \mathbf{A}\}_{\alpha < \kappa}$ as follows:

(i) If $\alpha = 0$, then $\mathcal{U}_\alpha = \mathcal{U}$.

(ii) If α is a nonzero limit ordinal, then $\mathcal{U}_\alpha = \bigcup_{\beta < \alpha} \mathcal{U}_\beta$.

(iii) If $\alpha = \beta + 1$, then $\mathcal{U}_\alpha$ is the full subcategory of $\mathbf{A}$ spanned by the following:

   (a) The objects which belong to $\mathcal{U}_\beta$.

   (b) The objects $F_\alpha^C(u) \in \mathbf{A}$, where $\alpha \in A$, $C \in \mathcal{C}_\alpha$, and $u : f \to g$ is a morphism from an object of $\mathcal{U}_\beta^{\mathcal{C}_\alpha}$ to a finite product of objects in $\mathcal{U}_\beta^{\mathcal{C}_\alpha}$.

   (c) The objects $G_\alpha^C(u) \in \mathbf{A}$, where $\alpha \in A$, $C \in \mathcal{C}_\alpha$, and $u : f \to g$ is a morphism from a finite coproduct of objects of $\mathcal{U}_\beta^{\mathcal{C}_\alpha}$ to an object in $\mathcal{U}_\beta^{\mathcal{C}_\alpha}$.

It is readily verified that the subcategory $\mathcal{V} = \bigcup_{\alpha < \kappa} \mathcal{U}_\alpha$ has the desired properties. $\square$

A.3.5 Homotopy Colimits of S-Enriched Categories

Our goal in this section is to give an explicit construction of (certain) homotopy colimits in the model category $\mathcal{C}\mathrm{at}_{\mathbf{S}}$, where $\mathbf{S}$ is an excellent model category. We begin with some general remarks concerning localization.

Notation A.3.5.1. Consider the canonical map $\bar{i} : [1]_{\mathbf{S}} \to [1]_{\mathbf{S}}^{\sim}$. We fix once and for all a factorization of $\bar{i}$ as a composition

$$[1]_{\mathbf{S}} \xrightarrow{i} \mathcal{E} \xrightarrow{i'} [1]_{\mathbf{S}}^{\sim},$$

where i is a cofibration and i' is a weak equivalence of $\mathbf{S}$-enriched categories. For every $\mathbf{S}$-enriched category $\mathcal{C}$ and every map of sets $W \to \mathrm{Hom}_{\mathcal{C}\mathrm{at}_{\mathbf{S}}}([1]_{\mathbf{S}}, \mathcal{C},$ we define a new $\mathbf{S}$-enriched category $\mathcal{C}[W^{-1}]$ by a pushout diagram

$$\begin{array}{ccc}
\coprod_{w \in W}[1]_{\mathbf{S}} & \longrightarrow & \mathcal{C} \\
\downarrow & & \downarrow \\
\coprod_{w \in W} \mathcal{E} & \longrightarrow & \mathcal{C}[W^{-1}].
\end{array}$$

Remark A.3.5.2. Since the model category $\mathcal{C}\mathrm{at}_{\mathbf{S}}$ is left proper, the construction $\mathcal{C} \mapsto \mathcal{C}[W^{-1}]$ preserves weak equivalences in $\mathcal{C}$.

We now characterize $\mathcal{C}[W^{-1}]$ by a universal property in $\mathrm{h}\mathcal{C}\mathrm{at}_{\mathbf{S}}$.

Lemma A.3.5.3. *Let $\mathcal{C}$ be a fibrant $\mathbf{S}$-enriched category and let f be a morphism in $\mathcal{C}$ classified by a map $j_0 : [1]_{\mathbf{S}} \to \mathcal{C}$. The following conditions are equivalent:*

(1) *The map f is an equivalence in $\mathcal{C}$.*

(2) *The extension problem depicted in the diagram*

$$\begin{array}{ccc}
[1]_{\mathbf{S}} & \xrightarrow{j_0} & \mathcal{C} \\
{\scriptstyle i}\downarrow & {\scriptstyle j}\nearrow & \\
\mathcal{E} & &
\end{array}$$

admits a solution.

Proof. The implication $(2) \Rightarrow (1)$ is clear since every morphism in $\mathcal{E}$ is an equivalence. For the converse, we observe that the desired lifting problem admits a solution if and only if the induced map $i' : \mathcal{C} \to \mathcal{C} \coprod_{[1]_{\mathbf{S}}} \mathcal{E}$ admits a left inverse. Since $\mathcal{C}$ is fibrant, it suffices to show that i' is a trivial cofibration. The map i' is a cofibration since it is a pushout of i, and a weak equivalence because of the invertibility hypothesis. $\square$

Lemma A.3.5.3 immediately implies the following apparently stronger claim:

Lemma A.3.5.4. *Let $f_0 : \mathcal{C} \to \mathcal{D}$ be an $\mathbf{S}$-enriched functor, where $\mathcal{D}$ is a fibrant $\mathbf{S}$-enriched category. Let $\psi : W \to \mathrm{Hom}_{\mathcal{C}\mathrm{at}_{\mathbf{S}}}([1]_{\mathbf{S}}, \mathcal{C})$ be a map of sets. The following conditions are equivalent:*

(1) *The map f_0 extends to a map $f : \mathcal{C}[W^{-1}] \to \mathcal{D}$.*

(2) *For each $w \in W$, f_0 carries the morphism $\phi(w)$ to an equivalence in $\mathcal{D}$.*

Proposition A.3.5.5. *Let* $\mathcal{C}$ *and* $\mathcal{D}$ *be* **S**-*enriched categories and let* $\psi :$ $W \to \text{Hom}_{\mathcal{C}\text{ats}}([1]_{\mathbf{S}}, \mathcal{C})$ *be a map of sets. Then the induced map*

$$\phi : \text{Hom}_{\text{h}\mathcal{C}\text{ats}}(\mathcal{C}[W^{-1}], \mathcal{D}) \to \text{Hom}_{\text{h}\mathcal{C}\text{ats}}(\mathcal{C}, \mathcal{D})$$

is injective, and its image is the subset $\text{Hom}_{\text{h}\mathcal{C}\text{ats}}^{W}(\mathcal{C}, \mathcal{D}) \subseteq \text{Hom}_{\text{h}\mathcal{C}\text{ats}}(\mathcal{C}, \mathcal{D})$ *consisting of those homotopy classes of maps which induce functors* $\text{h}\mathcal{C} \to \text{h}\mathcal{D}$ *carrying each element of* W *to an isomorphism in* $\text{h}\mathcal{D}$.

Proof. Without loss of generality, we may suppose that $\mathcal{C}$ is cofibrant and $\mathcal{D}$ is fibrant. The description of the image of ϕ follows immediately from Lemma A.3.5.4. It will therefore suffice to show that ϕ is injective. Suppose we are given a pair of maps $[f], [g] \in \text{Hom}_{\text{h}\mathcal{C}\text{ats}}(\mathcal{C}[W^{-1}], \mathcal{D})$ such that $\phi([f]) = \phi([g])$. Since $\mathcal{C}[W^{-1}]$ is cofibrant, we may assume that $[f]$ and $[g]$ are represented by actual **S**-enriched functors $f, g : \mathcal{C}[W^{-1}] \to \mathcal{D}$. Moreover, the condition that $\phi([f]) = \phi([g])$ guarantees that the restrictions $f| \mathcal{C}$ and $g| \mathcal{C}$ are homotopic. We wish to show that f and g are homotopic.

Invoking Proposition A.2.3.1, we deduce that g is homotopic to a map $g' : \mathcal{C}[W^{-1}] \to \mathcal{D}$ such that $g'| \mathcal{C} = f| \mathcal{C}$. Replacing g by g' if necessary, we may assume that $g| \mathcal{C} = f| \mathcal{C}$. It will now suffice to show that f and g are homotopic in the model category $(\mathcal{C}\text{ats})_{\mathcal{C}/}$. We observe that f and g determine a map

$$h : \mathcal{C}[(W \coprod W)^{-1}] \simeq \mathcal{C}[W^{-1}] \coprod_{\mathcal{C}} \mathcal{C}[W^{-1}] \to \mathcal{D}.$$

Using the invertibility hypothesis, we conclude that $\mathcal{C}[(W \coprod W)^{-1}]$ is a cylinder object for $\mathcal{C}[W^{-1}]$ in the model category $(\mathcal{C}\text{ats})_{\mathcal{C}/}$, so that h is the desired homotopy from f to g. $\quad\square$

Lemma A.3.5.6. *Let* $f : \mathcal{C} \to \mathcal{D}$ *be an* **S**-*enriched functor and let* $\mathcal{M}$ *be the categorical mapping cylinder of* f *defined as follows:*

(1) *An object of* $\mathcal{M}$ *is either an object of* $\mathcal{C}$ *or an object of* $\mathcal{D}$.

(2) *Given a pair of objects* $X, Y \in \mathcal{M}$, *we have*

$$\text{Map}_{\mathcal{M}}(X, Y) = \begin{cases} \text{Map}_{\mathcal{C}}(X, Y) & \text{if } X, Y \in \mathcal{C} \\ \text{Map}_{\mathcal{D}}(X, Y) & \text{if } X, Y \in \mathcal{D} \\ \text{Map}_{\mathcal{D}}(fX, Y) & \text{if } X \in \mathcal{C}, Y \in \mathcal{D} \\ \emptyset & \text{if } X \in \mathcal{D}, Y \in \mathcal{C}. \end{cases}$$

Here $\emptyset$ *denotes the initial object of* **S**.

We observe that there is a canonical retraction j *of* $\mathcal{M}$ *onto* $\mathcal{D}$ *described by the formula*

$$j(X) = \begin{cases} fX & \text{if } X \in \mathcal{C} \\ X & \text{if } X \in \mathcal{D}. \end{cases}$$

Let W *be a collection of morphisms in* $\mathcal{M}$ *with the following properties:*

(i) *For each $w \in W$, $j(w)$ is an identity morphism in $\mathcal{D}$.*

(ii) *For every object $C \in \mathcal{C}$, the morphism $C \to fC$ in $\mathcal{M}$ classifying the identity map from fC to itself belongs to W.*

Assumption (i) implies that the map j canonically extends to a map $\bar{j}$: $\mathcal{M}[W^{-1}] \to \mathcal{D}$. The map $\bar{j}$ is a weak equivalence of $\mathbf{S}$-enriched categories.

Proof. It will suffice to show that composition with $\bar{j}$ induces a bijection

$$\mathrm{Hom}_{h\mathcal{C}\mathrm{at}_\mathbf{S}}(\mathcal{D}, \mathcal{A}) \to \mathrm{Hom}_{h\mathcal{C}\mathrm{at}_\mathbf{S}}(\mathcal{M}[W^{-1}], \mathcal{A})$$

for every $\mathbf{S}$-enriched category $\mathcal{A}$. Equivalently, we must show that the map

$$t : \mathrm{Hom}_{h\mathcal{C}\mathrm{at}_\mathbf{S}}(\mathcal{D}, \mathcal{A}) \to \mathrm{Hom}_{h\mathcal{C}\mathrm{at}_\mathbf{S}}^{W}(\mathcal{M}, \mathcal{A})$$

is bijective, where $\mathrm{Hom}_{h\mathcal{C}\mathrm{at}_\mathbf{S}}^{W}(\mathcal{M}, \mathcal{A})$ is defined as in Proposition A.3.5.5. The map t has a section t' given by composition with the inclusion $\mathcal{D} \to \mathcal{M}$. It will therefore suffice to show that $t \circ t'$ is the identity on $\mathrm{Hom}_{h\mathcal{C}\mathrm{at}_\mathbf{S}}^{W}(\mathcal{M}, \mathcal{A})$.

Using Lemma A.3.4.6 and Corollary A.3.4.11, we can reduce to the case where $\mathcal{A} = \mathbf{A}^\circ$, where $\mathbf{A}$ is a combinatorial $\mathbf{S}$-enriched model category. Using Proposition A.3.4.12, we deduce that every element $[g] \in \mathrm{Hom}_{h\mathcal{C}\mathrm{at}_\mathbf{S}}(\mathcal{M}, \mathcal{A})$ can be represented by a diagram $g : \mathcal{M} \to \mathbf{A}^\circ$. We wish to prove that g and $g \circ i \circ j$ are homotopic. We observe that there is a canonical natural transformation $\alpha : g \to g \circ i \circ j$. Moreover, if g carries each element of W to an equivalence in $\mathbf{A}^\circ$, then assumption (ii) guarantees that α is a weak equivalence in the model category $\mathbf{A}^\mathcal{M}$. We now invoke Proposition A.3.4.12 to deduce that g and $g \circ i \circ j$ are homotopic, as desired. $\square$

Definition A.3.5.7. Let A be a partially ordered set. An *A-filtered $\mathbf{S}$-enriched category* is an $\mathbf{S}$-enriched category $\mathcal{C}$ together with a map r : $\mathrm{Ob}(\mathcal{C}) \to A$ with the following property: if $C, D \in \mathcal{C}$ and $r(C) \not\leq r(D)$, then $\mathrm{Map}_\mathcal{C}(C, D) \simeq \emptyset$, where $\emptyset$ denotes an initial object of $\mathbf{S}$.

If $\mathcal{C}$ is an A-filtered $\mathbf{S}$-enriched category and $a \in A$, then we let $\mathcal{C}_{\leq a}$ denote the full subcategory of $\mathcal{C}$ spanned by those objects $C \in \mathcal{C}$ such that $r(C) \leq a$.

Remark A.3.5.8. Let $\mathcal{C}$ be an A-filtered $\mathbf{S}$-enriched category and let ψ : $W \to \mathrm{Hom}_{\mathcal{C}\mathrm{at}_\mathbf{S}}([1]_\mathbf{S}, \mathcal{C})$ be a map of sets. For each $a \in A$, we let $W_a \subseteq W$ be the subset consisting of those elements $w \in W$ such that the morphism $\psi(w)$ belongs to $\mathcal{C}_a$. This data determines a diagram $\chi^{W} : A \to \mathcal{C}\mathrm{at}_\mathbf{S}$ described by the formula $a \mapsto \mathcal{C}_{\leq a}[W_a^{-1}]$. Moreover, we have a canonical isomorphism of $\mathbf{S}$-enriched categories $\mathcal{C}[W^{-1}] \simeq \varinjlim(\chi)$.

Using the small object argument, we easily deduce the following result:

Lemma A.3.5.9. *Let A be a partially ordered set and let $\mathcal{C}$ be an A-filtered $\mathbf{S}$-enriched category. Then there exists an $\mathbf{S}$-enriched functor $f : \mathcal{C}' \to \mathcal{C}$ with the following properties:*

(1) *The functor f is bijective on objects, and for every pair of objects $C, D \in \mathcal{C}'$, the map $\mathrm{Map}_\mathcal{C}(C, D) \to \mathrm{Map}_\mathcal{C}(fC, fD)$ is a trivial fibration in $\mathbf{S}$. In particular, f is a weak equivalence of $\mathbf{S}$-enriched categories.*

(2) *The A-filtration on $\mathcal{C}$ induces an A-filtration on $\mathcal{C}'$. In other words, if $C, D \in \mathcal{C}'$ and $r(fC) \not\leq r(fD)$, then $\mathrm{Map}_{\mathcal{C}'}(C, D)$ is an initial object of $\mathbf{S}$.*

(3) *The diagram $A \to \mathcal{C}\mathrm{at}_\mathbf{S}$ described by the formula $a \mapsto \mathcal{C}'_{\leq a}$ is projectively cofibrant.*

Proposition A.3.5.10. *Let A be a partially ordered set, let $\mathcal{C}$ be an A-filtered $\mathbf{S}$-enriched category, and let $\psi : W \to \mathrm{Hom}_{\mathcal{C}\mathrm{at}_\mathbf{S}}([1]_\mathbf{S}, \mathcal{C})$ be a map of sets. Let $\chi : A \to \mathcal{C}\mathrm{at}_\mathbf{S}$ be defined as in Remark A.3.5.8. Then the isomorphism $\varinjlim \chi \simeq \mathcal{C}[W^{-1}]$ exhibits $\mathcal{C}$ as the homotopy colimit of the diagram χ.*

Proof. Choose a map $\mathcal{C}' \to \mathcal{C}$ as in Lemma A.3.5.9 and a map $\psi' : W \to \mathrm{Hom}_{\mathcal{C}\mathrm{at}_\mathbf{S}}([1]_\mathbf{S}, \mathcal{C}')$ lifting ψ, and let $\chi' : A \to \mathcal{C}\mathrm{at}_\mathbf{S}$ be defined as in Remark A.3.5.8. Then we have a canonical map $\chi' \to \chi$, which is a cofibrant replacement for χ in the model category $\mathrm{Fun}(A, \mathcal{C}\mathrm{at}_\mathbf{S})$. It will therefore suffice to show that the induced map $\mathcal{C}'[W^{-1}] \simeq \varinjlim \chi' \to \varinjlim \chi \simeq \mathcal{C}[W^{-1}]$ is a weak equivalence of $\mathbf{S}$-enriched categories, which follows immediately from Remark A.3.5.2. $\square$

Definition A.3.5.11. Let A be a partially ordered set and let $p : A \to \mathcal{C}\mathrm{at}_\mathbf{S}$ be an A-indexed diagram of $\mathbf{S}$-enriched categories. Let us denote the image of $a \in A$ under p by $\mathcal{C}_a$.

The *Grothendieck construction on p* is a category $\mathrm{Groth}(p)$ defined as follows:

(1) The objects of $\mathrm{Groth}(p)$ are pairs (a, C), where $a \in A$ and $C \in \mathcal{C}_a$.

(2) Given a pair of objects $(a, C), (a', C')$ in $\mathrm{Groth}(p)$, we set

$$\mathrm{Map}_{\mathrm{Groth}(p)}((a, C), (a', C')) = \begin{cases} \mathrm{Map}_{\mathcal{C}_{a'}}(p_a^{a'}C, C') & \text{if } a \leq a' \\ \emptyset & \text{otherwise.} \end{cases}$$

Here $p_a^{a'}$ denotes the functor $\mathcal{C}_a \to \mathcal{C}_{a'}$ determined by p, and $\emptyset$ denotes an initial object of $\mathbf{S}$.

(3) Composition in $\mathrm{Groth}(p)$ is defined in the obvious way.

We observe that $\mathrm{Groth}(p)$ is A-filtered via the map $r : \mathrm{Ob}(\mathrm{Groth}(p)) \to A$ given by the formula $r(a, C) = a$. We let $W(p)$ denote the collection of all morphisms in $\mathrm{Groth}(p)$ of the form $\alpha : (a, C) \to (a', p_a^{a'}C)$, where $a \leq a'$ and α corresponds to the identity in $\mathcal{C}_{a'}$.

For each $a \in A$, there is a canonical functor $\pi_a : \mathrm{Groth}(p)_{\leq a} \to \mathcal{C}_a$ given by the formula $(C, a') \mapsto p_{a'}^a(C)$. We note that π carries each element of $W(p)_a$ to an identity map in $\mathcal{C}_a$, so that π_a canonically extends to a map $\overline{\pi}_a : \mathrm{Groth}(p)_{\leq a}[W(p)_a^{-1}] \to \mathcal{C}_a$. The maps $\overline{\pi}_a$ are functorial in a and therefore determine a map of diagrams $\chi(p) \to p$, where $\chi(p)$ is defined as in Remark A.3.5.8.

Lemma A.3.5.12. *Let $p : A \to \mathrm{Cat}_{\mathbf{S}}$ be as in Definition A.3.5.11. Then for each $a \in A$, the map $\overline{\pi}_a : \mathrm{Groth}(p)_{\leq a}[W(p)_a^{-1}] \to \mathcal{C}_a$ is a weak equivalence of $\mathbf{S}$-enriched categories.*

Proof. This is a special case of Lemma A.3.5.6.   □

Lemma A.3.5.13. *Let $p : A \to \mathrm{Cat}_{\mathbf{S}}$ be as in Definition A.3.5.11. Then there is a canonical isomorphism $\mathrm{Groth}(p)[W(p)^{-1}] \simeq \mathrm{hocolim}(p)$ in the homotopy category $\mathrm{hCat}_{\mathbf{S}}$.*

Proof. Combine Lemma A.3.5.12 with Proposition A.3.5.10.   □

Lemma A.3.5.14. *Let $\mathcal{C}$ and $\mathcal{D}$ be small $\mathbf{S}$-enriched categories. Let W be a collection of morphisms in $\mathcal{C}$ and let W' be the collection of all morphisms in $\mathcal{C} \otimes \mathcal{D}$ of the form $w \otimes \mathrm{id}_D$, where $w \in W$ and $D \in \mathcal{D}$. Then the canonical map*

$$(\mathcal{C} \otimes \mathcal{D})[W'^{-1}] \to \mathcal{C}[W^{-1}] \otimes \mathcal{D}$$

is a weak equivalence of $\mathbf{S}$-enriched categories.

Proof. It will suffice to show that for every $\mathbf{S}$-enriched category $\mathcal{A}$, the induced map

$$\phi : \mathrm{Hom}_{\mathrm{hCat}_{\mathbf{S}}}(\mathcal{C}[W^{-1}] \otimes \mathcal{D}, \mathcal{A}) \to \mathrm{Hom}_{\mathrm{hCat}_{\mathbf{S}}}((\mathcal{C} \otimes \mathcal{D})[W'^{-1}], \mathcal{A})$$

is bijective. Using Lemma A.3.4.6 and Corollary A.3.4.11, we can reduce to the case where $\mathcal{A} = \mathbf{A}^\circ$, where $\mathbf{A}$ is a combinatorial $\mathbf{S}$-enriched model category. We now invoke Propositions A.3.4.13 and A.3.5.5 to get a chain of bijections

$$\begin{aligned}
\mathrm{Hom}_{\mathrm{hCat}_{\mathbf{S}}}(\mathcal{C}[W^{-1}] \otimes \mathcal{D}, \mathbf{A}^\circ) &\simeq \mathrm{Hom}_{\mathrm{hCat}_{\mathbf{S}}}(\mathcal{C}[W^{-1}], (\mathbf{A}^{\mathcal{D}})^\circ) \\
&\simeq \mathrm{Hom}_{\mathrm{hCat}_{\mathbf{S}}}^W(\mathcal{C}, (\mathbf{A}^{\mathcal{D}})^\circ) \\
&\simeq \mathrm{Hom}_{\mathrm{hCat}_{\mathbf{S}}}^{W'}(\mathcal{C} \otimes \mathcal{D}, \mathbf{A}^\circ) \\
&\simeq \mathrm{Hom}_{\mathrm{hCat}_{\mathbf{S}}}((\mathcal{C} \otimes \mathcal{D})[W'], \mathbf{A}^\circ)
\end{aligned}$$

whose composition is the map ϕ.   □

Theorem A.3.5.15. *Let A be a partially ordered set and let $\mathcal{D}$ be an $\mathbf{S}$-enriched category. Then the functor $\mathcal{C} \mapsto \mathcal{C} \otimes \mathcal{D}$ commutes with A-indexed homotopy colimits. In other words, if $p : A \to \mathrm{Cat}_{\mathbf{S}}$ is a projectively cofibrant diagram and $p' : A \to \mathrm{Cat}_{\mathbf{S}}$ is defined by $p'(a) = p(a) \otimes \mathcal{D}$, then the canonical isomorphism $\varinjlim(p') \simeq \varinjlim(p) \otimes \mathcal{D}$ exhibits $\varinjlim(p) \otimes \mathcal{D}$ as a homotopy colimit of the diagram p'.*

Proof. In view of Lemma A.3.5.13, it will suffice to show that the canonical map $h : \mathrm{Groth}(p')[W(p')^{-1}] \to \mathrm{Groth}(p)[W(p)^{-1}] \otimes \mathcal{D}$ is a weak equivalence of $\mathbf{S}$-enriched categories. This is a special case of Lemma A.3.5.14.   □

A.3.6 Exponentiation in Model Categories

Let $\mathcal{C}$ be a category which admits finite products, containing a pair of objects X and Y. An *exponential of X by Y* is an object $X^Y \in \mathcal{C}$ together with a map $e : X^Y \times Y \to X$, with the following universal property: for every object $W \in \mathcal{C}$, the composition

$$\operatorname{Hom}_{\mathcal{C}}(W, X^Y) \to \operatorname{Hom}_{\mathcal{C}}(W \times Y, X^Y \times Y) \xrightarrow{\circ e} \operatorname{Hom}_{\mathcal{C}}(W \times Y, Z)$$

is bijective.

Our goal in this section is to study the existence of exponentials in the homotopy category of a model category $\mathbf{A}$. Suppose we are given a pair of objects $X, Y \in \mathbf{A}$ such that there exists an exponential of $[X]$ by $[Y]$ in the homotopy category $h\mathbf{A}$. We can then represent this exponential as $[Z]$ for some object $Z \in \mathbf{A}$. Without loss of generality, we may assume that X, Y, and Z are fibrant and cofibrant, so that we have a canonical identification $[Z] \times [Y] \simeq [Z \times Y]$. However, we encounter a technical difficulty: the evaluation map $[Z] \times [Y] \to [X]$ need not be representable by any morphism from $Z \times Y$ to X in the category $\mathbf{A}$ because $Z \times Y$ need not be cofibrant. We wish to work in certain contexts where this difficulty does arise (for example, where $\mathbf{A}$ is the category of simplicial categories). For this reason we are forced to work with the following somewhat cumbersome definition:

Definition A.3.6.1. Let $\mathbf{A}$ be a model category. We will say that a diagram

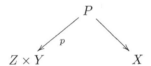

exhibits Z as a weak exponential of X by Y if the following conditions are satisfied:

(1) The map p exhibits P as a homotopy product of Z and Y; in other words, the induced map $[p] : [P] \to [Z] \times [Y]$ is an isomorphism in the homotopy category $h\mathbf{A}$.

(2) The composition $[Z] \times [Y] \xrightarrow{[p]^{-1}} [P] \to [X]$ exhibits $[Z]$ as an exponential of $[X]$ by $[Y]$ in the homotopy category $h\mathbf{A}$.

We will say that a map $Z \times Y \to X$ *exhibits Z as an exponential of X by Y* if the diagram

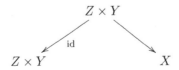

satisfies (1) and (2).

Remark A.3.6.2. Suppose we are given a diagram

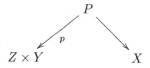

as in Definition A.3.6.1. We will say that this diagram is *standard* if $X, Y, Z \in$ **A** are fibrant, and the map p is a trivial fibration.

Suppose X and Y are fibrant objects of **A** such that there exists an exponential of $[X]$ by $[Y]$ in the homotopy category h**A**. Without loss of generality, this exponential has the form $[Z]$, where Z is a fibrant object of **A**. We can then choose a trivial fibration $P \to Z \times Y$, where P is cofibrant. The evaluation map $[Z \times Y] \simeq [Z] \times [Y] \to [X]$ is then representable by a map $P \to X$ in **A**, so that we obtain a *standard* diagram which exhibits Z as a weak exponential of X by Y.

Remark A.3.6.3. Suppose we are given a diagram

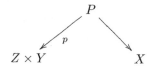

in a model category **A**. The condition that this diagram exhibits Z as a weak exponential of X by Y depends only on the image of this diagram in the homotopy category h**A**. We may therefore replace the above diagram by a weakly equivalent diagram when testing whether or not the conditions of Definition A.3.6.1 are satisfied.

Remark A.3.6.4. Let $\mathbf{A} \underset{G}{\overset{F}{\rightleftarrows}} \mathbf{B}$ be a Quillen equivalence of model categories. Suppose we are given a standard diagram

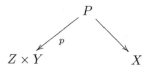

in $\mathcal{B}$. Then this diagram exhibits Z as a weak exponential of X by Y in **B** if and only if the associated diagram

exhibits GZ as a weak exponential of GX by GY in **A**.

To work effectively with weak exponentials, we need to introduce an additional assumption.

Definition A.3.6.5. Let **A** be a combinatorial model category containing a fibrant object Y. We will say that *multiplication by Y preserves homotopy colimits* if the following condition is satisfied:

($*$) Let A be a (small) partially ordered set, let $F : A \to \mathbf{A}$ be a projectively cofibrant diagram, and let $F' : A \to \mathbf{A}$ be another strongly cofibrant diagram equipped with a natural transformation $F'(a) \to F(a) \times Y$ which is weak equivalence for each $a \in A$. Then the induced map $\varinjlim F' \to (\varinjlim F) \times Y$ exhibits $\varinjlim F'$ as a homotopy product of Y with $\varinjlim F$ in **A**.

We will say that *multiplication in $\mathbf{A}$ preserves homotopy colimits* if condition ($*$) is satisfied for every fibrant object $Y \in \mathbf{A}$.

Remark A.3.6.6. Definition A.3.6.5 refers only to homotopy colimits indexed by partially ordered sets. However, every diagram indexed by an arbitrary category can be replaced by a diagram indexed by a partially ordered set having the same homotopy colimit. We formulate and prove a precise statement to this effect (in the language of ∞-categories) in §4.2. However, we will not need (or use) any such result in this appendix.

Remark A.3.6.7. Let $\mathbf{A} \underset{G}{\overset{F}{\rightleftarrows}} \mathbf{B}$ be a Quillen equivalence between combinatorial model categories and let $Y \in \mathbf{B}$ be a fibrant object. Then multiplication by Y preserves homotopy colimits in **B** if and only if multiplication by $G(Y)$ preserves homotopy colimits in **A**. Since the right derived functor RG is essentially surjective on homotopy categories, we see that multiplication in **B** preserves homotopy colimits if and only if multiplication in **A** preserves homotopy colimits.

Example A.3.6.8. Let **S** be an excellent model category with respect to the symmetric monoidal structure given by the Cartesian product in **S**. Then multiplication in $\mathcal{C}\mathrm{at}_\mathbf{S}$ preserves homotopy colimits. This is precisely the content of Theorem A.3.5.15.

Lemma A.3.6.9. *Let S be a collection of simplicial sets satisfying the following conditions:*

(i) *The simplicial set Δ^0 belongs to S.*

(ii) *If $f : X \to Y$ is a weak homotopy equivalence, then $X \in S$ if and only if $Y \in S$.*

(iii) *For every small partially ordered set A, if $F : A \to \mathcal{S}\mathrm{et}_\Delta$ is a projectively cofibrant diagram such that each $F(a) \in S$, then $\varinjlim F \in S$.*

Proof. Using (ii) and (iii), we deduce that if $F : A \to \mathcal{S}\mathrm{et}_\Delta$ is *any* diagram such that each $F(a)$ belongs to S, then the homotopy colimit of F belongs to S. In particular, S is closed under the formation of coproducts and homotopy pushouts.

We now prove by induction on n that every n-dimensional simplicial set X belongs to S. For this, we observe that there is a homotopy pushout diagram

$$\begin{array}{ccc} B \times \partial\,\Delta^n & \longrightarrow & B \times \Delta^n \\ \downarrow & & \downarrow \\ \mathrm{sk}^{n-1} X & \longrightarrow & X, \end{array}$$

where B denotes the set of n-simplices of X. The simplicial sets $B \times \partial\,\Delta^n$ and $\mathrm{sk}^{n-1} X$ belong to S by the inductive hypothesis. The simplicial set $B \times \Delta^n$ is weakly equivalent to the constant simplicial set B, which belongs to S in view of (i) and the fact that S is stable under coproducts. Since S is stable under homotopy pushouts, we conclude that $X \in S$, as desired.

An arbitrary simplicial set X can be written as the colimit of a projectively cofibrant diagram

$$\mathrm{sk}^0 X \subseteq \mathrm{sk}^1 X \subseteq \mathrm{sk}^2 X \subseteq \cdots$$

and therefore belongs to S by assumption (iii). $\qquad\square$

Proposition A.3.6.10. *Let* $\mathbf{A}$ *be a combinatorial simplicial model category containing a standard diagram*

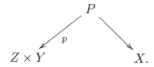

Assume further that multiplication by Y *preserves homotopy colimits in* $\mathbf{A}$. *The following conditions are equivalent:*

(i) *The above diagram exhibits* Z *as a weak exponential of* X *by* Y.

(ii) *Let* W *and* W' *be cofibrant objects of* $\mathbf{A}$ *and let* $W' \to W \times Y$ *be a map which exhibits* W' *as a homotopy product of* W *and* Y. *Then the induced map*

$$\mathrm{Map}_{\mathbf{A}}(W, Z) \times_{\mathrm{Map}_{\mathbf{A}}(W', Z \times Y)} \mathrm{Map}_{\mathbf{A}}(W', P) \to \mathrm{Map}_{\mathbf{A}}(W', X)$$

is a homotopy equivalence of Kan complexes.

Remark A.3.6.11. In the situation of part (ii) of Proposition A.3.6.10, the projection map $\mathrm{Map}_{\mathbf{A}}(W', P) \to \mathrm{Map}_{\mathbf{A}}(W', Z \times Y)$ is a trivial Kan fibration, so the fiber product $\mathrm{Map}_{\mathbf{A}}(W, Z) \times_{\mathrm{Map}_{\mathbf{A}}(W', Z \times Y)} \mathrm{Map}_{\mathbf{A}}(W', P)$ is automatically a Kan complex which is homotopy equivalent to $\mathrm{Map}_{\mathbf{A}}(W, Z)$.

Proof of Proposition A.3.6.10. First suppose that (ii) is satisfied. We wish to show that for every object $[W] \in \mathrm{h}\mathbf{A}$, the composition

$$\mathrm{Hom}_{\mathrm{h}\mathbf{A}}([W], [Z]) \to \mathrm{Hom}_{\mathrm{h}\mathbf{A}}([W] \times [Y], [Z] \times [Y])$$
$$\simeq \mathrm{Hom}_{\mathrm{h}\mathbf{A}}([W] \times [Y], [P])$$
$$\to \mathrm{Hom}_{\mathrm{h}\mathbf{A}}([W] \times [Y], [P])$$

is bijective. Without loss of generality, we may assume that $[W]$ is the homotopy equivalence class of a fibrant-cofibrant object $W \in \mathbf{A}$. Choose a cofibrant replacement $W' \to W \times Y$. We observe that the map in question can be identified with the composition

$$\pi_0 \operatorname{Map}_{\mathbf{A}}(W, Z) \to \pi_0 \operatorname{Map}_{\mathbf{A}}(W', Z \times Y)$$
$$\simeq \pi_0 \operatorname{Map}_{\mathbf{A}}(W', P)$$
$$\to \pi_0 \operatorname{Map}_{\mathbf{A}}(W', X),$$

which is bijective in view of (ii) and Remark A.3.6.11.

We now assume (i) and prove (ii). It will suffice to show that for every simplicial set K, the induced map

$$\operatorname{Hom}_{\mathrm{hSet}_\Delta}(K, \operatorname{Map}_{\mathbf{A}}(W, Z) \times_{\operatorname{Map}_{\mathbf{A}}(W', Z \times Y)} \operatorname{Map}_{\mathbf{A}}(W', P))$$
$$\downarrow$$
$$\operatorname{Hom}_{\mathrm{hSet}_\Delta}(K, \operatorname{Map}_{\mathbf{A}}(W', X))$$

is a bijection. Using Remark A.3.6.11, we can identify the left side with the set $\operatorname{Hom}_{\mathrm{hSet}_\Delta}(K, \operatorname{Map}_{\mathbf{A}}(W, Z)) \simeq \operatorname{Hom}_{\mathrm{h}\mathbf{A}}(W \otimes K, Z)$. Similarly, the right side can be identified with $\operatorname{Hom}_{\mathrm{h}\mathbf{A}}(W' \otimes K, X)$. In view of assumption (i), it will suffice to show that the map $\beta_K : W' \otimes K \to Y \times (W \otimes K)$ exhibits $W' \otimes K$ as a homotopy product of Y and $W \otimes K$. The collection of simplicial sets K with this property clearly contains Δ^0 and is stable under weak homotopy equivalence. The assumption that multiplication by Y preserves homotopy colimits guarantees that the hypotheses of Lemma A.3.6.9 are satisfied, so that the desired conclusion holds for *every* simplicial set K. $\square$

Lemma A.3.6.12. *Let $\mathbf{A}$ be a combinatorial model category and $i : B_0 \to B$ an inclusion of partially ordered sets. Suppose that there exists a retraction $r : B \to B_0$ such that $r(b) \leq b$ for each $b \in B$. Let $F : B \to \mathbf{A}$ be a diagram. Then a map $\alpha : X \to \lim(F)$ in $\mathbf{A}$ exhibits X as a homotopy limit of F if and only if α exhibits X as a homotopy limit of $i^* F$.*

Proof. Without loss of generality, we may assume that F is injectively fibrant. We have a canonical isomorphism $\lim(F) \simeq \lim(i^* F)$. It will therefore suffice to show that the functor i^* preserves injective fibrations. It now suffices to observe that i^* is right adjoint to r^* and that the functor r^* preserves weak trivial cofibrations. $\square$

Lemma A.3.6.13. *Let $\mathbf{A}$ be a combinatorial model category, $\mathcal{C}$ a small category, $F : \mathcal{C} \to \mathbf{A}$ a diagram, and $\alpha : X \to \lim(F)$ a morphism in the category $\mathbf{A}$. Suppose that*

(i) *For each $C \in \mathcal{C}$, the induced map $X \to F(C)$ is a weak equivalence in $\mathbf{A}$.*

(ii) *The category $\mathcal{C}$ has a final object C_0.*

Then α exhibits X as a homotopy limit of the diagram F.

Proof. Without loss of generality, we may assume that the diagram F is projectively fibrant. Let $F' : \mathcal{C} \to \mathbf{A}$ be defined by the formula $F'(C) = F(C_0)$. We observe that, for every $G \in \mathrm{Fun}(\mathcal{C}, \mathbf{A})$, we have $\mathrm{Hom}_{\mathrm{Fun}(\mathcal{C}, \mathbf{A})}(G, F') = \mathrm{Hom}_{\mathbf{A}}(G(C_0), F(C_0))$. In particular, we have a canonical map $\beta : F \to F'$. Condition (i) guarantees that β is a weak equivalence. Since $F(C_0) \in \mathbf{A}$ is fibrant, the diagram F' is injectively fibrant. It therefore suffices to show that the induced map $X \to \lim(F') \simeq F(C_0)$ is a weak equivalence, which follows from (i). $\qquad\square$

Lemma A.3.6.14. *Let $\mathbf{A}$ be a combinatorial model category, let A be a partially ordered set, and set $B = \{(a,b) \in A^{op} \times A : a \geq b\}$, regarded as a partially ordered subset of $A^{op} \times A$. Let $\pi : B \to A^{op}$ denote the projection onto the first factor.*

Suppose we are given diagrams $F : B \to \mathbf{A}$, $G : A \to \mathbf{A}$, and a natural transformation $\alpha : \pi^*(G) \to F$, which induces weak equivalences $G(a) \to F(a,b)$ for each $(a,b) \in B$. Then α exhibits G as a homotopy right Kan extension of F.

Proof. In view of Proposition A.2.8.9, it will suffice to show that for each $a_0 \in A$, the transformation α exhibits $G(a_0)$ as a homotopy limit of the diagram $F|\{(a,b) \in B : a \leq a_0\}$. Let $B_0 = \{(a,b) \in B : a = a_0\}$. In view of Lemma A.3.6.12, it will suffice to show that α exhibits $G(a_0)$ as a limit of the diagram $F_0 = F|B_0$. This follows immediately from Lemma A.3.6.13. $\qquad\square$

Proposition A.3.6.15. *Let $\mathbf{A}$ be a combinatorial model category and $A = A_0 \cup \{\infty\}$ a partially ordered set with a largest element ∞. Let $B = \{(a,b) \in A^{op} \times A : a \geq b\}$, regarded as a partially ordered subset of $A^{op} \times A$.*

Suppose we are given an object $X \in \mathbf{A}$ together with functors $Y : A \to \mathbf{A}$, $Z : A^{op} \to \mathbf{A}$, $P : B \to \mathbf{A}$, and diagrams $\sigma_{a,b}$:

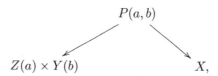

which depend functorially on $(a,b) \in B$. Suppose further that

(i) *Each diagram $\sigma_{a,b}$ exhibits $P(a,b)$ as a homotopy product of $Z(a)$ and $Y(b)$ in $\mathbf{A}$.*

(ii) *The diagrams $\sigma_{a,a}$ exhibit $Z(a)$ as a weak exponential of X by $Y(a)$.*

(iii) *Multiplication in $\mathbf{A}$ preserves homotopy colimits.*

(iv) *The diagram Y exhibits $Y(\infty)$ as a homotopy colimit of $Y_0 = Y|A_0$.*

Then the diagram Z exhibits $Z(\infty)$ as the homotopy limit of the diagram $Z_0 = Z|A_0^{op}$.

Proof. Making fibrant replacements if necessary, we may assume that each diagram $\sigma_{a,b}$ is standard. According to the main result of [19], there exists a Quillen equivalence $\mathbf{A}' \underset{G}{\overset{F}{\rightleftarrows}} \mathbf{A}$, where $\mathbf{A}'$ is a combinatorial *simplicial* model category. In view of Remark A.3.6.4, we may replace $\mathbf{A}$ by $\mathbf{A}'$ and thereby reduce to the case where $\mathbf{A}$ is a simplicial model category.

In view of Proposition A.3.3.12, it will suffice to prove the following: for every fibrant-cofibrant object $C \in \mathbf{A}$, if we define $G : A^{op} \to \mathrm{Set}_\Delta$ by the formula $G(a) = \mathrm{Map}_{\mathbf{A}}(C, Z(a))$, then G exhibits $G(\infty)$ as a homotopy limit of the diagram $G|A_0^{op}$.

Let $W : A \to \mathbf{A}$ be a cofibrant replacement for the functor $a \mapsto C \times Y(a)$. Let $G' : A^{op} \to \mathrm{Set}_\Delta$ be defined by the formula $G'(a) = \mathrm{Map}_{\mathbf{A}}(W(a), X)$.

Define $G'' : B \to \mathrm{Set}_\Delta$ by the formula

$$G''(a, b) = \mathrm{Map}_{\mathbf{A}}(C, Z(a)) \times_{\mathrm{Map}_{\mathbf{A}}(W(a), Z(a) \times Y(b))} \mathrm{Map}_{\mathbf{A}}(W(a), P(a, b)).$$

Let $\pi : B \to A^{op}$ denote projection onto the first factor, so that we have natural transformations of diagrams

$$\pi^* G \overset{\alpha}{\leftarrow} G'' \overset{\beta}{\to} \pi^* G'.$$

We observe that β induces a trivial Kan fibration $G''(a, b) \to G'(a)$ for all $(a, b) \in B$. In particular, for $a \leq b$ the induced map $G''(a, a) \to G''(a, b)$ is a homotopy equivalence. Condition (ii) guarantees that α induces a homotopy equivalence $G''(a, b) \to G(a)$ if $a = b$ and therefore for all $(a, b) \in B$.

Using Lemma A.3.6.14, we conclude that α and β exhibit G and G' as homotopy right Kan extensions of G'' along π. In particular, G and G' are equivalent in the homotopy category $\mathrm{hFun}(A^{op}, \mathbf{A})$. Assumptions (iii) and (iv) guarantee that W exhibits $W(\infty)$ as the homotopy colimit of $W|A_0$. Using Proposition A.3.3.12, we deduce that G' exhibits $G'(\infty)$ as the homotopy limit of $G'|A_0^{op}$. It follows that G exhibits $G(\infty)$ as the homotopy limit of $G|A_0^{op}$, as desired. $\square$

We conclude this section with an application of Proposition A.3.6.15.

Proposition A.3.6.16. *Let* $\mathbf{S}$ *be an excellent model category in which the monoidal structure is given by the Cartesian product. Let* $\mathbf{A}$ *be a combinatorial* $\mathbf{S}$*-enriched model category,* $A = A_0 \cup \{\infty\}$ *a partially ordered set with a largest element* ∞*, and* $\{\mathcal{C}_a\}_{a \in A}$ *a diagram of small* $\mathbf{S}$*-enriched categories indexed by* A*. Let* $\mathcal{U} \subseteq \mathbf{A}$ *be a chunk. For each* $a \in A$*, let* $\mathcal{U}_f^{\mathcal{C}_a}$ *denote the full subcategory of* $\mathcal{U}^{\mathcal{C}_a} \subseteq \mathbf{A}^{\mathcal{C}_a}$ *spanned by the projectively fibrant diagrams and let* W_a *denote the collection of weak equivalences in* $\mathcal{V}_a$*. Assume that*

(a) *For each* $a \in A$*, the* $\mathbf{S}$*-enriched category* $\mathcal{C}_a$ *is cofibrant and* $\mathcal{U}$ *is a* $\mathcal{C}_a$*-chunk of* $\mathbf{A}$*.*

(b) *The diagram* $\{\mathcal{C}_a\}_{a \in A}$ *exhibits* $\mathcal{C}_\infty$ *as a homotopy colimit of the diagram* $\{\mathcal{C}_a\}_{a \in A_0}$*.*

(c) *The chunk* $\mathcal{U}$ *is small.*

Then the induced diagram $\{\mathcal{U}_f^{\mathcal{C}_a}[W_a^{-1}]\}_{a\in A}$ *exhibits* $\mathcal{U}_f^{\mathcal{C}_\infty}[W_\infty^{-1}]$ *as a homotopy limit of the diagram* $\{\mathcal{U}_f^{\mathcal{C}_a}[W_a^{-1}]\}_{a\in A_0}$.

Before proving Proposition A.3.6.16, we need a simple lemma.

Lemma A.3.6.17. *Let* **S** *be an excellent model category,* **A** *a combinatorial* **S***-enriched model category, and* $\mathcal{U} \subseteq \mathbf{A}$ *a chunk. Let* $\mathcal{U}_f$ *denote the full subcategory of* $\mathcal{U}$ *spanned by those objects which are fibrant in* **A** *and let* W *denote the collection of weak equivalences in* $\mathcal{U}_f$. *Then the induced map* $\mathcal{U}^\circ \to \mathcal{U}_f[W^{-1}]$ *is a weak equivalence of* **S***-enriched categories.*

Proof. Let $W_0 = W \cap \mathcal{U}^\circ$. Since every weak equivalence in $\mathcal{U}^\circ$ is actually an equivalence, we conclude that the induced map $\mathcal{U}^\circ \to \mathcal{U}^\circ[W_0^{-1}]$ is a weak equivalence. It will therefore suffice to prove that the map $i : \mathcal{U}^\circ[W_0^{-1}] \to \mathcal{U}_f[W^{-1}]$ is a weak equivalence. Let F be an **S**-enriched fibrant replacement functor which carries $\mathcal{U}$ to itself, so that F induces a map $j : \mathcal{U}_f[W^{-1}]$ to $\mathcal{U}^\circ[W_0^{-1}]$. We claim that j is a homotopy inverse to i. To prove this, we observe that there is a natural transformation $\alpha : \mathrm{id} \to F$, which we can identify with a map

$$\overline{h} : \mathcal{U}_f \otimes [1]_{\mathbf{S}} \to \mathcal{U}^\circ.$$

Let W_0' be the collection of all morphisms in $\mathcal{U}_f \otimes [1]_{\mathbf{S}}$ of the form $e \otimes \mathrm{id}$, where e is an equivalence in $\mathcal{U}_f$, and let W_1' be the collection of all morphisms of $\mathcal{U}_f \otimes [1]_{\mathbf{S}}$ of the form $\mathrm{id} \otimes g$, where $g : 0 \to 1$ is the tautological morphism in $[1]_{\mathbf{S}}$. Let $W' = W_0' \cup W_1'$, so that $\overline{h}$ determines a map

$$h : (\mathcal{U}_f \otimes [1]_{\mathbf{S}})[W'^{-1}] \to \mathcal{U}^\circ[W_0^{-1}].$$

We will prove that h determines a homotopy from the identity to $j \circ i$, so that j is a left homotopy inverse to i. Applying the same argument to the restriction $\overline{h}|\,\mathcal{U}^\circ \otimes [1]_{\mathbf{S}}$ will show that j is a right homotopy inverse to i.

To prove that h gives the desired homotopy, it will suffice to show that the inclusions $\{0\}, \{1\} \hookrightarrow [1]_{\mathbf{S}}$ induce weak equivalences

$$\mathcal{U}_f[W^{-1}] \to (\mathcal{U}_f \otimes [1]_{\mathbf{S}})[W'^{-1}].$$

This follows immediately from Corollary A.3.4.11 and Lemma A.3.5.14. $\square$

Proof of Proposition A.3.6.16. Let $B = \{(a,b) \in A^{op} \times A : a \geq b\}$. For each $(a,b) \in B$, we define $P(a,b) = (\mathcal{U}_f^{\mathcal{C}_a} \times \mathcal{C}_b)[V_{a,b}^{-1}]$, where $V_{a,b}$ is the collection of all morphisms of $\mathcal{U}_f^{\mathcal{C}_a} \times \mathcal{C}_b$ of the form $e \otimes \mathrm{id}_C$, where $e \in W_a$ and $C \in \mathcal{C}_b$. We have an evident family of diagrams $\sigma(a,b)$:

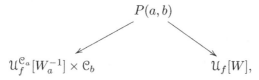

where $\mathcal{U}_f$ denotes the full subcategory of $\mathcal{U}$ spanned by the fibrant objects and W is the collection of weak equivalences in $\mathcal{U}_f \subseteq \mathbf{A}$. To complete the proof, it will suffice to show that the hypotheses of Proposition A.3.6.15 are satisfied. Condition (i) follows from Lemma A.3.5.14, condition (iii) from Theorem A.3.5.15, and condition (iv) from (b). To prove (ii), we observe that for each $a \in A$, the diagram $\sigma(a, a)$ is weakly equivalent to the diagram

This diagram exhibits $(\mathcal{U}^{\mathcal{C}_a})^{\circ}$ as a weak exponential of $\mathcal{U}^{\circ}$ by $\mathcal{C}_a$ by Corollary A.3.4.14. $\qquad \square$

Corollary A.3.6.18. *Let* **S** *be an excellent model category in which the monoidal structure is given by the Cartesian product. Let* **A** *be a combinatorial* **S**-*enriched model category, let* $A = A_0 \cup \{\infty\}$ *be a partially ordered set with a largest element* ∞, *and let* $\{\mathcal{C}_a\}_{a \in A}$ *be a diagram of small* **S**-*enriched categories indexed by* A.

For each $a \in A$, *let* $\mathbf{A}_f^{\mathcal{C}_a}$ *denote the collection of projectively fibrant objects of* $\mathbf{A}^{\mathcal{C}_a}$ *and let* W_a *denote the collection of weak equivalences in* $\mathbf{A}_f^{\mathcal{C}_a}$. *Assume that the diagram* $\{\mathcal{C}_a\}_{a \in A}$ *exhibits* $\mathcal{C}_\infty$ *as a homotopy colimit of the diagram* $\{\mathcal{C}_a\}_{a \in A_0}$. *Then the induced diagram* $\{\mathbf{A}_f^{\mathcal{C}_a}[W_a^{-1}]\}_{a \in A}$ *exhibits* $\mathbf{A}^{\mathcal{C}_\infty}[W_\infty^{-1}]$ *as a homotopy limit of the diagram* $\{\mathbf{A}_f^{\mathcal{C}_a}[W_a^{-1}]\}_{a \in A_0}$.

Proof. Without loss of generality, we may suppose that each $\mathcal{C}_a$ is cofibrant. The proof of Proposition A.2.8.2 shows that there exists a (small) regular cardinal κ such that the collection of homotopy limit diagrams in $\mathrm{Fun}(A, \mathcal{C}at_\mathbf{S})$ is stable under κ-filtered colimits. This cardinal depends only on A and $\mathbf{S}$ and remains invariant if we enlarge the universe. Using Lemma A.3.4.15, we can write $\mathbf{A}$ as a κ-filtered union of full subcategories $\mathcal{U} \subseteq \mathbf{A}$, where $\mathcal{U}$ is a $\mathcal{C}_a$-chunk for each $a \in A$. We now conclude by applying Proposition A.3.6.16. $\qquad \square$

A.3.7 Localizations of Simplicial Model Categories

Let $\mathbf{A}$ and $\mathbf{A}'$ be two model categories with the same underlying category. We say that $\mathbf{A}'$ is a *(Bousfield) localization* of $\mathbf{A}$ if the following conditions are satisfied:

(C) A morphism f of $\mathcal{C}$ is a cofibration in $\mathbf{A}$ if and only if f is a cofibration in $\mathbf{A}'$.

(W) If a morphism f of $\mathcal{C}$ is a weak equivalence in $\mathbf{A}$, then f is a weak equivalence in $\mathbf{A}'$.

It then also follows that

(F) If a morphism f of $\mathcal{C}$ is a fibration in $\mathbf{A}'$, then f is a fibration in $\mathbf{A}$.

Our goal in this section is to study the localizations of a fixed model category **A** and to relate this to our study of localizations of presentable ∞-categories (§5.5.4).

Let **A** be a simplicial model category. Let h**A** be the homotopy category of **A** obtained from **A** by inverting all weak equivalences. Alternatively, we can obtain h**A** by first passing to the full subcategory **A**° ⊆ **A** spanned by the fibrant-cofibrant objects and then passing to the homotopy category of the simplicial category **A**°. From the second point of view, we see that h**A** has a natural enrichment over the homotopy category $\mathcal{H}$: if $X, Y \in$ h**A** are represented by fibrant-cofibrant objects $\overline{X}, \overline{Y} \in$ **A**, then we let

$$\mathrm{Map}_{h\mathbf{A}}(X, Y) = [\mathrm{Map}_{\mathbf{A}}(\overline{X}, \overline{Y})].$$

Here $[K] \in \mathcal{H}$ denotes the object of $\mathcal{H}$ represented by a Kan complex K. In fact, this description is accurate if we assume only that $\overline{X}$ is cofibrant and $\overline{Y}$ fibrant.

Let S be a collection of morphisms in h**A**. Then

(*i*) We will say that an object $Z \in$ h**A** is *S-local* if, for every morphism $f : X \to Y$ in S, the induced map

$$\mathrm{Map}_{h\mathbf{A}}(Y, Z) \to \mathrm{Map}_{h\mathbf{A}}(X, Z)$$

is a homotopy equivalence. We say that an object $\overline{Z} \in$ **A** is S-local if its image in h**A** is S-local.

(*ii*) We will say that a morphism $f : X \to Y$ of h**A** is an *S-equivalence* if, for every S-local object $Z \in$ h**A**, the induced map

$$\mathrm{Map}_{h\mathbf{A}}(Y, Z) \to \mathrm{Map}_{h\mathbf{A}}(X, Z)$$

is a homotopy equivalence. We say that a morphism $\overline{f}$ in **A** is an *S-equivalence* if its image in h**A** is an S-equivalence.

If $\overline{S}$ is a collection of morphisms in h**A**, with image S in h**A**, we will apply the same terminology: an object of **A** (or h**A**) is said to be $\overline{S}$-*local* if it is S-local, and a morphism of **A** (or h**A**) is said to be an $\overline{S}$-*equivalence* if it is an S-equivalence.

Lemma A.3.7.1. *Let* **A** *be a left proper simplicial model category, let S be a collection of morphisms in* h**A**, *and let $i : A \to B$ be a cofibration in* **A**. *The following conditions are equivalent:*

(1) *The map i is an S-equivalence.*

(2) *For every fibrant object $X \in$ **A** which is S-local, the map i induces a trivial Kan fibration $\mathrm{Map}_{\mathbf{A}}(B, X) \to \mathrm{Map}_{\mathbf{A}}(A, X)$.*

Proof. Choose a trivial fibration $f : A' \to A$, where A' is cofibrant, and choose a factorization

$$A' \xrightarrow{i'} B' \xrightarrow{f'} B$$

of $i \circ f$, where i' is a cofibration and f' is a trivial fibration. We have a commutative diagram

$$
\begin{array}{ccc}
A' & \xrightarrow{\;i'\;} & B' \\
{\scriptstyle f}\downarrow & & \downarrow{\scriptstyle g} \qquad \searrow{\scriptstyle f'} \\
A & \xrightarrow{\;i\;} & A\coprod_{A'} B' \xrightarrow{\;j\;} B.
\end{array}
$$

Since f is a weak equivalence and i' is a cofibration, the left properness of $\mathbf{A}$ guarantees that g is a weak equivalence. It follows from the two-out-of-three property that j is also a weak equivalence.

Suppose first that (1) is satisfied. Let X be an S-local fibrant object of $\mathbf{A}$. The map $p : \mathrm{Map}_{\mathbf{A}}(B, X) \to \mathrm{Map}_{\mathbf{A}}(A, X)$ is a Kan fibration. We wish to show that p is a trivial Kan fibration. Our assumption that X is S-local guarantees that the map $q' : \mathrm{Map}_{\mathbf{A}}(B', X) \to \mathrm{Map}_{\mathbf{A}}(A', X)$ is a homotopy equivalence and therefore a trivial fibration (since i' is a cofibration). The map

$$
q : \mathrm{Map}_{\mathbf{A}}(A\coprod_{A'} B', X) \to \mathrm{Map}_{\mathbf{A}}(A)
$$

is a pullback of q' and therefore also a trivial fibration. To show that p is a trivial Kan fibration, it will suffice to show that for every $t : A \to X$, the fiber $p^{-1}\{t\}$ is a contractible Kan complex. Since the corresponding fiber $q^{-1}\{t\}$ is contractible, it will suffice to show that composition with j induces a homotopy equivalence

$$
p^{-1}\{t\} \to q^{-1}\{t\}.
$$

This is clear since j is a weak equivalence between cofibrant objects of the simplicial model category $\mathbf{A}_{A/}$.

Now assume that (2) holds. We wish to show that i is an S-equivalence. For this, it suffices to show that for every fibrant S-local object $X \in \mathbf{A}$, the map

$$
q' : \mathrm{Map}_{\mathbf{A}}(B', X) \to \mathrm{Map}_{\mathbf{A}}(A', X)
$$

is a trivial Kan fibration. The preceding argument shows that the fiber of q' over a morphism $t' : A' \to X$ is contractible, provided that t' factors as a composition

$$
A' \xrightarrow{f} A \xrightarrow{t} X.
$$

To complete the proof, it suffices to show that the same result holds for an arbitrary vertex t' of $\mathrm{Map}_{\mathbf{A}}(A', X)$. The map t' factors as a composition

$$
A' \xrightarrow{u} Y \xrightarrow{v} X,
$$

where u is a cofibration and v is a trivial fibration. We have a commutative diagram

$$
\begin{array}{ccc}
\mathrm{Map}_{\mathbf{A}}(B', Y) & \longrightarrow & \mathrm{Map}_{\mathbf{A}}(A', Y) \\
\downarrow & & \downarrow \\
\mathrm{Map}_{\mathbf{A}}(B', X) & \longrightarrow & \mathrm{Map}_{\mathbf{A}}(A', X)
\end{array}
$$

in which the vertical arrows are trivial Kan fibrations. It will therefore suffice to show that the fiber $\mathrm{Map}_{\mathbf{A}}(B', Y) \times_{\mathrm{Map}_{\mathbf{A}}(A', Y)} \{u\}$ is contractible. Choose a trivial cofibration $A' \coprod_A Y \to Z$, where Z is fibrant. We observe that the map $Y \to A' \coprod_A Y$ is the pushout of a weak equivalence by a cofibration and therefore a weak equivalence (since $\mathbf{A}$ is left proper). It follows that the map $Y \to Z$ is a weak equivalence between fibrant objects of $\mathbf{A}$. We have a commutative diagram

$$
\begin{array}{ccc}
\mathrm{Map}_{\mathbf{A}}(B', Y) & \longrightarrow & \mathrm{Map}_{\mathbf{A}}(A', Y) \\
\downarrow & & \downarrow \\
\mathrm{Map}_{\mathbf{A}}(B', Z) & \xrightarrow{q''} & \mathrm{Map}_{\mathbf{A}}(A', Z)
\end{array}
$$

in which the vertical maps are homotopy equivalences and the horizontal maps are Kan fibrations. It will therefore suffice to show that the fiber of q'' is contractible when taken over the composite map $t'' : A' \xrightarrow{u} Y \to Z$. We now observe that t'' factors through A, so that the desired result follows from the first part of the proof. $\qquad\square$

Corollary A.3.7.2. *Let $\mathbf{A}$ and $\mathbf{B}$ be simplicial model categories and suppose we are given a simplicial adjunction*

$$
\mathbf{A} \underset{G}{\overset{F}{\rightleftarrows}} \mathbf{B}.
$$

Assume that $\mathbf{B}$ is left proper. The following conditions are equivalent:

(1) *The adjunction between F and G is a Quillen adjunction.*

(2) *The functor F preserves cofibrations, and the functor G preserves fibrant objects.*

Proof. The implication $(1) \Rightarrow (2)$ is obvious. Conversely, suppose that (2) is satisfied. We wish to prove that F is a left Quillen functor. Since F preserves cofibrations, it will suffice to show that for every trivial cofibration $u : A \to A'$ in $\mathbf{A}$, the image Fu is a weak equivalence in $\mathbf{B}$. Applying Lemma A.3.7.1 in the case $S = \emptyset$, it will suffice to prove the following: for every fibrant object $B \in \mathbf{B}$, the induced map

$$
\mathrm{Map}_{\mathbf{B}}(FA', B) \to \mathrm{Map}_{\mathbf{B}}(FA, B)
$$

is a trivial Kan fibration. Since F and G are adjoint simplicial functors, this is equivalent to the requirement that the map $\mathrm{Map}_{\mathbf{A}}(A', GB) \to \mathrm{Map}_{\mathbf{A}}(A, GB)$ be a trivial Kan fibration, which follows from our assumption that u is a trivial cofibration in $\mathbf{A}$ and that $GB \in \mathbf{A}$ is fibrant. $\qquad\square$

Proposition A.3.7.3. *Let $\mathbf{A}$ be a left proper combinatorial simplicial model category and let S be a (small) set of cofibrations in $\mathbf{A}$. Let $S^{-1}\mathbf{A}$ denote the same category, with the following distinguished classes of morphisms:*

(C) *A morphism g in $S^{-1}\mathbf{A}$ is a cofibration if it is a cofibration when regarded as a morphism in $\mathbf{A}$.*

(W) *A morphism g in $S^{-1}\mathbf{A}$ is a weak equivalence if it is an S-equivalence.*

Then

(1) *The above definitions endow $S^{-1}\mathbf{A}$ with the structure of a combinatorial simplicial model category.*

(2) *The model category $S^{-1}\mathbf{A}$ is left proper.*

(3) *An object $X \in \mathbf{A}$ is fibrant in $S^{-1}\mathbf{A}$ if and only if X is S-local and fibrant in $\mathbf{A}$.*

Proof. Enlarging S if necessary, we may assume:

(a) For every morphism $f : A \to B$ in S and every $n \geq 0$, the induced map

$$(A \times \Delta^n) \coprod_{A \times \partial \Delta^n} (B \times \partial \Delta^n) \to B \times \Delta^n$$

belongs to S.

(b) The set S contains a collection of generating trivial cofibrations for $\mathbf{A}$.

It follows that an object $X \in \mathbf{A}$ is fibrant and S-local if and only if it has the extension property with respect to every morphism in S. Since $S \subseteq C \cap W$, we deduce that every fibrant object of $S^{-1}\mathbf{A}$ is S-local and fibrant in $\mathbf{A}$. The converse follows from Lemma A.3.7.1; this proves (3).

To prove (1), it will suffice to show that the classes C and W satisfy the hypotheses of Proposition A.2.6.8 (the compatibility of the simplicial structure on $S^{-1}\mathbf{A}$ with its model structure follows immediately from Proposition A.3.1.7). We observe that Lemma A.3.7.1 implies that $C \cap W$ is a weakly saturated class of morphisms in $\mathbf{A}$. The only other nontrivial point is to show that W is an accessible subcategory of $\mathbf{A}^{[1]}$.

Proposition A.1.2.5 implies the existence of a functor $T : \mathbf{A} \to \mathbf{A}$ and a natural transformation $\mathrm{id}_{\mathbf{A}} \to T$ having the following properties:

(i) For every $X \in \mathbf{A}$, the object $TX \in \mathbf{A}$ is fibrant and S-local.

(ii) For every $X \in \mathbf{A}$, the map $X \to TX$ belongs to the smallest weakly saturated class of morphisms containing S; in particular, it belongs to $W \cap C$ and is therefore an S-equivalence.

(iii) There exists a regular cardinal κ such that T commutes with κ-filtered colimits.

It follows that a morphism $f : X \to Y$ in $\mathbf{A}$ is an S-equivalence if and only if the induced map $Tf : TX \to TY$ is an S-equivalence. Since TX and TY are S-local, Yoneda's lemma (in the category h$\mathbf{A}$) implies that Tf is an S-equivalence if and only if Tf is a weak equivalence in $\mathbf{A}$. It follows that

W is the inverse image under T of the collection of weak equivalences in $\mathbf{A}$. Corollaries A.2.6.5 and A.2.6.6 imply that W is an accessible subcategory of $\mathbf{A}^{[1]}$, as desired. This completes the proof of (1).

We now prove (2). We need to show that the collection of S-equivalences in $\mathbf{A}$ is stable under pushouts by cofibrations. We observe that every morphism $f : X \to Z$ admits a factorization

$$X \xrightarrow{f'} Y \xrightarrow{f''} Z,$$

where f' is a cofibration and f'' is a weak equivalence in $\mathbf{A}$ (in fact, we can choose f'' to be a trivial fibration in $\mathbf{A}$). If f is an S-equivalence, then f' is an S-equivalence, so that $f' \in C \cap W$. It will therefore suffice to show that $C \cap W$ and the class of weak equivalences in $\mathbf{A}$ are stable under pushouts by cofibrations. The first follows from the assertion that $C \cap W$ is weakly saturated, and the second from the assumption that $\mathbf{A}$ is left proper. $\square$

Proposition A.3.7.4. *Let $\mathbf{A}$ be a left proper combinatorial simplicial model category. Then*

(1) *Every combinatorial localization of $\mathbf{A}$ has the form $S^{-1}\mathbf{A}$, where S is some (small) set of cofibrations in $\mathbf{A}$.*

(2) *Given two (small) sets of cofibrations S and T, the localizations $S^{-1}\mathbf{A}$ and $T^{-1}\mathbf{A}$ coincide if and only if the class of S-local objects of $\mathrm{h}\mathbf{A}$ coincides with the class of T-local objects of $\mathrm{h}\mathbf{A}$.*

Proof. The "if" direction of (2) is obvious, and the converse follows from the characterization of the fibrant objects of $S^{-1}\mathbf{A}$ given in Proposition A.3.7.3. We now prove (1). Let $\mathbf{B}$ be a combinatorial model category which is a localization of $\mathbf{A}$ and let S be a set of generating trivial cofibrations for $\mathbf{B}$. We claim that $\mathbf{B} = S^{-1}\mathbf{A}$. The cofibrations of $S^{-1}\mathbf{A}$ and $\mathbf{B}$ coincide. Moreover, the collection of trivial cofibrations in $S^{-1}\mathbf{A}$ is a weakly saturated class of morphisms which contains S and therefore contains every trivial cofibration in $\mathbf{B}$. To complete the proof, it will suffice to show that every trivial cofibration $f : X \to Y$ in $S^{-1}\mathbf{A}$ is a trivial cofibration in $\mathbf{B}$.

Choose a diagram

$$
\begin{array}{ccc}
X' & \xrightarrow{f'} & Y' \\
\downarrow & & \downarrow \\
X & \xrightarrow{f} & Y,
\end{array}
$$

where X' is cofibrant, f' is a cofibration, and the vertical maps are weak equivalences in $\mathbf{A}$. Then f' is a trivial cofibration in $S^{-1}\mathbf{A}$, and it will suffice to show that f' is a trivial cofibration in $\mathbf{B}$. For this, it will suffice to show that for every fibrant object $Z \in \mathbf{B}$, the map

$$\mathrm{Map}_{\mathbf{B}}(Y', Z) \to \mathrm{Map}_{\mathbf{B}}(X', Z)$$

is a trivial fibration. In view of Lemma A.3.7.1, it will suffice to show that Z is S-local and fibrant as an object of $\mathbf{A}$. The second claim is obvious, and the first follows from the fact that S consists of trivial cofibrations in $\mathbf{B}$. $\square$

Remark A.3.7.5. In the situation of Proposition A.3.7.4, we may assume that for every cofibration $f : A \to B$ in S, the objects A and B are themselves cofibrant. To see this, choose for each cofibration $f : A \to B$ in S a diagram

$$
\begin{array}{ccc}
A' & \xrightarrow{\ g_f\ } & B' \\
\downarrow{\scriptstyle u} & & \downarrow{\scriptstyle v} \quad \searrow{\scriptstyle w} \\
A & \xrightarrow[\ f'\ ]{} & A \coprod_{A'} B' \xrightarrow{\ f''\ } B
\end{array}
$$

as in the proof of Lemma A.3.7.1, so that u and w are trivial cofibrations, $f = f'' \circ f'$, and g_f is a cofibration between cofibrant objects. Then g_f is a trivial cofibration in $S^{-1}\mathbf{A}$. We claim that the localizations $S^{-1}\mathbf{A}$ and $T^{-1}\mathbf{A}$ coincide, where $T = \{g_f\}_{f \in S}$. To prove this, it will suffice to show that for each $f \in S$, every g_f-local fibrant object $X \in \mathbf{A}$ is also f-local.

Suppose that X is g_f-local. We wish to prove that the map

$$p : \mathrm{Map}_{\mathbf{A}}(B, X) \to \mathrm{Map}_{\mathbf{A}}(A, X)$$

is a trivial Kan fibration. Since p is automatically a Kan fibration, it will suffice to show that the fiber $p^{-1}\{t\}$ is contractible for every morphism $t : A \to X$. Since X is g_f-local, we deduce that the fiber $q^{-1}\{t\}$ is contractible, where q is the projection map $\mathrm{Map}_{\mathbf{A}}(A \coprod_{A'} B', X) \to \mathrm{Map}_{\mathbf{A}}(A, X)$. It will therefore suffice to show that f'' induces a homotopy equivalence of fibers

$$\mathrm{Map}_{\mathbf{A}_{A/}}(B, X) \to \mathrm{Map}_{\mathbf{A}_{A/}}(A \coprod_{A'} B', X).$$

This is clear because f'' is a weak equivalence between cofibrant objects of the simplicial model category $\mathbf{A}_{A/}$.

Proposition A.3.7.6. *Let $\mathcal{C}$ be an ∞-category. The following conditions are equivalent:*

(1) *The ∞-category $\mathcal{C}$ is presentable.*

(2) *There exists a combinatorial simplicial model category $\mathbf{A}$ and an equivalence $\mathcal{C} \simeq \mathrm{N}(\mathbf{A}^\circ)$.*

Proof. According to Theorem 5.5.1.1 and Proposition 5.5.4.15, $\mathcal{C}$ is presentable if and only if there exists a small simplicial set K, a small set S of morphisms in $\mathcal{P}(K)$, and an equivalence $\mathcal{C} \simeq S^{-1}\mathcal{P}(K)$. Let $\mathcal{D}$ be the simplicial category $\mathfrak{C}[K]^{op}$ and let $\mathbf{B}$ be the category $\mathrm{Set}_{\Delta}^{\mathcal{D}}$ of simplicial functors $\mathcal{D} \to \mathrm{Set}_{\Delta}$ endowed with the injective model structure. Proposition 4.2.4.4 implies that there is an equivalence $\mathcal{P}(K) \simeq \mathrm{N}(\mathbf{B}^\circ)$. Moreover, Propositions A.3.7.3 and A.3.7.4 imply that there is a bijective correspondence between accessible localizations of $\mathcal{P}(K)$ (as a presentable ∞-category) and combinatorial localizations of $\mathbf{B}$ (as a model category). This proves the implication

$(1) \Rightarrow (2)$. Moreover, it also shows that $(2) \Rightarrow (1)$ in the special case where $\mathbf{A}$ is a localization of a category of simplicial presheaves.

We now complete the proof by invoking the following result, proven in [19]: for every combinatorial model category $\mathbf{A}$, there exists a small category $\mathcal{D}$, a set S of morphisms of $\mathrm{Set}_{\Delta}^{\mathcal{D}^{op}}$, and a Quillen equivalence of $\mathbf{A}$ with $S^{-1}\,\mathrm{Set}_{\Delta}^{\mathcal{D}}$. Moreover, the proof given in [19] shows that when $\mathbf{A}$ is a *simplicial* model category, then F and G can be chosen to be simplicial functors. $\square$

Remark A.3.7.7. Let $\mathbf{A}$ and $\mathbf{B}$ be combinatorial simplicial model categories. Then the underlying ∞-categories $\mathrm{N}(\mathbf{A}^{\circ})$ and $\mathrm{N}(\mathbf{B}^{\circ})$ are equivalent if and only if $\mathbf{A}$ and $\mathbf{B}$ can be joined by a chain of simplicial Quillen equivalences. The "only if" assertion follows from Corollary A.3.1.12, and the "if" direction can be proven using the methods described in [19].

Proposition A.3.7.8. *Let $\mathbf{A}$ be a left proper combinatorial simplicial model category and let $\mathcal{C} = \mathrm{N}(\mathbf{A}^{\circ})$ denote its underlying ∞-category. Suppose that $\mathcal{C}^{0} \subseteq \mathcal{C}$ is an accessible localization of $\mathcal{C}$ and let $L : \mathcal{C} \to \mathcal{C}^{0}$ denote a left adjoint to the inclusion.*

Then there exists a localization $\mathbf{A}'$ of $\mathbf{A}$ satisfying the following conditions:

(1) *An object $X \in \mathbf{A}'$ is fibrant if and only if it is fibrant in $\mathbf{A}$ and the associated object of the homotopy category $\mathrm{h}\mathbf{A} \simeq \mathrm{h}\mathcal{C}$ belongs to the full subcategory $\mathrm{h}\mathcal{C}^{0}$.*

(2) *A morphism $f : X \to Y$ in $\mathbf{A}'$ is a weak equivalence if and only if the functor $L : \mathrm{h}\mathcal{C} \to \mathrm{h}\mathcal{C}^{0}$ carries f to an isomorphism in the homotopy category $\mathrm{h}\mathcal{C}^{0}$.*

Proof. According to Proposition A.3.7.6, the ∞-category $\mathcal{C}$ is presentable. The results of §5.5.4 imply that we can write $\mathcal{C}^{0} = S^{-1}\mathcal{C}$ for some small collection of morphisms S in $\mathcal{C}$. We then take $\widetilde{S}$ to be a collection of representatives for the elements of S as cofibrations between cofibrant objects of $\mathbf{A}$ and let $\mathbf{A}'$ denote the localization $\widetilde{S}^{-1}\mathbf{A}$. $\square$

We conclude this section by establishing a universal property enjoyed by the localization of a combinatorial simplicial model category.

Proposition A.3.7.9. *Suppose we are given a simplicial Quillen adjunction*

$$\mathbf{A} \underset{G}{\overset{F}{\rightleftarrows}} \mathbf{B}$$

between left proper combinatorial simplicial model categories and let $\mathbf{A}'$ be a Bousfield localization of $\mathbf{A}$. The following conditions are equivalent:

(1) *The adjoint functors F and G determine a Quillen adjunction between $\mathbf{A}'$ and $\mathbf{B}$.*

(2) *Let α be a morphism in $\mathbf{A}$ which is a weak equivalence in $\mathbf{A}'$. Then the left derived functor $LF : \mathrm{h}\mathbf{A} \to \mathrm{h}\mathbf{B}$ carries α to an isomorphism in the homotopy category $\mathrm{h}\mathbf{B}$.*

(3) *For every fibrant object $X \in \mathbf{B}$, the image GX is a fibrant object of $\mathbf{A}'$.*

Proof. The implication $(1) \Rightarrow (2)$ is obvious, and the implication $(3) \Rightarrow (1)$ follows from Corollary A.3.7.2. We will complete the proof by showing that $(2) \Rightarrow (3)$. According to Proposition A.3.7.4 and Remark A.3.7.5, we may suppose that $\mathbf{A}' = S^{-1}\mathbf{A}$, where S is a small collection of cofibrations between cofibrant objects of $\mathbf{A}$. Let X be a fibrant object of $\mathbf{B}$; we wish to show that GX is a fibrant object of $\mathbf{A}'$. Since GX is fibrant in $\mathbf{A}$, it will suffice to show that GX is S-local (Proposition A.3.7.3). In other words, we must show that if $\alpha : A \to B$ belongs to S, then the induced map $p : \operatorname{Map}_{\mathbf{A}}(B, GX) \to \operatorname{Map}_{\mathbf{A}}(A, GX)$ is a weak homotopy equivalence. Since F and G are simplicial functors, we can identify p with the map $\operatorname{Map}_{\mathbf{B}}(FB, X) \to \operatorname{Map}_{\mathbf{B}}(FA, X)$. To prove that p is a weak homotopy equivalence, it will suffice to show that $F(\alpha)$ is a weak equivalence between cofibrant objects of $\mathbf{B}$. This follows immediately from assumption (2) (because α is a cofibration between cofibrant objects of $\mathbf{A}$, we can identify $F(\alpha)$ with the left derived functor $LF(\alpha)$). $\qquad\square$

Corollary A.3.7.10. *Let $\mathbf{A}$ and $\mathbf{B}$ be left proper combinatorial simplicial model categories and suppose we are given a simplicial Quillen adjunction*

$$\mathbf{A} \underset{G}{\overset{F}{\rightleftarrows}} \mathbf{B}.$$

Then

(1) *There exists a new left proper combinatorial simplicial model structure $\mathbf{A}'$ on the category $\mathbf{A}$ with the following properties:*

   (C) *A morphism α in $\mathbf{A}'$ is a cofibration if and only if it is a cofibration in $\mathbf{A}$.*

   (W) *A morphism α in $\mathbf{A}'$ is a weak equivalence if and only if the left derived functor LF carries α to an isomorphism in the homotopy category $h\mathbf{B}$.*

   (F) *A morphism α in $\mathbf{A}'$ is a fibration if and only if it has the right lifting property with respect to every morphism in $\mathbf{A}'$ satisfying (C) and (W).*

(2) *The functors F and G determine a new simplicial Quillen adjunction*

$$\mathbf{A}' \underset{G'}{\overset{F'}{\rightleftarrows}} \mathbf{B}.$$

(3) *Suppose that the right derived functor RG is fully faithful. Then the adjoint pair (F', G') is a Quillen equivalence.*

Proof. The functors F and G determine a pair of adjoint functors

$$N \mathbf{A}^\circ \underset{g}{\overset{f}{\rightleftarrows}} N \mathbf{B}^\circ$$

between the underlying ∞-categories (see Proposition 5.2.4.6), which are themselves presentable (Proposition A.3.7.6). Let $\overline{S}$ be the collection of all morphisms u in $N \mathbf{A}^\circ$ such that $f(u)$ is an equivalence in $N \mathbf{B}^\circ$. Proposition 5.5.4.16 implies that $\overline{S}$ is generated (as a strongly saturated class of morphisms) by a small subset $S \subseteq \overline{S}$. Without loss of generality, we may suppose that the morphisms of S are represented by some (small) collection T of cofibrations between cofibrant objects of $\mathbf{A}$. Let $\mathbf{A}' = T^{-1}\mathbf{A}$. We claim that $\mathbf{A}'$ satisfies the description given in (1). In other words, we claim that a morphism α in $\mathbf{A}$ is a T-equivalence if and only if the left derived functor LF carries α to an isomorphism in $h\mathbf{B}$. Without loss of generality, we may suppose that α is a morphism between fibrant-cofibrant objects of $\mathbf{A}$, so that we can view α as a morphism in the ∞-category $N \mathbf{A}^\circ$. In this case, both conditions on α are equivalent to the requirement that α belong to $\overline{S}$. This completes the proof of (1). Assertion (2) follows immediately from Proposition A.3.7.9.

We now prove (3). Note that the homotopy category $h\mathbf{A}'$ can be identified with a full subcategory of the homotopy category $h\mathbf{A}$ and that under this identification the left derived functor LF restricts to the left derived functor LF'. It follows that for every fibrant object $X \in \mathbf{B}$, the counit map

$$(LF')(RG')X \simeq (LF')(GX) \simeq (LF)(GX) \simeq (LF)(RG)X \simeq X$$

is an isomorphism in $h\mathbf{B}$ (where the last equivalence follows from our assumption that RG is fully faithful). It follows that the functor RG' is fully faithful. To complete the proof, it will suffice to show that the left derived functor LF' is conservative. In other words, we must show that if $\alpha : X \to Y$ is a morphism in $\mathbf{A}'$, then α is a weak equivalence if and only if $LF(\alpha)$ is an isomorphism in $\mathbf{B}$; this follows immediately from (1). $\qquad\square$

Bibliography

[1] Adámek, J., and J. Rosicky. *Locally Presentable and Accessible Categories.* Cambridge University Press, Cambridge, 1994.

[2] Artin, M. *Théorie des Topos et Cohomologie Étale des Schémas.* SGA 4. Lecture Notes in Mathematics 269. Springer-Verlag, Berlin and New York, 1972.

[3] Artin, M., and B. Mazur. *Étale Homotopy.* Lecture Notes in Mathematics 100. Springer-Verlag, Berlin and New York, 1969.

[4] Baez, J., and M. Shulman. *Lectures on n-categories and cohomology.* Available for download at math.CT/0608420.

[5] Berkovich, V. *Spectral Theory and Analytic Geometry over Non-Archimedean Fields.* Mathematical Surveys and Monographs 33. American Mathematical Society, Providence, R.I., 1990.

[6] Beilinson, A., Bernstein, J., and P. Deligne. *Faisceaux pervers.* Astérisque 100, 1982.

[7] Bergner, J.E. *A model category structure on the category of simplicial categories.* Trans. Amer. Math. Soc. 359, 2007, 2043–2058.

[8] Bergner, J.E. *A survey of* $(\infty, 1)$*-categories.* Available at math.AT/0610239.

[9] Bergner, J.E. *Rigidification of algebras over multi-sorted theories.* Algebr. Geom. Topol. 6, 2006, 1925–1955.

[10] Boardman, J.M., and R.M. Vogt. *Homotopy Invariant Structures on Topological Spaces.* Lecture Notes in Mathematics 347. Springer-Verlag, Berlin and New York, 1973.

[11] Bourn, D. *Sur les ditopos.* C.R. Acad. Sci. Paris Ser. A 279, 1974, 911–913.

[12] Bousfield, A.K. *The localization of spaces with respect to homology.* Topology 14, 1975, 133–150.

[13] Breen, L. *On the classification of 2-gerbes and 2-stacks.* Asterisque 225, 1994, 1–160.

[14] Brown, K. *Abstract homotopy theory and generalized sheaf cohomology.* Trans. Amer. Math. Soc. 186, 1973, 419–458.

[15] Chapman, T. *Lectures on Hilbert Cube Manifolds.* American Mathematical Society, Providence, R.I., 1976.

[16] Cordier, J.M. *Sur la notion de diagramme homotopiquement cohérent.* Cah. Topol. Géom. Différ. 23 (1), 1982, 93–112.

[17] Cordier, J.M., and T. Porter. *Homotopy coherent category theory.* Trans. Amer. Math. Soc. 349 (1), 1997, 1–54.

[18] Dydak, J., and J. Segal. *Shape Theory.* Lecture Notes in Mathematics 688. Springer-Verlag, Berlin and New York, 1978.

[19] Dugger, D. *Combinatorial model categories have presentations.* Adv. Math. 164, 2001, 177–201.

[20] Dugger, D., S. Hollander, and D. Isaksen. *Hypercovers and simplicial presheaves.* Math. Proc. Camb. Phil. Soc. 136 (1), 2004, 9–51.

[21] Dwyer, W.G. and D.M. Kan. *Simplicial localizations of categories.* J. Pure Appl. Algebra 17, (1980), 267–284.

[22] Dwyer, W.G., and D.M. Kan. *Homotopy theory of simplicial groupoids.* Nederl. Akad. Wetensch. Indag. Math. 46 (4), 1984, 379–385.

[23] Dwyer, W.G., and D. M. Kan. *Realizing diagrams in the homotopy category by means of diagrams of simplicial sets.* Proc. Amer. Math. Soc. 91 (3), 1984, 456–460.

[24] Dwyer, W.G., P.S. Hirschhorn, D. Kan, and J. Smith. *Homotopy Limit Functors on Model Categories and Homotopical Categories.* Mathematical Surveys and Monographs 113. American Mathematical Society, Providence, R.I., 2004.

[25] Ehlers, P.J., and T. Porter. *Ordinal subdivision and special pasting in quasicategories.* Adv. Math. 217, 2007, 489–518.

[26] Eilenberg, S., and N.E. Steenrod. *Axiomatic approach to homology theory.* Proc. Nat. Acad. Sci. USA 31, 1945, 117–120.

[27] Engelking, R. *Dimension Theory.* North-Holland, Amsterdam, Oxford, and New York, 1978.

[28] Freitag, E., and R. Kiehl. *Étale Cohomology and the Weil Conjectures.* Springer-Verlag, Berlin and New York, 1988.

[29] Fresnel, J., and M. van der Put. *Rigid Analytic Geometry and Its Applications.* Boston, Birkhauser, 2004.

[30] Friedlander, E. *Étale Homotopy of Simplicial Schemes.* Annals of Mathematics Studies 104. Princeton University Press, Princeton, N.J.,1982.

[31] Giraud, J. *Cohomologie non abelienne.* Springer Verlag, Berlin and New York, 1971.

[32] Goerss, P., and J.F. Jardine. *Simplicial Homotopy Theory.* Progress in Mathematics, Birkhauser, Boston, 1999.

[33] Gordon, R., A.J. Power, and R. Street. *Coherence for Tricategories.* Mem. Amer. Math. Soc. 117(558), American Mathematical Society, Providence, R.I., 1995.

[34] Grothendieck, A. *Sur quelques points d'algebra homologique.* Tohoku Math. J. 9, 1957, 119–221.

[35] Grothendieck, A. *A la poursuite des champs.* Unpublished letter to D. Quillen.

[36] Günther, B. *The use of semisimplicial complexes in strong shape theory.* Glas. Mat. 27(47), 1992, 101–144.

[37] Haver, W. *Mappings between ANRs that are fine homotopy equivalences.* Pacific J. Math. 58, 1975, 457–461.

[38] Hirschhorn, P. *Model Categories and Their Localizations.* Mathematical Surveys and Monographs 99. American Mathematical Society, Providence, R.I., 2003.

[39] Hirschowitz, A., and C. Simpson. *Descente pour les n-champs.* Available for download at math.AG/9807049.

[40] Hovey, M. *Model Categories.* American Mathematical Society, Providence, R.I., 1998.

[41] Jardine, J.F. *Simplicial Presheaves.* J. Pure Appl. Algebra 47(1), 1987, 35–87.

[42] Johnstone, P. *Stone Spaces.* Cambridge University Press, Cambridge, 1982.

[43] Joyal, A. *Quasi-categories and Kan complexes.* J. Pure Appl. Algebra 175, 2005, 207–222.

[44] Joyal, A. *Theory of quasi-categories I.* In preparation.

[45] Joyal, A., and M. Tierney. *Strong Stacks and Classifying Spaces.* Lecture Notes in Mathematics 1488. Springer-Verlag, Berlin and New York, 1991, 213–236.

[46] Kashiwara, M., and P. Schapira. *Sheaves on Manifolds.* Fundamental Principles of Mathematical Sciences 292. Springer-Verlag, Berlin and New York, 1990.

[47] Lazard, D. *Sur les modules plats.* C.R. Acad. Sci. Paris 258, 1964, 6313-6316.

[48] Leinster, T. *A survey of definitions of n-category.* Theor. Appl. Categ. 10(1), 2002, 1–70.

[49] Leinster, T. *Higher Operads, Higher Categories.* London Mathematical Society Lecture Note Series 298. Cambridge University Press, Cambridge, 2004.

[50] Lurie, J. *Derived Algebraic Geometry.* In preparation.

[51] Lurie, J. *Elliptic Cohomology.* In preparation.

[52] MacLane, S. *Categories for the Working Mathematician.* 2nd ed. Graduate Texts in Mathematics 5. Springer-Verlag, Berlin and New York, 1998.

[53] MacLane, S., and I. Moerdijk. *Sheaves in Geometry and Logic.* Springer-Verlag, Berlin and New York, 1992.

[54] Makkai, M., and R. Pare. *Accessible Categories.* Contemporary Mathematics 104. American Mathematical Society, Providence, R.I., 1989.

[55] Mardesic, S., and J. Segal. *Shape Theory.* North-Holland, Amsterdam, 1982.

[56] May, J.P. *Simplicial Objects in Algebraic Topology.* Chicago Lectures in Mathematics. University of Chicago Press, Chicago, 1993.

[57] May, J.P., and J. Sigurdsson. *Parametrized Homotopy Theory.* Mathematical Surveys and Monographs 132, American Mathematical Society, Providence, R.I., 2006.

[58] Moerdijk, I., and J. Vermeulen. *Proper Maps of Toposes.* Memoirs of the American Mathematical Society 148(705). American Mathematical Society, Providence, R.I., 2000.

[59] Munkres, J. *Topology.* Prentice Hall, Englewood Cliffs, N.J., 1975.

[60] Nichols-Barrer, J. *On Quasi-Categories as a Foundation for Higher Algebraic Stacks.* Doctoral dissertation, Massachusetts Institute of Technology, Cambridge, Mass., 2007.

[61] Polesello, P., and I. Waschkies. *Higher monodromy.* Homology Homotopy Appl. 7(1), 2005, 109–150.

[62] Quillen, D. *Higher Algebraic K-theory I.* In "Algebraic K-theory I: Higher K-Theories," Lecture Notes in Mathematics 341. Springer, Berlin, 1973.

[63] Quillen, D. *Homotopical Algebra.* Lectures Notes in Mathematics 43. SpringerVerlag, Berlin and New York, 1967.

[64] Rezk, C. *A model for the homotopy theory of homotopy theory.* Trans. Amer. Math. Soc. 35(3), 2001, 973–1007.

[65] Rezk, C., Shipley, B., and S. Schwede. *Simplicial structures on model categories and functors.* Amer. J. Math. 123, 2001, 551–575.

[66] Rosicky, J. *On Homotopy Varieties.* Adv. Math. 214(2), 2007, 525–550.

[67] Schwartz, L. *Unstable Modules over the Steenrod Algebra and Sullivan's Fixed Point Set Conjecture.* Chicago Lectures in Mathematics. Chicago University Press, Chicago, 1994.

[68] Segal, G. *Classifying spaces and spectral sequences.* Inst. Hautes Etudes Sci. Publ. Math. 34, 1968, 105–112.

[69] Serre, J.P. *Cohomologie Galoisienne.* Lecture Notes in Mathematics 5. Springer-Verlag, Berlin and New York, 1964.

[70] Simpson, C. *A Giraud-type characterization of the simplicial categories associated to closed model categories as ∞-pretopoi.* Available for download at math.AT/9903167.

[71] Simpson, C. *A closed model structure for n-categories, internal Hom, n-stacks and generalized Seifert-Van Kampen.* Available for download at math.AG/9704006.

[72] Spaltenstein, N. *Resolutions of Unbounded Complexes.* Compos. Math. 65(2), 1988, 121–154.

[73] Stasheff, J. *Homotopy associativity of H-spaces I, II.* Trans. Amer. Math. Soc. 108, 1963, 275–312.

[74] Street, R. *Two-dimensional sheaf theory.* J. Pure Appl. Algebra 23(3), 1982, 251–270.

[75] Tamsamani, Z. *On non-strict notions of n-category and n-groupoid via multisimplicial sets.* Available for download at math.AG/9512006.

[76] Toën, B. *Vers une axiomatisation de la théorie des catégories supériures.* K-theory 34(3), 2005, 233–263.

[77] Toën, B. *Vers une interpretation Galoisienne de la theorie de l'homotopie.* Cah. topol. geom. différ. categ. 43, 2002, 257–312.

[78] Toën, B., and G. Vezzosi. *Segal topoi and stacks over Segal categories.* Available for download at math.AG/0212330.

[79] Toën, B. and G. Vezzosi. *Homotopical algebraic geometry I: Topos theory.* Advances in Mathematics 193 (2005), no. 2, 257-372.

[80] van den Dries, L. *Tame Topology and O-minimal Structures.* Cambridge University Press, Cambridge, 1998.

[81] Wall, C.T.C. *Finiteness conditions for CW-complexes.* Ann. Math. 81, 1965, 56–69.

[82] *Čech and Steenrod Homotopy Theories with Applications to Geometric Topology.* Lecture Notes in Mathematics 542. Springer-Verlag, Berlin and New York, 1976.

General Index

abelian category
    Grothendieck, 763
absolute neighborhood retract, 774
accessible
    adjoint functors, 450
    category, xiv
    coproducts, 448
    functor, 422
    functor categories, 433
    homotopy fiber products, 446
    ∞-category, 421
    ∞-category of sections, 455
    localization, 457
    overcategories, 447
    products, 448
    subcategory, 450, 812
    undercategories, 442
adjoint functor
    between categories, 331
    and (co)limits, 346
    and composition, 339
    existence of, 342
    between Ind-categories, 406
    between ∞-categories, 337
    Quillen, 350
    and unit transformations, 340
adjoint functor theorem, 465
adjunction, 337
    Quillen, 811
algebraic morphism, 628
anodyne, 53
    inner, 53
    left, 53, 63
    marked, 148
    right, 53

Berkovich space, 778
bicategory, 3
bifibration, 141
    associated to a correspondence, 143
    and smoothness, 235

canonical covering, 588
canonical topology, 589
Cartesian
    equivalence, 157

    locally, 121
Cartesian edge, 115, 118
    and composition, 117
    and simplicial categories, 120
Cartesian fibration, 123
    and categorical fibrations, 209
    classified by $f : S \to \mathcal{C}at_\infty^{op}$, 211
    and functor categories, 153
    and overcategories, 128
    and pullbacks, 205
    and right fibrations, 124
    and trivial fibrations, 134
    universal, 211
Cartesian model structure, 159
Cartesian transformation, 543
categorical equivalence, 25
    and products, 92
    weak, 94
categorical fibration, 90
    of ∞-categories, 139
category, 1, 781
    cofibered in groupoids, 56
    enriched, 790
    homotopy, 806
    model, 803
    monoidal, 786
    Reedy, 829
    of simplices, 254, 537
    simplicial, 18
    topological, 7
Čech nerve, 542
cell-like, 681
    map of ∞-topoi, 775
    map of topological spaces, 774
chunk, 879
classifying map
    for a (co)Cartesian fibration, 211
    for a collection of morphisms, 565
    for a left fibration, 212
    for objects, 565
    for relatively κ-compact morphisms, 569
    for a right fibration, 212
    for subobjects, 564
closed
    immersion, 754

Index of Notation

Printed and bound by CPI Group (UK) Ltd, Croydon, CR0 4YY

16/04/2025

14658360-0001